Control Theory for Physicists

Control theory, an interdisciplinary concept dealing with the behavior of dynamical systems, is an important but often overlooked aspect of physics. This book is the first broad and complete treatment of the topic tailored for physicists, one that goes from the basics right through to the most recent advances. Simple examples develop a deep understanding and intuition for the systematic principles of control theory, beyond the recipes given in standard engineering-focused texts. Up-to-date coverage of control of networks and complex systems and a thorough discussion of the fundamental limits of control, including the limitations placed by causality, information theory, and thermodynamics, are included. In addition, this text explores important recent advances in stochastic thermodynamics on the thermodynamic costs of information processing and control. For all students of physics interested in control theory, this classroom-tested, comprehensive approach to the topic with online solutions and further materials delivers both fundamental principles and current developments.

John Bechhoefer is Professor of Physics at Simon Fraser University and Fellow of the American Physical Society. His research is at the intersection of thermodynamics, statistical physics, control theory, and information theory.

Control Theory for Physicists

JOHN BECHHOEFER
Simon Fraser University

CAMBRIDGE
UNIVERSITY PRESS

CAMBRIDGE
UNIVERSITY PRESS

University Printing House, Cambridge CB2 8BS, United Kingdom

One Liberty Plaza, 20th Floor, New York, NY 10006, USA

477 Williamstown Road, Port Melbourne, VIC 3207, Australia

314–321, 3rd Floor, Plot 3, Splendor Forum, Jasola District Centre, New Delhi – 110025, India

79 Anson Road, #06–04/06, Singapore 079906

Cambridge University Press is part of the University of Cambridge.

It furthers the University's mission by disseminating knowledge in the pursuit of education, learning and research at the highest international levels of excellence.

www.cambridge.org
Information on this title: www.cambridge.org/9781107001183
DOI: 10.1017/9780511734809

First published 2021

Printed in the United Kingdom by TJ Books Ltd, Padstow Cornwall.

A catalogue record for this publication is available from the British Library.

ISBN 978-1-107-00118-3 Hardback

Additional resources for this publication at www.cambridge.org/bechhoefer

Contents

Preface

This book extends a tutorial I wrote on control theory (Bechhoefer, 2005). In both the article and this book, my goal has been "to make the strange familiar, and the familiar strange."[1] The *strange* is control theory – feedback and feedforward, transfer functions and minimum phase, $\mathcal{H}_\infty$ metrics and Z-transforms, and many other ideas that are not usually part of the education of a physicist. The *familiar* includes notions such as causality, measurement, robustness, and entropy – concepts physicists think they know – that acquire new meanings in the light of control theory. I hope that this book accomplishes both tasks.

If You Are an Experimental Physicist

Control techniques such as feedback can dramatically increase the performance of your experimental apparatus, and most experimentalists, by necessity, pick up a certain amount of control theory informally. The results are often ok, but with a bit of savvy, they can usually be made much better. In this book, you will learn how to set the parameters and structure of a control system to maximize – by your own measure – the performance of your experiment. This is *optimal control.*

If you have learned a bit about proportional-integral-differential (PID) control and think that is enough, this book is for you: Although PID control is indeed a good generic solution – especially if you know how to use the D – knowing something about the dynamics of the system can lead to much better control. Generic solutions are fine for the family sedan, but a race car needs to be tuned.

An instructive example is the atomic force microscope (AFM). The original instruments were built by experimental physicists and were based on PI control (or slight variations). Starting about 2005, people in the industry realized that more advanced techniques could be valuable, and the current commercial AFM controllers now employ the full panoply of techniques discussed in this book, with thousandfold increases in scanning rates.[2]

[1] Like many sayings, this one has a complicated history. Its origin is attributed to Novalis (Baron Friedrich von Hardenberg, 1772–1801), poet, philosopher, and a founder of German Romanticism. In an unpublished fragment, he wrote, "To romanticize the world is to make us aware of the magic, mystery and wonder of the world; it is to educate the senses to see the ordinary as extraordinary, the familiar as strange, the mundane as sacred, the finite as infinite" (translated in Beiser, 1998). Subsequent reexpressions have boiled this down and added the converse, ending up as the pithier phrase I quote.

[2] Interestingly, the involvement of control specialists in the AFM industry came from the "top down." The National Science Foundation, a leading source of support for control theory in the United States, was concerned about an overly mathematical bent to control-theory research and began actively encouraging researchers to explore new applications, of which AFM was a prominent example.

If You Are a Theoretical Physicist

From theoretical physicists, a remark about control theory that I have heard is, "That's just engineering and linear systems" – said, if not with a sneer, then with an implication that there is nothing fundamental to interest a theoretical physicist. I hope to convince you that such sentiments give short shrift to one of the great ideas of the twentieth century, worth knowing on its own and for the deeper understanding of physical systems that it leads to.

What is different about control theory? In a word, *purpose*. Control always has a goal, and control theory shows how to achieve that goal. By contrast, physical theories describe nature, and historically physicists have gone to great lengths to remove the subjective from their theories. Thus, to study control theory as a physicist is a bit like licking the forbidden fruit. We know we are not supposed to do it, but it does seem interesting. I hope to assuage such feelings of guilt, as much in the natural world is designed, either explicitly by people, or implicitly through evolution and natural selection. And even if you do not share this way of looking at the world, learning about a great idea and its implications is worthy in itself, and the contrast can teach you much about "nonpurposeful" dynamics.

Control theory also deepens our appreciation of physical systems. There has been a steady expansion of physics and physical methods to the study of complex systems and networks. In many of these systems (such as the earth's climate, as discussed in Chapter 1), the notion of feedback is important, and this book will help give perspective on what those feedbacks can mean (Chapter 14). And, as we will discuss in Chapter 15, there are fundamental limits set on control from thermodynamics and causality (as well as information theory), and the effort to understand the interplay among these topics has generated much recent effort, including in my own research. Notions such as Maxwell's demon and Szilard's engine have evolved from thought experiments to real ones, and those achievements have then stimulated much ongoing theoretical work where control theory plays an important role.

If You Are a Control Theorist or Engineer

If you have a professional interest in control, you have many fine books to choose from, ranging from those focused on particular applications for the engineer to more theoretical books written for the control theorist and applied mathematician. What a physicist can offer is a concise introduction that takes a broader view and devotes more attention to the foundations of topics than is usual in textbook engineering discussions. I have tried to sort through a vast field, epitomized by the 3,500-page, three-volume series by Levine (2011b,c,a), and produce a broad, coherent view of what control can mean. Also, physicists focus on foundational issues, and this book includes topics rarely seen in introductory control-theory texts, such as the relations between stochastic and worst-case approaches to robust control (Chapter 9) and the limitations of control due to causality, information theory, and thermodynamics (Chapter 15).

Keeping this book to a reasonable size led to a rather ruthless competition, and much valuable material had to be cut. I apologize for all the pretty and important topics – Nyquist stability criterion! – that are not present.

Why This Book?

University libraries have shelves full of control-theory books. Why write another? Existing control-theory books are mostly written either for engineers or for control theoreticians and mathematicians. For a physicist, neither type of book is quite right. The mathematical approach is usually concise but not great at shaping intuitions. The engineering approach uses different notation (j for – i, Laplace transforms rather than Fourier transforms, factors of 2π, etc.) that can render familiar expressions mysterious. Moreover, the engineering style is quite different, with more emphasis on numerical, practical examples. Physicists prefer scaled variables and examples that are stylized to illustrate concepts. Here, I follow physics conventions by default but compromise to connect to control-theory texts.

This book has broader scope than most engineering texts. Because control theory is an essential part of an engineering degree, introductory texts are broad but low level, while advanced texts tend to be narrow. A similar problem exists with mathematics: Most physics departments have one or more courses in "mathematics for physicists" that compress half-a-dozen courses (tensors, boundary-value problems, partial differential equations, differential geometry, algebra, group theory, etc.) into a single course. You can view this book similarly (hence its title). Our topics are distilled from what is often covered in courses on classical, modern, digital, optimal, stochastic, robust, adaptive, and nonlinear control. Indeed, each chapter can be – and often is – the subject of a semester-long course. What physicist has time to take so many courses on control? Yet the answer is not the de facto current situation, where physicists get little or no exposure to control ideas at all. This book attempts a rational compromise.[3]

Finally, this book includes topics of recent interest, such as the stochastic approach to robust control, network control, quantum control, and the connections between control and fundamental issues of thermodynamics and information theory. Indeed, the period of writing this book has coincided with a renewed interest in the foundations of control theory.[4] Notions such as controllability, which seemed settled 50 years ago, have been reopened, with sometimes surprising results. And issues such as Maxwell's demon that had befuddled researchers since the dawn of statistical physics 150 years ago have been greatly clarified. While this book focuses on basic theory in control, it also aims to show where new areas fit with more traditional ones and to be a bridge to such recent work.

[3] Åström and Murray (2008) have written an excellent control-theory book for general scientists. They seek a wider audience, including those less comfortable with mathematics. Although full of insights and a broad range of examples, the style, level, and choice of topics are collectively rather different from this book.

[4] The interest can be measured by the recent spate of control-theory articles in *Nature*, *Physical Review Letters*, and *PNAS*. These journals had published little on these topics for many years, until recently.

A word on approach: a little intuition can be worth a lot of formalism, and intuition can be formed through detailed analysis of simple representative examples. That was Einstein's sentiment when he stated that "Everything should be as simple as possible, but not simpler." Typically, I solve the simplest nontrivial example in the text (analytically, if possible) and then elaborate in the problems. There are only a few applications, but there is a thorough treatment of familiar generic examples (e.g., first- and second-order systems, the pendulum). Your job is to transfer acquired intuitions to your own applications!

Figures and Style

The gray-scale figures in this book have been drawn using a minimalist style inspired by the work of Edward Tufte and others (Tufte, 1990, 2001; Frankel and DePace, 2012). Similarly, my writing tries to follow William Strunk's famous advice to "Omit needless words":

> Vigorous writing is concise. A sentence should contain no unnecessary words, a paragraph no unnecessary sentences, for the same reason that a drawing should have no unnecessary lines and a machine no unnecessary parts. This requires not that the writer make all his sentences short, or that he avoid all detail and treat his subjects only in outline, but that every word tell.
>
> (Strunk, W., Jr., 1918)

Although the figure design tends to minimalism, there are hundreds of them: control is a discipline that takes well to illustrations.

Problems

"The problems are an essential part of the text." Such platitudes are often found in textbook prefaces, but here it really is true. I try to demystify the problems as much as possible, while still leaving something for you to do. The goal is that you should be able to know whether you have done the problem correctly. Sometimes, I ask you to "Show that," or to add decimals to an approximate answer, or to reproduce a graph. The main text tries to be "lean but friendly," focusing on the main ideas. Hopefully, this creates a clear narrative, while the problems fill in steps and explore further cases and applications.

Solving problems is the difference between being a tourist in a new land and trying to live there. And who knows? Perhaps you may even decide to stay awhile.

Acknowledgments

I am grateful for the support of many friends and colleagues: Albert Libchaber, Jean-Christophe Géminard, Normand Fortier, and Ray Goldstein first encouraged me to take my interest in control theory seriously. Simon Capelin proposed a book and was incredibly patient; Nicholas Gibbons, along with Róisín Munnelly and Henry Cockburn, saw it to completion. Suckjoon Jun hosted a stay at the Harvard Bauer Center, where the first part of this book was written; Massimiliano Esposito invited me to Luxembourg to finish it off; and Stanislas Leibler invited me for brief but very productive stays at the Simons Center for Systems Biology at the Institute for Advanced Study in Princeton. Thanks to all! I also thank Ben Adcock, Karl Åström, Leslie Ballentine, Jules Bloomenthal, Braden Brinkman, Jean Carlson, Laurent Daudet, Jean-Charles Delvenne, Michel Devoret, Steve Dodge, Max Donelan, Guy Dumont, Alex Fields, Mark Freeman, Barbara Frisken, Harvey Gould, Paul Haljan, Shalev Itzkovitz, Philipp Janert, Malcolm Kennett, John Milton, Elisha Moses, Dugan O'Neill, Rik Pintelon, Max Puelma Touzel, Riccardo Rao, Henrik Sandberg, Pablo Sartori, David Sivak, Philipp Strasberg, Juzar Thingna, Hugo Touchette, Anand Yethiraj, Joel Zylberberg, and the students of Physics 493/881 for suggestions, critiques, and insights. Raphaël Chétrite read the entire manuscript and caught many mistakes. The ones left are my fault! In Canada, NSERC and Simon Fraser University supported my work throughout the writing. I thank Guy Immega for supplying the cover image.

Most of all, I thank Adrienne, qui reste au coeur du tout.

PART I

CORE MATERIAL

Historical Introduction 1

Any sufficiently advanced technology is indistinguishable from magic.

Arthur C. Clarke[1]

This is a book about magic. Not the magic of wizardry and sorcerers, nor the magic of fairy tales or fables, but *real* magic: the magic that powers planes and runs computers, that keeps our investments high and our blood pressure low, our beer cold and our bodies warm. It is the magic behind physics experiments of extraordinary precision, from gravity-wave detectors that probe the cosmos to scanning probe microscopes that image atoms. It is the magic that regulates biological processes from the pupil size in our eyes to the gene expression in our cells. It is the magic made possible by control theory.

The study of control theory can lead to something of a culture shock for physicists. Of course, jargon, technical methods, and applications may all be new. But something more fundamental is at play: As physicists, we study the world as it is. We look for the fundamental laws that govern time and energy, fields and forces, matter and motion, at the level of individual particles and collective phenomena. We do this in settings that range from the very large scales of the cosmos to the very small scales of fundamental particles to the very complex systems that rule the human scale. But we do all of this on Nature's terms, content to describe the actual dynamics of real "physical" systems.

Control theorists ask, instead, what might be. They seek to alter the states and dynamics of a system to make it *better*. The word "better" already implies a human element, or at least an active agent that can influence its environment. The Ancient Greeks coined the notion of *teleology* to denote the purpose or end (*telos*) of an object. While science has moved away from endowing objects in themselves with purpose, engineers design machines or systems to accomplish predefined tasks. Control theory tells, in a precise way, how to accomplish these tasks and indicates what is possible or not. *Uncertainty* – about initial conditions, external disturbances, dynamical rules, etc. – can limit possibilities.

Since all systems are physical ones, ruled by the laws of physics, physics will play a role in our story. But in many ways, it will have a supporting role, as we seek to create "augmented" systems that perform in ways that seemingly ignore the laws of physics. Of course they do not. Even so, we will see that a larger, open, physical "supersystem" can give a subsystem effective dynamics with new laws and properties.

[1] *Profiles of the Future: An Inquiry Into the Limits of the Possible*, New York, Harper and Row, Rev. ed., 1973.

In this book, we will take a broad look at control, from both the fundamental point of view that seeks to understand what it can accomplish and what not, and how control in general meshes with other topics in physics such as thermodynamics and statistical physics. At the same time, we will also be interested in control for its practical applications. Just as control is fundamental to the technological devices of modern life, so too does it play a key role in the techniques an experimental physicist should know.

Sometimes called the "hidden technology," control is often invisible, despite its omnipresence in modern technology. We do not notice it until something fails. Planes are very safe, but occasionally they fall from the sky. Our bodies also depend on many control loops. To name one: to survive, we must maintain a core temperature within 27–44 °C, implying the need to keep maximum deviations to $< \pm 3\%$ and to regulate typical fluctuations to be $< \pm 0.3\%$. Again, we pay little attention to our body's temperature – except when it begins to deviate when we get sick or cold or hot. Our ability to ignore control under normal circumstances is a testament to its *robustness* to specific types of situations; our need to confront the often-drastic consequences of its failure is a consequence of its *fragility* to unforeseen circumstances. As we will see, the two aspects are linked.

In this introductory chapter, we present briefly the historical development of control and its theory, which gives some insight as to what "better" dynamics might mean. We then list some of these *goals* for control. Then we introduce, in an intuitive way, some of the principal *methods* of control, notably *feedback* and *feedforward*. We conclude with a discussion of the *types* of control systems.

1.1 Historical Overview

We can divide the development of control techniques and theory into five periods:

- Early control (before 1900)
- Preclassical period (1900–1940)
- Classical period (1930–1960)
- Modern control (1945–2000)
- Contemporary control (after 2000)

The overlaps are deliberate, as actual developments are not as well ordered chronologically as the classifications would imply. Although it seems logical to "begin at the beginning," this summary may be easier to follow after you have learned some of the material from later chapters. Partly for this reason, the discussion is relatively brief, with some aspects deferred to the relevant later chapters. Of course, a short exposition inevitably simplifies a complex story. The notes and references give pointers to more extensive presentations.

1.1.1 Early Control (before 1900)

The word *feedback* is of relatively recent origin, with the Oxford English Dictionary reporting its first use in 1920, in connection with an electrical circuit.[2] However, uses of feedback and the broader notion of control are far more ancient. Ktesibios (285–222 BC), a Greek working in Alexandria, Egypt, used feedback to improve the stability of water clocks, vessels that measure time by the outflow of water. However, as the fluid level in a vessel decreases, so too will its outflow rate. Keeping the level constant, or *regulating* it, stabilizes the rate of outflow. There are no original records of the device, but reconstructions based on Vitruvius's *De architectura* (~ 30–15 BC) and later Arab water clocks indicate that the mechanism was the same as that used in the modern flush toilet: a ball floating in the tank follows the water level. When the level is low, a float lets in more water, raising the level and increasing q_{out}; when high, the float shuts off the valve, decreasing q_{out} (see right).

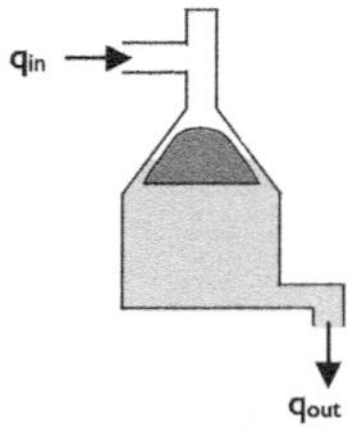

In the Middle Ages, *mechanical clocks* powered by falling weights or springs were developed, with various ratchets ("escapements") that translate oscillating into rotational motion. These clocks also have feedback mechanisms to ensure constant rotation rates.

Because fluid density depends on temperature, the level of a fluid can be used to regulate temperature. René-Antoine Ferchault de Réamur (1683–1757) invented such a device, based on the temperature sensor of Cornelius Drebbel (1572–1663), a Dutch engineer working in England. In France, Jean-Simon Bonnemain (1743–1830) patented in 1783 an improved temperature controller based on a bimetallic rod that flexed when the temperature changed. He used it to make practical hot-water central heating for buildings.

The beginning of the Industrial Revolution, centered on England in the second half of the eighteenth century, led to the first important applications of feedback. The most prominent was the *governor*, which was developed to keep windmills turning at a constant rate and then adapted to the steam engine for more general purposes by James Watt in the late 1780s.[3] The issue was that variable loads would alter the rotation rate of the engine. To keep it constant, Watt and his partner Matthew Boulton adapted a *flyball* sensor for rotation rates that had been patented by Thomas Mead in 1787. As illustrated in Figure 1.1, the sensor has two heavy balls that rotate with the engine shaft. If the engine rotates too quickly, centrifugal force pushes the balls out, pulling down a lever and shutting off the throttle valve that lets steam in, thus slowing the motor. If it rotates too slowly, the balls fall in, pushing up the lever, opening the value, letting more steam in, and speeding up the motor. If all goes well, the steam-engine rotation rate settles at a desired value.

The nineteenth century saw a steady improvement in the technology of governors. The 1868 paper *On governors* by James Clerk Maxwell gave the first theoretical analysis. A flaw of governors was their tendency to make the engine "hunt" for the right

[2] The related term *feedforward* was first used even more recently, in 1952 (also according to the OED).
[3] By the 1670s, Christiaan Huygens had invented a governor to regulate pendulum clocks (Bateman, 1945).

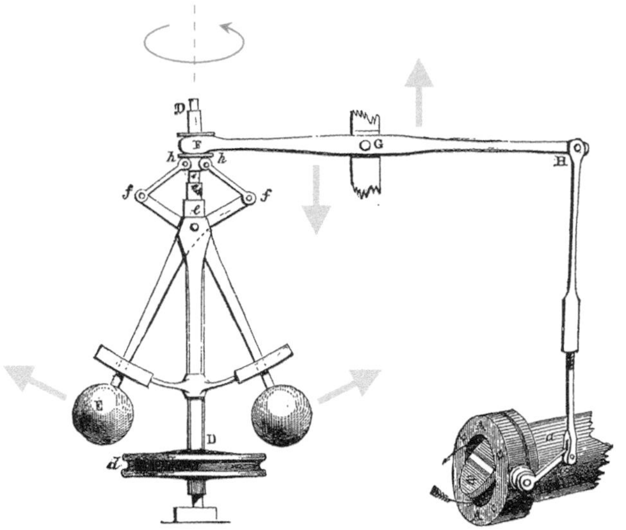

Fig. 1.1 Flyball governor and throttle valve, with rotation around the indicated axis. Flyballs move out, lever at F is pulled down, pivots about G, and pushes up at H, closing the throttle valve at lower right. Adapted from Routledge (1900).

rotation speed. In more modern language, there could be long-lived oscillatory transients before settling to a steady state. Even worse, the engine could become unstable and show erratic motion. Maxwell analyzed the conditions for stability of regulation against small perturbations using *linear stability analysis*. His stability conditions were generalized by Edward J. Routh and Adolf Hurwitz later in the nineteenth century. Although these early analyses of control systems eventually became part of the techniques of control theory in the mid twentieth century, they had little immediate impact on practical realizations, which was driven by the innovations of "tinkerers." Another emerging class of control applications concerned the position of a moving object. Thus, ships needed steering and missiles guiding to their target. In England, J. McFarlane Gray patented in 1866 a steering engine using feedback. In France, Jean Joseph Farcot introduced a range of position-control devices that he called *servomotors*. More generally, servomechanisms were used to *track* desired time-dependent trajectories, a generalization of the simpler goal of *regulation*, where the desired trajectory is simply a constant.

1.1.2 Preclassical Period (1900–1940)

Pre-1900 regulators were all *direct acting*: the elements that measured the quantity being regulated also had to change the system. The lever in a fluid-level regulator that moves in response to a change in level also opens the value that lets in more water. Of course, there is a "power source" (a high-pressure supply of water) that makes the response possible, but one "gadget" must still carry out two actions. Around 1900 a long process of abstraction began that led to distinct notions of *sensors*, *controllers*, and *actuators*. The sensor *measures* a quantity of concern, the controller *decides* how to respond, and the actuator *executes* the response. Each element can have its own, independent source of power. Such ideas, however, took several decades to become clear.

Meanwhile, the first decades of the twentieth century saw the beginnings of industrial process control. Applications included boiler control for steam generation, electric motor speed regulation, steering for ships and airplanes, temperature and pressure control, and more. A key development was of stand-alone controllers that could be added on to existing equipment. For example, around 1910, Elmer Sperry greatly improved the gyrocompass and designed a gyroscope autopilot to steer ships. The Sperry Gyroscope Company supplied the US Navy with navigational aids, as well as bomb sights and fire-control systems.

In 1922, Nicholas Minorsky gave a detailed analysis of such mechanisms, introducing the notion of three types of control. The first is *proportional* to the error between *set point* and actual signal, the second to the *integral* of that error, and the third to its *derivative*. Together, they form the three-term regulator, or proportional-integral-derivative (PID) control, which is discussed in Chapter 3. Although these ideas now seem very general, they were at first encountered separately in each domain of application. Thus, Minorsky's analyses were little known in the broader technical community for a number of years.

Another important development was the first airplane flight by Orville and Wilbur Wright. Others had built (and sometimes died testing) unsuccessful flying machines. The Wright brothers' success was based on their mastery of control, using flaps to alter yaw, pitch, and roll (three axes). Moreover, they recognized the advantages of an inherently *unstable* design stabilized by control (e.g., a human pilot). Unstable systems are more maneuverable than stable ones. They need active feedback to produce stable motion but can respond to disturbances (gusts of wind, abrupt change in terrain, etc.) much more quickly. The concept should be familiar: when we stand upright, we are unstable and must use (unconscious) small muscle movements to prevent ourselves from falling over. Indeed, the ability to walk on two legs is what distinguished the first hominids from other apes.[4]

Along with developments in mechanical control systems came parallel ones in electrical circuits. By the end of the nineteenth century, there was already a division between the power and signal applications of electricity. In both, the *amplifier* was a key element, allowing separation of the functions of sensor and actuator. Early high-power amplifiers took the form of relays and spring-based solenoids, which became the basis of many kinds of actuators.

For low-power electrical signals and their circuits, a key development was Lee de Forest's 1906 *grid audion*, a vacuum-tube amplifier that could boost the voltage level of a weak signal, compensating for signal losses in transmission and making possible

[4] What were the evolutionary advantages of walking upright? Darwin thought that it improved our ability to fight. But walking is also more efficient for traveling large distances on the ground (e.g., over grasslands). As with many evolutionary developments, the "why" is elusive (Wayman, 2012).

long-distance telephone networks. But the amplifiers had serious flaws: the signal gain was both nonlinear and prone to drifts, which led to distortion and volume variations.

Finally, there was a transformation in our view of living beings. Life in the nineteenth century was fixed on Newtonian, mechanical motion. Things alive *moved*, powered perhaps by the electric spark that jolted Frankenstein's monster to life or by some other unknown vital force. In the 1920s, the physiologist Walter Cannon introduced the term *homeostasis*, the ability to maintain conditions in the face of external perturbations. These conditions include the core temperature of the body and the concentrations of glucose, iron, oxygen, calcium, sodium, potassium, and other chemicals or ions. All these quantities are closely regulated, even when external conditions change dramatically: through hot or cold, our core temperatures are close to 37°C, our sodium levels stay between 135 and 145 milliequivalents per liter, and so on. The ability to regulate so many quantities in the body using multiple, hierarchical systems is one of the defining features of the modern view of life. Conversely, death is associated with a *failure cascade* that shuts down the essential functions of the body with its nested control loops, often one after the other. Understanding homeostasis was a goal of Wiener's influential book *Cybernetics*, a founding text of control theory, discussed below.

1.1.3 Classical Period (1930–1960)

At Bell Telephone Laboratories, a group of engineers was set up to address quality problems in the growing telephone network. Initial progress was slow, but on Tuesday morning, August 2, 1927, Harold Black had an epiphany while riding the Lackawanna Ferry across the Hudson to Manhattan to get to work. His idea, sketched out on a blank page of the *New York Times*, was that by taking a portion of the amplifier output signal and subtracting it from the input, one could reduce distortion, at the cost of a reduced gain. Thus was born the *negative-feedback amplifier*, which had the immediate effect of improving long-distance telephone calls and was a key development in the history of control. Its descendant, the *operational amplifier*, is described in Chapter 3. More broadly, efforts to understand what Black had created led to a "classical" formulation of control theory.

In the 1930s, Black's colleagues at Bell Labs, Harold Nyquist and Hendrik Bode, contributed theoretical analyses that put negative feedback and other ideas of *classical control* on a firmer footing. Work by Harold Hazen and Gordon Brown at MIT also was influential. In contrast to earlier studies based on solving ordinary differential equations in the time domain, they used frequency-domain methods based on the Laplace transform to derive a set of heuristic rules (often expressed in graphical form) for controllers of reasonable performance that work well for a relatively large class of systems. Bode's 1945 book *Network Analysis and Feedback Amplifier Design*, delayed because of the war, is perhaps the apotheosis of classical control. It considered *robustness* in depth, pointing out the fundamental compromises inherent in control: feedback that suppresses the response to disturbances at some frequencies will inevitably boost that response at other frequencies.

The next great impetus to the fledgling field of control engineering (and its counterpart, control theory) came with World War II. Engineers worked on a variety of control problems, notably the aiming of antiaircraft guns and automatic radar tracking. The *Radiation Lab* at MIT was a particularly important center for such research. To the scientists and engineers working at such centers, the war made particularly clear the need for unified, abstracted treatments of control based on concepts that were independent of specific applications. The classified results released en masse at the end of the war spurred rapid progress afterwards.

1.1.4 Modern Control (1945–2000)

After the end of World War II, control emerged as a distinct technical discipline. Engineering societies such as the American Society for Mechanical Engineers (AMSE), the Instrument Society of America (ISA), and the Institute of Radio Engineers (IRE, later the IEEE) all launched subgroups, and new professional societies such as the American Automatic Control Council (AACC) and International Federation of Automatic Control (IFAC) were created. Where MIT had stood almost alone as an academic center, many universities around the world added groups focusing on automatic control. The military-industrial complex took shape: think tanks such as the RAND corporation in Santa Monica, California and the Research Institute for Advanced Study in Baltimore and companies blurred military and industrial roles on scales larger than had been known before the War.[5] Prominent companies included IBM, General Electric, Hughes Aircraft, Bell Labs, Honeywell, Westinghouse, Leeds and Northrup in the United States, and Siemens (Germany), Schneider (France), ASEA (Sweden), and Yokogawa and Mitsubishi (Japan). Regular national and international conferences began: The first IFAC World Conference, in 1960 in Moscow at the height of the Cold War, marked the emergence of *modern control*.[6]

Modern control introduced state-space methods that marked a return to analysis in the time domain, in contrast to the frequency-domain methods characterizing classical control. The latter is fine for time-invariant, linear systems but cannot describe easily time-varying, nonlinear dynamics, which is omnipresent in applications. Although "modern," the state-space approach reaches back to the late nineteenth and early twentieth centuries. and includes figures such as Aleksandr Lyapunov in Russia and Henri Poincaré in France. A key insight was that knowing the system dynamics could improve performance spectacularly relative to the classical methods, which were developed assuming much less about the system under control. The resulting *optimal control* gave a systematic way to generate "the best" controller for a given task. With key contributions from Richard Bellman and Rudolph Kalman in the US and Lev Pontryagin in the Soviet Union, optimal control had spectacularly successful applications in the space program, particularly the Apollo moon-landing project.

[5] The RIAS was absorbed into the Martin Marietta Corporation, which survives as Lockheed Martin.

[6] Obviously, this use of "modern" is dated, as is "modern physics" (relativity and quantum mechanics), "modern art" (Impressionism, Dada, etc.), and "modern architecture" (Bauhaus, International Style, etc.).

The *digital computer* had a long gestation that was greatly advanced by war efforts – e.g., to formulate tables to aid in fire control. The history of computers is a separate story; in control, there was a gradual shift from *analog controllers* to *digital controllers*. The former had been implemented by external electrical, hydraulic, or mechanical circuits. Then came a long evolution to digital mainframes, minicomputers, microcomputers, laptops, and microcontrollers. In parallel came a shift from analog control methods for continuous-time dynamical systems to digital control methods for discrete dynamical systems.

At MIT in World War II, Norbert Wiener, introduced the stochastic analysis of control problems at roughly the same time and independently of efforts in the Soviet Union led by Andrei Kolmogorov. Wiener's primary technical publication, *The Extrapolation, Interpolation, and Smoothing of Time Series with Engineering Applications*, was circulated as a classified report in 1942 and eventually published in 1949.[7] His famous 1948 contribution, *Cybernetics: or Control and Communication in the Animal and the Machine*, showed that control theory applied not only to engineering systems but also to human, biological, and social systems. The book was inspired by the notion of homeostasis in organisms and by similar issues in controlling complex systems. Coined by Wiener from the Greek word for "governance," the word "cybernetics" was a tribute to Maxwell's 1868 governor paper.

The successes of optimal- and stochastic-control methods during the 1960s soon led to overconfidence, as it was forgotten how much the optimized performance depends on knowing system dynamics. In the 1970s and 1980s, control theory underwent something of an identity crisis. While state-space methods work well in the aerospace industry, where dynamics can be known accurately, they do poorly in industrial settings where the dynamics are more complex and harder to characterize (paper mills, chemical plants, etc.). This disenchantment led many practical engineers (and physicists) to avoid advanced techniques in favor of the tried-and-true PID controller. Indeed, academic research on control theory from 1960 through at least the 1970s had "negligible" impact on industrial applications. In response came a new subfield, *robust control*, to optimize the performance of systems whose underlying dynamics had at least moderate uncertainty. Its goal was to merge the robustness of classical control methods with the performance of modern control. In parallel came the subfields of *system identification* and *adaptive control*, where the goal was to *learn* better the system dynamics, either through independent or online measurements. Common applications of adaptive algorithms include noise-cancelling headsets, automobile cruise control, and thermostats.

1.1.5 Contemporary Control (after 2000)

Beginning around 2000, control theorists began to tackle increasingly complex systems. One notable example is the attempt to understand biological systems from

[7] Its nickname "Yellow Peril" came from the color of its cover, difficulty of its contents, and racism of its times.

an engineering perspective emphasizing control especially. The resulting field of *systems biology* contrasts, occasionally sharply, with a parallel effort in physics known as *biological physics*. Perhaps the most striking conclusion concerning control is strong empirical evidence that organisms have *internal models* of the world that allow them to anticipate and plan ahead. Recognizing the role of such planning and anticipation has been key to understanding how humans move and act in the world. See Example 3.4 for an application in connection with human balance – how we manage to stand upright without falling down. Later, in Chapter 10 we introduce *reinforcement learning*, a technique for learning how to plan and anticipate from repeated supervised trials.

Another example is the development of *autonomous*, self-driving vehicles. Indeed, the development of the automobile recapitulates the entire history of control. The early twentieth-century automobile was a mechanical device, like the governor and steam engine. In the 1970s, a variety of electrical control systems appeared, many based on microcontrollers that implemented feedback loops. By 2007, the typical automobile had 20–80 microprocessors, dealing with powertrain control to reduce emissions (e.g., by controlling the air-fuel ratio), performance optimization (e.g., variable cam timing), and driver assistance (e.g., cruise control and antilock brakes), and more. And, while the driver – the "human in the loop" – remains the ultimate controller for an automobile, many responsibilities are off-loaded (e.g., GPS and its associated navigational aids). At present, many companies seek to eliminate the driver from these control loops, a goal that must integrate many subproblems and use techniques from fields such as machine learning, big data, and wireless communications. Finally, control is expanding beyond the scale of single vehicles. Highways and smart phones already give real-time information on traffic for more efficient routing. In the future, platoons of trucks may travel in closely spaced groups that reduce traffic and increase fuel efficiency by controlling the collective air flow around the group (*drafting*).

Another application is to *climate science* and models of climate change, where it is crucial to understand feedbacks on both fast and slow timescales. On the one hand, water vapor is an effective greenhouse gas that is a "fast feedback" because the amount of water vapor in the air adjusts within days to changes in temperature. On the other hand, the area of land covered by glaciers and ice sheets adjusts much more slowly. (Glaciers melt, exposing darker surfaces, which absorb more sunlight.) Such *positive feedback* can lead to instability that will drive a dynamical system to another attractor (a new steady state or, sometimes, an oscillatory one). Negative feedbacks occur in climate models, too. As warmer temperatures lead to greater cloud cover, more light will be reflected away by the clouds, lessening absorption. Unfortunately, positive feedbacks seem likely to dominate.

The desire to understand complex systems has led to a discipline of *network science*. In the context of control, the goal is to understand collective network dynamics rather than individual dynamical systems. One focus has been to understand how the structure of a network affects one's ability to control it. Applications are widespread, as networks are everywhere, from the world-wide web to the proteins that control cellular processes, to control of the brain and neurological disorders, to ecological species interactions, to sets of friends and other social networks. See Chapter 14 for background on the principles.

In parallel with distributed systems came distributed control. The world is moving rapidly to an *Internet of things*, where everyday objects are networked together but also have autonomous controllers. The cities of today and homes of tomorrow are or will be filled with intelligent devices that communicate with and among each other.

These developments have been accompanied by an increasing awareness of the role of *information* in control. Information theory is an older topic, born in the work of Claude Shannon at Bell Labs, whose 1949 book *The Mathematical Theory of Communication* had an enormous impact, spread over many fields. Curiously, it did not have much initial impact on control theory. But by the 2000s, it was clear that many of the laws due to Bode, Nyquist, and others that limit what is possible in control have connections to information theory. Their complicated frequency-domain expressions turn out to have simpler time-domain versions when expressed in information-theoretic terms. See Chapter 15.

In parallel, there has been a renewed interest from physicists in control. In addition to his paper on governors, Maxwell actually made a second, nearly simultaneous contribution to control in 1867.[8] Asked to review a draft of an early treatise on thermodynamics by Peter Guthrie Tait, Maxwell responded with a letter outlining the thought experiment now known as *Maxwell's demon*. In Chapter 15, we discuss how this thought experiment led eventually to an understanding of some of the fundamental limits of control and to deep connections among thermodynamics, information theory, and feedback.

Finally, there has been great theoretical interest and practical success in controlling quantum systems. We give a brief introduction in Chapter 13.

1.2 Lessons from History

History can be very interesting on its own and can illuminate the sociology of a profession – where engineers are "coming from." But we can also learn lessons from history. From control applications, we can extract the types of goals or objectives that control can have. We can also attempt to define terms such as feedback and feedback loop that can be slippery when tested against the full range of situations where the terms are used. Finally, we can classify control – open and closed loop, autonomous and nonautonomous, simple and complex controller – to help make sense of the various historical examples.

[8] Maxwell's two contributions to control theory are considered to be among his "minor work," but only compared to those on kinetic theory and, especially, on the unified description of electricity and magnetism.

1.2.1 Goals of Control

From the historical development, we can abstract several goals for control systems. To best express these goals, we define an internal state **x**, a real-valued vector that evolves continuously in time t and obeys an ordinary differential equation, $\dot{x}(t) = \boldsymbol{f}(\boldsymbol{x})$, a case that we shall study a lot in this book.[9] The internal state might correspond to a point in phase space, as in classical mechanics, but the notion is more general in that any dynamical rule will do, including ones with phenomenological damping terms, and the like. We will discuss this notion of internal state in more detail beginning in Chapter 2. Here, we just list some *goals for control* and illustrate them in sketches at right:

1. *Regulation*. The internal state should be kept constant (e.g., $\mathbf{x}(t) = \mathbf{x}^*$). This was historically perhaps the first goal of control. The idea is to fix a system's state in the face of perturbations of various sorts. Thus, at home, we regulate temperature against cold and hot swings of the environment. Our body does much the same for our internal body temperature. And temperature control is often an important requirement in an experiment. Of course, many other quantities such as pressure, position, velocity, concentration – any kind of internal state or quantity related to such a state – can be regulated.

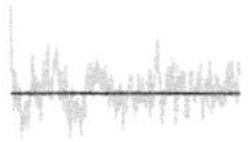

2. *Tracking*. The system state vector $\mathbf{x}(t)$ should approximate a desired time-dependent trajectory $\mathbf{x}^*(t)$ as closely as possible. That is, we generalize regulation about a constant to regulation about a curve in state space.

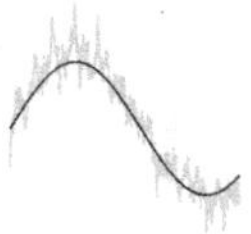

3. *Changing the attractor*. Rather than trying to generate a particular trajectory, we can change the *type* of motion that a dynamical system can generate. For example, a system that would normally oscillate might be altered to have a steady state. More complicated motion such as *chaos* can also be suppressed. Conversely, you might want to induce any of these types of motion.

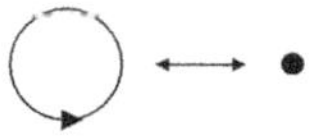

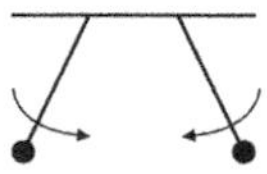

4. *Collective dynamics*. The goal is to induce multiple dynamical systems to work together to accomplish some task. An example is *synchronization*, where two or more systems evolve identically in time, even in the absence of "target" system dynamics. Temporal behavior can also be correlated according to more complex rules, as found in *musical ensembles* and *dancing*. By contrast, *swarming*, illustrated at right, refers to correlated spatial motion.

While not exhaustive, the above list gives a sense of the broad range of possibilities. Mostly, we will focus on regulation and tracking, as they involve most of the basic ideas and issues in control theory. In Chapter 11, we discuss some of the other types of goals.

1.2.2 What is a Feedback Loop?

Examining the historical development shows an evolution in the different types of control and controllers. Figure 1.2 shows four examples. In (a), we show the governor

[9] We shall also study many other types of dynamical systems, such as extended systems, discrete-time evolution, and discrete states.

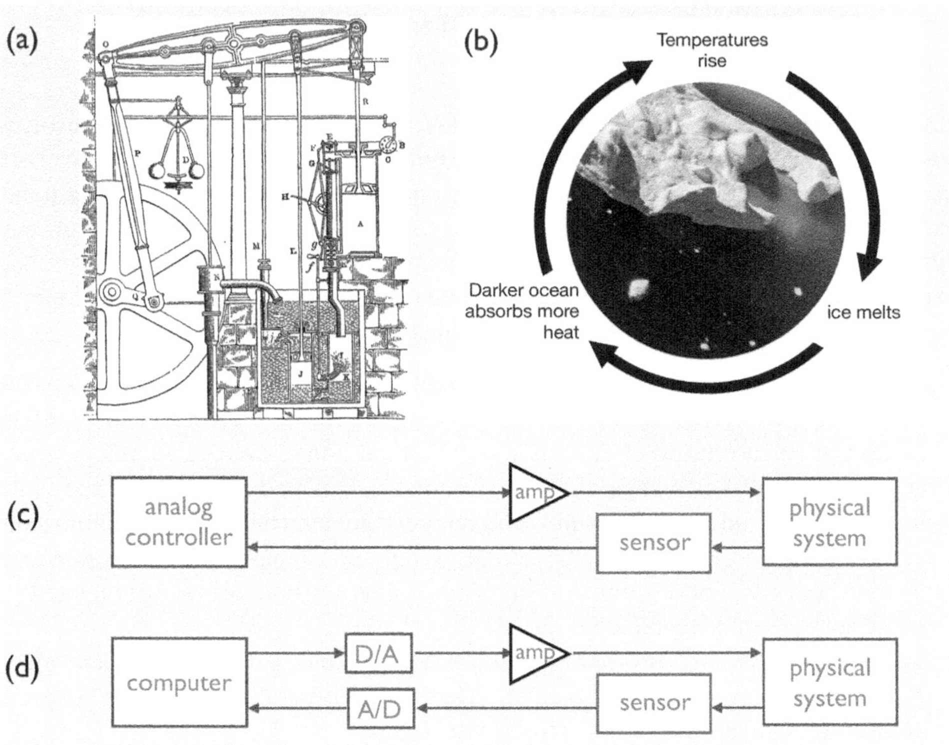

Fig. 1.2 Four examples of feedback systems. (a) Watt's steam engine with governor (labeled D). Adapted from Routledge (1900). (b) Positive feedback loop can reinforce climate change. Author's adaptation of photograph supplied by Monica Bertolazzi/Getty Images. (c) Analog feedback controller with accompanying sensor and power amplifier. (d) Digital feedback controller with digital-analog and analog-digital converters.

depicted previously in Figure 1.1 integrated into the full steam engine designed by Watt. The governor is indicated by "D" in the drawing. If you had not already seen a governor (e.g., in Figure 1.1), you might have difficulty identifying it amid all the clutter of parts. The whole "contraption" is one physical system consisting of many wheels, rods, and the like joined together in complicated ways, and nothing in particular distinguishes the pieces that form part of the engine from those that form part of the controller. Both are obviously physical subsystems.

In (b), we set history aside to consider a "natural system," where it is difficult to distinguish the controller from the system. The system is the earth, or at least its climate system, and control (in this case, an undesirable destabilizing control) is given by sea ice and ocean.

In (c), we show the kind of analog controller that evolved in the first half of the twentieth century. It consists of the physical system of interest, plus some add-ons: controller, sensor, power amplifier, and power supply (not shown). In contrast to the cases of (a) and (b), it is easy to identify all elements – they are connected by wires – and each is obviously a physical (sub)system on its own. The add-ons are analog circuits. The system of interest can range widely (e.g., a block of metal to keep at constant temperature). In Chapter 3, we will see how to make a controller circuit using operational amplifiers.

In (d), we have transitioned from analog to digital control. The controller is now implemented as a program on a computer (or microcontroller or other similar digital device). The computer communicates in 1s and 0s to the analog world via converters that take a number and output a corresponding voltage, and vice versa. After conversion to analog signals, the other elements are the same as in (c). See Chapter 5.

One of the most important concepts expressed in Figure 1.2 is the notion of a *feedback loop*, where the output is "fed back" to the input. The system thus affects the controller (via a measurement) and, conversely, the controller affects the system (by sending out a response). In the sketch at right, we show the structure of this basic, stripped-down notion of a feedback loop. Its two distinct parts, the system of interest $\boldsymbol{x}$ and controller $\boldsymbol{y}$, typically obey coupled dynamical equations, such as

Feedback loop

$$\dot{\boldsymbol{x}} = \boldsymbol{f}(\boldsymbol{x}, \boldsymbol{y}) \,, \qquad \dot{\boldsymbol{y}} = \boldsymbol{g}(\boldsymbol{x}, \boldsymbol{y}) \,. \tag{1.1}$$

The sketch illustrates two essential features of a feedback loop: *separability* and *causality*.

Separability implies that the dynamics can be decoupled into two coupled (vector) equations (here into $\boldsymbol{x}$ and $\boldsymbol{y}$) in a way that makes sense physically. That is, $\boldsymbol{x}$ and $\boldsymbol{y}$ should each describe physically distinct systems with a meaningful identity when considered in isolation. Mathematically, in the equation $\dot{\boldsymbol{x}} = \boldsymbol{f}(\boldsymbol{x}, \boldsymbol{y})$, we could freeze $\boldsymbol{y}$ at some constant value (e.g., $\boldsymbol{0}$) and then look at the "pure" $\boldsymbol{x}$ dynamics. Or we could do the same with the $\boldsymbol{y}$ dynamics. But physically, the separation should make sense. For example, the N identical molecules in a gas are typically described as a single N-particle system of interacting particles, whereas two large particles in a gas of N molecules are usefully described separately, and one can imagine the motion of one "controlling" the motion of the other.

Causality is linked with notions of time: The controller *first* measures and *then* responds. Or, $\boldsymbol{x}$ affects $\boldsymbol{y}$, which then affects $\boldsymbol{x}$. In Eq. (1.1), imagine that the coupled system settles to a fixed point $\{\boldsymbol{x}^*, \boldsymbol{y}^*\}$. If the response to fast variations goes to zero as the frequency characterizing the variation increases, then some kind of notion of causality is present. In Chapter 15, we will develop the formal links between causality and response functions, beginning with the classic Kramers–Kronig relations.

Not all coupled systems form feedback loops. Imagine a simpler set of equations, $\boldsymbol{f}(\boldsymbol{x}, \boldsymbol{y}) = \boldsymbol{0}$, $\boldsymbol{g}(\boldsymbol{x}, \boldsymbol{y}) = \boldsymbol{0}$, which also have solution $\{\boldsymbol{x}^*, \boldsymbol{y}^*\}$. But now if we perturb one of these equations, the equilibrium solution will be modified instantaneously, because we are just solving an algebraic set of equations. Even though it is still true that $\boldsymbol{x}$ affects $\boldsymbol{y}$ and vice versa, we would probably not call this a case of feedback – it would just be coupling.

We might try to formalize this notion by comparing the interaction time scales, the time it takes $\boldsymbol{x}$ to affect $\boldsymbol{y}$ or vice versa, with time scales within each system. Typically, the interaction times will be faster than most time scales in the system. If not, as in the "instantaneous" case from the previous paragraph, we would hesitate to call the interaction a feedback loop. I am sure that one can find violations of any specific requirement, making these criteria heuristic, not rigorous. But I think they can be useful in understanding the different types of systems that different communities call "feedback loops."

"Feedback loop" is thus a theoretical construct imposed on a system. In a modern version of a controlled system where each element (controller, amplifier, sensor, system, etc.) is housed in its own box and connected to other elements by wires, it is easy to see the loop. The governor and similar devices lack the wires, but we can still easily distinguish system from controller. In the climate-change example, still more imagination is required. Indeed, in a number of fields such as biophysics, identifying feedback loops is an important research topic. The cell is a complicated pile of proteins, but the proteins function as machines – of a different sort, since they live in a strongly fluctuating environment, but machines nonetheless. They are every bit as complex as the mechanical factories of the machine age, and there has been a long effort to identify functions such as gene regulation, where feedback plays a key role.

Finally, one aspect that is implicit in the notion of feedback loops but not readily apparent visually, is the notion of correcting against *uncertainty* of various types. We will see that the major – some would argue the only – reason to use feedback is to compensate for unknown disturbances and system dynamics. If there is no uncertainty, then other types of control such as feedforward (see below) will be more appropriate.

1.2.3 Progression of Control Types

In Figure 1.2, we see a progression in abstraction of control systems, from "contraptions" – machines that are constructed to improve a particular performance characteristic such as engine rotation rate – to an ensemble of physical subsystems to digital systems that seem quite divorced from the original physical phenomena, even though computers are, of course, physical systems in themselves. This increasing abstraction has been useful in understanding key issues: the need for feedback loops, the role of time delays and noise, and so on. It is at the cost of an analysis that strays far from any given particular system, and one must be careful in applications that too much has not been lost. For example, we will see that all controller design implies a model of the system, whether explicitly expressed or not. The quality of control will then be linked to the quality of the model.

How does feedback in natural systems fit into the above picture? For example, we have seen that in climate science, positive and negative feedback is an important part of understanding the earth's climate and its response to man-made changes. On the one hand, you might take the view that, since these are entirely natural systems that have not been engineered, feedback is not an appropriate concept. On the other hand, the terms are in common use and well accepted. They are best applied to situations where one can distinguish subsystems that operate independently, at least to some extent. In the example shown in Figure 1.2b, the different parts (air temperature, sea ice, and so on) each have distinct identities and can be measured independently. Thus, it acts very similarly to engineered systems such as the steam-engine governor, and using the term "feedback" seems reasonable. But even in climate science, identifying all such relevant feedbacks is far from clear, never mind quantifying their potential impacts.

1.2.4 Classifying Control and Controllers

We introduce a few distinctions that will be useful in organizing our understanding of control. In particular, we distinguish between open-loop and closed-loop designs, between feedforward and feedback control algorithms, and between autonomous and non-autonomous systems. We will not try for rigorous definitions because these terms are used inconsistently in different communities. Moreover, without learning more about the subject, it is difficult to appreciate fine points. Nonetheless, there are general ideas and useful distinctions that are worth making.

Open and Closed Loops

We have already met the concept of feedback and the feedback loop in Section 1.2.2. We view feedback as a technique for *closed-loop* control, which is a design where the control is based on the system's current (and past) states. So pervasive is closed-loop control that its alternative is known as *open-loop* control, which is a strange name when you think about it. The distinction is shown graphically in Figure 1.3. Open-loop designs use *feedforward*, where an input is "fed forward" to the system, and the control acts *independently* of the system state.[10] Although feedforward might seem just a limited version of feedback, its reliance on anticipated events can speed response relative to feedback, which reacts to perturbations after the fact. Feedforward also will not destabilize a system, a pitfall of feedback. We will see in Chapter 3 that the best control designs often incorporate elements of both feedforward and feedback.

The distinction between open- and closed-loop control makes sense – and indeed resulted from, human-engineered systems, where subsystems and wiring connections are often known. It is less clearly useful when applied to the feedbacks of natural systems such as the climate example given above.

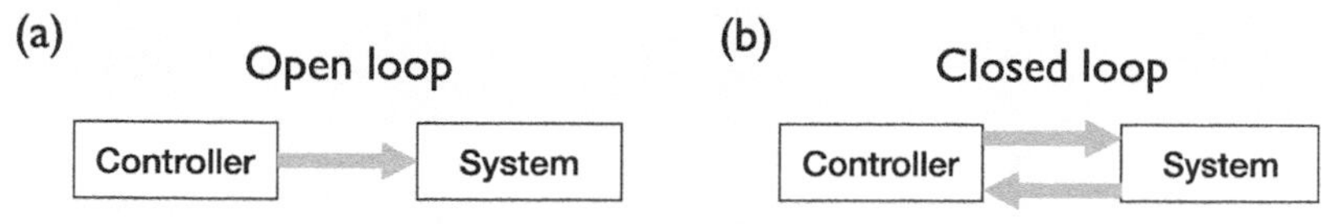

Types of control. (a) open loop; (b) closed loop.

Fig. 1.3

[10] Some people reserve the term "feedforward" for the special case where measurements of incoming disturbances allow a controller to act before the disturbance arrives. We will refer to that case as *disturbance feedforward* (Section 3.4.3) and here use "feedforward" more broadly.

Negative and Positive Feedback

Another basic distinction in control theory is between *negative* and *positive feedback*. To develop our intuition, consider a system with a static, *open-loop response*, where the output y is a static function of the input u. Since the response of a physical system is limited in range, let

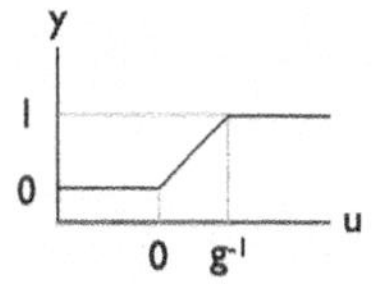

$$y = \mathrm{sat}_g(u)\,. \tag{1.2}$$

where the *saturation* function is defined at left: the output y goes from 0 to 1 as the input u goes from 0 to g^{-1}. The constant g is the linear *gain* of the response function.

Using the open-loop response as a reference (b), we can explore the impact of negative (a) and positive (c) feedbacks in Figure 1.4. In (a), we show a *block diagram* illustrating the signal flow for negative feedback, with $k < 0$. The output y is fed back and subtracted from the input, using a *feedback gain*, k. There is now a *closed-loop response*

$$y = \mathrm{sat}_g(u - |k|y) \equiv \mathrm{sat}_{g'}(u)\,, \tag{1.3}$$

where the new saturation function has gain

$$g' = \frac{g}{1 + |k|g}\,. \tag{1.4}$$

The linear-response range of the input has increased from g^{-1} to $g^{-1} + |k|$, as illustrated by the dotted lines in the response curve at lower left in Figure 1.4a. Notice that for large gain ($|k| \gg g^{-1}$), the linear range is approximately independent of the linear range of the original system. If the goal is to regulate the output y, we see that we can keep y in a given range over a wider range of inputs, which might come from a noisy environment. Thus, negative feedback can make the output less sensitive (more robust) to variations of the input.

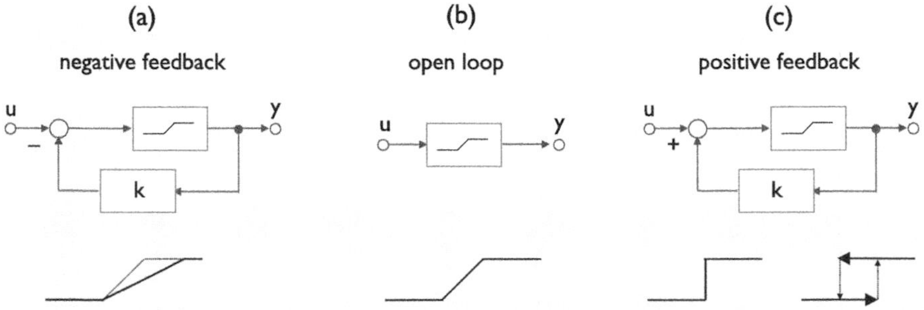

Fig. 1.4 Feedback affects response. (a) Negative feedback increases the linear range. (b) Open-loop response. (c) Positive feedback decreases the linear range, until it produces a switch (left) or memory (right). Adapted from Sepulchre et al. (2019).

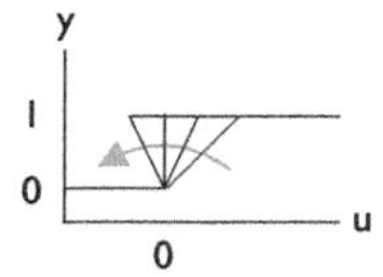

Figure 1.4c and the sketch at right show the effects of positive feedback, $k > 0$. For $k < g^{-1}$ the response is steepened, obeying

$$g' = \frac{g}{1 - kg}, \tag{1.5}$$

until at $k = g^{-1}$, the gain diverges and the response becomes a step (part c, lower left sketch, and vertical line at left). Such a response can be viewed as a *switch* or threshold detector. It implies extreme sensitivity: a finite response due to an infinitesimal input variation. For larger gains $k > g^{-1}$, the response has a backwards slope, and it is easy to show that there is *hysteresis* and a binary output. The output is $y = 0$ until it suddenly jumps to $y = 1$ when u is increased from negative values through 0. But if the output is $y = 1$ for positive u, it stays at that value until $u = g^{-1} - k < 0$, when it transitions back to $y = 0$. See the lower right sketch in Figure 1.4c. We have created a *memory*, which can exist in two states over some range of input values.

The system described here is very simple, as it is just a static response. Yet already, we see that negative feedback can regulate an output, making it less sensitive to variations in its input and that positive feedback can create a switch or memory. In Chapter 3, we shall explore how dynamics generalizes this picture, leading to gains that can be negative at some time scales (frequencies) but positive at others. And in Chapter 11, we will understand better how the positive-feedback response is shaped by nonlinearities in the dynamics. Here, the nonlinearity corresponds to the saturation of the output at 0 and 1.

Autonomous and Nonautonomous

To understand the various types of feedback systems, physicists interested in control have classified systems into *autonomous* and *nonautonomous* cases. The terminology comes from the analogous classification of ordinary differential equations:[11]

$$\underbrace{\dot{x} = f(x)}_{\text{autonomous}}, \qquad \underbrace{\dot{x} = f(x,t)}_{\text{nonautonomous}}, \tag{1.6}$$

where the distinction is that the dynamics of nonautonomous systems have explicit time dependence, whereas autonomous systems do not. Thus, Watt's combination of steam engine and governor would be an autonomous feedback system, as the ultimate power sources are simply two heat reservoirs at different temperatures. All the time dependence of motion and its regulation is generated through the internal dynamics. By contrast, digital control systems with their explicit measurements and decisions taken internally in a computer program would be the archetype of nonautonomous control.

One subtlety is that the distinction between autonomous and nonautonomous depends on the level of description. This is true for the theory of ordinary differential equations, where one can always rewrite an n-dimensional system with

[11] We define and review such equations in more detail in Chapter 2.

nonautonomous dynamics as the projection of a system that lives in an $n+1$-dimensional state space. That is,

$$\dot{\boldsymbol{x}} = \boldsymbol{f}(\boldsymbol{x}, t) \quad \rightarrow \quad \begin{cases} \dot{\boldsymbol{x}} &= \boldsymbol{f}(\boldsymbol{x}, x_t) \\ \dot{x}_t &= 1\,, \qquad\qquad y(0) = 0\,. \end{cases} \tag{1.7}$$

In Eq. (1.7), the $\boldsymbol{x}$ subsystem behaves nonautonomously in both cases. Because the second system has $x_t = t$, we would not be able to distinguish the variable x_t from time t.

Physically, the dynamics describing one system may be nonautonomous, but they can be part of a larger autonomous system. For example, a digital control loop (Figure 1.2d) implies that the system under control will seem nonautonomous, because the digital controller can drive the system in a time-dependent way. But the entire system – with computer, amplifiers, system, and the like – is autonomous because it is powered by fixed power supplies delivering a constant power level.

Simple and Complex Controllers

Let us define complex controllers as ones that are (equivalent to) *universal Turing machines*. In particular, they are capable of implementing any algorithm. Simple ones are all the others. This distinction will allow us to distinguish controllers depicted in Figure 1.2c from d. The analog controller in (c) can be very simple. We will see in Chapter 3 how to make a basic controller from an electrical circuit with just a small number of components. The controller in (d) is a digital computer, and we know that such machines can be programmed to give almost any output.

The distinction has the advantage of allowing us to include in the set of complex control systems *cybernetical* cases, where a human being is part of the control loop. (If you drive a car or ride a bike, you are the human in the control loop.)

As with the other distinctions, this definition is meant to be more instructive than rigorous. In particular, we can imagine a family of machines progressing in complexity from the simple circuits of basic analog controllers to the universal computers in digital controllers. Just where to draw the line between simple and complex may not be immediately obvious.

Other ambiguous cases would arise for natural systems such as the climate-change example shown in Figure 1.2b. Consider, too, a biophysical example, the various functions of a cell. Some actions – sensing the environment, deciding which gene-expression pathway to activate, translating the appropriate proteins – might end up be classified as simple. And the brain, as we have already noted, is complex in our terminology. But what about the metabolic control system, which is fiendishly complex? And immune response? Where to draw the line? Conceptually, the requirement to be a Turing machine seems reasonable, but understanding whether a given system is complex may not be an easy task.

1.3 Control and Information

We already mentioned briefly the recent interest in links between control theory and *information theory*. Historically, the two subjects grew up together. Wiener's book on cybernetics and Shannon's major papers on information theory both appeared in 1948, and the two authors knew each other well. Yet, until recently, there were very few explicit connections between the two topics. We will develop some of these connections in Chapter 15, but it is useful to summarize even now the heuristic perspective that comes with this view. In the information-theoretic version of control, the quality of control is limited by the information available about the system of interest. Control then consists of gathering information, making a decision about what to do, and then acting. How to act? A basic principle is

Exploit what you know; learn what you can.

These eight words summarize most of the field. If this book is long, it is because what you know or can learn and what is worth doing with that knowledge differ greatly from situation to situation. This book is a kind of catalog of situational responses incorporating these guiding ideas.[12]

In the next chapter, we begin with a discussion of continuous-time dynamical systems, which will be the focus of much of our discussion. In part a review, it introduces some differences in the way engineers, as opposed to physicists, view such systems. Read it even if you know the general topic. Then, in Chapters 3 and 4, we present core elements of classical and modern control theory. These chapters form the core of a brief course on control. Later chapters will develop more sophisticated and more specialized points.

1.4 Notes and References

The first historical book on control was published by Otto Mayr in German in 1969 and in English translation a year later (Mayr, 1970). At the time, feedback was viewed as largely a twentieth-century. development, with a few antecedents such as Watt's governor; Mayr showed that it had ancient historical roots. Mayr's work was summarized by Bennett (2002), who "[took] up the story where Mayr left off" and wrote a two-volume history of control covering 1800–1930 (Bennett, 1979) and 1930–1955 (Bennett, 1993). The water-clock description comes from Mayr, but the sketch is based on Lepschy (1992), who also gives a quantitative analysis of the clock dynamics and a description of medieval mechanical clocks, too.

[12] You might argue that it is better to learn before exploiting. But learning has a cost: Typically, we start with prior knowledge and see how far we get. Only when simple, generic solutions fail do we need to learn more.

Unfortunately, the three books by Mayr and Bennett are all out of print and relatively hard to find. Two useful and brief historical summaries are by Bennett (1996) and Bissell (2009). The presentation I give here also draws from a review by two influential control theorists (Åström and Kumar, 2014) that includes more modern developments. (The term "hidden technology" is due to Åström.) The review on quantum feedback by Zhang et al. (2017) has many good insights in its brief sections on history and on classical control.

The classification into periods given is a mixed scheme taken from Bennett (first four periods) and Åström and Kumar for the "contemporary" period, although the name is my own coinage. I also increased the overlap in dates and reclassified some material. For example, the work of Wiener and Kolmogorov on stochastic systems occurred prior to and during World War II but did not have an impact until well after.

The description of homeostasis and the view of life as a hierarchy of regulated systems and death as a failure cascade is taken from a beautiful, poignant essay by Siddhartha Mukherjee (2018) on the death of his father. Although the term "homeostasis" was coined by Walter Cannon in the 1920s, the notion itself was discussed in the 1860s by Claude Bernard, in work that had no influence until the early twentieth century (Gross, 1998).

The Harold Black epiphany that led to the negative-feedback amplifier is recounted in Black (1977), but David Mindell notes how, like most "origin myths," it gives short shrift to its precursors and the many contributions required to abstract the problem of "signals" well enough to formulate such ideas (Mindell, 2002). Mindell also presents a useful "prehistory" to *Cybernetics* (Wiener, 1961, originally published in 1948), drawing attention to the "human in the loop" in instrumentation developed in American industry (Ford Instrument Company, Sperry Gyroscope, Bell Telephone Laboratory) and how these control systems fed into US military efforts prior to and during World War II (Mindell, 2002).

For a review of control applications in the automotive industry, see Cook et al. (2007). The discussion of climate change is from Hansen et al. (2017). A key paper in the development of systems biology and control is Hartwell et al. (1999). Ingalls (2013) discusses how to model control of the metabolic system. And Germain (2001) reviews how similar ideas connect to the immune system. McNamee and Wolpert (2019) review, from the perspective of control, the hypothesis that animals (especially humans) use internal models of the world to make rational inferences about sensory data and plans for action. Tang and Bassett (2018) discuss efforts to control dynamics in brain networks.

The note at the end of Section 1.1.4 on the "negligible" impact of academic control methods prior to the development of robust control is from Morari and Zafirioiu (1989). For more discussion, see Chapter 9 and also Åström and Kumar (2014). The discussion of goals is adapted (and then slightly altered) from a discussion by Fradkov (2007), Chapter 2. The notion of feedback without loops has been emphasized (from a slightly different point of view) by Jacobs (2014). Finally, the discussion of negative and positive feedback is from Sepulchre et al. (2019).

O'Keeffe et al. (2017) contrast the notions of synchronization (temporal correlation) with swarming (spatial correlation) and go on to introduce dynamical systems with spatiotemporal correlations, which they dub *swarmalators*.

2 Dynamical Systems

> And now, reader …Bestir thyself …for, though we will always lend thee proper assistance in difficult places, as we do not, like some others, expect thee to use the arts of divination to discover our meaning, yet we shall not indulge thy laziness where nothing but thy own attention is required; for thou art highly mistaken if thou dost imagine that we intended, when we began this great work, to leave thy sagacity nothing to do; or that, without sometimes exercising this talent, thou wilt be able to travel through our pages with any pleasure or profit to thyself –
>
> Henry Fielding, *Tom Jones*

Control-theory techniques allow us to modify a dynamical system. Perhaps we want to regulate a physical variable such as temperature or pressure, linearize the response of a sensor, speed up a sluggish system or slow down a jumpy one, or stabilize an unstable system or destabilize a stable one (to create an oscillator or a switch). Whatever the reason, we alter a given dynamical system in hopes of creating a "better" one.

Before "improving" the dynamics of a system, one should understand it. In this chapter, we review ideas and tools about dynamics that mostly should be familiar from introductory and intermediate physics courses. We present just enough material to get started and will return to and extend the ideas introduced here later. The value of outlining this material in advance is to clearly separate the tools that are common to the analysis of all dynamical systems, whatever the application, from the tools that are important for controlling systems.

We start on familiar ground, showing how to write equations of motion in a compact form. We diverge, however, from traditional physics presentations by specifying carefully how inputs affect the dynamics and how outputs, or measurements, are made on the system. Such care about inputs and outputs will turn out to be crucial for successful control.

Dynamical systems are characterized by an internal *state vector* $\boldsymbol{x}(t)$, which can represent a position, a velocity, a concentration, or any other variable relevant to the system being considered. In this chapter and the next two, we will assume that both the state $\boldsymbol{x}$ and time t are continuous variables. In Chapter 5, we will focus on the case of discrete time and (approximately) continuous $\boldsymbol{x}$, which is relevant for computer-based controllers. Chapter 12 introduces discrete-state dynamics, with discrete (or continuous) observations.

Most dynamical systems obey nonlinear equations. While there are tools for understanding certain aspects of *nonlinear systems*, most must be studied case by case. In contrast, *linear systems* can be analyzed completely. We will review appropriate tools, such as the *Laplace* transform, a cousin of the more familiar *Fourier* transform. In traditional presentations of control theory, linear systems have pride of place. The cynic would say that such systems are favored because they are the only ones we understand in general. Although there is some truth to this notion,[1] the motivation is more subtle. We can often model the behavior of a nonlinear system near a point of interest using a linear approximation. If our model and our control algorithms are good, the *difference* between the real dynamics will stay small enough that a linear approximation is valid. For nonlinear systems, we focus on a few key concepts: the multiplicity of possible solutions, the *stability* of base solutions, and *bifurcations* (or switching) between different solutions.

Advice to students: If you are familiar with the analysis of dynamical systems, this chapter will remind you of ideas and will introduce notation. Since some of the language and applications are likely to be unfamiliar, you should at least skim these pages.

If you are less familiar with this material, you may not want to go through the chapter all at once but rather cover the material as needed. For example, Chapter 3 covers frequency-domain methods and depends largely on Section 2.3, which itself refers to the Appendix online, Section A.4. But other sections from this chapter are not needed.

2.1 Introduction: The Pendulum as a Dynamical System

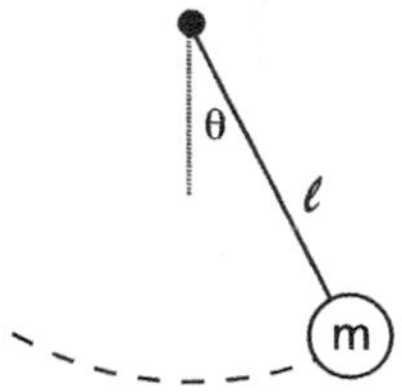

Before we consider a general formulation, let us start with a familiar dynamical system, the *pendulum*, illustrated at right. From Newton's laws (or using a Lagrangian), one can write the equation for torque dynamics,

$$m\ell^2\ddot{\theta} + mg\ell \sin\theta = 0\,, \tag{2.1}$$

where m is the mass of the pendulum bob, ℓ its length, and g the acceleration of gravity, and where we neglect any dissipation. We will use various notations to denote derivatives, $\dot{\theta}(t) = \mathrm{d}_t\theta = \mathrm{d}\theta/\mathrm{d}t$, and partial derivatives, $U_x(x,t) = \partial_x U = \partial U/\partial x$, at fixed t.

Although Eq. (2.1) seems to depend on three parameters (m, ℓ, and g), they can all be eliminated by a careful rescaling of variables. Divide by $m\ell^2$ and define $\omega_0^2 = g/\ell$, giving

$$\ddot{\theta} + \omega_0^2 \sin\theta = 0\,. \tag{2.2}$$

Next, define a dimensionless time $\bar{t} \equiv \omega_0 t$ and transform the angular acceleration as $\mathrm{d}_{tt}\theta \to (\omega_0^2)\,\mathrm{d}_{\bar{t}\bar{t}}\theta$. Finally, divide by ω_0^2 in Eq. (2.2) and write

$$\ddot{\theta} + \sin\theta = 0\,. \tag{2.3}$$

[1] It is a bit like looking for your lost keys under a streetlight, because that is where you can see.

In Eq. (2.3), we let the same variable denote the dimensional and dimensionless quantities (i.e., we drop the overbar and let $\bar{t} \to t$). This reduces notation clutter.

Equation (2.3) is a second-order, nonlinear *ordinary differential equation* (ODE). We can rewrite it as a pair of coupled first-order equations by defining $x_1(t) \equiv \theta(t)$ and $x_2(t) \equiv \dot{\theta}$. The two new variables, x_1 and x_2, can then be grouped into an *internal state vector* $\boldsymbol{x}(t)$ ("state vector" for short), and we can write Eq. (2.3) as a nonlinear, first-order *vector* ODE:

$$\frac{\mathrm{d}}{\mathrm{d}t}\begin{pmatrix} x_1 \\ x_2 \end{pmatrix} = \begin{pmatrix} x_2 \\ -\sin x_1 \end{pmatrix} \equiv \begin{pmatrix} f_1(\boldsymbol{x}) \\ f_2(\boldsymbol{x}) \end{pmatrix} \quad \Longrightarrow \quad \dot{\boldsymbol{x}} = \boldsymbol{f}(\boldsymbol{x}), \tag{2.4}$$

where $\boldsymbol{f}$ is a vector field (with two components here) that depends on the state vector $\boldsymbol{x}$. In short, we have gone from Eq. (2.1) to $\dot{\boldsymbol{x}} = \boldsymbol{f}(\boldsymbol{x})$. We can always put the models and equations of motion for dynamical systems that we consider into this general form.

Motion in Phase Space

Although it is possible to solve Eq. (2.3) analytically in terms of an elliptic integral, most nonlinear ODEs cannot be solved exactly.[2] One way to make further progress was pioneered by Poincaré and others at the beginning of the twentieth century. In their geometric approach, we analyze the trajectories of the state vector $\boldsymbol{x}(t)$ in a *phase-space* plot.[3] For a two-dimensional state vector, the result is a two-dimensional phase plot, which is shown for the pendulum in Figure 2.1. We can characterize the trajectories by their energy, because the latter is a conserved constant during motion. (Proof: integrate Eq. (2.3) to find $\frac{1}{2}\dot{\theta}^2 - \cos\theta + 1 = E$, where E is the total energy and satisfies $\dot{E} = 0$. The constant is chosen so that $E = 0$ when the pendulum is at rest and hanging down.)

For energies less than $E_c = 2$, the trajectory is a circle about the origin, corresponding to an oscillation of the bob about the downward position. For $E > E_c$, the pendulum swings end over end in a "running trajectory." For $E = E_c$, the pendulum starts upside down and makes one swing and, after an infinite amount of time,

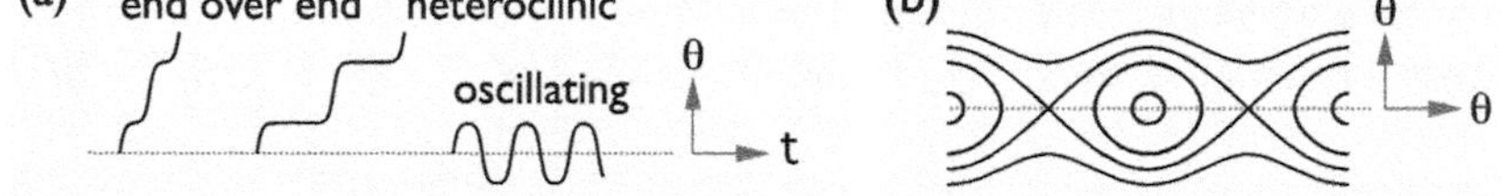

Fig. 2.1 State-space trajectories of a nonlinear simple pendulum without friction. (a) Oscillator angle versus time, showing end-over-end motion, an approximation to a heteroclinic connection, and oscillations. (b) Phase-space plot of angular velocity versus angle. Because angles are periodic in 2π radians, we should view the repeating plot as being wrapped around a cylinder.

[2] Even a solution in terms of an elliptic integral is not always interesting. Although knowing an analytic form can lead to useful asymptotic expressions, general evaluations must still be done numerically.

[3] The physics term *phase space* is just a particular example of a state space, where the internal state vector components are positions and momenta. More generally, the internal state $\boldsymbol{x}$ can be any dynamical quantity.

returns to the upright position, a trajectory known as a *heteroclinic connection*. The three classes of trajectory are illustrated (for energies $E = 2.001, 2.000$, and 1.999) in Figure 2.1a. Note that because of errors in the integration routine, the heteroclinic connection begins another swing after a finite time.

In phase space, trajectories cannot cross by the uniqueness of solutions to ODEs.[4] In two dimensions, the closed orbits corresponding to oscillatory motion divide the state space into two disjoint sets (inside and outside). As a consequence, a trajectory confined to a region for all time has at least one fixed point or limit cycle in that region (*Poincaré-Bendixson theorem*). Otherwise, it is impossible that trajectories not cross each other after an infinite amount of time. Of course, the phase space often has more than two dimensions. In higher dimensions, lines are generically *skew*: although they may seem to cross when projected down to a two-dimensional subspace, they do not cross in the full space. The motion can then have more exotic structure (strange attractors, chaos); see Chapter 11.

There are two general features of phase-space plots. One is that they shift the focus from understanding single solutions to understanding the properties of groups of solutions (that is, those originating from a region of phase space). Also, they suggest the use of geometrical and topological methods to analyze motion (Section 2.6).

Linearization

Another way to make progress is to consider motion that does not deviate much from an equilibrium solution. For example, Eq. (2.3) has two equilibrium solutions, $\theta_0 = 0$ and π, corresponding to a pendulum that hangs straight down or balances upside down, respectively. Small deviations about one of the equilibria, say $\theta_0 = 0$, can be studied by linearizing: $\sin\theta \approx \theta$, leading to a linear ODE,

$$\ddot{\theta} + \theta = 0 \quad \Longrightarrow \quad \frac{\mathrm{d}}{\mathrm{d}t}\begin{pmatrix} x_1 \\ x_2 \end{pmatrix} = \begin{pmatrix} 0 & 1 \\ -1 & 0 \end{pmatrix}\begin{pmatrix} x_1 \\ x_2 \end{pmatrix} \quad \Longrightarrow \quad \dot{\boldsymbol{x}} = \boldsymbol{A}\boldsymbol{x} \tag{2.5}$$

where the dynamics is encoded into a 2×2 matrix, $\boldsymbol{A}$. Solving Eq. (2.5) gives

$$\begin{aligned} &\theta(t) = \theta(0)\cos t + \dot{\theta}_0(0)\sin t \\ \Longrightarrow\ &\theta(t) = \theta(0)\cos\omega_0 t + \left(\dot{\theta}_0(0)/\omega_0\right)\sin\omega_0 t, \end{aligned} \tag{2.6}$$

with $\theta(0) = x_1(0)$ and $\dot{\theta}_0 = x_2(0)$ the initial angular position and velocity. Equation (2.6) is in dimensional units. We can thus interpret ω_0 as the angular frequency of the oscillations. Note how linearizing the equations allows us to solve for the evolution of the state variable $\boldsymbol{x}(t)$ in terms of combinations of (complex) exponential functions.

[4] In Figure 2.1, trajectories seem to cross at the unstable fixed points. Remember that these fixed points correspond to pointing up, with zero angular velocity. It takes an infinite amount of time for a trajectory to go from one such fixed point to another with $\Delta\theta = 2\pi$.

So far, we have not specified how the outside world interacts with our pendulum. Let us start with the input. One way to alter the motion of a pendulum is to exert an external torque $\tau(t)$. Equation (2.1) then becomes

$$m\ell^2\,\ddot{\theta} + mg\ell\,\sin\theta = \tau(t) \qquad \to \qquad \ddot{\theta} + \sin\theta = u(t)\,, \tag{2.7}$$

with $u = \tau/(m\ell^2\omega_0^2)$. The variable $u(t)$ is known as a control input. In state-vector notation, we have, for the nonlinear and linear (about 0) cases

$$\dot{\boldsymbol{x}} = \boldsymbol{f}(\boldsymbol{x}, u) \to \begin{pmatrix} 0 & 1 \\ -1 & 0 \end{pmatrix}\begin{pmatrix} x_1 \\ x_2 \end{pmatrix} + \begin{pmatrix} 0 \\ 1 \end{pmatrix} u \to \boldsymbol{A}\boldsymbol{x} + \boldsymbol{B}u\,. \tag{2.8}$$

The vector $\boldsymbol{B}$ couples the input $u(t)$ to the state-vector dynamics. That $B_1 = 0$ and $B_2 = 1$ follows since u is a *torque*.

Next, we consider observations, or system outputs. A common situation is that we measure the angular displacement. We might also measure the angular velocity directly, for example by using an accelerometer located at the pendulum bob.[5] Or we might measure both quantities. For control, it matters a great deal what information is available about the dynamical system. We introduce an explicit notation to specify what exactly is being measured, defining a (single) *output variable* $y(t)$ as a linear combination of state-vector elements:

$$y = \begin{pmatrix} C_1 & C_2 \end{pmatrix}\begin{pmatrix} x_1 \\ x_2 \end{pmatrix} \equiv \boldsymbol{C}\boldsymbol{x}\,. \tag{2.9}$$

If only the angular position is measured, $C_1 = 1$ and $C_2 = 0$; if only the angular velocity is measured, then $C_1 = 0$, $C_2 = 1$. If both are measured independently, then y is a two-component vector and $\boldsymbol{C}$ a 2×2 matrix – equal to the identity matrix in this special case, so that $y_1 = x_1$ and $y_2 = x_2$.

2.2 General Formulation

Motivated in part by the pendulum example, we will discuss dynamical systems more generally. We start by assuming that the dynamics can be written as[6]

$$\dot{\boldsymbol{x}} = \boldsymbol{f}(\boldsymbol{x}, \boldsymbol{u}) \qquad \boldsymbol{y} = \boldsymbol{h}(\boldsymbol{x}, \boldsymbol{u})\,, \tag{2.10}$$

where there are n linearly independent elements of the state vector $\boldsymbol{x}$, m independent inputs (driving terms) $\boldsymbol{u}$, and p independent outputs $\boldsymbol{y}$. The state vector generalizes the notion of phase space from classical mechanics: it contains all the information

5 The pendulum bob traces a circle, implying a radial acceleration of $a = \dot{\theta}^2\,\ell$, in addition to gravity.

6 Even more general forms of dynamics can be envisioned (e.g., $\boldsymbol{F}(\dot{\boldsymbol{x}}, \boldsymbol{x}, \boldsymbol{u}) = \boldsymbol{0}$). Such a form can represent differential-algebraic equations (DAE), differential equations supplemented by algebraic constraints.

necessary to predict the future state of the system. If ever you find that the dynamics of a "state vector" seems to depend on the history of the system, it is missing some elements. Once all the necessary information is included in an enlarged state vector, a measurement at the present time t suffices to predict the future (assuming future inputs, including disturbances, are known). In dynamics derived from partial differential equations describing spatially extended systems, the state vector is formally infinite dimensional, although a finite number of components often gives a sufficiently accurate representation. As we will see, values of state-vector elements depend on the choice of coordinate system.

The measurements $\boldsymbol{y}$ may be made in discrete or continuous time. The vector-valued function $\boldsymbol{f}$ represents the (nonlinear) dynamics and $\boldsymbol{h}$ converts the state $\boldsymbol{x}$ into observations. The dependence of $\boldsymbol{h}$ on $\boldsymbol{u}$ allows for static input-output relationships (see Section 2.2.1). The number of internal state variables (n) is often greater than the number of measured quantities (p). The important point is that Eq. (2.10) defines not only the system dynamics but also its inputs and outputs, giving rise to the *state-space* formulation of dynamics.

One feature *not* in Eq. (2.10) is an explicit time dependence in $\boldsymbol{f}$ or $\boldsymbol{h}$, which would create a *nonautonomous* dynamical system. A nonautonomous dynamical system for $\boldsymbol{x}$ in n dimensions can always be rewritten as a time-independent, *autonomous* system in $n+1$ dimensions by defining a new state variable, $x_{n+1} = t$, which obeys $\dot{x}_{n+1} = 1$. Thus, while it may be convenient to define nonautonomous systems – as we will sometimes do – there is no fundamental distinction between the two types.

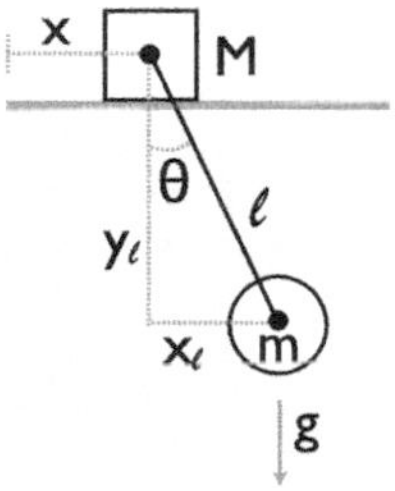

Example 2.1 (Cart-pendulum system) Let us consider the slightly more complicated pendulum shown at right. A cart, of mass M, is confined to a rail (gray bar) that allows horizontal motion. It carries a pivot for a pendulum of mass m (assumed concentrated at the end) of length ℓ. By pushing the cart back and forth by a horizontal force $u(t)$, you can excite motion of the pendulum. A typical task is to swing up and balance the pendulum in its unstable "up" equilibrium. In Problem 2.1, you will show that the scaled equations of motion, neglecting friction, can be written as

$$\ddot{x} + \left(\frac{m}{M+m}\right)\left(\ddot{\theta}\cos\theta - \dot{\theta}^2\sin\theta\right) = u\,, \qquad \ddot{\theta} + \sin\theta + \ddot{x}\cos\theta = d\,, \tag{2.11}$$

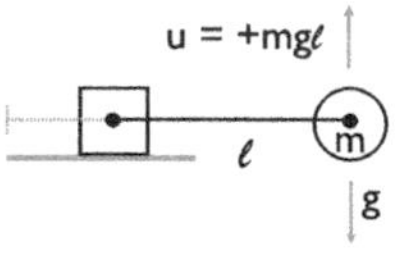

where $u(t)$ and $d(t)$ are the force and disturbing torque, with u scaled by $(M+m)g$, the total weight of the cart-pendulum combination, and d by $mg\ell$, the torque needed to hold the pendulum against gravity at $\theta = \pi/2$ (see right). The two second-order equations for $x(t)$ and $\theta(t)$ can be written as a single first-order, nonlinear equation for $\boldsymbol{x} = (x, \dot{x}, \theta, \dot{\theta})^{\mathrm{T}}$.

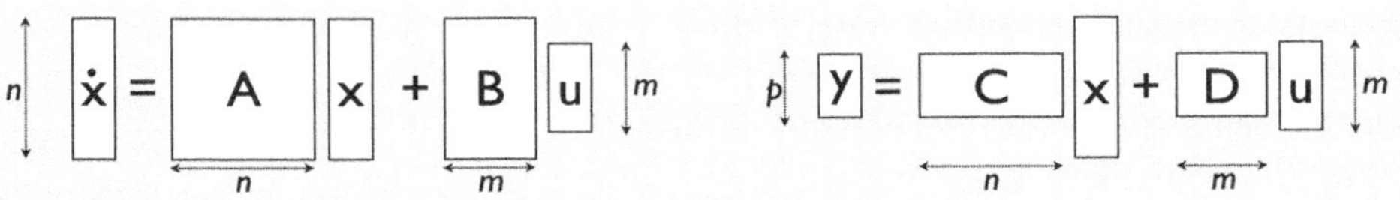

Fig. 2.2 Dimensions in a general linear dynamical system.

2.2.1 Linear Systems

We will often deal with linear systems, which can be written[7]

$$\dot{\boldsymbol{x}} = \boldsymbol{A}\boldsymbol{x} + \boldsymbol{B}\boldsymbol{u} \qquad \boldsymbol{y} = \boldsymbol{C}\boldsymbol{x} + \boldsymbol{D}\boldsymbol{u}\,. \tag{2.12}$$

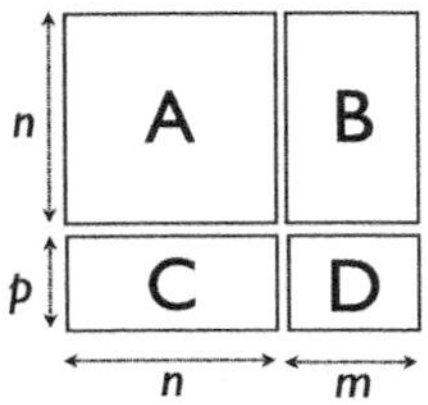

Here, the *dynamics* $\boldsymbol{A}$ are encoded by an $n \times n$ matrix, the *input coupling* $\boldsymbol{B}$ by an $n \times m$ matrix (n rows, m columns), the *output coupling* $\boldsymbol{C}$ by a $p \times n$ matrix, and the input-output *feedthrough* $\boldsymbol{D}$ by a $p \times m$ matrix.[8] If the system lacks explicit time dependence in $\boldsymbol{f}$ and $\boldsymbol{h}$, the matrices $\boldsymbol{A}$, $\boldsymbol{B}$, and $\boldsymbol{C}$ will all be constant, and the system is *linear time invariant* (LTI).[9] Figure 2.2 and the sketch at left show how the dimensions of the matrices couple together. Equation (2.12) defines the *state-space* representation of the linear dynamics.

A system with only a single input and a single output is a SISO system, while a system with multiple inputs and multiple outputs is a MIMO system.[10] We focus on SISO systems, but the formalism is designed to generalize easily to the MIMO case.

Example 2.2 (Low-pass filter) As a simple example of a dynamical system based on electrical circuits rather than mechanics, many sensors act as a low-pass filter and are equivalent to the electrical circuit shown in Example 2.2, for which we find

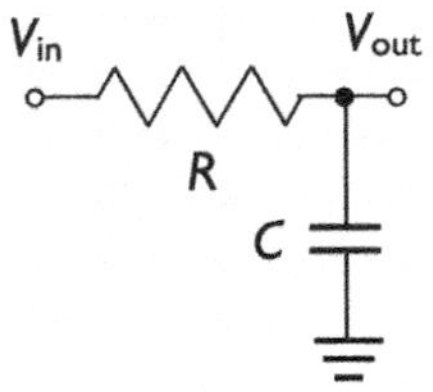

$$\dot{V}_{\text{out}}(t) = -\frac{1}{RC} V_{\text{out}}(t) + \frac{1}{RC} V_{\text{in}}(t)\,. \tag{2.13}$$

Here, $n = m = p = 1$, $x = y = V_{\text{out}}$, $u = V_{\text{in}}$, $A = -1/RC$, $B = 1/RC$, $C = 1$, and $D = 0$.[11] As before, we can simplify the equations by scaling. If we define the time scale $\tau = RC$ and rescale time to be dimensionless and let $u \equiv V_{\text{in}}$, we have

$$\dot{x} = -x + u\,, \qquad\qquad y = x\,. \tag{2.14}$$

In our scaled units, $A = -1$, and $B = 1$. All quantities are functions of the scaled time $\tilde{t} = t/\tau$. In practice, we drop the tildes in the scaled equations. Again, we start from

[7] For differential-algebraic equations (DAEs), the corresponding linear system is known as the *descriptor* representation and takes the form $\boldsymbol{E}\dot{\boldsymbol{x}} = \boldsymbol{A}\boldsymbol{x}+\boldsymbol{B}\boldsymbol{u}$, with $\boldsymbol{E}$ a possibly singular $n\times n$ matrix. The noninvertible subspace represents algebraic constraints. See Section 8.1.2 for an example. Software such as *Modelica*, a modeling language to describe complex systems, uses DAEs extensively.

[8] By defining a new output $\boldsymbol{y}' = \boldsymbol{y} - \boldsymbol{D}\boldsymbol{u}$, we can always consider the simpler case $\boldsymbol{y}' = \boldsymbol{C}\boldsymbol{x}$.

[9] If $\boldsymbol{u}$ is constant, then Eq. (2.12) is also homogeneous. For time-dependent $\boldsymbol{u}(t)$, it is LTI but not homogeneous.

[10] On occasion, SIMO and MISO are used for the mixed cases.

[11] Do not confuse the output coupling C with the capacitance.

an equation that depends on two parameters, R and C, and find an equation with no parameters.

Example 2.3 (Driven, damped harmonic oscillator) Consider a mass on a linear spring that exerts a force $= -kq$, with q the displacement of the mass from the equilibrium position set by the spring. The mass is connected in parallel to a dashpot to provide damping (force $= -bv = -b\dot{q}$) and is driven by an external force $F(t)$, as shown in the illustration on the right. The equations of motion are

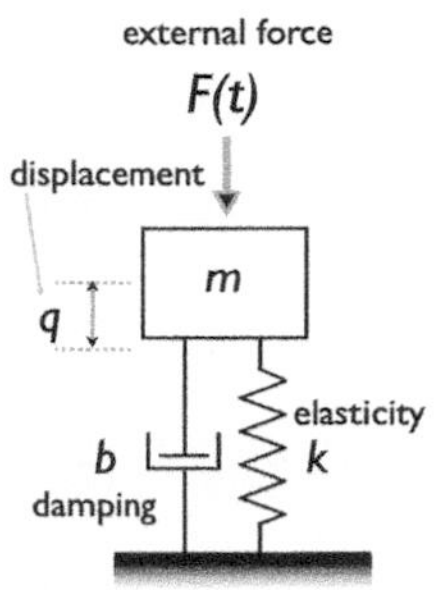

$$m\ddot{q} + b\dot{q} + kq = F(t)\,. \tag{2.15}$$

We can simplify Eq. (2.15) by defining $\omega_0^2 = k/m$ and then scaling time by $\tilde{t} = \omega_0 t$. We also define $2\zeta \equiv b/(m\omega_0) = b/\sqrt{mk}$, a dimensionless damping parameter, with $0 < \zeta < 1$ for an underdamped oscillator and $\zeta > 1$ for an overdamped system. (You may be more familiar with the quality factor $Q = 1/2\zeta$.) Then,

$$\ddot{q} + 2\zeta\,\dot{q} + q = \frac{1}{k}\,F(t)\,. \tag{2.16}$$

To put Eq. (2.16) in state-space form (Eq. 2.12), we let $x_1 = q$, $x_2 = \dot{q}$. Then $n = 2$ (second-order system), $m = p = 1$ (one input, $u = F/k$, and one output, q), and

$$\frac{d}{dt}\begin{pmatrix} x_1 \\ x_2 \end{pmatrix} = \underbrace{\begin{pmatrix} 0 & 1 \\ -1 & -2\zeta \end{pmatrix}}_{A}\begin{pmatrix} x_1 \\ x_2 \end{pmatrix} + \underbrace{\begin{pmatrix} 0 \\ 1 \end{pmatrix}}_{B} u(t)\,, \qquad y = \underbrace{\begin{pmatrix} 1 & 0 \end{pmatrix}}_{C}\begin{pmatrix} x_1 \\ x_2 \end{pmatrix}\,. \tag{2.17}$$

Note how scaling simplifies our original equation, with three dimensional parameters (m, b, and k), to an equation that has a single dimensionless parameter, ζ.

Why study linear systems? As suggested in the introduction to this chapter,

1. Linear systems can be analyzed easily, while most nonlinear systems cannot.
2. For systems near an equilibrium, small deviations obey linear dynamics.
3. If we use a nonlinear calculation to generate a trajectory for a nonlinear system, deviations may be small enough to obey linear dynamics. (See Section 7.6.1.)

We will illustrate the first point below, where we present Laplace-transform and time-domain methods. The second point is easily understood algebraically: Let $\boldsymbol{x}_0$, $\boldsymbol{u}_0$, and $\boldsymbol{y}_0$ be a *fixed point* of the dynamical system $\dot{\boldsymbol{x}} = \boldsymbol{f}(\boldsymbol{x},\boldsymbol{u})$, $\boldsymbol{y} = \boldsymbol{h}(\boldsymbol{x},\boldsymbol{u})$. By fixed point, we mean that $\boldsymbol{f}(\boldsymbol{x}_0,\boldsymbol{u}_0) = \boldsymbol{0}$ and $\boldsymbol{y}_0 \equiv \boldsymbol{h}(\boldsymbol{x}_0,\boldsymbol{u}_0)$. That is, for $\boldsymbol{x} = \boldsymbol{x}_0$ and $\boldsymbol{u} = \boldsymbol{u}_0$, the state vector $\boldsymbol{x}$ is unchanging. We saw an example of a fixed point for the pendulum, for which $\theta = 0$ and 2π were fixed points (for $u = 0$). Given a fixed point ($\boldsymbol{x}_0$, $\boldsymbol{u}_0$, $\boldsymbol{y}_0$), we can define

$$\boldsymbol{x} \equiv \boldsymbol{x}_0 + \boldsymbol{x}_1 \qquad \boldsymbol{u} \equiv \boldsymbol{u}_0 + \boldsymbol{u}_1 \qquad \boldsymbol{y} \equiv \boldsymbol{y}_0 + \boldsymbol{y}_1 \tag{2.18}$$

and substitute into Eq. (2.10). If the deviations $\boldsymbol{x}_1$, $\boldsymbol{u}_1$, $\boldsymbol{y}_1$ are all small, we can Taylor expand $\boldsymbol{f}$ and $\boldsymbol{h}$.

$$\dot{\boldsymbol{x}}_1 = \boldsymbol{f}(\boldsymbol{x}_0 + \boldsymbol{x}_1, \boldsymbol{u}_0 + \boldsymbol{u}_1) \approx \cancelto{0}{\boldsymbol{f}(\boldsymbol{x}_0, \boldsymbol{u}_0)} + \left.\frac{\partial \boldsymbol{f}}{\partial \boldsymbol{x}}\right|_{(\boldsymbol{x}_0,\boldsymbol{u}_0)} \boldsymbol{x}_1 + \left.\frac{\partial \boldsymbol{f}}{\partial \boldsymbol{u}}\right|_{(\boldsymbol{x}_0,\boldsymbol{u}_0)} \boldsymbol{u}_1 + \cdots \tag{2.19a}$$

$$\cancel{\boldsymbol{y}_0} + \boldsymbol{y}_1 = \boldsymbol{h}(\boldsymbol{x}_0 + \boldsymbol{x}_1, \boldsymbol{u}_0 + \boldsymbol{u}_1) \approx \cancel{\boldsymbol{h}(\boldsymbol{x}_0, \boldsymbol{u}_0)} + \left.\frac{\partial \boldsymbol{h}}{\partial \boldsymbol{x}}\right|_{(\boldsymbol{x}_0,\boldsymbol{u}_0)} \boldsymbol{x}_1 + \left.\frac{\partial \boldsymbol{h}}{\partial \boldsymbol{u}}\right|_{(\boldsymbol{x}_0,\boldsymbol{u}_0)} \boldsymbol{u}_1 + \cdots \tag{2.19b}$$

and identify the matrices for the linear system discussed in Eq. (2.12),

$$\boldsymbol{A} = \left.\frac{\partial \boldsymbol{f}}{\partial \boldsymbol{x}}\right|_{(\boldsymbol{x}_0,\boldsymbol{u}_0)} \qquad \boldsymbol{B} = \left.\frac{\partial \boldsymbol{f}}{\partial \boldsymbol{u}}\right|_{(\boldsymbol{x}_0,\boldsymbol{u}_0)} \qquad \boldsymbol{C} = \left.\frac{\partial \boldsymbol{h}}{\partial \boldsymbol{x}}\right|_{(\boldsymbol{x}_0,\boldsymbol{u}_0)} \qquad \boldsymbol{D} = \left.\frac{\partial \boldsymbol{h}}{\partial \boldsymbol{u}}\right|_{(\boldsymbol{x}_0,\boldsymbol{u}_0)} . \tag{2.20}$$

The matrices depend on the point of linearization, $\boldsymbol{x}_0$ and $\boldsymbol{u}_0$, and different reference points will lead to different matrices. Note that we will generally drop the subscript for the deviations $\boldsymbol{x}_1$ and $\boldsymbol{u}_1$. To be explicit, recall that $\boldsymbol{f}$ is an n-component vector, as is $\boldsymbol{x}$. Then the components of $\boldsymbol{A}$, for example, are given by $A_{ij} = \partial_{x_j} f_i$, evaluated at $\boldsymbol{x}_0$ and $\boldsymbol{u}_0$.

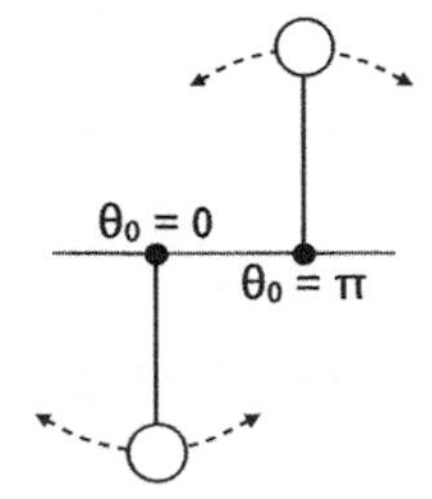

In our pendulum example, we had two fixed points, at $\theta = 0$ and π (see left). Linearizing about the $\theta = 0$ fixed point (pendulum down) gave $\boldsymbol{A}_0 = \left(\begin{smallmatrix} 0 & 1 \\ -1 & 0 \end{smallmatrix}\right)$. If we linearize about the inverted position $\theta = \pi$, we have $\boldsymbol{A}_\pi = \left(\begin{smallmatrix} 0 & 1 \\ 1 & 0 \end{smallmatrix}\right)$. Another important point is that the linear dynamics can be a good approximation to the full system only when the motion remains near the fixed point. For example, we might expect the approximation to be good for a while for small kicks to a pendulum that is "down" but not for one that is "up."

2.2.2 Scaling and Dimensional Analysis

We have seen that the equations of motion for a pendulum, a low-pass filter, and a linear oscillator can be simplified by identifying characteristic length and time scales and then scaling variables with respect to these quantities. If you are not used to scaling equations, details about the change of variables might seem mysterious at first. And even if you are used to scaling equations, it is worth thinking about the origins of the procedure and the questions it raises: Can we know in advance how to scale equations? How many dimensionless parameters will be left in the end?

About 100 years ago, Edgar Buckingham suggested a systematic approach.[12] Physical laws should not depend on the units chosen: $\boldsymbol{F} = m\boldsymbol{a}$, whether we use meters or

[12] Buckingham spent much of his career at the US National Bureau of Standards, now the National Institute of Science and Technology, or NIST, which seems appropriate.

feet.[13] In the early nineteenth century, Joseph Fourier had already realized an important consequence of this principle: terms that are added together in an equation must have the same units.[14]

If not, by changing units, we could systematically make one term bigger or smaller than the others, hence dominant or negligible. This freedom to alter the relative sizes of different terms of an equation contradicts the principle that the laws of physics should be the same for all systems of units. By contrast, if all terms have the same units, any rescaling will affect each term in the same way: the equation is *homogeneous* under rescaling. As a consequence, we can make any equation dimensionless by choosing appropriate scales and then using them to divide through each element of an equation.

To do this systematically, we classify the physical quantities that are relevant to a problem as *variables* or *parameters*. The former include time (an *independent* variable) and *dependent variables* that are functions of time, such as position, velocity, and force. In a spatially extended system, x, y, and z would all be additional dependent variables, and the dependent quantities (fields) would depend on both space and time.

Equations also have parameters, quantities that characterize the physical system. Some may be *fixed parameters*, such as the mass of Example 2.3; others may be *control parameters*, whose values may be varied at will. Here, the control parameter is the magnitude of the applied force $F(t)$. What is fixed and what is variable is somewhat arbitrary, and the distinction is made for clarity rather than any fundamental reason. In a different version of the experiment, the classification of fixed and control parameters might differ. In the harmonic-oscillator example, we might more typically fix the mass and vary the input frequency, but nothing stops us from fixing the frequency and varying the mass.

We will frame our discussion in terms of Example 2.3, with forcing function $F(t) = F\cos\omega t$, where F is the magnitude and ω the frequency of the external force. Then,

$$m\ddot{q} + b\dot{q} + kq = F\cos\omega t\,. \tag{2.21}$$

There are seven physical quantities in Eq. (2.21): two variables (q and t) and five parameters (m, b, k, F, and ω). Using brackets to denote the units (e.g., $[q] = L =$ length), we see that there are three fundamental units (M, L, and T), from which the units of the seven physical quantities may be *derived* (i.e., expressed as powers of the fundamental units). See Table 2.1. Note that we could equally well substitute force for mass to define the FLT class of units, as opposed to the MLT. What counts is the ability to express all other units in terms of the fundamental ones. More generally, there would be m fundamental units, k variables, and p parameters, and $N = k + p$ total physical variables.

[13] Although physical laws do not depend on the choice of units, politics seems to, as witnessed by controversies over the use of the metric system in some otherwise developed countries.

[14] "Chaque grandeur indéterminée ou constante a une dimension qui lui est propre, et ... les termes d'une même équation ne pourraient pas être comparés, s'ils n'avaient point le même exposant de dimension." (Each variable or constant has its own units, and terms of an equation cannot be compared if they have different dimensional exponents.) J.-B. J. Fourier, *Théorie Analytique de la Chaleur*, Section 160, Fermin Didot, Paris (1822).

Table 2.1 Oscillator scaling

Fundamental units		Derived units	
M	mass	m	$[M]$
L	length	b	$[MT^{-1}]$
T	time	k	$[MT^{-2}]$
		F	$[MLT^{-2}]$
		ω	$[T^{-1}]$
		q	$[L]$
		t	$[T]$

We first choose *scales* for the variables (length scale for q, time scale for t). Since the units of each parameter are of the form $M^\alpha L^\beta T^\gamma$, we look for powers of parameters that have the right units (L and T). In this case, we can choose a length scale $q_0 \equiv F/k$ and a time scale $t_0 \equiv \sqrt{m/k}$. The choice of scales is not unique. The time scale b/k would also work. Having chosen our scales, we define dimensionless variables

$$\bar{q} = \frac{q}{q_0}, \qquad \bar{t} = \frac{t}{t_0}, \tag{2.22}$$

Changing to these new variables then gives

$$\begin{aligned}
\frac{mq_0}{t_0^2}\,\bar{q}'' + \frac{bq_0}{t_0}\,\bar{q}' + kq_0\bar{q} &= F\cos\left(\omega\, t_0\, \bar{t}\right) \\
\frac{mF}{k}\cdot\frac{k}{m}\bar{q}'' + \frac{bF}{k}\cdot\sqrt{\frac{k}{m}}\bar{q}' + k\cdot\frac{F}{k}\bar{q} &= F\cos\left(\omega\sqrt{\frac{m}{k}}\,\bar{t}\right) \\
\bar{q}'' + \frac{b}{\sqrt{km}}\bar{q}' + \bar{q} &= \cos\left(\omega\sqrt{\frac{m}{k}}\,\bar{t}\right) \\
\bar{q}'' + 2\zeta\bar{q}' + \bar{q} &= \cos\left(\bar{\omega}\bar{t}\right) \\
\ddot{q} + 2\zeta\,\dot{q} + q &= \cos\omega t\,,
\end{aligned} \tag{2.23}$$

where $\bar{q}' = d\bar{q}/d\bar{t}$. As we progress through Eq. (2.23), we first divide through by the dimensional factors, then define dimensionless parameters ζ and $\bar{\omega}$. Finally, we drop the overbars, as it is clearer to have just one set of symbols. We might view a variable q as representing abstractly the displacement, in any set of units.

We could have predicted that we would end up with two dimensionless parameters. We started with seven physical quantities and have three basic units (M, L, and T). We can thus form three basic scales and use these to make the other four quantities dimensionless. We first made the variables q and t dimensionless, leaving two dimensionless parameters.

This logic is generalized in the Π *theorem* (Buckingham, 1914). For N physical quantities q_i (including variables and parameters) and m fundamental units, the Π

theorem states that an equation $f = 0$ depending on N dimensional physical quantities is equivalent to an equation $g = 0$ that depends on only $N - m$ dimensionless quantities:

$$f(q_1, \ldots, q_N) = 0 \quad \rightarrow \quad g(\Pi_1, \ldots, \Pi_{N-m}) = 0 \,, \tag{2.24}$$

where the Π_i are $N - m$ dimensionless quantities that are formed by dividing by the appropriate powers of fundamental scales. In the harmonic-oscillator example, f is the original functional relation, Eq. (2.21), while g is the scaled functional relation, Eq. (2.23).

Although we have used the Π theorem to understand how equations are scaled, it has even deeper applications, which go under the rubric of *dimensional analysis*. The Π theorem holds for any relation among physical quantities, not just equations of motion. Thus, for example, even if we did not know the equations of motion for an oscillator, we could relate, say, the period of oscillation to other physical quantities. We will simplify the discussion by dropping the forcing ($F = 0$) and the friction ($b = 0$). Then, introducing T_{osc} for the period, the relationship would be of the form $f(T_{\text{osc}}, m, k) = 0$.

We now have three physical quantities (of units T, M, and MT^{-2}) and two fundamental units (M and T). We thus have just one dimensionless variable, $\Pi \equiv T_{\text{osc}} \sqrt{k/m}$, and the Π theorem asserts that there is a function g such that $g(\Pi) = 0$. Because the function g depends on just one variable, its roots (solutions) will be isolated values of Π. If there is just one relevant value, then Π = constant. As a result,

$$T_{\text{osc}} = \text{const.} \sqrt{\frac{m}{k}} \,. \tag{2.25}$$

Thus, up to a constant, we can determine the period of a mass on a spring without knowing the equations of motion! In this sense, dimensional analysis is more fundamental. Knowing – or intuiting – the relevant physical quantities puts constraints on *any* model of the dynamics. These constraints lead to testable scaling laws such as period $\propto \sqrt{\text{mass}}$. A main point of this book is that successful control encodes, as much as possible, the knowledge one has about the dynamics into the control algorithm. In practice, we never know the dynamics perfectly. Even if we knew the correct form of the equations, we would not know the parameter values perfectly. By scaling and reducing the number of parameters, we limit the kinds of errors we can make about the dynamics.

You might be worrying about the choice of relevant variables. Have we found them all? In our analysis of the simple harmonic oscillator, we left out the initial conditions, $q(0)$ and $\dot{q}(0)$. And why stop there? We also left out a dependence on the radius of the proton (the mass is made of protons, which are certainly physical quantities). Of course, we intuitively believe that some quantities are certainly relevant (the seven that we originally included); some might be relevant [$q(0)$ and $\dot{q}(0)$]; and some seem clearly irrelevant. When can we exclude quantities that do not seem relevant?

We can formalize this question by noting that if we did have p extra quantities, say $q'_1, \ldots, q'_p$, then the Buckingham Π theorem would state that there is a g such that

$$g(\Pi_1, \ldots, \Pi_{N-k}; \Pi'_1, \ldots, \Pi'_p) = 0 \tag{2.26}$$

The extra dimensionless terms are formed using the p extra variables. When we scale them, we will typically find that the "relevant" dimensional terms are of order unity, and the "irrelevant" ones are much less than one.[15] Then, if some of our dimensionless parameters are very small and if the variation (derivative) of g with respect to those parameters also is small, we can just set them equal to zero and redefine our function to act on the remaining variables, justifying our choice of relevant physical variables.

Although such reasoning often works, it can fail in two ways. The first, more banal, is that some quantity assumed not to be irrelevant is. That is, we forgot to include a relevant parameter. Or, we may have assumed that a dimensionless constant is small, but it turns out to be of order unity. In either case, we have made a mistake!

The second, more subtle way to go astray is that even a small dimensionless parameter can have a big effect on the function g. Implicitly, we have assumed that the limit where the "irrelevant" parameter Π' goes to zero is *regular* – i.e., that $g(\Pi' \to 0) \approx g(\Pi' = 0)$. In other words, g tends to a constant value (when the other parameters are fixed) that is neither zero nor infinity. But such *singular* cases do arise, most famously at continuous phase transitions, where the result is that critical exponents can deviate from their mean-field values. For a detailed discussion, see Goldenfeld (1992) and Barenblatt (2003).

We conclude by summarizing the advantages of working with scaled equations:

1. We reduce the number of quantities that must be varied in an experiment.
2. Dimensionless quantities are required for numerical work.
3. Analytic derivations are easier to do using scaled equations.
4. Reducing the number of parameters reduces the number of possible modeling errors.
5. Measurements at one scale can predict the behavior at another scale.

The last point justifies using *scale models* in moderate-sized wind tunnels and wave tanks to study the dynamics of full-sized airplanes and ships.

2.3 Frequency Domain

Many of the systems we will be interested in are linear and time invariant (LTI). Such systems – the focus of much of practical control theory – are often much simpler when viewed in the frequency domain. There are a few basic ideas:

1. A sinusoidal input, $u(t) = u \cos \omega t$, gives an output that is a periodic steady state, which we write in the form $y(t) = y \cos(\omega t + \varphi)$. The ratio of amplitudes, y/u, and

[15] Or much greater than one, but then we use their inverse, which is much less than one.

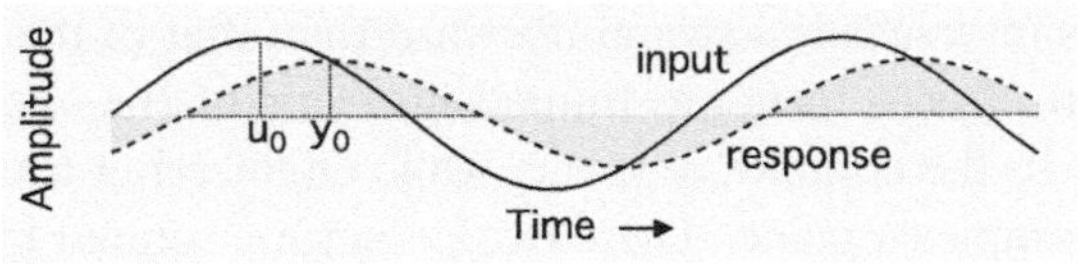

Fig. 2.3

Transfer function showing response to a sinusoidal input. Here, the relative amplitude $|G| = \left|\frac{y_0}{u_0}\right| = \frac{1}{\sqrt{2}}$, and the phase shift $\angle G = \angle u_0 - \angle y_0 = -45°$.

the phase shift φ between the two curves characterize the response at (angular) frequency $\omega = 2\pi f$, where f is the physical driving frequency. See Figure 2.3.

2. These relations are more conveniently expressed using complex numbers. Then the input $u(t) = u_0\, e^{i\omega t}$ gives rise to an output $y(t) = y_0\, e^{i\omega t}$, and the ratio of *complex* amplitudes y_0/u_0 gives the frequency response.
3. An *arbitrary* input $u(t)$ can be written as a sum or integral of sinusoids.
4. The *linearity* of the system implies that the response to $u(t)$ is just the sum or integral of the responses to each sinusoidal input.

Although these ideas are probably familiar to you, we review them briefly below, adding a few twists that are peculiar to control-theory applications.

2.3.1 Laplace vs. Fourier

Although physicists usually examine frequency response via the Fourier transform, control engineers almost invariably use the Laplace transform. Recall their definitions:

$$\text{Fourier transform: } f(\omega) = \int_{-\infty}^{\infty} dt\, f(t)\, e^{-i\omega t} \equiv \mathcal{F}[f]$$
$$\text{Laplace transform: } f(s) = \int_{0}^{\infty} dt\, f(t)\, e^{-st} \equiv \mathcal{L}[f] \tag{2.27}$$

For control applications, the Laplace transform has some minor advantages:

1. It is better suited for initial-value problems.
2. It applies to a wider class of functions:[16]
   a. The Fourier transform $\Longleftrightarrow$ $f(t)$ must decay to 0 at $t = \pm\infty$.
   b. The Laplace transform $\Longleftrightarrow$ $f(t)$ can grow as an arbitrary-order polynomial.
3. The Laplace transform is more natural for decaying exponential functions.

Often, the transforms can be used interchangeably. For example, we can often recover the Fourier transform from the Laplace by setting $s = i\omega$. One issue with the Laplace

[16] If – as is often the case in the physics literature – the Fourier transform variable ω is complex, then convergence of the transform is nearly the same as for Laplace transforms (up to rotation by 90° and infinite domains).

transform is that its inverse is less straightforward than that of the Fourier transform. Also, discussions in physics texts are framed in terms of the analytic properties of Fourier transforms in the complex ω plane, while engineering texts refer to Laplace transforms in the complex s plane. These two planes are rotated by 90 degrees. Even more confusing, the engineering literature usually defines the Fourier transform $f(\omega)$ using $e^{-\mathrm{i}\omega t}$ in the integral of $f(t)$, as we did in Eq. (2.27), above, while the physics literature mostly uses $e^{+\mathrm{i}\omega t}$. Thus, where a physicist considers the properties of functions in the upper and lower frequency plane, the engineer considers the left- and right-hand parts of the s plane.

In Eq. (2.27) we used the same symbol, f, for the time and transform domains, even though the two quantities are quite different functions of the arguments t, ω, and s. This "overloading" of notation makes it easier to keep track of variables. The elementary properties of Fourier and Laplace transforms are reviewed in the Appendix online, Section A.4.

2.3.2 Transfer Functions

One reason to use Laplace transforms is that they give an intuitive way to understand how a linear system transforms an input signal into an output signal. The Laplace transform converts an nth-order linear differential equation with constant coefficients to an nth-order algebraic equation in s.[17] For example, the first-order system and transform

$$\dot{y}(t) = -y(t) + u(t)$$
$$sy(s) = -y(s) + u(s)\,,$$

lead to

$$G(s) \equiv \frac{y(s)}{u(s)} = \frac{1}{1+s}\,, \tag{2.28}$$

where the *transfer function* $G(s)$ is the ratio of output to input, in Laplace transform space, assuming zero initial conditions. The frequency response is obtained from $G(s = \mathrm{i}\omega)$:

$$G(\mathrm{i}\omega) = \frac{1}{1+\mathrm{i}\omega}\,. \tag{2.29}$$

Because $G(\mathrm{i}\omega)$ is a complex number, we can write $G(\mathrm{i}\omega) = |G(\mathrm{i}\omega)|\,\mathrm{e}^{\mathrm{i}\varphi}$, with

$$|G(\mathrm{i}\omega)| = \frac{1}{\sqrt{1+\omega^2}}\,, \qquad \varphi = \tan^{-1}\left(\frac{\mathrm{Im}\,G(\mathrm{i}\omega)}{\mathrm{Re}\,G(\mathrm{i}\omega)}\right) = -\tan^{-1}\omega\,. \tag{2.30}$$

The magnitude $|G|$ of the transfer function G is the ratio of output to input amplitude, or *gain*. The argument of the transfer function, $\angle G$, gives the phase shift φ between the output and input sine waves, as illustrated in Figure 2.3.

[17] We assume that all boundary terms in the Laplace transforms are zero, which requires the relevant initial conditions to be zero. Occasionally, it will be important to keep the initial conditions (see, for example, Eq. (3.52)). See the Appendix online, Section A.4 for an example where we use Laplace transforms to solve an ODE with initial conditions.

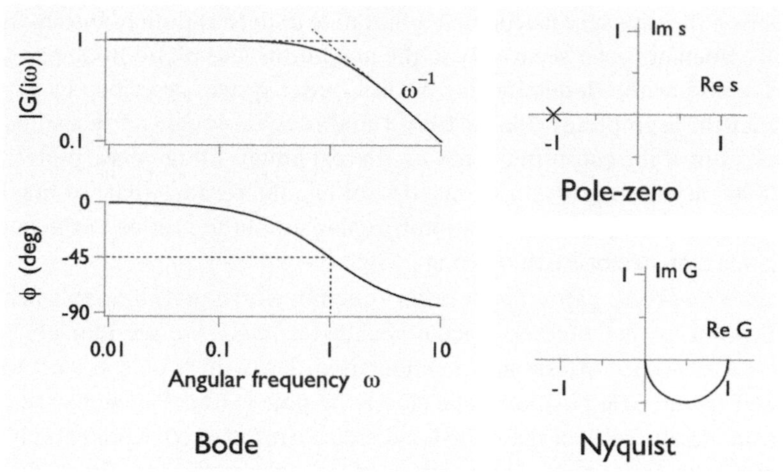

Bode, pole-zero, and Nyquist plots for first-order transfer function (low-pass filter). Fig. 2.4

Note that to return to dimensional units, we let $s \to s/\omega_0$ and $\omega \to \omega/\omega_0$, with $\omega_0 = 2\pi f_0$. We will see that f_0 can be interpreted as a *cutoff frequency* of the low-pass filter (see Figure 2.4, where the cutoff frequency is scaled to be unity).

When returning to dimensional units, the numerator may also acquire a constant value, G_0, known as the *DC gain*, which represents the system's response at zero frequency. In electrical engineering jargon, DC means *direct current* – the steady current at zero frequency – and AC or *alternating current* refers to a response at a frequency ω. The jargon derives from electrical circuits such as Example 2.2 and is commonly used even when the applications have nothing to do with electrical circuits. Note that if the input and output have different units, then G_0 is a dimensional conversion factor. For example, if the input is a current and the output a voltage, then G_0 has units of electrical resistance (ohms).

Graphical Tools

There are several types of graphs that can help to visualize the properties of transfer functions. We consider here three types of plots: *Bode*, *pole-zero*, and *Nyquist*. All of these are standard in control software. The plots are illustrated in Figure 2.4 for the first-order system of Eq. (2.28), so that we can start to develop an intuition for them.

1. *Bode plots.* Log-log graphs of the transfer function magnitude $|G(\mathrm{i}\omega)|$ and linear-log graphs of $\arg(G(\mathrm{i}\omega))$ are commonly known as *Bode plots*,[18,19] and we

[18] In the physics literature, the role of the transfer function $G(s)$ is played by the dynamical linear response function, $\chi(\omega) \equiv G(s = -\mathrm{i}\omega)$ (Chaikin and Lubensky, 1995). The negative sign arises because the engineering literature adopts a different sign for the Fourier transform than the physics literature. Another way to "convert" is to remember the mnemonic $\mathrm{j} \to -\mathrm{i}$.

[19] *Hendrik Bode* (1905–1982), was a pioneer in the use of transfer functions and Laplace transforms to analyze dynamics. Much of his career took place at Bell Telephone Laboratories. His name is pronounced "Boh-dah."

will see that they are very useful for a qualitative understanding of system dynamics. We will sometimes refer separately to the magnitude and phase Bode plots. In Figure 2.4, we show the Bode plots for the first-order system described by Eqs. (2.30). Note that the asymptotes $G(\omega \to 0) = 1$ and $G(\omega \to \infty) \sim \omega^{-1}$ in the magnitude plot intercept at the cutoff frequency ω_0 (in real units). In the phase plot, the phase lag is $-90°$ asymptotically, crossing $-45°$ at ω_0. Note, too, that we break partly from traditional engineering notation by using amplitude ratios in the magnitude plot rather than decibels (dB $= 20\log_{10}|G|$).

2. *Pole-zero / root-locus plots.* The transfer function $G(s)$ equals infinity when $s = -1$. Such a point in the complex s-plane is called a *pole*. We see that poles on the negative real s-axis correspond to exponential decay of impulses, with the decay rate fixed by the pole position. The closer the pole is to the imaginary s-axis, the slower the decay. Poles in the right-hand side correspond to exponentially increasing amplitudes. We can plot the positions of the poles in the complex s-plane, using an **×** by convention. Similarly, if $G = 0$ at some s, then we denote the position of this *zero* by a **O**.

   If the transfer function depends on a (dimensionless) control parameter, then the positions of poles and zeros will in general be a function of this parameter and thus move as the control parameter is varied. See the example of a second-order system, below. The locus of points traced out by the poles (and/or zeros) as a control parameter is varied is called a *root-locus* plot. Control software can animate the graph, so that the poles and zeros move in real time in response to a control parameter variation.

3. *Nyquist plots.* Also known to physicists as *Cole–Cole plots*,[20] Nyquist plots trace the imaginary part of $G(i\omega)$ along the y-axis against its real part along the x-axis. In other words, we make a 2D vector from G in the complex plane and then track the head of the vector as the frequency changes from 0 to ∞. In the present case of a first-order system, as $\omega \to 0$, we see that $G(i\omega) \to (1, 0)$. As $\omega \to \infty$, we similarly see that $G(i\omega) \to (0, 0)$. In Problem 2.2, you will show that the intermediate frequencies map to a semicircle.

Example 2.4 (Second-order system) We rewrite Eq. (2.16) in terms of y and u and then Laplace transform, ignoring the initial conditions.

$$\begin{aligned} \ddot{y} + 2\zeta\,\dot{y} + y &= u(t)\,, \\ \left(s^2 + 2\zeta\,s + 1\right)y(s) &= u(s)\,, \end{aligned} \tag{2.31}$$

so that the transfer function is

$$G(s) = \frac{1}{1 + 2\,\zeta\,s + s^2} \qquad \text{or} \qquad G(i\omega) = \frac{1}{1 + 2i\,\zeta\,\omega - \omega^2}\,. \tag{2.32}$$

[20] Cole–Cole plots were introduced to model the frequency response of dielectric media (Cole and Cole, 1941). The complex dielectric, $\epsilon(\omega)$, may be thought of as a transfer function between the electric field, $\boldsymbol{E}$ (the "input"), and the displacement field, $\boldsymbol{D}$ (the "output"), as described by the *constitutive relation* $\boldsymbol{D} = \epsilon\boldsymbol{E}$.

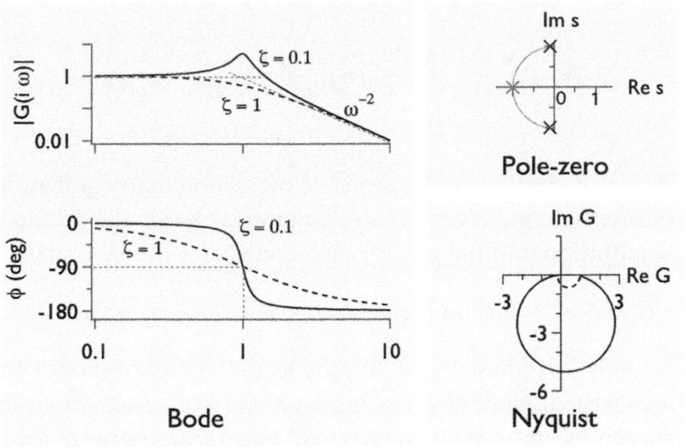

Fig. 2.5

Bode, pole-zero, and Nyquist plots for a second-order system. $\zeta = 0.1$ (solid line) gives underdamped and $\zeta = 1$ (dashed line) gives critically damped dynamics. In the pole-zero plot, the two black crosses denote the poles for $\zeta = 0.1$, while the gray cross shows the double pole at -1, for $\zeta = 1$.

Bode, pole-zero, and Nyquist plots for various damping ratios ζ are shown in Figure 2.5. (Recall that we have scaled $\omega_0 = 1$.) For a second-order system, there are two poles. In the underdamped case ($\zeta < 1$), they form a complex-conjugate pair $s = -\zeta \pm \mathrm{i}\sqrt{1-\zeta^2}$. In the overdamped case ($\zeta > 1$), the two poles are both on the real s-axis. Since $|s| = 1$ for $0 < \zeta < 1$, we see that, in the underdamped case, the poles lie on a circle of radius 1 (ω_0 is dimensional units) in the complex s-plane. The details are left to Problem 2.3.

Rational Transfer Functions

Both the low-pass filter and the second-order system have transfer functions $G(s)$ that are rational functions; that is, they can be written as $N(s)/D(s)$, where $N(s)$ and $D(s)$ are polynomials (numerator and denominator, respectively). Most of the linear dynamical systems that we will consider have rational transfer functions. In general, we can consider an nth-order ODE driven by an kth-order input, that is,

$$\frac{\mathrm{d}^n y}{\mathrm{d}t^n} + a_1 \frac{\mathrm{d}^{n-1} y}{\mathrm{d}t^{n-1}} + \cdots + a_n y = b_0 \frac{\mathrm{d}^k u}{\mathrm{d}t^k} + b_1 \frac{\mathrm{d}^{k-1} u}{\mathrm{d}t^{k-1}} + \cdots + b_k u \,. \tag{2.33}$$

Neglecting initial conditions, the Laplace transform is

$$(s^n + a_1 s^{n-1} + \cdots + a_n)\, y(s) = (b_0 s^k + b_1 s^{k-1} + \cdots + b_k)\, u(s)\,, \tag{2.34}$$

and the transfer function is given by

$$G(s) = \frac{y(s)}{u(s)} = \frac{b_0 s^k + b_1 s^{k-1} + \cdots + b_k}{s^n + a_1 s^{n-1} + \cdots + a_n} \equiv \frac{b(s)}{a(s)} . \quad (2.35)$$

The *order* of $G(s)$ is defined as the order of the denominator polynomial (n here). Because the transfer function is a ratio, we can always choose the coefficient $a_0 = 1$.

To gain some intuition into the general case, consider a simple transfer function,

$$G(s) \sim s^{-n} \implies \text{frequency response } G(\mathrm{i}\omega) = (-\mathrm{i})^n \, \omega^{-n} . \quad (2.36)$$

Since $-\mathrm{i}$ corresponds to a phase lag of 90°, the term $(-\mathrm{i})^n$ corresponds to a lag of $n \times 90°$. This lag is consistent with the phase lags at high frequencies we see in Figures 2.4 and 2.5 for first- and second-order systems (−90° and −180°, respectively). The second term, ω^{-n}, is a straight line of slope $-n$ on a log-log plot. Note the slopes of −1 and −2 in the high-frequency portions of Figures 2.4 and 2.5.

The behavior of the general $G(s)$ in Eq. (2.35) can be pieced together from regions where $G(s)$ is approximately s^p, where p is an integer. For example, if the order of the numerator $N(s)$ is k and that of the denominator $D(s)$ is n, then at high frequencies, we expect a phase lag of $(n - k)\pi/2$ and a slope of $n - k$ on the Bode magnitude plot. Likewise, the behavior at $\omega \to 0$ is dominated by a different exponent. Often, $G \to$ constant (the *DC gain*) at low frequencies. The phase lag is then zero.

Another way to understand the general $G(s)$ is to use *partial fractions* to write G as a sum of lower-order polynomials. (See Section A.4.4 online.) For example,

$$G(s) = \frac{1}{(s+2)(s+3)} = \frac{1}{s+2} - \frac{1}{s+3} . \quad (2.37)$$

Writing $G(s)$ as a sum of partial fractions is more accurate numerically for higher-order equations. (See the discussion surrounding Eq. (4.44).)

Proper Transfer Functions

The response of physical systems is small at high frequencies and vanishes in the limit that the input frequency tends to infinity. Physically, a system just cannot respond infinitely quickly. At high frequencies, the response must decay to zero. In the language of transfer functions, a vanishing response at high frequency means that $|G(\mathrm{i}\omega)| \to 0$ as $\omega \to \infty$. If $G(s)$ is given by a rational function, then the denominator order n must be larger than the numerator order k. Such transfer functions are called *strictly proper*.

Although physical transfer functions must be strictly proper, we will often use transfer functions with equal numerator and denominator orders. Such transfer functions are called *biproper*.[21] Many controller transfer functions, for example, are biproper (including the common proportional controller discussed in Chapter 3). This can

[21] Or, sometimes, *semiproper*.

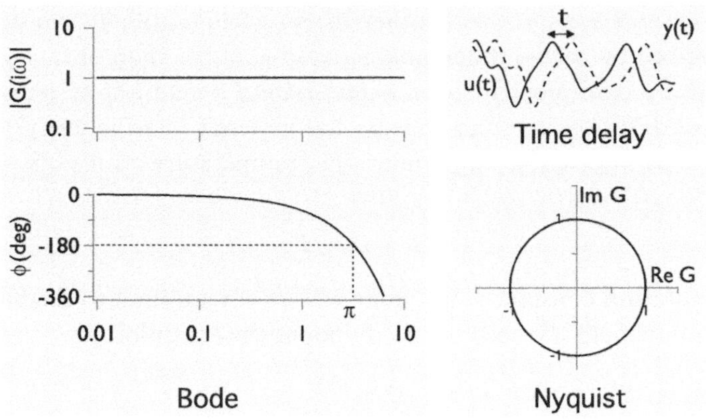

Fig. 2.6 Time series, Bode, and Nyquist plots for a time delay. Input (solid line) $u(t)$ is delayed by τ to produce output $y(t)$ (dashed line).

make sense if the overall transfer function of controller in series with a physical system is strictly proper.

More subtly, transfer-function *models* can be biproper. As models are valid only up to some high-frequency limit, the "constant" response at infinite frequency summarizes, in a crude way, what "remains" at the high-frequency limit. A transfer function is *proper* if it is either strictly proper or biproper. Otherwise, it is *improper*.

In the next section, we will discuss *extended systems* further. Problem 2.6 goes through a detailed derivation of a lumped-element approximation in the context of heat conduction.

Nonrational Transfer Functions: Delays and Extended Systems

Not every transfer function can be written as a rational function. For example, consider a sensor that faithfully records a signal, but with a time delay τ – i.e., $y(t) = u(t - \tau)$. From the shift theorem for Laplace transforms, the transfer function for such a sensor is $G(s) = \mathrm{e}^{-s\tau}$. Because Taylor's theorem shows that the exponential function is an infinite-order polynomial, the time dynamics are equivalent to an infinite-order ODE. Note that the magnitude of $G(s = \mathrm{i}\omega)$ is always 1 and that the phase lag $\varphi = \tan^{-1}(\mathrm{Im}\, G/\mathrm{Re}\, G) = -\omega\tau$ increases linearly with frequency. In contrast, in Examples 2.2 and 2.4, the phase lag tended to an asymptotic value (90° and 180°, respectively). In Figure 2.6, we plot the delayed signal, with its Bode and Nyquist plots. In contrast to Examples 2.2 and 2.4, there are no poles or zeros, as $|G| = 1$ for all ω. The Nyquist plot is the locus of $(\cos\omega\tau, \sin\omega\tau)$ as ω is varied from 0 to ∞ and is just a circle of radius 1, centered at the origin. Note how the phase in the Bode plot bends down, a consequence of the logarithmic frequency axis.

Spatially extended systems are another origin of nonrational transfer functions. They are described by partial differential equations rather than ordinary differential equations and are equivalent to *infinite-dimensional* dynamical systems. A simple example is heat diffusion in a one-dimensional material (see Figure 2.7). Let $T(x, t)$ be the temperature at position x and time t. The temperature profile obeys a diffusion equation,

$$\partial_t T = D\,\partial_{xx} T\,. \tag{2.38}$$

The thermal diffusion constant D has units of ℓ^2/t. In addition to the bulk equation for the temperature field, we need to specify boundary conditions:

$$T(x \to \infty, t) = T_\infty\,, \tag{2.39a}$$

$$-\lambda\,\partial_x T|_{(x=0,t)} = J(t) = P(t)/a\,, \tag{2.39b}$$

where λ is the thermal conductivity of the material and $J = P/a$ is the heat flux (power P divided by the heater area a). Equation (2.39b) is *Fourier's Law*: heat flux is proportional to the (negative) gradient of the temperature. We also need to specify the output,

$$y(t) = T(x = \ell, t)\,. \tag{2.40}$$

As usual, scaling can simplify the dynamical equations. Let $x \to x/\ell$, $t \to t/(\ell^2/D)$, $T \to (T - T_\infty)/T_\infty$, and $P \to P\ell/(\lambda a T_\infty) \equiv u$. Note that in addition to rescaling T, we also shifted the reference temperature T_∞ to 0. The simplified equations then become

$$\begin{aligned} \partial_t T &= \partial_{xx} T\,, & y &= T(1, t)\,, \\ T(x \to \infty, t) &= 0\,, & -\partial_x T|_{(x=0,t)} &= u(t)\,. \end{aligned} \tag{2.41}$$

We proceed by Laplace transforming in time,

$$sT = \partial_{xx} T \quad \Longrightarrow \quad T(s, x) = \overset{0}{\cancel{A}}\,\mathrm{e}^{\sqrt{s}x} + B\,\mathrm{e}^{-\sqrt{s}x}\,, \tag{2.42}$$

where we take $A = 0$ since $T(\infty, t) = 0$. The gradient condition then gives

$$\partial_x T|_0 = -\sqrt{s}B \cdot 1 = -u(s) \quad \Longrightarrow \quad u(s) = B\sqrt{s}\,. \tag{2.43}$$

The output equation then gives

$$y(s) = T(1, s) = B\,\mathrm{e}^{-\sqrt{s}\cdot 1} \quad \Longrightarrow \quad G(s) = \frac{y(s)}{u(s)} = \frac{B\,\mathrm{e}^{-\sqrt{s}}}{B\sqrt{s}} = \frac{\mathrm{e}^{-\sqrt{s}}}{\sqrt{s}}\,. \tag{2.44}$$

We have found the transfer function, $G = \mathrm{e}^{-\sqrt{s}} / \sqrt{s}$ for a case where the input is power oscillations at $x = 0$ and the output is the temperature at $x = 1$ (or ℓ in dimensional units). The form of the transfer function depends on where we choose the observation point. (Remember that the material is semi-infinite.)

In Figure 2.7, we show the Bode and Nyquist plots. Note that, as with the delay, the phase lag increases indefinitely with frequency. This behavior is in accord with the idea that the exponential contains all powers of its argument (as expressed in its Taylor expansion). In contrast to the case of rational functions, the magnitude plot

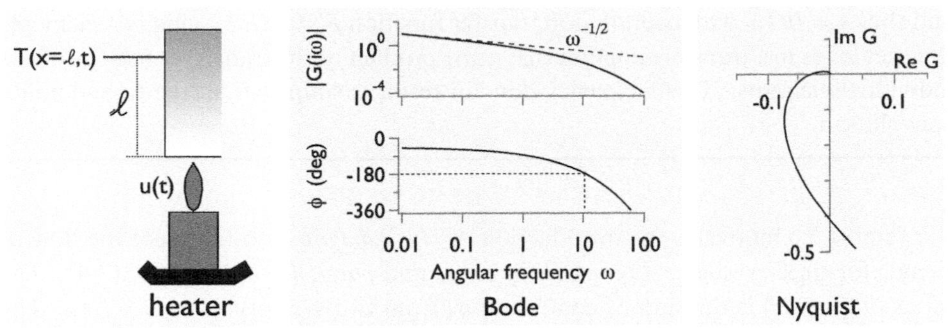

Fig. 2.7 Bode and Nyquist plots for a heated, semi-infinite, one-dimensional bar. The transfer function G relates the input at the flame, $u(t)$, to the sensor probe at $x = \ell$.

can have noninteger power laws (slopes on the log-log plot). For example, $G \sim \omega^{-1/2}$ at low frequencies. Further properties of Equation (2.44) are explored in Problem 2.4.

We conclude this section by noting that we have considered two nonrational transfer functions: $G = \mathrm{e}^{-s}$ for a delay (scaled), and $G = \mathrm{e}^{-\sqrt{s}} / \sqrt{s}$ for thermal diffusion in one dimension. While neither can be expressed in terms of finite-order polynomials, they are analytic and meromorphic functions of the complex variable s and $\sqrt{s}$, respectively.[22] Problem 2.4 also discusses how nonanalytic transfer functions can be approximated, over a finite range of frequencies, by rational polynomials in s (Padé approximants).

2.3.3 Convolution Theorem and Block Diagrams

The convolution theorem (Appendix A.4.3 online) allows important manipulations of transfer functions. Given two functions $G(t)$ and $H(t)$ and their convolution $F(t) = G * H \equiv \int_0^\infty \mathrm{d}t'\, G(t')\, H(t-t')$, then $\mathcal{L}[G * H] = G(s)\, H(s) = F(s)$. Convolution describes compound systems where the output of one element is fed into the input of the next element.

Example 2.5 (Sensor dynamics) Consider a first-order sensor element that reports the position of a second-order mechanical system. We have $\ddot{v} + 2\zeta\dot{v} + v = u(t)$ and $\dot{y} + y = v(t)$, where $u(t)$ drives the oscillator position $v(t)$, which drives the output $y(t)$. Taking the Laplace transform, we have

$$\begin{aligned} v(s) &= G(s)\, u(s) = \frac{1}{1 + 2\zeta s + s^2}\, u(s) \\ y(s) &= H(s)\, v(s) = \frac{1}{1+s}\, v(s) = \frac{1}{(1+s)(1 + 2\zeta s + s^2)}\, u(s) \equiv F(s)\, u(s), \end{aligned} \tag{2.45}$$

[22] *Meromorphic* functions are analytic (have a well-defined derivative with respect to their argument) everywhere in the complex plane, except at a set of isolated poles (here, at $\sqrt{s} = 0$).

and thus $y = HG\,u$, with overall loop transfer function $F = HG$. Having two elements in series leads to a transfer function that is the product of the transfer functions of the individual elements. This frequency-domain result is simpler than the time-domain convolution.

Series

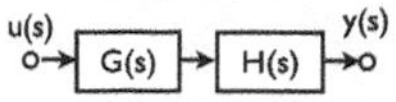

y(s) = [G H] u(s)

Parallel

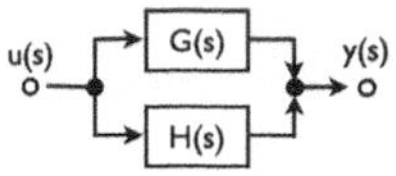

y(s) = [G+H] u(s)

Example 2.5 motivates the introduction of *block diagrams* to represent the flow of signals for linear systems. Dynamics in *series* and *parallel* are depicted at left. The series output $y(s)$ is the sum of contributions from two systems: $y = (G + H)\,u$. The arrows give the direction of signal flow and are omitted when the context makes that direction clear. Input and output ports are denoted by open circles, and summing junctions are denoted by small black disks. Later, a negative sign will indicate a difference junction. We will also see, in Chapter 3, that block diagrams are particularly useful for describing feedback loops.

2.4 Time Domain

Although the frequency domain can be intuitive, the time domain can be better for systems with multiple inputs and outputs (MIMO), stochastic effects, transients and other time inhomogeneities, nonlinearities, and combinations of the above. We start with the simplest case of linear time-invariant systems. Recall Eq. (2.12):

$$\dot{\boldsymbol{x}} = \boldsymbol{A}\boldsymbol{x} + \boldsymbol{B}\boldsymbol{u} \qquad y = \boldsymbol{C}\boldsymbol{x} + \boldsymbol{D}\boldsymbol{u}\,. \tag{2.46}$$

To solve Eq. (2.46), we introduce a change of variable, $\boldsymbol{x} = \mathrm{e}^{\boldsymbol{A}t}\,z$, where the *matrix exponential* is defined in Section A.1.4 online. Substituting into Eq. (2.46), we have

$$\dot{z} = \mathrm{e}^{-\boldsymbol{A}t}\,\boldsymbol{B}u \quad \Longrightarrow \quad z(t) = z(0) + \int_0^t \mathrm{d}t'\,\mathrm{e}^{-\boldsymbol{A}t'}\,\boldsymbol{B}u(t')\,. \tag{2.47}$$

Changing back to $\boldsymbol{x}$, we have

$$\dot{\boldsymbol{x}} = \boldsymbol{A}\boldsymbol{x} + \boldsymbol{B}\boldsymbol{u} \quad \Longrightarrow \quad \boldsymbol{x}(t) = \mathrm{e}^{\boldsymbol{A}t}\,\boldsymbol{x}(0) + \int_0^t \mathrm{d}t'\,\mathrm{e}^{\boldsymbol{A}(t-t')}\,\boldsymbol{B}\,\boldsymbol{u}(t')\,. \tag{2.48}$$

In Eq. (2.48), the first term, $\mathrm{e}^{\boldsymbol{A}(t-0)}$, propagates the initial condition $\boldsymbol{x}_0$ from time 0 to the present, time t. The second term, $\mathrm{e}^{\boldsymbol{A}(t-t')}$, propagates the input from time t' in the past to the present and sums over all the intermediate times t' between 0 and t.

If A can be diagonalized as $A = RD_0R^{-1}$, then its matrix exponential is (Appendix A.1.4 online)

$$\mathrm{e}^{At} = \mathrm{e}^{RD_0R^{-1}t} = R\,\mathrm{e}^{D_0t}\,R^{-1}\,, \qquad \mathrm{e}^{D_0t} = \begin{pmatrix} \mathrm{e}^{\lambda_1 t} & \cdots & 0 \\ \vdots & \ddots & \vdots \\ 0 & \cdots & \mathrm{e}^{\lambda_n t} \end{pmatrix}, \tag{2.49}$$

where we distinguish the diagonal matrix D_0 from the feedthrough matrix D.

The eigenvalues of the dynamical matrix determine the character of the time evolution, because the components of the state vector are linear combinations of the $\mathrm{e}^{\lambda_i t}$. Assuming no repeated values of λ, there are four basic types of behavior: When λ is real, the state shows exponential decay or growth, depending on whether λ is negative or positive. When λ is complex, the state oscillates, decaying for $\mathrm{Re}\,\lambda < 0$ and growing for $\mathrm{Re}\,\lambda > 0$. The angular frequency ω is given by $\omega = \mathrm{Im}\,\lambda$. The four cases are illustrated at right.

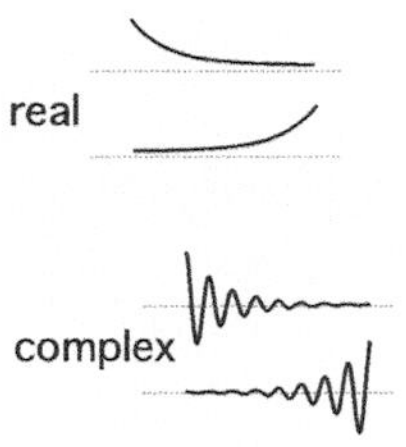

Example 2.6 (Driven harmonic oscillator) We consider the driven harmonic oscillator of Eq. (2.17), setting the damping $\zeta = 0$, for simplicity. We have

$$A = \begin{pmatrix} 0 & 1 \\ -1 & 0 \end{pmatrix} = \underbrace{\frac{1}{\sqrt{2}}\begin{pmatrix} 1 & 1 \\ \mathrm{i} & -\mathrm{i} \end{pmatrix}}_{R} \underbrace{\begin{pmatrix} \mathrm{i} & 0 \\ 0 & -\mathrm{i} \end{pmatrix}}_{D} \underbrace{\frac{1}{\sqrt{2}}\begin{pmatrix} 1 & -\mathrm{i} \\ 1 & \mathrm{i} \end{pmatrix}}_{R^{-1}}. \tag{2.50}$$

Thus,

$$\begin{aligned} \mathrm{e}^{At} &= \underbrace{\frac{1}{\sqrt{2}}\begin{pmatrix} 1 & 1 \\ \mathrm{i} & -\mathrm{i} \end{pmatrix}}_{R} \underbrace{\begin{pmatrix} \mathrm{e}^{\mathrm{i}t} & 0 \\ 0 & \mathrm{e}^{-\mathrm{i}t} \end{pmatrix}}_{\mathrm{e}^{Dt}} \underbrace{\frac{1}{\sqrt{2}}\begin{pmatrix} 1 & -\mathrm{i} \\ 1 & \mathrm{i} \end{pmatrix}}_{R^{-1}} = \begin{pmatrix} \cos t & \sin t \\ -\sin t & \cos t \end{pmatrix}, \\ x(t) &= x(0)\,\cos t + v(0)\sin t\,, \end{aligned} \tag{2.51}$$

where $x(0)$ and $v(0)$ are the initial position and velocity, respectively (that is, the components of x). We assume no inputs ($u = 0$), for simplicity. Equation (2.51) is the elementary direct solution of the second-order equation. Although this example is straightforward to do by hand, computers make these kinds of problems easy. See Problem 2.7.

2.4.1 Converting between Time and Frequency Domains

From State Space to the Frequency Domain

We can study linear systems equivalently in either the frequency (Laplace) or time domains, and it is useful to be able to go back and forth between them. Laplace transforming Eq. (2.46) and dropping the initial condition, we have,

$$\begin{aligned} s\,x(s) &= Ax(s) + Bu(s) & \qquad y(s) &= Cx(s) + Du(s) \\ (s\mathbb{I} - A)\,x(s) &= Bu(s) & x(s) &= (s\mathbb{I} - A)^{-1}\,Bu(s). \end{aligned} \tag{2.52}$$

The transfer function $\mathbf{G}(s)$, a $p \times m$ matrix between the m inputs $\mathbf{u}$ and p outputs $\mathbf{y}$, is then

$$\mathbf{G}(s) = \mathbf{C}\,(s\mathbb{I} - \mathbf{A})^{-1}\,\mathbf{B} + \mathbf{D}\,. \tag{2.53}$$

The *poles* of the transfer function, $\mathbf{G}(s)$, are given by the eigenvalues of the matrix $\mathbf{A}$:

$$\mathbf{A}\mathbf{v} = \lambda\,\mathbf{v} \quad \Longrightarrow \quad (\lambda\,\mathbb{I} - \mathbf{A})\,\mathbf{v} = \mathbf{0} \quad \Longrightarrow \quad \det\,(\lambda\,\mathbb{I} - \mathbf{A}) = 0\,. \tag{2.54}$$

The last term gives the nth-order *characteristic polynomial* of $\mathbf{A}$, whose n roots $(\lambda_1, \ldots, \lambda_n)$ are the n eigenvalues of $\mathbf{A}$. For a SISO system, the transfer function $G(s)$ is scalar, with singularities at each $s_i = \lambda_i$. The equivalence between the eigenvalues of $\mathbf{A}$ and the poles of G explains the interpretation we gave the pole-zero plots in Section 2.3.2. Finally, we see explicitly that the $\mathbf{D}$ matrix implies a constant term in the (biproper) transfer function. But since we typically know $\mathbf{u}(t)$ and $\mathbf{D}$, we can always define a new output $\mathbf{y}' = \mathbf{y} - \mathbf{D}\mathbf{u} = \mathbf{C}\mathbf{x}$, which corresponds to a strictly proper transfer function.

Example 2.7 (Second-order system) We return to the example of a damped, driven harmonic oscillator (Eq. 2.17). If we observe position,

$$G(s) = \underbrace{\begin{pmatrix}1 & 0\end{pmatrix}}_{\mathbf{C}}\left[s\underbrace{\begin{pmatrix}1 & 0\\ 0 & 1\end{pmatrix}}_{\mathbb{I}} - \underbrace{\begin{pmatrix}0 & 1\\ -1 & -2\zeta\end{pmatrix}}_{\mathbf{A}}\right]^{-1}\underbrace{\begin{pmatrix}0\\ 1\end{pmatrix}}_{\mathbf{B}} = \frac{1}{1 + 2\zeta s + s^2}\,. \tag{2.55}$$

If we observe *both* position and velocity,

$$\mathbf{G}(s) = \underbrace{\begin{pmatrix}1 & 0\\ 0 & 1\end{pmatrix}}_{\mathbf{C}}\left[s\underbrace{\begin{pmatrix}1 & 0\\ 0 & 1\end{pmatrix}}_{\mathbb{I}} - \underbrace{\begin{pmatrix}0 & 1\\ -1 & -2\zeta\end{pmatrix}}_{\mathbf{A}}\right]^{-1}\underbrace{\begin{pmatrix}0\\ 1\end{pmatrix}}_{\mathbf{B}} = \frac{1}{1 + 2\zeta s + s^2}\begin{pmatrix}1\\ s\end{pmatrix}. \tag{2.56}$$

That is, $\mathbf{G}$ is a 2×1 matrix of transfer functions. $G_{11}(s)$ is our familiar transfer function between the input force and position, and $G_{21}(s)$ is the transfer function between the input and the velocity. Not surprisingly, $G_{21} = s\,G_{11}$, because taking a derivative in the time domain is equivalent to multiplying by s, or $i\omega$, in the frequency domain.

You should check the various matrix multiplications (Problem 2.8).

And Back Again

Now we go back from a transfer function to a state-space formulation. Recall Eq. (2.35),

$$G(s) = \frac{y(s)}{u(s)} = \frac{b_0 s^k + b_1 s^{k-1} + \cdots + b_k}{s^n + a_1 s^{n-1} + \cdots + a_n} \equiv \frac{b(s)}{a(s)}\,. \tag{2.57}$$

We assume that $G(s)$ is strictly proper, with $k < n$. For state-space matrices $\{\boldsymbol{A}, \boldsymbol{B}, \boldsymbol{C}\}$, there is a unique transfer function $G(s)$, given by Eq. (2.53). (We neglect feedthrough, so that $\boldsymbol{D} = \boldsymbol{0}$.) But the converse does not hold: given a transfer function $G(s)$, there is not a unique way to specify $\{\boldsymbol{A}, \boldsymbol{B}, \boldsymbol{C}\}$. That is, although the inputs $\boldsymbol{u}$ and outputs $\boldsymbol{y}$ have physical significance, the individual components of the matrices $\boldsymbol{A}$, $\boldsymbol{B}$, and $\boldsymbol{C}$ do not depend on the coordinates (Section 2.4.3). Here is one way to construct such matrices. The idea is that, for a SISO system, an nth-order transfer function corresponds to an nth-order ODE for $y(t)$. Then, we write $\boldsymbol{x}$ as $\dot{x}_1 = x_2, \dot{x}_2 = x_3, \ldots, \dot{x}_{n-1} = x_n$, and

$$\dot{x}_n = -a_n x_1 - a_{n-1} x_2 - \cdots - a_1 x_n + u \,. \tag{2.58}$$

The state-space version of these equations is equivalent to Eq. (2.57). See Problem 2.10.

$$\dot{\boldsymbol{x}} = \underbrace{\begin{pmatrix} 0 & 1 & 0 & \cdots & 0 & 0 \\ 0 & 0 & 1 & \cdots & 0 & 0 \\ \vdots & \vdots & \vdots & & \vdots & \vdots \\ 0 & 0 & 0 & \cdots & 0 & 1 \\ -a_n & -a_{n-1} & -a_{n-2} & \cdots & -a_2 & -a_1 \end{pmatrix}}_{\boldsymbol{A}} \begin{pmatrix} x_1 \\ x_2 \\ \vdots \\ x_{n-1} \\ x_n \end{pmatrix} + \underbrace{\begin{pmatrix} 0 \\ 0 \\ \vdots \\ 0 \\ 1 \end{pmatrix}}_{\boldsymbol{B}} u \tag{2.59}$$

$$y = \underbrace{\begin{pmatrix} b_k & b_{k-1} & \cdots & b_1 & b_0 & 0 & \cdots & 0 \end{pmatrix}}_{\boldsymbol{C}} \boldsymbol{x} \,.$$

Equation (2.59) is the *controller canonical* form of the linear dynamical system $\{\boldsymbol{A}, \boldsymbol{B}, \boldsymbol{C}\}$. Because there is a chain of integrators ($u \to x_n \to x_{n-1} \to \ldots \to x_2 \to x_1$), a single control can affect all n states (Section 4.2). Note that we need to add $n - k$ zeros to the row vector $\boldsymbol{C}$ to make its total length equal to n, the dimension of the state vector $\boldsymbol{x}$. These zeros correspond to the "missing orders" in the numerator of the transfer function.

Example 2.8 The second-order system $G(s) = \frac{1}{1+2\zeta s+s^2}$ has $a_1 = 2\zeta,\ a_2 = 1,\ b_0 = 1$,

$$\boldsymbol{A} = \begin{pmatrix} 0 & 1 \\ -1 & -2\zeta \end{pmatrix}, \quad \boldsymbol{B} = \begin{pmatrix} 0 \\ 1 \end{pmatrix}, \quad \boldsymbol{C} = \begin{pmatrix} 1 & 0 \end{pmatrix}, \quad \boldsymbol{D} = 0 \,. \tag{2.60}$$

Notice how the matrix $\boldsymbol{C}$ is padded by a zero to have size 2, to match the state vector $\boldsymbol{x}$.

2.4.2 Feedthrough and Biproper Transfer Functions

In state space, the biproper transfer-function models introduced in Section 2.3.2 correspond to ones with a nonzero feedthrough matrix $\boldsymbol{D}$. That is, a nonzero $\boldsymbol{D}$ results from having dynamics with frequencies higher than the range covered by the system model *and* having an output proportional to some or all of the fast variables. Problem 2.11 shows how feedthrough can be created after eliminating fast modes using a quasistatic approximation.

These "esoteric" ideas are actually common. For Ohm's Law, $V = IR$, the voltage V across a resistor R is proportional to the current I flowing through it. Viewing Ohm's law as a dynamical model, its transfer function $G(s) = V(s)/I(s) = R$ is constant for all frequencies and is thus biproper. But Ohm's law is not valid for all frequencies. Signals need a finite time t_0 to propagate from one terminal to the other. If the terminals are separated by a distance L (see left), then $t_0 < L/c$, where c is the velocity of light. At frequencies $\omega_0 \sim t_0^{-1}$, we should use the full set of Maxwell's equations to model the response and would then find a response that vanishes $\omega \gg \omega_0$. But for $\omega \ll \omega_0$, the biproper Ohm's law can be a good *lumped-element* approximation to the dynamics.

2.4.3 Changing Coordinates

Although the observed quantities $\boldsymbol{y}$ generally have physical meaning, the components of the internal state vector $\boldsymbol{x}$ may not. In particular, we can show that it is always possible to change coordinates in ways that make the components of the state vector $\boldsymbol{x}$ look very different. We can use this freedom to simplify the equations in various ways. These coordinate changes are called *similarity transformations*.

We start from our familiar state-space equations (dropping the feedthrough),

$$\dot{\boldsymbol{x}} = \boldsymbol{A}\boldsymbol{x} + \boldsymbol{B}\boldsymbol{u} \qquad \boldsymbol{y} = \boldsymbol{C}\boldsymbol{x}\,, \tag{2.61}$$

and define a coordinate transformation $\boldsymbol{x}' = \boldsymbol{T}\boldsymbol{x}$, or $\boldsymbol{x} = \boldsymbol{T}^{-1}\boldsymbol{x}'$. Then

$$\begin{aligned} \boldsymbol{T}^{-1}\dot{\boldsymbol{x}}' &= \boldsymbol{A}\boldsymbol{T}^{-1}\boldsymbol{x}' + \boldsymbol{B}u \qquad \boldsymbol{y} = \boldsymbol{C}\boldsymbol{T}^{-1}\boldsymbol{x}' \\ \dot{\boldsymbol{x}}' &= \left(\boldsymbol{T}\boldsymbol{A}\boldsymbol{T}^{-1}\right)\boldsymbol{x}' + (\boldsymbol{T}\boldsymbol{B})\,\boldsymbol{u}\,. \end{aligned} \tag{2.62}$$

Defining

$$\begin{aligned} \boldsymbol{A}' &= \boldsymbol{T}\boldsymbol{A}\boldsymbol{T}^{-1} \qquad & \boldsymbol{A} &= \boldsymbol{T}^{-1}\boldsymbol{A}'\boldsymbol{T} \\ \boldsymbol{B}' &= \boldsymbol{T}\boldsymbol{B} & \boldsymbol{B} &= \boldsymbol{T}^{-1}\boldsymbol{B}' \\ \boldsymbol{C}' &= \boldsymbol{C}\boldsymbol{T}^{-1} & \boldsymbol{C} &= \boldsymbol{C}'\boldsymbol{T}\,, \end{aligned} \tag{2.63}$$

we conclude that

$$\dot{\boldsymbol{x}}' = \boldsymbol{A}'\boldsymbol{x}' + \boldsymbol{B}'\boldsymbol{u} \qquad \boldsymbol{y} = \boldsymbol{C}'\boldsymbol{x}'\,. \tag{2.64}$$

Thus, the state-space equations take exactly the same form in the new coordinate system, but the matrices $\boldsymbol{A}$, $\boldsymbol{B}$, and $\boldsymbol{C}$ are all transformed according to Eq. (2.63). Note that the input $\boldsymbol{u}$ and output $\boldsymbol{y}$ do not transform: you input what you input and you measure what you measure, regardless of the coordinates used to describe the state vector.

The transfer function $G(s)$ is also invariant under the coordinate transformation (Problem 2.12). In other words, the poles of G (or equivalently the eigenvalues of $\boldsymbol{A}$) are invariant under coordinate transformation. This equivalence makes sense: poles and eigenvalues correspond to physical movements and should not depend on internal coordinates.

2.4.4 Time-Domain Response

Just as we characterized dynamics in the frequency regime by the frequency response, we can do the same in the time domain by looking at the temporal response to a given input. And just as Bode and Nyquist plots were helpful in visualizing the response, so too are time-response plots of the state vector components. As in the frequency-response case, we restrict ourselves to linear time-invariant (LTI) systems.

Although we can choose any input and graph any output, certain inputs are particularly useful. Consider the *impulse* and the *step*. The impulse $\delta(t)$, a Dirac delta function at $t = 0$, is impossible to realize physically. But giving a system a "kick" (a large-amplitude pulse for a short duration) is often a reasonable approximation. In that case, for an LTI system, the transfer function $G(s)$ for a SISO system is the Laplace transform of the *impulse response function*, $G(t)$. To see this, assume that the input function $u(t) = \delta(t)$. Then

$$y(s) = G(s)\,u(s) \qquad \Longrightarrow \qquad y(t) = G(t) * u(t) = G(t) * \delta(t) = G(t)\,. \tag{2.65}$$

Here, $G * u$ represents the convolution of $G(t)$ with $u(t)$, and we have used the fact that the delta function is unity under convolution, since $\int \mathrm{d}t'\, G(t')\,\delta(t - t') = G(t)$. We can also compute $G(t)$ directly in the time domain:[23]

$$G(t) = \boldsymbol{C}\left[\int_0^t \mathrm{d}t'\, \mathrm{e}^{\boldsymbol{A}(t-t')}\,\boldsymbol{B}\,\delta(t')\right] = \boldsymbol{C}\,\mathrm{e}^{\boldsymbol{A}t}\,\boldsymbol{B}\,. \tag{2.66}$$

To complete the connection, we Laplace transform Eq. (2.66) and again find Eq. (2.53):

$$G(s) = \int_0^\infty \mathrm{d}t\,\boldsymbol{C}\,\mathrm{e}^{\boldsymbol{A}t}\,\boldsymbol{B}\,\mathrm{e}^{-ts} = \boldsymbol{C}\left(\int_0^\infty \mathrm{d}t\,\mathrm{e}^{-(s\mathbb{I}-\boldsymbol{A})t}\right)\boldsymbol{B} = \boldsymbol{C}\,(s\mathbb{I} - \boldsymbol{A})^{-1}\,\boldsymbol{B}\,. \tag{2.67}$$

$G(t)$ is known as the *Green function* of the differential equation relating $y(t)$ and $u(t)$.

Because it is impossible to produce a true delta-function input, it is worth exploring other inputs. The *step function*, $\theta(t)$ is 0 for negative t and 1 for positive t and corresponds to a sudden shift in input level. Although it is still impossible to produce a perfectly abrupt step, it is easier to get a closer approximation than with a delta function. The two are related: because the step function is the integral of the delta function, the response to a step is the integral of the impulse response, too. Because integration is a kind of averaging operation, we see that a step function is useful when the output is contaminated by measurement noise. Then, the step function helps to average out such noise while still providing information about lower frequencies. Later, we will explore more complicated inputs, such as *frequency chirps* (Problem 6.3) and *white noise*.

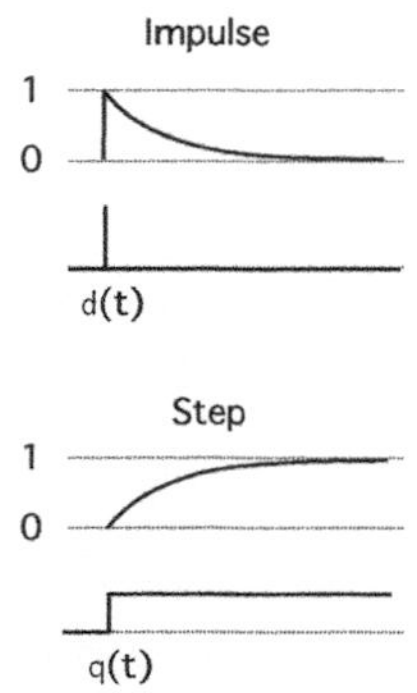

As illustrated at right, first-order systems relax exponentially. For impulses, they return to zero; for steps of unit amplitude, they converge to one. For the second-order system, depending on the damping, the response is underdamped, critically damped,

[23] Another way to compute the impulse response is to integrate over the time interval $(-\epsilon, +\epsilon)$ and take the limit $\epsilon \to 0$. Effectively, you replace the delta function by an equivalent initial condition.

or overdamped. The impulse responses for the second order system are just the familiar solutions for transient motion of a simple damped harmonic oscillator with initial conditions $y(0) = 0$, $\dot{y}(0) = 1$. This result is intuitive – it is what we mean by a "kick." More formally, we can show the result by applying $\lim_{\varepsilon\to 0}\int_{-\varepsilon}^{\varepsilon}$ to Eq. (2.16), which leads to the conclusion that the discontinuity in the velocity $\Delta\dot{y}(0) = 1$. That is, the object goes instantly from being at rest to having unit velocity at time 0. You can check that

$$y(t) = \begin{cases} \left(\frac{1}{\sqrt{1-\zeta^2}}\right) e^{-\zeta t} \sin\sqrt{1-\zeta^2}\, t & 0 < \zeta < 1 \text{ (underdamped)}, \\ t\, e^{-t} & \zeta = 1 \text{ (critically damped)}, \\ \left(\frac{1}{\sqrt{\zeta^2-1}}\right) e^{-\zeta t} \sinh\sqrt{\zeta^2-1}\, t & \zeta > 1 \text{ (overdamped)}. \end{cases} \tag{2.68}$$

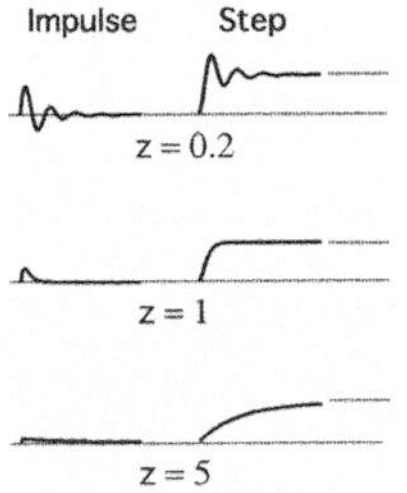

Mostly, these types of response are computed numerically. Loosely, we see that the temporal response is a *transient* response, while the transfer function is equivalent to the *steady-state* response of the system to a persistent sinusoidal input. Some other features to note in the illustration at left: damping decreases the amplitude of the impulse response; "ringing" or *overshoot* in the step response occurs in the underdamped case, with an associated *settling time*; and the critically damped response has the quickest decay to the new equilibrium. Critical damping is often a desirable response for a second-order system.

2.5 Stability

Things fall apart; the centre cannot hold.

W. B. Yeats, *The Second Coming*

Stability, a concept that is important for dynamical systems in general, is central to control theory. As we will see, feedback can make an unstable system stable, and vice versa. Intuitively, a stable solution is one that does not "run away." In Section 2.5.1, we will make this notion more precise. One subtlety is that the size of a perturbation to a solution can matter. Sometimes, a solution will be stable to small perturbations but unstable to larger ones. Proving that a solution is stable to arbitrarily large perturbations can be hard. In Section 2.5.2, we discuss Lyapunov stability, which is one of the few tools that can be used to analyze such questions of *global* stability. In Section 2.5.3, we will look at the effect of small perturbations, using a technique known as *linear stability* analysis.

2.5.1 Definitions

In order to focus on basic points, let us restrict our discussion to autonomous systems of the form $\dot{\boldsymbol{x}} = \boldsymbol{f}(\boldsymbol{x})$ that have an *equilibrium point* $\boldsymbol{x} = \boldsymbol{0}$. Equivalently, $\boldsymbol{f}(\boldsymbol{0}) = \boldsymbol{0}$

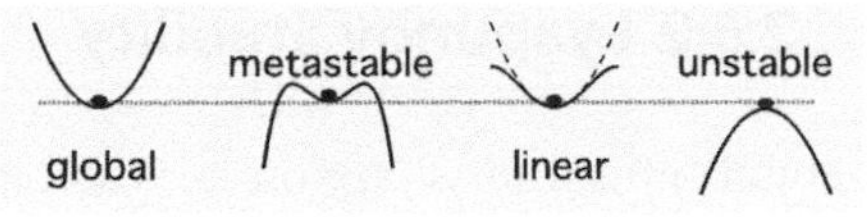

Fig. 2.8 Stability classified according to the size of the perturbation.

is a *critical point*. At right is an equilibrium *stable in the sense of Lyapunov*, depicted by the small circle: trajectories that start near the equilibrium point stay nearby. At bottom, the equilibrium is *asymptotically stable*: trajectories eventually converge to the equilibrium.

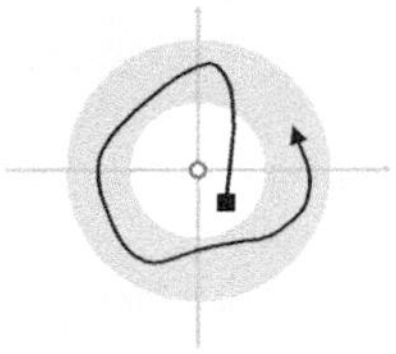

stability (Lyapunov)

In more detail, stability in the sense of Lyapunov means that it will always be possible to keep the solution (for $t > 0$) within a ball of radius ϵ (shaded disk) whose size we are free to choose. Usually, a smaller ϵ implies a smaller δ. Further, this confinement must work for any initial condition (the black square) located within a ball of radius δ (white disk). At right, the trajectory can start anywhere in the white disk of radius δ and must always stay within the shaded disk of radius ϵ. More formally, a system is stable in the sense of Lyapunov if,

$$\forall \epsilon > 0\,,\ \exists \delta > 0\,,\ \text{such that } ||\boldsymbol{x}(0)|| < \delta \implies ||\boldsymbol{x}(t)|| < \epsilon \text{ for all } t \geq 0\,. \tag{2.69}$$

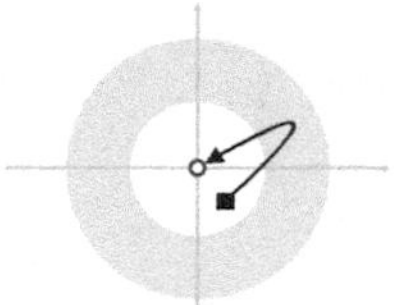

asymptotic stability

For asymptotically stable motion, the solution also converges to $\mathbf{0}$ – i.e., $\lim_{t\to\infty} \boldsymbol{x}(t) = \mathbf{0}$.

To give a more intuitive motivation for these definitions of stability, consider the down equilibrium point of a pendulum. If the pendulum is frictionless, then it oscillates about the downward equilibrium point after a small tap. We can always make a small enough tap (the δ) that keeps the pendulum oscillations within any given bound (the ϵ) on the motion in state space: the motion is stable. Asymptotic stability requires damping, which ensures that the pendulum returns to rest as $t \to \infty$. By contrast, a tap at the up equilibrium leads to motion that leaves a localized region: the equilibrium is *unstable*.

Another way to classify (asymptotic) stability is to look at the largest allowable size of perturbation. Figure 2.8 illustrate four cases in order of decreasing size of perturbation.

1 *Globally stable*: The solution is stable for *all* perturbations, of any size.
2 *Metastable*: The solution is stable for small perturbations but not for large perturbations.
3 *Linearly stable*: The solution is stable for infinitesimal perturbations but may or may not be globally stable, depending on (unexamined) dynamics far from the fixed point.
4 *Unstable*: If a solution is not stable by any of these criteria, it is unstable. A particle atop a mountain will fall to either side at the slightest push.

2.5.2 Lyapunov Stability

One way to investigate stability in a large region of state space is via Lyapunov's method,[24] which formalizes the intuition that in a stable mechanical system with friction, the energy always decreases as equilibrium is approached. To generalize this idea, we define a *Lyapunov function*, $V(\boldsymbol{x})$, that decreases (or is nonincreasing) when evaluated for any solution to the equation of motion and has a lower bound. A solution $\boldsymbol{x}(t)$ that achieves that lower bound is asymptotically stable. Physical examples of Lyapunov functions include energy or free energy, but the functions need not have an immediate physical interpretation.

Define a scalar continuous function $f(\boldsymbol{x})$ to be *positive definite*[25] if $f(\boldsymbol{x} \neq \boldsymbol{0}) > 0$ and $f(\boldsymbol{0}) = 0$ – i.e., a minimum at $\boldsymbol{x} = 0$. If $f(\boldsymbol{x}) \geq 0$ and $f(\boldsymbol{0}) = 0$, then f is *positive semidefinite*. The function f is *negative definite* if $(-f)$ is positive definite. These characteristics are *local* if they hold in a neighborhood of $\boldsymbol{x} = \boldsymbol{0}$ and *global* if they hold for all $\boldsymbol{x} \neq \boldsymbol{0}$.

Then a scalar function $V(\boldsymbol{x})$ is a *Lyapunov function* if, in a neighborhood of $\boldsymbol{x} = \boldsymbol{0}$,

- V is positive definite;
- $\dot{V}(\boldsymbol{x}) = \frac{\partial V}{\partial \boldsymbol{x}} \, \dot{\boldsymbol{x}} = \frac{\partial V}{\partial \boldsymbol{x}} \, \boldsymbol{f}$ is negative semidefinite.

As a consequence, $\boldsymbol{x} = \boldsymbol{0}$ is locally stable in the sense of Lyapunov. If $\dot{V}$ is actually *negative definite*, then the origin is locally asymptotically stable. Here, $\boldsymbol{x}(t)$ is a vector in state space that traces a solution to $\dot{\boldsymbol{x}} = \boldsymbol{f}(\boldsymbol{x})$. Then $\dot{V}$ is the change in V along a trajectory $\boldsymbol{x}(t)$ determined by the dynamics. We can choose coordinates to move the equilibrium point from $\boldsymbol{0}$ to $\boldsymbol{x}^*$, if desired. Note also that the lower bound of V is 0. Since adding a constant to V does not change anything, the lower bound can have any convenient value.

If the first two properties hold everywhere in state space and if also

- $V(\boldsymbol{x}) \to \infty$ as $\|\boldsymbol{x}\| \to \infty$,

then the origin is *globally* stable (and asymptotically stable if the stronger condition that $\dot{V}$ is negative definite holds). The condition that V be unbounded as $\|\boldsymbol{x}\| \to \infty$ is needed to make the contours of V closed curves. Otherwise, the state $\boldsymbol{x}$ can sneak off to infinity while always decreasing V by staying within a given contour. In the illustration at left, the dotted lines show contours of V about its global minimum, at the black dot. The Lyapunov function V decreases along the solid trajectory $\boldsymbol{x}(t)$, even as $\boldsymbol{x}$ heads out to infinity.

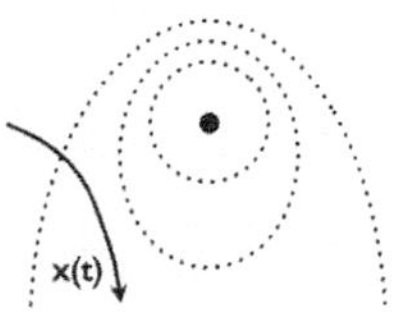

Example 2.9 Consider the first-order system, $\dot{x} = -x$, and define $V(x) = \frac{1}{2}x^2$. The reference solution is the fixed point $x^* = 0$. We then verify that

$$V(0) = 0\,, \qquad V(x \neq 0) > 0\,, \qquad \dot{V} = (\partial_x V)\,\dot{x} = x(-x) = -x^2 < 0 \text{ if } x \neq 0\,.$$

[24] Actually, Lyapunov's *second* method. His first relates to linear stability analysis (Section 2.5.3).

[25] If $\boldsymbol{M}$ is positive definite, then so is the quadratic form $f(\boldsymbol{x}) = \boldsymbol{x}^{\mathrm{T}} \boldsymbol{M} \boldsymbol{x}$ (Section A.1.1 online).

Thus, V is a global Lyapunov function with $\dot{V}$ negative definite, and we have proven that $x = 0$ is asymptotically stable globally for all perturbations $x(0)$. We could also reach this conclusion by noting that the solution $x(t) = x(0)\, e^{-t} \to 0$ as $t \to \infty$ for all starting conditions $x(0)$. But with the Lyapunov approach, we do not need to know the solution.

Although this example is linear, the Lyapunov method works with nonlinear equations, too. For example, for $\dot{x} = -x - x^3$ and $x^* = 0$, then $V = \frac{1}{2}x^2 + \frac{1}{4}x^4$. Check it!

Note that both cases in Example 2.9 may be written in the form

$$\dot{x} = f(x) = -(\partial_x V) \quad \Longrightarrow \quad \dot{V} = (\partial_x V)\,\dot{x} = (\partial_x V)\, f(x) = -(\partial_x V)^2 \le 0\,. \tag{2.70}$$

Thus, V is a Lyapunov function, with *relaxational*, or *potential* dynamics.[26] In higher dimensions, the derivative becomes a gradient, $\boldsymbol{f}(\boldsymbol{x}) = -\nabla V(\boldsymbol{x})$. In one dimension, we can write $V(x) = -\int^x dx'\, f(x')$ when x is defined on a line but not in general in higher dimensions. (A necessary condition is that $\boldsymbol{f}(\boldsymbol{x})$ be curl free: $\nabla \times \boldsymbol{f} = \boldsymbol{0}$.)

Example 2.10 Consider the second-order system, $\ddot{x} + 2\zeta\dot{x} + x = 0$, whose state-space representation is $\dot{x}_1 = x_2$, $\dot{x}_2 = -x_1 - 2\zeta x_2$. The reference solution is the fixed point $x^* = 0$. To test its stability, we use the total energy as a Lyapunov-function candidate: $V(\boldsymbol{x}) = \frac{1}{2}(x_1^2 + x_2^2)$. As before, $V(\boldsymbol{0}) = 0$ and $V(\boldsymbol{x} \ne \boldsymbol{0}) > 0$. But now,

$$\dot{V} = x_1(\dot{x}_1) + x_2(\dot{x}_2) = x_1x_2 - x_2x_1 - 2\zeta x_2^2 = -2\zeta x_2^2 \le 0\,. \tag{2.71}$$

Since $\dot{V}$ is only negative semidefinite, $x_1 = x_2 = 0$ may not be asymptotically stable.

Example 2.10 seems unsatisfactory: it is clear physically that all solutions continually lose energy and tend to the origin. The solution is asymptotically stable, yet Lyapunov's theorem implies only that it is stable in the sense of Lyapunov. Fortunately, a small extension, the *Krasovskii–LaSalle Invariance Principle*, covers such cases.

When $\dot{V} \le 0$ is negative semidefinite, define the set E of state-space vectors $\boldsymbol{x}$ such that $\dot{V}(\boldsymbol{x}) = 0$. Further, let $M \subset E$ be the positively invariant set, defined by requiring that $\boldsymbol{x}(0) \in M \implies \boldsymbol{x}(t) \in M$, for all $t > 0$. Then the invariance principle implies that all bounded solutions tend to M at long times. The requirement that the solution be bounded is needed to rule out cases where $|\boldsymbol{x}| \to \infty$, instead.

To apply this to the damped harmonic oscillator of Example 2.10, we note that imposing $\dot{V} = 0$ at a *single time* implies that

$$\dot{V} = 0 \implies x_2 = 0 \implies \dot{x}_1 = 0 \implies x_1 = \text{const}\,. \tag{2.72}$$

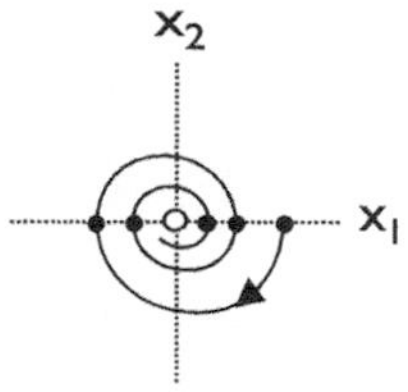

More physically, the velocity of the damped oscillator vanishes at each turning point (if underdamped). These turning points define the set E, denoted by black markers in the parametric (state-space) plot of velocity, $x_2(t)$ versus position, $x_1(t)$, at right. But

[26] The negative sign in the definition of V is a convention: dynamics are defined to flow *down* the gradient.

to be a member of the positively invariant set M means that $x_2 = 0$ for *all* later times. In this case,

$$\dot{V} = 0 \implies x_2 = 0,\ \forall t \implies \dot{x}_2 = 0 \implies x_1 = 0\,. \tag{2.73}$$

The set M is just the origin $x_1 = x_2 = 0$ (white marker), which is thus asymptotically stable (Krasovskii–LaSalle). Problem 2.16 extends this case to the damped nonlinear pendulum.

The main catch with the Lyapunov method is that there is no general way to construct a Lyapunov function. Expressions that resemble energies (or free energies) often work. Otherwise, many tricks will work in special cases. See Problem 2.17 for an example. But given a candidate V, you can easily check whether it is a Lyapunov function.

2.5.3 Linear Stability

Linear stability analysis is one strategy for understanding motion in the vicinity of a fixed point of a nonlinear equation. Here are some simple cases.

One Dimension

Consider a one-dimensional state that obeys $\dot{x} = f(x) = x - x^3$. There are three fixed points, since $f(x_0) = 0$ at $x_0 = 0, \pm 1$. We study small deviations about the fixed point by linearizing the dynamics. With $x = x_0 + x_1(t)$, the linear deviations x_1 obey $\dot{x}_1 = f'(x_0)\, x_1$. Thus, the sign of $f'(x) = 1 - 3x^2$, evaluated at a fixed point x_0, determines the stability of that fixed point. A positive $f'(x_0)$ means perturbations increase exponentially, and a negative one means that they decrease exponentially. It is easy to see that the fixed point at $x_0 = 0$ is unstable, while the two at $x_0 = \pm 1$ are stable:

$$\begin{array}{llll} x_0 = 0 & f'(x_0) = +1 & \dot{x}_1 = +x_1 & x_1(t) = x_1(0)\ \mathrm{e}^{+t} \\ x_0 = \pm 1 & f'(x_0) = -2 & \dot{x}_1 = -2x_1 & x_1(t) = x_1(0)\ \mathrm{e}^{-2t}\,. \end{array} \tag{2.74}$$

f(x) = x³- x

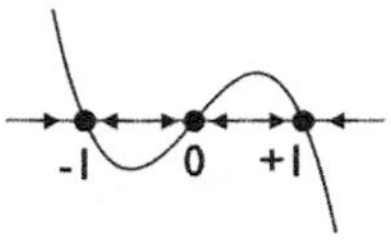

At left, we plot both the one-dimensional phase space for x, along with the graph of $f(x)$. The intersections of f with the horizontal axis give the three fixed points. The sketch also shows arrows toward the stable fixed points and away from the unstable fixed points. Because the phase space is one-dimensional and f is a continuous function, the arrows must keep their direction and change only when passing through fixed points. Thus, we know the global behavior of solutions (just follow the arrows), even though we have not explicitly integrated the equation of motion. Note that we can reach the same conclusions by defining a Lyapunov function $V(x) = -\frac{1}{2}x^2 + \frac{1}{4}x^4$.

Two dimensions

As an example of a two-dimensional state vector, consider the dynamics of a point particle in one spatial dimension, which is described by its position and velocity. If $\boldsymbol{x}_0$ is a fixed point of $\dot{\boldsymbol{x}} = \boldsymbol{f}(\boldsymbol{x})$, then small perturbations $\boldsymbol{x} = \boldsymbol{x}_0 + \boldsymbol{x}_1(t)$ obey

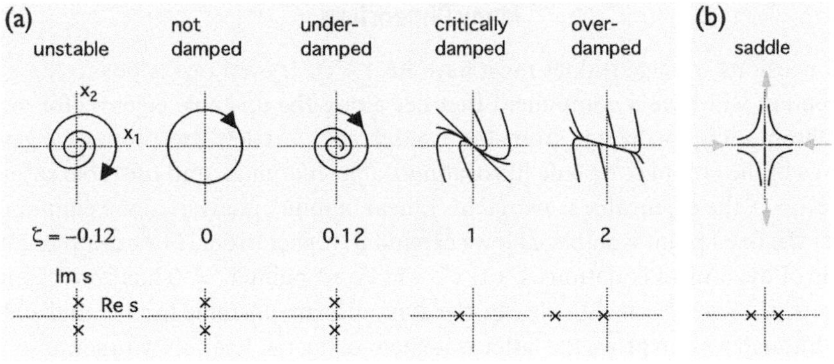

Fig. 2.9 Phase portraits and pole-zero plots for two-dimensional dynamics. (a) Harmonic oscillator for various damping parameters ζ. (b) Saddle-node dynamics.

$\dot{x}_1 = A x_1$. In two dimensions, A is a 2×2 matrix, with eigenvalues λ_1 and λ_2. Some of the different cases for the λ's – real or complex, positive or negative real parts, degenerate or distinct – are summarized in Figure 2.9a, where the left-hand group of phase space plots shows the behavior of a harmonic oscillator for different damping coefficients ζ. When the damping is negative (at left), oscillations grow exponentially, and the system is unstable. With no damping, there are sinusoidal oscillations, and the stability is uniform. We show the underdamped, critically damped, and overdamped cases. Note how the overdamped case has two time constants: a fast decay, corresponding to a pole at $s = -\zeta - \sqrt{\zeta^2 - 1} \approx -2\zeta$ (for $\zeta \gg 1$), and a slow decay, corresponding to $s = -\zeta + \sqrt{\zeta^2 - 1} \approx -1/(2\zeta)$. See the phase space plot, with the rapid decay to a line, followed by a slow decay along this line to the origin.

Figure 2.9b shows a different system, with one positive and one negative eigenvalue. This *saddle-node* case illustrates that for motion to be stable,[27] the real part of both eigenvalues must be negative: $\mathrm{Re}\,\lambda_{1,2} < 0$. Here, a set of initial conditions of measure zero (along the x-axis) converge to the fixed point at the origin. A generic initial condition will be off the *stable manifold*, at a point that does not fall on the x-axis. Such a point will initially move toward the origin but then diverge away near the *unstable manifold*. Near the origin, the stable and unstable manifolds are given by the eigenvectors corresponding to the stable (negative) and unstable (positive) eigenvalues. In the saddle shown in Figure 2.9b, the stable and unstable manifolds are the x- and y-axes, respectively. Farther away from the origin, these directions will usually vary – hence the need to describe the trajectory by a curved manifold rather than just a direction. Informally, the trajectory for the unstable manifold is the path starting from a point in phase space near the fixed point, along the unstable eigenvector. The stable manifold is defined analogously: start close to the fixed point along the stable eigenvector and run the dynamics *backwards* in time.

[27] "Saddle node" refers to the shape of a potential near a fixed point with one stable and one unstable eigenvalue. It curves up in one direction, down in the other, and looks a bit like a horse's saddle. See Strogatz (2014).

Many Dimensions

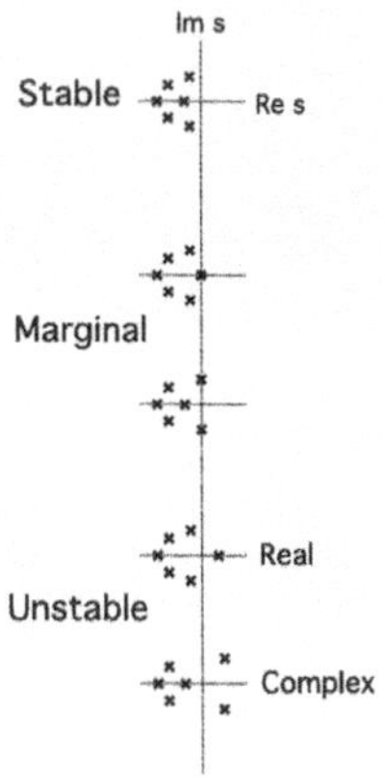

In n dimensions, *all* eigenvalues must have $\text{Re}\,\lambda < 0$. If even one is positive, a generic perturbation will have a component that lies along the unstable eigenvector and will cause the motion to diverge from the fixed point. At left are plots of eigenvalue locations in the complex s-plane illustrating stable, marginal, and unstable cases.

If $\text{Re}\,\lambda = 0$, the eigenvalue is *marginal*. Linear stability analysis alone cannot decide whether the fixed point is stable, and we expand to higher order. For example, consider the pair of dynamical equations $\dot{x} = \pm x^3$. The fixed point $x_0 = 0$ has zero eigenvalue, because $f'(0) = \pm 3x^2 = 0$. But clearly, the dynamics are unstable for $+x^3$ and stable for $-x^3$. Although we can prove the latter statement using the Lyapunov function $V = \frac{1}{4}x^4$, the main point is the need to analyze the effect of the nonlinear terms.

2.6 Bifurcations

So far, we have considered dynamical systems whose general properties such as stability and pole locations are fixed. More generally, we can imagine using a control parameter to vary the properties of a dynamical system. In the context of control theory, this will often relate to the amount of feedback gain applied. For most values of the parameter, the changes are only minor, but at special values, known as *bifurcation* points, the changes may be qualitative, and an unstable solution can become stable, and vice versa.

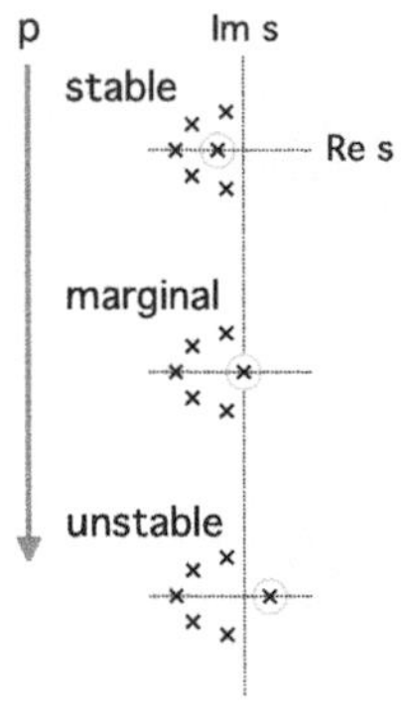

As a start, consider the matrix $\boldsymbol{A}(p)$, which describes dynamics near a fixed point that depends on a parameter p. In general, the positions of all the eigenvalues λ of $\boldsymbol{A}$ – poles of the transfer function $G(s)$ – will be functions of p, so that $\lambda = \lambda(p)$. Assume $\boldsymbol{A}$ to be stable, initially. As shown at left, as p is increased,[28] a real eigenvalue may cross the imaginary axis in the s-plane.[29] At this point, the system – according to linear analysis – is marginal and may be stable or unstable, depending on the nonlinear terms. As the control parameter is further increased, the system definitely becomes unstable.

At most values of the control parameter p, the stability of the dynamical system does not change: it is either linearly stable or is linearly unstable. This statement will typically be true even when the dynamics are perturbed, as such perturbations will shift the eigenvalues by only a small amount. This property is called *structural stability* and is a prerequisite for sensible modeling. In specifying a model, we always choose what is relevant and needs to be in the model and what may be "small" and thereby omitted.

However, at one special control-parameter value p_c, the stability properties change. At this *bifurcation point*, the fixed point becomes unstable. Perturbations grow exponentially (and sometimes oscillate). Of course, a physical solution cannot keep

[28] Here, instability occurs as p increases; in other cases, it can happen via a decrease.
[29] The relative change of the eigenvalue closest to zero will be much larger than the relative change of other eigenvalues. We thus neglect p-dependence in the other eigenvalues.

growing forever. Either nonlinear terms will saturate the solution, or the model will simply break down.

Although "bifurcation" may not be a familiar term, examples are. Phase transitions are a kind of bifurcation, as are instabilities, such as the buckling of an overloaded beam, the transition from conduction to convection for a liquid heated from below, and so on. In control theory, varying the feedback parameters can lead to instabilities and bifurcations, creating a new steady state or oscillation (*Hopf bifurcation*). Let us describe some of its properties.

Hopf Bifurcations

In a Hopf bifurcation, a pair of eigenvalues goes unstable, creating a spontaneous oscillation, whose angular frequency ω, at onset, is given $\omega = \text{Im}\,\lambda(p)$ (see right). Because the coefficients of the underlying equations are real, the complex eigenvalues will always be in complex-conjugate pairs. Notice that there are now two time scales. On the one hand, there are fast oscillations, on a time scale ω^{-1}. On the other hand, the amplitude of those oscillations will typically change at a rate $\epsilon \sim (p - p_c)$. If $z(t)$ denotes the time-dependent amplitude of fast oscillations, the state $x(t)$ can be described as $x(t) = z(t)\,\mathrm{e}^{-i\omega t} + z^*(t)\,\mathrm{e}^{+i\omega t}$, where, near the bifurcation point, z is slowly varying ($\dot{z} \ll \dot{x}$), changing over many periods of the fast oscillation, $\mathrm{e}^{-i\omega t}$. Because $z = z(t)$, it also obeys a differential equation,

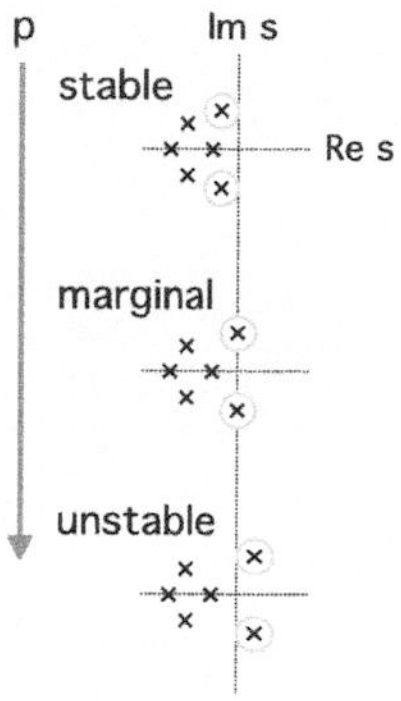

$$\dot{z} = \varepsilon z - |z|^2 z\,. \tag{2.75}$$

The form for Eq. (2.75) is a consequence of invariance under arbitrary time translation: the physical properties of the solutions to the equations of motion should depend only on time intervals and not on absolute time. If $t \to t + \Delta t$, the new trajectories should be solutions to the same equations of motion. Here, the solutions oscillate at angular frequency ω. Shifting time forward by Δt adds a phase lag $\Delta\varphi = -\omega\,\Delta t$, and $z \to z\,\mathrm{e}^{-i\omega\Delta t}$. The form of Eq. (2.75) is invariant under this transformation, and we can view the equation as a kind of Taylor expansion of the dynamics that respects the time-invariance symmetry.

The nonlinear $|z|^2 z$ term in Eq. (2.75) leads to a finite-amplitude oscillation (*limit cycle*), $|x| \sim \sqrt{\varepsilon}$, for $\varepsilon > 0$. For $\varepsilon \gtrsim 0$, the motion is nearly sinusoidal but typically distorts farther away from the bifurcation point, where nonlinearities excite ever-higher numbers of harmonics. Although the period typically changes as well, the motion remains periodic – at least until another bifurcation occurs. See Figure 2.10, which shows dynamics of the van der Pol oscillator, $\ddot{x} - \varepsilon(1 - x^2)\dot{x} + x = 0$, which has a Hopf bifurcation at $\varepsilon = 0$.

A limit cycle is distinguished from the oscillations of an undamped oscillator in that it is asymptotically stable: kick it, and it will come back to the same amplitude (albeit displaced in phase). In contrast, a kick will change the energy and thus amplitude of a linear, undamped oscillator, which is only neutrally stable. Figure 2.11a illustrates

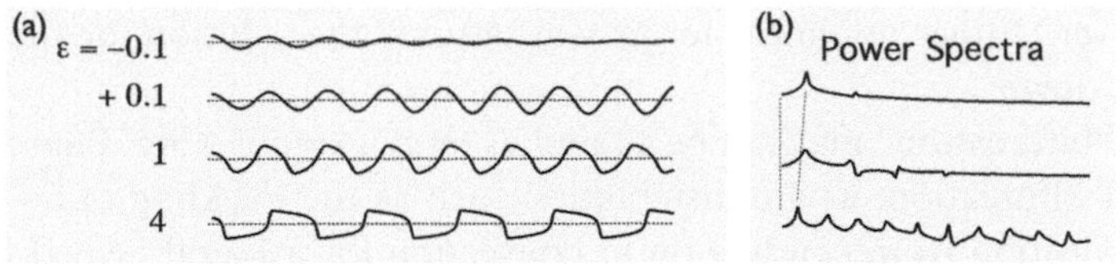

Fig. 2.10 Limit cycle for the van der Pol oscillator. (a) time series $x(t)$ for different values of the control parameter ε. (b) power spectra of the corresponding limit cycles. Note the higher harmonics and lower fundamental frequency as ε increases.

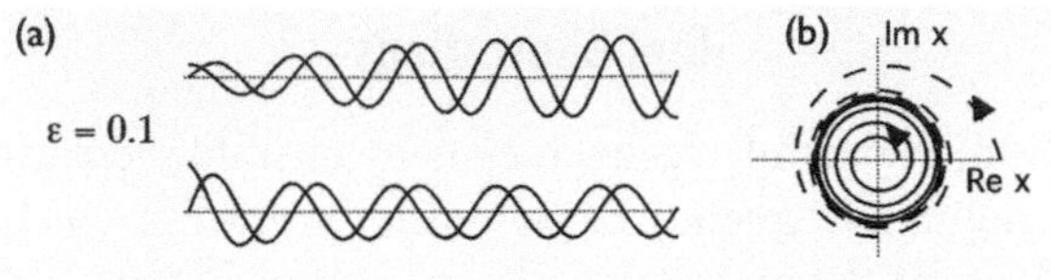

Fig. 2.11 Limit cycle for generic Hopf bifurcation dynamics (Eq. 2.75). (a) Time series for x and $\dot{x}$, for small- and large-amplitude initial conditions. (b) Parametric plot of the two components of $\boldsymbol{x}(t)$. Phase portrait for lower time series is dashed.

these cases. Notice how the change in amplitude occurs over several oscillations, as discussed above. Figure 2.11b shows anharmonicity and the shift in oscillation frequency with ε.

2.7 Summary

In this chapter, we have briefly reviewed key concepts in continuous dynamical systems that are relevant for our study. We began by noting that we can cast a wide variety of dynamical systems into the form

$$\dot{\boldsymbol{x}} = \boldsymbol{f}(\boldsymbol{x}, \boldsymbol{u}) \qquad \boldsymbol{y} = \boldsymbol{h}(\boldsymbol{x}, \boldsymbol{u}), \tag{2.76}$$

where $\boldsymbol{f}$ and $\boldsymbol{h}$ are nonlinear vector functions of the state vector $\boldsymbol{x}$ and control inputs $\boldsymbol{u}$, and $\boldsymbol{y}$ represents the system outputs. The equations are much simpler when scaled. Typically, we are interested in motion near a reference solution, often a fixed point $\boldsymbol{x}_0$ satisfying $\boldsymbol{f}(\boldsymbol{x}_0, \boldsymbol{u}) = \boldsymbol{0}$. Near the fixed point, the dynamics may be described by linear equations,

$$\dot{\boldsymbol{x}} = \boldsymbol{A}\boldsymbol{x} + \boldsymbol{B}\boldsymbol{u} \qquad \boldsymbol{y} = \boldsymbol{C}\boldsymbol{x} + \boldsymbol{D}\boldsymbol{u}, \tag{2.77}$$

where the dynamics of deviations $\boldsymbol{x}$ are given by the eigenvalues λ of $\boldsymbol{A}$. The components of the state vector and the elements of the matrices $\{\boldsymbol{A}, \boldsymbol{B}, \boldsymbol{C}, \boldsymbol{D}\}$ depend on the choice of coordinate system, but the physical properties – the inputs and outputs and the eigenvalues of $\boldsymbol{A}$, which give the dynamics, $e^{\lambda t}$ of each mode – are independent of the choice of coordinates. An advantage of the vector-matrix state-space notation is that it automatically handles multiple inputs and multiple outputs (the MIMO case).

Linear systems are also conveniently described in the frequency domain, using Laplace transforms to define the transfer function matrix

$$G(s) = C(s\mathbb{I} - A)^{-1} B + D. \tag{2.78}$$

where initial conditions are usually assumed to be zero. For a single input, single output (SISO) system, $G(s)$ is a scalar function, D is a constant, and $\boldsymbol{B}$ and $\boldsymbol{C}$ are column and row vectors, respectively. Evaluating G at $s = \mathrm{i}\omega$ in the complex s-plane gives the Fourier transform. Graphical (Bode) plots of the magnitude and phase of the frequency response give an intuitive way of understanding the dynamics.

In general, nonlinear equations can have multiple solutions, which leads one to ask under what conditions a given base solution (fixed point, limit cycle, etc.) is stable. If we can define a Lyapunov function $V[\boldsymbol{x}(t)]$ that is strictly decreasing within some region of phase space, then that base solution is asymptotically stable within that region. Unfortunately, while it is straightforward to verify that a proposed V is a Lyapunov function, there is no general algorithm to construct such a function. Mostly, we have to content ourselves with tests of linear stability, obtained by linearizing the dynamics about the base solution. If all eigenvalues of the resulting dynamical matrix $\boldsymbol{A}$ have negative real part, or if all poles lie in the left-hand side of the complex s-plane, the system is linearly stable.

If the dynamics depend on a control parameter, a base solution such as a fixed point or limit cycle may lose (or gain) stability at a bifurcation point. We focused on the Hopf bifurcation to a limit cycle, as it is commonly encountered in feedback systems.

Having developed basic notions of dynamical system, we can ask how to use the information provided by observations (system outputs) to design inputs that will "improve" the dynamics of the system. Explaining these words will occupy us for the rest of the book.

2.8 Notes and References

Dynamics as part of classical mechanics is part of the standard physics curriculum. Good intermediate references include Taylor (2005) (very clear) and Morin (2008) (wonderful problems). Classic advanced texts include the very concise Landau and Lifshitz (1976), the somewhat old-fashioned Goldstein (1980), and the challenging Arnold (1989). The pendulum – particularly when forced by the motion of a cart – is a standard control-theory example (Leong and Doyle, 2016).

The focus on inputs and outputs by engineers goes under the broad name of *(linear) systems theory* and is developed in many engineering texts (e.g., Kailath (1980)). Åström and Murray (2008) introduce systems ideas to a general scientific audience. For a survey of differential-algebraic equations and descriptor systems, see Sjöberg (2005).

Sometimes, the notions of *input* and *output* are not as clearcut as engineers would like, since they are imposed on physical systems "from the outside." A more general *behavioral* approach to defining dynamical systems focuses on the set of possible time trajectories for motion and defines notions such as *terminals* and *ports* that relate to energy input and connections between different physical systems (Willems, 2007, 2010).

For scaling and nondimensionalization, the paper by Edgar Buckingham (1914) is still worth reading. The book by Nobel laureate Percy Bridgman (1922) is a pleasant and straightforward read. Beginning in the 1950s in the Soviet Union, G. I. Barenblatt introduced the notion of *intermediate asymptotics* and discussed how setting a small dimensionless parameter to zero may result in physical quantities that diverge to infinity, as we briefly discussed here. The book *Scaling* (Barenblatt, 2003) is a good introduction. The connection between intermediate asymptotics and the renormalization group was pioneered by Nigel Goldenfeld and his collaborators (Goldenfeld, 1992).

For nonlinear dynamics – stability, bifurcations, and the like – see Strogatz (2014). Finding Lyapunov functions remains a topic of current research. For an interesting way to assign a standard physical interpretation to Lyapunov functions for dissipative dynamical systems, see Yuan et al. (2014). A popular book, *The Tipping Point* (Gladwell, 2002) lists numerous everyday examples of bifurcations in social experiences, such as the sudden resurgence in popularity of a brand of shoes after decades of neglect.

For background on mathematical methods (Laplace transforms, etc.), see the Appendices in this book, as well as the references given there. I am grateful to Karl Åström for pointing out Ångström's method for measuring thermal diffusivity and for formulating his results for copper in modern terms (cf. Problem 2.5).

Problems

2.1 Balancing a pendulum by moving a cart.

a. Using a Lagrangian, derive Eq. (2.11).

b. For the case of a uniform stick of mass m and length ℓ, show that

$$\ddot{x} + \tfrac{1}{2}\left(\ddot{\theta}\cos\theta - \dot{\theta}^2\sin\theta\right) = u\,, \qquad \ddot{\theta} + \sin\theta + \tfrac{3}{2}\ddot{x}\cos\theta = d\,,$$

where $u(t)$ acts directly on the stick and $d(t)$ is a torque on the pendulum.

c. Write these equations as an equivalent first-order equation of the form $\dot{\boldsymbol{x}} = \boldsymbol{f}(\boldsymbol{x}, \boldsymbol{u})$.

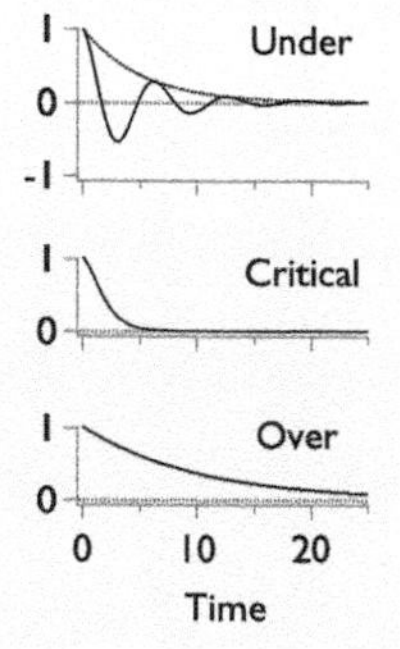

2.2 First-order systems, frequency domain. Derive analytically the curves in the graphs of the first-order dynamical system $G(s) = \frac{1}{1+s}$ (Bode plots, pole-zero, and Nyquist plots) shown in Figure 2.4. For the Nyquist plot, derive the geometric shape of the curve, not just the parametric form.

2.3 Second-order systems.

a. Consider a generic second-order system $G(s) = (1 + 2\zeta s + s^2)^{-1}$. As in Problem 2.2, derive analytic expressions for the Bode, pole-zero, and Nyquist plots (Figure 2.5), for $\zeta < 1$, $\zeta = 1$, and $\zeta > 1$, (underdamped, critically damped, and overdamped), respectively, for some parts of this problem. Find analytic approximations for $\zeta \ll 1$ and $\zeta \gg 1$. Give the Nyquist plots in parametric form.

b. A common control goal is to have critically damped closed-loop dynamics ($\zeta = 1$). Show that the damping time of the decay of an oscillator is shortest for critical damping. Confirm the time-decay plots at right ($\zeta = 0.2, 1, 5$) and the plot below.

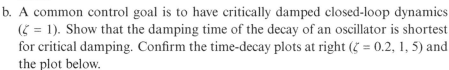

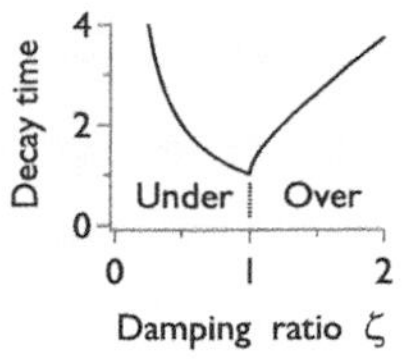

2.4 Transfer function for thermal conduction. We showed that a semi-infinite, one-dimensional thermal conductor gives rise to a transfer function between heater and thermometer of $G(s) = \mathrm{e}^{-\sqrt{s}} / \sqrt{s}$. See Eq. (2.44). The probe is at $x = \ell$ (scaled to 1).

a. Derive explicit expressions for the magnitude and phase of the frequency response. Use the results to reproduce the Bode and Nyquist plots.

b. How are Bode and Nyquist plots altered for a different sensor point ($x = 2$)?

c. We can approximate non-rational transfer functions by rational polynomials (Padé approximants; Cf. Section 3.6.4). The second-order Padé approximation to $G(s)$ about $s = 1$ is $G_{2,2}(s) = (\mathrm{e}^{-1})\,\frac{44-76s+8s^2}{5+26s+55s^2}$. Compare its Bode plot with that of $G(s)$. Use a computer-algebra program to compute and compare $G_{3,3}(s)$.

d. Show that if we use two temperature probes, at distances x_1 and x_2, the transfer function between the two probes is $G(s) = \mathrm{e}^{-\sqrt{s}\ell}$, where $\ell = x_2 - x_1$.

2.5 Ångström's method for measuring the thermal diffusion coefficient. Ångström (1861) derived a "remarkably simple" result for the thermal diffusion constant that led to far more accurate measurements of D. Distributed heat losses to the environment modify the diffusion equation to be $\partial_t T(x,t) = D\partial_{xx}T - \mu T$. Unfortunately, μ is difficult to measure and reflects all the details of the geometry of the experiment. Ångström found an expression for D that was *independent* of μ.

a. Show that the transfer function between two points on the rod that are separated by a distance $\Delta\ell$ is given by $G(s) = \exp\left[-\sqrt{\mu + s}\,(\Delta\ell / D^{1/2})\right]$. The temperature is the "input" at a point ℓ_1 and the "output" at point $\ell_2 = \ell_1 + \Delta\ell$.

b. Derive Ångström's result: $[\ln|G(\mathrm{i}\omega)|]\,\theta(\omega) = \frac{(\Delta\ell)^2\omega}{2D}$, which is indeed independent of μ. Hints: $G = |G|\,\mathrm{e}^{\mathrm{i}\theta} \implies \ln G = \ln|G| + \mathrm{i}\theta$. Then look at $(\ln G)^2$.

Thus, you simply oscillate one end of your material at ω and measure the log of the ratio of temperature-oscillation amplitudes at two different points and their phase difference. Ångström varied the temperature by alternating cold water and steam using a valve. He used the method to measure the thermal diffusivity of copper (expressed as a conductivity). His result of 382 W/m/K is 5% below the modern value of 401 W/m/K (Haynes, 2014). The best previous measurement, 80 W/m/K, was far too low because it did not account correctly for heat losses. Ångström's method became the standard one for measuring thermal diffusivity and was the first use of thermal "diffusion waves" to probe material properties (Mandelis, 2000).

2.6 From lumped-element circuits to infinite objects. A physical object, such as the one-dimensional conductor considered in Problem 2.4, can show three qualitatively different types of behavior depending on ω, the signal frequency (Frick et al., 2018).

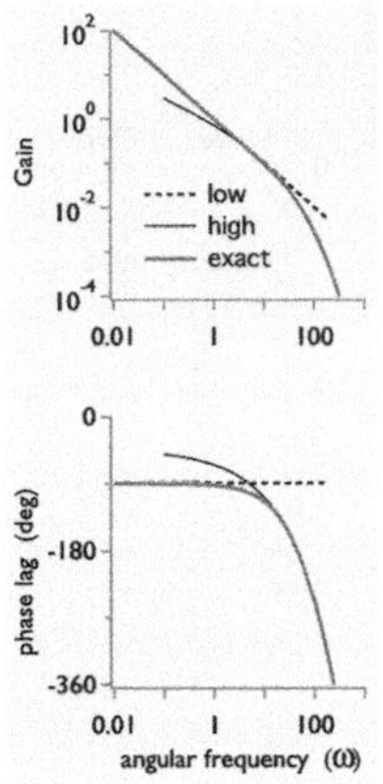

a. Derive the transfer function for thermal conduction of a one-dimensional material of finite length L, with insulating boundary conditions at $x = L$ and a probe at $x = l_0$. Following Problem 2.4 and scaling L to one, show that $G(s, \ell) = \frac{\cosh \sqrt{s}(1-\ell)}{\sqrt{s} \sinh \sqrt{s}}$.

b. Simplify the transfer function at low frequencies ($s \to 0$) and high frequencies ($s \to \infty$). Be precise about what sets the scale that defines low and high frequencies. For each limit, express your result in dimensional as well as dimensionless units, and interpret. (Hint: for dimensional units, use $D = \lambda/(\rho C_p)$.)

c. Reproduce the Bode plot at left of the exact solution and its two limits, which is plotted for $\ell = l_0/L = 0.5$. Explore other values of ℓ.

2.7 Example 2.6. Give the steps in the calculations.

2.8 Example 2.7. Check the matrix calculations.

2.9 Critically damped harmonic oscillator. Confirm Eq. (2.68)b that the response of a critically damped harmonic oscillator to a unit-velocity "kick" at time $t = 0$ is $y(t) = t\, e^{-t}$. Hint: Write the dynamical matrix $\boldsymbol{A} = \left(\begin{smallmatrix} 0 & 1 \\ -1 & -2 \end{smallmatrix}\right)$ as $(-1)[\left(\begin{smallmatrix} 1 & 0 \\ 0 & 1 \end{smallmatrix}\right) + \left(\begin{smallmatrix} -1 & -1 \\ 1 & 1 \end{smallmatrix}\right)]$ and exponentiate directly. See Problem 13.4a for a more sophisticated approach.

2.10 Converting transfer functions to canonical state-space form. Show:

a. State-space equations in control-canonical form, Eq. (2.59) correspond to the strictly proper transfer function, Eq. (2.57).

b. A "merely proper" transfer function can still be written in control-canonical form.

$$G(s) = \frac{b_0 s^n + b_1 s^{n-1} + \cdots + b_n}{s^n + a_1 s^{n-1} + \cdots + a_n} \quad \longrightarrow \quad \left(\begin{array}{c|c} \boldsymbol{A} & \boldsymbol{B} \\ \hline \boldsymbol{C} & \boldsymbol{D} \end{array}\right),$$

2.11 Feedthrough as unmodeled dynamics. Consider highly overdamped motion, with a small term multiplying the highest derivative (*singular perturbation*): $\varepsilon\ddot{x} + \dot{x} + x = u$ and $y = \dot{x}$, where $u(t)$ is the input, the output $y(t)$ is the *velocity*, and the mass $\varepsilon \ll 1$.

a. Transform to a two-dimensional state-space form with $\boldsymbol{x} = (x_1 \; x_2)^{\mathsf{T}}$. Find the two poles, to lowest order in ε. One mode should be slow, the other fast.

b. Solve for $x_2(t)$ in the quasistatic limit, $\varepsilon\dot{x}_2 \approx 0$, and show that the reduced, one-dimensional state-space equations have an output with feedthrough $D = 1$.

2.12 Invariance of the transfer function.

a. Show that the transfer function is invariant under coordinate transformation:

$$G(s) = \boldsymbol{C}\,(s\mathbb{I} - \boldsymbol{A})^{-1}\,\boldsymbol{B} = \boldsymbol{C}'\,(s\mathbb{I} - \boldsymbol{A}')^{-1}\,\boldsymbol{B}'\,.$$

b. Consider the second-order system $\boldsymbol{A} = \left(\begin{smallmatrix} 0 & 1 \\ -1 & -2\zeta \end{smallmatrix}\right)$, $\boldsymbol{B} = \left(\begin{smallmatrix} 0 \\ 1 \end{smallmatrix}\right)$, $\boldsymbol{C} = (1\ \ 0)$, and transformation $\boldsymbol{T} = \frac{1}{\sqrt{2}}\left(\begin{smallmatrix} 1 & -1 \\ 1 & 1 \end{smallmatrix}\right)$. Physically, $\boldsymbol{T}$ corresponds to a 45° rotation in the x_1-x_2 plane. Find new matrices $\{\boldsymbol{A}', \boldsymbol{B}'$, and $\boldsymbol{C}'\}$ and show that the corresponding transfer function calculated is the same as that calculated using $\{\boldsymbol{A}$, $\boldsymbol{B}$, and $\boldsymbol{C}\}$.

2.13 First-order systems, step and impulse response. Derive the step and impulse response from Section 2.4.4 for the first-order system $G(s) = (1 + s)^{-1}$. Do it directly in the time domain and also by Laplace transforms. For responses, the system is in equilibrium at $x = 0$ for $t < 0$. For the step response, see Eq. (2.66) or Footnote 22.

2.14 Lyapunov function for linear dynamics, 1. Let $\dot{\boldsymbol{x}} = \boldsymbol{A}\boldsymbol{x}$, with all the eigenvalues of $\boldsymbol{A}$ having a negative real part, and let $\boldsymbol{Q}$ be an arbitrary $n \times n$ positive definite matrix.

a. Show $V(\boldsymbol{x}) = \boldsymbol{x}^{\mathsf{T}}\boldsymbol{P}\boldsymbol{x}$ is a Lyapunov function, where $\boldsymbol{P} = \int_0^\infty \mathrm{d}t\, \mathrm{e}^{\boldsymbol{A}^{\mathsf{T}}t}\,\boldsymbol{Q}\,\mathrm{e}^{\boldsymbol{A}t}$ and $\boldsymbol{P}$ satisfies the *Lyapunov equation*, $\boldsymbol{A}^{\mathsf{T}}\boldsymbol{P} + \boldsymbol{P}\boldsymbol{A} = -\boldsymbol{Q}$.

b. If we can choose $\boldsymbol{Q}$, then Lyapunov functions are not unique. For one-dimensional dynamics $\dot{x} = -x$ and arbitrary $Q > 0$, construct P, show that it obeys the Lyapunov equation, and find V. Why does an arbitrary, positive Q work?

2.15 Lyapunov function for linear dynamics, 2. Consider two-dimensional linear dynamics, with $\dot{\boldsymbol{x}} = \boldsymbol{A}\boldsymbol{x}$ and Lyapunov equation $\boldsymbol{A}^{\mathsf{T}}\boldsymbol{P} + \boldsymbol{P}\boldsymbol{A} = -\boldsymbol{Q}$, with $\boldsymbol{A} = \left(\begin{smallmatrix} a_{11} & a_{12} \\ a_{21} & a_{22} \end{smallmatrix}\right)$, $\boldsymbol{Q} = \left(\begin{smallmatrix} q_1 & 0 \\ 0 & q_2 \end{smallmatrix}\right)$, $\boldsymbol{P} = \left(\begin{smallmatrix} p_a & p_b \\ p_b & p_c \end{smallmatrix}\right)$. Solve the linear matrix equation with a trick:

a. Define the *vector* $\boldsymbol{p}^{\mathsf{T}} = (p_a\ \ p_b\ \ p_c)$. Then write down and solve the corresponding 3×3 matrix version of the Lyapunov equation for $\boldsymbol{p}$.

b. Write $V(\boldsymbol{x})$ explicitly in terms of the elements of $\boldsymbol{A}$ and $\boldsymbol{Q}$ for $a_{12} = a_{21} = 0$.

2.16 Lyapunov and the damped pendulum. Some Lyapunov functions are more useful than others. Consider a damped pendulum whose angle $\theta(t)$ obeys $\ddot{\theta} + \dot{\theta} + \sin\theta = 0$.

a. Show that the total energy V_1 is a negative semidefinite Lyapunov function.

b. Use the Krasovskii–LaSalle Invariance Principle to conclude from the analysis of V_1 that the down position is locally asymptotically stable.

c. Show that another Lyapunov function $V_2 = \frac{1}{2}\dot{\theta}^2 + \frac{1}{2}(\dot{\theta} + \theta)^2 + 2(1 - \cos\theta)$ is locally negative definite. Discuss the local stability of the down equilibrium.

2.17 Krasovskii's method for Lyapunov functions. Although there are no general methods for finding Lyapunov functions, there are tricks. Here is one: let $\dot{\boldsymbol{x}} = \boldsymbol{f}(\boldsymbol{x})$ be an n-dimensional nonlinear dynamical system with $\boldsymbol{A} = \frac{\partial \boldsymbol{f}}{\partial \boldsymbol{x}}$ its $n \times n$ Jacobean matrix.

a. Show that if $\boldsymbol{A} + \boldsymbol{A}^{\mathsf{T}}$ is negative definite, then $V = \boldsymbol{f}^{\mathsf{T}}\boldsymbol{f}$ is a Lyapunov function.
b. Show that the origin is globally stable for $\dot{x}_1 = -2x_1 + x_2$, $\dot{x}_2 = x_1 - 2x_2 - x_2^3$.

3 Frequency-Domain Control

We begin our study of control by considering linear, time-invariant dynamics in the frequency domain, using Laplace transforms of signals and transfer functions. Chapter 1 introduced the idea of *closed-loop* control, where the input to a dynamical system depends on measurements of its output. We now look at these ideas in more detail.

To motivate feedback techniques, consider the *tracking problem* of making the output $y(t)$ of a system follow, as closely as possible, a desired reference signal $r(t)$.[1] The Laplace transform of the output $y(s)$ should then equal that of the reference $r(s)$. A first solution is to send the reference signal directly to the system input (see right). Unfortunately, the system dynamics (transfer function) usually alters the signal so that the output

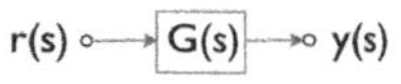

$$y(s) = G(s)\, r(s)\,, \tag{3.1}$$

differs from the input. A better, if still naive, strategy is to transform the reference using the inverse dynamics of the system. Loosely, the idea is to "shape" the input so that the system dynamics transforms the signal back to the original, desired form. Mathematically, the transfer function of the new dynamical element must invert the system transfer function:

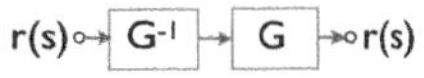

$$y(s) = G^{-1} G\, r(s) = r(s)\,. \tag{3.2}$$

The dynamical element G^{-1} is known as a *(reference) feedforward* controller, because it feeds the input signal (the reference) forward to the system. It is an example of open-loop control and gives the "perfect" solution, where the output tracks perfectly the input. We cannot do better!

The open-loop, inverse strategy sounds too good to be true – and usually is. We identify quickly some problems. Their full resolution will take the rest of the book.

- *Identification*: In order to invert $G(s)$, we need to know what $G(s)$ is. Inferring G is the problem of *system identification*, the subject of Chapter 6.
- *Robustness*: Even if we measure $G(s)$, it may change afterwards: Parameters may drift or jump abruptly, loads change, components fail, and so on.
- *Realizability*: Even with the right $G(s)$ and even if it is unchanging, the inverse may not be realizable physically. Even then, the required inputs may exceed actuator limits.

[1] The *regulation problem* corresponds to the special case where $r(t)$ is constant.

These problems with naive feedforward motivate the introduction of *feedback*, which can generate an effective inverse of the system, even if the transfer function is unknown, and even if it changes over time. There are three fundamental reasons for using feedback:

1. *Uncertain signals.* These can include disturbances, actuator noise, etc.
2. *Uncertain dynamics.* You do not know (or do not want to know) your system precisely.
3. *Unstable dynamics.* Your system is not stable unless you act on it.

We will elaborate on all of these ideas throughout this book.

3.1 Basic Feedback Ideas

We begin with a simplified single-input, single-output (SISO) feedback loop, as illustrated in the block diagram of Figure 3.1. We go through the different elements in the diagram and develop ideas that we introduced in Section 2.3.3.

1. All quantities are represented in the frequency domain by their Laplace transform and are functions of s. We use the same symbol to represent the time and the frequency domains, so that the output $y(s)$ is the Laplace transform of a signal $y(t)$, etc.
2. Transfer functions of the physical system $G(s)$ and controller $K(s)$ are denoted by white boxes. The signals $r(s)$, $e(s)$, $u(s)$, and $y(s)$ are represented by lower-case letters.[2]
3. Input and output ports are represented by small open circles.
4. Summing junctions, which add multiple inputs to produce one output, are represented by larger open circles. Inputs are positive unless marked by a negative sign.
5. Nodes, where a signal branches, are represented by small filled circles. For example, the output y is sent both to an output port and back to a summing junction (as feedback).
6. The *error signal* $e \equiv r - y$. The choice of sign is standard in control books. Feeding minus the output back into the controller input is termed *negative feedback*.

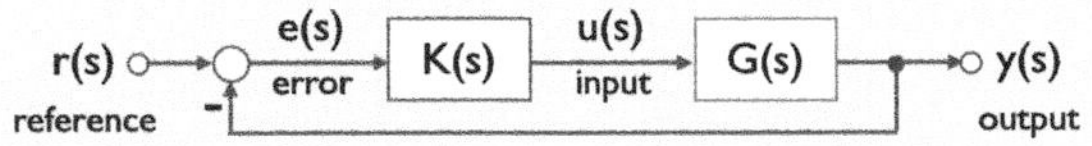

Fig. 3.1 Block diagram of simple feedback loop, with reference signal r, error e, controller K, input u, system G, and output y.

[2] Why is the symbol u often used for the control input? The convention seems to be a bow to early Soviet and Russian contributions to the discipline. In Russian, "control" is *upravlenie* (Gamkrelidze, 1999).

How does the feedback loop affect dynamics? Starting from a system whose *open-loop dynamics* are given by $G(s)$, we add a feedback loop and a controller $K(s)$ to produce a new, *closed-loop* dynamical system. From the block diagram, the closed-loop dynamics is

$$u = Ke = K(r-y) \quad \text{and} \quad y = Gu = GK(r-y) \quad \Longrightarrow \quad y = \left(\frac{GK}{1+GK}\right) r. \tag{3.3}$$

Adding the feedback loop and controller alters the transfer function of the physical system G ("open loop") into the "closed-loop" transfer function $\frac{GK}{1+GK}$.

Let us define some other terms for later use:

$$\left.\begin{aligned} L &= GK && \text{loop transfer function} \\ S &= \frac{1}{1+L} = \frac{1}{1+GK} && \text{sensitivity function} \\ T &= \frac{L}{1+L} = \frac{GK}{1+GK} = 1-S && \text{complementary sensitivity function} \end{aligned}\right\} \tag{3.4}$$

The loop transfer function L is just the product of transfer functions "going around the feedback loop." The complementary sensitivity function T is the closed-loop transfer function between r and y. We will discuss its name and relation to S below.

3.1.1 Proportional Control of a First-Order System

Consider a first-order physical system that acts as a low-pass filter with gain G_0 and cutoff frequency ω_0,

$$G(s) = \frac{G_0}{\omega_0 + s}, \tag{3.5}$$

The controller $K(s) = K_\text{p}$ defines a *proportional controller*, with *proportional gain* K_p and input $u(s) = K_\text{p}\, e(s)$. Errors are thus corrected in "proportion" to their size. Because $e = r - y$, the feedback is *negative*: the correction is opposite to the direction of deviation. When the output y exceeds the reference *set point* r, the input u decreases, and vice versa.

The closed-loop transfer function between the reference $r(s)$ and output $y(s)$ is then

$$T(s) = \frac{y(s)}{r(s)} = \frac{GK}{1+GK} = \frac{K}{K+G^{-1}} = \frac{K_\text{p}}{K_\text{p} + \frac{\omega_0 + s}{G_0}} = \left(\frac{G_0 K_\text{p}}{G_0 K_\text{p} + \omega_0 + s}\right), \tag{3.6}$$

which is again a first-order (low-pass) system, but now with gain $K_\text{p} G_0$ and with cutoff frequency $\omega_\text{c} = \omega_0 + G_0 K_\text{p}$. Notice that the controller transfer function $K(s)$ is only biproper, but the closed-loop response $T(s)$ is strictly proper.

Converting back to the time domain via the techniques of Section 2.4.1, we have

$$\begin{aligned} \dot{y} &= -\omega_0 y + G_0 u \qquad u = K_\text{p}(r-y) \\ &= -(\omega_0 + G_0 K_\text{p})\, y + G_0 K_\text{p}\, r. \end{aligned} \tag{3.7}$$

For a step-function command signal $r(t) = \theta(t)$ and initial conditions $y(0) = 0$, we have

$$y(t) = y_{ss}\left[1 - e^{-(\omega_0 + G_0 K_p)t}\right], \quad y_{ss} = \frac{G_0 K_p}{\omega_0 + G_0 K_p}. \tag{3.8}$$

In the limit of large gain $K_p \gg \omega_0/G_0$ and long times ($\gg (\omega_0 + G_0 K_p)^{-1}$), Eq. (3.8) implies that $y(t) \to y_{ss} = 1$. Consequently, the closed-loop transfer function $T(s) \to 1$, and $y = Tr \to r$. Thus, the feedback loop has generated the desired inverse controller. Even more remarkably, it has done so in a way that is independent of the details of $G(s)$. All that needs to be satisfied is the large-gain and long-time conditions and that the system actually be first order. Since $T \to 1$ independently of the exact parameters describing $G(s)$, the closed-loop behavior is *robust*.

We can also calculate the control signal u. For the step-function command $r(t) = \theta(t)$, we can immediately write, from Eq. (3.8)

$$\begin{aligned} u(t) = K_p e(t) = K_p[1 - y(t)] &= K_p\left[(1 - y_{ss}) + y_{ss}\, e^{-(\omega_0 + G_0 K_p)t}\right] \\ &= K_p\left[\left(\frac{\omega_0}{\omega_0 + G_0 K_p}\right) + \left(\frac{G_0 K_p}{\omega_0 + G_0 K_p}\right) e^{-(\omega_0 + G_0 K_p)t}\right]. \end{aligned} \tag{3.9}$$

This decreases from a maximum value at $t = 0$ given by $u(0) = K_p$. Thus, the speed-up of the response rate, from $\omega_0 \to \omega_0 + G_0 K_p$ has a cost: the actuator must be able to reach an amplitude of at least K_p.

3.1.2 Scaling the Transfer Function

Although the first-order low-pass filter discussed in Section 3.1.1 is simple, the physical constants obscure the main ideas. For this reason (and those discussed in Section 2.2.2), let us scale time and the units of input and output so that

$$G(s) = \frac{1}{1 + s}. \tag{3.10}$$

For proportional feedback control, $K(s) = K_p$, now K_p a dimensionless number, the *gain*. The closed-loop command response is

$$T(s) = \frac{GK}{1 + GK} = \frac{K}{K + G^{-1}} = \frac{K_p}{K_p + 1 + s} = \left(\frac{K_p}{1 + K_p}\right)\left(\frac{1}{1 + \frac{s}{1 + K_p}}\right), \tag{3.11}$$

which is a low-pass filter with cut-off frequency increased from 1 to $1 + K_p$. We recover the steady-state solution y_{ss} using the Final-Value Theorem (Section A.4.3 online):

$$y(t \to \infty) = \lim_{s \to 0} s\, T(s)\, r(s). \tag{3.12}$$

For the first-order case described above, the step command $r(t) = \theta(t) = 1$ for $t > 0$, so that

$$r(s) = \int_0^\infty dt\, (1) \cdot e^{-st} = \frac{1}{s}, \tag{3.13}$$

and

$$y_{ss} = \lim_{s \to 0} s\,T(s) \left(\frac{1}{s} \right) = \lim_{s \to 0} T(s) = \frac{K_p}{1 + K_p}, \tag{3.14}$$

which agrees with Eq. (3.8) when we take $G_0 = \omega_0 = 1$. Notice how simple the Final-Value Theorem is for a step command: $y_{ss} = T(s \to 0)$.

We can also evaluate the control signal, $u(s)$. In general (omitting the s dependence),

$$u = Ke = K(r - y) = K\left(r - \frac{KG}{1 + KG} r\right) = \left(\frac{K}{1 + KG}\right) r. \tag{3.15}$$

For $K(s) = K_p$ and $G(s) = \frac{1}{1+s}$, the control is

$$u(s) = \left(\frac{K_p}{1 + \frac{K_p}{1+s}} \right) r(s). \tag{3.16}$$

For a step command $r(t) = \theta(t)$, the Initial Value Theorem (Section A.4.3 online) gives

$$u(t \to 0) = \lim_{s \to \infty} s\,u(s)\,r(s) = u(s \to \infty) = \left. \frac{K_p}{1 + \frac{K_p}{1+s}} \right|_{s \to \infty} = K_p, \tag{3.17}$$

in agreement with Eq. (3.9). The "one" in the numerator of the scaled system $G(s)$ means that one unit of "u" will, for the open-loop system without control, lead to a unit change in the value of y. The gain of the closed-loop system has thus increased by a factor K_p.

In these scaled units, a "good" controller should not require control signals greatly exceeding one. Requiring a large control signal will limit the operating range relative to that of the open-loop system. This "cost" of control (for the proportional-control algorithm) partly motivates a search for better control algorithms, even for a first-order system.

3.1.3 Robustness and Sensitivity Functions

As we saw in examples such as Eq. (3.8), feedback can add robustness to a system's dynamical response. To understand this statement, we return to the naive feedforward strategy, where the controller is chosen to be the inverse, $G^{-1}(s)$, of the system, $G(s)$. Intuitively, any perturbation, or *disturbance* to the output will directly affect that output in the feedforward case but will be corrected in the feedback case.

To quantify "robustness," we introduce the dimensionless *sensitivity function* S_{QP}, the fractional change in a quantity Q due to a fractional change in a parameter P:

$$S_{QP} \equiv \frac{P}{Q}\frac{dQ}{dP} = \frac{d \ln Q}{d \ln P}. \tag{3.18}$$

If the meaning of Q and P is clear from the context of the discussion, we will drop the subscripts and just use S. Note that

$$Q \sim P^n \quad \Longrightarrow \quad S = n, \tag{3.19}$$

so that the sensitivity equals the exponent in the parameter dependence. If $Q \propto P$, a 1% change in P changes Q by 1%. If $Q \sim P^2$, a 1% change in P changes Q by 2%, and so on. For a more general function $Q(P)$, we can interpret S as the *local* exponent. For example,

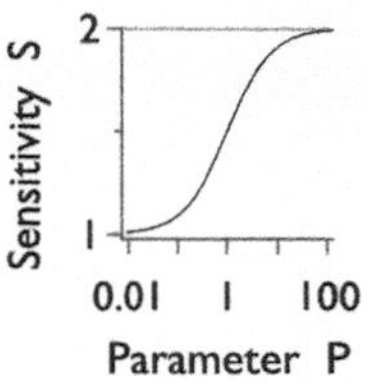

$$Q = P + P^2 \quad \Longrightarrow \quad S = \begin{cases} 1 & P \ll 1\,, \\ 2 & P \gg 1\,. \end{cases} \tag{3.20}$$

In other words, S interpolates between 1 for small P and 2 for large P (see left).

We use the sensitivity function of the system's transfer function to measure robustness. A large value of S implies that changes to a parameter produce large changes in the dynamical response, while a small value of S means that the system is insensitive to changes.

Let us see how to apply these notions of robustness to our first-order system. We assume that variations in $G(s)$ arise because of variations in its DC gain, G_0. Then we can compare the sensitivities of open- and closed-loop transfer functions:

$$\begin{aligned} T_{\text{open}} &= G(s)\,K_{\text{p}} & \Longrightarrow \quad S_{\text{open}} &= \left(\frac{G_0}{T_{\text{open}}}\right)\frac{\mathrm{d}T_{\text{open}}}{\mathrm{d}G_0} = 1 \\ T_{\text{closed}} &= \frac{G(s)\,K_{\text{p}}}{1+G(s)\,K_{\text{p}}} & \Longrightarrow \quad S_{\text{closed}} &= \left(\frac{G_0}{T_{\text{closed}}}\right)\frac{\mathrm{d}T_{\text{closed}}}{\mathrm{d}G_0} = \frac{1}{1+G(s)\,K_{\text{p}}}\,. \end{aligned} \tag{3.21}$$

The open-loop sensitivity S_{open} equals 1: disturbances affect the output directly, with no attenuation. By contrast, for large values of feedback gain, K_{p}, the closed-loop sensitivity function S_{closed} tends to zero. The scale K_{p}^* that divides small from large is obtained by balancing the terms in the denominator: $1 \approx G_0 K_{\text{p}}^*$, or $K_{\text{p}}^* \approx \frac{1}{G_0}$.

The closed-loop system is thus much less sensitive to variations in the gain G_0 than the open-loop system. The reduced sensitivity to parameter variations makes the closed-loop dynamics more robust than the open-loop dynamics.

It is no accident that S_{closed} is the sensitivity function S defined in Eq. (3.4). In general, we can write $T = \frac{L}{1+L}$, using the loop transfer function $L = GK$. Then

$$S_{TL} = \frac{L}{T}\frac{\mathrm{d}T}{\mathrm{d}L} = \frac{L}{T}\frac{\mathrm{d}}{\mathrm{d}L}\left(1 - \frac{1}{1+L}\right) = \frac{1}{1+L}\,. \tag{3.22}$$

That is, $S_{TL} = 1 - T$; hence the name of the latter. If the loop transfer function L is itself a function of a parameter, $L = L(P)$, we apply the *chain rule for sensitivities*:

$$S_{TP} = \frac{P}{T}\frac{\mathrm{d}T}{\mathrm{d}P} = \frac{P}{T}\left(\frac{\mathrm{d}T}{\mathrm{d}L}\frac{\mathrm{d}L}{\mathrm{d}P}\right) = \left(\frac{L}{T}\frac{\mathrm{d}T}{\mathrm{d}L}\right)\left(\frac{P}{L}\frac{\mathrm{d}L}{\mathrm{d}P}\right) = (S_{TL})\,(S_{LP})\,. \tag{3.23}$$

In our example of a first-order system with proportional control, $L(G_0) = K_{\text{p}}G_0/(1+s)$. Since $L \sim G_0$, we see that $S_{LP} = 1$, giving the result of Eq. (3.21).

We draw two general conclusions:

1. A high loop gain $|L|$ implies that the closed-loop transfer function will be less sensitive than the open-loop system to variations in a parameter P.

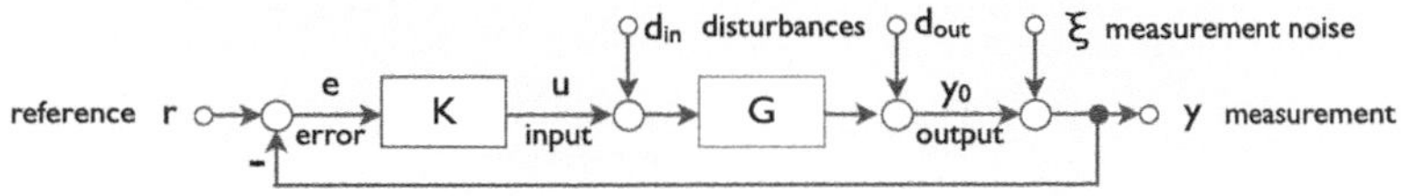

Block diagram of simple feedback loop, with disturbance d and measurement noise ξ. All elements are functions of s. **Fig. 3.2**

2. Because of linearity, the above conclusion holds frequency by frequency. That is, when $|L(s = i\omega)| \gg 1$, the dynamics of the system, at frequency ω, is robust to perturbations.

At this point, we are tempted to be greedy and let $K_p \to \infty$, implying $T = 1$. This is the perfect transfer function between input and output that effectively inverts the system dynamics. Unfortunately, what seems too good to be true, is:

- A very large gain K_p requires larger inputs u than are physically possible.
- A high gain can lead to instability.
- Measurement noise is multiplied by the feedback gain and "contaminates" the system dynamics, a kind of *back action* of the measurement on the system.

As a result, the simple feedback loop of Figure 3.1 is not the complete solution to the problem of designing a controller that will perfectly track a reference signal r. In other words, if we could take $K_p \to \infty$, we would indeed generate an effective inverse to the problem. But, for the reasons just mentioned, K_p must not be too large.

To understand these points, consider the more detailed block diagram in Figure 3.2, which shows how *disturbances*, $d(s)$, enter before or after the system. Figure 3.2 also allows for noise $\xi(s)$ from the measurement, distinguishing between the ideal output y_0 and the measured output y. The former refers to the true value of the output variable (e.g., temperature); the latter refers to the noisy measurement reported by a probe (e.g., a thermistor). Similarly $d_{out}(s)$ affects the physical system whereas $\xi(s)$ affects the measurement.

Using Figure 3.2, we derive the transfer function between the reference r and true output y_0. First, $u = K(r - y) = K(r - y_0 - \xi)$ and $y_0 = G(u + d_{in}) + d_{out}$, which implies that

$$y_0 = \underbrace{\left(\frac{GK}{1+GK}\right)}_{T}(r-\xi) + \underbrace{\left(\frac{G}{1+GK}\right)}_{GS} d_{in} + \underbrace{\left(\frac{1}{1+GK}\right)}_{S} d_{out}. \qquad (3.24)$$

Equation (3.24) shows that the effect of disturbances on the ideal system output is captured by the sensitivity function S, while the effect of measurement noise is captured by its complement T. Ideally both would be small, but the constraint $S + T = 1$ implies trade-offs, explored below.

Note a shortcut for calculating closed-loop transfer functions: the numerator is the transfer function of the direct path between input and output, while the denominator is always $1 + L$, with L the loop transfer function, equivalent to multiplication by S. Schematically,

$$\text{Closed loop} = \frac{\text{Direct}}{1 + \text{Loop}}.$$

We can then immediately write any other desired transfer function. For example, when noise and disturbances are present, Eq. (3.15) generalizes to $u = KS(r - \xi - d_{\text{out}}) - Td_{\text{in}}$. Referring to Figure 3.2, notice that the four inputs $(r, \xi, d_{\text{in}}, d_{\text{out}})$ are all summed after first multiplying by the direct-path transfer function.

As we saw previously for the simplified block diagram of Figure 3.1, the limit of high gain, $K \gg G^{-1}$, leads to appealing conclusions: From Eq. (3.24), we have $y \approx r$. Not only does the measurement y then follow the reference signal r, but disturbances of all kinds are suppressed. Similarly, the input $u \approx G^{-1}(r - \xi - d_{\text{out}}) - d_{\text{in}}$. Notice how u directly counteracts the input disturbance. Of course, these appealing properties hold only if the gain can be set arbitrarily high, which is usually not possible. Finally, note a drawback of high-gain feedback: the actual output $y_0 \to r - \xi$, meaning that measurement noise is injected into the physical output. As we will see later in this chapter and in Chapter 8, filtering the measurement noise can reduce that cost but will increase the effect of disturbances.

Figure 3.2 shows the disturbances entering both before and after the system. Both cases can be relevant. For example, consider the temperature of a block. Noise in the power amplifier that delivers current to an electrical heater would be a disturbance that enters at the input d. A cool breeze could generate a separate disturbance that affects the output via a temperature probe.

In Figure 3.3, the transfer function $G_{\text{d}}(s)$ allows for even more general disturbances while recovering input and output disturbances as special cases. For example, $G_{\text{d}} = G$ models an input disturbance (e.g., noise from the power amplifier), while $G_{\text{d}} = 1$ models an output disturbance (e.g., breeze on the temperature probe).

Finally, whereas in this chapter, we will think of $d(s)$ as representing some definite disturbance – an impulse, a step, a sine wave, etc. – another approach is to represent it as a random signal. Then $G_{\text{d}}(s)$ can be thought of as a kind of low-pass filter that "shapes" the disturbances in frequency space. Chapter 8 develops these ideas in detail.

The closed-loop transfer function between measurement noise and ideal output, T, captures another cost to feedback. Although a large gain is good for tracking the reference, it injects noise into the system. And if disturbances affect the output

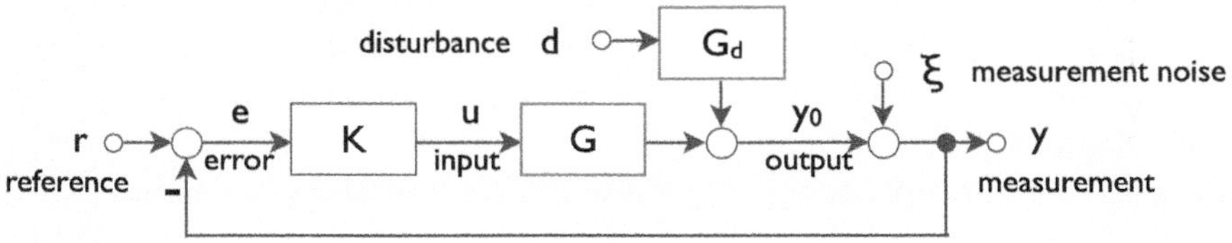

Fig. 3.3 Block diagram showing a disturbance d entering via an arbitrary path.

even without control, measurement noise does not. Thus, before adding control, we should ask whether the advantages of regulated dynamics outweigh the extra noise input into the system. A clever dodge is to use a high gain at low frequencies, where variations in r dominate, and a low gain at high frequencies, where the noise remains important. This strategy is the essence of classical control: because it is not possible to invert the system's transfer function exactly, we try to construct a partial inverse. For a linear system, the idea is to divide the dynamics by frequency, using large gains – approximately inverting the dynamics – over important frequency ranges while leaving nearly unchanged the dynamics at other frequencies. Thus, we will consider control algorithms $K(s)$ that have frequency-dependent gain.

Feedback loops are characterized by four transfer functions, the "Gang of Four" :[3]

$$\underbrace{\frac{1}{1+L}}_{\text{sensitivity}}, \quad \underbrace{\frac{G}{1+L}}_{\text{load sensitivity}}, \quad \underbrace{\frac{K}{1+L}}_{\text{noise sensitivity}}, \quad \underbrace{\frac{L}{1+L}}_{\text{complementary sensitivity}}, \tag{3.25}$$

where $L(s) = K(s)G(s)$ is the *loop transfer function*. It is important to understand the properties of these different transfer functions, especially their variation with frequency ($s = i\omega$). We begin by introducing the concept of *closed-loop feedback bandwidth*, defined to be the lowest frequency ω_c for which $\left|\frac{T(i\omega)}{T(0)}\right| = \frac{1}{\sqrt{2}}$. For a first-order system with proportional feedback, we found (Eq. 3.6) that T was given by a low-pass filter, with a characteristic frequency, in dimensional units, of $\omega_c = \omega_0(1 + G_0K_p)$. In other words, the complementary sensitivity function T is close to its DC value (≈ 1 for large K_p) for frequencies up to ω_c, and falls off for higher frequencies. Conversely, since $S + T = 1$, the sensitivity function is a high-pass filter: it is low throughout the feedback bandwidth, rising to one at higher frequencies. The behavior of S and T are summarized at right, where we see that the feedback "fades away" beyond ω_c.[4]

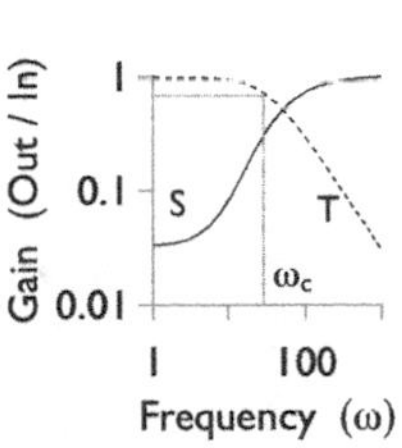

3.2 Two Case Studies

Before going further, we present two applications of high-gain proportional feedback. The first is taken from a relatively recent design of an optical interferometer by Gray et al. (1999). The second is a classic electronic device, the *operational amplifier*.

[3] Åström and Murray (2008). The Gang of Four, led by Chairman Mao's wife Jiang Qing, were influential during China's Cultural Revolution. Their "removal" in 1976, after Mao's death, marked the end of a dark period of Chinese history.

[4] Skogestad and Postlethwaite (2005) define feedback bandwidth as the lowest frequency for which $|S(\omega)| = \frac{1}{\sqrt{2}}$. They note that although $|S| \ll 1$ implies that $|T| \approx 1$, the converse is not true. Since S and T are complex, the relation $S + T = 1$ can be satisfied when S and T have magnitudes larger than 1 (for example, with $S = 10$ and $T = -9$). Thus, while a small $|S|$ guarantees both good disturbance rejection and tracking, $|T| \approx 1$ guarantees good tracking but not necessarily good disturbance rejection. Defining the closed-loop bandwidth as $\left|\frac{T(i\omega_c)}{T(0)}\right| = \frac{1}{\sqrt{2}}$ is closer to the usual definition of bandwidth and is often simpler analytically.

3.2.1 Michelson Interferometer

In Section 3.1, we looked at controlling a device that was merely a low-pass filter. Such a simple example might seem academic. Yet, as the following case study shows, simplicity can often be forced on a system, purchased at the price of possible performance.

We consider a Michelson interferometer proposed for the LIGO gravity-wave project (Gray et al., 1999). The detector (itself an interferometer) must isolate all its elements – which are several kilometers long! – from earth-induced vibrations, so that gravity-wave-induced distortions may be identified. Isolating the large masses in the LIGO apparatus requires knowing their position relative to the ground – hence the need for accurate displacement measurement. Here, we consider a simplified version of the interferometer that highlights the role of feedback.

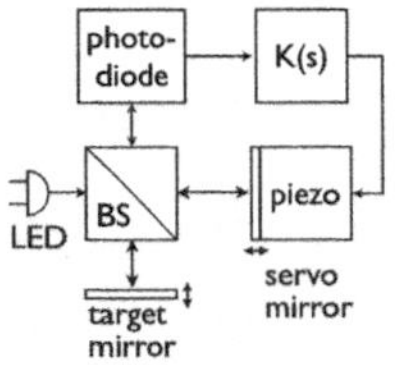

At left is a schematic diagram of the interferometer.[5] Without the control element $K(s)$ and the piezo actuator shown at bottom, it would depict just an ordinary Michelson interferometer. As such, its output is a sinusoidal function of the displacement of the target mirror. The relative displacement of actuator and target mirrors mimics the relative displacement of apparatus and ground in the LIGO experiment. In open-loop operation, the interferometer output is linear only over a small fraction of a wavelength. By adding the actuator, Gray et al. force the servo-mirror to track the moving target mirror. The actuator signal to the servo-mirror effectively becomes the sensor signal.

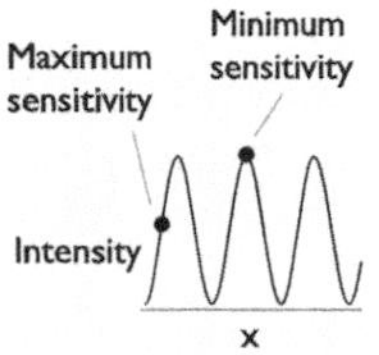

One immediate advantage of tracking a desired *set point* on the fringe is that a linear actuator can effectively *linearize* the original, highly nonlinear sensor signal.[6] A second advantage of using feedback is that measurements have maximum sensitivity if they are taken at a position where the variation of intensity with position is largest. Since the intensity $I(x)$ varies sinusoidally, it is clear that the measurement sensitivity will oscillate between a maximum and zero (where $\mathrm{d}I/\mathrm{d}x = 0$) – see left. By using feedback, we can always use a point of maximum sensitivity.[7] In open loop, the sensitivity would vary with the position being measured. A widely used application of such *feedback linearization* is the scanning tunneling microscope (STM), where the exponential dependence of tunneling current on the distance between conductors is linearized by feedback (Oliva et al., 1995).[8]

Gray et al. chose a band-limited, proportional-gain feedback law, $K(s) = \frac{K_0}{1+s/\omega_0}$, which looks just like our simple system $K(s)G(s)$ in Eq. (3.5) above. They assume that their system has *no* dynamics, $G(s) = 1$, up to the feedback bandwidth $\omega_c = \omega_0(1 + K_0)$. Of course, their system does have dynamics: the photodiode signal rolls

[5] A second detector (not shown) measures and compensates for intensity fluctuations of the LED.

[6] The actuator actually used piezoelectric ceramic stacks, which also give a nonlinear response. But that response can be linearized via another feedback circuit, which is itself based on a linear position sensor.

[7] A second beam splitter (BS) can capture the light directed *back* toward the source. The sum of the two intensities will be constant then for all displacements, and the contrast, $C = \frac{I_1 - I_2}{I_1 + I_2}$, will indicate displacement with uniform sensitivity (Bellon et al., 2002).

[8] Section 11.1 presents another kind of feedback linearization, where a nonlinear feedback law leads to linear closed-loop dynamics. Section 15.2.3 also discusses using feedback to improve measurements.

off at about 30 kHz, and the piezo actuator has a mechanical resonance at roughly the same frequency. But they chose $\omega_0/2\pi \approx 25$ Hz and $K_0 \approx 120$, so that the feedback bandwidth ($\omega_c/2\pi \approx 120 \times 25 = 3$ kHz) was much less than the natural frequencies of their dynamical system. The result is a closed-loop transfer function

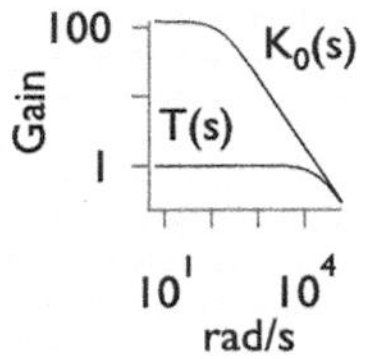

$$T(s) = \frac{1}{1+(KG)^{-1}} = \frac{1}{1+\frac{1+s/\omega_0}{K_0}} = \frac{K_0}{K_0+1+s/\omega_0}, \tag{3.26}$$

which is flat out to nearly $K_0\omega_0$, as shown at right.

The large DC gain of the design increases measurement precision of static displacements. The position noise, 2×10^{-14} m/$\sqrt{\text{Hz}}$, is only about ten times greater than the shot-noise limit imposed by the laser intensity used ($\approx$ 10 mW at $\lambda = 880$ nm). The response was linear over 2 μm of displacement, instead of $\lambda/4 \approx 0.2$ μm for the open-loop interferometer.[9] The lesson: using high-performance components can simplify the feedback design. Here, we are "killing the problem with bandwidth": we start with far more bandwidth than is ultimately needed, in order to simplify the design. Of course, we do not always have that luxury, which motivates the study of more sophisticated feedback algorithms.

3.2.2 Operational Amplifier

Another application of proportional feedback for first-order systems that most experimentalists will not be able to avoid is the operational amplifier – *op-amp* to its friends (Mancini, 2002). The op-amp is perhaps the most widely used analog device and is the basis of most modern analog electronic circuits, including amplifiers, filters, differentiators, integrators, multipliers, buffers, interfaces with sensors, and so forth. Almost any analog circuit will contain several of them. For example, you can use op-amps to implement analog controllers with electronic circuits (Problem 3.6). Here, we study the op-amp itself.

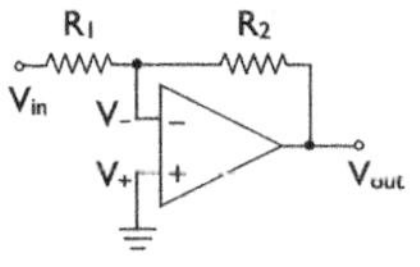

An op-amp is essentially a differential amplifier with very high gain that uses negative feedback to trade off that high gain for reduced sensitivity to drift. A typical circuit, an *inverting amplifier*, is shown at right. The op-amp acts as a differential amplifier:

$$V_{\text{out}} = G(s)\,(V_+ - V_-)\,. \tag{3.27}$$

The + and – inputs serve as an error signal for the feedback loop. In an ideal op-amp, the amplifier gain G is so high that we can neglect any difference between the V_+ and V_- inputs. In addition, the input impedance is so high that we can neglect any input currents. Applying these two rules to the circuit shown above, we note that since the V_+ input is at ground, then V_- is, too. And since the inputs draw no current,

$$\frac{V_{\text{out}}-0}{R_2} = \frac{0-V_{\text{in}}}{R_1} \implies \frac{V_{\text{out}}}{V_{\text{in}}} = -\left(\frac{R_2}{R_1}\right). \tag{3.28}$$

[9] The op-amp used to power the piezo has a signal-to-noise ratio of about 10^9. With a range of 2 μm, we expect a noise floor of $\approx$ 2 fm (2×10^{-15} m), about ten times smaller than was observed. Presumably, the laser power was chosen to make the shot noise comparable to the amplifier noise floor.

The minus sign here corresponds to the "inverted" gain and not to negative feedback.

Real op-amps differ in a number of ways from the above ideal. The most important difference is that, at higher frequencies, we cannot assume arbitrarily large gain. We thus relax the condition that $V_+ = V_-$ and use Eq. (3.27), with $G(s) = \frac{G_0}{1+s}$, where G_0 is the DC gain and the characteristic frequency ω_0 has been scaled out.[10] Summing the currents gives $\frac{V_{\text{out}}-V_-}{R_2} = \frac{V_- - V_{\text{in}}}{R_1}$. Also, $V_{\text{out}} = -G\,V_-$, since the positive input is grounded. Thus,

$$\frac{V_{\text{out}}}{V_{\text{in}}} = -\left(\frac{R_2}{R_1}\right) \frac{G_0}{G_0 + (1 + \frac{R_2}{R_1})(1 + s)} \,. \tag{3.29}$$

At low frequency ($s \ll 1$) and high gain ($G_0 \gg R_2/R_1$), we recover the gain of the ideal op-amp circuit, Eq. (3.28). In physical units, this approximation holds when $(\frac{R_2}{R_1})\,\omega \ll G_0\,\omega_0$, which can be phrased more generally:

$$\text{(Closed-loop gain) (circuit bandwidth)} \approx G_0\,\omega_0 \,, \tag{3.30}$$

where the bandwidth is the maximum frequency for which Eq. (3.28) is valid. The product $G_0\,\omega_0$ in Eq. (3.30), a characteristic of the op-amp itself, is the *gain bandwidth product*, also known as the *unity-gain bandwidth*. Its value typically ranges between 1 MHz and 1 GHz. Equation (3.30) implies a trade-off between the gain and bandwidth of an op-amp-based voltage amplifier: the higher the gain, the lower the bandwidth. It also is the bandwidth of an amplifier with unity gain ($V_{\text{out}} = -V_{\text{in}}$).[11]

The gain-bandwidth trade-off applies more generally. Indeed, the essential element is just having a frequency-dependent gain, $G(\omega) \approx \frac{G_0}{\omega/\omega_0} \approx \frac{G_0\,\omega_0}{\omega}$, with G_0 the DC gain and ω_0 the bandwidth of the amplifier response at maximum gain (G_0). Then $G\omega \approx G_0\omega_0$.

Op-amp circuits reduce sensitivity to variations in amplifier gain. For example,

$$S_{TG_0} = \frac{G_0}{(V_{\text{out}}/V_{\text{in}})} \left| \frac{\mathrm{d}}{\mathrm{d}(V_{\text{out}}/V_{\text{in}})} G_0 \right| = \left| 1 + \frac{G_0}{R_2/R_1} \right|^{-1} \approx \frac{(R_2/R_1)}{G_0} \ll 1 \,. \tag{3.31}$$

By contrast, an open-loop amplifier would have $S = 1$ for variations in its gain. Physically, we trade the large temperature dependence of the semiconductors in transistors, which affects G_0, for the much feebler temperature dependence of resistors.

3.3 Integral, Derivative, and PID

The closed-loop, DC gain $T_0 = T(s = 0)$ of a first-order system with proportional control is $T_0 = G_0 K_{\text{p}}/(1 + G_0 K_{\text{p}})$, with G_0 the open-loop DC gain of the physical system and K_{p} the controller's proportional gain. For finite K_{p}, the steady-state output $T_0 < 1$

[10] The first-order response of Eq. (3.27) is designed into the internal circuitry of the op-amp, which is considerably more complex than a low-pass filter.

[11] See Mancini (2002) for many examples of such "rough-and-ready" op-amp calculations.

is less than the input. Can we eliminate this *proportional droop* without using infinite proportional gain?

A nice trick, *integral control*, solves the problem. Let us send the integral of the error, rather than the error itself, to the feedback controller. The input signal $u(t)$ is then

$$u(t) = K_{\rm i} \int_0^t {\rm d}t' \, e(t') \,, \tag{3.32}$$

with $K_{\rm i}$ the *integral gain* and $t = 0$ an arbitrary starting time. Intuitively, if the error is, say, positive at $t = 0$, then the input will keep increasing until the error vanishes, at which point, $u(t)$ can be constant. But be aware that this argument does not preclude oscillations about the desired steady-state value, or even instability.

3.3.1 Integral Control

More formally, consider a first-order system with integral feedback. In scaled units,

$$\dot{y} = -y + K_{\rm i} \int_0^t {\rm d}t' \left[r_{\rm ss} - y(t') \right] \,, \tag{3.33}$$

where $r_{\rm ss}$ is the desired steady-state set point. Differentiating both sides of Eq. (3.33) gives

$$\ddot{y} = -\dot{y} + K_{\rm i} \left[r_{\rm ss} - y(t) \right] \,, \tag{3.34}$$

converting a first-order integrodifferential equation into an ordinary, second-order ODE that clearly has a steady-state solution $y_{\rm ss} = r_{\rm ss}$, as desired.

Let us use the frequency domain to generalize this analysis to an arbitrary LTI dynamical system. We can use the Final Value Theorem to find the steady-state value $y_{\rm ss}$ for a step input of amplitude $r_{\rm ss}$. The closed-loop transfer function is $T(s) = \frac{1}{1+L^{-1}}$. Here, $L = GK$, with $K = K_{\rm i}/s$ and $G(s)$ a transfer function whose only restriction is that the DC gain, $|G(s = 0)|$, is finite.[12] Then,

$$T(s) = \frac{1}{1 + (GK)^{-1}} = \frac{1}{1 + s\left(\frac{1}{K_{\rm i}}\right) G^{-1}(s)} \,. \tag{3.35}$$

and $\lim_{s\to 0} T(s) = 1$, meaning that $y_{\rm ss} = r_{\rm ss}$. This is a remarkable result: No matter what the time constant (here scaled to one), no matter what the value of $K_{\rm i}$, no matter what $G(s)$ is, the closed-loop dynamics will converge to the desired steady-state value.

When $G(s) = \frac{1}{1+s}$ is first order, Eq. (3.35) describes a second-order system:

$$T(s) = \frac{K_{\rm i}}{s^2 + s + K_{\rm i}} \,, \quad \text{with poles at} \quad s = -\tfrac{1}{2} \pm \sqrt{\tfrac{1}{4} - K_{\rm i}} \,. \tag{3.36}$$

Thus, the response for $K_{\rm i} < \frac{1}{4}$ is overdamped. For $K_{\rm i} = \frac{1}{4}$, it is critically damped, and for $K_{\rm i} > \frac{1}{4}$, underdamped. For $K_{\rm i} \gg 1$, we have $s \approx -\frac{1}{2} \pm {\rm i}\sqrt{K_{\rm i}}$.

12 Actually, the DC gain can be infinite. We can let $G(s) \sim s^a$ with $a < 1$ for small s.

The weakly damped oscillatory closed-loop response at high gain K_i illustrates a problem with feedback: although $K_i > 0$ eliminates proportional droop, too much gain gives a response whose amplitude decays over many oscillations. If the goal is to eliminate perturbations quickly, such a slowly decaying oscillatory response is not desirable.

3.3.2 Proportional-Integral (PI) Control

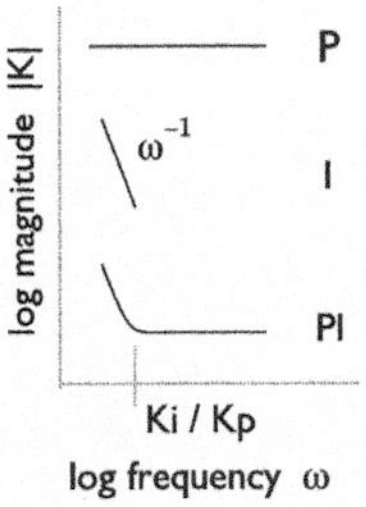

Although integral control solves the problem of set point offset, or proportional droop, it is not the solution to all control problems, as suggested by the collection of Bode magnitude plots for different controllers shown at left. At top is a representation of proportional feedback in the frequency domain. The response is constant for all frequencies. Next, we plot the frequency response of integral control. Since the Laplace transform of an integral is equivalent to multiplication by $1/s$, the magnitude of the frequency response is $1/\omega$, which drops off quickly at high frequencies. Pure integral control has a low closed-loop bandwidth, unless the coefficient can be made very large. But we saw in the previous section that the effect of large K_i is to make a first-order system have a rapidly oscillating, slowly decaying response, which may not be desirable. Indeed, (see Problem 3.1) for large K_i, the bandwidth $\omega_c \sim \sqrt{K_i/2}$, while the damping coefficient is $\zeta \sim 1/(2\sqrt{K_i})$. We seem forced to compromise between bandwidth and an undesirable oscillation in the response.

An obvious thought is to try to combine the advantages of proportional and integral control by summing their response, which leads to *PI control*:

$$u(t) = K_p\, e(t) + K_i \int_0^t dt'\, e(t') \qquad \Longrightarrow \qquad u(s) = \left(K_p + \frac{K_i}{s}\right) e(s). \tag{3.37}$$

The frequency response at bottom shows that the ω^{-1} response of the integral control dominates at low frequencies, while the constant response of proportional control dominates at high frequencies. The crossover frequency is set by the gains: $\omega^* \approx K_i/K_p$.

We now reanalyze our first-order system, $G(s) = \frac{1}{1+s}$, with our PI controller, $K(s) = K_p + K_i/s = (K_p s + K_i)/s$. The closed-loop transfer function between the reference signal r and the output y is

$$T(s) = \frac{1}{1 + \frac{s}{K_p s + K_i}(1+s)} = \frac{K_p s + K_i}{s^2 + (1+K_p)s + K_i}. \tag{3.38}$$

Since $T(s = 0) = 1$, the steady-state solution matches the set point, $y_{ss} = r_{ss}$ – no proportional droop. But, in contrast to pure integral control, in the limit $K_p \gg K_i$, the system looks approximately like our first-order system with proportional gain, where the closed-loop response is again first order, but with the bandwidth increased from ω_0 to $\approx \omega_0 K_p$. Thus, we combine the advantages of both types of control: the bandwidth is increased, with no oscillations in response and no steady-state error.

3.3.3 Proportional-Integral-Derivative (PID) Control

While PI control does a good job on a first-order system, it is less successful with second-order systems (and even worse at higher-order ones).[13] For second-order systems, it turns out that *derivative control* (D) can be very useful. Pure derivative control is given by

$$u(t) = K_{\mathrm{d}} \frac{\mathrm{d}e}{\mathrm{d}t} \qquad \Longleftrightarrow \qquad u(s) = K_{\mathrm{d}}\, s\, e(s)\,, \tag{3.39}$$

Intuitively, if we see the system moving at high "velocity," we "know" that the system state is changing rapidly. We can thus speed the feedback response greatly by anticipating this state excursion and taking counteraction immediately. For example, if the temperature of an object starts to fall, we can increase the heat immediately, counteracting a presumably large perturbation. The word "presumably" highlights a difficulty of derivative control. We infer a rapid temperature change by measuring the derivative of the system state. If the sensor is noisy, random fluctuations can lead to large spurious rates of change and to inappropriate controller response. Sadly, many experimentalists try derivative control, find out that it makes the system noisier, and then give up. But there are many benefits to derivative control, and the spurious response to noise can be minimized. I hope this book will give you the tools to work around the potential problems.[14]

To begin, consider derivative control of $G(s) = \frac{1}{s^2+2\zeta s+1}$. The closed-loop response is

$$T(s) = \frac{K}{K+G^{-1}} = \frac{K_{\mathrm{d}} s}{K_{\mathrm{d}} s + s^2 + 2\zeta s + 1} = \frac{K_{\mathrm{d}} s}{s^2 + 2\left(\zeta + \frac{K_{\mathrm{d}}}{2}\right) s + 1}\,. \tag{3.40}$$

For $K_{\mathrm{d}} > 0$, the derivative term increases the effective damping. This frequency dependence matches our intuition, stated above, that the derivative term alters the "velocity" of the system's response. Cf. Problem 3.2 to apply to the damping of disturbances.

The ability to change damping can be very useful. For example, transforming an underdamped system to an overdamped one can reduce mechanical vibrations (Problem 3.2). We can even imagine turning an unstable system with negative ζ into a stable one (Problem 3.4). Alternatively, a negative K_{d} could destabilize a stable system, to create an oscillator.

[13] For example, in Section 3.5, we will see that integral control can destabilize a stable oscillator.

[14] Derivative control is also interesting for deeper reasons. The fundamental equations of motion – both classical and quantum – are invariant under time reversal, $t \to -t$. Terms such as $\sim \dot{y}(t)$ break this invariance. They are not fundamental and arise only after partitioning the world into macroscopic, observable degrees of freedom and microscopic, unobservable degrees of freedom (heat) and then taking a thermodynamic limit. Fluctuations (cf. Chapter 8) arise from a similar partitioning and must be included to be consistent with the fluctuation-dissipation theorem. Thus, there is a fundamental distinction between feedback terms that are even under time reversal, such as proportional control, and those that are not. In particular, *velocity-dependent* feedback terms such as derivative control can reduce the entropy of a macroscopic system (Kim and Qian, 2004). See the discussion of stochastic thermodynamics in Section 15.3.

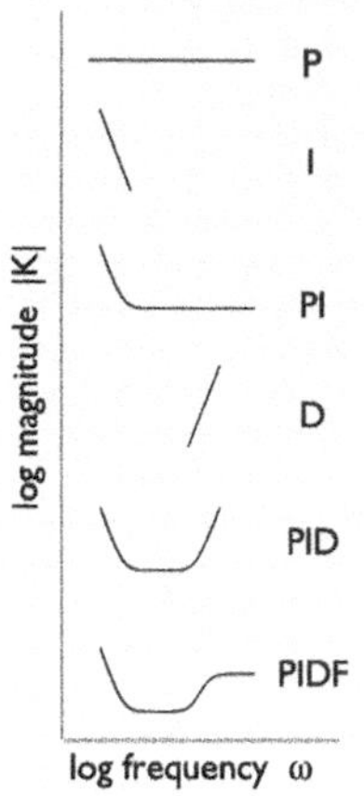

Derivative control is rarely used alone and is most often found in combination with one or both of proportional and integral control. The three-term combination is known as PID control and remains, some three-quarters of a century after its introduction, the "generic" controller for much industrial (and experimental-physics) equipment.[15]

We illustrate various combinations of P, I, and D magnitude plots at left. Note the PIDF control at bottom, which filters, or turns off, the derivative response at high frequencies. Since the feedback loop itself injects noise from the sensor into the physical system (Section 3.1.3), the linear s dependence of D control amplifies the high-frequency components of the noise. An obvious response is to limit the action of the derivative term by adding low-pass elements to the control law. The result is then a controller of the form

$$K(s) = K_{\mathrm{p}} + \frac{K_{\mathrm{i}}}{s} + \frac{K_{\mathrm{d}}\, s}{1 + \frac{s}{\omega_{\mathrm{d}}}}, \tag{3.41}$$

where ω_{d} gives the frequency where the derivative term begins to turn off. The low-pass filter flattens the gain of K at high frequencies. In the absence of filtering ($\omega_{\mathrm{d}} = 0$), the PID transfer function is improper, as the degree of its numerator exceeds that of its denominator, so that its magnitude diverges for $s \to \infty$. Such a transfer function can be realized only at low frequencies.

Another possibility would be to directly filter the sensor to reduce its noise at high frequencies. Indeed, many sensors have a built-in low-pass filter. Sometimes, these hardware features allow us to drop the low-pass filter in the derivative term, leading to the simpler form, $K(s) = K_{\mathrm{p}} + K_{\mathrm{i}}/s + K_{\mathrm{d}}\, s$. Finally, a sometimes-undesirable side effect of derivative control is that changing the reference by a step (a common task) generates an immediate pulse (a "kick") in the output. Problem 3.7 offers a simple workaround.

PID control, with proper precautions for the derivative term, is often a very effective type of controller, but we must still choose parameter values: K_{p}, K_{i}, and K_{d}, as well as parameters associated with any noise-filtering term. For a second-order system, the task is straightforward. The closed-loop response of a PID controller is

$$T(s) = \frac{K_{\mathrm{p}} s + K_{\mathrm{i}} + K_{\mathrm{d}} s^2}{\left(K_{\mathrm{p}} s + K_{\mathrm{i}} + K_{\mathrm{d}} s^2\right) + s\,(1 + 2\zeta s + s^2)}. \tag{3.42}$$

Note how we can use the three tunable parameters K_{p}, K_{i}, and K_{d} to turn the denominator into an *arbitrary* third-order system.[16] To fully understand the closed-loop dynamics, we need to look at the structure of the numerator, as well. Nonetheless, the PID controller leads to closed-loop dynamics with arbitrary poles (eigenvalues).

[15] One fun application: In 2001, Andy Schecter was the first to apply PID control to an espresso maker to stabilize its water temperature. Most high-end machines now use this innovation, which improves the likelihood of making a "good" shot of espresso! See https://home.lamarzoccousa.com/history-of-the-pid/.

[16] Since we are interested only in the position of the roots, the overall scale of the denominator polynomial for $T(s)$ is unimportant, and we can choose it so that the coefficient of the cubic term equals one. The three Ks then specify independently the quadratic, linear, and constant terms.

See Section 3.7.5 for further discussion of how to "tune" the parameters of the PID controller.

Although nonlinearities, neglected modes, and so on mean that a system is never truly second order, the PID controller is often a good choice. In a simplistic view, it represents a kind of lowest-order approach to incorporating information from the present (P), past (I), and future (D). But if a system deviates too much from ideal second-order behavior, the low-order solution will be less and less satisfactory. Before we proceed to more complex controllers, we consider some of the issues that can arise in more complex situations.

In Section 3.4.1, we return to the notion that there is a conflict between rejecting disturbances and tracking a reference signal. We show that feedforward, while not the simple solution to control problems, helps eliminate the conflict between disturbance rejection and reference tracking. In Section 3.5, we discuss how too much gain can lead to an instability caused by the feedback itself.

3.4 Feedforward

We began this chapter by presenting a naive feedforward algorithm for reference tracking by inverting the system dynamics, setting $F = G^{-1}$ so that $y = FG\,r = r$ (see Eq. 3.2). We argued that the naive approach is impractical. Issues include

- Robustness: if G changes but the inverse does not, the tracking will not be accurate.
- Feasibility: it may not be possible to realize the inverse controller.

Although naive feedforward is ineffective, we shall see that a more sophisticated approach combining feedforward with feedback can do better than either technique alone.

3.4.1 Control with Two Degrees of Freedom

In Section 3.1.3, we studied the trade-off between rejecting disturbances and injecting measurement noise into the dynamics of a controlled system. Feedforward can help resolve this trade-off. Consider the block diagram in Figure 3.4, which resembles Figure 3.2 but adds a filter $F(s)$ to the reference signal r. The actual output y_0 becomes

$$y_0 = T(Fr - \xi) + SGd\,, \tag{3.43}$$

where ξ is the amplitude of measurement noise, d the amplitude of (input) disturbances, $S = \frac{1}{1+L}$ and $T = \frac{L}{1+L}$ the sensitivity functions, with $L = GK$. Because $S + T = 1$, when $|S| \ll 1$, then $|T| \approx 1$. Suppressing disturbances (small S) adds measurement noise to the dynamics of the output (large T).

The above trade-offs hold for each frequency. But if disturbances dominate at low frequencies and measurement noise at high frequencies, we can make S small at low

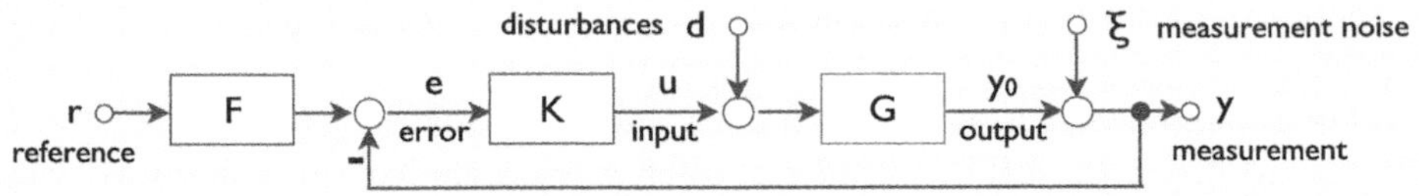

Fig. 3.4 Block diagram of controller with two degrees of freedom, with feedforward F.

frequencies and T small at high frequencies. In Chapter 8, we will see more precisely how to balance the competing demands of rejecting disturbances and limiting measurement noise.

Another trade-off is between limiting feedback bandwidth to limit measurement noise and following command signals. A step, for example, has high-frequency Fourier components. A solution is to add *reference feedforward* dynamics to the controller, in the form of the *prefilter* F in Figure 3.4. The controller then has *two degrees of freedom* for K and F and can simultaneously suppress measurement noise and track rapid set point variations. In this strategy, instead of inverting $G(s)$, we invert the closed-loop transfer function $T(s)$. Consequently, we can choose $T(s)$ to be a "nice" function to invert, such as a first-order low-pass filter with bandwidth ω_c. With similar logic, we realize that asking for a perfect inverse to T is also a mistake, as having the output follow the reference to all frequencies will require arbitrarily high inputs, which cannot be supplied. More modestly, we ask that FT be another first-order low-pass filter whose bandwidth ω_r is higher than ω_c. In other words, we would like to keep the feedback bandwidth ω_c relatively low to minimize noise in the output while still tracking r to higher bandwidths ω_r. Then,

$$F(s)\,T(s) = \frac{F(s)}{1+s/\omega_c} = \frac{1}{1+s/\omega_r} \quad \Longrightarrow \quad F(s) = \frac{1+s/\omega_c}{1+s/\omega_r}\,. \tag{3.44}$$

The *lead compensator* $F(s)$ increases the phase (the "lead") over a limited frequency range. Its Bode plot, at left, shows that the gain interpolates from 1, for frequencies $\omega \ll \omega_c$ to $\frac{\omega_r}{\omega_c} = 10$, for $\omega \gg \omega_r$. Between those frequencies, F resembles derivative control, while it "turns off" outside this frequency range. Lead compensators and variants are also useful transfer-function building blocks for feedback controllers (Section 3.7.3).

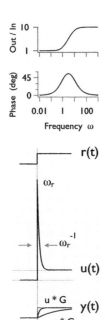

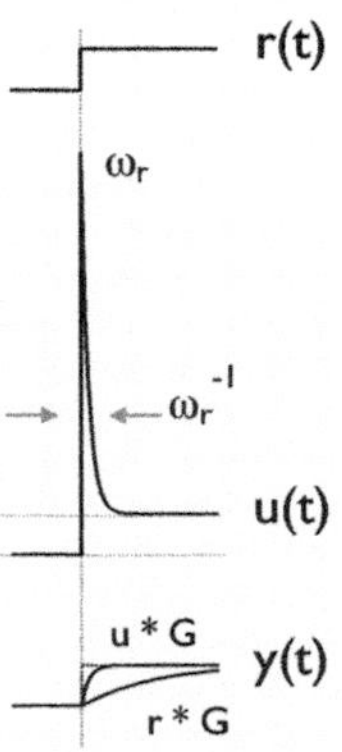

To look at the signals in the time domain, we could invert the Laplace transform of $F(s)$, or just solve directly for $u(t)$ in $\dot{y} = -y + u(t)$. For example, a desired first-order response $y_d(t) = 1 - e^{-\omega_r t}$ is generated by the feedforward input

$$u(t) = y_d + \dot{y}_d = \left(1 - e^{-\omega_r t}\right) + \left(\omega_r\, e^{-\omega_r t}\right) = 1 + (\omega_r - 1)\, e^{-\omega_r t}\,. \tag{3.45}$$

These time-domain signals are plotted at left. If the desired output is sufficiently differentiable (once, in the present case), we can readily generate the required feedforward input. As always, such feedforward neglects disturbances, which can be handled by feedback.

In the present case, we should limit $y_d(t)$ to functions with finite first derivative, such that $y_d + \dot{y}_d$ stays within the allowed input range of the system. Thus, we can choose

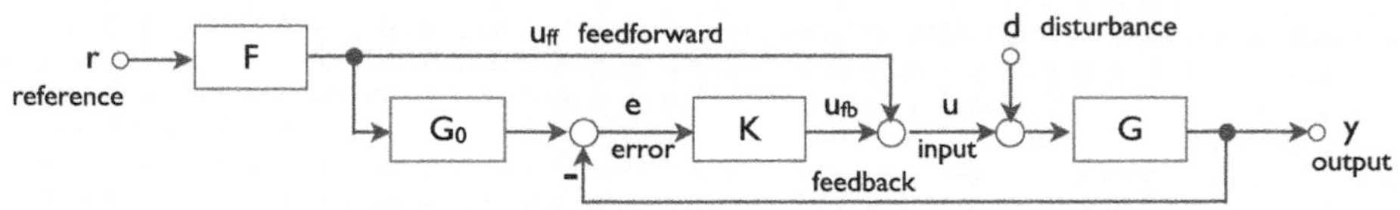

Fig. 3.5 Decoupling feedforward from feedback. F and K act independently if $G_0 = G$.

ω_r as desired, but the maximum amplitude of the actuator signal will be $u(t = 0) = \omega_r$. If ω_r is too large, the actuator will not be able to supply the needed signal. In addition, the range of accessible inputs is reduced by a factor of ω_r/ω_c. The trade-off between range and bandwidth is the price for increasing the speed of response. The important point is that the bandwidth limit is no longer set by the closed-loop bandwidth.

Reference feedforward is particularly effective if we can improve the estimate of the system transfer function by monitoring the performance of the controlled system. This approach is known as adaptive control and will be discussed in Section 10.3.2.

3.4.2 Decoupling Feedforward from Feedback

In the two-degrees-of-freedom controller presented in Figure 3.4, altering the controller K will change both the disturbance and command responses. (Equation 3.43 shows that the former is proportional to S, the latter to T.) But a slightly more elaborate control strategy can decouple the effects of the feedforward block $F(s)$ from the feedback controller $K(s)$. We can then design K to reject disturbances and F to improve command tracking.

Figure 3.5 illustrates the strategy. We send the reference feedforward signal, $u_{ff} = Fr$ directly to the input of the physical system G. The reference is also sent to the feedback loop as in standard feedback control. If our model G_0 of the system transfer function is perfect, and if we can invert it to form $F = G_0^{-1}$, then it is easy to show that, in the absence of disturbances d, the feedback signal $u_{fb} = 0$. If there is a disturbance d, it will generate a feedback signal without affecting the feedforward signal. The two contributions $u_{fb} + u_{ff}$ are thus decoupled. See Problem 3.8.

3.4.3 Disturbance Feedforward

Feedforward can sometimes help reject disturbances, too. If sensors give “advance warning” of a disturbance, we can use the information to compensate for it. For example, consider regulating the temperature y_2 of a sample that is located along a rod (black bar in Figure 3.6a). The device is placed in an insulating “can” that creates slow time constants for perturbations coming from all sides except the top. The lightly insulated top is a thermal link to the ambient environment that evacuates heat efficiently (e.g., to a cold bath) but also introduces an easy path for external disturbances $d(t)$.

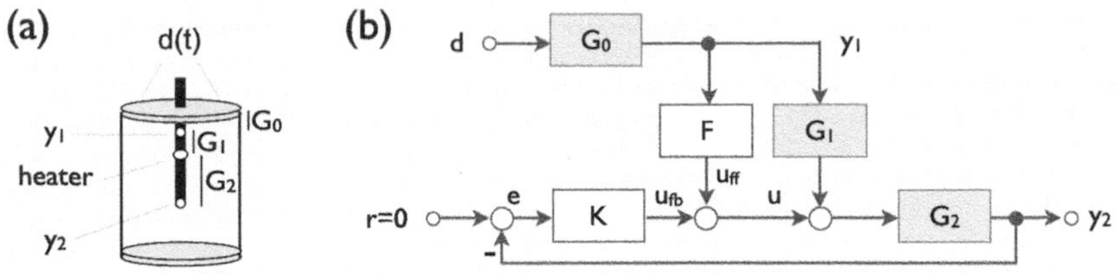

Fig. 3.6 Feedforward compensation of disturbances. (a) Sketch of a temperature-control apparatus showing the two temperature outputs, y_1 and y_2, along with the heater position. The insulation ensures that external disturbances, $d(t)$, arrive only through the top of the apparatus. (b) Block diagram showing signal flow, with feedback (K) and feedforward (F) controllers. The reference input $r = 0$ for this regulator.

In a standard control arrangement, the disturbances would propagate to y_2 (low-pass filtered by the thermal mass of the system) and would be measured and then corrected entirely by the feedback loop, at the same location. Here, we measure at y_1, which encounters the disturbance via G_0 before it reaches y_2 (or the heater). We can then send an appropriate correction "ahead" to the heater. As thermal disturbances are filtered by this transfer function, the probe at y_1 need have only modest bandwidth.[17] We can thus use a feedforward algorithm to send a signal that can compensate for the disturbance. To make sure that the actual temperature at the sample is correct, we add a feedback loop using integral control. If disturbances are well corrected, the gains in the feedback loop may be kept low, as we will see below. The feedback loop will correct for any modeling error in the feedforward element (see below) and also for any disturbances that enter through other paths. (By design, these extra disturbances, which pass through thick thermal insulation, should have low bandwidth and small amplitude.)

From the block diagram in Figure 3.6(b), the sample temperature is $y_2 = G_2(u + G_1 G_0 d)$, with $u = u_{\text{fb}} + u_{\text{ff}}$. The feedback signal is $u_{\text{fb}} = -K y_2$, assuming .that the reference $r = 0$. The feedforward signal is $u_{\text{ff}} = F y_1 = F G_0 d$. Putting all this together gives

$$y_2 = \left[\frac{G_2 G_0 (F + G_1)}{1 + K G_2} \right] d \,. \tag{3.46}$$

Notice that we can reduce the effect of disturbances two ways. Either we choose the feedforward gain to be $F = -G_1$ in the numerator, or we increase the gains in the controller K in the denominator. If the model for G_1 were perfect, we could use only feedforward, setting $K = 0$. The worse the model, the greater the feedback gains needed to compensate. But larger feedback gains will amplify the measurement noise injected into the control loop. Section 3.8.1 discusses further advantages of combining feedforward with feedback.

[17] If the bandwidth of the combined elements $G_0(s) G_1(s)$ is too low, it will take a long time to cool the system. But a higher bandwidth for $G_0 G_1$ increases the frequency range over which disturbances have to be corrected. Adjust the trade-off to fit the specific requirements of your application.

The above scheme needs an estimate of the transfer function G_l, to calculate the feedforward signal u_{ff}; however, the estimate need be accurate only for a limited range of frequencies. Since G_l is low pass, its bandwidth (and that of G_0) will limit the upper frequency range where accurate estimates are needed. Similarly, the feedback controller can typically do a good job of compensating for low-frequency drifts (via integral control). If the bandwidth of G_l is ω_l and that of the controller K is ω_K, then $F = -G_l$ should be accurate only over the frequency range $\omega_K < \omega < \omega_l$. If $\omega_K > \omega_l$, we can eliminate the active feedforward block F, reverting to passive shielding (G_0 and G_l) and feedback.

3.4.4 Applications of Feedforward

Feedforward has many applications.[18] A classic one of disturbance feedforward is to control the beam in particle accelerators based on circular storage rings. At CERN, the technique was key to increasing event rates of proton-antiproton collisions. A disturbance to the beam can be detected and a correction sent ahead. Since a chord that cuts across a circle is shorter than a path around it, signals that propagate at or near the speed of light will arrive earlier than the bunch of particles (see right). The technique led to the discovery of the W and Z bosons and to a Nobel prize to Simon van der Meer, in 1984.

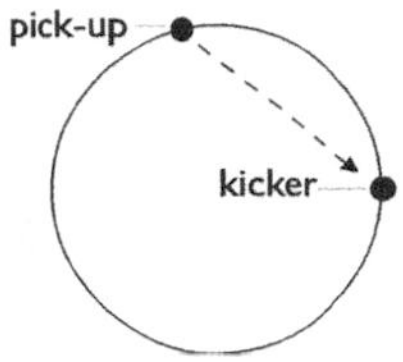

As outlined above, feedforward can compensate for an external disturbance in a temperature regulator. Also, adding a feedforward prefilter to the reference signal can speed up set point changes and ensure a first-order response that prevents overshooting the set point.[19] For example, a transient overshoot might trigger an undesired phase transition. Similarly, irreversible chemical reactions may have appreciable rates only above threshold temperatures. For example, in cooking, transient overshoots often lead to unwanted, irreversible denaturing of proteins (browning, burning, curdling, and so forth).

In one class of problems, instruments such as microscope and AFM stages that displace samples have dynamics that are approximately an underdamped, second-order system. Reference feedforward can transform the input to create a rapid displacement in such a stage without exciting the resonant mode (Problem 3.3; cf. Sections 5.4.2 and 9.1).

Finally, *coherent control*, which is central to nuclear magnetic resonance and to the control of chemical reactions and, more recently, to the control of micro- or nanomechanical oscillators, can be achieved using a feedforward prefilter, inputting a sequence of pulses that makes a target system behave as desired. Designing the pulse sequence is easy in the time domain (Section 5.4.2) or by optimal control (Chapter 7).

[18] *Reference* and *disturbance* feedforward denote the source of the signal that is to be compensated for. We will use the simpler term "feedforward" when either the source is obvious or its type does not matter.

[19] Just measure the appropriate heat capacity and calculate the amount of energy needed to raise the temperature a given amount. The energy needed is $\int_T^{T+\Delta T} \mathrm{d}T' \, C_\mathrm{p}(T')$, with T the temperature, ΔT the increment, and $C_\mathrm{p}(T)$ the measured heat capacity. Some energy is lost to the outside environment; a more refined calculation compensates for these losses, as well. Thanks to Seth Fraden for this suggestion.

In general, all information known in advance about the reference signal and the system itself should be incorporated into feedforward elements, with feedback left to handle the residual uncertainties. You just need a reasonable estimate of the dynamics (Chapter 6). If the estimate is less accurate, there are ways to make feedforward more robust (Chapter 9).

3.5 Stability of Closed-Loop Systems

Although adding feedback to a system can improve the dynamics, it can also lead to disaster (sometimes literally), in that feedback can turn a stable system into an unstable one or fail to stabilize an unstable system.[20] For example, a PID controller can change the dynamics of an arbitrary second-order system into an arbitrary third-order system, which can be unstable. For example, $G(s) = \frac{1}{(1+s)^2}$ and $K(s) = \frac{K_i}{s}$ correspond to a critically damped oscillator under integral control. The closed-loop transfer function is a third-order system,

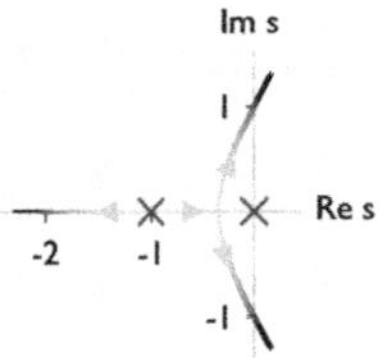

$$T(s) = \frac{1}{1+(GK)^{-1}} = \frac{1}{1+\frac{(1+s)^2 s}{K_i}} = \frac{K_i}{K_i + (1+s)^2 s} . \tag{3.47}$$

At left, a *root-locus* plot (Section 2.3.2) shows the closed-loop eigenvalues as a function of K_i. The lightest shade of gray represents $K_i = 0$, the darkest $K_i = 4$.

To derive the conditions for instability, note that the complementary sensitivity function $T = GK/(1+GK)$ becomes infinite when $GK \equiv L(s) = -1$. An infinite steady-state response corresponds to an instability, since the exponentially growing terms of unstable dynamics come to "steady state" only at infinite amplitude. The singularity occurs when

$$L(i\omega) = G(i\omega)\, K(i\omega) = -1 . \tag{3.48}$$

Here, it is simpler to write the equivalent condition $G^{-1} K^{-1} = -1$, which gives

$$\begin{aligned} (1+i\omega)^2 \frac{i\omega}{K_i} &= -1 \\ \left(1+2i\omega-\omega^2\right)(i\omega) &= -K_i \\ (K_i - 2\omega^2) + i\omega\left(1-\omega^2\right) &= 0 . \end{aligned} \tag{3.49}$$

[20] In his 1989 Bode Lecture, reprinted in 2003, Gunter Stein presented the sobering example of the Chernobyl nuclear disaster (Stein, 2003). In a simple version of the story, the "RBMK" boiling water reactor at Chernobyl used water both for cooling and to provide steam to the power turbines. The denser water absorbs neutrons better than steam, giving rise to a potential instability: as the fission rate is increased, more water turns to steam, increasing the fission rate further. In the RBMK design, overall stabilization is provided by boron control rods that absorb neutrons, by regulating the water flow rate, *and* by having a set of control and operations policies intended to ensure that the unstable regime is never encountered in practice. On the night of April 25, 1986, plant operators conducted a safety test in a way that violated many operational guidelines and exposed design flaws that led to an instability in the plant, whose core exploded. At least thirty people were killed directly; the number of subsequent deaths from cancer and related illnesses remains highly controversial, with estimates ranging from O(10) to O(10^5). Whatever the actual numbers, the point remains that for engineers, control theory can be a matter of life and death.

Both real and imaginary terms must vanish separately. The latter gives the frequency of oscillation at onset, $\omega^* = 1$, the real term the critical gain $K_{\rm i}^* = 2$ for instability. The instability is a Hopf bifurcation (Section 2.6). Although the third pole (−2) is stable at onset, it takes only a single unstable pole or complex-conjugate pair to have an instability.

With negative feedback, the typical instability is a Hopf bifurcation. To understand this observation, note that the condition for instability, $L = -1$ means that a perturbation that goes "once around the loop" and returns with the same magnitude but opposite sign. The negative sign arises from trying to act so fast that the system response slips 180° in phase with respect to the reference signal. This negative sign is then inverted by the feedback input and reenters the feedback loop at full amplitude. Thus, after a complete loop, the signal is identical and is thus maintained indefinitely. If the gain is now increased, amplitudes will grow each cycle (instability), while if it is decreased, the perturbation will decay each cycle (stability).

In a linear model, the amplitude of an unstable mode grows indefinitely. In a real system, its saturates at a finite value, for one of two reasons:

1. Nearly all systems are nonlinear when presented with large enough amplitudes. Linearity is just an approximation in a given region of state space.
2. The actuator corresponding has a finite range before it saturates, another nonlinear effect. Physically, the system can drive the instabilities only up to a maximum amplitude.

3.5.1 Graphical Tools and Stability Margins

Although it is good to be able to calculate the conditions for instability, it can be equally important to know how far from instability a stable system is. The Bode and Nyquist plots introduced in Section 2.3.2 can give such information via three useful quantities: the *gain margin*, the *phase margin*, and the *stability margin*, which measure how "close" or "far" the closed-loop transfer function, $L(s)$, is from the instability point, −1, using slightly different notions of distance. We begin with the gain and phase margins:

- *Gain Margin*: At the frequency where the phase delay is 180°, the gain margin (GM) is the factor by which the gain must be increased to have $|G| = 1$. In other words, we can increase the loop gain by a factor GM before instability results. More precisely,

$$GM = \frac{1}{|L(\mathrm{i}\omega_{180})|}\,, \qquad \omega_{180} = \text{frequency where phase lag} = 180^\circ\,. \tag{3.50}$$

- *Phase Margin*: At the frequency where $|G| = 1$, the phase margin PM is the phase shift required to reach 180°. In other words, we can add an additional phase shift of PM degrees before instability results. More precisely,

$$PM = \angle L(\mathrm{i}\omega_1) \qquad \omega_1 = \text{frequency where } |L| = 1\,. \tag{3.51}$$

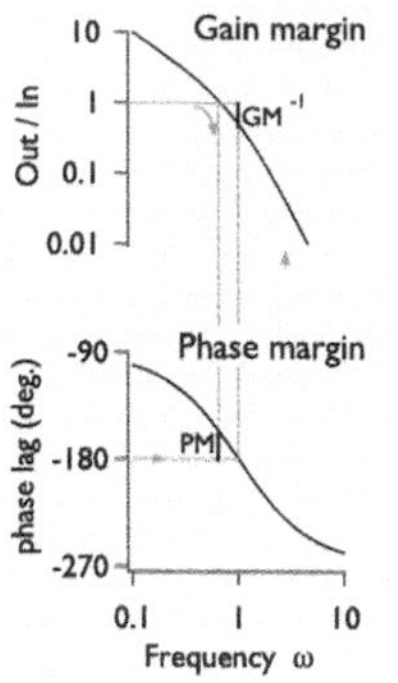

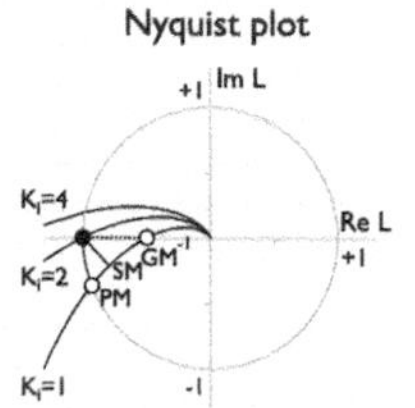

It is perhaps easier to illustrate these definitions pictorially. For the critically damped, second-order system with integral gain, we can see either graphically at left or algebraically that when $K_i = 1$, we have $\omega_{180} = 1$. Then $L(\mathrm{i}\cdot 1) = -1/2$, implying a gain margin of 2. The phase margin, $\approx 20°$, is more complicated algebraically (Problem 3.5).

Another interpretation of gain and phase margin comes from the Nyquist graph of L at left. Recall that the Nyquist plot is the image $L(s)$ in the complex L-plane for $s = \mathrm{i}\omega$ and $0 < \omega < \infty$ (see Section 2.3.2). On the plot, the instability point $L = -1$ is indicated by a filled circle. The $K_i = 2$ curve passes through this point, indicating that an instability occurs at that gain. Changing K_i just scales L, which means that the $K_i = 1$ and $K_i = 3$ curves that are also plotted intersect the x-axis at -0.5 and -1.5, respectively. Focusing on the $K_i = 1$ curve, the open circles on the x-axis indicate the gain margin (GM = factor to multiply along x-axis) and the phase margin (PM = degrees along the arc of the unit circle). Thus, on a Nyquist plot of $L(\mathrm{i}\omega)$, the gain margin indicates how much the L curve can be "inflated" by a uniform factor, while the phase margin indicates how much the L curve can be rotated before reaching the point $(-1, 0)$, which indicates instability.

We summarize the discussion of gain and phase margin by noting

1. One can have a good gain margin but poor phase margin, or vice versa.
2. The gain margin should be at least two, the phase margin at least 60°. Any closer to an instability, and the response will typically be very underdamped and decay slowly.

A third, perhaps better, indication of stability is given by the point on the Nyquist plot that is closest to $L = -1$. This *stability margin* (SM) is indicated by a short line drawn to -1 in the plot above at left. It is potentially a truer indication of the closeness to instability in that, in somewhat pathological cases, SM might be small even while GM and PM are large. (Think of a squared-off Nyquist contour whose corner is near $L = -1$.) While the stability margin is easy to evaluate graphically and numerically, it is harder to calculate analytically. Again, the best way to develop an intuition is via an interactive graph.[21]

3.5.2 Stabilizing the Unstable

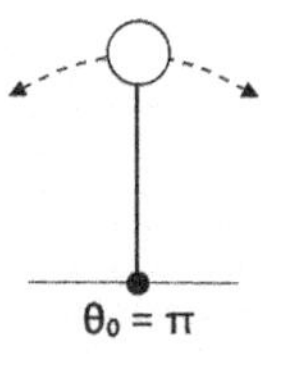

Although feedback can destabilize a stable system (perhaps unwittingly), it can also do the reverse. As a prototype, consider a nearly upside-down pendulum (see left), whose small-angle deviations $\theta(t)$ from the unstable vertical equilibrium are described by $\ddot{\theta} - \theta = u(t)$, with $y = \theta$. The control torque $u(t)$ is supplied by a motor. (We neglect its time constant, as we do any mechanical friction in the pendulum's motion.) In discussing derivative control, we noted that it changes the effective damping coefficient

[21] Given current software graphics, the tools developed here are more valuable for intuition than calculation. For similar reasons, we omit the Nyquist criterion for instability, based on a Nyquist plot for $L(s)$, where s runs along the entire imaginary axis in the complex s-plane and is closed in a "D" -shaped contour by an infinite-radius half-circle to the right (Åström and Murray, 2008).

of a second-order system. This will be part of our strategy for stabilizing the unstable oscillator.

Many practical engineering applications depend on stabilizing unstable dynamics. Military fighter planes, for example, are often designed to be open-loop unstable, as unstable planes can change orientation faster and maneuver more quickly than stable planes.[22] Also, to minimize radar reflections, *stealth fighters* are made of angled flat surfaces assembled into an overall polygonal hull. The flat surfaces deflect radar beams away from their source but make the aircraft unstable. Active control allows the pilot to fly the stabilized plane. Of course, a failure of the control system has dramatic consequences!

Unstable systems abound in physics. Charged particles cannot be confined to a stable equilibrium point by static fields (Earnshaw's Theorem) but can be stabilized using feedback or open-loop techniques (*Paul trap*). The commercial Segway personal transport device is another inverted pendulum. Rockets are also unstable and require active *vector thrusting*. A river kayak or canoe designed to run rapids is short and has a flat bottom, whereas a sea kayak or lake canoe is longer and has a keel for stability. Below, in Example 3.4, we consider balancing a stick in the palm of a hand and related questions of human balance. In Problem 3.4, we discuss in more detail how PD control can stabilize the prototypical unstable oscillator $G(s) = \frac{1}{s^2-1}$.

3.5.3 Canceling Unstable Poles: A Tempting, Bad Idea

At about this point in the discussion of unstable systems, an idea often pops up: If there is an unstable pole $p > 0$ in the system $G(s) \sim \frac{1}{s-p}$, why not "cancel" it using a corresponding *zero*? That is, why not use controller $K(s) \sim (s-p)$? This is such a simple, seductive, *bad* idea that it is worth understanding clearly why it will not work. Here are three reasons:

1. *Input disturbances blow up.* The full response of the dynamics is governed by the "Gang of Four" transfer functions (Eq. 3.25). Even though the transfer function between r and y, $T = \frac{KG}{1+KG}$ is stable, the transfer function between the system input u and the output y, which is $\frac{G}{1+KG} \sim \frac{1}{s-p}$, is not.
2. *Initial conditions blow up.* The usual representation of a transfer function in the Laplace domain drops the initial-value terms. With stable dynamics, those transients decay. With unstable dynamics, they do not. We can see this in a simple calculation:

$$
\begin{aligned}
\dot{y}(t) &= y(t) - u(t) + \dot{u}(t) \\
sy(s) - y_0 &= y(s) - u(s) + su(s) - \cancelto{0}{u_0} \\
\cancel{(s-1)}y(s) &= \cancel{(s-1)}u(s) + y_0 \\
y(s) &= u(s) + \frac{y_0}{s-1},
\end{aligned}
\tag{3.52}
$$

[22] The key technical advance that allowed the Wright brothers to make the first successful flight in 1903 was a design that made the aircraft itself less stable, but more maneuverable, and relied on the human pilot to stabilize the overall system (Tomayko, 2000).

where $y_0 \equiv y(t=0)$ and $u_0 \equiv u(t=0)$. We can set $u_0 = 0$, since we control $u(t)$, but not y_0. Thus, even though we can cancel the $s-1$ term in front of $y(s)$ and $u(s)$, we cannot do the same for the initial-value term. Any nonzero value of y_0 will grow exponentially. From the inverse Laplace transform of Eq. (3.52), $y(t) = u(t) + y_0\, e^t$.

3. *Perfect cancellation is unlikely.* In practice, the controller zero never matches exactly the unstable system pole. As the zero position approaches the pole, control becomes increasingly difficult, and finally impossible (Problem 3.17).

For all these reasons, do not try to cancel an unstable pole by placing a zero that vanishes in the right-hand part (RHP) of the complex s plane in your controller! By contrast, canceling a stable pole with a zero that is in the left-hand part (LHP) of the complex s plane can make sense and be part of a reasonable control strategy (Example 3.1). For the same reasons, do not try to stabilize an unstable system using reference feedforward with an RHP zero to cancel the RHP pole (Problem 3.4 and Section 3.7.1).

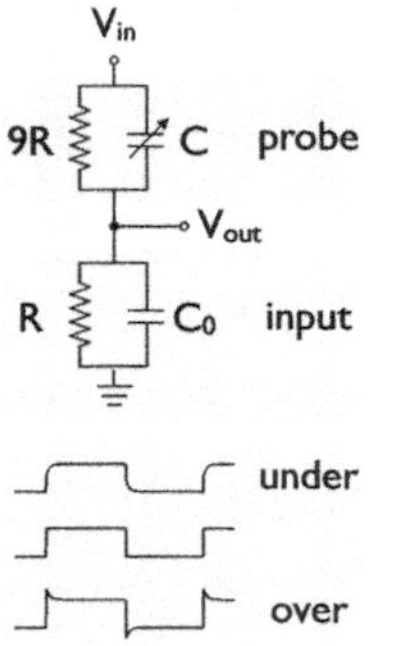

Example 3.1 (Oscilloscope probe) A common application of the idea of cancelling a stable pole via a zero is the passive oscilloscope test probe. Standard laboratory oscilloscopes have an input impedance of $R = 1$ MΩ. This resistance combines with capacitance in the cable and input, $C_0 \lesssim 100$ pF, to form a one-pole low-pass filter with time constant $RC_0 \approx 10^{-4}$ s. To avoid signal distortion by frequency components in the kHz range, one uses a compensated probe. Normally, the input impedance is also increased (e.g., by 10x) to reduce signal loading. The probe then adds a 9 MΩ resistor in parallel with an adjustable capacitance. The capacitance C is tuned by looking at the response to a square-wave input. The distortion is minimized when C produces a zero that cancels the RC pole. A simplified schematic and response is at left. You can easily show that the response $G(s) = \frac{V_{\text{out}}}{V_{\text{in}}} = \frac{1+s}{1+0.1s(\tau+9)}$, including a gain of 10 to compensate for the voltage-divider effect produced by the probe and scaling time so that $RC_0 = 1$.

3.6 Delays and Nonminimum Phase

In Section 3.5.1, we saw that when the loop transfer function $L = -1$ (phase lag of π), the closed-loop system is unstable. Since an nth order system has a phase delay of $-\frac{\pi}{2}n$, first- and second-order systems can never become unstable.[23] Nevertheless, a high-enough gain will make any system eventually become unstable. Why? As we will see, systems usually have delays or approximations to delays that add unwanted phase at higher frequencies. These phase delays then lead to oscillatory instability at high controller gain.

[23] For a second-order system, as $\omega \to \infty$, the phase delay $\varphi \to -\pi$. But the magnitude vanishes.

3.6.1 Time Delays

An explicit time delay, or *latency*, τ shifts the output of a function relative to its input:

$$h(t) = y(t-\tau) \qquad \Longrightarrow \qquad h(s) = \mathrm{e}^{-s\tau}\, y(s)\,. \tag{3.53}$$

One source of time delay is the A/D (analog-to-digital) converter, present in all computer-controlled feedback loops. The basic issue is shown at right. We will discuss sampling in detail in Chapter 5, but the main point is simple: because we update our information only once every T_s, we assign the value we measure at that time t to the midpoint of the interval, $t + \frac{1}{2}T_s$. In other words, a sampling interval T_s implies a delay $\tau = \frac{1}{2}T_s$, as seen at right.

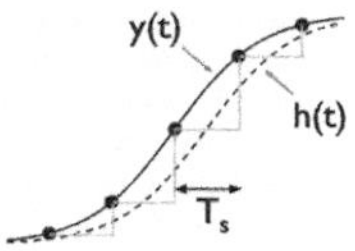

Turning up the gain of a first-order system with delay leads to a Hopf bifurcation (instability). Taylor expanding the delay, $y(t-\tau) = y(\tau) + \dot{y}(\tau)\,(t-\tau) + \frac{1}{2}\ddot{y}(\tau)\,(t-\tau)^2 + \cdots$, shows that it is equivalent to an infinite-order dynamical system, with arbitrarily large phase delays.

Example 3.2 (First-order system with small delay) Let $G(s) = \frac{1}{1+s}$, with proportional feedback, $K(s) = K_\mathrm{p}$. The A/D converter adds a small delay, represented by the transfer function $H(s) = \mathrm{e}^{-s\tau}$ shown in the block diagram in Figure 3.7.

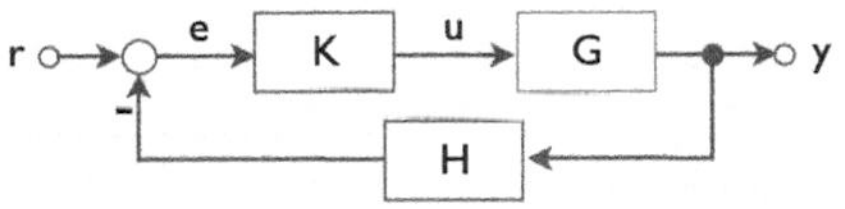

Fig. 3.7 Block diagram illustrating A/D converter delay in a feedback loop.

With the extra dynamics in the feedback path, the closed-loop transfer function between r and y is $T = \frac{GK}{1+GKH}$, and the condition for loop instability is $L(s) = KGH = -1$. If $\tau \ll 1$, the frequencies where τ is relevant are so high that we can approximate $G(s) \sim \frac{1}{s}$. Then

$$L(s) = K_\mathrm{p} \cdot \frac{1}{s} \cdot \mathrm{e}^{-s\tau}\Big|_{s=\mathrm{i}\omega} = \frac{K_\mathrm{p}\,\mathrm{e}^{-\mathrm{i}\omega\tau}}{\mathrm{i}\omega} = -\left(\frac{K_\mathrm{p}}{\omega}\right)\mathrm{e}^{-\mathrm{i}(\omega\tau-\pi/2)} = -1\,. \tag{3.54}$$

Thus, $K_\mathrm{p}^* = \omega^* = \pi/(2\tau)$. The instability is a Hopf bifurcation with oscillation frequency ω^*. Since $K_\mathrm{p}^* = \mathrm{O}(1/\tau)$, the sampling rate limits the maximum gain that can be applied. At higher frequencies, the phase delay increases and eventually reaches 180°, at which point the closed-loop system starts oscillating. If the gain is one, the oscillations do not decay. Finally, in physical units, $\tau \to \tau/\tau_0$, where τ_0 is the time scale for dynamics – i.e., $G(s) = \frac{1}{1+s\tau_0}$. Thus, $K_\mathrm{p}^* \sim \omega^* \sim \frac{\tau_0}{\tau}$.

The conclusions in the above example hold generally, as the source of the instability is the accumulated phase delay in the sensor, H, rather than the system, G. We can see an example in the Bode plots for G, H, and GH together (see Figure 3.8). The

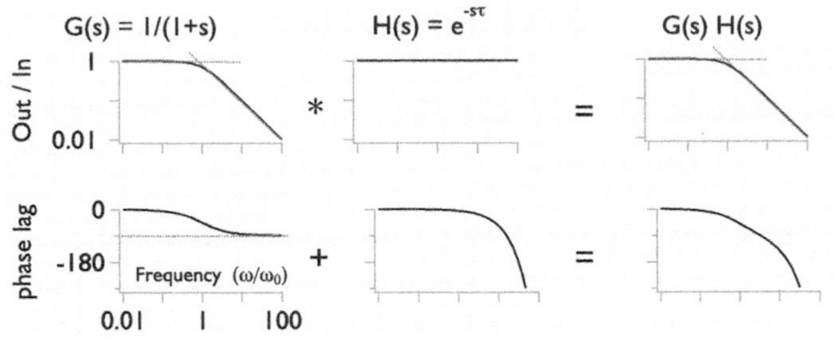

Fig. 3.8 Bode plots of a first-order system, $G(s)$, a delay, $H(s)$, and their combination.

delay has no effect on the magnitude plot but dominates at high frequencies on the phase plot.

As another example, consider the semi-infinite, one-dimensional rod of Section 2.3.2, with transfer function $G(s) = e^{-\sqrt{s}} / \sqrt{s}$ between a power source at $x = 0$ and a probe at $x = \ell$. Adding proportional feedback leads to an oscillatory instability at (Problem 3.9)

$$\omega^* = \frac{9}{8}\pi^2 \approx 11\,\frac{D}{\ell^2}, \qquad K_p^* = \sqrt{\omega^*}\, e^{\sqrt{\omega^*/2}} \approx 35\,\frac{\lambda}{\ell}, \tag{3.55}$$

where the numerical expressions are in physical units. Recall that K_p relates a temperature difference to a heater power and thus has units of [power × length / temperature]. The maximum gain is thus $K_p^* \sim \lambda/\ell$ in dimensional units, where λ is the thermal conductivity of the bar. At instability, the oscillations ω^* occur at a time scale of ℓ^2/D. The instability results from a delay at ω^* that is just enough to give a 180° phase shift at that frequency. Thus, this example is just a fancier version of the pure delay discussed above. Note a practical conclusion: To control temperature well, place the temperature probe near the heater. The distance between the two, ℓ, sets the maximum gain K_p^*, which, in turn, sets the robustness against disturbances. Recall that the transfer function between disturbances d and measurement y is $\frac{1}{1+GK} \sim (1/K_p)$. Thus, increasing K_p attenuates temperature perturbations. You should thus try to place sample and sensor in good thermal contact with the heater. Even better, add a second sensor closer to the sample (Section 3.8.1).

The effective delays that occur in the temperature-control example are present in the response of nearly any physical system, and adding proportional control with high-enough gain will usually lead to instability for the same reason. In Problem 2.6, we discussed how the temperature response of a rod has three regimes, depending on the frequency. At low frequencies, the temperature is essentially uniform over the rod, which behaves as a single *lumped element*. At intermediate frequencies, there are a countable infinity of modes. Often, the behavior is well-modeled by a small number of these modes. Finally, at high frequencies, the rod behaves as a semi-infinite medium, with a continuum of modes. The scale of low and high frequencies is determined by comparing the frequency of the (temperature) signal to the inverse time required for the signal to traverse the object. Applying proportional gain to such a system shortens the time constants by a factor $\sim K_p$. Eventually, the frequencies are high enough that the lumped-element approximation breaks down, and the continuum-system delays we have discussed become important. Thus, any physical system under proportional control with negative feedback will eventually undergo a Hopf bifurcation, unless another instability occurs first. Here, we have shown this for an explicit delay and for the implicit delay in a diffusive system. An example where the delay is due to a wave propagation speed is the familiar (and annoying) case of speaker and microphone, where too much microphone gain leads to a feedback instability.

Delays and Unstable Systems

In Example 3.2, we saw that a small delay τ, or *feedback latency*, limits the amount of proportional feedback gain one can apply to a stable first-order system, with $K^*_{\mathrm{max}} \lesssim \tau^{-1}$. This destabilizing effect of delay is even more important when the system is unstable. Consider the system $\dot{x} = ax + u$, with $u(t) = -K_p x(t-\tau)$. Its Laplace transform implies that

$$s - a + K_p\, e^{-s\tau} = 0\,. \tag{3.56}$$

If there is no delay ($\tau = 0$), then $s = a - K_p$ and $K_p > a$ stabilizes the system. Thus, instability implies a minimum gain K_{min}. But if delays imply a maximum gain K_{max} that decreases with τ, then a long delay can make $K_{\mathrm{min}} > K_{\mathrm{max}}$, so that there is no gain that can stabilize the system.

Examining the roots of s in Eq. (3.56) show that for $\tau > a^{-1}$, no value of K_p will lead to stable dynamics (Problem 3.10). At right is a situation where there is a range of possible gains. With a long delay, there may not be such an interval.

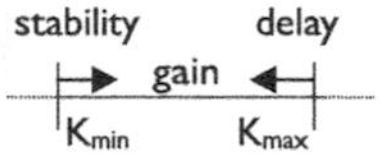

Is there any way to stabilize an unstable solution for large delay? Yes! But the feedback strategy changes radically: knowing the state of the system at time τ in the past and knowing $u(t)$ in the interval $(t-\tau, t)$, you can *predict* the current state of the system and use that in the standard feedback algorithm. We pass from an algorithm based solely on information available at one time t to an algorithm that depends on information over an interval of time. Thus, by keeping a *memory* of past outputs, along with an *internal model* of the dynamics, we can stabilize even in the presence of long delays. Of course, disturbances will still grow uncontrollably during the time τ before the feedback controller becomes aware of their existence. In a nonlinear system, the feedback may then fail if the disturbance grows too large, a scenario explored in Problem 3.15.

It is interesting that there seems to be a sharp transition in the need to keep a memory versus using single-time-point information, a topic we will return to in Chapter 12. Since predictive feedback algorithms are more naturally expressed in the time domain, and the calculations are simpler in discrete time, we defer further discussion to Section 5.4.2.

3.6.2 RHP Poles and Zeros

In Chapter 15, we will see that causality implies minimum phase delays in the response of a transfer function. The minimum phase delay for a given transfer function is captured by Bode's *gain-phase* relation, which generalizes the observation that systems whose dynamics are asymptotically s^{-n} have minimum phase lags of $n\frac{\pi}{2}$. We will state and prove this relation later in Section 15.1.2. Here, we make use of it.

Systems that obey the Bode relations are called *minimum phase* systems. However, nothing prevents extra phase lags, such as those arising from explicit time delays. Transfer functions with RHP poles or zeros also have *extra* phase delays. Such systems are *nonminimum phase* (NMP) transfer functions, with phase lags greater than the lower bound set by the Bode theorem. For example, consider $G(s) = \frac{1-s}{1+s}$, which has a pole at $s = -1$ but a zero at $s = +1$ (see left). The magnitude response is

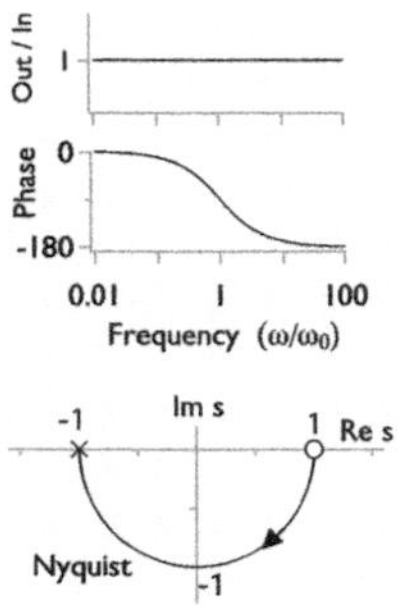

$$|G(s = i\omega)| = \left[\left(\frac{1 - i\omega}{1 + i\omega}\right)\left(\frac{1 + i\omega}{1 - i\omega}\right)\right]^{1/2} = 1\,, \tag{3.57}$$

while the phase lag $= -\pi$ as $\omega \to \infty$ [since $G(i\omega) \to -1$], as shown at left. Clearly, this is an NMP transfer function, as $G(s) = 1$ has the same gain relation but 0 phase lag.

Transfer functions are *all pass* if $|G(i\omega)| = 1$ for all ω: all sinusoidal inputs, whatever their frequency, are passed without attenuation. All-pass functions include true delays ($e^{-s\tau}$) and effective ones (Eq. 3.57) and can be realized by an electronic circuit (Problem 3.11). Their Nyquist plot is an arc of a circle of unit radius, centered on the origin.

We can decompose a stable transfer function into the product of a minimum-phase function times an all-pass function. For G_{mp} minimum phase and G_{ap} all pass,

$$G(s) = G_{\text{mp}}(s)\, G_{\text{ap}}(s)\,. \tag{3.58}$$

Proof: Since $G(s)$ is stable, it is either already minimum phase or it has zeros at $s = \{z_1, z_2, \ldots\}$. Let us define

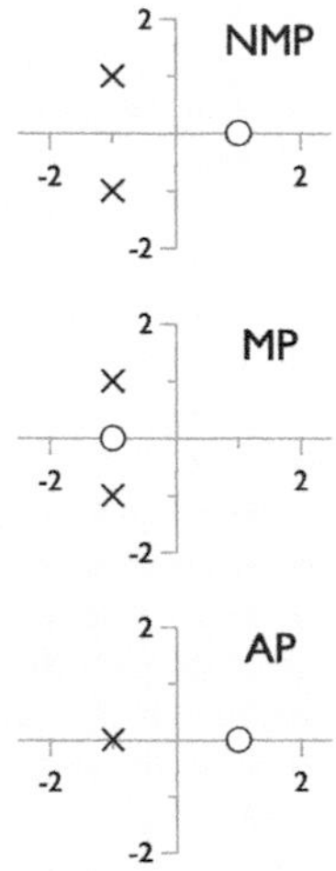

$$G_{\text{ap}}(s) = \left(\frac{z_1 - s}{z_1^* + s}\right)\left(\frac{z_2 - s}{z_2^* + s}\right) + \cdots\,, \tag{3.59}$$

which is all pass.[24] (To verify, substitute $s = i\omega$ and take the magnitude. Note the need for z^* in the denominator.) Then define $G_{\text{mp}} = G/G_{\text{ap}}$, which swaps the RHP zeros for their mirror reflection in the LHP. For example, at left is illustrated the decomposition

$$G(s) = \underbrace{\frac{1 - s}{s^2 + 2s + 2}}_{\text{nonminimum phase}} = \underbrace{\left(\frac{s + 1}{s^2 + 2s + 2}\right)}_{\text{minimum phase}} \underbrace{\left(\frac{1 - s}{s + 1}\right)}_{\text{all pass}}\,. \tag{3.60}$$

Note how the zero at $s = 1$ in $G(s)$ is assigned to the all-pass function, while the minimum-phase function has a "reflected" zero at $s = -1$. The poles at $-1 \pm i$ are untouched.

[24] The right-hand side of Eq. (3.59) is also known as a *Blaschke product* (Toll, 1956).

3.6.3 NMP Systems Are Hard to Control

Nonminimum phase systems are hard to control for two reasons:

- As the proportional gain is increased, a NMP system always goes unstable.
- The transient response to a disturbance initially "goes the wrong way."

We first illustrate these points by example and then explain them afterwards.

Example 3.3 (First-order response plus all pass filter) Consider

$$G_{\mathrm{mp}} = \frac{1}{1+s} \quad \text{and} \quad G_{\mathrm{nmp}} = \frac{1}{1+s}\left(\frac{1-s/2}{1+s/2}\right). \tag{3.61}$$

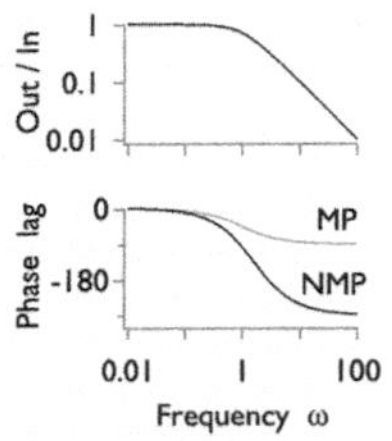

G_{mp} is a first-order, low-pass, minimum-phase transfer function. G_{nmp} is an all-pass transfer function having the form of Eq. (3.57). At right is the Bode plot of the two transfer functions. The magnitude responses are the same, but G_{NMP} has an extra 180° phase lag.

The step response is even more interesting. The minimum phase function shows the expected exponential rise. The nonminimum-phase function initially moves the wrong way and then recovers and follows the path of G_{mp} but with a delay of about one time unit (Problem 3.12). Adding proportional feedback, which acts immediately, will then lead to an instability, as it does generally for systems with delay (Problem 3.13).

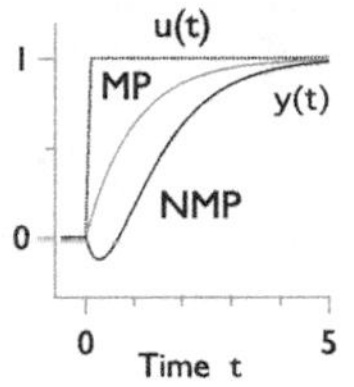

To understand why adding proportional control to an NMP system inevitably leads to instability at high gain, consider a stable system with RHP zeros. For a rational transfer function, $G(s) = N(s)/D(s)$, where the numerator $N(s)$ has some roots (zeros of G) in the RHP and the denominator $D(s)$ has roots (poles of G) in the LHP. Thus, G is stable but NMP. Now add proportional control, $K(s) = K_{\mathrm{p}}$:

$$T(s) = \frac{KG}{1+KG} = \frac{K_{\mathrm{p}}\,N/D}{1+K_{\mathrm{p}}\,N/D} = \frac{K_{\mathrm{p}}\,N}{D+K_{\mathrm{p}}\,N}. \tag{3.62}$$

Thus, the poles of the closed-loop system are the roots of $D(s) + K_{\mathrm{p}}N(s)$. For small K_{p}, the poles are nearly the same as the roots of $D(s)$, the open-loop dynamics. For large K_{p}, the poles approximate the roots of $N(s)$. On a root locus diagram:

> The poles of the closed loop system move toward the open-loop zeros as K_{p} increases.

If the open-loop zeros are RHP, then the closed-loop system *must* become unstable at high feedback gain. In effect, the control approximates an inverse, but the inverse of the transfer function of an NMP system is unstable. Note that the zeros do *not* vary with K_{p}.

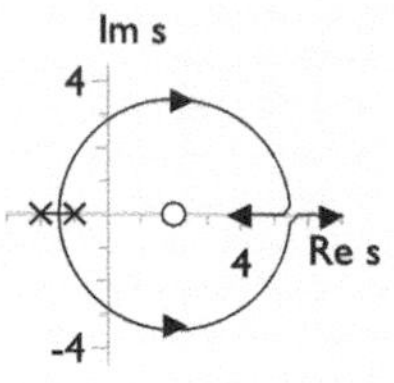

This movement of poles in response to open-loop zeros is illustrated at left for the dynamics of Example 3.3. At low gain, the two LHP poles collide, become complex conjugate pairs, cross over to RHP (where the system becomes unstable), and collide again. Then one of the poles goes toward the fixed RHP zero at $s = 2$, while the other goes off to infinity.

The inverse response to a step makes control difficult because the system first goes the wrong way before recovering. Overcorrection leads to an oscillatory instability.

3.6.4 Origins of NMP Systems

Both a pure delay and an all-pass filter can lead to oscillatory instabilities at high feedback gains. Indeed, all-pass filters resemble approximate delays. Physically, delays occur when inputs and outputs are spatially separated. The propagation time adds a phase delay relative to the case of a *collocated* input and output, where there is no spatial separation.

Padé Approximations to a Delay

Example 3.3 shows that the all-pass function $G_{\text{ap}} = \frac{1-s/2}{1+s/2}$ behaves similarly to a delay. Mathematically, G_{ap} is also the (1,1) rational fraction (Padé) expansion of the unit delay function, e^{-s}, about $s = 0$. The (1,1) Padé expansion matches the Taylor expansion through order 1+1=2 but differs at higher order. Unlike the latter, it has unit gain for $s = \mathrm{i}\omega$.

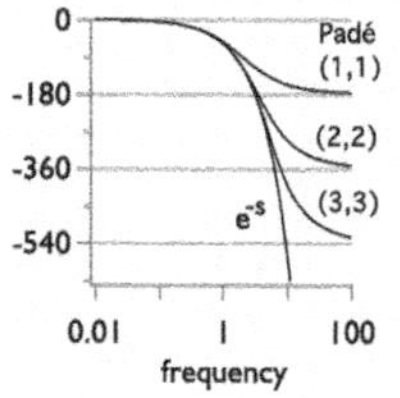

Thus, if a transfer function has a simple RHP zero, decompose it into a minimum-phase system times an all-pass system, as in Eq. (3.60). Analyze the minimum-phase part as usual. The remaining all-pass part is just a delay, up to a frequency of $O(\tau^{-1})$. Above that frequency, the phase delay of a real system continues to grow (as $\omega\tau$), while the (1,1) Padé approximation saturates at 180°. For higher orders, the (n, n) Padé expansion continues to be all-pass and better approximates the phase response of a delay (see left).

As we have seen, even though a single zero is just a low-frequency approximation to a delay, it is enough to limit the maximum feedback gain that can be applied. Physically, a high gain shortens the time constants of the system so that they become comparable to the approximate delay generated by the zero. Mathematically, the positions of the closed-loop poles are drawn toward the open-loop zeros (Eq. 3.62). In Problem 3.16, we will see that an RHP zero also limits the maximum possible feedback bandwidth.

Flexible and Multimode Objects

Another class of NMP systems includes flexible objects – ones whose dynamics show contributions from more than one mode. Often, the modes are lightly damped. As a toy model of a flexible system, consider a system whose output has additive contributions from two undamped modes, with frequencies scaled to 1 and ω. If the mode amplitudes are $\pm\alpha$ and β, the response in the Laplace domain is

$$G_\pm(s) = \pm\frac{\alpha}{1+s^2} + \frac{\beta}{1+\frac{s^2}{\omega^2}} = \frac{\pm\alpha+\beta+\left(\frac{\pm\alpha}{\omega^2}+\beta\right)s^2}{(1+s^2)\left(1+\frac{s^2}{\omega^2}\right)} \implies z^2 = -\frac{\pm\alpha+\beta}{\frac{\pm\alpha}{\omega^2}+\beta}. \quad (3.63)$$

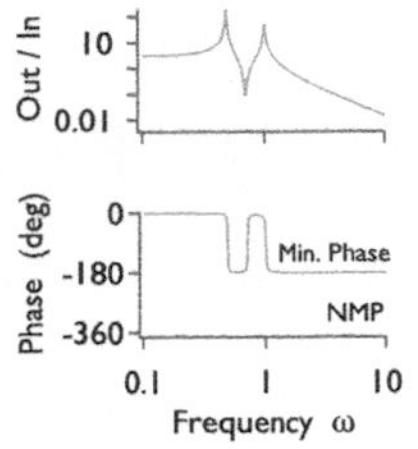

Note how adding two oscillatory modes creates two zeros whose locations depend on the mode amplitudes α and β. At right are Bode plots for a case where $\alpha = 1$, $\beta = 2$, and $\omega = \frac{1}{2}$ are chosen so that G_+ is minimum phase and G_- is NMP. Note how the minimum-phase function has an asymptotic phase lag of 180°, as expected for a system of relative order = 2, while that of the NMP system is larger (360°). We have added a small amount of damping ($\zeta = 0.01$ for each mode) to soften the phase jumps and to keep responses finite. The damping shifts the poles and zeros slightly to the left in the complex s-plane. The poles then are strictly LHP, as they must be for a stable system. With damping, the zeros give rise to finite-magnitude response minima, known as *antiresonances*.

The full analysis of flexible objects is complex. Problem 3.18 analyzes a vibrating string, whose infinite number of modes each has a pole. Depending on the observation point, there may or may not be a zero in between, and the response can be either MP or NMP.

The string is a prototype for robotic arms, atomic force cantilevers, and many other floppy objects. NMP can also arise in a nonmechanical system such as a column of boiling water. If you add cold water, the level will initially fall, as the boiling is suppressed, but will later rise as the water reheats. The inverse response signifies an NMP transfer function between liquid mass and water level.

Example 3.4 (Balancing a stick by moving your hand) Problem 2.1 introduced the problem of balancing a stick of mass m and length ℓ by moving a hand laterally. In controlling the stick angle near the vertical, you gaze at a point ℓ_0 partway along a stick, indicated at right by the white circle. Strangely, where you look determines the nature and location of zeros in the transfer function between the force input u and the measured position y (Problem 3.14). In particular, there are zeros at $s = \pm(1-\frac{3}{2}\ell_0)^{-1/2}$.

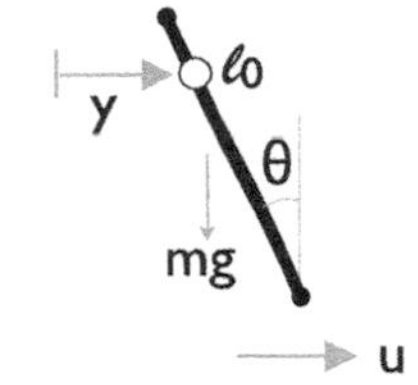

For $\ell_0 > 2/3$, there is a complex-conjugate pair of zeros on the imaginary axis. For $\ell_0 < 2/3$, there are stable and unstable real zeros. The latter makes the transfer function NMP. And for $\ell_0 = 2/3$, there are no zeros at all. Thus, a poor choice of observable makes a system NMP, while a good choice avoids the problem. It is much easier to balance a rod looking near the top than near the bottom. Note that in physical units, $\ell_0 \to \ell_0/\ell$.

Delay. Another obstacle to balancing a stick is the delay in applying a restoring force, due mostly to the processing of visual information. In Problem 3.15, we analyze the effect of delay τ on a proportional-derivative (PD) feedback algorithm to stabilize the up equilibrium. The calculation is similar to the first-order example of

Section 3.6.1 and leads to the conclusion that for $\ell < 3g\tau^2$, no choice of K_p and K_d can stabilize the stick.

A simpler calculation is also very instructive: as there is no way to correct for a disturbance within the delay interval, it can grow large enough before it is detected that it is impossible to supply a stabilizing correction, no matter what the internal control algorithm is. Following this logic leads to the conclusion that a stick will always fall for $\ell \lesssim g\tau^2$.

Both the PD and disturbance calculations imply that short sticks should be much harder to balance. Indeed, balancing a meter stick is easy, but a pencil is impossible. Try it! However, experiments imply that humans can balance sticks that are shorter than the PD feedback limit ($\tau \approx 0.23$ s, which corresponds to e *ll* > 1.5 m).[25] Indeed, it seems likely that humans use the kind of predictive model we introduce in Section 5.4.2.

Delays and unstable zeros are thus two different mechanisms that make sticks hard to balance. While delay is the more practical limitation, unstable zeros can be the more serious problem. Indeed, a very long stick that would be easy to balance if you look at the end can be impossible when you restrict your view to the lower half. Again, please try it! If you do, make sure not to cheat by peeking at the top. A baseball cap can help.

Finally, stick balancing is more than a game: when we stand, *we* are unstable. Close your eyes and balance on one foot to feel how we actively control our balance. As we age, reaction times slow and increase the risk of falling – a major health concern for seniors.

3.7 Designing the Control

So far, we have discussed desirable features for a controller (tracking, disturbance rejection) and pitfalls (instabilities, noise injection). Let us use these insights to design controllers. We have already shown that a PID controller is a good, generic control algorithm when not much is known about the system. More generally, applying high gains is a robust strategy that is independent of the details of the system under control.

Although high-gain control can work for poorly known systems, it does not take advantage of specific features of the system dynamics. If we knew the controller transfer function, could we do better? Here, we give some heuristic, intuitive design principles that show how to profit from knowledge of the system transfer function.

We thus consider four overall strategies to design ("synthesize") a controller:

1 *High-gain control* is effective when the system is poorly known.

[25] Accounting for the mass of the hand can lower this limit by as much as a factor of four. Humans can balance sticks smaller than this reduced limit, too! The conclusion remains that our central nervous systems use some kind of predictive control (Insperger and Milton, 2017).

2 *Pole placement* designs a controller via the position of closed-loop poles.
3 *Loop shaping* adds stability considerations to high-gain control.
4 *Internal model control* allows exact reference tracking or disturbance compensation.

Later, we will add more systematic design approaches to our repertoire of control-design strategies: In Chapter 7, we present *optimal control* for a perfectly known system $G(s)$. In Chapter 9, we discuss *robust control* for a partially known $G(s)$.

3.7.1 High-Gain Control

Recall that high-gain control at a particular frequency makes the closed-loop response track the reference well and reduces the effect of disturbances. That is,

$$K \gg G^{-1} \quad \Longrightarrow \quad T = \tfrac{KG}{1+KG} \approx 1 \quad \Longrightarrow \quad S = \tfrac{1}{1+KG} \approx 0\,. \tag{3.64}$$

As we saw in Section 3.6.3, the effect of high gain is to move poles:

$$K = \tfrac{N_K}{D_K} \quad \text{and} \quad G = \tfrac{N_G}{D_G} \quad \Longrightarrow \quad T = \tfrac{N_K N_G}{N_K N_G + D_K D_G}\,. \tag{3.65}$$

That is, for high gain ($N_K \gg D_K D_G$), we have $T \approx 1$ independently of N_G and D_G. This *robust high-gain* feedback is a general strategy for controlling poorly known systems. Of course, we have already seen its limitations: High gains may require a larger control input than is possible, may cause instabilities, and does not move the zeros. Nonetheless, high-gain control is a basic control strategy and the starting point for more sophisticated methods. Problem 3.4 shows how high-gain control can stabilize an unstable oscillator.

3.7.2 Pole Placement

From Eq. (3.65), the closed-loop dynamics has poles at the roots of $D_L = N_K N_G + D_K D_G = 0$. A tempting strategy is to choose the numerator $N_K(s)$ and denominator $D_K(s)$ of the controller $K(s)$ so that the poles of the closed-loop response have "good" positions. As we will see, giving poles a very large and negative real part makes the closed-loop dynamics rapid, but the required control input may be too large.

We give a brief example of pole-placement techniques here but defer most of our discussion to Section 4.2, as state-space algorithms are more straightforward. We return to the undamped oscillator $G(s) = \frac{1}{s^2+1}$, a demanding system to control, since natural damping cannot help bring about the desired response. Indeed, the only way to make the effects of a disturbance decay is to apply feedback.

Example 3.5 For $G(s) = \frac{1}{s^2+1}$, we have $N_G = 1$ and $D_G(s) = s^2 + 1$. We seek closed-loop dynamics with three poles at $s = -3$. That is, the denominators of S, T, etc. should be $(s+3)^3 = s^3 + 9s^2 + 27s + 27$. To match coefficients of the third-order polynomial, our controller should have four coefficients. Let us choose a *lead compensator*, which can be viewed as a modified PD controller:

$$K(s) = \frac{N_K(s)}{D_K(s)} = \frac{n_1 + n_2 s}{d_1 + d_2 s}, \tag{3.66}$$

where n_1, n_2, n_3, and n_4 are coefficients to be determined. Matching coefficients gives

$$N_K N_G + D_K D_G = (n_1 + n_2 s)(1) + (d_1 + d_2 s)(s^2 + 1) = (s+3)^3$$
$$\Longrightarrow \quad n_1 = 18,\ n_2 = 26,\ d_1 = 9,\ d_2 = 1\,. \tag{3.67}$$

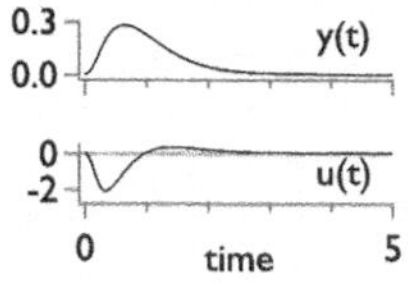

Thus, $K(s) = \frac{18+26s}{9+s}$ and $S = \frac{(s+9)(s^2+1)}{(s+3)^3}$ and $T = \frac{2(13s+9)}{(s+3)^3}$. The sensitivity functions indeed have the desired closed-loop dynamics. See the input-disturbance response at left.

One issue with pole placement is that we do not directly care where the poles are located; rather, what matters is the response to a perturbation, the amount of control effort required, and so on. Pole positions do influence these properties but not as directly as might be desired. The numerators of the various transfer functions also affect dynamics. Chapter 7 will show how to better incorporate control goals and how to directly use those criteria to design controllers. Problem 3.20 compares the modified PD controller to a standard PID one.

3.7.3 Loop Shaping

Loop shaping combines ideas of high-gain robust control with the stability insights gained from Bode's gain-phase relations for minimum-phase systems. The goal is to choose $K(s)$ so that the loop transfer function $L = GK$ has the "right" shape. The design is not optimal but can work well for simple systems.

From the block diagram of Figure 3.3, we see that perfect control without error would require $S = T = 0$. Since $S + T = 1$, there will be trade-offs. The premise of *loop shaping* is that if different inputs dominate at different frequencies, we can choose the best compromise in the trade-offs at each frequency. Possible objectives include

- Performance (disturbance rejection) $\Longrightarrow$ large gains (L large)
- Performance (tracking) $\Longrightarrow$ L large
- Stabilization of unstable G $\Longrightarrow$ L large
- Sensor noise $\Longrightarrow$ L small
- Reduce actuator wear $\Longrightarrow$ L small
- Physical controller must be proper $\Longrightarrow$ $K \to$ constant and $L \to 0$ as $\omega \to \infty$
- Nominal stability $\Longrightarrow$ L small (for RHP zeros, time delays)
- Robust stability $\Longrightarrow$ L small (where dynamics are uncertain or are neglected)

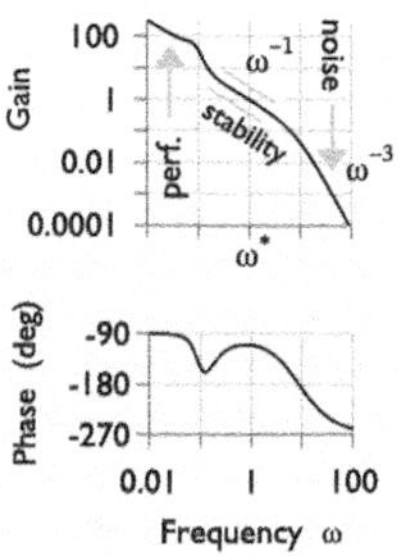

Thus, there are reasons to make L big and reasons to make L small. A typical compromise is to make L big at low frequencies and small at high frequencies, as illustrated at left.

The heuristic rules for loop shaping can be summarized as follows:

- **Performance:** We want $|L| \gg 1$ so that $S = \frac{1}{1+L} \approx 0$ and $T = \frac{L}{1+L} \approx 1$, implying a large $|K|$ at low frequencies. Integral control to eliminate offsets implies $L(s) \sim s^{-1}$ as $\omega \to 0$.
- **Stability:** When $|L| = 1$, we need a reasonable phase margin. If L is minimum phase, a dependence $L \sim s^{-1}$ for frequencies near where $|L| \approx 1$ implies a phase margin of $\approx 90°$. A steeper decrease would imply too large a phase lag (and too small a phase margin).
- **Noise suppression:** For $\omega \gg \omega_c$, $|L|$ should decrease rapidly (i.e., $|L| \sim s^{-a}$, with $a > 1$ to suppress injecting sensor noise).

We thus boost the gain $|L|$ at low frequencies for good input tracking; make $|L| \sim \omega^{-1}$ near the crossover ω^* where $|L| = 1$ for good phase margin; and suppress the gain more quickly when $\omega \gg \omega^*$ for good noise rejection.

Tweaking the Loop

Having seen criteria for a "good" shape for $L(s)$, we now describe transfer function building blocks that can "tweak" the shape of L by altering its form over a limited frequency range.

- **Lag compensation:** This element is a "soft" approximation to integral feedback:

$$K_{\text{lag}}(s) = \frac{a(1+s)}{1+as}\,, \qquad a > 1\,, \tag{3.68}$$

and its Bode and pole-zero plots are shown at right. The transfer function adds gain, at the cost of an extra 90° phase shift, over a frequency range $(\frac{\omega_0}{a}, \omega_0)$. Outside this range, the lag element is constant. By contrast, integral control adds a 90° phase lag for all frequencies. Another difference between lag compensation and integral control is that the DC gain of the former is finite (a), whereas integral control has infinite DC gain. A lag compensator by itself thus cannot completely compensate for a step perturbation.

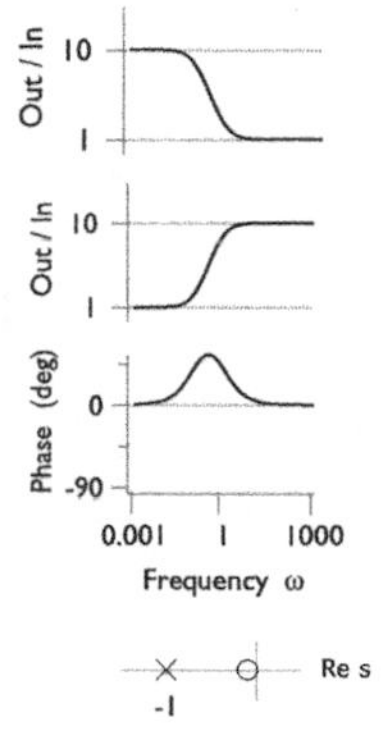

- **Lead compensation:** This element is a "soft" approximation to derivative control:

$$K_{\text{lead}}(s) = \frac{1+as}{1+s}\,, \qquad a > 1\,, \tag{3.69}$$

and its Bode and pole-zero plots are given at right. As with derivative control, a phase advance stabilizes the dynamics, but only over a finite range of frequencies. Unlike the pure derivative term, the high-frequency gain boost is limited to a finite value. Since the unbounded high-frequency gain magnifies the effect of noise at high frequencies, the cutoff in gain is an advantage.
- **Lag-lead compensation:** This element combines a lag with a lead compensator:

$$K_{\text{lag-lead}}(s) = \frac{\left(1+\frac{s}{\omega_1}\right)\left(1+\frac{s}{\omega_2}\right)}{\left(1+\frac{as}{\omega_1}\right)\left(1+\frac{s}{a\omega_2}\right)}\,, \qquad a > 1\,, \quad \omega_2 > \omega_1\,, \tag{3.70}$$

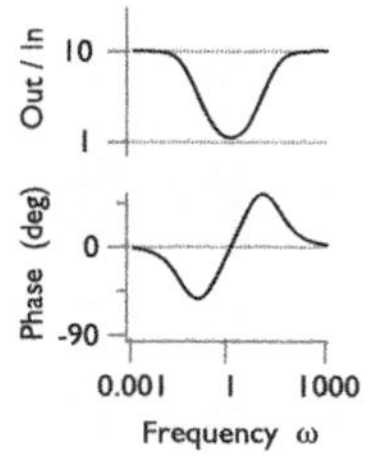

and its Bode and pole-zero plots given at right. Note the local amplitude "notch" between ω_1 and ω_2. The obvious lead-lag version has a form that is just the inverse,

$K_{\text{lead-lag}} \sim K^{-1}_{\text{lag-lead}}(s)$ and boosts the amplitude between ω_1 and ω_2. Although second-order elements produce sharper amplitude boosts or notches, they have the disadvantage of creating a 180° phase shift between the low- and high-frequency response. The lag-lead and lead-lag compensators limit their phase shifts to a finite range of frequencies.

These three types of transfer function can be used to "tweak" the shape of $K(s)$ to give the loop function the form that best accomplishes the desired trade-offs.

Example 3.6 (Loop shaping) Consider again $G(s) = \frac{1}{1+s}\left(\frac{1-s/2}{1+s/2}\right)$, the transfer function from Example 3.3 that represents a first-order system with an first-order approximation to a delay. Because G is nonminimum phase, it is hard to control. We apply the ideas of loop shaping step by step, aided by interactive graphical software.

We start with proportional control, with gain $K_p = 1$. In Figure 3.9a, the Bode plot of $L_p = GK_p$, with $K_p = 1$ has a gain margin of about 3 (at $\omega = 2.8$) and a phase margin of 180° (at $\omega = 0$). The exact result is $\text{GM} = 3$ at $\omega = 2\sqrt{2}$.

Following the rule of thumb that the gain margin should be at least 2 and the phase margin at least 60°, we increase the proportional gain to $K_p = 1.5$, for a gain margin of 2. The phase margin $\approx 73°$ meets our criteria. The resulting time-domain step response is shown in Figure 3.9b. As you see, there is a large overshoot, slowly decaying oscillations, and a proportional droop (factor of $\frac{K_p}{1+K_p} = \frac{3}{5} = 0.6$). Looking at the proportional gain, you might think it would be better to reduce K_p a little, even though the droop would be greater. In any case, the result encourages us to try a more complicated control architecture.

To eliminate the proportional droop, we could add integral control to make low-frequency loop transfer function $L \sim s^{-1}$. However, simply multiplying the controller

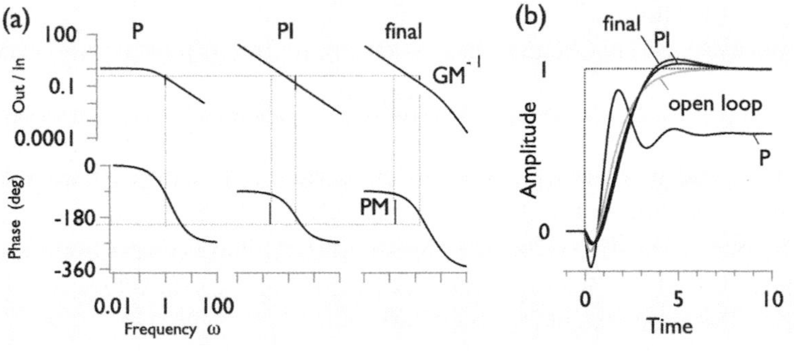

Fig. 3.9 Loop shaping. (a) Bode plots for proportional control with $K_p = 1.5$, proportional-integral control with $K_p = K_i = 0.5$, and additional low-pass filter at $\omega = 10\omega_0$. The inverse gain margins are indicated by thin black vertical lines (GM^{-1}) on the magnitude plot, and the phase margin is indicated by thin black vertical lines on the phase plots (PM). (b) Corresponding step responses, with ideal response (dotted line) and open-loop system response (gray line).

by $\frac{1}{s}$ (pure integral control) adds a 90° phase shift for all frequencies. Instead, we choose a controller of the form $K\frac{z+s}{s}$, with $z = -1$ to cancel the poles at $s = -1$. Leaving untouched the pole at $s = -\frac{1}{2}$ keeps the all-pass structure. The controller is now $K(s) = K_{\mathrm{pi}}\frac{1+s}{s}$, a PI controller with equal gains for the proportional and integral terms.

Removing the pole-zero pair at -1, the loop now is $L_{\mathrm{pi}} = K_{\mathrm{pi}}\frac{1-s/2}{s(1+s/2)}$. For $K_{\mathrm{pi}} = 0.5$, the gain margin is 4 and the phase margin $\approx 62°$. Note the pure $\frac{1}{s}$ form of $|L|$ (the other terms have all-pass response). The step response in Figure 3.9b is good, with no steady-state offset and a much-reduced overshoot, a consequence of the larger gain margin.

Because gain roll-off at higher frequencies improves the rejection of measurement noise, we multiply by a low-pass filter. The final controller and loop transfer functions are then

$$K_{\mathrm{final}} = 0.5\frac{(1+s)}{s\left(1+\frac{s}{10}\right)}, \qquad L_{\mathrm{final}} = 0.5\frac{\left(1-\frac{s}{2}\right)}{s\left(1+\frac{s}{2}\right)\left(1+\frac{s}{10}\right)}. \tag{3.71}$$

The cutoff frequency at $\omega = 10$ is arrived at by trial and error. Note that the final form for $K(s)$ is an integrator times a lead compensator. Adding the low-pass filter degrades performance, because its extra phase lag reduces the phase margin. The value chosen here reduces the gain margin of 3.4 and the phase margin to 59°. Again, the only reason to add the low-pass filter is to reduce the effects of measurement noise. The more noise, the more it is worth trading off performance, by lowering the bandwidth. Of course, one could also use a fancier low-pass filter, such as a higher-order Butterworth or Bessel filter (see Problem 5.1). In Chapter 8, we will explore such trade-offs systematically.

Shaping T and S

Instead of designing a controller based on the shape of the open-loop transfer function $L(s)$, we could specify the closed-loop transfer functions $T(s)$ and $S(s)$, a choice that highlights the difficulties of using a single controller to track a reference signal and reject disturbances.

To track a reference signal, we assume a block diagram of the form of Figure 3.3. Neglecting measurement and disturbances, we solve for the controller $K(s)$ in terms of $T(s)$:

$$T = \frac{KG}{1+KG} \quad \Longrightarrow \quad K = \frac{T}{1-T}G^{-1}. \tag{3.72}$$

For example, to track r up to a bandwidth $\omega_c \equiv 1$, we can ask that T be a low-pass filter:

$$T = \frac{1}{1+s} \quad \Longrightarrow \quad K_r = \frac{1}{s}G^{-1}, \tag{3.73}$$

where the controller is denoted K_r to remind us that we chose it to track r as well as we could. This controller has the structure of an integral controller times the inverse of the physical system. Choosing $T(0) = 1$ thus implies integral control.

The success of this approach depends on being able to invert $G(s)$. If the transfer function has right-hand-plane zeros (NMP), the inverse will be unstable. Also, if $G(s) \sim s^{-n}$ at high frequencies, with $n > 1$ (i.e., second- or higher-order system dynamics), the controller will be improper. We can add poles to the prefactor of K in that case.

The strategy for designing a controller to reject disturbances is different:

$$y = \frac{G_d}{1 + KG} d \equiv G_d S d \quad \Longrightarrow \quad \left(K_d = \frac{G_d}{S} - 1\right) G^{-1} . \tag{3.74}$$

To suppress disturbances up to a frequency $\omega_c \equiv 1$, we ask that S be a high-pass filter:

$$S = \frac{s}{1 + s} \quad \Longrightarrow \quad K_d = G_d \frac{1 + s}{s} G^{-1} = [G_d(1 + s)] \frac{1}{s} G^{-1} . \tag{3.75}$$

The most important point about Eqs. (3.74) and (3.75) is that the controller K_d is different from K_r in Eq. (3.73). Although they share the common elements $\frac{1}{s}G^{-1}$, their difference means that a controller that tracks a disturbance well may reject disturbances poorly, and vice versa. Again, we see the value of adding a second-degree of freedom in the controller by including reference feedforward elements (see Section 3.4).

Finally, note how designing $T(s)$ differs from the pole-placement method discussed above: here, we fix both the numerator and denominator of $T(s)$. In the pole-placement method, we choose a common denominator for all four transfer functions but leave their numerators unspecified.

3.7.4 Internal Model Principle

In Section 3.3, we saw that integral control makes a closed-loop system converge to an arbitrary constant reference level. Here, we generalize this result to track asymptotically time-dependent signals such as a ramp or oscillation and to reject similar disturbances. We will find that the controller must contain a copy of the dynamics that can produce those reference or disturbance signals.[26] This important result is known as the

> *Internal Model Principle*: To track a reference or to cancel a disturbance, a controller must have an internal model of the reference or disturbance dynamics.

Recall that a constant reference signal is tracked via an integral controller of the form $K(s) \sim \frac{1}{s}$. Recall, too, that the Laplace transform of the reference $r(t) = r_0\theta(t)$ is $r(s) = \frac{r_0}{s}$. Thus, $K(s) \sim r(s)$. To see why this observation is not a coincidence, let $r(s) = \frac{N_r(s)}{D_r(s)}$, where N_r and D_r are polynomials in s. Then define the controller

[26] The internal model could alternatively be contained in the system G. That is, it is the loop dynamics $L = KG$ that should contain the internal model. Usually, we have to add the internal model to the controller.

$K(s) = \frac{K_0(s)}{D_r(s)}$, with a denominator that contains that of $r(s)$. From the Final-Value Theorem,

$$e(s) = r(s) - y(s) = \frac{1}{1 + KG} r(s) \implies e(t \to \infty) = \lim_{s \to 0} s \left(\frac{N_r}{D_r + K_0 G} \right) = 0 \,, \quad (3.76)$$

assuming that K_0, G, N_r, and D_r are all finite as $s \to 0$ and that $D_r + K_0 G$ has no poles on the imaginary axis or with positive real s. We can choose $K_0(s)$ to ensure this last condition. Thus, a controller that contains a copy, or "model," of the poles of the input signal $r(s)$ can track the signal at long times. Conversely, a linear system that tracks a reference signal asymptotically must have $K \sim D_r(s)^{-1}$: A system that tracks a constant input *must* have an integral controller, independent of other details of the system. For example, Yi et al. (2000) showed that bacterial chemotactic systems track nutrient gradients independently of absolution concentration levels because they implicitly use integral control.

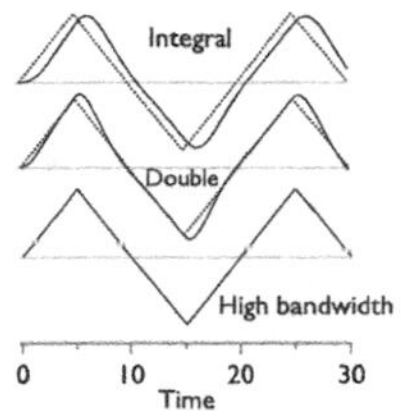

Example 3.7 (Tracking a ramp) For $G = \frac{1}{1+s}$, we seek to track a ramp command $r(t) = r_1 t$. Its Laplace transform $r(s) = \frac{r_1}{s^2}$, implying that we should choose $K(s) \sim r(s) \sim \frac{1}{s^2}$. At right are the time responses to a triangle wave input of closed-loop systems controlled by $K_i = \frac{1}{s}$, a pure integral controller (top), and by an "integrated PID" controller, $K_{ii} = \frac{(1+s)^2}{s^2} = \frac{1}{s}\left(\frac{1}{s} + 2 + s\right)$ (middle). The double integrator in K_{ii} includes extra terms to reduce the 180° phase lag at high frequencies.

As advertised, the K_i controller cannot track ramps and responds to a triangle input (alternating ramps) with a phase lag. By contrast, the K_{ii} controller tracks the signal, although there is a transient each time the ramp changes sign. The overshoot can be reduced by increasing the controller action at high bandwidth, for example with $K_{hb} = \frac{(1+10s)^2}{s^2}$ (bottom). But note that a controller with finite input power cannot track a triangle wave exactly, since instantaneously reversing velocity would require an infinite force. A better approach would be to track a rounded triangle wave whose maximum required acceleration is achievable. Do not ask for the impossible!

One application where tracking a ramp is desirable is for controlling the motion of scanning microscopes.[27] Usually, we want to scan back and forth at constant velocity, so that data acquired periodically is spaced evenly. Thus, the desired waveform is a triangle wave. An ordinary PID controller will always show an offset, while a "PI^2D" controller will not.[28]

We can suppress output disturbances $d(s)$ similarly (i.e., by including a copy of their signal in the controller). Defining a controller $K(s) = K_0(s)\, d(s) = K_0 \frac{N_d}{D_d}$ leads to

[27] A realistic mechanical-scanner model would include at least one resonance and possibly an RHP zero.
[28] The I^2 in PI^2D means that there is a double integrator.

$$e(s) = -y(s) = -\frac{d}{1+KG} \quad \Longrightarrow \quad e(t \to \infty) = -\lim_{s \to 0} s \left(\frac{N_d}{D_d + K_0 G} \right) = 0\,, \tag{3.77}$$

again assuming K_0, G, N_d, and D_d are all finite at $s = 0$ and that the denominator $D_d + K_0 G$ has all its poles in the LHP. You can pick $K_0(s)$ to give "nicely damped" dynamics for $y(t)$.

Another way to think about the internal model principle is that specifying the form of the reference $r(s)$ or the disturbance $d(s)$ is equivalent to saying that the signal is a member of signals that obey a differential equation.[29] For example, if we are interested in tracking a constant, then $r = \frac{1}{s}$ implies tracking the class of signals defined by $\dot{r}(t) = 0$. Alternatively, to suppress sinusoidal disturbances $d(t)$ of known frequency ω, we ask that $\ddot{d}(t) + \omega^2 d(t) = 0$. With the initial-condition terms, the Laplace transform is

$$d(s) = \left(\frac{s d(0) + \dot{d}(0)}{s^2 + \omega^2} \right) \equiv \frac{N_d(s)}{D_d(s)}\,. \tag{3.78}$$

We then design a controller $K(s) \sim D_d(s)^{-1}$ (see Problem 3.21).

One weakness of the above formulation of the Internal Model Principle is that it is robust only to a limited class of functions and not at all robust to functions outside that class. For example, in the above example we can suppress a sine wave with any amplitude and any phase – but not any frequency. Only ω will work. What if we do not know the frequency? Three strategies are to measure the disturbance in advance (Chapter 6), learn it online (Chapter 10), or use a phase-locked loop (Section 11.3.3).

Finally, while the Internal Model Principle guarantees asymptotic tracking or disturbance rejection (as $t \to \infty$), the resulting controller may perform poorly at finite times. Adding feedforward for the reference and, when possible, for the disturbance can improve transient responses, in keeping with the general rule to take advantage of all that we know about reference signals and disturbances. Even with inaccurate information, feedforward can "reduce the burden" of the feedback loop and improve overall performance.

3.7.5 (Auto) Tuning PID Controllers

We end this section on designing controllers by returning to the generic PID controller. We saw in Section 3.3.3 that a PID controller can transform one second-order system into another whose parameters may be set at will. But if the system under control is of higher order, tuning the three parameters K_p, K_i, and K_d requires more thought. You might measure carefully the dynamics and then choose the PID parameters rationally. But if you know $G(s)$ well enough to do this, you should probably choose a controller architecture that is better adapted to the system's dynamics.

29 The dynamical equations are sometimes referred to as the *exosystem* (Isidori and Byrnes, 1990).

Are there rules of thumb to tune PID parameters to control higher-order systems using only vague knowledge of the system's transfer function? Rather than try to survey all possibilities, we give a simple example of such rules and show how it, and its kin, are used in the now-ubiquitous "autotune" features of commercial PID controllers (and software).

The first step is to do a simple, easy characterization of the system's dynamics. One approach is to use pure proportional control and turn up the gain until instability occurs. Of course, we have seen that if $G(s)$ is first or second order, there will not be any instability. In practice, delays and unmodeled high-frequency dynamics almost always lead to instability at high gain. The measurement then consists of noting the critical gain K_c and the period T_c of limit-cycle oscillations that are observed just above onset. The classic *Ziegler–Nichols* tuning rules, from the 1940s, are usually given for PID parameters expressed in the form

$$K(s) = K_p \left(1 + \frac{1}{T_i s} + T_d s \right) . \tag{3.79}$$

The rules for a PI controller are $K_p = 0.4K_c$ and $T_i = 0.8T_c$. For a PID controller, they are $K_p = 0.6K_c$, $T_i = 0.5T_c$, and $T_d = 0.125T_c$. Note how adding a derivative term allows for an increase in proportional gain. An alternative version of Ziegler–Nichols is based on parameters measured from the open-loop step response.

Measuring T_c is easy, because it typically changes little for $K_p > K_c$, but finding K_c is tricky: near the instability, the growth and decay rates vanish at onset. A direct search for K_c can be slow. A better, faster method that gives both K_c and T_c uses *relay* feedback,

$$u(t) = u_r \operatorname{sign} e(t) . \tag{3.80}$$

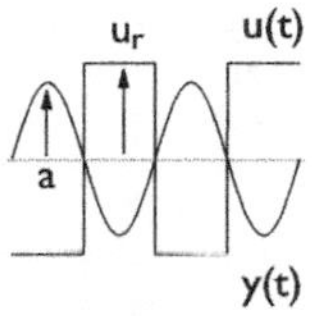

That is, $u = +u_r$ when the output is below the set point and $-u_r$ when it is above. The system typically oscillates, with an approximately sinusoidal response, as depicted at right. In Problem 3.22, you will estimate K_c and T_c from the output. Some commercial PID controllers have an "autotune" mode where they briefly implement relay feedback, infer K_c and T_c, and then automatically set either the PI or PID parameters accordingly.

Although the Ziegler–Nichols rules were historically influential, they often lead to rather mediocre results, and there are many competing schemes. But remember – no general rule can do as well as you can. If there is a specific property that you want to optimize (step response, disturbance response) for a particular system, and if you take the time to search the parameter space, you can do better than any of these rules. See Problem 3.22.

Even better, you can measure the transfer function (Chapter 6) and use optimal-control (Chapter 7) to design a controller that best satisfies the particular constraints that you care about. With adaptive-control techniques (Chapter 10), you tune and control simultaneously, which is particularly useful if the system transfer function drifts. Still, there is no denying the appeal of a simple solution: just push the button and don't worry (too much).

3.8 MIMO Systems

In this section, we discuss systems that have multiple inputs and multiple outputs. Although many ideas are more naturally discussed in the time domain, we introduce a few basic ideas. We have already seen examples that are "trivial" MIMO systems. All systems have multiple inputs: at the very least, the reference and disturbance can be thought of as two separate inputs.[30] And often, there is freedom in choosing the output, as well. For example, in the inverted-rod system (Example 3.4), we showed that putting the angle sensor near the pivot made control much harder than putting it near the center of mass. Here, multiple inputs or outputs will couple together in ways that alter system behavior.

Under the right circumstances, adding more sensors and actuators can significantly improve performance. However, the added complexity can also lead to problems, even when each individual element seems well designed. These problems motivate a deeper look at MIMO systems, and we introduce relevant mathematical tools and control techniques.

The main technical complication is the use of matrix techniques to handle the array of transfer functions that connect the ith input with the jth output. In the SISO case, at each frequency, a single transfer function multiplies the input by a complex gain. In the MIMO case, the inputs are vectors, and a transfer function matrix stretches them and rotates their direction. This new "twist" complicates the control problem.

3.8.1 Opportunities

Sometimes, "More is more" : extra degrees of freedom can enable control that would not otherwise be possible.

More Actuators

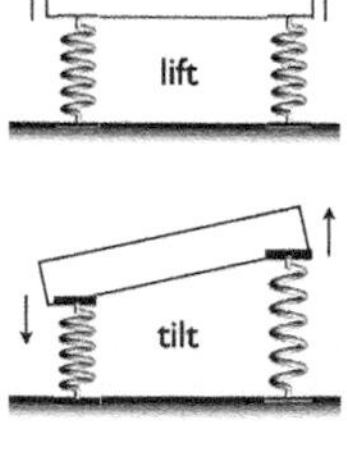

Consider a mass supported by two springs (see left). With one off-center actuator, lift and tilt motions are inevitably coupled. With a second actuator and with independent control over each, we can easily decouple those motions: symmetric combinations give pure lift, while antisymmetric combinations give pure tilt. Thus, with two actuators, we create and control motions far more easily. The important points here are the need to (1) use two actuators together to be effective and (2) recognize specific linear combinations, such

as $u_1 \pm u_2$, to get the maximum benefit of the added control capabilities. In general, controlling n outputs requires at least n independent inputs.

More Sensors

Multiple sensors can lead to similar qualitative improvements. Here, we return to the problem of temperature control in an extended system begun in Section 3.4.3.

[30] Of course, we can control the reference but not the disturbance. Even so, both are input signals.

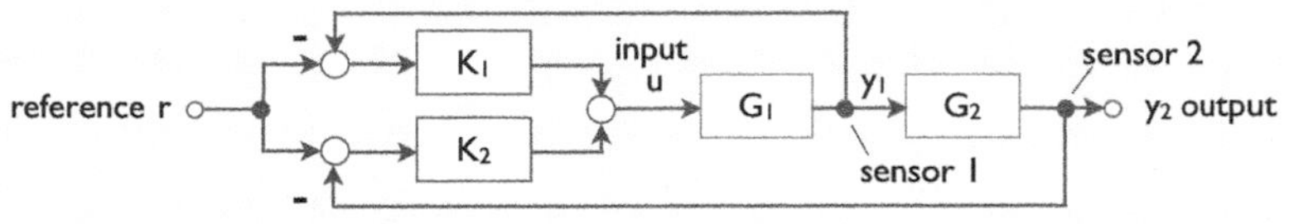

Fig. 3.10 Block diagram of an extended system with two sensors and reference $r(t)$. In Case I, we use Sensor 1. In Case II, we use Sensor 2. In Case III, we use both.

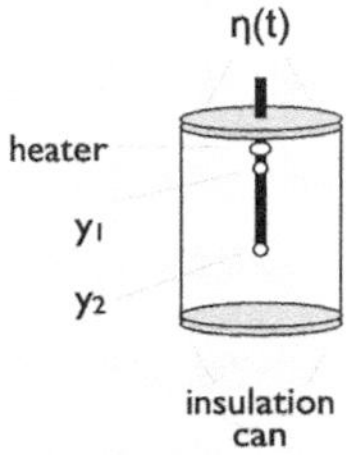

Consider a long rod, with the "sample" and a temperature probe at the end, y_2. The heater is near the top of the rod, and another temperature probe is at y_1. The apparatus is well insulated along its length, as it is inside a cylindrical can, with either vacuum or an insulating material such as styrofoam to isolate it from outside disturbances. However, power and other leads must still enter, and it is through this path, at top, that temperature disturbances $d(t)$ will primarily enter, as shown at right. Our previous discussion focused on the use of feedforward for disturbance compensation. Here, we look at a different strategy, splitting the control algorithm into two pieces, one associated with each sensor.

Is it better to put a probe near the heater (y_1) or near the experiment (y_2)? As we will see, either choice is problematic: using *both* sensors leads to better performance.

The heat equation leads to transfer functions that are not rational polynomials in s and must typically be approximated by rational polynomials. To simplify our discussion, we consider a simplified model with two low-pass filters that exhibit the same qualitative behavior.

Figure 3.10 shows a block diagram that represents the extended rod by two transfer functions, $G_1(s) = G_2(s) = \frac{0.5}{1+s}$. The two transfer functions are equal, implying that the probe at y_1 is in the middle of the rod, while the probe at y_2 is at the far end. The heater is at the other end. Having DC gains less than one reflects heat losses to the environment: the temperature falls to ambient (here, 0) as one goes along the bar (that is, from heater through G_1 through G_2). Their time constants reflect, crudely, the lag due to thermal diffusion.

In Case I, we use the sensor at y_1 with controller $K_1(s) = K_{\rm pi}(s) \equiv K_{\rm p} + K_{\rm i}/s$, with $K_{\rm p} = K_{\rm i} = 5$. The other sensor is ignored, and $K_2(s) = 0$. The feedback signal is then

$$u = -K_{\rm pi}(y_1 - r) \quad \Longrightarrow \quad y_1 = \frac{K_{\rm pi}G_1}{1 + K_{\rm pi}G_1} r, \quad y_2 = \frac{K_{\rm pi}G_1G_2}{1 + K_{\rm pi}G_1} r. \tag{3.81}$$

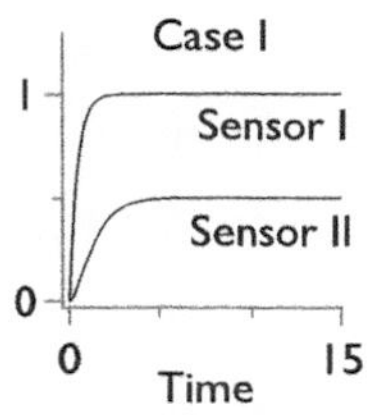

As shown at right, using the information from a single sensor, at position 1, you would conclude that the temperature is well regulated, with disturbances much attenuated and a steady-state value equaling the desired set point. But at the sensor at position 2 (at the sample), temperature gradients create an offset. Its value depends on the rate of heat loss to the environment and varies with ambient temperature changes. These temperature drifts would be invisible to someone who looked only at data coming from Sensor 1. Failing to recognize a "Case I" situation is a common mistake made by experimentalists.

In Case II, we use only the sensor at y_2, with $K_2 = K_{pi}(s) = K_p + K_i/s$ and $K_1 = 0$. Then

$$u = -K_{pi}(y_2 - r) \quad \Longrightarrow \quad y_2 = \frac{K_{pi}G_1G_2}{1 + K_{pi}G_1G_2} r . \tag{3.82}$$

Finally, in Case III, we use both sensors and divide the controller among them in a *split-PI* configuration where the first sensor uses proportional feedback and the second integral feedback: $K_1(s) = K_p$ and $K_2(s) = K_i/s$. Then

$$u = K_{pi}r - K_p y_1 - \frac{K_i}{s} y_2 \quad \Longrightarrow \quad y_2 = \frac{K_{pi}G_1G_2}{1 + K_pG_1 + \frac{K_i}{s}G_1G_2} r . \tag{3.83}$$

Notice that the transfer functions have the same numerator but different denominators.

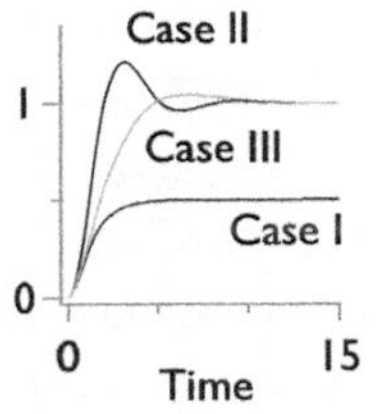

At left are the temperature step responses to all three transfer functions at y_2. Case I, sensor near heater, reproduces the response shown above. In Case II, sensor at sample, the longer delay between heater and sensor degrades the response, but integral control still removes the offset at the sample. Finally, in Case III, proportional control using the near sensor and integral control using the sample sensor combine to eliminate the offset and preserve response, giving the best of both worlds. See Problem 3.23.

The apparent trade-off between good disturbance rejection from a sensor near the actuator and good set point tracking from a sensor near the sample can thus be overcome by combining the information from the two. In the context of typical PID control of equipment, we could term this control algorithm *P$\bar{I}$D*. Adding feedforward can further improve performance by decoupling the command and disturbance responses. *P$\bar{I}$D* is easy to implement and deserves consideration in many applications.

3.8.2 Pitfalls

If adding inputs and outputs creates opportunities, it can also create pitfalls. For example, in the SISO case, a first-order system ($\frac{1}{1+s}$) is stable under proportional feedback for any positive gain K (see Section 3.1.1). Consider now the 2×2 generalization,

$$\boldsymbol{G}(s) = \begin{pmatrix} \frac{1}{1+s} & \frac{1}{1+\tau s} \\ \frac{1}{1+\tau s} & \frac{1}{1+s} \end{pmatrix} . \tag{3.84}$$

For a physical picture, think of a shower whose hot- and cold-water pipes mix to form a single stream. The shower output is characterized by an overall flow and temperature. Its two inputs set the flow of cold and hot water. The simplified form (and symmetry) in Eq. (3.84) is more for mathematical convenience than physical realism.

Let us now try to regulate the shower's temperature and flow by adjusting the hot and cold flows with proportional gains K that are assumed identical for each. The controller matrix $\boldsymbol{K}(s)$ is then $K\mathbb{I}$, with $\mathbb{I}$ the 2×2 identity matrix. The closed-loop stability is determined by the matrix analog of Eq. (3.4), $\boldsymbol{S} = (\mathbb{I} + \boldsymbol{GK})^{-1}$, where the closed-loop system is unstable when one of the eigenvalues of $\boldsymbol{S}$ has a positive real part for any value of $s = i\omega$, generalizing the ideas of Section 3.5. The eigenvalues of $\boldsymbol{S}$ are given by

$$\lambda_\pm(s) = \frac{1}{1 + K(\frac{1}{1+s} \pm \frac{1}{1+\tau s})}, \qquad \Longrightarrow \text{ poles when } \quad K = \frac{(1+s)(1+\tau s)}{s(1-\tau)}. \tag{3.85}$$

Setting $s = \pm i\omega$ and isolating real and imaginary parts shows that there is an instability (Hopf bifurcation) with $\omega^* = \frac{1}{\sqrt{\tau}}$ and $K^* = \frac{1+\tau}{1-\tau}$. For $\tau < 1$, the system will be unstable using only proportional gain when $K > K^*$. Even though the connections between input and output are all first-order systems – and hence always stable in isolation – the full, coupled system can nonetheless be unstable. Comfortable showers are not easy to achieve, as anyone who has been in the shower when someone else flushes the toilet can attest!

One reason that the system $\boldsymbol{G}(s)$ in Eq. (3.84) is difficult to control is that its determinant vanishes at DC, and the matrix cannot be inverted. Since most feedback algorithms amount to generating an approximate inverse of the system dynamics, we expect control to be difficult at low frequencies. More generally, a transfer function may not be invertible at some frequencies because of pathologies in the interactions among different components rather than because of a problem with any individual component.

3.8.3 Matrix Techniques for MIMO

Motivated by the opportunities that adding actuators and sensors can bring and cautioned by the pitfalls that can arise, we introduce formalism to understanding MIMO systems more deeply. The tools, while standard parts of linear algebra, are not often taught in the standard physics curriculum. Then, in Section 3.8.4, we apply these ideas to develop techniques for analyzing and controlling MIMO systems in the frequency domain.

Defining "Gain" for a MIMO System

In a MIMO system, the frequency response is given by a matrix of transfer functions. Let us generalize the notion of gain for a transfer function to the matrix case. We recall the definition of gain of a SISO transfer function in the frequency domain:

$$\text{SISO gain} = \frac{|y(s)|}{|u(s)|} = \frac{|G(s)\,u(s)|}{|u(s)|} = \frac{|G(s)|\,|u(s)|}{|u(s)|} = |G(s)|\,. \tag{3.86}$$

For a MIMO system, the situation is less simple: the input $\boldsymbol{u}(s)$ is a vector of length m, the output $\boldsymbol{y}(s)$ a vector of length p, and the transfer function $\boldsymbol{G}(s)$ is an $m \times p$ element matrix connecting the m inputs to the p outputs. The matrix $\boldsymbol{G}(s)$ stretches and rotates an input vector. In particular, the amount of stretching depends on the orientation of $\boldsymbol{u}$. As a first step, we define the "size" of $\boldsymbol{u}$ using the Euclidean norm:

$$\|\boldsymbol{u}\|_2 \equiv \sqrt{\sum_j |u_j|^2} = \sqrt{\boldsymbol{u}^\dagger \boldsymbol{u}}, \tag{3.87}$$

where the sum is over the p inputs and where † denotes the conjugate transpose (Hermitian conjugate). Thus, $\boldsymbol{u}^\dagger$ is a row vector whose entries are the complex conjugate of the entries of the corresponding column vector, giving a scalar for the product with $\boldsymbol{u}$. We will drop the subscript on the norm notation unless we need to distinguish it from other definitions of the norm. Then the effect of $\boldsymbol{G}$ on a particular input $\boldsymbol{u}$ is

$$\frac{\|y(s)\|}{\|u(s)\|} = \frac{\|G(s)\,u(s)\|}{\|u(s)\|} \tag{3.88}$$

Singular Value Decomposition

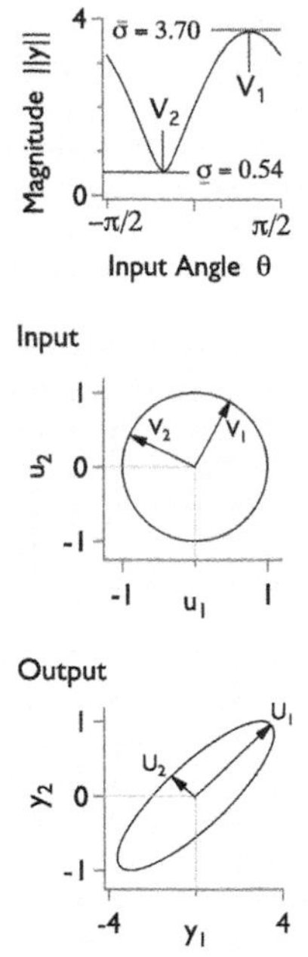

To understand better this notion of a direction-dependent gain, consider an example where we evaluate $\boldsymbol{G}$ at a particular frequency, so that it is just a matrix. For example, let

$$\boldsymbol{G}_0 = \begin{pmatrix} 2 & 3 \\ 0 & 1 \end{pmatrix}, \qquad \boldsymbol{u} = \begin{pmatrix} \cos\theta \\ \sin\theta \end{pmatrix}. \tag{3.89}$$

In Eq. (3.89), we choose $\boldsymbol{u}$ to have unit magnitude, since the linear relationship between $\boldsymbol{y}$ and $\boldsymbol{u}$ implies that any overall scaling factor in the input will cancel out in Eq. (3.88). At left, we plot the magnitude of $\boldsymbol{y} = \boldsymbol{G}_0\boldsymbol{u}$ as a function of θ, which parametrizes the direction of $\boldsymbol{u}$ over the range $\theta \in (0, \pi)$.[31] From the graph of $\|\boldsymbol{y}[\boldsymbol{u}(\theta)]\|$, we discover that the maximum gain, which we denote $\overline{\sigma}$, occurs at an angle $\theta \approx 58°$, while the minimum gain, denoted $\underline{\sigma}$, occurs at the orthogonal direction, $\approx -32°$.

That the gain varies with direction should not surprise you. However, you might be surprised that the values of maximum and minimum gain, 3.70 and 0.54, are not the same as the eigenvalues, which equal the diagonal elements, 2 and 1. To visualize what is going on, we also plot at left the two unit vectors, $\boldsymbol{V}_1$ and $\boldsymbol{V}_2$, that correspond to the input directions producing the maximum and minimum gains.[32] In the gains for each input vector, the principal input directions are mapped to another orthogonal pair of principal output directions, $\boldsymbol{U}_1$ and $\boldsymbol{U}_2$. Thus, the action of $\boldsymbol{G}_0$ is to rotate from the

[31] The magnitude of the response is periodic in π, and not 2π, because a rotation of $\boldsymbol{u}$ by π is equivalent to multiplying $\boldsymbol{u}$ by a factor of -1, which is canceled when the gain is calculated.

[32] Note the unfortunate clash of two notation conventions: the input vector is $\boldsymbol{u}$, but the SVD decomposition $\boldsymbol{G} = \boldsymbol{U\Sigma V}^\dagger$ denotes the *principal input directions* by columns of $\boldsymbol{V}$ and the *principal output directions* by columns of $\boldsymbol{U}$. Since $\boldsymbol{u}$ and $\boldsymbol{U}$ are different symbols, we have done nothing wrong, but it's easy to get confused. Fortunately, there will be only a few situations where the two symbols are used together.

input vector to a standard coordinate frame, stretch it, and then rotate it to the output direction. Explicitly, to two-digit precision,

$$G_0 = \begin{pmatrix} 2 & 3 \\ 0 & 1 \end{pmatrix} = \underbrace{\begin{pmatrix} 0.97 & -0.23 \\ 0.23 & 0.97 \end{pmatrix}}_{\text{rotate}} \underbrace{\begin{pmatrix} 3.70 & 0 \\ 0 & 0.54 \end{pmatrix}}_{\text{stretch}} \underbrace{\begin{pmatrix} 0.53 & 0.85 \\ -0.85 & 0.53 \end{pmatrix}}_{\text{rotate}} \equiv U \Sigma V^{\dagger} . \tag{3.90}$$

This way of writing G_0 is known as the *singular value decomposition*, or SVD. Compare to the more familiar decomposition into eigenvectors and eigenvalues, $G_0 = R\,D\,R^{-1}$.

In an eigenvalue decomposition, the eigenvalues form columns of the matrix R. The vectors are not necessarily orthogonal. Also, the diagonal matrix of eigenvalues, D, may have positive, negative, or complex values. By contrast, in the singular value decomposition of Eq. (3.90), the matrices U and V are unitary (so that $UU^{\dagger} = VV^{\dagger} = \mathbb{I}$). Also, the elements of Σ are real and nonnegative. Here, Σ is a square matrix, but its dimensions follow those of G. Further properties of the SVD are collected in Appendix A.1.7 online. Here, we focus on those properties that relate to the frequency-dependent gains of MIMO systems.

In the general situation (Eq. 3.88), the input and output are related by a frequency-dependent matrix. Scaling inputs so that $||u|| = 1$, we have,

$$y = Gu \quad \Longrightarrow \quad ||y||^2 = ||Gu||^2 = u^{\dagger} G^{\dagger} G u = u^{\dagger}\, W D_A W^{\dagger} u . \tag{3.91}$$

In Eq. (3.91), the last identity holds since the matrix $A \equiv G^{\dagger}G$ is square, symmetric, and has positive definite eigenvalues – even if G itself has none of these properties. Thus, we can write $A = WD_AW^{\dagger}$, with W unitary and D_A a diagonal matrix with elements σ_j^2, where j goes from 1 to k. These are the *singular values* of G. If a system has m inputs and p outputs, then G is an $m \times p$ matrix, and $k = \min(m, p)$. By convention, $\bar{\sigma}$ denotes the largest singular value and $\underline{\sigma}$ the smallest. We can then construct inequalities using $\bar{\sigma}\,\mathbb{I} \geq D_A \geq \underline{\sigma}\,\mathbb{I}$. Since u has unit norm and W is unitary, we conclude that

$$\bar{\sigma} \geq ||y|| \geq \underline{\sigma} . \tag{3.92}$$

That is, the magnitude of the output y is bounded by the largest and smallest of the singular values, as we saw in Eq. (3.90) in a special case. Using Eq. (3.92), we define $||G||$ by

$$||G(s)|| \equiv \sup_{u(s)} \frac{||G(s)\,u(s)||}{||u(s)||} = \bar{\sigma}[G(s)] . \tag{3.93}$$

Similarly, we have

$$\underline{\sigma}[G(s)] = \inf_{u(s)} \frac{||G(s)\,u(s)||}{||u(s)||} \tag{3.94}$$

If the transfer function matrix G is *normal*, so that it can be expressed as $G = UDU^{\dagger}$, the singular values coincide with eigenvalues, but usually, they are different. The example $G_0 = \left(\begin{smallmatrix}2 & 3\\ 0 & 1\end{smallmatrix}\right)$ is not normal (although it is typical!). Notice that the eigenvalues, 2 and 1, are independent of the value of the upper-right element. It could be 3 or 100 or whatever – the eigenvalues are the same. The singular values for $i = 1, \ldots, k$ are given by

$$\sigma_i(\boldsymbol{G}) = \sqrt{\mathrm{eig}_i(\boldsymbol{G}^\dagger \boldsymbol{G})} = \sqrt{\mathrm{eig}_i(\boldsymbol{G}\boldsymbol{G}^\dagger)} \tag{3.95}$$

and depend on all the elements. For example, the general 2×2 matrix $\boldsymbol{G} = \left(\begin{smallmatrix}a & b\\ c & d\end{smallmatrix}\right)$ has

$$\sigma_{1,2}(\boldsymbol{G}) = \frac{1}{\sqrt{2}}\sqrt{G^2 \pm \sqrt{G^4 - 4(\det)^2}} \quad \Longrightarrow \quad \overline{\sigma}(\boldsymbol{G}) \approx G\,, \quad \underline{\sigma}(\boldsymbol{G}) \approx \frac{\det}{G}\,,$$
$$G^2 \equiv a^2 + b^2 + c^2 + d^2\,, \quad \det \equiv ad - bc\,. \tag{3.96}$$

Notice that G is greater than the magnitude of the largest entry of $\boldsymbol{G}$. The approximations in Eq. (3.96) hold for $\det \ll G$. The condition number $\gamma \equiv \overline{\sigma}/\underline{\sigma} \approx G^2/\det$ (see below).

If $\boldsymbol{G}$ is a nonsquare, $m \times p$ matrix, then $\boldsymbol{G}^\dagger \boldsymbol{G}$ is a $p \times p$ matrix, while $\boldsymbol{G}\boldsymbol{G}^\dagger$ is an $m \times m$ matrix. The number of nonzero singular values then is at most $k = \min(m,p)$.

In short, for MIMO systems, singular values are better than eigenvalues:

1 They are the correct measure of gain.
2 The principal input and output directions are always orthogonal.
3 They can be defined for nonsquare matrices.

MIMO Frequency Response

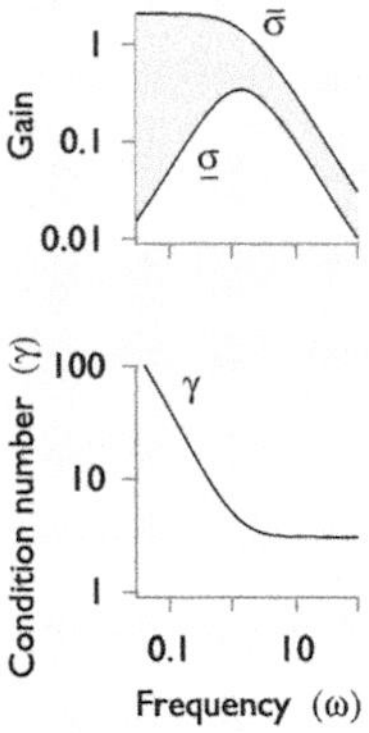

A simple and natural way to analyze MIMO systems in the frequency domain is to look at the *singular value plot*, a generalization of the Bode magnitude plot. We plot $\overline{\sigma}(s)$ and $\underline{\sigma}(s)$ for $s = \mathrm{i}\omega$, as a function of the angular frequency ω. As an example, at left is the singular value plot of the "shower" system of first-order transfer functions in Eq. (3.84), for $\tau = \frac{1}{2}$. The shaded area between $\overline{\sigma}$ and $\underline{\sigma}$ represents the set of possible responses of the dynamics. Note the divergence between the two curves $\overline{\sigma}(s)$ and $\underline{\sigma}(s)$.

Their ratio (bottom plot) is known as the *condition number* $\gamma(\omega)$, which is a measure of how easy or hard numerically it is to invert the transfer function matrix at the frequency ω. Matrices with $\gamma \gtrsim 1$ are easy to invert. Matrices that are nearly singular have $\gamma \gg 1$. The shower system is hard to control for low frequencies because $\lim_{\omega\to 0} \boldsymbol{G}(\mathrm{i}\omega) = \left(\begin{smallmatrix}1 & 1\\ 1 & 1\end{smallmatrix}\right)$, which is singular.

MIMO Zeros

In the shower system, Eq. (3.84), the diverging condition number at low frequencies and the difficulties for control that the divergence causes are reminiscent of the issues we saw in SISO systems when zeros were present in the response function. In MIMO systems, we define zeros to be values of the complex variable s where the transfer function matrix $\boldsymbol{G}(s)$ loses rank. (See Appendix A.1 online for a discussion of rank.)

It is easiest to understand the notion of a MIMO zero by example. Consider

$$G(s) = \frac{1}{(1+2s)^2}\begin{pmatrix} 1 & 1 \\ 1+2s & 2 \end{pmatrix}. \tag{3.97}$$

Although the individual matrix elements of $\boldsymbol{G}$ have no zeros, there is a MIMO zero in the right-hand plane (RHP), at $s = \frac{1}{2}$, where, as Eq. (3.97) shows, $\boldsymbol{G} = \frac{1}{4}\left(\begin{smallmatrix} 1 & 1 \\ 2 & 2 \end{smallmatrix}\right)$ has rank one. Alternatively, det $\boldsymbol{G}(s) = \frac{1-2s}{(1+2s)^4}$, obviously has a zero at $s = \frac{1}{2}$.[33] If the matrix $\boldsymbol{G}$ were not square, we could look at the singular values.[34]

Unlike zeros in a SISO system, MIMO zeros have a *direction*. For normal, square matrices, the direction is that of the eigenvector associated with the zero eigenvalue. More generally, it is the associated input vector in the singular value decomposition. Physically, it is the combination of inputs to the matrix that yields zero output. In the example of Eq. (3.97), the associated direction is $\left(\begin{smallmatrix} 1 \\ -1 \end{smallmatrix}\right)$. We can see this at right in the plots of step responses of both outputs for three different input combinations. The top plot is the response to $\left(\begin{smallmatrix} u_1 \\ u_2 \end{smallmatrix}\right) = \left(\begin{smallmatrix} 1 \\ 0 \end{smallmatrix}\right)$, the middle to $\left(\begin{smallmatrix} 0 \\ -1 \end{smallmatrix}\right)$, and the bottom to $\left(\begin{smallmatrix} 1 \\ -1 \end{smallmatrix}\right)$. The last case shows that when the input combination is along the zero direction, one of the outputs is zero. In addition, because this is an RHP zero, the other output displays the initial inverse response that we saw for RHP zeros in SISO systems (see Section 3.6.3).

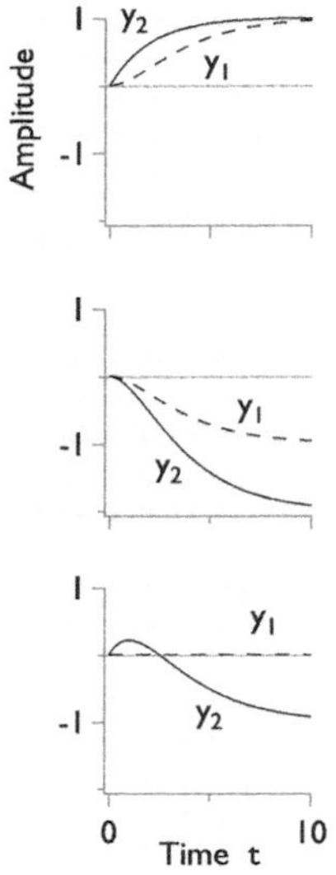

A "cure" for RHP zeros in a MIMO system is to add more inputs: more control means more ways for a signal to reach its output and less danger of the kind of cancellation we saw in this example. For example, a graphical plot of singular values shows that we can eliminate the RHP zero by adding an input to our example,

$$G_1(s) = \frac{1}{(1+2s)^2}\begin{pmatrix} 1 & 1 & 1 \\ 1+2s & 2 & 1 \end{pmatrix}. \tag{3.98}$$

3.8.4 MIMO Control Techniques

The most obvious way to control a MIMO system is to invert the transfer matrix $\boldsymbol{G}(s)$. In the following, we again assume we have m inputs and p outputs. Assuming for the moment $m = p$ and that it is possible to invert $\boldsymbol{G}$, then a controller can take the form

$$\boldsymbol{K}(s) = \boldsymbol{G}^{-1}(s)\,\boldsymbol{L}(s) \quad \Longrightarrow \quad \boldsymbol{G}(s)\,\boldsymbol{K}(s) = \boldsymbol{L}(s)\,, \tag{3.99}$$

where we need to remember that, in general, matrix multiplication does not commute ($\boldsymbol{GK} \neq \boldsymbol{KG}$) and that there are two clashing conventions: signals on block diagrams such as Figure 3.1 flow left to right, but matrices have their input on the right and output on the left. Thus, we write $\boldsymbol{GK}$ because $\boldsymbol{K}$ is applied to $\boldsymbol{r}$ and its output, $\boldsymbol{u}$, is sent to the system, $\boldsymbol{G}$. In Eq. (3.99), the $p \times p$ matrix $\boldsymbol{L}(s)$ is the loop transfer function,

[33] Be cautious in using the determinant to find zeros. Although a zero of the determinant is a zero of $\boldsymbol{G}$, the converse may not be true. A zero in one direction can cancel a pole in another direction. For example, $\boldsymbol{G} = \left(\begin{smallmatrix} 1/(1+s) & 0 \\ 0 & 1+s \end{smallmatrix}\right)$ has det = 1, even though there is a zero in the $\left(\begin{smallmatrix} 0 \\ 1 \end{smallmatrix}\right)$ direction and a pole in the $\left(\begin{smallmatrix} 1 \\ 0 \end{smallmatrix}\right)$ direction.

[34] Since even 2×2 matrix expressions for singular values are complicated, we depend on numerical calculations using standard software routines.

as measured starting from the output $\boldsymbol{y}$. (If we started from the system input $\boldsymbol{u}$, then the loop transfer function would be the $m \times m$ matrix $\boldsymbol{KG}$.) We can also derive, in analogy to the SISO case, sensitivity and complementary sensitivity matrices,

$$\boldsymbol{S}(s) = [\mathbb{I} + \boldsymbol{L}(s)]^{-1}, \qquad \boldsymbol{T}(s) = [\mathbb{I} + \boldsymbol{L}(s)]^{-1}\boldsymbol{L}(s) = \boldsymbol{L}(s)\,[\mathbb{I} + \boldsymbol{L}(s)]^{-1}, \tag{3.100}$$

where we note that $\boldsymbol{L}$ and $(\mathbb{I} + \boldsymbol{L})^{-1}$ commute.

If inversion is possible, then we can choose a simple $\boldsymbol{L}(s) = L(s)\mathbb{I}$ to give all p elements straightforward transfer functions. The choice of $L(s)$ (and even whether it should be the same for all elements) depends on the details of the given problem, for example, whether the goal is reference tracking or disturbance rejection, and so on. The polynomial degree of $L(s)$ should at least equal that of $\boldsymbol{G}^{-1}(s)$ so that the controller $\boldsymbol{K}(s)$ is realizable.

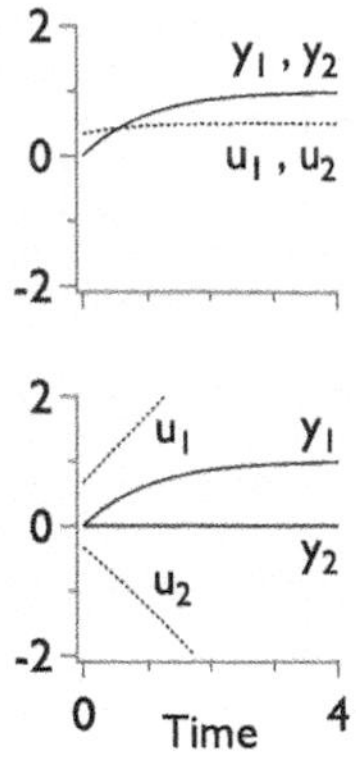

Example 3.8 (Controlling the shower) Let us try a controller that inverts the toy "shower" model of Eq. (3.84) with $\tau = 1/2$. A controller of the form

$$\boldsymbol{K}(s) = \frac{1}{s}\boldsymbol{G}_{\rm sh}^{-1}, \qquad \boldsymbol{G}_{\rm sh}(s) = \begin{pmatrix} \frac{1}{1+s} & \frac{1}{1+s/2} \\ \frac{1}{1+s/2} & \frac{1}{1+s} \end{pmatrix} \tag{3.101}$$

leads to "ideal" sensitivity and complementary sensitivity functions $S = \frac{s}{1+s}\mathbb{I}$ and $T = \frac{1}{1+s}\mathbb{I}$. Applying a unit step function simultaneously to both the $r_1(t)$ and $r_2(t)$ reference signals – i.e., $\boldsymbol{r}(t) = \theta(t)\left(\begin{smallmatrix}1\\1\end{smallmatrix}\right)$ – leads to well-behaved step responses (see left, top). Now, consider a unit step in just $r_1(t)$ that leaves $r_2(t) = 0$ – i.e., $\boldsymbol{r}(t) = \theta(t)\left(\begin{smallmatrix}1\\0\end{smallmatrix}\right)$. The output $y_1(t)$ rises to the desired unit output level, while $y_2(t)$ stays at 0. But there is a fatal problem: the inputs $\boldsymbol{u}(t)$ that are actually supplied to the system increase linearly in time! Since they cannot do so forever, the system cannot hold indefinitely the desired values ($y_1 = 1$ and $y_2 = 0$).

The difficulty we found with our inversion controller is fundamental. We saw in our singular value plot that the condition number γ of $\boldsymbol{G}_{\rm sh}(s)$ diverges as $s \to 0$. More simply, we observe that at $s = 0$, $\boldsymbol{G}_{\rm sh} = \left(\begin{smallmatrix}1&1\\1&1\end{smallmatrix}\right)$, which is clearly not invertible. Since, at DC, $y_1 = y_2 = u_1 + u_2$, there is simply no way to have constant u_1 and u_2 and set independent values for y_1 and y_2. The fault is not in our strategy, but in our system: it really has a pathology at $s = 0$. Of course, in this case, $\boldsymbol{G}_{\rm sh}$ was chosen to have this problem, but you should always look out for such issues. Checking the frequency-dependent condition number γ is a good way to assess how difficult it will be to control a MIMO system. Divergences in γ signal pathologies while near divergences, $\gamma \gg 1$, signal difficulties.

When confronted with a system that looks like $\boldsymbol{G}_{\rm sh}$, what to do? The easier, better strategy is to change the system to make it more amenable to control. Problem 3.24 shows how a small change in the DC gains of $\boldsymbol{G}$ makes an impossible problem straightforward.

As with SISO systems, an inversion controller may be impractical:

1. $\boldsymbol{G}$ may not be a square matrix and thus have no inverse.

2. At some, or all, values of s, the matrix Σ may have some zero elements along the diagonal. This situation arises, for example, when we have RHP zeros.
3. Even if G is invertible, the required inputs may be too large.
4. G may not be well known or may change over time, and the inverse of the nominal transfer function matrix may not be close to the true inverse.

The first issue, nonsquare G, can be partially dealt with by using a singular value decomposition $G = U\Sigma V^\dagger$, so that $K = (V\Sigma^{-1}U^\dagger)L$. If Σ has elements along the diagonal that equal zero, one can invert just the nonzero elements (giving the *pseudoinverse*).

The other issues can be dealt with by other kinds of selective inversion. For example, one might try to invert the system just at one frequency (perhaps at DC or at the feedback bandwidth). Another compromise is to try to reduce the system to an upper or lower triangular matrix, that is, one whose nonzero entries lie along the diagonal or entirely above or below it. Clearly, such a form is easier to achieve than a fully diagonal form.

Such a triangular matrix implies *hierarchical control*. For example, if $y = GKe$ and GK is lower triangular, then $GK = \left(\begin{smallmatrix} a & 0 \\ b & c \end{smallmatrix}\right)$, which implies $y_1 = ae_1$, $y_2 = be_1 + ce_2$. Thus, the control loop involving y_1 can be closed first. Tune it by choosing a, which sets the dynamic behavior of $e_1(t)$. Then tune Loop 2 by choosing b and c, considering $e_1(t)$ as fixed. If the upper-right element of K were nonzero, there would be no decoupling, as tuning loop 2 would change $e_2(t)$, which would then alter the signal in loop 1. Thus, triangular loop matrices lead to a straightforward way to design the controller as a sequence of loops.

MIMO Loop Shaping

Section 3.7.3 introduced the strategy of loop shaping for SISO systems. Large gains at low frequencies give tight control; loop gains $\sim \omega^{-1}$ near the crossover frequency where $|L(i\omega)| = 1$ insure stability; and rapid gain cut-off at higher frequency minimizes the injection of sensor noise into the closed-loop dynamics. The main tools were the standard PID elements, augmented by lag, lead, and related transfer function "tweaks." How do these ideas apply to MIMO systems? Although loop shaping is an intuitive strategy that is most effective on simple systems, it is still relevant for the more complex MIMO case. If we account for the spread in gains, as measured by the singular-value plot, the principles are still the same and provide qualitative insight into controller design.

We begin with a couple of simple observations. First, we remind the reader that the definition of loop transfer function in a MIMO system depends on where the loop is broken. We will always choose the system output, for which $L = GK$. Second, in cases where it is possible to invert a matrix and define a scalar loop function $L(s)$, then the shape of $|L(s)|$ may be chosen as for a SISO system. If you cannot invert G completely, try to minimize the condition number of the loop-transfer-function matrix.

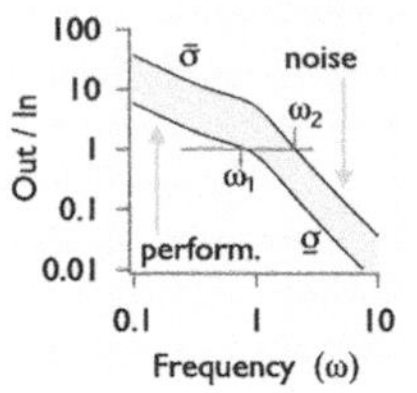

At left, we see that MIMO loop shaping is similar to the SISO case. Increasing $\underline{\sigma}$ increases the gain at low frequencies and suppresses disturbances. Decreasing $\overline{\sigma}$ at high frequencies suppresses sensor noise. In addition, the fall-off of the gain near the crossover frequency at unit gain controls the phase margin. In the SISO case, $|L(i\omega)| \sim \omega^{-1}$ near the crossover frequency leads to an adequate phase margin ($\approx 90°$); the system will be far enough from an instability to have good dynamical behavior. In the MIMO case, general relationships between the frequency dependence of singular values and stability margins are complicated. One strategy is to first ensure stability and then use loop shaping to improve controller performance and robustness (see Chapter 9).

Last Words on MIMO Control

The above is a taste of the opportunities, pitfalls, and techniques for MIMO control. Although we have emphasized many of the difficulties and pitfalls, we should not end on such a pessimistic note. As we observed in the introduction to this section, it is often true that "more is more" : adding more sensors and actuators allows more and better control, particularly when care is taken to avoid the pitfalls we have just discussed. In Chapter 15, we shall link control theory and information theory. In that perspective, the inputs and outputs of a control system are channels for exchanging information between a controller and a physical system, and the quality of control depends on the rates of information exchange. Adding more sensors and actuators then increases the information flow. Such speedup is of enormous practical importance in engineering.

To cite but one example, the recent wireless standard for communication, IEEE 802.11n, owes its increased information rate to the adoption of MIMO techniques, with transmitters and receivers using multiple antennas. If there are N transmitters and N receivers, information transmission can be increased by a factor of N relative to a single channel. More subtly, the pathologies of radio transmission – multiple signal paths, destructive interference, and the like – can be compensated by using information spread over the N^2 links among the transmitters and receivers. With only a single transmitter and receiver, a given path might have destructive interference – a zero – for a given arrangement of transmitter and receiver. Adding other receivers spaced a few wavelengths apart gives independent measurements: if one path has destructive interference, other paths will likely not. Thus, redundancy leads to both a speed-up of N and an increase in reliability. In a physics context, you could improve the quality of measurements by sending multiple excitations to a sample and recording multiple outputs from it.

Another way to interpret MIMO Bode plots is that the range between $\overline{\sigma}$ and $\underline{\sigma}$ represents the set of all possible outcomes, at a given frequency, in response to the set of all possible inputs. Since we do not know the actual input, it makes sense to design the control so that it works well – satisfies the various requirements of performance, stability, robustness, and so forth – for any possible input. Notice, though, that we would have precisely the same design task if our system were a simple SISO system but had an uncertain transfer function. In that case, $\overline{\sigma}$ and $\underline{\sigma}$ would represent the *uncertainty bounds* at a given frequency for the response. See the MIMO Bode plot in the shower example.

3.9 Summary

In this chapter, we have presented the basics of classical control, as developed mostly between 1945 and 1960. We began by considering the simplest approaches to controlling the output of a dynamical signal. Starting from the observation that the system's transfer function will alter the control input, we first tried an open-loop, reference-feedforward controller that inverts the system's dynamics. Unfortunately, the system dynamics may not be well known, and, even if known at one time, may change. The consequent lack of robustness of feedforward techniques motivates the introduction of feedback loops that approximate the inverse without knowing the full details about the system dynamics. A combination of feedback and feedforward techniques works even better. While feedback has many desirable properties, it can lead to unstable dynamics when too much feedback is used – when the gains are set too high. The root cause of instabilities is that delay in the system response that can invert the sign of the feedback correction and lead to runaway behavior. Because physical response functions are causal, there are constraints that imply a minimum phase delay given the frequency response of the magnitude of the transfer function. These delays trace back to the particular way actuators and sensors are connected to the system and are often linked to a physical separation between the two. The result is a transfer function that has poles or zeros in the right-hand part of the complex s-plane. In such a case, the approximate inverse generated by a feedback loop will always lead to instability when the gain is too high. Conversely, feedback can also stabilize an unstable system. Controller design must balance several competing constraints: the system (usually) should be stable; causality imposes minimum phase lags, and actuators can supply only a finite input. These constraints lead to informal heuristics (loop shaping, internal model control) for designing controllers. Finally, adding inputs and outputs (MIMO) lead to both opportunities and pitfalls. Singular value plots generalize the Bode magnitude plots and can be used to design controllers in ways that are similar to, but more complicated than, the SISO case.

In the next chapter, we will reformulate these ideas in the time domain, which clarifies and makes systematic the intuitive, heuristic frequency-domain methods just presented.

3.10 Notes and References

Much of the material in this chapter is influenced by the crisp, succinct summary of classical control theory by Skogestad and Postlethwaite (2005). For sensitivity functions, see Åström and Murray (2008); for PID control, Åström and Hägglund (2006). For numerous ways to tune PID parameters, see Skogestad and Grimholt (2012). Doyle et al. (1992) have good discussions of loop shaping and NMP functions and the various limitations on control that they impose. Bode's 1937 patent application on loop shaping was published as Bode (1940). For experiments that test various models based on the inverted pendulum, see Winter et al. (1998). For associated control strategies, see Kuo (1995).

Fleming and Leang (2014) give a comprehensive presentation of feedback and feedforward techniques, as applied to nanopositioning, where the system to be controlled typically consists of a series of lightly damped vibration modes. Much of the discussion focuses on piezoelectric positioning stages for scanning-probe microscopy, but the lessons are broader. For a description of van der Meer's work in stochastic cooling, see Marriner (2004).

Goodwin et al. (2001) discuss the Internal Model Principle (IMP) and the fundamental limitations imposed by NMP functions. For an analysis of the NMP response of an elastic cantilever used for atomic force microscopes, see Rubio-Sierra et al. (2006). The IMP was first introduced to control theory by Francis and Wonham (1975), using a state-space formalism. But Conant and Ashby (1970) developed essentially the same idea for state spaces with a finite number of elements.

For singular values and MIMO, see Skogestad and Postlethwaite (2005) and Goodwin et al. (2001). The MIMO-zero example is adapted from Skogestad and Postlethwaite (2005). The two-sensor example was studied experimentally by Znaimer and Bechhoefer (2014). For delays and stick balancing, see Milton et al. (2016). For the role of nonminimum-phase zeros, see Doyle et al. (1992) and Leong and Doyle (2016).

Problems

3.1 Bandwidth of integral control. For the first-order system with integral control discussed in Section 3.3.1, derive the feedback bandwidth for arbitrary K_i. Find simpler expressions for the limits $K_i \ll 1$ and $K_i \gg 1$. Also, rescale the closed-loop transfer function of Eq. (3.36) to find the damping coefficient ζ as a function of K_i.

3.2 Rejecting disturbances in undamped oscillators. Let $G(s) = \frac{1}{1+s^2}$ and consider how an input disturbance d modifies the block diagram in Figure 3.1.

a. *Response functions.* Show that the disturbance response of the output $y(s)$ is given by $y(s) = \frac{G}{1+KG}\, d(s)$. Show that the controller output is $u(s) = \frac{-KG}{1+KG}\, d(s)$.

b. *Proportional (P) control.* Find and plot the $y(t)$ and $u(t)$ disturbance impulse responses for the P-control algorithm, $K(s) = K_p$. What is wrong with P-control?

c. *Derivative (D) control.* Find and plot the $y(t)$ and $u(t)$ disturbance impulse responses for D control, $K(s) = K_d s$. Find the critical value of K_d^* where the response changes from oscillatory to damped. Using the final-value theorem, show that the initial control input is $u(t \to 0^+) = -K_d$. What is wrong with D control?

d. *Proportional-derivative (PD) control.* Consider the PD algorithm, $K = K_p + K_d s$. With K_p as a free parameter, find K_d^* for critical damping. Find and plot the y and u disturbance impulse responses for this choice of K_d. Find $u(0^+)$, and discuss the penalty for PD control relative to D control.

e. *Proportional-integral-derivative (PID) control.* For a step disturbance, $d(t) = \theta(t)$, show that PD control (d) results in a steady-state offset. Add integral control to make a full PID controller, $K = K_p + K_i/s + K_d s$. By matching coefficients of $\{1, s, s^2, s^3\}$ of the denominator polynomial (pole placement), show that choosing $K_p = 3a^2 - 1$, $K_i = a^3$, and $K_d = 3a$ leads to closed-loop dynamics with three degenerate poles at $s = -a$. Plot step and impulse responses for y and u, for $a = 2$.

3.3 Feedforward control for an undamped oscillator. In Problem 3.2, we discussed how a PID controller can damp impulse and step disturbances. Now let's make an undamped oscillator reject disturbances *and* track commands. Recall the system, $G(s) = \frac{1}{1+s^2}$, and the PID controller, $K(s) = K_p + K_i/s + K_d s$, with $K_p = 3a^2 - 1$, $K_i = a^3$, and $K_d = 3a$, chosen to make the closed-loop denominator $(1 + KG) \sim (1 + s/a)^3$.

a. Find $r \to \{y, u\}$ transfer functions and analytic time-domain formulas for the command step response, for $K_p = 3a^2 - 1$, $K_i = a^3$, and $K_d = 3a$. Plot for $a = 2$.

b. Since the above PID controller worked well for disturbances, we would like to keep it and add feedforward instead. Modify the command signal $r(s)$ by an element $F(s)$. The simple strategy $F \sim T^{-1}$, for $T = \frac{KG}{1+KG}$, does not lead to proper response. Try $F(s) = \frac{T^{-1}(s)}{(1+s/a)^2}$, which is proper. Find $r \to \{y, u\}$ transfer functions and analytic formulas for the step response. Plot for $a = 2$. Compare the output and control time responses with and without feedforward.

3.4 Stabilizing an unstable oscillator. Consider an unstable oscillator $G(s) = \frac{1}{s^2-1}$, subject to an impulse input disturbance, $d(t) = \delta(t)$.

a. *Proportional (P) control.* For $K(s) = K_p$, show that $K_p > 1$ stabilizes the closed-loop disturbance response $\frac{G}{1+KG}$. Describe the response and its problems. What is the minimum proportional gain if the oscillator has natural frequency ω?

b. *Proportional-Derivative (PD) control.* For $K(s) = K_p + K_d s$, use pole placement (matching denominator polynomials) to show that choosing $K_d = K_d^* \equiv 2\sqrt{K_p - 1}$ and $K_p > 1$ gives the desirable critical-damping response. Choose a to give two poles at $s = -a$. For $a = 1$, plot the response for y and u.

c. *Filtered Proportional-Derivative (PD) control.* Limit the derivative control by filtering: For $K(s) = K_p + \frac{K_d^* s}{1+s/\omega_d}$, find the $d \to \{y, u\}$ transfer functions and also the explicit time-domain differential equations for $y(t)$ and $u(t)$.

d. Matching denominators, choose K_p, K_d, and ω_d to get a disturbance response with three poles at $s = -a$. For $a = 1$, plot the disturbance response $y(t)$ and $u(t)$.

3.5 Integral-control instability. Using Eq. (3.47), we studied an instability in integral control for $G = \frac{1}{(1+s)^2}$ and $K = \frac{K_i}{s}$. Please verify that

a. At the onset of instability, $K_i = 2$, the stable root is at -2.

b. For $K_i = 0$, there is a double pole at -1 and a single pole at 0. Show that one of the -1 poles collides with the 0 pole at $K_i = \frac{4}{27}$. Find the location of that collision (and that of the other pole while you're at it).

c. Verify that the gain margin is $2/K_i$.

d. Plot the phase margin as a function of K_i over the parameter range $0 < K_i < 2$. Evaluate the phase margin for $K_i = 1$ numerically.

3.6 Analog circuits for PID control. Although less flexible than digital controllers, analog circuits can nonetheless be the easiest, cheapest solution. Simple algorithms such as PID control can be implemented with a single operational amplifier costing just a few cents, with bandwidths easily reaching 100 kHz.

a. Show how the circuit in Section 3.2.2 could be used as a proportional controller. If the controller output $u(t)$ is related to the error $e(t) = r(t) - y(t)$ by $u(t) = K_p e(t)$, what is K_p? It is important to get the signs correct. How is $V_{in}(t)$ related to $e(t)$?

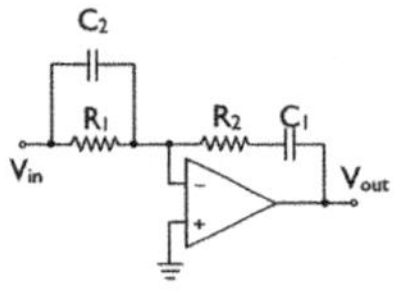

b. The op-amp circuit at left implements a PID controller of the form $u(t) = [K_p + \frac{K_i}{s} + K_d s]\, e(t)$. Find K_p, K_i, and K_d.

c. Draw circuits for PI, PD, and I control.

3.7 Eliminating derivative kick. If the basic PID algorithm is tuned to regulate against disturbances, it will tend to perform poorly when tracking a step command. A simple improvement is to apply the derivative term not to the error $e = (r - y)$ but to $-y$ directly. (The error still enters other terms, as usual.) This modification eliminates the output "kick" when changing the reference signal.

a. Derive this algorithm in the Laplace domain, and explain in simple terms why it works. Show that the corresponding block diagram has a new feedforward term.

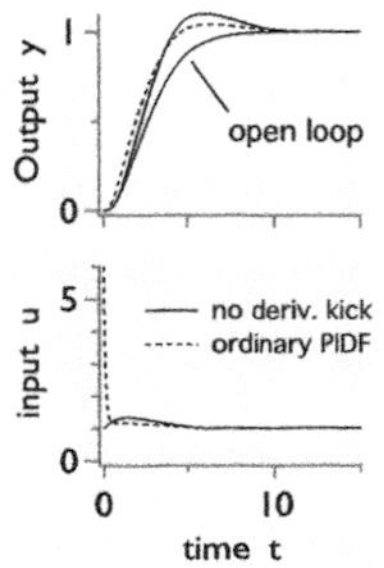

b. The system $G(s) = \frac{1}{(1+s)^3}$ has a "sluggish" response. Explore PIDF control (Eq. 3.41), with and without derivative kick. For $K_p = 1$, $K_i = 0.5$, $K_d = 0.5$, and $\omega_d = 10$, reproduce the plots at left. The dashed line (ordinary PIDF) shows a large spike in the signal $u(t)$ sent to the system. The solid line shows that applying the DF part of the algorithm to y eliminates the spike, with only minor deterioration of the step response.

3.8 Decoupling feedforward from feedback. Naive implementations of feedforward (e.g., Problem 3.3) lead to schemes where the feedforward filter F depends on the controller K. Figure 3.5 presented a scheme for combining feedforward and feedback that decouples the two transfer functions F and K.

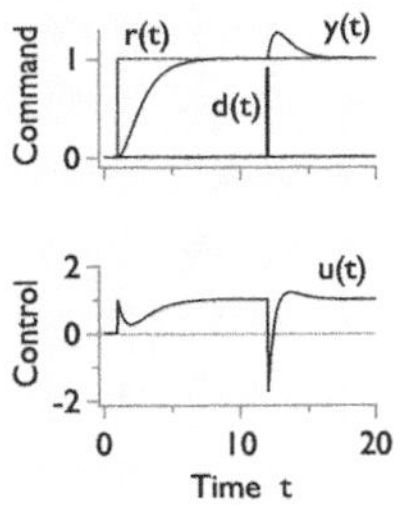

a. Solve the block diagram in Figure 3.5 for y. Show that when the model is perfect ($G_0 = G$) that $y = FG\,r + (\frac{G}{1+KG})\,d$. Thus, F acts only on r, and K only on d.

b. Plot command step and disturbance impulse response for the undamped oscillator $G = \frac{1}{s^2+1}$ (see right). Use a PD controller with $K_\mathrm{p} = 1$ and $K_\mathrm{d} = 2\sqrt{2}$ and a feedforward controller $F = G^{-1}/(1+s)^2$.

3.9 Proportional temperature control of an extended rod. In Section 2.3.2, we showed that the transfer function between a heater at $x = 0$ and a temperature probe at $x = \ell$ was $G(s) = \frac{1}{\sqrt{s}}\,\mathrm{e}^{-\sqrt{s}}$, with lengths scaled by ℓ and times by $\frac{\ell^2}{D}$. Here, D is the thermal diffusion constant. Find the maximum proportional gain K_p^* that can be applied without oscillation. Derive the frequency of oscillations at instability onset, ω^*.

3.10 Instability with a long delay. Consider the unstable system $\dot{x} = ax + u$, with $u(t) = -K_\mathrm{p}x(t-\tau)$. Show that $a < K_\mathrm{p} < \tau^{-1}$ for stability. Thus, when $\tau > a^{-1}$ no value of K_p can make all roots of s have a negative real part. Hint: When s is real, expand the exponential. Consider a possible Hopf bifurcation ($s = \pm\,\mathrm{i}\omega$), too.

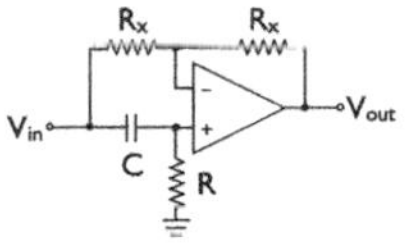

3.11 Op amp allpass. Consider the op-amp circuit shown at right. Derive the transfer function between V_in and V_out and show that it is all pass, with a zero in the RHP. You should find that the transfer function is independent of R_x.

3.12 Inverse response. NMP zeros can lead to *inverse* response: the initial response to a step input is in the opposite direction to the step. We saw such behavior in Example 3.3. Here, we explore this behavior analytically for that example, as well as for a slightly more general transfer function, $G_0(s) = \frac{1-s/z}{(1+s/p)(1+s/p^*)}$.

a. Show that the initial value of the derivative of the response $y(t)$ to a step function $\theta(t)$ is, for a general transfer function $G(s)$, given by $\dot{y}(t \to 0) = \lim_{s\to\infty} s\,G(s)$. Hint: see the discussion of the initial-value theorem in Appendix A.4.3 online.

b. Then show that the transfer function G_NMP in Ex. 3.3 has an inverse response.

c. Show that G_0 has an inverse response when there is an NMP zero.

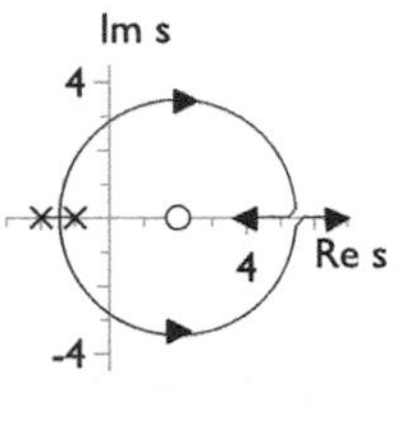

3.13 Response of a nonminimum phase (NMP) system. In Section 3.6.2, we analyzed the system $G_\mathrm{NMP}(s) = \frac{1}{1+s}\left(\frac{1-s/2}{1+s/2}\right)$, which has a zero in the right-hand plane (RHP). Reproduce the pole-zero plot at right and describe its main features analytically.

3.14 Balancing a stick by moving your hand. In Problem 2.1, we showed, neglecting any mass associated with the hand or arm, that $\ddot{x} + \frac{1}{2}\left(\ddot{\theta}\cos\theta - \dot{\theta}^2\sin\theta\right) = u$ and $\ddot{\theta} + \sin\theta + \frac{3}{2}\ddot{x}\cos\theta = 0$, where θ increases counterclockwise, from the bottom, and x is the horizontal displacement of the stick bottom, relative to a reference position.

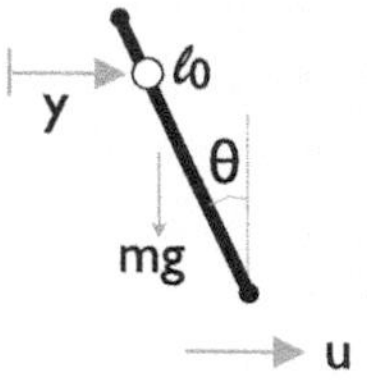

a. Linearize the equations of motion about the vertical equilibrium $\theta = \pi$.
b. Show that the transfer function from u to the *fixation point* $y = x + \ell_0 \sin\theta$ is $G(s) = [(1 - \frac{3}{2}\ell_0)s^2 - 1]/[s^2(\frac{1}{4}s^2 - 1)]$.
c. From Problems 3.16 and 3.17, show that one cannot balance a stick if $\ell_0 \leq \frac{1}{2}$.

3.15 Balancing a stick, with delay in applying feedback.

a. For small deviations of a stick from the vertical, show that the equation of motion with delayed PD feedback is $\ddot{\theta}(t) - (6g/\ell)\theta(t) = -K_{\mathrm{p}}\theta(t-\tau) - K_{\mathrm{d}}\dot{\theta}(t-\tau)$. As in Problem 3.14, neglect the mass of the hand in your calculation.
b. Inserting $\theta(t) \sim \mathrm{e}^{st}$, show that no choice of K_{p} and K_{d} stabilizes the stick for $\tau > \sqrt{\ell/(3g)}$. Thus, there is instability for $\ell < 3g\tau^2$. Hint: Write $s = \mathrm{i}\omega$; separate into real and complex equations; expand $\cos\omega\tau$; and look for ω roots in $0 < \omega\tau < \frac{\pi}{2}$. A careful argument about the roots is subtle (Stepan, 2009).
c. Whatever the controller, disturbances grow uncorrected over the delay time τ. Use this idea to argue that $\ell \lesssim g\tau^2$ implies instability. Experiments by Milton et al. (2016) suggest that, in humans, $\tau \approx 0.23$ s, while $\ell_{\mathrm{min}} \approx 0.32$ m. Show that these observations imply that uncontrolled disturbances grow by a factor ≈ 20. Humans thus seem to use memory and an internal model to predict motion.

3.16 Control of NMP systems, 1. Nonminimum phase systems restrict the possible feedback bandwidths. Consider an RHP zero and then an RHP pole (both real). Apply the Bode gain-phase relation to the minimum-phase part of the system. Decompose the loop transfer function $L(s) = G(s)\,K(s) = G_{\mathrm{mp}}(s)\,G_{\mathrm{ap}}(s)\,K(s)$, where G_{mp} is minimum phase, G_{ap} is all pass, and $K(s)$ is the controller. Let n be the slope of the gain curve at the crossover frequency ω^*, defined by $|L(\mathrm{i}\omega^*)| = 1$.

a. *Simple zero*. For $G_{\mathrm{ap}} = \frac{z-s}{z+s}$, show that we must choose $\omega^* < z \tan\frac{\varphi}{2}$, where z is the position of the zero and where the phase lag $\varphi \equiv \pi - \varphi_{\mathrm{m}} + n\frac{\pi}{2}$, with φ_{m} being the desired phase margin. For $\varphi_{\mathrm{m}} = 90° = \pi/2$, the bandwidth $\omega^* < z$.
b. *Simple pole*. For $G_{\mathrm{ap}} = \frac{s+p}{s-p}$, show that the *minimum* bandwidth to stabilize is $\omega^* > p/\tan\frac{\varphi}{2}$. For $\varphi_{\mathrm{m}} = 90°$, the bandwidth $\omega^* > p$. Intuitively, to stabilize an unstable system, the feedback must correct a perturbation faster than it grows.

3.17 Control of NMP systems, 2. If pole and zero are both unstable, you are caught between a rock and a hard place: the pole imposes minimum bandwidth requirements while the zero imposes maximum requirements. The sensitivity function $S \equiv \frac{1}{1+L}$ then cannot be small at all frequencies. For $L(s) = K(\frac{z-s}{s-p})$ and $p, z > 0$, show that

a. the system is stable if $\frac{p}{z} < K < 1$ or $1 < K < \frac{p}{z}$;
b. the largest stability margin is $s_{\mathrm{m}} = \frac{|p-z|}{p+z}$, for $K = \frac{2p}{p+z}$ (Hint: Look at the Nyquist plot of the loop transfer function as a function of K.);
c. the maximum magnitude of S equals or exceeds $|S| = \frac{p+z}{|p-z|}$.

Thus, poles near zeros make control difficult. Recall that S gives the sensitivity to disturbances, with $|S| = 1$ being open loop. If $|S(i\omega)| > 1$, then disturbances at that frequency are amplified. An RHP zero and pole guarantees such a frequency. And if they are close, you will do much worse (Åström and Murray, 2008).

3.18 Flexible string transfer function. A string supporting transverse waves $\psi(x,t)$ of unit velocity is driven at one end, $\psi(0,t) = u(t)$ and free at $x = 1$. Show that $G(s) = \frac{\psi(x,s)}{u(s)} = \frac{\cosh s(1-x)}{\cosh s}$ for an observation point $0 \le x \le 1$. Plot the magnitude of frequency response for $x = \{0, \frac{1}{2}, \frac{1}{4}, 1\}$, and discuss its structure.

3.19 Zero cancellation. To see why cancelling a zero is dangerous, consider $G(s) = \frac{s+z}{(s+p_1)(s+p_2)}$. Find a controller $K(s)$ such that $T(s) = \frac{KG}{1+KG} = \frac{1}{(s+p_1)(s+p_2)}$ is the closed loop transfer function. What goes wrong?

3.20 Synthesizing a PID controller. Example 3.5 used pole placement to synthesize a controller $K(s) = \frac{18+26s}{9+s}$ for the undamped oscillator $G(s) = \frac{1}{s^2+1}$. Repeat the controller synthesis using a PID form, $K(s) = \frac{K_i + K_p s + K_d s^2}{s}$.

a. Find K_p, K_i, and K_d.

b. Plot the input disturbance response and controller input for the two controllers.

3.21 Rejecting an output disturbance. For $G = \frac{1}{1+s}$ and $d(t)$ a sinusoid of frequency ω:

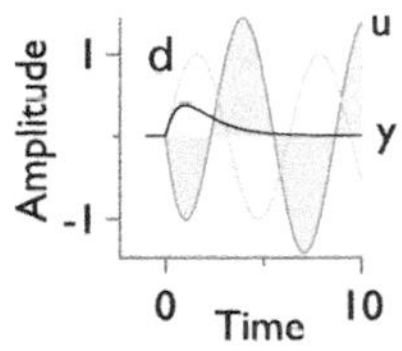

a. Design a controller $K(s)$ to reject $d(t)$. Try choosing the controller to make the output $y(s) = \frac{N_d(s)}{(s+\omega)^2}$, where $d(s) = \frac{N_d(s)}{D_d(s)}$. Why is this a "nice" form for $y(s)$?

b. Calculate the time response of the output, $y(t)$, to a sinusoidal input, $d(t)$, of the form described above. Plot $y(t)$, $d(t)$, and $u(t)$, as shown at right for $\omega = 1$.

c. Investigate the output when $d(t) = \sin \omega_d t$ has the "wrong" frequency. For $\omega_d = \omega(1+\epsilon)$, show that $y(t)$ converges to $\epsilon \cos \omega t + O(\epsilon^2)$.

3.22 Autotuning a PI controller.

a. Argue that relay feedback, Eq. (3.80), leads to output oscillations of period T_c, with critical gain $K_c = 4u_r/(\pi a)$. Here, u_r is the amplitude of the relay feedback and a the amplitude of the output oscillations. Hint: At instability, the loop gain $L(i\omega_c) = -1$. Why does only the first harmonic of the square-wave matter?

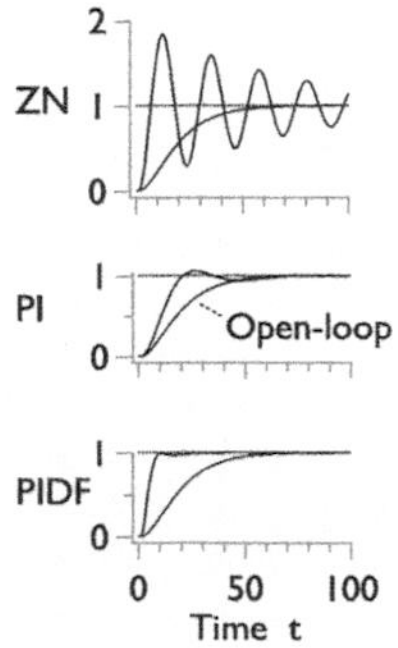

b. Consider a second-order system with delay, $G(s) = \frac{e^{-s}}{(10s+1)^2}$. Simulate the output responses $y(t)$ for both proportional control near the instability threshold and relay feedback. Show that both lead to $K_c \approx 20$ and $T_c \approx 14$.

c. Implement the PI Ziegler–Nichols rule and evaluate the closed-loop response for step command and for step input disturbance. Find PI parameters that are better than ZN. Then explore the full PIDF architecture. Reproduce the step commands at right and their associated $u(t)$. Plot the response to input step disturbances, too.

3.23 Analysis of a two-sensor system. For the split-PI example of Section 3.8.1:

a. Derive the closed-loop transfer functions of Eqs. (3.81), (3.82), and (3.83).

b. Reproduce the step responses for all three cases.
c. For $K_p = 5$, why is there no instability in Case 1 for any K_i? For Cases 2 and 3, find the instability threshold K_i^* and oscillation frequency ω^* at onset, again fixing $K_p = 5$. Do this part analytically or numerically.

3.24 Fixing the shower. Slightly altering the "shower" transfer function $\mathbf{G}_{sh}$ given in Eq. (3.101) turns an impossible control problem into a straightforward one. Consider $\mathbf{G}'_{sh} = \begin{pmatrix} 1/(1+s) & 1/(2+s) \\ 1/(2+s) & 1/(1+s) \end{pmatrix}$, which differs from the expression in Eq. (3.101) in that the DC cross gains equal $\frac{1}{2}$ rather than 1. The control goal is a good step response (e.g., to step rapidly the outputs from 0 to $y_1 = 1$ and $y_2 = \frac{1}{3}$, using inputs $(u_1, u_2) \in (-10, 10)$).

a. Make a singular value plot. Show that the condition number still diverges at $s \to \infty$ but not at $s \to 0$. Why is a high-frequency divergence allowable but not a low-frequency one? Can your computer program compute a step response in the time domain? Find a fix that leads asymptotically to the correct DC outputs.
b. Try to improve the controller by canceling a pole or zero in the inverse. Remember "tweaks" such as lag and lead.

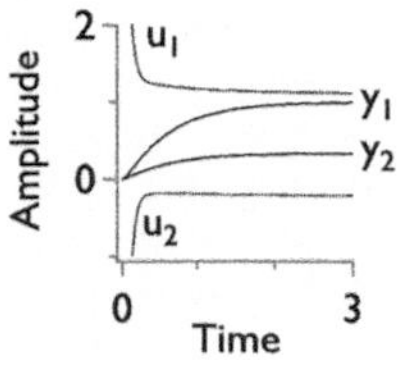

At left are step responses for the modified-shower problem. Reference signals r_1 and r_2 are given steps of $(1, \frac{1}{3})$. Asymptotically, the outputs $y_1 \to r_1$ and $y_2 \to r_2$.

Time-Domain Control 4

In this chapter, we focus on linear systems in the time domain, an approach that has advantages over the frequency-domain methods developed in the previous chapter. First, time-domain methods are more general. As we saw in Section 2.2, we can treat in the time domain both nonlinear systems and linear systems that are not time invariant. In this chapter, we continue to focus on linear, time-invariant dynamics. Beginning in Chapter 7, we will take advantage of the full power of state-space methods.

Second, time-domain dynamics introduce the useful distinction between observable quantities and the internal state of the system. By contrast, the frequency-domain formalism of Chapter 3 deals explicitly with inputs and outputs but is vague about the structure "within the black box." By formalizing more carefully the notion of a system, we will acquire tools that will tell us when control is possible and when it is destined to fail. We will also see that handling multiple inputs and outputs (MIMO systems) is very natural in the time domain, and the linear-algebra formalism is well adapted to computer calculations. Later, in Chapter 7, we will develop the topic of optimal control, a systematic way to design controllers that is best formulated using the state-space formalism of the time domain. With the state-space formalism, we will see that we can design controllers that take complete advantage of our knowledge of the system dynamics. By contrast, frequency-domain techniques such as loop shaping lead to heuristic design principles (e.g., choosing loop gains at high and low frequencies) but not a systematic way of designing a controller.

If there are advantages to working in the time domain and using the state-vector formalism, there are also drawbacks: Methods based on linear algebra and state space are less intuitive and more abstract. While well-adapted to computer calculations, they *require* such calculations for all but the simplest examples. Hopefully, the intuition developed in Chapter 3 for frequency-domain methods will make the time-domain formalism easier to digest. A more serious drawback is that time-domain methods tend to be more sensitive to modeling errors in the dynamics. In Chapter 9 (Robust Control), we address this problem.

We begin, in Section 4.1, by asking whether control of a given system is possible. More precisely, for a specified system dynamics and specified set of inputs, can we drive the system from one state to another in finite time? Conversely, given a system and its observations, can we infer the underlying state? We then go on to decompose the control problem into two tasks: The first develops control algorithms assuming that the full state vector is available to the controller (Section 4.2). The second provides a way to convert a time series of past observations into estimates of the full state vector (Section 4.3).

4.1 Controllability and Observability

As in Chapter 2, let the n-dimensional state vector $\boldsymbol{x}$ evolve in time according to

$$\dot{\boldsymbol{x}}(t) = \boldsymbol{f}(\boldsymbol{x}, u) , \qquad y = h(\boldsymbol{x}) . \tag{4.1}$$

To build intuition, we restrict ourselves to the single-input, single-output (SISO) case with no feedthrough (h independent of u). If we choose coordinates so that $\boldsymbol{f}(\boldsymbol{0}, 0) = \boldsymbol{0}$, then linearizing the functions $\boldsymbol{f}$ and h about the fixed point gives linear state-space equations:

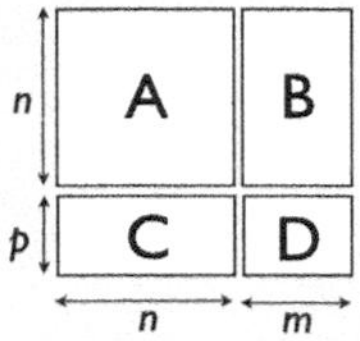

$$\dot{\boldsymbol{x}} = \boldsymbol{A}\boldsymbol{x} + \boldsymbol{B}u \qquad y = \boldsymbol{C}\boldsymbol{x} , \qquad \text{with} \qquad \boldsymbol{A} = \partial_{\boldsymbol{x}}\boldsymbol{f} , \quad \boldsymbol{B} = \partial_u \boldsymbol{f} , \quad \boldsymbol{C} = \partial_{\boldsymbol{x}} h . \tag{4.2}$$

All derivatives are evaluated at ($\boldsymbol{x} = \boldsymbol{0}, u = 0$). For the SISO case, $\boldsymbol{A}$ is an $n \times n$ matrix, $\boldsymbol{B}$ is an n-dimensional column vector ($m = 1$), and $\boldsymbol{C}$ is an n-dimensional row vector ($p = 1$), and we assume no feedthrough ($D = 0$). The dynamics are illustrated in Figure 4.1.

Recall that Eq. (4.2) can be solved explicitly for an arbitrary input $u(t)$:

$$\boldsymbol{x}(t) = \mathrm{e}^{\boldsymbol{A}t}\, \boldsymbol{x}(0) + \int_0^t \mathrm{d}t'\, \mathrm{e}^{\boldsymbol{A}(t-t')}\, \boldsymbol{B}\, u(t') . \tag{4.3}$$

By Laplace transforming Eq. (4.2) and neglecting the initial condition, we can go from the time domain to the frequency domain, with the transfer function $G(s)$ given by

$$G(s) = \boldsymbol{C}(s\mathbb{I} - \boldsymbol{A})^{-1}\, \boldsymbol{B} , \tag{4.4}$$

with $\mathbb{I}$ the $n \times n$ identity matrix. We can also start from a transfer function and derive a state-space representation such as Eq. (2.59). Recall from Section 2.4.3 that changing coordinates, $\boldsymbol{x}' = \boldsymbol{T}\boldsymbol{x}$, with $\boldsymbol{T}$ an $n \times n$ invertible matrix, leads to the same transfer function and dynamics. There are an infinite number of equivalent state-space representations.

Implementing feedback in the state-vector formalism amounts to choosing the control input $u(t)$. The "obvious" choice $u = u[y(t)]$ is known as *output feedback*. Yet a less-obvious, two-step process has advantages: First, we assume that from the present

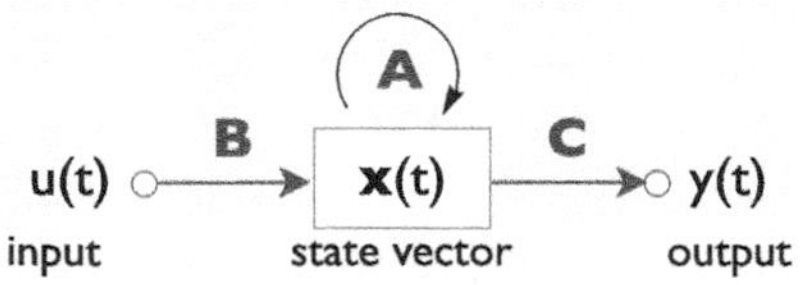

Fig. 4.1 Schematic block diagrams illustrating how $\boldsymbol{B}$ and $\boldsymbol{C}$ couple the input and output of a SISO system to the intrinsic dynamics given by $\boldsymbol{A}$.

and past values of the observations $y(t)$, we can somehow infer an estimate $\hat{\boldsymbol{x}}(t)$ of the state vector $\boldsymbol{x}(t)$. Then we construct a feedback scheme base on $u[\hat{\boldsymbol{x}}(t)]$. This two-step process is termed *state feedback*. Its conceptual advantages will compensate the extra complexity.

One difference between the state-space and frequency-domain representations is that, in state space, we work with an n-dimensional state vector, whereas in the frequency domain, the input and output, in a SISO system, are scalars. Now, if the ability to generate a single input $u(t)$ lets you access, or "control" all n elements of the state vector, then the system is *controllable*. Likewise, if the time series history from a single output $y(t)$ gives information about the n-dimensional vector $\boldsymbol{x}$, the system is *observable*. A failure of either condition implies that the input and output connections, illustrated in Figure 4.1, are incomplete and limit the ability to control the system, as we now will discuss.

4.1.1 Controllability

We begin by defining the notion of controllability of a dynamical system:

Controllable system: For every initial state $\boldsymbol{x}_0$ and final state $\boldsymbol{x}_\tau$, there exists a time τ and control $u(t)$ such that $\boldsymbol{x}(0) = \boldsymbol{x}_0$ and $\boldsymbol{x}(\tau) = \boldsymbol{x}_\tau$.[1]

The notion of controllability depends only on the (feedforward) input $u(t)$ and state $\boldsymbol{x}(t)$ and is independent of the output $y(t)$ and its coupling $\boldsymbol{C}$. Note that a single $u(t)$ must control *all* n states. Here are some simple examples.

Example 4.1 Consider the first-order system $\dot{x} = -\lambda x + u(t)$. Notice that we can "invert" the dynamics: A *desired trajectory* $x_{\mathrm{d}}(t)$ is produced by $u(t) = \dot{x}_{\mathrm{d}}(t) + \lambda x_{\mathrm{d}}(t)$. To show controllability, we design a trajectory that satisfies $x_{\mathrm{d}}(0) = x_0$ and $x_{\mathrm{d}}(\tau) = x_\tau$. The ability to control a trajectory in state space is a much stronger claim than controllability. For example, the family of trajectories $x_{\mathrm{d}} = t^n$ all satisfy $x(0) = 0$ and $x(1) = 1$ and can be generated by $u(t) = nt^{n-1} + \lambda t^n$. See right, for $\lambda = \tau = 1$ and $n = 1, 2$ (solid lines are $x_{\mathrm{d}} = t, t^2$, dashed are the corresponding inputs u). The ability to control a state trajectory is the exception, not the rule, and many systems are controllable in the technical sense defined here, even when arbitrary trajectories are not possible (Problem 4.2).

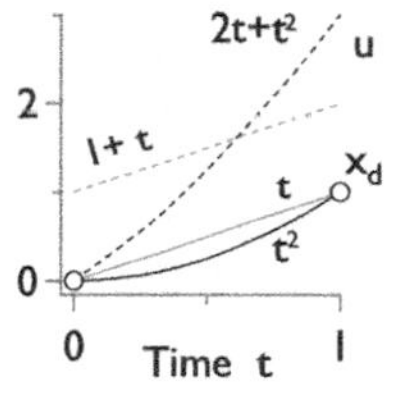

Example 4.2

$$\begin{pmatrix} \dot{x}_1 \\ \dot{x}_2 \end{pmatrix} = \begin{pmatrix} -\lambda_1 & 0 \\ 0 & -\lambda_2 \end{pmatrix} \begin{pmatrix} x_1 \\ x_2 \end{pmatrix} + \begin{pmatrix} 1 \\ 0 \end{pmatrix} u(t), \qquad \text{or} \qquad \begin{aligned} \dot{x}_1 &= -\lambda_1 x_1 + u, \\ \dot{x}_2 &= -\lambda_2 x_2 \,, \end{aligned} \tag{4.5}$$

Since x_2 is completely unaffected by u, the system is not controllable.

[1] The time required for control, τ, in general depends on the initial and final states, $\boldsymbol{x}_0$ and $\boldsymbol{x}_\tau$. For *linear* systems, it is always possible to find a control for any τ, for any initial and final states. See Problem 7.9.

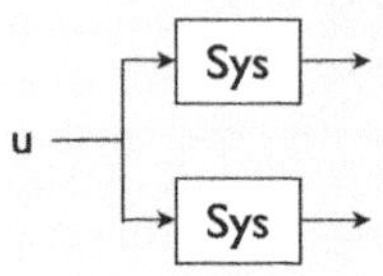

Another category of systems that are not controllable consists of systems where one input drives identical subsystems with identical coupling, as illustrated at left.

Example 4.3

$$\begin{pmatrix}\dot{x}_1\\ \dot{x}_2\end{pmatrix}=\begin{pmatrix}-\lambda & 0\\ 0 & -\lambda\end{pmatrix}\begin{pmatrix}x_1\\ x_2\end{pmatrix}+\begin{pmatrix}1\\ 1\end{pmatrix}u(t)\,, \qquad \text{or} \qquad \begin{aligned}\dot{x}_1 &= -\lambda x_1 + u\,,\\ \dot{x}_2 &= -\lambda x_2 + u\,,\end{aligned} \tag{4.6}$$

Since the two *identical* systems are driven by two *identical* inputs, we cannot independently control the two states. The system is not controllable.

Example 4.4 Determining controllability in Examples 4.2 and 4.3 was fairly obvious, but just looking at the equations is usually not enough. For example, consider

$$\begin{pmatrix}\dot{x}_1\\ \dot{x}_2\end{pmatrix}=\begin{pmatrix}-\lambda_1 & 0\\ 0 & -\lambda_2\end{pmatrix}\begin{pmatrix}x_1\\ x_2\end{pmatrix}+\begin{pmatrix}1\\ 1\end{pmatrix}u(t)\,, \qquad \text{or} \qquad \begin{aligned}\dot{x}_1 &= -\lambda_1 x_1 + u\,,\\ \dot{x}_2 &= -\lambda_2 x_2 + u\,,\end{aligned} \tag{4.7}$$

The only change relative to Example 4.3 is that the λ's are now different. We still feed the same input to each independent system. In Problem 4.1, you will find an input $u(t)$ that demonstrates explicitly that this system *is* controllable for $\lambda_1 \neq \lambda_2$.

A Test for Controllability

We have argued heuristically that very similar systems may or may not be controllable. Here, we develop a simple quantitative test for controllability of a linear system that depends on only the dynamics matrix $\boldsymbol{A}$ and input coupling $\boldsymbol{B}$.

For the SISO case, $u(t)$ is a scalar and $\boldsymbol{B}$ an n-component vector. Let the initial condition be $\boldsymbol{x}(0) = \boldsymbol{0}$. Since controllability means that the set of inputs can collectively excite all dynamical modes, we consider a delta-function input, whose white-noise spectrum will excite all accessible modes. From Eq. (4.3), the impulse response function, $\boldsymbol{x}_{\text{imp}}(t) = \mathrm{e}^{\boldsymbol{A}t}\,\boldsymbol{B}$.

In coordinates where $\boldsymbol{A}$ is diagonal, each normal mode is a basis vector of $\boldsymbol{x}$, and controllability loosely implies that a delta-function input leads to nonzero values of all components. If $\boldsymbol{A}$ is not diagonal, we can look at the *Gramian* matrix created by $\boldsymbol{x}_{\text{imp}}\,\boldsymbol{x}_{\text{imp}}^{\mathsf{T}}$.[2] Controllability implies that $\boldsymbol{x}_{\text{imp}}$ span the full state space and thus that the Gramian matrix have nonzero determinant and be invertible. We need to be careful, however, as a momentary vanishing of a component does not imply lack of controllability – think about an oscillating system. But vanishing over a finite time interval would imply loss of controllability. Thus we are led to consider the *controllability Gramian* matrix,

$$\boldsymbol{P}(\tau) \equiv \int_0^{\tau} \mathrm{d}t\, \boldsymbol{x}_{\text{imp}}(t)\,\boldsymbol{x}_{\text{imp}}^{\mathsf{T}}(t) = \int_0^{\tau} \mathrm{d}t\, \mathrm{e}^{\boldsymbol{A}t}\,\boldsymbol{B}\,\boldsymbol{B}^{\mathsf{T}}\,\mathrm{e}^{\boldsymbol{A}^{\mathsf{T}}t}\,. \tag{4.8}$$

[2] In bra-ket notation, the Gramian is simply $|\boldsymbol{x}_{\text{imp}}\rangle\langle\boldsymbol{x}_{\text{imp}}|$. Dirac notation is *not* standard in control texts.

Controllability over a time interval $[0, \tau]$ is equivalent to the invertibility of $\boldsymbol{P}(\tau)$, and controllability for all time is equivalent to asking that the symmetric, positive, semidefinite $\boldsymbol{P}_\infty \equiv \boldsymbol{P}(\tau = \infty)$ be invertible. Of course, $\boldsymbol{A}$ must be stable, if we are to integrate to $t = \infty$.

Although informative, Gramian matrices can be hard to compute. One trick is to show that $\boldsymbol{P}_\infty$ obeys the *Lyapunov equation*, $\boldsymbol{A}\boldsymbol{P}_\infty + \boldsymbol{P}_\infty \boldsymbol{A}^\mathrm{T} = -\boldsymbol{B}\boldsymbol{B}^\mathrm{T}$, which is often easier to solve than integrating Eq. (4.8). See Problem 2.14.

Gramian matrices have other applications, as well. In Chapter 7 (Problem 7.9), we will see that the control $u(t) = \boldsymbol{B}^\mathrm{T}\, \mathrm{e}^{\boldsymbol{A}^\mathrm{T}(\tau - t)}\, \boldsymbol{P}^{-1}(\tau)\, \boldsymbol{x}_\tau$ minimizes the *control effort* $\mathcal{E} = \int_0^\tau \mathrm{d}t\, u^2(t)$ to move a system between $\boldsymbol{x}(0) = \boldsymbol{0}$ and $\boldsymbol{x}(\tau) = \boldsymbol{x}_\tau$. We see a direct link between the invertibility of $\boldsymbol{P}(\tau)$ and the explicit expression for the required control $u(t)$. In Chapter 6, we will see that Gramians can also be used to obtain reduced dynamical systems of fewer states that accurately approximate the motion of the full dynamics.

There is a test for controllability that is even easier than solving the Lyapunov equation. From the Cayley–Hamilton theorem (see Appendix A.1.6 online), we recall that a matrix $\boldsymbol{A}$ obeys its own characteristic equation; as a consequence, $\boldsymbol{A}^n$ can be written as a linear combination of lower powers of $\boldsymbol{A}$. Likewise, we can express $\mathrm{e}^{\boldsymbol{A}t}$ as an n-element power series, giving

$$\boldsymbol{x}_{\mathrm{imp}}(t) = \mathrm{e}^{\boldsymbol{A}t}\, \boldsymbol{B} = \sum_{j=0}^{n-1} \alpha_j(t)\, \boldsymbol{A}^j\, \boldsymbol{B}\,. \tag{4.9}$$

The $\alpha_j(t)$ are the expansion coefficients at time t and are *not*, in general, $\sim t^j$.

Since controllability requires the impulse response function to span the full n-dimensional state space, the system will be controllable for any τ if the set of n vectors $\boldsymbol{B}, \boldsymbol{AB}, \boldsymbol{A}^2\boldsymbol{B}, \ldots \boldsymbol{A}^{n-1}\boldsymbol{B}$ are linearly independent. To check the linear independence of n column vectors, we form $\boldsymbol{W}_\mathrm{c}$, the $n \times n$ *controllability matrix*:[3]

$$\boldsymbol{W}_\mathrm{c} \equiv \begin{pmatrix} \boldsymbol{B} & \boldsymbol{AB} & \boldsymbol{A}^2\boldsymbol{B} & \cdots & \boldsymbol{A}^{n-1}\boldsymbol{B} \end{pmatrix}. \tag{4.10}$$

The invertibility of $\boldsymbol{W}_\mathrm{c}$ (determinant $\neq 0$) is equivalent to the controllability of the system with dynamics matrix $\boldsymbol{A}$ and input coupling $\boldsymbol{B}$. We can recap our previous examples:

Example 4.2:

$$\boldsymbol{A} = \begin{pmatrix} -\lambda_1 & 0 \\ 0 & -\lambda_2 \end{pmatrix} \; \boldsymbol{B} = \begin{pmatrix} 1 \\ 0 \end{pmatrix} \implies \boldsymbol{AB} = \begin{pmatrix} -\lambda_1 \\ 0 \end{pmatrix} \implies \boldsymbol{W}_\mathrm{c} = \begin{pmatrix} 1 & -\lambda_1 \\ 0 & 0 \end{pmatrix}, \tag{4.11}$$

and $\det \boldsymbol{W}_\mathrm{c} = 0$.

Examples 4.3 and 4.4:

$$\boldsymbol{A} = \begin{pmatrix} -\lambda_1 & 0 \\ 0 & -\lambda_2 \end{pmatrix} \; \boldsymbol{B} = \begin{pmatrix} 1 \\ 1 \end{pmatrix} \implies \boldsymbol{AB} = \begin{pmatrix} -\lambda_1 \\ -\lambda_2 \end{pmatrix} \implies \boldsymbol{W}_\mathrm{c} = \begin{pmatrix} 1 & -\lambda_1 \\ 1 & -\lambda_2 \end{pmatrix}, \tag{4.12}$$

[3] The span of the n columns of $\boldsymbol{W}_\mathrm{c}$ forms a vector space known as the *Krylov subspace* $\mathcal{K}_n(\boldsymbol{A}, \boldsymbol{B})$. Thus, controllability is equivalent to $\mathrm{Dim}[\mathcal{K}_n(\boldsymbol{A}, \boldsymbol{B})] = n$.

and det $W_c = \lambda_1 - \lambda_2$. The system is controllable, except when $\lambda_1 = \lambda_2$ (identical systems).

Multiple Inputs: When B Is a Matrix

If there are m inputs, B is an $n \times m$ matrix. Then each element $A^j B$ is also a $n \times m$ matrix and W_c is a rectangular $n \times (mn)$ matrix. If W_c is full rank (i.e., rank $= n$, see Appendix A.1.2 online), then $W_c v = w$ has a solution v, and the system is controllable. Thus, the condition for controllability for a linear system, whether SISO or MIMO, is given by the *Kalman rank condition*: that rank $W_c = n$. To summarize,

> A linear system $\{A, B\}$ is *controllable* if and only if the matrix W_c has full rank.

While the Kalman rank condition is straightforward for simple systems, it can be hard to apply to high-dimensional systems with many inputs. See Chapter 14.

Controllability in Practice

Having tried to convince you that the notion of controllability is important, let me point out ways in which it is not as important as you might hope.

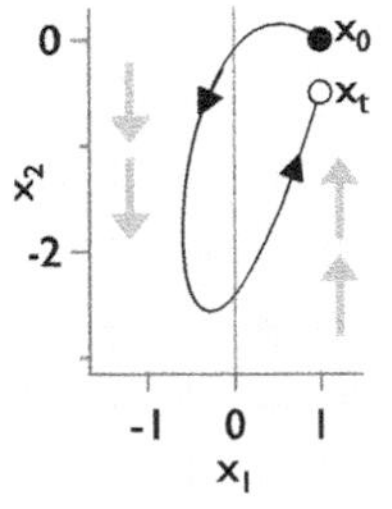

1. Too large a control input may be required. Any transition in finite time between any two states should be possible, no matter how far apart they are. But the control signal that is required will eventually exceed the possible limits of the input signal.
2. Control trajectories are often *nonlocal*: moving the system an arbitrarily small distance "against the flow" can require trajectories of finite length. See left and Problem 4.3. The nonlocality of control trajectories can alter the controllability of nonlinear systems (Chapter 11), because even though x_0 and x_τ may be close enough that the same local linearization governs both states, the intermediate trajectory may venture into a part of phase space where the linearization is completely different. In such a case, the controllability of a nonlinear system can differ from that of the linear approximation that holds near the start and end points.

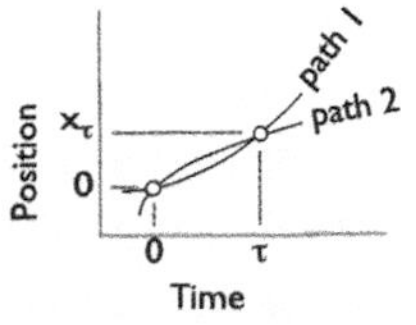

3. The notion of controllability does *not* imply that you can make a system follow a prescribed path from x_0 to x_τ. All that we have argued is that we can somehow get the system from x_0 to x_τ. In Problem 4.2, we consider the converse problem of finding a set of controls $u(t)$ given a prescribed path $x(t)$ for the state vector. As illustrated at left, many paths $x(t)$, each with its own input $u(t)$, can connect two points at $t = 0$ and τ.
4. Similarly, even though we can achieve a start x_τ at time τ, we may not be able to hold that state indefinitely. Reaching a state and holding it are two different tasks.
5. In a linear system, inputs have infinite range. If inputs have finite range, as they always do, the system is nonlinear. We will explore the consequences in more detail in Chapter 11, but clearly, the set of controllable states will be reduced. For example,

a space ship obeying Newtonian dynamics can go from point A to point B in a time τ by accelerating for $\frac{1}{2}\tau$ and decelerating for the second half. If the space ship cannot accelerate enough, the state will not be reachable in time τ.

6. Finally, you may not care about all the states. If so, then it is immaterial whether the whole system is controllable, as long as the subspace you care about is controllable.

In short, controllability – of the subspace you care about – is necessary but not sufficient to achieve typical performance goals.

4.1.2 Observability

Controllability asks whether an input u "couples to" the entire state vector $\boldsymbol{x}$. *Observability* asks whether the output y couples to the state vector. The two notions are closely related.

Consider the case of a single output, $y = \boldsymbol{Cx}$, and no input. Can the n-element state vector $\boldsymbol{x}$ be inferred from its observations $y(t)$? More precisely, a system is observable if one can determine the state $\boldsymbol{x}(\tau)$ given measurements $y(t)$ over the interval $(0, \tau)$.

For linear, time-invariant dynamics, $\dot{\boldsymbol{x}} = \boldsymbol{Ax}$ and $y = \boldsymbol{Cx}$. Recall that $\boldsymbol{A}$ is an $n \times n$ matrix and $\boldsymbol{C}$ is an $1 \times n$ row vector. At first glance, it would seem impossible to infer $\boldsymbol{x}$ from y. After all, $\boldsymbol{x}$ has n quantities, while y just has 1. However, we observe y over an interval of time, which gives information on the time derivatives of $y(t)$. For example, from $y = \boldsymbol{Cx}$, we have $\dot{y} = \boldsymbol{C}\dot{\boldsymbol{x}} = \boldsymbol{CAx}$. Differentiating again gives $\ddot{y} = \boldsymbol{CA}\dot{\boldsymbol{x}} = \boldsymbol{CA}^2\boldsymbol{x}$. We can stop at the $n-1$ derivative of $y(t)$ because the Cayley–Hamilton theorem tells us we can always express a higher power of $\boldsymbol{A}$ in terms of lower powers. Putting this all together,

$$\begin{pmatrix} y \\ \dot{y} \\ \ddot{y} \\ \vdots \\ y^{(n-1)} \end{pmatrix} = \begin{pmatrix} \boldsymbol{C} \\ \boldsymbol{CA} \\ \boldsymbol{CA}^2 \\ \vdots \\ \boldsymbol{CA}^{n-1} \end{pmatrix} \boldsymbol{x} \equiv \boldsymbol{W}_{\mathrm{o}}\, \boldsymbol{x}\,, \tag{4.13}$$

where the $n \times n$ *observability matrix* $\boldsymbol{W}_{\mathrm{o}}$ has n row vectors formed by $\boldsymbol{C}$, $\boldsymbol{CA}$, and so on. Formally, $\boldsymbol{Y} = \boldsymbol{W}_{\mathrm{o}}\boldsymbol{x}$, with $\boldsymbol{Y}$ the vector of y and its $n-1$ derivatives. Thus, if we know $\boldsymbol{Y}(t)$, then inverting $\boldsymbol{W}_{\mathrm{o}}$ gives us $\boldsymbol{x}(t)$. Generalizing to p outputs, the matrix $\boldsymbol{W}_{\mathrm{o}}$ becomes rectangular ($pn \times n$), and the condition for observability – the ability to solve $\boldsymbol{Y} = \boldsymbol{W}_{\mathrm{o}}\boldsymbol{x}$ for $\boldsymbol{x}$ – becomes the same as we saw for controllability: $\boldsymbol{W}_{\mathrm{o}}$ must have rank n (full rank).

Do we know $\boldsymbol{Y}(t)$? Well, we can always estimate a derivative from multiple observations using finite differences. For example,

$$\frac{\mathrm{d}y}{\mathrm{d}t} \approx \frac{y(t) - y(t - \Delta t)}{\Delta t}\,, \tag{4.14}$$

which can be applied repeatedly to get an estimate of any order of derivative we want. Since Δt is arbitrary, we need to know $y(t)$ in a small, finite interval about t. This gives a poor estimate, as finite differencing amplifies noise, a situation that becomes worse with each higher derivative. Still, the question here is one of principle, not practice. In Section 4.3, we will discuss a better way to recover the state $\boldsymbol{x}$ from noisy observations $y(t)$.

Finally, adding a known input $u(t)$ does not alter observability, as $u(t)$ merely changes the state $\boldsymbol{x}(t)$ relative to its natural evolution without the input. Since we have proven that we can observe any state $\boldsymbol{x}$, it does not matter whether an input was used to produce it or an initial condition. The issue is always how $\boldsymbol{x}$ couples to the output y. To summarize,

A linear system $\{\boldsymbol{A}, \boldsymbol{C}\}$ is *observable* if the matrix $\boldsymbol{W}_o$ has full rank.

Example 4.5 (Prototype of an unobservable system) Consider the system

$$\begin{pmatrix} \dot{x}_1 \\ \dot{x}_2 \end{pmatrix} = \begin{pmatrix} 1 & 0 \\ 0 & -1 \end{pmatrix} \begin{pmatrix} x_1 \\ x_2 \end{pmatrix}, \qquad y = \begin{pmatrix} 1 & 0 \end{pmatrix} \begin{pmatrix} x_1 \\ x_2 \end{pmatrix}. \tag{4.15}$$

Since knowing y tells nothing about x_2, we do not have an observable system. What if we had only the x_1 equation? This system would be observable, albeit unstable: $\dot{x}_1 = x_1$, $y = x_1$. Clearly, knowing $y(t)$ we know $x_1(t)$. Thus, the *system* defined in Eq. (4.15) is not observable, although the subspace spanned by x_1 is.

Observability can be subtle: Consider a system very similar to Example 4.5.

Example 4.6 (Observability of a linear oscillator) We have

$$\ddot{x} + x = 0 \quad y = x, \tag{4.16}$$

which, in state-space representation is, with $x_1 = x$ and $x_2 = \dot{x}$,

$$\begin{pmatrix} \dot{x}_1 \\ \dot{x}_2 \end{pmatrix} = \begin{pmatrix} 0 & 1 \\ -1 & 0 \end{pmatrix} \begin{pmatrix} x_1 \\ x_2 \end{pmatrix}, \qquad y = \begin{pmatrix} 1 & 0 \end{pmatrix} \begin{pmatrix} x_1 \\ x_2 \end{pmatrix}. \tag{4.17}$$

Then

$$\boldsymbol{C}\boldsymbol{A} = \begin{pmatrix} 1 & 0 \end{pmatrix} \begin{pmatrix} 0 & 1 \\ -1 & 0 \end{pmatrix} = \begin{pmatrix} 0 & 1 \end{pmatrix} \quad \Longrightarrow \quad \boldsymbol{W}_o = \begin{pmatrix} \boldsymbol{C} \\ \boldsymbol{C}\boldsymbol{A} \end{pmatrix} = \begin{pmatrix} 1 & 0 \\ 0 & 1 \end{pmatrix}, \tag{4.18}$$

which is obviously invertible. Thus, if you observe the position variable of an undamped oscillator, you can infer the full state vector (position and velocity).

What if we observe the velocity rather than the position, by choosing $\boldsymbol{C} = (0 \quad 1)$? You can check that the system $\{\boldsymbol{A}, \boldsymbol{C}\}$ is still observable.

How Observability Can Fail

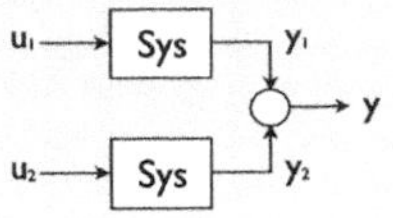

To see how observability might fail, we can consider two identical systems whose output is added (see right). Switching the inputs, u_1 and u_2, leaves the output exactly the same: the combined system is not observable. See also Example 4.7.

Example 4.7 (Identical systems)

$$\begin{pmatrix} \dot{x}_1 \\ \dot{x}_2 \end{pmatrix} = \begin{pmatrix} -\lambda & 0 \\ 0 & -\lambda \end{pmatrix} \begin{pmatrix} x_1 \\ x_2 \end{pmatrix}, \quad y = \begin{pmatrix} 1 & 1 \end{pmatrix} \begin{pmatrix} x_1 \\ x_2 \end{pmatrix}. \implies \boldsymbol{W}_{\mathrm{o}} = \begin{pmatrix} \boldsymbol{C} \\ \boldsymbol{CA} \end{pmatrix} = \begin{pmatrix} 1 & 1 \\ -\lambda & -\lambda \end{pmatrix}. \tag{4.19}$$

The observability matrix $\boldsymbol{W}_{\mathrm{o}}$ is clearly not invertible. If the individual systems were different (e.g., $\lambda_2 \neq \lambda_1$), then $\det \boldsymbol{W}_{\mathrm{o}} \neq 0$, and the system *would* be observable.

Observability Is the Dual of Controllability

You may have noticed a mathematical resemblance between observability and controllability. Recall that $\boldsymbol{W}_{\mathrm{c}} = (\boldsymbol{B}\ \boldsymbol{AB}\ \boldsymbol{A}^2\boldsymbol{B}\ \cdots\ \boldsymbol{A}^{n-1}\boldsymbol{B})$, which is the transpose of $\boldsymbol{W}_{\mathrm{o}}$ if $\boldsymbol{A}^i\boldsymbol{B} \to (\boldsymbol{CA}^i)^{\mathsf{T}}$. If you recall that $\boldsymbol{B}$ is the input coupling and $\boldsymbol{C}$ the output coupling, this *duality* makes sense. To push the analogy further, let us define a *dual dynamical system*:

$$\dot{\boldsymbol{x}} = -\boldsymbol{A}^{\mathsf{T}}\boldsymbol{x} - \boldsymbol{C}^{\mathsf{T}}\boldsymbol{u}\,, \qquad \boldsymbol{y} = \boldsymbol{B}^{\mathsf{T}}\boldsymbol{x}\,, \tag{4.20}$$

In comparison to the original system, we have

$$\boldsymbol{A} \leftrightarrow \boldsymbol{A}^{\mathsf{T}}\,, \quad \boldsymbol{B} \leftrightarrow \boldsymbol{C}^{\mathsf{T}}\,, \quad t \leftrightarrow \tau - t\,. \tag{4.21}$$

To understand the time reversal, note that the $\boldsymbol{u}(t)$ affects a state at *future* times, whereas the output $\boldsymbol{y}(t)$ is determined by the *past* states $\boldsymbol{x}(t)$. As Shannon observed long ago, the duality between forward and time-reversed dynamics is fundamental: "We may have knowledge of the past but cannot control it; we may control the future but have no knowledge of it." Putting it more pithily:

> We can know the past but not control it; we can control the future but do not know it.

Using duality, we can define an *observability Gramian* matrix,

$$\boldsymbol{Q}(\tau) = \int_0^{\tau} \mathrm{d}t\, \mathrm{e}^{\boldsymbol{A}^{\mathsf{T}}t}\, \boldsymbol{C}^{\mathsf{T}}\boldsymbol{C}\, \mathrm{e}^{\boldsymbol{A}t}\,. \tag{4.22}$$

To be observable, $\boldsymbol{Q}(\tau)$ must be invertible. We thus have two tests for observability. They are analogous to the two tests for controllability discussed above.

4.1.3 Putting It All Together: The Structure of State Space

Kalman Decomposition

Since controllability and observability are independent properties, it is clear that there are four cases. A system may be

controllable and observable

controllable but not observable

not controllable but observable

neither controllable nor observable.

Indeed, Kalman showed that the general linear system may be decomposed into four such subspaces. If the eigenvalues of the dynamical matrix $\boldsymbol{A}$ are all distinct, the *Kalman decomposition* is straightforward:

$$\begin{pmatrix} \dot{x}_{\mathrm{co}} \\ \dot{x}_{\mathrm{c\bar{o}}} \\ \dot{x}_{\mathrm{\bar{c}o}} \\ \dot{x}_{\mathrm{\bar{c}\bar{o}}} \end{pmatrix} = \begin{pmatrix} A_{\mathrm{co}} & 0 & 0 & 0 \\ 0 & A_{\mathrm{c\bar{o}}} & 0 & 0 \\ 0 & 0 & A_{\mathrm{\bar{c}o}} & 0 \\ 0 & 0 & 0 & A_{\mathrm{\bar{c}\bar{o}}} \end{pmatrix} \begin{pmatrix} x_{\mathrm{co}} \\ x_{\mathrm{c\bar{o}}} \\ x_{\mathrm{\bar{c}o}} \\ x_{\mathrm{\bar{c}\bar{o}}} \end{pmatrix} + \begin{pmatrix} B_{\mathrm{co}} \\ B_{\mathrm{c\bar{o}}} \\ 0 \\ 0 \end{pmatrix} u \; y = \begin{pmatrix} C_{\mathrm{co}} & 0 & C_{\mathrm{\bar{c}o}} & 0 \end{pmatrix} \begin{pmatrix} x_{\mathrm{co}} \\ x_{\mathrm{c\bar{o}}} \\ x_{\mathrm{\bar{c}o}} \\ x_{\mathrm{\bar{c}\bar{o}}} \end{pmatrix}, \tag{4.23}$$

where the subscripts c and $\bar{\mathrm{c}}$ represent controllable and uncontrollable, and so on. The important point is that the transfer function $G(s)$ represents just one of these subspaces:

> The transfer function represents the subspace that is *both* controllable and observable.

In general, the three other subspaces exist as well. If $\boldsymbol{A}$ has repeated eigenvalues, then an analysis similar to that which leads to the *Jordan canonical form* of a matrix adds some off-diagonal blocks to the representation of $\boldsymbol{A}$ given in Eq. (4.23).

Zeros in the Time Domain

In Chapter 3, we looked at zeros, complex values z in the s plane where $G(z) = 0$. We recall that, except for special cases such as extended systems, the complex transfer function $G(s)$ can be written as a rational polynomial,

$$G(s) = \frac{y(s)}{u(s)} = \frac{b_0 s^k + b_1 s^{k-1} + \cdots + b_k}{s^n + a_1 s^{n-1} + \cdots + a_n} \equiv \frac{b(s)}{a(s)}, \tag{4.24}$$

which, neglecting initial conditions, is the Laplace transform of the time-domain equation,

$$y^{(n)} + a_1 y^{(n-1)} + \cdots + a_n y = b_0 u^{(k)} + b_1 u^{(k-1)} + \cdots + b_k u, \tag{4.25}$$

where $y^{(n)}$ is the nth time derivative of $y(t)$, and so on. Just as poles are the roots of the transfer function denominator, $a(s)$, zeros are the roots of the numerator, $b(s)$.

In the time domain, we can define the *zero dynamics* as the dynamical equation governing $u(t)$:

$$b_0 u^{(k)} + b_1 u^{(k-1)} + \cdots + b_k u = 0 \,. \tag{4.26}$$

The solutions $u(t)$ to the zero dynamics of Eq. (4.26) are *blocked* from exciting the dynamics of the output, $y(t)$. Since the fundamental theorem of algebra allows us to write $b(s)$ as a product of real and complex roots, the set of blocked signals $u(t)$ consists of linear combinations of terms of the form e^{zt}, where z is a root of $b(s) = 0$. If z is complex, then $e^{z^* t}$ is also a solution: the roots come in complex-conjugate pairs, since the b_i are all real. We see, too, that the zero dynamics will be stable for LHP zeros (real part in the LHP part of the complex plane) and unstable for RHP zeros. These conclusions parallel our stability criteria for poles in characterizing the dynamics of $y(t)$.

Example 4.8 Consider the linear system $\{\boldsymbol{A}, \boldsymbol{B}, \boldsymbol{C}\}$, with $\boldsymbol{A} = \left(\begin{smallmatrix} -1 & 0 \\ 0 & -2 \end{smallmatrix}\right)$, $\boldsymbol{B} = \left(\begin{smallmatrix} 2 \\ 3 \end{smallmatrix}\right)$, and $\boldsymbol{C} = (1 \;\; -1)$. The transfer function representation is

$$G(s) = \boldsymbol{C}(s\mathbb{I} - \boldsymbol{A})^{-1}\boldsymbol{B} = \frac{1 - s}{(1+s)(2+s)} \,. \tag{4.27}$$

We see that there is a single (RHP) zero at $z = 1$, corresponding to e^t. Indeed, the zero dynamics are governed by $u - \dot{u} = 0$, which implies $u(t) = u_0\, e^t$. We can also verify in the state-space representation the signal-blocking properties claimed for $u(t)$. Returning to the time-dependent equations, starting from zero initial conditions, we choose the input $u(t) = e^t$. The response is then

$$\boldsymbol{x}(t) = \begin{pmatrix} x_1 \\ x_2 \end{pmatrix} = \begin{pmatrix} e^t - e^{-t} \\ e^t - e^{-2t} \end{pmatrix} \quad \Longrightarrow \quad y(t) = x_1(t) - x_2(t) = e^{-2t} - e^{-t} \,. \tag{4.28}$$

In Eq. (4.28), the states $x_1(t)$ and $x_2(t)$ diverge in response to the diverging input, yet $y(t)$ does not diverge. The signal $u(t) = e^t$ is blocked by the zero.

Finally, the observability and controllability matrices are $\boldsymbol{W}_c = \left(\begin{smallmatrix} 2 & -2 \\ 3 & -6 \end{smallmatrix}\right)$ and $\boldsymbol{W}_o = \left(\begin{smallmatrix} 1 & -1 \\ -1 & 2 \end{smallmatrix}\right)$. Both are rank 2: Despite blocking $u(t) = e^t$, the system is controllable and observable.

Pole-Zero Cancellation, Revisited

In Section 3.5.3, we discussed using a controller zero to cancel a system pole and showed that it was a bad idea to cancel an unstable pole using a RHP zero. Here, we revisit the notion of pole-zero cancellation in a state-space context. The basic insight is that the cancellation can hide a loss of controllability or observability in a system.

Example 4.9 (Pole-zero cancellation, two ways) We will look at the series connection of two simple systems,

$$G_1(s) = \frac{s-\alpha}{(s-p)^2} \quad \text{and} \quad G_2(s) = \frac{1}{s-\alpha} \tag{4.29}$$

Can we cancel the pole in G_2 at α by the zero in G_1? We include the denominator of $(s-p)^2$ to keep the systems strictly proper. As always, it is tempting to say that

$$G_1 G_2 = G_2 G_1 = \frac{1}{(s-p)^2} \,. \tag{4.30}$$

We know from Section 3.5.3 that there is more to the story: the cancellation neglects initial conditions that are commonly discarded from the transfer function, not to mention that in practice the zero will never exactly match the pole. Here, we look at what happens in state space. First, we convert the two systems into the state-space representation, using the canonical form given in Section 2.4.1. We find (Problem 4.4.),

$$\begin{aligned} G_1(s) &\Longrightarrow \begin{pmatrix} \dot{x}_1 \\ \dot{x}_2 \end{pmatrix} = \begin{pmatrix} 0 & 1 \\ -p^2 & 2p \end{pmatrix} \begin{pmatrix} x_1 \\ x_2 \end{pmatrix} + \begin{pmatrix} 0 \\ 1 \end{pmatrix} u_1 \,, \quad y_1 = \begin{pmatrix} -\alpha & 1 \end{pmatrix} \begin{pmatrix} x_1 \\ x_2 \end{pmatrix} \\ G_2(s) &\Longrightarrow \dot{x} = \alpha x + u_2 \,, \quad y_2 = x \,. \end{aligned} \tag{4.31}$$

We can then form the series connection $G_1(s)G_2(s)$ by taking the output $y_1(t)$ from G_1 and feeding it to the input u_2 of G_2. The new system has three states and is given by

$$G_1 G_2 \Longrightarrow \begin{pmatrix} \dot{x}_1 \\ \dot{x}_2 \\ \dot{x}_3 \end{pmatrix} = \left(\begin{array}{cc:c} 0 & 1 & 0 \\ -p^2 & 2p & 0 \\ \hdashline -\alpha & 1 & \alpha \end{array} \right) \begin{pmatrix} x_1 \\ x_2 \\ x_3 \end{pmatrix} + \begin{pmatrix} 0 \\ 1 \\ 0 \end{pmatrix} u_1, \quad y_2 = \begin{pmatrix} 0 & 0 & 1 \end{pmatrix} \begin{pmatrix} x_1 \\ x_2 \\ x_3 \end{pmatrix}. \tag{4.32}$$

Conversely, the series connection $G_2(s)G_1(s)$ formed by taking the output $y_2(t)$ from G_2 and feeding to the input u_1 of G_1 gives

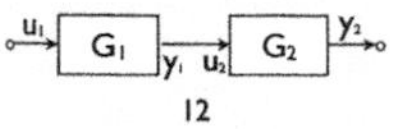

12

$$G_2 G_1 \Longrightarrow \begin{pmatrix} \dot{x}_1 \\ \dot{x}_2 \\ \dot{x}_3 \end{pmatrix} = \left(\begin{array}{c:cc} \alpha & 0 & 0 \\ \hdashline 0 & 0 & 1 \\ 1 & -p^2 & 2p \end{array} \right) \begin{pmatrix} x_1 \\ x_2 \\ x_3 \end{pmatrix} + \begin{pmatrix} 1 \\ 0 \\ 0 \end{pmatrix} u_2, \quad y_1 = \begin{pmatrix} 0 & -\alpha & 1 \end{pmatrix} \begin{pmatrix} x_1 \\ x_2 \\ x_3 \end{pmatrix}. \tag{4.33}$$

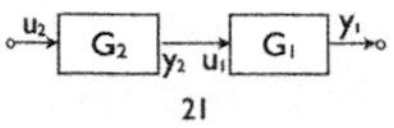

21

In Eqs. (4.32) and (4.33), the partitions are guides to show the structure of G_1 and G_2, with 12 and 21 connections, respectively. We illustrate these connections at left.

The two systems given in Eqs. (4.32) and (4.33) have different state-space representations. Although you might think the differences are related to a similarity coordinate transformation (Section 2.4.3), something else is going on. From the controllability and observability matrices, the 12 connection is not controllable (but is observable) while the 21 connection is not observable (but is controllable). The zero in $G_1(s)$ blocks signals (exponentials of the form $e^{\alpha t}$) and severs the 12 connection between input and dynamics, leading to a loss of controllability. The 21 connection severs the output from the dynamics, leading to a loss of observability. Finally, the transfer function

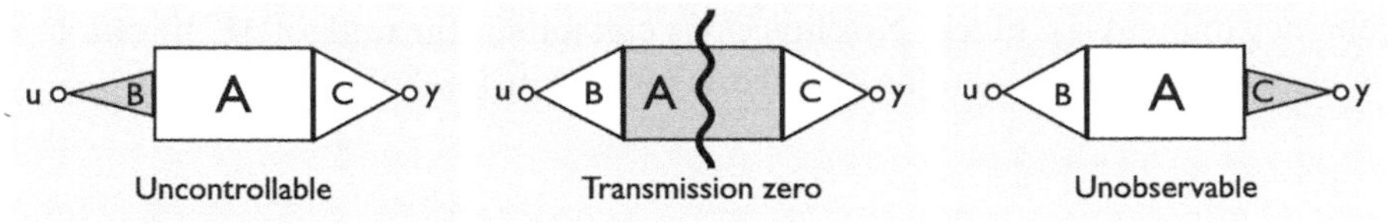

Three pathologies of a dynamical system: lack of controllability, transmission zero, and lack of observability.

Fig. 4.2

$G_1G_2 = G_2G_1 = \frac{1}{(s-p)^2}$ represents the controllable, observable subspace, which is a two-dimensional subspace that differs from the three-dimensional state spaces representing G_1G_2 and G_2G_1.

The conclusion that we can draw from Examples 4.8 and 4.9 and our previous discussion is that there are three closely related types of pathologies of dynamical systems, as illustrated graphically in Figure 4.2. In Example 4.8, the block was due to a zero in the transfer function. Such zeros are also known as *transmission zeros*. In Example 4.9, the input could not reach all the modes of the dynamics (uncontrollable) in one version and, in the other version, all the dynamical modes did not reach the output (unobservable). All three cases share the common theme of a pathology in the connections between the outside world and the dynamical system, and all three issues can be eliminated either by a physical rearrangement or by adding sensors and/or actuators.

Multiple Inputs and Outputs Can Eliminate Zeros

In Section 3.8.3, we saw that adding inputs or outputs can eliminate zeros. The intuition is that an extra input or output adds new paths for signals. When one signal path is "blocked" by a zero, another may yet transmit information. But adding inputs or outputs raises the possibility that a new linear combination of them may be blocked, creating a new kind of zero. Here, we explore such ideas more carefully.

For a zero at s, the input $u(t) = u_0\, \mathrm{e}^{st}$ leads to a state vector of the form $\boldsymbol{x}(t) = \boldsymbol{x}_0\, \mathrm{e}^{st}$ and an output $y(t) = 0$. Substituting into our state-space equations gives

$$s\, \mathrm{e}^{st}\, \boldsymbol{x}_0 = \mathrm{e}^{st}\, \boldsymbol{A}\boldsymbol{x}_0 + \mathrm{e}^{st}\, \boldsymbol{B}\boldsymbol{u}_0 \qquad 0 = \mathrm{e}^{st}\, \boldsymbol{C}\boldsymbol{x}_0\,, \tag{4.34}$$

or, in matrix-vector form,

$$\begin{pmatrix} \boldsymbol{A} - s\mathbb{I} & \boldsymbol{B} \\ \boldsymbol{C} & \boldsymbol{0} \end{pmatrix} \begin{pmatrix} \boldsymbol{x}_0 \\ \boldsymbol{u}_0 \end{pmatrix} \mathrm{e}^{st} \equiv \boldsymbol{M}\boldsymbol{v}_0\, \mathrm{e}^{st} = \boldsymbol{0}\,, \tag{4.35}$$

where the four elements of the matrix $\boldsymbol{M}$ in Eq. (4.35) are themselves matrices. To have a solution to Eq. (4.35) with nonzero $\boldsymbol{x}_0$ and $\boldsymbol{u}_0$, the matrix $\boldsymbol{M}$ must have less than full rank. Zeros occurs at values of s such that $\boldsymbol{M}$ loses rank, and, indeed, you can use Eq. (4.35) to find zeros numerically in a MIMO system.

Here, we point out that $\boldsymbol{M}$ cannot lose rank if either $\boldsymbol{B}$ or $\boldsymbol{C}$ is of rank n (the number of states). For if rank $\boldsymbol{B} = n$, there are n independent rows in $\boldsymbol{M}$, independent of any degeneracy in the $A - s\mathbb{I}$ block. No choice of s can reduce the rank of M. If rank $C = n$, there are n independent columns, and the same conclusion holds. In general:

n states and (n actuators or n sensors) $\Longrightarrow$ no transmission zeros

On the other hand, with n actuators but fewer sensors, the system may not be observable; and with n sensors but fewer actuators, the system may not be controllable.

What about our previous example in Eq. (3.97) where a MIMO system had zeros even though no element $G_{ij}(s)$ had an explicit zero? In that case, the 2×2 transfer matrix had a four-dimensional state space, whereas there were two inputs and two outputs. Thus, zeros are allowed (and were found). Here, we have proven that four inputs or four outputs guarantee that there are no zeros. See also Problem 4.5.

4.2 Control Based on the State

If a system is controllable, we can use a linear feedback law to modify the dynamics arbitrarily by moving the eigenvalues of A [poles of $G(s)$] to *arbitrary* positions. The strategy is equivalent to the pole-placement method discussed in the previous chapter (Section 3.7.2), but the state-space formulation will give us better collective control over the full set ("gang of 4") transfer functions.

A linear feedback law is of the form $u(t) = -Kx$. For a SISO system, K is a $1 \times n$ row vector; for m inputs, K is an $m \times n$ matrix. Here, we focus on the SISO case: if the linear system $\{A, B\}$ (*open-loop dynamics*) is controllable, then the *closed-loop dynamics*

$$\dot{x} = (A - BK)x \equiv A'x \tag{4.36}$$

will have n eigenvalues that can be chosen arbitrarily by choosing the n feedback gains in the row vector K. One caveat is that if any eigenvalue of A' is complex, we must include its complex conjugate. Note that even if some eigenvalues are complex, there are always n independent parameters in total: Each complex-conjugate pair has an independent real and imaginary part and "takes up" two elements among the eigenvalues.

We justify our claim by writing out the components of the system $\{A, B\}$ in the *controller canonical* form of Eq. (2.59),

$$A = \begin{pmatrix} 0 & 1 & 0 & \cdots & 0 & 0 \\ 0 & 0 & 1 & \cdots & 0 & 0 \\ \vdots & \vdots & \vdots & & \vdots & \vdots \\ 0 & 0 & 0 & \cdots & 0 & 1 \\ -a_n & -a_{n-1} & -a_{n-2} & \cdots & -a_2 & -a_1 \end{pmatrix}, \quad B = \begin{pmatrix} 0 \\ 0 \\ \vdots \\ 0 \\ 1 \end{pmatrix}. \tag{4.37}$$

In the SISO case, the row vector $\boldsymbol{K}$ is multiplied as an exterior product with the column vector $\boldsymbol{B}$ to form a $n \times n$ matrix,

$$\boldsymbol{K} = \begin{pmatrix} k_1 & k_2 & \cdots & k_n \end{pmatrix} \implies \boldsymbol{BK} = \begin{pmatrix} 0 & \cdots & 0 \\ \vdots & & \vdots \\ 0 & \cdots & 0 \\ k_1 & \cdots & k_n \end{pmatrix}, \tag{4.38}$$

and gives, for the closed-loop dynamics,

$$\boldsymbol{A}' = \begin{pmatrix} 0 & 1 & 0 & \cdots & 0 & 0 \\ 0 & 0 & 1 & \cdots & 0 & 0 \\ \vdots & \vdots & \vdots & & \vdots & \vdots \\ 0 & 0 & 0 & \cdots & 0 & 1 \\ -a_n - k_1 & -a_{n-1} - k_2 & -a_{n-2} - k_3 & \cdots & -a_2 - k_{n-1} & -a_1 - k_n \end{pmatrix}. \tag{4.39}$$

In Eq. (4.39), each element k_i can be used to define an arbitrary new dynamical element $a_i' = a_i + k_{n-i+1}$. Thus, we can choose the eigenvalues of $\boldsymbol{A}'$ arbitrarily, as claimed.

Example 4.10 (Controlling the harmonic oscillator) We illustrate full-state control of the harmonic-oscillator system, equivalent to a PD controller.

$$\begin{pmatrix} \dot{x}_1 \\ \dot{x}_2 \end{pmatrix} = \underbrace{\begin{pmatrix} 0 & 1 \\ -1 & -2\zeta \end{pmatrix}}_{\boldsymbol{A}} \begin{pmatrix} x_1 \\ x_2 \end{pmatrix} + \underbrace{\begin{pmatrix} 0 \\ 1 \end{pmatrix}}_{\boldsymbol{B}} u(t), \quad u = -\underbrace{\begin{pmatrix} k_1 & k_2 \end{pmatrix}}_{\boldsymbol{K}} \begin{pmatrix} x_1 \\ x_2 \end{pmatrix} = -k_1 x_1 - k_2 x_2 . \tag{4.40}$$

Then

$$\boldsymbol{BK} = \begin{pmatrix} 0 \\ 1 \end{pmatrix} \begin{pmatrix} k_1 & !!k_2 \end{pmatrix} = \begin{pmatrix} 0 & 0 \\ k_1 & k_2 \end{pmatrix} \implies \boldsymbol{A}' = \boldsymbol{A} - \boldsymbol{BK} = \begin{pmatrix} 0 & 1 \\ -1 - k_1 & -2\zeta - k_2 \end{pmatrix}, \tag{4.41}$$

whose eigenvalues are given by

$$\left| s\mathbb{I} - \boldsymbol{A}' \right| = \begin{vmatrix} s & -1 \\ 1 + k_1 & s + 2\zeta + k_2 \end{vmatrix} = s^2 + (2\zeta + k_2)s + (1 + k_1) = 0 . \tag{4.42}$$

It is clear that choosing arbitrary values of k_1 and k_2 suffices to put the roots of s anywhere. For example, to make a critically damped system with rate –2, we choose repeated eigenvalues $s = \{-2, -2\}$ (see left, for $\zeta = 0$). Our desired characteristic equation is

$$(s + 2)^2 = s^2 + 4s + 4 = 0 , \tag{4.43}$$

which implies setting $4 - 2\zeta = k_2$ and $1 + k_1 = 4$, or $k_1 = 3$. Thus $\boldsymbol{K} = (3 \;\; 4 - 2\zeta)$.

If, more generally, we wanted repeated eigenvalues of $s = \{-a, -a\}$, the characteristic equation would be $s^2 + 2as + a^2 = 0$ and $\boldsymbol{K} = (a^2 - 1 \;\; 2(a - \zeta))$. Note how moving the eigenvalues of $\boldsymbol{A}'$ a greater distance requires a larger gain $\boldsymbol{K}$ and concomitant control effort u. Mathematically, we can put the closed-loop eigenvalues anywhere we like, but there are practical limits as to how much we can and should move them.

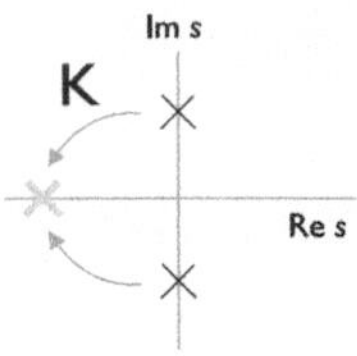

We can obtain these results in the frequency domain using the pole-placement method. For $G(s) = \frac{N_G}{D_G} = \frac{1}{1+s^2}$, we choose $K(s) = \frac{D_G}{(s+2)^2-N_G} = \frac{1+s^2}{(s+2)^2-1} = \frac{1+s^2}{s^2+4s+3}$, which gives a double pole at -2 for $T(s)$. See left for a sketch.

4.2.1 Choosing the Controller Gains

We have shown that, in principal, we can transform the dynamics $\boldsymbol{A}$ of a controllable system to an arbitrary $\boldsymbol{A}'$ by suitable choice of feedback gains $\boldsymbol{K}$. Although the algebra in Example 4.10 is easy, it usually is not. Fortunately, nearly all control software programs have routines that automatically find the controller gains given a system $\{\boldsymbol{A}, \boldsymbol{B}\}$ and a set of desired poles.[4] Since the numerics to solve for the feedback gains can be tricky, you should use the available routines.

Of course, the next question is, Where to put the poles? Freedom is good, but determining n poles is a lot of choice! The first temptation, perhaps, is to be greedy and imagine transforming our system to a set of nth-order low-pass filters with some high cutoff frequency, since the cutoff frequency is equivalent to the feedback bandwidth. However, we already saw that a high feedback bandwidth injects sensor noise into the feedback loop. In addition, moving poles requires control effort. Too much and the actuator will saturate, making the system effectively nonlinear. Thus, we temper our greed and ask, instead, What is the least amount that we can move the poles and still have good feedback performance? Of course, to answer this question, we need to think carefully about what constitutes "good performance." We will postpone a systematic answer to this question until our discussion of optimal control in Chapter 7. Here, we give some heuristic principles.

Before we start, we note that placing eigenvalues arbitrarily is analogous to the pole-placement method discussed in Section 3.7 and contrasts with heuristic methods such as high-gain and loop shaping. The state-space formulation is particularly easy to do on computer, and there are good numerical methods built into control software packages. An advantage of the state-space formulation is that is tools such as observability and controllability tests can identify problems due to an inadvertent cancellation of a pole by a controller zero. On the other hand, we have no systematic way of knowing whether the desired motion for $y(t)$ is possible, given actuator limits.

Given this limitation, here are some principles, based on the goal of ensuring a stable system controlled to a bandwidth given by $|\text{Re } s| = \omega^*$.

- Stabilize the system by moving all unstable poles to the LHP.
- If any stable poles have $|\text{Re } s| < \omega^*$, move them to ω^*.
- Leave the rest.

Implicit in these rules are several approximations. The first is that most of the uninteresting modes decay quickly, leaving one (or a small number) of slower modes. The second is that, in the case of several slow modes, the bandwidth remains approximately

[4] We use the terms "poles" and "eigenvalues" interchangeably, whether we refer to $\boldsymbol{A}$ or $G(s)$.

ω^*. (Be careful: the bandwidth of a series of cascaded first-order filters $G = \frac{1}{(1+s/\omega^*)^n}$ can be significantly less than ω^* for large n.) Check carefully the actual bandwidth of the closed-loop system to see whether it is close enough to the desired value.

The behavior is particularly easy to understand if there is a *dominant pole*: a single complex-conjugate pair that is significantly slower than the other poles. Note that "slower" is sometimes interpreted in two ways. It can refer to the decay rate Re s, as indicated above. Alternatively, it can be interpreted as at right, where the other poles lie in the "shadow" of the dominant pair of eigenvalues. Defining the "shadow line" is equivalent to using the dimensionless damping ratio $\zeta = -\frac{\mathrm{Re}\, s}{|s|}$ instead of Re s to order the eigenvalues. The latter looks at the absolute time for a mode to decay; the former at the number of oscillations (*Q factor*) during the decay. An individual application will dictate which is more important.

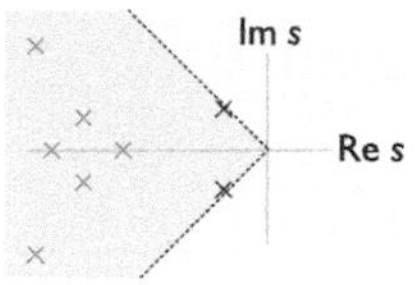

Robustness

The above procedure for placing eigenvalues of a closed-loop system starting from $\{\boldsymbol{A}, \boldsymbol{B}\}$ has a serious weakness: If the system is not known precisely – and, really, it never is – errors in the model for $\boldsymbol{A}$ and $\boldsymbol{B}$ lead to errors in the eigenvalues. Unfortunately, those errors may be bigger than you might like. Chapter 9 deals with such questions in detail. Here is just a hint of what can go wrong.

Consider the polynomial equation

$$\begin{aligned} p(s) &= (s+1)(s+2)(s+3)(s+4)(s+5) \\ &= 120 + 274s + 225\, s^2 + 85s^3 + 15s^4 + s^5 = 0\,. \end{aligned} \tag{4.44}$$

We imagine that $p(s) = 0$ is the characteristic equation of a dynamical matrix $\boldsymbol{A}$. Now change the s^2 coefficient from 225 to 226, a change of less than 0.5% in just one coefficient. To a precision of ± 0.1, the new roots are

$$\begin{pmatrix} -1 \\ -2 \\ -3 \\ -4 \\ -5 \end{pmatrix} \implies \begin{pmatrix} -1.1 \\ -1.6 \\ -3.4 + 1.1i \\ -3.4 - 1.1i \\ -5.5 \end{pmatrix} \tag{4.45}$$

That is quite a change in roots for such a small change in coefficient in the characteristic equation! What this extraordinary sensitivity means in practice is that when you calculate a feedback gain matrix $\boldsymbol{K}$ based on an experimentally determined $\boldsymbol{A}'$, the closed-loop eigenvalues of $\boldsymbol{A}' = \boldsymbol{A} - \boldsymbol{B}\boldsymbol{K}$ may not be what you had planned for. Be careful!

Fortunately, good numerical practices can help: if we regard $G(s) = \frac{1}{p(s)}$ as a transfer function, its expansion in terms of partial fractions (Section A.4.4 online),

$$G(s) = \frac{1}{p(s)} = \frac{1}{24}\left(\frac{1}{1+s} - \frac{4}{2+s} + \frac{6}{3+s} - \frac{4}{4+s} + \frac{1}{5+s}\right), \tag{4.46}$$

gives dynamics that are robust to parameter perturbations: Perturbing any numerical coefficient in Eq. (4.46) leads only to a proportional change in either the amplitude or the time constant of one of the exponential decay modes, in sharp contrast to the large, qualitative change in roots produced by the perturbation shown in Eq. (4.44).

Chapter 9 will offer a more sophisticated approach to achieving robustness. For now, check that the dominant closed-loop roots match their expected values, and remember that it is better numerically to write a high-order transfer function as a sum of low-order systems, using partial fractions. Similarly, a high-order characteristic polynomial is better specified by its roots than by its monomial coefficients.

4.2.2 Stabilizing a Constant Output

We have seen how to choose a controller input $u(t) = -\boldsymbol{Kx}$ to stabilize the state $\boldsymbol{x} = \boldsymbol{0}$, or, equivalently, $y = 0$. To stabilize the output at a nonzero constant, $y(t) = r$, there are two ways to proceed. The more straightforward way is to introduce *feedforward*. Indeed, (see Problem 4.6) choosing

$$u = -\boldsymbol{Kx} + k_r r\,, \qquad k_r = \frac{-1}{\boldsymbol{C}(\boldsymbol{A} - \boldsymbol{BK})^{-1}\boldsymbol{B}}\,, \tag{4.47}$$

we can drive the system to $y = r$ and hold it there. Note that k_r is just a number, as there is a single input and a single output. As always, the issues with feedforward are its lack of robustness: we must know $\boldsymbol{A}$, $\boldsymbol{B}$, and $\boldsymbol{C}$ accurately. *Integral feedback* provides a more robust solution.

Integral Feedback

Although integral feedback will stabilize an output $y = r$ on its own, we will combine it with feedforward, as we have seen before that the combination of feedback and feedforward outperforms either technique alone. As before, we define a control action proportional to $\int_0^t (r-y)\,dt'$. In state space, we need to define an extra state z that obeys

$$\dot{z} = y - r\,. \tag{4.48}$$

If the system dynamics force $z = 0$, then $y = r$. Thus, we consider an *augmented* system

$$\begin{pmatrix}\dot{\boldsymbol{x}}\\ \dot{z}\end{pmatrix} = \begin{pmatrix}\boldsymbol{Ax} + \boldsymbol{B}u\\ y - r\end{pmatrix} = \begin{pmatrix}\boldsymbol{A} & 0\\ \boldsymbol{C} & 0\end{pmatrix}\begin{pmatrix}\boldsymbol{x}\\ z\end{pmatrix} + \begin{pmatrix}\boldsymbol{B}\\ 0\end{pmatrix}u + \begin{pmatrix}\boldsymbol{0}\\ -1\end{pmatrix}r, \quad y = \begin{pmatrix}\boldsymbol{C} & 0\end{pmatrix}\begin{pmatrix}\boldsymbol{x}\\ z\end{pmatrix} = \boldsymbol{Cx}. \tag{4.49}$$

Here, the new state vector $\left(\begin{smallmatrix}\boldsymbol{x}\\ z\end{smallmatrix}\right)$ is an $n+1$ component vector (n for $\boldsymbol{x}$ and 1 for z). We just have a slightly bigger dynamical system, with an $(n+1)\times(n+1)$ element dynamical matrix.[5] If z is constant, then $y = r$. With this in mind, we choose

$$u = -\boldsymbol{Kx} - k_i z + k_r r\,, \tag{4.50}$$

[5] Strictly speaking, we should really treat our system as a MIMO system having two inputs, u and r. That is, we should define $\boldsymbol{u}' = \left(\begin{smallmatrix}u\\ r\end{smallmatrix}\right)$ and $\boldsymbol{B}' = \left(\begin{smallmatrix}\boldsymbol{B} & \boldsymbol{0}\\ 0 & -1\end{smallmatrix}\right)$. However, it will be simpler to abuse our SISO notation slightly.

where k_i is the integral gain and k_r the feedforward gain. The new dynamical system is

$$\begin{pmatrix} \dot{\boldsymbol{x}} \\ \dot{z} \end{pmatrix} = \begin{pmatrix} \boldsymbol{A} - \boldsymbol{BK} & -\boldsymbol{B}k_\mathrm{i} \\ \boldsymbol{C} & 0 \end{pmatrix} \begin{pmatrix} \boldsymbol{x} \\ z \end{pmatrix} + \begin{pmatrix} \boldsymbol{B}k_\mathrm{r} \\ -1 \end{pmatrix} r \,, \tag{4.51}$$

which has fixed points $z = \boldsymbol{Cx} - r = 0$, or $y = r$. Note that the output converges to $y = r$ independent of the values of $\boldsymbol{A}$, $\boldsymbol{B}$, $\boldsymbol{K}$, and $\boldsymbol{C}$, and even k_r, which means that the closed-loop dynamics are much more robust than before. One pitfall is that the closed-loop dynamics now depend on both $\boldsymbol{K}$ and k_i, so that if you have designed a feedback control based on $\boldsymbol{K}$, you may need to reduce (and perhaps redesign) $\boldsymbol{K}$ to compensate for the integral gain k_i. We reached a similar conclusion in Section 3.3.2, where going from proportional (P) control to proportional-integral (PI) control in the frequency domain implied that the proportional gain usually must be reduced after adding integral control.

Why use the feedforward term k_r? The closer the feedforward dynamics are to the "correct" value, the better the feedback dynamics will perform. Unlike feedback, there is no penalty in dynamical performance when using feedforward. Roughly, it pays to use all available information about a system to form feedforward elements whenever possible. Then feedback is used for the truly uncertain aspects such as disturbances. This idea, along with other aspects of feedforward and feedback, is illustrated in Example 4.11.

Example 4.11 (First-order system) For $\dot{x} = -ax + u$ and $y = x$, we seek $y = r$. Define $\dot{z} = y - r = x - r$ and write the augmented dynamics as

$$\begin{pmatrix} \dot{x} \\ \dot{z} \end{pmatrix} = \begin{pmatrix} -a & 0 \\ 1 & 0 \end{pmatrix} \begin{pmatrix} x \\ z \end{pmatrix} + \begin{pmatrix} 1 \\ 0 \end{pmatrix} u + \begin{pmatrix} 0 \\ -1 \end{pmatrix} r \,, \tag{4.52}$$

where

$$u = -(k \;\; k_\mathrm{i}) \begin{pmatrix} x \\ z \end{pmatrix} + k_\mathrm{r} r \,, \qquad k_\mathrm{r} = \frac{-1}{(1)(-a-k)^{-1}(1)} = a + k \,. \tag{4.53}$$

Then

$$\begin{pmatrix} \dot{x} \\ \dot{z} \end{pmatrix} = \begin{pmatrix} -a-k & -k_\mathrm{i} \\ 1 & 0 \end{pmatrix} \begin{pmatrix} x \\ z \end{pmatrix} + \begin{pmatrix} k_\mathrm{r} \\ -1 \end{pmatrix} r \,, \tag{4.54}$$

The eigenvalues of the augmented dynamical matrix are given by

$$\begin{vmatrix} s + a + k & k_\mathrm{i} \\ -1 & s \end{vmatrix} = s^2 + (a+k)s + k_\mathrm{i} = 0 \implies s = \frac{1}{2}\left[-(a+k) \pm \sqrt{(a+k)^2 - 4k_\mathrm{i}}\right] . \tag{4.55}$$

Note that the closed-loop modes s are independent of the feedforward gain k_r but not of the integral gain k_i, which makes the closed-loop modes more oscillatory.

If we set the integral gain ($k_\mathrm{i} = 0$) but keep the feedforward term, we have $\dot{x} = -ax+u$ and $u = -kx + k_\mathrm{r} r$. The closed-loop dynamics $\dot{x} = -(a+k)x + k_\mathrm{r} r$ has a steady-state solution

$$x_\mathrm{e} = \frac{k_\mathrm{r} r}{a+k} = \left(\frac{a+k}{a+k}\right) r = r \qquad \implies \qquad y_\mathrm{e} = r \,. \tag{4.56}$$

Table 4.1 Feedback types

state	$\boldsymbol{x} \to u(\boldsymbol{x})$
output	$y \to u(y)$
observer	$y \to \hat{\boldsymbol{x}} \to u(\hat{\boldsymbol{x}})$

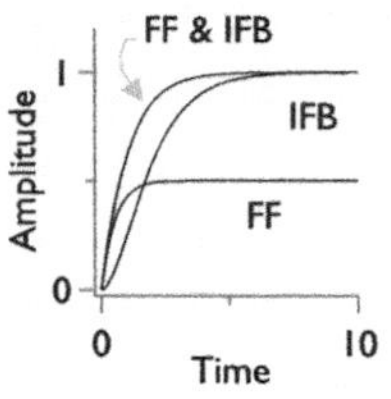

Thus, y does converge to $y_e = r$. But notice how the output follows the input only for these parameter values. Varying them – especially k_r – leads to a different steady-state value.

With the integral term $k_i > 0$, the steady-state value equals r for all values of k, a, and k_r. But changing these parameters does change the closed-loop dynamics. By contrast, the feedforward term k_r does *not* change the closed-loop dynamics. At left, we plot the response to a step in the reference $r(t) = \theta(t)$ for three cases:

Feedforward (FF):with incorrect gain $k_r = 1$. (Half the correct value.) Although the response is fast, the steady-state value is off by a factor of 2.

Integral feedback (IFB):$k_i = 1$. Correct steady-state value but sluggish response.

Both (FF & IFB):feedforward, $k_r = 1$, and integral feedback, $k_i = 1$. Despite the incorrect k_r, the steady-state value is correct, and the response is faster.

The best performance is obtained from a mix of feedforward and feedback.

4.3 Control Based (Indirectly) on the Output

For the state-vector control discussed in Section 4.2, we assume that the full n-dimensional state vector $\boldsymbol{x}$ is observable and design a feedback law, $u(\boldsymbol{x})$, accordingly. However, we often have access to only a smaller number $p < n$ of components $\boldsymbol{y}$. Indeed, here we assume that $p = 1$: only a single output $y(t)$ is available. In such a case, we can try to design a feedback law $u(y)$ directly, *output feedback*, or go through a more complicated, two-step procedure where we use present and past observations to form an estimate $\hat{\boldsymbol{x}}$ of the state vector $\boldsymbol{x}$ and then use that estimate in a state-vector control algorithm. Below, we will introduce a technique known as an *observer* to calculate such an estimate. Schematically, $y \to \hat{\boldsymbol{x}} \to u(\hat{\boldsymbol{x}})$. Although the procedure seems more complicated – there are now two steps rather than one – we can use our familiar state-vector control algorithms. If we were to base feedback directly on the output, $y \to u(y)$, we would need new design techniques. The three approaches to feedback are summarized in Table 4.1.

In Section 4.1.2, we showed that if a system is observable – if the observability matrix $\boldsymbol{W}_o$ has full rank – then estimating the state from observations is in principle possible. Of course, if, in the output equation $\boldsymbol{y} = \boldsymbol{C}\boldsymbol{x}$, the matrix $\boldsymbol{C}$ happens to be square and invertible, then we can trivially find the state vector as $\boldsymbol{x} = \boldsymbol{C}^{-1}\boldsymbol{y}$. In such a case, the number of observations equals the numbers of states. Most commonly, the number of observations is less than the number of states, and inversion is not possible. Here, we focus on the SISO case, where there is one observation y but n states $\boldsymbol{x}$.

A naive strategy is to use finite differences to estimate the first $n-1$ time derivatives of $y(t)$. In effect, from one variable measured over time, we generate n independent observations (y and its derivatives). Knowing the time derivatives and the $\boldsymbol{A}$ and $\boldsymbol{C}$ matrices, you can then solve Eq. (4.13) for the state vector $\boldsymbol{x}$. However, this method does not work well in practice, because estimating derivatives amplifies noise, a problem that becomes worse with each higher derivative estimated.

A better strategy is to simulate a "shadow" dynamical system, the *observer*, that has identical dynamics to the physical system and can *synchronize* to it. If so, then you can simply use the observer's synchronized state as an estimate for the physical state.

To construct a SISO observer, we try an exact copy of the physical system's dynamics,

$$\left.\begin{aligned} \dot{\boldsymbol{x}} &= \boldsymbol{A}\boldsymbol{x} + \boldsymbol{B}u \\ \dot{\hat{\boldsymbol{x}}} &= \boldsymbol{A}\hat{\boldsymbol{x}} + \boldsymbol{B}u \end{aligned}\right\} \quad \Longrightarrow \quad (\dot{\boldsymbol{x}} - \dot{\hat{\boldsymbol{x}}}) \equiv \dot{\boldsymbol{e}} = \boldsymbol{A}\boldsymbol{e}\,, \tag{4.57}$$

where $\boldsymbol{x}$ is the true state of the system, $\hat{\boldsymbol{x}}$ is the *estimate*, which obeys the same dynamics as the true system, and $\boldsymbol{e}$ is the *error* in the state. Note how the inputs, being identical for both systems, drop out of the expression for the difference. Then, at long times, the error decays exponentially to zero. One hidden assumption is that the equation for $\hat{\boldsymbol{x}}$ uses the "correct" $\boldsymbol{A}$ and $\boldsymbol{B}$. In other words, the observer equation is really $\dot{\hat{\boldsymbol{x}}} = \hat{\boldsymbol{A}}\hat{\boldsymbol{x}}+\hat{\boldsymbol{B}}u$, and we assume that somehow we know enough about the physical system that $\hat{\boldsymbol{A}} \approx \boldsymbol{A}$ and $\hat{\boldsymbol{B}} \approx \boldsymbol{B}$. Not only must we estimate the state, we also have to estimate the dynamics. Often, we can separate the two tasks: First, we learn the dynamics, for example by taking preliminary measurements. Then, knowing the dynamics, we can track the state $\boldsymbol{x}$ in real time. See in Chapter 6 how to estimate dynamics and in Chapter 9 the consequences of $\hat{\boldsymbol{A}} \neq \boldsymbol{A}$.

Although the estimator given in Eq. (4.57) might seem satisfactory, it is not. If $\boldsymbol{A}$ is unstable, the scheme obviously fails, as $\boldsymbol{e}(t) \to \infty$. Less obviously, even when $\boldsymbol{A}$ is stable, the error converges to zero at a rate that is set by the slowest eigenvalue of $\boldsymbol{A}$. For brevity, we say that this rate "sets the time scale" of $\boldsymbol{A}$. However, if the goal is to use the estimate $\hat{\boldsymbol{x}}(t)$ as an input for state-vector control, it must be accurate – the estimate must have converged – at the time scale of the system dynamics. But that time scale is also set by $\boldsymbol{A}$. Thus, Eq. (4.57) would not be useful in practice: the estimator and the dynamics are both at the time scale of $\boldsymbol{A}$. We need estimates that are *faster* than $\boldsymbol{A}$.

4.3.1 Observers

The naive observer given above can be improved by adding feedback based on the observations. If the observer is successful at tracking the dynamics, its output $\hat{y} = \boldsymbol{C}\hat{\boldsymbol{x}}$ should match the observed output y from the physical system. Any discrepancy between the two can be used as a feedback signal to correct the estimate. Thus,

$$\left.\begin{aligned} \dot{x} &= Ax + Bu\,, \quad y = Cx \\ \dot{\hat{x}} &= A\hat{x} + Bu + L(y - C\hat{x}) \end{aligned}\right\} \implies \dot{e} = (A - LC)e\,. \tag{4.58}$$

The n element column vector $\boldsymbol{L}$ is a set of *observer gains* that can be chosen to make the *error dynamics* $\boldsymbol{A}' = \boldsymbol{A} - \boldsymbol{LC}$ as desired. In particular, we can choose the time scale of $\boldsymbol{A}'$, its slowest eigenvalue, to be faster than the time scale of $\boldsymbol{A}$. Of course, if $\boldsymbol{A}$ is unstable, we should first make $\boldsymbol{A}'$ stable. Beyond that, a rule of thumb is that $\boldsymbol{A}'$ should be 5–10 times faster than $\boldsymbol{A}$. Any slower, and coupling with the system dynamics will degrade the accuracy of the estimator. Any faster, and the noise injected into the estimate by the feedback will degrade its accuracy. In Chapter 8, we will show how to make a systematic optimization that balances these considerations more properly.

The estimator error dynamics in Eq. (4.58), $\boldsymbol{A} - \boldsymbol{LC}$, closely resembles Eq. (4.36) for state vector control, $\boldsymbol{A} - \boldsymbol{BK}$. Indeed, the duality principle of Section 4.1.2,

$$A \to A^{\mathsf{T}}\,, \quad B \to C^{\mathsf{T}}\,, \quad K \to L^{\mathsf{T}}\,, \tag{4.59}$$

gives an exact correspondence: $(\boldsymbol{A} - \boldsymbol{LC}) \to (\boldsymbol{A}^{\mathsf{T}} - \boldsymbol{C}^{\mathsf{T}}\boldsymbol{L}^{\mathsf{T}}) \leftrightarrow \boldsymbol{A} - \boldsymbol{BK}$. This duality between controllability and observability allows us to use the same techniques to choose modified dynamics for the observer as for the controller. Mathematically, the mapping implies that just as controllability gives the ability to place poles at will, so, too, does observability give the ability to choose observer gains $\boldsymbol{L}$ to make arbitrary observer dynamics $\boldsymbol{A}' = \boldsymbol{A} - \boldsymbol{LC}$.

The duality between controllability and observability extends to the rules of thumb that we gave for choosing controller and observer gains. If the controller gains are too high, too much control effort (too much input) is required. In a practical system, the range of an actuator limits the control effort that can be applied. Analogously, if the observer gains are too high, too much noise enters the system. Unlike control effort, there is no limit to how much we can amplify the noise through large observer gains (Problem 4.7).

Example 4.12 (Harmonic oscillator observer) We observe the position x_1 of a harmonic oscillator and want an observer to infer the velocity $x_2 = \dot{x}_1$. We have

$$A = \begin{pmatrix} 0 & 1 \\ -1 & 0 \end{pmatrix}, \quad C = \begin{pmatrix} 1 & 0 \end{pmatrix}. \tag{4.60}$$

We set

$$L = \begin{pmatrix} \ell_1 \\ \ell_2 \end{pmatrix}, \quad LC = \begin{pmatrix} \ell_1 \\ \ell_2 \end{pmatrix}\begin{pmatrix} 1 & 0 \end{pmatrix} = \begin{pmatrix} \ell_1 & 0 \\ \ell_2 & 0 \end{pmatrix} \quad A' \equiv A - LC = \begin{pmatrix} -\ell_1 & 1 \\ -1-\ell_2 & 0 \end{pmatrix} \tag{4.61}$$

The eigenvalues of L are then determined by

$$|s\mathbb{I} - A'| = \begin{vmatrix} s+\ell_1 & -1 \\ 1+\ell_2 & s \end{vmatrix} = s^2 + \ell_1 s + \ell_2 + 1 = 0\,. \tag{4.62}$$

Assuming that the measurement noise is small, we would choose $\boldsymbol{L}$ to make $\boldsymbol{A}'$ about 10 times faster than $\boldsymbol{A}$. Since the time scale of $\boldsymbol{A}$ is set by $|s| = |\pm i| = 1$, we would place the eigenvalues of $\boldsymbol{A}'$ at $(-10, -10)$. Here, to visualize better the observer dynamics, we slow down its dynamics by choosing $(-2, -2)$. The eigenvalues of $\boldsymbol{A}' = \boldsymbol{A} - \boldsymbol{LC}$ should then be

$$(s+2)^2 = s^2 + 4s + 4 = 0 \quad \Longrightarrow \quad \boldsymbol{L} = \begin{pmatrix} 4 \\ 3 \end{pmatrix}, \tag{4.63}$$

where we find $\boldsymbol{L}$ by matching polynomial coefficients.

We illustrate this observer at right. Although the initial conditions of the physical system and observer are very different, the observer synchronizes its state to track the state of the physical system. Notice that the transient lasts a bit less than half an oscillator period, as expected given that the observer dynamics are only twice as fast as the system.

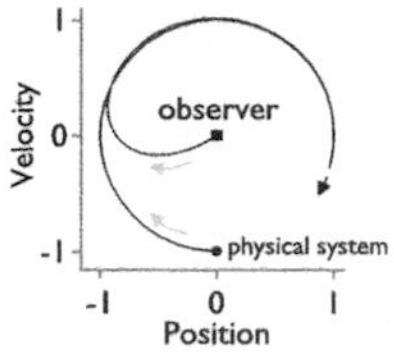

4.3.2 Observer-Controller System

The next step is to combine state-vector control and observers into a single observer-based feedback system. We simply use the estimate $\hat{\boldsymbol{x}}$ in place of the true (but unknown) state $\boldsymbol{x}$ in our feedback law. In other words, we set $u = -\boldsymbol{K}\hat{\boldsymbol{x}} + k_r r$, where we recall that the *feedforward* term $k_r r$ is put in so that the system will go to a desired state r. Then,

$$\begin{aligned} \dot{\boldsymbol{x}} &= \boldsymbol{A}\boldsymbol{x} + \boldsymbol{B}(-\boldsymbol{K}\hat{\boldsymbol{x}} + k_r r), \quad y = \boldsymbol{C}\boldsymbol{x} \\ \dot{\hat{\boldsymbol{x}}} &= \boldsymbol{A}\hat{\boldsymbol{x}} + \boldsymbol{B}(-\boldsymbol{K}\hat{\boldsymbol{x}} + k_r r) + \boldsymbol{L}(y - \boldsymbol{C}\hat{\boldsymbol{x}}) \\ \dot{\boldsymbol{e}} &= \boldsymbol{A}\boldsymbol{e} - \boldsymbol{LC}\boldsymbol{e}, \end{aligned} \tag{4.64}$$

where the last line uses $\boldsymbol{e} = \boldsymbol{x} - \hat{\boldsymbol{x}}$. The difference $\boldsymbol{e}$ is useful for analysis and for understanding the dynamics. In a real implementation, we have only $\hat{\boldsymbol{x}}$.

Next, we combine $\boldsymbol{x}$ and $\boldsymbol{e}$ to make a $2n \times 2n$ dimensional system:

$$\begin{pmatrix} \dot{\boldsymbol{x}} \\ \dot{\boldsymbol{e}} \end{pmatrix} = \underbrace{\begin{pmatrix} \boldsymbol{A} - \boldsymbol{BK} & \boldsymbol{BK} \\ 0 & \boldsymbol{A} - \boldsymbol{LC} \end{pmatrix}}_{\boldsymbol{M}} \begin{pmatrix} \boldsymbol{x} \\ \boldsymbol{e} \end{pmatrix} + \begin{pmatrix} \boldsymbol{B}k_r \\ 0 \end{pmatrix} r. \tag{4.65}$$

For the upper row of $\boldsymbol{M}$, we note that the original term is $-\boldsymbol{BK}\hat{\boldsymbol{x}} = -\boldsymbol{BK}(\boldsymbol{x} - \boldsymbol{e})$. Because $\boldsymbol{M}$ is *block triangular*, its eigenvalues s are given by the roots of the characteristic equation

$$\det(s\mathbb{I} - \boldsymbol{A} + \boldsymbol{BK})\,\det(s\mathbb{I} - \boldsymbol{A} + \boldsymbol{LC}) = 0. \tag{4.66}$$

In other words, the characteristic equation of the combined system is the product of two characteristic equations. One corresponds to the controller $(\boldsymbol{A} - \boldsymbol{BK})$, the other to the observer $(\boldsymbol{A} - \boldsymbol{LC})$. Since the coupling term (upper-right element of $\boldsymbol{M}$) does not contribute, the eigenvalues are the same as for the separate problems. Because the eigenvalue equations of the controller and observer decouple, the problem of designing a controller and an observer can be divided into two separate problems that are solved individually. This observation was formalized by Kalman as the *separation principle*:

The design of the observer and controller can done independently (separately).

As we will discuss in Chapter 8, the separation principle depends heavily on the assumption of linear dynamics and other assumptions about noise sources.

To assess the performance of a combined observer-controller, we can simulate the combined system on a computer:

$$\begin{aligned}\dot{\boldsymbol{x}} &= \boldsymbol{A}\boldsymbol{x} + \boldsymbol{B}u \quad y = \boldsymbol{C}\boldsymbol{x} \\ \dot{\hat{\boldsymbol{x}}} &= \boldsymbol{A}\hat{\boldsymbol{x}} + \boldsymbol{B}u + \boldsymbol{L}(y - \boldsymbol{C}\hat{\boldsymbol{x}}) = (\boldsymbol{A} - \boldsymbol{B}\boldsymbol{K} - \boldsymbol{L}\boldsymbol{C})\hat{\boldsymbol{x}} + \boldsymbol{L}y \\ u &= -\boldsymbol{K}\hat{\boldsymbol{x}} + k_r r\,. \end{aligned} \tag{4.67}$$

In terms of a combined state vector $\binom{\boldsymbol{x}}{\hat{\boldsymbol{x}}}$,

$$\frac{\mathrm{d}}{\mathrm{d}t}\begin{pmatrix}\boldsymbol{x}\\ \hat{\boldsymbol{x}}\end{pmatrix} = \begin{pmatrix}\boldsymbol{A} & -\boldsymbol{B}\boldsymbol{K}\\ \boldsymbol{L}\boldsymbol{C} & (\boldsymbol{A}-\boldsymbol{B}\boldsymbol{K}-\boldsymbol{L}\boldsymbol{C})\end{pmatrix}\begin{pmatrix}\boldsymbol{x}\\ \hat{\boldsymbol{x}}\end{pmatrix} + \begin{pmatrix}\boldsymbol{B}k_r\\ \boldsymbol{B}k_r\end{pmatrix} r + \begin{pmatrix}\boldsymbol{B}\\ \boldsymbol{0}\end{pmatrix} d,\ y = \begin{pmatrix}\boldsymbol{C} & \boldsymbol{0}\end{pmatrix}\begin{pmatrix}\boldsymbol{x}\\ \hat{\boldsymbol{x}}\end{pmatrix}. \tag{4.68}$$

In Eq. (4.68), we have allowed for a separate reference $r(t)$ and input disturbance $d(t)$. Note how the reference input couples to both $\boldsymbol{x}$ and $\hat{\boldsymbol{x}}$, since we know its value, whereas the disturbance, being unknown, enters only the physical system, through $\boldsymbol{x}$. Strictly speaking, we should couple the two inputs together and write $\left(\begin{smallmatrix}\boldsymbol{B}k_r & \boldsymbol{B}\\ \boldsymbol{B}k_r & \boldsymbol{0}\end{smallmatrix}\right)\binom{r}{d}$.

A good heuristic design goal is to make the observer dynamics (governed by $\boldsymbol{L}$) faster than the controller dynamics ($\boldsymbol{K}$), which in turn should be faster than the system dynamics ($\boldsymbol{A}$). In the time domain, this translates into $-\mathrm{Re}\lambda_L > -\mathrm{Re}\lambda_K > -\mathrm{Re}\lambda_A$, where $-\mathrm{Re}\lambda$ denotes the decay rate of the smallest (in magnitude) eigenvalues for observer, controller, and system dynamics. Of course, the maximum observer gain is also limited by measurement noise, while the maximum controller gain may be limited by the need for a gain margin.

Laplace transforming Eq. (4.67) gives the transfer function for output control ($y \to u$),

$$u(s) = -\left[\boldsymbol{K}\,(s\mathbb{I} - \boldsymbol{A} + \boldsymbol{B}\boldsymbol{K} + \boldsymbol{L}\boldsymbol{C})^{-1}\,\boldsymbol{L}\right] y(s) \equiv -K_{\mathrm{ob}}(s)\,y(s)\,. \tag{4.69}$$

One could use $K_{\mathrm{ob}}(s)$ as a controller that directly computes a system input given the measured output. The above state-space design has the advantage of making the design criteria (observer, shift of system dynamics) more transparent and systematic.

Example 4.13 (Observer-based feedback for the harmonic oscillator) We illustrate the combined observer-controller on the harmonic oscillator.

$$\boldsymbol{A} = \begin{pmatrix}0 & 1\\ -1 & 0\end{pmatrix}, \quad \boldsymbol{B} = \begin{pmatrix}0\\ 1\end{pmatrix} \quad \boldsymbol{C} = \begin{pmatrix}1 & 0\end{pmatrix} \tag{4.70}$$

We set

$$\boldsymbol{K} = \begin{pmatrix}k_1 & k_2\end{pmatrix}, \quad \boldsymbol{B}\boldsymbol{K} = \begin{pmatrix}0\\ 1\end{pmatrix}\begin{pmatrix}k_1 & k_2\end{pmatrix} = \begin{pmatrix}0 & 0\\ k_1 & k_2\end{pmatrix} \tag{4.71}$$

$$L = \begin{pmatrix} \ell_1 \\ \ell_2 \end{pmatrix}, \quad LC = \begin{pmatrix} \ell_1 \\ \ell_2 \end{pmatrix} \begin{pmatrix} 1 & 0 \end{pmatrix} = \begin{pmatrix} \ell_1 & 0 \\ \ell_2 & 0 \end{pmatrix} \quad A' \equiv A - BK - LC = \begin{pmatrix} -\ell_1 & 1 \\ -(1 + k_1 + \ell_2) & -k_2 \end{pmatrix}, \tag{4.72}$$

which can be substituted into Eqs. (4.68). Problem 4.8 analyzes this controller, focusing on the roles of the time constants for system dynamics, controller, and observer.

4.3.3 Disturbance Cancellation

As an application of observer-controller systems – and to test your understanding of the material in this chapter – we look at the problem of compensating for an unknown disturbance $d(t)$ that acts at the input $u(t)$ of our system. Assume that the reference $r = 0$ for simplicity. As in our discussion of integral feedback in Section 4.2.2, we create a new dynamical system obeyed by a new state vector $\boldsymbol{x}_\mathrm{d}$, whose output will be the disturbance $d(t)$ that affects the original dynamical system. Then we have

$$\begin{aligned} \dot{\boldsymbol{x}} &= \boldsymbol{A}\boldsymbol{x} + \boldsymbol{B}(u + d)\,, & y &= \boldsymbol{C}\boldsymbol{x}, \\ \dot{\boldsymbol{x}}_\mathrm{d} &= \boldsymbol{A}_\mathrm{d}\boldsymbol{x}_\mathrm{d}\,, & d &= \boldsymbol{C}_\mathrm{d}\boldsymbol{x}_\mathrm{d}\,. \end{aligned} \tag{4.73}$$

We combine $\boldsymbol{x}$ and $\boldsymbol{x}_\mathrm{d}$ into a single, augmented system:

$$\frac{\mathrm{d}}{\mathrm{d}t}\begin{pmatrix} \boldsymbol{x} \\ \boldsymbol{x}_\mathrm{d} \end{pmatrix} = \begin{pmatrix} \boldsymbol{A} & \boldsymbol{B}\boldsymbol{C}_\mathrm{d} \\ \boldsymbol{0} & \boldsymbol{A}_\mathrm{d} \end{pmatrix} \begin{pmatrix} \boldsymbol{x} \\ \boldsymbol{x}_\mathrm{d} \end{pmatrix} + \begin{pmatrix} \boldsymbol{B} \\ \boldsymbol{0} \end{pmatrix} u\,, \quad y = \begin{pmatrix} \boldsymbol{C} & \boldsymbol{0} \end{pmatrix} \begin{pmatrix} \boldsymbol{x} \\ \boldsymbol{x}_\mathrm{d} \end{pmatrix}. \tag{4.74}$$

Note that in this augmented system, we do not directly observe the disturbance $d(t)$. [Of course, you might have a way of measuring the disturbance independently, but that situation is too easy. Here, we assume that we can sense the effects of the disturbance only through the measured variable $y(t)$.] Nonetheless, in the technical sense of Section 4.3.1, the combined system, Eq. (4.74), is observable, but not controllable:

$$\boldsymbol{W}_\mathrm{o} = \begin{pmatrix} \boldsymbol{C} & \boldsymbol{0} \\ \boldsymbol{C}\boldsymbol{A} & \boldsymbol{C}\boldsymbol{B}\boldsymbol{C}_\mathrm{d} \end{pmatrix}, \qquad \boldsymbol{W}_\mathrm{c} = \begin{pmatrix} \boldsymbol{B} & \boldsymbol{A}\boldsymbol{B} \\ \boldsymbol{0} & \boldsymbol{0} \end{pmatrix}. \tag{4.75}$$

We see that $\boldsymbol{W}_\mathrm{o}$ can have full rank, since the columns are linearly independent, in general. (Observability may fail in special cases, for unfortunate choices of system matrices.) By contrast, $\boldsymbol{W}_\mathrm{c}$ does not have full rank, implying that the combined system is not controllable. Thus, we cannot control a disturbance but can detect its influence on other states.

Our control strategy is as follows: We create an observer to estimate the disturbance state vector $\boldsymbol{x}_\mathrm{d}$ (and also the dynamical state $\hat{\boldsymbol{x}}$). Then we introduce the control law,

$$u = -\boldsymbol{K}\hat{\boldsymbol{x}} - \hat{d} = -\begin{pmatrix} \boldsymbol{K} & \boldsymbol{C}_\mathrm{d} \end{pmatrix} \begin{pmatrix} \hat{\boldsymbol{x}} \\ \hat{\boldsymbol{x}}_\mathrm{d} \end{pmatrix}. \tag{4.76}$$

The $\hat{x}$ term is our usual term to control the dynamics of A. The $\hat{x}_d$ term tries to "cancel out" the disturbance $d(t)$. The observer equations are

$$\begin{aligned} \dot{\hat{x}} &= A\hat{x} + B(-K\hat{x} - \hat{d} + \hat{d}) + L(y - \hat{y}), \quad & \hat{y} &= C\hat{x} \\ \dot{\hat{x}}_d &= A_d\hat{x}_d + L_d(y - \hat{y}), & \hat{d} &= C_d\hat{x}_d, \end{aligned} \tag{4.77}$$

with separate observer gains L and L_d for the system and observer dynamics. Note that we replace d by $\hat{d}$ in the $\hat{x}$ equation as we do not know d. Similarly, note in the $\hat{x}_d$ equation that the observer gain L_d, multiplies the measurement error $(y - \hat{y})$ rather than the disturbance error $(d - \hat{d})$. Using the latter would make designing a controller simpler and the performance better, but that would be cheating: the whole point of this exercise is that we do not know $d(t)$. Thus, we have to make do with the measurement errors alone.

Putting Eqs (4.74) and (4.77) for the state vectors and their estimates together, we have

$$\begin{aligned} \dot{x} &= Ax - BK\hat{x} - BC_d\hat{x}_d + BC_dx_d, \quad & \dot{x}_d &= A_dx_d, \\ \dot{\hat{x}} &= A\hat{x} - BK\hat{x} + L(Cx - C\hat{x}), & \dot{\hat{x}}_d &= A_d\hat{x}_d + L_d(Cx - C\hat{x}). \end{aligned} \tag{4.78}$$

Collecting everything in vector-matrix notation (useful for computer simulation), we have

$$\frac{d}{dt}\begin{pmatrix} x \\ x_d \\ \hat{x} \\ \hat{x}_d \end{pmatrix} = \begin{pmatrix} A & BC_d & -BK & -BC_d \\ 0 & A_d & 0 & 0 \\ LC & 0 & (A - BK - LC) & 0 \\ L_dC & 0 & -L_dC & A_d \end{pmatrix}\begin{pmatrix} x \\ x_d \\ \hat{x} \\ \hat{x}_d \end{pmatrix} \tag{4.79}$$

We change variables from $\hat{x}$ and $\hat{x}_d$ to $e_x = x - \hat{x}$ and $e_d = x_d - \hat{x}_d$:

$$\begin{aligned} \dot{x} &= Ax - BK(x - e_x) + BC_de_d, \quad & \dot{e}_x &= Ae_x - LCe_x + BC_de_d, \\ \dot{x}_d &= A_dx_d, & \dot{e}_d &= A_de_d - L_dCe_x, \end{aligned} \tag{4.80}$$

or

$$\frac{d}{dt}\begin{pmatrix} x \\ x_d \\ e_x \\ e_d \end{pmatrix} = \left(\begin{array}{cc|cc} A - BK & 0 & BK & BC_d \\ 0 & A_d & 0 & 0 \\ \hline 0 & 0 & A - LC & BC_d \\ 0 & 0 & -L_dC & A_d \end{array}\right)\begin{pmatrix} x \\ x_d \\ e_x \\ e_d \end{pmatrix} \tag{4.81}$$

The dynamical matrix in Eq. (4.81) can be split into 2×2 blocks, as shown. We see that the eigenvalues of $A - BK$ and A_z are independent of the error dynamics given by the bottom right 2×2 block, implying that the separation principle – that you can design the controller and observer separately – still holds. We can understand the more complicated error dynamics by computing the transfer function $G_{uy}(s)$ between the observations $y(s)$, which function as the observer's input, and the controller signal $u(s)$, which is the observer's output. From Eqs. (4.76) and (4.77) and generalizing

Eq. (4.69), we have

$$u(s) = -\begin{pmatrix} K & C_d \end{pmatrix} \underbrace{\begin{pmatrix} s\mathbb{I} - A + BK + LC & 0 \\ L_d C & s\mathbb{I} - A_d \end{pmatrix}^{-1}}_{M^{-1}} \begin{pmatrix} L \\ L_d \end{pmatrix} y(s). \tag{4.82}$$

The eigenvalues of M form the denominator of the transfer function and explicitly contain $s\mathbb{I} - A_d$, which is the model of the dynamical system that generates the disturbance. We came to the same conclusion in Section 3.7.4 when we formulated the *internal model principle*. In the more complicated analysis just given, we gain some insight into the origin of the principle: we need the internal model to estimate the disturbance. Once we have the disturbance estimate, we use it in the control input to cancel the physical disturbance.

A similar analysis holds for tracking the reference trajectory, $r(t)$. Since the reference, unlike a disturbance, generally is known, feedforward techniques should also be used. One limitation of the internal model principle is that you have to know a fair bit about the disturbance or reference. The precise claim is that the signals are generated by a known linear dynamical system, with the only unknown being the initial condition. Thus, an unknown constant is generated by $\dot{x}_d = 0$, or $A_d = 0$; a periodic disturbance of known frequency but unknown amplitude and phase is generated by $\ddot{x}_d + \omega^2 x_d = 0$; and so on. In practice, we can use the internal model principle for disturbances and references that obey known dynamics and random *inputs* (not just initial conditions), as long as the inputs change at a rate that is slow compared to the observer dynamics. In Chapter 8, we will discuss more systematic ways of including the influence of random inputs.

Example 4.14 (Disturbance cancellation in 1d) The simplest case is 1d dynamics and a constant disturbance:

$$\begin{aligned} &\dot{x} = -x + u + d, \quad y = x, \quad A = -1, \quad B = C = 1 \\ &d = \text{const.} \Leftrightarrow \dot{x}_d = 0, \quad d = x_d, \quad A_d = 0, \quad C_d = 1 \\ &u = -K\hat{x} - \hat{d}, \end{aligned} \tag{4.83}$$

Thus,

$$\begin{aligned} \dot{x} &= -x - K\hat{x} - \hat{d} + d, & \dot{\hat{x}} &= -\hat{x} - K\hat{x} + L(y - \hat{y}), \\ \dot{x}_d &= 0, & \dot{\hat{x}}_d &= 0 + L_d(y - \hat{y}). \end{aligned} \tag{4.84}$$

Substituting $y = x$ and $\hat{y} = \hat{x}$, $\hat{d} = \hat{x}_d$, and plugging into Eqs. (4.79), we have

$$\frac{d}{dt}\begin{pmatrix} x \\ x_d \\ \hat{x} \\ \hat{x}_d \end{pmatrix} = \begin{pmatrix} -1 & 1 & -K & -1 \\ 0 & 0 & 0 & 0 \\ L & 0 & (-1-K-L) & 0 \\ L_d & 0 & -L_d & 0 \end{pmatrix} \begin{pmatrix} x \\ x_d \\ \hat{x} \\ \hat{x}_d \end{pmatrix} \tag{4.85}$$

The transfer function between $y(s)$ and $u(s)$ is

$$u(s) = -\begin{pmatrix} K & 1 \end{pmatrix} \begin{pmatrix} s+1+K+L & 0 \\ L_{\mathrm{d}} & s \end{pmatrix}^{-1} \begin{pmatrix} L \\ L_{\mathrm{d}} \end{pmatrix} y(s) = -\frac{L_{\mathrm{d}}(K+1) + (KL + L_{\mathrm{d}})s}{s(1+K+L+s)}\, y(s)\,. \tag{4.86}$$

The control law in Eq. (4.86) is of the form

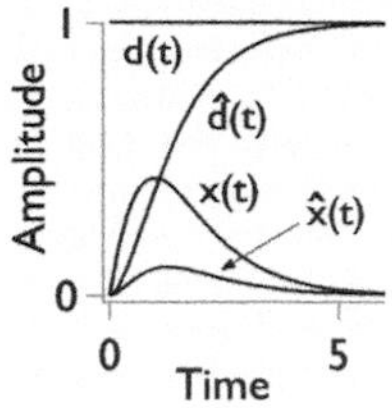

$$u(s) = -\frac{\alpha + \beta s}{s(\gamma + s)}\,, \tag{4.87}$$

which is the expected integral control $1/s$ times a lead controller. The state $x(t)$ and the disturbance d and their estimates are plotted at left, for $K = 1$, $L = L_{\mathrm{d}} = 2$, and $d = 1$. We confirm that, despite the disturbance, the state $x(t)$ returns to the desired value of 0 and that the estimates converge to the state and to the disturbance.

4.4 Summary

In this chapter, we have explored basic ideas of control in the time domain, for linear, time-invariant dynamics governed by state equations $\dot{\boldsymbol{x}} = \boldsymbol{A}\boldsymbol{x} + \boldsymbol{B}\boldsymbol{u}$ and $\boldsymbol{y} = \boldsymbol{C}\boldsymbol{x}$. Two key concepts are *controllability* and *observability*. The Kalman rank condition states that, for an n-dimensional state vector $\boldsymbol{x}$, if the matrix $\boldsymbol{W}_{\mathrm{c}} = (\boldsymbol{B} \;\; \boldsymbol{A}\boldsymbol{B} \;\; \cdots \;\; \boldsymbol{A}^{n-1}\boldsymbol{B})$ has rank n, then all n modes of the state space may be controlled by the inputs $\boldsymbol{u}$. If the rank is lower, then a lower-dimensional subspace can be controlled, with the other modes unaffected by the input. Similarly, if the system is observable and $\boldsymbol{W}_{\mathrm{o}}$ has rank n, then from the output signals $\boldsymbol{y}(t)$, you can infer the state vector $\boldsymbol{x}(t)$ even if, as is usually the case, the number of measurements is fewer than the number of state-vector elements.

If a system is controllable, then a linear control $\boldsymbol{u} = -\boldsymbol{K}\boldsymbol{x}$ can alter the eigenvalues of the dynamics matrix $\boldsymbol{A}$ as desired. But altering the dynamical modes requires control effort, and the maximum range of the inputs $\boldsymbol{u}$ is limited. Thus, you should change as few modes as possible: Unstable modes must be stabilized, and slow modes may need to decay faster.

Observers are a technique to estimate the state vector, using a shadow dynamical system that synchronizes to the physical system using feedback based on the discrepancy between observed and predicted outputs. The observer design is *dual* to the control-design problem. If a system is observable, the estimator dynamics can be set arbitrarily by varying the *observer gains*. But when gains are too high, too much measurement noise is injected into the system. Ideally, the observer dynamics should be faster than the controller dynamics.

If a system is controllable, observable, and linear, then the *separation principle* holds: the choice of controller gain does not affect the observer dynamics, and vice versa. The dynamics of controller and observer are independent and may be designed separately. A related idea, the *certainty equivalence principle*, holds that the design of feedback is the same whether the true state or an unbiased estimate is used.

As an example of state-space design, we considered the problem of *disturbance rejection*, where the goal is to keep the state vector near the origin when the system is perturbed by a disturbance that is generated by a known differential equation. One principle was the *internal model principle*: the controller must have a copy of the dynamics that generates the disturbance. For example, an observer can estimate a disturbance and then use the estimate to cancel out the physical disturbance at the input.

So far, we have given rules of thumb to design controllers and observers, but the time-domain methods allow a more systematic approach, developed in Chapters 7 and 8.

4.5 Notes and References

In this chapter, we have drawn often on presentations by Åström and Murray (2008), Dutton et al. (1997), and Goodwin et al. (2001). For more details on the Kalman decomposition, see (Åström and Murray, 2008). For controllability, see also Manneville (2004). A very thorough discussion of zeros in a state-space context is given by MacFarlane and Karcanias (1976). Equation (4.44) is a simpler version of *Wilkinson's polynomial* (Wilkinson, 1959), which went up to 20th order. The sensitivities increase with polynomial order. The presentation of the disturbance observer was adapted from Goodwin et al. (2001).

Curiously, Shannon's discussion of duality that we quote (Shannon, 1959) was published just before Kalman's 1960 paper, in a work concerned with information theory rather than control. Although the basic picture we present of controllability, observability, and duality dates to Kalman (1960b), the subject has taken on renewed interest recently, particularly in the context of controlling complex networks that have many states, inputs, and outputs. These recent developments will be discussed in Chapter 14.

Problems

4.1 **Controllability of nearly identical systems**. Consider two first-order systems with relaxation rates $\lambda_1 = 1$ and $\lambda_2 = 2$ that are driven by identical inputs (Eq. 4.7). Find an input $u(t)$ that takes the system from an initial state $\boldsymbol{x}_0 = \binom{0}{0}$ to a final state $\boldsymbol{x}_\tau = \binom{1}{1}$, for $\tau = 1$. Plot your solution. Hint: Try a step function with two parameters.

4.2 **Prescribing a path in state space**. A system may be controllable, but that does not mean we can make it follow a desired trajectory $\boldsymbol{x}(t)$ in state space.

a. Show that you cannot prescribe a path for the system defined in Example 4.4.
b. Consider the undamped oscillator with torque control, $\ddot{x} + x = u$. Following Example 4.1, find and plot $u(t)$ that leads to the desired trajectory $x_{\mathrm{d}}(t) =$

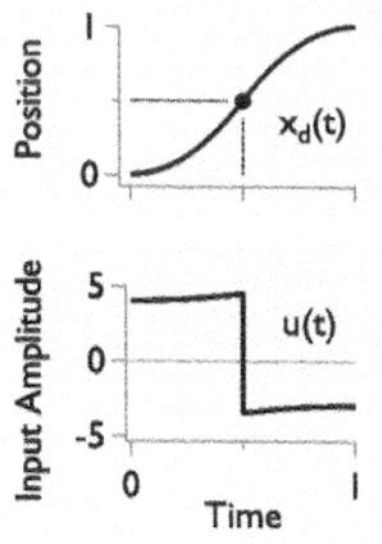

$2(t/\tau)^2 - [1 - 2(t/\tau)^2]\theta[2(t/\tau) - 1]$, which is sketched at left. Verify by integrating the differential equation numerically that your $u(t)$ produces the desired $x_d(t)$.

c. Comment on the required control effort for $\tau \to 0$, with fixed x_τ and ω_0.
d. Can you give any intuition about why the second case works but not the first?

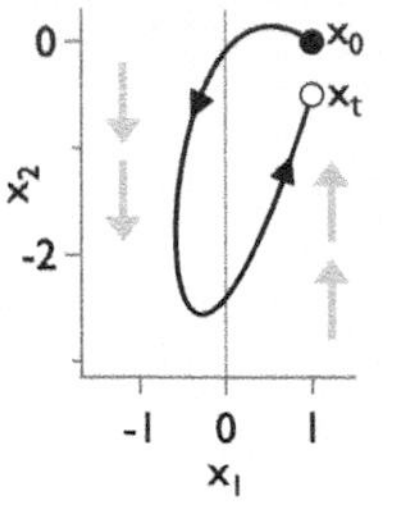

4.3 Nonlocal control. If there are fewer control nodes than state variables – and there usually are – then moving the system from an initial state $\boldsymbol{x}_0$ to a final state $\boldsymbol{x}_\tau$ may require a finite-length trajectory, even when $|\boldsymbol{x}_\tau - \boldsymbol{x}_0| = \varepsilon$, and $\varepsilon \to 0$. To illustrate this *nonlocality* of control trajectories, see the dynamics at left, which depict a kind of shear "flow" that is directed *down* for $x_1 < 0$ and *up* for $x_1 > 0$.

a. At left, $\dot{x}_1 = u$, $\dot{x}_2 = x_1 + u$. Write these equations in the form $\dot{\boldsymbol{x}} = \boldsymbol{A}\boldsymbol{x} + \boldsymbol{B}u$. Calculate analytically $e^{\boldsymbol{A}t}$, $e^{\boldsymbol{A}t}\boldsymbol{B}$, and the Gramian $\boldsymbol{P}(\tau) \equiv \int_0^\tau dt\, e^{\boldsymbol{A}t}\boldsymbol{B}\boldsymbol{B}^{\mathsf{T}} e^{\boldsymbol{A}^{\mathsf{T}}t}$.
b. Show, by substitution into the general solution, $\boldsymbol{x}(t) = e^{\boldsymbol{A}t}\boldsymbol{x}_0 + \int_0^t dt'\, e^{\boldsymbol{A}(\tau - t')}\boldsymbol{B}u(t')$, that the control $u(t) = \boldsymbol{B}^{\mathsf{T}} e^{\boldsymbol{A}^{\mathsf{T}}(\tau - t)}\boldsymbol{P}^{-1}(\tau)\Delta\boldsymbol{x}$ brings the initial state at $t = 0$, $\boldsymbol{x}_0$, to the final state at τ of $\boldsymbol{x}_\tau$. Here, $\Delta\boldsymbol{x} \equiv \boldsymbol{x}_\tau - e^{\boldsymbol{A}\tau}\boldsymbol{x}_0$. See also Problem 7.9.
c. Show the above formula gives $u(t) = 0.126(t - 5)$ and moves an initial state $\binom{1}{0} \to \binom{1}{-\varepsilon}$, with $\varepsilon = 0.5$ in a time $\tau = 10$. Reproduce the plot at left.

This problem is adapted from Sun and Motter (2013).

4.4 Pole-zero cancellation. In Example 4.9, we explored how the different input-output connections between two transfer functions can lead to different issues (loss of controllability vs. loss of observability). Verify the following:

a. Check the state-space forms for G_1 and G_2.
b. Show that the 12 and 21 series connections lead to different 3d systems.
c. Show that if you start from either the 12 or the 21 system, you derive the same transfer function ($= G_1G_2$ or G_2G_1).
d. Write down the observability and controllability matrices for the 12 and 21 systems, and verify that 12 is uncontrollable and 21 unobservable.

4.5 Zeros with more actuators and sensors. In Section 4.1.3, we saw that if the state vector is n-dimensional and there are either n independent inputs or outputs, the transfer function of the enlarged system cannot have a zero. Here, we verify this in a simple example. Consider the transfer function G with a single RHP zero,

$$G(s) = \frac{s-2}{(s-1)^2} \quad \Longleftrightarrow \quad \boldsymbol{A} = \begin{pmatrix} 0 & 1 \\ -1 & 2 \end{pmatrix} \quad \boldsymbol{B} = \begin{pmatrix} 0 \\ 1 \end{pmatrix} \quad \boldsymbol{C} = \begin{pmatrix} -2 & 1 \end{pmatrix}.$$

Now consider a second input or output, by taking $\boldsymbol{B}' = \left(\begin{smallmatrix} b & 0 \\ 0 & 1 \end{smallmatrix}\right)$ or $\boldsymbol{C}' = \left(\begin{smallmatrix} -2 & 1 \\ 0 & c \end{smallmatrix}\right)$.

a. Keeping the original $\boldsymbol{A}$ and $\boldsymbol{C}$, consider the new inputs $\boldsymbol{B}'$ and show that the 1×2 transfer function matrix has no zeros. Recall that in a MIMO transfer matrix, a zero is a value of s for which the transfer function matrix loses rank.
b. Repeat the calculation for the case where you keep $\boldsymbol{A}$ and $\boldsymbol{B}$ and use $\boldsymbol{C}'$.
c. Why do the above conclusions become invalid if b or $c = 0$?

4.6 **Feedforward gain for constant output**. For the SISO system $\dot{x} = Ax + Bu$, $y = Cx$, show that choosing $u = -Kx + k_r r$ leads to $y = r$ if $k_r = -[C(A - BK)^{-1}B]^{-1}$.

4.7 **Noise-tracking tradeoffs for observers**. If observer gains are too low, the observer states will not track the state vector well. If the gains are too high, too much measurement noise will be injected into the system. Here, we show this trade-off explicitly.

a. Add measurement noise $\xi(t)$ to the observer equation for the dynamical system:

$$\dot{x} = Ax + Bu, \quad y = Cx$$
$$\dot{\hat{x}} = A\hat{x} + Bu + L[y(t) + \xi(t) - \hat{y}(t)] .$$

Take the Laplace transform of the error dynamics ($e = x - \hat{x}$), keeping the initial value term to give $e(s)$ as the sum of two terms, one proportional to $e(t = 0) \equiv e_0$ and one proportional to $\xi(s)$. Argue that large values of L make the initial-value term decay quickly but will simultaneously keep the noise term large.

b. Specialize to the first-order system $\dot{x} = -x$ and $y = x + \xi \equiv x + \xi_0 \sin t$. The sinusoidal "noise" is a simple stand-in for stochastic noise, which would be the sum of sines of all frequencies and with random phases. Solve for the Laplace transform of the observer error, $e(s)$, in terms of the initial error e_0 and ξ_0.

The "best" value of observer gain ℓ balances the convergence rate of estimator errors against noise injection. One missing ingredient is the notion that disturbances continually inject new state-estimation errors, which the observer must try to remove. Here, an initial error e_0 will decay away for all values of ℓ so that an observer would not be necessary for long-time estimation. Continuously injecting new disturbances into the dynamics highlights the role of ℓ in balancing the rate that the observer removes disturbances against noise injection. See the *Kalman filter* in Chapter 8.

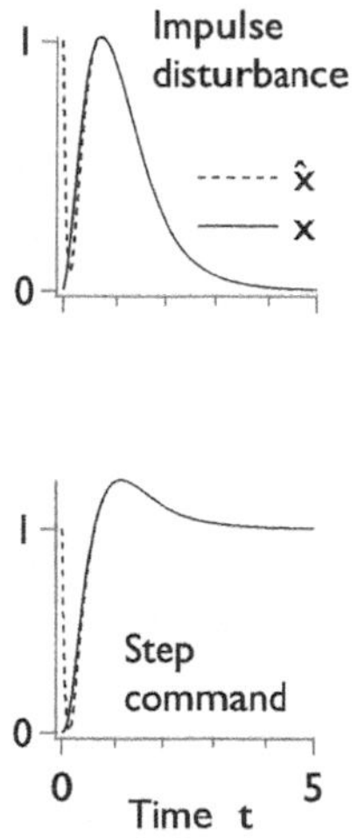

4.8 **Observer-based feedback for the harmonic oscillator**. In Example 4.13, we set up a structure for an observer-based control of a harmonic oscillator.

a. Write down the coupled system plus observer. Include a feedforward gain to make a step command go to the right value. Design the controller to have poles $(-2, -2)$ and the observer to have poles $(-10, -10)$. Give numerical values for the controller (K), observer (L), and feedforward (k_r) gains.

b. Reproduce the numerical responses at right for an impulse disturbance and step command. Use discordant initial conditions: $\hat{x}(0) = \binom{1}{0}$, but $x(0) = \binom{0}{0}$

4.9 **Stabilizing an inverted pendulum forced by a torque**.

a. For small displacements about the vertical, show that the scaled equations of motion have $A = \left(\begin{smallmatrix} 0 & 1 \\ 1 & 0 \end{smallmatrix}\right)$ and $B = \binom{0}{1}$.

b. Check that $\{A, B\}$ is controllable.

c. Assuming a known state vector, design a feedback law $u = -\boldsymbol{Kx}$ to stabilize the vertical fixed point with eigenvalues $(-1, -1)$. Find the gains K_1 and K_2.
d. If you measure only the position, $\boldsymbol{C} = (1\ \ 0)$. Show that $\{\boldsymbol{A}, \boldsymbol{C}\}$ is observable.
e. Design an observer with dynamics $(-2, -2)$.
f. Design a combined observer-controller that regulates the system about $\binom{0}{0}$. This *regulator* is a four-dimensional system. Find the state-space matrices $\{\boldsymbol{A}_{\text{oc}}, \boldsymbol{B}_{\text{oc}}, \boldsymbol{C}_{\text{oc}}\}$ using an input torque disturbance $d(t)$ as input and the angle $\theta(t)$ as output.
g. Find the transfer function $K_{\text{ob}}(s)$ for output control $(y \to u)$.
h. Plot $\{\theta, \hat{\theta}\}$ and $\{\dot{\theta}, \hat{\dot{\theta}}\}$ for an impulse input disturbance $[u(t) = \delta(t)]$.

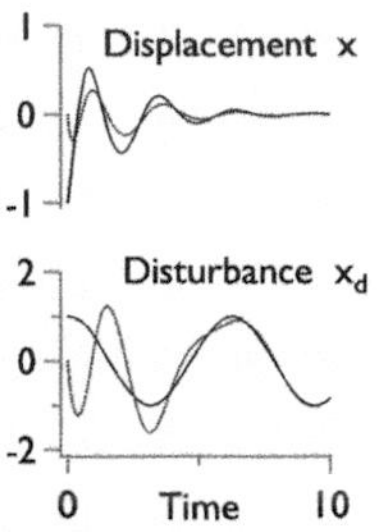

4.10 Canceling a sinusoidal disturbance. Following Section 4.3.3, explore a sinusoidal disturbance that affects a first-order system. Calculate the displacement x and the disturbance position x_{d} and their estimates $\hat{x}$ and $\hat{x}_{\text{d}}$. Remember that Eq. (4.79) describes a 6×6 matrix. Reproduce the plot at left, using $\boldsymbol{A} = -1$, $\boldsymbol{B} = \boldsymbol{C} = 1$, $\boldsymbol{A}_{\text{d}} = \begin{pmatrix} 0 & 1 \\ -1 & 0 \end{pmatrix}$, $\boldsymbol{B}_{\text{d}} = \binom{0}{0}$, $\boldsymbol{C}_{\text{d}} = (1\ \ 0)$ and $K = 2$, $L = 4$. Choose $\boldsymbol{L}_{\text{d}}$ so that the poles of the disturbance observer are at $(-4, -4)$. Initial conditions are $x(0) = -1$, $\boldsymbol{x}_{\text{d}} = \binom{1}{0}$. The initial conditions for the estimators $\hat{x}$ and $\hat{\boldsymbol{x}}_{\text{d}}$ are zero.

Discrete-Time Systems 5

The earliest control loops were analog and were implemented using mechanical, electronic, or hydraulic circuits. As a result, analog control techniques predominated through the 1960s. Even today, analog electronic control circuits can sometimes still be the cheapest, easiest solution (Problem 3.6). Nonetheless, control is now almost always implemented digitally, for reasons of flexibility: very complicated control algorithms are easily programmed (and reprogrammed!), and performance is easy to assess. We thus introduce digital signals and controllers, focusing on the effects of discretizing time. A computer, of course, also uses discrete state variables. Mostly, we will assume that this state discretization is at a fine enough scale that a continuous approximation suffices.[1] Chapter 12 discusses discrete-state systems with a small number of states (even two).

A second, less obvious motivation for discussing discrete-time dynamics is that advanced control algorithms such as those used in optimal and stochastic control are easier to present and understand in their discrete form. The basic point is that continuous functions are elements of infinite-dimensional vector spaces, whereas discrete functions live in finite-dimensional spaces. Finite-dimensional spaces are much simpler to work with.

Our plan will be as follows: First, we discuss how to turn a continuous-time signal into a discrete-time signal (Section 5.1). There are subtleties. The easiest approach, *sampling*, records the value of the continuous signal periodically. The danger is that we do not know explicitly what the signal does in between samples, and, indeed, the unobserved behavior may be completely different from a naive interpolation between measurements, a phenomena known as *aliasing*. Because this subject touches on issues of practical implementation, we will discuss more technical details than normal.

We then introduce mathematical tools for analyzing discrete-time dynamical systems (Section 5.2). In the frequency domain, we introduce the Z-transform, a discrete analog of the Laplace transform. In the time domain, we introduce finite impulse response (FIR) and infinite impulse response (IIR) filters, which implement using different approaches the digital equivalent of analog filters (low pass, high pass, bandpass, and the like).

With the appropriate tools in hand, we discuss how to discretize continuous-time dynamical systems (Section 5.3). If a continuous system is fed the staircase-like signal that is output from a digital controller, and if the continuous output is then sampled,

[1] Section 5.1.2 briefly discusses state quantization.

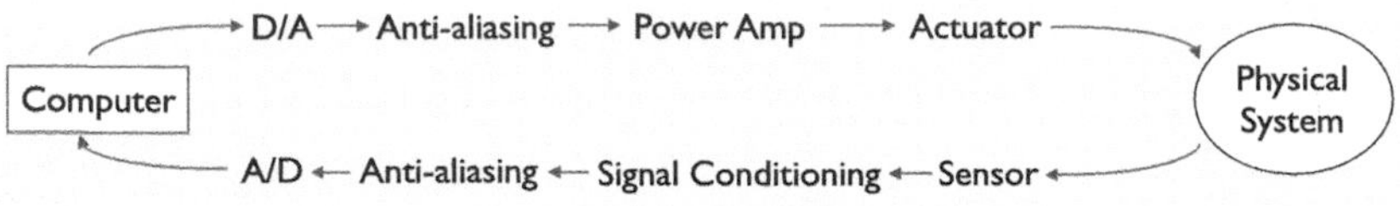

Fig. 5.1 Schematic of a digital control loop.

it turns out that we can give an exact expression for the discrete response at sampling times. In other cases, where the continuous system is fed a continuously variable input, an exact expression is not possible, and we consider various approximations.

Finally, we discuss digital control systems (Section 5.4). In order of increasing complexity, the three approaches are as follows:

- *Emulation*: Design an appropriate continuous controller, and then translate the continuous controller into a discrete approximation of the continuous controller. The strategy is simple and works well if the sampling rate is high enough.
- *Direct digital design*: Derive a discrete approximation to the continuous physical system and then design a digital controller directly. This method potentially allows for a lower sampling frequency.
- *Hybrid dynamics*: Model the controller as discrete but the physical system as continuous. Only this last approach can reveal the "hidden" behavior of the closed-loop system at times between sampling intervals.

Before delving into details, let us consider the schematic illustration of a practical digital control system, shown in Figure 5.1. Starting at the physical system, we take measurements using a *sensor*. Since sensors typically generate small signals, we include *signal conditioning*, such as a voltage amplifier or a transimpedance amplifier (to convert currents to voltage). Then, as we will discuss, the signal needs to be low-pass filtered by an *antialiasing* filter, before being converted to a sequence of numbers by an *analog-to-digital*, or A/D converter. The computer processes the sequence of sensor signals and sends out a sequence of control outputs. These outputs are converted to a voltage by a *digital-to-analog*, or D/A converter and then smoothed by an antialias filter. The filter is sometimes omitted if the system is well damped. The analog output is then boosted in strength by a *power amplifier* and sent to the *actuator*, which acts on the physical system. The loop is thus closed.

For those who are impatient to "get going," the obvious naive approach – first-order approximations to the integrals and derivatives of a continuous controller – is often perfectly adequate. For example, you can replace a continuous PID controller with

$$u(t) = K_{\rm p} e(t) + K_{\rm i} \int^t {\rm d}t'\, e(t') + K_{\rm d} \frac{{\rm d}e}{{\rm d}t}$$

$$\rightarrow \quad u_{k+1} = K_{\rm p} e_k + K_{\rm i} T_{\rm s} I_k + K_{\rm d} \frac{(e_k - e_{k-1})}{T_{\rm s}}, \qquad I_{k+1} = I_k + e_k\,. \tag{5.1}$$

The last term is a running integral value, and subscripted values reflect the sampled digital value of a signal at time kT_S. If you sample fast enough and are not too picky, such a simple algorithm may be good enough. But there are pitfalls, and a modest effort can sometimes produce significantly better results. Read on!

5.1 Discretizing Signals

Our first task is to turn a continuous signal that typically comes from a measurement into a digital signal, a sequence of numbers, represented by a finite number of bits and sampled periodically, with period T_S (or, equivalently, rate $f_S \equiv 1/T_S$ and angular sampling frequency $\omega_S = 2\pi f_S$). Among the important questions: How best to sample a signal? Is the sampled signal "close" to the original?

Answering these questions will lead us to some subtle points. Consider Figure 5.2. If you measure the points denoted by filled circles, your natural instinct is to "connect the dots" and draw the dashed line interpolating between them. However, the same measured points could equally well result from one of an infinite number of higher-frequency curves, such as the one depicted by a light solid curve. This confusion of a higher-frequency sine wave with a lower-frequency one reflects the phenomonon of *aliasing* and is a common occurrence. The wagon wheels that rotate backwards at the movies, the shimmering Moiré patterns seen through a screen door, the false detail seen when looking at periodic structures in a microscope – all these are due to aliasing. But when does it occur?

Intuitively, we can exclude the higher-frequency curve in Figure 5.2 if we know in advance that the continuous signal has no components at high frequencies. That is, the dashed sine wave is the lowest-frequency curve that interpolates through the measured points. If we use the *antialiasing filter* in Figure 5.1 to remove higher frequencies, then the naive interpretation is correct. To be more quantitative, we consider a continuous signal $f(t)$ sampled periodically every T_S at times kT_S. The sampled signal $f_S(t)$ is

$$f_s(t) = f(t)\left[\sum_{k=-\infty}^{\infty} \delta\,(t - kT_S)\right], \tag{5.2}$$

where δ is the Dirac δ-function. The δ-function results from an implicit averaging of the original signal over a very short time that is a necessary part of sampling. From

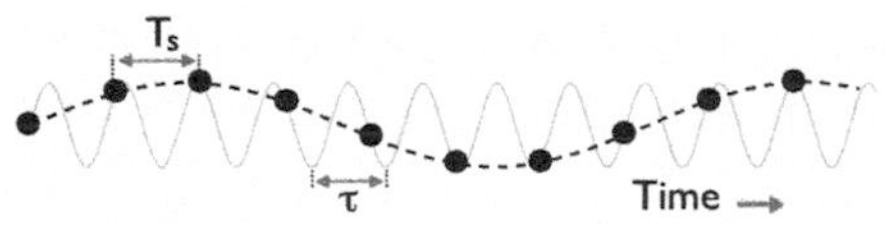

Fig. 5.2 Aliasing: Do the filled circles come from a low- or from a high-frequency curve?

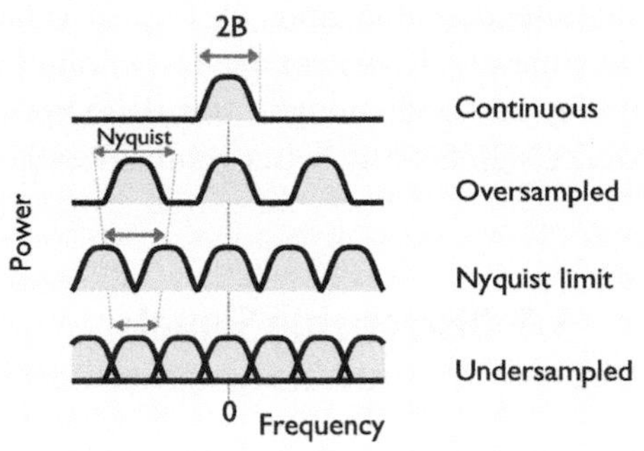

Fig. 5.3 Nyquist limit, in frequency space. The continuous signal has a bandwidth B and spans frequencies from $-B$ to $+B$. Oversampled: the sampling rate exceeds the Nyquist limit $f_k = 2B$. Undersampled: the sampling rate is below $2B$. The Nyquist limit is indicated by left-right arrows.

Problem A.4.4 online and the second form of the Convolution Theorem, Eq. (A.83 online), the Fourier transform of the sampled signal is

$$f_s(\omega) = \left(\frac{1}{2\pi}\right) f(\omega) * \left[\sum_{n=-\infty}^{\infty} \left(\frac{2\pi}{T_s}\right) \delta\left(\omega - n\omega_s\right)\right]$$
$$= \frac{1}{T_s} \sum_{n=-\infty}^{\infty} \int_{-\infty}^{\infty} \mathrm{d}w\, f(w)\, \delta\left(\omega - n\omega_s - w\right) = \frac{1}{T_s} \sum_{n=-\infty}^{\infty} f(\omega - n\omega_s). \qquad (5.3)$$

In words, the power spectrum $|f_s(\omega)|^2$ of the sampled signal $f_s(t)$ is the sum of an infinite number of copies of the power spectrum $|f(\omega)|^2$ of the continuous signal $f(t)$. The copies are evenly spaced in frequency, with a separation $\omega_s = \frac{2\pi}{T_s}$.

Next, consider signals that are *bandlimited*, limited to frequencies in the range $(-B, B)$, or angular frequencies in the range $(-\omega_B, \omega_B)$, where $\omega_B = 2\pi B$ (see Figure 5.3). If we sample such a signal, there are three cases, depending on the sampling frequency f_s:

Oversampled: $f_s > 2B$, and the copies in frequency space are well separated;

Nyquist limit: $f_s = 2B$, and the copies just touch;

Undersampled: $f_s < 2B$, and the copies overlap.

In this last case, the observed power spectrum, which is the sum of the individual copies, will differ from that of the continuous signal in the areas of overlap. In Figure 5.3, the overlap is small compared to B, but as the sampling rate slows down, the overlap increases until there is overlap at all frequencies. The overlap implies *aliasing*, a distortion of the power spectrum of the signal that is introduced by a sampling rate that is too slow. Thus, signals should be oversampled to have, in the range $(-B, B)$, the same spectrum as the continuous signal. Equivalently, we should measure at least two points per period of the highest-frequency sine wave. The above observations lead to the *sampling theorem*:

A function of bandwidth B is completely determined by a sequence of points sampled at intervals $(2B)^{-1}$.

Following Shannon, we have

$$f(t) = \int_{-\infty}^{\infty} \frac{d\omega}{2\pi} f(\omega)\, e^{i\omega t} = \int_{-\omega_B}^{\omega_B} \frac{d\omega}{2\pi} f(\omega)\, e^{i\omega t}, \tag{5.4}$$

since $f(\omega) = 0$ for $|\omega| > \omega_B$. Evaluating Eq. (5.4) at the sampling interval $T_s = \frac{1}{2B}$ gives

$$f_k \equiv f(t = kT_s) = \int_{-\omega_B}^{\omega_B} \frac{d\omega}{2\pi} f(\omega)\, e^{i\omega k T_s}, \tag{5.5}$$

with k an integer. The sampled values f_k thus equal the kth-order Fourier-series coefficient of the periodic function $f(\omega)$, taking $(-B, B)$ as the fundamental period. These Fourier-series coefficients determine $f(\omega)$ completely, since it is zero outside of $(-B, B)$.

We can also derive an explicit interpolation formula. We write

$$\begin{aligned} f(t) &= \mathcal{F}^{-1}[f(\omega)] = \mathcal{F}^{-1}\left[f_s(\omega)\, \text{rect}\left(\frac{\omega}{\omega_s}\right)\right] \\ &= \left[\sum_{k=-\infty}^{\infty} f(t)\, \delta(t - kT_s)\right] * \text{sinc}\left(\frac{t}{T_s}\right) = \sum_{k=-\infty}^{\infty} f_k\, \text{sinc}\left(\frac{t-kT_s}{T_s}\right), \end{aligned} \tag{5.6}$$

using the relation for the Fourier transform of the rectangle function given in Example A.8 online. Equation (5.6) shows how to compute the continuous function $f(t)$ given the samples f_k. We see that the continuous signal $f(t)$ can be thought of as the sum of sinc functions. Note that the kth sinc function is 1 at time kT_s and 0 at all other integer multiples of T_s. In theory, we could implement the interpolation in a D/A converter by sending the values f_k through a perfect, "brick-wall" low-pass filter. Since practical low-pass filters have a softer roll-off, the general strategy is to set the filter bandwidth $\omega_0 = \omega_B$ in order to pass nearly all the signal and to sample at $\omega_s > 2\omega_B$. We will see below that it is wise to choose $\omega_s \gtrsim 10\omega_B$. Under these conditions, the continuous signal is approximately equal to the ideal continuous signal implied by Eq. (5.6).

Figure 5.3 implies a simple test of sampling quality: If there is significant power at the Nyquist frequency (relative to the noise floor), then aliasing will be important.[2]

5.1.1 Antialiasing Filter

The Nyquist sampling theorem derived above states that frequencies higher than $\frac{1}{2}f_s$ will be erroneously interpreted as lower frequencies by the digitizer because of aliasing.

[2] The requirement to sample frequencies below the Nyquist limit is equivalent to restricting wavenumbers to the first Brillouin zone for a one-dimensional crystal in condensed-matter physics (Marder, 2010).

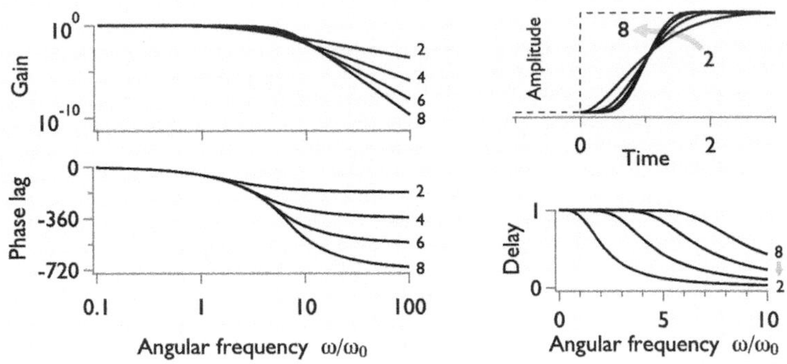

Fig. 5.4 Bode plots, step response, and group delay for Bessel filters of order 2–8.

Thus, you should sample at least twice per period in order to reconstruct the signal present at that period. You should also add an *analog* low-pass filter with cutoff frequency no higher than $\frac{1}{2}f_s$. In principle, a simple, passive "RC" filter would suffice, but a steeper frequency cutoff increases the usable bandwidth. For this reason, the first-order RC circuit is replaced by a circuit of higher order (n = 6 or 8 is typical). Such circuits are usually implemented using active components. Commercial solutions are readily available. As will be discussed in Section 15.1.2, circuit designs trade off the steepness of the frequency cutoff against the amount of phase lag at high frequencies, with $\varphi \approx -n\frac{\pi}{2}$.

Beyond the choice of order, different filter circuits make subtler trade-offs, as discussed in Problem 5.1. *Bessel* filters have, for a given order, the most linear phase-delay curve; they distort time-domain signals least, at the cost of a slower roll-off in frequency than Butterworth or Chebyshev filters. For control problems, the time-domain signal shape is often important. Figure 5.4 shows Bode plots and step responses of Bessel filters of orders n = 2, 4, 6, and 8. Increasing the order makes the gain fall off more quickly (asymptotically, as ω^{-n}) at the cost of phase lag and time delay (the response is initially $\sim t^n$). At bottom right, we show the defining feature of the Bessel filter, its uniform delay $\tau = 1+O(\omega^{2n})$ of time-domain signals. Alternatively, the system transfer function is multiplied by $\approx e^{-\tau s}$.

Some engineering applications sample close to the Nyquist limit, using filters with steep cutoffs, such as the inverse Chebyshev filter. For example, the audio-industry standard samples compact disk recordings at 44.1 kHz, with a bandwidth of 20 kHz ($\omega_s/\omega_B \approx 2.2$). The file sizes are smaller, but the phase and time-domain response is distorted; fortunately, the ear is usually not sensitive to such distortions. For many physics applications, accuracy in the time domain is important, and sampling $> 10B$ is then advised (about ten points are needed for a waveform to "look like" a sine wave). Since digitizing hardware has become cheap, you should usually "buy the board": pay a bit more to increase the sampling rate rather than worry about the subtleties of sampling too close to minimum rates.

Practice safe sampling!
Use an antialiasing filter before converting an analog signal into a digital one.
Sample at a rate at least ten times the signal bandwidth.

5.1.2 Quantization and Dithering

So far, we have sampling artifacts such as aliasing that may arise when the sampling rate is too low compared to the bandwidth of the signal. Another important consideration is that computers *quantize* signals, representing them by a finite number of bits. A typical A/D converter uses 16 bits, which means that it can represent $2^{16} = 65\,536$ levels.[3] The result is a nonlinear transformation of the signal by the converter's *staircase* response, as shown in Figure 5.5. If the quantization step is less than the noise level of your signal (after the antialias filter), the nonlinearity is not important. But when the bit resolution is lower, a strategy known as *dithering* can help. The idea is to deliberately *add* noise to the signal to be digitized so that the noise level is comparable to the quantization step size and then oversample the noisy signal. Why does adding noise improve the accuracy of quantization? Consider digitizing a signal that has a value of 2.3 (in units of the quantization step). In the digitizer shown at left in Figure 5.5, the digitizer will always return 2. Now add noise and measure (center). You will measure "2" about 2/3 of the time and "3" about 1/3 of the time. Averaging the measurements then gives a result very close to the true value of 2.3.

We can think of dithering as a nonlinear *stochastic process* $x = Q(x_0 + \xi)$ that takes a number x_0 and adds to it a random variable ξ. Here $Q(y)$ is the *rounding* function, which rounds its argument y up or down to the nearest integer. When we dither, we measure many realizations of x, which can vary because the random variable inside takes on different values in each trial. The heavy dots in the center plot of Figure 5.5 show a sequence of such realizations. Thus, we can consider x to be a random variable, too, and ask how its statistical properties depend on the statistical properties of the dither variable ξ.

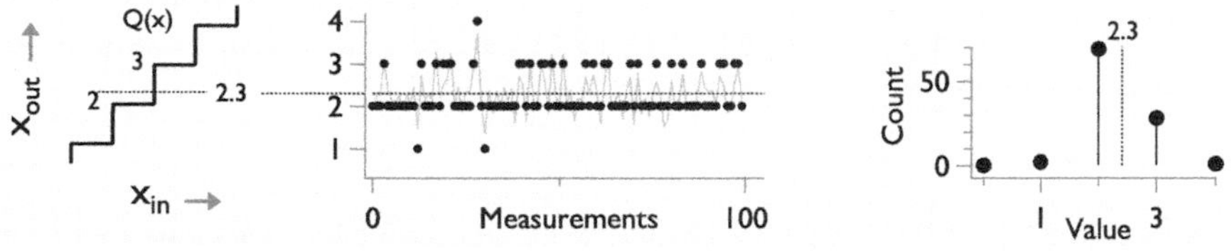

Fig. 5.5 Quantization and dithering. Left: an A/D converter has a nonlinear, "staircase" response, $x_{out} = Q(x_{in})$. Middle: adding a small amount of noise to an intermediate value (here, 2.3) and taking repeated measurements can recover the value. Right: in 100 trials, about $^2/_3$ give 2, about $^1/_3$ give 3, and there are a few trials that give 1 or 4. The noise is Gaussian, with $\sigma = {}^1/_2$.

[3] Commonly encountered A/D converters range from 8 bits (256 levels) to 24 bits (16 777 216 levels).

The "obvious" choice for ξ is a uniformly distributed random variable between $(-\frac{1}{2}, \frac{1}{2})$. Intuitively, this choice should, and does, lead to an unbiased x. That is, $\langle x \rangle = x_0$. Unfortunately, though, the estimation error will depend on x_0. To see this, assume x_0 to be an integer. Then, since the random number ξ is in the range $(-\frac{1}{2}, \frac{1}{2})$, we will have $Q(x_0 + \xi) = x_0$: we always round back to the original x_0, and our error is zero! But consider the case where x_0 is an integer + 0.5. In that case, the quantizer will round up for all $\xi > 0$ and down for all $\xi < 0$. The standard deviation for x is then clearly 0.5. Thus, we have an awkward situation, where the standard deviation σ of the noise depends on the level of the signal.

Problem 5.2 explores several solutions to this problem: instead of letting ξ be a uniform distribution, add noise drawn from a *triangular distribution*, $p(\xi) = 1 - |\xi|$ between -1 and 1. Then not only is $\langle x \rangle = x_0$, but also the noise level turns out to be independent of x_0.[4] In practice, simple analog noise sources are often Gaussian. If the standard deviation of this noise is chosen to be approximately half the spacing between D/A levels, then the bias and x_0 dependence of the dithered signal will be small (albeit not quite zero). Such a Gaussian dither is a good compromise between practicality and performance.

Pursuing this idea to its logical extreme, we can imagine measuring a signal with a *one-bit* converter. We would add noise and oversample by a large factor in order to be able to report an accurate value. This is the strategy behind *delta-sigma* A/D converters (Gershenfeld, 2000). Because of the oversampling, a simple low-pass filter, often built into the conversion chip itself, suffices to prevent aliasing. The disadvantage of delta-sigma conversion is a relatively long *latency time* – the lag between the signal and the digital output – which can be 10–100 times the sampling interval. Still, if the lag is small compared to the system timescales, this kind of conversion may be an attractive option.

Dithering can be used for output, too. For example, you can reduce the quantization nonlinearity of a slowly varying output by adding a random number to the output and, as rapidly as possible, changing the random number for the output. In other words, you merely implement Figure 5.5 backwards. For example, to output 2.3, add a random number drawn from a triangular distribution with standard deviation of range $(-1, +1)$ and round to the nearest D/A level.[5] Of course, this procedure assumes that you can cycle the output at a rate much faster than the system dynamics so that the extra noise is spread over a large frequency range, most of which is damped by the system itself.

[4] As Gray and Stockham (1993) recount, the dependence of the standard deviation of the dithered signal on the original level was discovered in 1980 while recording the album *Tusk* by Fleetwood Mac, which was one of the first digital rock and roll recordings. During a long fadeout of the University of Southern California Marching Band, the sound engineers could hear the quantization noise fading in and out as the signal decreased. They then figured out the triangular-noise trick, which eliminated the problem.

[5] If you can generate uniform random noise, then it is easy to generate noise with a triangular distribution: just add two independent trials of a uniformly distributed number from the range $(-\frac{1}{2}, \frac{1}{2})$. From Section A.6.1 online, the distribution of the sum of two random variables is their convolution. Here,
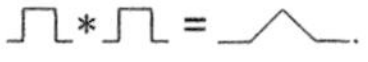

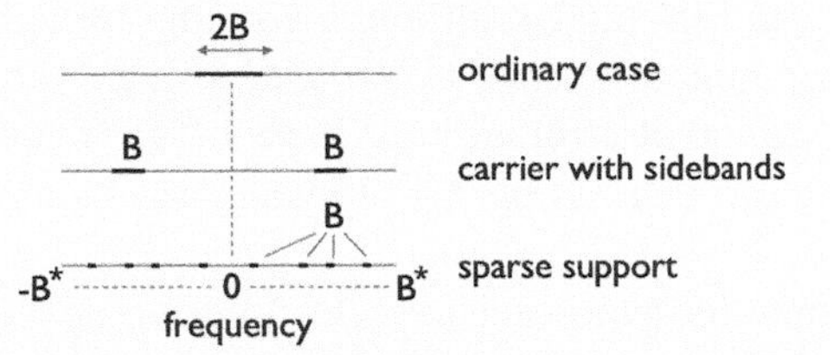

Fig. 5.6

The support of a signal with bandwidth B is usually measured from DC, or 0 frequency (top), but the sampling theorem holds when the bandwidth is shifted to higher frequencies (middle) or spread over a bandwidth B^* (bottom).

Stochastic output dithering is useful for temperature control, where updates are often at about 10 Hz (and system time constants are seconds or longer). If your D/A converter is faster, put the extra capacity to work to improve the resolution of your output! Since most D/A converters have onboard cyclic buffers that hold thousands of points, you can download those numbers and let them cycle (or just stream random numbers).

5.1.3 Beyond Nyquist

Although the Nyquist limit is fundamental, there are ways of "getting around" it. These ways do not violate the sampling theorem or the statements above, but they do force us to clarify the underlying assumptions. The sampling theorem states that you need to sample a signal of bandwidth B at a rate $2B$ in order not to have aliasing effects, *assuming* that you know nothing else about the signal. If you do know more, then the statement can change. For example, the sampling theorem requires a frequency range of $2B$ but does *not* require that the range be $(-B, B)$. In one case, the bandwidth may be centered on a nonzero *carrier* frequency ω_0. In another case, the bandwidth may be broken up into little disjoint pieces whose sum adds up to B. The first situation is amenable to a technique known as *downmixing*, the second to *compressed sensing*. See Figure 5.6.

Undersampling and Downmixing

A common type of signal consists of a carrier of angular frequency ω_0 and modulation in the range $(\omega_0 \pm \frac{1}{2}\omega_B)$. Because the signal is real, the support must also include the corresponding negative region $(-\omega_0 \pm \frac{1}{2}\omega_B)$. See Figure 5.6. There are two ways to sample such a signal using a data-acquisition system with a throughput of $2B$ symbols/time.

Undersampling. The naive approach is to undersample the high-frequency system and use aliasing to move the spectrum down into the $(-B, B)$ region. The time base of the sampling electronics needs to be stable on the time scale of the high-frequency signal. Thus, for a 1 GHz signal with a bandwidth of 10 kHz, timing must be stable to a fraction of a nanosecond. Similarly, although you digitize at roughly 20 kHz, your electronics must be fast enough to *sample and hold* a signal on the nanosecond time scale.[6] That is, the front-end electronics must have 2 GHz bandwidth. Despite these caveats, undersampling can be a simple and effective way to measure high-frequency signals.

Downmixing. Imagine that you want to measure a physical quantity that is noisy at low frequencies. One approach is to reduce measurement noise by choosing to measure the quantity in a "quiet" part of the noise spectrum. You then mix, or multiply, the signal of interest with a stable reference of frequency ω_0. If the signal of interest is $y(t) = A\cos[\omega_0 t + \varphi(t)]$ and the reference is $y_r(t) = \cos\omega_0 t$, then a standard trigonometric identity gives

$$y(t)\,y_r(t) = A\cos\omega_0 t \cdot \cos[\omega_0 t + \varphi(t)] = \tfrac{1}{2}A\left[\cos\varphi(t) + \cancel{\cos[2\omega_0 t + \varphi(t)]}\right], \qquad (5.7)$$

where we eliminate the $\cos 2\omega_0 t$ term with a low-pass filter. (If ω_0 is much higher than $\dot\varphi(t)$, then a simple low-pass filter is enough.) The low-frequency, downmixed signal corresponding to $\cos\varphi(t)$ then has bandwidth B and can be recorded by a conventional A/D converter. This procedure is often implemented using a *lock-in amplifier*, which supplies the reference and low-pass filter, or with separate analog mixing and filtering elements, especially for working with microwave signals.[7] Techniques to stabilize laser oscillators are also based on such ideas (Problem 11.20).

Compressed Sensing

A second situation where we can exceed the Nyquist limit is known as *compressed sensing* (or, sometimes, as *compressive sampling*). Figure 5.6 illustrates a situation where the signal bandwidth B is broken up into little pieces of width ΔB that are spread over a much larger bandwidth B^*. If we knew in advance which frequency intervals had nonzero elements, then we could apply band-pass filters centered on those intervals and measure the amplitude and phase of each signal component. Thus, we would measure a signal of bandwidth B^* using an A/D system with bandwidth $B \ll B^*$. Of course, we usually do not know in advance which frequency components will be present. Remarkably, compressed sensing allows one to reconstruct such signals using a sampling rate of order $B\log\frac{B^*}{B}$, which is still much lower than B^*. Since its inception in 2004, the algorithm and the thinking behind it have triggered a revolution in signal processing and applied mathematics.[8] Here, we give a few basic ideas about this rapidly developing area.

Sparseness: The idea that a signal has just a few frequency components can be generalized. A signal is *sparse* in a given basis B if it has just a few components in B. Thus, a signal with just a few frequency components is sparse over the basis of sines and cosines (used for Fourier series). Alternatively, a signal with a few isolated pulses

6 *Sampling* A/D converters include sample-and-hold electronics and specify the minimum sampling time.

7 *Heterodyne detection* is a related idea where the measured frequency ω_0' is offset slightly from that of the carrier, ω_0. The high-frequency term is again filtered, leaving a beat frequency $\sim \cos[(\omega_0 - \omega_0')t]$.

8 Candès et al. (2006) and Donoho (2006) have been together cited over 36,000 times.

that is otherwise zero is sparse over the basis set of discrete pulse functions (Dirac basis).

More formally, if a signal $f(t)$ is S-sparse over a set of complete or overcomplete basis functions $E = e_n(t)$, there are at most S nonzero amplitudes. That is, if

$$f(t) = \sum_{n=1}^{N} f_n \, e_n(t) , \tag{5.8}$$

then $f(t)$ is sparse if the vector $\boldsymbol{f}$ made from the components f_n has at most S nonzero elements, with $S \ll N$. If the $e_n(t)$ represent pulses, then $f(t)$ is zero except during a small number of time intervals. If they represent sinusoids, then $f(t)$ consists of just a few frequencies, and the expansion in Eq. (5.8) forms a Fourier series (Eq. A.63 online). In mathematical jargon, the set $\mathcal{S}$ of nonzero basis elements is the *support* of $\boldsymbol{f}$.

If we knew in advance the support $\mathcal{S}$, we could determine $f(t)$ from just S measurements. For a pulse basis, we would simply measure $f(t)$ during the few time-intervals when $f(t) \neq 0$. For a Fourier basis, we could use preconfigured band-pass filters whose center frequencies match those of the nonzero Fourier components. Then, having measured the S components, we could exactly reconstruct $f(t)$:

$$f(t) = \sum_{s \in \mathcal{S}} f_s \, e_s(t) . \tag{5.9}$$

Of course, we do not normally know which elements of $\boldsymbol{f}$ are nonzero without measuring all of them. But this is where the magic of compressed sensing begins. We illustrate the basic principle in Example 5.1, which is from Bryan and Leise (2013).

Example 5.1 (Finding a counterfeit coin) Imagine that you have seven coins, which all look identical. You have been told that six are genuine, and one is a fake. The coins look identical, but the fake is made of a cheaper metal and weighs a different amount from the other six, which all have mass m_0. Your task is to find out which coin is the fake using the fewest possible number of weight measurements. You might think that, in the worst case, you would have to weigh all seven coins. In fact, three measurements suffice.

The trick is to weigh *combinations* of coins. First, we weigh coins 1, 3, 5, and 7, which have a combined mass $M_1 = m_1 + m_3 + m_5 + m_7$. Next, coins 2, 3, 6, and 7 have mass $M_2 = m_2 + m_3 + m_6 + m_7$. Finally, coins 4 through 7, have mass $M_3 = m_4 + m_5 + m_6 + m_7$.

With a little thought, we can see that the three measurements $M_{1,2,3}$ are enough. For example, if coin 1 is the fake, then $M_1 \neq M_2 = M_3$. This is the only scenario that yields this particular combination of measurement results. In Problem 5.3, you will confirm that you can use the three values of M to uniquely determine which mass is fake.

To make contact with our sparseness notation, let $f_i = m_i - m_0$ be the deviation of the mass of coin i from the "genuine" value, m_0. Thus, the vector $\boldsymbol{f}$ of mass deviations is zero, except for one unknown element. It is sparse in the "basis" of coins. Next, we summarize the measurements by introducing a *measurement matrix* $\boldsymbol{\Phi}$ and a *measurement vector* $\boldsymbol{y}$. The latter is defined by $y_1 = f_1 + f_3 + f_5 + f_7$, and so on. Then $\boldsymbol{\Phi f} = \boldsymbol{y}$, with

$$\boldsymbol{\Phi} = \begin{pmatrix} 1 & 0 & 1 & 0 & 1 & 0 & 1 \\ 0 & 1 & 1 & 0 & 0 & 1 & 1 \\ 0 & 0 & 0 & 1 & 1 & 1 & 1 \end{pmatrix}. \tag{5.10}$$

See Problem 5.3 for details. Note that if the fake mass deviates from the others by one unit, then each measurement will give 0 or 1. With three measurements, there are $2^3 = 8$ possibilities, enough to pin down which of the seven coins is fake.

We can thus find the fake coin in three measurements, in contrast to the naive procedure, which needs one to seven measurements (average = 3.5). To generalize this example to a sparse vector $\boldsymbol{f}$ of N elements, of which at most S are nonzero, we need to understand

1. how many measurements to make (to choose M, the dimension of the $\boldsymbol{y}$ vector);
2. how to choose the measurement matrix (elements of $\boldsymbol{\Phi}$);
3. how to find $\boldsymbol{f}$ given $\boldsymbol{y}$ and $\boldsymbol{\Phi}$.

Here, we simply state some of the basic results.

Requirement 1: Naively, if we knew which elements of a particular S-sparse vector $\boldsymbol{f}$ were nonzero, we would need to make only S measurements. But not knowing which elements of $\boldsymbol{f}$ are nonzero beforehand, we need more measurements. For instance, in Example 5.1, we needed three measurements for $S = 1$. And if we allow measurement noise, still more measurements are required. The remarkable result, due to Candès, Tao, Donoho, and others, is that there are robust, practical algorithms that work for $M \gtrsim 2S \log N/S$ measurements, for a properly chosen measurement matrix $\boldsymbol{\Phi}$.

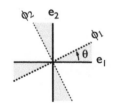

Requirement 2: The measurement matrix $\boldsymbol{\Phi}$ selects the particular linear combinations for each measurement. Intuitively, each measurement is the dot product between the desired unknown vector $\boldsymbol{f}$ and a row of $\boldsymbol{\Phi}$. The M rows of $\boldsymbol{\phi}_m$ of $\boldsymbol{\Phi}$ then form a set of vectors in $\mathbb{R}^N$. Because $M \ll N$, the vectors cannot form a complete basis, but they can be part of a basis. Intuitively, the $\boldsymbol{\phi}_m$ need to be as "incoherent" as possible from the basis $\boldsymbol{e}_n$. Each element $\boldsymbol{e}_n$ should project as evenly as possible onto the components $\boldsymbol{\phi}_m$. (See left).

More technically, the coherence μ between two sets of orthonormal basis vectors is

$$\mu(\boldsymbol{\phi}, \boldsymbol{e}) \equiv \sqrt{N} \max_{m \le M, n \le N} |\boldsymbol{\phi}_m \cdot \boldsymbol{e}_n| \tag{5.11}$$

To illustrate the notion of coherence, consider two 2d basis sets. The first has $\boldsymbol{e}_1 = \binom{1}{0}$ and $\boldsymbol{e}_2 = \binom{0}{1}$. The second is rotated at an angle θ, with $\boldsymbol{\phi}_1 = \binom{\cos\theta}{\sin\theta}$ and $\boldsymbol{\phi}_2 = \binom{-\sin\theta}{\cos\theta}$. Performing the various dot-products, we see that

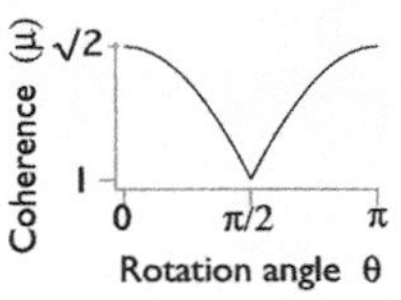

$$\mu(\boldsymbol{\phi}, \boldsymbol{e}) = \mu(\theta) = \sqrt{2} \max\left(|\cos\theta|, |\sin\theta|\right), \tag{5.12}$$

which varies from 1 to $\sqrt{2}$ (see left). The two bases are maximally incoherent at $\theta = \frac{\pi}{4}$, where a basis vector from one basis projects equally onto the other basis vectors.

With the normalization in Eq. (5.11), incoherent basis sets will have $\mu \approx 1$. (Perfect alignment, the worst case, gives $\mu = \sqrt{N}$; maximal incoherence corresponds to $\mu = 1$.) We can generalize our example to rotations in N-dimensional spaces. For the "pulse basis" in the time domain, it turns out that the Fourier basis is maximally incoherent (and vice versa). Schematically, ⊥ ↔ ∿: a delta function in the time domain contains all frequencies, while a single sine wave extends over all times. More strikingly, for $N \gg 1$, a randomly chosen basis also has $\mu \to 1$. This statement has been proven for matrices whose elements are drawn from independent Gaussian distributions and, empirically, holds even when the matrix elements are drawn from a Bernoulli distribution over two values (e.g., 0 and 1).

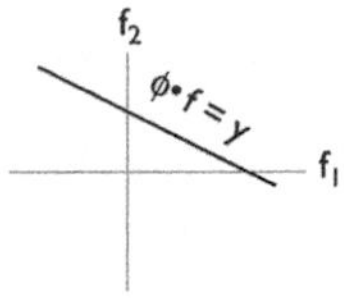

Requirement 3: We now know how many measurements M to make and how to choose the elements of the measurement matrix $\mathbf{\Phi}$. We can then proceed to take the measurements $\mathbf{y}$. The last step is to find the correct sparse $\mathbf{f}$ that satisfies $\mathbf{\Phi f} = \mathbf{y}$. By assumption, the problem is always *underdetermined*, with an infinite number of vectors $\mathbf{f}$ that satisfy the measurement constraints. But almost all of these vectors are not sparse in the basis $\mathbf{e}$. Schematically, the situation resembles the top sketch at right, where the constraint $\boldsymbol{\phi} \cdot \mathbf{f} = y$ defines a one-dimensional subspace in the two-dimensional basis set f_1, f_2. The middle plot shows that the smallest $\mathbf{f}$, as measured in the ℓ_2 norm, generically includes nonzero elements f_1 and f_2. The bottom plot then shows that if distances are measured in the ℓ_1 norm, the minimum generically consists of either one or the other component, but not both.

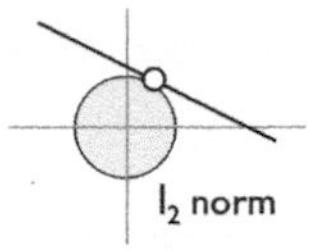

The problem of finding a sparse vector $\mathbf{f}$ can thus be stated as solving[9]

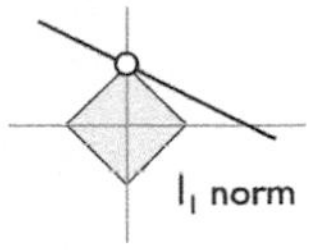

$$\min \|\mathbf{f}\|_1 , \quad \text{subject to} \quad \mathbf{\Phi f} = \mathbf{y} . \tag{5.13}$$

The problem defined by Eq. (5.13) has been well studied. Viewed as a constrained optimization (Section A.5.1 online), the problem is convex, meaning that there is a unique solution. Equally important, there are efficient numerical algorithms that can find this solution.

Finally, there are important generalizations. So far, we have assumed that the signal $f(t)$ is measured over a basis that is sparse. But, often, the signal is only sparse in a different basis. For example, images, although usually not sparse in either delta-function (pixel) or Fourier bases, typically are sparse over a *wavelet* basis, a fact used by the JPEG-2000 compression algorithm. To formalize these ideas, assume that there is a basis $\mathbf{e}'(t)$ where $f(t)$ is sparse, $f(t) = \sum_n f_n' \mathbf{e}'(t)$, but that we measure $f(t)$ in a different basis $\mathbf{e}(t)$ that is not sparse. Knowing the sparse basis, we can find the orthogonal transformation $\mathbf{\Psi}$ that transforms the Fourier coefficients $\mathbf{f}'$ into $\mathbf{f}$. We write $\mathbf{f} = \mathbf{\Psi f}'$. Then $\mathbf{\Phi f} = \mathbf{\Phi \Psi f}' = \mathbf{y}$. Thus, in terms of the measurement matrix $\mathbf{\Phi}' \equiv \mathbf{\Phi\Psi}$, the problem reduces to the original one of minimizing $\|\mathbf{f}'\|_1$ subject to $\mathbf{\Phi}' \mathbf{f}' = \mathbf{y}$. Remarkably, at least in some cases, the fact that $\mathbf{\Phi}'$ can be chosen randomly also implies that $\mathbf{\Phi}$ can be similarly chosen. In other words, the same matrix that

[9] It would be more logical to minimize $\mathbf{f}$ over the ℓ_0 norm, which just counts the number of nonzero amplitudes of basis vectors. Unfortunately, the ℓ_0 analog to Eq. (5.13) gives a nonconvex optimization problem that often has many local minima. Using the ℓ_1 norm gives a convex relaxation of the original problem, with a unique minimum that is generically sparse. But it may not be the sparsest solution.

works for the f basis will very likely work for the f' basis. Then, having found f' and knowing $\mathbf{\Psi}$, you can find the desired f.

Another generalization is from sparseness to *compressibility*. A sparse signal has a basis with a small number S of nonzero basis vectors. A *compressible* signal relaxes this assumption by allowing the amplitude of the other elements to be small, rather than zero.[10] This generalization is important, because any noisy signal cannot be sparse in the strict sense defined above. In addition, signals often result from dynamics that produces a spectrum of modes whose amplitudes decreases as some power law (Waterfall et al., 2006). Reconstruction by minimizing the ℓ_1 norm is typically robust to such noisy signals, and its performance degrades gracefully. Similar considerations show that compressed sensing will be robust to measurement noise, too. With noise, the reconstruction problem becomes

$$\min \|\boldsymbol{f}\|_1 , \quad \text{subject to} \quad \|\mathbf{\Phi} \boldsymbol{f} - \boldsymbol{y}\|_2 \leq \epsilon , \tag{5.14}$$

where ϵ reflects the amount of measurement noise. Note that while we continue to enforce sparseness by finding the smallest f according to the ℓ_1 norm, we quantify measurement errors using the usual ℓ_2 norm. We explore this extension briefly in Problem 5.4, whose surprising conclusion is that there can be phase transitions in reconstructions: For a given noise level and given sparseness, the probability of successful reconstruction rises abruptly from zero at a critical number of measurements M^*.

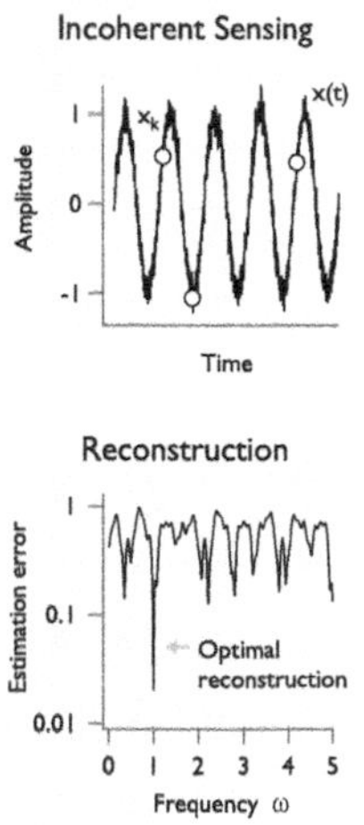

To see some of these ideas, we illustrate at left a very simple reconstruction problem that we solve by brute force (trying all possible frequencies). Using brute force allows us to bypass the techniques for ℓ_1 norm minimization. The figure shows a single, noise-corrupted sine wave. The signal is sparse in the frequency domain (but not in the time domain). We see that we can find the frequency of the sine wave from the widely spaced point measurements whose average separation clearly violates the Nyquist sampling limit (two points per period of the sine wave). The three sampling times have been chosen randomly, to assure incoherence with respect to the frequency basis.

Applications: One of the exciting applications of compressed sensing concerns imaging. Typically, a CCD camera records a two-dimensional array of pixel values. File compression saves space, eliminating information that was acquired but interpreted as noise. What if it were possible to acquire the useful information directly? Such an A/I (analog-to-information) converter would bypass the need to record a large initial file and would acquire directly the useful information that the compression algorithm preserves. In principle, compressed sensing algorithms can do this, in real time. But these algorithms are inevitably tied to hardware (in the way that an anti-alias filter is for ordinary sampling).

Here are a few preliminary projects:

- *Single-pixel camera*. Duarte et al. (2008) use a digital micromirror array to select and then sum onto a single detector a subset of image "pixels." Each set of pixel

[10] More precisely, there is an upper bound on the decay of coefficients when sorted in decreasing order, typically a power law. We can then upper bound the error made by truncation.

choices then corresponds to a row $\boldsymbol{\phi}_m$ in the $\boldsymbol{\Phi}$ matrix, with a 1 when the pixel is selected by a micromirror and a 0 if not. The usual algorithms are applied to M such measurements.

- *White-paint "lens."* Liutkus et al. (2014) pass coherent light through a thin, multiply scattering medium (a layer of white paint). Each camera pixel records light from a set of unique light paths, allowing reconstruction. Placing a film of white paint in front of a camera can actually *improve* its imaging resolution! Of course, the analog "measurement matrix" created by the paint layer is not known and must be calibrated carefully.
- *Xampling*. This is a general-purpose A/D board that can record GHz bandwidth signals while sampling at 240 MHz (Eldar and Kutyniok, 2012). The board rapidly and randomly modulates the signal, integrating over some time and recording only the slower, averaged output. The reconstruction is done rapidly by an FPGA processor.

5.2 Tools for Discrete Dynamical Systems

Before we tackle discrete dynamics and digital control, we need tools to handle such systems. We present mathematical background for three types: *digital filter*, the analog of a time-domain filter for continuous dynamics; *Z-transform*, an analog of the Laplace transform; and *discrete time Fourier transform*, in order to discuss the frequency domain.

5.2.1 Digital Filters

Consider the discrete dynamical system

$$y_k = f(y_{k-1}, y_{k-2}, \cdots, y_{k-n}; u_k, u_{k-1}, \cdots, u_{k-n}), \tag{5.15}$$

with y_k an output at discrete time k and u_k its corresponding input. The number of coefficients is $2n+1$. In general, f is a nonlinear function and can lead to very complex dynamics.[11] Here, we focus on linear discrete dynamics of the form

$$y_k = A_1 y_{k-1} + A_2 y_{k-2} + \cdots + A_n y_{k-n} + B_0 u_k + B_1 u_{k-1} + \cdots + B_n u_{k-n}. \tag{5.16}$$

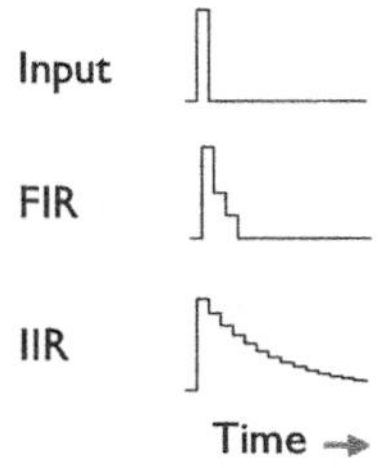

If all the A coefficients in Eq. (5.16) are zero, the dynamics give a *Finite-Impulse-Response* (FIR) filter.[12] If any of the A coefficients are nonzero, then the filter is an *Infinite-Impulse-Response* (IIR) filter. The names are self-explanatory: for FIR filters, the response to a pulse is zero after a finite number of time steps, as shown at right. For IIR filters, the response takes an infinite number of steps to decay to zero. (Typically, though, the magnitudes decay exponentially, so that after a finite number of time steps, the response can be very small.) IIR filters are often approximations to familiar continuous filters, which are also IIR. For example, the discrete approximation to the

[11] A famous example is the *logistic equation* $y_k = r y_k(1 - y_k)$, with r a control parameter (Strogatz, 2014).

[12] FIR filters are also called *feedforward*, *nonrecursive*, and *transversal* filters, as well as *tapped delay lines*.

familiar continuous low-pass filter (first-order system whose response decays exponentially) has a response that decays geometrically, taking an infinite number of steps to reach zero.

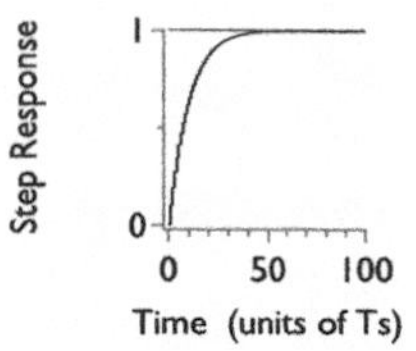

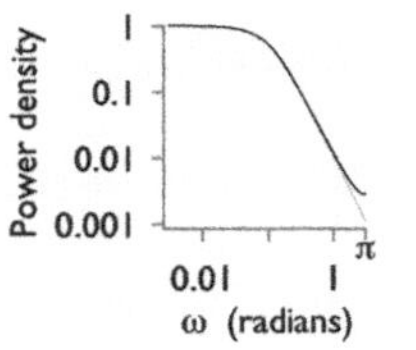

Example 5.2 (IIR vs. FIR low-pass filter) To compare the IIR and FIR approaches, consider the discrete equivalent of a one-pole, low-pass filter. The IIR filter is

$$y_k = a y_{k-1} + (1-a) u_{k-1}, \qquad 0 < a < 1, \tag{5.17}$$

where $A_1 = a$ and $B_0 = 1-a$, and all other coefficients equal 0. The step and frequency responses for $a = 0.9$ are shown at left. The light gray trace on the frequency-response plot is the response of the equivalent continuous filter.

Now consider the FIR equivalent:

$$y_k = \left(\frac{1-a}{1-a^{n+1}}\right)(u_k + a u_{k-1} + a^2 u_{k-2} + \cdots + a^n u_{k-n}). \tag{5.18}$$

For large n, the step and frequency responses of the two filters are indistinguishable (Problem 5.6). You can always approximate an IIR filter by a high-order FIR filter.

In Example 5.2, the IIR filter has two coefficients, whereas the FIR equivalent typically uses roughly a hundred. IIR filters are thus much faster and simpler to compute.[13] Nonetheless, FIR filters offer many practical advantages: with no feedback, they are always stable.[14] Because calculations on the input are done only once, they can use lower-precision arithmetic. There is no degradation of precision, which can result from feeding back the results of a calculation. Perhaps most important, FIR filters can have properties that no analog or IIR approximant can have, such as linear phase response (Problem 5.7). There, each frequency ω has its amplitude component changed and is then simply delayed by a time τ that is independent of ω (thus, $\phi = \omega\tau$). By contrast, minimum-phase IIR filters distort the phase response, changing the shape of the initial waveform.

Although the IIR structure is common in control applications, FIR filters are used, for example, as adaptive feedforward controllers (Chapter 10).

Example 5.3 (Fibonacci sequence) The *Fibonacci sequence* is generated by the IIR dynamical system (recursion relation)

$$y_k = y_{k-1} + y_{k-2}, \qquad y_0 = y_1 = 1. \tag{5.19}$$

Directly iterating forward in time gives the well-known sequence,

$$y_k = 1,\ 1,\ 2,\ 3,\ 5,\ 8,\ 13,\ 21,\ 34,\ \ldots. \tag{5.20}$$

We can also find y_k using a method that is reminiscent of solving a continuous second-order linear differential equation. Equation (5.19) is also second order because y_k

[13] FIR filters of length 2^n, with n integer, can be sped up using Fast Fourier Transforms.

[14] Choosing $|a| > 1$ in Eq. (5.17) leads to an instability, whereas Eq. (5.18) is stable for all a.

depends on two time-differenced values (discrete analogs of first and second derivatives). Note that we give two initial conditions, which is again what we expect for a second-order system. We assume a solution of the form $y_k = Az^k$ and substitute:

$$Az^k = Az^{k-1} + Az^{k-2} \quad \Longrightarrow \quad 1 = z^{-1} + z^{-2} \quad \Longrightarrow \quad z^2 - z - 1 = 0. \tag{5.21}$$

Solving the quadratic equation, we find $z_\pm = \frac{1}{2}(1 \pm \sqrt{5})$. Thus, $y_k = A_+ z_+^k + A_- z_-^k$. The initial conditions imply that $y_0 = A_+ + A_- = 1$ and $y_1 = A_+ z_+ + A_- z_- = 1$. In matrix form,

$$\begin{pmatrix} 1 & 1 \\ z_+ & z_- \end{pmatrix}\begin{pmatrix} A_+ \\ A_- \end{pmatrix} = \begin{pmatrix} 1 \\ 1 \end{pmatrix} \Rightarrow \begin{pmatrix} A_+ \\ A_- \end{pmatrix} = \frac{1}{z_- - z_+}\begin{pmatrix} z_- & -1 \\ -z_+ & 1 \end{pmatrix}\begin{pmatrix} 1 \\ 1 \end{pmatrix} = \frac{1}{\sqrt{5}}\begin{pmatrix} \frac{\sqrt{5}+1}{2} \\ \frac{\sqrt{5}-1}{2} \end{pmatrix}, \tag{5.22}$$

$$\Longrightarrow \quad y_k = \frac{1}{\sqrt{5}}\left(\frac{(1+\sqrt{5})^{k+1} - (1-\sqrt{5})^{k+1}}{2^{k+1}}\right). \tag{5.23}$$

Substituting $k = 1, 2, 3, \ldots$ gives, amazingly, the sequence given in Eq. (5.20). Note that $|z_-| = \left|\frac{1-\sqrt{5}}{2}\right| < 1$ and $z_+ = \frac{1+\sqrt{5}}{2} > 1$. The latter term grows geometrically (exponentially).

The steps used to solve the difference equation in Example 5.3 parallel those used to solve linear differential equations by Laplace transforms (Section A.4.5 online). Indeed,

$$y_k = A_1 y_{k-1} + A_2 y_{k-2} + \cdots, \quad \Longrightarrow \quad z^k - A_1 z^{k-1} - A_2 z^{k-2} - \cdots A_k = 0, \tag{5.24}$$

where $y_k \to Az^k$. If all k roots of the characteristic polynomial have $|z_k| < 1$, then the solutions are stable. The requirement $|z| < 1$ is thus analogous to the stability requirement Re $s < 0$ for solutions e^{st} to continuous, constant-coefficient, linear differential equations.

5.2.2 The Z-Transform

For ordinary differential equations, the properties of solutions of the form e^{st} motivated introducing the Fourier and Laplace transforms. For discrete dynamical systems, as already seen in Example 5.3, solutions of the form z^k play a similar role. We are thus led to define the Z-transform, a discrete analog of the Laplace transform.

Let $f(t)$ be a continuous signal that "starts at $t = 0$" and is sampled at times kT_s ($k = 0, 1, \ldots$). We can take the Laplace transform of the sampled signal f_s:

$$f_s(t) = \sum_{k=0}^{\infty} f(t)\,\delta(t - kT_s)$$

$$\mathcal{L}[f_s] = \int_0^{\infty} dt \sum_{k=0}^{\infty} f(t)\,\delta(t - kT_s)\, e^{-st} = \sum_{k=0}^{\infty} f(kT_s)\, e^{-s(kT_s)} \tag{5.25}$$

With $f_k \equiv f(kT_s)$ and $z \equiv e^{sT_s}$, we write

$$\mathcal{L}[f_s] = \sum_{k=0}^{\infty} f_k \, z^{-k} \equiv \mathcal{Z}[f_k] \equiv f(z) \,. \tag{5.26}$$

In Eq. (5.26), the *Z-transform* of the sampled function f_k is a function of the complex variable z. In our "operator overloading" convention, $f(z)$ denotes the Z-transform of f. We think of $f(\cdot)$ as representing one object with different representations in the t, s, and z domains. Note that the Z-transform operates on sequences of numbers, the f_k, whose values depend on the sampling time T_s. Thus, the function $f(z)$ depends on the sampling time indirectly through its dependence on the f_k. The Z-transform is closely related to the method of *generating functions*, a commonly used technique in statistical physics.

Example 5.4 (Elementary examples) Let $f_k = 0$ for $k < 0$. Then

$$f_k = \delta_0 \qquad f(z) = 1 \tag{5.27a}$$

$$f_k = \theta_k \qquad f(z) = \sum_{k=0}^{\infty} (1) \, z^{-k} = 1 + z^{-1} + z^{-2} + \cdots = \frac{1}{1 - z^{-1}} = \frac{z}{z-1} \tag{5.27b}$$

$$f_k = \alpha^k \qquad f(z) = \sum_{k=0}^{\infty} \alpha^k \, z^{-k} = \sum_{k=0}^{\infty} \left(\frac{\alpha}{z}\right)^k = \frac{1}{1 - \frac{\alpha}{z}} = \frac{z}{z - \alpha} \tag{5.27c}$$

$$f_k = k \qquad f(z) = \sum_{k=0}^{\infty} k z^{-k} = -z \sum_{k=0}^{\infty} \frac{\mathrm{d}}{\mathrm{d}z} z^{-k} = -z \frac{\mathrm{d}}{\mathrm{d}z} \left(\frac{z}{z-1}\right) = +\frac{z}{(z-1)^2} \,. \tag{5.27d}$$

Here, $\delta_0 = 1$ for $k = 0$ and equals 0 for $k = 1, 2, \ldots$, and the step function $\theta_k = 1$ for $k \geq 0$.

Many of the properties of Laplace and Fourier transforms also have analogs. For example, assuming $f_\ell = 0$ for $\ell < 0$ gives the *shift theorem*,

$$\mathcal{Z}[f_{k-m}] = \sum_{k=0}^{\infty} f_{k-m} \, z^{-k} = \sum_{\ell=-m}^{\infty} f_\ell \, z^{-(\ell+m)} = \sum_{\ell=0}^{\infty} f_\ell \, z^{-\ell} \, z^{-m} = z^{-m} f(z) \,. \tag{5.28}$$

Then, defining *convolution* as $x * y[k] = \sum_{m=0}^{\infty} x_m y_{k-m}$ leads to the *convolution theorem*,

$$\mathcal{Z}[f * g] = \sum_{k=0}^{\infty} \sum_{m=0}^{\infty} f_m \, g_{k-m} \, z^{-k} = \sum_{m=0}^{\infty} f_m \sum_{k=0}^{\infty} g_{k-m} \, z^{-k} = \sum_{m=0}^{\infty} f_m \, z^{-m} g(z) = f(z) \, g(z) \,. \tag{5.29}$$

Finally, using the shift theorem for $\Delta < 0$ gives the Z-transform of a difference,

$$\mathcal{Z}[f_{k+1}] = \sum_{k=0}^{\infty} f_{k+1} \, z^{-k} = z \sum_{k=0}^{\infty} f_{k+1} \, z^{-(k+1)} = z \sum_{m=1}^{\infty} f_m \, z^{-m} = z \, [f(z) - f_0] \,. \tag{5.30}$$

Compare Eq. (5.30) to the Laplace transform of a derivative. Note also the *initial-* and *final-value* theorems, $f_0 = \lim_{z\to\infty} f(z)$ and $\lim_{k\to\infty} f_k = \lim_{z\to 1}[(z-1)f(z)]$ (Problem 5.5).

Example 5.5 (First-order equation) Let $f_{k+1} = \alpha f_k$, with initial condition $f_0 = 1$ and $\alpha < 1$. Then, Z-transforming the equation, we have

$$z\,[f(z) - f_0] = \alpha f(z) \implies (z-\alpha)f(z) = z \implies f(z) = \frac{z}{z-\alpha} \implies f_k = \alpha^k \tag{5.31}$$

The *inverse* Z-transform is calculated most simply by manipulating $f(z)$ into a standard form and looking up the inverse from known forward transforms, as we did in the above example. Partial-fraction expansions are often useful and are more accurate numerically for high-order systems (Eq. 4.46). But sometimes the explicit inverse Z-transform is useful:

$$f_k = \mathcal{Z}^{-1}[f(z)] = \frac{1}{2\pi\mathrm{i}}\oint_C \mathrm{d}z\, f(z)\, z^{k-1}\,, \tag{5.32}$$

where the integral follows the closed path C counterclockwise and must encircle all poles of $f(z)$. To prove Eq. (5.32), we write

$$\frac{1}{2\pi\mathrm{i}}\oint_C \mathrm{d}z \left(\sum_{m=0}^{\infty} f_m\, z^{-m}\right) z^{k-1} = \frac{1}{2\pi\mathrm{i}}\sum_{m=0}^{\infty} f_m \oint_C \mathrm{d}z\, z^{-m+k-1} = \frac{2\pi\mathrm{i}}{2\pi\mathrm{i}}\sum_{m=0}^{\infty} f_m\, \delta_{mk} = f_k\,. \tag{5.33}$$

As a simple application, we return to Example 5.5. If $f(z) = \frac{z}{z-\alpha}$, with $\alpha < 1$, then

$$f_k = \frac{1}{2\pi\mathrm{i}}\oint_C \mathrm{d}z\, \frac{z}{z-\alpha}\, z^{k-1} = \frac{1}{2\pi\mathrm{i}}\oint_C \mathrm{d}z\, \frac{z^k}{z-\alpha} = \alpha^k\,, \tag{5.34}$$

where we use the Residue Theorem, Eq. (A.47 online), for the simple pole at $z = \alpha$.

5.2.3 The Discrete-Time Fourier Transform

Just as the Laplace and Fourier transforms are intimately related (Section 2.3.1), so, too, is the Z-transform related to a discrete analog of the Fourier transform, which characterizes the frequency response. To connect our previous expression for the Fourier transform of a sampled signal, Eq. (5.3), with the Z-transform, we return to the definition of the sampled signal, $f_s(t) = \sum_k f(t)\delta(t - kT_s)$, and write, with $f_k \equiv f(kT_s)$,

$$f_s(\omega_0) = \int_{-\infty}^{\infty} \mathrm{d}t\, f_s(t)\,\mathrm{e}^{-\mathrm{i}\omega_0 t} = \int_{-\infty}^{\infty} \mathrm{d}t \sum_{k=-\infty}^{\infty} f(t)\,\delta(t-kT_s)\,\mathrm{e}^{-\mathrm{i}\omega_0 t} = \sum_{k=-\infty}^{\infty} f_k\,\mathrm{e}^{-\mathrm{i}\omega_0 T_s k}\,. \tag{5.35}$$

Rescaling frequency by T_s leads to the *discrete time Fourier transform* (DTFT),

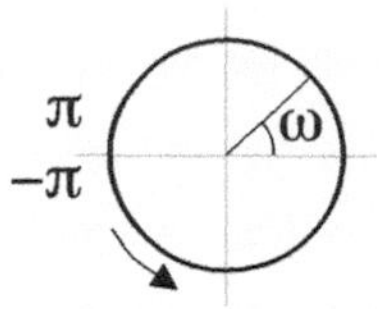

$$f(\omega) \equiv \sum_{k=-\infty}^{\infty} f_k \, \mathrm{e}^{-\mathrm{i}\omega k} \qquad \Longleftrightarrow \qquad f_k = \int_{-\pi}^{\pi} \frac{\mathrm{d}\omega}{2\pi} f(\mathrm{e}^{\mathrm{i}\omega}) \, \mathrm{e}^{\mathrm{i}\omega k} \, . \tag{5.36}$$

The integration path is illustrated at left. In Eq. (5.36), the integral resembles the definition of the Z-transform, with $z = \mathrm{e}^{\mathrm{i}\omega}$, except that the sum is over positive and negative k (in the same way that the Fourier transform is over positive and negative frequencies). When evaluated at real frequencies, the inverse DTFT is a special case of the inverse Z-transform, with the contour C being the unit circle. All poles are assumed to be inside the unit circle.

Another important relation is *Parseval's Theorem* (Problem 5.8), which equates signal energy in the time and frequency domains. For a real, discrete-time signal f_k,

$$\sum_{k=-\infty}^{\infty} f_k^2 = \int_{-\pi}^{\pi} \frac{\mathrm{d}\omega}{2\pi} \, |f(\omega)|^2 \, . \tag{5.37}$$

Figure 5.7 illustrates the different types of "Fourier" transforms. The *discrete Fourier transform* (DFT) is used for a *finite* time series, for example N evenly sampled points $f_k = f(kT_s)$, $k = 0, 1, \ldots, N-1$. The DFT is usually implemented using a *Fast Fourier Transform* (FFT) and has a discrete spectrum whose lowest-frequency component is $\Delta f = \frac{f_s}{N} = \frac{1}{NT_s}$. The signal implicitly repeats with a period NT_s. By contrast, the discrete-time Fourier transform (DTFT) applies to an infinitely long time series sampled at T_s and has a spectrum measured by a *continuous* variable ω. Like the DFT, the DTFT is periodic in frequency, with period ω_s. Unlike the DFT, the DTFT can describe motion of arbitrarily low frequency. As $N \to \infty$, the DFT $\to$ DTFT, analogous to the relation between the Fourier Transform and Fourier Series. Alternately, the DTFT is dual to the Fourier Series:

Fourier Series: periodic and continuous in time; discrete and aperiodic in frequency.

DTFT: periodic and continuous in frequency; discrete and aperiodic in time.

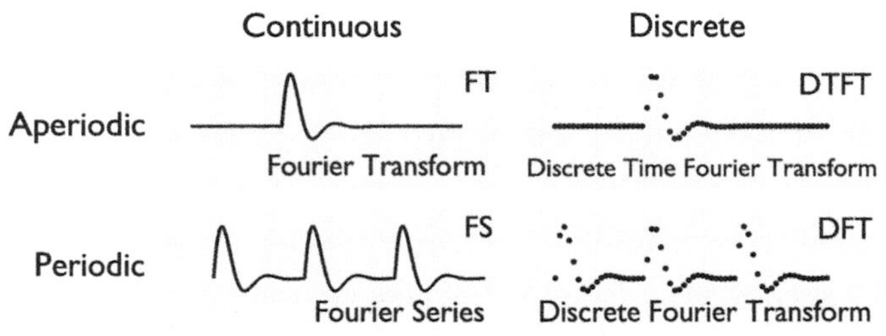

Fig. 5.7 Four different types of Fourier transformations. Cf. Smith (1999), Fig. 8-2.

5.3 Discretizing Dynamical Systems

Section 5.1 discussed how to use sampling to discretize a continuous signal. We focused on how to avoid pitfalls such as aliasing, which can make the sampled signal, a poor representation of the continuous signal. Now, using the tools from Section 5.2, we ask how to discretize a dynamical system so that it is "close" to its continuous counterpart. Because discretization methods lose the information between samples, different schemes are better at preserving different aspects.

5.3.1 State Space, with Zero-Order Hold

When the output of a D/A converter is sent to a continuous system, it often has a "staircase" form.[15] The continuous signal $u(t)$ becomes a *zero-order hold* (ZOH) signal,

$$u_k(t) = u(kT_s) \qquad kT_s < t < (k+1)T_s \,. \tag{5.38}$$

For a linear, time-invariant system with state-space matrices $\{\boldsymbol{A}, \boldsymbol{B}, \boldsymbol{C}\}$, we can then calculate the response exactly. If $\boldsymbol{A}$ can be inverted,[16] we have, with $\boldsymbol{x}(0) = \boldsymbol{x}_0$ and $u(0) = u_0$,

$$\boldsymbol{x}_1 \equiv x(T_s) = e^{\boldsymbol{A}T_s}\, \boldsymbol{x}_0 + \boldsymbol{A}^{-1}\left(e^{\boldsymbol{A}T_s} - \mathbb{I}\right)\boldsymbol{B}\, u_0 \quad \equiv \quad \boldsymbol{A}_d \boldsymbol{x}_0 + \boldsymbol{B}_d \boldsymbol{u}_0 \,. \tag{5.39}$$

Iterating then gives a linear, time-invariant, discrete dynamical system $\boldsymbol{x}_{k+1} = \boldsymbol{A}_d \boldsymbol{x}_k + \boldsymbol{B}_d u_k$ and $y_{k+1} = \boldsymbol{C}_d \boldsymbol{x}_{k+1}$. The system matrices for this *ZOH discretization* are given by

$$\boldsymbol{A}_d = e^{\boldsymbol{A}T_s}\,, \qquad \boldsymbol{B}_d = \boldsymbol{A}^{-1}\left(\boldsymbol{A}_d - \mathbb{I}\right)\boldsymbol{B}\,, \qquad \boldsymbol{C}_d = \boldsymbol{C}\,. \tag{5.40}$$

Example 5.6 (ZOH discretization of a first-order system) The simplest example is a first-order system, with $A = -1$, $B = C = 1$. The ZOH discretization rules give $A_d = e^{-T_s}$, $B_d = 1 - A_d$, $C_d = 1$. Recall that our scaling of time has absorbed a constant – call it τ – so that $e^{-T_s} \to e^{-T_s/\tau}$ in physical units. We illustrate the time dependence of the response to two inputs, for $T_s = 1$. At right is a step response. Notice that the discrete response exactly matches that of the continuous system, delayed by one step (the light dotted line). (It is *step invariant*.) By contrast, a sine-wave input also leads to a slightly distorted, delayed response, since the discretization of a sine-wave input differs from a continuous sine wave.

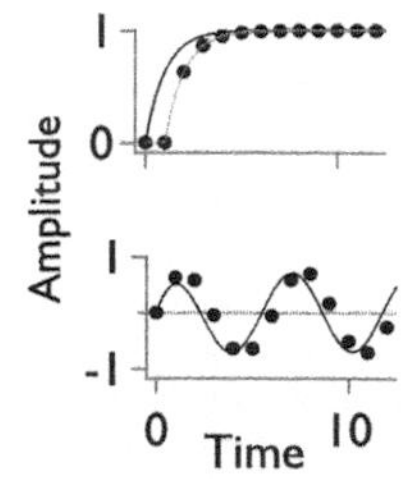

[15] Some D/A converters include an anti-aliasing filter that smooths the controller output. (See Figure 5.1.)

[16] If $\boldsymbol{A}$ is singular, we can still discretize by expanding the exponential. See Problem 5.9.

5.3.2 Frequency Domain and Transfer Functions

It is particularly instructive to compare the frequency responses of discrete and continuous systems. In order to find the frequency response of the linear system $\boldsymbol{x}_{k+1} = \boldsymbol{A}_{\mathrm{d}}\boldsymbol{x}_k + \boldsymbol{B}_{\mathrm{d}}u_k$ and $y_{k+1} = \boldsymbol{C}_{\mathrm{d}}\,\boldsymbol{x}_{k+1}$, we take the Z-transform of both sides and solve for the ratio of output to input, which we define as the ZOH (step-invariant) discrete transfer function:

$$G_{\mathrm{d}}(z) = \frac{y(z)}{u(z)} = \boldsymbol{C}_{\mathrm{d}}(z\mathbb{I} - \boldsymbol{A}_{\mathrm{d}})^{-1}\boldsymbol{B}_{\mathrm{d}} = \left(1 - z^{-1}\right) \mathcal{Z}\left\{\mathcal{L}^{-1}\left[\frac{G(s)}{s}\right]\right\}. \tag{5.41}$$

In the second expression, we convert directly from $G(s)$ to $G_{\mathrm{d}}(z)$ by first taking the inverse Laplace transform from s to t and then the Z-transform from $t = kT_{\mathrm{s}}$ to z. The term $\frac{1-z^{-1}}{s}$ corresponds to performing the ZOH, illustrated at left (cf. Problem 5.10).

Figure 5.8 illustrates the frequency response for the first-order system of Example 5.6. With $A_{\mathrm{d}} = \mathrm{e}^{-T_{\mathrm{s}}}$, $B_{\mathrm{d}} = 1 - A_{\mathrm{d}}$, $C_{\mathrm{d}} = 1$, the explicit transfer function is

$$G_{\mathrm{d}}(z) = C_{\mathrm{d}}\frac{1}{z - A_{\mathrm{d}}}B_{\mathrm{d}} = \frac{1 - \mathrm{e}^{-T_{\mathrm{s}}}}{z - \mathrm{e}^{-T_{\mathrm{s}}}}, \tag{5.42}$$

and the discrete Bode plots are generated by substituting $z = \mathrm{e}^{\mathrm{i}\omega T_{\mathrm{s}}}$ and letting ω range over several periods $\frac{2\pi}{T_{\mathrm{s}}}$ of frequency. Figure 5.8 (left) shows the standard first-order response as a solid black line and the discrete system, sampled at $T_{\mathrm{s}} = 0.1$, as a dotted line. The agreement between the two is reasonable in magnitude until about $\omega = 10$ and in phase until about $\omega = 5$ – about 16% and 8%, respectively, of the sampling frequency, $\omega_{\mathrm{s}} = 20\pi \approx 63$. At right in Figure 5.8 are the same plots with a linear frequency scale, which illustrate the periodicity in ω_{s}. The three different sampling intervals show that a faster sampling extends the agreement between the continuous system and the discrete approximation.

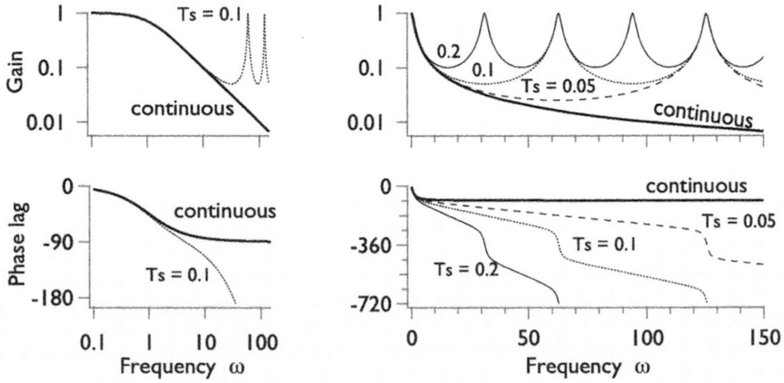

Fig. 5.8 First-order frequency response for continuous (heavy line) and discrete dynamics. Left: Bode plots, with curves for T_{s} in dotted line. Right: The same Bode plots with a linear frequency scale. Three discrete curves, for $T_{\mathrm{s}} = 0.2, 0.1$, and 0.05.

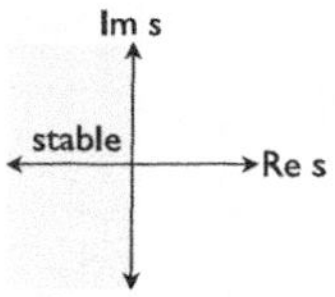

Continuous

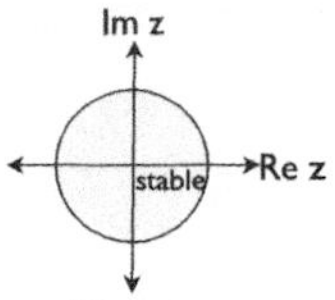

Discrete

Another important connection between continuous and discrete transfer functions is the relation between stability boundaries. Having all poles of a transfer function in Re $s < 0$ implies stable motion of a continuous linear system and corresponds to the condition that all poles have $|z| < 1$ for discrete transfer functions. To understand why, recall that in a continuous system $\dot{x} = sx$ implies $x(t) \sim \mathrm{e}^{st}$, which decays only if Re $s < 0$. Similarly, $x_{k+1} = zx_k$ implies $x_k \sim z^k$, which decays only if $|z| < 1$. Less obviously, the change of variables $z = \mathrm{e}^{sT_\mathrm{s}}$ maps the stable part of the s plane (Re $s < 0$) to the stable part of the z plane ($|z| < 1$). Accordingly, $\omega = 0$ is mapped to $z = 1$. See right and Problem 5.11.

5.3.3 Converting to Difference Equations

Discrete transfer functions can be used to generate time-domain difference equations. For example, let $K(z)$ be the transfer function between the error $e(z)$ and the controller output $u(z)$, with a form that is a rational polynomial in z, as in Eq. (5.16):

$$K(z) \equiv \frac{u(z)}{e(z)} = \frac{B_0 + B_1 z^{-1} + \cdots B_k z^{-k}}{1 + A_1 z^{-1} + A_2 z^{-2} + \cdots A_k z^{-k}} \,. \tag{5.43}$$

Since z^{-1} means "delay by T_s," we can write down the difference equation by inspection:

$$u_k = A_1 u_{k-1} + A_2 u_{k-2} + \cdots B_0 e_k + B_1 e_{k-1} + \cdots \,, \tag{5.44}$$

where the discrete error sequence e_k input to the controller leads to the output sequence u_k, which becomes a zero-order-hold "staircase" signal $u_k(t) \equiv$ , which is then input into the continuous system.

5.3.4 Euler and Tustin Discretization

Although ZOH discretization is appropriate for modeling a physical system, a discrete controller is different, as there is no underlying continuous physical system. We thus approximate the continuous dynamics directly. Two common methods are identified with the names of *Euler* and *Tustin*. The latter is also known as the *bilinear*, or *trapezoidal*, approximation. These methods are simply two different ways to approximate $z = \mathrm{e}^{sT_\mathrm{s}}$, with the former being a first-order and the latter a second-order approximation.

Euler Approximation

The Euler approximation has two versions, *forward* and *backward*. We used the former in our naive conversion of the PID control law into a discrete dynamical system in Eq. (5.1). Recall that $s \leftrightarrow \frac{\mathrm{d}}{\mathrm{d}t}$ and $z \leftrightarrow$ "advance by T_s." Expanding $\mathrm{e}^{sT_\mathrm{s}}$ to first order in sT_s, we have

$$z = \mathrm{e}^{sT_\mathrm{s}} \approx (1 + sT_\mathrm{s}) \quad \Longrightarrow \quad s \approx \frac{z-1}{T_\mathrm{s}} \quad \Longrightarrow \quad \dot{y} \approx \frac{y_{k+1} - y_k}{T_\mathrm{s}} \,. \tag{5.45}$$

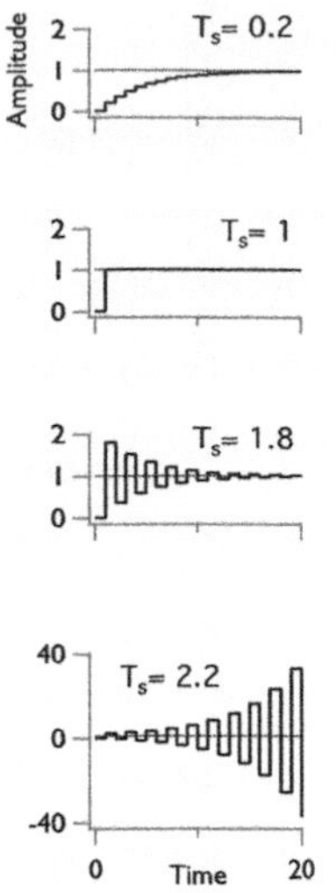

The forward Euler method is simple and often used in naive conversions of continuous controllers. It turns out to be a good choice for the PID control law. One problem with the forward Euler approximation is that it can turn a stable dynamical system into an unstable one if the discretization time step is too large. For example, consider the simple first-order system $\dot{y} = -y(t) + u(t)$. The forward Euler approximation gives

$$y_{k+1} = (1 - T_s)\, y_k + T_s u_k \,, \tag{5.46}$$

where we recall that in real units for a decay time τ, the sampling time $T_s \to T_s/\tau$. At left, we plot the step response for three values of T_s. From the plots and from Eq. (5.46), there is monotonic relaxation to the final value for $0 < T_s < 1$. The value $T_s = 1$ is a special case analogous to critical damping in a second-order system. For $1 < T_s < 2$, there is an oscillatory relaxation, or "ringing pole." Finally, $T_s = 2$ marks an instability threshold: For $T_s > 2$, the response is unstable (bottom graph). Of course, an unstable response is a problem, but ringing is usually also not desirable. In closed-loop systems, it can lead to wild cycling of the actuator between large positive and negative values.

The backward Euler method is similar to the forward but uses the current input u_k to estimate the current output y_k.

$$z = e^{sT_s} \approx \frac{1}{1 - sT_s} \quad \Longrightarrow \quad s \approx \frac{1 - z^{-1}}{T_s} \quad \Longrightarrow \quad \dot{y} \approx \frac{y_k - y_{k-1}}{T_s} \,. \tag{5.47}$$

For example, the discretization of $\dot{y} = -y(t) + u(t)$ becomes $y_{k+1} = (1 - T_s)\, y_k + T_s u_{k+1}$. As we shall see in Example 5.7, the backward Euler is stable for all T_s. The approximations are examples of what, in numerical analysis, are known as *explicit* (forward Euler) and *implicit* (backward Euler) integration schemes. For nonlinear equations, implicit methods require solving a nonlinear equation for y_k, but for linear equations they also lead to a linear equation for y_k. The backward Euler method is seldom used, losing out to either the simpler forward Euler method or the more accurate Tustin approximation.

Tustin Transformation

The second-order Tustin approximation, or bilinear transformation, corresponds to the (1,1) Padé rational-function approximation to the exponential that we discussed in Section 3.6.4:

$$z = e^{sT_s} \approx \frac{1 + \frac{1}{2} sT_s}{1 - \frac{1}{2} sT_s} \quad \Longrightarrow \quad s = \frac{2}{T_s}\left(\frac{1 - z^{-1}}{1 + z^{-1}}\right) = \frac{2}{T_s}\left(\frac{z - 1}{z + 1}\right) . \tag{5.48}$$

The Tustin transformation is usually more accurate than the Euler method and preserves stability, mapping Re $s < 0$ to $|z| < 1$ (Problem 5.11), although it distorts frequencies significantly at longer sampling times (Problem 5.12). It also does not work well with a derivative term in the controller and hence with PID control (Problem 5.13).

Table 5.1 Discretization of the first-order system $G(s) = \frac{1}{1+s}$.

Approximation scheme	s	G_d	z_p	Time domain
Forward Euler	$\frac{z-1}{T_s}$	$\frac{T_s}{z-1+T_s}$	$1 - T_s$	$y_{k+1} = (1 - T_s) y_k + T_s u_k$
Backward Euler	$\frac{1-z^{-1}}{T_s}$	$\frac{T_s}{T_s+1-z^{-1}}$	$\frac{1}{1+T_s}$	$y_{k+1} = \frac{1}{1+T_s} y_k + \frac{T_s}{1+T_s} u_{k+1}$
Tustin (bilinear)	$\frac{2}{T_s}\frac{1-z}{1+z}$	$\frac{T_s(1+z)}{(T_s-2)+(T_s+2)z}$	$\frac{2-T_s}{2+T_s}$	$y_{k+1} = \frac{2-T_s}{2+T_s} y_k + \frac{T_s}{T_s+2}(u_k + u_{k+1})$

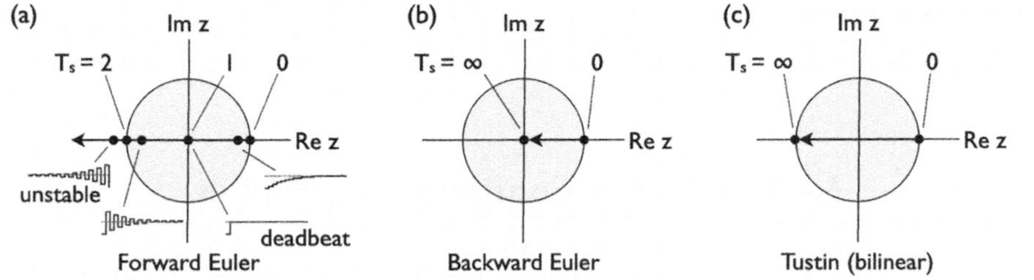

Fig. 5.9

Discretization of a first-order system by (a) forward Euler, (b) backward Euler, and (c) Tustin transformations. Black disks denote position of the pole z_p for different discretization times T_s. For the forward Euler approximation, the dynamics becomes unstable at $T_s = 2$, when the pole leaves the shaded unit circle at $z = -1$. Step responses are shown for $T_s = 2.2, 1.8, 1,$ and 0.2. For the backward Euler and Tustin schemes, the dynamics are stable for all $T_s > 0$.

Example 5.7 We compare the forward/backward Euler and Tustin discretizations of the first-order system $G(s) = \frac{1}{1+s}$ (Table 5.1 and Figure 5.9). The pole of the forward Euler scheme is at $z_p = 1 - T_s$, which exits the unit disk at -1 when $T_s = 2$. The backward Euler and Tustin schemes have z_p within the unit disk (and are thus stable) for all values of T_s. In the time domain, the value of y_{k+1} depends only on u_k for the forward Euler approximation but on u_{k+1} for the backward Euler and Tustin approximations.

5.3.5 Discretization of Nonlinear Equations

In the above sections, we focused on the discretization of linear equations. Many of the same ideas extend to nonlinear systems. Part of the field of *numerical analysis* is devoted to finding reasonable approximation schemes for integrating, in discrete time, ordinary differential equations, and those techniques – Runge Kutta methods and the like – can be used for control systems, too. However, the complexity of nonlinear equations leads to new discretization schemes, as well. In Chapter 11 on nonlinear systems, we introduce the notion of a *Poincaré section*, which can lead to discretization schemes where the times between sampled points are *not* uniform. See Section 11.4.1 for more discussion.

5.4 Design of Digital Controllers

We are finally ready to discuss the design of a digital controller. As discussed in the introduction, there are three approaches: (i) design a continuous controller for a physical system and then discretize a controller that emulates it; (ii) discretize a continuous transfer-function model of a physical system and then design the digital controller; (iii) design a controller that considers the full hybrid dynamics of digital controller and analog physical system. To picture these alternatives, we recall, in Figure 5.1, that a digital controller interacts with the continuous physical system. As a result, we can distinguish between discrete and continuous parts of the overall system, as shown in Figure 5.10.

5.4.1 Emulate a Continuous Controller

In our first approach, we start with a physical system, $G(s)$, design a continuous controller $K(s)$, and then convert it into an approximate $K_d(z)$. Because the continuous controller receives a continuous signal, we use either the forward or backward Euler or the Tustin conversion. For implementation, this conversion is all that is required.[17] To simulate the closed-loop dynamics – always a good idea – discretize the continuous dynamics using the zero-order hold (ZOH) procedure discussed in Section 5.3.1. Because the ZOH is included in the discretization of the model, do not again include it in the controller.

Example 5.8 (Euler vs. Tustin) Let $G(s) = \frac{1}{1+s^2}$, and let us focus on the response to an impulse disturbance exerted at input. We first design a continuous controller using a modified PD control law $K(s) = \frac{K_p + K_d s}{1+\tau s}$. From Example 3.5, the proportional (P) term increases the natural frequency, and the derivative (D) term adds damping. The

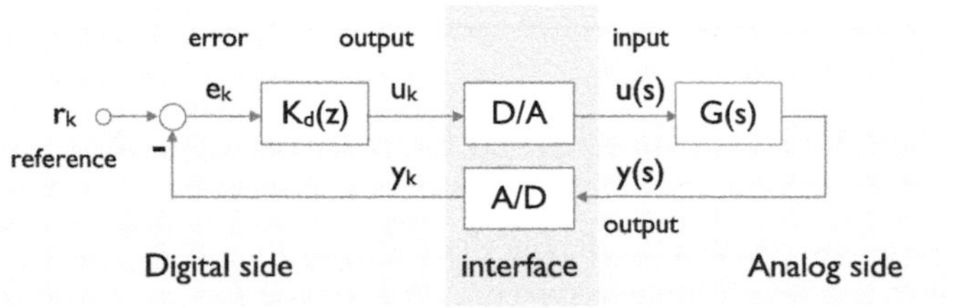

Fig. 5.10 A digital control system has three parts. Left: digital signals for the control algorithm. Center: the A/D and D/A interfaces convert between digital and analog signals. Right: $G(s)$ contains the system being controlled, plus all auxiliary analog elements, such as amplifiers, sensors, actuators, and anti-alias filters.

[17] One could model the delay $\tau = T_s$ in designing the controller, but here, we assume T_s is small enough that its effects are negligible. If not, it is simpler to digitize the system with a ZOH and then design the controller.

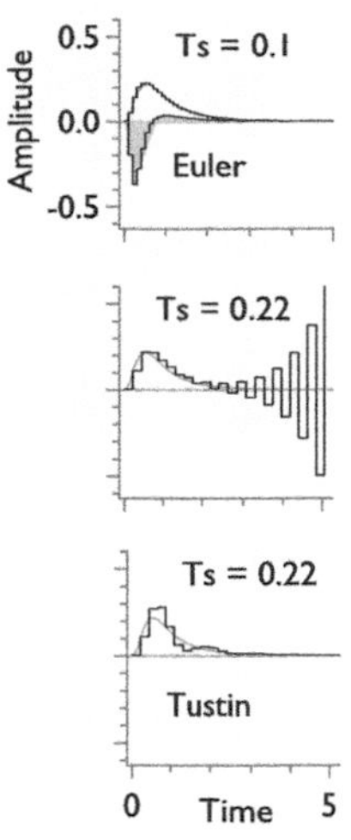

values $K_p = 3$, $K_d = 4$, and $\tau = \frac{1}{10}$ produce the response at right (solid black lines).[18] The denominator term filters the derivative, which is necessary so that the controller's response to high-frequency commands (such as a step) does not exceed the available range. The top plot at right shows both the output and the controller's response (divided by ten). An ideal response would be the negative of the gray bar and would leave the output undisturbed, $y(t) = 0$. Although the response here seems sluggish compared to the disturbance, remember that the disturbance is a pulse of width T_s. In general, a control system should have a sampling rate that is faster than all relevant time scales, including that of disturbances. In this case, the disturbance relaxes with a time constant of ≈ 1, which is only ten times slower than T_s. Lower-frequency disturbances would be even better controlled.

We also show the digital response, where the continuous controller has been replaced by a digital approximation based on the forward Euler method, with $T_s = 0.1$. See Eq. (5.49). The discrete response is a good approximation to the continuous response.

$$K(s) = \underbrace{\frac{3+4s}{1+s/10}}_{\text{continuous}} \to \underbrace{\frac{-37+40z}{z}}_{\text{Euler, } T_s = 0.1} \to \underbrace{\frac{-15.2+18.2z}{0.55+0.45z}}_{\text{Euler, } 0.22} \to \underbrace{\frac{-7.34+8.66z}{0.02+0.42z}}_{\text{Tustin, } 0.22} \tag{5.49}$$

Increasing the sampling interval to $T_s = 0.22$ (center plot) destabilizes the response. The Tustin transformation (bottom plot) stabilizes the output, but with worse tracking than the Euler discretization with $T_s = 0.1$.

Example 5.9 (Discretization of a PID controller) The forward Euler method, $\dot{x} \approx (x_{k+1} - x_k)/T_s$, works well if the sampling time is much shorter (by a factor of at least 10, but preferably 20 to 30) than the fastest dynamics that need to be modeled in the system. If we also differentiate and discretize the control, the PID law becomes

$$u_{k+1} = u_k + B_0 e_{k+1} + B_1 e_k + B_2 e_{k-1} \,, \tag{5.50}$$

with $B_0 = K_p + \frac{1}{2}K_i T_s + \frac{1}{T_s}K_d$, $B_1 = -K_p + \frac{1}{2}K_i T_s - \frac{2}{T_s}K_d$, and $B_2 = \frac{1}{T_s}K_d$. Here K_p, K_i, and K_d are the proportional, integral, and derivative terms, T_s is the sampling time, and u_k and e_k are the actuator and error signals at time kT_s. Equation (5.50) has the advantage relative to Eq. (5.1) that no integral explicitly appears and also the next value of the actuator signal u_{k+1} depends on its previous value u_k. This is useful to go from *manual mode* to closed-loop control without large transients in the actuator signal, since u_{k+1} is based on the actual previous value u_k. Note that e_{k+1} appears in Eq. (5.50), implying that the calculation time is much less than the sampling time. If not, account for the delay by shifting back the indices of the error terms by one.

[18] Compared to Example 3.5 (Eq. 3.67), the controllers have the same form but slightly different coefficients.

5.4.2 Direct Digital Design

In Section 5.4.1, the approach to digital design was to first design a continuous controller $K(s)$ and then transform s to z to derive an approximate controller $K_d(z)$. Now, consider the complementary tactic: approximate the continuous system $G(s)$ by a discrete $G_d(z)$, using the ZOH procedure, and then design the controller $K_d(z)$ directly in the digital domain. Such a direct digital design can give more precise control of the closed-loop response.

The good news about digital design is that the state-space formalism developed in Chapter 4 mostly carries over to the digital domain. For example, the test for controllability of a linear system based on the rank of the matrix

$$\boldsymbol{W}_c \equiv \begin{pmatrix} \boldsymbol{B} & \boldsymbol{AB} & \boldsymbol{A}^2\boldsymbol{B} & \cdots & \boldsymbol{A}^{n-1}\boldsymbol{B} \end{pmatrix} \tag{5.51}$$

continues to apply when the matrices describe discrete dynamics (Problem 5.14). Indeed, if $\{\boldsymbol{A}, \boldsymbol{B}\}$ is a controllable continuous system, its discretization $\{\boldsymbol{A}_d, \boldsymbol{B}_d\}$ is, too, *except* for T_s corresponding to pathological frequencies, such as that of a normal mode. The issues trace back to aliasing. If you sample significantly faster than system time scales, this situation should not arise. Similar considerations hold for observability.

Timing

The timing of signals for discrete-time dynamics becomes important. In simple terms, "What do we know, and when do we know it?"[19] In a continuous controller, a control law such as $u(t) = K_p e(t)$ responds instantly to the error signal. In a digital controller, we read in a sensor signal, process the new information to compute a new controller output, and finally send the output to the physical system. Each step – anti-alias filter, A/D conversion of the sensor signal, computation, and D/A conversion for the actuator – takes time, and the sum of these times, τ, represents a delay, or *feedback latency*, between the arrival of information from the experiment and the response from the controller.

For stability, the latency τ should be shorter than open- or closed-loop dynamical time scales. But what should be it be relative to the sampling time T_s? If $\tau \ll T_s$, just ignore the delay. If $\tau \lesssim T_s$, wait until the next cycle starts, so that $\tau = T_s$. And if $\tau \gg T_s$, average the "redundant" readings to have a lower-noise sampling with $\tau = T_s$. Note that it is usually not worth modeling any fractional-delay effects, as every little system modification will change τ; rather, just add extra delay so that $\tau = T_s$.[20]

If we can measure the full discrete state vector $\boldsymbol{x}_k$, then either $u_k = \boldsymbol{K}\boldsymbol{e}_k$ or $\boldsymbol{K}\boldsymbol{e}_{k-1}$, where $\boldsymbol{K}$ is a feedback vector (for a SISO system), $\boldsymbol{e}_k$ is the error signal at time kT_s,

[19] From the phrase "What did the president know, and when did he know it? by Howard Baker, from the US Senate Watergate Committee hearings, June 29, 1973.

[20] Because personal computers multitask, software-time control loops are inherently irregular. Special-purpose machines such as *digital signal processors*, *microcontrollers*, and *FPGAs*, with differing trade-offs among cost, speed, and ease of programming, can make timing predictable.

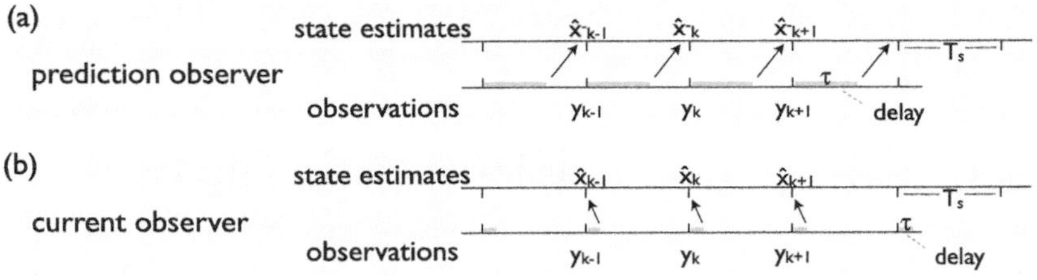

Fig. 5.11 Timing in a discrete system with feedback latency τ. (a) If $\tau \lesssim T_s$, the state estimate $\hat{x}_{k+1}^-$ at time $k+1$ is based on the observation y_k at time k (prediction observer). (b) If $\tau \ll T_s$, we may neglect it and base the state estimate at time $k+1$ on the observation y_{k+1} (current observer).

and u_k is the controller output sent to the D/A converter. The backward Euler and Tustin methods use $\boldsymbol{e}_k$ for formulating u_k, while the forward Euler method uses $\boldsymbol{e}_{k-1}$. The instability of the Euler method results from the delay between the input at time $k-1$ and the output at time k.

If we cannot measure the full state vector, we need to use an observer to infer it, as discussed in Chapter 4. Recall that for a continuous SISO system with state vector $\boldsymbol{x}(t)$, we create a dynamical system for an estimate $\hat{\boldsymbol{x}}(t)$. See Eq. (4.58). Again, depending on whether $\tau \ll T_s$ or $\tau \lesssim T_s$, we have two types of observer, as illustrated in Figure 5.11:

1 *Prediction observer*: $\hat{\boldsymbol{x}}_{k+1}^-$ uses the observation y_k.

2 *Current observer*: $\hat{\boldsymbol{x}}_{k+1}$ also uses the observation y_{k+1}. (An alternate notation is $\hat{\boldsymbol{x}}_{k+1}^+$.)

The minus sign in the prediction observer $\hat{\boldsymbol{x}}_{k+1}^-$ reflects that it is based on the prior information available just *before* the observation $\boldsymbol{y}_{k+1}$ (but incorporating $\boldsymbol{y}_k$). The current observer estimate, $\hat{\boldsymbol{x}}_{k+1}$, differs by incorporating the latest observation as a kind of posterior.[21]

The *prediction observer* has dynamical equations that are exactly analogous to Eq. (4.58):

$$\boldsymbol{x}_{k+1} = \boldsymbol{A}\boldsymbol{x}_k + \boldsymbol{B}u_k\,, \quad y_k = \boldsymbol{C}\boldsymbol{x}_k \tag{5.52a}$$

$$\hat{\boldsymbol{x}}_{k+1}^- = \boldsymbol{A}\hat{\boldsymbol{x}}_k^- + \boldsymbol{B}u_k + \boldsymbol{L}(y_k - \hat{y}_k)\,, \qquad \hat{y}_k = \boldsymbol{C}\hat{\boldsymbol{x}}_k^-\,. \tag{5.52b}$$

In Eq. (5.52), the estimate $\hat{\boldsymbol{x}}_{k+1}^-$ depends only on the observation made at the previous time step, y_k, which enters through the corrector term that is proportional to the estimator gain $\boldsymbol{L}$. If we want to use the observer in a feedback loop, we use u_k and y_k in Eq. (5.52b) to produce the state-vector estimate $\hat{\boldsymbol{x}}_{k+1}^-$. This estimate may then be used in a feedback law such as $u_{k+1} = -\boldsymbol{K}\hat{\boldsymbol{x}}_{k+1}^-$ to form the next controller output.

[21] Some texts use $\hat{\boldsymbol{x}}_{k+1|k}$ for the prediction observer and $\hat{\boldsymbol{x}}_{k+1|k+1}$ for the current observer.

To simulate system and observer together, we rewrite Eq. (5.52) in block-matrix form:

$$\begin{pmatrix} x \\ \hat{x}^- \end{pmatrix}_{k+1} = \begin{pmatrix} \boldsymbol{A} & \boldsymbol{0} \\ \boldsymbol{LC} & \boldsymbol{A}-\boldsymbol{LC} \end{pmatrix} \begin{pmatrix} x \\ \hat{x}^- \end{pmatrix}_k + \begin{pmatrix} \boldsymbol{B} \\ \boldsymbol{B} \end{pmatrix} u_k \quad \begin{pmatrix} y \\ \hat{y} \end{pmatrix}_{k+1} = \begin{pmatrix} \boldsymbol{C} & \boldsymbol{0} \\ \boldsymbol{0} & \boldsymbol{C} \end{pmatrix} \begin{pmatrix} x \\ \hat{x}^- \end{pmatrix}_{k+1} . \tag{5.53}$$

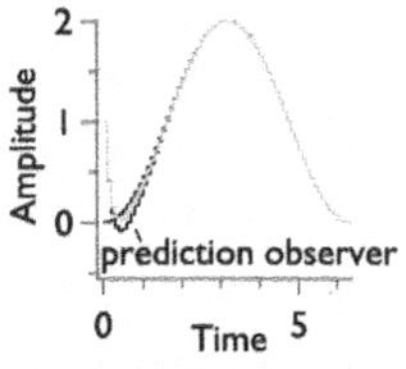

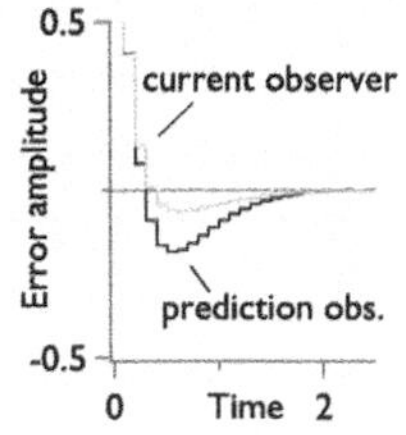

The combined system for $(\begin{smallmatrix} x \\ \hat{x}^- \end{smallmatrix})_{k+1}$ is driven by a common input, u_k. The difference between the system output y_k and its prediction $\hat{y}_k$ gives feedback for the state estimate. The term $\boldsymbol{A} - \boldsymbol{LC}$ appears in the observer dynamics, as for the continuous system, Eq. (4.58), although here the system matrices are the ZOH discretizations (Problem 5.15).

The *current observer* is similar but uses y_{k+1} to estimate $\boldsymbol{x}_{k+1}$. We then distinguish the *prediction*, $\hat{\boldsymbol{x}}_{k+1}^-$ (the value of $\hat{\boldsymbol{x}}$ that can be predicted for time $k+1$) from the *estimate* $\hat{\boldsymbol{x}}_{k+1}$. The resulting dynamical equations follow Eq. (5.53), with $\boldsymbol{LC} \to \boldsymbol{LCA}$ (Problem 5.16).

The sketch at left illustrates the two types of observer for an undamped harmonic oscillator, discretized by a zero-order hold. The initial conditions of the observer, $\hat{\boldsymbol{x}}_0 = (\begin{smallmatrix} 1 \\ 0 \end{smallmatrix})$, differ significantly from the actual state vector, $\boldsymbol{x}_0 = (\begin{smallmatrix} 0 \\ 0 \end{smallmatrix})$, as would be typical in practice. The current observer (lighter thin line) tracks the oscillator's position better than the prediction observer because it uses information that is more up to date.

Delays and Predictive Control

In the continuous domain, linear, finite-dimensional ordinary differential equations (ODEs) give rise to transfer functions $G(s)$ that are rational polynomials in s. Delays, however, are represented by $e^{-\tau s}$, which must be approximated by rational polynomials.

For a discrete system, a delay by the time step T_s is just z^{-1}, and a delay of kT_s is just z^{-k}. The behavior of the state vector then depends on only a finite number of values (e.g., the input u_k at time steps k_{max} into the past). However, a zero-order hold involves substituting $s = \frac{1}{T_s} \ln z$, which turns rational polynomials in s into transcendental functions that are approximated by the s-to-z conversion techniques discussed in Section 5.3.3. The upshot is that delays are easier to represent in the discrete formalism than in the continuous one. In the discrete case, we simply convert $G(s)$ to $G_d(z)$ ignoring the delay. Then we multiply by z^{-k} to include a delay of kT_s. Problem 5.17 discusses how to accommodate, in linear systems, fractional delays that are not equal to a multiple of the sampling time T_s.

In Chapter 3, we showed that long delays can make it difficult or impossible to control systems. For example, for the unstable first-order system $\dot{x}(t) = x(t) + u(t - \tau)$, with $u(t) = -K_p x(t)$, there is no value of K_p that can stabilize $x = 0$ when $\tau > 1$. We claimed, though, that a more sophisticated, *predictive* algorithm can work, even for long delays. The basic idea is that given an initial state in the past, $x(t - \tau)$, and a record of the controls we applied over the interval $(t - \tau, t)$, we can predict the current state $x(t)$:

$$x_{\text{pred}}(t) = e^{\tau}\, x(t-\tau) + \int_{t-\tau}^{t} dt'\, e^{(t-t')}\, u(t') \,. \tag{5.54}$$

The feedback algorithm based on $u(t) = -K_p x_{\text{pred}}(t)$ can stabilize the up solution for any τ.

Let us explore these issues using an equivalent discrete-time problem. For $a > 1$, let

$$x_{k+1} = a x_k + u_k\,, \qquad u_k = -K_p x_k\,. \tag{5.55}$$

The discrete open-look dynamics ($u_k = 0$) is unstable, but a negative feedback K_p can stabilize $x = 0$. Now consider feedback applied with unit delay, $u_k = -K_p x_{k-1}$. For $a > 2$, no value of K_p can stabilize $x = 0$. However, a predictive linear feedback can work: Problem 5.18 will show that if $u_k = -K_p x_k^{\text{pred}}$, with $x_k^{\text{pred}} = a x_{k-1} + u_{k-1}$, the dynamics can be stable for any a. Note that at time k, both x_{k-1} and u_{k-1} are known.

Digital Design and Deadbeat Control

The design of controllers in the digital domain closely follows that of design in the continuous domain and uses the same methods – high gain, pole placement, loop shaping, internal model principle, and so on. Here, we apply the pole-placement method to a case that has unique features relative to continuous systems.

Consider an overdamped particle in one dimension. The discrete-time system is $x_{k+1} = x_k + u_k$, and we seek a step response that goes to one in a single time step. If $x_0 = 0$, we want $x_k = 1$ for $k \geq 1$. We can achieve our goal in this simple example by setting $u_0 = 1$ and $u_k = 0$ for $k \geq 1$. This control "strategy" is known as *deadbeat control.* But rather than being a special virtue of discrete control, deadbeat control reflects *limitations* of discrete control. To understand this point, compare the equivalent continuous equivalent system, $\dot{x} = u(t)$, with $x(0) = 0$. For $x(t) = 1$ when $t \geq t_0$, with t_0 arbitrary, we can set $u(t) = \frac{1}{t_0}$ for $0 < t < t_0$. Of course, the shorter t_0, the larger the required u, and practical deadbeat control requires a careful choice of T_s. Thus, because the control can only be changed once every T_s in a discrete system, there is a *minimum* time for zeroing out a disturbance. By contrast, there is no such fundamental limit for a continuous system.

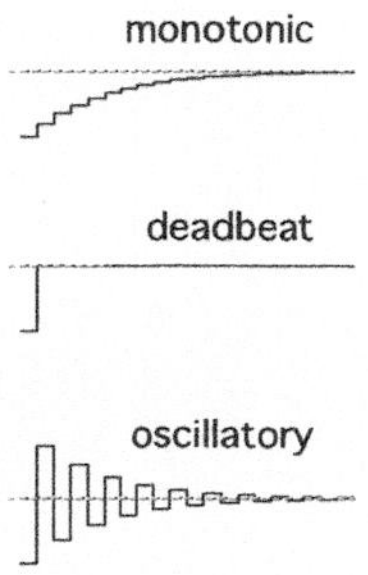

At right, we see that the fastest decay in a discrete first-order system is indeed one time step. Although a decay in one step seems fast, the result also means that perturbations cannot decay any more quickly. For a continuous system, there is no analogous limitation. On the other hand, deadbeat control achieves *complete* recovery in a single time step, while the continuous systems that we have looked at all require an infinite time (e^{-t} is zero only at $t = \infty$). Although this difference is sometimes described as being a special feature of digital control, it is actually a feature of *nonlinear* dynamics. (Staircase inputs are nonlinear.)

To generalize these ideas to an nth-order, SISO linear system $\boldsymbol{x}_{k+1} = \boldsymbol{A}\boldsymbol{x}_k + \boldsymbol{B}u_k$ and $y_k = \boldsymbol{C}\boldsymbol{x}_k$, observe that fixing the n components of the state vector requires n inputs and hence n time steps. Assume that the system is initially in the state $\boldsymbol{x}_0 = \boldsymbol{0}$, with $y_0 = 0$, as well. We seek a controller that, in response to a step reference command, creates an output $y_k = 1$ and $u_k = \text{constant}$, for $k \geq n$. The first step will be to convert the state-space problem to a transfer function $G(s)$ and then, using a zero-order hold, to a discrete transfer function $G_{\mathrm{d}}(z) \equiv N(z)/D(z)$. Then, in Problem 5.19, you will show the following:

- Fixing the reference step response $y(z) = T_{\mathrm{d}}(z)\, r(z)$ implies $T_{\mathrm{d}}(z) = \frac{f(z)}{z^n}$, with $f(1) = 1$.
- Making the input constant after n steps implies that $f(z) = N(z)/N(1)$.
- The corresponding controller then has the form $K_{\mathrm{d}}(z) = \frac{D(z)}{z^n N(1) - N(z)}$.

Because the closed-loop transfer function in deadbeat control has a denominator z^n, the method is equivalent to moving all the poles of the dynamical matrix $\boldsymbol{A}$ to the origin.

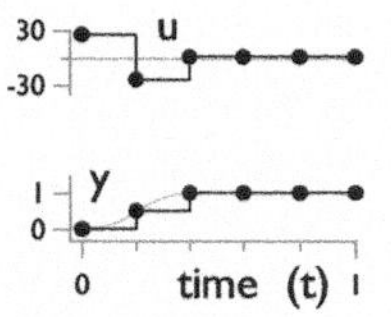

Example 5.10 (Deadbeat response of a second-order system) Consider the undamped oscillator, $G(s) = \frac{1}{1+s^2}$, digitized to $G_{\mathrm{d}}(z)$ using a zero-order hold and $T_{\mathrm{s}} = 0.2$ (Problem 5.10). At left is the deadbeat step response y_k, with its corresponding control signal u_k. The light trace gives the continuous response $y(t)$ due to the ZOH input $u(t)$.

Pole Placement and State-Space Design

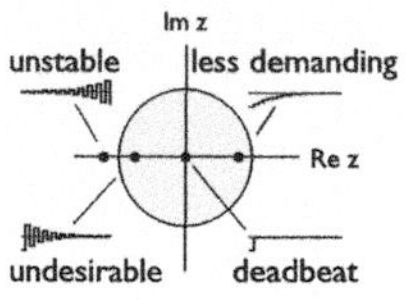

A practical issue with deadbeat control is that the only adjustable parameter is the sampling time, T_{s}. In particular, the required control effort increases rapidly for small sampling times. Often, it is easier to fix the sampling rate to be small and then move the open-loop poles away from the origin (but still inside the unit circle, to have stable response) (see left). Note that poles near $z = -1$ are to be avoided, as they cause oscillations at half the sampling frequency. Otherwise, the design ideas are similar to the continuous case.

Digital-controller design is often more straightforward using the state-space formalism, with an observer if need be. In analogy with Eq. (4.47), the feedback law is then

$$u_k = -\boldsymbol{K}\hat{\boldsymbol{x}}_k + k_{\mathrm{r}} r_k\,, \qquad {}_{\mathrm{r}} = [\boldsymbol{C}(\mathbb{I} - \boldsymbol{A} + \boldsymbol{B}\boldsymbol{K})^{-1}\boldsymbol{B}]^{-1}\,, \tag{5.56}$$

where the feedforward term k_r is chosen that $y_k \to r_k$ (Problem 5.20).

The discrete observer-controller system then obeys equations analogous to Eq. (4.68):

$$\begin{pmatrix} \boldsymbol{x} \\ \hat{\boldsymbol{x}}^- \end{pmatrix}_{k+1} = \begin{pmatrix} \boldsymbol{A} & -\boldsymbol{B}\boldsymbol{K} \\ \boldsymbol{L}\boldsymbol{C} & (\boldsymbol{A} - \boldsymbol{B}\boldsymbol{K} - \boldsymbol{L}\boldsymbol{C}) \end{pmatrix} \begin{pmatrix} \boldsymbol{x} \\ \hat{\boldsymbol{x}}^- \end{pmatrix}_k + \begin{pmatrix} \boldsymbol{B}k_{\mathrm{r}} & \boldsymbol{B} \\ \boldsymbol{B}k_{\mathrm{r}} & \boldsymbol{0} \end{pmatrix} \begin{pmatrix} r \\ d \end{pmatrix}_k, \quad y_k = \begin{pmatrix} \boldsymbol{C} & \boldsymbol{0} \end{pmatrix} \begin{pmatrix} \boldsymbol{x} \\ \hat{\boldsymbol{x}}^- \end{pmatrix}_k. \tag{5.57}$$

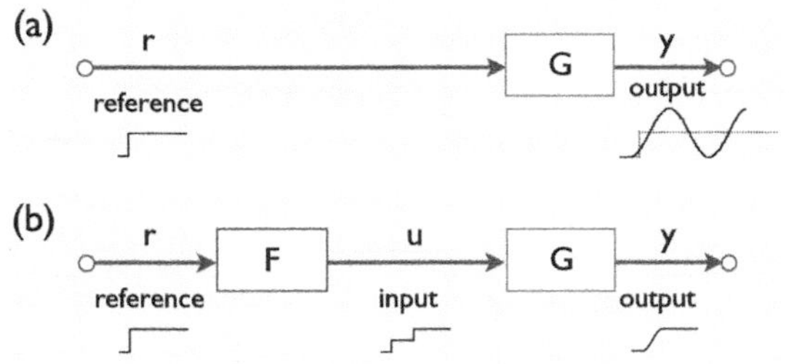

Fig. 5.12 Feedforward design for a digital controller. (a) A step sent to an oscillator leads to an oscillating response $y(t)$. (b) The reference signal r_k is transformed by a feedforward filter $F(z)$ into a distorted input u_k. The ZOH $u(t)$ of the discrete input u_k leads to an output that has a step but no oscillations.

Equation 5.57 has two inputs, the reference r_k and input disturbance d_k, and one output, the measurement y_k. The reference is fed to both the physical state $\boldsymbol{x}_k$ and the estimate $\hat{\boldsymbol{x}}_k^-$, whereas the disturbance affects only the physical state.

Feedforward Control

As with continuous systems, if dynamics are accurately known, feedforward control can improve a controller by adding another "degree of freedom" to its design.[22] We again consider an undamped oscillator driven by a digital controller and design a feedforward controller that modifies the system input to track reference changes. A feedback controller such as the discretized PD controller discussed in Example 5.8 can then deal with disturbances independently of the feedforward element.[23] Schematically, the feedforward design of Figure 5.12 follows the discussion in Section 3.4.

The design of a digital feedforward controller is straightforward:

1. Discretize the continuous system using a zero-order hold: $G(s) \xrightarrow{\text{ZOH}} G_{\text{d}}(z) = \frac{N(z)}{D(z)}$.
2. Define a scale factor $\lambda = D(1)$, to match reference and output amplitudes.
3. Let the feedforward cancel the system poles and add deadbeat response: $F(z) = \frac{D(z)}{\lambda z^2}$.
4. Invert the Z-transform to find the time domain equations.

In the above scheme, $F(z)$ has a denominator $\sim z^k$, where $k = 2$ corresponds to a second-order system. As discussed above, the deadbeat response in a kth-order system requires k time steps to reach the zero state. Notice that F is designed to have relative degree zero.

Figure 5.13 shows the result of this procedure, for the undamped harmonic oscillator $G(s) = \frac{1}{1+s^2}$. The left column represents different reference commands r_k, the middle column the shaped inputs $u(t)$, and the right column the continuous output

22 Feedforward can correct disturbances if there is "advance warning." For example, you can measure ambient changes to temperature and then "warn" the controller to "expect" a temperature perturbation.
23 See Problem 3.8 for an analogous continuous-time design.

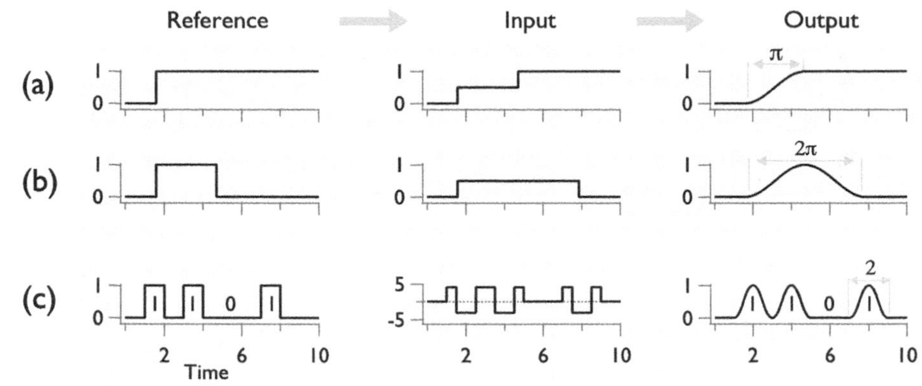

Fig. 5.13 Feedforward control of an undamped oscillator. The reference sent to the feedforward element distorts the input, which is sent to the oscillator, which "undistorts" the input. (a) Step input, $T_s = \frac{\pi}{2}$. (b) Pulse input, $T_s = \frac{\pi}{2}$. (c) Pulse sequence for "1101," with $T_s = \frac{1}{2}$. Note the larger input scale.

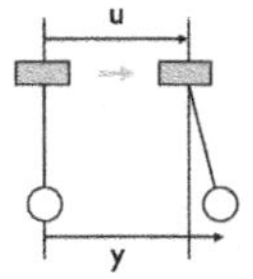

$y(t)$. The first case (a) shows that breaking the step into two halves leads to a smooth response. To understand why reshaping the step into two parts helps, picture a construction crane that picks up a load (see left). To move it from one point to another without swinging, quickly move half the distance, wait half an oscillator period, and – just when the load has reached its maximum swing but before it starts to fall back – move the second half of the distance.

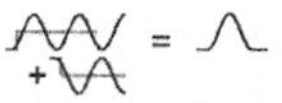

Figure 5.13b shows a pulse lasting one period of the oscillator. As seen here at left, each edge gives rise to oscillations. If we wait one period, and then apply a negative edge of the same amplitude, we generate the same oscillatory response – except that it is 180° out of phase. The sum cancels, and the output stays constant after its one-period pulse.

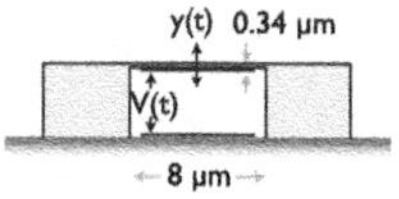

Liu et al. (2008) have implemented the single-pulse control depicted in Figure 5.13 at the nanoscale. Their oscillator – also known as a *resonator*[24] – illustrated at left, was an example of a NEMS (nanoelectromechanical system), consisting of a silicon cantilever clamped at both ends and bridging an 8 μm gap. It was 0.5 μm wide by 0.34 μm thick. Its lowest bending mode of ≈ 250 MHz could be excited electrically by applying a voltage $V(t)$ between the metal-coated cantilever and sample substrate. A single-cycle pulse could be excited, with possible applications to computation and information storage.

Toward the latter end, imagine the more complicated control pattern illustrated in Figure 5.13c, where a reference sequence corresponding to "1101" is sent to the feedforward element. The resulting sequence of pulses looks complicated but is nothing more than several copies of the pattern for an isolated pulse. To speed up information

[24] Liu et al. (2008) distinguish between *resonator* and *oscillator*. The former is a highly underdamped ($\zeta \ll 1$) but passive second-order system; the latter is capable of self-sustained oscillation. We use "oscillator" more loosely, to denote any underdamped, second-order system, whether the oscillations are self-sustained or not.

transmission, we have shortened T_s from $\frac{\pi}{2}$ to $\frac{1}{2}$. Of course, shorter pulses require a larger force and both negative and positive forces. Problem 5.21 shows that the required amplitude diverges as T_s^{-2}. Periods much shorter than the resonant frequency are thus likely to be limited by the actuator range. But you *can* generate pulses shorter than the natural oscillator period.

One problem with the feedforward designs discussed here is that they assume perfect knowledge of the system under control. For example, if the frequency of the oscillator differs from your model, then an unwanted oscillatory response will result. Chapter 9 shows that feedforward can be made robust to such parameter variations.

The time-domain *coherent control* of an oscillator discussed here has wide application, especially to quantum systems, where radiofrequency pulses are used in nuclear magnetic resonance (NMR) and short laser pulses control chemical reactions (Chapter 13). All spring from the same idea: feedforward control of a lightly damped oscillating system that is sufficiently isolated from its surrounding environment that the motion is coherent over long periods of time, relative to the period of the oscillator. The initial pulse from the feedforward element starts a local clock, and subsequent actions are timed relative to it.

5.4.3 Hybrid Dynamics

Most of our discussion has assumed that the sampled output of the dynamical system closely approximates the underlying continuous output. If the sampling time T_s is short compared to all dynamical time scales, this is a reasonable assumption. However, if the output of the digital controller is a staircase function, each edge injects a range of frequencies into the physical system and raises the possibility of aliasing and other artifacts.

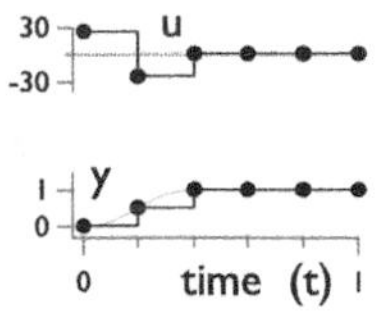

To see an example, consider again Example 5.10, with its deadbeat control algorithm for an oscillator that stabilized the response to a step command in two time steps. See right. The sampled signal is a good approximation to the continuous response (light curve).

Now let us try an algorithm that attempts to achieve deadbeat control after only *one* time step (Problem 5.22). The output does indeed go to one, and it stays constant after a delay of just one time step. However, the sampled output completely misses the oscillation seen in the continuous response to the ZOH input shown in the top plot. A moment's thought shows that this result is to be expected: With a single control, we can fix only one component of the state vector (the position), leaving the velocity undetermined. Indeed, the input is not constant after a single time step, as it should be in deadbeat control, a further clue. More abstractly, the step discontinuity in $u_k(t)$ has excited the resonant frequency of the oscillator. Because the control algorithm unfortunately sets T_s to be exactly half the sampling frequency, the output motion is aliased to DC and appears constant.

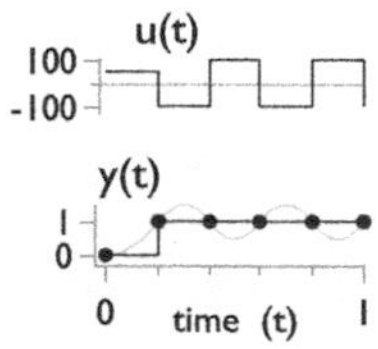

There are several responses to this kind of issue. The first is that it can be important to simulate the continuous *hybrid dynamics*. Although some software programs have built-in routines, you can easily make your own. The basic idea is to convert the discrete controller inputs u_k into a continuous *staircase* function $u_k(t)$, defined analogously to the zero-order hold of Eq. (5.38). In pseudocode, the hybrid dynamics algorithm is as follows:

1. Simulate the discrete recursion relation to generate (u_k, y_k) for all desired k.
2. Construct the *staircase* function $u_k(t)$ from u_k • • •, as in Eq. (5.38).
3. Use $u_k(t)$ in an ODE integrator as the forcing function for the continuous system.

You can then check explicitly for aliasing. For the attempt at deadbeat control given here, the issue is that the controller has a pole that cancels the *sampling zero* of $G_d(z)$ at $z = -1$ (Problem 5.22). Don't do that! Although theories of hybrid dynamics show how to avoid such problems, the most important point is to be aware of them and take corrective action when needed. This can involve modifying the control algorithm or adding an anti-alias filter to the D/A output, to ensure that only smooth, bandwidth-limited signals are injected into the physical system. Of course, you then have to account for the extra delay.

5.5 Summary

In our digital age, discrete controllers for continuous systems are ubiquitous. In this chapter, we introduced signal-processing tools for discrete systems. For sampling a continuous signal, an analog filter can set the signal bandwidth and eliminate the spurious frequencies caused by aliasing. From the Nyquist limit, a signal of bandwidth B needs to be sampled at a rate $> 2B$ (preferably $\approx 10B$). A priori knowledge about the signal (e.g., that its bandwidth B is centered on a carrier frequency or that the signal is sparse over some basis set) can allow spectacular improvements over the classic Nyquist rate.

Many of the tools for continuous equations have discrete counterparts. Thus, ordinary differential equations become difference equations, the Fourier transform becomes the discrete-time Fourier transform, and the Laplace transform becomes the Z-transform. Using these tools, we can convert a continuous linear system into a discrete one that accurately approximates the continuous system. If the signal fed to the system is a discrete signal that is converted into a continuous staircase equivalent by a sample-and-hold converter, then we can exactly discretize the response of the continuous system using the zero-order-hold technique. The ZOH is usually appropriate for modeling the continuous system (when simulating a control loop, for example). Since the controller itself receives a continuous signal from the outside world, any discretization will be approximate, and other methods, such as forward or backward Euler or the Tustin bilinear transformation, are more appropriate. The simplest method, forward Euler, can make the closed-loop system unstable if the time step T_s is too large. To test for hidden instabilities, the surest method is a hybrid simulation that first calculates the digital response at the sampling intervals and then inputs the staircase version of that signal into a simulation of the continuous system.

With appropriate tools, digital controllers are relatively straightforward. Most of the techniques and design principles for continuous control apply with small modifications. The frequency-domain techniques shift $s \to z$ plane (whose stability domain is the unit circle). We can have deadbeat response – stabilization after n time steps, for a nth-order system – which is not possible for a continuous linear controller and linear system. Delays are particularly straightforward for digital controllers since z^{-1} already represents a delay (and since implementing a delay in the time domain means simply using a later value of a signal). Of course, the gain limitations due to the presence of delays remain the same.

5.6 Notes and References

Much of this chapter would be considered signal processing rather than control theory, and there are many books dedicated to this topic. For the physicist, the book *Numerical Recipes* (Press et al., 2007) is good and easy to digest but not that systematic. Among the engineering books, Smith (1999) uses a minimum of mathematics and can be browsed online, while Oppenheim and Schafer (2010) is a standard text for engineering students.

The history of sampling theory is complicated. Most engineering books credit Nyquist for stating that a system of bandwidth B can transmit (or is characterized by) $2B$ symbols/time and Shannon (1949) (highly recommended) for the sampling theorem itself. However, there were many precursors, most notably work by K. Küpfmuller, V. A. Kotelnikov, H. Raabe, and E. T. Whittaker. Lüke (1999) nicely sums up the history: "first the practitioners put forward a rule of thumb, then the theoreticians develop the general solution, and finally someone discovers that the mathematicians have long since solved the mathematical problem which it contains, but in 'splendid isolation.' "

The history of compressed sensing is also complicated and, in some ways, refutes Lüke's pattern. In the 1970s, geophysicists discovered "rules of thumb" for seismic waves to probe the structure of the earth's interior at a resolution higher than what the Nyquist limit and Sampling Theorem suggest is possible. There were also isolated mathematical results on sampling at random intervals (e.g., by H. J. Landau in 1967). Nonetheless, the field was born only when mathematicians suddenly showed how and proved why compressed sensing in general cases was possible. Since then, the explosion of activity has seen many simultaneous contributions from both more practical and more theoretical and mathematical sources. See Candès and Wakin (2008) for an introduction, Bryan and Leise (2013) for the coin-weighing example, and both Eldar and Kutyniok (2012) and Foucart and Rauhut (2013) for book-length treatments.

For dithering, beware of naive discussions. Gray and Stockham (1993) tell the full story, carefully. Etchenique and Aliaga (2004) is easy to read and gives good examples of applications but makes inaccurate statements about the noise statistics. (Since their actual dither signal is Gaussian, their experiments are okay.) Stochastic dithering is widely used as a means to improve DAC resolution. I learned about it from Paul Dixon. Dithering is also closely related to *stochastic resonance*, where adding noise to a measurement increases the amount of information obtained (McDonnell et al., 2008). For audibility of phase distortions, see Lipshitz et al. (1982). (I thank Jules Bloomenthal for discussions on this.)

There are many techniques for designing time-domain FIR and IIR filters (Oppenheim and Schafer, 2010). For digital control itself, Dutton et al. (1997) is brief but does all the algebra explicitly for some simple examples. Franklin et al. (1998) and Åström and Wittenmark (1997) are book-length presentations. The latter discusses practical implementation (e.g., timing). Goodwin et al. (2001) is briefer and well organized. But much of its presentation is based on the discrete δ-transform, an alternative to the Z-transform that, while logical, has yet to win over the community at large. Its Chapter 14 in Goodwin et al. (2001) has a good discussion of hybrid control.

For feedforward, see Åström and Murray (2008) and also the MIT group (Singer and Seering, 1990). The work of the latter has been recently summarized in Singh (2010). Atomic force microscopy (AFM) has been another fruitful area of application (Clayton et al., 2009). My own introduction to feedforward was via an attempt to improve the scanning speed of a piezoelectric stage for an AFM (Li and Bechhoefer, 2007).

The topic of delays and predictive feedback is discussed in great generality in Krstic (2009). We have omitted the *Smith compensator*, an early (1959) frequency-domain compensator that can also incorporate delays. It is less clear physically than predictive feedback and much less general, being limited to stable linear systems. But it was the first algorithm to show that a delay must somehow be incorporated in the controller.

Problems

5.1 Analog low-pass filters. Low-pass filters with faster fall-off than a first-order system can remove high-frequency components of a signal before digitization. Recall the form $G(s) = (1 + 2\zeta s + s^2)^{-1}$ of the scaled transfer function of an arbitrary second-order system. Let us explore properties of different choices for the damping ζ.

a. *Butterworth filter.* For a given filter order, the Butterworth filter has the flattest magnitude response. Specifically, the nth-order Butterworth filter is defined as

$$|G(i\omega)| = \frac{1}{\sqrt{1+\omega^{2n}}} \approx 1 - \tfrac{1}{2}\omega^{2n} + O(\omega^{4n}) .$$

The lack of low-order terms leads to a flat response. Show that, for $n = 2$, the Butterworth filter corresponds to $\zeta = \frac{1}{2}\sqrt{2} \approx 0.71$.

b. *Bessel filter.* For a given filter order, the Bessel filter has the closest approximation to a linear phase response, $\varphi = -\tau\omega$, which corresponds to a delay in the signal by τ. At order n, the best approximation is to have $\varphi(\omega) = -\tau^*\omega + O(\omega^{2n})$, with τ^* the approximate delay. Show that the $n = 2$ Bessel filter has $\tau^* = \sqrt{3}$ and $\zeta = \frac{1}{2}\sqrt{3} \approx 0.87$, which is slightly more damped than the Butterworth filter.

c. Make Bode plots of the frequency response of the $n = 2$ Butterworth and Bessel filters. Compare the response of a naive cascade of first-order elements, $G_{\text{naive}}(s) = (1 + s)^{-2}$, corresponding to $\zeta = 1$. For the Bessel filter, plot also phase versus frequency on a linear scale, to see how well it approximates a linear phase response. Finally, plot the step response of all three filters. Why is G_{naive} not a great choice?

The Bessel filter has a "nicer" time response, the Butterworth a nicer frequency response. The differences are more striking for higher-order filters. Software packages can generate these, as well as variants such as elliptic and Chebyshev filters (Smith, 1999). Some Bessel filters are defined using a frequency scale that sets $\tau^* = 1$.

5.2 Dithering details. Why are some choices for random dither ξ better than others? Define $\langle x \rangle \equiv \int_{-\infty}^{\infty} d\xi\, p(\xi)\, Q(x_0 + \xi)$ and $\sigma \equiv \sqrt{\text{Var}}$, with $\text{Var} = \int_{-\infty}^{\infty} d\xi\, p(\xi)\, [Q(x_0 + \xi) - \langle x \rangle]^2$. Here, $p(\xi)$ is the probability density function of the added dither, and $Q(x)$ is the quantization nonlinearity, defined as rounding x to the nearest integer.

a. Consider uniform noise, $p(\xi) = 1$ for $(-\frac{1}{2}, +\frac{1}{2})$, or . Show that $\langle x \rangle = x_0$, so that there is no bias. Then show that $\text{Var} = (\delta x_0)(1 - \delta x_0)$, where δx_0 is the fractional part of x_0. (That is, if $x_0 = 3.1$, then $\delta x_0 = 0.1$.)

b. For triangular noise, $p(\xi) = 1 - |\xi|$, for $|\xi| < 1$ (and 0 for $|\xi| > 1$), or , show that $\langle x \rangle = x_0$ and $\text{Var} = \frac{1}{4}$. There is again no bias, and $\sigma = \frac{1}{2}$ is independent of x_0.

c. Investigate a Gaussian dither of standard deviation σ_0 numerically. Plot both the bias of $\langle x \rangle$ (deviation from the mean) and its variance as a function of x_0. Investigate for $\sigma_0 = 0.4$, 0.5, and 0.6. Is there an optimal value for σ_0?

d. *Subtractive dithering.* Let $x \equiv Q(x_0 + \xi) - \xi$. Show that $\langle x \rangle = x_0$ and $\text{Var} = \frac{1}{12}$ for this new x. Thus, with uniform noise, the standard deviation is not only independent of x_0, it is $\sqrt{3}$ lower than using triangular noise. Why doesn't everyone use subtractive dithering? The catch is the need to subtract the exact analog noise value added to the analog signal. Usually, this value is hard to know.

5.3 Compressed sensing and the counterfeit coin.

a. For the seven-coin / one-fake problem with the three measurements given, verify explicitly that each possibility leads to a unique pattern of measurements. To guide our intuition, change "coordinates" in the manner suggested in the text, so that each genuine mass has $f_i = 0$, while the fake coin has mass $f_j = 1$.

b. Now assume that there are one or two (identical) fakes. Find an explicit counterexample where inferring which masses are fakes is impossible.

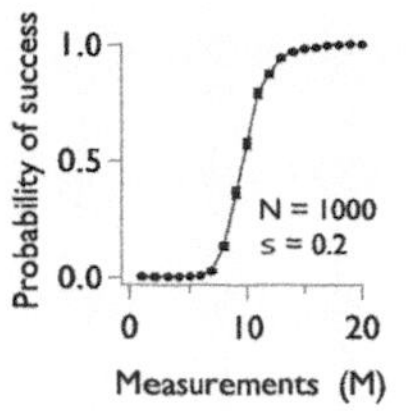

5.4 **Phase transition in compressed sensing**. Consider the N-coin / 1-fake problem. Generate the $M \times N$-dimensional measurement matrix $\mathbf{\Phi}$ numerically by letting each element be 0 or 1 with 50% probability. Then write a code to do the reconstruction numerically. Use a brute-force algorithm that examines each of the N possibilities explicitly, predicts the outcome $\mathbf{y}$ based on the choice of $\boldsymbol{f}$, and calculates the ℓ_2 norm of the error, $\|\mathbf{y} - \mathbf{y}_0\|_2$. Then select the $\boldsymbol{f}$ that minimizes this error. For a given m, repeat enough times to estimate the probability P of identifying the correct nonzero element of $\boldsymbol{f}$. Then vary M to estimate $P(M)$. You should find something resembling the graph at left. Add measurement noise, with $\mathbf{y}_0 = \mathbf{\Phi} \boldsymbol{f} + \boldsymbol{\xi}$, where $\boldsymbol{\xi} \sim \mathcal{N}(0, \sigma^2 \mathcal{I})$. Confirm that the reconstruction algorithm is robust against moderate noise levels.

Surprisingly, the probability for successful reconstruction rises sharply at $M^* \approx \ln N$: there is a *phase transition* in reconstruction probability, controlled by the relative measurement number M/N and sparseness S/N, with a qualitative difference between a low-data "phase" where reconstruction is impossible and a high-data "phase" where it succeeds almost always. This phase transition is universal: many choices for $\mathbf{\Phi}$ give the same reconstruction thresholds, or phase boundaries. See Donoho and Tanner (2009) and Krzakala et al. (2012).

5.5 **Final Value Theorem for Z-transform**. Show that $\lim_{k\to\infty} f_k = \lim_{z\to 1}[(z-1)f(z)]$. Hint: Take $\mathcal{Z}(f_{k+1} - f_k)$, and write the infinite sum as a limit $k \to \infty$ of a finite sum.

5.6 **IIR vs. FIR low-pass filter**. In Example 5.2, we claimed that the IIR filter

$$y_k = a y_{k-1} + (1-a) u_{k-1}\,, \qquad 0 < a < 1$$

is equivalent to the IIR filter

$$y_k = A(u_k + a u_{k-1} + a^2 u_{k-2} + \cdots + a^n u_{k-n})\,,$$

where A is a normalization constant chosen to make the DC gain equal to 1. For large n, the two transfer functions become very similar. To see this:

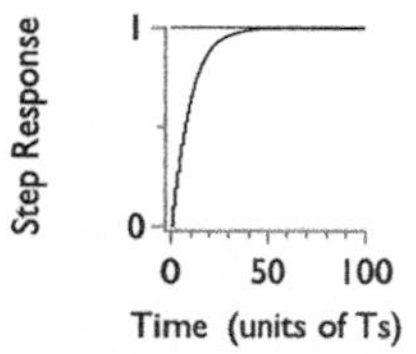

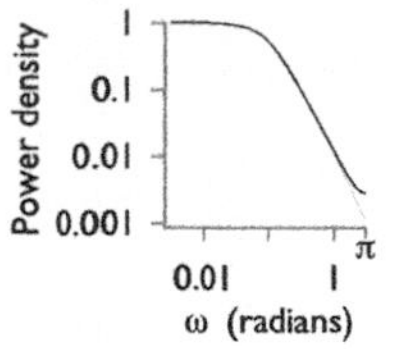

a. By Z-transformation, derive the form of the transfer function for both filters.
b. Show that the IIR and FIR transfer functions are identical for $n \to \infty$.
c. Show that the corner frequency of the equivalent continuous system is $\omega_0 = \frac{1-a}{\sqrt{a}}$.
d. Reproduce the step and frequency response graphs shown at left.

5.7 **FIR filter with linear phase response**. For $n = 2N + 1$ odd, consider the FIR filter

$$y_k = B_0 u_k + B_1 u_{k-1} + \cdots + B_{n-1} u_{k-n+1}\,,$$

a. Show that if $B_m = B_{n-1-m}$, then the complex frequency response $y(\omega)$ has a linear phase. That is, show that $y(\omega) = \tilde{y}(\omega)\, \mathrm{e}^{-\mathrm{i}\omega\tau}$, where $\tilde{y}(\omega)$ and τ are real.
b. An ideal low-pass filter would have a frequency response that is 1 for $\omega < \omega_c$ and 0 for $\omega_c < \omega < (\pi/T_s)$. Show we can realize the filter via FIR coefficients

$$(B_m)_{\text{acausal}} = \left(\frac{\omega_c}{\pi}\right)\left(\frac{\sin m\omega_c T_s}{m\omega_c T_s}\right), \qquad -\infty < m < \infty\,,$$

where m is integer. For negative m, the ideal low-pass filter is *acausal* and cannot be implemented in real time, since it needs future information.

c. To make a realizable filter, truncate to $n = 2N + 1$ terms and then delay each component to make it causal. The resulting filter has

$$B_m = \left(\frac{\omega_c}{\pi}\right)\left(\frac{\sin(m-N)\omega_c T_s}{(m-N)\omega_c T_s}\right), \qquad 0 < m < 2N,$$

Verify that this filter is linear phase and plot the magnitude of the frequency response for $n = 101$ and 1001. Note and explain the *Gibb's phenomenon.*

d. Discuss the effects of multiplying the FIR coefficients by a *Hamming window*,

$$B_m \to B_m \times \left[0.54 - 0.46\cos\left(\frac{\pi m}{N}\right)\right], \qquad 0 < m < 2N.$$

5.8 Discrete Parseval's theorem.

a. Using the definitions of the discrete time Fourier transform (DTFT) given in Section 5.2.3, derive Parseval's Theorem, Eq. (5.37).

b. By integrating around the unit circle, derive an alternate form of the theorem,

$$\sum_{k=0}^{\infty} f_k^2 = \frac{1}{2\pi \mathrm{i}} \oint \frac{\mathrm{d}z}{z} f(z)\, f(z^{-1}).$$

c. Show that Parseval's Theorem works explicitly for the transform pair $f_k = a^k \theta_k$, with $|a| < 1$ and $f(z) = \frac{z}{z-a}$, with $z = \mathrm{e}^{\mathrm{i}\omega}$. Here, $\theta_k = 1$ for $n \geq 0$ and 0 otherwise.

5.9 Discretization of a zero-order hold. To find the discrete matrices $\boldsymbol{A}_\mathrm{d}$ and $\boldsymbol{B}_\mathrm{d}$ from Eq. (5.40) in one step and without inverting $\boldsymbol{A}$, show that $\exp[T_s \left(\begin{smallmatrix} \boldsymbol{A} & \boldsymbol{B} \\ \boldsymbol{0} & \boldsymbol{0} \end{smallmatrix}\right)] = \left(\begin{smallmatrix} \boldsymbol{A}_\mathrm{d} & \boldsymbol{B}_\mathrm{d} \\ \boldsymbol{0} & \mathbb{I} \end{smallmatrix}\right)$.

5.10 ZOH discretization. For a continuous function $u(t)$, its zero-order-hold staircase function $u_k(t)$ is defined in Eq. (5.38). Show that

a. The Laplace transform of the zero-order hold is given by $\mathcal{L}[u_{\mathrm{ZOH}}] = (\frac{1-\mathrm{e}^{-sT_s}}{s})\mathcal{Z}[u]$, where $\mathcal{Z}[u]$ is the Z-transform of the sequence $u_k = u(kT_s)$.

b. The ZOH discrete transfer function $G_\mathrm{d}(z) = (1 - z^{-1})\,\mathcal{Z}\{\mathcal{L}^{-1}[\frac{G(s)}{s}]\}$, with $G(s)$ a continuous system transfer function and $\mathcal{Z}\{\mathcal{L}^{-1}[\cdot]\}$ the Z-transform of the time domain signal from the inverse Laplace transform, sampled at times kT_s.

c. The 1st-order system $G(s) = \frac{1}{1+s}$ implies that $G_\mathrm{d}(z) = \frac{1-\mathrm{e}^{-T_s}}{z-\mathrm{e}^{-T_s}}$ (cf. Eq. 5.42).

d. The 2nd-order system $G(s) = \frac{1}{1+s^2}$ implies that $G_\mathrm{d}(z) = \frac{(1-\cos T_s)(z+1)}{(z-\mathrm{e}^{\mathrm{i}T_s})(z-\mathrm{e}^{-\mathrm{i}T_s})}$. The *sampling zero* at $z = -1$ arises solely from the sampling process; $G(s)$ has no zero.

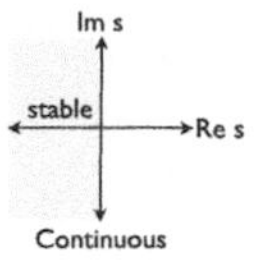

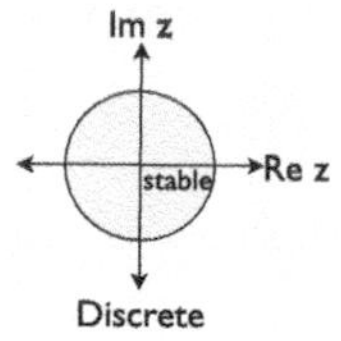

5.11 Mapping s to z. Show the following:

a. The change of variable $z = e^{sT_s}$ maps Re $s < 0$ to $|z| < 1$ (see left).
b. The same mapping is valid for the Tustin transformation: $s = \frac{2}{T_s}\frac{z-1}{z+1}$. Thus, if the continuous system is stable, so too is its Tustin discretization.
c. The backward Euler rule for $s \to z$ gives the *Euler* approximation to an integral, while the Tustin transformation gives the *trapezoidal* algorithm.

5.12 Tustin transformation and frequency warping.

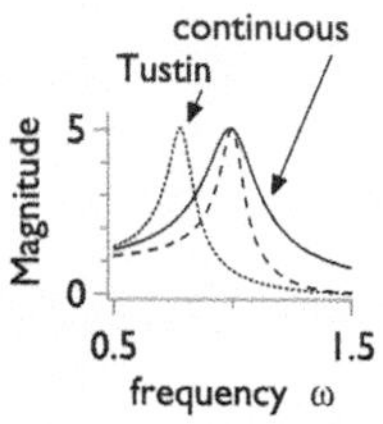

a. Show that that the Tustin transformation, $s \to \frac{2}{T_s}\frac{z-1}{z+1}$ distorts frequencies so that a frequency ω in the continuous system maps to a frequency $\omega' = \frac{2}{T_s}\tan(\omega T_s/2)$.
b. Find a value λ to rescale, or "prewarp," the Tustin transformation ($s \to s' = \lambda s$), so that its frequency response matches the continuous system at $\omega = \omega'$.
c. Plot $|G(s)| = \left|\frac{1}{1+2\zeta s+s^2}\right|$ with $\zeta = 0.1$, its discrete Tustin approximation $|G_{\text{Tustin}}(z)|$ for $T_s = 2$, and its prewarped version, matched at $\omega = \omega' = 1$. At left, the dashed line represents the prewarped approximation.

5.13 PID discretization: simpler can be better. Sometimes, the simple backward Euler discretization works best. Consider PI control of the continuous system $G(s) = \frac{1}{1+s}$, with $K(s) = 1 + \frac{1}{s}$. For $T_s = 0.1$, find the ZOH discretization $G_d(z)$.

a. Discretize the PI controller using backward Euler, $s \to \frac{1-z^{-1}}{T_s}$, and Tustin, $s \to \frac{2}{T_s}\frac{1-z^{-1}}{1+z^{-1}}$. Plot the step response for all three closed-loop systems.
b. Now add derivative control, $K(s) \to 1 + \frac{1}{s} + 0.1s$. Show that the backward Euler controller is little changed, but something goes wrong for the Tustin controller.

5.14 Controllability of a discrete system. Section 4.1.1 for continuous systems mostly carries over to discrete systems. But let us distinguish the *reachable set* of states $\mathcal{R}_k$ that may be reached from $\boldsymbol{x}_0$ in k steps from the *controllable set* of states C_k, the $\boldsymbol{x}_k$ that may be brought to $\boldsymbol{0}$ in k steps. A system is *reachable* if $\mathcal{R}_k = \mathbb{R}^n$ for all $k \geq n$.

a. Prove that a discrete SISO system is reachable if $\boldsymbol{W}_c = \begin{pmatrix} \boldsymbol{B} & \boldsymbol{AB} & \boldsymbol{A}^2\boldsymbol{B} & \cdots & \boldsymbol{A}^{n-1}\boldsymbol{B} \end{pmatrix}$ is invertible. As part of the proof, show that reachability requires n time steps (*deadbeat control*). Hint: Look explicitly at a sequence of iterates of $\boldsymbol{x}_0$.
b. Show that controllability is equivalent to reachability if $\boldsymbol{A}^{-1}$ exists. Thus, not all reachable discrete systems are controllable. Contrast with continuous systems.
c. For the undamped oscillator $\boldsymbol{A} = \begin{pmatrix} 0 & 1 \\ -1 & 0 \end{pmatrix}$, $\boldsymbol{B} = \begin{pmatrix} 0 \\ 1 \end{pmatrix}$, show that the continuous system and its ZOH are controllable, except at $T_s = m\pi$, for positive integer m. Why does controllability fail at these values of T_s? Why is it harder to control the oscillator when $T_s = 2\pi, 4\pi, \ldots$ than when $T_s = \pi, 3\pi, \ldots$?

5.15 Prediction observers. For the prediction observer:

a. Let the estimation error $\boldsymbol{e}_k^- \equiv \boldsymbol{x}_k - \hat{\boldsymbol{x}}_k^-$. Show that $\boldsymbol{e}_{k+1}^- = (\boldsymbol{A} - \boldsymbol{LC})\boldsymbol{e}_k^-$.

b. Show that the dynamics of the observer error and the physical system decouple (*separation principle*), in analogy with Eq. (4.65).

5.16 Current vs. prediction observers. The prediction observer state vector, $\hat{x}_{k+1}^-$, is based on observations up to time k (Eq. 5.52). Here, we construct an estimator that is based on observations up to time $k+1$. First: Given the old estimate $\hat{x}_k$, we predict the next state: $\hat{x}_{k+1}^- = A\hat{x}_k + Bu_k$. Then we correct the estimate using the difference between the new observation y_{k+1} and its predicted value, $\hat{y}_{k+1} = C\hat{x}_{k+1}^-$. Thus, $\hat{x}_{k+1} = \hat{x}_{k+1}^- + L(y_{k+1} - \hat{y}_{k+1})$, with L the observer gain. For the current observer,

a. Show that $\begin{pmatrix} x \\ \hat{x} \end{pmatrix}_{k+1} = \begin{pmatrix} A & 0 \\ LCA & A - LCA \end{pmatrix} \begin{pmatrix} x \\ \hat{x} \end{pmatrix}_k + \begin{pmatrix} B \\ B \end{pmatrix} u_k$.

b. Define the error $e_k \equiv x_k - \hat{x}_k$, and show that $e_{k+1} = (A - LCA)e_k$.

c. Reproduce the margin plots for y_k and $\hat{y}_k$ in Section 5.4.2 using the parameters given in the caption. Plot $\hat{y}_k$ for both current and prediction observers. Explore the output behavior for different estimator gains L. Why are there more problems with large gains for the prediction observer than for the current observer?

5.17 Fractional delays. For a linear system, fractional delays affect two neighboring time points. For example, consider $\dot{x} = -x(t) + u(t - \tau)$, with $0 < \tau < T_s$. Let the input $u_{k-\tau}(t)$ be a staircase ZOH signal delayed by τ with respect to the state x_k.

a. Draw a timing diagram for $x(t)$ and $u(t)$. Indicate x_k and $u_{k-\tau}$.

b. Show that the discrete dynamics have the form $x_{k+1} = Ax_k + B_1 u_{k-1} + B_0 u_k$. Find A, B_1, and B_0. Check the limits $\tau \to 0$ and $\tau \to T_s$. Hint: Split the integral.

c. Redo (b) assuming $T_s < \tau < 2T_s$. (Hint: The coefficients are almost the same.)

Other delays can be treated similarly (e.g., a delay between two subsystems can be analyzed as a delayed input to the second subsystem). Finally, another approach uses the *modified Z-transform*, $F(z, m) \equiv \sum_{k=0}^{\infty} f[(k + m - 1)T_s]z^{-k}$, with $0 < m < 1$.

5.18 Delays and predictive feedback.

a. For $x_{k+1} = ax_k + u_k$, with $u_k = -K_p x_k$, find the range of K_p that stabilizes $x = 0$.

b. For delayed proportional feedback $u_k = -K_p x_{k-1}$, show that $a > 2$ implies that no value of K_p can stabilize $x = 0$.

c. Show that $u_k = -K_p x_k^{\text{pred}}$, with $x_k^{\text{pred}} = ax_{k-1} + u_{k-1}$, can stabilize $x = 0$ for all a.

5.19 Deadbeat control of an undamped oscillator. Derive the deadbeat controller $K_d(z)$ described in Section 5.4.2 and Example 5.10. Reproduce the graphs in the example. Hint: For a step, $r(z) = \frac{z}{z-1}$. Use the final value theorem for y_k (or inverse transform).

5.20 Feedforward gain. For a steady-state output $y = r$, you can add an offset $k_r r$ to the input u. Show that $k_r = [C(\mathbb{I} - A + BK)^{-1}B]^{-1}$ for a discrete system. See Eq. 4.47.

5.21 Feedforward control of an oscillator. For the feedforward filter in Figure 5.13,

a. Implement numerically the feedforward control and reproduce the nine graphs. Design a feedforward filter by inverting the denominator and adding poles at zero. Produce the continuum response using the hybrid procedure of Section 5.4.3. For $T_s = \frac{\pi}{2}$ and $\frac{1}{2}$, find $G_d(z)$, the parameter λ, and the feedforward $F(z)$.

b. The z^n in the denominator of a feedforward filter means that we can write it as an FIR filter with delay: $F(z) = F_0 + F_1 z^{-1} + F_2 z^{-2} \cdots$. In the time domain, this is $u_k = F_0 r_k + F_1 r_{k-1} + F_2 r_{k-2} \cdots$. Put the transfer functions from (a) in this form, and show that they transform the reference r_k into the desired "shaped-input" u_k.

c. Show that the input amplitude $\sim [4\sin^2(\omega T_s/2)]^{-1}$, where $\omega = 1$ is the angular frequency of the oscillator. The factor is $\sim (\omega T_s)^{-2}$ as $T_s \to 0$. The ω^{-2} dependence mirrors the high-frequency response of the original transfer function.

d. Investigate numerically the impact of oscillator damping. Plot the shaped inputs and both discrete and continuous outputs for a reference step, for $\zeta = \{0, 0.4, 1\}$.

5.22 Deadbeat control of an undamped oscillator in one time step? Try to make a one-step deadbeat controller for $G(s) = \frac{1}{s^2+1}$ by enforcing $T_d(z) = \frac{1}{z}$.

a. Show that the required controller has the form $K_d(z) = \frac{1}{1-\cos T_s} \frac{z^2 - 2\cos T_s z + 1}{(z+1)(z-1)}$.

b. Reproduce the margin figure in Section 5.4.3.

c. The oscillatory step response arises because a controller pole cancels the sampling zero at -1. To see the problem in a simpler context, compare transfer functions $G_1(z) = 1$ and $G_2(z) = \frac{z-a}{z-a}$. Compute the output y_k given an initial condition y_0.

6 System Identification

If the key to improving the performance of a control system is to incorporate knowledge about the system dynamics, how can we learn the dynamics? An *offline* approach is to measure dynamical properties in preliminary experiments, determine model equations, and then design the controller. Once implemented, the controller is fixed. An alternative, *online* approach is to start with a nominal model of the system, base a controller on that model, monitor actual performance, and use "errors" to improve both model and controller. Such *adaptive control* techniques are the subject of Chapter 10. Here, we consider the offline approach. It is straightforward, simple, and leads to easy, practical advice for those who want to apply the ideas in this book as quickly as possible.

Offline approaches exist in both "physics" and "engineering" traditions. In brief: **Physicists want to understand; engineers need to predict.** It will take a chapter to unpack this brief sentence, but it can guide what might seem an overly technical discussion. Physicists use prior knowledge about a system and its setting to write down models based on physical principles. The models usually have a small number of parameters whose values must be fixed by measurements, although their orders of magnitude are often known a priori. Simplicity and clarity can matter as much as the ability to fit data. Engineers start with measured input and output signals and attempt to find a phenomenological dynamical model that maps one to the other, with the ability to predict used as the key measure of success. In control theory, the approach is termed *system identification* – hence the title of this chapter – but it is also known as *black-box* or *data-driven* modeling.

In this chapter, in Section 6.1, we compare the physics and engineering approaches in the simpler setting of determining a static functional relation. Section 6.2 discusses the choice of inputs and outputs for linear, time-invariant dynamical systems; Section 6.3, the fitting of models of known structure to input-output data; Section 6.4, the selection of the best model from several contenders; and Section 6.5, the "reduction" of complicated models to simpler approximations that are faster to compute, an important attribute for real-time control. Spoiler alert: although I tend to "root for the home team" and favor physics-based approaches, both have advantages and are worth knowing.

6.1 Physics or Phenomenology?

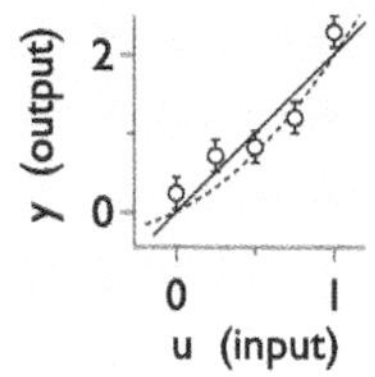

We can understand many issues in learning system dynamics by first considering the simpler problem of measuring a static function $y = f(u)$. In the spirit of this chapter, the function might represent the DC (zero-frequency) response of the system: the input has a constant value u and the output settles to a constant value y. At left, we sketch a possible measurement of some data, with error bars ± 0.2 representing the uncertainty in each measurement. The dashed line is $u + u^2$, which represents the (always unknown) "true" function that generated the data (plus noise of standard deviation 0.2). The solid line is a linear fit to $y = au$, with a a parameter estimated via the fit to be 2.0 ± 0.1.

To connect with the coming discussion, imagine that a physical model of the process predicts the linear relation $y = au$. Unfortunately, "extra" unmodeled and unknown physics leads to the quadratic term. In that case, which is all too typical, the fit function leads to systematic errors that depend on input levels – e.g., slightly too high for the central values of u and much too low for values of u extrapolated beyond the fitting range of (0,1).

The *data-driven* engineering approach is phenomenological: its goal is to fit data accurately. If we were to try $y = a_1 u$ and then $y = a_1 u + a_2 u^2$, we would find that the quality of the fit improves when the second order is added.[1] The phenomenological fits are thus usually more accurate than physics-based fits, which tend to neglect minor but real physical effects, especially at extreme ranges of parameters. And often, in control problems, it is more important that the algorithm perform well than to understand the dynamics well.

So why choose a physics-based approach nonetheless? A few reasons:

- *Portability*. Systems change, and a new phenomenological fit is needed each time. By contrast, if we understand the physical change (different values of physical parameters, for example), then we can often immediately adapt the model to reflect the new situation. Similarly, a good model will often extrapolate and apply to new situations outside the ones used to formulate the model. Data-driven approaches tend to fail completely when used in situations that differ from those used to formulate the model.
- *Robust control algorithms*. Good control algorithms have robustness against a reasonable level of error, which can be due to external disturbances but also from modeling discrepancies. As a rule of thumb, a reasonable control algorithm should tolerate mismatches of the order of 10%. Your model may need not to be more accurate, and a simple physical model may be enough to reach that level of accuracy.
- *Better controller intuition*. Physical modeling may suggest physical control algorithms (e.g., controlling the energy or damping of a system), giving insight as to why some control approaches are more successful than others.

[1] The fit continues to improve at still higher orders; at order four (for five data points), the fit would be perfect. However, including too many free parameters leads to *overfitting* – fitting to the noise rather than real dynamics, motivating the issue of *model selection* (Section 6.4)

- *Holistic design of system and controller*. Physical insights can improve the design of both system and controller. A system designed to be easy to control may not need accurate modeling (or fancy control algorithms). Such a system-controller combination will usually outperform a hard-to-control system, even one with perfectly known dynamics.

With no clear winner in the debate between physics and engineering approaches, we look at both in the following sections.

Example 6.1 (Holistic design and atomic force microscopes) Atomic force microscopes are sensitive to external vibrations. To minimize their effects, one should make the mechanical-resonance frequencies of the instrument as high as possible. Thus, an important part of designing and using an instrument is to both know the resonant frequencies and understand what sets their values. (See Problem 6.14.) Such understanding may then suggest a design improvement. For example, since the stiffness of a beam goes as the cube of its thickness, thickening a beam or plate-like structure by even a small amount can increase resonance frequencies significantly. Designing an experiment can be an iterative process that cycles through modifications of the physical system and measurements of the transfer function. The cycling can involve either physical prototypes or computer models that incorporate the important physical elements.

6.2 Measuring Dynamics

We discuss how to design system inputs that produce system outputs that can be used to infer the system dynamics for linear, time-invariant, single-input-single-output (SISO). After outlining some preliminary steps to determine how big an input can be without leading to nonlinearities, what the basic time scales are, and so on, we consider different types of inputs, focusing on frequency-domain techniques.

6.2.1 Preliminaries

Before making detailed measurements of a system's linear dynamics, you need to determine the range of inputs over which the system is well approximated as being linear. Then, as a quick start to determining dynamics, a simple step input leads to quick estimates of system time scales and can indicate the presence of delays or integrating poles, and so on.

Determine a Linear Range of Inputs

All dynamical systems are nonlinear when presented a large input. But what does "large" mean? Some nonlinearities are strong enough that inputs comparable to noise levels create a nonlinear response. In other systems, the limiting nonlinearity may be the saturation limits of the actuator. Since transfer functions assume linear response, it is essential to determine the range of inputs over which a system actually has a (locally) linear response.

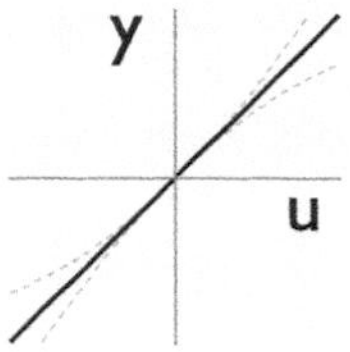

To start, look at the DC response (static input, static output). Ideally, the output y should vary linearly with the input u, as shown in the thick line at left (top). The dashed lines denote static nonlinearities from the dynamics itself or from the input or output couplings. For example, a (control) voltage V across a heater of resistance R produces a power $P \propto V^2/R$. Similarly, a thermistor sensor has a resistance that is a highly nonlinear function of temperature. Both of these relations are easy to compensate for.

Often, restricting the range of the input u will lead to an approximately linear system, and we can speak of an operating range – a range of u values – over which the system is well approximated by a linear response. For short, such a situation is *locally linear*.

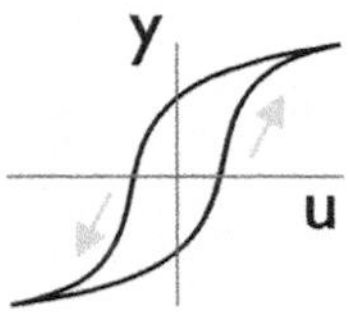

When *hysteresis* is present, the above procedures are insufficient, as the value of the measured variable y will depend on the system's history (see left). There are classic physical examples, such as the relation between magnetization and applied field. In mechanics, the *backlash* in a geared motor gives rise to hysteresis when changing direction, for example when using a machine lathe. Another example is *static friction*, where there can be a discontinuous, hysteretic transition between a stationary and a moving state. In electronics, hysteresis is often the basis for a switch: the two solutions can represent two states (e.g., one and zero in digital circuits such as a *flip flop*).

Step Function Input

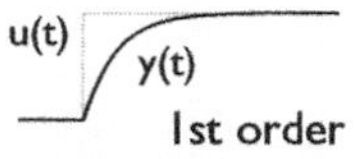

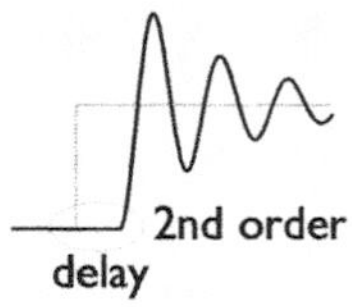

Step functions are a good "first step" in probing an unknown system. They are easy to apply: you simply make a sudden change in your control input and record the output. In particular, you can see how long it takes for the output to go to a constant, or to a linear slope for an integrating system. For systems with long time constants, this can be an efficient way to estimate low-frequency behavior. Likewise, the maximum slope gives a rough estimate of the upper limit of input frequencies. Steps are also convenient for detecting delays. At left are step responses of a first- and an underdamped, delayed second-order system (Section 2.4.4).

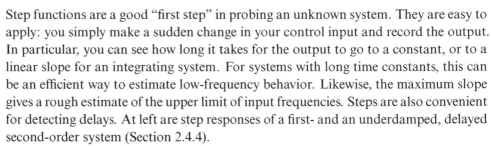

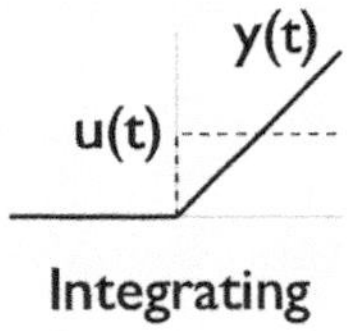

Step inputs are also useful for diagnosing the presence of *integrating* systems, which have a transfer function $G(s) \sim s^{-1}$ and a linearly increasing output (see left). For example, at short times, applying power to a heater in a well-insulated box leads to a linear temperature ramp. At high temperatures, losses to the environment dominate, and the temperature reaches a steady-state value where the power input balances the heat loss.[2]

[2] Problem 2.5 shows that a distributed loss μ to the environment, as described by a modified thermal diffusion equation, $\partial_t T(x,t) = D\partial_{xx}T - \mu T$, the (scaled) transfer function from $G(s) = \exp\left(-\sqrt{s}\right)/\sqrt{s}$ becomes $G(s) = \exp\left(-\sqrt{s+\mu}\right)/\sqrt{s+\mu}$. That is, $1/\sqrt{s}$ becomes $1/\sqrt{s+\mu}$, which "cuts off" after a time $\approx \mu^{-1}$. On time scales shorter than μ^{-1}, the system will appear to be integrating.

A step function thus allows us to estimate the range of relevant input frequencies and to test for delays and integrating response. Unfortunately, it gives only noisy information about high-frequency response. To understand why, let $u(t) = \theta(t)$, so that $u(s) = \frac{1}{s}$. Then, from Eq. (A.89 online), we write

$$G(s) = \frac{y(s)}{u(s)} = s\,y(s) \quad \Longrightarrow \quad y(t) = \mathcal{L}^{-1}\left[\frac{1}{s}G(s)\right] = \int_0^t dt'\, G(t') \,. \tag{6.1}$$

The step response is thus the integral of the impulse response, which can be recovered by differentiation. But differentiation amplifies noise at higher frequencies. Stating the problem in the frequency domain, a step input has little energy (and thus the output little information) at high frequencies. The input energy – and hence the signal-to-noise ratio of the output – decreases as ω^{-2}.

A similar difficulty renders impractical another seemingly desirable input, $u(t) = \delta(t)$, the Dirac delta function, or *impulse*, which has $u(s) = 1$. Thus, $y(s) = G(s)u(s) = G(s)$ directly! But again, a delta-function input – or its approximation by a short pulse – has little power overall and still less in any small frequency range. Trying to compensate by increasing the amplitude may lead to an undesired nonlinear response.

6.2.2 Sinusoidal Input

Measuring a transfer function accurately requires inputs with significant power, small amplitude, and wide frequency range. An intuitively appealing choice is a sequence of single sine waves, one for each desired measurement frequency.

In more detail, an input signal $u_r(t) = u_r \cos \omega t$ leads to a steady-state output $y_r(t) = y_r \cos(\omega t + \varphi)$. In terms of complex signals, we write $u(t)$ and $y(t)$, with real parts $u_r(t)$ and $y_r(t)$, respectively. Then, $u(t) = u_0\, e^{i\omega t}$ and $y(t) = y_0\, e^{i\omega t}$. We take $u_0 = u_r$ as real, but $y_0 = y_r\, e^{i\varphi}$ is in general complex. The real part of $y(t)$ gives the *in-phase* response (proportional to $\cos \omega t$), and the imaginary part gives the *out-of-phase* response (proportional to $\sin \omega t$). The transfer function at frequency ω is then $G(s = i\omega) = y_0/u_0$ (see Bode plot at right).

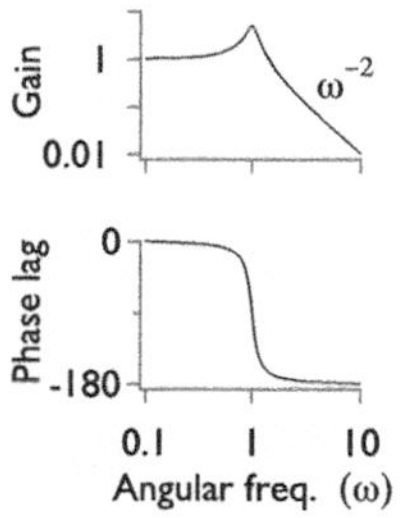

The full dynamical response has both a transient and periodic steady-state contribution. Recall from Eq. (2.48), for an initial condition $\boldsymbol{x}(0) = \boldsymbol{0}$, that $\boldsymbol{x}(t) = \int_0^t dt'\, e^{\boldsymbol{A}(t-t')}\, \boldsymbol{B}\, u(t')$. Even with a zero initial condition, the driving term can also produce transient response. To see this for a sinusoidal driving term, let $u(t) = e^{i\omega t}$. Then

$$\begin{aligned} \boldsymbol{x}(t) &= \int_0^t dt'\, e^{\boldsymbol{A}(t-t')}\, \boldsymbol{B}\, e^{i\omega t'} = e^{\boldsymbol{A}t}(i\omega\mathbb{I} - \boldsymbol{A})^{-1}\left[-\mathbb{I} + e^{(i\omega\mathbb{I}-\boldsymbol{A})t}\right]\boldsymbol{B} \\ &= \underbrace{-\,e^{\boldsymbol{A}t}(i\omega\mathbb{I} - \boldsymbol{A})^{-1}\boldsymbol{B}}_{\text{transient}} + \underbrace{(i\omega\mathbb{I} - \boldsymbol{A})^{-1}\boldsymbol{B}\, e^{i\omega t}}_{\text{periodic steady state}} \,. \end{aligned} \tag{6.2}$$

The transient term proportional to $e^{\boldsymbol{A}t}$ in Eq. (6.2) arises because of the mismatch between the initial condition, $\boldsymbol{x}(0) = \boldsymbol{0}$ and the periodic solution. For $t \gg \tau_{max}$ (longest relaxation time), only the periodic steady-state contribution will be present.

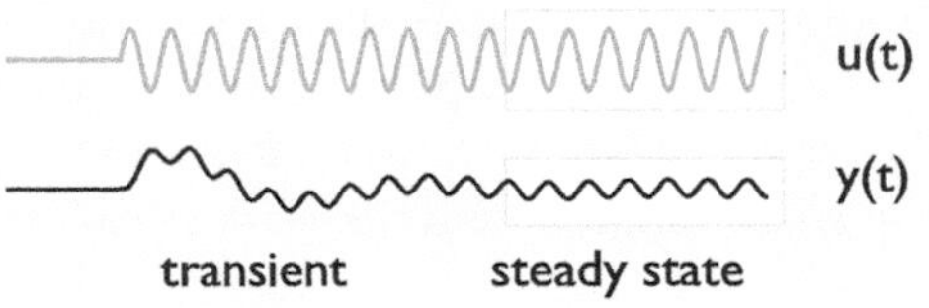

Fig. 6.1 Transient and periodic steady-state response to a sinusoidal input for a harmonic oscillator $\ddot{x} + 2\zeta\dot{x} + \omega_0^2 x = u(t)$, with $\zeta = 0.3$ and $\omega/\omega_0 = 10$. y scale ten times u scale.

Figure 6.1 illustrates these features. When we first apply a sinusoidal waveform to a damped harmonic oscillator with initial conditions $\boldsymbol{x}(0) = \mathbf{0}$, there is a transient, at the natural frequency ω_0. The transient decays after a time of order $(\omega_0\,\zeta_0)^{-1}$, leaving the periodic steady-state solution outlined in white on the figure. Notice that the decay time for a transient of a linear system is independent of the forcing amplitude. After each change in input, you still need to wait the same amount of time before taking a measurement.

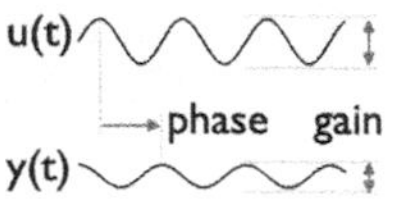

Although you could use transients to make inferences about the transfer function, non-periodic input signals are tricky to deal with properly. The simpler, safer course is to wait until transients die away before acquiring data.[3] Once you have waited long enough to reach a steady state, the only important parameters are the gain (relative amplitude) and phase difference. These two numbers, illustrated at left, give the magnitude and phase of the complex number $G(i\omega)$. In conclusion, the single-sine method for frequency-domain estimates of transfer functions is simple, safe, but sometimes slow:

- *Simple*: The input-output relation for sine waves can even be determined graphically.
- *Safe*: At each frequency, you can check nonlinearities and the signal-to-noise ratio.
- *Sometimes slow*: Low damping can lead to long transients.

Practical Issues

In implementing a sinusoidal input, there are a few more practical issues to consider:

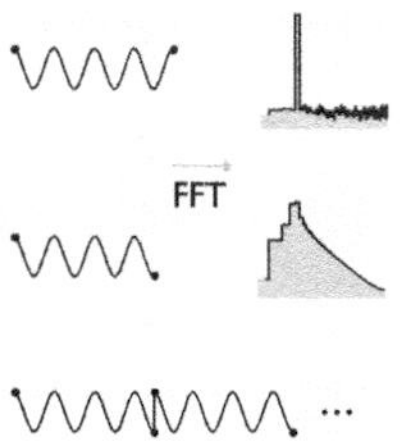

- *Timing*. Use the clock signal from the signal generator to set the acquisition timing. Alternatively, measure input and output synchronously or with known time offset. In either case, it is important to measure the input and output using the same clock (t).
- *Measure whole periods*. Record an integer number of periods of the particular frequency being tested. Remember that the discrete Fourier transform implicitly turns your time record of length τ into a periodic signal of period τ. Discontinuities between the point at $t = 0$ and $t = \tau$ lead to *spectral leakage*, as illustrated at left: The top sine wave is measured for exactly 4 periods, and all of its spectral energy

[3] A test for transients is to extend the last measured period to a periodic curve $y_p(t)$ and subtract from $y(t)$.

is concentrated in a single bin, as the log-log plot of the magnitude squared of the FFT shows. The other sine wave is sampled for 3.5 periods, and its spectral energy is spread over a wide range of frequencies. Since the waveform is in effect repeated, noninteger period lengths lead to jump discontinuities illustrated in the bottom graph and, hence, to broadened spectra. Another advantage of using periodic waveforms is that we can average the response in several blocks to reduce noise and to estimate the noise spectrum. See Section 6.2.4.

- *Lower frequency bound.* The finite time τ of a measurement implies that you probe to angular frequencies $\omega_{\min} \approx 2\pi/\tau$. Ideally, this frequency will be lower than any dynamics produced by your system. This may not be true if this system is approximately integrating, s^{-1}, with a pole at or near zero frequency. Also, disturbances at low frequencies lead to "drifts." Here are two strategies to deal with such situations:
  - *Decoupling*: A high-pass analog filter can remove low-frequency signals.[4] The filter may distort in the time domain at frequencies that are too close to the filter's cutoff.
  - *Detrending*: The effects of drifts can be removed digitally after acquisition by fitting to a "background function" such as a low-order polynomial. The choice of polynomial order is somewhat ad hoc, as it does not map cleanly to frequency cutoff. Still, if the separation between drifts and signal frequencies is big enough, this is an easy fix.
- *Upper frequency bound.* We need $\omega_{\max} \gg \omega_c$, a desired closed-loop bandwidth. Typically, there is no well-defined "asymptotic limit" to the form of the transfer function. Going to higher frequencies usually just exposes some new dynamical effect. But the response will eventually be very small, and the neglected dynamics at higher frequencies may be lumped in with external disturbances. Use *anti-aliasing filters* to limit the bandwidth of the measured analog signals to ω_c (Chapter 5).
- *Frequency step.* Choose intervals fine enough to resolve basic features of the transfer function. Strong resonances with small damping coefficient ζ need several measurements to infer the associated parameters accurately. Otherwise, choose a fixed number of frequencies per decade, so that the density of points on a Bode plot is constant.
- *Phase jumps.* Rapid phase variations can produce jumps in the response curve. Most data analysis packages or programming languages can "unwrap" such jumps, removing discontinuities that exceed some threshold, as illustrated at right.

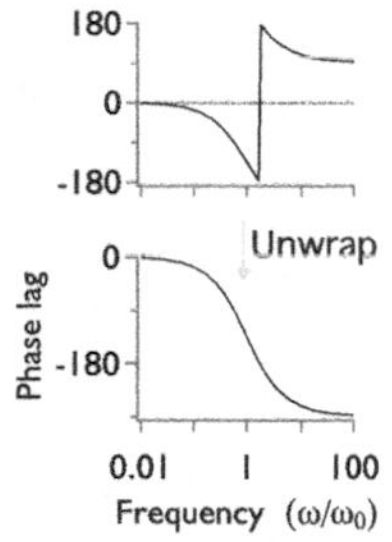

Example 6.2 (Harmonic oscillator with delay) To illustrate the above points, we measure the transfer function of a simulated harmonic oscillator with a delay in the signal path. For each frequency, we wait 50 s to let the transient die away and then take data for another 50 s. We then fit a sine wave to the last 50 s of data and extract an amplitude and relative phase. We compile 30 frequencies and then do a global fit to

[4] Oscilloscopes often have an "AC coupling" button that is just such a high-pass filter. Usually, the rationale is to remove large DC offsets from small AC signals. But it also eliminates drifts below the cutoff frequency.

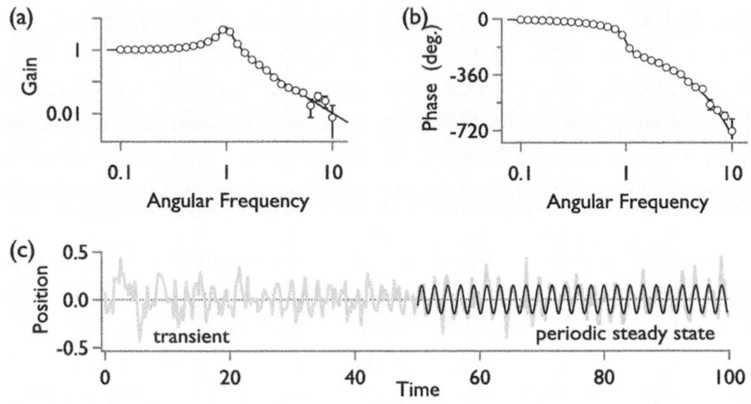

Fig. 6.2 Simulated measurement of a transfer function of a second-order system with delay, with $\omega_0 = 1$ rad/s, delay = 1 s, and damping $\zeta = 0.1$. We measure 30 frequencies each for 100 s, sampled at $f_s = 4$ Hz, a forcing amplitude of $A = 1$, and noise $\sigma = 0.1$. (a,b) Bode gain and phase plots. Solid lines are global fit, with $\omega_0 = 1.0003 \pm 0.0003$ rad/s, $\zeta = 0.1001 \pm 0.0003$, and delay $= 1.002 \pm 0.002$ s. The $\chi^2 = 56$ for 55 degrees of freedom. (c) Time response and fit to sine for $\omega = 3$.

the amplitude and phase response.[5] Figure 6.2 shows that even when the noise level is 10% of the nominal signal level (for gain = 1), the results are excellent. The parameters agree with their input values to better than a part in 1000, and the χ^2 statistic indicates a good fit. The drawback is that the total measurement time is 100 s / frequency, which for 30 frequencies is 3000 s, nearly 500 times the period corresponding to the natural frequency. Because we wait for transients to die away, the measurement time can be reduced only slightly before systematic errors caused by fitting to transients appear.

Finite-Frequency Tests of Nonlinearity

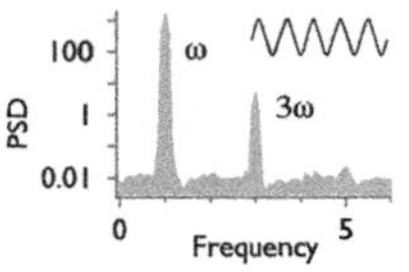

Static (DC) tests of linearity can be problematic. Systems may drift or be integrating. Nonlinearities such as slew-rate limits may increase with frequency. For these reasons, finite-frequency tests of linearity are important. For example, the nonlinear oscillator $\ddot{x} + 2\zeta\dot{x} + x + \varepsilon x^3 = u(t)$ has a nonsinusoidal response to a sinusoid input $u(t)$ (see left).

6.2.3 Multisine Inputs

In this section, we develop an idea championed by Pintelon and Schoukens (2012): using multifrequency periodic (*multisine*) inputs of controlled bandwidth can increase the efficiency of measurements relative to single-sine inputs. Consider signals $u_N(t)$

[5] An alternative strategy is to do a global fit to the real and imaginary parts of the response function. This strategy makes particular sense when a lock-in is used to directly measure real and imaginary responses.

of period τ that are the sum of N harmonics. With $\omega_n = 2\pi n f_0$ and $f_0 = 1/\tau$, we define

$$u_N(t) = \sum_{n=1}^{N} A_n \cos(\omega_n t + \varphi_n)\,, \tag{6.3}$$

where A_n and φ_n represent the amplitude and phase of the nth harmonic component. A single-sine waveform has $A_n = u_0 = 1$ and $A_{m \neq n} = 0$. Below in Section 6.2.4, we will also see that using a periodic waveform can help assess and then reduce the effects of noise. But first, let us see more clearly why multisines can be more efficient and how to choose the amplitudes, phases, and frequencies of the sine components.

Multisines are More Efficient

While you might think that by using a multisine with N sine waves you can estimate transfer functions N times faster, you should "keep the game fair" and keep the power injected the same for both cases. The root-mean-square (RMS) power of sine components add incoherently (Problem 6.2), with $u_{\text{rms}}^2 = \frac{1}{2}\sum_{n=1}^{N} A_n^2$. The single-sine excitation with $u_0 = 1$ has $u_{\text{rms}}^2 = \frac{1}{2}$. At equal RMS power, $u_N(t)$ should have amplitudes $A_n = a_n/\sqrt{N}$, with $a_n = \mathcal{O}(1)$. The multisine would seem to have no advantage, as the information gained depends on the total power injected, no matter what the frequency distribution. However, transients tip the balance in favor of the multisine: N single-frequency signals have N transients, while an N-frequency multisine has but one.

How to Choose the Amplitudes

Having fixed the overall amplitude scaling at $1/\sqrt{N}$, you might be tempted to choose the $a_n = 1$. The problem is that the high-frequency components will usually have a very small response and hence poor signal-to-noise ratio (SNR). A better strategy is to quickly estimate (either by a preliminary measurement, theoretical argument, or whatever) the approximate form of the transfer function and to choose the a_n to be proportional to the approximate inverse. The goal is to have the SNR of the response approximately equal at all frequencies. Of course, to do this perfectly, you would need to know the result of the measurement beforehand! But only a crude approximation is needed.

How to Choose the Phases

The maximum amplitude u_{max} of the input waveform $u(t)$ should be small, to minimize nonlinear response. To be more precise, we define the *crest factor* (Cr),

$$\text{Cr}[u] = \frac{u_{\text{max}}}{u_{\text{rms}}}\,, \qquad u_{\text{max}} = \max_{0\le t\le \tau} |u(t)| \qquad u_{\text{rms}} = \sqrt{\frac{1}{\tau}\int_0^{\tau} \mathrm{d}t\, u^2(t)}\,, \tag{6.4}$$

which balances the need to keep the RMS power high (good SNR ratios) with the need to keep the maximum amplitude low (smaller nonlinear response). In general, then,

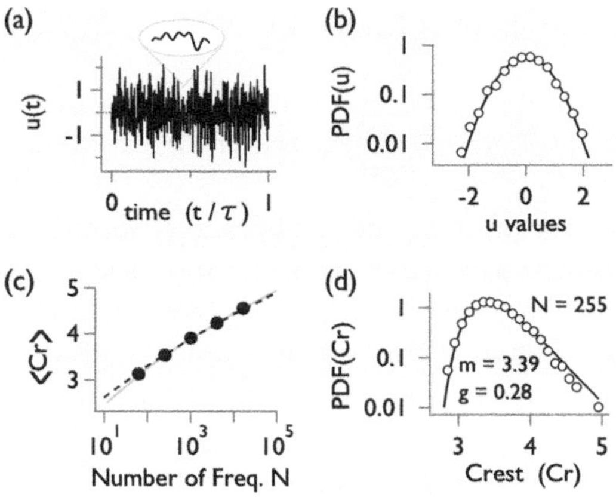

Fig. 6.3 N harmonics with random phases. (a) One period of a time series generated from 255 harmonics (sampled at 4096 points per period), with a blow-up of the oversampled wave above. (b) Histogram of values in (a); solid curve is $\mathcal{N}(0, \frac{1}{\sqrt{2}})$. (c) Markers are simulations of average crest factor as a function of the number of harmonics N. Dashed line is numerical solution to the transcendental equation from extreme-value theory, with $M = 5N$. Light-gray line is an empirical formula that approximates the asymptotic expansion of the erfc function, $\langle \text{Cr} \rangle = \sqrt{2\ln(2N)}$. (d) Gumbel distribution calculated from the underlying Gaussian and M draws.

we should design input signals that have low crest factors. Problem 6.2 explores crest factors. Here are a few conclusions:

1. Because the waveforms have $u_{\text{rms}}^2 = \frac{1}{2}$, the crest factor is determined entirely by $u_{\max}$.
2. For a single sine, $\text{Cr} = \sqrt{2} \approx 1.4$.
3. For N sine waves with $A_n = 1/\sqrt{N}$, Cr depends on the phases φ_n.
4. If all phases are the same ($\varphi_n = 0$), then $\text{Cr} = \sqrt{2N}$. This is the worst possible choice for measuring a transfer function.
5. Choosing uniform, random phases (Figure 6.3a) leads to Gaussian signal values (Figure 6.3b) and $\langle \text{Cr} \rangle \approx \sqrt{2\ln 2N}$, which is 3–4 for N between 100 and 1000 (Figure 6.3c).
6. Random phases lead to a distribution of input values $p(u) \sim \mathcal{N}(0, \frac{1}{2})$ – i.e., Gaussian with mean 0 and variance $\frac{1}{2}$. The crest factor is Gumbel distributed: positively skewed, with an exponential tail for large values and a sharp cutoff for low values (Figure 6.3d). That is, smaller crest factors are rarer than larger ones.
7. Nonrandom choices of φ_n can lead to lower crest factors. An *ansatz* by Schröder leads to $\text{Cr} \approx 1.66$ for signals whose harmonics have equal amplitude (see Figure 6.4c). Numerical optimization of the phases can lower Cr to ≈ 1.2–1.4.

Figure 6.4 illustrates multifrequency inputs with equal phases (a), random phases (b), and Schröder phases (c). Part (d) shows the magnitude for all cases; only the phases differ. The higher crest factor of the random-phase case results mostly from rare spikes that may lead to small nonlinear distortions but have only minor effect on

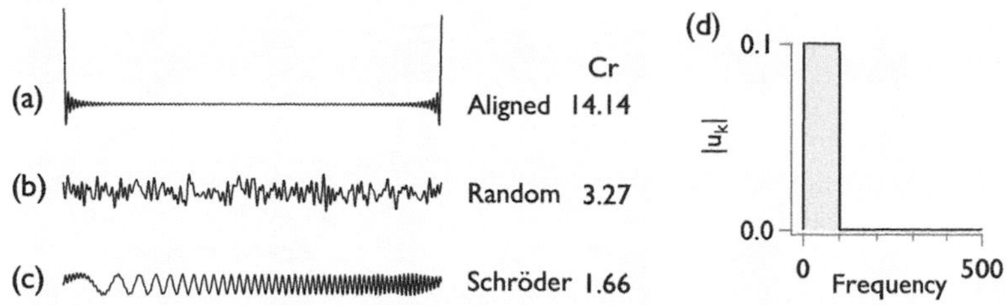

Fig. 6.4 Periodic multifrequency waveforms with 100 harmonics and 1000 samples per period. All three input waveforms have equal power and crest factors as indicated. (a) Equal phases. (b) Random phases. (c) Schröder phases, $\varphi_n = -n(n+1)\pi/N$. (d) Magnitude spectrum for all three waveforms.

the response function. The random-phase multifrequency waveform is convenient and practical for many applications.

How to Choose the Frequencies

Finally, we choose the multisine frequencies ω_n. Because $u(t)$ is periodic, the waveforms must be harmonics of the fundamental frequency $\omega_0 = 2\pi/\tau$. The highest harmonic is given by the Nyquist frequency, π/T_s, which limits the number of possible harmonics. Naively, you might want to use all of them, to maximize the frequency resolution. But the frequency response is typically plotted on a logarithmic frequency scale, so that you might opt for harmonics that are approximately logarithmically spaced (Example 6.3).

Example 6.3 (Harmonic oscillator with delay, multifrequency measurement) We repeat Example 6.2, using a multifrequency input $u(t)$ consisting of 30 harmonics, approximately logarithmically spaced. The period of the waveform is 50 s, and we measure 37 periods, for a total of 1850 s. Subtracting 50 s for the transient, we average over $M = 36$ periods, for a total of 1800 s of measurement time. The global fit gives parameters whose accuracy is comparable to that obtained before.

Another reason not to use all harmonics is to diagnose nonlinearities and unexpected transients. We discussed this for single-sine inputs above, but the idea works for multisines with a sparse set of harmonics.

6.2.4 Frequency-Domain Transfer Functions

So far, we have discussed how to measure the output response to single- and multisine inputs. The transfer function is the ratio of the two signals. If, as is common, we measure both input and output to a system, we have to take the ratio of Fourier transforms to estimate the frequency response. Here, we review some of the subtleties involved.

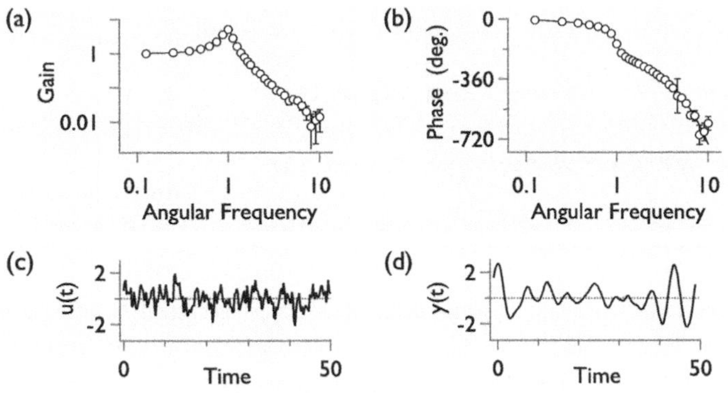

Fig. 6.5 Simulated multifrequency measurement of harmonic oscillator with delay, with $\omega_0 = 1$ rad/s, damping $\zeta = 0.1$, and delay = 1 s. (a, b) Bode gain and phase plots. Solid lines denote global fit, with $\omega_0 = 0.9994 \pm 0.0002$ rad/s, $\zeta = 0.0998 \pm 0.0002$, and delay = 0.997 ± 0.002 s. $\chi^2 = 78$ for 56 degrees of freedom. (c, d) Multifrequency input and corresponding output, sampled at $f_s = 4$ Hz.

To fix notation, denote the sampled input signal $u_k = u(kT_s)$ and output $y_k = y(kT_s)$, where T_s is the sampling period and $k = 0, 1, \ldots, N-1$ for one period. The period $\tau \equiv NT_s$ of the excitation waveform should be longer than the slowest important dynamical time scale, including any noise correlation time. Then, for each period, fluctuations in the input-output waveform are approximately independent. We have

$$u_k = (u_0)_k + (\xi_u)_k\,, \qquad y_k = (y_0)_k + (\xi_y)_k\,, \tag{6.5}$$

where u_0 is the "true" value of the input signal and ξ_u is the noise. Similarly, y_0 is the true output signal and ξ_y its noise. We do not distinguish here between different noise types (disturbance or measurement noise).

Now collect M periods of such data (after waiting for transients to decay). Make M an integer, to avoid leakage due to discontinuity between the end of one period and the beginning of another. Curve fits to the transfer function will be biased if such leakage is not taken into account since it alters the input amplitude at a given frequency.

The discrete Fourier transform (DFT) (see Chapter 5) of the input signal is denoted

$$u(\ell) = \sum_{k=0}^{N-1} u_k\, \mathrm{e}^{-\frac{2\pi i k\ell}{N}}\,, \qquad u_k = \frac{1}{N}\sum_{\ell=0}^{N-1} u(\ell)\, \mathrm{e}^{\frac{2\pi i k\ell}{N}}\,. \tag{6.6}$$

Because u_k is stochastic, so are the DFT components $u(\ell) = u_0(\ell) + \xi_u(\ell)$. If noise is white in the time domain, then in the frequency domain the noise $\langle \xi_u(\ell)\, \xi_u(\ell')^* \rangle = N\xi_u^2\, \delta_{\ell\ell'}$. That is, the fluctuations of individual Fourier components are independent complex Gaussian variables.[6] The expressions for y are similar. See Problem 6.5.

[6] More precisely, each frequency component is an independent *circular symmetric* complex Gaussian random variable. That is, real and imaginary parts are independently distributed as $\mathcal{N}(0, \frac{1}{2}\sigma^2)$, so that contours of the joint probability distribution function in the xy plane are circular. Even when the time-domain noise is correlated, the frequency-domain components are independent random variables.

If the fundamental period τ of the input waveform is much longer than the correlation time of the noise, the Fourier amplitudes of each period are independent random variables. We then estimate $u(\ell)$ and $y(\ell)$ by averaging:

$$\hat{u}(\ell) = \frac{1}{M}\sum_{m=1}^{M} u(\ell)_m \,, \qquad \hat{y}(\ell) = \frac{1}{M}\sum_{m=1}^{M} y(\ell)_m \,, \tag{6.7}$$

where $u(\ell)_m$ is the mth DFT measurement. We can then estimate the variance of each Fourier component and the covariance between input and output:

$$\hat{\sigma}_u^2(\ell) = \frac{1}{M-1}\sum_{m=1}^{M} |u(\ell) - \hat{u}(\ell)|_m^2 \,, \qquad \hat{\sigma}_y^2(\ell) = \frac{1}{M-1}\sum_{m=1}^{M} |y(\ell) - \hat{y}(\ell)|_m^2$$
$$\hat{\sigma}_{yu}^2(\ell) = \frac{1}{M-1}\sum_{m=1}^{M} [y(\ell) - \hat{y}(\ell)]_m \, [u(\ell) - \hat{u}(\ell)]_m^* \,. \tag{6.8}$$

Note that the (co)variances estimates are for a single sample. Divide them by M to obtain the corresponding estimates for the mean values. From the DFT averages $\hat{u}$ and $\hat{y}$, the nonparametric estimate of the complex frequency response is given by

$$\hat{G}(s = \mathrm{i}\omega_\ell) = \frac{\hat{y}(\ell)}{\hat{u}(\ell)} \,. \tag{6.9}$$

For *high signal-to-noise* ratios, where measurement noise and other disturbances are small, Eq. (6.9) estimates the transfer function $G(\mathrm{i}\omega_\ell)$ with negligible bias. Fluctuations about the mean are also approximately Gaussian. In Problem 6.8, we will see that the variance of $|G(\mathrm{i}\omega_\ell)|$ is

$$\hat{\sigma}_{|G|}^2 \approx \frac{|\hat{G}|^2}{M}\left[\frac{\hat{\sigma}_y^2}{|\hat{y}|^2} + \frac{\hat{\sigma}_u^2}{|\hat{u}|^2} - 2\,\mathrm{Re}\left(\frac{\hat{\sigma}_{yu}^2}{\hat{y}\hat{u}^*}\right)\right] \approx \frac{\hat{\sigma}_y^2}{M|\hat{u}|^2} \,, \tag{6.10}$$

where $\hat{G} = \hat{y}/\hat{u}$ and M is the number of repetitions of the input waveform, and where all other quantities depend on the frequency ω_ℓ. The latter identity holds if the input noise is negligible. The variance (in radians) of the phase is then $\hat{\sigma}_{\varphi}^2 = \hat{\sigma}_{|G|}^2/|\hat{G}|^2$ (Problem 6.9). For small $|G|$, the phase tends to a uniform distribution on $[0, 2\pi)$.

For *low signal-to-noise ratios* ratios, the situation is considerably more complicated. In Problem 6.7, we will see that the estimate in Eq. (6.9) is biased, with

$$\frac{\langle \hat{G}(\mathrm{i}\omega_\ell)\rangle}{G(\mathrm{i}\omega_\ell)} = 1 - \exp\left(-\frac{|u(\ell)|^2}{\sigma_u^2(\ell)}\right) , \tag{6.11}$$

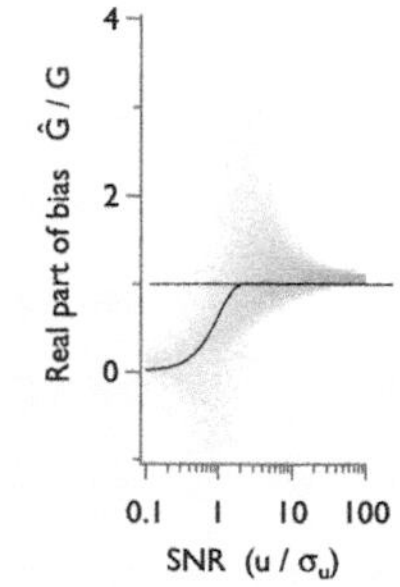

The bias of the transfer-function estimate $\langle\hat{G}\rangle$ is small when the input signal-to-noise ratio $u/\sigma_u \gtrsim 2$ (solid line at right). Even when the bias is moderate, the distribution of fluctuations can be very broad, as illustrated at right by the scattering of light-gray points. The points are calculated from 10^6 Monte Carlo realizations where the signal-to-noise ratio (SNR) u/σ_u is chosen randomly (uniform-in-log from 0.1–100). There is a huge scatter for $u/\sigma_u < 10$.

The root cause of the wild fluctuations is that whenever u fluctuates to a value near 0 (in the complex plane), the ratio y/u has a large magnitude. The large-noise limit is equivalent to the ratio $G = y/x$ of two real Gaussian variables $x, y \sim \mathcal{N}(0,1)$. The change-of-variables formula for probability distributions shows the ratio to be Cauchy distributed, with $p(G) = (\frac{1}{\pi})(\frac{1}{1+G^2})$ (Problem A.6.4 online). The fat tails (G^{-2}) imply that the mean and variance diverge and that one must use the median and full width at half maximum to quantify the distribution's location and scale. For complex G, the extra "phase space" of the complex plane keeps the mean in Eq. (6.11) finite, but the variance still diverges. These awkward features of $\hat{G}$ can be avoided by an *errors-in-variables* least-squares fit to a parametrized model for $G(i\omega)$ that avoids an explicit division by noisy variables (Section 6.3.1). When there are several noise sources, calculating the bias becomes even more complex. Surprisingly, adding noise can sometimes *decrease* bias (Problem 6.10).

Given the complexities of the low-SNR case, try to use large signal strengths whenever possible. Sometimes, this can be difficult: *Nonlinearities* in the system response can limit the allowable size of the input signal. That is, using larger signals will alter the transfer function you are trying to measure. At *high frequencies*, the physical response can be greatly attenuated, so that even large inputs generate very small outputs. The output measurement noise is likely to be significant. Finally, because of *weakly damped resonances*, even a small input leads to a large output (high Q factor). Thus, you are likely to be limited to small input signals, which may have poor SNR relative to either actuator noise or measurement noise. Thus, despite best efforts, you may sometimes need to think about the issues of bias and diverging variance that characterize the low-SNR case.

A couple more issues: First, you should average the complex DFT rather than its magnitude. Doing the latter biases the transfer function estimate and limits the averaging of noise (Problem 6.11). Averaging also decreases frequency resolution: the DFT of a time series of duration MT is $\Delta f = 1/(M\tau)$ while the frequency resolution of the averaged time series is $1/\tau$, coarser by a factor M. Of course, if we record the full time series, we can Fourier transform both with and without averaging.

Problem 6.12 explores many of the issues of frequency-domain estimates in the simplest possible setting, a static transfer function with no dynamics at all. The particular example examines different ways to estimate a resistance from voltage and current measurements. The best estimate turns out to be $\hat{R} = \overline{V}/\overline{I}$, where $\overline{V}$ is the average of the voltage measurements. Intuitively, averaging reduces fluctuations and thus the problems of bias and large variance noted above. The lesson immediately extends to transfer function estimates, where an effective strategy is to use a periodic input such as a multisine and then to estimate $\hat{G} = \hat{y}/\hat{u}$ at each frequency, with $\hat{u}$ and $\hat{y}$ the average of M measurements.

Transfer Functions for Systems with Feedback

Another situation where the measurement of a transfer function is prone to bias concerns systems under feedback control, where noise in the output is coupled back to the input. Although the most straightforward measurement strategy is to break open the feedback loop and record the open-loop response, sometimes this is not possible (or desirable):

- *Unstable systems* usually cannot be measured unless stabilized by feedback. Dynamics will drive the system to a region of state space whose local behavior is completely different from that near the operating point of the stabilized system.
- *Interacting subsystems*. You may be interested in one piece of the system but lack the ability to isolate it from the whole. For example, in biology, the expression of one gene can enhance or repress that of another gene. Studying one gene in isolation may require a mutant system that is difficult to create and may have further unwanted consequences.
- *Economics*. It may cost too much to shut down an experiment (or factory) just to characterize its dynamics.

As an example of the issues raised by measuring a subsystem, consider the closed-loop system in the block diagram at right. You impose r and measure u and y_m. If we can neglect measurement noise for u, the transfer function from u to y_m is $G_m \equiv \frac{y_m}{u} = \frac{KGr+\xi}{K(r-\xi)}$ (Problem 6.13). With no applied reference, $G_m \to -1/K$; that is, you measure the feedback dynamics rather than the system G. For large r, $G_m \to G$. For finite reference levels, we can estimate G by $\hat{G} = \langle y_m \rangle / \langle u \rangle$. In Problem 6.13, we explore how to minimize the bias and variance of $\hat{G}$ by using a multisine reference input.

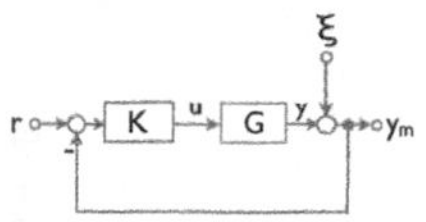

Sometimes the input $u(t)$ is not accessible. Examples include gene transcription in biological systems where the general interactions are known but where measuring all reaction constants may not be feasible. In these cases, we can measure a closed-loop transfer function $T = KG/(1 + KG)$, but unless we know K, we cannot estimate G. Even when K is known, the inversion from T to G may be very noisy.

6.2.5 Time-Domain Transfer Functions

Although the frequency-domain transfer function is physically appealing, we can also estimate the time-domain transfer response (Green function) $G(t)$. One method is known as *correlation testing* and uses an input-output pair $[u(t), y(t)]$. We include it for three reasons: First, it is the basis for widely used phenomenological methods in engineering. Second, it will lead to a deeper understanding of the notion of causality. Third, it leads to a practical test for the presence of feedbacks in a system. Even so, frequency-domain methods for estimating $G(s)$ are usually more intuitive and less prone to artifacts.

We begin by defining the correlation function between output and input, assuming that the dynamics and signals are statistically stationary and real (see Section A.9 online):

$$R_{uy}(\tau) \equiv \langle u(t)\, y(t+\tau) \rangle \equiv \lim_{T\to\infty} \frac{1}{2T} \int_{-T}^{T} \mathrm{d}t\, u(t)\, y(t+\tau) = \lim_{T\to\infty} \frac{1}{2T} \int_{-T}^{T} \mathrm{d}t\, u(t-\tau)\, y(t)\,, \tag{6.12}$$

where we use the stationarity assumption to shift both limits of integration by τ and then change variables. Substituting for $y(t)$ then gives

$$\begin{aligned}
R_{uy}(\tau) &= \lim_{T\to\infty} \frac{1}{2T}\int_{-T}^{T} \mathrm{d}t\, y(t)\, u(t-\tau)\,, \qquad \text{with} \qquad y(t) = \int_0^{\infty} \mathrm{d}t'\, u(t-t')\, G(t') \\
&= \lim_{T\to\infty} \frac{1}{2T}\int_{-T}^{T} \mathrm{d}t \int_0^{\infty} \mathrm{d}t'\, u(t-t')\, u(t-\tau)\, G(t') \\
&= \lim_{T\to\infty} \int_0^{\infty} \mathrm{d}t' \left[\frac{1}{2T}\int_{-T}^{T} \mathrm{d}t\, u(t-t')\, u(t-\tau)\right] G(t') \\
&= \int_0^{\infty} \mathrm{d}t'\, R_{uu}(t'-\tau)\, G(t') = R_{uu} * G\,.
\end{aligned} \tag{6.13}$$

Laplace transforming both sides of Eq. (6.13) gives a simple, elegant result:

$$G(s) = \frac{R_{uy}(s)}{R_{uu}(s)}\,. \tag{6.14}$$

Frequency chirp

Discrete random binary

White noise

Now that we know how to extract the transfer function from an arbitrary input signal, we again need to decide on a suitable signal. We have argued that a multi-sine with constant amplitude and random phases is a good choice. The engineering literature often focuses on others such as the frequency chirp, random binary, and white-noise signals at left. These all have at least approximately flat power spectra and are discussed in Problems 6.3 and 6.4. Note that if $u(t)$ has a white-noise (flat) power spectrum, then $R_{uu}(t) = \delta(t)$ and $R_{uy}(t) = G(t)$, the impulse response function. In the frequency domain, $R_{uu}(s) = \text{const.}$ and $G(s) \propto R_{uy}(s)$. A disadvantage of nonperiodic signals, though, is that they inject power beyond the Nyquist frequency and require anti-alias filters.

Although time-domain signals would seem not to have issues with transients – the input signal, not being periodic in general, is itself a kind of transient – that is not quite right. Implicitly, the time-domain method assumes that at $t = 0$ the system is in a steady state for the input $u = 0$. The effects of any previous input need to have died away beforehand.

Despite their disadvantages, stochastic input signals have been put to good use in physical settings. Using an argument equivalent to the fluctuation-dissipation theorem, Twiss (1955) showed that measuring the cross-correlation between two "ports" of a network whose components are in thermal equilibrium gives the real part of the impedance between the two points. The impedance can be related to the Green's function; a similar result holds for two points in a continuous field. Weaver and Lobkis (2001) applied these techniques to ultrasonics, and Apalkov et al. (2004) used the correlations between speckle intensities at two points in a disordered medium to deduce details of a photonic band gap. In the latter case, the random input signal is due to light that is multiply scattered by refraction-index variations at fixed random locations in the medium rather than thermal fluctuations.

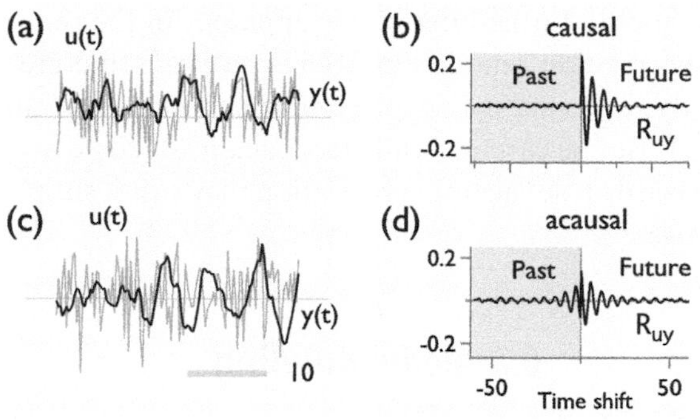

Fig. 6.6

Input-output correlations give the impulse-response function. (a) Input (light gray) and output (black) for a lightly damped harmonic oscillator with $\zeta = 0.1$ and $\omega = 1$. A small portion of the 3600 s time series, sampled at $f_s = 4$ Hz, is shown. (b) Cross-correlation of output to input shows the causal structure and the impulse response function. (c,d) Adding feedback to the input leads to an acausal response. The input $u(t) = \xi(t) - 0.1y(t)$, where $\xi(t) \sim \mathcal{N}(0, 3^2)$.

Example 6.4 (Cross correlation for a harmonic oscillator) Figure 6.6 applies these ideas to a harmonic oscillator, $G(s) = (1+2\zeta s+s^2)^{-1}$, with $\zeta = 0.1$ (lightly damped) and $f_s = 4$ Hz. We analyze 3600 s of data with a noise of standard deviation $\xi = 0.1$. In Figure 6.6b, the correlation function is a damped oscillation, as we expect. From two noisy time series, a hidden deterministic relationship emerges.

Causal Structure

Cross-correlation can reveal *causal structure*: In Figure 6.6b, the correlation function $R_{uy}(\tau) = 0$ for $\tau < 0$, implying that the output comes *after* the input, and the future has no effect on the past. We choose words carefully here. From just the measurements $u(t)$ and $y(t)$, we cannot conclude that $u(t)$ *causes* $y(t)$. Perhaps an earlier signal $s(t)$ influences both $u(t)$ and $y(t)$, as shown at right. Nonetheless, what we can conclude is that $y(t)$ cannot have caused $u(t)$.[7] The cross-correlation technique generalizes in a statistical sense, statements such as "Event B always follows event A."[8] We discuss causality further in Chapter 15.

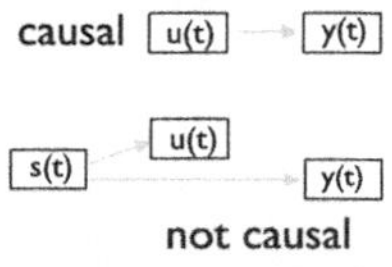

We can also use cross-correlation techniques to test for the presence of feedback loops in a dynamical system. In Figure 6.6(c,d), we repeat the analysis of (a,b) but

[7] Granger (1980) starts from N. Wiener's similar-in-spirit axiom that "The past and present may cause the future, but the future cannot cause the past." Granger's influential analysis relating causality to the statistics of time series received the Nobel Memorial Prize in Economics in 2003. As we discuss in Chapter 15, acting now on an *anticipated* future value $y(t)$ can also create correlations that seemingly violate causality. Distinguishing anticipation from causality is tricky (Hahs and Pethel, 2011; Pearl, 2009).

[8] In special relativity, the terms "before" and "after" lose their absolute meaning. By introducing concepts such as *light cones*, one can make statements about causality that are valid in all inertial reference frames.

add feedback to the input. Little difference is apparent in the time series themselves (a,c), but the cross-correlation function in (d) has *acausal* structure at negative times. We can understand this as resulting from the influence a previous output $y(t-\tau)$ has on the current input $u(t)$. But be careful. while feedback leads to acausal correlation, the converse is not necessarily true: acausal correlation may result simply from a common signal, as shown above.

6.3 Model Building

Having measured the input and output signals and constructed a nonparametric estimate of the transfer function in either the frequency or time domain, the next step is to find a model that describes the data. In a physics-based approach, we use physical reasoning to specify the model class and need only fix the parameters. In the phenomenological engineering approach of system identification, we assume a transfer function model of the form

$$G(s) = \mathrm{e}^{-s\tau} \frac{N(s)}{D(s)} = \mathrm{e}^{-s\tau}\, G_{\mathrm{DC}} \frac{\prod_{k=1}^{N} (1 - s/z_k)}{\prod_{\ell=1}^{M} (1 - s/p_\ell)} \tag{6.15}$$

or, for digital control, its discrete counterpart in the z-plane. If z_k or p_ℓ is complex, its complex conjugate also appears. The task is to find the various parameters: τ (delay), G_{DC} (DC gain), zeros $\{z_k\}$, and poles $\{p_\ell\}$ up to some order n.

Physics-based models are often but not always of the same form as their phenomenological counterpart. One exception is for spatially extended systems. For example, we saw that transfer functions of thermal systems have forms such as $G(s) \sim \mathrm{e}^{-\sqrt{s}} / \sqrt{s}$. For control algorithms, we typically will need to approximate such transfer functions by finite-order rational polynomials (over a given frequency range), so that the control response depends on a finite history of previous outputs and controls.

6.3.1 Physical Approach to Modeling

A physics-based approach should begin informally with quick initial estimates. Start by visually inspecting the Bode plot of your transfer function. In general, we expect transfer functions to consist of combinations of simple elements: low- and high-pass filters, resonances, antiresonances, and so on. Try to identify limits, including the asymptotic behavior at DC ($\omega \to 0$) and at high frequency ($\omega \to \infty$). Try, too, to understand the physical origin of each feature. As well, make back-of-the-envelope physical estimates of at least a few of the lower-order parameters. That way, when you get parameters from curve fits, you will have a sense as to whether the results are reasonable or not.

As a rule of thumb, rough estimates such as those given in the examples below can be accurate to within a factor of 10. A more careful calculation (that accounts for factors of 2, π, etc.) can decrease the error to a factor of ≈ 2. Using finite-element software to better model complicated geometries can decrease the error to about 10%. Of course, more refined techniques require more time and care to achieve the stated level of accuracy.

Example 6.5 (Estimating a time constant in a thermal system) Recall that temperature obeys a diffusion equation, $\partial_t T = D\partial_{xx} T$. From dimensional analysis (Chapter 2, Section 2.2.2), we then expect, for an object of diffusion constant D and size scale ℓ, that $\tau \approx \ell^2/D$. The quadratic dependence on ℓ implies that the time scales of small objects will be dramatically faster than for larger objects. This can be used to advantage in design, if you are trying to increase a control bandwidth. Keeping the heater close to the sensor is another implication. Alternatively, if you passively isolate an object from thermal fluctuations by adding insulation, the time constant will give the approximate pole position of the low-pass thermal filter. Note that the thermal diffusion constants can vary widely: metals such as copper have $D \approx 10^{-4}$ m^2/s, while plastics and water have $D \approx 10^{-7}$ m^2/s.

Example 6.6 (Estimating a mechanical resonance frequency) Experimental equipment is often sensitive to vibrations, and understanding their cause can help improve the design. We care primarily about the lowest resonance frequency, as it is typically the one most liable to be excited by external vibrations, which usually are strongest at lower frequencies. Naively, we might expect the frequency $f \approx c_{\mathrm{L}}/\ell$, where c_{L} is the sound speed in the material and ℓ is its largest dimension. In Problem 6.14, we show that objects with large aspect ratio a, defined as the ratio of largest to smallest length scale, have lower resonances of order $f \approx c_{\mathrm{L}}/(\ell a)$. A plate, for example, has $a = \ell/h$, where ℓ is the lateral plate size and h the thickness. Putting in numbers for an aluminum plate of size $\ell = 10$ cm and $h = 1$ mm, we estimate $f \approx (5000\ \mathrm{m/s})/[0.1\mathrm{m} \times (10\ \mathrm{cm}/0.1\ \mathrm{cm})] = 500$ Hz. To increase the lowest resonance, make ℓ or a smaller, or both.

Starting from a physical model, we can estimate its parameters by a least-squares "curve fit," taking care of some subtle points that can lead to bias if not attended to.

1. Least-squares fitting assumes independent Gaussian-distributed noise, and you should design and process your data to meet this condition. One trick is to apply the Central Limit Theorem: the average of N independent random variables each of variance σ is a random variable that tends to a Gaussian distribution with variance σ/N as $N \to \infty$. The average of M DFTs of a periodic signal will tend to have Gaussian statistics.
2. Given data with Gaussian-distributed noise, use the estimated variance of each data point to properly weight the least-squares fit. Then use the resulting χ^2 statistic to assess the quality of the curve fit. See Appendix Section A.8.2 online for more details.

Since the transfer function is complex, there are actually *two* fits to do. Physically, it makes sense to fit both the magnitude and phase. Since the parameters are the same for both curves, you should do a simultaneous, *global fit*. One issue is that the units for magnitude and phase are different, so that any errors in estimating the uncertainties can bias the fit to favor one or the other response. An easy alternative is to separate G into its real and imaginary components. These have the same units and thus same uncertainties.

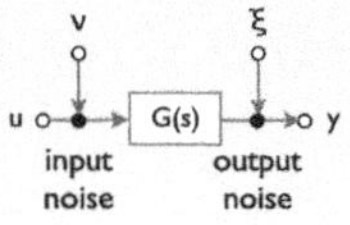

One issue with fitting to $\hat{G}$ is that the estimate is made by dividing a noisy output by a possibly noisy input, $\hat{G} = \hat{y}/\hat{u}$. (Recall the measurement schematic, illustrated at left.) As we have discussed, such estimates have bias and, potentially, fat-tailed distributions that seriously violate the Gaussian assumptions of least-squares fits. However, you can avoid such problems by doing a complex fit directly in terms of the input and output data sets.

The noise contributions from measurements of the input and output signals are usually independent. However, input noise can also result from disturbances such as amplifier noise or thermal fluctuations, which are physically input into the system and thus correlate the output and input noise terms. These noise terms and any correlations can be estimated at each frequency via the $\hat{\sigma}_u^2$, $\hat{\sigma}_y^2$, and $\hat{\sigma}_{yu}^2$ defined in Eq. (6.8).

In Problem 6.15, we argue that when inputs are noisy, we can avoid bias by fitting directly the complex transfer function, choosing model parameters $\boldsymbol{\theta}$ to minimize a modified *errors-in-variables* version of the usual χ^2 statistic:

$$\chi^2(\boldsymbol{\theta}) = \sum_{\ell=1}^{N} \frac{|\hat{y} - G\,\hat{u}|^2}{\hat{\sigma}_y^2 + |G|^2\,\hat{\sigma}_u^2 - 2\,\mathrm{Re}\left[\hat{\sigma}_{yu}^2\,G^*\right]}\,, \tag{6.16}$$

where all quantities are functions of the measured frequencies ω_ℓ and where the complex transfer function model $G = G(i\omega_\ell, \boldsymbol{\theta})$, with $\boldsymbol{\theta}$ the parameters that characterize the transfer function model. The shaded terms result from errors in u and their correlations with y. Notice that in Eq. (6.16), we do not take the quotient $\hat{y}/\hat{u}$ and that we find the usual χ^2 when input errors vanish. The discussion in Problem 6.15 parallels that for curve fitting with errors in the independent variable.

Equation (6.16) is more complicated than a standard curve fit, for two reasons. First, the quantities are complex, rather than real, as discussed above. Second, the presence of G in the denominator means that the χ^2 statistic is not quadratic in $\boldsymbol{\theta}$, even if G is linear in $\boldsymbol{\theta}$. The model parameters $\boldsymbol{\theta}$ must then be found by iterating a nonlinear-equation solver from initial guesses for $\boldsymbol{\theta}$. If input variable errors are small, ignore them, minimize the standard χ^2, and use the resulting $\boldsymbol{\theta}_0$ as an initial guess for the full problem. If input errors are large, then $\boldsymbol{\theta}_0$ may not be good enough. It is usually better to reduce the errors by averaging power spectra from more blocks of a periodic input than by improving initial guesses.

6.3.2 Direct Time-Domain Identification

In the previous sections, we have discussed frequency-domain methods for measuring parameters of physically motivated transfer functions. In the engineering approach, where the transfer function form is just a rational polynomial resulting from a Padé expansion, the parameters will not have obvious physical meaning. In such a case, there is no real advantage to transforming the data into a Bode plot and then fitting: you can do an equivalent fit to find the parameters directly from the time-domain input and output signals.

We thus discuss briefly a direct method for determining transfer functions from input-output data that skips all the intermediate steps we have discussed so far. Rather than give an exhaustive description, we illustrate some basic ideas on a simple one-dimensional example in discrete time and refer the reader to more specialized discussions and software packages for the general case. Let

$$y_{k+1} = a y_k + b u_k + \nu_k \,, \qquad \langle \nu_k \, \nu_\ell \rangle = \nu^2 \, \delta_{k\ell} \,, \tag{6.17}$$

which corresponds to a discrete first-order system with input noise. The goal: Given N pairs of input-output data, $\{u_k, y_k\}$, the structure of the first-order model, and known noise amplitude ν^2, estimate the parameters a and b.

Before proceeding further, we should point out what *not* to do: you might think that a reasonable approach would be to "fit to the solution of the differential equation." That is, you could assume values for the parameters and initial conditions (perhaps also treated as parameters), solve the differential equation, computer the χ^2 statistic for the predicted output relative to the observed, and then adjust the parameters to reduce the χ^2. Such a method would be horribly inefficient, as you would have to recompute a solution to the equations of motion each time. Fortunately, there is a much more efficient algorithm, based on fitting directly to the structure of the dynamical equations.

In the system-identification approach, we use Bayes' Theorem and notions of maximum likelihood to estimate the parameters (Section A.8.2 online). The procedure is equivalent to a curve fit where different data points have correlated fluctuations. Because the model structure in Eq. (6.17) is linear in the unknown parameters, the result is analogous to the *normal equations* from ordinary least squares (Section A.8.4 online):

$$\begin{pmatrix} \hat{a} \\ \hat{b} \end{pmatrix} = \begin{pmatrix} \sum y_k^2 & -\sum u_k \, y_k \\ -\sum u_k \, y_k & \sum u_k^2 \end{pmatrix}^{-1} \begin{pmatrix} -\sum y_{k+1} \, y_k \\ \sum y_{k+1} \, u_k \end{pmatrix} , \tag{6.18}$$

where the sums are over the N input-output pairs. (See Problem 6.16.)

At right, we simulate the identification process, for an input $u_k \sim \mathcal{N}(0, 1)$, $a = 0.9$, $b = 1$, and $\nu = 1$. The estimates are given by Eq. (6.18), which, for $N = 1000$ data points and 10 000 simulations, yield $\hat{a} = 0.8971 \pm 0.0001$, $\hat{b} = 0.9990 \pm 0.0003$. Single-sample standard deviations are $\sigma_a = 0.01$ and $\sigma_b = 0.03$. The standard deviations decrease with SNR $\equiv \sqrt{\langle u^2 \rangle / \nu^2}$. Both parameters are estimated with a small bias resulting from the nonlinear relation between log-likelihood and parameters. In Problem 6.16, we will see that the bias $\sim N^{-1}$ and is thus unimportant when a large

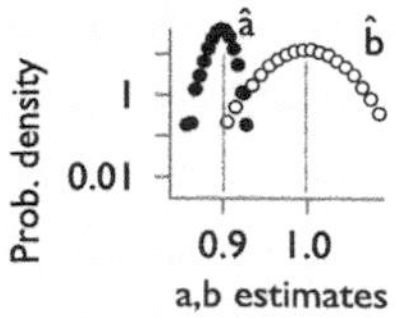

number of input-output pairs are probed, a property termed *consistent*. Time-domain methods have further pitfalls that can bias parameter estimates, even for large N:

1. *Persistence*. If the input $u(t)$ has limited bandwidth, the transfer function may be biased at other frequencies. Above, the input noise excites the system at all frequencies.
2. *Observation noise*. We must integrate over both noise sources.
3. *Closed-loop identification*. The noise terms are correlated with the system state.
4. *Initial conditions*. We have implicitly assumed that $x_0 = 0$. If $x_0 \neq 0$ (but known) or is stochastic, the likelihood functions are different.[9]

Obtaining accurate results from time-domain identification methods is therefore more subtle than it is in the frequency domain. You can avoid inaccurate inferences by using software packages that do the fits correctly. (See the Notes for suggestions.) You can also do *Monte Carlo simulations*, a procedure that is almost always worth the trouble. Use the extracted parameters from a model fit to generate surrogate data sets; repeat the analysis on each surrogate data set; and histogram each inferred parameter. You can then assess the degree of bias (compare the mean ± standard error with the known seed value for the data sets) and also the expected errors. You can even generate the covariance matrix that describes the coupling between estimates of parameter pairs.

Finally, time-domain methods have advantages if you need to learn about and control your system at the same time. Such *adaptive control* is the subject of Chapter 10. The least-squares methods discussed in this section are there formulated *recursively*, with running estimates of model fit parameters updated as each new measurement is made. With updated parameter estimates, we can update feedback controller parameters. Thus, although we favor frequency-domain methods for off-line determination of transfer functions, time-domain methods are preferred for online applications.

6.3.3 Nonlinear Models

So far, we have focused on modeling linear dynamical systems. Identifying the equations of motion for nonlinear dynamics is, unsurprisingly, much harder than for linear systems. Except for small nonlinearities, the frequency domain is less useful, and one must work directly in the time domain. For parametric uncertainty, one can use approaches that extend linear ideas, but the task is harder when the form of the equations is not known in advance. In the worst case, finding equations of motion is a computationally hard (NP hard) problem for which no polynomial-time algorithm is believed to exist (Cubitt et al., 2012). Fortunately, typical cases can be solved much

[9] A similar situation exists for frequency-domain fits. The easiest case is the periodic input, where the final state of the data $x_N = x_0$. Alternatively, $x_N - x_0$ may be nonzero and can be either deterministic or stochastic. This last occurs if the deterministic part of the input is periodic (and transients are left to decay) but stochastic noise is present at the input, too. See Agüero et al. (2010) for a discussion and a proof that maximum-likelihood estimates in the time and frequency domains are equivalent when done correctly.

faster, and there has been significant recent progress using different machine-learning techniques to develop phenomenological equations (Daniels and Nemenman, 2015; Brunton and Kutz, 2019). In Section 10.5, we briefly introduce one of these techniques, based on *neural networks*.

6.4 Model Selection

So far, we have assumed that the model structure is known. In the physical approach, this means knowing the correct physical model. In the engineering approach, this means selecting the proper order for both numerator and denominator of the transfer function. But if we are less certain about the physics or do not know the right order, as is usually the case, we need to decide whether adding phenomenological terms to a physical model or higher orders to a Padé expansion is warranted.

Loosely, *model selection* attempts to balance saying too little against making things up. "Saying too little" – *underfitting* – means that the model is too simple to adequately describe the data. "Making things up" – *overfitting* – means that the model has so many free parameters that it fits to the noise. Interestingly, physics-based approaches lead to strategies that differ from those used in phenomenological approaches. The first asks for the simplest model consistent with the data, where "simplest" will be defined by the *minimum description length* (MDL), and measured by the *Bayesian Information Criterion* (BIC). Alternatively, one can assess a model's ability to predict new observations via *cross-validation*, as measured via the *Akaike Information Criterion* (AIC).

Minimum Description Length and BIC

We start with notions of *algorithmic complexity* developed independently by Kolmogorov, Chaitin, and Solomonoff: The "complexity" of a set is determined by the length of computer code required to generate the set.[10] That length is equivalent to the minimum size of a reversibly compressed file storing the data. For example, a random sequence of N 1s and 0s can only be represented by the numbers themselves and thus requires N bits of information, with no possibility of compression. Indeed, the vast majority of binary sequences have this property. By contrast, the sequence 1010... can be represented by two bits (for 10) plus the binary coding for N.

In the context of data, we can think of a model as a means of compressing the data. That is, we can represent the data as random residuals from the model predictions (hopefully smaller and thus requiring fewer bits than the original values), along with the rules for generating the model itself. Loosely, a simple model that takes little information to specify will lead to a poor fit and thus large residuals. A complex model will fit the data very well but will then require a longer description to specify. We are

[10] As noted by Chaitin (2006), the core idea was expressed already by Leibniz, whose 1686 essay *Discours de Métaphysique* argued that a theory should be simpler than the data it explains.

thus led to the notion of the best model as one requiring the *minimum description length* (MDL) of a set of data.

Although Kolmogorov proved that no general algorithm can compute the MDL exactly in all cases, a simple, approximation often works well, for $N \gg 1$ data points: Assume that the data have independent additive Gaussian noise of standard deviation σ. For a model of K parameters and N data points, the *Bayesian Information Criterion* BIC is

$$\mathrm{BIC} = \chi^2 + K \ln N \tag{6.19}$$

and will approximate the MDL for large N. The "best" model then minimizes BIC over the model parameters $\boldsymbol{\theta}$ and the model class K, adding to the usual χ^2 statistic a term $K \ln N$ that penalizes models with more free parameters. Since $\chi^2 \sim N$, BIC favors simple models when there are few data points but more complex models for large N.

To understand the origin of the BIC criterion in Eq. (6.19), we generalize Eq. (A.211 online) to the case where the number of parameters K is also a random variable. For simplicity, we take the N points x_i to be fixed. Dropping them from the explicit notation, we write

$$p(K, \boldsymbol{\theta}|Y) \propto p(Y|K, \boldsymbol{\theta})\, p(K, \boldsymbol{\theta}) = p(Y|K, \boldsymbol{\theta})\, p(\boldsymbol{\theta}|K)\, \cancel{p(K)} . \tag{6.20}$$

In the last step in Eq. (6.20), we assume a uniform prior for K (no prejudice about what order of model to pick). Next, we integrate out $\boldsymbol{\theta}$:

$$p(K|Y) = \int \mathrm{d}^K\boldsymbol{\theta}\, p(K, \boldsymbol{\theta}|Y) \propto \int \mathrm{d}^K\boldsymbol{\theta}\, p(Y|K, \boldsymbol{\theta})\, p(\boldsymbol{\theta}|K) \propto \int \mathrm{d}^K\boldsymbol{\theta}\, \mathrm{e}^{-Nf(\boldsymbol{\theta})} , \tag{6.21}$$

where the "free energy" $f(\boldsymbol{\theta})$ is

$$f(\boldsymbol{\theta}) \equiv \frac{1}{2N}\chi^2 - \frac{1}{N}\ln p(\boldsymbol{\theta}) . \tag{6.22}$$

We have dropped the $(\cdot|K)$ from the various conditional expressions. The free energy has two terms: an "energy" χ^2/N that tries to force the fit into the "ground state" of perfect accord with the data, an "entropy" $\ln p(\boldsymbol{\theta})$ that is proportional to the "volume" of parameter space occupied by the model structure, and a "temperature" proportional to $1/N$. Recall that χ^2/N is of order 1, whereas $p(\boldsymbol{\theta})$ is independent of N. The entropy term measures the logarithm of the number of equivalent models in the function space of possible models (here a countable set indexed by integer K but in general an uncountable set). For small N – high temperatures – the entropy term dominates, and the free energy will be minimized by models with small K. For large N – low temperatures – the energy term dominates, and the free energy will be minimized by having a low χ^2 per data point.

Why not simply pick the model that fits the data best? Although "just so" models account perfectly for each data point, the number of equivalent models is very high. In another experiment with another set of data, a new model will fit those points.

Accounting for the entropy of equivalent models, the penalty for such a fit would be very high.[11]

To understand these ideas from statistical physics more quantitatively, we use the saddle-point method to approximate Eq. (6.21) for large N:

$$p(K|Y) \propto \int \mathrm{d}^K\boldsymbol{\theta}\, \mathrm{e}^{-Nf(\boldsymbol{\theta})} \approx \mathrm{e}^{-Nf(\boldsymbol{\theta}^*)}(2\pi)^{K/2} \exp\left[-\tfrac{1}{2}\ln\det(N\mathcal{H})\right], \qquad (6.23)$$

where the Hessian matrix $\mathcal{H}_{ij}$ of second derivatives with respect to θ_i and θ_j is given by $\partial_{ij}f$, evaluated at $\boldsymbol{\theta} = \boldsymbol{\theta}^*$. Taking the logarithm of $p(K|Y)^{-1}$ and noting that the determinant of the $K \times K$ matrix $\mathcal{H}$ implies that $\ln\det(N\mathcal{H}) = N^K \det\mathcal{H}$, we have

$$-\ln p(K|Y) = \tfrac{1}{2}\chi^2 + \tfrac{1}{2}K\ln N + \text{terms that do not increase with } N, \qquad (6.24)$$

Choosing the most probable value of K then amounts to minimizing $\mathrm{BIC} = \chi^2 + K\ln N$, as claimed. Intuitively, the $\frac{K}{2}\ln N = K\ln\sqrt{N}$ term in Eq. (6.24) is the reduction in information required to specify the K parameters. Imagine that the parameter θ_i has an initial uncertainty $\Delta\theta_i$ before the data are taken. Afterwards, we determine θ_i^* to a precision that is $\approx (\Delta\theta_i)/\sqrt{N}$. Just as the data are compressed from the information that is required to specify the initial data points y_i to that required to specify the (smaller) residuals $\varepsilon_i = y_i - f(x_i;\boldsymbol{\theta})$, the parameters are compressed from $\Delta\boldsymbol{\theta}$ to $\Delta\boldsymbol{\theta}/\sqrt{N}$. It thus makes sense to choose $K = K^*$ to minimize the sum of the total description length.

Cross-Validation and AIC

A different way to assess a model structure is to ask how well it predicts "new" data, the strategy used in system identification.[12] Note that "predict" need not have a temporal implication, although it often does. To assess such a property, we partition the data into *training* and *validation* sets. While there are many ways to make partitions, the one that is usually most relevant to control problems is to ask how well the *next* data point is predicted. We focus on predicting the next data point because, in principle, we can update our model after each data point (cf. Chapter 10).

We thus omit one data point, fit according to the remaining $N - 1$ points, and ask how well we can predict the missing data point.[13] Averaging over all N points (i.e., leaving out and then predicting each point, in turn) leads to *leave-one-out* cross-validation.[14]

[11] The statistical theory of model selection quantifies the intuition of Sir Karl Popper (1959), who introduced the notion of *falsifiability* to distinguish science from pseudoscience. Theories that are not falsifiable can always "explain" a particular set of observations.

[12] Our discussion of the AIC is framed in terms of leave-one-out cross validation (LOOCV). For data generated by a distribution $p(\boldsymbol{x})$, the actual AIC criterion is to pick the model $q(\boldsymbol{x})$ that minimizes the Kullback–Leibler divergence $D(p\|q)$ (see Appendix Section A.10.2 online). For large sample size N, the asymptotic evaluation of the KL divergence leads to the same expression for AIC as does the asymptotic evaluation of LOOCV.

[13] Omitting one point and adding one point are equivalent here.

[14] In other applications of cross-validation, it is usual to consider the effect of leaving out multiple points.

Cross-validation errors can be evaluated via an asymptotic analysis by Stone (1977), which shows that minimizing leave-one-out cross-validation for N data points and K model parameters requires minimizing

$$\mathrm{AIC} = \chi^2 + 2K\,. \tag{6.25}$$

Equation (6.25) is commonly known as the *Akaike Information Criterion* (AIC). It penalizes adding parameters more than does MDL or its BIC approximation.[15] Problem 6.17 shows how to derive AIC from cross-validation for a simple case. As an alternative, let us take as a model structure a complete, orthonormal basis set in function space,[16] with $e_k \cdot e_\ell \equiv \frac{1}{b-a}\int_a^b \mathrm{d}x\, e_k(x)\, e_\ell(x) = \delta_{k\ell}$, for $a < x < b$. Then

$$f(x, \boldsymbol{\theta}) = \sum_{k=1}^{\infty} \theta_k\, e_k(x)\,, \tag{6.26}$$

We also assume that we order the basis set so that $\theta_k \geq \theta_{k+1}$.

In Problem 6.18 (cf. Section A.8.5 online), we show that, if we define a model structure by taking the first K terms in Eq. (6.26), the average value of χ^2 is given by

$$\langle \chi^2 \rangle = \underbrace{\frac{N}{\sigma^2}\left(\sum_{k=K+1}^{\infty} \hat{\theta}_k\, e_k(x) \right)^2}_{\text{model mismatch}} + \underbrace{N - K}_{\text{noise}}\,. \tag{6.27}$$

The first term is the projection of the function onto the space that is the orthogonal complement to the K-dimensional model subspace. (It is over dimensions $K + 1$ to ∞.) Note that the parameters $\hat{\theta}_k$ are the optimal values (minimizing χ^2). The second term is the same as the usual term for a perfect model structure (see Appendix Section A.8.5 online). It is then straightforward to write $\langle\mathrm{AIC}\rangle = \langle\chi^2\rangle + 2K$ and $\langle\mathrm{BIC}\rangle = \langle\chi^2\rangle + K \ln N$.

Minimizing the AIC criterion by taking a (discrete) derivative with respect to K then gives an optimal K^* that satisfies

$$\frac{N}{\sigma^2}\,\hat{\theta}^2_{K^*+1} = 1\,. \tag{6.28}$$

We can interpret the AIC then as asking that the "power" of each basis coefficient θ_i be greater than the noise level, σ^2/N. In other words, Eq. (6.28) defines a signal-to-noise ratio (SNR) for each coefficient, and AIC asks that we keep parameters whose SNR > 1.

15 A "finite N" correction to AIC is sometimes used: $AIC_c = \chi^2 + 2K + [2K(K+1)]/(N - K - 1)$. Except when $K \propto N$, it also generally leads to a smaller penalty on extra parameters than does BIC.

16 Function sets may be orthogonal with respect to a weight. For example, Chebyshev polynomials of the first kind, a Fourier cosine expansion with a change of variable $x = \cos\theta$, satisfy $\int_{-1}^{1} \mathrm{d}x\, T_m(x)\, T_n(x)\, w(x) = \frac{\pi}{2}\delta_{mn}$ (except for $m = n = 0$, where the integral equals π). The weight $w(x) = \frac{1}{\sqrt{1-x^2}}$. The formalism discussed here carries over, once it is understood that inner products must always include the weight (Boyd, 2000).

AIC vs. BIC

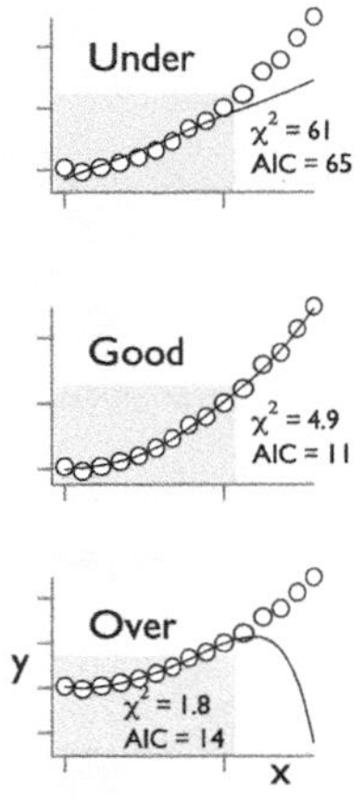

Although AIC, by design, does a good job of predicting new data points whose x value falls within the range used to evaluate the curve fits used to select the model (the *interpolated* case), it tends to select slightly overfit models. In effect, the criterion assumes that the penalty for overfitting occasionally is less than for underfitting. But while this statement is reasonable for interpolated data, it does not hold for extrapolation.

The simple example at right uses data generated from $f(x) = x^2 + v$, a quadratic model plus noise. As denoted by the darker shaded area, we fit to the first 10 points, with linear, quadratic, and 5th-order polynomials (from top to bottom). As expected, the χ^2 statistic decreases as the model order increases. We then extrapolate each fit to "predict" the next 5 points at right. We see that the true quadratic model structure minimizes the "prediction error" relative to both under- and overfitting. The underfit model does not have the freedom to follow the major trends of the data. The overfit model has so much freedom that it starts to adapt to the noise itself. Notice that the extrapolation error is much larger for the overfit model than for the underfit model.

To further illustrate these points, let us analyze, in two ways, noisy measurements of

$$f(t) = f_0(1 - 2t), \qquad 0 < t < 1. \tag{6.29}$$

In the first, we understand the problem well and pick a model with a small number of parameters that includes the true function f. In the second, we pick a poor set of basis functions that requires an infinite number of parameters to describe the true function.

Example 6.7 (AIC vs. BIC: Good model parametrization.) Consider a series of fits to Eq. (6.29) of orders $n = 0, 1, 2, \ldots$ (constant, linear, quadratic, ...), up to order 20. Record the χ^2, AIC, and BIC statistics for each fit and determine the fit order that minimizes the AIC and BIC statistics. Repeating for 1000 runs, we compile the histogram at right. The BIC statistic selects the correct order 99.5% of the time, whereas the AIC statistic selects the correct order only about 70% of the time. Indeed, AIC can spectacularly overfit the data, selecting orders as high as 13. When the correct model corresponds to a low-order fit, the BIC statistic does a better job of identifying that order. Remember that only if the order is "correct" will extrapolation be possible.

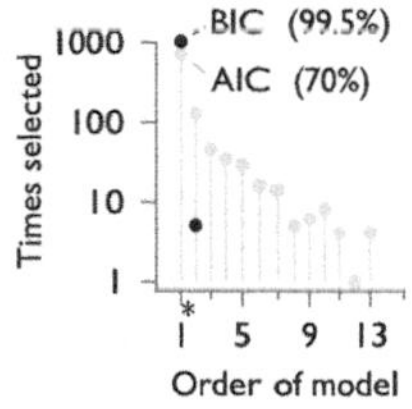

Example 6.8 (AIC vs. BIC: Bad model parametrization.) If we were unable to "guess" that a straight line would be a good possibility for Eq. (6.29), we might try a Fourier decomposition. Set $f_0 = \pi^2/8$ for simplicity. From Problem A.4.1 online,

$$f(t) = \cos \pi t + \tfrac{1}{9} \cos 3\pi t + \tfrac{1}{25} \cos 5\pi t + \cdots = \sum_{j=1}^{\infty} \frac{1}{j^2} \cos j\pi t\,. \tag{6.30}$$

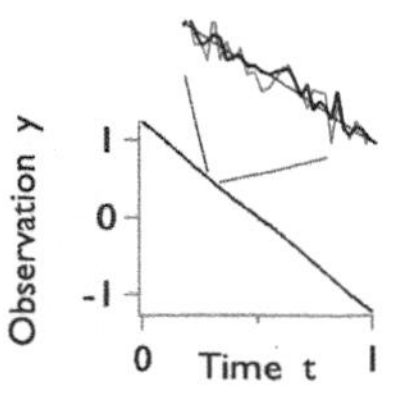

At right, we plot $f(t)$ with noise $\sigma = 0.01$, along with a Fourier-expansion fit of 500 coefficients. The inset shows overfitting: the fit follows much of the random noise.

We then evaluate χ^2, AIC, and BIC as a function of the number of Fourier coefficients (fit parameters), for 1000 data points and up to 500 parameters.

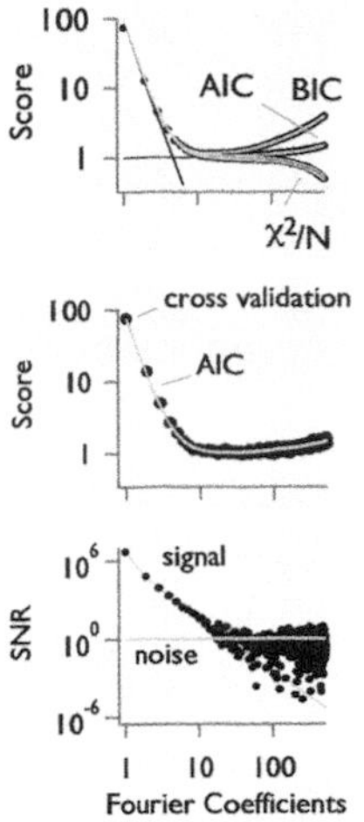

At left, we see that χ^2 decreases monotonically, as it must for a set of nested models. Both BIC and AIC have minima, at $K = 15$ and 24, respectively. BIC still selects a smaller model than AIC, but now the correct model requires an infinite number of parameters. As a result, all finite-n models will have poor extrapolation properties, removing a key advantage of BIC. Since the performance on interpolation is all that remains, AIC is thus more appropriate.

To see that the AIC criterion matches the cross-validation error, we evaluate a further 1000 randomly chosen points in the interval $t = (0, 1)$. For each t, we simulate a new measurement and calculate its normalized residual, based on a model of K parameters (left). Thus, minimizing AIC also minimizes the one-point prediction error for new points in the fit domain (interpolation). This case is common in control applications.

Finally, we show the signal-to-noise ratio (SNR), with lower Fourier orders having a high SNR and orders higher than $n = 24$ lost in the noise. Note that the SNR statistic is itself "noisy": the standard deviation of a coefficient of SNR $= 1$ is $\sqrt{2}$. The result is the wide spread of measured SNR values in the noise-limited Fourier coefficients shown at left (bottom). See Problem 6.19 for details.

To summarize, BIC tends to pick the simplest model consistent with the data and is best for extrapolation, while AIC tends to pick the model that is best for interpolation. To understand the physics or other fundamental properties of the system, try BIC. Given the right – or nearly right – model class and given that the best model requires few parameters, then BIC will likely find it. For complex models or cases that are poorly understood (and hence poorly parametrized), the parameter SNR values "march down into the noise." AIC will then identify those parameters that have reasonable SNR and lead to slightly lower prediction errors. In Example 6.8, the AIC error is $\approx 2\%$ lower than BIC.

Finally, remember that both AIC and BIC are asymptotic formulas, derived for large numbers of data points. For small data sets, a direct analysis based on predictive errors or Bayes' theorem may give better results.

Experimental physicists aim to understand the physics of an experiment and try to design their apparatus to be amenable to modeling. By contrast, constrained by time and money, the engineer focuses on equipment that works and whose behavior is predictable. We can see the difference in perspective when comparing consumer and scientific versions of similar equipment. For example, a 1 megapixel digital camera for consumers may cost \$10 while a scientific version may cost more than \$10 000. What do you get for the money? A scientific camera usually has a linear gain and a simple noise model (e.g., shot noise plus dark noise plus read-out noise), while the industrial camera may have a nonlinear gain (to "look good"), use lossy compression algorithms, and contain "hot pixels."

To illustrate the lengths physicists will go to understand an apparatus, consider the recent discovery of the Higgs boson at CERN. The boson was detected indirectly by looking for products of its decay that had the correct energy, momentum, and angular distribution. The relevant particle detection events are a small fraction of the total number of detection events – a bump on a background – and can be reliably identified only through an exquisite understanding of the physics of the Standard Model and of the detector itself. The characterization of the ATLAS detector was a ten-year project, summarized in a paper of over 400 pages with nearly 3000 authors (ATLAS, 2008).

6.5 Model Reduction

In a chapter devoted to finding the best models of transfer functions consistent with experimental data, it might seem odd to find a section devoted to replacing good models of transfer functions with poorer approximations. But the "best" controller transfer function can still have drawbacks. In particular, it may take too long to compute a response at the desired sample rate for the controller. Such situations can occur when the model time scales are very fast or when there are extended spatial scales.

In the latter case, models are finite-element approximations to a partial-differential equation on a three-dimensional grid, where even modest resolution leads rapidly to very high-order systems whose states can take a long time to compute. Further, controlling optimally an nth-order system typically leads to an nth-order controller, again slow to compute.

One easy strategy is to reduce the model order. But there are subtleties: a good approximation to the frequency response will not necessarily lead to a good time-domain response and vice versa. In general, different approximations to the higher-order model optimize different features of the dynamics. Simple *truncation* does well at high frequencies and short times but poorly at low frequencies and long times. Conversely, *residualization* (also known as *adiabatic elimination*) does well at low frequencies and long times but poorly at high frequencies and short times.

Both naive truncation and residualization often give poor results, since a complete system model must account not only for the dynamics but also the input and output coupling that a transfer function measures. "Balancing" how inputs and outputs couple to the dynamics gives better results – but at the cost of losing intuitive understanding of the reduced dynamical system, a recurrent theme of this chapter.

6.5.1 Naive Truncation and Residualization

Consider a two-mode system,

$$\dot{x}_1 = -\lambda_1 x_1 + b_1 u\,, \qquad \dot{x}_2 = -\lambda_2 x_1 + b_2 u\,, \qquad y = c_1 x_1 + c_2 x_2\,, \tag{6.31}$$

whose transfer function is $G(s) = \frac{c_1 b_1}{s+\lambda_1} + \frac{c_2 b_2}{s+\lambda_2}$.

Truncation implies keeping the slower modes (x_1) and dropping the faster modes (x_2):

$$\dot{x}_1 = -\lambda_1 x_1\,, \qquad y = c_1 x_1\,, \qquad G_{\rm tr}(s) = \frac{c_1 b_1}{s + \lambda_1}\,. \tag{6.32}$$

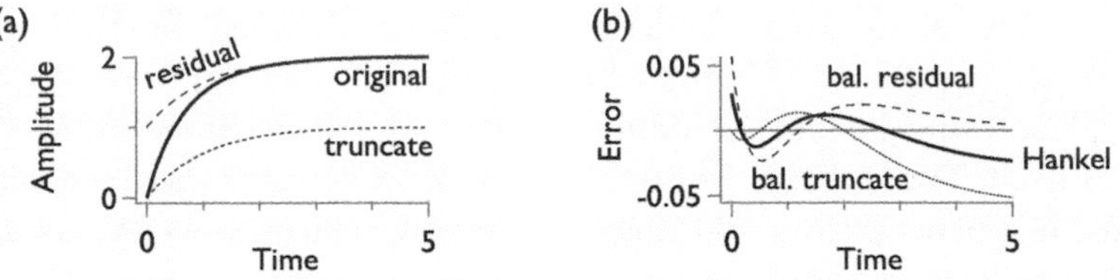

Fig. 6.7 Model reduction from two states to one. (a) Naive reduction. Solid line shows the step response of the original system $G(s)$, Eq. (6.31). Lower dashed line represents the truncation approximation, Eq. (6.32), while the upper dashed line represents the residualization approximation, Eq. (6.34). (b) Difference between original system and approximations using balanced coordinates. Dotted line denotes the difference for the truncated system; dashed line for the residualization; and solid line for the optimal Hankel approximant.

The error magnitude between the original and truncated systems is given by

$$|G(\mathrm{i}\omega) - G_{\mathrm{tr}}(\mathrm{i}\omega)| = \frac{c_2 b_2}{\sqrt{\omega^2 + \lambda_2^2}}\,, \tag{6.33}$$

which indeed is zero as $\omega \to \infty$, albeit with different DC gains: $\frac{c_1 b_1}{\lambda_1}$ versus $\frac{c_1 b_1}{\lambda_1} + \frac{c_2 b_2}{\lambda_2}$.

Residualization is a complementary strategy, where we assume that faster modes relax rapidly to quasistatic "local equilibrium," which is enforced by setting their time derivatives equal to zero. By adiabatic elimination, $\dot{x}_2 \approx 0$, and $x_2(t) \approx (b_2/\lambda_2)\,u(t)$. Then,

$$\dot{x}_1 = -\lambda_1 x_1\,, \qquad y = c_1 x_1 + \frac{c_2 b_2}{\lambda_2} u\,, \qquad G_{\mathrm{res}}(s) = \frac{c_1 b_1}{s + \lambda_1} + \frac{c_2 b_2}{\lambda_2}\,. \tag{6.34}$$

The error is now

$$|G(\mathrm{i}\omega) - G_{\mathrm{res}}(\mathrm{i}\omega)| = c_2 b_2 \left(\frac{1}{\lambda_2} - \frac{1}{s + \lambda_2} \right), \tag{6.35}$$

which is zero at $\omega \to 0$ but goes to a constant, $c_2 b_2/\lambda_2$, at $\omega \to \infty$, rather than 0. These results are illustrated in Figure 6.7a, for $\lambda_1 = 1$, $\lambda_2 = 2$, $\boldsymbol{B} = \left(\begin{smallmatrix}1\\1\end{smallmatrix}\right)$, and $\boldsymbol{C} = (1\ 2)$.

6.5.2 Balanced Coordinates

The approximations in Figure 6.7a, while accurate at $t = 0$ for truncation and $t \to \infty$ for residualization, fail at intermediate times because they do not account for the input and output couplings, $\boldsymbol{B} = \left(\begin{smallmatrix}b_1\\b_2\end{smallmatrix}\right)$ and $\boldsymbol{C} = (c_1\ c_2)$. These couplings determine the prefactors for the transfer function $G(s)$ and are as important as the time constants for finite-time behavior.

One way to improve model reduction is to transform $\boldsymbol{x} = \boldsymbol{T}\boldsymbol{x}'$ to *balanced* coordinates. To find the appropriate transformation $\boldsymbol{T}$, we note a result that will be proven in Chapter 7, Problem 7.9: to move a system state from $\boldsymbol{x}(0) = \boldsymbol{0}$ to $\boldsymbol{x}(\tau) = \boldsymbol{x}_\tau$, the control $u(t)$ that requires the least control effort $\|u\|^2 = \int_0^\tau \mathrm{d}t\, u^2(t)$ is given by

$$u(t) = \boldsymbol{B}^{\mathsf{T}}\, \mathrm{e}^{\boldsymbol{A}^{\mathsf{T}}(\tau - t)}\, \boldsymbol{P}^{-1}(\tau)\, \boldsymbol{x}_\tau\,, \tag{6.36}$$

where $\boldsymbol{P}(\tau) = \int_0^\tau \mathrm{d}t\, \mathrm{e}^{At}\, \boldsymbol{B}\,\boldsymbol{B}^{\mathsf{T}}\, \mathrm{e}^{A^{\mathsf{T}}t}$ is the *controllability Gramian* matrix defined in Chapter 4. In a basis in which $\boldsymbol{P}$ is diagonal, its eigenvalues indicate the "coupling" between the input and the corresponding eigenstate. States with smaller eigenvalues require more effort to alter, while states with larger eigenvalues need less effort. If the input is projected roughly equally onto the different modes, those modes associated with higher eigenvalues will have larger amplitudes and will thus dominate the state dynamics.

In Chapter 4 (Section 4.1.2), we introduced the notion that inputs and outputs have a dual relation to each other, which implies that the above discussion of controllability has a parallel one for observability. In particular, let us consider the output $y(t)$ over the time interval $(0, \tau)$ for a system with $\boldsymbol{x}(0) = \boldsymbol{x}_0$. The norm of y is given by

$$||y||^2 = \frac{1}{\tau}\int_0^\tau \mathrm{d}t\, y(t)^2 = \frac{1}{\tau}\int_0^\tau \mathrm{d}t\, \boldsymbol{x}_0^{\mathsf{T}}\, \mathrm{e}^{A^{\mathsf{T}}t}\, \boldsymbol{C}^{\mathsf{T}}\boldsymbol{C}\, \mathrm{e}^{At}\, \boldsymbol{x}_0 \equiv \frac{1}{\tau}\, \mathrm{d}t\, \boldsymbol{x}_0^{\mathsf{T}}\, \boldsymbol{Q}(\tau)\, \boldsymbol{x}_0\,, \tag{6.37}$$

where the *observability Gramian* matrix $\boldsymbol{Q}(\tau) \equiv \int_0^\tau \mathrm{d}t\, \mathrm{e}^{A^{\mathsf{T}}t}\, \boldsymbol{C}^{\mathsf{T}}\boldsymbol{C}\, \mathrm{e}^{At}$ measures the contribution of the state $\boldsymbol{x}_0$ to the output over the time interval $(0, \tau)$. In coordinates where $\boldsymbol{Q}$ is diagonal, its eigenvalues signify how much $\boldsymbol{x}_0$ excites the associated eigenvector.

Let us now consider the controllability and observability Gramians together. In Problem 6.20, you will see how to choose a coordinate transformation $\boldsymbol{T}$ that makes the infinite-time controllability and observability Gramians *simultaneously* equal and diagonal, with $\boldsymbol{P}(\infty) = \boldsymbol{Q}(\infty) = \boldsymbol{\Sigma}$. Here, $\boldsymbol{\Sigma}$ is a diagonal matrix whose entries $\sigma_1 \geq \sigma_2 \geq \ldots \geq \sigma_n > 0$. Such coordinates are known as *balanced*.

If a system $G = \{\boldsymbol{A}, \boldsymbol{B}, \boldsymbol{C}\}$ is balanced, then the eigenvectors measure the effect of the input on the output. In other words, $\boldsymbol{P}$ tells how an input $u(t)$ leads to a final state, whereas $\boldsymbol{Q}$ tells how an initial state leads to an output $y(t)$. If we take infinite-time Gramians, we consider the input over $(-\infty, 0)$ and the output over $(0, \infty)$. Schematically,

$$\underbrace{u(t)}_{-\infty < t < 0} \longrightarrow \boldsymbol{x}(t = 0) \longrightarrow \underbrace{y(t)}_{0 < t < \infty}\,.$$

The eigenvalues σ_i of $\boldsymbol{P}(\infty)$ and $\boldsymbol{Q}(\infty)$ are known as *Hankel singular values*, and their largest value, σ_1, gives the *Hankel norm* of the linear system, which is, equivalently,

$$||G||_H^2 \equiv \sup_{u(t)} \frac{\int_0^\infty \mathrm{d}t\, y(t)^2}{\int_{-\infty}^0 \mathrm{d}t\, u(t)^2}\,, \tag{6.38}$$

where the "sup" (*supremum*, or least upper bound) is evaluated over all square-integrable $u(t)$. The connection with Hankel singular values makes sense since the minimum $||u||$ to achieve a given end state can be written in terms of the controllability Gramian $\boldsymbol{P}(\infty)$, and there is a similar relation for the output $y(t)$ in terms of the observability Gramian.

6.5.3 Balanced Truncation and Residualization

The Hankel singular values (eigenvalues of $\boldsymbol{P}$ and $\boldsymbol{Q}$ in balanced coordinates) indicate how strongly a mode couples inputs to outputs. But that is also what a transfer function measures. Using truncation or residualization in balanced coordinates may then be expected to better approximate the transfer function than a naive procedure in the coordinate system that merely diagonalizes $\boldsymbol{A}$. Going back to Figure 6.7(b), we illustrate truncation and residualization using balanced coordinates. Since the results are good enough that it would be hard to distinguish the step responses in the time-domain plot shown in (a), we show the differences instead. We see that truncation continues to be exact at $t = 0$ and residualization at $t = \infty$, while the approximations in both cases are much better at finite times. In general, the worst-case errors have been reduced from $O(1)$ with the naive approximations to less than 3% for the same approximations in balanced coordinates.

As mentioned, the success of truncation or residualization using balanced coordinates comes at some cost of intuitive and physical understanding of the reduced system. In the present case, the original dynamics had two modes, with poles at $s = -1$ and -2. Truncation and residualization in the original coordinates meant keeping $s = -1$ and dropping the -2 mode. For balanced coordinates, truncation gives a single-mode system with a pole at $s = -\frac{3}{2}$, while residualization gives a pole at $s = -\frac{4}{3}$. The exact pole positions now depend in a complicated way on the components of $\boldsymbol{B}$ and $\boldsymbol{C}$. We have traded off physical insight for better input-output modeling.

Truncation and residualization are two ways to find reduced dynamics. Another method that works well is to minimize the Hankel norm of the subspace that is orthogonal to the reduced dynamics. The solid line in Figure 6.7(b) shows the result, whose worst error is reduced by about a factor of two. While the solution is no longer exact at $t = 0$ or ∞, it is overall a better approximation. The pole of the single-mode approximation is now at $-\sqrt{2}$, in between that of truncation ($-\frac{3}{2}$) and residualization ($-\frac{4}{3}$).

6.6 Summary

Controllers that take advantage of known system dynamics can outperform generic ones such as the PID algorithm. Estimating system dynamics is thus an important preliminary task. Here, we have presented both physical and phenomenological approaches. The former incorporate knowledge of the physical system to formulate equations of motion, with free parameters that may be determined from experiments. The latter (system identification) construct models based on relating outputs to inputs, approximating the transfer function as a low-order rational polynomial using a Padé expansion.

Since we focus on linear systems, we have emphasized frequency-domain methods using sine-wave inputs to estimate the transfer function, while briefly discussing corresponding time-domain methods based on correlation methods. Among many possible input signals used to infer dynamics, multisines use the linearity of the dynamics to speed up transfer-function measurements by probing behavior at many frequencies simultaneously.

A second focus concerned model building: formulating analytic models that capture experimental trends. When comparing models, different criteria lead to different conclusions. In particular, the Akaike Information Criterion (AIC) minimizes prediction errors for data that is taken under conditions similar to the training data. By contrast, the Bayesian Information Criterion (BIC) is more likely to identify the correct model and thereby minimize prediction errors for data that requires extrapolation from training data to new regimes.

Finally, we briefly discussed several ways to reduce high-order models with many free parameters to nearly equivalent lower-order models that have fewer parameters and are faster to compute. A useful definition of "nearly equivalent" for control problems is balanced coordinates, where each input-output combination is equally sensitive.

6.7 Notes and References

The advantages of periodic, multisine inputs have been championed for many years by Pintelon and Schoukens (2012). This chapter is heavily influenced by their presentation and may be seen as a physicist's introduction to their work, which includes many other cases (e.g., MIMO systems) and also gives a more rigorous analysis. Rare examples of the multisine method in physics include Pérez-Aparicio et al. (2015), who measure the dielectric response of a medium over four orders of magnitude in frequency. Using material from this chapter, Frick et al. (2018) measured the temperature response of an object of complex geometry at frequencies as low as 10^{-3} Hz. Geri et al. (2018) advocate a chirp with exponential frequency increase for probing rheological response in soft materials.

For background on Fourier transforms of stochastic processes, Davenport and Root (1958) is still a readable, straightforward introduction. Our derivation of cross-correlation between input and output and the impulse response follows Doyle et al. (1992). Ljung (1999) is a more traditional engineering presentation that focuses on discrete time-domain methods with nonperiodic inputs. Isermann and Münchhof (2011) discuss methods for both time and frequency domains and for both linear and nonlinear systems.

The first-order time-domain example follows a discussion by Åström and Wittenmark (1997), whose chapter on system identification is short and insightful. Our discussion of model reduction follows Skogestad and Postlethwaite (2005). The original work on balanced representations, Moore (1981), was inspired by *principal component analysis*, which has found increasing application in physics research.

Several software packages are available to aid system identification. The book by Pintelon and Schoukens (2012) has supporting exercises (Schoukens et al., 2012) with Matlab routines that may be freely downloaded from the publisher's website.

See also the Matlab toolbox (FDIDENT). The Matlab System Identification toolbox follows Ljung (1999) and has a wide range of available algorithms. A graphical interface guides the user through the required steps. The publicly available CONTSID toolbox, which requires the System Identification and Control toolboxes, focuses on continuous systems. The open-source Scilab has some functions for discrete-time identification.

Our "statistical-physics" derivation of minimum description length (MDL) follows Bialek (2012), which adapts the original arguments by Schwarz (1978) and Rissanen (2007) (whose original work is also from 1978). Wallace and Boulton (1968) did an earlier, more specialized calculation. As discussed by Rissanen, the BIC approximation to the MDL criterion can be greatly improved upon in specific settings, particularly when the number of data points is modest. A popular approach is to use *Bayesian nonparametric* data analyses that replace "parametric" models with finite numbers of parameters with models that sample from an infinite-dimensional function space of potential distributions (Müller et al., 2015).

The AIC criterion is due to Akaike (1974). Machta et al. (2013) have a different view of the relation between predictions and models that focuses on the Fisher information matrix and connections to the renormalization group. Lamont and Wiggins (2016) show that particular classes of models (e.g., segmentation of a time series into a set of N distinct segments) have singularities in the parameter dependence that cause AIC and BIC to give bad results. Their *Frequentist Information Criterion* performs well in such situations.

Skogestad and Postlethwaite (2005) discuss model reduction and optimal Hankel norm approximations. The significance of balanced coordinates and their Hankel singular values has only begun to be appreciated in the physics literature. Reynolds (2009) shows how to use balanced coordinates for a renormalization-group analysis where the Hankel singular values are used to *coarse grain* nonequilibrium dynamical systems in a way that captures the relevant physics of a problem automatically. An earlier preprint, Reynolds (2003), includes much valuable background that was deleted from the published version.

And, lastly, concerning the statistical subtleties of measurements, model construction, and selection, do not forget the advice of Press et al. (2007) (end of their Section 15.6.1):

> Offered the choice between mastery of a five-foot shelf of analytical statistics books and middling ability at performing statistical Monte Carlo simulations, we would surely choose to have the latter skill.

Problems

6.1 **Timing jitter.** Fluctuations in sampling a signal lead to low-pass filtering and distort the apparent transfer function. The phenomenon is similar to

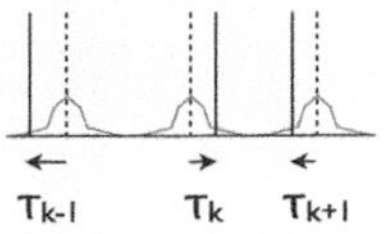

the Debye-Waller factor for X-ray scattering from crystal lattices at finite temperatures. To see this, we follow Souders et al. (1990) and consider timing fluctuations $\tau_k \sim p(\tau)$ at time step k, as illustrated at right. Define the jitter signal $f_j(t) \equiv \langle f(t+\tau) \rangle$, where the angle brackets denote an ensemble average over $p(\tau)$, which we assume to be even in τ.

a. Show that jitter acts as a convolution and hence that the continuous-time Fourier transform $f_j(\omega) = f(\omega)\,\varphi_\tau(\omega)$, where $f(\omega)$ is the Fourier transform of the original signal and $\varphi_\tau(\omega)$ is the characteristic function (Fourier transform) of $p(\tau)$.

b. Consider a measurement with sampling at nominal times kT_s. Because of jitter, the actual measurement times are at $kT_s + \alpha_k T_s$, where $\alpha_k \sim \mathcal{N}(0, \alpha^2)$. Show that timing jitter limits the bandwidth to $\omega_b = \omega_s/(\sqrt{2}\pi\alpha)$, with $\omega_s = 2\pi/T_s$.

6.2 Crest factor. The *crest factor* $\mathrm{Cr}[u(t)]$ measures the maximum amplitude of a signal for a given RMS power. For a waveform of period τ, it is defined as

$$\mathrm{Cr}[u] = \frac{u_{\max}}{u_{\mathrm{rms}}}, \qquad u_{\max} = \max_{0 \le t \le \tau} |u(t)| \qquad u_{\mathrm{rms}} = \sqrt{\frac{1}{\tau}\int_0^\tau \mathrm{d}t\, u^2(t)}\,.$$

A good input signal should have a small crest factor, to inject power into a system while keeping the maximum amplitude low enough to avoid a nonlinear response.

a. Elementary cases: Show that a square wave has Cr = 1, a single sine has Cr = $\sqrt{2}$, and a Dirac delta function has Cr = ∞.

b. Multisine signals. Consider periodic signals $u_N(t)$ of period T that are the sum of N harmonics. With $\omega_n = 2\pi n f_0 = 2\pi n/\tau$, we have $u_N(t) = \sum_{n=1}^N A_n \cos(\omega_n t + \varphi_n)$. Show that $u_{\mathrm{rms}}^2 = \frac{1}{2}\sum_{n=1}^N A_n^2$, independent of the value of the phases φ_n. Set $A_n = 1/\sqrt{N}$, so that $u_{\mathrm{rms}} = 1/\sqrt{2}$, and the crest function depends only on $u_{\max}$. Set also $\varphi_1 = 0$ by overall translational invariance and $f_0 = \tau = 1$ for convenience.

c. Fast numerical calculations. Show that you can vastly speed up the explicit sum for $u_N(t)$ by defining the signal in the Fourier domain and taking the inverse Fourier transform. Choose N_s, the total number of points in a period of the waveform, to be a power of 2. Why is the Fourier-method much faster than the time-domain sum? Write a program to calculate the waveform using the two methods. The two waveforms should agree to within machine precision. For $N_s = 1024$ and $N = 255$, show that the speedup is $\approx$ 100-fold.

d. N cosines, done wrong. For N cosines, the worst choice is to set $\varphi_n = 0$. Plot the $N = 2, 3, 4$ cases and show that Cr = $\sqrt{2N}$.

e. Small number of harmonics. We can use brute-force numerics to find the optimum phases. Search an $N-1$ dimensional grid for all possible phase values. Obviously, the time to solve the problem grows exponentially, but small problems can be solved easily on a laptop. Your solutions for $N \le 4$ should resemble the graphs at right. Note how, for $N \ge 2$, the optimal

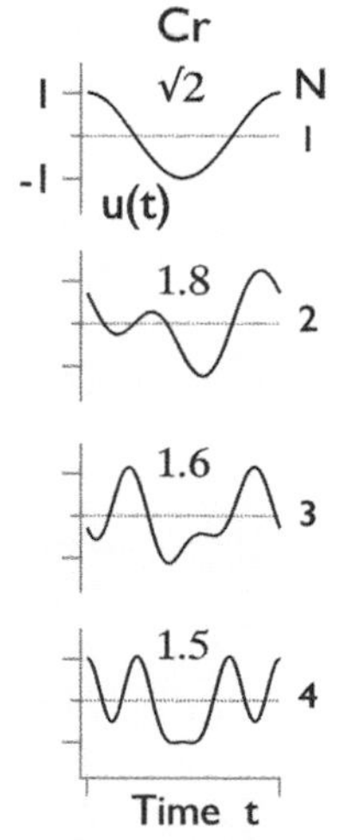

Cr decreases with N. Give the phases of the waves and state Cr with more precision.

f. Random phases: In the limit of a large number of harmonics, $N \to \infty$, one idea is to choose phases randomly from a uniform distribution between 0 and 2π. In this part, we investigate the properties of the average crest factor. In particular, we will see that $\langle \mathrm{Cr} \rangle \approx \sqrt{2 \ln(2N)}$, a number that is 3–4 for typical values N (say 100 to 1000) and varies little with N in this range.

   i. The $N = 1$ case corresponds to a single cosine. Clearly, for $N = 1$ and $A_1 = 1$, we have $u_{\max} = 1$ for all choices of φ_1. But if we choose φ_1 randomly and look at a random time t, what is the probability density $p_1(u)$? Argue that this is equivalent to picking a random angle θ from $(0, 2\pi)$. Use the change-of-variables formalism for probability distributions to show $p_1(u) = \frac{1}{\pi}\frac{1}{1-u^2}$. Verify that $\langle u \rangle = 0$ and $\langle u^2 \rangle = \frac{1}{2}$, consistent with the result in Figure 6.3b.

   ii. For N harmonics of amplitude $1/\sqrt{N}$, use the Central Limit Theorem (Appendix Section A.7.3 online) to show that $\lim_{N\to\infty} p_N(u) \sim \mathcal{N}(0, \frac{1}{2})$ (Figure 6.3b).

   iii. *Extreme Value Statistics.* If we draw M times from a probability distribution $p(u)$, what is the typical value for $u_{\max}$? Let $F(u) = \int_{-\infty}^{u} du'\, p(u')$ be the cumulative distribution for $p(u)$. The probability to draw a value greater than u from $p(u)$ is $1 - F(u)$. Argue that the typical largest absolute value $u_{\max}$ in M draws from $p(u)$ is given by the solution to the equation $2M[1 - F(u_{\max})] = 1$. Derive the transcendental equation for M (involves erfc). What value to take for M? Since $u_N(t)$ has frequencies up to Nf_0, we must sample several times the fastest period in order to see the maximum (16 samples per period determines the maximum to better than 1%). But if we sample too often, the maximum value will saturate, meaning that not all draws are independent. Empirically, the maximum number of effectively independent draws is about $M = 5N$.

   iv. Confirm via simulation the results of these calculations shown below. In Figure 6.3d, we simulate, using 10 000 runs, a histogram of the distribution of crest factors (for $N = 255$ and $N_s = 4096$). The histogram follows a *Gumbel* distribution, as expected for the maximum of M draws from a parent distribution that decays exponentially or faster. The theorem is very much analogous to the Central Limit Theorem. The black curve in Figure 6.3d is calculated given the normal distribution from this problem (Gumbel, 1958).

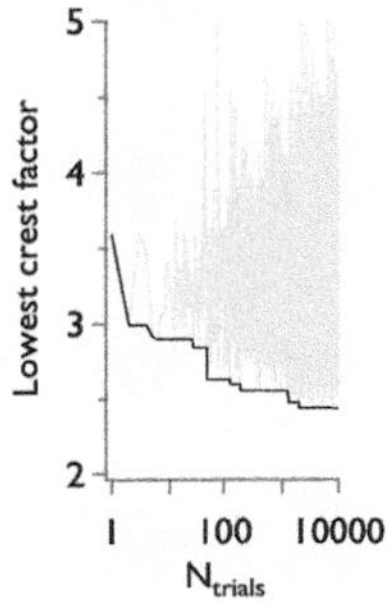

g. Not-so-random phases. Do random phases produce the lowest crest factor for large N? No! Figure 6.4c shows a deterministic choice that gives a crest factor of ≈ 1.66 but works only when all harmonics up to a maximum value are selected. Here is a simple way to lower the crest factor that works for any choice of harmonics: Generate N_{trials} random-phase multisine waveforms and select the one with the lowest crest factor. For 100 harmonics and 1000 points per fundamental period, your plot should resemble the one at left. Crest factors $\lesssim 2.5$ are readily obtained by this method. Using numerical optimization techniques to adjust systematically the phases can further lower the crest factor to ≈ 1.4 (Schoukens et al., 2012).

6.3 **Frequency chirp**. As illustrated at right, a frequency chirp consists of a sinusoid of continuously varying frequency. It "runs together" the individual sinusoids of the frequency-domain method. We retain the advantage of probing (almost) frequency by frequency but set aside the need to wait for the transients to die away. We also easily control the amplitude of each frequency, boosting in regions where the output is weak, if needed.

Frequency chirp

a. Find an analytic form for a chirp that sweeps from a frequency f_1 to f_2 in a time τ.

b. One disadvantage is that the power spectrum differs from the ideal "brick-wall" of a multisine. Plot the power spectrum numerically for a chirp that goes from 1 to 2 Hz over a time $\tau = 10$, 100, and 1000 s.

c. Compute the fraction of power that falls outside the 1–2 Hz range, as a function of τ. How does the error decrease with τ?

6.4 **Discrete random binary sequence (DRBS)**. Binary signals have a crest factor of 1 (see Problem 6.2) and thus inject the most power for a given input range. To define a DRBS, at every time interval kT_s, choose $\pm u_0$, with equal probability for $+u_0$ and $-u_0$, as shown at right. Here, we explore a number of properties of these signals.

Discrete random binary

a. Show – perhaps handwaving is good enough – that the autocorrelation function $R_{uu}(\tau) = 1 - \frac{\tau}{T_s}$ for $|\tau| < T_s$ and that it vanishes for larger $|\tau|$.

b. From the autocorrelation function, find the power spectrum.

c. Write a program to input a DRBS to a harmonic oscillator. Extract the transfer function and compare to the expected form. Be careful about aliasing. Show that oversampling – sampling the input and output at rates that are integer multiples of the original sampling frequency – helps.

Another binary signal variant is the *pseudo-random binary sequence* (PRBS), a *deterministic* sequence generated by a combination of shift registers and XOR logical operations. Its spectrum again is close to white. Its chief advantages are (i) because it is deterministic, its autocorrelation function may be calculated exactly, with no extra statistical uncertainty due to finite lengths of records. (ii) You can measure the signal repeatedly and average the output, reducing the effects of measurement noise. (iii) Because the sequence is periodic, the amount of power that "leaks" outside the desired band is much less than a DRBS signal.

Finally, because they probe just two values, binary sequences do *not* help to detect nonlinearity. It is then better to use an input that explores all levels.

6.5 **Noise and Fourier transforms**. Consider a sampled time series of observation noise ξ_k that is white, with $\langle \xi_k \rangle = 0$ and $\langle \xi_k \, \xi_{k'} \rangle = \xi^2 \delta_{kk'}$ and with $0 \le k \le N-1$.

a. Show that the discrete Fourier transform $\xi(\ell)$ satisfies $\langle \xi(\ell) \rangle = 0$ and $\langle \xi(\ell)\, \xi(\ell')^* \rangle = N\xi^2 \delta_{\ell\ell'}$. Here, $0 \le \{\ell, \ell'\} \le N-1$.

b. Why is each Fourier component statistically independent, even though it is built up from the entire time series?
c. The time series has N components, but the DFT has $2N$ components, since $\xi(\ell)$ is complex. Further, the real and imaginary parts of $\xi(\ell)$ are statistically independent. How can N independent noise components lead to $2N$ Fourier components?
d. Show that Parseval's theorem is satisfied and explain the physical significance:

$$\sum_{k=0}^{N-1} \langle \xi_k^2 \rangle = \frac{1}{N} \sum_{\ell=0}^{N-1} \langle |\xi(\ell)^2| \rangle$$

Note: For colored time-domain noise, $\xi_k = \sum_n h_{k-n}\, e_n$, for white noise $e_n \sim \mathcal{N}(0,1)$. Then a similar argument to (a) gives $\langle \xi(\ell)\, \xi(\ell')^* \rangle = N|H(e^{-2\pi i\ell/N})|^2\, \delta_{\ell\ell'}$, where $H(\cdot)$ is the Z-transform of h_k. The Fourier components remain independent complex, zero-mean Gaussian variables, but with frequency-dependent variances.

6.6 **Aliasing, two ways**. We can calculate the power spectrum of a sampled signal via the sampling theorem (Chapter 5) or directly from the discrete dynamics (Chapter 6). In a simple case, the two approaches give the same answer: Consider noise-free observations of a 1d Brownian particle, where $\gamma\dot{x}(t) = \xi_F(t)$. The noise $\langle \xi_F(t)\, \xi_F(t') \rangle = \sqrt{2D}\,\gamma\, \delta\,(t - t')$, and the power spectral density is $\langle |x|^2 \rangle(\omega) = 2D/\omega^2$. Now sample $x(t)$ at intervals T_s, giving x_k. Calculate its power spectrum two ways:

a. Discretize the continuous equations by integrating over T_s. Take the Z-transform and calculate the magnitude.
b. Use the sampling theorem, Eq. (5.3), and $\sum_{n=-\infty}^{\infty} \frac{1}{(\omega - n\omega_s)^2} = \frac{\pi^2}{\omega_s^2} \csc(\pi\omega/\omega_s)^2$. Derive the required identity by applying Parseval's Theorem to $f(t) = e^{i\omega t}$, assuming the function to be periodic with period T_s and to have jump discontinuities (Stone and Goldbart, 2009, Section 2.2.3).

Verify that as $\omega \to 0$, your discrete power spectrum approaches the continuous one.

6.7 **Bias of transfer function estimates**. The simple estimate of a transfer function as the ratio of two noisy DFT variables is biased. That is, $\langle G \rangle = \left\langle \frac{y+\xi_y}{u+\xi_u} \right\rangle \neq \frac{y}{u}$, where we drop the frequency dependence on all quantities. Assuming that $\langle \xi_y\, \xi_u^* \rangle = 0$, the bias arises entirely from the fluctuations in the input, ξ_u. Thus, we simplify by setting $\xi_y = 0$. Scaling by y/u and setting $z = \xi_u/u$, we can study the bias by comparing $\langle b(z) \rangle \equiv \left\langle \frac{1}{1+z} \right\rangle$ to 1. Here, $z = x + iy$, with $x, y \sim \mathcal{N}(0, \sigma^2/2)$ and $\sigma = 1/\mathrm{SNR}_u$.

a. Taylor expand to show that $\langle |b(z)|^2 \rangle = 1 + \sigma^2 + O(\sigma^4)$. Intuitively, negative fluctuations increase $|b|^2$ more than positive fluctuations decrease it. Indeed, if z can take the value ≈ -1, the corresponding fluctuation in b will be very large.
b. Since z is complex, its statistics are tricky. Show that $\langle z^2 \rangle = 0$.

c. Expand $\langle b(z)\rangle = \left\langle \frac{1}{1+z} \right\rangle$ in a full Taylor series and use the result from the previous part to conclude, incorrectly, that $\langle b(z)\rangle = 1$. Where is the flaw in the argument?

d. Calculate the bias directly, by integrating $b(z)$ over the probability distribution for z. By evaluating the integral first in terms of the real and imaginary components (x and y) and then converting to polar coordinates, show that $\langle b(z)\rangle = 1 - e^{-1/\langle |z|^2\rangle}$.

6.8 Variance of transfer function estimate. For high SNR, Problem 6.9 shows that fluctuations about the mean value of a complex number are approximately Gaussian. Using this idea and the result in Problem 6.7b, Taylor expand the transfer function to estimate $\delta G = (y_0 + \delta y)/(u_0 + \delta u) - G_0$ to derive Eq. (6.10). First derive the result in terms of the unknown true values u_0 and $\sigma_u^2 = \langle|\delta u|^2\rangle$, and so on. and then express in terms of the estimated means and variances given by Eqs. (6.7) and (6.8), for M periods of the input function $u = u_0 + \delta u$. All quantities are functions of the frequency ω_ℓ.

6.9 Amplitude and phase noise. Complex Gaussian random variables $z = x + iy$ with nonzero mean result from taking a discrete Fourier transform of a finite-amplitude signal. Let $x \sim \mathcal{N}(x_0, \sigma^2)$ and $y \sim \mathcal{N}(y_0, \sigma^2)$ be independent Gaussian random variables. Define fluctuations $\delta x = x - x_0$ and $\delta y = y - y_0$, and define magnitude and phase variables $r = r_0 + \delta r$ and $\theta = \theta_0 + \delta\theta$. Define the signal-to-noise ratio as $\text{SNR}_x = x_0/\sigma$ and $\text{SNR}_y = y_0/\sigma$.[17]

a. For high SNR ($\gg 1$), find δr and $\delta\theta$ to first order. Calculate the mean and variance of each and show that $\langle \delta r\, \delta\theta\rangle = 0$. Interpret the result geometrically.

b. Describe the zero SNR case ($r_0 = 0$). Derive (or guess) the radial and angular distributions for this case. Then describe how the low SNR case (δx and δy comparable to r_0) interpolates between the high-SNR and zero-SNR cases. Illustrate three cases (high, low, and zero SNR) by Monte Carlo simulations. For each case, plot $p(r,\theta)$ and the marginal plots $p(r)$ and $p(\theta)$.

c. One subtlety is that the mean of the magnitude is biased. An exact calculation gives $\langle r\rangle = \sqrt{r_0^2 + \sigma^2}$. Interpret this result physically. Derive it approximately by continuing the Gaussian expansion for r to second order in the noise.

This problem asks you to think physically about the distributions $p(r)$ and $p(\theta)$, but they are straightforward to investigate analytically, as well (Goodman, 2007). For reference, if $\boldsymbol{r}_0 = (r_0\cos\theta_0, r_0\sin\theta_0)$ and $\boldsymbol{r} = (r\cos\theta, r\sin\theta)$, then

$$p(\boldsymbol{r}) = \frac{1}{2\pi\sigma^2}\exp\left\{\left[-\frac{1}{2\sigma^2}|\boldsymbol{r} - \boldsymbol{r}_0|^2\right]\right\}$$

$$p(r,\theta) = \frac{r}{2\pi\sigma^2}\exp\left\{\left(-\frac{1}{2\sigma^2}\left[r^2 + r_0^2 - 2rr_0\cos(\theta-\theta_0)\right]\right)\right\}.$$

- Integrating out θ leads to $p(r) = \int_0^{2\pi} d\theta\, p(r,\theta) = \frac{r}{\sigma^2}\exp\left\{\left(\frac{-(r^2+r_0^2)}{2\sigma^2}\right)\right\} I_0\left(\frac{rr_0}{\sigma^2}\right)$ (*Rice distribution*), with $I_0(\cdot)$ a modified Bessel function of the first kind of order zero.

[17] Physicists often define signal-to-noise ratios in terms of amplitudes, engineers in terms of power.

- Integrating out r by $p(\theta) = \int_0^\infty \mathrm{d}r\, p(r,\theta)$ gives, with $\bar{r} = r_0/\sigma$,

$$p(\theta) = \frac{1}{2\pi} \exp\left\{\left(-\tfrac{1}{2}\bar{r}^2\right)\right\} \left[1 + \left(\frac{\bar{r}\cos\theta}{\sqrt{2}}\right) \exp\left\{\left(\tfrac{1}{2}\bar{r}^2\cos^2\theta\right)\right\} \int_{-\infty}^{\bar{r}\cos\theta} \mathrm{d}r'\, \exp\left\{\left(-\tfrac{1}{2}r'^2\right)\right\}\right].$$

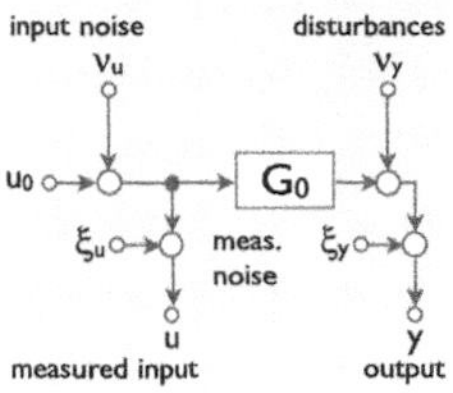

6.10 Transfer function bias and multiple noise sources. Consider a system $G_0 = y_0/u_0$ that is probed by noisy signals. In particular, let the input be given by $u = u_0 + v_u + \xi_u$, where v_u represents input noise (e.g., from a power amplifier) to the system and ξ_u the input measurement noise (see the block diagram at left). Let the output be given by $y = y_0 + v_y + \xi_y$, where v_y represents output noise (e.g., thermal fluctuations) and ξ_y the output measurement noise. Assume that the four noise sources are independent. All quantities are frequency-dependent, complex Fourier coefficients.

a. Show that the measured transfer function $G = y/u = G_0/[1 + \frac{\xi_u}{u_0+v_u}] + \frac{v_y+\xi_y}{u_0+v_u+\xi_u}$.
b. Why is G biased? Which noise source is responsible?
c. Explain why increasing input noise can reduce bias.

6.11 Do not average the magnitude of a Fourier Transform. Averaging the magnitude of multiple Fourier transforms is a poor strategy for reducing noise:

a. Write a program to generate a multifrequency sine wave of period 1 s, sampled 1000 times per period, and repeated for 1000 periods. Let the multifrequency sine wave have harmonics with amplitude $\propto 1/f^2$, with f the frequency, and choose the phases randomly. Calculate the magnitude of the power spectrum three ways: (i) Average each period in the time domain and then compute the DFT magnitude of the time-averaged waveform. (ii) Compute the DFT of each waveform, average the complex waves, and then find the magnitude. (iii) Take the DFT of each waveform, compute the magnitude, and then average. Plot all three magnitudes on one graph. Why is averaging the magnitude spectra wrong?
b. A less obvious issue is that when the input is noise dominated, the magnitude estimate is biased. To see this, consider an input $x \sim \mathcal{N}(0,1)$ and a transfer function =1. That is, show, both by Monte Carlo simulation and by analytic calculation, that the output has $\sqrt{\langle x^2 \rangle} = 1$ while $\langle |x| \rangle = \sqrt{2/\pi} \approx 0.80$.

6.12 Noisy resistor measurements. Noisy measurements can bias estimates even without dynamics. Let us estimate a resistance by applying a series of noisy currents I_k and measuring noisy voltages V_k (Pintelon and Schoukens, 2012). Let $I_k = 1 + \delta I_k$ and $V_k = 1 + \delta V_k$, with δI_k and $\delta V_k \sim \mathcal{N}(0,1)$ (resistance $R = 1$). Consider three estimators $\hat{R}$ for the resistance, each minimizing a different cost function J.

a. $J_1(R) \equiv \frac{1}{2}\sum_{k=1}^{N}(R_k - R)^2$, where $R_k = V_k/I_k$. Show that the value of R that minimizes J_1 is given by $\hat{R}_1 = \frac{1}{N}\sum_k R_k \equiv \overline{R_k}$, which is biased (Problem 6.7).
b. $J_2(R) \equiv \frac{1}{2}\sum_k(V_k - RI_k)^2$. Show that choosing $\hat{R}_2 = \overline{V_k I_k}/\overline{I_k^2}$ minimizes J_2. Show that $\langle \hat{R}_2 \rangle = 1/(1+\sigma_I^2)$, which implies that estimator is biased.

c. $J_3(R, I, V) \equiv \frac{1}{2}\sum_k \frac{(V_k-V)^2}{\sigma_V^2} + \frac{(I_k-I)^2}{\sigma_I^2}$, with the constraint that $V = RI$. Here, V and I are unknown "true values." Show that choosing $\hat{R}_3 = \overline{V_k}/\overline{I_k}$ minimizes J_3. Hint: Use the constraint to eliminate V; then differentiate with respect to both R and I.

d. Write a simulation for $\sigma_I = \sigma_V = 1$ and $N = 500$. Repeat for 10^4 trials and plot histograms of the values $\hat{R}_1$, $\hat{R}_2$, and $\hat{R}_3$. Explain why $\hat{R}_1$ is pathological here but not $\hat{R}_3$. How large a value of N is needed for $\hat{R}_3$ to be okay?

The first estimator is sometimes used in elementary physics laboratory courses, the second in intermediate courses, and the third (hopefully) in more advanced courses. The second is the common, unweighted least squares, assuming no error in the "input" variable (current). The third is the weighted least-squares estimate, taking into account errors in both variables.

6.13 Measuring a closed-loop transfer function.

a. For the block diagram at right, show that $G_m \equiv \frac{y_m}{u} = \frac{KGr+\xi}{K(r-\xi)}$.

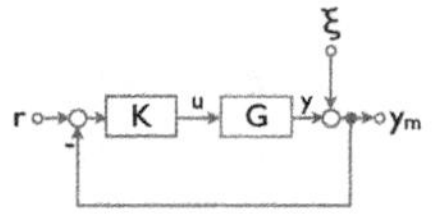

b. Simulate an unstable discrete first-order system that is stabilized by proportional feedback. The dynamics are given by $(y_m)_k = y_k + \xi_k$ and $u_k = -K[r_k - (y_m)_k]$, with $y_{k+1} = (1 + T_s)y_k + T_s u_k$, where the observational noise $\xi_k \sim N(0, \xi^2)$ and T_s is the sampling time. Use $K = 2$, $\xi^2 = 1$, $T_s = 0.05$ s, and scan frequencies from 0.01 to 10 Hz. Try three different reference signals: $r_k = 0$ (no reference), $r_k \sim N(0, 1)$ (white noise), and r_k a random-phase multisine of RMS amplitude = 1. Run the simulations for 10 periods of 100 s. Discard the first response to eliminate large transients. Your plot should resemble the figure at right, where the solid line is the transfer function of the noiseless discrete system (u_k to y_k), the triangles the no-reference case, the filled markers the random-phase case, and the white markers the multisine case. Explain the results.

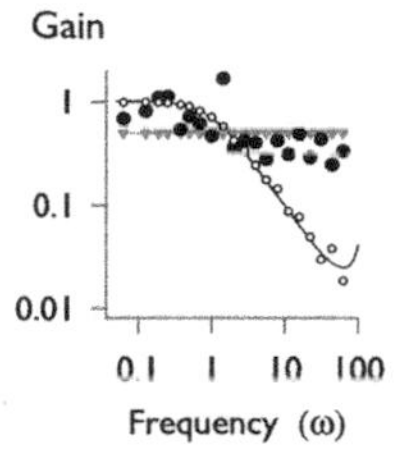

6.14 Resonance frequency of a thin plate We fill in some details of Example 6.6. For a careful approach, see Landau et al. (1986) and also Rossing and Russell (1990). Here, in the spirit of making rough approximations, feel free to use handwaving arguments. Consider just one transverse dimension, for simplicity.

a. Let $\psi(x, t)$ be a component of the elastic displacement in a material. Argue that the local kinetic energy per volume is $T = \frac{1}{2}\rho(\psi_t)^2$ and that the local elastic potential energy per volume is $U = \frac{1}{2}E(\psi_x)^2$. Here, ρ is the density and E the *Young's modulus*. The partial derivatives are $\psi_t = \partial_t\psi$ and $\psi_x = \partial_x\psi$.

b. From the Lagrangian $L = T - U$ and the Euler-Lagrange equations for a field, derive a wave equation for ψ. Find the dispersion relation, and show that the expected (longitudinal) sound speed is $c_L = \sqrt{E/\rho}$. For an object of size ℓ, the expected lowest resonance frequency is $f \approx c_L/\ell$.

c. Anisotropic objects such as plates can bend with a radius that is much greater than the plate thickness h, leading to lower resonance frequencies. Argue that the bending energy per unit area is of order $U_{\text{bend}} \sim Eh^3(\psi_{xx})^2$, where h is the

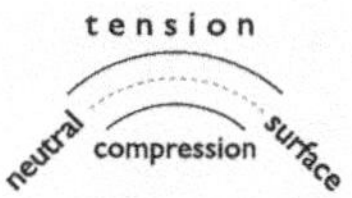

plate thickness. Hint: As shown at left, a bent plate of thickness h has one surface under tension and the other under compression. You can also use symmetry, or even a ball-and-spring model to derive the energy.

d. Use the higher-order version of the Lagrangian argument above to show that the resonance is lowered to $f \approx c_L/(\ell a)$, where the aspect ratio $a = \ell/h$.

6.15 Measuring a transfer function with noisy inputs. We explore the implications of input noise on transfer function measurements. The noisy input at frequency ω_ℓ is $u = u_0 + \nu$, where $\nu \sim \mathcal{N}(0, \sigma_u^2)$ and where u_0 is the (unobservable) true value of the input. Similarly, the noisy output is $y = y_0 + \xi$, where $\xi \sim \mathcal{N}(0, \sigma_y^2)$. From M input periods, we can use Eq. (6.7) to estimate the averages $\hat{u}$ and $\hat{y}$ and Eq. (6.8) to estimate the (co)-variances $\hat{\sigma}_u^2$, $\hat{\sigma}_y^2$, and $\hat{\sigma}_{yu}^2$. Note that we have simplified the notation by dropping the ω_ℓ dependence from all quantities.

a. Estimating G via $y = Gu$ is equivalent to fitting data to a straight-line relation between u and y with errors in both variables. Generalizing the Bayesian derivation of the χ^2 statistic in Appendix, Eq. A.210 online, define $z = \begin{pmatrix} \hat{y}-y_0 \\ \hat{u}-u_0 \end{pmatrix}$ and $\boldsymbol{\Sigma} = \begin{pmatrix} \hat{\sigma}_y^2 & \hat{\sigma}_{yu}^2 \\ \hat{\sigma}_{yu}^2 & \hat{\sigma}_u^2 \end{pmatrix}$ and show that the best estimate of the transfer function is given by minimizing the *errors-in-variables* cost function $\chi^2 = \sum_{\ell=1}^{N} z^\dagger \boldsymbol{\Sigma}^{-1} z$. If input-output correlations can be neglected, show that the general χ^2 simplifies to $\chi^2 = \sum_{\ell=1}^{N} \left(\frac{|\hat{y}-y_0|^2}{\hat{\sigma}_y^2} + \frac{|\hat{u}-u_0|^2}{\hat{\sigma}_u^2} \right)$. In both cases, we minimize χ^2 with respect to the unknown true input and output values u_0 and y_0, subject to the constraint (at each frequency ω_ℓ) that $y_0 = G\,u_0$. Recall that the transfer function model $G = G(\mathrm{i}\omega, \boldsymbol{\theta})$, with $\boldsymbol{\theta}$ the fit parameters.

b. We can minimize the above χ^2 with respect to the transfer function parameters $\boldsymbol{\theta}$ and the *nuisance parameters* u_0 and y_0 using Lagrange multipliers to enforce the constraints. Here, in a Bayesian approach that confirms that all distributions are Gaussian, we *marginalize* (integrate out) u_0 and y_0, again limiting ourselves to the $\hat{\sigma}_{yu}^2 = 0$ case for simplicity. To carry this out, assume a uniform prior on the u_0 and y_0 and impose the constraint $y_0 = G\,u_0$. Integrate the probability density $p(G|u_0, y_0)$ over the u_0 to show that the maximum-likelihood solution minimizes $\chi^2 = \sum_{\ell=1}^{N} \frac{|\hat{y}-G\hat{u}|^2}{\hat{\sigma}_y^2+|G|^2\hat{\sigma}_u^2}$ with respect to the parameters $\boldsymbol{\theta}$ in $G(\mathrm{i}\omega_\ell, \boldsymbol{\theta})$. Interpret intuitively the denominator in this expression. The required u_0 integral is a messy version of the identity in Problem A.7.1 online. You can use a computer-algebra program or try the real case, which has a similar but simpler structure.

6.16 Time-domain identification. We go through the example presented in Section 6.3.2. Consider the first-order system $y_{k+1} = a y_k + b u_k + \nu_k$, with $\langle \nu_k\, \nu_\ell \rangle = \nu^2\, \delta_{k\ell}$.

a. Starting from the Bayesian and maximum-likelihood ideas formulated in Section A.8.2 online, show that the best estimate for a and b is the one that minimizes $\chi^2 = \frac{1}{\nu^2} \sum_k (y_{k+1} + a y_k - b u_k)^2$.

b. Minimize χ^2 to show that the best estimates for a and b are given by Eq. (6.18).

c. To show that the parameter estimates are biased but *consistent*, plot the relative bias $|\hat{a} - a|/a$, against the number of data pairs N. Confirm that the bias scales as N^{-1}, rather than the $N^{-1/2}$ scaling that is characteristic of stochastic errors.

6.17 AIC and cross-validation. Consider N measurements of $y = \theta^* x + \eta$, with independent scalar variable x, observed variable y, and parameter θ^*. The measurement noise $\eta \sim N(0, \sigma^2)$, and the log likelihood is $L(\theta) = -\frac{1}{2\sigma^2} \sum_{j=1}^{N} (y_j - \theta x_j)^2$.

a. Show that the maximum-likelihood (ML) estimate is $\hat{\theta} = \sum_{j=1}^{N}(x_j y_j)/\sum_{j=1}^{N}(x_j^2)$.
b. Now consider the same data set, but without point i. Show that, to $O(1/N)$, the ML estimate of θ is $\hat{\theta}_{-i}$, where $\hat{\theta}_{-i} = \hat{\theta} - \frac{x_i}{\sum x_j^2}(y_i - \hat{\theta} x_i)$.
c. In one-point cross-validation, we calculate the likelihood of a missing point using $\hat{\theta}_{-i}$ and then average over all points. Define $A \equiv -\frac{1}{2\sigma^2} \sum_{i=1}^{N} (y_i - \hat{\theta}_{-i} x_i)^2$ as an assessment value and show that $A = L(\hat{\theta}) - 1 + O(1/N)$.

The case with K parameters proceeds similarly and leads to $A = L(\hat{\theta}) - K + O(1/N)$.

6.18 $\langle \chi^2 \rangle$ for an orthonormal basis and model mismatch. Consider a model function with an orthonormal basis set $\{e_k\}$, with $y^* = \sum_{k=1}^{\infty} (y^* \cdot e_k) e_k$ and $e_k \cdot e_\ell \equiv \frac{1}{N} \sum_{i=1}^{N} e_k(x_i) e_\ell(x_i) = \delta_{k\ell}$. For N points and K parameters, let $y = y^* + \xi$, with $\langle \xi^2 \rangle = \sigma^2$, and show that $\langle \chi^2 \rangle = \langle \frac{N}{\sigma^2} \|\boldsymbol{y} - \hat{\boldsymbol{y}}\|^2 \rangle = (N-K) + \frac{N}{\sigma^2} \sum_{\ell=K+1}^{\infty} (y^* \cdot e_\ell) e_\ell$. Hint: Subtract and add the true vector $\boldsymbol{y}^*$. The N in the definition of χ^2 is traditional.

6.19 AIC vs. BIC example. We work through the details of Example 6.8.

a. Write code to reproduce the plots in Example 6.8. First, simulate the data set itself from the true function $f(t) = \frac{\pi^2}{8}(1 - 2t)$, $0 < t < 1$, adding Gaussian noise of $\sigma = 0.01$. Then calculate the first 500 Fourier coefficients and the corresponding χ^2, AIC, and BIC statistics for each order.
b. Calculate the signal-to-noise ratio (SNR) of Example 6.8. Show that for large K, we have $p(\text{SNR}) = \frac{1}{\sqrt{2\pi \text{SNR}}} e^{-\text{SNR}/2}$, where $\text{SNR} = \theta_j^2/(\sigma^2/2N)$. The extra factor of 2 comes from the normalization of the basis vectors.
c. By approximating a sum by an integral, show that an asymptotic, large N, analytic approximation for the model-mismatch term, accurate enough for $N > 2$, is given by $\chi^2_{\text{mm}} = \frac{1}{96\sigma^2 (K+1)^3}$. Add the result to the stochastic contribution, $1 - K/N$, to generate the solid curves in the bottom three plots in the example.

6.20 Balanced coordinates. Show that, for a system $G = \{\boldsymbol{A}, \boldsymbol{B}, \boldsymbol{C}\}$, we can choose a coordinate transformation $\boldsymbol{T}$, with $\boldsymbol{x} = \boldsymbol{T}\boldsymbol{x}'$ such that $\boldsymbol{P}' = \boldsymbol{Q}' = \boldsymbol{\Sigma}$. Here, the Gramians are over $t = (0, \infty)$ and $\boldsymbol{\Sigma}$ is the diagonal matrix of Hankel singular values.

a. For the coordinate transformation $\boldsymbol{x} = \boldsymbol{T}\boldsymbol{x}'$, show that $\boldsymbol{A}' = \boldsymbol{T}^{-1}\boldsymbol{A}\boldsymbol{T}$, $\boldsymbol{B}' = \boldsymbol{T}^{-1}\boldsymbol{B}$, $\boldsymbol{C}' = \boldsymbol{C}\boldsymbol{T}$. Then show that $\boldsymbol{P}' = \boldsymbol{T}^{-1}(\boldsymbol{P}\boldsymbol{T}^{-1})^{\mathsf{T}}$ and $\boldsymbol{Q}' = \boldsymbol{T}^{\mathsf{T}}\boldsymbol{Q}\boldsymbol{T}$.
b. Decompose $\boldsymbol{P} = \boldsymbol{R}\boldsymbol{R}^{\mathsf{T}}$ using Cholesky decomposition (Section A.1.5 online), and write $\boldsymbol{R}^{\mathsf{T}}\boldsymbol{Q}\boldsymbol{R} = \boldsymbol{U}\boldsymbol{\Sigma}^2\boldsymbol{U}^{\mathsf{T}}$, and $\boldsymbol{T} = \boldsymbol{R}\boldsymbol{U}\boldsymbol{\Sigma}^{-1/2}$. Show that $\boldsymbol{P}' = \boldsymbol{Q}' = \boldsymbol{\Sigma}$.
c. Consider the example from Figure 6.7: $\boldsymbol{A} = \begin{pmatrix} -1 & 0 \\ 0 & -2 \end{pmatrix}$, $\boldsymbol{B} = \begin{pmatrix} 1 \\ 1 \end{pmatrix}$, and $\boldsymbol{C} = \begin{pmatrix} 1 & 2 \end{pmatrix}$.

i. Find the infinite-time Gramians $\boldsymbol{P}$, $\boldsymbol{Q}$. Construct a balanced representation.

ii. Find symbolically or numerically $\boldsymbol{R}$, $\boldsymbol{U}$, $\boldsymbol{T}$, Σ and also $\boldsymbol{A}'$, $\boldsymbol{B}'$, and $\boldsymbol{C}'$. You will want to use a computer-algebra program for this part and the next.

iii. In the new coordinate system, verify that $\boldsymbol{P}' = \boldsymbol{Q}' = \Sigma$, with the diagonal elements of Σ being $\sigma_{\pm} = \frac{1}{2} \pm \frac{\sqrt{2}}{3}$. (If you do not have access to symbolic-manipulation software, do this part numerically.)

PART II

ADVANCED IDEAS

7 Optimal Control

Optimal control is an approach to designing a controller that takes advantage of knowing a good dynamical model for a system. In Chapter 4, we introduced the notion of *controllability* – whether a system's input can determine all of its internal states – and showed that a linear MIMO system is controllable if the matrix $\boldsymbol{W}_c \equiv \begin{pmatrix} \boldsymbol{B} & \boldsymbol{AB} & \boldsymbol{A}^2\boldsymbol{B} & \cdots & \boldsymbol{A}^{n-1}\boldsymbol{B} \end{pmatrix}$ has full rank. If a linear system is controllable, then we can always pick a controller $K(s)$ that will place arbitrarily the eigenvalues of the closed-loop linear system. But where to put the poles? In Section 4.2.1, we gave some rules of thumb – move as few as possible, as little as possible – that work well for simple systems but are difficult to implement in more complex situations. In optimal control, we define a *cost function*, a function that assigns a numerical score to each choice of control input and then choose the control algorithm that minimizes this cost.[1] This approach can work well for higher-order SISO and MIMO systems and can be generalized to nonlinear systems. One pitfall is that the dynamics of the system need to be known accurately, and disaster can arise when optimizing for the wrong dynamics. But when the dynamics are well known, optimal control can be a powerful design method. Historically, many of the ideas were first developed and then applied to the space program. Indeed, the Apollo mission to the moon would not have been possible without the ideas considered in this chapter and the next.

We first formulate the *cost function* J. Often, the goal is to balance performance – for example, a small control error – against the effort, or cost, of control. But the trade-off among competing objectives is usually very specific to a given application. This arbitrariness in choosing the cost function is a distinguishing feature of control theory and other human-based endeavors, including not just technology but also economics, sociology, and so on. In physics, cost functions such as the action or thermodynamic potentials are typically dictated by Nature. In biology, objective functions often reflect survivability – hence the adage "Nothing in biology makes sense except in the light of evolution" (Dobzhansky, 1973). And yet, despite the vastly differing origins of cost functions in all these different fields, there is a common theme of optimization under constraints.

Having chosen a cost function, we then search for the dynamics (parametrized, for example, by the feedback parameters) that minimizes it. For background, see Appendix A.5.2 online. In this chapter proper, we start with a one-dimensional example that illustrates some of the basic ideas

[1] Optimists might prefer instead to maximize a *reward function*, such as $-J$.

of optimal control. We generalize to n-element state vectors with continuous and then discrete dynamics. We next discuss how to optimize controllers that take into account the constraint of having finite control *authority* – that is, a limited range for the input variable $u(t)$. We end with a synthesis that combines nonlinear feedforward control with linear feedback for basic robustness, augmented by the possibility of recalculating the feedforward solution whenever discrepancies between the desired state trajectory and the observed one differ too much. The combination offers a reasonable solution to the problem of controlling nonlinear systems: it works well if the model is "accurate" and perturbations are either "small" or "rare." The terms in quotes will be discussed in this chapter.

7.1 One-Dimensional Example

We start with a simple example that can be analyzed completely. Let

$$J\left([u]_0^\infty, x(0)\right) \equiv \int_0^\infty \mathrm{d}t\, L(x,u) \equiv \int_0^\infty \mathrm{d}t\, \tfrac{1}{2}\left(Qx^2(t) + Ru^2(t)\right),$$
$$\dot{x} = -ax + u(t), \qquad x(0) = x_0, \qquad a > 0. \tag{7.1}$$

The *cost function* J integrates the *running cost* $L(x,u)$ over the time interval $[0,\infty)$. It is a *functional* of $[u]_0^\infty$, the *path* of the control $u(t)$, and a function of the initial state $x(0) = x_0$. Notice that J does *not* depend on the state trajectory $[x]_0^\infty$ because the dynamics determine $x(t)$ once the initial condition and control law $u(t)$ are specified. The value of J should also be bounded from below to prevent "run-away" solutions tending to $J \to -\infty$.[2]

The running cost $L(x,u)$ is a quadratic function of $x(t)$ and $u(t)$. The coefficient Q penalizes deviations of the state $x(t)$ from the set point 0, while R penalizes the *control effort* $u^2(t)$. Choosing a quadratic form for L leads to linear equations for determining the minimum, just as a quadratic potential leads to linear dynamical equations in mechanics.

The requirement that $J \geq 0$ implies $Q \geq 0$ and $R \geq 0$. Physically, either Q or R must be positive. If $R = 0$, we seek to minimize excursions in the state $x(t)$ whatever the costs of control (limited to a finite range of values in the real world but assumed unlimited in this problem). If $Q = 0$, we seek to minimize control usage, without caring about the states $x(t)$. The result is uncontrolled dynamics, $u = 0$. To simplify the notation in this one-dimensional problem, we set $Q = 1$ and vary R. This does not change the problem, as we can always rescale $J \to J/Q$ and $R \to R/Q$.

Let us next assume that the form of the control law that minimizes J is a linear, negative feedback of the form $u(t) = -Kx(t)$, a simplification justified in Problem 7.1. Then, the *closed-loop* equation for $x(t)$ becomes $\dot{x} = -(a + K)\,x$, a one-

[2] The condition $a > 0$ implies that the uncontrolled dynamics are stable. See Problem 7.2 for the unstable case.

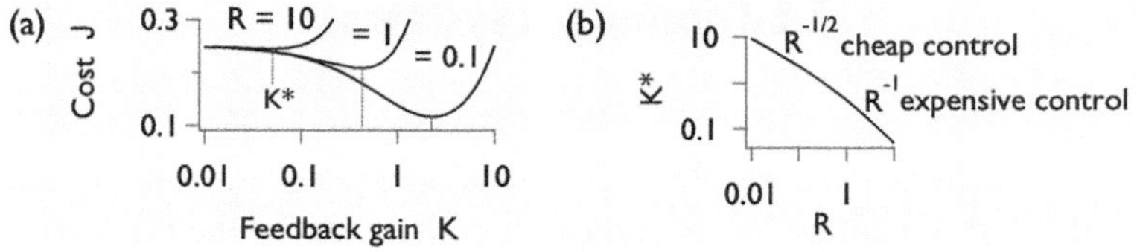

Optimal control in one dimension. (a) Performance index J for three values of R, with $a = 1$. (b) The optimal feedback gain K^* decreases with R. Fig. 7.1

dimensional differential equation that we can solve explicitly for $x(t)$ and then $u(t)$ and $J(K,R)$:

$$x(t) = x_0 \, \mathrm{e}^{-(a+K)t} \quad \Longrightarrow \quad u(t) = -K \, x_0 \, \mathrm{e}^{-(a+K)t} \, , \quad J = x_0^2 \left(\frac{1 + RK^2}{4(a+K)} \right) . \tag{7.2}$$

Equation (7.2) shows that J is also a function of the initial condition x_0; however, the linearity of the equations of motion implies that we can eliminate it by scaling $J \to J/x_0^2$. In Figure 7.1a, we plot J as a function of K for different values of R. By taking the derivative of J with respect to K, we see that there is a unique minimum for $K = K^*$:

$$\partial_K J = 0 \quad \Longrightarrow \quad K^2 + 2aK - \frac{1}{R} = 0 \quad \Longrightarrow \quad K^* = -a \pm \sqrt{a^2 + \frac{1}{R}} \, . \tag{7.3}$$

We choose the positive root $K > 0$, since we impose negative feedback. We also check that $(\partial_{KK} J)(K^*) = \frac{R^2}{2\sqrt{R(1+a^2R)}} > 0$, implying that K^* is a local minimum. Consider two limits:

$$\begin{aligned} Ra^2 &\ll 1 : \textit{cheap control}, & J &\approx \int_0^\infty \mathrm{d}t \, \tfrac{1}{2} x^2 & &\Rightarrow & K^* &\approx R^{-1/2} \to \infty \, , \\ Ra^2 &\gg 1 : \textit{expensive control}, & J &\approx \int_0^\infty \mathrm{d}t \, \tfrac{1}{2} R u^2 & &\Rightarrow & K^* &\approx \frac{1}{2Ra} \to 0 \, . \end{aligned} \tag{7.4}$$

Figure 7.1b shows how cheap control leads to large values of K^* and hence large values of $u(t)$ while expensive control leads to the opposite. We can draw some general observations:

- Rather than choosing K, we choose R.
- The utility of optimal control is based on the hope that we have more intuition for choosing R than we have for choosing K or the eigenvalue position.
- "Optimal" may not imply "good": A poor choice for R leads to poor control.

In this simple, one-dimensional setting, there is not much point to optimal control – choosing K directly is probably at least as intuitive as choosing R. However (see Section 7.3), when $\boldsymbol{x}$ has n components, the gain $\boldsymbol{K}$ becomes an $n \times n$ matrix. Then, optimal control fixes all n^2 components of $\boldsymbol{K}$ via a single parameter, R, that reflects the cost of the control input $u(t)$. Optimal control can greatly simplify parameter tuning for such control problems.

7.2 Continuous Systems

In Section 7.1, we discussed a simple, one-dimensional problem with a quadratic cost function and linear dynamics. We postulated a control law of the form $u(t) = -Kx(t)$ and optimized by taking a derivative with respect to the feedback gain K. Here, we approach optimal control more generally, inspired by variational calculus and the Lagrangian approach to mechanics.

Choosing the Cost Function

A broad class of cost functionals can be expressed in the form

$$J = \varphi[\boldsymbol{x}(\tau)] + \int_0^\tau \mathrm{d}t\, L(\boldsymbol{x}, \boldsymbol{u}), \tag{7.5}$$

where the running cost L puts a value on the instantaneous state $\boldsymbol{x}(t)$ and control $\boldsymbol{u}(t)$, and J is again a scalar functional of the control path $[\boldsymbol{u}]_0^\tau$. The *terminal cost* φ puts a value on the final state at $t = \tau$. Alternatively, we can replace the soft end-time constraint φ with a hard constraint on $\boldsymbol{x}(\tau)$, which simply fixes the value of $\boldsymbol{x}$ at time τ via a boundary condition.

We seek to minimize J, subject to the constraint that the nonlinear dynamical equations $\dot{\boldsymbol{x}} = \boldsymbol{f}(\boldsymbol{x}, \boldsymbol{u})$ are obeyed for $t \in [0, \tau]$. For problems with an infinite horizon ($\tau \to \infty$), as in Section 7.1, we omit $\varphi(\boldsymbol{x})$.

Our problem belongs to the calculus of variations, as we have to consider continuous paths $[\boldsymbol{u}]_0^\tau$ that lie in a *function space*. The control signal $\boldsymbol{u}(t)$ is here unconstrained in this function space. Since actuators usually have physical limits, we will consider the implications of bounding the set of allowable $\boldsymbol{u}(t)$ in Section 7.5. Appendix A.5 online reviews the mathematical background on the calculus of variations with constraints.

What to choose for L? There are a few "common-sense" restrictions: L should be a scalar function, bounded from below, and twice differentiable to facilitate analysis. A quadratic form obeys these rules and, in Section 7.1, led to simple results. However, quadratic forms have some less desirable features: Because the immediate cost function L is a convex function of $\boldsymbol{x}$, it is unbounded for large deviations in the system state. But maybe the worst outcome is not so bad. We can then cap the penalty for a bad outcome, perhaps replacing $L = (x/x_0)^2$ with $L' = 1 - \mathrm{e}^{-(x/x_0)^2}$, which $\sim L$ for small errors but is bounded by 1 for large errors (see left). Of course, a nonconvex L' is harder to deal with analytically.

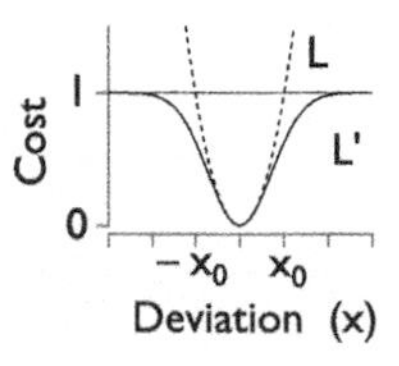

The end-point constraint φ puts a special weight on the end state at time τ. If $L \neq 0$ and $\varphi = 0$, then the final state has no special significance, and costs are summed over the interval $[0, \tau]$. Example 7.4, below, is such a case. Alternatively, if $\varphi \neq 0$ and $L(\boldsymbol{x}, u, t) = 0$ for $[0, \tau]$, then only the final state $\boldsymbol{x}(\tau)$ matters ("the end justifies the means"). Problem 7.16, below, is such a case. In general, both L and φ are nonzero (Problem 7.11) and there may be final conditions on $\boldsymbol{x}(\tau)$, as well (Example 7.12).

Euler–Lagrange, with Constraints

To recap, our problem is to find the control path $[\boldsymbol{u}]_0^\tau = \boldsymbol{u}(t)$ over $t \in [0, \tau]$ that minimizes the cost function $J = \varphi[\boldsymbol{x}(\tau)] + \int_0^\tau \mathrm{d}t\, L(\boldsymbol{x}, \boldsymbol{u})$ subject to the dynamical constraint $\dot{\boldsymbol{x}} = \boldsymbol{f}(\boldsymbol{x}, \boldsymbol{u})$. There are several possible approaches we might take:

1. Solve $\dot{\boldsymbol{x}} = \boldsymbol{f}(\boldsymbol{x}, \boldsymbol{u})$ for $\boldsymbol{x}[\boldsymbol{u}(t)]$ and eliminate the dynamics from our cost function, leaving $J[\boldsymbol{u}(t)] = \int_0^\tau \mathrm{d}t\, L'(\boldsymbol{u})$ as an optimization over the arbitrary control $\boldsymbol{u}(t)$, which we can solve using calculus of variations. This was our strategy in Section 7.1, except that we "cheated" by assuming a special feedback structure, $u(t) = -Kx(t)$, which simplified the solution for $x(t)$ greatly. Indeed, in most practical problems, it will be impossible to solve dynamical equations and cost functions analytically for arbitrary inputs.
2. Extract $\boldsymbol{u} = \boldsymbol{u}(\boldsymbol{x}, \dot{\boldsymbol{x}})$ from the equation $\dot{\boldsymbol{x}} = \boldsymbol{f}(\boldsymbol{x}, \boldsymbol{u})$, substitute into $L(\boldsymbol{x}, \boldsymbol{u}) \to \tilde{L}(\boldsymbol{x}, \dot{\boldsymbol{x}})$, and optimize over variations in $\boldsymbol{x}(t)$ alone. In many cases (e.g., the affine form $\dot{\boldsymbol{x}} = \boldsymbol{f}(\boldsymbol{x}) + \boldsymbol{g}(\boldsymbol{x})\boldsymbol{u}$), this is easy. Indeed, we use this method in Problem 7.1 to solve the example in Section 7.1 without assuming linear feedback a priori. In general, the Euler–Lagrange variational equations for $\boldsymbol{x}$ become second order, with an $\ddot{\boldsymbol{x}}$ term that also makes them hard to solve. This method does, however, justify choosing L as our notation for running cost, as $\tilde{L}(\boldsymbol{x}, \dot{\boldsymbol{x}})$ has the same structure as the Lagrangian of classical mechanics.[3]
3. Use the method of *Lagrange multipliers* and augment (or *adjoin*) the cost function. The minimization is now done over *both* the state $\boldsymbol{x}$ and control $\boldsymbol{u}$ simultaneously. Not only docs this method avoid the need to solve explicitly for either $\boldsymbol{x}$ or $\boldsymbol{u}$, it also maintains the structure of the control problem, which adds conceptual clarity, as we will see below.

To implement the Lagrange-multiplier strategy, we define

$$J' \equiv \varphi[\boldsymbol{x}(\tau)] + \int_0^\tau \mathrm{d}t \left(L(\boldsymbol{x}, \boldsymbol{u}) + \boldsymbol{\lambda}^\mathsf{T}(t)[\boldsymbol{f}(\boldsymbol{x}, \boldsymbol{u}) - \dot{\boldsymbol{x}}] \right), \tag{7.6}$$

where the Lagrange multiplier $\boldsymbol{\lambda}(t)$ is an *adjoint vector* (or *costate vector*) that provides the time-dependent "force" that ensures $\boldsymbol{x}(t)$ obeys the equations-of-motion constraint for each component at each time t. (See Example A.14, in A.5.1 online.) We now solve the *unconstrained* minimization of J' by allowing functional variations with respect to $\delta\boldsymbol{x}(t)$, $\delta\boldsymbol{u}(t)$, and $\delta\boldsymbol{\lambda}(t)$:

$$\delta J' = (\partial_{\boldsymbol{x}}\varphi)\,\delta\boldsymbol{x}(\tau) + \int_0^\tau \mathrm{d}t \left[\underbrace{(\partial_{\boldsymbol{x}}L)\,\delta\boldsymbol{x} + (\partial_{\boldsymbol{u}}L)\,\delta\boldsymbol{u}}_{\delta L} + \boldsymbol{\lambda}^\mathsf{T}\Big(\underbrace{(\partial_{\boldsymbol{x}}\boldsymbol{f})\,\delta\boldsymbol{x} + (\partial_{\boldsymbol{u}}\boldsymbol{f})\,\delta\boldsymbol{u}}_{\delta\boldsymbol{f}} - \delta\dot{\boldsymbol{x}} \Big) + \delta\boldsymbol{\lambda}^\mathsf{T}(\boldsymbol{f} - \dot{\boldsymbol{x}}) \right]$$

[3] The analogy is not perfect: optimal control differs in several ways from more familiar problems in classical mechanics. First, physics dictates that the Lagrangian L is $T - V$, the difference between kinetic and potential energies. In optimal control, by contrast, we can choose the running cost $L(\boldsymbol{x}, \boldsymbol{u})$ relatively freely. Also, while in classical mechanics the end points are fixed, here they may not be. In addition, the equations of motion are imposed as a constraint for optimal-control problems. Finally, the basic Lagrangian formalism in classical mechanics does not accommodate dissipative forces. There is no such restriction in the control formulation.

$$
\begin{aligned}
&= (\partial_{\boldsymbol{x}}\varphi)\,\delta\boldsymbol{x}(\tau) - \boldsymbol{\lambda}(\tau)^{\mathsf{T}}\,\delta\boldsymbol{x}(\tau) + \boldsymbol{\lambda}^{\mathsf{T}}(0)\delta\boldsymbol{x}(0) + \int_0^{\tau} \mathrm{d}t\left[(\partial_{\boldsymbol{x}}L) + \boldsymbol{\lambda}^{\mathsf{T}}\,(\partial_{\boldsymbol{x}}\boldsymbol{f}) + \dot{\boldsymbol{\lambda}}^{\mathsf{T}}\right]\delta\boldsymbol{x}(t) \\
&\quad + \int_0^{\tau} \mathrm{d}t\left[(\partial_{\boldsymbol{u}}L) + \boldsymbol{\lambda}^{\mathsf{T}}\,(\partial_{\boldsymbol{u}}\boldsymbol{f})\right]\delta\boldsymbol{u}(t) + \int_0^{\tau} \mathrm{d}t\,(\boldsymbol{f} - \dot{\boldsymbol{x}})^{\mathsf{T}}\,\delta\boldsymbol{\lambda}(t)\,.
\end{aligned}
\tag{7.7}
$$

For the second equality, we integrate the $\boldsymbol{\lambda}^{\mathsf{T}}(t)\delta\dot{\boldsymbol{x}}(t)$ term by parts and isolate terms belonging to each type of variation. We also took the transpose of the last (scalar) term to make it more resemble the other terms. Inside the integrals, all functions are evaluated at time t. Since $\boldsymbol{x}(0)$ is fixed, its variation $\delta\boldsymbol{x}(0)$ vanishes.[4] The condition that $\delta J' = 0$ for unconstrained small variations of $\boldsymbol{x}(t)$, $\boldsymbol{u}(t)$, and $\boldsymbol{\lambda}(t)$ then gives,

$$
\begin{aligned}
0 \to \tau, &\quad \dot{\boldsymbol{x}} = \boldsymbol{f}(\boldsymbol{x},\boldsymbol{u})\,, && \boldsymbol{x}(0) = \boldsymbol{x}_0\,, \\
\tau \to 0, &\quad \dot{\boldsymbol{\lambda}} = -\,(\partial_{\boldsymbol{x}}\boldsymbol{f})^{\mathsf{T}}\,\boldsymbol{\lambda} - (\partial_{\boldsymbol{x}}L)^{\mathsf{T}}\,, && \boldsymbol{\lambda}(\tau) = (\partial_{\boldsymbol{x}}\varphi|_{\tau})^{\mathsf{T}}\,, \\
t, &\quad \boldsymbol{0} = (\partial_{\boldsymbol{u}}\boldsymbol{f})^{\mathsf{T}}\,\boldsymbol{\lambda} + (\partial_{\boldsymbol{u}}L)^{\mathsf{T}}\,.
\end{aligned}
\tag{7.8}
$$

Equation (7.8) contains dynamical equations for $\boldsymbol{x}$ and $\boldsymbol{\lambda}$, with boundary conditions at $t = 0$ and τ, respectively.[5] The notation $0 \to \tau$ for the dynamical equation denotes that the equation obeys an initial condition at $t = 0$ and can be integrated forward in time. The notation $\tau \to 0$ denotes that the adjoint equation has a final condition at $t = \tau$ and can be integrated backwards in time, *knowing* the state $\boldsymbol{x}(t)$. The bottom line gives an *algebraic* equation relating the control $\boldsymbol{u}$ to the state and adjoint at each time t and has no associated boundary condition. If there is only one input, then the algebraic equation for u is scalar. The last two sets of equations transpose the terms in Eq. (7.7), putting all equations in column-vector form. (Recall that the gradient is a row vector – Section A.1.9 online.)

As with all minimization techniques based on derivatives, solutions to the variational equations (7.7) give *necessary conditions* for an optimal solution, which could, for example, still be a local maximum in cost (analogous to Point 1 in the sketch at left). One must check that second variations are positive definite (2) to ensure a *local cost minimum*. For multiple local minima, one must determine which is the *global minimum* (3).

[4] Contrast the situation for classical mechanics: given equations of motion with no external input ($\boldsymbol{u} = \boldsymbol{0}$), both $\boldsymbol{x}(0)$ and $\boldsymbol{x}(\tau)$ are fixed: the former by initial conditions, the latter by the determinism of the equations of motion. Thus, a variation $\delta\boldsymbol{x}(t)$ must vanish at both $t = 0$ and $t = \tau$, and there are no boundary terms in the Euler–Lagrange equations. In optimal control, one may not wish to impose final conditions on $\boldsymbol{x}$, giving more freedom to choose $\boldsymbol{u}(t)$. Since we then do not know $\boldsymbol{x}(\tau)$, variations associated with the final time do not vanish but rather appear as boundary terms in the Euler–Lagrange equations.

[5] The boundary conditions on the adjoint $\boldsymbol{\lambda}$ are for cases where the final state $\boldsymbol{x}(\tau)$ is free (but penalized for "bad" values). If we impose instead hard boundary conditions on the final state, there will be no boundary conditions at all for $\boldsymbol{\lambda}$, which is then fixed by the algebraic equation relating $\boldsymbol{u}$ to $\boldsymbol{x}$ and $\boldsymbol{\lambda}$.

Interpretation of the Adjoint Vector $\lambda(t)$

That the equation for state dynamics requires an initial condition $\boldsymbol{x}(0) = \boldsymbol{x}_0$ is no surprise, as it was an input to our problem. But the *final conditions* imposed at time τ for $\lambda(t)$ may be unexpected. To understand their origin, let us ignore any dependence on λ in the term $\partial_{\boldsymbol{x}}\boldsymbol{f}$. Then the adjoint equation, $\dot{\boldsymbol{\lambda}} = -(\partial_{\boldsymbol{x}}\boldsymbol{f})^{\mathrm{T}}\,\boldsymbol{\lambda} - (\partial_{\boldsymbol{x}}L)^{\mathrm{T}}$, is linear in $\boldsymbol{\lambda}$. We further specialize to a one-dimensional case, where $\lambda \propto \partial_x L$. That is, $|\lambda|$ is proportional to the *sensitivity* of the running cost L to changes in the state. The adjoint dynamics evolve with a time constant $\sim -(\partial_x f)^{-1}$. This time constant represents a *planning horizon*, the "lead time" needed to control the system state in the future. To reach a desired state at a given time t requires planning control signals in advance; the system cannot change state in an arbitrary way instantaneously. "Foreseeable" future events requiring advance planning can be created by the penalty on the end state φ at time τ. Alternatively, an $\boldsymbol{x}$ dependence in the running cost $L(\boldsymbol{x}, \boldsymbol{u})$ can act as a source term for the adjoint equation. The adjoint equation shows exactly how this planning should be done, with the $\boldsymbol{x}$ in its coefficients reflecting that differing locations in state space require different planning horizons.

With no end-time constraint, $\varphi = 0$ and $\boldsymbol{\lambda}(\tau)^{\mathrm{T}} = \boldsymbol{0}$: the adjoint would not constrain the state at time τ because there are no future consequences to worry about. Similarly, if the control function is independent of the system state, then that "source" of adjoint will not be present and the adjoint will tend to decay to zero as one integrates backward in time.

Finally, in Section 7.5.2, we will see that we can also interpret the adjoint as a kind of canonical momentum in a Hamiltonian description of dynamics. But just as the Lagrangian analogy made here is only partial, so too will be the Hamiltonian analogy.

Example 7.1 (One-dimensional control, with initial and final state conditions) Consider again the example from Section 7.1, for $L = \frac{1}{2}(x^2 + u^2)$ and $\dot{x} = -x + u$, but let the initial condition be the equilibrium, $x(0) = 0$, and require instead that $x(\tau) = x_\tau$. Rather than looking for a feedback solution, we can solve the Euler–Lagrange equations,

$$\dot{x} = -x + u\,, \qquad \dot{\lambda} = +\lambda - x\,, \qquad u = -\lambda\,. \tag{7.9}$$

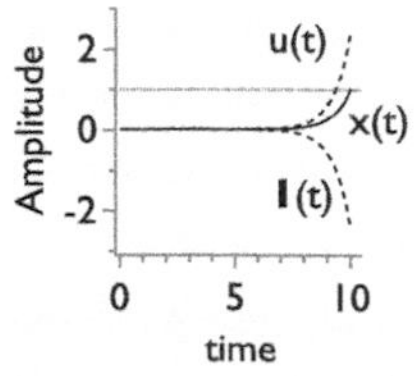

Because λ is linked to anticipated future events, the positive sign in the right-hand side of its evolution equation is appropriate, as it implies that a finite amount of planning is needed for such events. Note that there are two boundary conditions for x and none for λ. With two conditions on the second-order system of $\{x(t), \lambda(t)\}$, the equations are well defined (and can be solved analytically; see Problem 7.3). The plot at right shows that x, λ, and u are all nearly zero until shortly before the end time $\tau = 10$, which keeps L and hence J small. Then $\lambda(t)$ starts to increase so that x can reach its final value, $x_\tau = 1$. The adjoint λ and control u need to "wake up" a time of order $-(\partial_x f)^{-1} = 1$ before the constraint. The magnitude of the adjoint $\lambda(t) \propto x_\tau$: that is, the target value of the state sets the scale of $\lambda(t)$.

Nature of the Control Signal

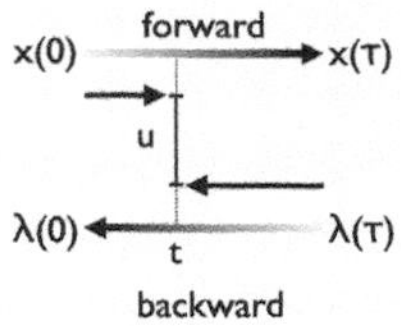

The control $\boldsymbol{u}$ at time t is influenced by past state values $\boldsymbol{x}(t)$ from the interval $[0, t)$ and also by future values of the adjoint $\boldsymbol{\lambda}(t)$, from the interval $(t, \tau]$. That is, the control at a given time t is based on the actual history of the system, as summarized by $\boldsymbol{x}(t)$ and by the projected future consequences, as summarized by $\boldsymbol{\lambda}(t)$. We thus have a *boundary-value* problem, with an initial condition for $\boldsymbol{x}(t)$ and a final condition for $\boldsymbol{\lambda}(t)$, coupled at time t through the control $\boldsymbol{u}(t)$, which must also be determined (see left).

Solving the Euler–Lagrange equations in Eq. (7.8) gives the optimal control $\boldsymbol{u}(t)$, which can be used to drive the system to the desired goal. The $\boldsymbol{\lambda}$ dependence of $\boldsymbol{u}$ reflects the influence of the future on present control. Similarly, the $\boldsymbol{x}$ dependence of $\boldsymbol{u}$ reflects knowledge gained from the past. Recall that we usually do not have direct access to the full state but instead must infer it from past observations $\boldsymbol{y}(t)$. Section 4.3 showed how the observer structure allows using the *history* of observations up to time t, the path $[\boldsymbol{y}]_0^t$, to find an estimator $\hat{\boldsymbol{x}}(t)$ of the current state $\boldsymbol{x}$. Thus, $\hat{\boldsymbol{x}}(t)$ summarizes knowledge of the past.

Usually, we will have to solve all the equations in Eq. (7.8) in order to produce the control path $[\boldsymbol{u}]_0^\tau$. Our basic strategy is to implement this nominal optimal control as an open-loop, feedforward scheme. By solving the equations offline (numerically) before implementation, we skirt the issue that $\boldsymbol{u}$ depends on $\boldsymbol{\lambda}$ and hence the future. Further, since off-line analysis always has access to the full state, we do not need to use an observer.

For the special case of a linear dynamical system and quadratic cost function, the adjoint $\boldsymbol{\lambda} \propto \boldsymbol{x}$, so that $\boldsymbol{u} = \boldsymbol{u}(\boldsymbol{x})$, allowing for a simple closed-loop, feedback control. See the simple example in Section 7.1 and also the general linear dynamics case of Section 7.3.1.

Example 7.2 (Swing up a pendulum) To illustrate feedforward control of a nonlinear system, consider swinging up a pendulum from its stable to its unstable equilibrium by applying torque (Section 2.1). The goal will be to arrive at the state $(\theta = \pi, \dot{\theta} = 0)$ in a fixed time τ while minimizing work.[6] The optimal-control problem is then to minimize $J = \int_0^\tau \mathrm{d}t\, \frac{1}{2} u^2(t)$, subject to the conditions

$$\ddot{\theta}(t) + \sin\theta(t) = u(t)\,, \qquad \theta(0) = \dot{\theta}(0) = 0\,, \qquad \theta(\tau) = \pi\,, \dot{\theta}(\tau) = 0\,. \tag{7.10}$$

In Problem 7.5, you will convert these equations to the standard state-space form used in Eq. (7.6) and find the corresponding Euler–Lagrange variational equations. Converting back to our original notation leads to Eq. (7.10) and also

$$\ddot{\lambda}(t) + \lambda(t)\cos\theta(t) = 0\,, \qquad u(t) = -\lambda(t)\,. \tag{7.11}$$

[6] The torque τ produced by a DC motor $\sim I$, the current supplied to its armature coils. If we neglect back EMF and coil inductance, the control signal $u(t) \sim I(t)$. Driving the coils dissipates a power $P = I^2 R$, with R the coil resistance. Integrating the power gives the total energy supplied to the motor as it delivers torque to the pendulum. Thus, the work to swing up the pendulum $\sim \int \mathrm{d}t\, u^2(t)$.

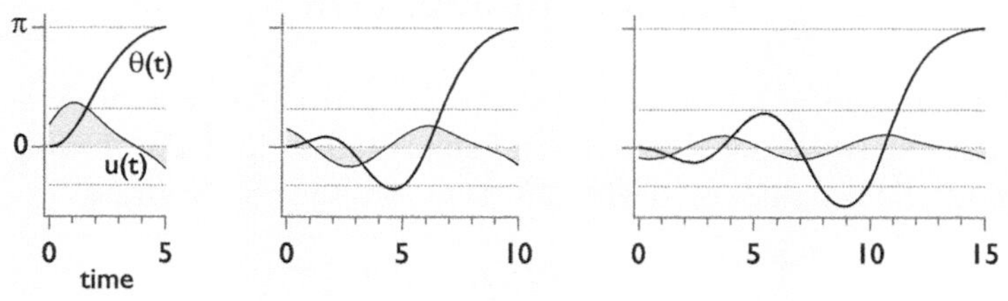

Fig. 7.2 Optimal control for pendulum swing up, for protocol times $\tau = 5$, 10, and 15. Thin solid lines with fill to zero depict the optimal control, $u(t)$. Solid lines show the pendulum angle, $\theta(t)$. Dotted lines at ± 1 delimit the weak-actuator regime, $|u| < 1$.

Figure 7.2 illustrates the solution for three values of protocol time τ. We defer discussion of the numerical methods used until Section 7.7. Notice how the optimal strategy shifts with the amount of time given for the task. The longer the protocol, the more it pays to reduce $u(t)$ and achieve the swing-up through several passes. For $\tau = 15$, the best strategy is to swing the pendulum initially in the "wrong" direction.

The dotted lines at ± 1 in each plot indicate the *weak-actuator* regime $|u| < 1$, where the magnitude of the external torque $mg\ell$, in physical units for a pendulum of mass m and length ℓ in gravity g, is less than that needed to counter the gravitational torque. (See right.) If we confine ourselves to torques less than 1, then we need a more sophisticated strategy than local feedback control, which accounts only for the current value of the state, $\boldsymbol{x} = (\theta \ \ \dot{\theta})^{\mathsf{T}}$.

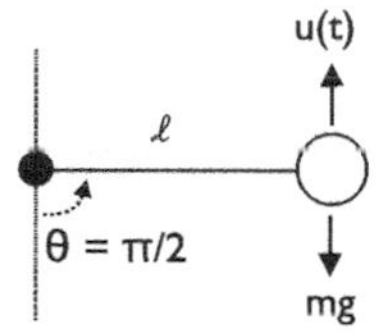

In Figure 7.2, the optimal control for $\tau = 10$ and 15 always lies in the region $|u| < 1$, but the signal briefly exits the region for $\tau = 5$. Swinging up in shorter times requires more control effort (torque). If our motor were limited to $|u| < 1$, then the optimal solution that we found for $\tau = 5$ would not be realizable. One solution to this problem is to learn how to optimize in the presence of hard constraints (Section 7.5). The solution found will have a higher value of J but will respect the limit on u. Or, we could simply increase τ by a small amount, so that the solution for $u(t)$ remains in the region $|u| < 1$.

7.3 Linear Quadratic Regulator

One of the few optimal-control problems that can be solved analytically is the case where the dynamics $\dot{\boldsymbol{x}} = \boldsymbol{f}(\boldsymbol{x}, \boldsymbol{u})$ is given by a *linear* function $\boldsymbol{f}$, the cost function is a *quadratic* function L, and the control has a goal of *regulation*: LQR. A nice feature of state-space notation is that the solution for the full multiple-input-multiple-output (MIMO) problem is formally the same as for the single-input-single-output (SISO) case. We begin with continuous dynamics and then quickly consider the discrete case, which is similar.

7.3.1 Continuous LQR

Consider regulating an n-dimensional state vector $\boldsymbol{x}(t)$ about $\boldsymbol{x} = \boldsymbol{0}$ by applying m inputs $\boldsymbol{u}(t)$. Leaving situations where the end-state matters to the problems, we define L to place quadratic penalties on deviations from zero for both $\boldsymbol{x}$ and $\boldsymbol{u}$:

$$J = \int_0^\tau \mathrm{d}t\,\tfrac{1}{2}\left(\boldsymbol{x}^\mathsf{T}\boldsymbol{Q}\boldsymbol{x} + \boldsymbol{u}^\mathsf{T}\boldsymbol{R}\boldsymbol{u}\right), \tag{7.12}$$

where $\boldsymbol{Q}$ and $\boldsymbol{R}$ are symmetric matrices of cost functions, usually chosen to be diagonal. We assume a positive-definite matrix $\boldsymbol{R}$, to keep control efforts finite, but require $\boldsymbol{Q}$ to be only positive semidefinite – only a single nonzero diagonal element is needed.[7] From Eq. (7.8), the equations of motion are

$$\dot{\boldsymbol{x}} = \boldsymbol{A}\boldsymbol{x} + \boldsymbol{B}\boldsymbol{u}\,, \qquad \dot{\boldsymbol{\lambda}} = -\boldsymbol{Q}\boldsymbol{x} - \boldsymbol{A}^\mathsf{T}\boldsymbol{\lambda}\,, \qquad \boldsymbol{u} = -\boldsymbol{R}^{-1}\boldsymbol{B}^\mathsf{T}\boldsymbol{\lambda}\,. \tag{7.13}$$

In Eq. (7.13), the controls $\boldsymbol{u}(t)$ depend on the adjoint $\boldsymbol{\lambda}(t)$. Notice that the dynamics of $\boldsymbol{x}$ is governed by the matrix $\boldsymbol{A}$ while that of $\boldsymbol{\lambda}$ is governed by its negative adjoint, $-\boldsymbol{A}^\mathsf{T}$. The great simplification of linear dynamics is that the optimal-control solution has $\boldsymbol{\lambda} \propto \boldsymbol{x}$, so that we can write the optimal control law as a linear feedback relation $\boldsymbol{u} = -\boldsymbol{K}\boldsymbol{x}$, with the gain $\boldsymbol{K}$ a time-dependent, $n \times m$ matrix. Optimal control gives a simple way of choosing these nm parameters based on meaningful criteria (as expressed in J).

For our quadratic-cost, linear-dynamics problem, we can decouple the state and adjoint equations by introducing the $n \times n$ matrix $\boldsymbol{S}(t)$, defined by $\boldsymbol{\lambda}(t) = \boldsymbol{S}(t)\,\boldsymbol{x}(t)$. Then[8]

$$\boldsymbol{u} = -\boldsymbol{R}^{-1}\boldsymbol{B}^\mathsf{T}\boldsymbol{S}\boldsymbol{x} \equiv -\boldsymbol{K}\boldsymbol{x}\,. \tag{7.14}$$

If we differentiate $\boldsymbol{\lambda} = \boldsymbol{S}\boldsymbol{x}$ with respect to time, the adjoint equation becomes

$$\dot{\boldsymbol{\lambda}} = -\boldsymbol{Q}\boldsymbol{x} - \boldsymbol{A}^\mathsf{T}\underbrace{\boldsymbol{S}\boldsymbol{x}}_{\boldsymbol{\lambda}} = \dot{\boldsymbol{S}}\boldsymbol{x} + \boldsymbol{S}\underbrace{\left(\boldsymbol{A}\boldsymbol{x} - \boldsymbol{B}\boldsymbol{R}^{-1}\boldsymbol{B}^\mathsf{T}\boldsymbol{S}\boldsymbol{x}\right)}_{\dot{\boldsymbol{x}}}\,. \tag{7.15}$$

Factoring $\boldsymbol{x}(t)$ and remembering that Eq. (7.15) must hold for all times, we have

$$\dot{\boldsymbol{S}} = -\boldsymbol{Q} - \boldsymbol{A}^\mathsf{T}\boldsymbol{S} - \boldsymbol{S}\boldsymbol{A} + \boldsymbol{S}\boldsymbol{B}\boldsymbol{R}^{-1}\boldsymbol{B}^\mathsf{T}\boldsymbol{S}\,, \tag{7.16}$$

Transposing Eq. (7.16) and recalling that $\boldsymbol{Q}$ and $\boldsymbol{R}$ are symmetric matrices implies that the symmetry of $\boldsymbol{S}(\tau)$ is conserved. Thus, $\boldsymbol{S}(t)$ is symmetric for $t < \tau$, since $\boldsymbol{S}(\tau) = \boldsymbol{0}$ is a symmetric matrix. Equation (7.16) is known as the *continuous time matrix Riccati equation.*[9] Because Eq. (7.16) is independent of the states $\boldsymbol{x}(t)$, it can be computed ahead of time, if desired.

[7] And if we added an end cost $\varphi[\boldsymbol{x}(\tau)]$, we could even set $\boldsymbol{Q} = \boldsymbol{0}$, as we did in the pendulum swing-up problem.

[8] In the SISO case, the matrix $\boldsymbol{K}$ becomes a row vector.

[9] Equation (7.23) is named for *Jacopo Riccati* (1676–1754), a Venetian scholar (and Count) who studied its continuous, one-dimensional version, $\dot{s}(t) = a(t) + b(t)s + s^2$. The Riccati equation is a nonlinear first-order equation but may be reduced to a linear second-order equation by the substitution $s = -\dot{u}/u$, which implies $\ddot{u} - b\dot{u} + au = 0$. Note that a term $c(t)s^2$ can be changed to s^2 by rescaling t and s. Similar

Alternatively, if you do not care about transients, let $\tau \to \infty$ and solve for the steady-state matrix $\boldsymbol{S}$, which obeys the *algebraic Riccati equation*,

$$-\boldsymbol{Q} - \boldsymbol{A}^{\mathsf{T}}\boldsymbol{S} - \boldsymbol{S}\boldsymbol{A} + \boldsymbol{S}\boldsymbol{B}\boldsymbol{R}^{-1}\boldsymbol{B}^{\mathsf{T}}\boldsymbol{S} = \boldsymbol{0} \tag{7.17}$$

Because Eq. (7.17) is quadratic, it has multiple solutions; however, only one will lead to the positive-definite gains $\boldsymbol{K}$ needed for a minimum in the cost function. Conversely, such a solution will always exist, and the LQR solution thus leads to stable closed-loop dynamics. Unfortunately, when the state is estimated via an observer and the system differs slightly from its model, the controller may be unstable (Skogestad and Postlethwaite, 2005).

For n states and m inputs, the LQR solution for the optimal feedback $\boldsymbol{K}$ is an $m \times n$ matrix. This result gives us another perspective on the PID control introduced in Chapter 3. Since that controller has three terms, it can be optimal only for a system with three states and one input. When the dynamics are more complex, the optimal controller is, too.

Example 7.3 (One-dimensional control, revisited) As a quick check on the formalism, we generalize the one-dimensional problem of Section 7.1 to finite-time protocols. Recall that $\boldsymbol{A} = -a$, $\boldsymbol{B} = \boldsymbol{Q} = 1$, and $\boldsymbol{R} = R$. Then Eq. (7.16) becomes

$$\dot{S} = -1 + 2aS + \frac{S^2}{R}, \qquad S(\tau) = 0, \tag{7.18}$$

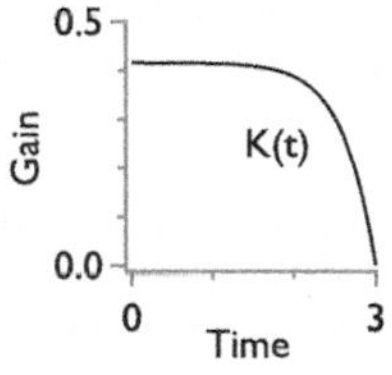

whose steady-state solution is $S = -aR \pm \sqrt{a^2R^2 + R}$. This solution holds for times far enough in the past from the final condition at $t = \tau$. The corresponding gain $K = R^{-1}(1)S = -a + \sqrt{a^2 + \frac{1}{R}}$ agrees with our previous result, Eq. (7.3). At right, for $a = R = 1$, we calculate the full time-dependent gain. Integrating back from the final time ($\tau = 3$), where $K = 0$, we see that K quickly reaches its steady-state value, $\sqrt{2} - 1$.

As before, the time constant for $K(t)$ reflects the time the system must plan ahead in order to meet the constraint at the final time. Intuitively, K must go to 0 as $t \to \tau$: the benefits of feedback are a *future* reduction in the error signal, but the costs are billed *now* in the function J. Thus, close to $t = \tau$, there will not be enough time left to benefit from the control, and K should be reduced. As a practical matter, it is common to use the steady-state gain as a simple suboptimal approximation. Whether this is reasonable depends on how much you are concerned with final-state costs relative to the integrated running costs.

tricks can simplify the matrix Riccati equation. The substitution $\boldsymbol{S}(t) = \boldsymbol{N}\boldsymbol{D}^{-1}$ leads to coupled linear equations for $\boldsymbol{N}(t)$ and $\boldsymbol{D}(t)$, but in general, they still have to be solved numerically.

7.3.2 Discrete LQR

The analysis for continuous systems has direct analogs for discrete dynamics. Indeed, the mathematics, while more complicated algebraically, is simpler conceptually. In the continuous case, we need to identify functions $\boldsymbol{u}(t)$ that are elements of an infinite-dimensional function space, whereas in the discrete case, we choose control inputs $\boldsymbol{u}_k$ at a finite number of instants in time. Thus, while the continuous problem requires the calculus of variations, the discrete equivalent uses only multivariable calculus. Because the analysis is similar, we state the main results here, leaving the derivations as an exercise.

In the discrete case, the analog of Eq. (7.6) for the augmented cost function J' is

$$J' = \frac{1}{2}\sum_{k=0}^{N}\left(\boldsymbol{x}_k^{\mathsf{T}}\boldsymbol{Q}\boldsymbol{x}_k + \boldsymbol{u}_k^{\mathsf{T}}\boldsymbol{R}\boldsymbol{u}_k\right) + \boldsymbol{\lambda}_{k+1}^{\mathsf{T}}(-\boldsymbol{x}_{k+1} + \boldsymbol{A}\boldsymbol{x}_k + \boldsymbol{B}\boldsymbol{u}_k)\,, \tag{7.19}$$

where the Lagrange multipliers $\boldsymbol{\lambda}_{k+1}$ enforce the linear dynamics $\boldsymbol{x}_{k+1} = \boldsymbol{A}\boldsymbol{x}_k + \boldsymbol{B}\boldsymbol{u}_k$ at every time step k.[10] The matrices $\{\boldsymbol{A}, \boldsymbol{B}, \boldsymbol{Q}, \boldsymbol{R}\}$ can be approximated from continuous versions (see Section 5.3) or specified directly.

We optimize J' by differentiating with respect to $\boldsymbol{x}_k$, $\boldsymbol{u}_k$, and $\boldsymbol{\lambda}_k$. From the derivatives $\partial_{\boldsymbol{x}_k} J' = \partial_{\boldsymbol{\lambda}_{k+1}} J' = \partial_{\boldsymbol{u}_k} J' = \boldsymbol{0}^{\mathsf{T}}$, we find the analog of Eq. (7.13):

$$\boldsymbol{x}_{k+1} = \boldsymbol{A}\boldsymbol{x}_k + \boldsymbol{B}\boldsymbol{u}_k\,, \qquad \boldsymbol{\lambda}_k = \boldsymbol{Q}\boldsymbol{x}_k + \boldsymbol{A}^{\mathsf{T}}\boldsymbol{\lambda}_{k+1}\,, \qquad \boldsymbol{u}_k = -\boldsymbol{R}^{-1}\boldsymbol{B}^{\mathsf{T}}\boldsymbol{\lambda}_{k+1}\,. \tag{7.20}$$

The equation for $\boldsymbol{x}_k$ has initial condition $\boldsymbol{x}_0$. As before, there is a final condition for $\boldsymbol{\lambda}_N$:

$$u_N = 0 \qquad \Rightarrow \qquad \boldsymbol{\lambda}_{N+1} = 0 \qquad \Rightarrow \qquad \boldsymbol{\lambda}_N = \boldsymbol{Q}\boldsymbol{x}_N\,, \tag{7.21}$$

$x_0 \rightarrow \rightarrow \rightarrow \rightarrow \rightarrow \rightarrow x_N$
$\lambda_0 \leftarrow \leftarrow \leftarrow \leftarrow \leftarrow \leftarrow \lambda_N$

Again, we have a mixed-boundary problem, with the equation for $\boldsymbol{x}_k$ having an initial condition and that for $\boldsymbol{\lambda}_k$ having a final condition, as shown schematically at left.

Again, we decouple the adjoint equation via the matrix $\boldsymbol{S}_k$, with $\boldsymbol{\lambda}_k \equiv \boldsymbol{S}_k\boldsymbol{x}_k$:

$$\boldsymbol{u}_k = -\underbrace{\left(\boldsymbol{B}^{\mathsf{T}}\boldsymbol{S}_{k+1}\boldsymbol{B} + \boldsymbol{R}\right)}_{\boldsymbol{R}^*}{}^{-1}\boldsymbol{B}^{\mathsf{T}}\boldsymbol{S}_{k+1}\boldsymbol{A}\boldsymbol{x}_k \quad \equiv -\boldsymbol{R}^{*-1}\boldsymbol{B}^{\mathsf{T}}\boldsymbol{S}_{k+1}\boldsymbol{A}\boldsymbol{x}_k\,. \tag{7.22}$$

Then $\boldsymbol{S}_k$ obeys the *discrete Riccati equation*, $\boldsymbol{K}_k$ is determined from $u_k = -\boldsymbol{K}_k\boldsymbol{x}_k$, and both are evaluated recursively, going backward from final conditions:

$$\begin{aligned} \boldsymbol{S}_k &= \boldsymbol{A}^{\mathsf{T}}\left[\boldsymbol{S}_{k+1} - \boldsymbol{S}_{k+1}\boldsymbol{B}(\boldsymbol{R}^*)^{-1}\boldsymbol{B}^{\mathsf{T}}\boldsymbol{S}_{k+1}\right]\boldsymbol{A} + \boldsymbol{Q}\,. \\ \boldsymbol{K}_k &= (\boldsymbol{R}^*)^{-1}\boldsymbol{B}^{\mathsf{T}}\boldsymbol{S}_{k+1}\boldsymbol{A}\,, \\ \boldsymbol{S}_N &= \boldsymbol{Q}\,, \quad \boldsymbol{K}_N = \boldsymbol{0}\,. \end{aligned} \tag{7.23}$$

[10] We choose the index of λ to be $k+1$ rather than k for convenience. Remember that we are free to choose the index as we wish, as long as we are careful about the boundary conditions.

For long-time problems with $N \to \infty$, the gains $\boldsymbol{K}_k \to \boldsymbol{K}$, a constant, determined from the steady-state solution to the Riccati equation for $\boldsymbol{S}$. With $\boldsymbol{S}_k = \boldsymbol{S}_{k+1} \equiv \boldsymbol{S}$, we find the *discrete algebraic Riccati equation* and the discrete LQR feedback gain $\boldsymbol{K}$:

$$\boldsymbol{S} = \boldsymbol{A}^{\mathsf{T}}\left[\boldsymbol{S} - \boldsymbol{S}\boldsymbol{B}(\boldsymbol{R}^*)^{-1}\boldsymbol{B}^{\mathsf{T}}\boldsymbol{S}\right]\boldsymbol{A} + \boldsymbol{Q}$$

$$\boldsymbol{K} = (\boldsymbol{R}^*)^{-1}\boldsymbol{B}^{\mathsf{T}}\boldsymbol{S}\boldsymbol{A}\,. \tag{7.24}$$

As with the continuous case, control software programs have built-in routines to solve for the time-independent gains $\boldsymbol{K}$ given a linear system $\{\boldsymbol{A}, \boldsymbol{B}\}$ and weights $\{\boldsymbol{Q}, \boldsymbol{R}\}$, making the computation of optimal gain easy in practice. Again, the Riccati equation will have multiple solutions, but $\boldsymbol{S}$ is positive definite for only one of them.

7.4 Dynamic Programming

Sections 7.2 and 7.3.2 show that the solution to the optimal control problem leads to a backwards differential (or discrete recurrence) equation backwards for the adjoint, $\lambda(t)$: Starting from a final condition at $t = \tau$, we integrate back to $t = 0$. But why backward dynamics? We suggested previously that the adjoint incorporates information from the future and reflects the need for *planning*. To understand these claims more deeply, we consider some basic ideas from *dynamic programming*, a technique pioneered by Richard Bellman in the 1950s at the Rand Corporation. Beyond the insight into connections with planning, dynamic programming also leads to a fourth strategy for solving optimal control problems, in addition to the three introduced in Section 7.2. This fourth strategy will involve solving the *Bellman equation*, the fundamental relation of dynamic programming.

7.4.1 Shortest Path between Cities

To understand the role of planning, we introduce an apparently unrelated problem: determining the *shortest path* between cities. Consider the set of possible journeys from Start → End, passing through various "cities." The towns are labeled a_1 and a_2 for the first day and b_1 and b_2 for the second. They are connected by roads, with distances indicated in Figure 7.3. The goal is to find the shortest path going from Start → End.

The strategy for working out the shortest path is based on the idea that if you know the best trajectory and are partway along it, the rest of the trajectory is the best trajectory starting from where you are. This is Bellman's *principle of optimality*:

> For any point on an optimal trajectory, the remaining trajectory is optimal, starting at that point.

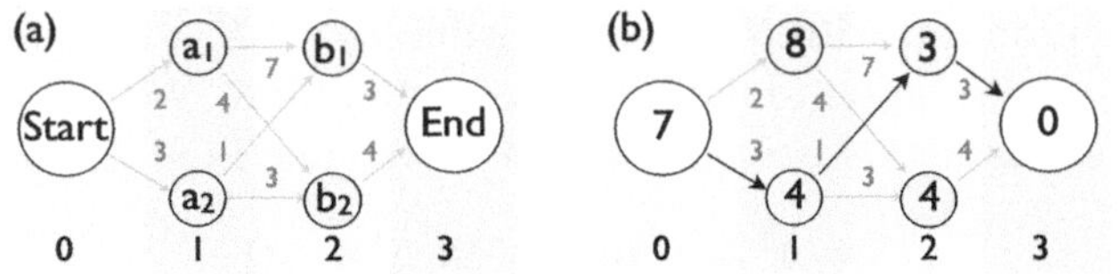

Fig. 7.3 Dynamic programming can find the shortest path from Start to End. (a) Graph of cities (nodes) and distances (edges), not to scale. (b) Minimum cost-to-go for each city is found by backwards recursion and is indicated by the corresponding node. The path segment used in each stage is in black; the unused path segments are in gray. The optimal path is read out by tracing forward the pointer states. Minimum cost $J^*(\text{Start}) = 3 + 1 + 3 = 7$.

To find the best route in Figure 7.3, we define $J(x)$ to be the *cost-to-go function*, the distance from town x to the destination *End* along a specified route (graph edges). Applying the principle of optimality, we can deduce an optimal path, working back from the end:

- $J^*(End) = 0$. (You are already there.)
- $J^*(b_1) = 3$ and $J^*(b_2) = 4$. (There is only possibility.)
- Going back one step in time, we have (shading light gray the actual minimum)
  - $J^*(a_1) = \min\{7 + J^*(b_1),\ 4 + J^*(b_2)\ \} = 8.$
  - $J^*(a_2) = \min\{\ 1 + J^*(b_1)\ ,\ 3 + J^*(b_2)\} = 4.$
- Finally, $J^*(Start) = \min\{2 + J^*(a_1),\ 3 + J^*(a_2)\ \} = 7.$

Having enumerated the value (optimal cost-to-go) for each node via the *backward sweep* from *End* $\to$ *Start*, we find the minimum cost to *End*, beginning from *Start* (equal to 7). To find the optimal path itself, we carry out a *forward sweep* from *Start* $\to$ *End*, choosing the option at each stage corresponding to the minimum cost to go, using the gray shading. Here, the path is from *Start* $\to a_2 \to b_1 \to$ *End*, with a cost $3 + 1 + 3 = 7$.

7.4.2 Bellman Equation: Discrete Case

Let us map the travel-planning problem back to our optimal-control language, for a discrete, one-dimensional dynamics with a single control:

- x_k = city at time k, for example a_1 or a_2.
- u_k = road choice at time step k. For example, if the state at time 1 is a_1, then u_1 is either "choose b_1" or "choose b_2".
- $L(x_k, u_k) \equiv L_k$ = running cost of the current step, given x_k and u_k.

We then see that the shortest-path problem is really just a simple optimal-control problem. To understand what to do more generally, the first insight is that cost functions that are the sum of independent running costs can be formulated recursively. The total cost is then

$$J = \sum_{n=0}^{N-1} L(\boldsymbol{x}_n, \boldsymbol{u}_n) + \varphi(\boldsymbol{x}_N), \tag{7.25}$$

where we consider a finite time horizon $[0, N]$ with terminal costs imposed on the state at time N.[11] Let J_k be the *cost-to-go*, or *future cost* starting at time k. We write

$$J_k = \sum_{n=k}^{N-1} L(\boldsymbol{x}_n, \boldsymbol{u}_n) + \varphi(\boldsymbol{x}_N) = L(\boldsymbol{x}_k, \boldsymbol{u}_k) + \underbrace{\sum_{n=(k+1)}^{N-1} L(\boldsymbol{x}_n, \boldsymbol{u}_n) + \varphi(\boldsymbol{x}_N)}_{J_{k+1}} . \tag{7.26}$$

Thus, the additive structure of the cost-to-go function in conjunction with the terminal cost $\varphi(\boldsymbol{x}_N)$ together lead to a *backwards* recursion relation. The full problem then combines the backward recursion for the cost with the forward relation for the state:

$$\begin{aligned} J_k &= L(\boldsymbol{x}_k, \boldsymbol{u}_k) + J_{k+1}, & J_N &= \varphi(\boldsymbol{x}_N) \\ \boldsymbol{x}_{k+1} &= \boldsymbol{f}(\boldsymbol{x}_k, \boldsymbol{u}_k), & \boldsymbol{x}_0 &\text{ given}. \end{aligned} \tag{7.27}$$

The second insight is that the recursive structure of the cost function J suggests a recursive strategy for solving the optimal-control problem. Imagine that we have already solved the problem starting at time $k+1$, finding an optimal sequence of controls $\boldsymbol{u}_{k+1}, \boldsymbol{u}_{k+2}, \ldots$ and corresponding states along the optimal path, $\boldsymbol{x}_{k+1}, \boldsymbol{x}_{k+2}, \ldots$. Let J_k^* denote the cost-to-go function calculated using the optimal controls and resulting optimal states. For brevity, we refer to $J_k^* = J^*(\boldsymbol{x}_k)$ as the *value* function at time k.[12] The value J^* is a function of $\boldsymbol{x}_k$, which we can think of as an initial condition for the reduced problem starting at time k. Then, to find the best course of action at time k, we need only look at each *admissable* control choice $\boldsymbol{u}_k$ and consider the sum of the current cost it imposes plus the *precalculated* optimal future costs (value) associated with the state $\boldsymbol{x}_{k+1}$ that results from the choice $\boldsymbol{u}_k$:

$$J^*(\boldsymbol{x}_k) = \min_{\{\boldsymbol{u}_k\}} \left[L(\boldsymbol{x}_k, \boldsymbol{u}_k) + J^*(\boldsymbol{x}_{k+1}) \right] . \tag{7.28}$$

Equation (7.28) is the (discrete) Bellman equation. Its meaning is illustrated schematically at right. The dashed lines indicate the value $J(\boldsymbol{x})$ for two possible states at time 1, $\boldsymbol{x}_1$, which would result from the control $\boldsymbol{u}_0$, and the state $\boldsymbol{x}_1'$, which would result from the control $\boldsymbol{u}_0'$. The Bellman equation tells us to choose the control $\boldsymbol{u}_0$: although its running cost is higher, the value of the state $\boldsymbol{x}_1$ is low enough that the overall optimal choice is $\boldsymbol{u}_0$, with a total cost of $2 + 1 < 1 + 3$. Such situations capture the essence of planning: it can be better to choose a higher-cost activity now that will lower overall future costs.

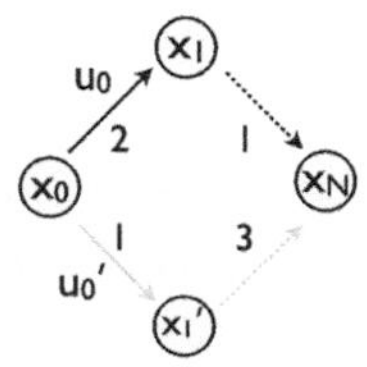

[11] The sum goes to $N-1$ and not N because $u_N = 0$, since u_N only affects u_{N+1}.

[12] The best value has the lowest cost.

Notice that in passing from the backwards recursion relation in Eq. (7.26) to the Bellman equation, Eq. (7.28), we implicitly assume a *state structure* for the dynamics. That is, knowing the state $\boldsymbol{x}_k$ and the control $\boldsymbol{u}_k$, we can use the dynamical rule $\boldsymbol{f}(\boldsymbol{x}_k, \boldsymbol{u}_k)$ to generate the next state, $\boldsymbol{x}_{k+1}$. If the dynamical model is incorrect, then the combination $(\boldsymbol{x}_k, \boldsymbol{u}_k)$ does not determine $\boldsymbol{x}_{k+1}$ and the minimization in Eq. (7.28) would have to be over not just $\boldsymbol{u}_k$ but also over future control choices, with $J^*(\boldsymbol{x}_{k+1})$ replaced by J_k.

To find the best path to the final time N, we thus determine the optimal control $\boldsymbol{u}_k^*$ backwards step by step to the present, as we did in Section 7.3.2 in terms of the adjoint vector $\boldsymbol{\lambda}_k$. We then apply the $\boldsymbol{u}_k^*$ to realize the optimal path $\boldsymbol{x}_k^*$ by integrating forwards. As the 19th-century Danish philosopher Søren Kierkegaard observed,[13]

> Life must be understood backwards but lived forwards.

The most important lesson of dynamic programming is the value of planning: the optimal control path is not the *greedy solution* that minimizes each stage's running costs. What is better in the short run may not be good in the long run. Planning ahead is essential.

The dynamic programming algorithm is also much more efficient than a naive search. To estimate the number of operations involved, assume that there are (on average) n nodes at each step. Each node has n possibilities for the next layer, implying $O(n^2)$ operations per time step. With N time steps, the total is $O(Nn^2)$. By contrast, a naive search from the starting point would examine $O(n^N)$ possible paths. For large N, the difference is huge!

The concepts behind dynamic programming were discovered independently in several fields, making a big impact each time. For example, the *Viterbi* algorithm,[14] an error-correction scheme for noisy digital communications links, made cell-phone communications possible. The *Needleman–Wunsch* and *Smith–Waterman* algorithms are used in bioinformatics to align (genetic) sequences. In economics, dynamic programming is used to develop policies for managing finances, resource extraction, budgeting, and so forth.

Dynamic programming assumes that the problem is decomposable into stages, with no loops in dependencies. Dynamic programming is natural for control problems, because time leads to natural stages. The structure of local cost functions permits replacing a simultaneous optimization over all variables with a sequential optimization (Problem 7.12). The recurring theme is to replace a single hard problem by a sequence of easier ones.

[13] Although this phrase is commonly attributed to Kierkegaard, he apparently never stated the idea in quite such a pithy form. In an early journal, he wrote, "Philosophy is perfectly right in saying that life must be understood backwards. But then one forgets the other clause – that it must be lived forwards." See *Journals* IV A 164 *n. d.*, 1843.

[14] In 1985, Andrew Viterbi became a founder of the Qualcomm Corporation. His 1967 algorithm is used in billions of cell phones, as well as satellite TV receivers, cable TV decoders, and the like.

7.4.3 Bellman Equation: Continuous Case

We can generalize the previous arguments to continuous dynamics with n-dimensional state space $\boldsymbol{x}(t)$ and m-dimensional inputs $\boldsymbol{u}(t)$. Define the scalar cost-to-go function,

$$J(\boldsymbol{x},\boldsymbol{u},t) = \varphi[\boldsymbol{x}(\tau)] + \int_t^{\tau} \mathrm{d}t'\, L[\boldsymbol{x},\boldsymbol{u}]\,, \quad \text{and} \quad J^*(\boldsymbol{x},t) = \inf_{\boldsymbol{u}(t\le t'\le\tau)} J(\boldsymbol{x},\boldsymbol{u},t)\,, \tag{7.29}$$

from time t to final time τ (including the end-state penalty φ), where $J^*(\boldsymbol{x})$ is the value function, which we get by choosing the optimal $\boldsymbol{u}^*(t)$ over the time interval from t to τ. Applying the principle of optimality over a small time interval Δt, we can write

$$\begin{aligned}
J^*(\boldsymbol{x},t) &= \inf_{\boldsymbol{u}(t\le t'\le\tau)} \left[\varphi[\boldsymbol{x}(\tau)] + \int_t^{\tau} \mathrm{d}t'\, L[\boldsymbol{x},\boldsymbol{u}]\right] \\
&= \inf_{\boldsymbol{u}(t\le t'\le\tau)} \left[\int_t^{t+\Delta t} \mathrm{d}t'\, L[\boldsymbol{x},\boldsymbol{u}] + \varphi[\boldsymbol{x}(\tau)] + \int_{t+\Delta t}^{\tau} \mathrm{d}t'\, L[\boldsymbol{x},\boldsymbol{u}]\right] \\
&= \inf_{\boldsymbol{u}(t\le t'\le t+\Delta t)} \left[L(\boldsymbol{x},\boldsymbol{u})\,\Delta t + \inf_{\boldsymbol{u}(t+\Delta t\le t'\le\tau)} [J(\boldsymbol{x}+\dot{\boldsymbol{x}}\,\Delta t, \boldsymbol{u}(t+\Delta t), t)]\right], \\
&= \inf_{\boldsymbol{u}(t\le t'\le t+\Delta t)} [L(\boldsymbol{x},\boldsymbol{u})\,\Delta t + J^*(\boldsymbol{x}+\boldsymbol{f}\,\Delta t, t)]\,, \\
&\approx \inf_{\boldsymbol{u}(t\le t'\le t+\Delta t)} [L(\boldsymbol{x},\boldsymbol{u})\,\Delta t + J^*(\boldsymbol{x},t) + (\partial_t J^*)\,\Delta t + (\partial_{\boldsymbol{x}} J^*)\,\boldsymbol{f}\,\Delta t]\,,
\end{aligned} \tag{7.30}$$

where $\dot{\boldsymbol{x}} = \boldsymbol{f}(\boldsymbol{x},\boldsymbol{u})$. Notice that we have decomposed the minimization over $\boldsymbol{u}$ into two nested minimizations: we need to find $\boldsymbol{u}$ going backwards from $t=\tau$, with each earlier value depending on the future evolution of $\boldsymbol{u}$. Canceling $J^*(\boldsymbol{x},t)$ and letting $\Delta t \to 0$, we find

$$\partial_t J^*(\boldsymbol{x},t) + \inf_{\boldsymbol{u}} [L(\boldsymbol{x},\boldsymbol{u}) + (\partial_{\boldsymbol{x}} J^*)\, \boldsymbol{f}(\boldsymbol{x},\boldsymbol{u})] = 0\,, \tag{7.31}$$

the *Hamilton–Jacobi–Bellman* equation[15] (HJB), which generalizes the *Hamilton–Jacobi* equations of classical mechanics. The inf (minimum) is over $\boldsymbol{u}(t)$, or, equivalently, over the values of $\boldsymbol{u}$ directly. We integrate HBJ *backwards* in time from the final condition $J^*[\boldsymbol{x}(\tau),\tau] = \varphi(\boldsymbol{x}(\tau))$. Recall that the gradient is a row vector, so that $(\partial_{\boldsymbol{x}} J^*)\boldsymbol{f}$ is a scalar quantity. Finally, having worked out the optimal control $\boldsymbol{u}^*[\boldsymbol{x}(t)]$ at each time, we compute the state vector by integrating *forward* in time from $\boldsymbol{x}_0$: $\dot{\boldsymbol{x}} = \boldsymbol{f}[\boldsymbol{x}, \boldsymbol{u}^*[\boldsymbol{x}(t)]]$.

[15] In Section 7.5.2, we define the Hamiltonian function $H(\boldsymbol{x},\boldsymbol{u},\boldsymbol{\lambda}) = L(\boldsymbol{x},\boldsymbol{u},t) + \boldsymbol{\lambda}^T \boldsymbol{f}(\boldsymbol{x},\boldsymbol{u})$, so that we can also write $\partial_t J^*(\boldsymbol{x}) + \inf_{\boldsymbol{u}} H(\boldsymbol{x},\boldsymbol{u},\partial_{\boldsymbol{x}} J^*) = 0$. This is a common form of the HJB equation.

Example 7.4 (One-dimensional control, yet again.) Assume a cost-to-go function $J = \int_t^\tau \mathrm{d}t' \, \frac{1}{2}(x^2 + u^2)$, with $\dot{x} = -x + u$. The HJB equation is

$$\partial_t J^* = -\inf_u \left[\underbrace{\tfrac{1}{2}(x^2 + u^2)}_{L(x,u)} + (\partial_x J^*) \underbrace{(-x + u)}_{f(x,u)} \right] . \tag{7.32}$$

With no restriction on $u(t)$, we minimize by differentiating the right-hand side of Eq. (7.32) with respect to u, which gives $u^* = -\partial_x J^*$. Substituting u^* into the HJB equation, we have

$$\partial_t J^* = -\tfrac{1}{2}x^2 + \tfrac{1}{2}(\partial_x J^*)^2 + x(\partial_x J^*) . \tag{7.33}$$

We solve Eq. (7.33) by substituting $J^*(x,t) = \frac{1}{2}x^2 S(t)$, which leads to $\dot{S} = -1 + S^2 + 2S$, with $S(\tau) = 0$, as in Eq. (7.18). We deduce $u^* = -\partial_x J^* = -Sx \equiv -Kx$. We substitute $u^*[x(t)]$ into $\dot{x} = -x + u^*$ and integrate to find $x(t)$, $J^*[x(t), t]$, u^*, and x.

Why are there two seemingly different methods for solving the same optimal control problem? One uses the calculus of variations to solve for an optimal trajectory $\boldsymbol{x}(t)$ and adjoint $\lambda(t)$, with Lagrange multipliers enforcing the dynamical equations. The other, dynamical programming, solves the HJB equation for $J^*(\boldsymbol{x}, t)$. To understand more intuitively why there are two versions of optimal control, we can appeal to an optical analogy.

Recall that *geometrical optics* describes the propagation of light in two ways. One solves for the trajectories of light rays, which are lines in free space but curve in a variable-index medium $n(\boldsymbol{r})$. The trajectory of rays is determined by making stationary the *optical path length*, $\int \mathrm{d}s \, n(\boldsymbol{r})$, which is conceptually similar to our task of finding a control trajectory $u(t)$ that makes stationary a cost function. The other describes the propagation of light by describing the *wavefronts*, contours of equal phases of the complex electric (or magnetic) field. If we introduce a function $S(\boldsymbol{r}, t)$ such that $S(\boldsymbol{r}, t) = 0$ describes a wavefront at time t that is locally perpendicular to light rays that traverse the surface, then the Hamilton–Jacobi equation is a nonlinear partial differential equation for S describing the evolution of the shape of these wavefronts.[16] For example, in optics, a point source emits rays that diverge from a geometrical point and has spherical wavefronts, as shown at left.

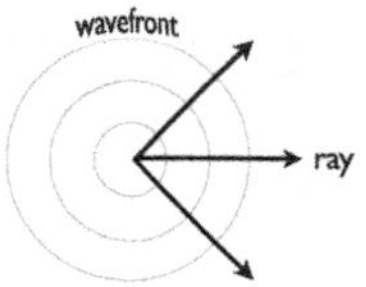

In the optimal-control problem, finding specific controls $\boldsymbol{u}(t)$ given an initial condition $\boldsymbol{x}_0$ is analogous to following the rays, while the HJB equation describes the wavefronts. An advantage of finding the wavefront surface is that we can then understand the properties of trajectories from *any* initial condition.[17] But, as you can appreciate already from Example 7.4, finding the wavefront surface is usually harder than finding an individual trajectory. Still, in any given problem, one or the other method might work better.

[16] In optics, the analog of the HJB equation is known as the *eikonal* equation.

[17] In Rolf Landauer's memorable phrase, the Hamilton–Jacobi equation for the action $S(\boldsymbol{r}, t)$ describes the paths of a "swarm" of particles (Landauer, 1952).

Finally, a global solution to the HJB equation gives necessary and sufficient conditions for an optimum, whereas the Euler–Lagrange method gives only necessary conditions, as it tests only a single trajectory. HJB tests all trajectories from all initial conditions.

7.5 Hard Constraints

So far, our discussion of optimal control has centered on the *soft constraints* imposed by cost functions. These constraints penalize unwanted behavior but do not forbid it. But some constraints are *hard*: they impose bounds that may not be crossed. In a sense, the dynamical equations are hard constraints and are imposed via a Lagrange multiplier. Another class of constraints places hard bounds, in the form of inequalities. The system may or may not actually be constrained by these inequalities, but any design must respect them.

One type of hard constraints concerns the values of the states themselves. For example, state variables should not exceed a threshold, to prevent damage. In this section, for simplicity, we will consider a different class of constraint, on the input function $\boldsymbol{u}(t)$.

A constant theme in our discussion of controller design has been the need to consider the required magnitude of control effort. Control algorithms that seem lovely on paper often turn out to require control signals well beyond the range that can be supplied. Our general strategy has been to limit the gains or shape the frequency response of the controller so that the required signals will usually not surpass the maximum available range of $\boldsymbol{u}(t)$.

Before suggesting more sophisticated strategies, do not forget *Method Zero*: if a controller saturates often, consider increasing the range of its actuator – by substituting a larger amplifier, a bigger motor, or whatever. In the indelicate language of control engineers, this tactic is known as *increasing control authority*. For example, simple temperature control uses resistive heaters and depends on losses to the environment for cooling. In controlling temperatures around ambient, the losses are very slow, and the controller will often saturate at $u = 0$ – no power to the heater – as the controller waits for the system to cool. A solution is to actively cool using cold water or Peltier elements.[18] These allow negative values of u.

Of course, money or technology will always limit the actuator range. Likewise, if the goal is to do something as fast as possible, the solution will usually be limited by the actuator range. We deal with such cases below. But many difficulties controlling systems trace back to an undersized actuator whose range can be extended. **Brute force can work!**

In this section, we will design controllers taking saturation into account. But adding hard constraints complicates the analysis and numerics significantly. Assess carefully

[18] Peltier elements are semiconductor devices that can cool as well as heat.

whether constraints are important before trying to implement the methods below. As a useful guide:

- Solve the optimal control problem in the absence of constraints.
- For typical time series coming from the unconstrained solution, identify all variables, whether states or controls, that violate the constraints.
- Classify the constraint violations into three categories:
  - The constraint is only rarely violated and by small amounts.
  - The constraint is always violated.
  - There is a mix of constrained and unconstrained behavior.
- If constraint violations occur, consider increasing the penalties in your cost function on states and control variables (and iterate back to the first step).

You can simply ignore rarely violated constraints, as they are unlikely to influence significantly the solution. Check this statement by running the solution computed without the constraint using a simulation that does implement the constraints. For always-violated constraints, the solution will typically always live at the constraint boundary for the variable. If you are lucky, the variable always "wants" to be higher (or lower) than the limit and you can simply fix it at that limiting value. In Section 7.5.2, we treat cases where it alternates between minimum and maximum limits a finite number of times. Problems where the goal is to do something as fast as possible (minimum-time problems) often have this feature. Finally, the mixed case requires a more careful treatment, and we outline some procedures below. The important point is to apply constraints sparingly, focusing on those with the most influence on behavior relative to the unconstrained problem.

Saturation of the Control Variable

Saturation of a control variable is a nonlinear effect, always present, and has nothing to do with the physical system itself, which might be linear. Moreover, the nonlinearity itself – a hard saturation when the input $\boldsymbol{u}$ exceeds upper or lower limits – differs from the kinds of nonlinearity typically found in the natural world, where the nonlinear function $\boldsymbol{f}(\boldsymbol{x}, \boldsymbol{u})$ that governs the dynamics usually has a smoother (differentiable) functional form. We thus study this particular kind of nonlinearity on its own.

For special situations such as Problem 7.16, we can use the HJB equation with constraints. We simply choose, at each time, the value of u that minimizes the relevant part of Eq. (7.31). Unfortunately, solving the resulting nonlinear partial differential equation is hard, and only limited classes of problems may be approached this way. In this section, we present two other approaches to the problem of accommodating input saturation in control design. The first is *anti-windup* control, a heuristic modification to the PID controller. Although not based on optimization, it is easy to implement and often works well.

A second approach generalizes optimal control using the *Pontryagin Minimum Principle* (PMP). The key point is that the constraint on the input variable is an *inequality*

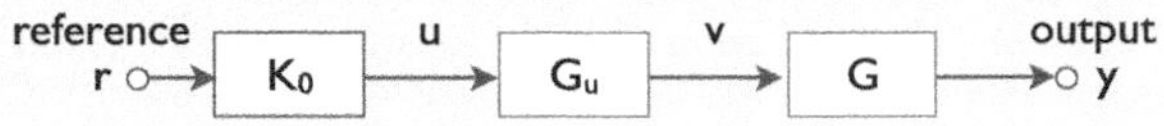

Fig. 7.4 Control signal u can change discontinuously. The u dynamics has transfer function G_u and generates the "slew-rate limited" signal v sent to the system G.

constraint – $\boldsymbol{u}$ lies within an allowed set of values (Appendix A.5.1 online). We will see that its value can change discontinuously. Typically, the switching is from one limit of $\boldsymbol{u}$ to another. Since there is no reason for $\boldsymbol{u}(t)$ to be a continuous function, taking advantage of the freedom to switch it abruptly can lead to better performance. Such discontinuous switching is known as *bang-bang control*: Bang! You go to one limit. Bang! You go to the other.

Discontinuities in the Control Variable

Discontinuities in optimal control, such as those appearing in bang-bang control, may seem curious in physics, where smooth evolution is the rule. In fact, control signals can have jump discontinuities. Unlike a physical system, which would require infinite energy to change its state discontinuously, the $\boldsymbol{u}$ signal simply reflects our "instructions" to the system. These can vary discontinuously, even if the system cannot respond instantly. Although engineers often refer to "slew rate" limitations to the control signal,[19] they really mean only that the signal electronics has a finite response time. But those electronics form a different system, which can be lumped together with the physical system under study (Figure 7.4).

7.5.1 Anti-Windup Schemes

Anti-windup is best illustrated by a simple example, based on the first-order system $\dot{x} = -ax + u$, with decay rate $a = 0.1$ and transfer function $G(s) = \frac{1}{s+a}$ (Section 2.3.2). Figure 7.5a shows the response to a step of amplitude $r = 3$, using a proportional control with gain $K_p = 1$. The signal $x(t)$ does not quite reach the reference value because of "proportional droop" (Section 3.3). Adding integral control, $1/s$, forces $x(t)$ to $r = 3$ (b). The controls in both cases, however, exceed the nominal limit $|u| = 1$. Enforcing that limit (c) leads to a large overshoot. Implementing an anti-windup scheme tames the overshoot (d).

To understand what goes wrong in (c), recall that an integral controller produces a controller signal $u(t) \propto \int_0^t \mathrm{d}t'\, e(t')$, where the error $e(t) = r - x(t)$. When the PI algorithm implies a value for $u(t)$ that exceeds its limits, the integral term keeps accumulating a large, positive value, $v(t)$, in Figure 7.5c. To switch the sign for $u(t)$, the state $x(t)$ must not only reach the set point (making $e = 0$), it must exceed it until the integral term is "erased." The integral term then switches sign and, aided by the proportional term, starts to return x to the set point. The nonlinear saturation in the controller signal has

19 The voltage slew rate at unity gain is a common specification for an op-amp integrated circuit.

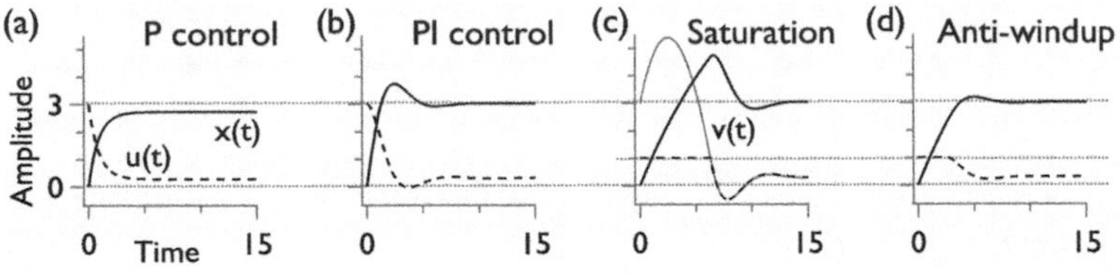

Fig. 7.5 Anti-windup control eliminates the overshoot due to saturation. (a) Proportional control. (b) Proportional-integral control. (c) Saturation makes the PI algorithm overshoot. (d) Anti-windup improves control. Thick solid line is state $x(t)$. Thick dashed line is control $u(t)$. Thin solid line in (c) is control before saturation, $v(t)$. Dotted lines show reference value $r = 3$, control saturation $u_{\max} = 1$, and zero.

led to a difference between the value of v called for by the PI algorithm and the value that is supplied to the system, $u(t)$. In effect, the control loop has been broken, and the system is operating in open loop.

The solution (d) is to "turn off the integral" term whenever the value of $v(t)$ would exceed the limits. Do not update the integral term – stop integrating – whenever the PI algorithm implies a control signal outside of the possible range for u (Problem 7.14).

7.5.2 Pontryagin Minimum Principle

Although anti-windup control can be effective, it is still a heuristic method. In this section, we introduce some basic ideas of optimal control with the inequality constraints for the range of the control inputs. To start, we assume that each input $u_i(t)$ $(i = 1, \cdots, m)$ is constrained so that $|u_i(t)| \leq 1$. That is, at any time t, $\boldsymbol{u}$ lies within an m-dimensional hypercube, $\mathcal{U}$, the set of allowed functions.[20] To get a feel for the kinds of behavior that we might expect, we first reconsider the one-dimensional example of Section 7.1.

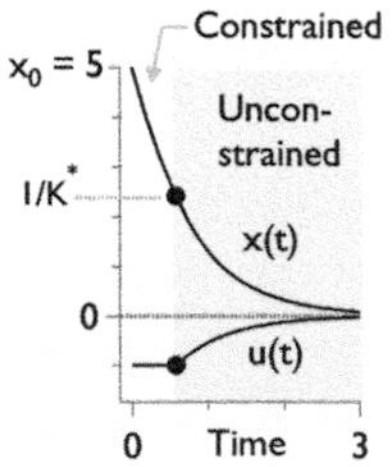

Example 7.5 (One-dimensional dynamics, with input constraints) As before, the dynamics $\dot{x} = -x + u$, with initial condition $x(0) = x_0$, running cost $L = \frac{1}{2}(x^2 + u^2)$, and total cost $J = \int_0^{\infty} \mathrm{d}t\, L(x, u)$. Now, we require, in addition, that $|u(t)| \leq 1$.

Recall the solution to the unconstrained problem (Problem 7.1, with $a = R = 1$). The optimal control $u(t)$ could be expressed as a linear feedback, $u = -Kx$, with optimal steady-state gain $K^* = \sqrt{2} - 1$. The state $x(t) = x_0\, \mathrm{e}^{-\sqrt{2}t}$, with input $u(t) = -K^*\, x_0\, \mathrm{e}^{-\sqrt{2}t}$.

When $x_0 > (K^*)^{-1} = \sqrt{2} + 1$, the control $u(0) < -1$, so that the constraint $u = -1$ is active at short times. Note the intuitive optimization result, at left: For short times, $u^* = -1$. At a time $t = \tau$, the solution switches to the unconstrained one (Problem 7.15).

[20] The exact shape – cube, sphere, or whatever – of the region defining the allowable $u_i(t)$ is not important. What counts is that it is bounded.

Hamiltonian Formalism

The control literature describes optimization by a formalism analogous to Hamiltonian mechanics. Recall that the Lagrangian formalism for n-dimensional coordinates $\boldsymbol{q}(t)$ and $\dot{\boldsymbol{q}}(t)$ uses a Lagrangian $L(\boldsymbol{q}, \dot{\boldsymbol{q}}) = T - U$, where T is the kinetic energy and U the potential. In the Hamiltonian formalism, we change variables from n generalized velocities $\dot{\boldsymbol{q}}$ to n generalized momenta $\boldsymbol{p} = \partial_{\dot{q}} L$, using a *Legendre transformation* $H = \boldsymbol{p} \cdot \dot{\boldsymbol{q}} - L$. The *Hamiltonian*, $H(\boldsymbol{q}, \boldsymbol{p})$ obeys $2n$ first-order equations,

$$\dot{\boldsymbol{q}} = \left(\partial_{\boldsymbol{p}} H\right)^{\mathsf{T}}, \qquad \dot{\boldsymbol{p}} = -\left(\partial_{\boldsymbol{q}} H\right)^{\mathsf{T}}, \tag{7.34}$$

with $H = T + U$ the total energy. If the Lagrangian does not depend explicitly on time, then neither does H, which is constant along the solution trajectories $\boldsymbol{x}(t)$ and $\boldsymbol{p}(t)$:

$$\mathrm{d}_t H = \left(\partial_{\boldsymbol{p}} H\right) \dot{\boldsymbol{p}} + \left(\partial_{\boldsymbol{q}} H\right) \dot{\boldsymbol{q}} = 0 . \tag{7.35}$$

In the control-theory version of the Hamiltonian formalism, the cost function is analogous to the Lagrangian, and $\boldsymbol{\lambda}$ plays the role of the canonical momenta. The analogy is imperfect, as the $\boldsymbol{\lambda}$ are already present in the Lagrangian $L + \boldsymbol{\lambda}^{\mathsf{T}}[-\dot{\boldsymbol{x}} + \boldsymbol{f}(\boldsymbol{x}, \boldsymbol{u})]$ as a constraint for the system dynamics $\dot{\boldsymbol{x}} = \boldsymbol{f}(\boldsymbol{x}, \boldsymbol{u})$. Also, the control version of the Hamiltonian formalism can accommodate dissipative forces, unlike the classical Hamiltonian. We then define the control Hamiltonian $H(\boldsymbol{x}, \boldsymbol{\lambda}, \boldsymbol{u})$,[21]

$$H = L + \boldsymbol{\lambda}^{\mathsf{T}} \dot{\boldsymbol{x}} = L + \boldsymbol{\lambda}^{\mathsf{T}} \boldsymbol{f} \tag{7.36}$$

and transform the Euler–Lagrange equations into their equivalents in terms of H:

$$\dot{\boldsymbol{x}} = (\partial_{\boldsymbol{\lambda}} H)^{\mathsf{T}} = \boldsymbol{f}, \qquad \dot{\boldsymbol{\lambda}} = -(\partial_{\boldsymbol{x}} H)^{\mathsf{T}}, \qquad (\partial_{\boldsymbol{u}} H)^{\mathsf{T}} = \boldsymbol{0} . \tag{7.37}$$

The adjoint states $\boldsymbol{\lambda}(t)$ play the role of the conjugate momenta in the classical-mechanics formulation.[22] In general, the state and adjoint vectors each have n components while the control $\boldsymbol{u}$ has m components. In Example 7.5, we have $H = \frac{1}{2}(x^2 + u^2) + \lambda(-x + u)$.

Equation (7.37) gives necessary conditions for an optimal solution. The first two resemble Hamilton's equations, but the inputs $\boldsymbol{u}(t)$ need to be specified. For unconstrained inputs, the equation $(\partial_{\boldsymbol{u}} H)^{\mathsf{T}} = \boldsymbol{0}$ is consistent with controls $\boldsymbol{u}(t)$ that minimize H. Indeed, even when $\boldsymbol{u}(t)$ are constrained and the constraints active, the control $\boldsymbol{u}(t)$ continues to minimize H. Thus, we define a quantity formally analogous to the Hamiltonian,

$$\mathcal{H}(\boldsymbol{x}, \boldsymbol{\lambda}) = \inf_{\boldsymbol{u} \in \mathcal{U}} H(\boldsymbol{x}, \boldsymbol{\lambda}, \boldsymbol{u}) \qquad \text{with} \quad \dot{\boldsymbol{x}} = (\partial_{\boldsymbol{\lambda}} \mathcal{H})^{\mathsf{T}}, \quad \dot{\boldsymbol{\lambda}} = -(\partial_{\boldsymbol{x}} \mathcal{H})^{\mathsf{T}}, \tag{7.38}$$

with mixed boundary conditions holding at initial ($t = 0$) and final ($t = \tau$) times:

$$\boldsymbol{x}(0) = \boldsymbol{x}_0 , \qquad \boldsymbol{\lambda}(\tau) = (\partial_{\boldsymbol{x}} \varphi)^{\mathsf{T}}\big|_{t=\tau} . \tag{7.39}$$

[21] The sign of L is opposite that of the Lagrangian from classical mechanics. The original work in this field (Pontryagin et al., 1964) used the traditional definition and led to what they called the *Maximum Principle*. Most control texts change the sign of L so that H is minimized when the cost function J is minimized.

[22] In an economics interpretation, $\boldsymbol{\lambda}^{\mathsf{T}}(t)$ plays the role of a current *price* of a resource $\boldsymbol{x}$ (Whittle, 1996).

The final condition at $t = \tau$ includes a penalty $\varphi(\boldsymbol{x})$, which is useful when you care about the final state of the system and how it got there. The minimization over $\boldsymbol{u}$ in Eq. (7.38) is done at each time, respecting the constraints on the allowed set of controls $\boldsymbol{u}(t)$. Bowing to common usage, we also refer to the unminimized control Hamiltonian H in Eq. (7.37) as the Hamiltonian. As before, $\dot{\boldsymbol{\lambda}}$ equation is integrated *backward* from $t = \tau$, and then the $\dot{\boldsymbol{x}}$ equation is integrated *forward* in time from $t = 0$.

Equation (7.38) expresses the famous *Pontryagin Minimum Principle* (PMP):[23] at each time t, choose the control $\boldsymbol{u}(t)$ that minimizes $H[\boldsymbol{x}(t), \boldsymbol{\lambda}(t), \boldsymbol{u}(t)]$. The condition for a local stationary point of H, given in Eq. (7.37), is replaced by the requirement for a global minimum. Intuitively, because the $\boldsymbol{u}$ variables do not have dynamics, choosing values that minimize H also minimize J. One feature of the solution is that the optimal $\boldsymbol{u}(t)$ are in general only piecewise continuous. That is, of course, admissible, as we are free to send any value we like to the controller, as long as it is within the high and low limits defined by $\mathcal{U}$. In previous examples (e.g., Example 7.5), we have seen that discontinuities in $\boldsymbol{u}(t)$ arise when the control switches from one constraint to another.

The PMP allows us to solve problems that would have no solution within the traditional analytical framework of calculus or the calculus of variations. (Cf. Section A.5.1 online.) For example, the derivative of the function $f(u) = u$ is never zero; however, if u is constrained to be positive, then f has a well-defined minimum at $u^* = 0$ (see left). The PMP provides a necessary condition for how the control variables $\boldsymbol{u}$ must be selected on the boundary set of their allowed domain. When the constraints on $\boldsymbol{u}$ are active, it is generally best to first impose the minimization condition by choosing $\boldsymbol{u}$ appropriately and then to solve the remaining Hamilton equations. If the constraints on $\boldsymbol{u}$ are not active, then there is no general rule to indicate which equation is best to solve first. Some final points:

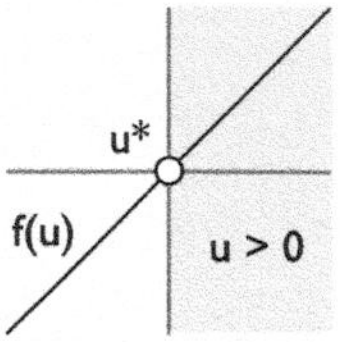

1. Since the PMP still involves first derivatives, it is a necessary, but not sufficient requirement for optimal control.
2. If the cost $L[\boldsymbol{x}, \boldsymbol{u}]$ and the dynamical function $\boldsymbol{f}(\boldsymbol{x})$ have no explicit time dependence, then H is constant (cf. Eq. 7.35). Imposing a constant H (that meets the boundary condition at the final time) is an alternative way to solve the control problem.
3. The PMP and Hamilton–Jacobi–Bellman (HJB) equations are, in principle, equivalent methods, but the PMP applies more widely, as $u(t)$ is not assumed to be continuous.
4. The PMP leads to $2n$ ordinary differential equations, whereas HJB leads to a partial differential equation in n space and 1 time dimension. Whereas discretizing $n + 1$

[23] The Soviet mathematician Pontryagin was blinded by an accident at the age of fourteen and was thereafter tutored by his mother, who had no education in mathematics and read and described the mathematical symbols as they appeared to her (Naidu, 2002). Nonetheless, Pontryagin became one of the leading mathematicians of the twentieth century, making major contributions to topology. Later in life, he turned his attention to engineering problems and was asked to solve a problem that arose in the context of the trajectory control of a military aircraft. The basic insights required to solve such problems with constraints on the control variables came after three consecutive, sleepless nights (Gamkrelidze, 1999).

dimensions for the HJB is expensive numerically relative to the 1 time dimension of the PMP equations, the mixed boundary conditions of the PMP can be challenging.

Example 7.6 (Moving in minimal time) As an example where the control is always at the boundary of allowed values (i.e., that the constraints are active), consider a free particle ("in space") that obeys Newton's laws with no friction and subject only to a controllable force, in one dimension. The equations of motion are $\dot{x}_1 = x_2$ and $\dot{x}_2 = u$, where x_1 is the particle's position, x_2 its velocity, and $u = F/m$ is the external force divided by the particle's mass. The goal is to bring the particle from an arbitrary initial state $\binom{x_1(0)}{x_2(0)}$ to the state $\binom{0}{0}$ as fast as possible. Assume that the applied force has limits that correspond to $|u(t)| \le 1$. Moving in minimal time then implies minimizing the cost $J = \int_0^\tau dt\,(1)$, where the final time τ is not fixed, with the equations of motion and $|u(t)| \le 1$ as constraints.

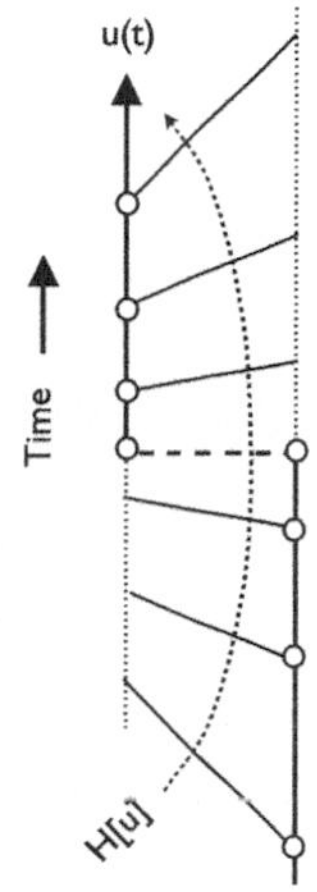

From Eqs. (7.36) and (7.37) and using $L = 1$, the control Hamiltonian is

$$H = 1 + \lambda_1 x_2 + \lambda_2 u\,, \quad \Longrightarrow \quad \dot{\lambda}_1 = 0\,,\ \dot{\lambda}_2 = -\lambda_1 \quad \Longrightarrow \quad \lambda_2(t) = -c_1 t + c_2\,. \tag{7.40}$$

Naively minimizing H by setting $\partial_u H = 0$ implies that $\lambda_1 = \lambda_2 = u = 0$ and hence that there is no nontrivial solution. But the linear function H *does* have a minimum if u is constrained to ± 1. Thus, $u = -\text{sign}(\lambda_2)$, implying that the applied force will always be pegged to its maximum value. Since $\lambda_2(t)$ varies linearly in time, it and thus $u(t)$ can make a (single) sign change, as illustrated at right.[24] Solving the equations of motion separately for $u = \pm 1$ shows that the system's dynamics follow a family of parabolas described by

$$\underbrace{x_1 = \tfrac{1}{2}x_2^2 + c_p\,,}_{u=+1} \qquad \underbrace{x_1 = -\tfrac{1}{2}x_2^2 + c_n\,,}_{u=-1} \tag{7.41}$$

with c_p and c_n determined by initial conditions.

The full solution is illustrated in the phase-space plot shown at right. Curves with downward-pointing arrows are generated with $u = -1$; those with upwards arrows are generated with $u = +1$. The dashed curve AOB passing through the end target point, the origin (with $c_p = c_n = 0$), has special significance. The left-hand segment AO is defined by motion with $u = -1$ and will end up at the target point (the origin). The right-hand segment, OB, is defined by a similar segment with $u = 1$. The optimal motion is then determined by evaluating the location of the system in phase space. If the state is below the curve AOB, choose $u = 1$, wait until the system state hits the curve AO, and then impose $u = -1$. If the state is above AOB, follow the reverse recipe. The curve AOB is thus known as the *switching curve*. The heavy black curve denotes the optimal solution for a particle initially at rest. The best strategy is to accelerate as fast as possible until halfway to the goal and then to decelerate as fast as possible the rest of the way, a strategy that is perhaps obvious (in retrospect). Note that formulating the problem in phase space leads to a local rule for the control – essential

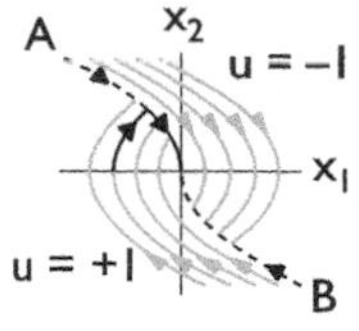

[24] An nth-order system linear in $\boldsymbol{x}$ and $\boldsymbol{\lambda}$ leads to n linear equations for the n-element vector $\boldsymbol{\lambda}$. By similar logic, there can be at most n sign changes and thus n discontinuities in the control.

for implementing a practical optimal *feedback* controller – based on the geometry of motion in phase space, as determined by the switching curve. This law is nonlinear even though the original problem seems to have linear dynamics. The nonlinearity has crept in through the saturation constraint.

Bang-Bang Control

In the previous section, we saw that bang-bang control is the optimal solution for control problems where the Hamiltonian is linear or independent of u, a situation that occurs particularly when you want to do something as fast as possible. Bang-bang control can also be cheap: a controller with two states (e.g., ON and OFF) can use simple relays, which can easily handle high powers. For industrial plants, there may be a production level that is most efficient, and it is then better to alternate between zero and that optimal level. Just remember: you can get "more bang-bang for the buck-buck."

7.6 Feedback

LQR, the linear quadratic regulator, leads to feedback control $\boldsymbol{u}(t)$ that is a function of the state $\boldsymbol{x}(t)$ only. In general, the optimal control will depend on both state and adjoint $\lambda(t)$ and cannot be calculated without considering future outcomes. Given an initial state $\boldsymbol{x}_0$, such a control can be calculated offline and implemented as feed-forward, which is simpler to calculate and implement than feedback, does not lead to instabilities, does not need an observer to supply missing state-vector components, and is not affected by measurement noise. But pure feedforward control is not robust: with no corrections, disturbances and modeling errors will cause the actual state to deviate quickly from the desired path. Some kind of feedback is essential! In this section, we present two complementary approaches: local linear feedback and model predictive control (MPC). The former is based on a Taylor expansion around the nominal feedforward dynamics. It is easy to formulate and solve, for it reduces to a slight generalization of the LQR theory of Section 7.3. The latter is based on the simple of idea of re-solving the full optimal-control problem at *each* time step, starting from the current state $\boldsymbol{x}(t)$. Unfortunately, MPC is computationally heavy and may not be stable. In a final subsection, we unite the two approaches, using local linear feedback supplemented by *event-triggered* MPC: whenever the system state deviates more than some critical amount away from its nominal value, we recompute the feedforward trajectory, starting from the current state.

Here, we focus on deviations that arise from external disturbances. While the feedback techniques used to counter such disturbances are also effective against modeling errors, we defer a systematic discussion of the latter to Chapter 9, on robust control.

7.6.1 Local Linear Feedback

Assume that we have solved, probably numerically, the nominal optimal-control problem defined in Section 7.2, where $J = \varphi[\boldsymbol{x}(\tau)] + \int_0^\tau \mathrm{d}t\, L(\boldsymbol{x},\boldsymbol{u})$, subject to dynamics $\dot{\boldsymbol{x}} = \boldsymbol{f}(\boldsymbol{x},\boldsymbol{u})$ for $t \in [0,\tau]$, with initial condition $\boldsymbol{x}(0) = \boldsymbol{x}_0$ and, possibly, final conditions at $t = \tau$. That is, we have found a nominal feedforward control $\boldsymbol{u}_{\mathrm{ff}}(t)$, along with the state evolution $\boldsymbol{x}_{\mathrm{ff}}(t)$ and adjoint $\boldsymbol{\lambda}_{\mathrm{ff}}(t)$. The feedforward control $\boldsymbol{u}(t)$ is independent of the state $\boldsymbol{x}(t)$, although it is calculated assuming a known initial condition $\boldsymbol{x}(0)$ (cf. Section 3.4). However, if there are external disturbances or if the model differs slightly from the actual dynamics, then small deviations will appear for $t > 0$. Let us therefore define

$$\boldsymbol{x}(t) = \boldsymbol{x}_{\mathrm{ff}}(t) + \delta\boldsymbol{x}(t)\,, \qquad \boldsymbol{u}(t) = \boldsymbol{u}_{\mathrm{ff}}(t) + \delta\boldsymbol{u}(t)\,, \qquad \boldsymbol{\lambda}(t) = \boldsymbol{\lambda}_{\mathrm{ff}}(t) + \delta\boldsymbol{\lambda}(t)\,. \tag{7.42}$$

Because the optimal solution is defined by making the first variation $\delta J = 0$, the lowest-order contribution will come from the second variation:

$$\delta^2 J = \tfrac{1}{2}\,\delta\boldsymbol{x}^{\mathrm{T}}\,(\partial_{\boldsymbol{xx}}\varphi)\,\delta\boldsymbol{x}\Big|_{t=\tau} + \int_0^\tau \mathrm{d}t \begin{pmatrix}\delta\boldsymbol{x}^{\mathrm{T}} & \delta\boldsymbol{u}^{\mathrm{T}}\end{pmatrix}\begin{pmatrix}\boldsymbol{Q} & \boldsymbol{M}\\ \boldsymbol{M} & \boldsymbol{R}\end{pmatrix}\begin{pmatrix}\delta\boldsymbol{x}\\ \delta\boldsymbol{u}\end{pmatrix}, \tag{7.43}$$

where $\boldsymbol{Q}$, $\boldsymbol{M}$, and $\boldsymbol{R}$ are all evaluated along the nominal optimal trajectory, $[\boldsymbol{x}_{\mathrm{ff}}(t), \boldsymbol{u}_{\mathrm{ff}}(t)]$:

$$\boldsymbol{Q}(t) = \partial_{\boldsymbol{xx}}L\,, \qquad \boldsymbol{M}(t) = \partial_{\boldsymbol{xu}}L\,, \qquad \boldsymbol{R}(t) = \partial_{\boldsymbol{uu}}L\,. \tag{7.44}$$

The augmented cost perturbation is then

$$\delta^2 J' = \delta^2 J + \delta\boldsymbol{\lambda}^{\mathrm{T}}\,(\boldsymbol{A}\,\delta\boldsymbol{x} + \boldsymbol{B}\,\delta\boldsymbol{u} - \delta\dot{\boldsymbol{x}})\,, \qquad \boldsymbol{A}(t) = \partial_{\boldsymbol{x}}\boldsymbol{f}\,, \quad \boldsymbol{B}(t) = \partial_{\boldsymbol{u}}\boldsymbol{f}\,. \tag{7.45}$$

where derivatives are again evaluated along the nominal optimal trajectory, $[\boldsymbol{x}_{\mathrm{ff}}(t), \boldsymbol{u}_{\mathrm{ff}}(t)]$.

The Euler–Lagrange equations for the perturbed cost are

$$\delta\dot{\boldsymbol{\lambda}} = -\boldsymbol{Q}\,\delta\boldsymbol{x} - \boldsymbol{M}\,\delta\boldsymbol{u} - \boldsymbol{A}^{\mathrm{T}}\delta\boldsymbol{\lambda}\,, \qquad \delta\boldsymbol{u} = -\boldsymbol{R}^{-1}\left(\boldsymbol{M}\,\delta\boldsymbol{x} + \boldsymbol{B}^{\mathrm{T}}\delta\boldsymbol{\lambda}\right). \tag{7.46}$$

Thus, to first order, small variations obey the same kind of linear equations we solved for the LQR problem, with two minor differences: The cost function and dynamical matrices all depend on time, and the cost function can have a cross term $\boldsymbol{M}(t)$.

Notice that a quadratic cost function arises, even when the original cost function $L(\boldsymbol{x},\boldsymbol{u})$ is not quadratic itself. Neither the time dependence nor the cross term leads to any fundamental changes. We will see this in a particular example, but it is easy to understand why: Because the problem is linear (and costs are quadratic), the control perturbation $\delta\boldsymbol{u}(t)$ depends only on $\delta\boldsymbol{x}(t)$ and does not require computations into the future. Thus, all that counts is the values of the LQR matrices at time t.

More formally, let $\delta\boldsymbol{\lambda}(t) = \boldsymbol{S}(t)\,\delta\boldsymbol{x}(t)$. With $\boldsymbol{S}(\tau) = \partial_{\boldsymbol{xx}}\varphi(\tau)$, we have

$$\dot{\boldsymbol{S}} = -\boldsymbol{Q} + \boldsymbol{M}\boldsymbol{R}^{-1}\boldsymbol{M}^{\mathrm{T}} - \boldsymbol{A}'^{\mathrm{T}}\boldsymbol{S} - \boldsymbol{S}\boldsymbol{A}' + \boldsymbol{S}\boldsymbol{B}\boldsymbol{R}^{-1}\boldsymbol{B}^{\mathrm{T}}\boldsymbol{S}\,, \qquad \boldsymbol{A}' = \boldsymbol{A} - \boldsymbol{B}\boldsymbol{R}^{-1}\boldsymbol{M}^{\mathrm{T}}\,. \tag{7.47}$$

Again, the matrices $\boldsymbol{A}$, $\boldsymbol{B}$, $\boldsymbol{Q}$, $\boldsymbol{M}$, and $\boldsymbol{R}$ are all time dependent. When the matrices are constant and the cross term $\boldsymbol{M} = \boldsymbol{0}$, Eq. (7.47) reduces to the Riccati equation of Eq. (7.16). Defining a time-dependent gain $\boldsymbol{K}(t)$ gives an optimal perturbative feedback law,

$$\boldsymbol{K}(t) = \boldsymbol{R}^{-1}\left[\boldsymbol{B}^{\mathsf{T}}\boldsymbol{S} + \boldsymbol{M}^{\mathsf{T}}\right], \qquad \boldsymbol{u}_{\mathrm{fb}}(t) = -\boldsymbol{K}(t)\,\delta\boldsymbol{x}(t). \tag{7.48}$$

The final control signal, then, is a sum of the feedforward term optimizing the nominal problem and the feedback term correcting for moderate disturbances and modeling errors. Changing the sign of the error term to follow our earlier conventions, we write

$$\boldsymbol{u}(t) = \boldsymbol{u}_{\mathrm{ff}}(t) + \underbrace{\boldsymbol{K}(t)\,[\boldsymbol{x}_{\mathrm{ff}}(t) - \boldsymbol{x}(t)]}_{\boldsymbol{u}_{\mathrm{fb}}(t)}. \tag{7.49}$$

This relation reduces to the case discussed in Chapter 3 when the underlying dynamics is linear. As a result, Figure 3.5 can describe the nonlinear case equally well once it is understood that the nonlinear dynamics are used to calculate the feedforward solution in the block F and that the controller K is based on the local linearization.

Example 7.7 (Swing up a pendulum, with feedback) We extend Example 7.2 to swing up a pendulum and balance it upside down afterwards. First, we find the matrices of the linearized dynamics for the state $\boldsymbol{x}^{\mathsf{T}} = (\theta\ \dot{\theta})^{\mathsf{T}}$:

$$\boldsymbol{A}(t) = \begin{pmatrix} 0 & 1 \\ -\cos\theta_{\mathrm{ff}}(t) & 0 \end{pmatrix}, \qquad \boldsymbol{B} = \begin{pmatrix} 0 \\ 1 \end{pmatrix}. \tag{7.50}$$

The cost function J requires some thought. For $t < \tau$, we can set $\boldsymbol{Q} = 0$ because there is a final condition at $t = \tau$. The need to arrive at the balance state by $t = \tau$ excludes the possibility of doing nothing ($u = 0$). Once the pendulum is balanced ($t > \tau$), this can no longer work: we must place a penalty on state deviations. As a consequence, for $t > \tau$, we add the cost[25]

$$J_\tau^\infty = \frac{1}{2}\int_\tau^\infty \mathrm{d}t\left[Q\left(\theta^2(t) + \dot{\theta}^2(t)\right) + Ru^2(t)\right]. \tag{7.51}$$

For numerical work, we take $Q = R = 1$. Because J is already quadratic, the linear-feedback does not generate an $\boldsymbol{M}$ term. The feedback is applied to the swing-up problem that we solved in Example 7.2 to find $\theta_{\mathrm{ff}}(t)$ for $0 \le t \le \tau$ (with $Q = 0$). We will refer to this as the *nominal* solution, as it holds in the absence of disturbances. Here, we extend this nominal solution past $t = \tau$ by setting for $t > \tau$ the nominal control to be $u_{\mathrm{ff}} = 0$, corresponding to the upright state of the pendulum, $\theta = \pi$ and $\dot{\theta} = 0$.

Using the extended nominal solution to the swing-up problem, we then find $\boldsymbol{A}(t)$. The other matrices are all constant or zero. Then we can write down and solve numerically the differential Riccati equation, either as a matrix differential equation or as three coupled, nonlinear ordinary differential equations for the matrix elements of

[25] In practice, we interpret the infinite end time of the protocol to be some very large time τ_1, with J representing the long-time average $\lim_{\tau_1\to\infty}(\tau_1-\tau)^{-1}\int_\tau^{\tau_1}\mathrm{d}t\,L(\boldsymbol{x},\boldsymbol{u})$. However, because the optimal control signal depends on future consequences as well as the immediate state, we cannot simply minimize the running cost L. As we saw in our discussion of the trip-planning example for dynamic programming, such "greedy" minimization is generally not optimal.

S, denoted s_{11}, s_{12}, and s_{22}. (Because S is symmetric, $s_{21} = s_{12}$.) More simply, we can invoke a *quasistatic approximation* by solving the time-*independent* Riccati equation (with $\dot{S} = 0$ in Eq. 7.47). The time dependence of S then is directly generated by the time dependence of $\theta_{\text{ff}}(t)$. The resulting gains differ only slightly from the full time-dependent solution (Problem 7.19).

At right, we illustrate the role of feedback in stabilizing the pendulum. The top plot shows the nominal feedforward solution for $u_{\text{ff}}(t)$ (extended to $u_{\text{ff}} = 0$ for $t > \tau = 5$), applied to the "real" system (in simulation). Because of numerical error in the integration, the pendulum is not exactly balanced and eventually falls over. The bottom plot adds feedback control that both corrects for numerical error and stabilizes against substantial perturbation "kicks" at $t = 3$ (while swinging up) and $t = 8$ (while balancing). Just before the disturbance at $t = 13$, the feedback is turned off, and the disturbance rapidly knocks the pendulum over.

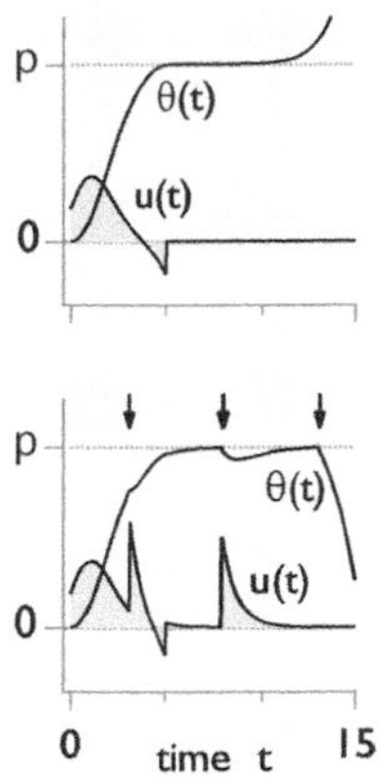

To summarize: adding feedback to the feedforward solution is essential to providing robustness to the control. Given a nonlinear optimal-control problem, there is a systematic way to generate local linear feedback that remains locally optimal relative to the original nonlinear problem. Simpler feedback schemes that trade performance against simplicity and speed of calculation may also be satisfactory. In descending order of performance,

1. *Time-dependent LQR*: This is the best strategy but requires solving the time-dependent Riccati equation, which usually has only numerical solutions. If the state-space dimension is high and if the dynamics takes the system all over state space, one may need to integrate these differential equations online, at every time step.
2. *Quasistationary LQR*: We neglect the $\dot{S}$ term in the Riccati equation but allow time dependence in the matrices A, B, etc. An off-line solution is possible if you can pre-calculate solutions numerically and interpolate. If the state-space dimension is too high, this may not be feasible and online implementation would be necessary.
3. *Basic LQR*: Pick a "typical" point in state space x_0 and calculate the LQR gain matrix K for that point, and use it for all state-space locations.
4. *Heuristic control*: Pick the gains K using one of the methods discussed in Chapters 3 and 4. For example, the basic LQR control of an inverted pendulum about its upright balance state is just proportional-derivative (PD) control, which you might have guessed without solving the LQR problem. (You still have to tune the gains.)

Depending on the needs of the control problem – the time scales, the computer power available to the controller, your time for off-line calculations, etc. – one or another of these solutions may be best. See also Problem 7.19.

7.6.2 Nonlinear Observers

So far, our discussion of local linear feedback has assumed that the full state is available. Often, only some components (or functions of components) are measured, and measurement noise may be significant for the observed components. Thus, we need a way to go from observations to estimates of the underlying system state.

In Section 4.3.1, we introduced, for linear systems, the notion of an *observer*, a shadow dynamical system that synchronizes to the original system. The approach generalizes easily to nonlinear systems and can be integrated into our optimal-control framework of nonlinear feedforward coupled with linear feedback. We begin by constructing the observer for a dynamical system. After an example where we reconstruct the angular velocity of a pendulum from observations of its angle, you will get a chance to integrate an observer into the nonlinear-feedforward, local-linear-feedback control scheme.

Consider an n-dimensional dynamical system, $\dot{\boldsymbol{x}} = \boldsymbol{f}(\boldsymbol{x})$ with p outputs $\boldsymbol{y} = \boldsymbol{h}(\boldsymbol{x})$. The observer is a parallel dynamical system whose estimator state $\hat{\boldsymbol{x}}(t)$ is driven by the error between the observables $\boldsymbol{y}(t)$ and their estimates $\hat{\boldsymbol{y}}(t) = \boldsymbol{h}(\hat{\boldsymbol{x}})$. Thus,

$$\dot{\hat{\boldsymbol{x}}} = \boldsymbol{f}(\hat{\boldsymbol{x}}) + \boldsymbol{L}(\boldsymbol{y} - \hat{\boldsymbol{y}}) , \tag{7.52}$$

where $\boldsymbol{L}$ is an n-dimensional vector of observer gains that we need to choose.[26] As before, we define the error $\boldsymbol{e} = \boldsymbol{x} - \hat{\boldsymbol{x}}$ and Taylor expand about $\hat{\boldsymbol{x}}$. The error dynamics obey

$$\dot{\boldsymbol{e}} = \boldsymbol{f}(\boldsymbol{x}) - \boldsymbol{f}(\hat{\boldsymbol{x}}) - \boldsymbol{L}[\boldsymbol{h}(\boldsymbol{x}) - \boldsymbol{h}(\hat{\boldsymbol{x}})] \approx (\partial_{\boldsymbol{x}}\boldsymbol{f})\,\boldsymbol{e} - \boldsymbol{L}\,(\partial_{\boldsymbol{x}}\boldsymbol{h})\,\boldsymbol{e} \equiv (\boldsymbol{A} - \boldsymbol{L}\boldsymbol{C})\boldsymbol{e} , \tag{7.53}$$

with $\boldsymbol{A}(t)$ an $n \times n$ matrix and $\boldsymbol{C}(t)$ a $p \times n$ matrix, both based on gradients evaluated at $\hat{\boldsymbol{x}}(t)$ and thus time. Equation (7.53) has the same form as our linear observer, and we can choose the np observer gains $\boldsymbol{L}(t)$ as in Section 4.3.1. Thus, if two dynamical systems are "close" enough in state space, their difference can obey stable linear dynamics. The delicate point is the initial condition: the state and its estimate must also start close enough together.

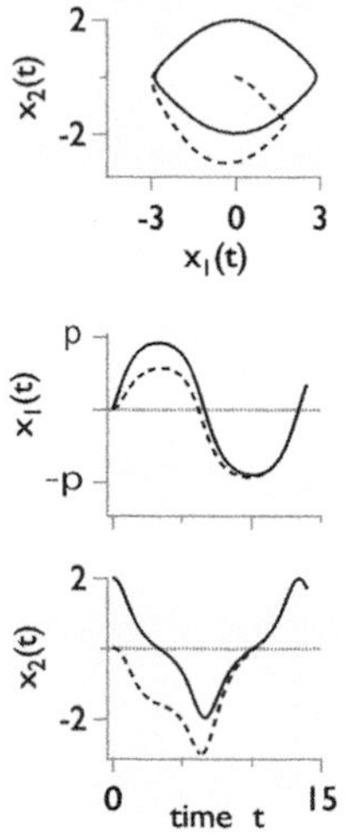

Example 7.8 (Pendulum observer) Consider an undamped pendulum with position observations: $\dot{x}_1 = x_2$, $\dot{x}_2 = -\sin x_1$, $y = x_1$. The Jacobians for the observer are

$$\boldsymbol{A} = \begin{pmatrix} 0 & 1 \\ -\cos\hat{x}_1 & 0 \end{pmatrix}, \quad \boldsymbol{C} = \begin{pmatrix} 1 & 0 \end{pmatrix} \quad \Longrightarrow \quad \boldsymbol{A} - \boldsymbol{L}\boldsymbol{C} = \begin{pmatrix} -\ell_1 & 1 \\ -\cos\hat{x}_1 - \ell_2 & 0 \end{pmatrix} . \tag{7.54}$$

The eigenvalues of $\boldsymbol{A}-\boldsymbol{L}\boldsymbol{C}$ are given by $\lambda = \frac{1}{2}\left(-\ell_1 \pm \sqrt{\ell_1^2 - 4(\ell_2 + \cos\hat{x}_1)}\right)$. If we choose $\ell_2 = \ell_2^* = \frac{1}{4}\ell_1^2 - \cos\hat{x}_1$, then both eigenvalues equal $-\frac{1}{2}\ell_1$. The matrix $\boldsymbol{A}$ and observer gain ℓ_2 both depend on time since they are functions of $\hat{x}_1(t)$. At left, we illustrate the system (solid lines) and observer (dashed lines) for $\ell_1 = 1.5$ and $\ell_2 = \ell_2^*$. The initial velocity $x_2(0) = 1.98$. Since the observer does not know $x_2(0)$, its initial condition is set arbitrarily to $\hat{x}_2(0) = 0$. Increasing ℓ_1 (and hence ℓ_2) will speed convergence but also add noise to the state estimate. You can tune ℓ_1 and $\ell_2 = \ell_2^*$ to balance these trade-offs.

---

[26] Recall that for one observable, $y(t)$ is a scalar, $\boldsymbol{L}$ a column vector, and $\boldsymbol{C}$ a $1 \times n$ row vector.

In Chapter 8, we describe how the optimal observer gains for $\boldsymbol{L}$ depend on noise levels. However, heuristic rules such as those used in Example 7.8 often work well. In Problem 7.17, you will apply these ideas to the van der Pol equation.

In control problems, the feedback is based on observations that must be translated into state estimates. The nonlinear observer presented here works well in the feedforward-feedback scheme presented in this section. The control signal applied to the observer will be $\boldsymbol{u} = \boldsymbol{u}_{\text{ff}}(t) + \boldsymbol{u}_{\text{fb}}(t)$, the sum of feedforward and feedback components. Previously, we took $\boldsymbol{u}_{\text{fb}}(t)$ to be a function of the state $\boldsymbol{x}(t)$. Here, it is a function of the state estimate $\hat{\boldsymbol{x}}(t)$ and hence of the observations $\boldsymbol{y}(t)$. Because the observer dynamical system has its own time constants, we implicitly incorporate the history of observations, the path $[\boldsymbol{y}]_0^t$, into the construction of the current state estimate $\hat{\boldsymbol{x}}(t)$.

Although the computation of the feedforward control $\boldsymbol{u}_{\text{ff}}$ does not need an observer, we do need to know the initial state $\boldsymbol{x}(0)$, at least approximately. We then solve the boundary-value problem to find $\boldsymbol{u}_{\text{ff}}(t)$ and integrate $\dot{\boldsymbol{x}} = \boldsymbol{f}(\boldsymbol{x}, \boldsymbol{u}_{\text{ff}})$ forward a desired amount. The feedback $\boldsymbol{u}_{\text{fb}}[\hat{\boldsymbol{x}}(t)]$, however, does use the observer. Example 7.9 explores the integration of an observer into control combining nonlinear feedforward and linear feedback.

Example 7.9 (Pendulum swing up, with observer) We choose observer gains as in Example 7.8 and feedback gains $\boldsymbol{K} = (k_1 \ k_2)$, with $k_1 = k_2 = 1 + \sqrt{2}$. In Figure 7.6(a)–(b) show the angle and angular velocity. Without feedback, the pendulum falls, as before, but the observer easily tracks the system. The feedforward control $u_{\text{ff}}(t)$ in (c) is the same as our previous calculation. In (d)–(f), we add feedback and a disturbance at $t = 8$. As before, the system recovers, but notice how the need to estimate the velocity degrades the control. In particular, the spike in $u(t)$ generated near $t = 0$ by the mismatch of initial conditions might require more control effort (larger u^2) than can be supplied. It is easy to check that limiting the control to ± 2 does not significantly change the response. The initial spike in u can also be reduced by letting the observer converge with the system in the stable down equilibrium before starting the swing up.

7.6.3 Model Predictive Control

Plans are of little importance, but planning is essential.

Winston Churchill[27]

Model predictive control (MPC) is based on two key ideas: The first is to approximate an optimization over an indefinite time period by evaluating the cost of an *open-loop* control strategy $\boldsymbol{u}(t)$ a finite time ahead and then optimizing over that finite

[27] Similar statements have been attributed to Carl von Clausewitz and Helmuth von Moltke the Elder.

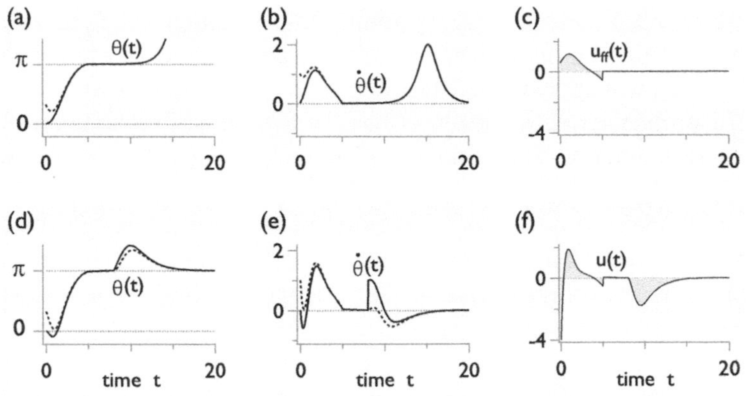

Fig. 7.6 Pendulum swing up and balance, with observer. Solid line shows physical system; dashed lines show the observer, which converges to the physical state. (a)–(c) Without feedback, the pendulum falls; (c) shows the feedforward control $u_{\text{ff}}(t)$. (d)–(f) With feedback, the pendulum recovers from a large disturbance.

time.[28] In the presence of disturbances and noise, we take only one step of the optimal path and then, having remeasured or reestimated the state vector, we repeat the optimization process. In effect, rather than greedily optimizing the next move of the controller, we take a longer view and optimize over a finite set of moves. Because we take only one step, we can calculate the optimum as an *open-loop* control – explicit instructions for $\boldsymbol{u}$ that are independent of the state of the system – rather than as a closed-loop control. In effect, we introduce feedback implicitly by a repeated open-loop optimization based on the current state of the system. MPC is also known as *receding horizon control*.

The second key idea in MPC, which we have already seen in our calculations of feedforward control, is that it can be much easier to solve the optimal control problem for a *specific* initial condition than for *all* initial conditions. Dynamic programming and the Hamilton–Jacobi–Bellman (HJB) Equation (7.31) do the latter. For the MPC algorithm, we need only solve the optimization given the present state. For linear problems, the solution to the HJB is a set of feedback gains $\boldsymbol{K}$ that are independent of $\boldsymbol{x}$, but that is not true for nonlinear problems in general – hence the simplification. In effect, in MPC we solve a simpler problem repeatedly (at each time step), whereas the full optimal control problem solves a much harder problem just once.

MPC is particularly effective in dealing with constraints on the control inputs, states, or outputs. Unlike anti-windup control, which merely prevents saturation from doing silly things to your system and makes no attempt to consider the future consequences

[28] More sophisticated MPC algorithms estimate the cost over an infinite time horizon. They first optimize over N steps, as described here, and then use a fixed-gain approximation to estimate the cost of the remaining stages, $[N+1, \infty)$. The basic idea is that if future consequences are sufficiently remote, a crude estimate will suffice. This more complicated scheme leads to proofs of the stability of the algorithm, in some cases.

of controller output, MPC tries to optimize constrained control sequences over a longer horizon.

The usefulness – and limitations – of MPC arise because there are highly efficient numerical methods to deal with the *linear-quadratic* case, where the dynamics are linear, the cost function quadratic, and there are constraints on the inputs and possibly outputs. The main point is that because the cost function $L(\mathbf{x}, \mathbf{u})$ is a positive-definite quadratic form, it must be a *convex* function. Such functions are easy to optimize, even with constraints, because they typically have just one minimum (see Appendix A.5 online). Specialized *quadratic programming* routines can easily solve such problems, even with many variables. See the discussion of software options in Section 7.9.

Despite its complexity and limitations to problems with dynamical time scales that are slow enough to compute the feedforward control fast enough, MPC is a widely used alternative to PID control in more complex industrial situations, particularly when control constraints are important. In the next section, we present a simplified version of MPC that is much easier to implement.

7.6.4 Event-Triggered Control

In event-triggered control, we activate a special control algorithm conditionally, typically when an observation (or inferred state) takes on "bad" values. As long as the system is in a "good" region of state space, there is no such control. Often, the goal is to minimize the costs of sensing or actuating. Here, we propose a slightly different view: We start from the scheme discussed in this chapter, which combines a nonlinear feedforward computation of a nominal trajectory coupled with local linear feedback to regulate against small deviations of the actual inferred state from the nominal one. However, we also define a threshold: when the system deviates from its nominal state beyond this threshold, we recompute the feedforward control from the current (bad) state back to the desired state. We can view this threshold condition as forming a kind of *tube in state space* around the basic feedforward solution. See right, where the dark line represents the nominal feedforward solution and the dark-gray tube about it is the domain inside which linear feedback is effective.[29]

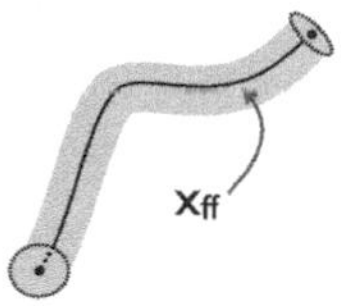

If the system ever exits the tube, a new feedforward solution replaces the old, and the local linear feedback, always present, now computes deviations relative to the new nominal trajectory. Such an algorithm adds robustness, particularly against occasional large perturbations that push the system into a different region of state space where the previous local linearization is no longer even qualitatively correct. Similarly, it can compensate for model disagreements that are somewhat localized to particular regions of state space. In principle, the threshold to activate feedforward computation can vary over state space.

Event-triggered feedforward control is a conditional version of the MPC discussed in the previous section. Perhaps its most important benefit is the reduced computational burden. The reduction does not come because controls are computed only

[29] In principle, the radius of the tube can vary along the feedforward solution, as the domain over which linear response is effective may differ in different regions of the state space.

occasionally; rather, linear feedback typically allows a longer delay between the detection of an anomaly and the application of revised feedforward control. Local linear feedback also reduces the required accuracy of the feedforward solution and hence its computational time (Section 7.7). Thus, the relevant measure of computational burden is how much time is required for a new solution to be computed relative to dynamical time scales.

Example 7.10 (GPS navigation) Before doing a technical example, let us start with a qualitative example from everyday life where we are part of the control system. The Global Positioning System (GPS) and its "apps" have revolutionized the way we navigate in the world, whether by foot, by car, or other types of transport.[30] When we drive, for example, we can set a destination and the GPS will decide upon a route. This is a nonlinear feedforward calculation. When small deviations are required, the system issues routine instructions to turn here or there. These are analogous to a linear feedback mechanism about the nominal feedforward trajectory. The app issues the instructions in advance, so that we can prepare to act at the proper time. This is equivalent to basing control $\boldsymbol{u}$ on the adjoint $\boldsymbol{\lambda}$, so that information from future requirements is acted upon at the present time.

The most interesting part, though, is when the driver makes a mistake. Then we may hear the dreaded "Recalculating ... " announcement, which means that the system is recomputing the feedforward. This is just what we mean by event-triggered control.

Example 7.11 (Plan B) Another everyday example is the expression "Plan B." Colloquially, we use it when the original plan (Plan A) has failed because of some significant change in circumstance. (Implicitly, small perturbations can be handled within Plan A.) A new plan is needed, and we formulate it taking into account the current state of affairs.

Below, we continue the pendulum-swing-up example, showing that event-triggered control can increase robustness to perturbations. It turns out that our previous version of a pendulum is too simple to show the limitations of linear feedback stabilization. In order for linear feedback to fail (absent a limitation on $|u|$), it must be possible for a perturbation to push the system into a region of state space that is so far from the nominal location that local feedback makes a qualitatively wrong decision. For a pendulum with torque as the forcing, the system can always recover by having u directly counter a perturbation. That is, if θ is too large, a negative u always corrects the perturbation in the "right" direction.

[30] The term "revolutionize" is overused and often trivialized but here is truly merited. Consider the many apps that take advantage of GPS information to offer services. Or, more broadly, think how the notion of being lost has changed: arguably, it is no longer possible on earth to be lost in the way it has always been, until recently.

In order to have an example where local linear feedback can be qualitatively wrong, consider a pendulum on a cart (Example 2.1). Scaling the equations and considering a cart that is much heavier than the pendulum leads to a simplified equation of motion,

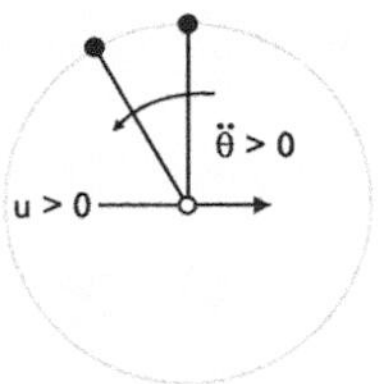

$$\ddot{\theta} + \sin\theta = -u\cos\theta\,, \tag{7.55}$$

where the control $u(t)$ is now the horizontal force applied to the cart. Recall that we define u to be positive when the force is to the left and that θ increases in the counterclockwise direction. The sign of the feedback now depends on the angle of the pendulum (via $\cos\theta$). When the pendulum is hanging down, $u > 0$ decreases $\ddot{\theta}$; however, when the pendulum is upside down, $u > 0$ increases $\ddot{\theta}$. The feedback algorithm must therefore choose the sign of the feedback according to the angle $\theta(t)$. Interestingly, for $\theta = \pm\pi/2$, the pendulum is horizontal and the u term is zero: the system is not controllable.

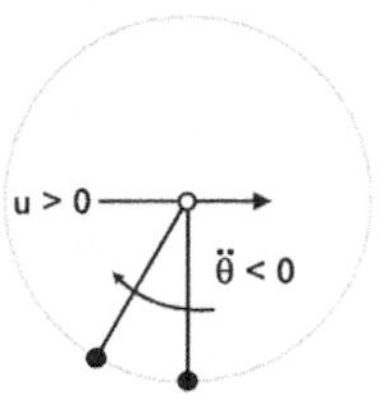

Example 7.12 (Swing up a pendulum: Event-triggered control) Figure 7.7 shows the behavior of a horizontally forced pendulum. The swing-up part of the protocol is executed over $0 \leq t \leq \tau = 3$, and the system balances thereafter. At $t = 5$, there is a large disturbance. Linear feedback (dashed line) is unable to stabilize the pendulum, and the solution runs away. The threshold is set at $\Delta\theta = 1$ radian, which represents a significant perturbation, yet still below the change in sign of the coupling at $\Delta\theta = \pi/2$. Using the feedforward solution that is recalculated from the threshold, the system can stabilize.[31]

Notice that the form of the optimal force $u_{\text{ff}}(t)$ for swing up differs qualitatively from the optimal torque seen (e.g., in Figure 7.6c). Here, u is small when $\theta \approx \pi/2$: since control is ineffective for those angles, there is cost but little benefit from applying force.

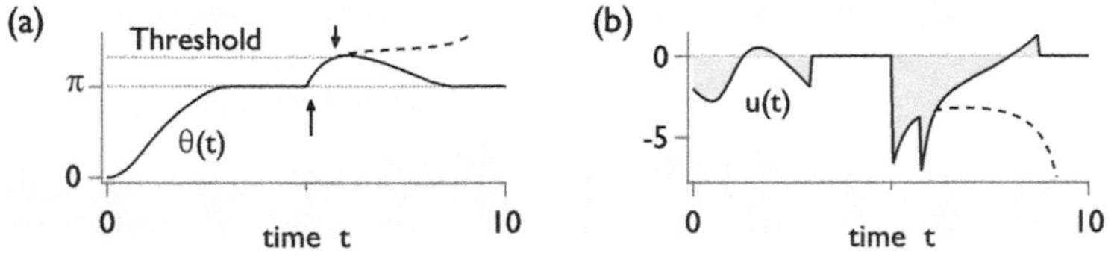

Horizontally forced pendulum tries to swing up during $0 \leq t \leq \tau = 3$ and balance afterwards. (a) Pendulum angle $\theta(t)$. Event triggered at a 1 radian deviation threshold at $t = 5$ (up arrow), with response at $t \approx 5.7$ (down arrow). Dashed line shows behavior without triggering; solid line with solid line shows behavior with triggering. (b) Control force $u(t)$. **Fig. 7.7**

[31] The choice of threshold, $\Delta\theta = 1$, is appropriate assuming that perturbations occur when the nominal system is already balanced upside down. More generally, the threshold should involve deviations in both θ and $\dot{\theta}$.

As a final point, the combination of nonlinear feedforward plus local linear feedback is not guaranteed to always work (nor is model predictive control). A perturbation might force the system into a "trap," a state from which no escape is possible or from which it is impossible to reach the target state. Think of driving a car on a road hugging a cliff. A bad perturbation can make the car go over the cliff. But for systems with states that are always *reachable*, the combination will always work, at least in the sense of reaching a target state.

7.6.5 Feedback with Input Saturation

Our scheme of starting with a feedforward control and adding robustness via local linear feedback has not considered constraints on the size of controls, $\boldsymbol{u}(t)$. It is easy to check that moderate and occasional saturations do not generally pose much of a problem. In effect, they become another kind of modeling error. We can, of course, calculate the feedforward command using the Pontryagin minimum principle of Section 7.5.2, which typically leads to intervals where the control is constrained by the boundary of allowed values.

In such time intervals, we cannot apply local linear feedback when the required response is beyond the constraint boundary. In effect, a system is open loop if the control is saturated, at least for perturbations in the "wrong" direction. This issue makes bang-bang control inherently less robust than the feedforward-feedback scheme presented above. What to do? If your control problem needs to operate in regimes where constraints are active, then MPC (Section 7.6.3) is a good approach. The main problem issue is likely to be whether the computations can be done within one time step. Alternatively, if the system is only occasionally operating against constraint limits, the local linear feedback plus event-triggered nonlinear feedforward calculation can be much faster and more robust. We will return to the problem of balancing performance against robustness in Chapter 9.

7.7 Numerical Methods

So far, we have not emphasized numerical methods for finding the nonlinear feedforward solution, and for good reason: these relatively standard boundary-value problems can be solved using widely available programs and numerical packages. For off-line computation, there may be no need to worry about the specifics of a numerical method, much as there is little reason to worry about other standard numerical tasks (eigenvalue decomposition, integration of ordinary differential equations, etc.). However, we have seen that we can increase robustness by invoking "event-triggered" recomputation of the feedforward solution whenever the system deviates too far from

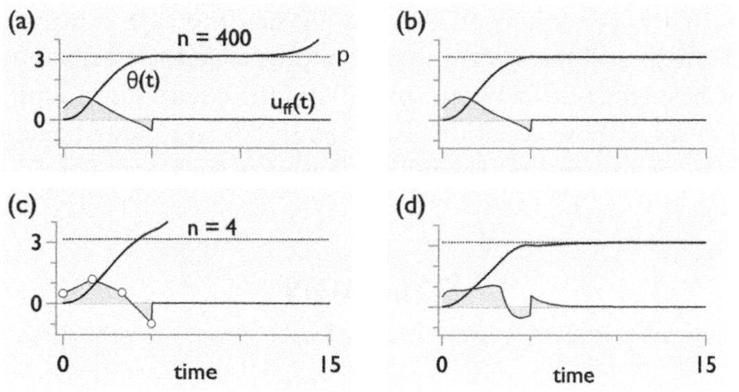

Fig. 7.8

Numerical solution to swing up and balance a pendulum using torque. (a) Angle $\theta(t)$ and control signal $u_{\mathrm{ff}}(t)$ computed using $n = 400$ grid points. (b) Same, with linear feedback. (c) Solution using 4 grid points. (d) Same, with linear feedback.

its intended path. Obviously, such a strategy can work only when computations are fast.[32]

A key point is that when you recompute the new feedforward solution, the local linear feedback should be evaluated relative to this new solution. A pleasant implication is that there is no need to find a highly accurate feedforward solution, because the feedback can correct for numerical inaccuracies. A solution accurate to 10% is probably good enough. The situation argues for a different kind of numerical analysis than usually practiced, one that values speed over accuracy. A nominal solution need only be accurate enough that linear feedback creates qualitatively the right correction. With tongue in cheek, we term this new approach *cowboy numerics*: fast, crude, and (usually) effective.

Figure 7.8 illustrates these ideas by comparing two numerical calculations of the feedforward control, $u_{\mathrm{ff}}(t)$ for the torque required to swing up a pendulum in $\tau = 5$, while minimizing the work. Part (a) shows a nominal calculation that uses $n = 400$ grid points to discretize the time domain, linearly interpolating in-between. The response, $\theta(t)$, diverges after a relatively long time. In comparison, Part (c) shows the same response calculated using only 4 grid points. Just four! Not surprisingly, the pendulum overshoots the upright position. Yet feedback readily stabilizes both solutions against numerical errors (parts b and d). As long as there are enough points to qualitatively represent the shape of $u_{\mathrm{ff}}(t)$, numerical errors may not be important. Because numerical calculations almost always scale faster than linearly with the number of grid intervals, cowboy numerics can lead to an enormous speedup that allows real-time computations.

Another way to speed up computations is to simplify the model. For example, the equations of motion of the cart-pendulum system of Example 2.1 include the effect

[32] Sometimes, the feedforward input can be computed very quickly, by direct differentiation, starting from the desired state trajectory (cf. Problem 4.2). More generally, the theory of *differential flatness* gives conditions for such calculations to be possible (Åström and Murray, 2008; Lévine, 2009).

of back action of the pendulum on the cart. When the cart is heavy, back action can be a small effect, and the system can be approximately described by the simpler Eq. (7.55). The best trade-off between simplifying the equations of motion and using a coarser grid to solve those equations is straightforward to investigate in individual cases. Whether there are general principles to optimize such a trade-off is not yet clear.

7.8 Summary

> We would often be sorry if our wishes were gratified.
>
> Moral from "The Old Man and Death," Aesop[33]

In optimal control, we introduce a positive, scalar cost function $J(\boldsymbol{x}, \boldsymbol{u})$ that hopefully captures our intuitive goals for control. The problem is then to find the control $\boldsymbol{u}(t)$ that minimizes J, under the constraint that $\boldsymbol{x}(t)$ must obey some known equations of motion, $\dot{\boldsymbol{x}} = \boldsymbol{f}(\boldsymbol{x}, \boldsymbol{u})$, perhaps with further constraints on the allowed values of the inputs and states. We introduced a scheme that combines feedforward computation of the command signal for the nonlinear dynamics, coupled with local linear feedback and event-triggered feedforward to add robustness. The scheme takes advantage of the idea that it is easier to find a single trajectory starting from a known state $\boldsymbol{x}_0$ than it is to find a general feedback rule that works for any state $\boldsymbol{x}(t)$. Also, the accompanying local linear feedback can be solved explicitly, although simpler, suboptimal approximations also work well. A pleasant surprise is that adding robustness simplifies the numerical calculation of the feedforward command, since feedback can correct for many inaccuracies. Another way to look at feedback is that it helps the feedforward calculation in two ways: First, it prolongs the time before which perturbations (due to external disturbances or modeling inaccuracies) limit the accuracy of the feedforward solution. Second, it reduces the need for accuracy in the feedforward solution (or model) itself. Both these aspects greatly broaden the ability to compute feedforward solutions in real time.

As powerful as the optimal-control methods developed here are, they have important limitations. The most serious is that the quality of results depends on the wisdom of the choice of cost function. Never confuse "optimal" with "good": a poor cost function will lead to poor performance. An optimal-control design is not necessarily better than a design based on other principles, if that design somehow better reflects what is important to you than does your cost function. What optimal control can do, however, is give a systematic way to translate from desires to actions. It is up to you to know what you really want.

A second important limitation is that we have assumed reasonable knowledge of the system dynamics. If the dynamics are too uncertain, then it can be very dangerous to optimize the response for a model system that differs significantly from reality.

[33] *The Fables of Aesop*, ed. J. Jacobs, Macmillan and Co., Ltd., St. Martin's St., London, 1894, pp. 164–165.

Historically, in the 1960s, optimal control swept in on a swell of optimism that soon dissipated once it was realized how poorly many optimally designed controllers perform in the "real world." Adding local linear feedback helps, and Chapters 9 and 10 offer further strategies.

The general lesson of dynamical programming is to be careful not to optimize too locally.[34] The idea applies to more than just the computation of $u(t)$. A classic mistake of control engineers is to start with a specified system and to ask what kind of control optimizes a given cost function. But if the quality of control limits performance, you should step back and ask whether the system itself can be redesigned to make better control possible. In Section 3.8.1, we saw that adding actuators and sensors could lead to fundamental improvements in possible performance. In designing an experiment, the physical system should be planned and built in a way that leads to easier control. For example, since signal delays limit the amount of feedback gain that may be applied, make sure that paths between actuator and sensor are as short as possible. In any case, do not separate the design of the physical system from the design of the control system.

7.9 Notes and References

Morin (2008) has an entertaining, elementary introduction to the calculus of variations and its role in classical mechanics. The book's website links to an unpublished chapter on Hamiltonian dynamics that is clear and thorough and does a good job of explaining Legendre transformations. The first chapter of Stone and Goldbart (2009) gives a more advanced treatment of the calculus of variations that also considers boundary variations. There are several mathematics texts that develop calculus of variations and then go on to connect to optimal control. Gelfand and Fomin (1963) give a detailed, authoritative discussion that is straightforward to read. Further recent texts include Mesterton-Gibbons (2009) (easier) and Liberzon (2012) (more advanced).

The connections between the calculus of variations, Hamilton–Jacobi equations, and optimal control were first pointed out by Kalman in 1960 [see his summary, Kalman (1963)]. Our approach here draws on Franklin et al. (1998) for the discrete-dynamics case and Stengel (1994) for the continuous case. Bryson and Ho (1975) is a classic text by two authors at the origin of many methods in the field. Lewis et al. (2012) has much of the same spirit and has been updated to include broader discussions of robustness. Both have nice discussions of the relation between optimal control and Hamiltonian dynamics. The online appendix to Goodwin et al. (2001) details the properties of matrix Riccati equations. Our discussion of local linear feedback follows Stengel (1994), who refers to it as *neighboring optimal control*. Finally, Whittle (1996) is full of deep insights into the conceptual structure of the field. If you want to read only one book on optimal control, make it this one.

34 The adage "Penny wise, pound foolish" reminds us that this is an old and familiar idea.

For an overview of the history of optimal control and its relations with both the calculus of variations and with classical mechanics, see Sussmann and Willems (1997). There are many interesting examples of applications of optimal control to ecology and population biology. Lenhart and Workman (2007) give a detailed exposition, and Alexander (1996) has fun examples. As one of the few analytic solutions to optimal control, the LQR has had a huge influence. To give just one application, a discounted version of LQR (see Problem 7.7) was used by the Federal Reserve in the early 1970s to make policy suggestions involving unemployment-inflation trade-offs in the US economy (Athans and Kendrick, 1974).

The problem of swinging up and balancing an inverted pendulum has a long history as a prototypical control task. Åström and Furuta (2000) discuss swing-up methods based on controlling the total energy of the pendulum. The book by Block et al. (2007), designed for a laboratory course on control, has an extensive and very practical discussion of the nonlinear pendulum. Our treatment of the nonlinear observer follows this source. One topic they present well that is omitted here is how to deal with the unpleasant mix of friction and stiction that exists in real DC motors. Graichen et al. (2007) use a generalization of the nonlinear feedforward and local linear feedback idea presented here to swing up a double pendulum mounted on a cart. The combination of horizontal forcing (with its momentary lack of controllability when the pendula are horizontal) with the two pendula makes the system more reliant on an accurate feedforward calculation. Cowboys beware! Finally, a fascinating, recent hardware implementation is the "Cubli," a small cube that uses reaction-wheel motors to provide internal torque and thereby jump up and balance on a corner of the cube (Gajamohan et al., 2012). See, especially, the associated YouTube videos!

Geering (2007) discusses dynamic programming and the Hamilton–Jacobi–Bellman equation, as does the encyclopedic work of Bertsekas (2005a). As mentioned, Bellman's insights were made independently by many other researchers (Lester Forde, Edward Moore, Edsger Dijkstra, Andrew Viterbi, etc.).

Small and Lam (2011) give a wonderfully simple derivation of the Hamilton–Jacobi equations of classical mechanics and relate them to the eikonal equations of optics. The connection between the Hamilton–Jacobi equation and the Lagrangian formulation, analogous to that between wavefronts and rays in geometrical optics, is nicely discussed in the second edition of Goldstein (1980) but is unfortunately missing from the third edition. Very few discussions of optimal control mention the analogous relation between the HJB equation and the minimum principle. The book by Gelfand and Fomin (1963) is a welcome exception: See Remark 3 in Appendix II. See, too, Schulz (2006), who also extends optimal control to the control of fields.

The original work by Pontryagin et al. (1964) is still an excellent reference on the Minimum Principle and its consequences. Stengel (1994) and Naidu (2002) are also good; Liberzon (2012) gives a careful yet readable mathematical approach. A recent pedagogical article by Andresen et al. (2011) studies in detail the optimal control of a parametric oscillator, which is an example that is more involved and subtle than the ordinary oscillator.

Our presentation of anti-windup control follows in part Åström and Murray (2008). Goodwin et al. (2001) discusses input constraints and is a strong advocate of model predictive control (MPC). For MPC, see the books by Rawlings et al. (2017) or Kouvaritakis and Cannon (2016). There are many open-source and commercial implementations of MPC. The Matlab toolbox is one starting point. Open-source solutions include μ AO-MPC for linear and ACADO for linear and nonlinear problems. The PSOPT (pseudospectral optimization) package takes a different approach to solving the nonlinear equations: instead of direct discretization, functions are expanded over an orthonormal basis set (Legendre or Chebyshev polynomials), an approach that is particularly good for representing smooth functions. See Boyd (2000) and also the PSOPT manual (Becerra, 2010).

The idea of using local linear feedback to create a moderate degree of robustness to feedforward optimal control dates back to the early 1960s and is described in Bryson and Ho (1975). Our version follows Stengel (1994) and is similar to "gain-scheduled linear controller with feedforward" (Åström and Murray, 2008). Event-triggered control is reviewed by Åström (2008). The idea of computing the feedforward control only when linear feedback fails is a type of *event-triggered* control. See also discussions of combined feedback and MPC (in particular, *tube-based* robust MPC [Rawlings et al., 2017]). Similar schemes are used in robotics to compute torques that make robotic arms move along desired trajectories (Lewis et al., 2004). The idea that only very crude numerics is needed to compute the feedforward solution seems relatively unexplored.

The control literature focuses on complicated shooting and iteration numerical methods, which are well adapted to problems with control constraints (Stengel, 1994). I thank Paul Tupper for pointing out how well standard finite-difference methods can work on problems without such constraints. The method used here (Problem 7.18) is adapted from the `FindRoot` documentation of *Mathematica*. For a guide to the many available optimization packages, see http://plato.asu.edu/sub/nlores.html.

Problems

7.1 One-dimensional optimization, without shortcuts. Redo the example in Section 7.1 without assuming that $u(t) = -Kx(t)$. Substitute the equation of motion $u = \dot{x} + ax$ into $L = \frac{1}{2}(x^2 + Ru^2)$, and find $x(t)$ directly. Confirm the assumed form of $u(t)$.

7.2 Unstable system. Repeat the example in Sec. 7.1 for an unstable system, $\dot{x} = +ax + u$, with $a > 0$. Compare with the stable case. Discuss the cheap ($R \to 0$) and expensive ($R \to \infty$) control limits. Show that expensive control leads to a gain that replaces the unstable eigenvalue a with its stable "mirror image" at $-a$. This is a general result.

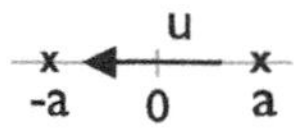

7.3 One-dimensional control, with initial and final state conditions. Solve the coupled linear equations for $x(t)$ and $\lambda(t)$ in Example 7.1. Show, in particular, that $x(t) = x_\tau\left(\frac{\sinh\sqrt{2}t}{\sinh\sqrt{2}\tau}\right)$ and $\lambda(t) = -(1 + \sqrt{2}\coth\sqrt{2}t)\,x(t)$, implying that $u(t) =$

$-\lambda(t) = +K(t)\,x(t)$ can be expressed as a positive feedback with time-dependent gain. Explain the behavior of λ and K for $\tau \to 0$ and $\tau \to \infty$.

7.4 **Move a harmonic oscillator**. For an undamped, simple harmonic oscillator, let the goal be to move from rest states $x = 0$ to $x = 1$ using the least control effort:

$$\ddot{x} + x = u\,, \quad x(0) = \dot{x}(0) = 0\,, \quad x(\tau) = 1\,,\ \dot{x}(\tau) = 0\,, \qquad J = \int_0^\tau \mathrm{d}t \left(\tfrac{1}{2}u^2(t)\right).$$

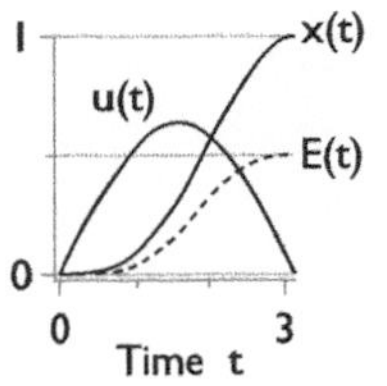

a. Solve the problem analytically, and find expressions for $x(t)$ and $u(t)$. Find also $J(\tau)$ and show that $J_{\text{short}}(\tau) \sim 12/\tau^3$ and $J_{\text{long}}(\tau) \sim 2/\tau$.

b. Plot $x(t)$ and $u(t)$ for $\tau = 1, 2, \pi, 2\pi, 3\pi$, and 10π and discuss. A sample plot for $\tau = \pi$ is shown at left, along with the energy $E(t) = \frac{1}{2}(x^2 + \dot{x}^2)$. What if $\tau < \pi$?

c. Plot $J(\tau)$, with its short- and long-protocol limits, and discuss.

7.5 **Pendulum swing up**. Fill in the missing steps from Example 7.2. In particular:

a. Identify $\boldsymbol{x}$, $\boldsymbol{u}$, and $\boldsymbol{\lambda}$, along with the functions L and $\boldsymbol{f}$.

b. Compute the various derivative terms: $\partial_{\boldsymbol{x}}L$, $\partial_{\boldsymbol{u}}L$ and $\partial_{\boldsymbol{x}}\boldsymbol{f}$, $\partial_{\boldsymbol{u}}\boldsymbol{f}$.

c. Write the Euler-Lagrange equations as two sets of equations for the two-vectors $\boldsymbol{x}$ and $\boldsymbol{\lambda}$. Then rewrite as two coupled second-order equations for $\theta(t)$ and $\lambda(t)$.

d. Derive a single fourth-order, nonlinear differential equation for $\theta(t)$.

e. In Figure 7.2, why is $u(\tau) < 0$? Hint: Relate the pendulum energy E to u.

7.6 **Discrete, one-dimensional dynamics**. Consider applying a controllable force to a free, overdamped particle. The forward Euler method for the continuous dynamics $\dot{x} = u$ gives $x_{k+1} = x_k + T_s u_k$. Let the cost function be $J = \frac{1}{2}\sum_{k=0}^{N}\left(x_k^2 + R u_k^2\right)$.

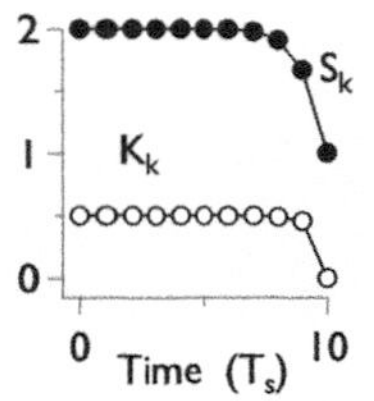

a. Form the one-dimensional augmented cost function J' that respects the constraint that x_k and u_k must obey the equations of motion.

b. Optimizing J' directly, show that $K_{N-1} = \frac{T_s}{R+T_s^2}$. Show, too, that this expression agrees with the result derived from general formula, Eq. (7.23).

c. Show that, in steady state, $S = \frac{1}{2} + \sqrt{\frac{1}{4} + \frac{R}{T_s^2}}$ and $K = \frac{S\,T_s}{R+S\,T_s^2}$.

d. For $T_s = 1$ and $R = 2$, plot S_k and K_k (see left).

e. For finite T_s, take the limit $R \to 0$. Why must $R > 0$ in the continuous case?

7.7 **Discounted LQR**. The cost function $J = \frac{1}{2}\int_0^\infty \mathrm{d}t\, \mathrm{e}^{-2\alpha t}\left(\boldsymbol{x}^{\mathsf{T}}\boldsymbol{Q}\boldsymbol{x} + \boldsymbol{u}^{\mathsf{T}}\boldsymbol{R}\boldsymbol{u}\right)$ is particularly popular in economics. The parameter $\alpha > 0$ *discounts*, or reduces the influence of future costs exponentially, on a time scale $(2\alpha)^{-1}$. This type of cost function has a steady-state solution, even though it is effectively a finite-horizon control problem.

a. By defining new variables $\tilde{\boldsymbol{x}} = \mathrm{e}^{-\alpha t}\boldsymbol{x}$ and $\tilde{\boldsymbol{u}} = \mathrm{e}^{-\alpha t}\boldsymbol{u}$, show that the problem reduces to solving a time-independent LQR problem with modified dynamics $\tilde{\boldsymbol{A}}$.

b. Show that the optimal control of the discounted problem has the form $\boldsymbol{u} = -\tilde{\boldsymbol{K}}\boldsymbol{x}$, and find $\tilde{\boldsymbol{K}}$ in terms of the solution to a steady-state Riccati equation.

c. Find $\tilde{K}(\alpha)$ for the one-dimensional problem of Section 7.1. You should get the plot at right, for $a = R = 1$. Intuitively, why does $\tilde{K}$ decrease with α?

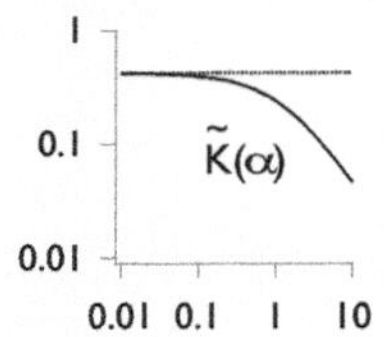

7.8 Optimal control of an undamped harmonic oscillator. In Chapter 4, we studied the PD strategy for regulating a harmonic oscillator against input disturbances. In Example 4.10, we found that for $G = \frac{1}{1+s^2}$ that the PD controller $K = k_1 + k_2 s$ gave good results for $k_1 = 3$ and $k_2 = 4$. Here, design a similar controller using optimal control. Fix the weights $\boldsymbol{Q}$ to be the 2×2 identity matrix and vary R.

a. Find, numerically or algebraically the "LQR" gains $k_1(R)$ and $k_2(R)$.

b. By plotting the system output $y(t)$ and controlled input $u(t)$, show that $R \approx 0.08$ gives a response similar to the PD controller. (Plot should resemble one at right.)

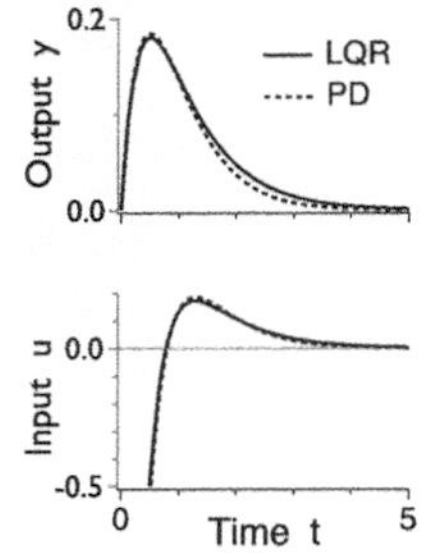

c. Show that the only value of R giving critical damping is $R = 1/8$.

d. Why is the LQR controller, in general, not critically damped?

7.9 Minimum-effort control. Consider our canonical SISO linear system, $\dot{\boldsymbol{x}} = \boldsymbol{A}\boldsymbol{x} + \boldsymbol{B}u$. Assume that we want to move the state $\boldsymbol{x}(0) = \boldsymbol{0}$ to $\boldsymbol{x}(\tau) = \boldsymbol{x}_\tau$. We do not care what the intermediate path $\boldsymbol{x}(t)$ is; rather, our goal is to choose the function $u(t)$ so as to minimize the required *control effort* $\mathcal{E} \equiv \int_0^\tau \mathrm{d}t\, u^2(t)$ (often an energy-like quantity).

a. Show that the optimal input is given by $u = \boldsymbol{B}^\mathsf{T} \mathrm{e}^{\boldsymbol{A}^\mathsf{T}(\tau - t)}\, \boldsymbol{P}^{-1}(\tau)\, \boldsymbol{x}_\tau$, where the $n \times n$ dimensional *controllability Gramian* matrix is $\boldsymbol{P}(\tau) = \int_0^\tau \mathrm{d}t\, \mathrm{e}^{\boldsymbol{A}t}\, \boldsymbol{B}\, \boldsymbol{B}^\mathsf{T}\, \mathrm{e}^{\boldsymbol{A}^\mathsf{T} t}$. Hint: This problem is just LQR with a boundary condition on $\boldsymbol{x}(\tau)$.

b. Deduce the normalized minimum control effort, $\mathcal{E}_\mathrm{n} = \frac{\boldsymbol{x}_\tau^\mathsf{T} \boldsymbol{P}(\tau)^{-1} \boldsymbol{x}_\tau}{\boldsymbol{x}_\tau^\mathsf{T} \boldsymbol{x}_\tau} = \hat{\boldsymbol{n}}^\mathsf{T}\, \boldsymbol{P}(\tau)^{-1}\, \hat{\boldsymbol{n}}$, assuming that the target $\boldsymbol{x}_\tau = \hat{\boldsymbol{n}}$ is a vector on the unit n-sphere.

c. Show that if $\boldsymbol{x}(0) = \boldsymbol{x}_0 \neq \boldsymbol{0}$, then $\boldsymbol{x}_\tau \to \Delta\boldsymbol{x} \equiv \boldsymbol{x}_\tau - \mathrm{e}^{\boldsymbol{A}\tau} \boldsymbol{x}_0$. Please interpret.

d. For $\dot{x} = -x + u$, with $x(0) = 0$ and $x(\tau) = x_\tau$, find the input $u(t)$ with minimum effort and the corresponding state $x(t)$. Show that the corresponding normalized minimum effort is $\mathcal{E}_\mathrm{n} = 2(1 - \mathrm{e}^{-2\tau})^{-1}$. Physically, the system corresponds to moving a Brownian particle in a harmonic potential, and we ask for the minimum effort to push a particle "up the potential." Discuss the limits $\tau \to \infty$ and $\tau \to 0$.

7.10 Minimum-energy control. In Problem 7.9, we discussed the minimum-effort control; however, the relation of "effort" to thermodynamic work is not entirely obvious. In this problem, we explore operations that minimize the heat dissipated into the surrounding fluid bath. Consider an overdamped particle in a harmonic potential, with equations of motion $\dot{x} = -x + u$, with $x(0) = 0$ and $x(\tau) = x_\tau$. For simplicity, ignore thermal fluctuations, which would add a stochastic term.

a. Define the heat dissipated into the bath as $Q = -\int_0^\tau \mathrm{d}t\, \dot{x}^2$, where the negative sign denotes that energy is lost by the particle and reappears as heat in the bath. Generalize the cost function in Eq. (7.12) to allow for a cross term proportional to $u(t)\, x(t)$; deduce the trajectory $x(t)$ and control $u(t)$ that minimizes Q.

b. Calculate $Q_{\min}$ for this minimum-dissipation trajectory. Discuss the limits of large and small τ (assuming x_τ to be fixed and finite).
c. The work done on the particle is $W = \int_0^\tau \mathrm{d}t\, u(t)\, \dot{x}(t)$, where $u(t)$ is interpreted as the applied "force." Calculate $W_{\min}$, and interpret for large and small τ.
d. Verify that the first law of thermodynamics holds, in the form of $\Delta U = W + Q$.

7.11 Soft end-time constraints. Consider one-dimensional motion of a Newtonian particle with $\ddot{x} = u$ and $x(0) = \dot{x}(0) = 0$. The goal is to move the particle close to $x(\tau) = x_\tau$, while minimizing the fuel cost. We impose no constraint on the velocity at $t = \tau$, and the dynamics-constrained cost functional is given by Eq. (7.6). The soft end-time constraint leads to the usual Euler–Lagrange equations (7.8), which hold for $t \in (0, \tau]$; however, the boundary condition at time τ becomes $\partial_x \varphi + \partial_{\dot{x}} L = 0$. If $\dot{\boldsymbol{x}}$ enters only in the constraint on the dynamical equation, the boundary condition becomes $\boldsymbol{\lambda}^{\mathsf{T}}(\tau) = \partial_x \varphi(\tau)$. We choose $L = \frac{1}{2}u^2$ (no x dependence) and $\varphi = \frac{1}{2}S(\bar{x} - 1)^2$. Scaling x by x_τ, t by τ, and defining $\bar{x} = x(\tau)$, we have,

$$J' = \tfrac{1}{2}S(\bar{x} - 1)^2 + \tfrac{1}{2}\int_0^1 \mathrm{d}t\, u^2(t) + \int_0^1 \mathrm{d}t\, \boldsymbol{\lambda}^{\mathsf{T}}(-\dot{\boldsymbol{x}} + \boldsymbol{A}\boldsymbol{x} + \boldsymbol{B}u).$$

where $(\bar{x} - 1) = x(\tau) - x_\tau$ in unscaled units. The parameter S balances accuracy of the end state against control effort ("fuel consumption").

a. Write the equations of motion in the standard form $\dot{\boldsymbol{x}} = \boldsymbol{A}\boldsymbol{x} + \boldsymbol{B}u$.
b. Write the adjoint equations for $\boldsymbol{\lambda}(t)$, with boundary conditions at $t = \{0, 1\}$.
c. Find $x_1(t)$, $x_2(t)$, $\lambda_1(t)$, $\lambda_2(t)$, and $u(t)$, as well as a relation between $\bar{x}$ and S and an expression for $J^*(S)$. Show that $u(t) = \frac{3S}{S+3}(1 - t)$.
d. Discuss the limits $S \to \infty$ and $S \to 0$.
e. For $S = 1$ and $S = \infty$, plot u, x, and $\dot{x}$ over the interval $t = (0, 1)$.

7.12 Sequential optimization and the Bellman equation. The "magic" of the Bellman equation arises because the optimization over N variables has a special "sequentially coupled" form. Consider optimizing $L(x_1, x_2, x_3) = f_0(x_0, x_1) + f_1(x_1, x_2) + f_2(x_2, x_3)$, where x_0 is given (the "initial condition" for a dynamical problem). Although one could solve the three coupled equations $\partial_{x_1} L = \partial_{x_2} L = \partial_{x_3} L = 0$, the special "intertwined" structure of the problem suggests a simpler, sequential solution.

a. Solve the equations sequentially, starting from $\partial_{x_3} f_2 = 0$.
b. For $f_0 = \frac{1}{2}(x_0 - x_1)^2 - x_1$, $f_1 = \frac{1}{2}(x_1 - x_2)^2 - x_2$, $f_2 = \frac{1}{2}(x_2 - x_3)^2 - x_3$, find the minimizing set $\{x_1^*, x_2^*, x_3^*\}$ by naive "global" optimization and by the easier "sequential" optimization. See Rawlings et al. (2017), Section 1.3.2.

7.13 HJB for LQR. Using Example 7.4 and the *ansatz* $J^* = \frac{1}{2}\boldsymbol{x}^{\mathsf{T}}\boldsymbol{S}\boldsymbol{x}$, derive the steady-state Linear Quadratic Regulator (Eq. 7.17) by starting from the Hamilton–Jacobi–Bellman equation (Eq. 7.31). The running cost is $L = \frac{1}{2}(\boldsymbol{x}^{\mathsf{T}}\boldsymbol{Q}\boldsymbol{x} + \boldsymbol{u}^{\mathsf{T}}\boldsymbol{R}\boldsymbol{u})$.

7.14 Anti-windup control. In Section 7.5.1, we showed that we could improve a PI controller's performance by ensuring that the integrator does not update if the controller value would exceed its physical limits. Here, the goal is to reproduce Figure 7.5.

a. Start with parts (a) and (b). Although you could use the step response function of standard control packages, write a simple forward Euler routine to integrate the equations, $\dot{x} = u$, $u(t) = K_0\left(e + \int_0^t \mathrm{d}t'\, e(t')\right)$ directly. Here, $e(t) = x_r - x(t)$, $x_r = 5$, and $K_0 = 1$. Recall that for forward Euler, $\dot{x} \approx \frac{1}{T_s}(x_{k+1} - x_k)$, with T_s the time step. Find a time step T_s that is small enough that numerical accuracy is good.
b. To reproduce (c), impose saturation, $|u| \leq 1$. In your code, distinguish the signal $v(t)$ that the controller would send to the system from $u(t)$, the signal actually sent.
c. To reproduce (d), add anti-windup control: whenever $|v(t)| > 1$, freeze the integral value by disabling the update.

7.15 One-dimensional constrained optimization. Finish the analysis in Example 7.5.

a. Show that the Pontrayagin minimum principle implies that $u(t) = -\mathrm{sat}[\lambda(t)]$, where the saturation function sat limits $\lambda(t)$ to ± 1 (see right).
b. By minimizing the Hamiltonian $H(x, \lambda, u)$, show that the the crossover between constrained and unconstrained dynamics occurs at $\tau = \ln[(1+x_0)/(2+\sqrt{2})]$. Or, make a less-rigorous argument by assuming that $u(t)$ is continuous at $t = \tau$.
c. Generate the plot shown in the example, with $x_0 = 5$ and $\tau \approx 0.56$.

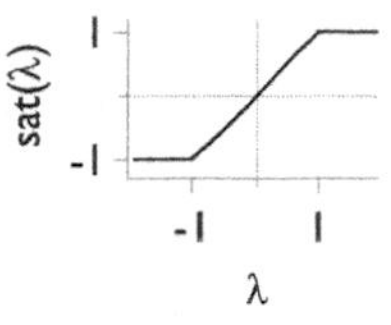

7.16 Bang-bang control of a harmonic oscillator. Consider an undamped harmonic oscillator, $\ddot{x} + x = u$, where the piecewise-continuous forcing $u(t)$ is restricted to the range $|u(t)| \leq 1$. Starting from initial conditions $x(0) = \dot{x}(0) = 0$, find the piecewise-continuous control that maximizes $x(\tau)$. Encode this goal in J by setting the penalty $L(x, u, t) = 0$ for $0 \leq t < \tau$ and the end-time penalty $\varphi[x(\tau)] = -x(\tau)$.

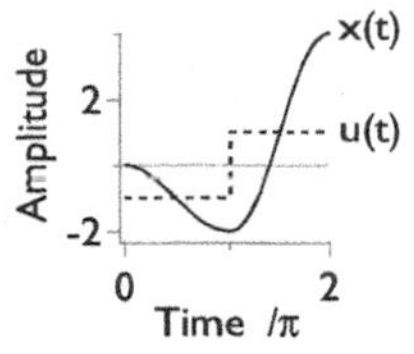

a. Solve for $u^*(t)$ using the Hamilton–Jacobi–Bellman equation.
b. Solve again using the Pontryagin Minimum Principle.
c. Show that if $\tau = 2\pi$, then $x(\tau) = 4$. Plot $u^*(t)$ and $x^*(t)$ for $0 \leq t \leq 2\pi$.
d. If you "knew" that $u(t)$ switched from -1 to $+1$ at an unknown time τ_0, then you could compute $J(\tau_0, \tau) \equiv -x(\tau)$ and minimize J directly. Do this analytically for this problem and confirm the result of part (c). See Kappen (2011).

7.17 Observer for van der Pol oscillator. Design an observer for the van der Pol equation, $\dot{x}_1 = x_2$, $\dot{x}_2 = \epsilon(1 - x_1^2)x_2 - x_1$, with output $y = x_1$.

a. Derive a time-dependent, linear equation for the error $\boldsymbol{e} = \boldsymbol{x} - \hat{\boldsymbol{x}}$.
b. Find gains $\boldsymbol{L}$ that give critically damped observer dynamics.
c. Plot $\dot{\theta}$ and $\dot{\hat{\theta}}$, θ and $\hat{\theta}$ versus time and each other for $\epsilon = 1$ and $\ell_1 = 2$. (See right.)

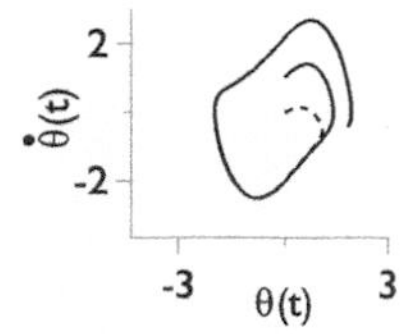

7.18 Pendulum swing up: numerics. A simple numerical method to solve the pendulum swing-up boundary-value problem is to discretize time functions.

a. Rewrite the equations of motion given in Example 7.2 to eliminate $u(t)$. Write them as a four-component vector $\dot{z} = \theta, \dot{\theta}, \lambda, \dot{\lambda}$ obeying $\dot{z} = \boldsymbol{h}(z)$.
b. Use a standard numerical routine to solve the pendulum swing-up boundary-value problem for a given value of τ. Confirm the plots shown in Figure 7.2.
c. Define n time intervals $\Delta t = \tau/n$ and denote z_k the values of the four components at time $t = k\,\Delta t$, with $k \in \{0, \dots, n\}$. From the trapezoidal rule of integration,

$$z_{k+1} = z_k + \tfrac{1}{2}\Delta t\,(\boldsymbol{h}(z_k) + \boldsymbol{h}(z_{k+1}))\ , k = 0, \dots\ n-1\,,$$

and adding the four boundary conditions, write down a coupled set of $4(n+1)$ nonlinear algebraic equations for the $4(n+1)$ variables z_k. Express your equations in the form $\boldsymbol{h}(z_k) = \mathbf{0}$. For coding, write them out explicitly, too.
d. Solve these equations using a standard numerical root finder. You can try Newton's method, which requires calculating the Jacobian matrix for $\boldsymbol{h}(z_k)$ (tricky). The secant method, which approximates the Jacobian using finite differences, is simpler and also works. Confirm Figures 7.8 and 7.2.

7.19 Pendulum swing up: adding feedback. Add linear feedback to the swing-up-and-balance protocol. First, calculate the nominal optimal control $u_{\text{ff}}(t)$ and $\theta_{\text{ff}}(t)$ (Problem 7.18). To find a linear feedback law for small deviations, assume a cost function where the weight Q on each state deviation matches the weight R on control effort.

a. Calculate the linear feedback $\boldsymbol{K} = (k_1\ \ k_2)$ gains three ways:
   i. Use the time-independent LQR gains for the upright, balanced state, $k_1 = k_2 = 1 + \sqrt{2}$. Standard LQR routines will give this result numerically. Find it analytically by solving the algebraic Riccati equations. (Cf. Problem 7.8.)
   ii. Find $k_1[\theta_{\text{ff}}(t)]$ and $k_2[\theta_{\text{ff}}(t)]$ assuming the quasistationary approximation.
   iii. Find the optimal $k_1(t)$ and $k_2(t)$ by solving the time-dependent Riccati equation assuming the dynamical matrix $\boldsymbol{A}(t) = \boldsymbol{A}[\theta_{\text{ff}}(t)]$.

   Plot and discuss $k_1(t)$ and $k_2(t)$ for the three cases.
b. Add feedback to your numerical code and produce plots resembling those in Example 7.7. Recall that a "kick" to $\dot{\theta}(t)$ imposes a slope discontinuity on $\theta(t)$. Show that all three feedback schemes give very nearly the same response.

Stochastic Systems 8

In a world without perturbations, there would be no need for feedback: A feedback law $u[x(t)]$ would be equivalent to its "feedforward replacement" $u(t)$, obtained by just substituting a previously computed $x(t)$ into $u(\cdot)$. Only the uncertainty created by random disturbances, measurement noise, and modeling errors justifies the use of feedback. Since probability and statistics are a natural language for addressing uncertainty, we need to develop analytic tools to incorporate such *stochasticity* into our treatment of control. Thus, this chapter revisits control theory from the point of view of stochastic systems.

A major focus will be *state estimation*. The state-space formulation of control of Chapter 4 distinguishes between a hidden, internal state $x(t)$ and accompanying (noisy) observations $y(t)$. There, we saw that a good way to estimate the hidden state is via an *observer*, a copy of the system's dynamics that is synchronized to the physical dynamics via feedback based on the difference between actual observations y and predictions $\hat{y}$ based on observer dynamics. The internal estimated state $\hat{x}$ can then be used in feedback laws. We suggested informally that the associated observer feedback gains should be chosen to minimize the effects of various kinds of noise. Here, a stochastic analysis will lead to systematic rules for choosing observer gains.

For linear dynamics and Gaussian, "white" noise, the optimal observer is known as the *Kalman filter* (Section 8.1). Once we have formulated an estimator, we can combine the optimal observer with the optimal control of Chapter 7. For linear dynamics, quadratic cost function, and Gaussian noise, the method is known as LQG (*Linear Quadratic Gaussian*) control (Section 8.2). Section 8.3 reformulates state estimation using notions of *Bayesian inference*. In Section 8.4, we use this broader, Bayesian approach to generalize observers to estimate states for weakly nonlinear dynamical systems driven by non-Gaussian noise. Sections 8.5 and 8.6 consider novel features of state estimation and control in the presence of more strongly nonlinear dynamics. We conclude with a discussion of smoothing, where one incorporates information from the "future" (Section 8.7).

The early parts of this chapter assume a background in stochastic effects that is roughly at the level of an undergraduate statistical physics course, in combination with the notions of probability and statistics taught in undergraduate laboratory courses. As we progress, the sophistication of the material rises somewhat. We introduce elementary ideas about Langevin equations, Fokker–Planck equations, and stochastic processes for discrete dynamical systems (see also Appendix Sections A.6 and A.8

online). The mathematics of stochastic processes for continuous dynamics is more subtle than that for discrete processes and leads to notions such as the *Itō* or *Stratonovich* calculus (van Kampen, 2007; Gardiner, 2009), whose subtleties we avoid. We will quote some relevant results, as needed.

8.1 Kalman Filter

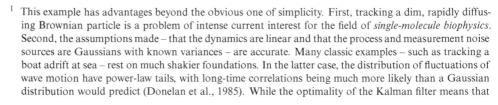

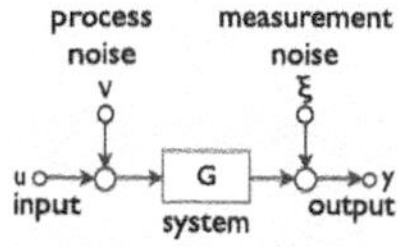

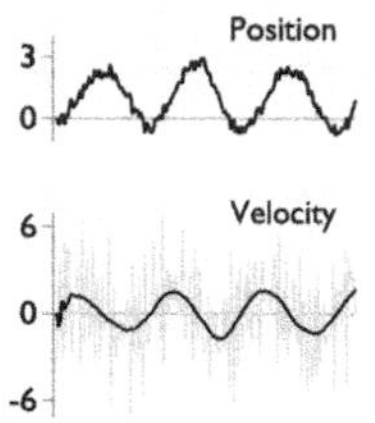

We begin by seeking an optimal observer in situations where noise is important. We will consider two types of noise in a system: *measurement noise* $\xi(t)$ at the output and *process noise* $\nu(t)$ at the input. The latter can result from a variety of sources but can often be traced back to thermal fluctuations. For example, the electrical output of a power amplifier (which supplies the "muscle" to the actuator) adds noise. In other words, you may think you are sending a command $u(t)$ to your system, but the actual command, $u(t) + \nu(t)$, has a noise term added.

To motivate a heavier chapter, let us consider briefly again the undamped harmonic oscillator, adding noise both to the system's input and to the measurement. The graphs at left are typical. On top is a simulated measurement of oscillator position. On bottom is a corresponding plot for the velocity. The naive velocity estimate (light gray line) oscillates wildly, as numerical differencing of nearest neighbors amplifies noise greatly. By contrast, the true state and its estimate (heavy dark lines) have comparatively little noise. Indeed, it is impossible to see the difference between the velocity estimate and its true value on the scale of the graph, as printed. As you can imagine, feedback based on the velocity state variable will be much more successful using the Kalman filter to estimate the velocity rather than a naive finite difference (Problem 8.7).

Section 8.1.1 begins with one-dimensional examples that already illustrate the most important features of the optimal estimator. The first is a simple physical example, tracking a colloidal particle or even a single molecule that diffuses in a liquid. The second is even simpler: estimating a fixed quantity with noisy measurements. The goal there is to relate the Kalman filter to the more familiar idea of reducing noise by averaging.

8.1.1 One-Dimensional Examples

Tracking a Diffusing Particle

Our first example is that of a Brownian particle that diffuses in a liquid but is subject to random thermal shocks from the surrounding medium.[1] If we assume that the

[1] This example has advantages beyond the obvious one of simplicity. First, tracking a dim, rapidly diffusing Brownian particle is a problem of intense current interest for the field of *single-molecule biophysics*. Second, the assumptions made – that the dynamics are linear and that the process and measurement noise sources are Gaussians with known variances – are accurate. Many classic examples – such as tracking a boat adrift at sea – rest on much shakier foundations. In the latter case, the distribution of fluctuations of wave motion have power-law tails, with long-time correlations being much more likely than a Gaussian distribution would predict (Donelan et al., 1985). While the optimality of the Kalman filter means that

diffusion takes place in one dimension, the state vector will have only one component. In an unbounded, isotropic medium, the diffusion in each dimension may be treated independently, and the full three-dimensional diffusion reduces to three uncoupled one-dimensional problems. This very simple, one-dimensional calculation nonetheless captures the essential aspects of the full estimation problem, just as a one-dimensional, optimal-control example captured many of the features of the full problem in Chapter 7. If we understand the observer in this simple scalar setting, generalizing to n dimensions will be straightforward.

Our problem then is to estimate the particle's position as well as possible, given noisy measurements and fluctuating force perturbations from solvent molecules at finite temperature. However, we also assume that we know other aspects of the system: the equations of motion (e.g., diffusion equation) and the values of all relevant parameters (diffusivity and measurement noise variance). Such knowledge of the dynamics and noise statistics can improve our estimate of the particle's position.

Notice that this thought experiment has revealed another reason to use an observer. Our original motivation was that an observer could "fill in" the missing components of the state vector. In the present case, there is one observed quantity and a one-dimensional state vector. But because observations are noisy, an observer can "filter out" the noise and return an estimate of the state that is more accurate than the raw observation. For this reason, observers for stochastic systems are often called *filters*.

Adapting the observer, Eq. (4.58), to a stochastic one-dimensional SISO system gives

$$\dot{x} = Ax + Bu + \nu\,, \qquad y = Cx + \xi, \tag{8.1a}$$

$$\dot{\hat{x}} = A\hat{x} + Bu + L(y - \hat{y})\,, \qquad \hat{y} = C\hat{x}\,. \tag{8.1b}$$

In Eq. (8.1a), we have added the process noise due to thermal fluctuations, $\nu(t)$, and the measurement noise, $\xi(t)$. Note that in one dimension, all matrices are just simple numbers. Equation (8.1b) then is the standard form of an observer, and the *Kalman filter* is just a particular choice for the *observer gain L*. Notice that we do not include the noise terms ν and ξ from Eq. (8.1a) in the estimator, Eq. (8.1b), since we do not know what their values are. In the spirit of Chapter 7, we will choose L to minimize a cost function (defined below) that penalizes deviations of the estimate $\hat{x}(t)$ from the true state $x(t)$.

Process noise can affect the observed position quite differently from measurement noise. Here, a viscous fluid acts as a low-pass filter that attenuates high-frequency movements. By contrast, measurement noise fluctuations occur at all frequencies. Thus, low-frequency fluctuations are mostly thermal noise, high-frequency fluctuations mostly measurement noise. The Kalman filter accounts for and takes advantage of this difference.

it often performs well even in situations where it should not, it is nice to start from a case where the assumptions really do hold.

The Dynamics

Assume that the particle is large compared to the solvent molecules (even a single protein or DNA molecule is large compared to a water molecule). Diffusion is then a continuous stochastic process on experimentally accessible time scales, but the modeling is simpler when framed in terms of discrete dynamics. This is reasonable, as measurements of the particle's position are taken at regular intervals T_s. Thus, the position of the particle is described by a sequence of observations $y_k \equiv y(t = kT_s)$. Discretizing the dynamics from $x(t) \to x_k$ and $y(t) \to y_k$ and scaling so that $T_s = 1$, we have the linear SISO system,

$$x_{k+1} = x_k + v_k\,, \qquad y_k = x_k + \xi_k\,, \tag{8.2}$$

where x_k represents the actual one-dimensional position of the particle at time k (its *state*) and y_k represents the measurement at that time. The dynamics are very simple, as there are no forces on the particle other than the fluctuating forces due to thermal motion of the solvent molecules. Also, there is no input u_k, as we only *track* the particle. Later, we will add a control input in order to *trap* the particle. At left, the light gray line graphs the continuous trajectory $x(t)$, and white markers represent x_k, the particle's position at time k. The filled black markers represent measurements y_k. The error bars on the y_k give the standard deviation ξ of measurement errors, while the typical displacement $x_{k+1} - x_k$ is v.

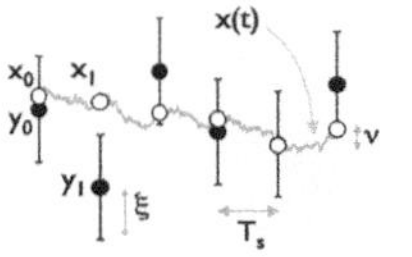

The position fluctuation due to diffusion during the previous interval T_s at time k is given by v_k, and the measurement error at time k is given by ξ_k. The noise distributions at each time k are independent Gaussian random variables, with mean and variance given by

$$\begin{aligned} \langle v_k \rangle = \langle \xi_k \rangle = 0\,, \qquad & \langle v_k \xi_k \rangle = \langle v_k x_k \rangle = \langle \xi_k x_k \rangle = 0\,, \\ \langle v_k v_\ell \rangle = v^2\, \delta_{k\ell}\,, \qquad & \langle \xi_k \xi_\ell \rangle = \xi^2\, \delta_{k\ell}\,, \end{aligned} \tag{8.3}$$

where the Kronecker delta symbol $\delta_{k\ell}$ is 1 for $k = \ell$ and 0 otherwise. The angle brackets $\langle \cdots \rangle$ denote ensemble averages over the relevant Gaussian probability distributions. An alternate notation is $v_k \sim \mathcal{N}(0, v^2)$ and $\xi_k \sim \mathcal{N}(0, \xi^2)$. (See Section A.9 online.)

The measurement and process noise terms are characterized by their standard deviations v and ξ. If the measurements are taken, for example, by imaging the particle through a microscope, the observation noise ξ in a simple setup will be proportional to the wavelength of light, λ, used to illuminate the particle. Different imaging techniques simply give different proportionality constants. The process noise from thermal shocks has variance

$$v^2 = 2DT_s = 2\left(\frac{k_B T}{\gamma}\right) T_s\,, \tag{8.4}$$

where D is the diffusion coefficient (units of ℓ^2/t, with ℓ a length scale and t a time scale), k_B is Boltzmann's constant, T is the absolute temperature, and γ is the hydrodynamic friction constant. For a sphere of radius R immersed in a fluid of viscosity

η, the *Stokes–Einstein* relation gives $\gamma = 6\pi\eta R$.[2] Physically, if the sphere moves at a steady-state velocity $\boldsymbol{v}$, the force resisting it is $\boldsymbol{F} = -\gamma \boldsymbol{v}$.

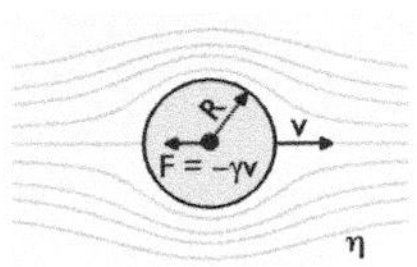

The Observer

As discussed in Section 5.4.2, there are two types of observer for estimating the state of the dynamical system, x_{k+1}, given a sequence of observations (see Figure 5.11). The *current observer* assumes that the latency in applying feedback is much less than T_s and can be ignored. Its estimate for x_k thus includes information from observations y_k. The *prediction observer* uses only observations through time $k - 1$. In contrast to Chapter 5, we focus here on the current observer. Although its equations are slightly more complicated, they give a clearer picture in the case of *hybrid dynamics*, where the system dynamics are taken to be continuous while the controller dynamics are taken to be discrete (Section 5.4.3).

We thus begin by reviewing two closely related quantities associated with the "true" state x_{k+1} and used by the current observer. Recalling that the "hat" notation ($\hat{x}_{k+1}$) is used to denote a statistical estimate (Section A.8 online.), we have

1. The *prediction*, $\hat{x}^-_{k+1}$. This estimate of x_{k+1} uses the dynamical equations to predict the state at the next time step. Starting from the estimate, $\hat{x}_k$, at time k, we use the dynamical equations to propagate forward one step in time. Thus, $\hat{x}^-_{k+1}$ uses the observations up to time k to predict the state at time $k + 1$ just prior to the observation y_{k+1}.
2. The *estimate*, $\hat{x}_{k+1}$. After the measurement y_{k+1} is acquired, we *update* the prediction to incorporate this latest information. The estimate $\hat{x}_{k+1}$ of x_{k+1} is thus conditioned on all the information available at time $k+1$, including the observation y_{k+1}.

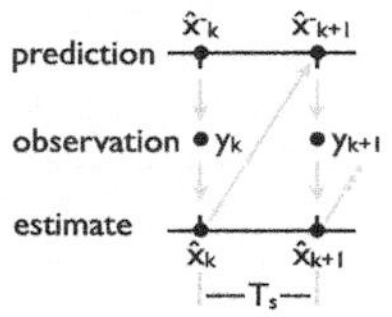

Using the equations of motion, Eqs. (8.2), we thus write for this simple case

$$
\begin{aligned}
\hat{x}^-_{k+1} &= \hat{x}_k\,, \qquad\qquad \hat{y}_{k+1} = \hat{x}^-_{k+1}\,, \\
\hat{x}_{k+1} &= \hat{x}^-_{k+1} + L\,(y_{k+1} - \hat{y}_{k+1}) = (1 - L)\hat{x}_k + L y_{k+1}\,,
\end{aligned}
\tag{8.5}
$$

where $\hat{y}_{k+1}$ is the predicted observation based on $\hat{x}_k$, so that $\hat{y}_{k+1} = \hat{x}_k = \hat{x}^-_{k+1}$. The latter differs from the actual observation $y_{k+1} = x_{k+1} + v_{k+1}$. Note that $\hat{x}^-_{k+1} = \hat{x}_k$ because diffusion is unbiased. If our best estimate of a position at time k is $\hat{x}_k$, that position

[2] To calculate γ, use low-velocity hydrodynamics (Stokes equations) to compute the fluid flow field around a solid sphere moving at a small, constant velocity v and sum the forces exerted at each point on the sphere. The fluid is assumed not to slip at the sphere's surface. The total force on the sphere resisting its motion then has magnitude γv. It is easy to forget the many assumptions that go into the derivation of the Stokes resistance. The numerical coefficient $6\pi \approx 19$ in γ holds only when the sphere is immersed in an infinite fluid. Nearby solid surfaces will tend to increase this coefficient, as the enhanced shear near the walls adds to the viscous resistance to thermal kicks. Conversely, a nearby free surface such as an air-water interface will decrease the coefficient. Since the Stokes relation holds only for steady-state motion, changes in the velocity $v(t)$ alter the resistance, as do inertial effects related to the particle's mass. Both effects are more important at relatively high frequencies and are neglected here. See Berg-Sørensen and Flyvbjerg (2004) for a fuller discussion.

continues to be the best estimate at a future time $k + 1$, in the absence of any new observation: the particle will move because of diffusion, as often to the right as to the left.

The Cost Function

Our goal is now to choose the "best" observer gain L, a problem that parallels the problem of optimal choice of feedback gain K discussed previously. In the present context, the optimal choice for L is also known as the *Kalman gain*. Here, our cost function will be the error between the estimated state, $\hat{x}_{k+1}$ and the true state, x_{k+1}. The error results from both the position measurement and thermal fluctuations. To be more precise, let us define

$$e_k \equiv x_k - \hat{x}_k \tag{8.6}$$

to be the difference between the true state and the best estimate, including the observation y_k. Because there is noise in the observations and dynamics, e_k is also a stochastic process and thus is not itself a good cost function. However, we can use its variance $\langle e_k^2 \rangle$ as a measure of performance, equivalent to the cost function J for optimal control. Note that this variance may change over time (k) and that the ensemble averaging is over different realizations at fixed time k. Following a notation commonly used in control theory, we denote the variance by P_k. Then, at time $k + 1$, our cost function is

$$P_{k+1} \equiv \left\langle e_{k+1}^2 \right\rangle = \left\langle (x_{k+1} - \hat{x}_{k+1})^2 \right\rangle , \tag{8.7}$$

We can also predict the variance at time $k + 1$ just *before* the observation y_{k+1}, in analogy with the distinction between prediction and estimate for the state:

$$P_{k+1}^- \equiv \left\langle \left(e_{k+1}^-\right)^2 \right\rangle \equiv \left\langle \left(x_{k+1} - \hat{x}_{k+1}^-\right)^2 \right\rangle . \tag{8.8}$$

To evaluate the cost function, we first evaluate the predicted error e_{k+1}^-:

$$e_{k+1}^- = x_{k+1} - \hat{x}_{k+1}^- = (x_k + \nu_k) - \hat{x}_k = e_k + \nu_k . \tag{8.9}$$

Then we have

$$P_{k+1}^- = \left\langle \left(e_{k+1}^-\right)^2 \right\rangle = P_k + \nu^2 , \tag{8.10}$$

since the cross term, $\langle e_k \nu_k \rangle = 0$. Equation (8.10) shows that in the time interval T_s ending just prior to the observation y_{k+1}, the variance increases because of the process noise, ν^2. In our particle-tracking example, this increase is $2DT_s$.

Next, to estimate the variance *after* measuring y_{k+1}, we write

$$\begin{aligned} e_{k+1} &= x_{k+1} - \left[\hat{x}_{k+1}^- + L(y_{k+1} - \hat{y}_{k+1})\right] \\ &= e_{k+1}^- - L(x_{k+1} + \xi_{k+1} - \hat{x}_{k+1}^-) \\ &= (1 - L)e_{k+1}^- - L\xi_{k+1} . \end{aligned} \tag{8.11}$$

The variance is then

$$P_{k+1} = \left\langle e_{k+1}^2 \right\rangle = (1 - L)^2 P_{k+1}^- + L^2 \xi^2 , \tag{8.12}$$

which shows that the new observation y_{k+1} alters the variance in the error estimate in two ways: The first term shows that new information reduces estimation variance while the second term shows that the accompanying measurement noise increases it.

Optimizing the Observer

We optimize P_{k+1} by differentiating with respect to the observer gain:

$$\frac{dP_{k+1}}{dL} = -2(1-L)P^-_{k+1} + 2L\xi^2 = 0 \quad \Longrightarrow \quad L^*_{k+1} = \frac{P^-_{k+1}}{P^-_{k+1} + \xi^2} = \frac{P_k + \nu^2}{P_k + \nu^2 + \xi^2}, \tag{8.13}$$

where L^*_{k+1} is the optimal gain at time $k+1$. Taking a second derivative gives

$$\left.\frac{d^2 P_{k+1}}{dL^2}\right|_{L=L^*_{k+1}} = 2\left(P^-_{k+1} + \xi^2\right) > 0\,, \tag{8.14}$$

confirming that the optimal gain L_{k+1} indeed minimizes P_{k+1} in Eq. (8.12).[3]

Equations (8.12)–(8.13) give the minimum variance:

$$P^*_{k+1} = \frac{\xi^4 P^-_{k+1}}{\left(P^-_{k+1} + \xi^2\right)^2} + \frac{\xi^2 \left(P^-_{k+1}\right)^2}{\left(P^-_{k+1} + \xi^2\right)^2} = \xi^2 L^*_{k+1}\,. \tag{8.15}$$

Equations (8.13) and (8.15) give two coupled recursion relations for the optimal gain L^*_{k+1} and the optimal estimator variance (the "error") P^*_{k+1}. The recursion relations start with an estimate and variance for x at $k = 0$ and propagate forward in time ($k \to k+1$).

Since the statistics are stationary (ξ^2 and ν^2 do not depend on k), the $L^*_k \to L^*$ and $P^*_k \to P^*$ at long times ($k \to \infty$). Thus, whatever the initial estimates for the Kalman gain and variance, they quickly approach L^* and P^*.[4] Equations (8.13) and (8.15) then lead to

$$L^* = \frac{P^* + \nu^2}{P^* + \nu^2 + \xi^2}\,, \qquad P^* = \xi^2 L^*\,. \tag{8.16}$$

Eliminating P^*, we see that L^* satisfies a quadratic equation,

$$L^{*2} + \alpha L^* - \alpha = 0\,, \qquad \alpha \equiv \frac{\nu^2}{\xi^2}\,, \tag{8.17}$$

with $\alpha = \mathrm{SNR}^2$ a signal-to-noise ratio for power. The relevant solution (Figure 8.1a) is

$$L^* = \tfrac{1}{2}\left[-\alpha + \sqrt{\alpha^2 + 4\alpha}\right]. \tag{8.18}$$

[3] Equation (8.12) is quadratic in L and can thus have only one extremum, which is either a minimum or a maximum. Hence our *local* minimum must also be the *global* minimum. The same logic applies to the more general calculation given in Section 8.1.2 and, indeed, to all *convex* functions. See Appendix A.5 online.

[4] Again, we assume that we know the correct variances for the noise statistics. With incorrect variances, the algorithm will converge to an L^* and P^* that do not minimize state-estimation errors.

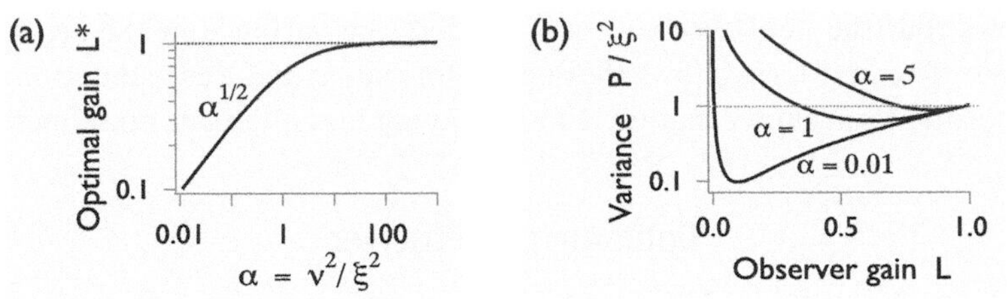

Fig. 8.1 One-dimensional Kalman filter. (a) Optimal gain L^* vs. $\alpha = \nu^2/\xi^2$. (b) Scaled variance P/ξ^2 vs. (nonoptimal) observer gain L, for three different values of α.

We can understand our result better by considering the implication of the limits of large and small α for the observer equation, $\hat{x}_{k+1} = \hat{x}^-_{k+1} + L(y_{k+1} - \hat{y}_{k+1})$:

$$\begin{array}{llll} \alpha \gg 1: & L^* \to 1 & P^* \approx \xi^2 & \text{trust the measurements} \\ \alpha \ll 1: & L^* \to \sqrt{\alpha} & P^* \approx \nu\xi & \text{trust the model} \end{array} \tag{8.19}$$

These results are what we might expect: When input noise dominates ($\alpha \gg 1$), we ignore the observer structure and simply use the observations. The variance is that of the raw observations, ξ^2. However, when the process itself has little noise driving it, we can do much better (by a factor of ξ/ν) using the Kalman filter. For diffusion, the standard deviation of the position estimate, $\delta x = \sqrt{P}$, in the $\alpha \ll 1$ limit, is

$$\delta x \approx (\nu\xi)^{1/2} = (2DT_s)^{1/4}\,\xi^{1/2}\,. \tag{8.20}$$

This error estimate depends rather weakly on D and T_s: increasing the particle diffusion constant by 10 000 increases the tracking error by a factor of only 10.

What are the consequences of *not* choosing the optimal gain? Solving Eqs. (8.10) and (8.12) for the steady-state variance P as a function of arbitrary Kalman gain L gives

$$\frac{P(L)}{\xi^2} = \frac{(1-L)^2\alpha + L^2}{1-(1-L)^2}\,. \tag{8.21}$$

This is illustrated in Figure 8.1b, which shows that the minimum in the estimator variance is rather shallow and tolerates considerable error in the estimate of the optimal gain L, especially for larger-than-optimal values of L. For small L, the variance diverges as L^{-1}. Thus, it is better to err by choosing L too large than too small. Put another way, if in doubt, put more weight on the observation than on the prediction. On the other hand, we also see from Figure 8.1 that there is not much point to the Kalman filter unless you can use low values of L (or small α), as the minimum in the variance is hardly lower than the naive value ($L = 1$). The usefulness in the Kalman filter is precisely in being able to trust the prediction well enough to value it more than the observation. If you cannot trust the prediction, then you may as well just use the observations. If you can – if the noise on the observations exceeds that on the input – then a Kalman filter can lead to spectacular improvements, particularly when used in a control loop (Problem 8.7).

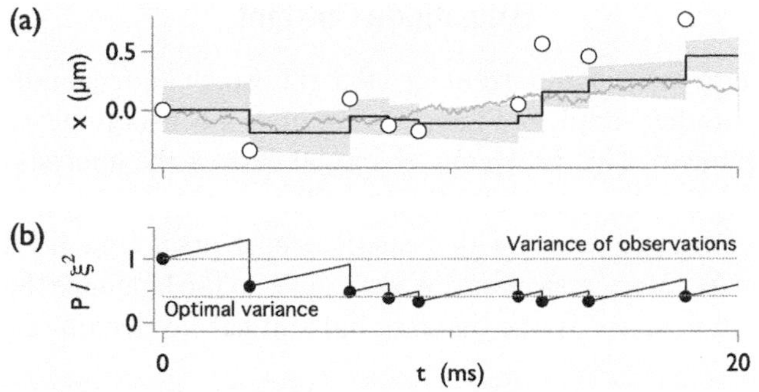

Fig. 8.2

Optimal estimate of position of a diffusing particle. (a) Jagged line shows the true position of the diffusing particle. Open circles show observations of position. Horizontal segments show $\hat{x}_k$. Shaded areas show ± 1 standard deviation for the estimation error. (b) Solid line denotes the time-dependent predicted variance $P(t)$. Filled circles show the optimal variance just after a new observation. Horizontal dashed lines show the normalized observation variance of observations – unity without the Kalman filter, and ≈ 0.4 with the optimal gain. Simulation parameters: $D = 2$ μm²/s; photon flux = 350 Hz; $\xi = 0.2$ μm; $\alpha = 0.28$.

Hybrid Dynamics

Figure 8.2b simulates continuous, one-dimensional particle diffusion (jagged gray line). Observations are photon limited (open circles), and the estimate of position is updated after each detected photon. Photon arrival times are Poisson distributed, with exponentially distributed time intervals between photon arrival times. The average interval $\langle \tau \rangle = \Gamma_0^{-1}$, where Γ_0 is the average rate of photon detections. Between observations, the particle freely diffuses, with a probability density function that obeys a diffusion equation (a simple version of the *Fokker–Planck equation*). That is if $p(x,t)$ is the probability density function for finding a particle at position x and time t, we have

$$\partial_t p = D\,\partial_{xx} p \quad \Longrightarrow \quad p(x,t) = \frac{1}{\sqrt{2\pi[P_k + 2D(t-t_k)]}} \exp\left[-\frac{(x-\hat{x}_k)^2}{2[P_k + 2D(t-t_k)]} \right], \tag{8.22}$$

where the initial conditions, at time $t = t_k$, give a mean $\hat{x}_k$ and variance P_k, the variance immediately after observation k. In Figure 8.2b, the variance increases continuously during the interval between photon counts. Between observations, the system dynamics are continuous, and the variance increases linearly (Eq. 8.22) as $P_k(t) = P_k + 2D(t - t_k)$, with P_k the variance after observation k and $P_k(t)$ the variance after the observation at time t_k but prior to the new observation at time t_{k+1}. The previous discrete-time calculation is then modified by letting the observation times t_k be arbitrary (that is, not $= kT_s$). In the Kalman-filter calculation, this is taken care of by letting the variances ν depend on time, with $\langle \nu_k \nu_\ell \rangle = 2D(t_k - t_{k-1})\,\delta_{k\ell}$. The dynamics are thus *hybrid*, as discussed in Section 5.4.3: observations arrive at discrete times t_k, but the dynamics evolve continuously. In practice, it is simpler – and little performance is lost – to use the steady-state gain L from Eq. (8.18).

Estimating a Constant

The Kalman filter is such an important idea that it is worth understanding from several points of view. Another simple problem is to estimate a constant measured repeatedly by a noisy instrument. This problem is a special case of the analysis for a diffusing particle, but it clarifies the action of a Kalman filter.

Estimating a constant quantity with measurement errors is arguably the most important elementary laboratory measurement technique. In the language that we have been developing, the state vector is one-dimensional and has no dynamics and no process noise. Thus,

$$x_{k+1} = x_k\,, \qquad y_k = x_k + \xi_k\,, \qquad \langle \xi_k \xi_\ell \rangle = \xi^2 \delta_{k\ell}\,. \tag{8.23}$$

This is just our one-dimensional problem of a diffusing particle in the limit ν^2 and $\alpha \to 0$. From Eqs. (8.13) and (8.15), we have

$$L^*_{k+1} = \frac{P^*_k}{P^*_k + \xi^2}\,, \qquad P^*_{k+1} = \xi^2 L^*_{k+1}\,. \tag{8.24}$$

If the initial estimate $\hat{x}_0$ has variance $P_0 = \xi^2$, then iterating Eq. (8.24) gives

$$L^*_{k+1} = \frac{1}{k+1}\,, \qquad P^*_{k+1} = \frac{\xi^2}{k+1}\,. \tag{8.25}$$

The result for P is just the usual statement that taking $k+1$ measurements reduces the uncertainty of each meaurement by $\sqrt{k+1}$. The update equation for the state estimate is

$$\hat{x}_{k+1} = (1 - L^*_{k+1})\hat{x}_k + L^*_{k+1} y_{k+1} = \left(\frac{k}{k+1}\right)\hat{x}_k + \left(\frac{1}{k+1}\right) y_{k+1}\,. \tag{8.26}$$

Although complicated, Eq. (8.26) is just the average, in disguise. Indeed, the average, usually calculated in a *batch* mode from all $k+1$ measurements, can also be calculated *recursively*, combining its previous estimate with the new measurement:

$$\begin{aligned}
\hat{x}_{k+1} &= \left(\frac{1}{k+1}\right)\sum_{i=1}^{k+1} x_i\,, && \textit{batch}\text{ algorithm} \\
&= \left(\frac{1}{k+1}\right)\sum_{i=1}^{k} x_i + \frac{1}{k+1}\,x_{k+1} \\
&= \left(\frac{k}{k+1}\right)\left(\frac{1}{k}\right)\sum_{i=1}^{k} x_i + \frac{1}{k+1}\,x_{k+1} \\
&= \left(\frac{k}{k+1}\right)\hat{x}_k + \left(\frac{1}{k+1}\right) x_{k+1}\,, && \textit{recursive}\text{ algorithm} \\
&= \hat{x}_k + \underbrace{\left(\frac{1}{k+1}\right)}_{L^*_{k+1}} (x_{k+1} - \hat{x}_k) && \text{Kalman gain form}\,.
\end{aligned} \tag{8.27}$$

If we identify $\hat{x}_k$ with the estimate of the true state x after k measurements, then we see that the Kalman estimate is just the average of the k measurements. Note that in formulating the Kalman filter, we are careful to distinguish between the measurement y and the actual state x. The usual notation in Eq. (8.27) is more casual.

The time dependence of $L^*_{k+1} = \frac{1}{k+1}$ is interesting. For small k, the gain L^*_{k+1} puts a large weight on the newest observation. Loosely, when the current estimate of the average is not very good (because not many observations have been made), it is better to be influenced strongly by each new data point. But after many points have been observed, you should weigh the previous estimate more heavily. The time dependence of L^*_{k+1} adjusts the weights of new and old observations appropriately. Notice that the steady-state value of the variance $P^* = 0$: after an infinite number of measurements of an unchanging quantity, the average converges to the true value. By contrast, when the quantity we estimate is itself changing, averaging too long will smooth the system dynamics. In those cases, there is an optimal amount of averaging and P^* tends to a finite positive value.

Both the ordinary average and the Kalman filter are equivalent ways to calculate the best estimate of a constant given N measurements. Batch algorithms are straightforward but require that the whole calculation be repeated each time a new point is measured. The calculation time grows, and you need to store all previous values. Recursive algorithms such as the Kalman filter are faster, and their storage requirements do not grow with time.

8.1.2 Kalman Filter: General Case

We now extend the above ideas on optimal observers to n-dimensional state vectors and multiple inputs and outputs (MIMO). Generalizing Eq. (8.1), we have

$$\begin{aligned} &\boldsymbol{x}_{k+1} = \boldsymbol{A}\boldsymbol{x}_k + \boldsymbol{B}\boldsymbol{u}_k + \boldsymbol{\nu}_k\,, \qquad \boldsymbol{y}_k = \boldsymbol{C}\boldsymbol{x}_k + \boldsymbol{\xi}_k\,, \\ &\langle \boldsymbol{\nu}_k \rangle = \langle \boldsymbol{\xi}_k \rangle = \left\langle \boldsymbol{\xi}_k \boldsymbol{\nu}_\ell^{\mathrm{T}} \right\rangle = \left\langle \boldsymbol{x}_k \boldsymbol{\nu}_\ell^{\mathrm{T}} \right\rangle = \boldsymbol{0}\,, \quad \left\langle \boldsymbol{\nu}_k \boldsymbol{\nu}_\ell^{\mathrm{T}} \right\rangle = \boldsymbol{Q}_\nu\, \delta_{k\ell}\,, \quad \left\langle \boldsymbol{\xi}_k \boldsymbol{\xi}_\ell^{\mathrm{T}} \right\rangle = \boldsymbol{Q}_\xi\, \delta_{k\ell}\,, \end{aligned} \tag{8.28}$$

where the state vector $\boldsymbol{x}$ has n components and where there are m deterministic inputs $\boldsymbol{u}$ and p observations $\boldsymbol{y}$. The matrix $\boldsymbol{A}$ is $n \times n$, and $\boldsymbol{B}$ is $n \times m$, and $\boldsymbol{C}$ is $p \times n$. The noise is characterized by the $m \times m$ and $p \times p$ covariance matrices $\boldsymbol{Q}_\nu$ and $\boldsymbol{Q}_\xi$, respectively, and is again assumed to be distributed as a multivariate Gaussian. The discrete Kronecker delta, $\delta_{k\ell}$, implies that the fluctuations at different times are independent.

Note that the covariance matrices $\boldsymbol{Q}_\nu$ and $\boldsymbol{Q}_\xi$ encode the input and output coupling of the noise. For example, an amplifier that adds noise at the system input would correspond to a noise term $\boldsymbol{\nu}_k = \boldsymbol{B}\nu_k$ with covariance matrix $\boldsymbol{Q}_\nu = \boldsymbol{B}\boldsymbol{B}^{\mathrm{T}}\nu^2$, where $\nu^2 = \left\langle \nu_k^2 \right\rangle$.

We follow the previous section and define the predicted state and updated estimate as

$$\begin{aligned} \hat{\boldsymbol{x}}^-_{k+1} &= \boldsymbol{A}\hat{\boldsymbol{x}}_k + \boldsymbol{B}\boldsymbol{u}_k\,, \\ \hat{\boldsymbol{x}}_{k+1} &= \hat{\boldsymbol{x}}^-_{k+1} + \boldsymbol{L}\left(\boldsymbol{y}_{k+1} - \hat{\boldsymbol{y}}_{k+1}\right)\,, \qquad \hat{\boldsymbol{y}}_{k+1} = \boldsymbol{C}\hat{\boldsymbol{x}}^-_{k+1}\,, \end{aligned} \tag{8.29}$$

where L is an $n \times p$ observer matrix, and $\hat{y}_{k+1}$ is the vector of predicted observations. Adding feedback $u_k = -K\hat{x}_k$, we can write these coupled equations as

$$\underbrace{\begin{pmatrix} \mathbb{I} & 0 & 0 \\ 0 & \mathbb{I} & 0 \\ 0 & 0 & 0 \end{pmatrix}}_{E'} \underbrace{\begin{pmatrix} x \\ \hat{x}^- \\ \hat{x} \end{pmatrix}_{k+1}}_{x'_{k+1}} = \underbrace{\begin{pmatrix} A & 0 & -BK \\ 0 & 0 & A - BK \\ LC & \mathbb{I} - LC & -\mathbb{I} \end{pmatrix}}_{A'} \underbrace{\begin{pmatrix} x \\ \hat{x}^- \\ \hat{x} \end{pmatrix}_k}_{x'_k} + \underbrace{\begin{pmatrix} \mathbb{I} & 0 \\ 0 & 0 \\ 0 & L \end{pmatrix}}_{B'} \underbrace{\begin{pmatrix} \nu \\ \xi \end{pmatrix}_k}_{u'_k} , \tag{8.30a}$$

$$y_k = \underbrace{\begin{pmatrix} C & 0 & 0 \end{pmatrix}}_{C'} \underbrace{\begin{pmatrix} x \\ \hat{x}^- \\ \hat{x} \end{pmatrix}_k}_{x'_k} + \underbrace{\begin{pmatrix} 0 & \mathbb{I} \end{pmatrix}}_{D'} \underbrace{\begin{pmatrix} \nu \\ \xi \end{pmatrix}_k}_{u'_k} . \tag{8.30b}$$

Equation (8.30) has the form of a descriptor state-space system $\{A', B', C', D', E'\}$ (Section 2.2.1) that simulates the combined observer-feedback system: the inputs are the two noise terms ν_k and ξ_k and the output is the measured state $y_k = Cx_k + \xi_k$. The update equation for $\hat{x}_k$ is entered as a constraint enforced by the singular descriptor matrix E'. Note that the identity matrix $\mathbb{I}$ is n-dimensional in Eq. (8.30a) and p-dimensional in Eq. (8.30b) .

Our goal is to choose the components of L to minimize the expected variance (error) in the estimates of each component of x. For reasons that will become clear below, we will first define the *covariance matrix* for state estimation errors,

$$P_k \equiv \langle e_k e_k^{\mathsf{T}} \rangle , \text{ with } e_k = x_k - \hat{x}_k . \tag{8.31}$$

Note that P_k is an $n \times n$ matrix, not a scalar. ($e^{\mathsf{T}}e$ is a scalar, but ee^{T} is a matrix.) Each element of the matrix P_k is a covariance. Dropping subscripts for simplicity, we have, at each time k, matrix elements $P_{ij} = \langle e_i e_j \rangle$. The diagonal elements $P_{ii} = \langle e_i^2 \rangle$ give the variance of component i of e, while the off-diagonal P_{ij} gives the covariance of fluctuations between components i and j of the state-vector estimate. In general, we expect nonzero covariances. For example, if state-vector components correspond to position and velocity, then a nonzero covariance between these components merely implies that the errors in the estimate of velocity are correlated with errors in the estimate of position. Finally, because the noise is Gaussian and the system dynamics linear, the components of $\hat{x}_k$ are each Gaussian random variables, as linear combinations of Gaussian random variables are themselves Gaussian (Appendix A.9 online). Here, $\hat{x}_k$ may be written in terms of the previous estimate $\hat{x}_{k-1}$, which is itself Gaussian, and so on.

We also define the covariance of the predicted state $\hat{x}_{k+1}^-$, just before an observation, as

$$P_{k+1}^- \equiv \langle e_{k+1}^- e_{k+1}^{-\,\mathsf{T}} \rangle , \text{ with } e_{k+1}^- = x_{k+1} - \hat{x}_{k+1}^- , \tag{8.32}$$

and follow the same steps as before, in Eq. (8.9):

$$e_{k+1}^- = x_{k+1} - \hat{x}_{k+1}^- = Ax_k + \cancel{Bu_k} + \nu_k - A\hat{x}_k - \cancel{Bu_k} = Ae_k + \nu_k . \tag{8.33}$$

From Problem A.7.8 online and the independence of v_k and $\hat{x}_k$, the covariance is

$$P_{k+1}^- = A P_k A^\mathsf{T} + Q_v \,. \tag{8.34}$$

Like Eq. (8.10), Eq. (8.34) implies that during the time interval between k and $k+1$, the uncertainty in the state is propagated by the dynamics A and increases as new noise Q_v is introduced via B. Note that the expectation angle brackets "pull through" A and A^T to give P_k. Also, P_{k+1}^- is independent of B and of u_k. This observation is important when using $\hat{x}$ in place of the unknown true state x in a feedback loop (Section 8.2).

Next, we calculate the covariance after the observations y_{k+1} at time $k+1$ are made. To keep the indices under control, we will write simply L for the observer gain.

$$\begin{aligned} e_{k+1} &= x_{k+1} - \hat{x}_{k+1} \\ &= x_{k+1} - \hat{x}_{k+1}^- - L\left(y_{k+1} - \hat{y}_{k+1}\right) \\ &= e_{k+1}^- - L\,\varepsilon_{k+1} \,, \end{aligned} \tag{8.35}$$

where $\varepsilon_{k+1} \equiv y_{k+1} - \hat{y}_{k+1}$ is the *innovation*, the deviation between the observation and its prediction.[5] Compare Eq. (8.35) with Eq. (8.11). The updated covariance is then

$$\begin{aligned} P_{k+1} &= \left\langle e_{k+1}\, e_{k+1}^\mathsf{T} \right\rangle \\ &= \left\langle \left(e_{k+1}^- - L\,\varepsilon_{k+1}\right)\left(e_{k+1}^- - L\,\varepsilon_{k+1}\right)^\mathsf{T} \right\rangle \\ &= P_{k+1}^- - L\,P_{k+1}^{xy\,\mathsf{T}} - P_{k+1}^{xy}\,L^\mathsf{T} + L\,P_{k+1}^{y}\,L^\mathsf{T} \,, \end{aligned} \tag{8.36}$$

where P_{k+1}^y is the covariance of the innovations,

$$\begin{aligned} P_{k+1}^y &\equiv \left\langle \varepsilon_{k+1}\,\varepsilon_{k+1}^\mathsf{T} \right\rangle \\ &= \left\langle \left(C x_{k+1} + \xi_{k+1} - C\hat{x}_{k+1}^-\right)\left(C x_{k+1} + \xi_{k+1} - C\hat{x}_{k+1}^-\right)^\mathsf{T} \right\rangle \\ &= \left\langle \left(C e_{k+1}^- + \xi_{k+1}\right)\left(C e_{k+1}^- + \xi_{k+1}\right)^\mathsf{T} \right\rangle \\ &= C P_{k+1}^- C^\mathsf{T} + Q_\xi \,, \end{aligned} \tag{8.37}$$

and P_{k+1}^{xy} is the predicted state errors and the innovations,

$$P_{k+1}^{xy} \equiv \left\langle e_{k+1}^-\,\varepsilon_{k+1}^\mathsf{T} \right\rangle = \left\langle e_{k+1}^- \left(C e_{k+1}^- + \xi_{k+1}\right)^\mathsf{T} \right\rangle = P_{k+1}^- C^\mathsf{T} \,. \tag{8.38}$$

A scalar for the SISO case, P^y has two independent contributions, one from propagating the uncertainty in the predicted state P^- through the observation matrix and the other due to observation noise. Finally, since the terms on the right-hand side of Eq. (8.36) are all explicitly symmetric and positive definite, so is P_{k+1}.

Next, we pick L to minimize the sum of the mean-square estimation errors at a given time step. The cost function is the scalar $P_k \equiv \langle e_k^\mathsf{T} e_k \rangle = \mathrm{Tr}(P_k)$, the trace of Eq. (8.36). Differentiating P_{k+1} with respect to L then gives (see Problems A.1.4 and A.1.5 online)

$$\frac{\mathrm{d}}{\mathrm{d}L} \mathrm{Tr}\, P_{k+1} = -2 P_{k+1}^{xy\,\mathsf{T}} + 2 P_{k+1}^y L^\mathsf{T} = 0 \quad \Longrightarrow \quad L_{k+1}^* = P_{k+1}^{xy}\left(P_{k+1}^y\right)^{-1} . \tag{8.39}$$

[5] See Section 10.2.4 for more on innovations.

Differentiating Eq. (8.39) again gives $2\boldsymbol{P}^{y}_{k+1}$, which is positive definite. Thus, choosing $\boldsymbol{L}_{k+1}$ gives a local minimum in the covariance – and a global one, too, since Eq. (8.36) is quadratic in $\boldsymbol{L}$ and thus has only one critical point.

Substituting $\boldsymbol{P}^{xy} = \boldsymbol{L}\boldsymbol{P}^{y}$ into Eq. (8.36) gives the covariance of the optimal state estimate:

$$\boldsymbol{P}^{*}_{k+1} = \boldsymbol{P}^{-}_{k+1} - \boldsymbol{L}^{*}_{k+1}\boldsymbol{P}^{y}_{k+1}\boldsymbol{L}^{*\mathrm{T}}_{k+1}\,. \tag{8.40}$$

We summarize the Kalman filter recursion relations:

$$\left.\begin{aligned}
\hat{\boldsymbol{x}}^{-}_{k+1} &= \boldsymbol{A}\hat{\boldsymbol{x}}_k + \boldsymbol{B}\boldsymbol{u}_k && \text{state mean}\\
\hat{\boldsymbol{y}}_{k+1} &= \boldsymbol{C}\hat{\boldsymbol{x}}^{-}_{k+1} && \text{observation mean}\\
\boldsymbol{P}^{-}_{k+1} &= \boldsymbol{A}\boldsymbol{P}_k\boldsymbol{A}^{\mathrm{T}} + \boldsymbol{Q}_{\nu} && \text{state covariance}\\
\boldsymbol{P}^{y}_{k+1} &= \boldsymbol{C}\boldsymbol{P}^{-}_{k+1}\boldsymbol{C}^{\mathrm{T}} + \boldsymbol{Q}_{\xi} && \text{observation covariance}\\
\boldsymbol{P}^{xy}_{k+1} &= \boldsymbol{P}^{-}_{k+1}\boldsymbol{C}^{\mathrm{T}} && \text{state-observation cov.}
\end{aligned}\right\}\ \text{predict} \tag{8.41a}$$

$$\left.\begin{aligned}
\boldsymbol{L}^{*}_{k+1} &= \boldsymbol{P}^{xy}_{k+1}\left(\boldsymbol{P}^{y}_{k+1}\right)^{-1} && \text{observer gain}\\
\hat{\boldsymbol{x}}_{k+1} &= \hat{\boldsymbol{x}}^{-}_{k+1} + \boldsymbol{L}^{*}_{k+1}(\boldsymbol{y}_{k+1} - \hat{\boldsymbol{y}}_{k+1}) && \text{state mean}\\
\boldsymbol{P}^{*}_{k+1} &= \boldsymbol{P}^{-}_{k+1} - \boldsymbol{L}^{*}_{k+1}\boldsymbol{P}^{y}_{k+1}\boldsymbol{L}^{*\mathrm{T}}_{k+1} && \text{state covariance}
\end{aligned}\right\}\ \text{update} \tag{8.41b}$$

Equation (8.41) gives the update scheme for the Kalman filter. Note several points:

1. The optimal observer gain $\boldsymbol{L}^{*}$ is proportional to the state-observation correlation $\boldsymbol{P}^{xy}$: When observations tell you more about the state, place more weight on the current observation than on predictions based on past information.
2. The simplified covariance-update equation in Eq. (8.41) is valid only when using the optimal $\boldsymbol{L}^{*}$, whereas the longer version in Eq. (8.36) holds for any choice of $\boldsymbol{L}$.
3. To save time, all of the various predictions (states, observations, covariances) can be done during the time interval $(k, k+1)$ – i.e., *after* the observation $\boldsymbol{y}_k$ and *before* $\boldsymbol{y}_{k+1}$.
4. Since $\boldsymbol{L}^{*}_{k+1}$ and $\boldsymbol{P}^{*}_{k+1}$ in Eq. (8.41b) do not couple to the observations $\boldsymbol{y}$, they can also be computed ahead of time, leaving as update only the $\hat{\boldsymbol{x}}_{k+1}$ equation. But for nonlinear dynamics and observations, the optimal gain and covariance will, in general depend on the most current observation. We thus include them here in the update step.
5. Often, the Kalman filter is written using fewer but messier equations that substitute the covariance expressions into the main relations. Equation (8.41) is clearer conceptually and motivates the more sophisticated filters used for nonlinear dynamics.

6. Variants of the Kalman filter equations may be better for numerical computation, as the straightforward version given in Eq. (8.41) can lead to round-off error.[6]

To gain further insight into the rather complicated Eq. (8.41), we can write the update equations for the state covariance matrix as

$$P^*_{k+1} = \underbrace{AP^*_kA^\mathsf{T}}_{\text{dynamics}} + \underbrace{Q_\nu}_{\text{disturbances}} - \underbrace{L^*_{k+1}P^y_{k+1}L^{*\mathsf{T}}_{k+1}}_{\text{observations}} \,. \tag{8.42}$$

The first term describes how the dynamics A propagates the covariance matrix P^* from k to $k+1$. The second term gives the increase in covariance due to disturbances. The third term gives the decrease in covariance due to observations, which increase our knowledge.

Also from Eq. (8.41), we can write the dynamics of the predicted covariance matrix P^-,

$$P^-_{k+1} = A\left(P^-_k - P^-_kC^\mathsf{T}(CP^-_kC^\mathsf{T} + Q_\xi)^{-1}CP^-_k\right)A^\mathsf{T} + Q_\nu\,, \tag{8.43}$$

a discrete Riccati equation with a structure that is dual to Eq. (7.23), reflecting the duality principle (Eq. 4.59). Thus, $A \to A^\mathsf{T}$, $B \to C^\mathsf{T}$, and $K \to L^\mathsf{T}$ maps the forward observer-Riccati equation to the backward controller-Riccati equation. We estimate states by propagating forward our knowledge of the state after each observation and determine control by working backwards from where we want to end up.

If we iterate Eq. (8.43), the covariance P^- (and hence P and L) quickly converge to steady-state values (assuming a constant linear system $\{A, B, C\}$ and stationary noise, meaning that Q_ν and Q_ξ are time-independent matrices). Then, from the optimal steady-state Kalman gain L^*, we can estimate the state via

$$\hat{x}_{k+1} = (\mathbb{I} - L^*C)(A\hat{x}_k + Bu_k) + L^*y_{k+1}\,. \tag{8.44}$$

The steady-state limit of the Kalman filter is commonly used, as the improvement from using time-dependent updates is often small.[7] Standard control software

[6] Because of round-off error, the updates to the P^* matrix may not be symmetric, or even positive definite. Using an explicitly symmetric version of the update such as Eq. (8.42) helps, as does solving the stationary equations rather than the time-dependent ones. You can also increase the precision of the arithmetic (e.g., from integer to floating point, or from single to double precision), which may be a problem in a fast, real-time application. Since the root of the problem is that the condition number, the ratio of largest to smallest eigenvalue magnitude, $\kappa(P^*)$ is too large, scaling the working variables so that all measurements and state-vector components have comparable magnitudes is a more fundamental fix. If rescaling is not possible, another option is to rewrite the update equations for P^* in terms of a *matrix square-root* $P^* = CC^T$, where C is uniquely specified by defining it to be lower triangular (*Cholesky decomposition*; see Appendix A.1.5 online). The equations for C are more complicated but require only half the number of digits or bits to represent since the condition number $\kappa(C) = \sqrt{\kappa(P^*)}$. Updating C recursively also has the advantage that P^*, by construction, is always symmetric and positive semidefinite. See Anderson and Moore (2005) and Maybeck (1979).

[7] Well before Kalman's 1960 paper, Norbert Wiener in the United States and Andrei Kolmogorov in the Soviet Union developed the frequency-domain equivalent of the steady-state Kalman filter in the early 1940s. The secrecy of the Second World War meant that neither work was immediately available. The *acausal* version of the *Wiener filter*, where information from the future is available, along with the past, is easy to formulate (Press et al. [2007]; cf. Problem 8.23). The causal filter requires solving complicated *Wiener–Hopf* integral equations (Stone and Goldbart, 2009). The Kalman filter is much simpler than the Wiener filter and can also deal with time-varying estimates and dynamics. It was a notable advance.

packages have built-in commands to solve directly for the optimal steady-state Kalman-gain and estimation-error matrices $\{\boldsymbol{L}^*, \boldsymbol{P}^*\}$, given the system matrices $\{\boldsymbol{A}, \boldsymbol{B}, \boldsymbol{C}\}$ and noise covariance matrices $\{\boldsymbol{Q}_\nu, \boldsymbol{Q}_\xi\}$.[8, 9] A final reminder: if the transient estimation errors from the steady-state Kalman filter are large, use the fully time-dependent equations instead.

8.1.3 Continuous and Hybrid Dynamics

So far, we have focused on discrete dynamics, mentioning the continuous case only in passing. As discussed in the introduction to Chapter 5, one reason is most practical controllers are digital, with discrete-time inputs and outputs. Another issue is the difficulty and subtlety of the mathematics of stochastic differential equations that are used. On the other hand, even though observations are discrete, the physical systems they refer to are usually continuous. Recall, for example, the diffusing particle discussed in Section 8.1.1. For continuous dynamics with $\boldsymbol{A}$ replaced by its continuous counterpart $\boldsymbol{A}_c$, etc.,

$$\frac{\mathrm{d}\boldsymbol{x}}{\mathrm{d}t} = \boldsymbol{A}_c\boldsymbol{x} + \boldsymbol{B}_c\boldsymbol{u} + \nu_c\,, \qquad \boldsymbol{y} = \boldsymbol{C}\boldsymbol{x} + \xi_c\,, \tag{8.45}$$

where the noise terms

$$\langle \nu_c(t)\,\nu_c^{\mathrm{T}}(t')\rangle = \boldsymbol{Q}_\nu^c\,\delta(t-t')\,, \qquad \langle \xi_c(t)\,\xi_c^{\mathrm{T}}(t')\rangle = \boldsymbol{Q}_\xi^c\,\delta(t-t') \tag{8.46}$$

for continuous, uncorrelated Gaussian white noise are related to the discrete ones by[10]

$$\boldsymbol{Q}_\nu = \frac{1}{T_s}\boldsymbol{Q}_\nu^c\,, \qquad \boldsymbol{Q}_\xi = \frac{1}{T_s}\boldsymbol{Q}_\xi^c\,, \tag{8.47}$$

which reflects the averaging of the white noise over T_s: If the power-spectral density of the noise is $\boldsymbol{Q}_\nu^c$ (in units of position^2/Hz for the diffusing-particle example), then averaging over a time T_s is equivalent to "allowing" frequencies up to T_s^{-1}, which gives Eq. (8.47).

Since the system is continuous, the observer (*Kalman–Bucy filter*) has the form

$$\frac{\mathrm{d}\hat{\boldsymbol{x}}}{\mathrm{d}t} = \boldsymbol{A}_c\hat{\boldsymbol{x}} + \boldsymbol{B}_c\boldsymbol{u} + \boldsymbol{L}(\boldsymbol{y}-\hat{\boldsymbol{y}})\,, \qquad \hat{\boldsymbol{y}} = \boldsymbol{C}\hat{\boldsymbol{x}}\,, \tag{8.48}$$

and the optimal observer gain is

$$\boldsymbol{L}^* = \boldsymbol{P}^{xy}(\boldsymbol{P}^y)^{-1} = \boldsymbol{P}^*\boldsymbol{C}^{\mathrm{T}}\left(\boldsymbol{Q}_\xi^c\right)^{-1}. \tag{8.49}$$

The optimal observer gain in the discrete case, $\boldsymbol{L}^* = \boldsymbol{P}^-\boldsymbol{C}^{\mathrm{T}}(\boldsymbol{C}\boldsymbol{P}^-\boldsymbol{C}^{\mathrm{T}} + \boldsymbol{Q}_\xi)^{-1}$, is lower because extra variance $\boldsymbol{P}^-$ accumulates during the interval T_s between measurements.

In Eq. (8.48), the observations are fed *continuously* to the estimator. This would be possible, for example, if implemented by analog electronics. Here, the issue is less one of practical implementation and more of understanding conceptually the continuum

[8] Deterministic and noise inputs can be grouped together using an overall input-coupling matrix $\boldsymbol{B}$.

[9] Some packages give the current observer, others the prediction observer (Problem 8.1).

[10] Equation (8.46) implicitly assumes that positive and negative frequencies are used. Power spectra defined for only positive frequencies must be multiplied by a factor of 2 to account for the negative frequencies. For example, the one-sided power spectral density for a diffusing particle sampled every T_s is $4DT_s^2$.

limit. The optimal covariance matrix $\boldsymbol{P}^*(t)$ can be shown to evolve continuously as (with $\mathrm{d}_t \equiv \frac{\mathrm{d}}{\mathrm{d}t}$)

$$\frac{\mathrm{d}\boldsymbol{P}^*}{\mathrm{d}t} = \underbrace{\boldsymbol{A}_\mathrm{c}\boldsymbol{P}^* + \boldsymbol{P}^*\boldsymbol{A}_\mathrm{c}^\mathrm{T}}_{\text{dynamics}} + \underbrace{\boldsymbol{Q}_\nu^\mathrm{c}}_{\text{disturbances}} \underbrace{- \boldsymbol{P}^*\boldsymbol{C}^\mathrm{T}\left(\boldsymbol{Q}_\xi^\mathrm{c}\right)^{-1}\boldsymbol{C}\boldsymbol{P}^*}_{\text{observations}} . \tag{8.50}$$

Equation (8.50) is the continuous analog of Eq. (8.42), as may be seen by writing the last term as $-\boldsymbol{L}^*\boldsymbol{Q}_\xi^\mathrm{c}\boldsymbol{L}^{*\mathrm{T}}$. Again, the terms on the right-hand side have a clear physical origin, reflecting contributions from system dynamics, the increase in variance due to disturbances (process noise), and the decrease due to observations. As before, if the system matrices are time independent, then Eq. (8.50) evolves to a stationary covariance matrix $\boldsymbol{P}^*$, with a corresponding stationary observer gain $\boldsymbol{L}^*$.

In a hybrid system, the observations $\{y_k\}$ are discrete, but the dynamics are continuous. Using the discrete update, Eq. (8.41) with $\boldsymbol{A} = \exp\{\boldsymbol{A}_\mathrm{c}T_\mathrm{s}\} \approx \mathbb{I} + \boldsymbol{A}_\mathrm{c}T_\mathrm{s}$ implies that

$$\boldsymbol{P}_{k+1}^- = \boldsymbol{A}\boldsymbol{P}_k^*\boldsymbol{A}^\mathrm{T} + \boldsymbol{Q}_\nu , \tag{8.51}$$

becomes

$$\begin{aligned} \boldsymbol{P}_{k+1}^- &\approx (\mathbb{I} + \boldsymbol{A}_\mathrm{c}T_\mathrm{s})\boldsymbol{P}_k^*(\mathbb{I} + \boldsymbol{A}_\mathrm{c}^\mathrm{T}T_\mathrm{s}) + T_\mathrm{s}\left(\frac{1}{T_\mathrm{s}}\boldsymbol{Q}_\nu^\mathrm{c}\right)T_\mathrm{s} \\ &= \boldsymbol{P}_k^* + T_\mathrm{s}\left[\boldsymbol{A}_\mathrm{c}\boldsymbol{P}_k^* + \boldsymbol{P}_k^*\boldsymbol{A}_\mathrm{c}^\mathrm{T} + \boldsymbol{Q}_\nu^\mathrm{c}\right] + O(T_\mathrm{s}^2) , \end{aligned} \tag{8.52}$$

which gives, on taking the limit $T_\mathrm{s} \to 0$,

$$\mathrm{d}_t\boldsymbol{P}_k^* = \boldsymbol{A}_\mathrm{c}\boldsymbol{P}_k^* + \boldsymbol{P}_k^*\boldsymbol{A}_\mathrm{c}^\mathrm{T} + \boldsymbol{Q}_\nu^\mathrm{c} , \qquad t_k < t < t_{k+1} , \quad \text{with } \boldsymbol{P}_k^*(t_k) = \boldsymbol{P}_k^* . \tag{8.53}$$

This is simply Eq. (8.50) *without* the term due to observations. That is, we use the continuous Kalman–Bucy equations to propagate the covariance between observations. In practice, we would again use steady-state dynamics so that, for each time interval, we replace $\boldsymbol{P}_k(t)$ with $\boldsymbol{P}'(t)$, $\boldsymbol{P}_k$ with $\boldsymbol{P}$, and $\boldsymbol{L}_k$ with $\boldsymbol{L}$. Note, finally, that this is not at all rigorous, in that we have glossed over many details, such as justifying $\frac{1}{T_\mathrm{s}}(\boldsymbol{P}_{k+1}^- - \boldsymbol{P}_k) \to \mathrm{d}_t\boldsymbol{P}_k$.

For our one-dimensional particle-tracking example, $A_\mathrm{c} = 0$ and $\nu_\mathrm{c}^2 = 2D$ (units of velocity^2/Hz). Then $\mathrm{d}_tP_k^* = 2D$, as we saw previously. In the particle-tracking problem, we also considered the case where the update occurred at each random arrival of a photon. Thus, the times k need not be evenly spaced. You simply integrate Eq. (8.53) from t_k to t_{k+1}, whatever the interval happens to be.

8.2 Linear Quadratic Gaussian Control

At this point, we can put together a complete solution for the optimal control of a linear system with noise at the inputs (process noise) and where the control is based on the measurements (system outputs), which are contaminated by measurement noise. We combine our previous discussion in Chapter 7 on optimal control and the discussion here on optimal estimation and find what is known, in its steady-state version, as *linear quadratic Gaussian* (LQG) control. To give the punchline away, the main result is a reaffirmation of the *separation principle* found in Chapter 4. The estimate $\hat{\boldsymbol{x}}_k$ is used in place of the true (but unknown) state $\boldsymbol{x}_k$, just as we did in our discussion of nonoptimal observers.

Even more striking, we will find that LQG control obeys the *certainty equivalence principle*, as well: the optimal choice of feedback gain $\boldsymbol{K}$ is independent of the noise variance.[11] The optimal strategy uses the Kalman filter discussed above to estimate the state, $\hat{\boldsymbol{x}}_k$, and the linear quadratic regulator theory (LQR) of Section 7.3.2 to choose the controller gains $\boldsymbol{K}$. The optimal control is then simply $\boldsymbol{u}_k = -\boldsymbol{K}\hat{\boldsymbol{x}}_k$. Conversely, the optimal observer gains of the Kalman filter are independent of the cost function ($\boldsymbol{Q}$, $\boldsymbol{R}$, and φ). Problem 8.5 gives a very simple example of these principles in a "one step" version of LQG. Below, we discuss a one-dimensional example that can be solved analytically.

8.2.1 Trapping a Diffusing Particle

We begin with a one-dimensional example. In Section 8.1.1, we discussed how to *track* a diffusing particle. Now let us add feedback and *trap* it. Biophysicists have trapped "particles" as small as individual DNA, protein, and fluorescent dye molecules for several seconds. Using long observation times, one can calculate statistics that are sensitive to different internal conformations of the molecules.

To set up our problem, we first recall the discrete, one-dimensional dynamics:

$$x_{k+1} = x_k + u_k + \nu_k\,, \qquad y_k = x_k + \xi_k\,, \tag{8.54}$$

with $\nu_k \sim \mathcal{N}(0, \nu^2)$ and $\xi_k \sim \mathcal{N}(0, \xi^2)$ independent Gaussian random variables denoting measurement noise and process noise ($\nu^2 = 2DT_s$), respectively.

We now want to generalize the notion of optimal control to stochastic systems such as Eq. (8.54). We begin with the stochastic generalization of a quadratic cost function

$$J = \sum_{k=0}^{N} \left\langle x_k^2 + R u_k^2 \right\rangle = \sum_{k=0}^{N} \left\langle x_k^2 \right\rangle . \tag{8.55}$$

With $R = 0$, the cost function penalizes only position fluctuations, implying *minimum-variance control*.[12] We average running costs because the cost at any single time step is a random variable. If N is large (or if we conduct many short protocols), then the law of large numbers guarantees that the total cost will converge to the expected value of J times N. Under these assumptions, N is large enough that we can focus on the steady-state cost per time step, $\Delta J = \langle x_k^2 \rangle$. Because the feedback gain K obeys a final

[11] The separation and certainty equivalence principles are sometimes presented as a single theorem.

[12] A minimum-variance cost function is valid for discrete but not continuous problems, where it would imply infinite control effort. See Problem 7.6.

condition while the observer gain L obeys an initial condition, the gains are constant for an intermediate range of times, as illustrated at right.

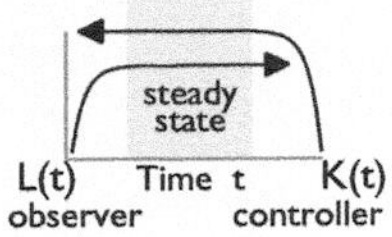

Let us explore numerically three control strategies:

1. *Perfect information*: $u_k = -Kx_k$. We know the state perfectly and use it.
2. *Naive observations*: $u_k = -Ky_k$. The feedback is based on the observations $y_k = x_k + \xi_k$.
3. *Observer*: $u_k = -K\hat{x}_k$. The feedback is based on an estimate of the state.

At right is the steady-state position variance $\langle x^2 \rangle$ as a function of feedback gain K for the three different feedback strategies. For all three, increasing the feedback gain first lowers the variance and then increases it. The decrease represents the direct action of the feedback. The increase results from the sampling delay T_s and from the measurement noise injected into the feedback loop. Below, we discuss the three strategies more quantitatively. Derivations are deferred to Problem 8.6.

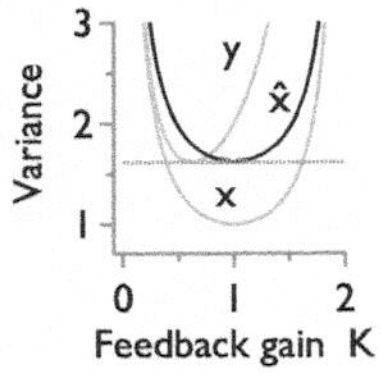

Perfect information: The feedback $u_k = -Kx_k$ implies

$$\langle x^2 \rangle = \frac{\nu^2}{1-(1-K)^2}, \tag{8.56}$$

which is plotted as the gray line marked "x." The minimum of $\langle x^2 \rangle$ occurs at $K^* = 1$ and has value $\Delta J^* = 1$ for $\nu^2 = 1$. Intuitively, this *deadbeat control*, $u_k = -x_k$, tries to "zero out" the fluctuation at each time step. Between observations, the particle is effectively open loop and diffuses an average distance $\nu = 1$. For the closed-loop system to be stable, the eigenvalues of its dynamics must be inside the unit circle (i.e., $0 < K < 2$).

Naive observations: The feedback $u_k = -Ky_k = -K(x_k + \xi_k)$, implying a variance

$$\langle x^2 \rangle = \frac{K^2\xi^2 + \nu^2}{1-(1-K)^2}, \tag{8.57}$$

which is plotted as the gray line marked "y." For $\xi^2 = \nu^2 = 1$, the minimum is at $K^* = \frac{1}{2}(\sqrt{5}-1) \approx 0.62$ and implies a variance (cost per step) of $\Delta J^* = \frac{1}{2}(\sqrt{5}+1) \approx 1.62$. The price for using noisy observations is a smaller feedback gain and larger variance.

Observer: We estimate the state and set $u_k = -K\hat{x}_k$. The observer is updated using

$$\begin{aligned} \hat{x}_{k+1}^- &= \hat{x}_k + u_k = (1-K)\hat{x}_k && \text{predictor step}, \\ \hat{x}_{k+1} &= (1-L)\hat{x}_{k+1}^- + Ly_{k+1} && \text{corrector step}, \end{aligned} \tag{8.58}$$

which leads to a complicated expression for the variance. Minimizing over K and L leads to the same optimal gains, $K^* = 1$ and $L^* = \frac{1}{2}[-\alpha + \sqrt{\alpha^2 + 4\alpha}] = \frac{1}{2}(\sqrt{5} - 1)$ for $\alpha \equiv \nu^2/\xi^2 = 1$, which we found before for the perfect state (LQR) and for the uncontrolled Kalman filter, Eq. (8.18). The variance can then be written for general K as

$$\langle x^2 \rangle = L^*\xi^2 + \frac{\nu^2}{K(2-K)}, \tag{8.59}$$

whose optimum $K^* = 1$ leads to $\Delta J^* = L^*\xi^2 + \nu^2 = \frac{1}{2}(\sqrt{5}+1)$ for $\nu^2 = \xi^2 = 1$, the same result as for naive observations.

The important point here is the separation principle: Using the observer structure, we design an optimal observer (Kalman filter, with $L = L^*$) independent of the control problem. Then, using the observer estimate, we design an optimal regulator (LQR, with $K = K^*$). Although the solution is no better than that based on naive observations (notice the horizontal dotted line in the above figure), it is much easier to calculate.

8.2.2 LQG in General

The results we established in Section 8.2.1 about Linear Quadratic Gaussian (LQG) control generalize in a straightforward way to the n-dimensional case:

1. The separation principle continues to decouple the optimal-control and state-estimation problems. From the Kalman gain matrix $\boldsymbol{L}$, we estimate the optimal state $\hat{\boldsymbol{x}}$, design the optimal feedback matrix $\boldsymbol{K}$ assuming full knowledge of the state $\boldsymbol{x}$, and combine the estimate and the gain in the control law $\boldsymbol{u} = -\boldsymbol{K}\hat{\boldsymbol{x}}$.
2. Each source of stochasticity adds its own penalty to J.

Let us then summarize the problem, first for the fully continuous case and then for the discrete case. For simplicity we restrict our attention to steady-state problems. Collecting our results for both optimal control and optimal state estimation, we have (dropping the *s)

$$\left.\begin{aligned}
&\dot{J} = \left\langle \boldsymbol{x}^{\mathrm{T}} \boldsymbol{Q} \boldsymbol{x} \right\rangle + \left\langle \boldsymbol{u}^{\mathrm{T}} \boldsymbol{R} \boldsymbol{u} \right\rangle \\
&\dot{\boldsymbol{x}} = \boldsymbol{A}\boldsymbol{x} + \boldsymbol{B}\boldsymbol{u} + \boldsymbol{\nu}, \qquad \boldsymbol{y} = \boldsymbol{C}\boldsymbol{x} + \boldsymbol{\xi} \\
&\quad \langle \boldsymbol{\nu}(t)\, \boldsymbol{\nu}^{\mathrm{T}}(t') \rangle = \boldsymbol{Q}_{\nu}^{\mathrm{c}}\, \delta(t-t'), \qquad \langle \boldsymbol{\xi}(t)\, \boldsymbol{\xi}^{\mathrm{T}}(t') \rangle = \boldsymbol{Q}_{\xi}^{\mathrm{c}}\, \delta(t-t'), \\
&\boldsymbol{u} = -\boldsymbol{K}\hat{\boldsymbol{x}}, \qquad \boldsymbol{K} = \boldsymbol{R}^{-1}\boldsymbol{B}^{\mathrm{T}}\boldsymbol{S}, \quad \text{with} \quad \boldsymbol{S}\boldsymbol{B}\boldsymbol{R}^{-1}\boldsymbol{B}^{\mathrm{T}}\boldsymbol{S} - \boldsymbol{S}\boldsymbol{A} - \boldsymbol{A}^{\mathrm{T}}\boldsymbol{S} - \boldsymbol{Q} = \boldsymbol{0} \\
&\boldsymbol{0} = \boldsymbol{A}\boldsymbol{P} + \boldsymbol{P}\boldsymbol{A}^{\mathrm{T}} + \boldsymbol{Q}_{\nu}^{\mathrm{c}} - \boldsymbol{P}\boldsymbol{C}^{\mathrm{T}}(\boldsymbol{Q}_{\xi}^{\mathrm{c}})^{-1}\boldsymbol{C}\boldsymbol{P} \\
&\boldsymbol{L} = \boldsymbol{P}\boldsymbol{C}^{\mathrm{T}}(\boldsymbol{Q}_{\xi}^{\mathrm{c}})^{-1} \\
&\dot{\hat{\boldsymbol{x}}} = \boldsymbol{A}\hat{\boldsymbol{x}} + \boldsymbol{B}\boldsymbol{u} + \boldsymbol{L}(\boldsymbol{y} - \hat{\boldsymbol{y}}), \qquad \hat{\boldsymbol{y}} = \boldsymbol{C}\hat{\boldsymbol{x}}.
\end{aligned}\right\} \tag{8.60}$$

The calculation of the cost function rate $\dot{J}$ for the optimal solution is lengthy and not presented here. The result is

$$\dot{J} = \underbrace{\mathrm{Tr}\, \boldsymbol{S}\boldsymbol{Q}_{\nu}^{\mathrm{c}}}_{\text{disturbances}} + \underbrace{\mathrm{Tr}\, \boldsymbol{P}\boldsymbol{K}^{\mathrm{T}}\boldsymbol{R}\boldsymbol{K}}_{\text{state-estimation error}}, \tag{8.61}$$

Similarly, the steady-state discrete case for the current observer is

$$\left.\begin{aligned}
&\Delta J = \langle x^{\mathsf T} Q x\rangle + \langle u^{\mathsf T} R u\rangle \\
&x_{k+1} = Ax_k + Bu_k + \nu_k\,, \qquad y_{k+1} = Cx_{k+1} + \xi_{k+1}\,, \\
&\quad \nu_k \sim \mathcal{N}(0, Q_\nu)\,, \quad \xi_k \sim \mathcal{N}(0, Q_\xi)\,, \qquad \nu_k\,, \xi_k \text{ i.i.d.}\,, \\
&u_k = -K\hat{x}_k\,, \qquad K = (R')^{-1}B^{\mathsf T}S\,, \qquad R' = R + B^{\mathsf T}SB\,, \\
&\qquad\qquad S = A^{\mathsf T}SB(R')^{-1}B^{\mathsf T}SA - A^{\mathsf T}SA - Q\,, \\
&\hat{x}^-_{k+1} = A\hat{x}_k + Bu_k\,, \qquad \hat{y}_{k+1} = C\hat{x}^-_{k+1}\,, \\
&P^- = APA^{\mathsf T} + Q^{\mathrm c}_\nu\,, \qquad L = P^{xy}(P^y)^{-1}\,, \\
&P^y = CP^-C^{\mathsf T} + Q^{\mathrm c}_\xi\,, \qquad \hat{x}_{k+1} = \hat{x}^-_{k+1} + L(y_{k+1} - \hat{y}_{k+1})\,, \\
&P^{xy} = P^-C^{\mathsf T}\,, \qquad P = P^- - LP^yL^{\mathsf T}\,.
\end{aligned}\right\} \tag{8.62}$$

The cost function increment then has the same overall form as Eq. (8.61). Note that R' appears instead of R. The extra $B^{\mathsf T}SB$ reflects the penalty that arises because the control between measurement updates is open loop. The cost function increment is

$$\Delta J = \underbrace{\operatorname{Tr} SQ_\nu}_{\text{disturbances}} + \underbrace{\operatorname{Tr} PK^{\mathsf T}R'K}_{\text{state-estimation error}}\,. \tag{8.63}$$

We can now make contact with our one-dimensional problem by noting that $S = B' = K = R' = 1$, $R = 0$, and $P = \frac{1}{2}(-\nu^2 + \sqrt{\nu^4 + 4\nu^2\xi^2})$.

Although the expressions collected in Eqs. (8.60) and (8.62) are lengthy, every term has a clear physical significance. (To test your understanding, try to identify where each term comes from.) In practice, the separation theorem simplifies calculations tremendously: we simply solve the optimal-control and optimal-state-estimation problems separately and combine the solution. The certainty equivalence principle is inferred from the expression for the optimal feedback gain K, which is independent of the variance P. In other words, whatever the level of certainty we have in our estimates $\hat{x}$, as expressed by P, the same feedback gain minimizes the cost function. Of course, as Eq. (8.63) shows, the performance *does* deteriorate in the face of disturbances and measurement noise.

The expressions for the increase in cost functions have mainly conceptual value: they imply that each stochastic input adds a cost to the optimization problem. Note that the above expressions are for time-independent problems. Adding time dependence and a starting and stopping times means that the total cost should be integrated, as the contribution at each time step will be different (although of the same form). At the initial time, there will be an extra cost in J associated with the uncertainty of the initial state. At the final time, there may be an extra cost associated with performance goals placed on the final state.

Also, remember that the matrices in the continuous and discrete equations differ if the latter approximate the former. In that case, we could distinguish $\boldsymbol{A}_c$ from $\boldsymbol{A}_d$, and so on.

Example 8.1 (Vibration isolation in the Advanced LIGO experiment) A recent physics application that illustrates all the techniques discussed so far in this chapter is given by Beker et al. (2014), who discuss the design of an active vibration-isolation system for the Advanced LIGO detector for gravitational waves. The experiment uses interferometers with arms four kilometers long and must detect minute strains of $O(10^{-21})$. Although there are many ways to rule out spurious signals – the most important is to have two independent instruments 3000 km apart – isolating the interferometer mirrors from ground vibrations is crucial. The original LIGO experiment, which depended on passive isolation, failed to detect gravity waves. The Advanced LIGO experiment added active vibration cancellation and reported the first direct detection of gravity waves, believed to be induced by the collision of two black holes. Beker et al. (2014) present an LGQ controller for vibration isolation that models the apparatus by two coupled modes driven by colored-noise sources. All the transfer functions are measured, using techniques discussed in Chapter 6. Compared to a reference PID-control design, ground-induced displacements were reduced by a factor of two in the crucial low-frequency ranges below 1 Hz.

Limitations of the Separation Principle

The separation principle, certainty equivalence principle, and the linearity of the optimal control law greatly simplify the analysis of stochastic optimal control problems; however, their validity depends heavily upon assumptions such as the linearity of the dynamics and measurement relations and the Gaussian character of the noise.

As a quick example of how the separation principle can break down when the noise level depends on the state of the system, consider trapping a diffusing Brownian particle that is hidden near the set point $x = 0$ but visible outside an obstruction (see left). Let

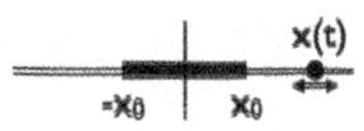

$$\dot{x} = -ax + u + \nu\,, \qquad y = \begin{cases} x & |x| > x_0 \\ 0 & |x| < x_0 \end{cases}. \tag{8.64}$$

Thus, $y(t) = x(t)$ for $x(t)$ outside the range $\pm x_0$ and 0 when inside that range. Clearly, the right strategy for control is to do nothing when the particle is hidden and then to control as usual when it is visible. The point here is that if the signal-to-noise ratio depends on the state of the system, then the optimal control policy is modified. The problems of estimating the state and controlling it are coupled.

More subtly, the separation principle, certainty equivalence principle, and linearity of the optimal control law also depend on the *information structure* of the problem.

If parameters (such as a) are unknown or observations "forgotten," then nonlinear control algorithms do better than the linear ones discussed here (see Chapter 10).

8.3 Bayesian Filtering

In Section 8.1, we followed the historical development in presenting the principles of state estimation for linear systems subject to Gaussian noise. Here, we show the same results can be recovered from a very different point of view, rooted in the Bayesian view of probability theory. This alternate approach has the advantage that it more easily generalizes to nonlinear dynamics and non-Gaussian noise.

We begin with a mathematical curiosity: if we rewrite slightly the Kalman filter update, Eq. (8.41b),

$$\hat{\boldsymbol{x}}_{k+1} = \hat{\boldsymbol{x}}^-_{k+1} + \boldsymbol{P}^{xy}_{k+1}\left(\boldsymbol{P}^{y}_{k+1}\right)^{-1}(\boldsymbol{y}_{k+1}-\hat{\boldsymbol{y}}_{k+1}), \quad \boldsymbol{P}_{k+1} = \boldsymbol{P}^-_{k+1}-\boldsymbol{P}^{xy}_{k+1}\left(\boldsymbol{P}^{y}_{k+1}\right)^{-1}\left(\boldsymbol{P}^{xy}_{k+1}\right)^{\mathrm{T}}$$

and let $\boldsymbol{x}^-_{k+1} \to \boldsymbol{x}_1$ and $\boldsymbol{y}_{k+1} \to \boldsymbol{x}_2$, then the equations have precisely the same form as the expressions for the mean and covariance of conditional Gaussian variables (Eq. A.197 online):

$$\boldsymbol{\mu}_{1|2} = \boldsymbol{\mu}_1 + \boldsymbol{\Sigma}_{12}\boldsymbol{\Sigma}_{22}^{-1}(\boldsymbol{x}_2 - \boldsymbol{\mu}_2), \qquad \boldsymbol{\Sigma}_{1|2} = \boldsymbol{\Sigma}_{11} - \boldsymbol{\Sigma}_{12}\boldsymbol{\Sigma}_{22}^{-1}\boldsymbol{\Sigma}_{21}. \tag{8.65}$$

To appreciate this "coincidence" better, we reformulate the state-estimation problem using notions of Bayesian probability and inference. We start with a toy model of the state-estimation problem and then derive general filtering equations based on Bayesian ideas. Although the Bayesian filter equations are easy to derive, they are hard to solve. One exception, of course, is for linear dynamics and Gaussian noise, where the Bayesian solution reduces to the Kalman filter.

8.3.1 A Toy Model for State Estimation

Perhaps the simplest state estimation problem is one without dynamics. Consider the noisy measurement of a scalar $y = x + \xi$. We observe y and want to know x. The unknown "state" x that we hope to measure is drawn from a Gaussian distribution, with mean 0 and variance σ_x^2, which we denote $\mathcal{N}(x; 0, \sigma_x^2)$, a slightly more explicit version of our usual notation for a Gaussian distribution. Because of the measurement noise ξ, the values of the measurement y are distributed about the true state x as $\mathcal{N}(y; x, \sigma_\xi^2)$.

To estimate x after making the measurement y, we use Bayes' Theorem and write for the posterior probability distribution,

$$p(x|y) \propto p(y|x)\, p(x) = \mathcal{N}(y; x, \sigma_\xi^2)\, \mathcal{N}(x; 0, \sigma_x^2) \propto \exp\left[-\frac{(y-x)^2}{2\sigma_\xi^2}\right] \exp\left[-\frac{x^2}{2\sigma_x^2}\right].$$

$$\propto \exp\left[-\frac{(x-\hat{x})^2}{2\sigma_0^2}\right], \quad \text{with} \quad \hat{x} = \left(\frac{\sigma_x^2}{\sigma_x^2+\sigma_\xi^2}\right) y \quad \text{and} \quad \frac{1}{\sigma_0^2} = \frac{1}{\sigma_x^2} + \frac{1}{\sigma_\xi^2}. \tag{8.66}$$

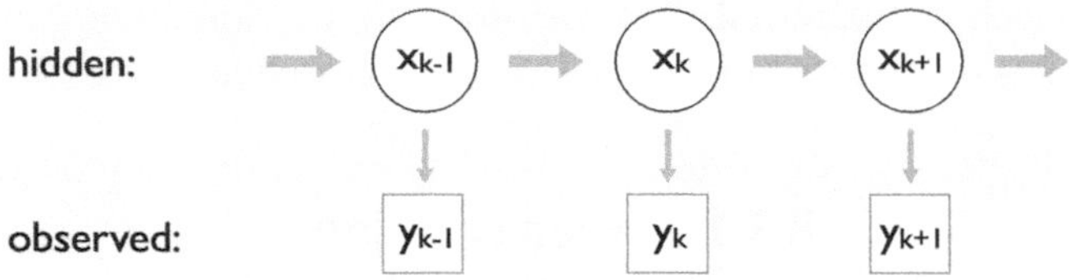

Fig. 8.3 Probabilistic state space model structure. The states $\boldsymbol{x}_k$ form a "hidden" Markov process that is not directly observable. The observations $\boldsymbol{y}_k$ depend only on $\boldsymbol{x}_k$.

In the last line, we complete the square and drop terms that do not involve x. Thus, $p(x|y)$ is also a Gaussian distribution, with variance σ_0. Its mean $\hat{x}$ is the "best estimate."[13] Notice for high signal-to-noise ratios ($\sigma_x \gg \sigma_\xi$) that $\hat{x} \to y$, while for low signal-to-noise ratios $\hat{x} \to 0$. That is, when the noise is low, we take the measurement at face value. When the noise is high, we assume that the measurement is dominated by noise and fall back on the expected value of the prior. Notice, too, that the variance of the optimal estimate, σ_0^2, is smaller than either σ_x^2 or σ_ξ^2. In Problem 8.9, you will show that taking a large number N of independent measurements leads to the usual measurement expressions: $\hat{x} \approx \bar{y}$ with an uncertainty whose standard deviation is $\sigma_\xi / \sqrt{N}$. Here, $\bar{y}$ is the average of the measurements. Thus, with enough data, the result is independent of the prior.

8.3.2 Bayesian Foundations

For a stochastic dynamical system, we cannot predict exact quantities such as the state vector, because various inputs into the system are not known. Following the lead from our toy model, we will track the probability distributions for different quantities. We begin by introducing the notion of a *probabilistic state space model*, which is illustrated schematically in Figure 8.3 and defined more formally by three probability density functions,[14]

$$\underbrace{\boldsymbol{x}_1 \sim p(\boldsymbol{x}_1)}_{\text{initial state}}, \qquad \underbrace{\boldsymbol{x}_{k+1} \sim p(\boldsymbol{x}_{k+1}|\boldsymbol{x}_k)}_{\text{dynamics}}, \qquad \underbrace{\boldsymbol{y}_k \sim p(\boldsymbol{y}_k|\boldsymbol{x}_k)}_{\text{observation}}. \tag{8.67}$$

In words, we specify the probability distribution of the initial state, $\boldsymbol{x}_1$, the dynamics that transform $\boldsymbol{x}_k$ into $\boldsymbol{x}_{k+1}$, and the observations $\boldsymbol{y}_k$ based on the current state $\boldsymbol{x}_k$. We assume Markov dynamics: the state $\boldsymbol{x}_{k+1}$ is determined solely by $\boldsymbol{x}_k$, independent of all previous states. That is, the future is determined entirely by the present state

[13] The mean of a Gaussian is the natural "typical value." Appendix A.8.7 online explains why.

[14] Graphical structures with more intricate topology can also be studied (Murphy, 2012). For example, for *trees*, a state can have multiple *parents*, which are states that connect to x_k. In a Markov chain, every state has just one parent. If we denote those states $\boldsymbol{x}_{\mathrm{pa}(k)}$, then the Markov rule becomes $p(\boldsymbol{x}_k|\boldsymbol{x}^{k-1}) = p(\boldsymbol{x}_k|\boldsymbol{x}_{\mathrm{pa}(k)})$. We can use the notion of parents to define an ordering, which makes trees similar to chains. Networks with *cycles* and with connecting *arcs* that lack directionality (*Markov random fields*) are still more complicated.

of the system – not on its history, nor on its observations. This is the notion of *conditional independence*: conditioning on $\boldsymbol{x}_k$"blocks" the influence of all other variables (Section A.9.2 online). More formally,

$$p(\boldsymbol{x}_{k+1}|\boldsymbol{x}^k, \boldsymbol{y}^k) = p(\boldsymbol{x}_{k+1}|\boldsymbol{x}_k)\,, \tag{8.68}$$

where $\boldsymbol{x}^k \equiv \{\boldsymbol{x}_1, \boldsymbol{x}_2, \ldots, \boldsymbol{x}_k\}$.[15] Similarly, we assume that observations are a memoryless function of the state alone.[16] That is,

$$p(\boldsymbol{y}_k|\boldsymbol{x}^k, \boldsymbol{y}^{k-1}) = p(\boldsymbol{y}_k|\boldsymbol{x}_k)\,. \tag{8.69}$$

In this probabilistic notation, the explicit functions $\boldsymbol{f}(\cdot)$ and $\boldsymbol{h}(\cdot)$ that describe state space dynamics (e.g., Eq. 2.10) are subsumed into the different probability distributions. The probabilistic state space model generalizes to discrete dynamics the $p(x|y)$ of Section 8.3.1.

Although the state $\boldsymbol{x}_{k+1}$ is influenced directly by the state $\boldsymbol{x}_k$, our knowledge of $\boldsymbol{x}_{k+1}$ is affected by observations – present, past, and even future (which might be available in an off-line analysis). This motivates considering various conditional probability distributions $p(\boldsymbol{x}_k|\cdot)$, which depend on differing sets of observations $\boldsymbol{y}_k$. In analogy with the distinction between current and prediction observers, we define (and distinguish between) two conditional probabilities for the state $\boldsymbol{x}$ at time $k+1$,

$$p(\boldsymbol{x}_{k+1}|\boldsymbol{y}^k)\,, \qquad \text{and} \qquad p(\boldsymbol{x}_{k+1}|\boldsymbol{y}^{k+1})\,. \tag{8.70}$$

The first $p(\cdot)$ is conditioned on the observations made between times 1 and k. The second includes also the observation $\boldsymbol{y}_{k+1}$. We will soon connect these quantities to $\hat{\boldsymbol{x}}^-_{k+1}$ and $\hat{\boldsymbol{x}}_{k+1}$.

As with the Kalman filter, we seek a recursive algorithm that has two stages: we *predict* the next state of the system, based on the present information, and then *update* our state estimate after receiving the new information (measurement). These two stages capture the idea that, as time passes, we can use all previous information to extrapolate our expectations into the future, but, when new information arrives, we should revise these expectations in the light of the new measurement. Schematically,

$$p(\boldsymbol{x}_k|\boldsymbol{y}^k) \xrightarrow[\text{predict}]{} p(\boldsymbol{x}_{k+1}|\boldsymbol{y}^k) \xrightarrow[\text{update}]{} p(\boldsymbol{x}_{k+1}|\boldsymbol{y}^{k+1})\,. \tag{8.71}$$

The *prediction step* uses the *Chapman–Kolmogorov* relation for Markov dynamics:

$$p(\boldsymbol{x}_{k+1}|\boldsymbol{y}^k) = \int \mathrm{d}\boldsymbol{x}_k\, p(\boldsymbol{x}_{k+1}|\boldsymbol{x}_k)\, p(\boldsymbol{x}_k|\boldsymbol{y}^k)\,. \tag{8.72}$$

See Section A.9.2 online. The dynamical equations are incorporated in the distribution $p(\boldsymbol{x}_{k+1}|\boldsymbol{x}_k)$, discussed below. We assume the dynamics is well described

[15] This attractively compact notation is common in the field of information theory. Control-theory books often use $\boldsymbol{x}_{1:k}$ and write the Markov assumption as $p(\boldsymbol{x}_{k+1}|\boldsymbol{x}_{1:k}, \boldsymbol{y}_{1:k}) = p(\boldsymbol{x}_{k+1}|\boldsymbol{x}_k)$. Often, the initial state is "far in the past" and need not be specified precisely. When the initial condition is important, we will write $\boldsymbol{x}^k_1$.

[16] In other words, the sensor has instantaneous dynamics. This assumption is not as restrictive as it might appear to be, as any sensor dynamics can be included in the system itself.

by discrete update equations, deferring discussion of hybrid dynamics to the next subsection.

To evaluate $p(\boldsymbol{x}_{k+1}|\boldsymbol{x}_k)$, we need to invoke a dynamical law, such as $\boldsymbol{x}_{k+1} = \boldsymbol{f}(\boldsymbol{x}_k, \boldsymbol{u}_k, \boldsymbol{v}_k)$, where the dynamics evolve according to a nonlinear function $\boldsymbol{f}(\cdot)$ that depends on the state $\boldsymbol{x}_k$, on deterministic controls $\boldsymbol{u}_k$, and on various noise sources $\boldsymbol{v}_k$. To keep things simple, we will assume that the input noise is independent of the state vector. Then

$$\begin{aligned} p(\boldsymbol{x}_{k+1}|\boldsymbol{x}_k) &= \int \mathrm{d}\boldsymbol{v}_k \, p(\boldsymbol{x}_{k+1}|\boldsymbol{x}_k, \boldsymbol{v}_k)\, p(\boldsymbol{v}_k|\boldsymbol{x}_k) && \text{marginalization} \\ &= \int \mathrm{d}\boldsymbol{v}_k \, \delta\left[\boldsymbol{x}_{k+1} - \boldsymbol{f}(\boldsymbol{x}_k, \boldsymbol{u}_k, \boldsymbol{v}_k)\right] p(\boldsymbol{v}_k) && \boldsymbol{v}_k \text{ independent of } \boldsymbol{x}_k \\ &= p(\boldsymbol{v}_k^*)\,, \end{aligned} \tag{8.73}$$

where $\boldsymbol{v}_k^*$ is determined by setting the argument of the delta function to zero. That is, $\boldsymbol{v}_k^*$ solves the equation $\boldsymbol{x}_{k+1} - \boldsymbol{f}(\boldsymbol{x}_k, \boldsymbol{u}_k, \boldsymbol{v}_k) = \boldsymbol{0}$. In the middle line of Eq. (8.73), the delta function arises because $\boldsymbol{x}_{k+1} = \boldsymbol{f}(\boldsymbol{x}_k, \boldsymbol{u}_k, \boldsymbol{v}_k)$ is *deterministic*, once we specify the noise term, $\boldsymbol{v}_k$. Also, since the process noise $\boldsymbol{v}_k$ is independent of the state $\boldsymbol{x}_k$, we have $p(\boldsymbol{v}_k|\boldsymbol{x}_k) = p(\boldsymbol{v}_k)$. Thus, $p(\boldsymbol{x}_{k+1}|\boldsymbol{x}_k)$ is simply the noise distribution function for $\boldsymbol{v}_k$, evaluated at $\boldsymbol{v}_k = \boldsymbol{v}_k^*$.

The *update step* uses *Bayes' Theorem* of probability (Section A.6 online) to write

$$p(\boldsymbol{x}_{k+1}|\boldsymbol{y}^{k+1}) = \frac{1}{Z}\, p(\boldsymbol{y}_{k+1}|\boldsymbol{x}_{k+1}, \boldsymbol{y}^k)\, p(\boldsymbol{x}_{k+1}|\boldsymbol{y}^k)\,, \tag{8.74}$$

where Z is a normalization constant, discussed below. Because the observation at $k+1$ depends *only* on the state $\boldsymbol{x}_{k+1}$, and nothing else, $p(\boldsymbol{y}_{k+1}|\boldsymbol{x}_{k+1}, \boldsymbol{y}^k) = p(\boldsymbol{y}_{k+1}|\boldsymbol{x}_{k+1})$.

By similar logic, $p(\boldsymbol{y}_k|\boldsymbol{x}_k) = p(\boldsymbol{\xi}_k^*)$, which is the noise distribution for $\boldsymbol{\xi}_k$, evaluated for given $\boldsymbol{y}_k$ and $\boldsymbol{x}_k$ by solving the equation $\boldsymbol{y}_k = \boldsymbol{h}(\boldsymbol{x}_k, \boldsymbol{\xi}_k^*)$ for $\boldsymbol{\xi}_k^*$. Here, $\boldsymbol{h}(\cdot)$ is the (possibly nonlinear) function relating the current observation to the current state $\boldsymbol{x}_k$ and noise $\boldsymbol{\xi}_k$. Finally, the normalization constant Z, equivalent to the *partition function* of statistical mechanics, is fixed by requiring $\int \mathrm{d}\boldsymbol{x}_k\, p(\boldsymbol{x}_k|\boldsymbol{y}^k) = 1$ for each k, which leads to

$$Z = \int \mathrm{d}\boldsymbol{x}_{k+1}\, p(\boldsymbol{y}_{k+1}|\boldsymbol{x}_{k+1})\, p(\boldsymbol{x}_{k+1}|\boldsymbol{y}^k) = p(\boldsymbol{y}_{k+1}|\boldsymbol{y}^k)\,. \tag{8.75}$$

Note that the partition function depends on time (i.e., $Z = Z_k$); however, we will always understand that Z represents whatever normalization is required for the distribution at hand and not bother to give it a different name in each case.

We collect the *Bayesian filtering equations* for convenience:

$$p(\boldsymbol{x}_{k+1}|\boldsymbol{y}^k) = \int \mathrm{d}\boldsymbol{x}_k\, p(\boldsymbol{x}_{k+1}|\boldsymbol{x}_k)\, p(\boldsymbol{x}_k|\boldsymbol{y}^k) \qquad \text{predict} \tag{8.76a}$$

$$\downarrow \qquad\qquad\qquad\qquad \downarrow$$

$$p(\boldsymbol{x}_{k+1}|\boldsymbol{y}^{k+1}) = \frac{1}{Z}\, p(\boldsymbol{y}_{k+1}|\boldsymbol{x}_{k+1})\, p(\boldsymbol{x}_{k+1}|\boldsymbol{y}^k) \qquad \text{update} \tag{8.76b}$$

$$p(\boldsymbol{x}_{k+1}|\boldsymbol{x}_k) \quad \longleftrightarrow \quad \boldsymbol{x}_{k+1} = \boldsymbol{f}(\boldsymbol{x}_k, \boldsymbol{u}_k, \boldsymbol{v}_k) \qquad \text{dynamics} \tag{8.77a}$$

$$p(\boldsymbol{y}_k|\boldsymbol{x}_k) \quad \longleftrightarrow \quad \boldsymbol{y}_k = \boldsymbol{h}(\boldsymbol{x}_k, \boldsymbol{\xi}_k) \qquad \text{measurement} \tag{8.77b}$$

Problem 8.10 shows more concretely how these relations work in a simple case. Together, Eqs. (8.76) and (8.77) solve, in principle, the state-estimation problem. Unfortunately, the simplicity of the equations hides considerable complexity. One issue is that the integrals in Eq. (8.76a) and in Z are over the n-dimensional state space. Still, although the general solution turns out to be intractable in practice, we will see how to apply the Bayesian approach exactly in special cases and approximately elsewhere.

8.3.3 Discrete State Spaces

The Bayesian formalism is particularly simple and natural when the values of states are restricted to a finite set. So far, we have considered the control of systems whose state spaces have elements $\boldsymbol{x}$ that are continuous sets. This was true whether the dynamical evolution in time was continuous or discrete. But often it is natural to think about systems whose internal states are discrete. In the simplest case, there are just two configurations. Historically, such systems have been considered more by operations research, artificial intelligence, and other computer-science disciplines than by the control community. As a result, they often seem like distinct topics, when in fact they are closely related.

We present the basics in Chapter 12. Beyond the direct applications, a motivation is that the Bayesian formalism of Eqs. (8.76) and (8.77) is easily applied to cases with arbitrary dynamics $P(\boldsymbol{x}_{k+1}|\boldsymbol{x}_k)$ and arbitrary observation relations $P(\boldsymbol{y}_{k+1}|\boldsymbol{x}_{k+1})$. By contrast, generalizing Bayesian filtering to nonlinear equations and observation relations on continuous state spaces is difficult and possible only in special cases.

8.3.4 Hybrid Dynamics

The above discussion is for discrete dynamics and discrete measurements. The generalization to the hybrid case of continuous dynamics and discrete measurements parallels the example shown in Figure 8.2. In the prediction step, the one-step-ahead prediction $p(\boldsymbol{x}_{k+1}|\boldsymbol{y}^k)$ is replaced by the continuous-time prediction $p[\boldsymbol{x}(t)|\boldsymbol{y}^k]$, where the state $\boldsymbol{x}$ is evaluated at an arbitrary time $t > t_k$, with t_k the time of the previous observation. When the next observation arrives at time t_{k+1}, we update using Bayes' theorem, Eq. (8.76b).

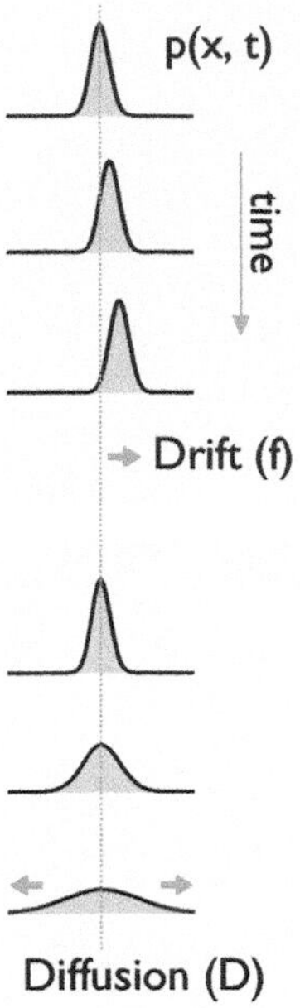

The probability density function $p[\boldsymbol{x}(t)|\boldsymbol{y}^k]$ obeys the *Fokker–Planck* equation. Indeed, writing down Eq. (8.76a) is an early step in the derivation of the Fokker–Planck equation. For continuous dynamics of the form

$$\dot{\boldsymbol{x}} = \boldsymbol{f}(\boldsymbol{x}) + \boldsymbol{g}(\boldsymbol{x})\boldsymbol{\nu}, \tag{8.78}$$

the Fokker–Planck equation for the probability distribution $p[\boldsymbol{x}(t)|\boldsymbol{x}(t=0)]$ is given by

$$\partial_t p = -\boldsymbol{\nabla} \cdot [\boldsymbol{f}(\boldsymbol{x})\, p] + \nabla_{\boldsymbol{x}}^2(\boldsymbol{D}\, p), \tag{8.79}$$

with $\boldsymbol{D} = \frac{1}{2}\boldsymbol{g}\,\boldsymbol{g}^{\mathsf{T}}$ and $\nabla_{\boldsymbol{x}}^2 \equiv \sum_{i,j} \partial_{x_i}\partial_{x_j}$ the Laplacian operator. In the example of a freely diffusing particle, $f = 0$, and the diffusion tensor $\boldsymbol{D}$ was a scalar diffusion D, implying that the probability distribution p obeys a simple diffusion equation. The diffusion equation implies that the variance increases linearly with time as $2Dt$. Thus, prediction over the time interval T_s, from k to $k+1$, is handled by the continuous Fokker–Planck equation, but periodic, instantaneous measurements are incorporated as before. At time $t_k = kT_s$, we use the Bayes rule to update the prior $p(\boldsymbol{x}_k)$ as described by Eq. (8.76b).

The Fokker–Planck equation is easy to understand qualitatively. The $-\boldsymbol{\nabla}\cdot(\boldsymbol{f}\,p)$ term forces the distribution to drift along the direction of $\boldsymbol{f}$ in state space. If $\boldsymbol{f}$ is independent of $\boldsymbol{x}$, then the whole probability distribution is translated uniformly. If $\boldsymbol{f}(\boldsymbol{x})$ is state dependent, then different parts of $p(\boldsymbol{x}, t)$ drift by different amounts and directions. The $\nabla_{\boldsymbol{x}}^2(\boldsymbol{D}p)$ term, as we have already seen, gives rise to a diffusive spreading of $p(\boldsymbol{x}, t)$. In higher dimensions, the spreading can be anisotropic – different in different directions.

Analytically solving the Fokker–Planck equations for state-dependent forcing and diffusion is usually impossible. Even numerically, there are difficulties. For example, since p is a probability density, it cannot be allowed to become negative.

8.3.5 Continuous Measurements

For completeness, we write down the filtering equations for the case with additive noise and continuous measurements. The dynamics and measurement equations become

$$\dot{\boldsymbol{x}} = \boldsymbol{f}(\boldsymbol{x}) + \boldsymbol{g}(\boldsymbol{x})\boldsymbol{\nu}, \qquad \boldsymbol{y}(t) = \boldsymbol{h}(\boldsymbol{x}) + \boldsymbol{\xi}, \tag{8.80}$$

with covariances $\boldsymbol{Q}_\eta$ and $\boldsymbol{Q}_\xi$ for the noise terms. We then consider the probability distribution $p(\boldsymbol{x}, t|\boldsymbol{y})$, which is the probability density for the state $\boldsymbol{x}$, given the time series of observations $\boldsymbol{y}(t)$ over the time interval $(0, t)$. It obeys the *Kushner–Stratonovich equation*,

$$\partial_t p = \underbrace{-\boldsymbol{\nabla} \cdot [\boldsymbol{f}(\boldsymbol{x})\, p] + \nabla_{\boldsymbol{x}}^2(\boldsymbol{D}\, p)}_{\text{Fokker-Planck}} + \underbrace{\left[\boldsymbol{h}(\boldsymbol{x}) - \langle \boldsymbol{h}(\boldsymbol{x})\rangle_y\right]^{\mathsf{T}} \boldsymbol{Q}_\xi^{-1} \left[\boldsymbol{y} - \langle \boldsymbol{h}(\boldsymbol{x})\rangle_y\right] p}_{\text{measurements}}\,. \tag{8.81}$$

Because the averages $\langle \cdots \rangle_y$ are with respect to the probability distribution $p(\boldsymbol{x}, t, |\boldsymbol{y})$, the Kushner–Stratonovich equation is both nonlinear in p and stochastic (because the stochastic measurements $y(t)$ appear). Although the equations are typically impossible to solve analytically, we can see explicitly how the stochastic observations enter as a driving term for $p(\boldsymbol{x}, t)$. Not surprisingly, in the limit of linear dynamics and Gaussian noise, Eq. (8.81) reduces to the Kalman–Bucy filter, Eqs. (8.48) and (8.50).

8.3.6 From Bayes to Kalman

From the general prediction-update relations, Eqs. (8.76) and (8.77), we can recover our previous results for the Kalman filter. The dynamics and measurement relations Eq. (8.77) become the familiar

$$\begin{aligned} &\boldsymbol{x}_{k+1} = \boldsymbol{f}(\boldsymbol{x}_k, \boldsymbol{u}_k, \boldsymbol{\nu}_k) = \boldsymbol{A}\boldsymbol{x}_k + \boldsymbol{B}\boldsymbol{u}_k + \boldsymbol{\nu}_k\,, \qquad \boldsymbol{y}_k = \boldsymbol{h}(\boldsymbol{x}_k, \boldsymbol{\xi}_k) = \boldsymbol{C}\boldsymbol{x}_k + \boldsymbol{\xi}_k\,, \\ &\langle \boldsymbol{\nu}_k \rangle = \langle \boldsymbol{\xi}_k \rangle = \boldsymbol{0}\,, \qquad \langle \boldsymbol{\nu}_k \boldsymbol{\nu}_\ell^{\mathsf{T}} \rangle = \boldsymbol{Q}_\nu \, \delta_{k\ell}\,, \quad \langle \boldsymbol{\xi}_k \boldsymbol{\xi}_\ell^{\mathsf{T}} \rangle = \boldsymbol{Q}_\xi \, \delta_{k\ell}\,, \quad \langle \boldsymbol{\nu}_k \boldsymbol{\xi}_\ell^{\mathsf{T}} \rangle = \boldsymbol{0}\,. \end{aligned} \tag{8.82}$$

Since the noise terms are all drawn from Gaussian distributions, they are fully characterized by their mean and covariance matrix. The prediction and update relations, Eqs. (8.76), then turn into rules for propagating the mean and variances. The key point is that since the linear dynamical system is driven by white noise, the state vector will also be a vector of Gaussian random variables (Appendix A.6.5 online).

We now sketch how the familiar Kalman update rules emerge from the general Bayesian formulation. Since the algebra can be unpleasant, we just quote some of the steps here. In Problem 8.10, you can work through a simple, one-dimensional example that illustrates the essential features of the general case.

The goal is to show by recurrence that $p(\boldsymbol{x}_k|\boldsymbol{y}^k) = \mathcal{N}(\boldsymbol{x}_k;\, \hat{\boldsymbol{x}}_k,\, \boldsymbol{P}_k)$, with the state estimate $\hat{\boldsymbol{x}}_k$ and covariance $\boldsymbol{P}_k$ defined as in Section 8.1.2. The first step is to use the Chapman–Kolmogorov relation to show that the distribution of the predicted state is Gaussian:

$$p(\boldsymbol{x}_{k+1}|\boldsymbol{y}^k) \sim \mathcal{N}(\boldsymbol{x}_{k+1};\, \hat{\boldsymbol{x}}_{k+1}^-,\, \boldsymbol{P}_{k+1}^-)\,, \tag{8.83}$$

where $\mathcal{N}(\boldsymbol{x};\, \boldsymbol{\mu}, \boldsymbol{P})$ denotes a multivariate Gaussian distribution for $\boldsymbol{x}$ with mean $\boldsymbol{\mu}$ and covariance matrix $\boldsymbol{P}$ (see Section A.7.5 online). Here, the predicted mean $\hat{\boldsymbol{x}}_{k+1}^-$ and covariance matrix $\boldsymbol{P}_{k+1}^-$ are given by Eq. (8.41). With similar logic, the two other terms in the Bayesian update step are also Gaussian:

$$\begin{aligned} p(\boldsymbol{y}_{k+1}|\boldsymbol{x}_{k+1}) &\sim \mathcal{N}(\boldsymbol{y}_{k+1} - \boldsymbol{C}\boldsymbol{x}_{k+1};\, \boldsymbol{0},\, \boldsymbol{Q}_\xi) \\ p(\boldsymbol{y}_{k+1}|\boldsymbol{y}^k) &\sim \mathcal{N}(\boldsymbol{y}_{k+1};\, \boldsymbol{C}\hat{\boldsymbol{x}}_{k+1}^-,\, \boldsymbol{C}\boldsymbol{P}_{k+1}^-\boldsymbol{C}^{\mathsf{T}} + \boldsymbol{Q}_\xi)\,. \end{aligned} \tag{8.84}$$

The last, and algebraically most unpleasant, step is to show that the result of the Bayes relation is indeed Gaussian:

$$p(\boldsymbol{x}_{k+1}|\boldsymbol{y}^{k+1}) = \frac{p(\boldsymbol{y}_{k+1}|\boldsymbol{x}_{k+1})\, p(\boldsymbol{x}_{k+1}|\boldsymbol{y}^k)}{p(\boldsymbol{y}_{k+1}|\boldsymbol{y}^k)} \sim \mathcal{N}(\boldsymbol{x}_{k+1};\, \hat{\boldsymbol{x}}_{k+1},\, \boldsymbol{P}_{k+1})\,, \tag{8.85}$$

with the updated mean $\hat{\boldsymbol{x}}_{k+1}$ and covariance matrix $\boldsymbol{P}_{k+1}$ given by Eq. (8.41).

Although the derivation involves complicated algebra, the result is pleasing and intuitive. At each time step, the state vector $\boldsymbol{x}_k$ is a Gaussian random variable. Since Gaussian distributions are completely characterized by their mean and covariance matrix, the evolution equations for $\hat{\boldsymbol{x}}_k$ and $\boldsymbol{P}_k$ completely specify the posterior distribution at each time k. To use the jargon of Bayesian statistics, for linear dynamics with Gaussian noise, a Gaussian initial condition is a *conjugate prior*: its posterior pdf belongs to the same family of distributions (Gaussian, here). Only the parameters change.

8.4 Nonlinear Filtering

The Bayesian formulation of the state-estimation problem opens the door to generalizing the Kalman filter to situations beyond linear dynamics and Gaussian white noise. Unfortunately, the general equations (Eqs. 8.76 and 8.77 for discrete-time and Eq. 8.81 for continuous-time dynamics) are hard to solve. There are two overall strategies:

- *Local approaches* (Sections 8.4.1 and 8.4.2). These methods "project" distributions onto a Gaussian form at each time step and work well for weak nonlinearities.
- *Global approaches* (Sections 8.4.3 and 8.4.4). These solve approximately, using deterministic or Monte Carlo numerical techniques, the general Bayesian filter equations.

8.4.1 Extended Kalman Filter (EKF)

The "obvious" extension of the Kalman filter, the EKF, uses the nonlinear equations to predict the mean values of state and observation and a local linearization to propagate the covariances. The dynamics are given by $\boldsymbol{x}_{k+1} = \boldsymbol{f}(\boldsymbol{x}_k, \boldsymbol{u}_k, \boldsymbol{\nu}_k)$ and $\boldsymbol{y}_k = \boldsymbol{h}(\boldsymbol{x}_k, \boldsymbol{\xi}_k)$. If nonlinearities are weak enough that distributions stay close to Gaussians, tracking means and covariances is enough. The goal is still to estimate the mean value of the state $\hat{\boldsymbol{x}}_k$ and its covariance matrix $\boldsymbol{P}_k$.

To predict the evolution of mean values, we use the full nonlinear equations:[17]

$$\hat{\boldsymbol{x}}^-_{k+1} = \boldsymbol{f}(\hat{\boldsymbol{x}}_k, \boldsymbol{u}_k), \qquad \hat{\boldsymbol{y}}_{k+1} = \boldsymbol{h}(\hat{\boldsymbol{x}}^-_{k+1}), \tag{8.86}$$

where, in a slight abuse of notation, $\boldsymbol{f}(\hat{\boldsymbol{x}}_k, \boldsymbol{u}_k) \equiv \boldsymbol{f}(\hat{\boldsymbol{x}}_k, \boldsymbol{u}_k, \boldsymbol{0})$ and $\boldsymbol{h}(\hat{\boldsymbol{x}}^-_{k+1}) \equiv \boldsymbol{h}(\hat{\boldsymbol{x}}^-_{k+1}, \boldsymbol{0})$, with zeroes representing the mean values of the noise.

The EKF then uses the linearized dynamics, Eq. (8.41), to predict and update the covariance matrix $\boldsymbol{P}_k$. The local linearization for propagating covariances has matrices

$$\boldsymbol{A}_k = \left.\frac{\partial \boldsymbol{f}}{\partial \boldsymbol{x}}\right|_{\hat{\boldsymbol{x}}_k, \boldsymbol{u}_k, \boldsymbol{0}} \qquad \boldsymbol{B}_k = \left.\frac{\partial \boldsymbol{f}}{\partial \boldsymbol{u}}\right|_{\hat{\boldsymbol{x}}_k, \boldsymbol{u}_k, \boldsymbol{0}} \qquad \boldsymbol{C}_{k+1} = \left.\frac{\partial \boldsymbol{h}}{\partial \boldsymbol{x}}\right|_{\hat{\boldsymbol{x}}^-_{k+1}, \boldsymbol{0}}. \tag{8.87}$$

[17] Even though Eq. (8.86) uses the nonlinear equations, it is nonetheless an approximation as, in general, $\langle f(x)\rangle \neq f(\langle x\rangle)$. For example, if $f(x) = x^2$ and $x \sim \mathcal{N}(\mu = 0, \sigma^2)$, then $\langle x\rangle = 0$ but $\langle x^2\rangle = \sigma^2 \neq 0$. But when $\sigma \ll \mu$, the approximation is reasonable. See also the discussion of Jensen's inequality in Section A.6.5 online.

The matrices $(\boldsymbol{A}_k, \boldsymbol{B}_k, \boldsymbol{C}_k)$ depend on time because the linearization base $(\boldsymbol{x}_k, \boldsymbol{u}_k)$ keeps changing. The covariance of the noise terms also depend on k:

$$\boldsymbol{Q}_k^{\nu} = \boldsymbol{M}_k \boldsymbol{Q}_{\nu} \boldsymbol{M}_k^{\mathsf{T}} , \qquad \boldsymbol{Q}_k^{\xi} = \boldsymbol{N}_k \boldsymbol{Q}_{\xi} \boldsymbol{N}_k^{\mathsf{T}}$$
$$\text{where} \qquad \boldsymbol{M}_k = \left. \frac{\partial \boldsymbol{f}}{\partial \boldsymbol{\nu}} \right|_{\hat{\boldsymbol{x}}_k, \boldsymbol{u}_k, \boldsymbol{0}} , \qquad \boldsymbol{N}_k = \left. \frac{\partial \boldsymbol{h}}{\partial \boldsymbol{\xi}} \right|_{\hat{\boldsymbol{x}}_k, \boldsymbol{0}} . \tag{8.88}$$

If noise is additive, with $\boldsymbol{x}_{k+1} = \boldsymbol{f}(\boldsymbol{x}_k, \boldsymbol{u}_k) + \boldsymbol{\nu}_k$ and $\boldsymbol{y}_k = \boldsymbol{h}(\boldsymbol{x}_k) + \boldsymbol{\xi}_k$, then $\boldsymbol{M}$ and $\boldsymbol{N}$ are identity matrices.

Notice that our previous Kalman-filter (KF) derivation is also valid when the dynamical matrices $(\boldsymbol{A}, \boldsymbol{B}, \boldsymbol{C})$ and covariances $(\boldsymbol{Q}_{\nu}, \boldsymbol{Q}_{\xi})$ depend on time; simply replace $\boldsymbol{A} \to \boldsymbol{A}_k$, and so on. Of course, if the dynamical matrices change, the covariance and Kalman gain matrices will not be stationary. Unlike the ordinary Kalman filter, the EKF gains and covariances usually do not reach a steady state: they reflect the history of motion in state space.

These tweaks to the KF affect only the predictions of mean and covariance. The form of update equations is unaltered. The EKF algorithm is then (dropping *s in the notation),

$$\left.\begin{aligned}
\hat{\boldsymbol{x}}_{k+1}^- &= \boldsymbol{f}(\hat{\boldsymbol{x}}_k, \boldsymbol{u}_k) && \text{state mean} \\
\hat{\boldsymbol{y}}_{k+1} &= \boldsymbol{h}(\hat{\boldsymbol{x}}_{k+1}^-) && \text{observation mean} \\
\boldsymbol{P}_{k+1}^- &= \boldsymbol{A}_k \boldsymbol{P}_k \boldsymbol{A}_k^{\mathsf{T}} + \boldsymbol{Q}_k^{\nu} && \text{state covariance} \\
\boldsymbol{P}_{k+1}^{y} &= \boldsymbol{C}_{k+1} \boldsymbol{P}_{k+1}^- \boldsymbol{C}_{k+1}^{\mathsf{T}} + \boldsymbol{Q}_k^{\xi} && \text{observation covariance} \\
\boldsymbol{P}_{k+1}^{xy} &= \boldsymbol{P}_{k+1}^- \boldsymbol{C}_{k+1}^{\mathsf{T}} && \text{state-observation cov.}
\end{aligned}\right\} \text{predict} \tag{8.89a}$$

$$\left.\begin{aligned}
\boldsymbol{L}_{k+1} &= \boldsymbol{P}_{k+1}^{xy} \left(\boldsymbol{P}_{k+1}^{y} \right)^{-1} && \text{observer gain} \\
\hat{\boldsymbol{x}}_{k+1} &= \hat{\boldsymbol{x}}_{k+1}^- + \boldsymbol{L}_{k+1} (\boldsymbol{y}_{k+1} - \hat{\boldsymbol{y}}_{k+1}) && \text{state mean} \\
\boldsymbol{P}_{k+1} &= \boldsymbol{P}_{k+1}^- - \boldsymbol{L}_{k+1} \boldsymbol{P}_{k+1}^{y} \boldsymbol{L}_{k+1}^{\mathsf{T}} && \text{state covariance}
\end{aligned}\right\} \text{update} \tag{8.89b}$$

For weak nonlinearities, the EKF can work well. In the 1960s, it guided astronauts to the Moon and back. Extended Kalman filters are also standard in Global Positioning System (GPS) devices, which integrate noisy timing signals from satellites into a position estimate. In this section, we consider two one-dimensional examples, one successful and the other not. The failure will motivate further approaches.

Example 8.2 (Soft spring) We explore the motion of a Brownian (overdamped) particle in a *soft-spring* potential. Specifically, we let $U(x) = -\mathrm{e}^{-\frac{1}{2}x^2}$, which implies a local force $f(x) = -\partial_x U = -x\,\mathrm{e}^{-\frac{1}{2}x^2}$. The discrete dynamics has the form

$$x_{k+1} = x_k \left(1 - a\,\mathrm{e}^{-\frac{1}{2}x_k^2} \right) + \nu_k , \qquad y_k = x_k + \xi_k , \tag{8.90}$$

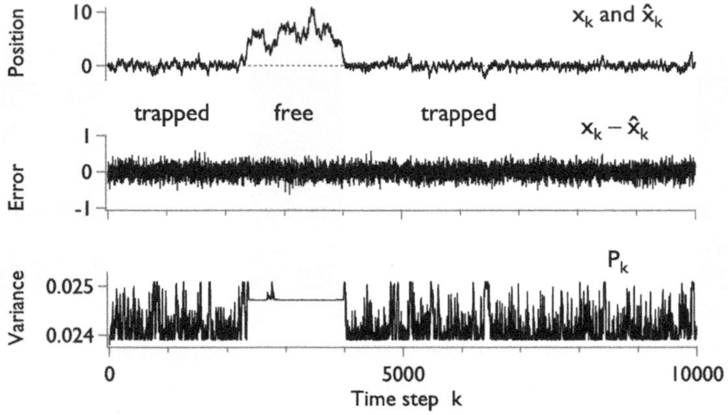

Fig. 8.4 Extended Kalman filter for particle motion in a soft potential, . The well-depth parameter $a = 0.1$. The noise standard deviations are $\nu = \xi = 0.2$. The true state x_k and estimate $\hat{x}_k$ are indistinguishable when plotted together (top).

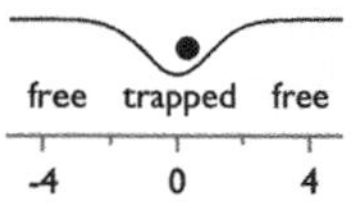

where the noise terms are both independent Gaussian random variables of mean zero and variance ν^2 and ξ^2. The constant a is proportional to T_s and the overall scale of the force. The linearization gives $A_k = 1 - \left(1 - x_k^2\right) e^{-\frac{1}{2}x_k^2}$, with $B_k = 0$ and $C_k = M_k = N_k = 1$. The noise terms are ν^2 and ξ^2. In the sketch at left, motion is *trapped* in the central potential well and *free* outside, where the particle undergoes Brownian motion.

Figure 8.4 illustrates EKF performance in Example 8.2, showing trapped and free motion.[18] Because the EKF constantly recalculates the local dynamical constant A_k, it can handle both types of motion without a glitch, as indicated by the time series of the estimation error. The variance also settles down to a steady state. During the periods of free diffusion, the variance "freezes" because the dynamical constant $A_k \to 1$ away from the trapping region. Finally, because the motion is accurately described by the extended Kalman filter, the estimate of the steady-state variance, $P = 0.0243$, is very close to the observed variance of the estimation error, 0.0245. Thus, the filter performs better than just using the observations (variance $\xi^2 = 0.04$) and accurately assesses its own performance.

The next example resembles the previous, but the outcome is less successful.

Example 8.3 (Double-well potential) Consider noisy observations of a Brownian particle in a double-well potential, $U(x) = -\frac{1}{2}x^2 + \frac{1}{4}x^4$, with force $f(x) = -\partial_x U = x - x^3$ and

[18] In this one-dimensional example, a freely diffusing particle always returns to the origin. Not so in three dimensions. For optical tweezers, for example, once a particle escapes, it is unlikely to be trapped again.

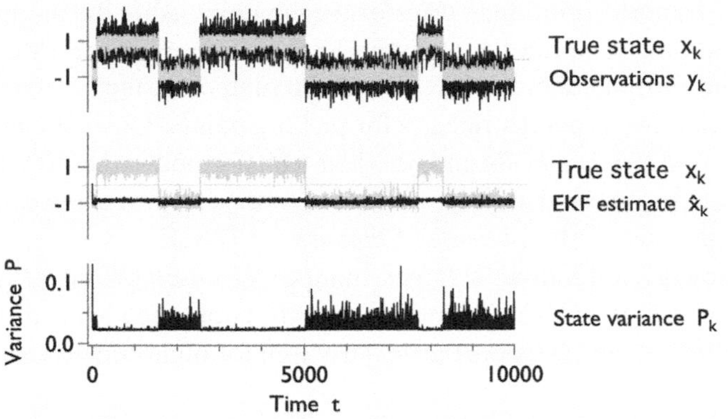

EKF for particle motion in a double-well potential. The noise terms are $\nu = 0.15$ and $\xi = 0.5$, and $a = 0.8$. Fig. 8.5

local equilibria at ± 1, as illustrated at right. The discrete equations are

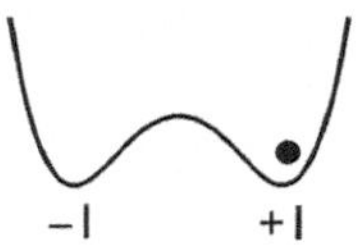

$$\begin{aligned} x_{k+1} &= x_k + a(x_k - x_k^3) + \nu_k\,, \qquad y_k = x_k + \xi_k\,, \\ &= 2x_k - x_k^3 + \nu_k\,, \end{aligned} \tag{8.91}$$

where, for convenience, we choose the time constant and potential scale so that $a \propto T_s = 1$. Although using such a large time step T_s means that the motion will deviate from the continuous dynamics, the qualitative behavior is similar. The additive noise terms have $\nu_k \sim \mathcal{N}(0, \nu^2)$ and $\xi_k \sim \mathcal{N}(0, \xi^2)$.

Figure 8.5 illustrates the EKF in Example 8.3. Notice that the estimate gets stuck in one well regardless of which well the particle is actually in. Even worse, the variance P_k actually *decreases* when the particle is in the wrong well: the filter is unaware of its own failings. To understand why, note that near the ± 1 local equilibria $A \approx 1$, so that $\hat{x}_{k+1}^- \approx \hat{x}_k$, even if the particle is in the wrong well. The variance P is then roughly constant and thus pays little attention to the observation. More fundamentally, near the top of the potential barrier, the predicted state distribution is not close to a Gaussian but rather is a double-peaked function that has one peak over each well (Section 8.5.2).

How serious are the limitations of the EKF? If the goal is to estimate the state of a continuous system using discrete measurements, then the nonlinearity of the dynamics, the function $f(\cdot)$, depends on the time interval T_s. Measuring more frequently (reducing T_s) will make the dynamics more linear and improve the EKF, even if the "global" nonlinearity of the dynamics is significant (Problem 8.13).

8.4.2 Ensemble Approaches

Example 8.3 shows that the EKF fails when nonlinearities are too strong. To do better, we can retain the local-Gaussian approximation but try to improve the accuracy of transformed moments. One idea is to represent the distribution by an ensemble of points "drawn" from the prior, then transform each point using the nonlinear dynamics and observations, compute the mean and covariance of the transformed points, and then use these estimates in the standard Kalman update step.

We shall explore two general strategies for picking points. One is to choose a small number of points *deterministically* and use their transformed values to estimate posterior means and covariances. This approach leads to the *cubature Kalman filter* (CKF), as well as variants such as the *unscented Kalman filter* (UKF).

A second strategy is to draw a larger number of ensemble elements randomly from the prior distribution. This Monte Carlo (MC) approach leads to the *ensemble Kalman filter* (EnKF), which works particularly well for high-dimensional state-space models.

Cubature Kalman Filter (CKF)

To understand the basic idea behind deterministic-sampling approaches such as the CKF, we recall that one of the limitations of the extended Kalman filter (EKF) is that it predicts the new state $\hat{\boldsymbol{x}}_{k+1}^-$ by propagating the old estimate $\hat{\boldsymbol{x}}_k$ through the nonlinear equation describing the dynamics: $\hat{\boldsymbol{x}}_{k+1}^- = \boldsymbol{f}(\hat{\boldsymbol{x}}_k, \boldsymbol{u}_k)$. The problem is that, as noted in the footnote to Eq. (8.86), $\langle f(x)\rangle \neq f(\langle x\rangle)$. To see this in a simple scalar-variable context, let us assume that $x \sim \mathcal{N}(\mu, \sigma^2)$ is a Gaussian random variable. Then we can improve our estimate of $\langle f(x)\rangle$ by Taylor expanding the function about $x = \langle x\rangle = 0$, which gives,

$$\langle f(x)\rangle \approx f(\mu)\,(1) + f'(\mu)\,\underbrace{\langle x-\mu\rangle}_{0} + \frac{1}{2}f''(\mu)\,\underbrace{\langle (x-\mu)^2\rangle}_{\sigma^2} = f(\mu) + \frac{1}{2}f''(\mu)\,\sigma^2\,. \tag{8.92}$$

Implementing an algorithm based on Eq. (8.92) is problematic, as it depends on taking the second derivative of the function $f(x)$. Any uncertainty in the functional form would be greatly amplified by the two derivatives. However, it is easy to find an approximate equivalent expression that needs no derivatives. Consider the sum,

$$\begin{aligned}\frac{1}{2}\left[f(\mu+\sigma) + f(\mu-\sigma)\right] &= \frac{1}{2}\Big[f(\mu) + f'(\mu)\,\sigma + \frac{1}{2}f''(\mu)\,\sigma^2 + \frac{1}{6}f'''(\mu) + \cdots \\ &\quad + f(\mu) - f'(\mu)\,\sigma + \frac{1}{2}f''(\mu)\,\sigma^2 - \frac{1}{6}f'''(\mu) + \cdots\Big] \\ &= f(\mu) + \frac{1}{2}f''(\mu)\,\sigma^2 + \cdots\,,\end{aligned} \tag{8.93}$$

which matches $\langle f(x)\rangle$ in Eq. (8.92) to 4th order. For example, consider $f(x) = x^2$, with $x \sim \mathcal{N}(0, \sigma^2)$. The transformed variable has $\langle x^2\rangle = \sigma^2$, which matches the approximation, $\frac{1}{2}[\sigma^2 + (-\sigma)^2]$. However, the variance is less well approximated (Problem 8.14).

To improve the estimate of quantities such as $\langle f(x)\rangle$, we can use more than two points, an idea that connects to classic approaches in numerical analysis. Indeed, in

Gauss–Hermite quadrature, we write, to order m,

$$\int_{-\infty}^{\infty} \mathrm{d}x\, \mathrm{e}^{-x^2} f(x) \approx \sum_{i=1}^{m} w_i f(x_i), \tag{8.94}$$

where the *node points* x_i and *weights* w_i are chosen so that the integrals are exact for monomial functions $f(x) = x^i$, for $i \in (0, 2m-1)$.[19]

The recipe given above corresponds to $m = 2$, and the results improve rapidly with increasing m. However, n-dimensional state vectors require m^n evaluations, which can be a large number. We will discuss this kind of approach further in Section 8.4.3, which considers the full numerical solution of the Bayesian filtering equations.

The *cubature Kalman filter* (CKF) uses the lowest-order (two-point) approximation, which allows larger values of n. For an n-dimensional state vector $\boldsymbol{x} \sim \mathcal{N}(\hat{\boldsymbol{x}}, \boldsymbol{P})$, the $2n$ *cubature points* are chosen symmetrically about each coordinate,[20] with $\hat{\boldsymbol{x}}_i = \hat{\boldsymbol{x}} + \left(\sqrt{n}\sqrt{\boldsymbol{P}}\right) \boldsymbol{e}_i$. Here, the variance $\boldsymbol{P} = \sqrt{\boldsymbol{P}}(\sqrt{\boldsymbol{P}})^{\mathrm{T}}$ uses Cholesky decomposition (Section A.1.5 online), and $\boldsymbol{e}_i$ is a positive ($i = 1, \cdots, n$) or negative ($i = n+1, \cdots, 2n$) unit vector:

$$\{\boldsymbol{e}_i\} = \left\{ \underbrace{\begin{pmatrix}1\\0\\0\\\vdots\\0\end{pmatrix}, \begin{pmatrix}0\\1\\0\\\vdots\\0\end{pmatrix}, \cdots,}_{i=1,\cdots,n} \underbrace{\begin{pmatrix}-1\\0\\0\\\vdots\\0\end{pmatrix}, \begin{pmatrix}0\\-1\\0\\\vdots\\0\end{pmatrix}, \cdots}_{i=n+1,\cdots,2n} \right\}. \tag{8.95}$$

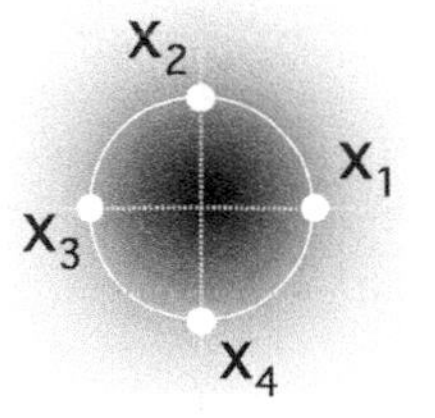

The $n = 2$ case is sketched at right.

Using the cubature ensemble for $\hat{\boldsymbol{x}}_k$ and $\hat{\boldsymbol{x}}^-_{k+1}$ gives the CKF algorithm:

$$\left.\hat{\boldsymbol{x}}_{i,k} = \hat{\boldsymbol{x}}_k + \left(\sqrt{n}\sqrt{\boldsymbol{P}_k}\right)\boldsymbol{e}_i, \qquad \hat{\boldsymbol{x}}^-_{i,k+1} = \hat{\boldsymbol{x}}^-_{k+1} + \left(\sqrt{n}\sqrt{\boldsymbol{P}^-_{k+1}}\right)\boldsymbol{e}_i \right\} \text{ensemble} \tag{8.96a}$$

[19] The matching works because there are $2m$ equations and $2m$ unknowns (m weights and m nodes).

[20] The term *cubature* has origins in numerical analysis, where it can refer to numerical integration in $n > 1$ dimensions. The term *quadrature* usually refers to integrals in one dimension. In addition, the cubature points are given by the intersection between an n-dimensional sphere and the coordinate axes.

$$\left.\begin{aligned}
&\hat{x}^-_{i,k+1} = f(\hat{x}_{i,k}, u_k)\,, \qquad \hat{x}^-_{k+1} = \frac{1}{2n}\sum_i \hat{x}^-_{i,k+1} && \text{state mean}\\
&\hat{y}_{i,k+1} = h(\hat{x}^-_{i,k+1})\,, \qquad \hat{y}_{k+1} = \frac{1}{2n}\sum_i \hat{y}_{i,k+1} && \text{obs. mean}\\
&P^-_{k+1} = \frac{1}{2n}\sum_i \left(\hat{x}^-_{i,k+1} - \hat{x}^-_{k+1}\right)\left(\hat{x}^-_{i,k+1} - \hat{x}^-_{k+1}\right)^{\mathrm{T}} + Q_\nu && \text{state cov.}\\
&P^y_{k+1} = \frac{1}{2n}\sum_i \left(\hat{y}_{i,k+1} - \hat{y}_{k+1}\right)\left(\hat{y}_{i,k+1} - \hat{y}_{k+1}\right)^{\mathrm{T}} + Q_\xi && \text{obs. cov.}\\
&P^{xy}_{k+1} = \frac{1}{2n}\sum_i \left(\hat{x}^-_{i,k+1} - \hat{x}^-_{k+1}\right)\left(\hat{y}_{i,k+1} - \hat{y}_{k+1}\right)^{\mathrm{T}} && \text{state-obs. cov.}
\end{aligned}\right\}\ \text{predict} \tag{8.96b}$$

$$\left.\begin{aligned}
L_{k+1} &= P^{xy}_{k+1}\left(P^y_{k+1}\right)^{-1} && \text{observer gain}\\
\hat{x}_{k+1} &= \hat{x}^-_{k+1} + L_{k+1}\left(y_{k+1} - \hat{y}_{k+1}\right) && \text{state mean}\\
P_{k+1} &= P^-_{k+1} - L_{k+1} P^y_{k+1} L^{\mathrm{T}}_{k+1} && \text{state covariance}
\end{aligned}\right\}\ \text{update} \tag{8.96c}$$

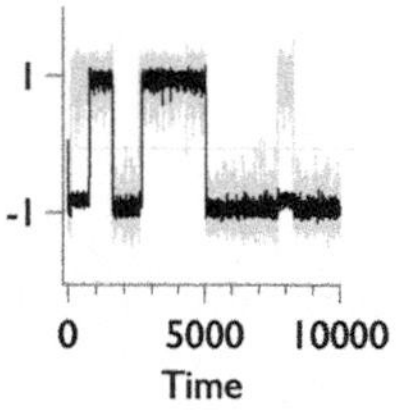

In Eq. (8.96c), the update equations are unchanged from the KF and EKF: All we are doing is (trying to) improve the predictions of means and covariances.

We can use the CKF algorithm to redo the double-well potential example. The plot at left has $a = 0.8$ and noise terms $\nu = 0.15$ and $\xi = 0.5$. For better comparison, we use the same noise realization as Figure 8.5. The result is better than the EKF but still not satisfactory. (To tweak CKF to improve the result, see Problem 8.14.)

There are many variations of the CKF. One particularly popular one, the *unscented Kalman filter* (UKF), uses a different set of weights and includes the original mean value as an extra ensemble element. (See Problem 8.15.) Unlike the CKF, it does not assume that the original distribution is Gaussian, but rather is merely symmetric about the mean. In general, as with many ad hoc, suboptimal methods, it works well in some situations but not in others.

Ensemble Kalman Filter (EnKF)

In the EnKF, we simply draw n_{E} ensemble elements from the prior distribution. Including random elements from the noise ensembles for ν and ξ allows us to generalize from the additive-noise case with very little additional effort. The EnKF algorithm is given by

$$\hat{x}_{i,k} \sim \mathcal{N}(\hat{x}_k, P_k)\,, \qquad \nu_{i,k} \sim \mathcal{N}(0, Q_\nu)\,, \qquad \xi_{i,k} \sim \mathcal{N}(0, Q_\xi) \quad \big\}\ \text{ensemble} \tag{8.97a}$$

$$
\left.
\begin{aligned}
&\hat{\boldsymbol{x}}^-_{i,k+1} = \boldsymbol{f}(\hat{\boldsymbol{x}}_{i,k}, \boldsymbol{u}_k, \boldsymbol{v}_{i,k}), \qquad \hat{\boldsymbol{x}}^-_{k+1} = \frac{1}{n_\mathrm{E}} \sum_i \hat{\boldsymbol{x}}^-_{i,k+1} && \text{state mean} \\
&\hat{\boldsymbol{y}}_{i,k+1} = \boldsymbol{h}(\hat{\boldsymbol{x}}^-_{i,k+1}, \boldsymbol{\xi}_{i,k}), \qquad \hat{\boldsymbol{y}}_{k+1} = \frac{1}{n_\mathrm{E}} \sum_i \hat{\boldsymbol{y}}_{i,k+1} && \text{obs. mean} \\
&\boldsymbol{P}^-_{k+1} = \frac{1}{n_\mathrm{E}-1} \sum_i \left(\hat{\boldsymbol{x}}^-_{i,k+1} - \hat{\boldsymbol{x}}^-_{k+1}\right)\left(\hat{\boldsymbol{x}}^-_{i,k+1} - \hat{\boldsymbol{x}}^-_{k+1}\right)^\mathrm{T} && \text{state cov.} \\
&\boldsymbol{P}^y_{k+1} = \frac{1}{n_\mathrm{E}-1} \sum_i \left(\hat{\boldsymbol{y}}_{i,k+1} - \hat{\boldsymbol{y}}_{k+1}\right)\left(\hat{\boldsymbol{y}}_{i,k+1} - \hat{\boldsymbol{y}}_{k+1}\right)^\mathrm{T} && \text{obs. cov.} \\
&\boldsymbol{P}^{xy}_{k+1} = \frac{1}{n_\mathrm{E}-1} \sum_i \left(\hat{\boldsymbol{x}}^-_{i,k+1} - \hat{\boldsymbol{x}}^-_{k+1}\right)\left(\hat{\boldsymbol{y}}_{i,k+1} - \hat{\boldsymbol{y}}_{k+1}\right)^\mathrm{T} && \text{state-obs. cov.}
\end{aligned}
\right\} \text{predict} \qquad (8.97\text{b})
$$

$$
\left.
\begin{aligned}
\boldsymbol{L}_{k+1} &= \boldsymbol{P}^{xy}_{k+1}\left(\boldsymbol{P}^y_{k+1}\right)^{-1} && \text{observer gain} \\
\hat{\boldsymbol{x}}_{k+1} &= \hat{\boldsymbol{x}}^-_{k+1} + \boldsymbol{L}_{k+1}\left(\boldsymbol{y}_{k+1} - \hat{\boldsymbol{y}}_{k+1}\right) && \text{state mean} \\
\boldsymbol{P}_{k+1} &= \boldsymbol{P}^-_{k+1} - \boldsymbol{L}_{k+1}\boldsymbol{P}^y_{k+1}\boldsymbol{L}^\mathrm{T}_{k+1} && \text{state covariance}
\end{aligned}
\right\} \text{update} \qquad (8.97\text{c})
$$

Notice that we normalize covariance elements by $n_\mathrm{E}-1$. The -1 factor is the familiar $n-1$ in sample estimates of the variance or standard deviation and reflects the idea that using an empirically determined mean "uses up" one degree of freedom in the data. Problem 8.16 shows that the EnKF can handle the double-well case of Example 8.3, without the ad hoc fixes needed by the CKF. Below, in Section 8.4.4, we develop a more general MC method that can handle non-Gaussian distributions.

8.4.3 Direct Numerical Solution

The second approach for estimation in systems with nonlinear dynamics or non-Gaussian statistics is to solve approximately for the full probability distribution densities. Approximating a continuous probability density function by $p(\boldsymbol{x}) \approx \sum_i w^i\, \delta\,(\boldsymbol{x}-\boldsymbol{x}^i)$, with $\sum_i w^i = 1$, we can replace the continuous integrals of Bayes' theorem with discrete, finite sums.[21] The trick is to choose wisely the points $\boldsymbol{x}^i$ and their weights w^i.[22] Here, we discuss *grid methods* that place the reference points on regular lattices in state space. In the next section, we will discuss random, Monte Carlo choices.

To apply the grid method to state estimation, we write

$$
p(\boldsymbol{x}_k|\boldsymbol{y}^k) \approx \hat{p}(\boldsymbol{x}_k|\boldsymbol{y}^k) \equiv \sum_{i=1}^{N} w^i_k \;\; \delta\left(\boldsymbol{x}_k - \boldsymbol{x}^i\right), \qquad (8.98)
$$

[21] Although we describe the grid method as an approximation, it can be exact when the state space for $\boldsymbol{x}$ is finite, so that a finite number of grid points can cover all possible $\boldsymbol{x}$ values.

[22] The superscript i here denotes a summation index and not an exponent!

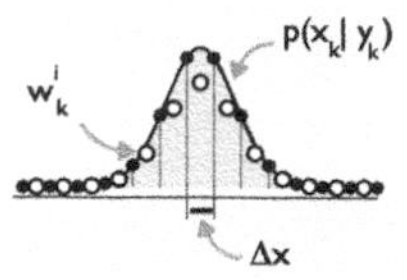

where $\hat{p}$ estimates the true density p and where the weights w^i_k give the probability for a state $\boldsymbol{x}$ to lie in the range $\boldsymbol{x}^i \pm \frac{1}{2}\Delta\boldsymbol{x}$. The approximation is illustrated at left for one dimension, with the horizontal black line indicating the box size (Δx) for the grid point x^i. Note that the grid values w^i (white disks) do not fall on the curve of the distribution function $p(x)$, as the normalization fixes their sum $\sum_i w^i = 1$, while probability densities are normalized so that $\int \mathrm{d}x\, p(x) = 1$. Thus, $w^i(x^i) \approx p(x^i)\Delta x$. Above, $\Delta x \approx 0.8$.[23]

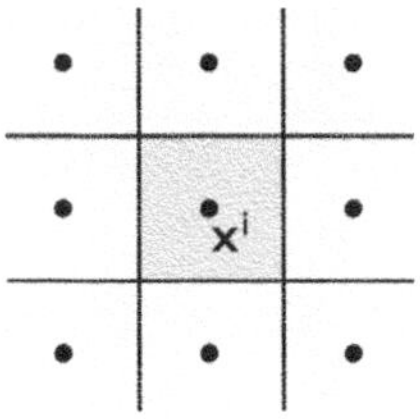

For the n-dimensional case, the probabilities are an integral of the probability density over the cell centered on the grid point, equivalent to the first *Brillouin zone* in condensed-matter physics (see shaded area in the 2-dimensional case at left). In general, we assume that the grid spacing is tight enough that we can approximate the integral over the hypercube as the value of P at the grid point multiplied by the grid volume.

Similarly, we represent the other conditional distributions by

$$\hat{p}(\boldsymbol{x}_{k+1}|\boldsymbol{y}^k) = \sum_{i=1}^{N} \left(w^i\right)^-_{k+1} \delta\left(\boldsymbol{x}_{k+1} - \boldsymbol{x}^i\right), \quad \hat{p}(\boldsymbol{x}_{k+1}|\boldsymbol{y}^{k+1}) = \sum_{i=1}^{N} w^i_{k+1}\, \delta\left(\boldsymbol{x}_{k+1} - \boldsymbol{x}^i\right). \tag{8.99}$$

The state-estimation problem is then to estimate the new weights w^i_{k+1} given the old ones w^i_k and the dynamics. Substituting Eqs. (8.98) and (8.99) into Eqs. (8.76) gives

$$\left(w^i\right)^-_{k+1} = \sum_{j=1}^{N} w^j_k\, p\left(\boldsymbol{x}^i|\boldsymbol{x}^j\right), \qquad w^i_{k+1} = \frac{\left(w^i\right)^-_{k+1} p\left(\boldsymbol{y}_{k+1}|\boldsymbol{x}^i\right)}{\sum_{j=1}^{N} \left(w^j\right)^-_{k+1}\, p\left(\boldsymbol{y}_{k+1}|\boldsymbol{x}^j\right)} \tag{8.100}$$

To evaluate Eqs. (8.100), we need to know $p\left(\boldsymbol{x}^i|\,\boldsymbol{x}^j\right)$ and $p\left(\boldsymbol{y}_{k+1}|\,\boldsymbol{x}^i\right)$. However, these are just related to the nonlinear dynamics and measurement equations, as in Eqs. 8.77. From those equations, the probabilities are easy to calculate for additive noise:

$$\begin{aligned} \boldsymbol{x}_{k+1} &= \boldsymbol{f}(\boldsymbol{x}_k) + \boldsymbol{\nu}_k \quad &\Longrightarrow \quad p\left[\boldsymbol{x}^i|\boldsymbol{x}^j\right] &= \mathcal{N}\left[\boldsymbol{x}^i - \boldsymbol{f}\left(\boldsymbol{x}^j\right);\, \boldsymbol{0},\, Q_\nu\right] \\ \boldsymbol{y}_k &= \boldsymbol{h}(\boldsymbol{x}_k) + \boldsymbol{\xi}_k \quad &\Longrightarrow \quad p\left[\boldsymbol{y}_{k+1}|\boldsymbol{x}^i\right] &= \mathcal{N}\left[\boldsymbol{y}_{k+1} - \boldsymbol{h}\left(\boldsymbol{x}^i\right);\, \boldsymbol{0},\, Q_\xi\right]. \end{aligned} \tag{8.101}$$

In Figure 8.6, we show the results of the grid filter for the double-well potential discussed in Example 8.3. In that one-dimensional case, the plot of $p(x_k|y^k)$ denoted by "optimal Bayes" in the figure is an image not a graph, as can be seen in the blowup in the middle of the figure. If we wanted to extract a single number to characterize $p(x_k|y^k)$ at a given time k – our version of $\hat{x}_k$, we could choose the conditional mean $\langle x_k\rangle_{y^k} = \int \mathrm{d}x_k\, x_k\, p(x_k|y^k)$. The resulting graph is hard to distinguish against the image and so is not shown. The main point is that the optimal Bayes estimate matches very well the true state. There are no mistakes, as there were with the extended Kalman filter (EKF).

Figure 8.6 illustrates the virtues of the Bayesian approach. Because we estimate the entire probability distribution function for the posterior distributions $p(x_k|y^k)$, we

[23] The weights w are dimensionless numbers, with the dimension of the density carried by the delta function.

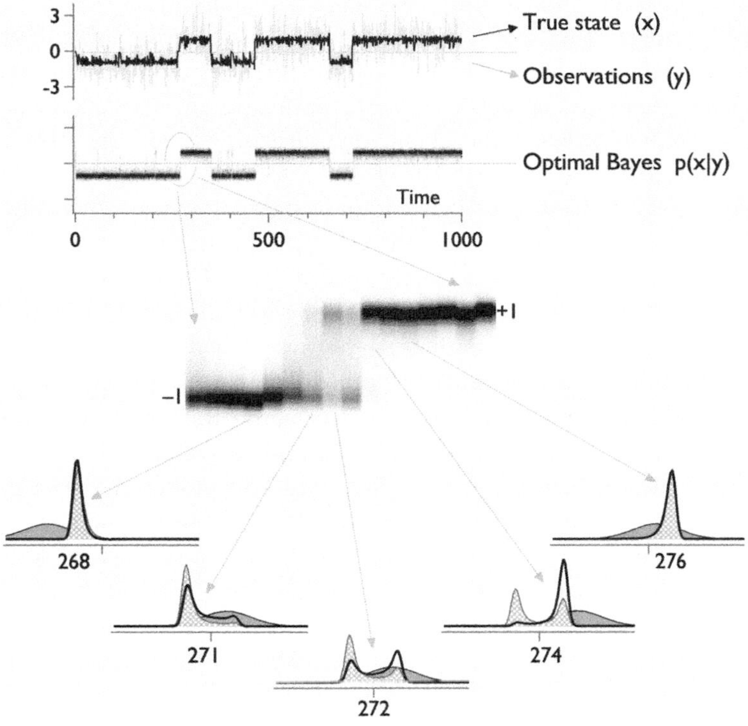

Fig. 8.6

Optimal Bayes estimation for particle motion in a double-well potential, ⩊. Parameters and noise are the same as for Figure 8.5, with $\nu = 0.15, \xi = 1$, and 100 grid points. Top graphs show the observations y_k, the underlying true state x_k, and an image plot of the optimal Bayes estimation $p(x_k|y_k)$. At the middle is a blowup of a portion of the Bayes estimation showing a transition from one well to another. Individual probability distributions for the Bayes estimate $p(x_{k+1}|y^{k+1})$ is shown as a black curve in the time slices 268–272. The Bayes estimate is the product of the prediction distribution $p(x_{k+1}|y^k)$ (hatched pattern) and the measurement $p(x_{k+1}|y_{k+1})$ (solid gray shading).

can capture all the nuances of our knowledge of the system. For example, look at the estimates for times 268–276, which capture the transition from −1 to +1. The sequence starts at 268, which concludes a long "run" of states that fluctuate about −1. The estimate distribution is thus firmly centered on −1. At this point, a succession of observations centered on +1 begin. Because the uncertainty associated with each observation is relatively large – the distribution of the observations is roughly twice that of the prediction – each +1 observation merely decreases the −1 peak of the optimal estimate and increases that of the +1 peak. By 274, a new consensus has been reached, that the particle has switched to the +1 well, and by 276 the optimal estimate is unambiguously centered on +1. At intermediate times, the initial observations in the other well "raise doubts," and the distribution becomes bimodal, "leaning to" −1 at 271, "slightly favoring" +1 at 272, and "firmly decided" by 274. Notice how easily everyday language adapts to describe the subtle differences in the probability distribution for the optimal estimate. There is evidence that human beings use Bayesian inference at many levels, from abstract reasoning to unconscious control of our movements. The language of Bayesian reasoning may be so intuitive and so rich because it is such a familiar part of life. Indeed, current theories of the brain argue that it creates internal, *generative* models of the external world and uses Bayesian methods (the *Bayesian brain hypothesis*) to evaluate how surprising events are relative to those models (Friston, 2010).

In principle, the grid method can give optimal state estimates with arbitrarily complex dynamics and noise statistics. But there is a catch: the method works only if we can "cover" the state space with a closely spaced grid. The first worry is that state spaces often have infinite extent. In practice, physical states tend to have appreciable support only over finite ranges that can always be scaled to be of order unity.[24] But what grid resolution do we need? For our double-well potential, 100 grid points gave a reasonably accurate solution, but that was mainly to show pretty pictures. Had we merely wanted a good estimate of the mean and variance, 10 points would have sufficed. Alternatively, instead of expanding over grid points, we can expand over orthogonal functions, and a cleverly chosen basis set may give a good approximation with only a small number of elements. Again, 10 might be a reasonable number to hope for. In both cases, for an n-dimensional state space, we would then need roughly 10^n grid points. This rapid growth is an example of the *curse of dimensionality* that Richard Bellman called attention to in the 1950s. Thus, although the grid method can work for low-dimensional state spaces, it surely fails in higher dimensions, motivating techniques that scale more gracefully with n.

8.4.4 Monte Carlo Techniques

If the dimension of state space (and hence number of required grid points) is too large for direct numerical solution, then Monte Carlo techniques, which involve a random sampling of the state space, are an alternative. In control theory and state estimation, Monte Carlo techniques go under the rather obscure name of *particle filters*. The individual samples from the probability distribution are pictured as *particles*, and the goal is to use them to *filter* the noisy observations. In any case, the relevant techniques are classic ones relevant to almost all areas of physics: they require averaging over and drawing random numbers from complicated, high-dimensional probability distribution functions. Such tasks are a major focus of computational and statistical physics, along with machine learning, neuroscience, and economics. Some twists, however, are peculiar to applications to state estimation.

In the Bayesian approach to state estimation, each time step requires the calculation of integrals over domains that have the dimension of the state-space vector $\boldsymbol{x}_k$. We can see this in the Chapman–Kolmogorov law, Eq. (8.76a), and in the partition function Z, Eq. (8.75). In addition, we often want to reduce the number of variables we track by marginalizing, or integrating out degrees of freedom in $\boldsymbol{x}$ that we do not care about. Alternatively, we may want to summarize the full probability distribution by simpler

[24] If not, $x \to x' = \tanh[(x - x_0)/\ell]$ maps $(-\infty, \infty)$ to $(-1, 1)$ and $1 - e^{-x/\ell}$ maps $(0, \infty)$ to $(0, 1)$. The parameters x_0 and ℓ give the typical location and scale of the dynamics in the original coordinates.

quantities, such as the (conditional) mean or variance. Such tasks lead us to consider integrals of the form

$$\langle \varphi(\boldsymbol{x}) \rangle = \int \mathrm{d}\boldsymbol{x}\, p(\boldsymbol{x})\, \varphi(\boldsymbol{x}) \,, \tag{8.102}$$

where, for example, $\varphi(\boldsymbol{x}) = \boldsymbol{x}$ when calculating the mean.

Recall that in the grid method, we approximate such integrals by discretizing the various continuous probability distributions. For a generic $p(\boldsymbol{x})$, the grid approximation is

$$p(\boldsymbol{x}) \approx \hat{p}(\boldsymbol{x}) \equiv \sum_{i=1}^{N} w_i\, \delta\left(\boldsymbol{x} - \boldsymbol{x}^i\right) \quad \Longrightarrow \quad \langle \varphi(\boldsymbol{x}) \rangle \approx \hat{\varphi} \equiv \sum_{i=1}^{N} w_i\, \varphi_i \,. \tag{8.103}$$

where $w_i = p(\boldsymbol{x}^i)$ and $\varphi_i = \varphi(\boldsymbol{x}^i)$, as may be seen by integrating over the Brillouin zone centered on $\boldsymbol{x}^i$, assuming that the grid spacing is fine enough that $p(\boldsymbol{x}) \approx$ constant. Both $\hat{p}$ and $\hat{\varphi}$ are estimators that approximate p and $\langle \varphi \rangle$, respectively. The difficulty is that an accurate approximation in n dimensions requires M^n points, using M grid points to resolve features in each linear dimension. As n grows, the number of required grid points becomes very large, very fast. Monte Carlo methods offer a way out of this impasse.

Particle Filters

The general strategy is to sample from $p(\boldsymbol{x}_k|\boldsymbol{y}^k)$, using methods from Section A.8.9 online,

$$\hat{p}(\boldsymbol{x}_k|\boldsymbol{y}^k) = \frac{1}{N} \sum_{i=1}^{N} \delta\left(\boldsymbol{x}_k - \boldsymbol{x}_k^i\right), \tag{8.104}$$

and then generate an approximation for the predicted distribution $p(\boldsymbol{x}_{k+1}|\boldsymbol{y}_k)$. When a new measurement arrives, we update the estimate of the predicted distribution to form a posterior distribution, giving a recursive way of going from one Monte Carlo estimate of $p(\boldsymbol{x}_k|\boldsymbol{y}^k)$ to another Monte Carlo estimate of $p(\boldsymbol{x}_{k+1}|\boldsymbol{y}^{k+1})$. Unfortunately, a naive iteration of this strategy fails: a single weight will eventually dominate over all the others. An idea is to periodically resample using the weighted bootstrap strategy (Section A.8.9 online), which gives rise to the *Sampling Importance Resampling* (SIR) algorithm.

To begin, we combine Chapman–Kolmogorov and the dynamics (Eqs. 8.72 and 8.73):

$$p(\boldsymbol{x}_{k+1}|\boldsymbol{y}^k) = \iint \mathrm{d}\boldsymbol{\nu}_k\, \mathrm{d}\boldsymbol{x}_k\, \delta\left[\boldsymbol{x}_{k+1} - \boldsymbol{f}(\boldsymbol{x}_k, \boldsymbol{\nu}_k)\right] p(\boldsymbol{\nu}_k)\, p(\boldsymbol{x}_k|\boldsymbol{y}^k) \,. \tag{8.105}$$

Next, we insert the Monte Carlo approximation for the prior given in Eq. (8.104):

$$\begin{aligned} p(\boldsymbol{x}_{k+1}|\boldsymbol{y}^k) &= \frac{1}{N} \sum_{i=1}^{N} \iint \mathrm{d}\boldsymbol{\nu}_k\, \mathrm{d}\boldsymbol{x}_k\, \delta\left[\boldsymbol{x}_{k+1} - \boldsymbol{f}(\boldsymbol{x}_k, \boldsymbol{\nu}_k)\right] p(\boldsymbol{\nu}_k)\, \delta\left(\boldsymbol{x}_k - \boldsymbol{x}_k^i\right) \\ &= \frac{1}{N} \sum_{i=1}^{N} \int \mathrm{d}\boldsymbol{\nu}_k\, \delta\left[\boldsymbol{x}_{k+1} - \boldsymbol{f}(\boldsymbol{x}_k^i, \boldsymbol{\nu}_k)\right] p(\boldsymbol{\nu}_k) \,. \end{aligned} \tag{8.106}$$

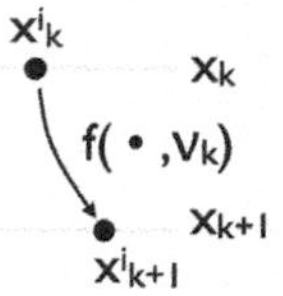

Similarly, we approximate the input-noise distribution by a *single* realization v_k^i for each particle i, which is acceptable because there are $N \gg 1$ particles. That is, $p(v_k) \approx \delta(v_k - v_k^i)$. Substituting into Eq. (8.106), we arrive at

$$p(x_{k+1}|y^k) \approx \frac{1}{N}\sum_{i=1}^{N} \delta\left[x_{k+1} - f(x_k^i, v_k^i)\right]. \tag{8.107}$$

Intuitively, we draw particles from the prior $p(x_k|y^k)$ and then propagate them forward using the dynamics and a noise realization to form particles for the prediction $p(x_{k+1}|y^k)$.

Our next task is to integrate the new information provided by the observation with our approximation to the prediction, using Bayes' Theorem. Our discussion of importance sampling (Section A.8.9 online) tells us how to do this: Given particles $x_{k+1}^i \sim p(x_{k+1}|y^k)$, we can draw from the posterior using a weighted sum, with weight

$$w_{k+1}^i = p(y_{k+1}|x_{k+1}^i), \qquad \tilde{w}_{k+1}^i = \frac{w_{k+1}^i}{\sum_i w_{k+1}^i}. \tag{8.108}$$

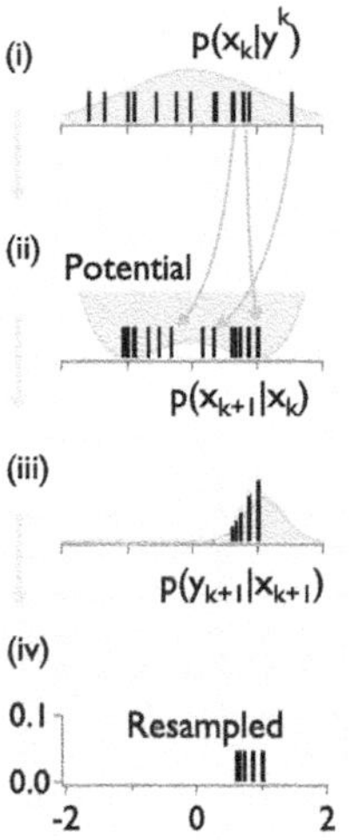

The SIR algorithm is summarized at left for the one-dimensional example of Brownian motion in a double-well potential. Part (i) shows, in the background, the prior distribution $p(x_k|y^k)$. Here, we assume a Gaussian that is centered on 0. We then show 20 particles drawn from this distribution as vertical black lines, whose height equals their relative weight. Initially, all particles have weight 0.05. In Part (ii), the system evolves under a double-well potential, shown in the second graph. Here, we show how particle positions move during a discrete update, under the influence of the potential. Note that because each particle receives a separate thermal noise realization, the order of particles can change, as illustrated by the crossed trajectories.[25] Notice how, under the influence of the double-well potential, about half of the particles end up near the well at -1 and half near $+1$.

In Part (iii), we show the probability density $p(y_{k+1}|x_{k+1})$ associated with the observation, y_k, which happens to be near $+1$. Then we use Eq. (8.108) to reweight the particles. Notice that more than half of the particles (the ones that ended up near the -1 well) have negligible weight after this operation and, as a result, the surviving particles have acquired greater weight. The weighted particles then form the Monte Carlo estimate of the posterior distribution, $p(x_{k+1}|y^{k+1}) = \sum_i w_i \, \delta(x_{k+1} - x_{k+1}^i)$.

In principle, we could iterate the above algorithm by adopting the weights of the posterior as a new prior for the measurement at time $k+2$. However, the number of effective particles has been reduced in going from k to $k+1$. Indeed, while subsequent losses may be less drastic (e.g., if the prior and the measurement are in the same well), the number of particles with substantial weight can only decrease. Ultimately, we

[25] For a continuous system, the trajectories between measurements would be continuous curves in state space. In a one-dimensional case, the lines cannot cross, but if a multidimensional case is projected for viewing convenience on one dimension, they can. In the discrete case shown here, the position of the particles hops, and the trajectories are merely guides for the eye. The crossing is then merely a reordering.

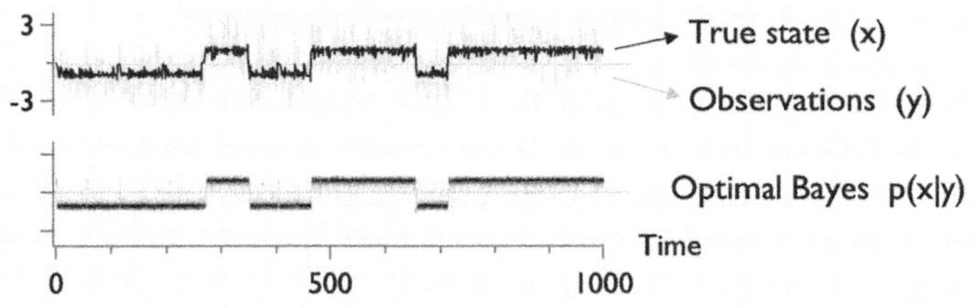

Fig. 8.7 Particle filter for Brownian motion in a double-well potential, . Parameters and noise values are the same as for Figure 8.6.

will be left with a single particle that has essentially unit weight, leading to a huge variance.

One solution to this problem, which is illustrated in Part (iv), is to resample the posterior distribution. The idea is simply to draw N particles from the posterior, as represented by its previous samples, including their weights. Thus a particle with weight w_i will be resampled with probability w_i. In effect, "strong" particles with high weights will have several representatives, while the particles with negligible weight are unlikely to be represented after the resampling. The strong thus "reproduce" while the weak are eliminated. The algorithm is indicated schematically at right. We first construct an empirical estimate of the cumulative density function (CDF), as shown by the heavy black line. Then we approximate N samples of the uniform distribution between 0 and 1 by a single sample between 0 and $1/N$ and $N-1$ others uniformly displaced by $1/N$. These are illustrated by horizontal lines at right. The particles selected are then determined by the index of the corresponding CDF value (white circles) Note that we take a commonly used shortcut in that only the first interval is random. A more proper resampling would choose N numbers uniformly from the interval (0,1). For particle-filtering algorithms, the simplification has been shown not to matter. See Kitagawa (1996) for a complete discussion.

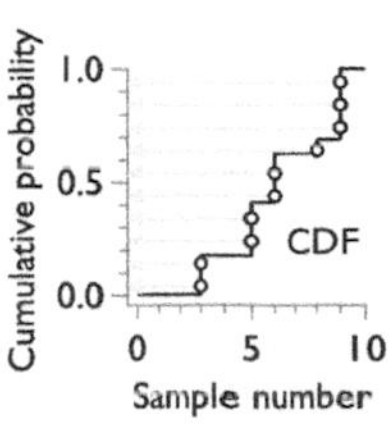

The degeneracy of the many particles after sampling is not necessarily cause for worry. As shown at right, any degeneracy in the predicted distribution disappears when each x_k^i is propagated forward with a different noise realization. Still, if the number of effective particles after the observation reweighting is very small, the input noise ν will not be enough to explore the prediction distribution well. We return to this question below.

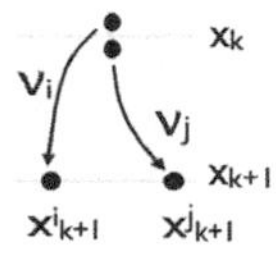

Finally, we apply the above ideas to one-dimensional Brownian motion in a double-well potential. Figure 8.7 thus reprises Figure 8.6. The latter was computed with a grid filter, while the present version was computed with the particle filter (SIR) algorithm discussed here. As we might expect, both methods work well. In one dimension, the grid method is actually more efficient and runs faster; higher dimensions favor particle filters.

Particle Filters with Smarter Particles

The Sequential Importance Resampling (SIR) algorithm described above works well for simple problems but fails for higher-dimensional problems with complex, multimodal probability distributions (Section 8.5). The main reason is that the number of particles (samples) required grows exponentially with state-space dimension in complex cases. At first glance, this seems surprising, as the whole motivation for Monte Carlo methods is their freedom from the curse of dimensionality. Alas, the curse is not so easily exorcized.

To understand why, consider once again the double-well example discussed in the previous section. Notice that in each step of the SIR, we lost more than half the particles, as the predicted distribution naturally put half the particles in one well and half in the other. Because the observation likelihood density is localized to one well, the particles that were propagated to the other well by the prediction step were effectively useless for estimating the posterior. In one dimension, of course, the inefficiency is not significant. But in high-dimensional state spaces where the support of distributions is tiny, efficiency is crucial. If we lose a fraction f of particles in each dimension in each time step, we lose f^n in n dimensions. And when the number of observed quantities is much smaller than the number of state-vector components, the problem is even worse.

An alternate approach is based on the idea that, for a sample with high weight, corresponding to a region of high probability, nearby locations may also have high probability. By wandering around the neighborhood, we can find out more of this structure than is probed by the noise realizations in the update step. Such thoughts are at the origin of the *Markov chain* approach to Monte Carlo simulation (MCMC), better known in physics as the *Metropolis algorithm*. In the context of state estimation, the algorithm is called the *Resample Move* algorithm, since it moves the particles after resampling them (Gilks and Berzuini, 2001). Many algorithms try to integrate these ideas to allow efficient estimation for high-dimensional state spaces (Andrieu et al., 2010).

8.5 Why State Estimation Can Be a Hard Problem

In the previous sections, we have discussed various ways to find approximate solutions to the Bayesian filtering equations. We saw that if nonlinearities lead only to small deviations from Gaussian posterior distributions, then all the methods can lead to a good solution. But stronger deviations pose considerable problems. For example, while a unimodal distribution  implies a straightforward state estimate, a bimodal one  implies that a system is in one of two coarse-grained states, a situation that we term *ambiguity*.

In this section, we consider the state-estimation problem from a different point of view, asking *why* state estimation is hard. In particular, what is the origin of ambiguity (multimodal posterior distributions)? What can we do to avoid hard situations?

8.5.1 Ambiguity from Nonlinear Dynamics

Although nonlinear dynamics can lead to great complexity, that complexity can often be tamed by simply taking more frequent and more accurate measurements. In Figure 8.8, we illustrate this idea on the double-well potential, Example 8.3.

Figure 8.8a depicts the worst case in terms of producing ambiguity: a state estimate centered on top of the unstable local maximum at $x = 0$. We start with $p(x_k|y^k) = \mathcal{N}(0, \sigma^2)$, a Gaussian distribution of width σ centered on 0. The image plot then tracks the evolution of this state forward in time, showing $p \equiv p(x_{k+1}|y^k)$, the predicted evolution. The curves are calculated by solving the Fokker–Planck equation using $\mathcal{N}(0, \sigma^2)$ as an initial condition,

$$\partial_t p = -\partial_x \left[\left(-x + x^3 \right) p \right] + D \partial_{xx} p \,, \tag{8.109}$$

and we see individual traces at right. The distribution stays reasonably close to a Gaussian up to times about 0.2 (the second graph from the bottom, at right) but becomes bimodal at longer times: the particle could be in one well or the other, and we do not know which.

Figure 8.8b shows a "less ambiguous" case: if the prior distribution is slightly offset to the right, then most of the probability ends up in the right well while some ends up in the left well. After a very long time, hopping over the barrier can equalize the probabilities and return the system to the Boltzmann distribution. But it is easy to make barriers so high that they make this final equilibration too slow to observe in practice.

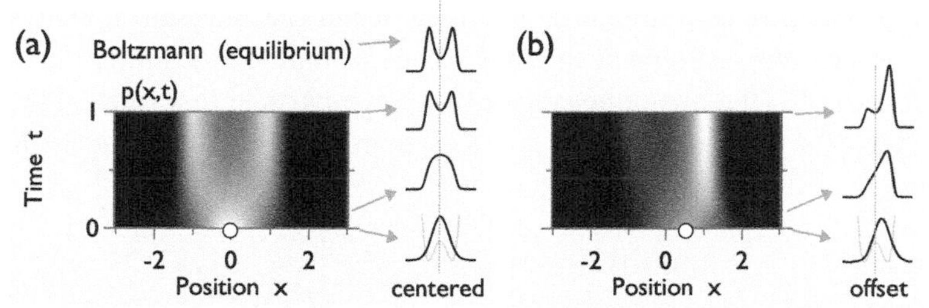

Fig. 8.8 Evolution of $p(x, t) = p[x_{k+1}(t) \mid y^k]$ for an overdamped particle in a double-well potential, ⌵. (a) Gaussian initial distribution centered on $x = 0$ (white dot). The particle goes to either well with equal probability. Individual profiles at $t = 0$, $t = 0.2$, $t = 1$, and $t = \infty$ (Boltzmann solution). (b) Initial condition offset to $x = 0.5$, with profiles at $t = 0$, $t = 0.2$, and $t = 1$ shown. The particle is likely to stay on the same side of the barrier. $D = 0.2$; $\sigma^2 = 0.5$.

8.5.2 Ambiguity from Nonlinear Measurements

Complexity can also result from the nonlinear measurement relation $\boldsymbol{y} = \boldsymbol{h}(\boldsymbol{x}, \boldsymbol{\xi})$, which appears in the measurement likelihood relation $p(\boldsymbol{y}_{k+1}|\boldsymbol{x}_{k+1})$. From Bayes' Theorem,

$$p(\boldsymbol{y}_{k+1}|\boldsymbol{x}_{k+1}) = \frac{1}{Z} p(\boldsymbol{x}_{k+1}|\boldsymbol{y}_{k+1}), \tag{8.110}$$

where we assume a uniform prior $p(\boldsymbol{x}_{k+1})$ (no knowledge of $\boldsymbol{x}$ before we do the measurement). Note that prior information is present in the full update relation, $p(\boldsymbol{x}_{k+1}|\boldsymbol{y}^k)$, which encodes all of the information obtained from the previous k measurements. To evaluate the probability distributions in Eq. (8.110), we use the inverse relation, $\boldsymbol{x}_k = \boldsymbol{h}^{-1}(\boldsymbol{y}_k, \boldsymbol{\xi}_k)$: given an observation (and noise), we infer the state.

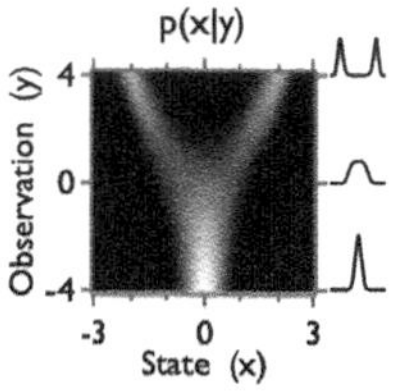

The figure at left illustrates the simple measurement relation $y = h(x, \xi) = x^2 + \xi$. It superficially resembles Figure 8.8 but has a quite different origin. Notice that for $y < 0$, the distribution is unimodal but for $y > 0$, it develops two modes. This makes sense: if you observe a negative value of y, then x is most likely at the origin, as a negative value of y can result only from a negative fluctuation of ξ. But $y = 4$ can come from $x \approx 2$ or $x \approx -2$: that is, a positive value of y can come from one of two "regions" of x.

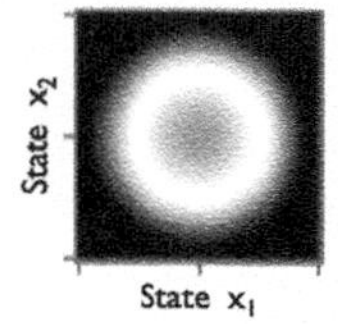

Ambiguity arising from measurement nonlinearity is unfortunately common. In crystallography, you measure the intensity of X-ray diffraction patterns, but inferring crystal structure requires knowing the phase. In quantum mechanics, you measure the modulus of a wavefunction $|\psi|^2$, but the dynamics of ψ are governed by the Schrödinger equation. These *phase retrieval* problems present an even greater ambiguity, in that a positive value for the modulus leads to a continuous ring of possibilities of phase, as illustrated at left in two dimensions for the measurement $p(y|\boldsymbol{x}) = \exp\left[-(|\boldsymbol{x}|^2 - y)^2/2\right] + \xi$ for $y = 1$ and $\xi^2 = 1$.

8.5.3 Ambiguity from Long-Tailed Distributions

Even when the dynamics are linear or otherwise unambiguous and even when the same is true for the measurement function, ambiguity may result when the PDFs of the prior and measurement are multiplied together in the Bayesian update step. For example, consider a measurement that has a background. For example, let us modify the measurement for the one-dimensional Brownian particle that freely diffuses, with $x_{k+1} = x_k + \nu_k$ and $y_k = x_k + \xi_k$. We add a background by modifying the likelihood function to

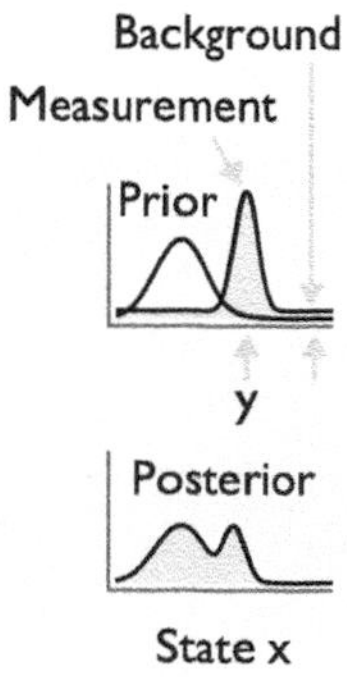

$$p(y_k|x_k) \propto \exp\left[-\frac{(x_k - y_k)^2}{2\xi^2}\right] + b \tag{8.111}$$

where b is a constant that is independent of the state x_k. (This requirement defines the term *background*.) The background extends over a range that is large compared to ξ but finite, so that normalization is possible. Turning the likelihood around to be a function of the observation y_k, we have the situation illustrated at left. The shaded distribution in the top graph gives the probability for the particle to be at

x given that the measurement was at y. We can see that while most of the density is centered on the particle position, there is a chance that the measurement of y could come from the background, meaning that the state could be anywhere else. That possibility is represented by the thin band of probability mass that stretches across the horizontal axis. More visually, imagine detecting the particle position photon by photon (e.g., from background fluorescence in an optical setup). At the same time, let the prior estimate be actually centered on a different position, as illustrated. This situation would correspond to one where the measurement really did come from the background rather than the object. When we multiply the two together and normalize, per Bayes' Theorem, we get the posterior shown at bottom, which is now multimodal: we do not know whether to trust the prior expectation or the new measurement.

The general issue is that the measurement probability distribution function (PDF) has a long tail. The heavier the tail, the more the overlap of the prior and measurement PDFs; hence multimodal posteriors. In addition to background counts, complexity in the dynamics or the measurement can lead to long tails. An example is motion in a turbulent medium.

Example 8.4 (Estimating a frequency) As an example of the issues created by a background, consider measuring a signal with a single unknown frequency. For simplicity, assume its amplitude and phase are known – inferring them does not change the problem significantly. Let us measure N time points of a noisy sinusoid of unit amplitude and zero phase. It turns out that the estimation is easier to implement on complex signals, which could be the in-phase and out-of-phase components of a bandpass-filtered noisy real signal such as the output of a lock-in amplifier. The measurements are then

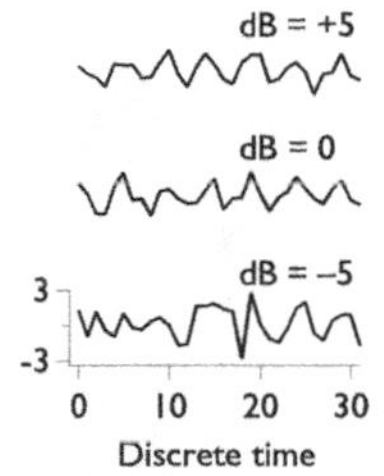

$$y_k = e^{i\omega k} + \xi_k\,, \qquad \xi_k \sim \mathcal{CN}(0, \sigma^2)\,, \tag{8.112}$$

with $k = 1, 2, \ldots$ the time index (kT_s, in physical units). The complex noise ξ_k is distributed as a *circular complex Gaussian*, with real and imaginary parts independent Gaussian distributions (uncorrelated with each other), $\sim \mathcal{N}(0, \frac{1}{2}\sigma^2)$. Explicitly, $p(\xi) = \frac{1}{\pi\sigma^2}\, e^{-|\xi|^2/\sigma^2}$. Defining the signal-to-noise ratio (SNR) by $\text{dB} = 20\log_{10}\sigma^{-1}$, we show time traces of the real part of the signal for $N = 32$ points with $\omega = 0.21$ at right, for different noise levels. The ability to "see" the sinusoid decreases notably when passing from +5 to 0 to −5 dB.

If we use maximum likelihood (uniform prior) for the frequency over a possible range of $(-\pi, \pi)$ (in units of T_s^{-1}), then $\ln p(\omega|y^k) \sim p(y^k|\omega)$. We can then solve numerically for the ω that maximizes the likelihood, as a function of SNR. The result at right shows the standard deviation σ_ω of the estimated frequency $\hat{\omega}$ (from the true value ω), averaging the results from 10^5 Monte Carlo trials per noise level. The dashed line shows the *Cramér–Rao Bound*, the lower limit for any unbiased estimator. The striking

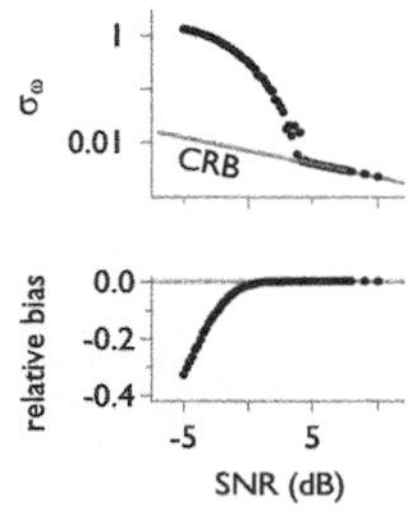

observation is that at a SNR of $\approx$ 3 dB, there is a threshold below which the estimation error and relative bias increase sharply.[26] The deviation occurs when the Fourier transform of the signal gives a peak value at an incorrect frequency (Problem 8.22). In other words, if, in a given trial, the background for some frequency has a higher value than the signal frequency, the maximum-likelihood algorithm will choose the wrong frequency. When the local maximum in the log-likelihood function is correctly chosen, the algorithm need only vary ω locally around the peak to refine the frequency estimate. The relative bias $(\hat{\omega} - \omega)/\omega$ is negative, as $\hat{\omega}$ is a mixture of signal at $\omega = 0.21$ and background, whose expected value is 0.

The argument in this last example is very general and has nothing to do with frequency estimation per se. Indeed, you might worry that it is too general: Why is there not always a sharp threshold near SNR $\approx$ 1? The answer is that textbooks dwell on the case of a Gaussian-distributed signal added to Gaussian-distributed noise. As we saw in Section 8.3, the distribution of the measurement then remains Gaussian, with a mean that smoothly interpolates between the prior and likelihood values. But typically, measurements are not linear, noise is not Gaussian, and thresholds are the rule, not the exception.

8.5.4 Dealing with Ambiguity

By far the best way to deal with ambiguity is to prevent it from arising in the first place. Perhaps surprisingly, the easiest source of ambiguity to eliminate is that associated with the nonlinear dynamics. The rather obvious idea is to measure quickly, to follow states so closely that ambiguity has no time to develop. In Figure 8.8, measuring at intervals less than 0.2 keeps the predicted distribution unimodal. And we recall that part (a) is the worst-case scenario, where the initial distribution is completely symmetric about $x = 0$. Any offset or other asymmetry will reduce the tendency to develop modes of equal weight, as shown in (b). More generally, dynamical systems evolve approximately linearly over short time intervals, and linear transformations will stretch and shift peaks in probability distributions but not create new ones: they will not add to ambiguity. Measuring rapidly is thus a general strategy for avoiding ambiguity produced by dynamics.

Measurement-induced ambiguity is probably the most difficult to deal with. You can plot the function $p(\boldsymbol{x}_k|\boldsymbol{y}_k)$ to see whether there is a problem and try to change your measurement function if possible. Unfortunately, in addition to straightforward situations such as the quadratic nonlinearity discussed above, ambiguities can also arise when there are fewer measured variables than state variables. As with issues related to observability (Chapter 4), increasing the number of independent measured quantities can then help.

[26] The relative bias is shown on a linear scale and fluctuates about zero for high SNR. On a log plot, there is an abrupt increase in relative bias magnitude at the same threshold as the increase in σ.

The ambiguity from long-tailed distributions depends on the weight of the PDF in the tails. Often, the criterion reduces to the integral of the signal relative to the background (SNR). For the diffusion example discussed above, ambiguity arose because a detected photon could arise either "from" the state or "from" the background. Intuitively, increasing the probability that a measurement reflects the signal reduces ambiguity. Thus, you should minimize the background signal (maximize the SNR). Perhaps this is a good time to remind the reader that everything we do in this chapter assumes that we know the dynamics and noise statistics. In this case, we should measure the background signal (which is possibly state dependent) before trying to estimate states. Of course, if the background is weaker in some local region, you should try to do your measurements there.

If ambiguity and its resultant complexity are truly unavoidable, then you will need to learn about more sophisticated state-estimation strategies discussed here, which range from the various extensions of Kalman filters to grid and Monte Carlo approaches.

8.6 Stochastic Optimal Control

> The race is not always to the swift, nor the battle to the strong; but that is the way to bet.
>
> Hugh E. Keough, 1864–1912.

In Section 8.5, we considered some of the complications in estimating the state of a strongly nonlinear system. Here, we consider the control of strongly nonlinear stochastic systems, which may have nonquadratic cost functions and non-Gaussian noise. We will consider a variety of approaches to this much more difficult problem in this chapter and later in Chapter 11. As a start, we extend the optimal-control techniques of Section 7.4 (Hamilton–Jacobi–Bellman, HJB equation). In the simplest case, we neglect measurement noise and consider a perfectly observed, noisy nonlinear dynamics:

$$\dot{\boldsymbol{x}} = \boldsymbol{f}(\boldsymbol{x},\boldsymbol{u}) + \boldsymbol{\nu}(t)\,, \qquad \langle \boldsymbol{\nu}(t)\rangle = \boldsymbol{0}\,, \qquad \langle \boldsymbol{\nu}(t)\,\boldsymbol{\nu}^{\mathrm{T}}(t')\rangle = \boldsymbol{Q}_{\nu}\delta\,(t-t')\,, \tag{8.113}$$

where $\boldsymbol{f}(\boldsymbol{x})$ can be nonlinear and $\boldsymbol{\nu}(t)$ is Gaussian white noise. The inputs $\boldsymbol{u}(t)$ and noise couple linearly. Let $L(\boldsymbol{x},\boldsymbol{u})$ be the cost function for "immediate costs" at time t. With an end-time penalty $\varphi[\boldsymbol{x}(\tau)]$, the cumulative cost-to-go over the time interval (t,τ) is

$$J(\boldsymbol{x},\boldsymbol{u},t) = \left\langle \varphi[\boldsymbol{x}(\tau)] + \int_t^{\tau} \mathrm{d}t'\, L[\boldsymbol{x}(t'),\boldsymbol{u}(t')] \right\rangle_{\boldsymbol{x}(t)=\boldsymbol{x}}\,, \tag{8.114}$$

where the only difference from Eq. (7.29) is that we now consider the expected value of the cost, as averaged over state trajectories $[\boldsymbol{x}]_t^{\tau}$ that begin at $\boldsymbol{x}(t)=\boldsymbol{x}$. We now proceed as in Section 7.4, first defining

$$J^*(\boldsymbol{x},t) = \inf_{\boldsymbol{u}(t\le t'\le\tau)} J(\boldsymbol{x},\boldsymbol{u},t)\,, \tag{8.115}$$

then breaking the integral into a piece from $(t, t+\Delta t)$ and a second one from $(t+\Delta t, \tau)$, and then Taylor expanding in Δt. This time, we include a second-order contribution in the expansion of $\langle J(\boldsymbol{x}+\boldsymbol{\Delta x})\rangle$ that captures the diffusion contribution from the noise term, $\langle \Delta x^2 \rangle \sim \nu^2 \Delta t$. Repeating the steps in Eq. (7.30) then leads to an HJB equation that accounts for stochasticity:

$$\partial_t J^*(\boldsymbol{x}, t) + \inf_{\boldsymbol{u}} \left[L(\boldsymbol{x}, \boldsymbol{u}) + (\partial_{\boldsymbol{x}} J^*)\, \boldsymbol{f}(\boldsymbol{x}, \boldsymbol{u}) + \tfrac{1}{2} \mathrm{Tr}\, (\boldsymbol{Q}_\nu \nabla^2 J^*) \right] = 0\,. \tag{8.116}$$

The final condition is $J^*(\boldsymbol{x}, \tau) = \varphi[\boldsymbol{x}(\tau)]$. Note the new $\frac{1}{2}\mathrm{Tr}\,(\boldsymbol{Q}_\nu \nabla^2 J^*)$ term relative to the HJB for a deterministic system, Eq. (7.31).

For linear dynamics and Gaussian white noise, the HJB equation for a stochastic system leads to LQG. (The derivation is just a slight generalization of that for deterministic LQR, explored in Problem 7.13.) Here, we explore instead an example of the new phenomena that can arise from the combination of nonlinearity and noise.

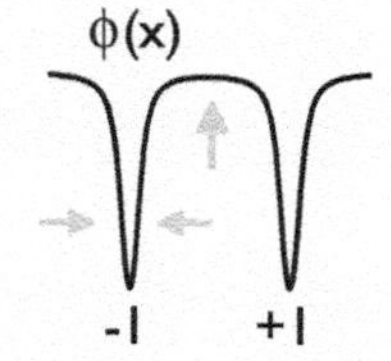

Example 8.5 (Delayed choice) Consider the motion of a particle that drifts at constant velocity but whose transverse position $x(t)$ is affected by noise. The particle starts at $x_0 = 0$ and must pass through narrow slits at $x = \pm 1$ at time τ. This final condition is effectively nonlinear: think of it as the limiting case of a terminal cost $\varphi(x)$ with local minima at $x = \pm 1$ that become infinitely deep (see left).[27] The equations of motion are

$$\dot{x} = u + \nu\,, \qquad x(0) = 0\,, \qquad x(\tau) = \pm 1\,, \tag{8.117}$$

with control input $u(t)$ and Gaussian white noise $\langle \nu(t)\, \nu(t')\rangle = 2D\,\delta\,(t-t')$ corresponding to a diffusion coefficient D. The goal is to minimize control effort, with running cost $L = \frac{1}{2}Ru^2$ and cost-to-go $J(x, u, t) = \int_t^\tau \mathrm{d}t'\, L[u(t')]$. The HJB equation for $J^*(x, t)$ is

$$\partial_t J^*(x, t) + \inf_u \left[\tfrac{1}{2} R u^2 + (\partial_x J^*)\, u + D \partial_{xx} J^* \right] = 0\,. \tag{8.118}$$

We minimize over the control at each time t by taking the derivative with respect to u in Eq. (8.118), which gives $u^* = -R^{-1}\partial_x J^*$. The HJB becomes

$$\partial_t J^*(x, t) - \frac{1}{2R} (\partial_x J^*)^2 + D \partial_{xx} J^* = 0\,. \tag{8.119}$$

A trick for solving Eq. (8.119) is to define the change of variable $J^*(x, t) = -\lambda \log \psi(x, t)$.[28] In Problem 8.8, you will show that choosing $\lambda = 2RD$ leads to

$$-\partial_t \psi = D\, \partial_{xx} \psi\,, \tag{8.120}$$

[27] By contrast, $x(\tau) = x_\tau$ is the limit $k \to \infty$ of $\phi(x) = k(x - x_\tau)^2$, which falls within the class of LQR problems.

[28] This is a simple example of a *Cole–Hopf* transformation and can be used for a class of HJB equations, as well as other situations, such as the study of the viscous Burgers equation modeling turbulence.

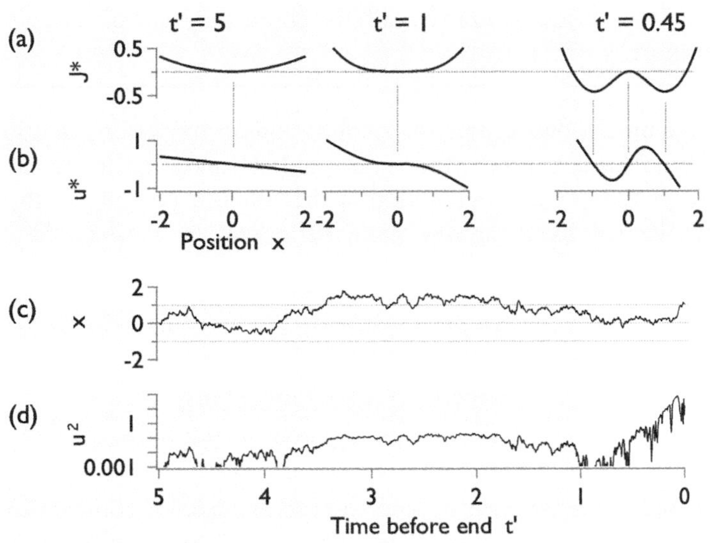

Fig. 8.9

Delayed-choice problem. (a) Cost-to-go $J^*(x, t')$ for $t' = 5, 1$, and 0.45. (b) corresponding optimal control $u^*(x, t')$. (c) Typical time series $x(t)$ and (d) control effort $u^2(t)$. Symmetry breaking occurs at $t'_c = 1$. We use $R = \nu^2 = x_0 = 1$.

a diffusion equation in *negative time*. At the end of the protocol, the particle must be at $x(\tau) = \pm 1$, implying a final condition for $\psi(x, t)$,

$$\psi(x, \tau) = \tfrac{1}{2}\left[\delta\left(x - 1\right) + \delta\left(x + 1\right)\right] . \tag{8.121}$$

Solving the diffusion equation then leads to (see Problem 8.8)

$$\begin{aligned} J^*(x, t') &= 2RD\left[\frac{x^2}{4Dt'} - \ln\cosh\left(\frac{x}{2Dt'}\right)\right] + f(t'), \\ u^*(x, t') &= \left(\frac{1}{t'}\right)\left(\tanh\frac{x}{2Dt'} - x\right), \end{aligned} \tag{8.122}$$

where $t' \equiv \tau - t$ is the time before the end and $f(t')$ collects all terms independent of x. Note that the optimal control $u^*(x, t') = -R^{-1}\partial_x J^*(x, t')$ is independent of $f(t')$.

Figure 8.9a shows J^* for $t' = 5$, 1, and 0.45. Its behavior suggests a free energy undergoing a phase transition (Kappen, 2005). Figure 8.9b shows the optimal control as a function of x for the three regimes; (c) shows a typical time series for $x(t)$; and (d) shows the control effort $u^2(t)$. The strategy is to do little before the transition at $t'_c = x_0^2/2D$, when the control "wakes up" and steers the particle to the nearest slit. There is spontaneous symmetry breaking between plus and minus x.

To understand qualitatively why there is a transition, we note that t'_c is also the time it takes to diffuse from $x = 0$ to one of the slits at $\pm x_0$. If the control acts much earlier than t'_c before the end, then it wastes effort regulating against diffusion. If the control acts much later, there will need to be a brief but large effort at the end to make sure the particle is directed through a slit. Because the cost $\sim u^2$ and not $|u|$, the trade-off is unfavorable. At t'_c, diffusion will typically be about the right distance, potentially saving control effort.

The above example shows a noise-dependent "phase transition" in strategy that results from the stochastic aspects – the noise $\nu(t)$ affects the particle – in combination with nonlinearity: the boundary condition at the slits is a potential with two delta-function wells. We will discuss another example in Section 10.4.1 based on discrete dynamics.

8.7 Smoothing and Prediction

The discussion so far has assumed that we want to estimate or control the state $\boldsymbol{x}$ at the present time k. But sometimes it is useful to project farther out into the future, or *predict*, and sometimes to reach back into the past, a task termed *smoothing*.

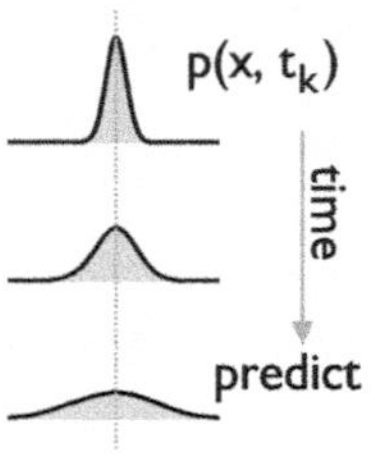

Prediction is straightforward:[29] we simply step the discrete equations (8.76a) or integrate the continuous Fokker–Planck equations (8.79), writing $\hat{x}^-_{k+\ell}$ for the discrete case and $\hat{x}_k(t)$ for the continuous case. Recall the freely diffusing particle, whose Fokker–Planck equation is just the diffusion equation for $p(x,t)$. Its Gaussian solution $p(x,t) = [2\pi P_k(t)]^{-1/2} \exp[-\frac{(x-\hat{x}_k)^2}{4P_k(t)}]$ is characterized for times $t \geq t_k$ entirely by its constant mean $\hat{x}_k$ and by its increasing variance $P_k(t) = P_k + 2D(t - t_k)$ (see left and Section 8.1.1).

Smoothing is slightly more complicated to implement than filtering. But why bother to estimate a state in the past? The main motivation is to improve state estimates by adding subsequent observations. You might want to do this for *data processing* after an experiment. It can also be useful to see how extra information improves an estimate. Finally, sometimes we *do* have information about the future (Chapter 15). This is not as strange as it sounds. If you want to control the temperature of a building, you know that nights are usually cool, and so you can anticipate, to some extent, future disturbances. Even partial information about the future can improve knowledge about and hence control of a current state.[30]

The goal of smoothing is to evaluate $p(\boldsymbol{x}_k|\boldsymbol{y}^N)$, which represents our knowledge of $\boldsymbol{x}$ at time k, given *all* the observations, $\boldsymbol{y}^N = \{y_1, y_2, \ldots, y_N\}$. By contrast, the filtering problem focused on $p(\boldsymbol{x}_k|\boldsymbol{y}^k)$, which is conditioned only on observations up to and including the present time. The basic trick is to split the estimation problem into two independent parts. Applying Bayes' rule to the future observations

[29] Or not. The old saying "Prediction is hard, especially about the future" – variously attributed to Niels Bohr, Yogi Berra, and many others – reminds us not to be too smug.

[30] Imagine what you could do in the stock market if you knew about the future. Or, think about how many details you notice when you read a book or see a movie for the second time. Knowing the future evolution of the plot – the "state" of the artistic work – reveals details in the artistic "signal," the words on the page or the images on the screen, that seem to be "noise" the first time around.

$y_{k+1}^N \equiv \{y_{k+1}, y_{k+2}, \ldots, y_N\}$, we have

$$p(x_k|y^N) \propto p(y_{k+1}^N|x_k, y^k)\, p(x_k|y^k) = \frac{1}{Z}\left(\frac{p(x_k|y_{k+1}^N)\, p(x_k|y^k)}{p(x_k)}\right). \tag{8.123}$$

In the first expression, a Markov process needs only x_k to predict future observations – past observations are of no further help. In the second expression, we use Bayes' rule again. Equation (8.123) implies that our knowledge of the state x_k arises from two components. One is the familiar Bayesian filter term, $p(x_k|y^k)$, which we compute by iterating in the forward direction (Eq. 8.76). The other is a similar term that can be computed *backwards* in precisely the same way, assuming that the dynamics can be inverted. In the Bayesian context, this implies the ability to compute $p(x_k|x_{k+1})$ in the "prediction" step of the backward dynamics $k+1 \to k$,

$$p(x_k|y_{k+1}^N) = \int \mathrm{d}x_{k+1}\, p(x_k|x_{k+1})\, p(x_{k+1}|y_{k+1}^N)\,. \tag{8.124}$$

Notice that the backward dynamics term is *not* $p(x_k|y_k^N)$, which is the best estimate of x_k including the observation y_k. That would double count the contribution of y_k. (See right.)

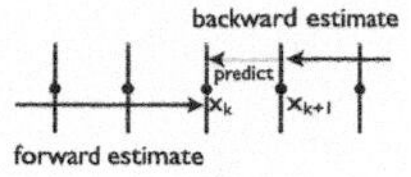

In summary, we carry out the following steps:

1. Estimate $p(x_k|y^k)$ using the observations y from time 1 to the present, k.
2. Starting from the last observation N (in the future) and working back to the present, construct $p(x_k|y_{k+1}^N)$ by prediction from $p(x_{k+1}|y_{k+1}^N)$.
3. Combine these two estimates into a single optimal estimate $p(x_k|y^N)$.

For linear systems and Gaussian noise, Step (1) is the Kalman filter. The reverse direction also leads to a Kalman filter, with *backwards dynamics*, $x_k = f^{-1}(x_{k+1})$. Thus,

$$\underbrace{(x^{\mathrm{f}})_{k+1} = A(x^{\mathrm{f}})_k + \nu_k}_{\text{forward}}\,, \qquad \underbrace{(x^{\mathrm{b}})_k = A^{-1}(x^{\mathrm{b}})_{k+1} - A^{-1}\nu_{k+1}}_{\text{backward}}\,, \tag{8.125}$$

where the backward dynamical matrix and input are modified to $A' = A^{-1}$ (assuming A is invertible).

Using the forward dynamics and the past and current observations y^k, we calculate the forward estimate $(\hat{x}^{\mathrm{f}})_k$ as before, from the Kalman filter recursion relations, Eq. (8.41). We proceed similarly with the backwards estimate, which leads to $(\hat{x}^{\mathrm{b}})_{k+1}$ and to the *backwards prediction* for x_k based on y_{k+1}^N, denoted $(\hat{x}^{\mathrm{b}})_k^+$. Combining these two independent estimates of x_k gives a single, best *smoother estimate*, $(\hat{x}^{\mathrm{s}})_k$ that uses all the observations, y^N.

Combining the past and future with the appropriate covariance weightings is equivalent to drawing two measurements from Gaussian distributions of the same mean but different variances. More prosaically, it is the problem of combining two independent estimates of the same quantity with different precisions (Problem A.8.1 online). The probability distribution for the state is again Gaussian, with variance and mean determined by weighted sums,

$$(P^{\mathrm{s}})^{-1} = (P^{\mathrm{f}})^{-1} + (P^{\mathrm{b},+})^{-1}\,, \quad \hat{x}^{\mathrm{s}} = P^{\mathrm{s}}\left[(P^{\mathrm{f}})^{-1}\hat{x}^{\mathrm{f}} + (P^{\mathrm{b},+})^{-1}\hat{x}^{\mathrm{b},+}\right], \tag{8.126}$$

where all quantities are evaluated at time k and where $(\boldsymbol{P}^{\mathrm{b},+})_k$ is the predicted covariance of the backwards dynamics at time step k, based on the future observations, $\boldsymbol{y}_{k+1}^N$.

The backwards dynamics are initialized by requiring the smoother estimate $\boldsymbol{x}^{\mathrm{s}}$ match the forward estimate $\boldsymbol{x}^{\mathrm{f}}$ at $k = N$. At that time, $(\boldsymbol{P}^{\mathrm{b},+})_N$ is infinite (and its inverse zero), as no further information is available to improve the forward estimate.

Example 8.6 (Smoothing for a diffusing particle) The forward equations, familiar from Section 8.1.1, are

$$x_{k+1} = x_k + \nu_k\,, \qquad y_k = x_k + \xi_k\,, \tag{8.127}$$

with $\langle \nu_k^2 \rangle = \nu^2$ and $\langle \xi_k^2 \rangle = \xi^2$. To estimate the state going forward, we recall the Kalman filter equations that we previously derived:

$$\begin{aligned} &(\hat{x}^{\mathrm{f},-})_{k+1} = (\hat{x}^{\mathrm{f}})_k\,, \quad (\hat{x}^{\mathrm{f}})_{k+1} = (\hat{x}^{\mathrm{f},-})_{k+1} + (L^{\mathrm{f}})_{k+1}\left[y_{k+1} - (\hat{x}^{\mathrm{f},-})_{k+1}\right], \\ &(P^{\mathrm{f},-})_{k+1} = (P^{\mathrm{f}})_k + \nu^2\,, \quad (L^{\mathrm{f}})_{k+1} = \frac{(P^{\mathrm{f},-})_{k+1}}{(P^{\mathrm{f},-})_{k+1} + \xi^2}\,, \quad (P^{\mathrm{f}})_{k+1} = \xi^2 (L^{\mathrm{f}})_{k+1}\,. \end{aligned} \tag{8.128}$$

The backward dynamics are

$$x_k = x_{k+1} - \nu_k\,, \qquad y_k = x_k + \xi_k\,, \tag{8.129}$$

with

$$\begin{aligned} &(\hat{x}^{\mathrm{b},+})_k = (\hat{x}^{\mathrm{b}})_{k+1}\,, \quad (\hat{x}^{\mathrm{b}})_k = (\hat{x}^{\mathrm{b},+})_k + (L^{\mathrm{b}})_k\left[y_k - (\hat{x}^{\mathrm{b},+})_k\right], \\ &(P^{\mathrm{b},+})_k = (P^{\mathrm{b}})_{k+1} + \nu^2\,, \quad (L^{\mathrm{b}})_k = \frac{(P^{\mathrm{b},+})_k}{(P^{\mathrm{b},+})_k + \xi^2}\,, \quad (P^{\mathrm{b}})_k = \xi^2 (L^{\mathrm{b}})_k\,. \end{aligned} \tag{8.130}$$

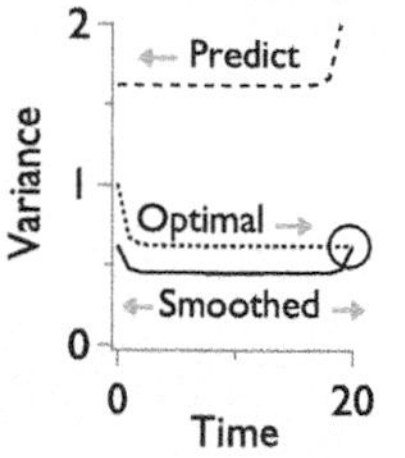

The variances are found by iterating these equations. For $\xi^2 = \nu^2 = 1$, the results are shown at left. The forward filter, shown in the fine-dotted line, starts with a variance of 1, which is the uncertainty of the state after the first observation. Following a brief transient, the variance goes to the stationary value derived before, of $\frac{1}{2}(\sqrt{5} - 1) \approx 0.61$. We do not show the backwards optimal estimate, but it would be just the mirror of the forward estimate, since the system is statistically time invariant. The backwards prediction variance (longer dashed line at top) is larger by ν^2; its information is independent and improves our forward estimate. We initialize the variance at time N by setting it to a large value, so that the smoother variance coincides with the forward variance (highlighted in white). The backward predicted variance then settles to an asymptotic value of $\frac{1}{2}(\sqrt{5} + 1) \approx 1.61$. Then, using Eqs. (8.126), we form the weighted average of $(\hat{x}^{\mathrm{f}})_k$ and $(\hat{x}^{\mathrm{b},+})_k$, and also find the steady-state smoother variance. Here, $P^{\mathrm{s}} = \frac{1}{\sqrt{5}} \approx 0.45$, or 72% of the forward variance.

Defining $\alpha = \nu^2/\xi^2$ and using Eq. (8.18), we see that the steady-state smoother gain is

$$P^{\rm s} = \frac{\xi^2}{2}\left(\frac{1}{\sqrt{\alpha^2+4\alpha}-\alpha} + \frac{1}{\sqrt{\alpha^2+4\alpha}+\alpha}\right)^{-1} = \xi^2 \frac{\alpha}{\sqrt{\alpha^2+4\alpha}}. \tag{8.131}$$

and thus

$$\frac{P^{\rm s}}{P^{\rm f}} = \frac{2\alpha}{\sqrt{\alpha^2+4\alpha}\,(\sqrt{\alpha^2+4\alpha}-\alpha)}. \tag{8.132}$$

Equation (8.132) implies that when input noise dominates ($\alpha \gg 1$), $P^{\rm f} = P^{\rm s} = \xi^2$. When measurement noise dominates ($\alpha \ll 1$), then $P^{\rm s} = \frac{1}{2}P^{\rm f}$. Both results make sense: with little observation noise, the best estimate is simply the measurement, with no need to use the dynamics, whether forward or backward. By contrast, when the input noise is small, forward and backward extrapolation contribute equally and the observation itself adds little. Thus, the variance combining the two is just half that of either alone, which is what we expect when we combine two equally accurate, independent measurements.

A factor of two in variance might seem meager reward for the investment in learning the smoothing concepts, but in higher-dimensional state spaces, the gain can be larger. Problem 8.21 explores filtering and smoothing for inertial motion (the "random rocket"), where variance gains can be up to a factor of four for the smoother relative to the filter.

The explicit forwards-backwards algorithm for smoothing that we have discussed here has many practical difficulties. The most serious is that it is not always possible to formulate a backwards dynamics, as the dynamical matrix $\mathbf{A}$ may not be invertible. Even so, future observations can improve present estimates. Problem 8.20 explores an alternate Bayesian smoother that uses only forward dynamics. Its linear form leads to the more efficient *Rauch–Tung–Striebel* (RTS) smoother algorithm (Problem 8.21).

The *fixed-interval smoothing* discussed above assume data from times 1 to N and seek smoothed estimates for all the time points. A variant, *fixed-lag smoothing*, seeks the smoothed estimate of a point that lags a fixed interval behind the current measurement, or, equivalently, assumes knowledge of a fixed *preview* of the future. We would then consider $P(\mathbf{x}_k|\mathbf{y}^{k+\ell})$, which uses information ℓ steps into the future (Särkkä, 2013). Finally, for steady-state problems, the frequency-domain approach of the *Wiener filter* is equivalent to Kalman smoothing but often much simpler to implement. See Problem 8.23.

Of course, every filtering algorithm that we have studied in this chapter – extended, cubature, ensemble Kalman, grid methods, particle filters – has its counterpart for smoothing.

8.8 Summary

This chapter began with a sharply focused question: how to choose the optimal gain for an observer for a linear system, as a complement to our studies in optimal control. Since sensible optimization criteria involve measurement and input noise, we introduced stochastic dynamical systems. For linear dynamics and Gaussian noise, we can offer a complete, closed solution, the Kalman filter. That solution combines in a wonderfully simple way with optimal control to form Linear Quadratic Gaussian (LQG) control, where each aspect may be addressed separately. The optimal state estimate is used in an optimal controller, as if it were the true state. To go beyond the restrictions of linear models and Gaussian noise, we reformulated the state estimation problem using notions of Bayesian inference as a problem to estimate the conditional probability distribution $p(\boldsymbol{x}_k|\boldsymbol{y}^k)$. The general estimation equations for this distribution are simple to formulate but hard to solve. Ad hoc methods such as extended and cubature Kalman filters project $p(\boldsymbol{x}_k|\boldsymbol{y}^k)$ onto Gaussian distributions and can be effective in problems with modest deviations from linearity and Gaussian statistics. When the deviations are more serious, numerical methods such as the grid method or the Monte Carlo techniques of particle filtering or path integrals are better, with the deterministic grid method suitable for smaller problems and Monte Carlo methods preferable for larger problems. Of course, for any real-time control implementation, computations must be done faster than the relevant time scales in the dynamics.

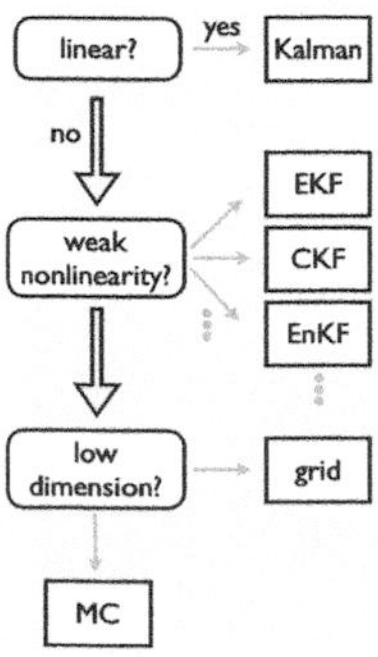

There are many other state-estimation methods that we have not discussed, such as different sampling strategies for Monte Carlo techniques. Not only do we lack the space to give each its due, there is a danger in overloading the reader with too many similar approaches, each with its own set of technical difficulties to master. But once you have understood the general ideas expressed here, you will be in a good position to understand the various approaches that have been proposed in the literature. We conclude with a flowchart, at left, that summarizes the different methods and when to use them.

8.9 Notes and References

For background on stochastic methods, many texts take a highly mathematical point of view. Among those that are more physical, Gardiner (2009) is a standard reference and van Kampen (2007) is particularly good on the conceptual background. Both of these books derive the Fokker–Planck equation, which is also discussed accessibly and comprehensively in Risken (1989). Miller et al. (1999) is a more recent discussion of issues for its numerical solution. For Kalman filters and stochastic control, Stengel (1994) gives a straightforward development of the theory. Åström (2006a) has a careful and thorough development of the mathematical background, especially for continuous dynamics. The original paper by Kalman (1960a) is worth reading, too. The first chapter of Maybeck (1979) has a very intuitive introduction to the Kalman filter. The remainder and its two companion volumes illustrate most aspects of the Kalman filter, including its use in hybrid dynamics. The relations between the problem of estimating a mean value and the Kalman filter, along with other applications of the Kalman filter to signal processing, are discussed by Cooper (1986). Haykin (2001) includes an overview of smoothing as an extension of Kalman filters and details the RTS smoother. The 1/4 power law for the precision of estimating (or controlling) a fluctuating quantity, Eq. (8.20), has been rediscovered many times in different contexts (e.g., by Lestas et al. [2010]; Mora and Nemenman [2019]).

The Bayesian formulation of state estimation originated, in the control literature, with Ho and Lee (1964), and we follow loosely their approach. Earlier but more complicated formulations include those by Stratonovich (1960) and Kushner (1962). Our discussion of Bayesian smoothing follows Särkkä (2013), who also discusses how the Bayesian formulation unifies the various state-estimation methods. Like Arulampalam et al. (2002), he gives a unified presentation of the extended Kalman filter, grid, and particle filters. Many aspects of assumed density filters are described by Maybeck (1982). The unscented transformation was originally proposed by Julier and Uhlmann (1997), and our presentation follows mostly Wan and van der Merwe (2001). The ensemble Kalman filter, blending assumed-density and Monte Carlo methods, is reviewed by its originator in Evensen (2009). For the grid method of state estimation and related numerical techniques, Miller et al. (1999) showed that it can be effective for a variety of problems, including the double-well potential, the Lorenz equations, and a higher-dimensional problem whose dominant dynamics projects down to an effectively two-dimensional subspace.

Although Monte Carlo methods for state estimation were attempted in the 1970s and 1980s, the explosion of recent interest began with the first practical implementation, the SIR given by Gordon et al. (1993). (They termed their algorithm the *bootstrap filter*.) For smoothing, the grid approximation leads to hidden Markov models (see Chapter 12). Our presentation of particle filters owes much to Kitagawa (1996). Crisan and Rozovskii (2011) is a 1000-page summary of the mathematical approach to nonlinear filtering. Daum (2005) gives a refreshingly candid overview and assessment of different extensions of the Kalman filter, noting several places where methods from theoretical physics may advance the state of the art. Daum and Huang (2016) advocate with verve particle filters whose particles "flow" to the relevant regions of state space. See also van Leeuwen (2010).

The separation principle for linear quadratic Gaussian control arose independently in closely related forms in several studies from the early 1960s. Wonham (1968) is an influential early summary. The certainty equivalence principle had been derived previously, in the context not of control theory but of economics, by Simon (1956). A polymath who made contributions to economics, mathematics, artificial intelligence, psychology, philosophy, and more, Herbert Simon won the Nobel Prize in Economics in 1978 for this and other results on *bounded rationality*.

Theoretical physicists have begun to show some interest in these problems and have started to apply the full arsenal of statistical physics methods – variational mean-field formulations, path integrals, and more – to the estimation problem. Attempts to make path-integral and other field-theory descriptions include Abarbanel (2013), Enßlin (2013), Restrepo (2008), Rasmussen and Williams (2006), Kappen (2005), and Lemm (2003). It is perhaps too early to judge the significance of this work. Many of these authors seem to be unaware of other similar approaches, and a broader assessment would be helpful.

The above path-integral methods can be generalized to include the kinds of information-theoretic concepts developed in Chapter 15.2. Williams et al. (2018) develop this approach and then apply it to the problem of autonomous driving. The implementation on a robot vehicle shows the state of the art of modern control theory: the authors' information-theoretic version of model predictive control uses GPUs to evaluate in parallel 1200 Monte Carlo trajectories at 40 Hz, making 4.8 million nonlinear function calls per second. The nonlinear function modeling the car is a seven-dimensional state space with two inputs. The dynamics are expressed either as a linear expansion over 25 nonlinear basis functions or, alternatively, as a neural network with two hidden layers of 32 neurons (1412 parameters). The resulting controller was able to drive "aggressively" the autonomous car around a dirt race track. Like the path-integral approach, this work is too much in flux to include here.

For more on why state estimation can be a hard problem, see Antenucci et al. (2019), who discuss *hard phases* of inference that have an exponentially large number of local minima. They apply methods from the study of spin glasses to inference.

Experimental physicists and chemists, especially those concerned with the noisy world of single-molecule biophysics and related problems, have used ideas of state estimation in their work. The ABEL (Anti-Brownian motion ELectrokinetic) trap for controlling the motion of a diffusing particle and its theoretical analysis was developed by Cohen (2005). Wang and Moerner (2010) proposed an optimal way to convert photon-limited observations into a standard Kalman-filtering problem and implemented an extended version of the scheme in Wang and Moerner (2011). Using a slightly different extension of Kalman filtering, Fields and Cohen (2011) could trap a single dye molecule in aqueous solution for about ten seconds. The frequency-estimation example is from Van Trees et al. (2013).

Many different fields have shown interest in state estimation. Unfortunately, each has its own notation and language, creating many barriers to communication, so that not all methods are in use in all communities. In addition to control theory, two fields in which state estimation plays a particularly strong role are meteorology and machine learning.

The meteorological literature seems especially divorced from engineering discussions. In that field, filtering is termed *data assimilation* and the method of choice is the ensemble Kalman filter. Forecasting the weather requires estimating the state of very high-dimensional models (e.g., $n = 10^6$), since the atmosphere is a spatially extended system in three dimensions. On the positive side, massive computer resources are generally available, and the update rate is slow (minutes to hours).

The machine-learning community grew out of the computer science and artificial intelligence fields. The past few years have seen many applications to physical problems, especially ones where pattern recognition plays an important role. For physicist-friendly introductions, see Chapter 16 of Press et al. (2007) and Mehta et al. (2019); see Murphy (2012) for a view from within the field.

Problems

8.1 Kalman filter for prediction observer. In the text, we have formulated Kalman filters in terms of the current observer, which uses observations up to y_{k+1} in the estimate $\hat{x}_{k+1}$. The prediction observer uses only up to y_k and is appropriate for cases where sensor and computational delays are on the order of T_s (Section 5.4.2).

a. Redo the 1d-Kalman filter calculations for the prediction observer, defining

$$e_k = x_k - \hat{x}_k \qquad x_{k+1} = x_k + \nu_k \qquad y_k = x_k + \xi_k$$
$$P_k = \langle e_k^2 \rangle \qquad \hat{x}_{k+1} = \hat{x}_k + L(y_k - \hat{y}_k) \qquad \hat{y}_k = \hat{x}_k \,.$$

Calculate the recurrence relations for the optimal L_k^* and P_k^*, as well as the steady state P^* and L^*. You should find that $P^* = L^*\xi^2 + \nu^2$ (cf. the previous result, $P^* = L^*\xi^2$). Comment on L^* and P^* in the limits $\alpha \ll 1$ and $\gg 1$, where $\alpha \equiv \nu^2/\xi^2$.

b. Generalize to an n-dimensional MIMO system, by defining

$$\boldsymbol{e}_k = \boldsymbol{x}_k - \hat{\boldsymbol{x}}_k \qquad \boldsymbol{x}_{k+1} = \boldsymbol{A}\boldsymbol{x}_k + \boldsymbol{B}\boldsymbol{u}_k + \boldsymbol{\nu}_k \qquad \boldsymbol{y}_k = \boldsymbol{C}\boldsymbol{x}_k + \boldsymbol{\xi}_k$$
$$\boldsymbol{P}_k = \langle \boldsymbol{e}_k \boldsymbol{e}_k^{\mathsf{T}} \rangle \qquad \hat{\boldsymbol{x}}_{k+1} = \boldsymbol{A}\hat{\boldsymbol{x}}_k + \boldsymbol{B}\boldsymbol{u}_k + \boldsymbol{L}(\boldsymbol{y}_k - \hat{\boldsymbol{y}}_k) \qquad \hat{\boldsymbol{y}}_k = \boldsymbol{C}\hat{\boldsymbol{x}}_k \,,$$

Show that the recurrence relations for the time-dependent Kalman filter become

$$\boldsymbol{P}_{k+1}^{y} = \boldsymbol{C}\boldsymbol{P}_k\boldsymbol{C}^{\mathsf{T}} + \boldsymbol{Q}_{\xi} \,, \quad \boldsymbol{P}_{k+1}^{xy} = \boldsymbol{P}_k\boldsymbol{C}^{\mathsf{T}} \,, \qquad \boldsymbol{L}_{k+1}^{*} = \boldsymbol{A}\boldsymbol{P}_{k+1}^{xy}\left(\boldsymbol{P}_{k+1}^{y}\right)^{-1} \,,$$
$$\boldsymbol{P}_{k+1}^{*} = \boldsymbol{A}\boldsymbol{P}_k^{*}\boldsymbol{A}^{\mathsf{T}} + \boldsymbol{Q}_{\nu} - \boldsymbol{L}_{k+1}^{*}\boldsymbol{P}_k^{y}\boldsymbol{L}_{k+1}^{\mathsf{T}} \,.$$

Show that, in steady state, $\boldsymbol{P}^*$ obeys an algebraic Riccati equation that maps precisely onto Eq. (7.24) from the discussion on LQR optimal control in Chapter 7. Comment on the different values of $\boldsymbol{L}^*$ for the two forms of Kalman filter.

8.2 Estimating unstable dynamics. Consider 1d *deterministic* dynamics with noisy observations. Let $x_{k+1} = ax_k$, with $y_k = x_k + \xi_k$ and x_0 unknown. The noise $\xi_k \sim \mathcal{N}(0, \xi^2)$. When $a > 1$, the dynamics are unstable. Using the prediction observer from Problem 8.1, show that the steady-state variance of the optimal estimate is $P^* = (a^2 - 1)\xi^2$ for $|a| > 1$, and 0 otherwise. Interpret the two cases.

8.3 Diffusion with continuous measurements. There are subtleties:

a. Start by formulating the steady-state Kalman–Bucy filter for one-dimensional diffusion. The equations of motion are $\gamma \dot{x} = \nu_c$, with $\langle \nu_c(t)\, \nu_c(t') \rangle = 2D\gamma^2\, \delta(t - t')$. The measurement relation is $y = x + \xi_c$, with $\langle \xi_c(t)\, \xi_c(t') \rangle = \xi_c^2\, \delta(t - t')$. Find the optimal Kalman gain, and show that the variance is $P = \sqrt{2D}\, \xi_c$.

b. Contrast the above results with those found for the discrete case: for signal-to-noise ratio $\alpha = \nu^2/\xi^2$ (for power), the limit $\alpha \gg 1$ implies $L \to 1$ and $P \approx \xi^2$. By contrast, the limit $\alpha \ll 1$ implies $L \to \sqrt{\alpha}$ and $P \approx \nu\xi$. Reconcile these different behaviors. Hint: Connect the discrete quantities ν^2 and ξ^2 with our continuous versions, $\nu_c^2 = 2D$ and ξ_c^2. Notice that the units of ν_c and ξ_c are different. Write α for the discrete case in terms of continuous quantities and the time step T_s.

c. Laplace transform the equations of motion to show that the Kalman filter acts as a first-order, low-pass filter between the observations, $y(t)$, and the estimate, $\hat{x}(t)$. What is the cutoff frequency? Argue (justify) that the filter "trusts the measurements" $y(t)$ at frequencies where the signal dominates over the noise. At higher frequencies, noise is important, and the filter attenuates the measurements.

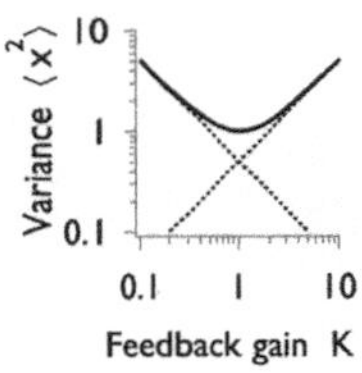

d. Add negative feedback, $u(t) = -K\, y(t)$, to stabilize the diffusing particle near the origin. Show that $\langle x^2 \rangle = (K^2\xi_c^2 + \nu_c^2)/2K$. See at left, for $\xi_c = \nu_c = 1$. Interpret.

8.4 Discretizion of a continuous stochastic system. We discretize a harmonic oscillator driven by thermal noise, following Nørrelykke and Flyvbjerg (2011).

a. Integrate the linear, time-invariant system $\dot{\boldsymbol{x}} = \boldsymbol{A}\boldsymbol{x} + \boldsymbol{B}\boldsymbol{\nu}$, with $\langle \boldsymbol{\nu}(t) \rangle = 0$ and $\langle \boldsymbol{\nu}(t)\, \boldsymbol{\nu}(t')^{\mathsf{T}} \rangle = \delta(t - t')$ over a time T_s to find discrete dynamics $\boldsymbol{x}_{k+1} = \boldsymbol{A}_d \boldsymbol{x}_k + \boldsymbol{\nu}_k$, with $\boldsymbol{A}_d = e^{\boldsymbol{A}T_s}$ and $\boldsymbol{\nu}_k = \int_0^{T_s} dt'\, \boldsymbol{\nu}(t')\, e^{\boldsymbol{A}(T_s - t')}\, \boldsymbol{B}$ a Gaussian random vector of mean $\boldsymbol{0}$ and covariance $\langle \boldsymbol{\nu}_k\, \boldsymbol{\nu}_\ell^{\mathsf{T}} \rangle = \delta_{k\ell} \int_0^{T_s} dt'\, e^{\boldsymbol{A}(T_s - t')}\, \boldsymbol{B}\boldsymbol{B}^{\mathsf{T}}\, e^{\boldsymbol{A}^{\mathsf{T}}(T_s - t')}$.

b. For the noisy critically damped harmonic oscillator, $\ddot{x} + 2\dot{x} + x = \sqrt{8D}\, \nu(t)$, show that $\boldsymbol{A} = \begin{pmatrix} 0 & 1 \\ -1 & -2 \end{pmatrix}$ and $\boldsymbol{A}_d = e^{-T_s} \begin{pmatrix} 1+T_s & T_s \\ -T_s & 1-T_s \end{pmatrix}$.

c. Show that the covariance matrix $\langle \boldsymbol{\nu}_n\, \boldsymbol{\nu}_n^{\mathsf{T}} \rangle = \begin{pmatrix} \sigma_{xx}^2 & \sigma_{xv}^2 \\ \sigma_{xv}^2 & \sigma_{vv}^2 \end{pmatrix}$, with $\sigma_{xx}^2 = 2D[1 - e^{-2T_s}(1 + 2T_s + 2T_s^2)]$, $\sigma_{xv}^2 = 4D\, e^{-2T_s}\, T_s^2$, and $\sigma_{vv}^2 = 2D[1 - e^{-2T_s}(1 - 2T_s + 2T_s^2)]$.

Notice that although the original physical system has only a single noise source (thermal fluctuations) that drives only the velocity, the sampled system is driven by *two* uncorrelated noise sources. The sources then become correlated by the input coupling, leading to a structure for the discrete equations that is quite different from that of the original continuous system. In the limit $T_s \to 0$, we see that $\sigma_{vv}^2 = 8DT_s + O(T_s^2)$, while σ_{xx}^2 and σ_{xv}^2 are higher order in T_s. We then recover the continuum situation.

8.5 One-step LQG. Consider one-dimensional, deterministic dynamics $x_1 = x_0 + u_0$, with cost function $J_{\text{det}} = \frac{1}{2}(Qx_1^2 + Ru_0^2)$. Only a single control u_0 is applied.

a. Find the optimal value of u_0, and show it has the form $u_0^* = -K^* x_0$, a linear feedback. Show that $K^* = Q/(R+Q)$, and evaluate the optimal cost $J_{\text{det}}^*(K^*)$.
b. Add a stochastic disturbance: $x_1 = x_0 + u_0 + \nu_0$, with $\nu_0 \sim \mathcal{N}(0, \nu^2)$. Show that u_0^* is unaltered and $\langle J^* \rangle = J_{\text{det}}^* + \frac{1}{2} Q \nu^2$ (certainty equivalence principle).
c. Consider a noisy observation $y_0 = x_0 + \xi_0$. Why is the optimal control gain still K^* (separation principle)? Why is $u_0 = -K y_0$ not optimal?
d. In formulating the cost function, you might have expected to see a total cost $J = \frac{1}{2} \sum_k \left(Q x_k^2 + R u_k^2 \right)$, with $k \in \{0, 1\}$. Why ignore the x_0 and u_1 terms?

8.6 Variance of observer control for a 1d Brownian particle. Observer-based feedback can lead to a minimum-variance control strategy (Section 8.2.1):

a. Find the variance $\langle x^2 \rangle$ and K^* for feedback based on perfect state information, naive observations, and observer. Find L^* for the observer case.
b. Write code to simulate all three cases; check the results from (a).

8.7 LQG for undamped, noisy oscillator. Let $\ddot{x} + x = u(t) + \nu(t)$, with $y(t) = x(t) + \xi(t)$.

a. What is the state-space representation of the continuous system?
b. For sampling at $T_s = 0.1$, what is the ZOH discrete state-space representation?
c. Assuming that the standard deviations $\nu = \xi = 0.3$ for the process and measurement noise (when sampled at intervals T_s) and assuming state-space weights of $\mathbf{Q} = \left(\begin{smallmatrix} 1 & 0 \\ 0 & 1 \end{smallmatrix}\right)$ and input weighting $r = 0.1$, derive the LQG controller. As a check, you should find an optimal observer (Kalman) gain of $\mathbf{L}^{\mathsf{T}} \approx (0.09, 0.03)$ and optimal control gain of $\mathbf{K} \approx (1.7, 3.3)$. (Give some more digits, please!)
d. Using the above observer and controller gains and adding process and measurement noise, plot the disturbance response (right, top graph, for $x(0) = 0$, $\dot{x}(0) = 1$).
e. Simulate the controller and plot the disturbance response (right, middle graph). Show that the difference between the position and its optimal (Kalman) estimate is the same for both the closed- and open-loop cases (dark lines in the bottom plot at right). Their difference is nearly zero, as shown by the gray trace.
f. Show that, after a transient, the standard deviation of both state and estimate are well below that of the measurement errors.

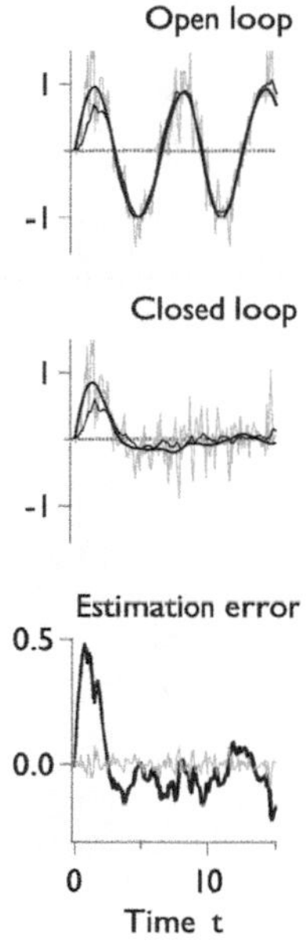

8.8 Delayed choice. Fill in the details from Example 8.5. Reproduce the plots in the example, using $R = 2D = x_0 = 1$.

8.9 Toy model of state estimation, part 2. Redo the calculations of Section 8.3.1, allowing for N noisy, independent measurements $y_i = x + \xi_i$. Find the posterior $p(x|y^N)$, where $y^N = \{y_i\}$ for $i = 1, \ldots, N$. Show that $\hat{x} = \frac{\sigma_x^2}{\sigma_x^2 + \sigma_\xi^2/N} \bar{y}$, where $\bar{y} = \frac{1}{N} \sum_i y_i$ is the average of the N measurements. Find the corresponding variance.

8.10 From Bayes to Kalman, in 1d. Equations (8.76) and (8.77) describe one-dimensional diffusion, $x_{k+1} = x_k + \nu_k$, with $y_k = x_k + \xi_k$, with i.i.d. Gaussian random noise terms $p(\nu_k) = \mathcal{N}(\nu_k; 0, \nu^2)$ and $p(\xi_k) = \mathcal{N}(\xi_k; 0, \xi^2)$.

a. Show that the PDF of $p(x_{k+1}|x_k)$ is $\mathcal{N}(x_{k+1} - x_k; 0, \nu^2) \equiv \frac{1}{\nu\sqrt{2\pi}} \exp\left[-\frac{(x_{k+1}-x_k)^2}{2\nu^2}\right]$.

b. Making the *ansatz* $p(x_k|y^k) = \mathcal{N}(x_k; \hat{x}_k, P_k)$ and marginalizing over the proper variable, show that $p(x_{k+1}|y^k) = \mathcal{N}(x_{k+1}; \hat{x}_k, P_k + \nu^2) \equiv \mathcal{N}(x_{k+1}; \hat{x}^-_{k+1}, P^-_{k+1})$.

c. The Bayesian update step, $p(x_{k+1}|y^{k+1}) = \frac{p(y_{k+1}|x_{k+1})\, p(x_{k+1}|y^k)}{p(y_{k+1}|y^k)}$ requires three probability distributions. We know one. Derive the other two:

$$p(y_{k+1}|x_{k+1}) = \mathcal{N}(y_{k+1} - x_{k+1}; 0, \xi^2), \quad p(y_{k+1}|y^k) = \mathcal{N}(y_{k+1}; \hat{x}_k, P_k + \nu^2 + \xi^2).$$

d. Finally, evaluate the update step using the three distributions to show that the conditional distribution $p(x_{k+1}|y^{k+1}) = \mathcal{N}(x_{k+1}; \hat{x}_{k+1}, P_{k+1})$, where

$$\hat{x}_{k+1} = \hat{x}_k + L_{k+1}(y_{k+1} - \hat{x}_k) \qquad L_{k+1} = \frac{P_k + \nu^2}{P_k + \nu^2 + \xi^2}$$

$$P^{-1}_{k+1} = (P_k + \nu^2)^{-1} + (\xi^2)^{-1} \qquad \text{or} \qquad P_{k+1} = L_{k+1}\,\xi^2,$$

Hint: Complete the square or use a computer-algebra program.

8.11 Instability with a hard-spring potential. Consider an overdamped particle in a hard-spring potential with input noise. If we ignore the nonlinear term, a quick analysis shows that the system is stable. Now add a nonlinear term that, in the absence of stochasticity, is stabilizing. Surprisingly, the state amplitude will eventually diverge, no matter how weak the noise. As shown at left, along with a normal quadratic potential (thin line), the *hard-spring potential* $V(x) = \frac{1}{2}x^2 + \frac{1}{4}x^4$, with force $F(x) = -\partial_x V = -(x + x^3)$. The spring is "hard" because the local stiffness $\mathrm{d}F/\mathrm{d}x$ increases away from equilibrium. In a first-order numerical integration, the system state evolves as $x_{k+1} = f(x_k) = x_k - T_s(x_k + x_k^3) + \nu_k$, with $\nu_k \sim \mathcal{N}(0, \nu^2)$.

a. Simulate the above equation. Generate various times series for x_k for different values of ν_k, using $T_s = 0.5$. What happens as you increase the input noise?
b. Show that the linearized system is stable for $0 < T_s < 2$, for arbitrary ν.
c. Fix $\nu = 0.5$, and modify your code to track the *lifetime* of the state, the typical time before x_k goes unstable. Run your code many times (≈ 1000), measuring the lifetime in each case. Plot a histogram. What distribution does it follow, and why? You should find that the average lifetime ≈ 700.
d. Show that the state dynamics go unstable for $|x_k| > x_0 = \sqrt{3}$. Hint: Look at the conditions $f(x) > x$ and $f(x) < -x$. Why do these lead to instability?
e. To estimate the average lifetime, assume $x_k \sim \mathcal{N}(0, \nu^2)$, which is only approximately true because of the x_k^3 term. Assume, too, that each time step brings an independent perturbation, which implies that perturbations relax in one time step. The probability of instability then reduces to the probability $P(|x| > x_0)$ for a Gaussian distribution. If, in either tail of the distribution, a point lies in the shaded area $|x| > x_0$, then instability will likely result (see left). ("Likely" because a fluctuation just larger than x_0 might come back.) Then, either numerically or by an asymptotic expansion of the erfc(·) function, estimate the lifetime of the state x, in units of the time step k. You should find $\approx$ 1000–2000, slightly > 700.

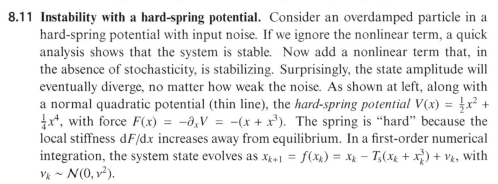

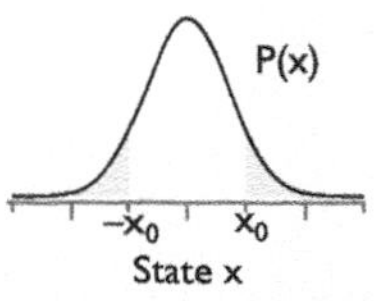

f. For the linearization $x_{k+1} = \frac{1}{2}x_k + \nu_k$, show that the variance of x_k is $\frac{4}{3}\nu^2$. Compare to the numerically estimated variance for the full nonlinear equation.

8.12 Conditioning matters! For the example in Figure 8.6, discuss and contrast

$$p(x_k), \quad p(x_k|y_k), \quad p(x_k|y^k), \quad p(x_k|y^{k-1}).$$

8.13 Extended Kalman filter (EKF). Code the EKF algorithm for one-dimensional dynamics such as Examples 8.2 and 8.3. Explore in particular the double-well potential. Show, for example, that the EKF fails for the parameters in Figure 8.4 but succeeds for smaller time steps (smaller a), even when the thermal noise ν is increased. Explain, and confirm for $a = 0.7$, $\nu = 0.2$, and $\xi = 0.5$.

8.14 Cubature Kalman filter (CKF). Code the CKF algorithm for one-dimensional dynamics such as Examples 8.2 and 8.3. Explore in particular the double-well potential.

a. Show that the $n = 1$ CKF algorithm matches the variance to $[\frac{1}{2}f''(0) + \frac{2}{3}f'(0)f'''(0)]$.
b. Plot true and estimated states for the CKF (see text and right).
c. Show that pushing out the two cubature points by a factor a can improve the CKF. In particular, find a value for a that works as well as the plot at right. The true state is shown in gray and the CKF estimate in black.

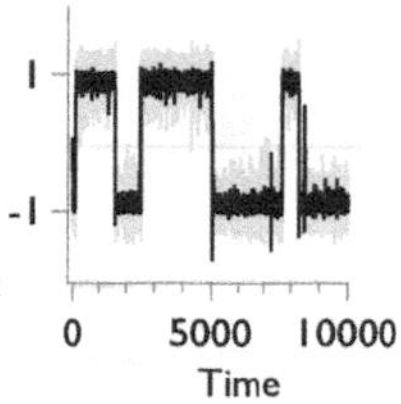

8.15 Unscented transform (UT). An alternative to the cubature Kalman filter (CKF) is the unscented Kalman filter (UKF), which is based on the "unscented" transform. It resembles the CKF but with different weights and an extra element. In particular, for a smooth function $y = f(x)$, with $x \sim \mathcal{N}(\mu, \sigma^2)$, there are three ensemble elements (*sigma points*), $x_0 = \mu$ and $x_{\pm 1} = \mu \pm a\sigma$ and weights w_0 and $w_{\pm 1}$.

a. Show that $a = \sqrt{3}$, $w_0 = \frac{2}{3}$, and $w_{\pm 1} = \frac{1}{6}$ matches the mean to fourth order.
b. Show that the variance is matched exactly to first order and partly to second order.
c. Find the mean $\overline{y}$ and variance $\overline{P}_y$ for $y = x^2$ and for $y = x^4$, assuming $x \sim \mathcal{N}(0, 1)$.

8.16 Ensemble Kalman filter (EnKF). Show that the EnKF can track motion in the double-well potential from Example 8.3. What is the effect of varying the number of elements n_{E} in the ensemble? (Remember that you choose n_{E} states and an equal number of noise elements ν and ξ.)

8.17 Grid method. We explore numerically the full Bayesian filtering solution for the double-well potential for various noise strengths ν and ξ.

a. Code the grid method and reproduce the equivalent of Figure 8.6.
b. Fix the input noise at $\nu = 0.15$ and study the state estimation problem as the measurement noise ξ goes from 0 (no noise) to large values. Discuss qualitatively.

c. Now fix the measurement noise $\xi = 1$ and vary the input noise from zero to large values. Again, describe qualitatively the different regimes.
d. In our problem of free diffusion, we found that the behavior of the Kalman filter depended only on the ratio $\alpha = \nu^2/\xi^2$ and not on the absolute values of the two noise strengths. Why does that conclusion not hold in this problem?

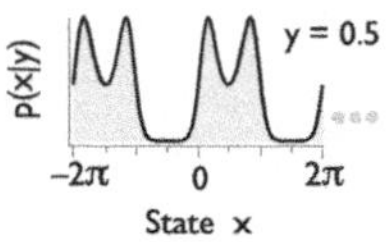

8.18 Sinusoidal nonlinearity in measurement function. The nonlinear measurement relation $y = \sin x + \xi$ can occur in interferometry experiments (Section 3.2.1). Note that a given y corresponds to an infinite number of possible x states. The graph at left for $P(x|y)$ was generated using $\xi^2 = 0.4^2$ and $y = 0.5$. Why the funny double bump? Explore the consequences of different noise strengths and observations, and explain what is going on in the different cases. Explain, in words, a strategy for dealing with the infinite number of possibilities. As usual, $\xi \sim \mathcal{N}(0, \xi^2)$.

8.19 Fat tails. To understand how non-Gaussian noise can affect state estimation, consider a somewhat artificial example where both system and observation noise are drawn from Lorentz distributions whose "fat" tails ($\sim 1/x^2$ for $|x| \to \infty$) imply that large fluctuations are vastly more probable than with a normal distribution.

a. Show that $x \sim \mathrm{Lor}(x_0, \nu) = \frac{1}{\pi}\frac{\nu}{(x-x_0)^2+\nu^2}$ is normalized, but the mean and variance diverge. Show that the median equals x_0, $\mathrm{Prob}(x_0 - \nu, x_0 + \nu) = \frac{1}{2}$ (thus connecting ν with a notion of width), and the characteristic function is $\varphi_x(k) = \mathrm{e}^{ikx_0 - \nu|k|}$.
b. Consider a toy state-estimation problem where we have a prediction x and an observation y that are both unbiased estimates of the true state of the system. As with the Kalman filter, we seek the best linear combination $\hat{x}$ of the two. Here, both prediction and observation are "Lévy-flights" that obey $x \sim \mathrm{Lor}(x_t, \nu)$ and $y \sim \mathrm{Lor}(x_t, \xi)$, with x_t the true state value. Let $\hat{x} = (1 - K)x + Ky$ and match characteristic functions to show that $\hat{x} \sim \mathrm{Lor}(x_t, \gamma)$,with width $\gamma = |1 - K|\nu + |K|\xi$.
c. Conclude that the "optimal" choice of Kalman gain K that minimizes γ is 0 if $\nu < \xi$ and 1 if $\xi < \nu$. That is, unlike the ordinary Kalman filter, "blending" the prediction x with the observation y does not improve the accuracy of estimation. Because the fluctuations in x and y are so "wild," the best action is to select at each time step whichever variable has the smaller distribution width. For fixed ν and ξ, this means ignoring all observations if $\nu < \xi$. For $\xi < \nu$, we would use only the naive observation. In other words, the Kalman reduces to a trivial course of action. Fat tails that go as $|x|^{-(1+\mu)}$, with $1 < \mu < 2$ (so that the mean but not the variance is defined), lead to a nontrivial gain K that minimizes the width γ. Sornette and Ide (2001) dub the result the *Kalman-Lévy* filter.

Unfortunately, the "stable" property of the Lorentz distribution (sum of two Lorentzians is also Lorentzian) holds for only a few distributions (extensively studied by Paul Lévy). The more realistic case where the system dynamics has a fat tail but observations are Gaussian does not lead to simple analytic results.

8.20 Bayesian RTS smoothing. In the text, we give a naive algorithm for smoothing that combines information from the forward and backwards dynamics to improve the estimate of a current state, $\boldsymbol{x}_k$. Here, we explore a more efficient, better-behaved smoother algorithm that does not need an explicit backwards dynamics.

a. By introducing the state $\boldsymbol{x}_{k+1}$ and applying causality, show that

$$p(\boldsymbol{x}_k|\boldsymbol{y}^N) = p(\boldsymbol{x}_k|\boldsymbol{y}^k) \int \mathrm{d}\boldsymbol{x}_{k+1} \, \frac{p(\boldsymbol{x}_{k+1}|\boldsymbol{x}_k)\, p(\boldsymbol{x}_{k+1}|\boldsymbol{y}^N)}{p(\boldsymbol{x}_{k+1}|\boldsymbol{y}^k)} \,,$$

which goes from $p(\boldsymbol{x}_{k+1}|\boldsymbol{y}^N)$ to $p(\boldsymbol{x}_k|\boldsymbol{y}^N)$. To apply it, first use forward Bayesian filtering, Eqs. (8.76), to find the $p(\boldsymbol{x}_k|\boldsymbol{y}^k)$ and their associated predictions $p(\boldsymbol{x}_{k+1}|\boldsymbol{y}^k)$.

b. Assuming linear dynamics and Gaussian probability distributions leads to the *Reich–Tung–Striebel* (RTS) smoother equations. Let $P(\boldsymbol{x}_k|\boldsymbol{y}^N) = \mathcal{N}\left[(\boldsymbol{x}^{\mathrm{s}})_k \,, (\boldsymbol{P}^{\mathrm{s}})_k\right]$ define the smoother state estimate $\boldsymbol{x}^{\mathrm{s}}$ and covariance matrix $\boldsymbol{P}^{\mathrm{s}}$. Show that these quantities may be found via the backwards recurrence relation

$$\begin{aligned}
(\boldsymbol{G}^{\mathrm{s}})_k &= \boldsymbol{P}_k \boldsymbol{A}^{\mathsf{T}} \, (\boldsymbol{P}^-)^{-1}_{k+1} \\
(\boldsymbol{P}^{\mathrm{s}})_k &= \boldsymbol{P}_k + (\boldsymbol{G}^{\mathrm{s}})_k \left[(\boldsymbol{P}^{\mathrm{s}})_{k+1} - \boldsymbol{P}^-_{k+1} \right] (\boldsymbol{G}^{\mathrm{s}})_k^{\mathsf{T}} \\
(\boldsymbol{x}^{\mathrm{s}})_k &= \boldsymbol{x}_k + (\boldsymbol{G}^{\mathrm{s}})_k \left[(\boldsymbol{x}^{\mathrm{s}})_{k+1} - \boldsymbol{x}^-_{k+1} \right] ,
\end{aligned}$$

where $\boldsymbol{x}$, $\boldsymbol{P}$, and $\boldsymbol{P}^-$ are first calculated using the forward Kalman filter. The recurrence relation starts at $k = N$, where $\boldsymbol{x}^{\mathrm{s}} = \boldsymbol{x}$ and $\boldsymbol{P}^{\mathrm{s}} = \boldsymbol{P}$ (Särkkä, 2013). Hint: Use Eq. (A.197 online) for conditional Gaussian distributions and computer algebra.

c. Using the RTS smoother equations, verify that the steady-state smoother variance of the one-dimensional diffusing particle is given by $P^{\mathrm{s}} = \xi^2\alpha / \sqrt{\alpha^2 + 4\alpha}$, in agreement with our previous result, Eq. (8.131).

8.21 The random rocket. In one dimension, a rocket with random forcing obeys $\ddot{x} = \nu(t)$. Assume that the measured positions are noisy: $y(t) = x(t) + \xi(t)$.

a. Discretize the system exactly for a time step that is scaled to $T_{\mathrm{s}} = 1$.

b. Write a numerical code for a Kalman filter and then a Kalman smoother. Explore the case where the discrete-noise variances are $\nu^2 = \xi^2 = 1$.

c. Scale all variances by ξ^2 and define $\alpha = \nu^2/\xi^2$. Confirm that the ratio of filter to smoother variances is given by the plot at right.

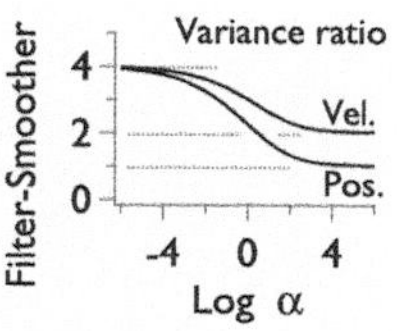

d. Plot time series of position, measurements, and both filter and smoother estimates.

Hints: $\boldsymbol{A}_{\mathrm{d}} = \begin{pmatrix} 1 & 1 \\ 0 & 1 \end{pmatrix}$. For $\alpha = 1$ and steady state, $\boldsymbol{P} = \begin{pmatrix} 3/4 & 1/2 \\ 1/2 & 1 \end{pmatrix}$ and $\boldsymbol{P}^{\mathrm{s}} = \begin{pmatrix} 1/3 & 0 \\ 0 & 1/3 \end{pmatrix}$.

8.22 Estimating a frequency. Consider the problem of estimating the frequency ω of a complex signal $y_k = \mathrm{e}^{\mathrm{i}\omega k} + \xi_k$, with $\xi_k \sim C\mathcal{N}(0, \sigma^2)$, for N data points.

a. Show that $\hat{\omega} = \text{argmax}_\omega\, F(\omega) \equiv \text{Re}\,(\frac{1}{N}\sum_{k=0}^{N-1} y_k\, e^{-i\omega k})$ is the maximum-likelihood estimator of ω. If you are not familiar with complex Gaussian distributions, start by considering the real and imaginary parts of the measurement equation, assuming that Re ξ_k and Im ξ_k are drawn from i.i.d. Gaussian distributions, $\mathcal{N}(0, \frac{1}{2}\sigma^2)$.

b. Use an FFT algorithm to find $\hat{\omega}$. Refine the estimate via a local-optimization routine. Plot $F(\omega)$ for good and bad SNR, and reproduce the figures in Example 8.4.

8.23 Wiener filtering. The Wiener filter is a frequency-domain technique equivalent to (and predating) the Kalman filter for time-invariant problems. We focus on the much-simpler smoothing case, where information is available for all times ($-\infty < t < +\infty$). Consider a signal $u(t)$ measured by an instrument with dynamical response $\mathcal{L}x(t) = u(t)$, where $\mathcal{L}$ is a differential operator. Its inverse in the Fourier domain is the transfer function $G(\omega)$, with $x(\omega) = G(\omega)u(\omega)$. The measured response $y(t) = x(t) + \xi(t)$ adds white noise $\xi(t)$ with spectral density ξ^2. The goal is to find an optimal linear "filter" that minimizes the mean-square estimation error. (Notice that the goal of estimating u rather than x is slightly different from that of a Kalman filter. But given $\hat{u}$, we have $\hat{x} = G\hat{u}$.) We thus define $\hat{u} = \frac{W}{G}y$, where all quantities are functions of ω. The mean-square estimation error is $E = \int_0^\infty dt\, [\hat{u}(t) - u(t)]^2 = \int_{-\infty}^{\infty} \frac{d\omega}{2\pi} |\hat{u}(\omega) - u(\omega)|^2$, where the signal $u(t)$ is taken to be white noise with spectral density u^2.

a. Show that $\hat{u} = [G^{-1}\,(1 + \frac{1}{|G|^2 \text{SNR}^2})^{-1}]\, y$ minimizes the mean-square error, with $\text{SNR} \equiv u/\xi$. The *whitening filter* G^{-1} compensates for the instrumental response.

b. Assume that $G = \frac{1}{1+i\omega}$, a first-order, low-pass filter. For SNR $\gg 1$, show that $(1 + \frac{1}{|G|^2\text{SNR}^2})^{-1}]$ is a low-pass filter with cut-off frequency $\omega_c \approx$ SNR. Thus, the Wiener filter cuts off the naive estimator $u(\omega) = G^{-1}(\omega)\, y(\omega)$ at ω_c.

The optimal filter contains the term $|G|^2$, making it acausal. Finding a causal Wiener filter that does not depend on future values of the signal turns out to be a harder problem and is solved more easily by the Kalman filter.

9 Robust Control

The frequency-domain methods from Chapter 3 show that high loop gains lead to *robustness*: reasonable closed-loop performance for a large class of systems. When control is robust, the system need not be known precisely, as the same controller will work well for systems with similar dynamics. The "generic" PID controller of Section 3.3 is the most prominent example. Robustness is important for industrial applications, where one controller needs to work for a batch of slightly different products. But generic controllers, even with tunable parameters, sometimes produce mediocre results, motivating control-design techniques that take into account a system's dynamics. For frequency-domain methods, heuristic tools such as sensitivity functions (Section 3.1.3) and loop shaping (Section 3.7.3) can improve control *performance*, as well as robustness. But when systems become complicated (high-dimensional, MIMO, etc.), heuristic approaches can be hard to implement.

By contrast, the time-domain, state-space methods from Chapter 4 incorporate MIMO structure from the outset and lead to controllers designed around specific properties of a system that can perform much better than generic controllers. In Chapters 7 and 8, we used optimal control to push this idea further: given the controller dynamics and a well-defined objective, there is a systematic way to derive the best-possible controller, even in the face of noisy observations and disturbances. Along the way (Chapter 6), we discussed how to measure dynamics and transfer functions experimentally. Combined, these ideas lead to a powerful control strategy: measure the dynamics and then use optimal control and optimal observers to *synthesize* the controller.

Optimal control and its stochastic extensions were developed in the 1960s and worked well in applications such as space flight where good models were available. It failed in "messy" industrial settings such as chemical plants, which are hard to model accurately. The many modeling issues – parameters differ from setup to setup, fast modes may be neglected, components age, and so on – are compounded when systems are as large and complex as modern industrial plants. In such situations, controllers optimized for a nominal system may perform poorly in practice.

Figure 9.1 summarizes the situation: Lacking any explicit model of the system to be controlled, we use techniques such as PID control. At the other extreme, optimal control based on a perfect model gives the best possible performance. In between, we know a little about the system dynamics and seek performance that is in between the modest levels of a PID controller and the unachievable performance of the optimal controller.

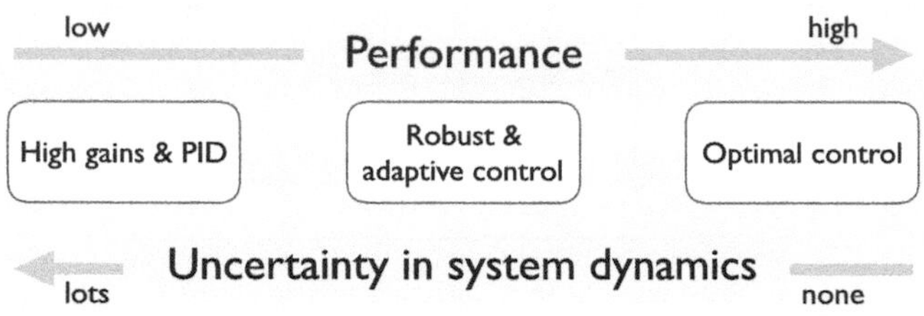

Fig. 9.1 Schematic of the trade-off between modeling uncertainty and control performance. At left: a PID controller gives modest performance but can handle considerable system uncertainty. At right: with perfect knowledge of system dynamics, optimal controllers give the best possible performance. In between, robust methods seek better performance in the face of moderate system uncertainty.

As Figure 9.1 suggests, there are two basic approaches to finding effective control laws in the face of such uncertainties. In Chapter 10, we discuss *adaptive control*, where we reduce uncertainty progressively by learning more and more about the system while it is under control. Here, we discuss *robust control*, whose goal is find controllers that explicitly take into account the partially known dynamics. In our discussion of optimal control, we focused on one approach, nonlinear feedforward coupled with linear feedback, that already incorporates some robustness (Section 7.6). But in that section, the emphasis was on correcting external disturbances. Here, we account more explicitly for model uncertainty.

Robust control might seem merely the poor cousin to adaptive control. Why design a controller for a range of systems but ignore what you find out about the actual system you are controlling? Adaptive control, which learns as it goes, would appear to be the better approach. But a robust design can be simpler and safer, as adaptive controllers are susceptible to instabilities (Chapter 10). There is always a tension between learning and doing, and a stable adaptive controller must learn relatively slowly, in order to distinguish between errors that result from disturbances (uncertain inputs) and previous controls (uncertain dynamics): If a system behaves unexpectedly, was there an external disturbance or was the model wrong? To know requires information from repeated attempts to control a system. As a result, at any given time, no matter how good the adaptive routine, the transfer function (or nonlinear dynamics) of a system will be known only imperfectly, and the controller, which always acts NOW, must still choose the best action given imperfectly known dynamics. This need to decide, at a given time with a given level of uncertainty, motivates the search for robust-control algorithms, perhaps combined with adaptive algorithms.

We begin, in Section 9.1, with the task of making feedforward commands more robust. We focus on *parametric uncertainty*, where a parameter of the model is poorly known. We start with a heuristic technique known as *input shaping* that is commonly used to avoid exciting unwanted mechanical vibrations. We then generalize the idea to nonlinear systems.

In Section 9.2, we introduce a probabilistic approach that modifies the results of optimal control to make feedback control of disturbances more robust. The strategy is to minimize the expected value of a cost function, $\langle J \rangle$, an approach already introduced in Sections 8.2 and 8.6. There, the focus was on uncertainty due to external disturbances and measurement errors. Here, we broaden our view to include uncertainty in the dynamical model itself, via a "low-noise" perturbative approach.

Probabilistic approaches to control focus on minimizing costs that are averaged over realizations of a control protocol; however, if the cost of an outcome is too high or its probability to occur too low, then calculating an average over realizations may be difficult. It then makes sense to consider more broadly the risks of control. One approach is to consider the entire probability distribution of costs, $p(J)$, especially its tails (Section 9.3).

The statistical description of risk leads to a surprising conclusion: in many cases, we can replace a probabilistic description with an equivalent deterministic one based on a *worst-case* scenario: We seek the best controller assuming that Nature selects the worst-possible system from a specified set of possible systems. In effect, we play a *game* with Nature. In Section 9.4, we apply this approach to systems where the dynamical uncertainties are *unstructured*: the form of the equations is uncertain, not just the parameters.

9.1 Robust Feedforward

In Chapter 7, we advocated controlling nonlinear systems by using nonlinear feedforward supplemented by local linear feedback, which adds robustness against disturbances and modeling errors. Here, we show that the feedforward commands themselves can be made more robust against parametric uncertainty. The main idea is to choose a control to make the first derivative of a performance index (or cost function) vanish with respect to the uncertain parameter. In Section 9.1.1, we present a heuristic version of this idea, which was developed as a way to change the state of weakly damped mechanical systems without exciting resonances. Then, in Section 9.1.2, we generalize to nonlinear systems.

9.1.1 Input Shaping

Consider a single control $u(t)$ and a system modeled by equations that depend on a single uncertain parameter, θ, which may differ by an amount $\delta\theta$ from the "true" value θ_0 that best describes the system. The feedforward optimal-control problem seeks to find the control path $[u]_0^\tau$ that minimizes a cost functional $J([u]_0^\tau, \boldsymbol{x}_0, \theta)$ for some predefined task over the time interval $[0, \tau]$, given a starting state $\boldsymbol{x}(0) = \boldsymbol{x}_0$. If the model is "perfect," then $\theta = \theta_0$ and $J(u_0, \theta_0)$ is the lowest possible cost of control, where $u_0 \equiv [u]_0^\tau$ denotes the optimal control path $u(t)$ for this "base" problem. (We omit the dependence of J on $\boldsymbol{x}_0$ for brevity.)

If the model is not perfect, then

$$J(u_0, \theta) = J(u_0, \theta_0) + \partial_\theta J|_{u_0,\theta_0}(\delta\theta) + \tfrac{1}{2}\partial_{\theta\theta} J|_{u_0,\theta_0}(\delta\theta)^2 + \cdots. \tag{9.1}$$

The basic idea of robust feedforward control is to take advantage of the infinite number of degrees of freedom of the control $u(t)$ to choose a new control $u^*(t)$ that makes $\partial_\theta J = 0$. If that is not good enough, then you find a $u^{**}(t)$ that makes $\partial_\theta J = \partial_{\theta\theta} J = 0$, and so on. For a linear system, we will find that we can design a prefilter block F that, when convoluted against the "naive" reference command, gives the desired "shaped input" that is fed into the system. The system then responds with a closer approximation to the desired command. Because of the system's linearity, the same prefilter works for all initial states $\boldsymbol{x}_0$ and all commands $r(t)$. This prefilter structure is the origin of the name "input shaping."

Feedforward Control of a Construction Crane

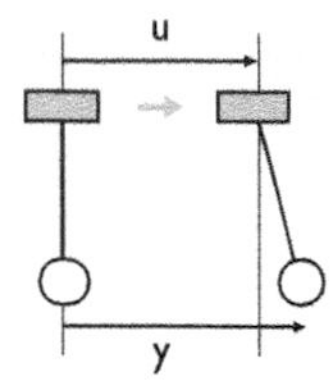

To illustrate these ideas in a concrete setting, we reprise the construction-crane problem of Section 5.4.2, illustrated at left. Recall that the goal is to use a pendulum-like crane to move a load from one location to another.[1] For simplicity, we ignore the pendulum's damping and nonlinearity. (We consider the nonlinear case below.) The input $u(t)$ will correspond to the position of the pendulum pivot, so that its acceleration is $\omega^2 u(t)$. The output $y(t)$ corresponds to the position of the pendulum bob relative to its starting point (see left). We also assume the desired motion is along a one-dimensional straight path (i.e., no complications due to swinging in a circle). With initial conditions corresponding to a stationary crane, the dynamics are $\ddot{y} + \omega^2 y = \omega^2 u(t)$, with $y(0) = \dot{y}(0) = 0$.

We wish to find a control $u(t)$ that makes $y = 1$ and $\dot{y} = 0$ for $t > \tau$. The state $(y, \dot{y})^T$ can vary arbitrarily for $0 < t < \tau$. The frequency ω has a nominal value ω_0 but is uncertain in the sense that a given system frequency will be fixed and near ω_0.

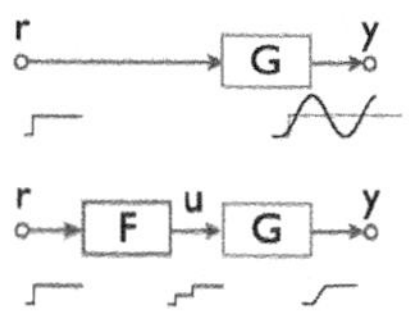

For a known frequency ω, we solved this problem in Section 5.4.2 (see left). Here, the oscillator frequency ω is unknown. If the dynamics are uncertain, then no solution will exactly satisfy our goals. But we can ask, instead, that $y(t) \approx 1$ and $\dot{y}(t) \approx 0$ for $t > \tau$.

A further simplification is that our goal is *point-to-point control*, from one point in state space to another, without caring about the intermediate states along the way. Since the end state is specified by a finite number of conditions (two numbers for the state, two more to make the first derivative of the state vanish, etc.), the control signal can depend on only a finite number of parameters. Then we can replace the infinite-dimensional function $u(t)$ with a set of steps $\theta(t - t_i)$. (Here θ represents a step function and not the unknown parameter ω.) We will also define deviations as relative variations, $\varepsilon \equiv (\frac{\omega-\omega_0}{\omega_0})$. The *input shaping* form for the control signal is then defined as

[1] Or a waiter might want to deliver a bowl of soup to a table without sloshing.

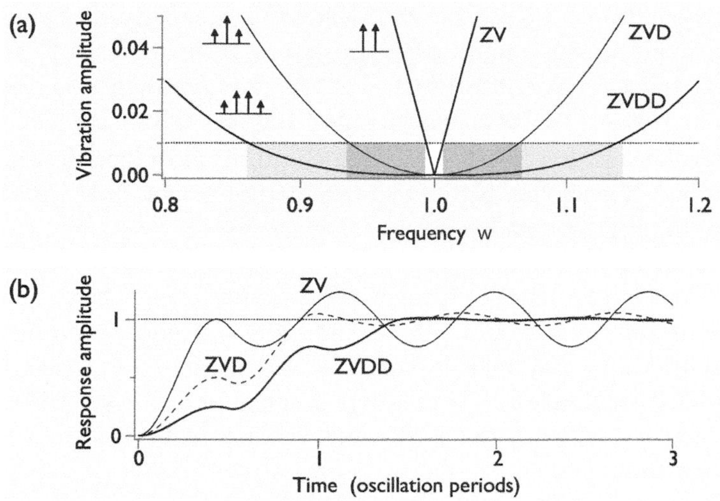

Fig. 9.2

Input shaping for step response of an undamped oscillator. (a) Residual relative vibration amplitudes for ZV, ZVD, and ZVDD protocols. Dotted horizontal line shows 1% residual amplitude. Pulses A_i are shown graphically for each protocol. (b) Step responses for $\omega = 1.15$ when designed for $\omega_0 = 1$.

$$u(t) = \sum_{i=0}^{n} A_i\,\theta(t - t_i)\,, \qquad \sum_{i=0}^{n} A_i = 1\,, \tag{9.2}$$

which is a sequence of n intervals (and $n+1$ jumps) of amplitude A_i applied at time t_i. We can view Eq. (9.2) as representing the convolution of a prefilter $F(t)$ with an ideal command $r(t) = \theta(t)$. The prefilter $F(t) \equiv \sum_i A_i\,\delta(t - t_i)$ is effectively the impulse-response function of the command signal, and $u = F * r$ is the shaped input sent to the system G.

As cost function, let J_n be the amplitude of residual oscillations for $t > t_n$,[2]

$$J_n \equiv J(\{A_i\},\{t_i\}) = \sqrt{\left(\sum_{i=0}^{n} A_i \cos\omega t_i\right)^2 + \left(\sum_{i=0}^{n} A_i \sin\omega t_i\right)^2}\,, \tag{9.3}$$

where $t_0 \equiv 0$ (Problem 9.1a). For $n = 0$ (single step function), $J_0 = A_0 = 1$. That is, a unit step naively applied will leave the system swinging with amplitude $J = 1$ (Figure 5.12a).

We start with the two-step protocol of Figure 5.12b. This *zero vibration* (ZV) protocol is illustrated in Figure 9.2. Choosing $A_0 = A_1 = \frac{1}{2}$ and $t_1 = \pi$ will make $J_1 = 0$ for a nominal frequency $\omega_0 = 1$ (Problem 9.1b). For small frequency deviations, $J_1 \sim \frac{\pi}{2}|\varepsilon|$. Although $J_1(0) = 0$, the linear increase in J with deviations of the actual frequency from the nominal one used to design the feedforward control implies the design is not very robust.

[2] Alternatively, J^2 is the energy (minus a static contribution due to the offset produced by the finite acceleration of the base). But engineers prefer to work with residual vibration amplitudes rather than energies.

To do better, let $J(0) = J'(0) = 0$, the ZVD (zero-vibration-derivative) protocol. Satisfying the extra constraint requires adding a step ($n = 3$) and leads to amplitudes $\boldsymbol{A} = \frac{1}{4}\{1, 2, 1\}$ and $\boldsymbol{t} = \{0, \pi, 2\pi\}$ (Problem 9.1c). The resulting $J_2 \sim (\frac{\pi}{2}|\varepsilon|)^2$ and is notably more robust. We can similarly cancel further derivatives. The next level (ZVDD protocol) leads to $\boldsymbol{A} = \frac{1}{8}\{1, 3, 3, 1\}$ and $\boldsymbol{t} = \{0, \pi, 2\pi, 3\pi\}$ and $J_3 \sim \left(\frac{\pi}{2}|\varepsilon|\right)^3$ and is still more robust (Figure 9.2). Notice that each derivative canceled "costs" an extra half period in the protocol duration.

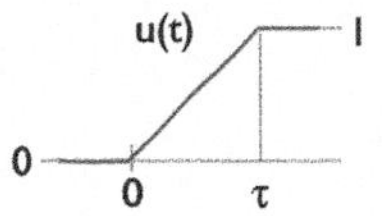

It is easy to see that n iterations leads to $J_n \sim \left(\frac{\pi}{2}|\varepsilon|\right)^n$, for a protocol that lasts a time $\tau = n\pi$. This large-n limit leads to an *adiabatic transition*, where *any* input $u(t)$ satisfying the boundary conditions $u(0) = 0$ and $u(\tau) = 1$ gives a transition that is arbitrarily robust. As an easy analytical example, a slow ramp over a time τ (see left) leads to a residual amplitude $\approx 1/(\omega\tau)$, which is small for any ω provided the ramp time is long enough (Problem 9.1g). The general lesson is that slowing a protocol can increase its robustness.

9.1.2 Robust Feedforward for Nonlinear Systems

We can generalize the perturbative input-shaping procedure described in Section 9.1.1 in two ways: First, we choose controls $\boldsymbol{u}(t)$, converting a finite number of control parameters to a continuous infinity. Second, we consider nonlinear dynamics.

Let us briefly recall the optimal-control scenario from Chapter 7. The cost functional $J = \varphi[\boldsymbol{x}(\tau)] + \int_0^\tau \mathrm{d}t\, L(\boldsymbol{x}, \boldsymbol{u})$, with L the running cost and φ the terminal cost. The goal is to choose the control path $[\boldsymbol{u}]_0^\tau$ that minimizes J, subject to the constraint that, at each time t during the protocol, the state $\boldsymbol{x}$ obeys the dynamics given by $\dot{\boldsymbol{x}} = \boldsymbol{f}(\boldsymbol{x}, \boldsymbol{u})$, with initial state $\boldsymbol{x}(0) = \boldsymbol{x}_0$. The constraint is enforced by adding a Lagrange multiplier, the adjoint $\boldsymbol{\lambda}(t)$, whose dynamics run backwards in time. The Euler–Lagrange equations for $(\boldsymbol{x}, \boldsymbol{\lambda}, \boldsymbol{u})$ are

$$\dot{\boldsymbol{x}} = \boldsymbol{f}(\boldsymbol{x}, \boldsymbol{u}), \qquad \dot{\boldsymbol{\lambda}} = -(\partial_{\boldsymbol{x}}\boldsymbol{f})^{\mathrm{T}}\boldsymbol{\lambda} - (\partial_{\boldsymbol{x}}L)^{\mathrm{T}}, \qquad \boldsymbol{0} = (\partial_{\boldsymbol{u}}\boldsymbol{f})^{\mathrm{T}}\boldsymbol{\lambda} + (\partial_{\boldsymbol{u}}L)^{\mathrm{T}}, \tag{9.4}$$

with final conditions for the adjoint, $\boldsymbol{\lambda}(\tau) = \left(\partial_{\boldsymbol{x}}\varphi|_\tau\right)^{\mathrm{T}}$. Solving equations (9.4) as a boundary value problem over the interval $[0, \tau]$ leads to a feedforward command $\boldsymbol{u}(t)$ that may be applied to the system. Here, we replace the terminal cost φ by final conditions $\boldsymbol{x}(\tau) = \boldsymbol{x}_\tau$ placed on the state. The goal is then to "move" the state from $\boldsymbol{x}_0$ to $\boldsymbol{x}_\tau$ by the end of the protocol, using the least "control effort" $\int \mathrm{d}t\, L(\boldsymbol{u})$.

To add robustness to the feedforward solution, we advocated adding linear feedback in Chapter 7. Here, we try to make the feedforward solution itself more robust, using a technique known as *forward sensitivity analysis*. The basic principle is to define new, additional state variables that represent those quantities that should be minimized and then to include them in the optimization. Specifically, let the dynamics include unknown parameters $\boldsymbol{\theta}$. Then all quantities in Eq. (9.4) carry a $\boldsymbol{\theta}$ dependence, as well. Next, we define *sensitivity states* $\boldsymbol{x}_\theta(t) \equiv \frac{\mathrm{d}\boldsymbol{x}}{\mathrm{d}\theta}$, which obey dynamics

$$\dot{\boldsymbol{x}}_\theta(t) = \mathrm{d}_\theta\left[\boldsymbol{f}(\boldsymbol{x}, \boldsymbol{u}, \boldsymbol{\theta})\right] = \partial_\theta \boldsymbol{f} + (\partial_{\boldsymbol{x}}\boldsymbol{f})\, \boldsymbol{x}_\theta\,. \tag{9.5}$$

If $\boldsymbol{x}_\theta(t)$ is small, then the cost of control has reduced sensitivity to parameter variations. Since we focus on the end state, we impose $\boldsymbol{x}_\theta(\tau) = \boldsymbol{0}$, which ensures that the terminal cost is independent of parameter variations, to first order.

If we start with n states in $\boldsymbol{x}$ and if there are n_θ parameters $\boldsymbol{\theta}$, then the total number of sensitivity states will be $n\,n_\theta$. We can then group the original states and the new sensitivity states into a giant state vector $\boldsymbol{X}$ that has $n_X \equiv n(n_\theta + 1)$ elements and obeys a dynamical equation $\dot{\boldsymbol{X}} = \boldsymbol{F}(\boldsymbol{X}, \boldsymbol{u})$, where $\boldsymbol{F}$ combines both $\boldsymbol{f}$ and the right-hand side of Eq. (9.5). We then treat this expanded system as a new optimal-control problem, with n_X Lagrange multipliers $\boldsymbol{\Lambda}$, which obey Eqs. (9.4). The boundary-value problem then involves solving $2n_X = 2n(n_\theta+1)$ nonlinear ordinary differential equations, subject to $2n_X$ boundary conditions that are distributed at $t = 0$ and τ, as appropriate. The new sensitivity states should vanish at both initial and final times. The final-time condition is to make states at $t = \tau$ insensitive to parameter variations. The initial-time condition means that we know the initial states, even if we do not know the parameters. These $2n_X$ equations are supplemented by m algebraic equations for the m control functions grouped into $\boldsymbol{u}(t)$.

Swing up a Pendulum Robustly

Let us apply these ideas to the problem of swinging up a pendulum, which we treated extensively in Chapter 7 and which is the nonlinear version of the input-shaping example of Section 9.1.1. To recap, the problem is to find a torque command with minimal effort that swings up a pendulum in time τ.[3] More precisely, the cost $J = \frac{1}{2}\int_0^\tau \mathrm{d}t\, u^2(t)$, and

$$\ddot{\theta}(t) + \omega^2 \sin\theta(t) = u(t)\,, \qquad \underbrace{\theta(0) = \dot{\theta}(0) = 0}_{\text{down}}\,, \qquad \underbrace{\theta(\tau) = \pi\,, \dot{\theta}(\tau) = 0}_{\text{up}}\,. \tag{9.6}$$

We define a two-dimensional state vector $\boldsymbol{x} = (\theta\ \dot{\theta})^{\mathsf{T}}$. The frequency ω is uncertain, with nominal value $\omega_0 = 1$. We define two sensitivity states $\boldsymbol{x}_\omega(t) = \mathrm{d}_\omega \boldsymbol{x}(t)$, which describe how the state elements vary with the frequency ω. The augmented system $\boldsymbol{X}$ has a four-dimensional state, and the boundary condition problem for $(\boldsymbol{X}\ \boldsymbol{\Lambda})^{\mathsf{T}}$ is eight-dimensional.

Problem 9.2 will take you through the details, and Figure 9.3 shows the results. We calculate two feedforward protocols, $u_0(t)$ and $u_{\text{rob}}(t)$. The first, $u_0(t)$ was calculated using standard optimal-control theory, while "rob" means that it was calculated using sensitivity states for robustness. The two controls were then applied to three different systems, with frequencies, $\omega/\omega_0 = 1$ and $\omega/\omega_0 \pm \varepsilon$, with ε corresponding to a 10% frequency variation. In (a), we see that nominal system arrives at the desired up state and can balance for at least 3 time units beyond the protocol. The two other curves for $\omega/\omega_0 = 1 \pm \varepsilon$ have significantly diverged, one with too much energy and the other with two little. In (b), the corresponding curves are much closer to the nominal response curve, reflecting the added robustness. For $t \geq \tau$, the higher and lower frequencies

[3] As before, the goal is a point-to-point trajectory. By adding a term $\propto \int_0^\tau \mathrm{d}t\, \frac{1}{2}(\boldsymbol{x} - \boldsymbol{x}_{\mathrm{d}})^2(t)$ to the cost function, we could also force the trajectory to lie near a proscribed path, $\boldsymbol{x}_{\mathrm{d}}(t)$.

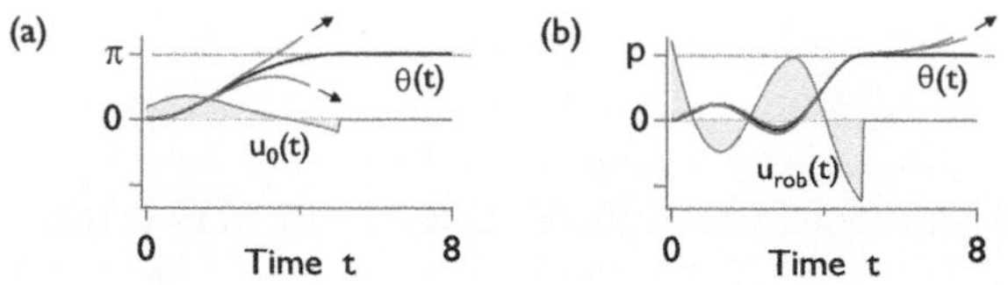

Fig. 9.3 Robust swing-up of a pendulum. (a) Feedforward protocol, with $u_0(t)$ (shaded fill) designed by ordinary optimal control. Three responses are shown: heavy line shows u applied to system with $\omega = \omega_0 = 1$. The lighter curves with arrows show the response of the same $u_{\text{nom}}(t)$ applied to systems with $\omega/\omega_0 = 1 \pm 0.1$ (10% variation). (b) Corresponding curves for a robust feedforward control $u_{\text{rob}}(t)$.

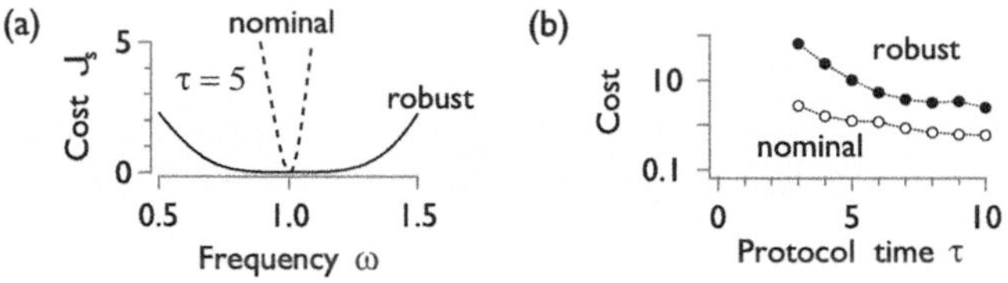

Fig. 9.4 Benefits and costs of robust control. (a) Smaller deviations of final state J_s for $\tau = 5$ for u_{rob}. (b) Greater integrated control costs $\frac{1}{2}\int_0^\tau \mathrm{d}t\, u(t)^2$ for u_{rob}.

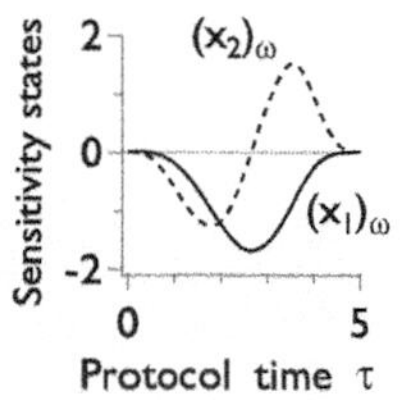

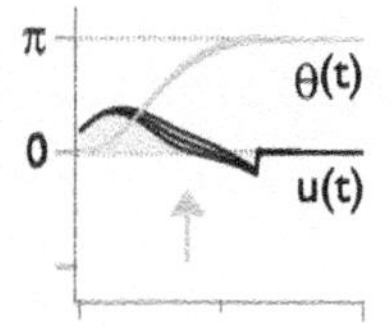

nominal

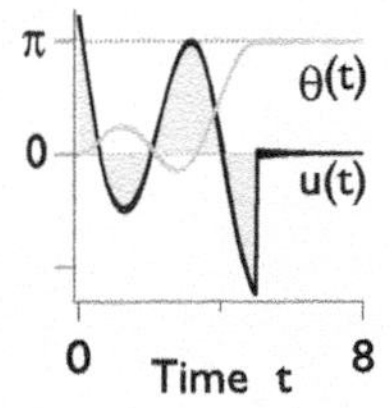

robust

both overshoot equilibrium point.[4] This behavior reflects the higher-order correction in the state variation at time τ, which is now $\sim \varepsilon^2$ and not ε. At left, we illustrate the sensitivity-state dynamics. Because they are zero at $t = \tau$, the states are insensitive to ω variations at the end of the protocol.

Figure 9.4 illustrates the benefits and costs of robust feedforward. Part (a) shows the deviations from the desired final state, $J_s \equiv \frac{1}{2}[(\theta-\pi)^2 + \dot{\theta}^2]$. The robust feedforward command, $u_{\text{rob}}(t)$, leads to deviations that are higher order in ω-deviations than the nominal feedforward command, $u_0(t)$. By contrast, (b) shows that the energy costs are higher for u_{rob}, which has an amplitude range four times greater than u_0. (Note the logarithmic scale for u.) Alternatively, we can keep the control effort constant but lengthen the protocol time τ, as in the input-shaping example. Either way, robustness has costs.

To conclude, adding robustness to feedforward is straightforward, even for nonlinear systems. But how does this method compare to adding local linear feedback? In fact, poorly: at left, we see that adding feedback to the nominal and robust feedforward commands stabilizes both, with minor additions to the command signals. Notice the spread in $u(t)$ commands (arising from the PD feedback $K_p = K_d = 1 + \sqrt{2}$, applied to systems that have 10% differences in frequency). This spread, in response to errors arising from the nominal control, is significantly larger than what arises when the robust command is used as a reference. See the arrow, which indicates the large

[4] Looking carefully at Figure 9.3b shows that $\delta x \sim \varepsilon^2$ only for $t \geq \tau$. At shorter times, $\delta x \sim \varepsilon$, and the trajectories are arrayed above and below the nominal response curve. This observation makes sense, as we impose the condition at time τ (and 0). See also the sensitivity states, graphed above at left.

spread produced relative to the spread in $u(t)$ for the robust protocols. That is, using robust feedforward command has lowered the feedback response; however, those gains are small relative to the extra costs of using feedforward in the first place. Estimating the cost via the control effort $J_u = \frac{1}{2}\int_0^T \mathrm{d}t\, u^2(t)$, we find that accommodating a 10% spread in frequencies by linear feedback increases J_u by $\approx$ 30% for the nominal case but only $\approx$ 10% for the robust case. But the base "cost" of the robust feedforward command is nearly ten times that of the nominal case. On this scale, the extra cost of feedback in the nominal case is trivial. In short, less control effort is required to respond to one specific disturbance with feedback than to respond adequately to an entire set of possible disturbances with feedforward.

Trade-offs

The example of swinging up a pendulum shows that there are trade-offs in strategies. More control effort can make a protocol more robust. Alternatively, we can fix the control effort but lengthen the protocol. Adding feedback improves these trade-offs but adds the cost of continuous monitoring and decision making. When applied to small-scale systems, such *energy-speed-accuracy* trade-offs lead to fundamental limits on control (Chapter 15). But it is interesting to see here that these trade-offs exist in the macroscopic world, too, where the specific strategy may be decided by the particular situation. In recent years, the costs of sensors and microcontrollers has plummeted, making feedback control a much more attractive option than it once was. The *Internet of things* with its ubiquitous feedback loops is a consequence of these developments, with everyday items from household appliances to clothing to cars all linked together in an intelligent network (Chapter 14).

9.2 Robust Feedback

A natural way to treat uncertainty is via a probabilistic approach where uncertainty is modeled as stochastic variables or processes.[5] In such a scenario, the uncertain dynamics is a *quenched* disorder: it varies from system to system but is constant, at least over the time scale of the control protocol (quasistatic), on any given system. Because it is constant or quasistatic for any given system, the cost of control, J, can be computed, conditional on a given system parameter, using a deterministic calculation. The stochasticity reduces to a transformation of variables, from the uncertain variables describing the range of possible systems to the distribution of cost outcomes, the probability density function $p(J)$. This important technical simplification allows us to treat many types of situations.

[5] Statistician Dennis V. Lindley takes an extreme stand: "The only satisfactory description of uncertainty is probability. ... [A]nything that can be done by alternative methods for handling uncertainty can be done better by probability" (Lindley, 1987). Our discussion of risk, below, offers a different view.

In this section, we develop a "low-noise," perturbative version of stochastic optimal control, where the system is assumed to be near some kind of nominal, simple model. Why low noise? As we discussed in the introduction, robust control makes sense only for a moderate amount of uncertainty. Otherwise, adaptive control is better. The perturbative approach works particularly well with parametric uncertainty and optimal control. When the uncertainty is small, we can Taylor expand the cost function in deviations of the parameters, compute the average expected correction, and alter the control accordingly.

9.2.1 Robust Feedback for Parametric Uncertainty

We now develop a method to respond to moderate parametric uncertainties in the system dynamics. Recall that parameters come in two types: *system parameters* $\boldsymbol{\theta}$, which are defined by the design of the system, and *control parameters* $\boldsymbol{K}$, which can changed at will and help define the control signals $\boldsymbol{u}(t)$. Because useful feedback rules are simple, we focus on feedback algorithms defined by a small number of control parameters.

Let us start with the case of a SISO (single input, single output) system whose dynamics contain a single uncertain system parameter θ. There may be other parameters, but their values are "certain enough" that they will play no role in our discussion. Similarly, the control law is assumed to have a single adjustable parameter K, the "feedback gain."

We thus define an optimal-control problem $J(\theta, K)$ for a scalar cost function J given a model having a known system parameter θ and single control parameter K. For a nominal value of $\theta \equiv \theta_0$ and smooth cost function, we assume that we can find an optimal gain $K = K_0$ by solving $\partial_K J(\theta_0, K) = 0$.

Now let the system parameter θ be a quenched random variable with mean $\langle\theta\rangle \equiv \theta_0$ and variance $\langle(\theta - \theta_0)^2\rangle \equiv \langle\,\delta\theta^2\rangle$. By "quenched," we mean that there is an ensemble of dynamical systems, each with a fixed value of θ, drawn from a probability density $p(\theta)$. For each of these systems, we use the same value K to execute a control protocol and measure the resulting value of J. How does the uncertainty of θ influence our choice of K?

Our general strategy will be to minimize a perturbative approximation to $\langle J\rangle$, by Taylor expanding about the nominal deterministic model. We thus expand the cost function $J(\theta, K)$ about the nominal (mean) system parameter $\theta_0 \equiv \langle\theta\rangle = \int d\theta\, p(\theta)\,\theta$:

$$J(\theta, K) = J(\theta_0, K) + \left.\frac{\partial J}{\partial \theta}\right|_{\theta_0, K} (\,\delta\theta) + \left.\frac{1}{2}\frac{\partial^2 J}{\partial \theta^2}\right|_{\theta_0, K} (\,\delta\theta)^2 + \cdots . \tag{9.7}$$

Next, we take an expectation over the probability density $p(\theta)$. Since $\langle\,\delta\theta\rangle = 0$,

$$\langle J\rangle(K) \approx J(\theta_0, K) + \left.\frac{1}{2}\frac{\partial^2 J}{\partial \theta^2}\right|_{\theta_0, K} \left\langle\,\delta\theta^2\right\rangle . \tag{9.8}$$

Notice that $\langle J \rangle$ is a function only of K, as the θ dependence is removed by taking the expectation. To find how the optimum value of K shifts K_0, we Taylor expand $\langle J \rangle$ about $K = K_0$ for both terms and define $\delta K = K - K_0$,

$$\begin{aligned}\langle J \rangle(K) \approx J(\theta_0, K_0) &+ \left.\frac{\partial J}{\partial K}\right|_{\theta_0,K_0}^{\nearrow 0} (\delta K) + \frac{1}{2}\left.\frac{\partial^2 J}{\partial K^2}\right|_{\theta_0,K_0} (\delta K)^2 \\ &+ \frac{1}{2}\left.\frac{\partial^2 J}{\partial \theta^2}\right|_{\theta_0,K_0} \left\langle \delta\theta^2 \right\rangle + \frac{1}{2}\left.\frac{\partial^3 J}{\partial K \partial \theta^2}\right|_{\theta_0,K_0} \left\langle \delta\theta^2 \right\rangle (\delta K)\,. \end{aligned} \tag{9.9}$$

We optimize $\langle J \rangle$ with respect to K by differentiating one more time:

$$\frac{\mathrm{d}\langle J \rangle}{\mathrm{d}K} = \left.\frac{\partial^2 J}{\partial K^2}\right|_{\theta_0,K_0} (\delta K) + \frac{1}{2}\left.\frac{\partial^3 J}{\partial K\, \partial \theta^2}\right|_{0} \left\langle \delta\theta^2 \right\rangle = 0\,. \tag{9.10}$$

Solving for δK and defining a dimensionless constant β gives the optimal gain shift δK^*,

$$\left(\frac{\delta K^*}{K_0}\right) = \beta \left(\frac{\left\langle \delta\theta^2 \right\rangle}{\theta_0^2}\right), \qquad \beta \equiv -\frac{1}{2}\left(\frac{\theta_0^2 \left.\frac{\partial^3 J}{\partial K\, \partial \theta^2}\right|_{\theta_0,K_0}}{K_0 \left.\frac{\partial^2 J}{\partial K^2}\right|_{\theta_0,K_0}}\right). \tag{9.11}$$

Scaling variables as $\vartheta \equiv \frac{\theta}{\theta_0}$ and $\varepsilon^2 \equiv \langle \delta\vartheta^2 \rangle = \frac{\langle \delta\theta^2 \rangle}{\theta_0^2}$ and $k \equiv \frac{K}{K_0}$, we can simplify Eq. (9.11),

$$\delta k^* = \beta \varepsilon^2\,, \qquad \beta \equiv -\frac{1}{2}\left(\frac{\partial_{k\vartheta\vartheta} J}{\partial_{kk} J}\right). \tag{9.12}$$

This is a nice result! Equation (9.12) states that you should change the relative feedback gain in proportion to the relative variance of the uncertain parameter. In many cases, the dimensionless parameter β will be of order unity. Note, however, that the sign of β depends on the sign of the derivative $\frac{\partial^3 J}{\partial k\, \partial \vartheta^2}$. (The denominator, $\partial_{kk} J$, is positive because it is evaluated at the optimal gain, where J has its minimum.) We can interpret β as a kind of sensitivity (or susceptibility) between control and uncertainty. Intuitively, we can understand the role of β by considering situations where it corresponds to a gain correction that is to be applied to an ensemble of systems with different parameters. Then $\beta > 0$ implies that it is costlier to use gains that are too low than too high, and $\beta < 0$ implies the converse. We will see some examples below and in the problems.

The theory developed here differs from that of Section 9.1 on robust feedforward control. There, we found control schemes that successively nulled derivatives of J with respect to parameter variations, at the cost of a progressively longer protocol. Here, we fix the number of control parameters but modify their values to minimize $\langle J \rangle$.

9.2.2 Priors and Probabilities

One subtle issue is the choice of probability density function for the scaled system parameter ϑ. A Gaussian, or normal distribution, $\vartheta \sim \mathcal{N}(1, \varepsilon^2)$ might seem natural, but this is often not the right choice. If changing the parameter can lead to instability and if we *know* that the uncontrolled system is stable, then the domain of possible values for ϑ is limited to a semi-infinite interval. For example, if systems are stable only for $\vartheta \geq 0$, then $\vartheta \sim \mathcal{N}(1, \varepsilon^2)$ implies a nonzero probability to have a system with negative ϑ, whatever the value of ε.

The problem is that the normal distribution conflicts with our prior knowledge that the uncontrolled system is stable. To incorporate this prior, we should choose a distribution that confines ϑ to its stable interval, $[0, \infty)$. It actually makes little difference what distribution is chosen, as long as it is single peaked and has the proper support, mean, and variance. A distribution that satisfies these criteria is the *lognormal*, where $\ln \vartheta \sim \mathcal{N}(\mu, \sigma^2)$, where μ and σ can be related to the mean and standard deviation of ϑ. The distribution is shown at left for relative uncertainties of 20% and 50%. Notice the rightward skew, which is more pronounced at higher uncertainties.

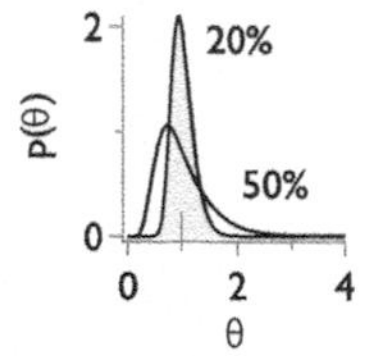

The lognormal distribution has semi-infinite support, specified mean and variance, and is easy to work with analytically. However, there are deeper reasons for choosing it: among all possible distributions with semi-infinite support and specified mean and variance for the order of magnitude (i.e., the log), it makes the fewest other assumptions about the nature of the distribution. All other distributions implicitly impose other criteria. In the jargon of Bayesian inference, it is the *maximum entropy* distribution satisfying the given constraints.

We can have different priori knowledge about a parameter. If there are no restrictions on sign, the maximum-entropy prior is a Gaussian distribution. If only the magnitude is known, then it is the exponential distribution. If the parameter is confined to an interval, it is the uniform distribution. And so on. The Bayesian viewpoint allows the incorporation of even vague prior knowledge.

Undamped, Uncertain Oscillator

To understand these ideas more concretely, consider an undamped simple harmonic oscillator with uncertain frequency ω. We can regulate the oscillator position about $x = 0$ against disturbances by using a simple derivative control. Modeling an impulse disturbance as the initial condition $x(0) = 0$, $\dot{x}(0) = 1$ gives

$$\ddot{x}(t) + \omega^2 x(t) = u(t)\,, \qquad u(t) = -K\dot{x}(t)\,, \tag{9.13}$$

which corresponds to simple derivative feedback. The physical time $\tilde{t}$ has been scaled to the dimensionless quantity $t = \omega_0 \tilde{t}$ using the nominal value of the oscillator. The frequency of the oscillator, ω, is fixed as far as the dynamics is concerned but modeled as a random variable with mean $\langle \omega \rangle = 1$ and variance $\langle\, \delta\omega^2 \rangle \equiv \varepsilon^2$ in the cost function. As discussed above, the randomness is quenched: ω is a random variable over the ensemble of systems, but any particular system is deterministic, with a fixed frequency.

Since the uncontrolled system is stable, $\omega > 0$ and we assume a lognormal distribution for $p(\omega)$.

The cost function $J(\omega, K)$, weighting state and control equally, is given by

$$J(\omega, K) = \int_0^{\infty} \mathrm{d}t \left(x^2(t) + u^2(t) \right) = \frac{1}{2} \left(\frac{1}{K\omega^2} + K \right) \tag{9.14}$$

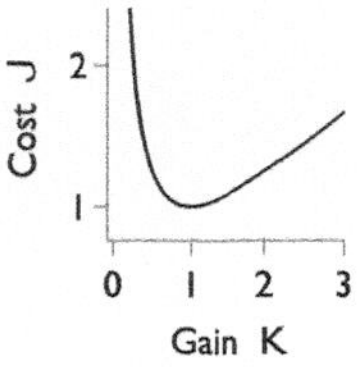

where the second identity comes from solving the equation of motion, Eq. (9.13), for fixed ω and then integrating (Problem 9.3). The result is plotted at right for $\omega = 1$. Differentiating J and setting ω to its nominal value 1, we find the optimal feedback gain is $K_0 = 1$, corresponding to a minimum cost $J_0 = 1$.

The optimal closed-loop impulse response is slightly underdamped, which is the best balance between the desire to minimize deviations (the x^2 term) and the desire to minimize the control effort (the u^2 term). A more general running cost of $x^2 + Ru^2$ leads to $K_0 = 1/\sqrt{R}$. Depending on the value of R, the "best" response can be either over- or underdamped.

We now consider ω as a lognormal stochastic variable with mean = 1 and variance ε^2. Equation (9.12) then leads to (Problem 9.3)

$$\delta k \equiv \left(\frac{\delta K}{K_0} \right) = +\tfrac{3}{2}\,\varepsilon^2 \,. \tag{9.15}$$

That is, the "susceptibility" $\beta = +\frac{3}{2}$. As a practical matter, guessing $\beta = 1$ would not change the quality of the control much, but getting the sign right is important. Here it reflects the sign of the curvature of cost with respect to parameter variations ($\partial_{\vartheta\vartheta} J$). Remarkably, the calculations in this example can be done exactly for lognormal $p(\omega)$. The result, $k^*_{\text{exact}} = (1 + \varepsilon^2)^{3/2}$, is consistent with the perturbative calculation.

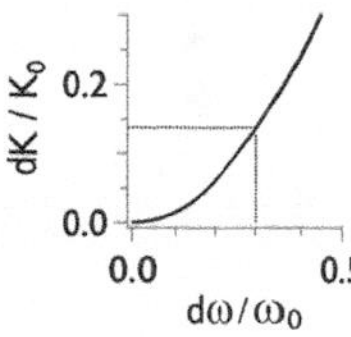

The plot at right shows that a 30% uncertainty in the relative frequency implies a shift in optimal gain of nearly 15%, a significant change. But a 10% uncertainty shifts it less than 2%, which is often not significant. This result goes a long way to explaining the success of naive optimal control and the certainty equivalence principle, which assume perfectly known parameters. If errors are relatively small, the optimum will not shift much.

The perturbative result in Eq. (9.15) makes more precise the notion of "middle ground" sketched in Figure 9.1. For a relative frequency uncertainty $\lesssim 20\%$, corrections are likely not important. For uncertainties $\gtrsim 60\%$, the perturbative approach will break down. While higher-order calculations are possible, it is typically so much easier to estimate a parameter to 20% that adaptive control is likely a better approach. But in that middle "sweet spot" of roughly 20–60%, the perturbative approach can yield real benefits.

9.2.3 More Parameters

We can generalize straightforwardly to cases with n_K control parameters $\boldsymbol{K}$ and n_θ uncertain system parameters $\boldsymbol{\theta}$. Defining scaled, dimensionless versions $\boldsymbol{\vartheta}$

and $\boldsymbol{k}$ of the nominal system and control parameters, we have the vector equation

$$\delta \boldsymbol{k} = -\tfrac{1}{2}(\partial_{\boldsymbol{kk}} J)^{-1}\, \partial_{\boldsymbol{k}} \mathrm{Tr}\,[\boldsymbol{\Sigma}\,(\partial_{\boldsymbol{\vartheta\vartheta}} J)]\ , \tag{9.16}$$

where the $n_\vartheta \times n_\vartheta$ covariance matrix for parameter variations $\boldsymbol{\Sigma}$ has elements $\Sigma_{ij} = \langle\, \delta\vartheta_i\, \delta\vartheta_j \rangle$ and is $O(\varepsilon^2)$. The gradient operator is $\partial_{\boldsymbol{k}}$ denotes partial derivatives with respect to the vector of scaled control parameters $\boldsymbol{k}$. The second-derivative matrices of the scalar cost function J are evaluated at the nominal values of control and system parameters. These *Hessian* matrices can be interpreted as curvature matrices for $\boldsymbol{k}$ and $\boldsymbol{\vartheta}$ surfaces. Note that the number of control parameters can differ from the number of uncertain system parameters.

Problem 9.4 extends the example of derivative control of an undamped harmonic oscillator, discussed in Section 9.2.2, against disturbances to PD control, which has two parameters to modify. There is still a single uncertain system parameter, the frequency.

Conversely, Problem 9.5 is a case with one control parameter and two uncertain parameters, the frequency and damping coefficient of an oscillator. For n uncertain system parameters with independent uncertainties, the trace operator leads to sums of the form $(\beta_1\varepsilon_1^2 + \beta_2\varepsilon_2^2 + \cdots)$. If a single uncertainty dominates, we can usually neglect the others.[6]

9.3 Risk

> There are known knowns.
> There are things we know we know.
> We also know there are known unknowns; that is to say
> we know there are some things we do not know.
> But there are also unknown unknowns –
> the ones we don't know we don't know.
> [They] tend to be the difficult ones.
>
> Donald Rumsfeld[7]

We have seen how to choose feedback to minimize the expected cost $\langle J \rangle$, given uncertain dynamical models. But is minimizing the expected cost what you really want to do?

Consider scenarios where $\langle J \rangle$ is dominated by large losses that occur with small probability. Since the product of a large number by a small number is prone to error, you might not trust expectation values that have a substantial contribution from the high-cost tail of a distribution. For example, how do we estimate a small probability?

[6] The same idea applies to classical error analysis, a fact that elementary discussions often regrettably underemphasize. The first step in an error analysis should be to identify the dominant uncertainty. And the ultimate goal should be to balance the important uncertainties, to have n comparable contributions.

[7] US Secretary of Defense; quoted from US Department of Defense news briefing, February 12, 2002.

In textbook situations, we can confidently calculate the probability of rare events. For example, the probability of flipping a coin N times and finding "heads" each time is 2^{-N}. More often, though, rare events result from a sequence of intermediate unlikely events. Moderate uncertainty in the intermediate steps will cascade into a large uncertainty in the probability of the outcome. Even worse, the event may be an "unknown unknown," an unanticipated scenario – a risk. How can you estimate the probability of event you cannot imagine?

Events due to unknown unknowns are also known as *black swans*: Before the discovery of black swans in Australia, Europeans had assumed that all swans were white. The term has become a metaphor for the limitations of induction: just because all the swans described for two thousand years were white does not mean that tomorrow a black swan cannot be encountered.[8] No evidence for black swans is not evidence for no black swans.

Nassim Taleb has discussed the considerable historical impact of *black-swan events*, outliers with major impact that are rationalized or "explained" after the fact. He warns in particular against the *ludic fallacy* of assuming that events are governed by ordinary probability whose odds you can calculate as in a casino: If someone flips a fair coin a hundred times and it always comes up heads, what is the probability that the next flip will be tails? Before you say "50%," is it not more likely that the "fair" coin is fake?

The other difficulty is that large costs can be hard to measure. What seems reasonable for small costs may not be sensible for large costs. For example, consider a trip across town. For every route you take, for every method of transport, there are scenarios leading to a fatal accident. The usual cost functions we employ (time of the trip, transit fare) seem inadequate to assign a value for one's life. We might be tempted to assign an infinite cost, but then all expected costs would be infinite, too. Yet we still take trips!

Leaving further discussion of such questions to the economists and psychologists,[9] we take away the point that large costs can be hard to measure on the same scale as small costs and thus it can make sense to design control in a way that considers rare disasters along with typical outcomes. With such a view in mind, let us return to the perturbative, low-noise approach to robust control begun in the previous section.

9.3.1 Perturbation Theory Breaks Down

In the examples of robust feedback control that we have so far considered, it was not possible for feedback to destabilize a system. But typically, it can. For example, consider disturbance rejection for an inverted, unstable pendulum of uncertain frequency constant, ω. For proportional-derivative (PD) control, the equations of motion are

[8] Engraved image from *Friends in Feathers and Fur*, J. Johannot, New York: D. Appleton and Co., 1885.
[9] See the Notes and References section at the end of the chapter.

$$\underbrace{\ddot{x} - \omega^2 x = -K_\mathrm{p} x - K_\mathrm{d}\dot{x}}_{u(t)}\,, \quad \Longrightarrow \quad \ddot{x} + K_\mathrm{d}\dot{x} + (K_\mathrm{p} - \omega^2)x = 0\,, \tag{9.17}$$

with an initial condition corresponding to a disturbance, $x(0) = 0$, $\dot{x}(0) = 1$. Except for the sign of ω^2, this is just the same dynamics as Eq. (9.13). But the sign matters! Indeed, if ω is known, then choosing $K_\mathrm{p} > \omega^2$ stabilizes the system. However, if ω can take on any positive value, then a fixed K_p cannot stabilize all systems.

Even if the probability of instability is very small, it has important consequences. For the cost function, assuming $K_\mathrm{p} > \omega^2 + \frac{1}{4}K_\mathrm{d}^2$,

$$J(\omega, K_\mathrm{p}, K_\mathrm{d}) = \frac{1}{2}\int_0^\infty \mathrm{d}t\,(x^2 + u^2) = \frac{1}{4}\left[K_\mathrm{d} + \frac{1 + K_\mathrm{p}^2}{K_\mathrm{d}(K_\mathrm{p} - \omega^2)}\right], \tag{9.18}$$

we optimize costs by choosing, for $\omega = 1$, the feedback gains $K_\mathrm{p}^{(0)} = 1 + \sqrt{2} \approx 2.4$ and $K_\mathrm{d}^{(0)} = \sqrt{2(1 + \sqrt{2})} \approx 2.2$. The minimum cost $J^{(0)} = \frac{1}{2}K_\mathrm{d}^{(0)} \approx 1.1$.

Now assume that ω is uncertain, again with the ensemble of possible systems having a lognormal distribution $p(\omega)$, with mean 1 and standard deviation ε. Because ω is unbounded from above, there will always be a finite fraction of systems that are not stabilized for *any* choice K_p of proportional gain. For these systems, the cost J is infinite, and thus so is $\langle J\rangle$. The conclusion holds for all $\varepsilon > 0$, no matter what the choice of control.

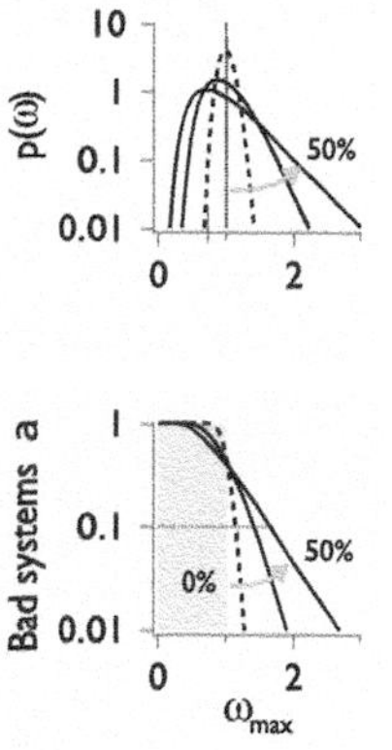

Our previous approach to robust control, based on minimizing $\langle J\rangle$, thus fails. To make progress, let us partition the set of possible systems into two classes: the *good* (stable) and the *bad* (unstable). In particular, we will choose a maximum frequency, ω_max, such that $P(\omega > \omega_\mathrm{max}) = \alpha$. At left are oscillator-frequency distributions $p(\omega)$ for different relative uncertainty levels ($\varepsilon = 0$, 10, 30, and 50%), all with average frequency $\langle\omega\rangle = 1$. The lower plot shows, for four values of ε, the fraction α of bad systems as a function of the dividing frequency, ω_max. For example, the thicker line, for $\varepsilon = 0.5$, shows that roughly 1% of systems will have frequencies ≥ 2.68. Thus, $\omega_\mathrm{max} \approx 2.68$ for $\alpha = 0.01$.

The good systems are thus defined to be those with frequencies ω in the range $(0, \omega_\mathrm{max})$. For that subset, we can use any convenient method to find "optimal" feedback parameters. Here, we minimize numerically the conditional expectation cost $\langle J\rangle_\mathrm{good}$,[10]

$$\langle J\rangle_\mathrm{good} = \int_0^{\omega_\mathrm{max}} \mathrm{d}\omega\, p(\omega)\, J(\omega)\,, \tag{9.19}$$

with respect to K_p and K_d (giving K_p^* and K_d^*). Figure 9.5 shows the results for a lognormal distribution with $\varepsilon = 50\%$ and a significance level $\alpha = 0.01$. For 99% of the time, control is attempted on a good system, while 1% of the time, it is on a bad system. The dividing frequency $\omega_\mathrm{max} \approx 2.68 < \omega^* = \sqrt{K_\mathrm{p}^*} \approx 2.73$, the maximum frequency that is stabilized.

[10] We omit a normalization of $\frac{1}{1-\alpha}$ because we focus on relative costs.

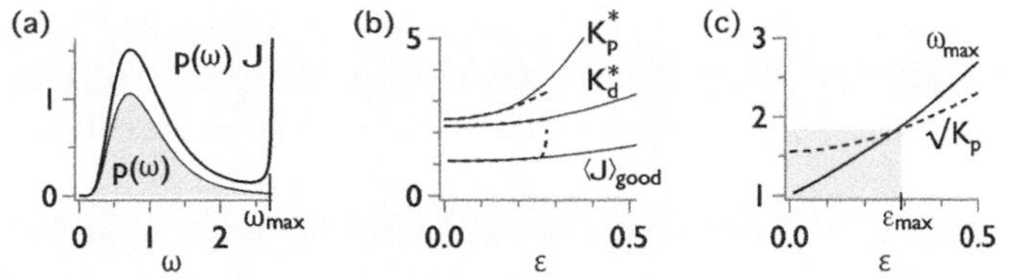

Fig. 9.5

Optimization of control for unstable oscillator. (a) Probability density of frequencies $p(\omega)$ and the divergent product $p(\omega)\,J(\omega, K_\mathrm{p}^*, K_\mathrm{d}^*)$ for $\varepsilon = 0.5$ and $\alpha = 0.01$. The optimal gains $K_\mathrm{p}^* \approx 7.4$ and $K_\mathrm{d}^* \approx 3.1$, and the minimized cost $\langle J \rangle_\mathrm{good} \approx 1.6$. (b) The optimal parameters K_p^* and K_d^* and $\langle J \rangle_\mathrm{good}$ as a function of relative uncertainty in the frequency, ε. Perturbation results indicated by dashed lines. (c) Crossover between ω_max and $\sqrt{K_\mathrm{p}}$ gives the upper limit, $\varepsilon_\mathrm{max} \approx 0.28$, of the perturbation theory (see shaded area).

Figure 9.5a shows the frequency dependence of $p(\omega)$ and of $p(\omega)J(\omega, K_\mathrm{p}^*, K_\mathrm{d}^*)$ (thicker line). This last curve is interesting, because its integral over frequency determines $\langle J \rangle_\mathrm{good}$. Notice how it neatly separates into two regimes, a well-behaved one for lower frequencies and a diverging peak as the instability frequency is approached.[11]

In Figure 9.5a, the optimal proportional gain $K_\mathrm{p}^* \approx 7.4$ is considerably higher than the corresponding deterministic result, $1 + \sqrt{2} \approx 2.4$. This makes sense: the proportional gain is responsible for stabilizing the system and needs to be large enough to stabilize beyond ω_max. By contrast, the optimal derivative gain $K_\mathrm{d}^* \approx 3.1$ is much closer to the deterministic value, ≈ 2.2. The derivative damping does not change much as K_p increases.

Figure 9.5b shows the general trend of the numerically determined optimal gain constants K_p^* and K_d^* with uncertainty ε (solid lines). The proportional gain increases much more sharply than does the derivative gain. Reassuringly, the average cost increases only moderately. The dashed lines are based on the low-noise perturbation theory of Section 9.2. The theory is valid up to a maximum uncertainty, $\varepsilon_\mathrm{max} \approx 0.28$, but then it fails spectacularly: the calculated $\langle J \rangle_\mathrm{good}$ diverges as $-\ln(\sqrt{K_\mathrm{p}} - \omega_\mathrm{max})$.

Figure 9.5c shows what is going on: the perturbation theory result $\sqrt{K_\mathrm{p}(\varepsilon)}$ has crossed the curve for $\omega_\mathrm{max}(\varepsilon)$. At larger ε, perturbation theory gives a gain that is too low to maintain $\alpha = 0.01$. It could be applied to larger ε if we were willing to allow more "bad" outcomes. The factor α would then be higher.

9.3.2 Diverging Cost Function

How should we interpret a diverging cost function? As defined, J makes sense for stable systems but not for unstable ones, where it becomes infinite. Partitioning the set of possible systems allows us to define a subset whose costs are well-defined, and where optimizing makes sense. As for the rest (a fraction α), we need to account for them via a process that is *outside* the original scope of the problem. In the example of the unstable oscillator, we can assign a cost J_bad to failures (instances where the

[11] You could choose gains to place ω_max at the minimum of $p(\cdot)J(\cdot)$, but $\langle J \rangle_\mathrm{good}$ would be slightly higher.

system is not stabilized). Here, we give a single cost to all possible failure events, as the main issue is whether the system is stabilized or not, not how fast it stabilizes. In other cases, a more nuanced cost function could account for different levels of "disaster."

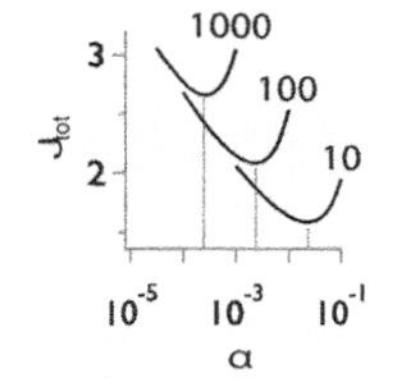

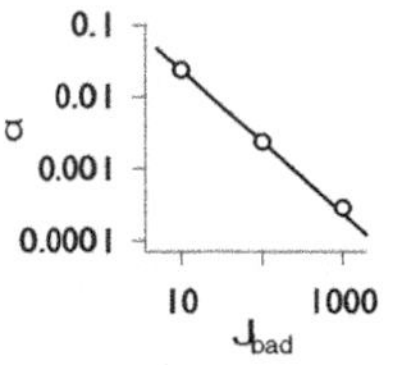

If we can assign a cost to failure, then we can globally optimize to fix the partition coefficient α. The total cost, accounting for both good and bad events, is then[12]

$$J_{\rm tot}(\alpha) = (1-\alpha)\langle J \rangle_{\rm good} + \alpha\, J_{\rm bad}\,. \tag{9.20}$$

Note that $\langle J \rangle_{\rm good}$ is a function of α, since we set the gains as a function of α and ε. The resulting $J_{\rm tot}$ is a function of α with a simple minimum. In the plot at left for three values of $J_{\rm bad}$, the relative parameter uncertainty $\varepsilon = 0.5$. We find that $\alpha^* \sim (J_{\rm bad})^{-1}$, as might be expected from Eq. (9.20): the relatively weak α dependence of $\langle J \rangle_{\rm good}$ means that we should expect $\alpha J_{\rm bad}$ to be approximately constant and that we need fix α or $J_{\rm bad}$ to only about an order of magnitude. The important point is that choosing the partition coefficient α is equivalent to placing a cost $J_{\rm bad}$ on all the events that lie outside of the "good" events.

9.3.3 The Partition Principle

The above examples suggest a general way to handle risk: partition the set of possible events into two groups. The *good* subset consists of all those events whose costs are low enough and probabilities for occurrence high enough that we can sensibly apply all the tools of probability theory and optimization. The *bad* subset consists of those events that we ignore as regards the optimization but possibly prepare for. This subset can actually be divided further. First, there are those events that are the extrapolation of a known probability tail, perhaps a fat tail: these are the known knowns of Rumsfeld or the gray swans of Taleb. The instabilities discussed in the previous section fall into this class. But unknown unknowns, the black swans that we have not anticipated, can also be present and are significant when they occur. However, they can be dealt with in the same way as the first category, by plans separate from those used to optimize over the "good" set.

I conjecture that essentially *all* strategies for managing risk employ this kind of partitioning, although sometimes the partitions can seem hidden.

Partition Principle: Strategies for managing risk partition the set of possible events into *good* events with finite costs and well-characterized probabilities, and *bad* events with large, possibly infinite costs and ill-defined probabilities. One needs to optimize costs over the good set, and plan for rare, bad events.

As an everyday example, the design of large buildings must take the possibilities of earthquakes into account. Since earthquake size is effectively unbounded, buildings

[12] Recall that we dropped a factor, $(1-\alpha)^{-1}$, in our definition of $\langle J \rangle_{\rm good}$. It will cancel the term in Eq. (9.20).

are designed not to collapse in earthquakes up to a specified size. Bigger earthquakes can occur, but they are not part of the design itself. Architects and engineers then try to minimize the costs of the building, subject to the earthquake design constraints. Here, partitioning is implemented via the choice of maximum earthquake magnitude, which itself might be selected by estimating frequency of event occurrences (e.g., "once a century").

Given that we have to choose a partition, there are several ways to proceed. Above, we chose the partition parameter α by directly fixing the fraction of bad events. Alternatively, we can use the distribution of costs, $p(J)$, for events to fix α:

$$\alpha \equiv P(J > J_{\text{max}}) = \int_{J_{\text{max}}}^{\infty} \mathrm{d}J\, p(J)\,, \tag{9.21}$$

which is illustrated as the shaded area at right.[13] The shaded area corresponds to the probability that an event occurs whose cost J exceeds a threshold value J_{max}. The attractive feature of this definition is its direct link to known costs. But finding the distribution $p(J)$ is not always easy. Here we give a simple case where this kind of analysis is possible.

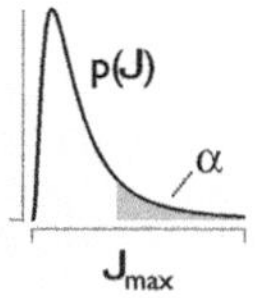

One-Step LQR

Consider one-decision control, introduced in Problem 8.5, without observation noise:

$$x' = x - u + v\,, \qquad J = \tfrac{1}{2}\left(x'^2 + Ru^2\right) = \tfrac{1}{2}\left((x-u+v)^2 + Ru^2\right), \tag{9.22}$$

where $v \sim \mathcal{N}(0, \sigma^2)$ is a random disturbance, x is the state (e.g., position of a particle), u is the control "move," and J is the cost function. Without disturbances (setting $v = 0$), the optimal-control problem can be solved by

$$\partial_u J = 0 \quad \Longrightarrow \quad u_{\text{det}} = \left(\frac{1}{1+R}\right) x \equiv K_{\text{det}} x\,, \quad J_{\text{det}} = J(u_{\text{det}}) = \frac{x^2}{2}\left(\frac{R}{1+R}\right). \tag{9.23}$$

If we now allow fluctuations v, the expected cost becomes

$$\langle J \rangle = J_{\text{det}} + \tfrac{1}{2}\sigma^2\,. \tag{9.24}$$

Since the change in mean value is independent of u and x, the optimal choice of control is unchanged: $K_{\text{mean}} = K_{\text{det}} = 1/(1+R)$, in accordance with the certainty equivalence principle. Fluctuations have simply increased the expected cost.

Figure 9.6a shows empirical distributions of Monte Carlo simulations (based on random variates of the Gaussian noise v). We can calculate explicitly the distributions

[13] Economists often define a related quantity, the *Value at Risk* (VaR): the value J_{max} such that $P(J > J_{\text{max}}) = \alpha$. VaR can even be expressed in dollars. Assigning a dollar value is psychologically dangerous because it encourages you to believe that this is the money at risk. It is not! Events beyond the "good" set, both from the tail of $p(J)$ and from black swans, can lead to effectively infinite losses. Likewise, we avoid a related quantity, the normalized complement $\frac{1}{\alpha}\int_{\omega_{\text{max}}}^{\infty} d\omega\, p(\omega) J(\omega)$, which is equivalent to the *expected shortfall* in economics (Acerbi and Tasche, 2002). Again, the tail of $p(\omega)$ is often poorly known or decays slowly enough that the conditional expectation diverges.

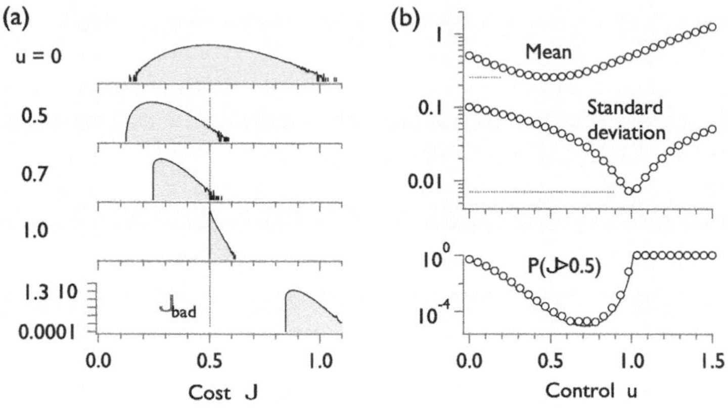

Fig. 9.6 Probability density functions $p[J(u)]$ for one-step LQR control, with $R = x = 1$ and $\sigma = 0.1$ (a) (top to bottom): No control ($u = 0$); control of $u = 0.5$ minimizes the mean $\langle J \rangle$; $u = 0.7$ approximately minimizes $\alpha = P(J > 0.5)$, shown by a thin dotted line; $u = 1$ minimizes the standard deviation σ_J; while $u = 1.3$ leads to larger mean and standard deviation. Monte Carlo simulations based on 10^6 samples per pdf. (b) Mean, standard deviation, and $\alpha = P(J > J_{\max})$, with $J_{\max} = 2J_{\text{det}} = 0.5$.

$p(J)$ for different values of u using the change of variable formalism for probabilities (Problem 9.6). We compare simulations to calculations for the mean, variance, and $P(J > J_{\max} = 0.5)$ in Figure 9.6b. Notice how all three quantities, $\langle J \rangle$, σ_J, and $P(J > J_{\max})$, all decrease with J before increasing after an optimum. But the optima are reached for different values of J.

The one-step LQR calculation readily generalizes to multistep, nth-order LQR processes, since the stochastic variables always appear as linear or quadratic terms. As another example, controlling the disturbance response of a stable oscillator with uncertain frequency (Problem 9.3) can be reformulated in terms of $p(J)$. See Problem 9.7.

9.3.4 Risk-Sensitive Control

In much of the statistical and economic literature, risk-sensitive statistics focuses on minimizing what, at first sight, might seem a strange quantity,

$$\tilde{J}(\gamma) \equiv \frac{1}{\gamma} \ln \left\langle e^{\gamma J} \right\rangle . \tag{9.25}$$

To see why $\tilde{J}$ might be an interesting statistic, let us Taylor expand, assuming $\gamma \ll 1$. Then, using $\ln(1 + x) \approx x - \frac{1}{2}x^2 + \cdots$ and applying the expectation operation,

$$\tilde{J}(\gamma) = \frac{1}{\gamma} \ln\left(1 + \gamma \langle J \rangle + \tfrac{1}{2}\gamma^2 \langle J^2 \rangle + \cdots\right) = \langle J \rangle + \tfrac{1}{2}\gamma\sigma_J^2 + O(\gamma^2) , \tag{9.26}$$

where σ_J^2 is the variance of J. Thus, $\tilde{J}$ is just a linear combination of mean and variance, for small γ. Choosing $\gamma > 0$ implies that fluctuations in the outcome of the control protocol are undesirable. The rational behavior is then to be *risk averse*, to accept a higher average nominal cost in return for reduced variation in that cost. Indeed, the language we used to describe Figure 9.6 draws on just such a picture. By contrast, $\gamma < 0$ means that you believe fluctuations on average are desirable: the rational behavior is to be *risk seeking*.

If the probability distribution for J happens to be Gaussian, then the higher-order terms vanish, and the mean-variance trade-off is exact for all γ. More generally, $\ln\langle e^{\phi}\rangle$ is the *cumulant generating function* (Section A.6.3 online). Non-Gaussian distributions lead to higher-order terms whose coefficients are the skewness, excess kurtosis, and higher cumulants. The quantity $\tilde{J}(\gamma)$ thus penalizes *all* the cumulants of $p(J)$.

In addition, *Jensen's inequality* (Section A.6.5 online) implies that $\tilde{J} \geq \langle J\rangle$ for $\gamma > 0$:

$$\tilde{J} = \gamma^{-1} \ln\left\langle e^{\gamma J}\right\rangle \geq \gamma^{-1} \ln e^{\gamma\langle J\rangle} = \langle J\rangle \,. \tag{9.27}$$

The risk-averse cost $\tilde{J}$ exceeds the expected cost $\langle J\rangle$ if there is any uncertainty.[14]

Although $\tilde{J}$ has attractive statistical features, its motivation is even stronger, as we can relate it to the tail probability, the statistic we are directly interested in. Notice that

$$P(J > J_{\max}) = \int_{J_{\max}}^{\infty} dJ\, p(J) = \int_{0}^{\infty} dJ\, p(J)\,\theta(J - J_{\max}) = \langle\theta(J - J_{\max})\rangle \,. \tag{9.28}$$

From the sketch at right,[15] $e^{\gamma(J-J_{\max})} \geq \theta(J - J_{\max})$ for $\gamma \geq 0$. Ensemble averaging then gives

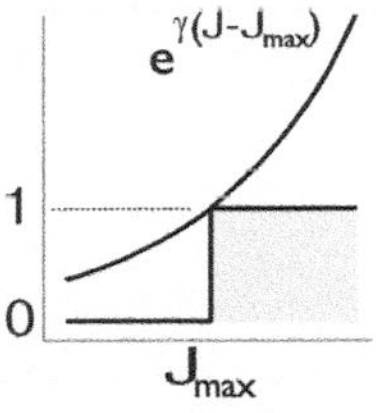

$$\left\langle e^{\gamma(J-J_{\max})}\right\rangle \geq \langle\theta(J - J_{\max})\rangle = P(J > J_{\max}) \implies P(J > J_{\max}) \leq e^{-\gamma J_{\max}}\left(e^{\gamma\tilde{J}}\right). \tag{9.29}$$

The last step uses Eq. (9.25) in the form $e^{\gamma\tilde{J}} \equiv \langle e^{\gamma J}\rangle$. Thus, a control that minimizes $\tilde{J}$ also bounds $P(J > J_{\max})$, the quantity we directly care about. As can be shown using *large-deviation theory*, the bound in Eq. (9.29) can actually be tight (tending to an equality) in the limit of protocols of long duration τ. In effect, such protocols sum many independent random variables, a situation that is common in many control problems.

One-Step LQR, Revisited

How does optimal control change if we seek to minimize $\tilde{J}(u)$ rather than $\langle J\rangle$? Recall the simple one-decision example with "dynamics" given by $x' = x - u + \nu$ and cost $J = \frac{1}{2}(x'^2 + Ru^2)$, with $\nu \sim \mathcal{N}(0, \sigma^2)$. We can calculate $\tilde{J}$ directly:

$$\tilde{J} = \gamma^{-1} \ln \frac{1}{\sqrt{2\pi\sigma^2}} \int_{-\infty}^{\infty} d\nu\, e^{-\frac{\nu^2}{2\sigma^2}}\, e^{\gamma J} \,. \tag{9.30}$$

[14] Equality in Jensen's theorem corresponds to $\gamma \to 0$, the *risk neutral* case.

[15] The argument is similar to that used to derive Markov and Chernoff bounds in probability (Whittle, 2000).

Equation (9.30) suggests we define the *stress* $S = J - v^2/(2\gamma\sigma^2)$ and rewrite the integral as $\sim \int \mathrm{d}v\, \mathrm{e}^{\gamma S}$. Since $S = S(u, v)$ is a quadratic form of u and v, it is easy to evaluate (Section A.7.2 online). Let v^* be the value of S that solves $\partial_v S = 0$. Then, with $\delta v = v - v^*$,

$$S(u, v) = S(u, v^*) + \tfrac{1}{2}(\delta v)^2\, \partial_{vv} S|_{v^*}\,. \tag{9.31}$$

Inserting this expression into the integral, we have

$$\int_{-\infty}^{\infty} \mathrm{d}v\, \mathrm{e}^{\gamma S(x,u,v)} = \mathrm{e}^{\gamma S(x,u,v^*)} \sqrt{\frac{2\pi}{-\gamma\, \partial_{vv} S|_{v^*}}}\,. \tag{9.32}$$

Notice that because S is a quadratic form in u and v, the second-derivative term $\partial_{vv} S|_{v^*}$ is independent of u. Using Eq. (9.32), we can evaluate $\tilde{J}$:

$$\tilde{J} = \gamma^{-1} \ln\left(\frac{1}{\sqrt{2\pi\sigma^2}} \int_{-\infty}^{\infty} \mathrm{d}v\, \mathrm{e}^{\gamma S}\right) = \gamma^{-1} \ln\left(\mathrm{e}^{\gamma S(u,v^*)} + \cdots\right) = S(u, v^*) + \cdots\,, \tag{9.33}$$

where the terms that are dropped are all independent of u. Up to such terms, $\tilde{J}(u) = S(u, v^*)$. Going back to the optimal-control problem, we now should choose $u = u^*$ to minimize $\tilde{J}$ and also $S(u, v^*)$. In general, we need to solve for u^* and v^* simultaneously:

$$\begin{aligned} \partial_u S &= -(x - u + v) + Ru = 0 \\ \partial_v S &= (x - u + v) - \frac{v}{\gamma\sigma^2} = 0\,. \end{aligned} \tag{9.34}$$

Adding the two equations implies $v^* = \gamma\sigma^2 R u^*$. Substituting back then gives

$$u^* = \left(\frac{1}{1 + R - \gamma\sigma^2 R}\right) x \equiv K^* x\,, \qquad K^* = \frac{1}{1 + R'}\,, \tag{9.35}$$

with $R' \equiv R(1 - \gamma\sigma^2)$. The stress is then

$$S(u^*, v^*) = \frac{1}{2}\left[(x - u^* + v^*)^2 + R(u^*)^2\right] - \frac{1}{2}\left(\frac{(v^*)^2}{\gamma\sigma^2}\right) = \frac{x^2}{2}\left(\frac{R}{1 + R'}\right). \tag{9.36}$$

Interpretations

Equation (9.22) reproduces the results found when minimizing $\langle J \rangle$ for $\gamma = 0$, where $R' = R$. This is the *risk-neutral* limit of $\tilde{J}$. For the risk-averse case, $R' < R$ and the gain $K^* > K^*_{\mathrm{det}}$. That is, a risk-averse actor will use larger feedback gains in a noisier situation (larger σ), in contrast to one who minimizes $\langle J \rangle$, where the optimal action is independent of σ. See Eq. (9.23). This represents a breakdown of *certainty equivalence* (Section 8.2), as now the optimal feedback gain does depend on the level of uncertainty.[16] Interestingly, such a dependence is observed in psychology experiments on humans asked to use control to keep a randomly perturbed disk on a computer screen on course to hit a target, with a penalty for control use. Each person seems to have an individual value of γ (Nagengast et al., 2010).

[16] Whittle (1996) shows that a new kind of certainty equivalence exists if one uses the stress S rather than the cost J to determine the optimal control. He uses this property to formulate a "risk-sensitive" LQG.

Notice that $\partial_{uu} S = 1 + R > 0$ for any value of v, so that u^* minimizes $S(u, v^*)$. As for v^*, from Eq. (9.32), we need $\gamma\, \partial_{vv} S < 0$ in order for the integral to converge. Thus, v^* *maximizes* S for $\gamma > 0$ (risk aversion) and *minimizes* S for $\gamma < 0$ (risk seeking).

To summarize, for the risk-averse case, we evaluate $\tilde{J}$ by choosing the value of u that minimizes S while assuming that v has been chosen to maximize S. In Peter Whittle's poetic expression, Nature is like a *phantom other* that picks the *worst* value of the noise v. You then pick the best control accordingly. In effect, you play a kind of *game* against the phantom other. In the language of the one-step LQR problem discussed above, the minimum risk-averse cost, $\tilde{J}^* = \min_u \max_v S(u, v)$, and we have turned the stochastic problem of risk aversion into a deterministic *minimax* problem for the stress.

Recall that we have not specified the value of γ, which, indeed, would seem to vary from person to person, reflecting psychology as much as rationality. Indeed, if we increase γ to $\gamma^* = 1/\sigma^2(1 + 1/R)$, then $R' \to -1$ and the stress $S \to \infty$ (as does $\tilde{J}$). This limit signals a *neurotic breakdown*: Nature is seen as so malicious that no response will be adequate. Better to do nothing![17]

Generalizations

It is possible to generalize the above calculations to n-state LQR. The stress S remains a quadratic form, now over a set of vectors $\boldsymbol{u}_k$ and perturbations $\boldsymbol{v}_k$. The exponential integral over the noise still leads to the minimax game described above where, at each time, Nature selects the worst value $\boldsymbol{v}_k^*$, and the optimal control chooses the best value $\boldsymbol{u}_k^*$ to counteract the phantom other. The case of imperfect, partial observations and a hidden state (the LQG problem) can also be treated, using tricks to decouple state estimation from control.

The second generalization is that the exponential integral can be approximated in cases where the noise is exponential but not Gaussian. One seeks limits where the higher-order terms in the saddle-point expansion resulting in Eq. (9.32) are small and can be dropped.

The third generalization is in two parts. First, we interpret v as arising from an uncertainty in the dynamics rather than a simple disturbance. Second, we analyze the case of linear, time-invariant dynamics and quadratic costs in the frequency domain. The analysis leads to a similar frequency-by-frequency game, where Nature chooses the worst value for a frequency-response function (within a specified set of possibilities) and the goal of the control is to optimize performance and guard against instability in the face of such a malevolent phantom other. In particular, the frequency with highest amplitude must still be small enough that the value of γ converges for it, too. Because this case so dominates the control literature, we discuss it in detail in the next section.

[17] In the risk-seeking case $\gamma < 0$, an actor believes that noise will tend to have values that help to reach the control goal. The extreme case, analogous to γ^*, is the *euphoric* limit.

9.4 Worst-Case Methods: The $\mathcal{H}_\infty$ Min-Max Approach

In this section, we move beyond parametric uncertainty in system dynamics and consider a *nonparametric* description of dynamical uncertainty that does not limit the nature or state-space dimension of the unknown, "true" dynamics. As motivated above, we will take a worst-case approach where we try to do the best we can in the face of the "worst" system. We focus on the analysis of linear, time-invariant systems in the frequency domain.

We develop the analysis in four parts: In Section 9.4.1, we show how controller design implicitly assumes a model of the system being controlled and suffers when that model is wrong. In Section 9.4.2, we discuss how to quantify a system's uncertainty. In Section 9.4.3, we show how to test whether an uncertain nominal system is stable. In Section 9.4.4, we ask how well a controller can perform given uncertain control disturbances, balancing the competing objectives of robust stability and performance.

9.4.1 Internal Model Control

To better appreciate how knowledge of system dynamics influences control, we introduce another way to parametrize control systems, *Internal Model Control* (IMC). The concept is related to the Internal Model Principle of Section 3.7.4: to track a reference or cancel a disturbance, the loop dynamics must include an internal model of the corresponding dynamics. Usually, the internal model is put into the controller. In a similar spirit, we now include a model $G_0(s)$ of the *system* dynamics, $G(s)$. This leads to the block diagram of Figure 9.7, which gives

$$y = \frac{GQ}{1 + (G - G_0)Q}\, r\,. \tag{9.37}$$

Notice that the feedback signal is $v = (G - G_0)u$. This shows explicitly that with a perfect model and no disturbances, there would be no need for feedback. Feedback is required *only* because of imperfect knowledge of the model system and its disturbances.

The IMC formulation of controllers has another advantage: For a perfect model, $G_0 = G$, choosing a stable transfer function $Q(s)$ *guarantees* that the overall controller $K(s)$ will be stable. That is, we no longer have to worry whether a particular parameter choice in a controller design may accidentally lead to unstable response!

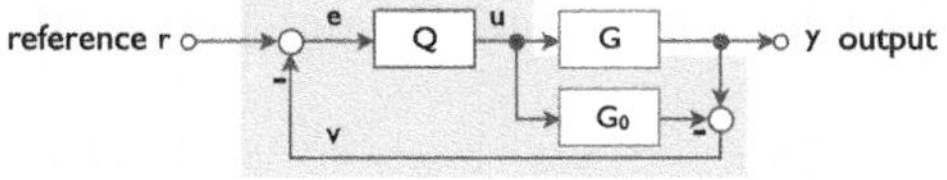

Fig. 9.7 Block diagram of an IMC controller. The darker shaded area is implemented either in a computer program or in control electronics.

To understand this claim, note that Eq. (9.37) implies, with $G_0 = G$, that $y = GQ\,r$. Since the physical system $G(s)$ is assumed stable and since $Q(s)$ is chosen stable, then the overall transfer function $G(s)\,Q(s)$ is obviously stable, too.

When the model, $G_0(s)$ does not match the actual system, $G(s)$, we can relate the IMC controller $Q(s)$ to the "classic" controller $K(s)$ from Chapter 3:

$$K = \frac{Q}{1 - G_0 Q}. \tag{9.38}$$

Because the denominator in Eq. (9.37) is $1 + (G - G_0)Q$, the feedback system will become unstable only if $Q = -\frac{1}{G-G_0}$, which will be large when $G \approx G_0$. More formally, the set of all stable $Q(s)$ generates all stable controllers $K(s)$ for a stable system $G(s)$, if you have a perfect model of the system. When the model is good but not perfect, IMC (Eq. 9.38) is still a reasonable place to start in the search for an appropriate controller.

Another advantage of the IMC structure is that the above remarks about stability carry forward to cases where the system G and model G_0 are nonlinear. If $G \approx G_0$, then their difference may be well approximated by a linear function, even when G and G_0 are themselves strongly nonlinear. See also Section 7.6.1.

If the model is perfect, with $G_0 = G$, then the sensitivity functions are $S = 1 - QG$ and $T = 1 - S = QG$, with $K = Q/S$. That is, the various sensitivity functions, which determine performance and stability, are linear in Q, even while they are nonlinear in the complete controller K. Problem 9.9 explores this IMC parametrization for control systems with feedforward and feedback (two degrees of freedom). Finally, our discussion has assumed that the underlying system $G(s)$ is stable. In such a case, a perfect model leads to effectively open-loop control. A more sophisticated approach is needed to handle unstable systems.

Example 9.1 (Tracking a step-function command) IMC can help design controllers in situations where simpler approaches may lead to instability. The internal model principle implies that including an integrating factor $\frac{1}{s}$ in the controller allows the system to track asymptotically a step-function command. The danger is that the new controller might be unstable. Starting from the Q parametrization of IMC avoids this problem.

Consider a system $G = \frac{1}{(1+s)^2}$ that is known accurately. To track a step-function command signal, we could use an integral-feedback controller, $K(s) = K_{\text{i}}/s$. But a quick analysis (Problem 3.5) shows that the closed-loop system becomes unstable for $K_{\text{i}} > 2$.

Let us design a controller that is guaranteed stable, no matter how fast the desired response. To set the latter, the controller should move the poles from $p = -1$ to $-1/\tau$, where we get to choose τ. To make the output track the reference as $t \to \infty$, we need $T(s \to 0) = 1$. Thus, our design goal is

$$T(s) = \frac{1}{(1+\tau s)^2} \quad \Longrightarrow S = 1 - T = \frac{\tau s(2 + \tau s)}{(1 + \tau s)^2}, \tag{9.39}$$

which satisfies $T(0) = 1$ and has two poles at the desired position. Notice that the sensitivity function S is proportional to s, the inverse of the signal $r = 1/s$ that we desire to track. Since $T = QG$ and $K = Q/S$, we have

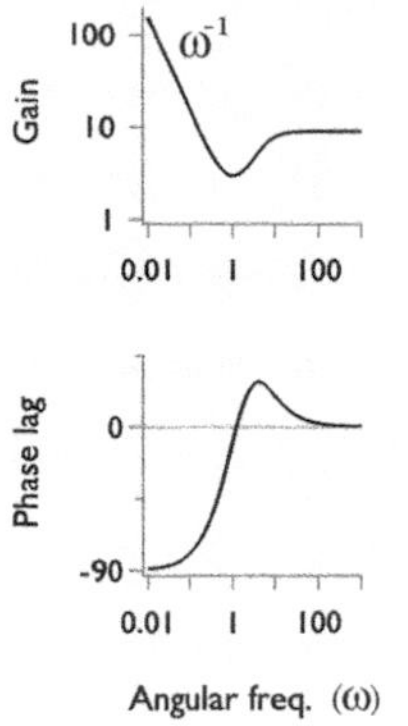

$$Q = TG^{-1} = \left(\frac{1+s}{1+\tau s}\right)^2 \quad \Longrightarrow \quad K(s) = \frac{(1+s)^2}{\tau s(2+\tau s)}, \tag{9.40}$$

whose Bode plot is sketched at left for $\tau = 1/3$. The controller $K(s)$ is realizable (biproper), since $K(s \to \infty) = \frac{1}{\tau^2}$ is finite. Notice that $K(s) \sim \frac{1}{s}$, showing that it indeed uses integral control. Notice, too, that the loop gain $L = KG = \frac{1}{\tau s(2+\tau s)}$ is stable for all τ. Of course, for small τ, the control response $K \sim \tau^{-1}$ will be large, leading to large responses to disturbances and possible saturation of the actuator. Problem 9.11 asks you to design a controller that tracks a ramp, using the same principles.

9.4.2 Quantifying Model Uncertainty

The IMC structure highlights the role of the internal system model and is a first step to analyzing the effects of having the wrong model. In earlier sections in this chapter, we have focused on parametric uncertainty, also known as *structured uncertainty*, since a certain model (or "structure") is assumed. While there are indeed many situations where uncertainty in a particular parameter dominates the model uncertainty, there are other situations where unknown dynamics dominate. For example, in Chapter 6, we saw that models are usually constructed from data over a limited frequency range. Typically, nonparametric uncertainty dominates at frequencies near the limits used to construct the system model.

For nonparametric, *unstructured uncertainty*, the actual system is unknown but assumed to be a member of a set of systems $\mathcal{G}(s)$ "near" a nominal system $G_0(s)$:

$$\mathcal{G}(s) \equiv [1 + \Delta(s)W(s)]G_0(s), \tag{9.41}$$

where we assume *multiplicative uncertainty*. In Eq. (9.41), the transfer function $\Delta(s)$ has a "size" ≤ 1 and is assumed stable, along with $W(s)$. (We define "size" below.) The entire family of transfer functions $\mathcal{G}(s)$ will then share the same unstable poles. Although this technical requirement can be avoided, it makes the analysis simpler. The transfer function $W(s)$ "weights" (or bounds) the uncertainty at different frequencies.

To define the "size" of a transfer function $\Delta(s)$, we use $||\Delta(s)||_\infty$, where the ∞ subscript on the norm refers to the "$\mathcal{H}_\infty$" norm. For stable $\Delta(s)$, it is given by[18]

$$||\Delta||_\infty \equiv \sup_\omega |\Delta(i\omega)|. \tag{9.42}$$

Intuitively, imagine supplying all possible frequencies, with unit amplitude, to the dynamical system represented by the transfer function $\Delta(s)$. We then define the size of $\Delta(s)$ to be the largest output amplitude, over all the frequencies. The $\mathcal{H}_\infty$ norm captures the idea of worst-case response and can be computed by finding the maximum of $|\Delta(i\omega)|$.

[18] More generally, the $\mathcal{H}_\infty$ space is the set of all matrix-valued functions that are analytic and bounded in the open right half of the complex plane, $\mathrm{Re}(s) > 0$. The "sup" then is over the largest singular value, $\overline{\sigma}(\omega)$.

By contrast, the more common Euclidean, or $\mathcal{H}_2$ norm is defined in frequency space for stable Δ by

$$||\Delta||_2 \equiv \left[\int_{-\infty}^{\infty} \frac{d\omega}{2\pi} |\Delta(i\omega)|^2 \right]^{1/2} . \tag{9.43}$$

For more on the different norms, see Section A.1.1 online and Problem 9.10. In brief, the $\mathcal{H}_2$ norm measures the "power" in a signal, while the $\mathcal{H}_\infty$ norm describes the "worst" response.

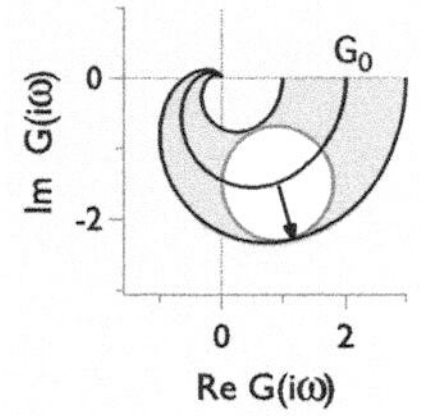

A good way to picture multiplicative uncertainty is in the complex s-plane. For example, the thick line in the plot at right shows the Nyquist plot of G_0, $\text{Im}[G_0(i\omega)]$ versus $\text{Re}[G_0(i\omega)]$ over $0 \le \omega < \infty$. (See Section 2.3.2.) Also shown are the superposition of the frequency-dependent multiplicative perturbations. At each frequency, the system is located within a disk of radius of radius $|W(i\omega)\, G_0(i\omega)|$. The union of all the disks gives the two bounding curves shown. The multiplicative bound expresses the uncertainties relative to $G_0(i\omega)$.

We could also parametrize the set of possible transfer functions by using *additive uncertainty*, defined by $\mathcal{G}(s) \equiv G_0(s) + \Delta(s) W(s)$, but this definition is easily reexpressed in terms of multiplicative uncertainty. In fact, multiplicative uncertainties arise naturally in control problems. For example, consider an actuator (a part of the system) that has a temperature-dependent DC gain A. If used in rooms of differing temperatures, then the set of loop transfer functions would range from $[A_0 - \delta A]\, G_0(s)$ to $[A_0 + \delta A]\, G_0(s)$, implying a fractional uncertainty $W = \frac{\delta A}{A_0}$. In general, both A_0 and δA depend on frequency.

Example 9.2 (Parametric vs. nonparametric uncertainties) To get a feel for the different types of uncertainties, let us consider a family of first-order systems with uncertain gain and delay. The systems are defined parametrically as

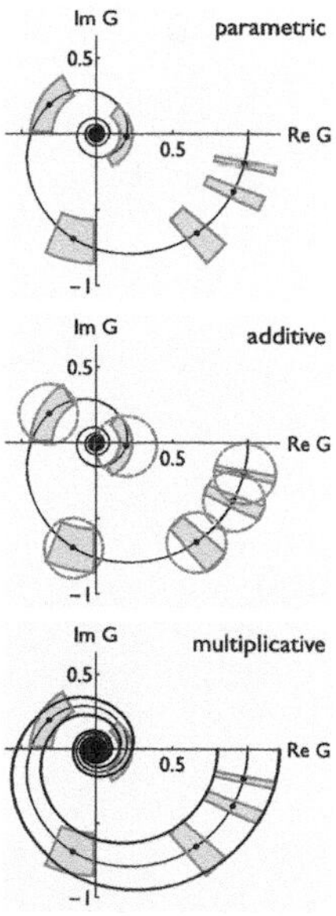

$$\mathcal{G}(s) = \frac{k}{1+s}\, e^{-\tau s}, \qquad 0.8 < k, \tau < 1.2 .$$

At right, the thick, black spiral curve shows the Nyquist plot of the nominal system ($k = \tau = 1$). The top plot shades the parametric uncertainty regions with dark fill. These sector-shaped regions are formed, for a given frequency ω, by varying k and τ over their allowed values. Their union is a set that can be complicated to compute. The middle plot adds dashed circles to show the approximate additive uncertainty bound. The explicit bound is

$$\mathcal{G}(s) = G_0(s) + \Delta(s)\, W(s), \qquad G_0(s) = \frac{e^{-s}}{1+s},$$

with $W(s) = 0.2$, independent of frequency. As the plot shows, it is a poor bound for low and high frequencies because it uses a circle to where the actual region is closer to a skinny rectangle. At low frequencies, however, the bound is better than it might seem because the union of all the circles is very similar to the union of all the parametric uncertainty regions. At high frequencies, though, the parametric-uncertainty rectangle is stretched mainly along the nominal response, and circular bounds cover a far larger region in the complex plane than does the union of the parametric uncertainty regions.

Finally, in the bottom plot, thin lines approximate the multiplicative uncertainty bound formed by $\mathcal{G}(s) = G_0(s)[1 + \Delta(s) W(s)]$, with $W(s) = 0.2$, again a constant. The multiplicative form makes the bounding lines "spiral in" and approximately match the union of the parametric uncertainty sets. (The bounds are slightly overconfident at higher frequencies.) In all three plots, the markers and regions correspond to $\omega = 0.1$, 0.2, 0.4, 1, 2.5, and 5.

In Example 9.2, we applied bounds graphically to the Nyquist plot. We can also use the Bode magnitude response plot to derive a multiplicative bound for nonparametric uncertainty, as the next example shows.

Example 9.3 (System with uncertain time delay) Consider a system with an unknown time delay $t \in [0, t_{max}]$. If the nominal system is $G_0(s)$ and known, then, at each frequency ω, we can bound the uncertainty by $|W(i\omega)|$, which must be greater than

$$\left| \frac{e^{-i\omega t} G_0}{G_0} - 1 \right| = \left| e^{-i\omega t} - 1 \right| , \qquad (0 \le t \le t_{max}) . \tag{9.44}$$

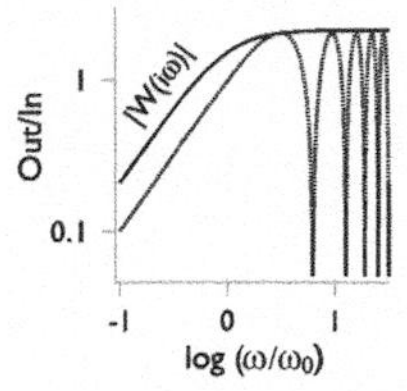

As shown at left, a suitable bound is $W(s) = 2.1st/(1 + st_{max})$, which can be derived graphically by noting the asymptotic behavior for $\omega \to 0$ and $\omega \to \infty$ of Eq. (9.44) and increasing the amplitude slightly to avoid "cutting the corner." This example is adapted from Doyle et al. (1992). While it treats a parametric uncertainty (the time delay), the way we derive a bound would also be appropriate for unstructured uncertainty.

The robust approach is particularly useful for dealing with *systematic* errors in transfer functions. Systematic errors in modeling system dynamics can arise for many reasons:

- Only part of a system's dynamics is modeled. The effects of higher-order modes may be neglected or projected away, and there may not be enough separation from the lower-order term to model the effect of those neglected modes by a white-noise term in a Langevin equation, which is the usual physics approach.
- The dynamics may be nonlinear. Because the control algorithms that we have been discussing are based heavily on linear techniques and because most real systems are at least somewhat nonlinear, the parameters and even the form of the effective linear dynamics will vary with the operating point. Relative to a fixed nominal system, there will be unmodeled dynamics. We discuss control of nonlinear systems in Chapter 11.
- A system may interact with a larger system whose dynamics are not modeled. Thus, an experiment in a room may show different behavior as a function of temperature. A biochemical network (Chapter 14) may function in a cell whose environment changes significantly in different conditions, and so on.

Typically, uncertainties in physical system dynamics are small at low frequencies and large at high frequencies. The bound $W(s)$ should also have these qualities. If its overall scaled amplitude is set to 1 and similarly for its frequency scale, it should satisfy $|W| \ll 1$ for $\omega \ll 1$ and $|W| \gg 1$ for $\omega \gg 1$. The analytical form is often taken to be a lead compensator (Eq. 3.44), which interpolates from a small constant value at low frequencies to a larger constant value at high frequencies. For example, we can use a lead compensator such as $W(s) = \frac{1+10s}{10+s}$, whose magnitude is plotted at right. Problem 9.15 generalizes this form to arbitrary low- and high-frequency limits and arbitrary frequency scales.

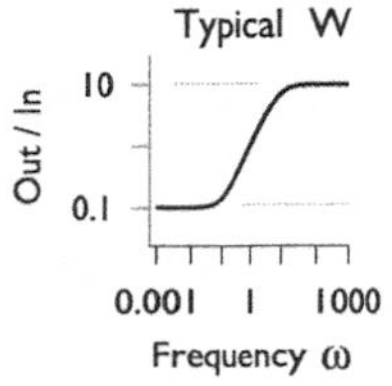

9.4.3 Stability

In Section 9.4.2, we quantified the uncertainty in a model of system dynamics by assuming that a particular system $G(s)$ belongs to a family of systems $\mathcal{G}(s)$. Having defined the family $\mathcal{G}(s)$, we would like to choose a controller $K(s)$ so that, at a minimum, the feedback loop is stable for all $G \in \mathcal{G}(s)$. From the example of an unstable oscillator in Section 9.3.1, we know this typically will not be possible. Still, we can proceed, in the spirit of separating the set of possible systems into "good" and "bad" and then stabilizing the set of "good" systems. The property of stability for all expected realizations is known as *robust stability*.

Unlike the system, we can assume that we know the controller $K(s)$ exactly.[19] Then, the loop gain $L = KG$ will be a member of a set $\mathcal{L} = K\mathcal{G}$. Since we have seen that a system goes unstable when the denominator of the transfer functions T and S equal zero, we must have that $1 + L(s) \neq 0$ for all Re $s > 0$, for all values of any parameters used in K, and for all systems G. The last requirement can be restated succinctly as $1 + \mathcal{L} \neq 0$.[20]

To analyze stability more precisely, define the set of systems $\mathcal{G}(s) = G_0(s)[1 + \Delta(s)W(s)]$, with W the uncertainty bound and Δ an arbitrary stable transfer function with magnitude ≤ 1. The set of loop transfer functions is then $\mathcal{L}(s) = L_0(s)[1 + \Delta(s)W(s)]$, with $L_0(s) = K(s)G_0(s)$. In terms of these quantities, a necessary condition for robust stability is

$$|1 + \mathcal{L}(i\omega)| = |1 + L_0(i\omega) + \Delta(i\omega)W(i\omega)L_0(i\omega)| \neq 0\,, \qquad \forall\omega\,. \tag{9.45}$$

The dark-shaded region in the diagram at right depicts a family of loop transfer functions $\mathcal{L}$. If the system is closed in a feedback loop, then the robust-stability condition expressed by Eq. (9.45) implies that the set of loop transfer functions cannot touch the point -1, indicated by the large dot. At each ω, $\Delta(i\omega)$ is a complex number

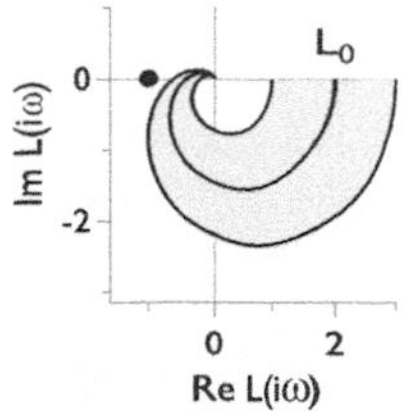

[19] Analog controllers would have uncertain component values and nonideal frequency response. But since robust algorithms generate fairly complex controllers, they are in practice always digital. A controller might also be connected to circuits such as power amplifiers that inject noise. But we can treat such noise as one kind of disturbance, perhaps referred to the output.

[20] As in Section 3.5.2, we are simplifying somewhat. For a given controller $K(s)$, we should invoke the Nyquist criterion to see whether any system in the Nyquist plot of $\mathcal{L}(i\omega)$ circles -1 an appropriate number of times to be unstable. In practice, it is more important to know the instability threshold, given by $1 + \mathcal{L} = 0$.

with magnitude ≤ 1. From the entire set of possible transfer functions, the worst case for the $\Delta(i\omega)$ minimizes the left-hand side of Eq. (9.45). We thus choose $\Delta(s)$ so that its magnitude $|\Delta(i\omega)| = 1$ and so that its phase $\phi(\omega)$ implies that

$$|1 + L_0(i\omega)| - |W(i\omega)L_0(i\omega)| > 0\,, \qquad \forall\omega\,. \tag{9.46}$$

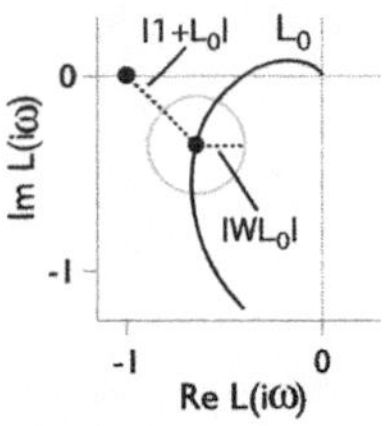

We illustrate this condition graphically at left, zooming in on the frequency response curve. The white disk shows the domain of uncertainty for one frequency. Its center is a distance $|-1 - L_0| = |1 + L_0|$ from the instability point, -1. This must be larger than the radius of the circular uncertainty domain, $|WL_0|$, in order to guarantee stability for all possible transfer functions. This condition, $|1 + L_0| > |WL_0|$, can be rewritten as[21]

$$\left|\frac{W(i\omega)L_0(i\omega)}{1 + L_0(i\omega)}\right| < 1\,, \qquad \forall\omega\,. \tag{9.47}$$

Since the complementary sensitivity function of the nominal system is $T = \frac{L_0}{1+L_0}$, we have

$$\|\, WT \,\|_\infty < 1\,, \tag{9.48}$$

where the $\mathcal{H}_\infty$ norm here serves simply as a shorthand for $|W(i\omega)T(i\omega)| < 1\ \forall\omega$. The use of the $\mathcal{H}_\infty$ norm arises from the desire to be conservative, to be stable for the worst possible realization of the system. Equation (9.48) also implies that T must be less than the inverse of the uncertainty bound, W^{-1}, at all frequencies for robust stability to hold.

Example 9.4 (First-order system with uncertain delay) To see how robust stability works, let us again consider a first-order system with a time lag that is constrained to lie within the interval $[0, t_{max}]$. The nominal system is $G_0(s) = (\frac{1}{1+s\tau})\, e^{-st_{max}/2}$. Since the maximum expected lag is t_{max}, we follow Example 9.3 and take $W(s) = \frac{2.1 s t_{max}}{1+s t_{max}}$. Given a controller $K(s) = K$, $\tau = 1$, and $t_{max} = 0.1$, a numerical calculation gives a maximum gain $K_{max} \approx 6.4$. This "robust" gain limit is significantly less than the stability limit for a first-order system with known time delay $t_0 = \frac{1}{2}t_{max}$, where, using the techniques from Example 3.2 (but not assuming $\tau \gg t_0$), we find $K_{max} \approx 32$. The robust-stability criterion thus leads to a smaller maximum gain than does a calculation based on a precise model of the system (with delay). Uncertainty makes the controller design more conservative.

Even if $t_0 = t_{max}$, we have $K_{max} \approx 16$, which exceeds the robust limit of 6.4. That is, the robust algorithm leads to a lower gain than the actual stability limit for the greatest possible delay, t_{max}. The lower value of K_{max} traces back to the imprecise bounding function for $W(i\omega)$. Notice the blank spaces "below" the worst-case function W, which

[21] Recall, for complex numbers a and b, that $|a/b| = |a|/|b|$. Prove this using polar form, $a = r e^{i\theta}$, etc.

was arbitrarily chosen to be first order. A higher-order approximation that "hugs" the function $|e^{-i\omega\tau} - 1|$ more tightly would increase the upper bound K_{max}. But such precision is misguided, as other errors (e.g., in setting the upper limit for t_{max}) are usually present (Problem 9.13).

9.4.4 Performance

Section 9.4.3 focused on stability, but performance counts, too. One goal is disturbance rejection, and previously, we have studied the response to a specific form of disturbance. Here, in the spirit of unstructured uncertainty, we consider a set of disturbance signals specified by a bounding function $W_1(s)$, the *performance weight*.

We thus characterize the set of possible $d(s)$ signals by $d(s) = \Delta(s)W_1(s)$, where $\Delta(s)$ is an arbitrary stable transfer function with $|\Delta| \leq 1$. Physical disturbances are generally filtered through physical systems whose response dies off at high frequencies. They are larger at low frequencies and smaller at high frequencies. The bound $W_1(s)$, which represents the largest disturbance at each frequency, should also have these qualities. If the amplitude and frequency scales are both set to 1, it should satisfy $|W_1| \gg 1$ for $\omega \ll 1$ and $|W_1| \ll 1$ for $\omega \gg 1$. The analytical form is often taken to be a lag compensator (Eq. 3.68). For example, we can use $W_1(s) = \frac{10+s}{1+10s}$, whose magnitude is plotted at right. Again, W_1 represents the largest expected disturbance at each frequency. If the loop transfer function $L(s)$ is known, the controller output $y(s)$ will depend on output disturbances as $y = S\,d$.[22] The output error is then bounded by

Typical W_1

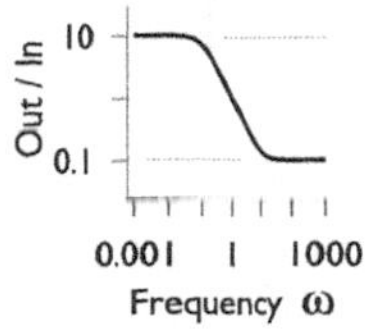

$$|y(i\omega)| = |dS| \leq \sup_\omega |W_1(i\omega)S(i\omega)| = \|W_1 S\|_\infty\,. \tag{9.49}$$

We can reasonably ask that the worst possible output deviation y resulting from the most dangerous expected output disturbance be bounded below the characteristic scale:

$$\|W_1 S\|_\infty < 1\,. \tag{9.50}$$

To understand this condition graphically, we rewrite it as $|W_1| < |1 + L|$, for all ω. This condition implies that a circle of radius $|W_1(i\omega)|$ centered on the instability point $(-1, 0)$ should not touch the loop transfer function $L(i\omega)$. See right.

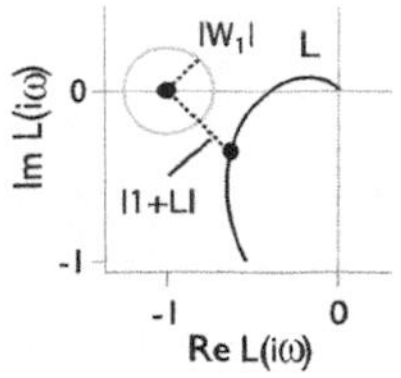

Uncertain disturbance

The next step is to allow for uncertainties in the system. Assuming multiplicative disturbances are bounded by $W_2(s)$, we expand the nominal loop transfer function $L_0(s)$ to a band defined by $L(s) = L_0(s)[1 + \Delta(s)\,W_2(s)]$. Graphically, this adds a second disk domain to the Nyquist plot, as shown at right. To guarantee robust performance, these domains – the disks – should be disjoint, as shown at right. The uncertainty bound $W_2(s)$, referred to as $W(s)$ in our earlier discussion, typically is modeled as a lead compensator that is nominally the inverse of $W_1(s)$ (nominal because the frequency and amplitude scales can be different, implying that the two transfer function bounds are no longer inverses of each other).

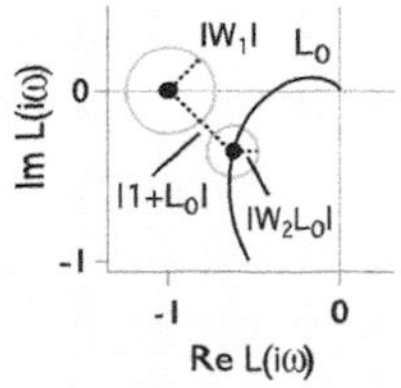

Uncertain disturbance and system

[22] We can similarly consider input disturbances, whose response is $y = GS\,d$, and reference functions, whose controller error $e = S\,r$.

The graphical condition leads to $|W_1| + |W_2 L_0| < |1 + L_0|$, or $|W_1 S| + |W_2 T| \leq 1$. To derive this algebraically from Eq. (9.50), we replace S by $\mathcal{S} = (1 + \mathcal{L})^{-1}$, the sensitivity function family corresponding to the possible loops $\mathcal{L} = K\mathcal{G} = KG_0(1 + \Delta W_2)$. Then

$$|W_1 \mathcal{S}| = \frac{|W_1|}{|1 + L_0 + \Delta W_2 L_0|} = \frac{|W_1 S|}{|1 + \Delta W_2 T|} < 1 \,. \tag{9.51}$$

Multiplying through gives $|W_1 S| < |1 + \Delta W_2 T| < 1 - |W_2 T|$, for all frequencies. Thus,

$$\| \, |W_1 S| + |W_2 T| \, \|_\infty < 1 \,. \tag{9.52}$$

Equation (9.52) is an instance of the *robust-performance problem*. As above, the $\mathcal{H}_\infty$ norm results from the requirement that the relation hold for all frequencies. In Eq. (9.52), all quantities are evaluated at $s = i\omega$, and S and T refer to the nominal system G_0.

From our point of view, the formulation of the robust-performance problem is perhaps more important than its solution. Equation (9.52) may be thought of as another type of optimal-control problem, where the goal is to find a controller K that minimizes $\| \, |W_1 S| + |W_2 T| \, \|_\infty$, given performance and stability weights W_1 and W_2 and given a nominal system G_0 with sensitivity functions S and T. The direct solution to this problem requires concepts from functional analysis that are beyond the scope of this book. Easy-to-use software, however, is available and, while the methods themselves require some background, they are simply a means to find $K(s)$.

Another approach uses loop shaping, a heuristic method introduced in Section 3.7.3. Assuming that W_1 is large at low frequencies and small at high frequencies and that W_2 is the reverse, we can rewrite Eq. (9.52) in terms of L_0 and easily derive that $|L_0| > |W_1|$ when $\omega \ll 1$ and $|L_0| < |W_2|^{-1}$ when $\omega \gg 1$ (Problem 9.16). From the loop shape, we can then derive the controller $K(s) = L_0(s)/G_0(s)$.

The above discussion of robust control methods neglects sensor noise. In Chapter 8 on state estimation, we saw how to use a Kalman filter to estimate the system state in the presence of noise. Such estimates assumed that one had accurate knowledge of the system's dynamics. Robust methods can be reformulated in the state-space formalism and integrated with state estimation. Risk-averse control can also accommodate state estimation.

Rather than developing further the mathematical techniques for solving the worst-case control problems defined above, we end with a simple example reprising the input-shaping technique discussed in Section 9.1.1. The calculations can be done by inspection, and the results illustrate well the advantages and problematic aspects of worst-case approaches.

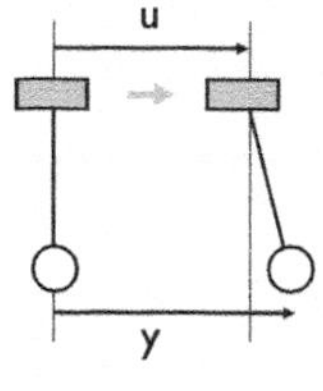

Example 9.5 (Input shaping by minimax) Recall that we seek the response $u(t)$ to a step command $\theta(t)$ that minimizes vibration amplitudes of $y(t)$ over a range of frequencies. If the class of command responses is given by

$$u(t) = A_0\theta(t) + A_1\theta(t - t_1) + A_2\theta(t - t_2)\,, \qquad \sum_{i=0}^{n} A_i = 1\,, \tag{9.53}$$

then the vibration amplitude after the step, the cost, is

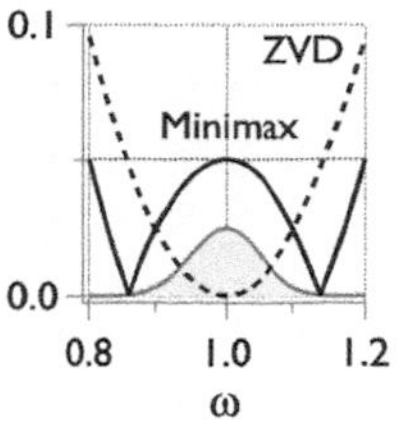

$$J_2 = \sqrt{(A_0 + A_1 \cos \pi\omega + A_2 \cos 2\pi\omega)^2 + (A_1 \sin \pi\omega + A_2 \sin 2\pi\omega)^2}\,. \tag{9.54}$$

In Eq. (9.54), the free parameters are $\{A_0, A_1, A_2, t_1, t_2\}$. The ZVD solution $\{A_i\} = \frac{1}{4}\{1, 2, 1\}$ and $\{t_i\} = \{0, \pi, 2\pi\}$ is shown as the dashed curve at right. Its amplitude $J_2(\omega = 1)$ vanishes at the nominal frequency $\omega_0 = 1$, as does its frequency derivative $J_2'(\omega = 1)$.

As an alternative approach, the solid line is the minimax solution, which minimizes the maximum amplitude of $J_2(\omega)$ over the frequency range $0.8 < \omega < 1.2$. The solution details are explored in Problem 9.18. It has contrasting aspects:

- The maximum vibration amplitude can be restricted to $\approx 5\%$, which is significantly better than the worst case of the ZVD solution over that frequency range, $\approx 10\%$.
- Near the nominal frequency, there is the same $\approx 5\%$ response amplitude. For the typical system this is much worse than the ZVD solution, which is nearly zero in that range.[23]

To illustrate the last point, consider a distribution of system frequencies with $\langle\omega\rangle = 1$ (margin plot above). The expected cost will be higher for the minimax control than for ZVD. If the distribution spreads outside the frequency range (0.8, 1.2), a small fraction of cases will give outcomes that exceed the worst-case limit, ≈ 0.05 (dotted horizontal line). But the minimax solution does guarantee a certain level of performance within the range. And adjusting the range can alter the fraction of "bad" cases that fall outside.

9.5 Summary

Optimal control is the "best," given a cost function and perfect model of system dynamics, but can fail spectacularly when the actual system encountered differs from the nominal design system. Control algorithms thus need to be robust in the face of uncertain dynamics. In this chapter, we have taken a broad view of robust-control problems. We began with a straightforward approach to making feedforward commands more robust, by nulling their lowest-order sensitivity to parameter variations. Typically, robustness requires either more control effort or longer protocols. Using probabilistic methods, we reached similar conclusions for feedback. Generically, the

[23] Adding another time-amplitude pair to $u(t)$ lengthens the protocol but allows the response have zero amplitude at $\omega = 1$, too. This reduces but does not eliminate the advantage of the ZVD for the expected cost.

lowest-order corrections to optimal control are proportional to the variance of the parameter distributions.

We then developed a simple perturbative, low-noise theory for uncertain dynamical parameters obeying probability distributions with known mean, variance, and support. We considered cases with finite numbers of system and control parameters, for linear and nonlinear control problems, including cases with potential instability. Because the method is based on the optimized cost for the certain case, it can be evaluated numerically for nonlinear problems. And because only derivatives are needed, they, too, could be calculated numerically by finite differences. The perturbative method should thus be widely useful. It would also be interesting to generalize to nonparametric (unstructured) uncertainties, which formally can be viewed as a limit with an infinite number of unknown parameters. Hopefully, a finite subset of important uncertainty "modes" will dominate.

Analyzing unstable systems led us to recognize that there can be costs so high that they need to be measured on a different scale and accounted for by a different planning process. In the case of unstable systems, the cost function is infinite when control is attempted on a system where the gain is too low to stabilize. Costs for this case would need to be based on different criteria. These considerations led us to partition the cases into "good" and "bad" and to use optimal control for the former and contingency plans for the latter.

Once we accept the need to partition systems into normal cases and disasters, there are many possible strategies for dealing with the normal cases. These include linear combinations of the mean and variance of a cost function,[24] minimization of the tail probability of bad events for some fixed threshold defining high costs, or fixing the percentage of bad systems from the start and trying to optimize performance over the "good" set.

One statistical approach is based on the risk-sensitive cost function $\tilde{J} = \gamma^{-1}\ln\langle\exp\gamma J\rangle$, a strange quantity that nonetheless has many attractive features. The parameter $\gamma > 0$ quantifies the degree of pessimism taken toward risk. It has analytic advantages that lead to exact solutions for linear dynamics and quadratic costs and, using advanced techniques, can evaluate more general systems in the low-noise or long-time limits. The long-time average of $\tilde{J}$ connects to the concept of "worst case" control, where we imagine that Nature will choose the worst member of a set of allowable uncertainties, and we respond with the best control given that unfortunate fact. The resulting $\mathcal{H}_\infty$ formalism is actually the dominant paradigm within the control community.

If I have emphasized the statistical approach and the links to statistical-physics methods at the expense of worst-case methods, it is because it seems natural for many

24 The problem of optimizing competing objectives is sometimes referred to as *multiobjective optimization*, for which a key notion is *Pareto optimality* (Seoane and Solé, 2016). The basic idea is to search for combinations of objectives, say mean and variance of an optimal protocol, where improving one objective is necessarily at the expense of the other. Solon and Horowitz (2018) give an example where the result is a phase transition in optimal control strategy as a function of the relative weight of one objective with respect to the other. These ideas are well worth developing further.

physics problems. The $\mathcal{H}_\infty$ approach is effective for linear time-invariant (LTI) systems, even ones that are relatively high order with many inputs and outputs. But it comes with heavy mathematical baggage from functional analysis to deal with the infinity norms, and that baggage can obscure the physics of a situation. And it is relevant only in the LTI, long-time limit. More obviously, the notion of bounded sets of uncertain transfer functions implicitly requires a partition into good and bad systems. There is danger in believing the formalism too much: a fraction of systems will fall outside the allotted uncertainty bounds, and control can fail in such cases, often spectacularly. John Doyle has termed this issue the *robust yet fragile* trade-off: robustness to perturbations within a prescribed set goes hand in hand with fragility to perturbations that exceed the limits assumed in "worst case" design (Alderson and Doyle, 2010). We will return to these ideas in Chapters 14 and 15. Here, it is a kind of "corollary" to the partition principle: often, too little attention is paid to the partitioning criterion. For example, most $\mathcal{H}_\infty$ analyses simply give frequency weights for unstructured uncertainty limits defining the "good" family of transfer functions representing a system or disturbance. Such choices should be carefully made, connecting explicitly the empirical measurements used to define the set of good functions and explicitly defining the fraction of bad systems α. I have yet to see a satisfactory discussion of these issues *anywhere*.

Why do engineers prefer the $\mathcal{H}_\infty$ approach to robustness over stochastic approaches? One guess is that it can help to assign blame when disaster strikes. As a profession, engineering needs rules that apportion legal responsibility for projects. Thus, a civil engineer signs off that a bridge will not collapse, an aerospace engineer that a plane will not fall from the sky, a chemical engineer that a factory will not release poisonous gases into the surrounding community. $\mathcal{H}_\infty$ techniques can *guarantee* that a control system will perform within specified limits, as long as the systems that are encountered lie within a prescribed set. If control fails on systems that are beyond those limits, no blame is assigned: such "disasters" are *acts of God*, a legal term. And while the law of large numbers *guarantees* that disasters will occur, each misfortune might lead to an argument, or lawsuit: Was the design at fault, or was the system too extreme? At any rate, engineers in practice must also have contingency plans for systems "beyond the worst case."

Informally, physicists also use worst-case thinking. The idea that "Nature will conspire against you" is just what good experimentalists assume when trying to decide whether to believe an unexpected experimental observation: You go through all possible scenarios that could conspire to produce it before you begin to believe that it may represent new physics.

Finally, as stated at the beginning of the chapter, robust techniques, whether statistical or deterministic, are most appropriate for moderate levels of uncertainty. Systems with small modeling errors can typically be well handled using standard optimal control supplemented by linear linear feedback. And for systems with larger uncertainties, it is better to learn about the individual system by studying its response to actual open- and closed-loop control. This is the approach of adaptive control, Chapter 10.

9.6 Notes and References

Input shaping began with Otto Smith's *Posicast* control, inspired by a fly-fishing maneuver where the fly is dropped in the water at the farthest reach of the cast, when its velocity vanishes (Smith, 1957). Singer and Seering (1990) extend the idea to accommodate damping and uncertainty in the mode frequency. Singh (2010) summarizes more recent work and also presents – but just in passing! – the idea of how to extend the approach to feedforward commands for nonlinear systems via forward sensitivity analysis (FSA). A good, general-purpose code for solving large-scale FSA problems numerically is the open-source package CVODES. Some of the techniques stemming from input shaping have been patented and commercialized by Convolve, Inc. (convolve.com). The construction-crane problem is formally similar to the problem of reducing reflections at lens surfaces in optics, and the multistage input shaping solutions (ZV, ZVD, etc.) are similar to *multilayer thin-film antireflection coatings* (Macleod, 2017).

For a quick overview on the maximum-entropy principle of inferring probability distributions, which we used in Section 9.2.2 on choosing priors, see Jaynes (2003).

Decision making under uncertainty is treated in economics via expected-utility theory (von Neumann and Morganstern, 1944; Fishburn, 1970). The *utility function* is an increasing, concave function of money, reflecting the idea that your first dollar is worth more than your millionth. Costs can be measured by similar convex *disutility* functions, as we explore in Section 9.3.4. Our discussion of risk-sensitive control is inspired by Whittle (2002), a deep article whose use of plain English to convey precise mathematical results is remarkable. Whittle (1996) gives details and discusses how the theory of *large deviations* can extend results beyond the standard LQR and LQG calculations. The discussion of black swans and ludic fallacies is from Taleb (2010). *Risk management* is a field unto itself, and we touch here only on aspects that seem relevant to simple control problems. In more complex settings, the range of options increases. For example, in ISO 31000-2018, the International Standards Organization identifies seven options for *risk treatment*: avoid it, take it as an opportunity, remove it, change the likelihood, change the consequence, share it (insurance), or retain it by informed consent. Finally, the way *humans* empirically make decisions under uncertainty differs from both the assumptions of expected utility theory and the discussion here. Daniel Kahneman and Amos Tversky developed *prospect theory* starting in 1979 to understand and characterize human decision making. That work and its extensions to *behavioral economics* led to a Nobel Prize in 2002.

A standard discussion of how to make optimal control more robust is given by Stengel (1994), who went on to explore Monte Carlo methods for assessing the affects of parametric uncertainty. The low-noise perturbative approach developed here is simpler.

Robust control dates from the late 1970s and early 1980s. For discussions and history about the problems in applying state-space design to "real-world" problems, see the introduction to Morari and Zafirioiu (1989), as well as Leigh (2004). A classic motivating example is John Doyle's paper on "Guaranteed margins for LQG controllers," with its memorable abstract: "There are none" (Doyle, 1978). For a game-theoretic approach to control, see Schulz (2006), chapter 9. On Internal Model Control, see Morari and Zafirioiu (1989), Goodwin et al. (2001), and chapter 13 of Dutton et al. (1997). Our discussion of robust control follows mainly Doyle et al. (1992) and Özbay (2000). Skogestad and Postlethwaite (2005) focus on robust techniques for MIMO systems. For an interesting discussion of "experimentalist paranoia" – the physicists' version of worst-case analysis for systematic errors – see Bailey (2017, 2018).

Because of space limitations, we have not presented several other approaches to robust control. Two notable ones that attempt nonparametric definitions of a "family of dynamical systems" (for describing unstructured uncertainty) are *Gaussian process regression* (MacKay, 2003; Diesenroth et al., 2015) and *polynomial chaos* (Kim et al., 2013).

Problems

9.1 Input shaping: Moving a load of uncertain mass. From Section 9.1.1, we consider a transfer function $G(s) = \frac{1}{1+(s/\omega)^2}$, with ω an unknown oscillation frequency of nominal value $\omega_0 = 1$. The goal is to move from y from 0 to 1 in finite time using the input-shaping protocol: $n + 1$ steps of amplitude $\mathbf{\Lambda} = \{A_i\}$, applied at times $\mathbf{t} = \{t_i\}$.

a. Show that the amplitude of residual oscillations is given by Eq. (9.3).

b. Zero Vibration (ZV): Show that $\boldsymbol{A} = \left\{\frac{1}{2}, \frac{1}{2}\right\}$ and $\mathbf{t} = \{0, \pi\}$ satisfies $J(\omega) = 0$ for $\omega = \omega_0 = 1$, which thus solves the control problem exactly if the system is known perfectly. Find the exact expression for $J_1(\omega)$. Here and below, set $t_0 \equiv 0$.

c. Zero Vibration Derivative (ZVD): Show that $\boldsymbol{A} = \left\{\frac{1}{4}, \frac{1}{2}, \frac{1}{4}\right\}$ and $\mathbf{t} = \{0, \pi, 2\pi\}$ satisfies $J_2 = J_2' = 0$ at $\varepsilon \equiv \omega - 1 = 0$.

d. Zero Vibration Double Derivative (ZVDD): Show that $\boldsymbol{A} = \left\{\frac{1}{8}, \frac{3}{8}, \frac{3}{8}, \frac{1}{8}\right\}$ and $\mathbf{t} = \{0, \pi, 2\pi, 3\pi\}$ satisfies $J_3 = J_3' = J_3'' = 0$ at $\varepsilon = 0$.

e. Show that the Taylor expansions of ZV, ZVD, and ZVDD solutions about $\omega = 1$ give $\left(\frac{\pi}{2}|\varepsilon|\right)^n$, with $n = 1, 2, 3$, respectively.

f. *Adiabatic limit.* Show that the ramp at right leads to residual oscillations whose typical amplitude is $(\omega\tau)^{-1}$, which is small for $\tau \gg \omega^{-1}$.

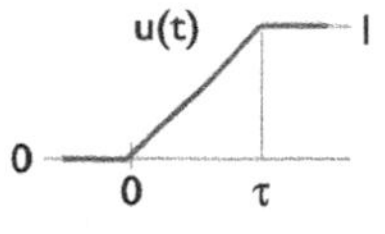

9.2 Swing up a pendulum robustly.

a. Derive the equations of motion for the four-dimensional augmented dynamics for $\boldsymbol{X} = (\boldsymbol{x}\ \boldsymbol{x}_\omega)^{\mathrm{T}}$ that augment Eqs. (9.6). Express them in the form $\dot{\boldsymbol{X}} = \boldsymbol{F}(\boldsymbol{X}, \boldsymbol{u})$.

b. Write the eight-dimensional equations for the combined state and adjoint $(\boldsymbol{X}\ \boldsymbol{\Lambda})^{\mathrm{T}}$.

c. Write a boundary-value code to solve the eight-dimensional equations of motion and make plots similar to the ones given in the text.

9.3 Robust rejection of disturbances for harmonic oscillator.

a. Derive the perturbative result that $\delta k \equiv (\frac{\delta K}{K_0}) = +\frac{3}{2}\,\varepsilon^2$.
b. Let $p(\omega)$ be lognormal, with $\langle\omega\rangle = 1$ and $\langle(\delta\omega)^2\rangle = \varepsilon^2$. Show that $\langle\omega^n\rangle = e^{n\mu+(n\sigma)^2/2}$ and thus $\ln\omega \sim \mathcal{N}(\mu,\sigma^2)$, with $\mu = -(1/2)\ln\left(1+\varepsilon^2\right)$ and $\sigma^2 = -2\mu$.
c. Find exact expressions for the scaled optimal cost $\langle j^*\rangle$ and gain k^* as functions of ε. Taylor expand to confirm $\beta = \frac{3}{2}$.

9.4 Harmonic oscillator with PD control. Redo Prob. 9.3 using PD control (two gains – proportional gain K_p and derivative gain K_d).

a. Find the solution $x(t)$ for the disturbance response. Evaluate the cost function J. Find the optimal values of the derivative gain K_d and proportional gain K_p.
b. Write out more explicitly Eq. (9.16) for the present case of two control parameters and one uncertain system parameter.
c. Find the scaled gain shifts, δk_d and δk_p as a function of ε^2.
d. Evaluate the cost function $\langle J\rangle(\varepsilon, K_\mathrm{d}, K_\mathrm{p})$ numerically by computing the expectation over the lognormal distribution for ω and minimizing over gains $K_\mathrm{d} = K_\mathrm{d}^*$ and $K_\mathrm{p} = K_\mathrm{p}^*$ to find $\langle J\rangle^*(\varepsilon)$. Plot and compare to the perturbative result. Do the same with the cost function and plot three curves (left): a *naive* control where you fix $K_\mathrm{d} = K_{\mathrm{d},0}$ and $K_\mathrm{p} = K_{\mathrm{p},0}$, the optimal values for the nominal model, a *perturbative* control using the techniques of this book, and an *exact* control using the gains determined using the numerically determined K_d^* and K_p^*.

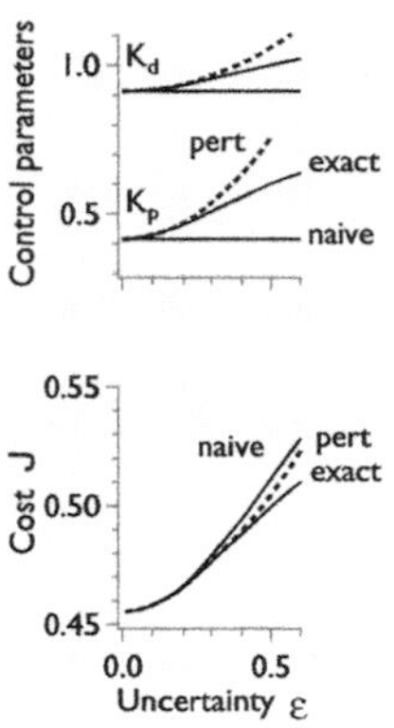

9.5 Harmonic oscillator with uncertain frequency and damping. Analyze the disturbance response of a harmonic oscillator $\ddot{x} + 2\zeta\omega\dot{x} + \omega^2 x = u$, with $u = -K\dot{x}$, for two uncertain parameters $\omega \approx 1$ and $\zeta \approx 0.1$ and cost function $J = \int_0^\infty \mathrm{d}t\,(x^2+u^2)$.

a. Show that the nominal optimal control problem leads to $K_0 = J_0 \approx 0.82$.
b. Write out more explicitly Eq. (9.16) for the present case of one control parameter and two uncertain system parameters.
c. Find the shift in relative gain $\delta K/K$ in terms of $\sqrt{\langle\,\delta\omega^2\rangle}/\langle\omega\rangle$ and $\sqrt{\langle\,\delta\zeta^2\rangle}/\langle\zeta\rangle$. Are uncertainties in both parameters important?

9.6 One-step LQR. The problem considered in Section 9.3.3 can be solved analytically.

a. One striking feature of the probability distributions is that they are zero when the cost $J < J_\mathrm{min}$. Explain why (without calculation) and find $J_\mathrm{min}(u)$.
b. By changing variables, transform the normal distribution for v and show that

$$p(J) = \left(\frac{1}{\sigma^2}\right)\frac{1}{\sqrt{\pi(\,\delta J)}}\,e^{-\left(\delta J+\frac{1}{2}(\,\delta x)^2\right)}\cosh\left(\sqrt{2(\,\delta J)}\,\delta x\right)\theta(\,\delta J)\,,$$

where $\theta(\cdot)$ is the step function, $\delta J = (J - J_\mathrm{min})/\sigma^2$, and $\delta x = (x-u)/\sigma$.

c. Find analytic expressions for the mean $\langle J \rangle$, standard deviation σ_J, and tail probability $P(J > J_{\max})$. Can you find a simpler approximation to this last quantity?

9.7 Harmonic oscillator statistics. We analyze the cost distribution $p(J)$ for an undamped harmonic oscillator with uncertain frequency, continuing Problem 9.3.

a. Derive the analytic form for $p(J)$ and plot for $K = 1.4, 2, 3$.
b. Derive analytic expressions for $\langle J \rangle$, σ_J, and the tail probability $p(J > J_{\max})$.
c. Show that the gain $K^{**} = 2J_{\max}$ minimizes the tail probability.
d. Plot $p(J)$ for $K = 1.4, 2, 3$; mean $\langle J \rangle$ and σ_J versus gain K; gain K^* that minimizes $\langle J \rangle$ versus ε; and tail probability $P(J > J_{\max} = 1)$ versus K.

9.8 Unstable first-order system. Consider an unstable first-order system,

$$\dot{x} = ax + u\,, \qquad u = -Kx\,, \qquad x(0) = 1\,, \qquad J = \int_0^{\infty} \mathrm{d}t \left(x^2 + u^2\right). \tag{9.55}$$

The closed-loop system, $\dot{x} + (K - a)x = 0$, is stable for feedback gain $K > a$. Now assume an uncertain a with lognormal distribution, as in Problem 9.3: $\ln a \sim \mathcal{N}(\mu, \sigma^2)$, with $\mu = -\frac{1}{2}\ln\left(1 + \varepsilon^2\right)$ and $\sigma = \sqrt{-2\mu}$. Here, ε^2 is the variance of a, and $\langle a \rangle = 1$.

a. Show that the optimal control is $K_0 = \sqrt{2} + 1$ and that $J_0(K_0) = K_0$, for $a = 1$.
b. Use perturbation theory to find the optimal feedback gain to $O(\varepsilon^2)$, ignoring the possibility of instability. Show that $\beta = 1$ and, thus, $K^*(\varepsilon) = K_0(1 + \varepsilon^2)$.
c. For $\alpha = 0.01$, show that perturbation theory is limited to $\varepsilon < \varepsilon_{\max} \approx 0.74$.
d. Minimize $\langle J \rangle_{\text{good}}(K)$ for $\alpha = 0.01$ and $\varepsilon < \varepsilon_{\max}$. Plot $K(\varepsilon)$ and $\langle J \rangle_{\text{good}}$ versus ε for both the numerical minimization and the perturbation theory.

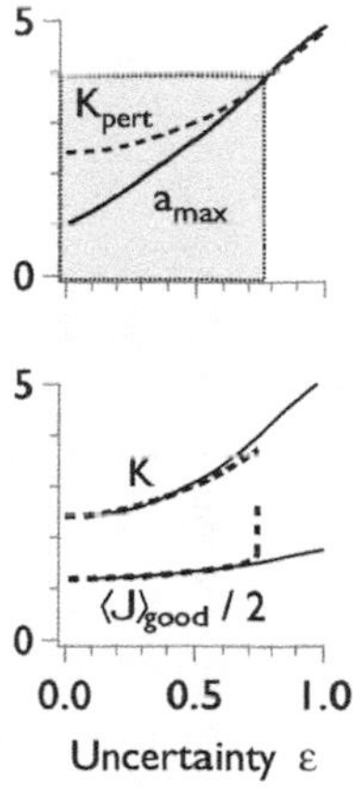

For the graphs at right, dashed quantities are from the perturbation theory. Intuitively, to stabilize systems up to $a = a_{\max}$, we need to set the gain K a "little bit" higher than $a_{\max}$. At $\varepsilon = \varepsilon_{\max}$, for example, $a_{\max} \approx 3.7$ for $\alpha = 0.01$. The perturbation-determined gain $K_{\text{pert}} \approx a_{\max}$ is too low and leads to the divergence of average cost. The numerically determined value that optimizes $\langle J \rangle_{\text{good}} \approx 3.0$ is $K^* \approx 4.0$.

9.9 Internal Model Control (IMC) and feedforward. Consider the "two degrees of freedom" variant of IMC shown below, with system model and two controller transfer functions Q_d and Q_r. The latter is the feedforward filter defined in Section 3.4.1.

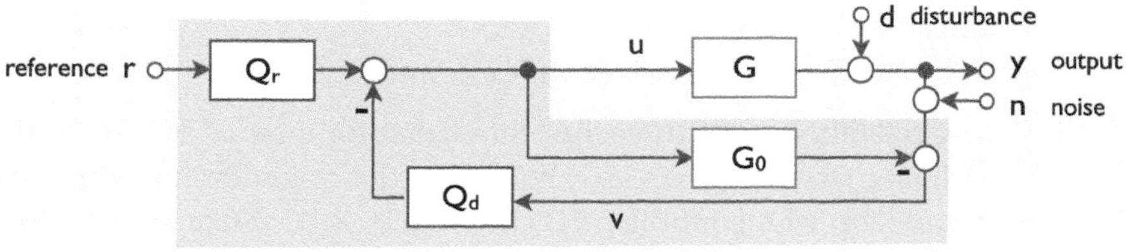

Show that the transfer function of the error signal $e = r - y$ is given by

$$e = \left(1 - \frac{GQ_r}{1 + Q_d(G - G_0)}\right) r - \left(\frac{1 - G_0 Q_d}{1 + Q_d(G - G_0)}\right) d + \left(\frac{GQ_d}{1 + Q_d(G - G_0)}\right) n .$$

For a perfect model, $G_0 = G$, this reduces to $e = (1 - GQ_r)r - (1 - GQ_d)d + GQ_d\, n$, which shows very clearly the role of Q_r in tracking the reference and Q_d in rejecting disturbances. Without feedforward ($Q_r = 1$), we cannot, in general, do both. We also see that rejecting disturbances generally adds noise to the error signal.

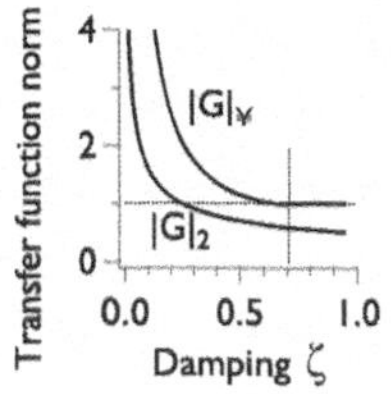

9.10 Transfer function norms. Define the 2 and ∞-norms of a transfer function $G(i\omega)$ as $\|G\|_2 \equiv \left[\int_{-\infty}^{\infty} \frac{d\omega}{2\pi} |G(i\omega)|^2\right]^{1/2}$ and $\|G\|_\infty \equiv \sup_\omega |G(i\omega)|$.

a. *2-norm.* Using Parseval's theorem (Problem A.4.3 online) and representing $G(s)$ in statespace via $\{\boldsymbol{A}, \boldsymbol{B}, \boldsymbol{C}\}$, show that $\|G\|_2 = \sqrt{\boldsymbol{C}\boldsymbol{P}\boldsymbol{C}^\mathsf{T}}$, where $\boldsymbol{P} = \int_0^\infty dt\, e^{\boldsymbol{A}t} \boldsymbol{B}\boldsymbol{B}^\mathsf{T} e^{\boldsymbol{A}^\mathsf{T} t}$ is the Gramian matrix introduced in Example 4.4. Recall that you can compute $\boldsymbol{P}$ directly or solve the Lyapunov equation, $\boldsymbol{A}\boldsymbol{P} + \boldsymbol{P}\boldsymbol{A}^\mathsf{T} = -\boldsymbol{B}\boldsymbol{B}^\mathsf{T}$ (Problem 2.15).

b. For $G(s) = \frac{1}{1+\tau s}$, show that $\|G\|_2 = 1/\sqrt{2\tau}$ and $\|G\|_\infty = 1$.

c. Show that the 2- and ∞-norms for $G(s) = \frac{1}{1+2\zeta s+s^2}$ give the curves at left.

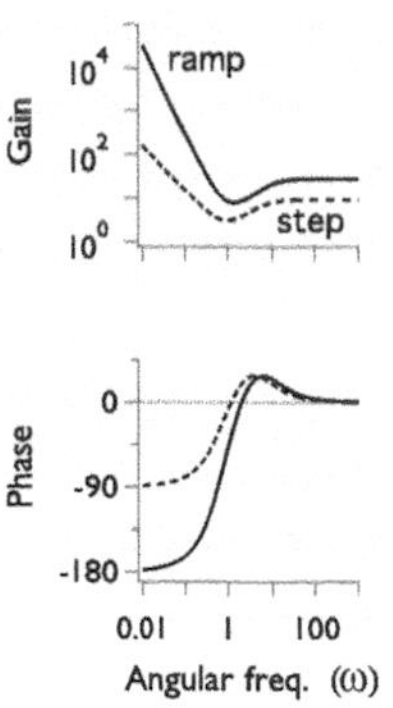

9.11 Tracking a ramp. For the system discussed in Example 9.1, use Internal Model Control to design a controller to track a ramp. Put all closed-loop poles at $s = -1/\tau$, and choose $\tau = 1/3$ for plots. Make a time-domain plot to show that the output does track a ramp. Make a Bode plot for $K(s)$. At left we add a dashed line showing the controller from Example 9.1, which tracks a step input but not a ramp.

9.12 Norms and transfer functions. Why is a transfer function with finite ∞-norm proper but one with finite 2-norm strictly proper?

9.13 First-order system with uncertain delay. Consider a nominal $G_0(s) = \frac{1}{1+s} e^{-s t_{\max}/2}$ and multiplicative uncertainty $G(s) = G_0[1 + \Delta(s)W(s)]$, with $W = \frac{2.1 s\, t_{\max}}{1+s\, t_{\max}}$ and $|\Delta| \le 1$. The controller $K(s) = K$, for $t \in [0, t_{\max}]$, with $t_{\max} = 0.1$ and $\tau = 1$.

a. From the robust stability limit $\|WT\|_\infty = 1$, find (numerically) the maximum allowable gain for which stability is guaranteed. Hint: Find the frequency ω^* and gain $K_{\max}$ such that $|WT| = 1$ and $\frac{d}{d\omega}|WT| = 0$. Here, $T = \frac{L}{1+L}$ and $L = KG$.

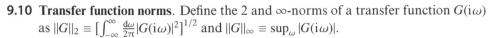

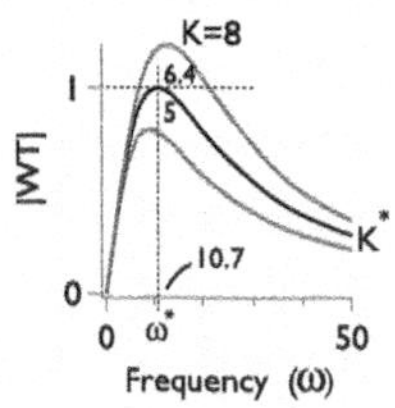

b. Plot $K(\omega)$ to confirm the values of $K_{\max}$ and ω^* (left). Next, make a Bode magnitude plot of $T(i\omega)$ and $W^{-1}(i\omega)$. Finally, plot in the complex plane the circles of possible loop transfer functions $L(i\omega)$ for a few different frequencies and a gain $K_{\max} \approx 6.4$. Highlight the circular domain corresponding to $K_{\max}$ and ω^*.

c. Calculate $K_{\max}$ for known delays $\frac{1}{2} t_{\max}$ and $t_{\max}$.

9.14 Robust stability for additive noise. Show that for a set of systems with additive noise limits, $\mathcal{G}(s) = G_0(s) + \Delta(s)\, W(s)$, that the condition for robust stability in a control loop with controller $K(s)$ is $\|WKS\|_\infty \le 1$, where $S = (1 + L_0)^{-1}$ is the sensitivity function of the nominal loop dynamics, $L_0 = K(s)\, G_0(s)$.

9.15 Bounding functions. The bounding functions $W_1(s)$ and $W_2(s)$ of Section 9.4.4 are typically lag and lead compensators, respectively. Find forms that give good approximations to arbitrary low- and high-frequency limits.

9.16 Loop-shaping criteria. Robust performance requires $\Gamma \equiv \| \, |W_1 S| + |W_2 T| \, \|_\infty < 1$. Typically, W_1 is large at low frequencies and small at high frequencies and W_2 the reverse. Here, $S = \frac{1}{1+L}$ and $T = \frac{L}{1+L}$. (Notice that $S + T = 1$.)

a. Show that $\Gamma < 1$ implies that $\text{Min}(|W_1|, |W_2|) < 1$ at all frequencies.
b. Show that $|L| > |W_1|$ at low frequencies and $|L| < |W_2|^{-1}$ at high frequencies.

9.17 Feedforward with model uncertainty (Devasia, 2002). What happens when an actual transfer function deviates from its model $G_0(s)$? Let $G(s) = G_0(s) + \Delta G(s)$.

a. Assuming an invertible model and a feedforward block $F = G_0^{-1}$, show that the tracking error $e(s) = r(s) - y(s)$ to a command signal $r(s)$ is $e(s) = -\frac{\Delta G/G_0}{1+KG}\, r(s)$.
b. Argue that feedforward helps only for frequencies where $\Delta G/G_0 < 1$.
c. Let $G_0 = \frac{1}{1+s}$ with uncertainty $\Delta G = \varepsilon$, a constant. What does such an uncertainty represent physically? Design a feedforward filter $F(s) = G_0^{-1}(s)\, G_{\text{lp}}(s)$, where $G_{\text{lp}}(s)$ is a low-pass cutoff whose frequency respects the criterion derived in (b).

9.18 Input shaping by minimax. Go through Example 9.5.

a. Argue that symmetry dictates that the command response function is symmetric under time reversal, implying that J_2 reduces to $J_2 = |1 - 2A_0 + 2A_0 \cos \omega t_1|$.
b. Conclude that the minimax solution has $A_0 = [2(1 + (\cos \frac{\pi}{2}\varepsilon)^2)]^{-1}$ and $t_1 = \pi$.
c. Generate the minimax plot of Example 9.5.

10 Adaptive Control

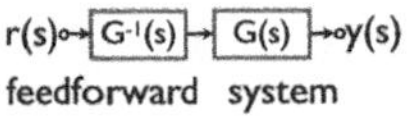

Control systems work better when you know the dynamics of the system being controlled. This statement is most obvious for the simple reference feedforward control $K(s) = G^{-1}(s)$ at left. As we discussed in Chapter 3, errors in estimating $G(s)$ alter the design of the controller $K(s)$ and output with unit sensitivity. Although feedback controllers with high loop gain can reduce the sensitivity of the output, algorithms such as the ever-popular PID still have parameters that should be optimized for a particular system $G(s)$.

In Chapter 6, we discussed off-line, system-identification techniques for inferring the dynamics of systems. When available, they work well if the system has stationary properties. Yet the characteristics of dynamical systems often do change, perhaps because parameters slowly "drift" or because operating conditions (e.g., the mass of a sample) suddenly change. In Chapter 9, we discussed how to make feedback perform well even when the dynamics of the system to be controlled is uncertain. This approach of robust control is intrinsically conservative: Control that works well for a wide range of systems is bound to fare worse than control optimized for one particular dynamics.

Here, we take the less-conservative approach of adaptive control: we monitor the performance of the control system and adjust the dynamics online (e.g., by changing the values of control parameters). In effect, we add an outer feedback loop to the control problem.

Adaptive algorithms almost always lead to nonlinear, time-varying dynamics, even when the underlying system is linear and time invariant. Consider, for example, a simple proportional control algorithm, $u = Ke$, with proportional gain K. An adaptive algorithm that adjusts the gain $K \to K[e(t)]$ implies the nonlinear control law $u = K[e(t)]\, e(t)$.

Often, K is adjusted on time scales that are *slow* compared to the time scales of the closed-loop dynamics. The separation of time scales simplifies the analysis, since the fast and slow dynamics are, separately, simpler than the full coupled system. Even so, proving stability and convergence is difficult, and many of our results will be heuristic. In Chapter 11, we will discuss nonlinear control more generally.

A Cautionary Example

To understand the lure and pitfalls of adaptive control, we consider the simple example of an unstable first-order system with linear feedback,

$$\dot{x} = ax + u\,, \qquad u = -kx\,, \qquad x(0) = x_0\,, \tag{10.1}$$

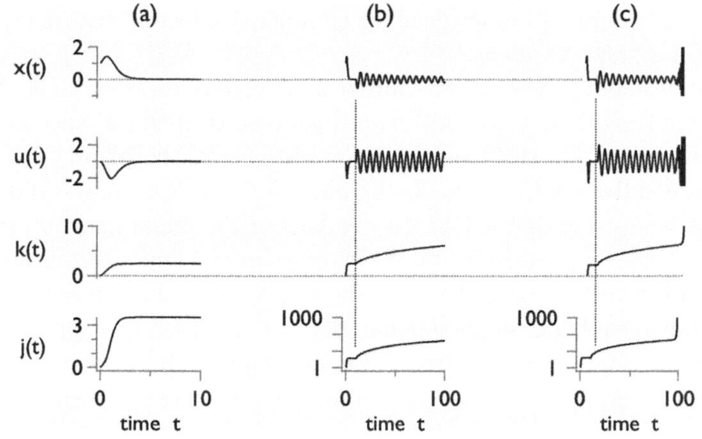

Fig. 10.1

Adaptive control of an unstable, first-order system. Plots of state x, control u, gain k, and cumulated cost j. (a) Adaptation stabilizes initial condition $x_0 = 1$, growth rate $a = 1$. (b) Disturbance $\sin t$ starts at $t = 10$. (c) Delay $\tau = 0.265$ included.

with $a > 0$. If a is known, then any $k > a$ will stabilize the system. Given an integrated cost function $j(t) = \int_0^t dt' \, (x^2 + u^2)$ that balances performance against control effort, we can use optimal-control methods to select the gain $k_{\text{opt}} = \sqrt{a^2+1} + a$ that minimizes $j(t)$ for $t \to \infty$ (see Problem 7.2).

If a is not known but is drawn from a known distribution, then we can use robust control, partitioning the set of possible a values into a finite interval of normal cases ("good values") and a semi-infinite interval of rare disasters ("bad values"). We can then optimize by assigning a large but finite cost to events in the bad subset (Problem 9.8).

To try to do better, we can monitor performance and adapt the control law to take into account the behavior of the specific system. In Eq. (10.1), let the gain depend on time:

$$\dot{k} = \gamma x^2 \,, \qquad k(0) = 0 \,, \tag{10.2}$$

with $\gamma > 0$ the *learning rate*. Intuitively, k increases until $x = 0$, no matter what the value of a. As Figure 10.1a shows, this approach seems to solve our problem.

Now let us add an external disturbance, $\sin t$, starting at $t = 10$. Nominally, the adaptive scheme seems satisfactory. But because $x(t)$ is almost always nonzero, the gain $k(t)$ and integrated cost $j(t)$ steadily rise; see Figure 10.1b. The real problem of the scheme becomes apparent in Figure 10.1c, where we add the effect of an unmodeled delay. That is, the actual relation between control and measured state is $u(t) = -k(t)x(t - \tau)$, even though the nominal model assumes $u(t) = -k(t)x(t)$. As we saw in Example 3.2, a first-order system with delay becomes unstable when the gain is too big. Here, the gain keeps increasing, until there is a sharp instability near $t = 100$. The time to instability depends sensitively on the delay, but *any* $\tau > 0$ eventually leads to instability. Indeed, we havė created a kind of bomb: once the cumulative effect of disturbances reaches a threshold, the system "explodes."

A similar instability mechanism caused a notorious disaster. The X-15 rocket-powered aircraft used adaptive feedforward gains to control roll and pitch. In a test flight on November 15, 1967, over the Nevada desert, the plane went into a spin, causing the adaptive control to make the aircraft "seem squirrelly" and go unstable, beginning a limit cycle that led to the breakup of the plane and the death of pilot Michael Adams. The crash ended the X-15 program and limited applications of adaptive control for nearly two decades, until the subject was on a firmer theoretical basis, with algorithms that better guarantee stability.

The above example is a truly pernicious instability. The system seems to behave normally until it explodes, diverging to infinity at a moment that depends on the number and size of perturbations and differs in every realization. But what is perhaps most remarkable is its sensitivity to "information structure." When a is known, the problem is simple and linear. But if a follows an unbounded probability distribution – even a Gaussian – the problem becomes a nonlinear one whose full solution remains unknown, to this day.

Outline

Cautioned by the above example, we outline some different approaches to adaptive control.

- Section 10.1 presents *direct methods*, which update *control parameters* to improve performance. For *model reference adaptive control* (MRAC), the goal is to make the response to a command follow that of a desired reference dynamical system. The difference between actual and desired response can serve as an error signal for parameter feedback, leading to adaptation laws (see above example). The motion can be unstable.
- Section 10.2 introduces *indirect methods*, which first estimate *system parameters* and then use these estimates to update the control parameters. We reformulate the least-squares fitting techniques of Chapter 6 into *recursive* algorithms that update parameter estimates and controller at each time step. With correct system models, prediction errors for measurements, or *innovations*, look like white noise. Modeling errors create dynamical correlations, which serve as an error signal for adjusting the controller dynamics. Unfortunately, the "natural dynamics" of the feedback system may not generate enough information to make accurate parameter estimates.
- Section 10.3 outlines *adaptive inverse control*, an approach based on *adaptive filtering*, a signal-processing technique.[1] The approach replaces the "inner" feedback loop of direct and indirect methods with feedforward control, leaving the "outer" adaptive feedback loop. The inherently stable feedforward has the structure of a Finite Impulse Response (FIR) filter, a structure that can approximate arbitrarily well the Infinite Impulse Response (IIR) filters created by linear feedback. The

[1] Historically, signal processing has been a distinct discipline from control engineering.

approach is favored on systems with fast dynamics, where high-bandwidth feedback would be difficult or expensive. But tuning the FIR coefficients adaptively can still lead to instability.

- Section 10.4 considers adaptation as a problem in optimal control, an approach also known variously as *dual control* or *reinforcement learning*. In particular, we show optimizing over a long time frame requires trading off the immediate cost of nonoptimal *exploration* and *learning* for the long-term benefit gained by *exploiting* better knowledge. The core insight is that it pays to actively *probe* the environment rather than passively accept the signals that "come naturally." These methods can resolve the issues of direct and indirect methods but are usually too hard to compute.
- Section 10.5 introduces *neural networks*, which are based on the premise that a high-dimensional nonlinear dynamical system can approximate arbitrary dynamics. We show how to train the network, off-line in a batch mode or online in an adaptive mode. When they work, these *data-driven* methods come close to a control engineer's dream: state your desires, turn on the controller, and let it figure out what to do.

10.1 Direct Methods

Adaptive control began in the 1950s in the aeronautics industry, where autopilots needed to work for varying values of speed, altitude, and wind. The basic problem is that the optimal parameter values for a controller (for example, a PID controller) may vary greatly as the set point and disturbances change. Such a scenario holds, for example, in regulating a nonlinear system: linearizing about the set point gives different dynamics – requiring different controllers – at different set points.

Gain scheduling is a simple solution: choose the gain as a function of the set point to achieve good control in the neighborhood of that particular point in state space. As a feedforward method, gain scheduling can quickly adjust to operating conditions. But it lacks robustness if the system itself changes. We treat gain scheduling in Chapter 11. Here, we focus on more truly adaptive techniques.

10.1.1 Model Reference Adaptive Control (MRAC)

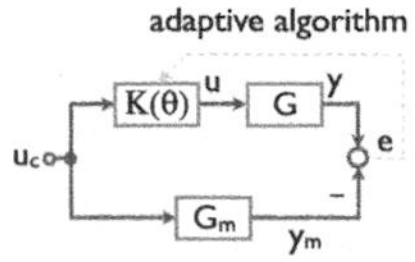

The goal of model reference adaptive control is to find control parameters that make a physical system behave similarly to a desired model system that serves as a reference. The sketch at right illustrates the basic setup: a command signal u_c is sent to a controller K that acts on a system G and produces an output y. Here, the control is feedforward, but feedback is also used. The output $y(t)$ should equal the output y_m produced by a fictitious model system G_m. In an adaptive scheme, we use the difference between the observed and desired outputs to adjust a control parameter θ to make $y \rightarrow y_m$.

The MIT Rule

The MIT rule was developed for the aerospace industry in the late 1950s and early 1960s at the Massachusetts Institute of Technology, in its Instrumentation Lab (now the Draper Laboratory), home to much of the early development of adaptive control. The idea is to reduce a cost function $J(\theta)$ by changing a control parameter θ. Let

$$\dot{\theta} = -\gamma J'(\theta) \quad \Longrightarrow \quad \dot{J}(\theta) = J'(\theta)\dot{\theta} = -\gamma J'(\theta)^2 \le 0\,, \tag{10.3}$$

where $\gamma > 0$ is the *learning rate* and $J'(\theta) = \mathrm{d}J/\mathrm{d}\theta$, and where $\dot{J} = 0$ at a (local) optimum value of θ. The idea generalizes immediately to multiple parameters $\boldsymbol{\theta}$ by taking $\dot{\boldsymbol{\theta}} = -\gamma\nabla J$: "walk downhill" along the $J(\boldsymbol{\theta})$ surface.

In the MRAC formalism with system $y(t,\theta)$ and model reference output $y_{\mathrm{m}}(t)$, the cost function $J(\theta)$ is typically a running cost such as the square of an error term, $e(t,\theta) = y(t,\theta) - y_{\mathrm{m}}(t)$, and not an integral. Then Eq. (10.3) becomes

$$J = \tfrac{1}{2}e^2(\theta) \quad \Longrightarrow \quad \dot{\theta} = -\gamma J'(\theta) = -\gamma\, e(\theta)\, \partial_\theta y\,. \tag{10.4}$$

Equation (10.4) is a straightforward rule for adjusting a control parameter. For a first-order problem, it leads to the adaptation law presented in the preliminary example to this chapter (Problem 10.3). But first, we consider even simpler cases.

Example 10.1 (Adaptive control for unknown gain) Consider a linear system with transfer function $G(s) = k\,G_0(s)$, with $G_0(s)$ known and k unknown. We wish to make this system follow the model system $G_{\mathrm{m}}(s) = k_{\mathrm{m}}\, G_0(s)$, with k_{m} specified. Multiplying the *command input* $u_{\mathrm{c}}(t)$ by a feedforward gain θ to create a *control input* $u(t) \equiv \theta\, u_{\mathrm{c}}(t)$ would solve our problem, if we knew k and could thereby set $\theta = \theta^* \equiv k_{\mathrm{m}}/k$. However, using the MIT rule, we can make the feedforward gain $\theta(t)$ approach θ^*. The controller K is simply $K(\theta) = \theta$. Since $y = Gu = (kG_0)(\theta u_{\mathrm{c}})$ and the output error $e(\theta) = y(\theta) - y_{\mathrm{m}}$, the MIT rule gives adaptive MRAC dynamics

$$\dot{\theta} = -\gamma e(\theta)\, y'(\theta) = -\gamma e(\theta)\, kG_0 u_{\mathrm{c}}\,. \tag{10.5}$$

For simplicity, let $G_0(s) = 1$ be a static transformation. Then

$$\begin{aligned} e &= y - y_{\mathrm{m}} = ku - k_{\mathrm{m}}u_{\mathrm{c}} = (k\theta - k_{\mathrm{m}})u_{\mathrm{c}} = k(\theta - \theta^*)u_{\mathrm{c}}\,, \\ \dot{\theta} &= -\gamma k^2 u_{\mathrm{c}}^2(\theta - \theta^*)\,. \end{aligned} \tag{10.6}$$

Finally, integrating the $\theta(t)$ equation gives

$$\theta(t) = \theta^* + [\theta(0) - \theta^*]\,\mathrm{e}^{-\gamma k^2 I(t)}\,, \qquad I(t) = \int_0^t \mathrm{d}t'\, u_{\mathrm{c}}^2(t')\,. \tag{10.7}$$

The long-time ($t \to \infty$) behavior of the feedforward gain $\theta(t)$ in Eq. (10.7) is very interesting. There are two main cases:

- $I(t) \to \infty$ (e.g., $u_{\mathrm{c}}(t) = t^{-\alpha}$ with $0 < \alpha \le \frac{1}{2}$). Then $\theta(t) \to \theta^*$ and, from Eq. (10.6), the output $y(t) \to y_{\mathrm{m}}(t)$.

- $I(t) \to I_0 < \infty$ (e.g., $u_c(t) = t^{-\alpha}$ with $\alpha > \frac{1}{2}$). Then $\theta(t) \not\to \theta^*$. Even so, Eq. (10.6) implies that the vanishing of u_c at long times makes $e \to 0$ and the output $y(t) \to y_m(t)$.

In other words, the output approaches that of the desired model even when the control-parameter estimate (and, implicitly, the estimate of the system gain k) is not the "correct" one. Good adaptive control does not require correct control parameters.

As Example 10.1 illustrates, the choice of control signal $u_c(t)$ determines whether control parameters converge to ideal values. In particular, the excitation needs to be *persistent* to make $\theta \to \theta^*$. The origin of the name is obvious in the above example. More generally, the signal needs to have a rich-enough frequency content. Loosely, for a "simple" command signal, there may be many sets of parameter values that give good control, whereas for a "complex" command signal, only the correct parameter values work.

MRAC Stability (Local Rules)

"Walk downhill" (Eq. 10.3) is a local adaptation rule that can work for convex cost functions with a unique minimum. It can also work for cost functions that have local minima, if the initial guesses are close enough (small values of the relative parameter uncertainty ε). However, the robust control of Chapter 9 also works well when relative parameter uncertainties are small. Adaptive control is of greater interest for *large* uncertainties; yet this is precisely the regime where a local rule such as MRAC can fail.

Indeed, problems with MRAC arise for a slightly more complicated system whose relative degree (order of denominator minus order of numerator) is not one, but two. For the underdamped oscillator $G_0(s) = (s^2 + s + 1)^{-1}$, the MRAC system dynamics are

$$\mathcal{L} y_m = k_m u_c , \qquad \mathcal{L} y = k u , \qquad \dot{\theta} = -\gamma y_m (y - y_m) , \tag{10.8}$$

with $\mathcal{L} = (\mathrm{d}_{tt} + \mathrm{d}_t + 1)$ and $u = \theta u_c$ and with y_m, y, u_c, u, and θ all functions of time. Note that the learning rate γ (chosen empirically) has absorbed a factor of $1/\theta^*$.

Figure 10.2 shows that for small learning rates ($\gamma = 0.1$) the system response stably adapts to follow the reference model, while for large gains ($\gamma = 1.1$) the response is unstable. Holding $\gamma = 0.1$ and increasing the square-wave amplitude to $u_0 > \sqrt{10} \approx 3.16$ also leads to an instability. A linear stability analysis of stationary solutions (Problem 10.1) shows that the response is stable when $0 < \gamma k k_m \left(u_c^0\right)^2 < 1$. Instability can thus result when either the learning rate γ or the operating level u_c^0 is too large. As we saw in our initial example, a "stable" system can all of a sudden become unstable if an unusual operating condition is encountered.

MRAC Stability (Lyapunov)

There are several strategies to reduce the tendency for adaptive systems to become unstable. Problem 10.2 alters the basic MIT rule of Eq. (10.5) by using the signal level

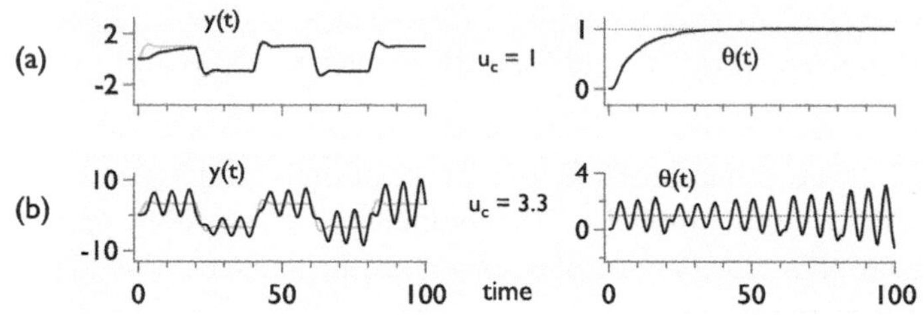

Fig. 10.2 Adaptation of feedforward gain for an underdamped oscillator with $k = k_{\mathrm{m}} = 1$ and learning rate $\gamma = 0.1$. (a) For $u_{\mathrm{c}} = 1$, response is stable (left) and the feedforward gain θ converges to the correct value 1 (right). (b) Unstable response for $u_{\mathrm{c}} = 3.3$. Dark line denotes system response, gray line the model.

to "normalize" the learning rate. The fix does not work for all input signals u_{c} but helps when u_{c} is nearly constant, with occasional abrupt variations.

Another approach is to use Lyapunov stability theory (Section 2.5.2). The trick is to find a Lyapunov function for the closed-loop system, which then will be guaranteed stable. For example, consider MRAC for a system with known matrices $\boldsymbol{A}$ and $\boldsymbol{B}$ but unknown input coupling k.[2] Let the reference model be the same, but with input gain k_{m}:

$$\begin{aligned} \dot{\boldsymbol{x}} &= \boldsymbol{A}\boldsymbol{x} + \boldsymbol{B}ku\,, & y &= \boldsymbol{C}\boldsymbol{x}, \\ \dot{\boldsymbol{x}}_{\mathrm{m}} &= \boldsymbol{A}\boldsymbol{x}_{\mathrm{m}} + \boldsymbol{B}k_{\mathrm{m}}u_{\mathrm{c}}\,, & y_{\mathrm{m}} &= \boldsymbol{C}\boldsymbol{x}_{\mathrm{m}}\,. \end{aligned} \tag{10.9}$$

Defining errors $\boldsymbol{e}_x = \boldsymbol{x} - \boldsymbol{x}_{\mathrm{m}}$ and $e_y = y - y_{\mathrm{m}}$, we have

$$\dot{\boldsymbol{e}}_x = \boldsymbol{A}\boldsymbol{e}_x + k\boldsymbol{B}(\theta - \theta^*)u_{\mathrm{c}}\,, \qquad e_y = \boldsymbol{C}\boldsymbol{e}_x\,, \tag{10.10}$$

where $\theta^* = k_{\mathrm{m}}/k$ and and $u = \theta\, u_{\mathrm{c}}$. A candidate Lyapunov function V is

$$V = \frac{1}{2}\left[\left(\boldsymbol{e}_x^{\mathsf{T}}\boldsymbol{P}\boldsymbol{e}_x\right) + \frac{k}{\gamma}(\theta - \theta^*)^2\right], \tag{10.11}$$

with $\boldsymbol{P}$ a symmetric, positive-definite matrix. The function V is bounded below by 0, which is achieved for zero state error ($\boldsymbol{x} = \boldsymbol{x}_{\mathrm{m}}$) and proper feedforward gain ($\theta = \theta^*$). We then need to make sure that $V(t)$ is nonincreasing. Differentiating,

$$\dot{V} = \frac{1}{2}\left[\left(\dot{\boldsymbol{e}}_x^{\mathsf{T}}\boldsymbol{P}\boldsymbol{e}_x + \boldsymbol{e}_x^{\mathsf{T}}\boldsymbol{P}\dot{\boldsymbol{e}}_x\right)\right] + \frac{k}{\gamma}(\theta - \theta^*)\dot{\theta}\,. \tag{10.12}$$

From Problem 2.14, a linear system $\dot{\boldsymbol{e}} = \boldsymbol{A}\boldsymbol{e}$ has a Lyapunov function $\frac{1}{2}\left(\boldsymbol{e}^{\mathsf{T}}\boldsymbol{P}\boldsymbol{e}\right)$ and is globally stable about $\boldsymbol{e} = \boldsymbol{0}$ if there are symmetric, positive-definite matrices $\boldsymbol{P}$ and $\boldsymbol{Q}$ satisfying the Lyapunov equation, $\boldsymbol{A}^{\mathsf{T}}\boldsymbol{P} + \boldsymbol{P}\boldsymbol{A} = -\boldsymbol{Q}$. From Eq. (10.10) for $\dot{\boldsymbol{e}}_x$, we have

$$\dot{V} = -\frac{1}{2}\left(\boldsymbol{e}_x^{\mathsf{T}}\boldsymbol{Q}\boldsymbol{e}_x\right) + \frac{k}{\gamma}(\theta - \theta^*)\left(\dot{\theta} + \frac{\gamma}{2}2\boldsymbol{B}^{\mathsf{T}}\boldsymbol{P}\boldsymbol{e}_x\, u_{\mathrm{c}}\right). \tag{10.13}$$

[2] For example, the input $u(t)$ might be sent to an amplifier whose gain is not known.

Choosing $\dot{\theta} = -\gamma \boldsymbol{B}^{\mathsf{T}} \boldsymbol{P} \boldsymbol{e}_x u_c$ then implies that $\dot{V} = -\frac{1}{2}\left(\boldsymbol{e}_x^{\mathsf{T}} \boldsymbol{Q} \boldsymbol{e}_x\right)$, which is negative definite. In order to implement the feedback law for $\dot{\theta}$, we need to formulate it in terms of the observable error e_y, not the unobservable state error $\boldsymbol{e}_x$. If we can solve $\boldsymbol{B}^{\mathsf{T}} \boldsymbol{P} = \boldsymbol{C}$, then

$$\dot{\theta} = -\gamma \boldsymbol{C} \boldsymbol{e}_x u_c = -\gamma e_y u_c , \tag{10.14}$$

which differs only slightly from the MIT law of Eq. (10.8) where $\dot{\theta} = -\gamma e_y y_m$. The change $y_m \to u_c$, however, guarantees stability for all γ and u_0.

To summarize the above construction, we need both to find symmetric, positive-definite matrices $\boldsymbol{P}$ and $\boldsymbol{Q}$ that satisfy the Lyapunov equation and to solve $\boldsymbol{B}^{\mathsf{T}} \boldsymbol{P} = \boldsymbol{C}$. We can do this when $G_0(s)$ is a first-order system.

Example 10.2 (First-order system with unknown input gain) Let $G_0(s) = (1 + s)^{-1}$. The state-space matrices are scalar ($A = -1$, $B = C = 1$), and the error is

$$\dot{e}_x = -e_x + k(\theta - \theta^*) u_c . \tag{10.15}$$

With $P = 1$, we define $V = \frac{1}{2}[e_x^2 + (k/\gamma)(\theta - \theta^*)^2]$, and $\dot{V} = -e_x^2 + (k/\gamma)(\theta - \theta^*)(\dot{\theta} + \gamma u_c e_x)$. Letting $\dot{\theta} = -\gamma u_c e_x = -\gamma e_y u_c$ gives stable dynamics. Note $\boldsymbol{B}^{\mathsf{T}} \boldsymbol{P} = \boldsymbol{C}$ is satisfied trivially.

Unfortunately, the construction fails when $G_0(s)$ is higher order.

Example 10.3 (Second-order system) Let $G_0(s) = (1 + s + s^2)^{-1}$. Then $\boldsymbol{A} = \left(\begin{smallmatrix} 0 & 1 \\ -1 & -1 \end{smallmatrix}\right)$, $\boldsymbol{B} = \left(\begin{smallmatrix} 0 \\ 1 \end{smallmatrix}\right)$, $\boldsymbol{C} = (1\ 0)$. The equation $\boldsymbol{B}^{\mathsf{T}} \boldsymbol{P} = \boldsymbol{C}$ then implies $\boldsymbol{P} = \left(\begin{smallmatrix} p_1 & 1 \\ 1 & 0 \end{smallmatrix}\right)$, with eigenvalues $\lambda = \frac{1}{2}(p_1 \pm \sqrt{p_1^2 + 4})$. Since one of the eigenvalues λ is positive and the other negative, $\boldsymbol{P}$ cannot be positive definite and our construction fails. Simulating the system confirms that Eq. (10.14) continues to be unstable for large γ. Nevertheless, if the full state vector $(y\ \dot{y})^{\mathsf{T}}$ is observed, then we can find a stable adjustment law (Problem 10.4).

MRAC Stability (Unmodeled Dynamics)

As we saw in our preliminary example, adaptive laws can fail in the face of unmodeled dynamics. This can happen even if a Lyapunov function guarantees stability for the model system. In Example 10.2, we had $G_0 = \frac{1}{1+s}$ and a feedforward adaptation rule of $\dot{\theta} = -\gamma e_y u_c$. Now consider what happens if the actual system is

$$G_1(s) = \left(\frac{1}{1+s}\right)\left(\frac{1}{1+s/\alpha}\right), \tag{10.16}$$

where $\alpha \gg 1$: Although we assume that the system is first order, there is actually a neglected, high-frequency pole. In Problem 10.5, you will show that the adaptive "Lyapunov" controller is unstable when a sinusoidal command signal $u_c(t) = u_0 \cos \omega t$ has $\omega > \sqrt{\alpha}$. Even more disturbing, the instability occurs for all values of γ and u_0.[3]

[3] The instability occurs for all γ and u_0, but the growth rate depends on the values of those parameters.

That is, a nominally "safe" but noisy input (with high-frequency components) can provoke an instability.

Thus, finding a Lyapunov function for a closed-loop system does not guarantee stability if there are unmodeled dynamics. In practice, adaptation is often stable for small learning rates γ. Because many scenarios can affect the stability, you have to work out empirically the maximum "safe" learning rate for each case.

10.1.2 Extremum-Seeking Control

Since unmodeled dynamics can lead to poor performance and instability, it is interesting to consider techniques that are *model free*. Recall the MIT rule $\dot{\theta} = -\gamma J'(\theta)$ from Eq. (10.3), which implicitly assumes you know the form of a cost function $J(\theta)$. The approach of *extremum-seeking control* is to estimate the gradient of $J(\theta)$ by perturbing the system, adding a deliberately introduced sinusoid to θ (see left).

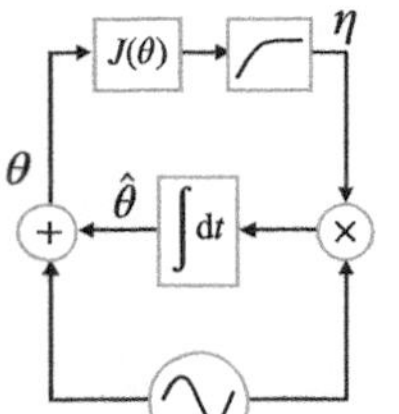

To see how the scheme works, assume that we have an estimate $\hat{\theta}$ that is not too far from the true optimal parameter θ_0 that minimizes $J(\theta)$. Taylor expanding about the minimum,

$$J(\theta) \approx J_0 + \tfrac{1}{2} J''(\theta_0)(\theta - \theta_0)^2 \,, \quad \Longrightarrow \quad J'(\hat{\theta}) \approx J''(\theta_0)(\hat{\theta} - \theta_0) \,, \tag{10.17}$$

and the MIT rule $\dot{\hat{\theta}} = -\gamma J'(\hat{\theta})$ leads to $\dot{\hat{\theta}} \sim -(\hat{\theta} - \theta_0)$, implying that $\hat{\theta} \to \theta_0$ at long times.

Previously, we assumed there was a model to evaluate $J'(\hat{\theta})$. In extremum-seeking control, we estimate the gradient from the dynamics by perturbing θ periodically. We start with an estimate $\hat{\theta}$ and then slowly modulate $\theta \to \hat{\theta} + a \cos \omega_0 t$ to see the quasistatic response.[4] The modulated $J(t)$ signal is then high-pass filtered to form a signal $\eta(t)$, with the DC frequency components of $J(t)$ removed. The cutoff frequency $\omega_h \ll \omega_0$, so that the $\cos \omega_0 t$ modulations are unaffected. The η signal is then *demodulated* by further multiplying by $a \cos \omega_0 t$ and integrated. The integration filters out all the modulated terms, leaving just a slowly varying signal, which becomes the estimate $\hat{\theta}$. More formally,

$$\begin{aligned} J(\theta) &= J(\hat{\theta} + a \cos \omega_0 t) && \text{modulate} \\ \dot{\eta} &= -\omega_h \eta + \dot{J} && \text{high pass} \\ \dot{\hat{\theta}} &= -k\,(a \cos \omega_0 t)\,\eta && \text{demodulate}\,. \end{aligned} \tag{10.18}$$

In Problem 10.6, we will see that these equations lead approximately to the MIT rule $\dot{\hat{\theta}} = -\gamma J'(\hat{\theta})$, with $\gamma = \tfrac{1}{2} k a^2 J''(\theta_0)$. A typical solution is shown at left, where J and $\hat{\theta}$ converge to the correct values (dotted lines). Note how the modulation (small rapid oscillations) vanishes near the extremum.

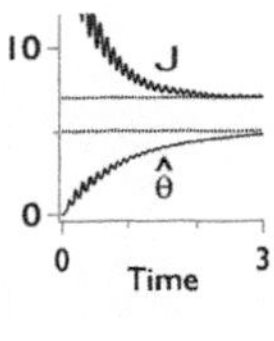

Problem 10.6 also shows that the solution can track slowly varying parameters, for example $\theta_0 \to \theta_0(t)$. The technique thus can maintain optimal conditions in the

[4] Here, *quasistatic* means that at each instant in time, $J \approx J(\theta)$, where $\theta = \theta(t)$. We note that $J(\theta)$ can be the time-dependent performance of a dynamical system governed by the parameter θ, which could be a feedback gain, for example. The modulation frequency ω_0 should be slower than any system time scales.

face of drifts and other slow disturbances. Applications include solar-cell and battery power management, where there is an optimal voltage-current combination to maintain that is a function of the sun intensity or user load. In physics, similar modulation techniques are used to measure derivative signals such as local conductance, dI/dV, using a scanning tunneling microscope. Problem 11.20 uses a similar idea to stabilize the frequency of a laser. The modulation-demodulation scheme is also similar to amplitude-modulation (AM) radio, and there are readily available circuits to implement those operations in hardware.

Extremum seeking readily extends to multiple parameters. The trick is to modulate with one frequency per parameter. Since modulation at frequency ω_0 produces a response at all harmonics $n\omega_0$ (albeit with an amplitude that decreases with n), the modulation frequencies should be incommensurate: their ratios should approximate irrational numbers.

10.1.3 Robust, Adaptive Control

The control community has devoted much effort to improving adaptive algorithms to make them robust in the face of unmodeled dynamics. A simple strategy is to introduce a *dead zone* in the adaptive routine, so that the update law for $\theta(t)$ is applied only for large variations. For example, you can set $\dot{\theta} = 0$ for $|e| < e_0$, where e_0 is a threshold value. Of course, you need to figure out how to choose e_0 in a given setting.

More generally, combining the adaptive techniques of this chapter with the robust techniques of the previous remains a significant challenge. One route has been to (implicitly) use the partition principle from Section 9.3.3: One classifies dynamics into a "normal" set, where good control and stability may be assured, and the rest, the "disasters" that need to be planned for and, in general, dealt with through a separate process.

In any case, one focus of effort in the adaptive control community has been to develop algorithms that are stable when applied to a bounded set of dynamical systems. Unfortunately, the techniques are complicated, the proofs technical, and the set of allowable dynamics not as large as one might hope. For all these reasons, description of this kind of robust, adaptive control is beyond the scope of the book.

10.2 Indirect Methods

Indirect methods for adaptive control first estimate system parameters and then use those estimates to choose control parameters. We focus on algorithms for parameter estimation that are equivalent to the least-squares method for system identification discussed in Chapter 6 but use the recursive techniques of Chapter 8. A dynamical system of known form has unknown parameter values collected in a vector $\boldsymbol{\theta}$. We update the estimate $\hat{\boldsymbol{\theta}}$ at each time step and then use the estimate $\hat{\boldsymbol{\theta}}$ in an indirect control algorithm. This naive substitution of $\hat{\boldsymbol{\theta}}$ for $\boldsymbol{\theta}$ in an otherwise unchanged control algorithm is again the *certainty equivalence principle* from Chapters 4 and 8. We will explore here its limitations, too.

10.2.1 Least-Squares Parameter Estimation

Consider models that are linear in any unknown parameters but possibly nonlinear in the equations themselves. Following Appendix A.8.4 online, we write the dynamical equations as

$$y_k = \boldsymbol{\varphi}_k^{\mathsf{T}} \boldsymbol{\theta} + \xi_k \,, \qquad \text{or} \qquad \boldsymbol{Y} = \boldsymbol{\Phi}\,\boldsymbol{\theta} + \boldsymbol{\xi} \,, \tag{10.19}$$

where the y_k are quantities that do not involve unknown parameters, $\boldsymbol{\theta}$ is a vector of parameters, $\boldsymbol{\varphi}_k$ are proportionality constants of the unknown parameters, and ξ_k is Gaussian white noise of zero mean and unit variance. That is, $\langle \xi_k \rangle = 0$ and $\langle \xi_k \xi_\ell \rangle = \delta_{k\ell}$. To make the noise term have unit variance, we use the linearity of the equations to scale inputs and outputs by ξ, the square root of the physical variance.

Equation (10.19) holds for k from 1 to N. As suggested by the second version, we can collect all N equations into a single vector relation by defining $\boldsymbol{Y} = (y_1 \; y_2 \; \dots y_N)^{\mathsf{T}}$, the *design matrix* $\boldsymbol{\Phi} = (\boldsymbol{\varphi}_1^{\mathsf{T}} \; \boldsymbol{\varphi}_2^{\mathsf{T}} \; \dots \boldsymbol{\varphi}_N^{\mathsf{T}})^{\mathsf{T}}$, and $\boldsymbol{\xi} = (\xi_1 \; \xi_2 \; \dots \xi_N)^{\mathsf{T}}$.

Example 10.4 Consider the linear dynamical system $y_k = a y_{k-1} + b u_{k-1} + \xi_k$.

- If a and b are not known, set $y_k = y_k$, $\boldsymbol{\varphi}^{\mathsf{T}} = (y_{k-1} \; u_{k-1})$, and $\boldsymbol{\theta} = \binom{a}{b}$.
- If a is known but not b, then set $y_k = y_k - a y_{k-1}$, $\boldsymbol{\varphi}^{\mathsf{T}} = (u_{k-1})$, and $\boldsymbol{\theta} = (b)$.
- If $\langle \xi_k^2 \rangle = \xi^2$, then scale $u_k \to u_k/\xi$ and $y_k \to y_k/\xi$.

For a and b unknown, the design-matrix version of the equations is

$$\begin{pmatrix} y_1 \\ \vdots \\ y_N \end{pmatrix} = \begin{pmatrix} y_0 & u_0 \\ \vdots & \vdots \\ y_{N-1} & u_{N-1} \end{pmatrix} \begin{pmatrix} a \\ b \end{pmatrix} + \begin{pmatrix} \xi_1 \\ \vdots \\ \xi_N \end{pmatrix} . \tag{10.20}$$

If there is no prior information about the parameters, Bayes' Theorem implies that the best estimate $\hat{\boldsymbol{\theta}}$ maximizes the likelihood function $p(\boldsymbol{\theta}|\boldsymbol{Y},\boldsymbol{\Phi})$ (Appendix A.8.2 online). For Gaussian noise, $p \propto \exp\{[-\frac{1}{2}\chi^2(\boldsymbol{\theta})]\}$, and the χ^2 statistic serves as a cost function:

$$\chi^2(\boldsymbol{\theta}) = (\boldsymbol{Y} - \boldsymbol{\Phi}\,\boldsymbol{\theta})^{\mathsf{T}}(\boldsymbol{Y} - \boldsymbol{\Phi}\,\boldsymbol{\theta}) = \boldsymbol{E}^{\mathsf{T}}\boldsymbol{E} = \sum_{k=1}^{N} \varepsilon_k^2 \,, \tag{10.21}$$

where $\boldsymbol{E} = \boldsymbol{Y} - \boldsymbol{\Phi}\,\boldsymbol{\theta}$ and $\varepsilon_k = y_k - \boldsymbol{\varphi}_k^{\mathsf{T}}\boldsymbol{\theta}$. Then $\nabla_{\boldsymbol{\theta}}(\chi^2) = \boldsymbol{0}^{\mathsf{T}}$ gives the *normal equations* (least-squares estimate) for $\hat{\boldsymbol{\theta}}$:

$$\hat{\boldsymbol{\theta}} = (\boldsymbol{\Phi}^{\mathsf{T}}\boldsymbol{\Phi})^{-1}\boldsymbol{\Phi}^{\mathsf{T}}\boldsymbol{Y} \equiv \boldsymbol{P}\,\boldsymbol{\Phi}^{\mathsf{T}}\boldsymbol{Y} \,, \tag{10.22}$$

where $\boldsymbol{P} \equiv (\boldsymbol{\Phi}^{\mathsf{T}}\boldsymbol{\Phi})^{-1}$ is the $N_p \times N_p$ covariance matrix of the least-squares estimate $\hat{\boldsymbol{\theta}}$, which has N_p components when N_p parameters are estimated (Problem A.8.2 online).

Finally, we often do not know the noise strength ξ^2 (and thus cannot scale by it). In such a case $\chi^2 \to \chi^2/\xi^2$, and we can estimate ξ^2 from the data, assuming that the model $y_k = \boldsymbol{\varphi}_k^{\mathsf{T}}\boldsymbol{\theta} + \xi_k$ has the correct structure.[5] Recall from Eq. (A.221 online) that the $\langle \chi^2 \rangle = N - N_p$, where N is the number of time steps used for the estimate. Then

$$\hat{\xi}^2 = \frac{(\boldsymbol{Y} - \boldsymbol{\Phi}\,\hat{\boldsymbol{\theta}})^{\mathsf{T}}(\boldsymbol{Y} - \boldsymbol{\Phi}\,\hat{\boldsymbol{\theta}})}{(N - N_p)} = \frac{\boldsymbol{Y}^{\mathsf{T}}(\boldsymbol{Y} - \boldsymbol{\Phi}\hat{\boldsymbol{\theta}})}{(N - N_p)}\,, \tag{10.23}$$

using the *orthogonality relation* $\boldsymbol{\Phi}^{\mathsf{T}}(\boldsymbol{Y} - \boldsymbol{\Phi}\,\hat{\boldsymbol{\theta}}) = \boldsymbol{0}$, which follows from Eq. (10.22).

Example 10.5 (One-parameter LS) Write the normal equations for the single-parameter estimate of the scalar equation, $y_k = \theta u_k + \xi_k$. The noise model is $\langle \xi_k \rangle = 0$ and $\langle \xi_k\, \xi_\ell \rangle = \xi^2 \delta_{k\ell}$, with ξ^2 not known. Assume N measurements, with $k \in (1, N)$.

Unpacking the notation of Eq. (10.22), we have $\varphi_k = u_k$ and

$$\hat{\theta} = \frac{\sum y_k\, u_k}{\sum u_k^2}\,, \qquad \hat{\xi}^2 = \frac{\sum y_k^2 - \hat{\theta} \sum u_k y_k}{N - 1}\,, \qquad P = \frac{\hat{\xi}^2}{\sum u_k^2}\,. \tag{10.24}$$

If ξ^2 were known, the last equation would be $P = \frac{\xi^2}{\sum u_k^2} \to \frac{1}{\sum u_k^2}$ after scaling u_k and y_k by ξ. Problem 10.7 explores the further simplification that occurs when $u_k = 1$.

10.2.2 Recursive Least-Squares Parameter Estimation

Equation (10.22), involving N data points, must be reevaluated each time a new point $k \to k + 1$ becomes available. Fortunately, there is a *recursive* version of the least-squares algorithm (RLS), where only the current result at step k need be recorded in order to process the new information at step $k + 1$ and form an updated, step $k + 1$ estimate of the parameters. The number of calculations is reduced by a factor of N. Although we could derive the recursive form directly from the standard least-squares equations, it is actually a special case of the Kalman filter derived in Chapter 8.

To make contact with our discussion of the Kalman filter, we view the parameters $\boldsymbol{\theta}$ as the state vector $\boldsymbol{x}_k$ to be estimated. If the parameters are constant, then their dynamics are trivial ($\boldsymbol{A} = \boldsymbol{I}$) and the full system is

$$\boldsymbol{\theta}_{k+1} = \boldsymbol{\theta}_k\,, \qquad y_k = \boldsymbol{\varphi}_k^{\mathsf{T}}\boldsymbol{\theta}_k + \xi_k\,. \tag{10.25}$$

Comparing with the Kalman filter, Eq. (8.28), we see that RLS is a special case, with $\boldsymbol{A} \to \boldsymbol{I}$, $\boldsymbol{B} \to \boldsymbol{0}$, $\boldsymbol{Q}_v \to 0$, and $\boldsymbol{C} \to \boldsymbol{\varphi}_k^{\mathsf{T}}$. The analog of $\boldsymbol{C}$ is now time dependent; in Chapter 8, it was typically a constant. Fortunately, the Kalman-filter derivation carries through identically, substituting time-dependent matrices into the previous results. The Kalman recurrence relations Eq. (8.41) then simplify to become the RLS update equations,

[5] The caution is worth emphasizing: Fitting to a bad model, you cannot independently estimate parameter errors, as the deviations mix both statistical and systematic errors.

$$\boldsymbol{L}_{k+1} \equiv \frac{\boldsymbol{P}_k \boldsymbol{\varphi}_{k+1}}{1+\boldsymbol{\varphi}_{k+1}^{\mathsf{T}} \boldsymbol{P}_k \boldsymbol{\varphi}_{k+1}}\,, \quad \boldsymbol{P}_{k+1} = \left(\boldsymbol{I} - \boldsymbol{L}_{k+1}\boldsymbol{\varphi}_{k+1}^{\mathsf{T}}\right)\boldsymbol{P}_k\,, \quad \hat{\boldsymbol{\theta}}_{k+1} = \hat{\boldsymbol{\theta}}_k + \boldsymbol{L}_{k+1}\varepsilon_{k+1}\,, \tag{10.26}$$

where the *innovations* $\varepsilon_{k+1} = y_{k+1} - \hat{y}_{k+1} = y_{k+1} - \boldsymbol{\varphi}_{k+1}^{\mathsf{T}}\hat{\boldsymbol{\theta}}_k$. In specializing from Eq. (8.41) to Eq. (10.26), we let $\hat{\boldsymbol{x}}_{k+1}^{-} = \hat{\boldsymbol{x}}_k \equiv \hat{\boldsymbol{\theta}}_k$ and $\boldsymbol{P}_{k+1}^{-} = \boldsymbol{P}_k$. Also, $\boldsymbol{Q}_\xi = \xi^2 \to 1$ in the denominator for $\boldsymbol{L}_{k+1}$. Note that the inverse is a scalar only for a single-output system.

Finally, we can also estimate the overall noise variance recursively. Since Eq. (10.23) is proportional to the average of ε_k^2, we can adapt Eq. (8.27) to find (Problem 10.8)

$$\hat{\xi^2} = \frac{1}{N-N_p}\sum_{k=1}^{N}\varepsilon_k^2 \quad \to \quad \hat{\xi_k^2} = \left(\frac{k-N_p-1}{k-N_p}\right)\hat{\xi_{k-1}^2} + \left(\frac{1}{k-N_p}\right)\varepsilon_k^2\,, \tag{10.27}$$

where N is the total number of time steps in the batch algorithm and N_p is the number of parameters determined. Initialize the recurrence relation with $\hat{\xi}^2_{N_p+1} = \varepsilon_1^2 + \cdots + \varepsilon_{N_p+1}^2$.

Example 10.6 (One-parameter RLS) We return to $y_k = \theta u_k + \xi_k$ from Example 10.5. The recursive update equations are all scalar, since there is only one parameter to determine. Using $\varphi_k = u_k$ and scaling the noise variance to one, they are

$$\begin{aligned} L_{k+1} &= \frac{P_k u_{k+1}}{1+u_{k+1}^2 P_k}\,, \qquad P_{k+1} = (1-L_{k+1})P_k = \frac{P_k}{1+u_{k+1}^2 P_k}\,, \\ \hat{\theta}_{k+1} &= \hat{\theta}_k + L_{k+1}(y_{k+1} - \hat{\theta}_k u_k)\,. \end{aligned} \tag{10.28}$$

Although these equations seem to differ from Eq. (10.24). Problem 10.9 shows that they are equivalent – if the initial conditions, P_1 and $\hat{\theta}_1$, are chosen correctly. If the initial conditions are unknown, the equations are still equivalent asymptotically for large k.

Another way to derive the RLS equations is to discretize the continuous LS equations. In Problem 10.10, we thus consider the continuous-time version of ordinary least squares. As with the Kalman–Bucy filter from Section 8.1.2, the continous-time equations are simpler than their discrete-time counterparts and are thus worth knowing for that reason alone.

10.2.3 Pitfalls

Least-squares estimation works when its assumptions are respected. Some pitfalls:

- *Persistence.* To find the optimal parameter estimates $\hat{\boldsymbol{\theta}}$, the matrix $\boldsymbol{P}^{-1} = \boldsymbol{\Phi}^{\mathsf{T}}\boldsymbol{\Phi}$ must be invertible and thus not have a zero determinant. Since $\boldsymbol{P}$ is the parameter covariance matrix, a small determinant for $\boldsymbol{\Phi}^{\mathsf{T}}\boldsymbol{\Phi}$ is also problematic, as it implies large parameter errors. As $\boldsymbol{\Phi}$ depends on the input, we just need to properly choose the u_k signal.

- *Colored noise*. The derivation of least squares assumes that the stochastic terms are white noise. Correlations will bias the estimates. Correcting for noise with known correlations turns out to be straightforward. If correlation strengths are not known, we can set up an extended recursion relation to estimate the noise model and other parameters. The resultant *extended recursive least squares* usually, but not always, converges.

- *Time-dependent parameters*. We have assumed that the parameters we determine are fixed and unchanging. If parameters drift or jump occasionally, the least-squares algorithm needs minor modification.

Persistence

We have already encountered briefly the notion of *persistent excitation*, both in the context of MRAC (see Eq. 10.7) and RLS parameter identification. The intuitive idea is that the input signal u_k must be "rich" enough to fully probe the system dynamics. As Problems 10.9d and 10.11 show, the signal should not be zero for too long a time and should have enough frequencies (or DC levels) to match the number of parameters to infer. More technically, a persistent input u_k is one that makes the inverse covariance matrix $\boldsymbol{P}^{-1} - \sum \varphi_k \varphi_k^{\mathrm{T}}$ positive definite, and hence invertible. Recall that the elements of φ_k are determined by u_k.

Another common way for an input to lose persistence is in closed-loop control, where u_k is chosen to be a function of the output y_k. Let us first look at a quick example.

Example 10.7 (Persistence in a closed-loop system) Consider the discrete system $y_{k+1} = -ay_k + bu_k + \xi_k$, with $\langle \xi_k \, \xi_\ell \rangle = \delta_{k\ell}$, with the following feedback rules:

- $u_k = -\kappa y_k$. Then $y_{k+1} = -(a + b\kappa)y_k + \xi_k$. Clearly, we can estimate only the combination $a + b\kappa$. If a and b are unknown, we will not be able to identify them individually.

- $u_k = -\kappa y_{k-1}$. Then $y_{k+1} = -ay_k - b\kappa y_{k-1} + \xi_k$. Now a and b affect different terms and can be separately identified. See Problem 10.12.

- $u_k = -\kappa_k y_k$, where κ_k is chosen from $\{\kappa_1, \kappa_2\}$, with equal probability. Can a and b be identified individually? Again, see Problem 10.12.

Example 10.7 shows that feedback can create an unexpected degeneracy in the identification problem. But once understood, the problem is relatively easy to solve. Other solutions are to add a square wave to the feedback signal, to add a stochastic component, and so on. In fact, the degeneracy is often delicate and thus easy to break.

Colored Noise

If the noise has correlations between values at different times (*colored noise*), the least-squares estimate $\hat{\boldsymbol{\theta}}$ is biased. From Eq. (10.22), the bias $\hat{\boldsymbol{\theta}} - \boldsymbol{\theta}$ is

$$(\boldsymbol{\Phi}^{\mathrm{T}}\boldsymbol{\Phi})^{-1}\boldsymbol{\Phi}^{\mathrm{T}}\boldsymbol{Y} - (\boldsymbol{\Phi}^{\mathrm{T}}\boldsymbol{\Phi})^{-1}(\boldsymbol{\Phi}^{\mathrm{T}}\boldsymbol{\Phi})\boldsymbol{\theta} = (\boldsymbol{\Phi}^{\mathrm{T}}\boldsymbol{\Phi})^{-1}\boldsymbol{\Phi}^{\mathrm{T}}(\boldsymbol{Y} - \boldsymbol{\Phi}\boldsymbol{\theta}) = (\boldsymbol{\Phi}^{\mathrm{T}}\boldsymbol{\Phi})^{-1}(\boldsymbol{\Phi}^{\mathrm{T}}\boldsymbol{E}), \tag{10.29}$$

which is nonzero unless $\boldsymbol{\Phi}^{\mathrm{T}}\boldsymbol{E} = \sum_k \varphi_k^{\mathrm{T}}\xi_k = 0$. In Example 10.5, this bias is

$$\hat{\theta} - \theta = \frac{\sum y_k u_k}{\sum u_k^2} - \frac{\theta \sum u_k^2}{\sum u_k^2} = \frac{\sum u_k(y_k - u_k\theta)}{\sum u_k^2} = \frac{\sum u_k \xi_k}{\sum u_k^2}. \tag{10.30}$$

The bias is nonzero if the noise, ξ_k, correlates with the input, u_k. This can happen either because the noise is colored (ξ_k depends on past noise values) or because there is feedback (u_k depends on past observations, which are influenced by the past noise). Here, we explore a very simple example of the complications of colored noise, in a setting where we can calculate everything explicitly. See Problem A.8.2 online for more general considerations.

Example 10.8 (Colored noise and bias) Consider $x_k = \theta x_{k-1} + \xi_k + a\xi_{k-1}$, where the a term implies that the noise is colored. By multiplying the equation of motion by each of $\{\xi_k, \xi_{k-1}, x_k, x_{k-1}\}$, we can show that

$$\langle x^2 \rangle = \frac{1 + 2a\theta + a^2}{1 - \theta^2}, \qquad \langle x\, x_{-1} \rangle = \frac{(\theta + a)(1 + \theta a)}{1 - \theta^2}. \tag{10.31}$$

Equation (10.30) for $\hat{\theta} - \theta$ then gives, in the limit of infinite data ($N \to \infty$),

$$\hat{\theta} - \theta = \frac{\langle x\, x_{-1} \rangle - \theta \langle x^2 \rangle}{\langle x^2 \rangle} = \frac{a(\theta^2 - 1)}{1 + 2a\theta + a^2}. \tag{10.32}$$

Thus, $\hat{\theta}$ is biased unless $a = 0$ (white noise) or $\theta = \pm 1$. See Problem 10.13 for details.

If we know the structure of the colored noise, then compensating for it is straightforward. In Example 10.8, if we know a, we can measure $\hat{\theta}$ and invert Eq. (10.32) for $\theta(\hat{\theta}, a)$. However, there is an easier, more systematic method: The Z-transform of the equation of motion in Example 10.8 is of the form

$$A(z^{-1})x(z) = C(z^{-1})\xi, \tag{10.33}$$

with $A = 1 - \theta z^{-1}$ and $C = 1 + az^{-1}$. Because ξ_k is white noise, its Z-transform ξ is constant (and often scaled to one). We can then "divide through" by C and define the filtered signal $x^{\mathrm{f}}(z) = C^{-1}x(z)$. Then $Ax^{\mathrm{f}} = \xi$. The noise term is now white, and the time-domain equation can be solved by the usual least-squares (or RLS) methods.

Example 10.9 (Colored noise: known correlations) We again consider the one-dimensional system $x_k = \theta x_{k-1} + \xi_k + a\xi_{k-1}$, with a known. Then

$$C(z)x^{\mathrm{f}}(z) = x(z) \quad \Longrightarrow \quad x_k^{\mathrm{f}} + a x_{k-1}^{\mathrm{f}} = x_k. \tag{10.34}$$

RLS amounts to solving the equations $x_k^f = \theta x_{k-1}^f + \xi_k$ for $\hat{\theta}$. The numerical simulation (for $\theta = -a = 0.5$) shows that with the "raw" signal x_k, the estimate $\hat{\theta}$ goes to the wrong value (dashed line). But with the "filtered" signal x_k^f, it goes to the true value (dotted line).

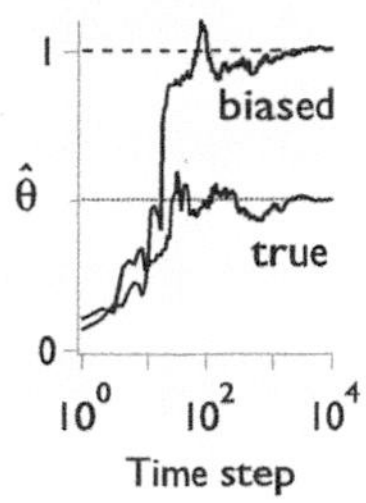

Problem 10.14 illustrates, for the case of a Brownian particle in a fluid, both how to measure the parameters of a system with colored noise adaptively and how to use those estimates to control the system adaptively, via the separation principle.

In Example 10.9, the division by $C(z^{-1})$ raises further issues (see Problem 10.15):

- A correlated noise source can be represented as the output of ideal white noise filtered by a stable dynamical system.
- Even if C has a zero *outside* the unit circle (meaning that its inverse has a pole outside the circle and thus corresponds to unstable dynamics), we can replace the unstable zero at a by a stable one at a^{-1} and keep the power spectrum the same.

We can generalize the "filter trick" to cases with explicit input signals u_k – we just filter it, too – but we run into problems if we do not know beforehand the noise correlations. In the above example, if we do not know a, we cannot implement the filtering. We can deal with this situation by adding a to our list of parameters to estimate. There is an important hitch: the coefficients of the usual terms in θ are quantities (y_k, u_k, etc.) that we measure directly, but the coefficient of a is ξ_{k-1}, which we do not measure directly. Nonetheless, we can estimate it using the $\hat{\theta}_k$. This is the *extended RLS* algorithm.

Example 10.10 (Colored noise: unknown correlations) Again we consider $x_k = \theta x_{k-1} + \xi_k + a\xi_{k-1}$, with unknown a and θ. Let us rewrite the equation as

$$\xi_k = x_k - \begin{pmatrix} x_{k-1} & \xi_{k-1} \end{pmatrix} \begin{pmatrix} \theta \\ a \end{pmatrix} . \qquad (10.35)$$

If we knew θ and a, we could estimate ξ_k at each step in terms of the previous noise estimate ξ_{k-1}, known recursively. In extended RLS, we estimate θ, a, and ξ_k. See Problem 10.13c.

In summary, to deal with colored noise, remember these points:

- Unaccounted-for noise correlations can lead to large biases for inferred parameters.
- The size of the covariance matrix $\boldsymbol{P}$ is not a good indicator of parameter error if there is bias. The "error bars" for $\boldsymbol{\theta}$ generated by $\boldsymbol{P}$ will be too small.
- To test for problems with bias, check whether the innovations $\varepsilon_k = y_k - \varphi_k^{\mathsf{T}} \hat{\theta}_k$ are white (e.g., by testing whether $\sum_k (\varepsilon_k \, \varepsilon_{k+m})$ is statistically equal to zero for $m \neq 0$).

Time-Dependent Parameters

The least-squares parameter-estimation algorithms that we have discussed so far, in both batch and recursive versions, tacitly assume that the estimated parameters are unknown but constant. However, adaptive control should also be able to deal with changing parameters. How they change – whether they drift or jump – leads to different strategies.

For slowly drifting parameter values, a good strategy is to limit the least-squares analysis to effectively a finite number of time steps. The χ^2 statistic then becomes,

$$\chi^2(\boldsymbol{\theta}) = \sum_{i=1}^{k} \lambda^{k-i}(y_i - \boldsymbol{\varphi}_i^{\mathsf{T}}\boldsymbol{\theta})^2\,, \tag{10.36}$$

where the *forgetting factor* $\lambda \in (0, 1)$ limits the sum to approximately the last $\frac{1}{1-\lambda}$ terms. From $\partial_{\boldsymbol{\theta}}\chi^2$, we find the normal equations for $\hat{\boldsymbol{\theta}}_k$,

$$\hat{\boldsymbol{\theta}}_k = \left(\sum_{i=1}^{k} \lambda^{k-i}\boldsymbol{\varphi}_i\boldsymbol{\varphi}_i^{\mathsf{T}}\right)^{-1}\left(\sum_{i=1}^{k} \lambda^{k-i}\boldsymbol{\varphi}_i\, y_i\right). \tag{10.37}$$

These differ from the constant-parameter case only in that the sum is dominated by the most recent terms instead of counting them all equally. We can derive a recursive version of the update equations by adapting the Kalman filter equations or by starting from the continuous-time equations (Problem 10.10):

$$\boldsymbol{L}_{k+1} \equiv \frac{\boldsymbol{P}_k\boldsymbol{\varphi}_{k+1}}{\lambda + \boldsymbol{\varphi}_{k+1}^{\mathsf{T}}\boldsymbol{P}_k\boldsymbol{\varphi}_{k+1}}, \quad \boldsymbol{P}_{k+1} = \left(\frac{1}{\lambda}\right)\left(\boldsymbol{I} - \boldsymbol{L}_{k+1}\boldsymbol{\varphi}_{k+1}^{\mathsf{T}}\right)\boldsymbol{P}_k, \quad \hat{\boldsymbol{\theta}}_{k+1} = \hat{\boldsymbol{\theta}}_k + \boldsymbol{L}_{k+1}\varepsilon_{k+1}. \tag{10.38}$$

We recover Eq. (10.26) when $\lambda \to 1$.

Example 10.11 (One-parameter LS with time-varying parameters) For $y_k = \theta u_k + \xi_k$, the normal equations are $\hat{\theta}_k = (\sum_i \lambda^{k-i} y_i u_i)/(\sum_i \lambda^{k-i} u_i^2)$. If $u_k = 1$, the estimate $\hat{\theta}$ can be written recursively as a simple *running average*,

$$\hat{\theta}_k = \lambda\hat{\theta}_{k-1} + (1 - \lambda)y_k\,, \tag{10.39}$$

which confirms that the forgetting factor limits the number of terms to $\approx \frac{1}{1-\lambda}$, with recent information counting more than older information (Problem 10.7).

More generally, parameters vary stochastically, and we can use a full Kalman filter to follow their "state." Recall that the LS update equations are a special form of the full Kalman filter. The static-parameter equation $\boldsymbol{\theta}_{k+1} = \boldsymbol{\theta}_k$ can be generalized by adding dynamics, $\boldsymbol{\theta}_{k+1} = \boldsymbol{A}\,\boldsymbol{\theta}_k + \boldsymbol{B}\boldsymbol{\nu}_k$. If parameters are driven by filtered white noise, then $\boldsymbol{A}$ sets the time scales of variation, and the covariance $\boldsymbol{\nu}\boldsymbol{\nu}^{\mathsf{T}}$ sets the associated random variation.

The full Kalman filter has one advantage over the simpler forgetting-factor algorithm: In the absence of a persistent input, the variance $\boldsymbol{P}$ in Eq. (10.38) diverges as λ^{-k}, or $e^{\lambda' t}$ for the equivalent continuous case (Problem 10.10c). The large variance can lead to large swings in the estimate $\hat{\theta}$. In the Kalman-filter formulation, such variations are prevented by fixing the relative variances of the θ dynamics and the observation noise. Alternatively, in the forgetting-factor algorithm, keep the updated variance below some maximum level.

In addition to smooth parameter variations, there may be occasional parameter jumps. For example, you control a device where a sample is changed from time to time. To deal with such a situation, monitor the innovations and reset the RLS algorithm whenever a large deviation in the innovations is detected. (You can filter the innovation signal over a short time scale to avoid confusing noise with actual system alterations.) Of course, you can use a forgetting factor *and* reset, if you have both parameter drift and jumps.

10.2.4 Innovations Approach

The innovations approach to adaptive control is based on the state-estimation formalism developed in Chapter 8. Let us first recall that formalism and the Kalman filter. For simplicity, we will restrict ourselves to a one-dimensional state space, with

$$x_{k+1} = Ax_k + \nu_k\,, \qquad y_k = x_k + \xi_k\,, \qquad \langle \nu_k\, \nu_\ell \rangle = \nu^2 \delta_{k\ell}\,, \qquad \langle \xi_k\, \xi_\ell \rangle = \xi^2 \delta_{k\ell}\,. \tag{10.40}$$

As before, the prediction $\hat{x}_{k+1}^-$ uses the observations $y^k \equiv \{y_k, y_{k-1}, \ldots, y_1\}$. By contrast, $\hat{x}_{k+1}$, includes the most recent observation y_{k+1}. We define the estimators by

$$\hat{x}_{k+1}^- = A\hat{x}_k\,, \qquad \hat{y}_k = \hat{x}_k\,, \qquad \hat{x}_{k+1} = \hat{x}_{k+1}^- + L\varepsilon_{k+1}\,, \qquad \varepsilon_{k+1} \equiv y_{k+1} - \hat{y}_{k+1}\,. \tag{10.41}$$

The *innovations*, or residuals ε_{k+1} are the difference between the predicted value of the measurement and the actual value and play a key role: We will show that when the observer gain L is chosen to minimize the mean-square estimation error $\left\langle (x_k - \hat{x}_k)^2 \right\rangle$ (the Kalman gain value), then the innovation sequence will be white, with $\langle \varepsilon_k\, \varepsilon_\ell \rangle = 0$ for $k \neq \ell$.

Intuitively, innovations are white when the correct dynamical model is used to predict the observations, because any nonrandom pattern in predictions $\hat{y}_{k+1}$ relative to the observations y_{k+1} means that some deterministic information has not been included in the estimate. The statement is true for general linear systems driven by colored Gaussian noise and even for nonlinear stochastic systems whose observations have additive Gaussian noise. Conversely, when innovations are not white, correlations can serve as an error signal in an adaptive control loop that alters the parameter estimates to *whiten* the innovations.

We start by establishing that the innovations form a white-noise sequence when the optimal observer gain is used. Note that a mismatch in parameters between model and physical system will lead to choosing the wrong gain L. We restrict ourselves to linear estimators, so that $\hat{x}_{k+1}^-$ is a linear combination of the observations y^k. For a linear dynamical system, one can prove that the best estimator must be linear. Then, as a lemma, we show that

$$\langle e_{k+1}^-\, \varepsilon_\ell \rangle = 0\,, \quad (1 \leq \ell \leq k)\,, \qquad e_{k+1}^- = x_{k+1} - \hat{x}_{k+1}^-\,, \tag{10.42}$$

where e_{k+1}^- is the state estimation error of the prediction.

We can interpret $\langle e_{k+1}^{-}\, \varepsilon_\ell \rangle = 0$ as a statement of orthogonality in a vector space. We first define a k-dimensional space $\mathrm{Sp}\{y_\ell\}$, the span (set of linear combinations) of the k observations y_ℓ, for $\ell \in (1,k)$. We also define an inner product $y_i \cdot y_j$ as the covariance $\langle y_i\, y_j \rangle$. The state that we wish to predict, x_{k+1}, is part of a higher-dimensional space (typically infinite dimensional). Since the prediction $\hat{x}_{k+1}^{-}$ is a linear combination of the y_ℓ, it is an element of the subspace spanned by the y_ℓ, and the best estimate minimizes the mean-square error, $\|x_{k+1} - \hat{x}_{k+1}^{-}\|^2 = \|e_{k+1}^{-}\|^2$.

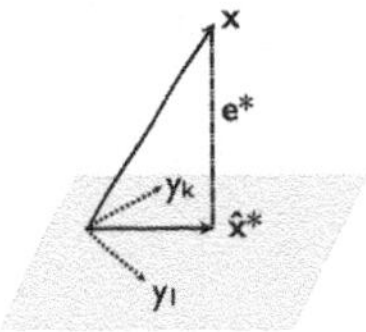

In the intuitive geometrical picture at left, $\hat{x}_{k+1} \equiv \hat{x}^*$ minimizes the distance $e_{k+1}^{-} \equiv e^*$ between $x_{k+1} \equiv x$ and $\hat{x}^*$ makes e^* orthogonal to $\mathrm{Sp}\{y_\ell\}$. The span of the $\{y_\ell\}$ is denoted by the darker plane in the figure. It is also clear that the minimum estimation error e^* is uniquely determined. More formally, consider a different estimate $\hat{x} \in \mathrm{Sp}\{y_\ell\}$. Then,

$$\|x - \hat{x}\|^2 = \|\underbrace{(x - \hat{x}^*)}_{e^*} - (\hat{x} - \hat{x}^*)\|^2 = \|x - \hat{x}^*\|^2 - \cancelto{0}{2e^* \cdot (\hat{x} - \hat{x}^*)} + \|\hat{x} - \hat{x}^*\|^2 \geq \|e^*\|^2 .$$

The middle term is zero by assumption, because e^* is orthogonal to *all* unbiased estimators $\hat{x}$. Orthogonality thus implies that the mean-square error is minimized by choosing $\hat{x} = \hat{x}^*$. Conversely, the estimate $\hat{x}_{k+1}^{-}$ that minimizes $\|e_{k+1}^{-}\|^2$ is orthogonal to each y_ℓ. Since the innovations ε_ℓ lie in $\mathrm{Sp}\{y_\ell\}$, we also have $\langle e_{k+1}^{-}\, \varepsilon_\ell \rangle = 0$ for $1 \leq \ell \leq k$, as claimed.

Returning to the innovations and our full notation, we have

$$\begin{aligned} \varepsilon_k &= y_k - x_k^{-} = x_k + \xi_k - x_k^{-} = e_k^{-} + \xi_k , \\ \langle \varepsilon_k\, \varepsilon_\ell \rangle &= \langle (e_k^{-} + \xi_k)\, \varepsilon_\ell \rangle = \langle e_k^{-}\, \varepsilon_\ell \rangle + \langle \xi_k\, \varepsilon_\ell \rangle . \end{aligned} \tag{10.43}$$

For $k > \ell$, the relation $\langle e_k^{-}\, \varepsilon_\ell \rangle = 0$ follows from the orthogonality lemma, and $\langle \xi_k\, \varepsilon_\ell \rangle = 0$ because noise in the future does not affect the present. Similarly, $\langle \varepsilon_k\, \varepsilon_\ell \rangle = 0$ for $k < \ell$.

For $k = \ell$, we have

$$\langle \varepsilon_k^2 \rangle = \langle (e_k^{-})^2 \rangle + 0 + 0 + \xi^2 \equiv P^{-} + \xi^2 , \tag{10.44}$$

where $P^{-} \equiv \left\langle (e_k^{-})^2 \right\rangle$ is the prediction variance, given by its Kalman value if the optimal observer gain is chosen and if dynamics are time invariant. If we put together all three cases, $\langle \varepsilon_k\, \varepsilon_\ell \rangle = (P^{-} + \xi^2)\, \delta_{k\ell}$ is a white-noise sequence, as claimed.

The above argument that the innovations are a white-noise sequence may seem abstract. We can make it more concrete by calculating explicitly the correlation $\langle \varepsilon_{k+1}\, \varepsilon_k \rangle$, allowing for an arbitrary observer gain L. First, we derive a recursion relation for e_{k+1}^{-}:

$$\begin{aligned} e_{k+1}^{-} &= x_{k+1} - \hat{x}_{k+1}^{-} \\ &= A x_k + \nu_k - A \hat{x}_k \\ &= A x_k + \nu_k - A[\hat{x}_k^{-} + L(e_k^{-} + \xi_k)] \\ &= A(1 - L) e_k^{-} - A L \xi_k + \nu_k . \end{aligned} \tag{10.45}$$

From Eq. (10.45), we see that $\langle e_{k+1}^{-}\, e_k^{-} \rangle = A(1-L)P^{-}$ and $\langle e_{k+1}^{-}\, \xi_k \rangle = -AL\xi^2$. Going back to the innovation correlations, we have $\langle \varepsilon_{k+1}\, \varepsilon_k \rangle = \langle e_{k+1}^{-}\, e_k^{-} \rangle + \langle e_{k+1}^{-}\, \xi_k \rangle$. Expanding, we have $A(1-L)P^- - AL\xi^2 = A(P^- - LP^- - L\xi^2) = A[P^- - L(P^- + \xi^2)] = A\left(\frac{P^-}{P^- + \xi^2} - L\right)(P^- + \xi^2)$. Thus, the correlations vanish for $L = L^* = P^-/(P^- + \xi^2)$. The value L^* is just the Kalman gain derived earlier and, indeed, the innovations approach provides an alternate derivation of the Kalman-filter equations. For $L \approx L^*$, we see that $\langle \varepsilon_{k+1}\, \varepsilon_k \rangle \propto (L^* - L)$. Thus, the correlations vary linearly, changing sign about the optimal value, and can serve as an error signal in a feedback loop that converges to the correct parameter values.

By iterating the recursion relation for e_{k+1} given in Eq. (10.45), we can show that the correlations $\langle \varepsilon_{k+\ell}\, \varepsilon_k \rangle$ for longer lags $\ell > 1$ are multiplied by the factor $A(1 - L)^{\ell-1}$. For stable observer dynamics, $|A(1 - L)| < 1$, and the correlations will decrease geometrically with ℓ. See Example 10.12 and its associated Problem 10.16.

Although our calculation is for a one-dimensional state space, the general case is similar and leads to the same qualitative conclusions: $\langle \boldsymbol{\varepsilon}_{k+\ell}\, \boldsymbol{\varepsilon}_k^{\mathsf{T}} \rangle$ vanishes when $\boldsymbol{L} = \boldsymbol{L}^*$ and decreases exponentially with ℓ when the wrong gain is chosen.

Example 10.12 (Estimate mobility of a trapped particle) Consider trapping a Brownian particle with unknown mobility μ_0 but known diffusion constant. If information acquired at step k is available for the output u_k, we have

$$x_{k+1} = x_k + \mu u_k + \nu_k\,, \qquad u_k = -\alpha \hat{x}_k\,, \quad \langle \nu_k\, \nu_\ell \rangle = \nu^2 \delta_{k\ell}\,, \tag{10.46}$$

where $\mu = \mu_0 T_s$ and T_s is the time interval between measurements. In our modeling, though, we use a different mobility $\mu' = \mu + \Delta\mu$. Neglecting observational noise ($\xi_k = 0$), we will see in Problem 10.16 that the innovation correlation function is

$$\langle \varepsilon_{k+\ell}\, \varepsilon_k \rangle = (\Delta\mu)\, \alpha \nu^2 (1 - \alpha\mu)^{\ell-1} + O(\Delta\mu^2)\,. \tag{10.47}$$

In Eq. (10.47), the correlation is linear in the mismatch $\Delta\mu$ and decreases geometrically with ℓ (solid curves at right top, for $\nu^2 = 1$ and $\alpha = 0.2$), as confirmed by simulation (markers). The plot at bottom right shows the standard deviation of lag-1 correlation measurements relative to expected value, as a function of the number of time steps N, for $\Delta\mu = 0.1$. Solid line is $\sim N^{-1/2}$.

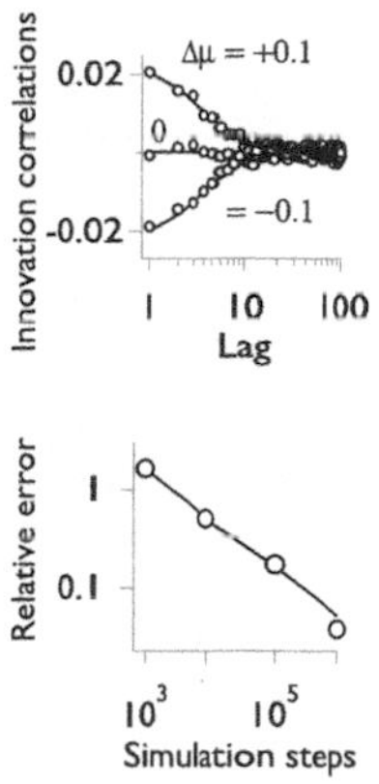

Example 10.12 shows that correlation-function matching works but is not very efficient. The plots required 10^6 data points and the calculation of multiple correlation lags (to extract all available information for the estimate).

We conclude by comparing recursive least squares (RLS) with correlation techniques:

- *RLS is efficient*: Recursive least-squares methods need less data, converge rapidly, and need relatively simple calculations. (Section 10.3 introduces RLS approximations that require even fewer calculations, at the cost of convergence speed.)
- *Correlation methods are robust*: Model mismatch generally leads to correlations, whereas a correct model leads to white noise. When least-squares methods are fooled (e.g., by an input signal that is not a persistent excitation), correlation methods may still work.

10.3 Adaptive Feedforward Control

Another approach to adaptive control focuses on finite-impulse-response (FIR) filters. From Section 5.2.1, we can always approximate a feedback system that has an infinite impulse response (IIR) by an equivalent FIR system. For long time constants, the FIR filter may require many terms (100 is common). Even so, FIR filters work well for some applications. In particular, because they require no measurements, they can be used when the update time T_s is so short that measuring and calculating a response is impractical. Also, FIR filters are inherently stable, with no possibility of a feedback-induced instability.

Both of these advantages are particularly helpful for adaptive control. First, the techniques introduced in Sections 10.1 and 10.2 require calculations that may take too long to compute for processes with short sampling times. Second, adapting FIR coefficients via a feedback loop creates only a single feedback loop, in contrast to the two loops that exist when adjusting IIR coefficients. Two loops increase the possibility of instability.

Below, in Section 10.3.1, we introduce adaptive FIR filters and apply them to system identification. Then, in Section 10.3.2, we apply adaptive FIR filters to two typical problems: feedforward control and disturbance estimation (noise cancellation).

10.3.1 Adaptive FIR Filters

We begin with the task of identifying, or estimating, the coefficients of an FIR filter given the input and the measured output time series. The output $y(kT_s) \equiv y_k$ at time kT_s is

$$y_k = \sum_{i=0}^{n} b_i\, u_{k-i} \equiv \boldsymbol{\varphi}_k^{\mathsf{T}} \boldsymbol{\theta}\,, \qquad \boldsymbol{\theta} \equiv \begin{pmatrix} b_0 \\ \vdots \\ b_n \end{pmatrix}, \qquad \boldsymbol{\varphi}_k \equiv \begin{pmatrix} u_k \\ \vdots \\ u_{k-n} \end{pmatrix}. \tag{10.48}$$

where we have grouped the parameters and inputs into vectors, as in Eq. (10.19).

Finding the coefficients $\boldsymbol{\theta}$ is just a special case of the general linear system considered in Section 10.2.1. Here, we seek a simple, computationally cheap approximation to the full least-squares solution. We use the prediction-observer form of Eq. (10.26),[6]

[6] The prediction observer updates the system state (parameter estimate) using y_k and not the y_{k+1} of the current observer. Since FIR filters often have very short sampling times, updates based on y_{k+1} are impractical.

$$\hat{\boldsymbol{\theta}}_{k+1} = \hat{\boldsymbol{\theta}}_k + \frac{\boldsymbol{P}_k \boldsymbol{\varphi}_k}{1 + \boldsymbol{\varphi}_k^{\mathsf{T}} \boldsymbol{P}_k \boldsymbol{\varphi}_k} \left(y_k - \boldsymbol{\varphi}_k^{\mathsf{T}} \hat{\boldsymbol{\theta}}_k \right) \approx \hat{\boldsymbol{\theta}}_k + \gamma\, \boldsymbol{\varphi}_k \left(y_k - \boldsymbol{\varphi}_k^{\mathsf{T}} \hat{\boldsymbol{\theta}}_k \right) . \tag{10.49}$$

In the second relation, we introduce the *Least Mean Square* (LMS) approximation, which replaces the matrix $\boldsymbol{P}_k/(1 + \boldsymbol{\varphi}_k^{\mathsf{T}} \boldsymbol{P}_k \boldsymbol{\varphi}_k)$ by the scalar γ.

To see why this drastic simplification is reasonable, let us derive LMS via another route. In the least-squares techniques of Section 10.2.1, we sought the $\boldsymbol{\theta}$ that minimized the sum of squared prediction errors, $\sum \varepsilon_k^2 = (y_k - \boldsymbol{\varphi}_k^{\mathsf{T}} \boldsymbol{\theta})^2$. For stationary statistics and sufficient data, we can approximate the sums by correlation functions. That is, we can alter our goal from minimizing a sum of squares to minimizing the mean-square error, $\langle \varepsilon^2 \rangle = \langle (y - \boldsymbol{\varphi}^{\mathsf{T}} \boldsymbol{\theta})^2 \rangle$, where we drop the k subscript by stationarity. Then we minimize $\langle \varepsilon^2 \rangle$ by *steepest descent*, following the gradient of $\langle \varepsilon^2 \rangle$ with respect to $\boldsymbol{\theta}$. The gradient is $\nabla_{\boldsymbol{\theta}} \langle \varepsilon^2 \rangle = -2 \langle \boldsymbol{\varphi}^{\mathsf{T}} \, (y - \boldsymbol{\varphi}^{\mathsf{T}} \boldsymbol{\theta}) \rangle$, and the steepest-descent update is given by

$$\hat{\boldsymbol{\theta}}_{k+1} = \hat{\boldsymbol{\theta}}_k - \frac{\gamma}{2} \left(\frac{\partial \langle \varepsilon^2 \rangle}{\partial \boldsymbol{\theta}} \right)^{\mathsf{T}} = \hat{\boldsymbol{\theta}}_k + \gamma \left\langle \boldsymbol{\varphi} \, (y - \boldsymbol{\varphi}^{\mathsf{T}} \hat{\boldsymbol{\theta}}_k) \right\rangle , \tag{10.50}$$

where the gradient is evaluated using the parameter estimate at time k and the factor of 2 is absorbed by convention into the learning rate γ. For estimates of the covariances $\langle \boldsymbol{\varphi}\, y \rangle$ and $\langle \boldsymbol{\varphi} \boldsymbol{\varphi}^{\mathsf{T}} \rangle$, the crude LMS approximation uses only the most recent measurement. That is, $\langle \boldsymbol{\varphi}\, y \rangle \approx \boldsymbol{\varphi}_k\, y_k$ and $\langle \boldsymbol{\varphi}^{\mathsf{T}} \boldsymbol{\varphi} \rangle \approx \boldsymbol{\varphi}_k^{\mathsf{T}} \boldsymbol{\varphi}_k$, which are just the expressions used in Eq. (10.49).

For n parameters, the LMS algorithm requires $O(n)$ calculations per update. Since the least-squares algorithm updates $\boldsymbol{P}$, it needs $O(n^2)$ calculations. FIR filters typically use large n, implying significant computational speedup.[7] We can also understand why the approximation is reasonable: although the variance of a single-point estimation of a mean value is large, the estimator is unbiased. Thus, although the correction to $\boldsymbol{\theta}_k$ will have a large error, it will, on average, point in the right direction, gradually decreasing the mean-square estimation error. For the fast processes where FIR filters are used, the trade-off between accuracy and speed of the LMS relative to the RLS algorithm can be favorable.

The LMS algorithm is unstable if γ is too large. Rewriting the LMS algorithm as

$$\hat{\boldsymbol{\theta}}_{k+1} = \left(\boldsymbol{I} - \gamma\, \boldsymbol{\varphi}_k \boldsymbol{\varphi}_k^{\mathsf{T}} \right) \hat{\boldsymbol{\theta}}_k + \gamma\, \boldsymbol{\varphi}_k\, y_k \tag{10.51}$$

shows there is an instability when an eigenvalue of $\boldsymbol{I} - \gamma\, \boldsymbol{\varphi}_k \boldsymbol{\varphi}_k^{\mathsf{T}}$ has magnitude greater than 1. But $\boldsymbol{\varphi}_k \boldsymbol{\varphi}_k^{\mathsf{T}}$ has only one nonzero eigenvalue,[8] the scalar $\boldsymbol{\varphi}_k^{\mathsf{T}} \boldsymbol{\varphi}_k$. Stability thus

[7] Other FIR filter architectures (e.g., lattice filters) lie between LMS and RLS for complexity and update speed.

[8] This is easy to prove. Consider the matrix $\boldsymbol{M} = \boldsymbol{\varphi}\boldsymbol{\varphi}^{\mathsf{T}}$, where we drop the k subscript to simplify notation. Then $M_{ij} = \varphi_i \varphi_j$, and the eigenvalue relation $\boldsymbol{M}\boldsymbol{v} = \lambda \boldsymbol{v}$ is $\varphi_i \varphi_j v_j = \lambda v_i$, with implicit summation over j. Choosing $v_i = \varphi_i$, we have $\varphi_i \varphi_j \varphi_j = \lambda \varphi_i$: we have chosen an eigenvector with eigenvalue $\lambda = \varphi_j \varphi_j = \varphi^{\mathsf{T}} \varphi$. Next, we choose $n-1$ vectors of the form $\boldsymbol{v}^{\mathsf{T}} = (-\varphi_k, 0, \ldots 0, \varphi_1, 0, \ldots 0)$, where the φ_1 entry is in position $k > 1$. Then $M_{ij} v_j = -\varphi_k \varphi_1 \varphi_j + \varphi_1 \varphi_j \varphi_k = 0$. We have thus found n linearly independent eigenvectors, with one eigenvalue $\varphi^{\mathsf{T}} \varphi$ and the rest equal to zero. Alternatively, since $\boldsymbol{M}$ is formed from a single vector, it must have rank 1 and thus only one nonzero eigenvalue, which must equal the trace: $\mathrm{Tr}\, \boldsymbol{M} = \varphi^{\mathsf{T}} \varphi$.

requires $0 < \gamma < \frac{2}{\varphi_k^T \varphi_k}$. For an FIR filter, $\varphi_k^T = \{u_k \; u_{k-1} \; \dots, \; u_{k-n}\}$, and the stability limit is

$$0 < \gamma < \frac{2}{\sum_{i=0}^{n} u_{k-i}^2} \approx \frac{2}{(n+1)\langle u^2 \rangle} . \tag{10.52}$$

Instability thus results if γ or the inputs are too large. The same reasoning suggests that the fastest convergence is given by $\gamma = \frac{1}{(n+1)\langle u^2 \rangle}$. In practice, fluctuations in small-n filters reduce these gain limits further. Problem 10.17 explores the LMS routine and a normalized variant that adjusts γ to compensate for different input amplitudes.

Example 10.13 (LMS system identification) Consider an equally weighted, four-coefficient moving-average (*boxcar*) filter of the form

$$y_k = b\,(u_k + u_{k-1} + u_{k-2} + u_{k-3}) + \xi_k . \tag{10.53}$$

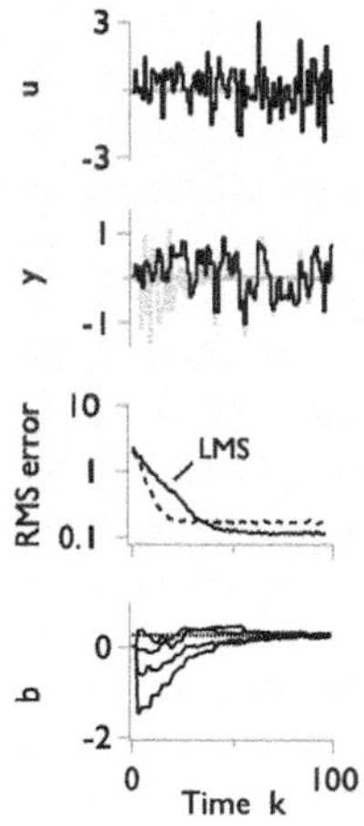

Simulating input-output data and applying the LMS algorithm leads to the plots at left. The top two show typical input and outputs (in black) and the predicted output based on the adapted system model (gray). With $\gamma = 0.1$, the LMS algorithm converges (solid line), while the dashed line shows the n-LMS variant, with $\gamma = 1.0$ and $\alpha = 10^{-4}$. At bottom, all four estimated FIR coefficients converge to the correct value, 0.25, as shown by the RMS error plots (averaged over 1000 simulations). See Problem 10.17.

In Example 10.13 and Problem 10.17, the parameters converge more quickly with a larger value of γ but have a higher steady-state error as a consequence. If the input statistics are stationary, you can use a larger value of γ initially to speed convergence and a smaller one at the end to minimize the steady-state error.

10.3.2 Control Using Adaptive Filters

Adaptive FIR filters can serve as building blocks for feedforward algorithms that perform many of the same tasks as the usual feedback-based control systems. Here, we introduce two basic examples: approximating the inverse to a dynamical system for feedforward control and disturbance compensation.

Adaptive Feedforward Control

We began our discussion of feedback in Chapter 3 by considering the *naive inverse* feedforward controller illustrated in Table 10.1. The idea is that if we can set $K = G^{-1}$, then y will follow r exactly. Although, for reasons summarized in Table 10.1, naive feedforward is unlikely to work well, adaptive feedforward can help address the issues.

The adaptive strategy outlined in Table 10.1 can work, but not if the LMS algorithm is used to adjust the filter weights for K. The LMS algorithm requires access

Table 10.1 Naive and adaptive inverse feedforward control

	Naive Feedforward	Adaptive Inverse Feedforward
	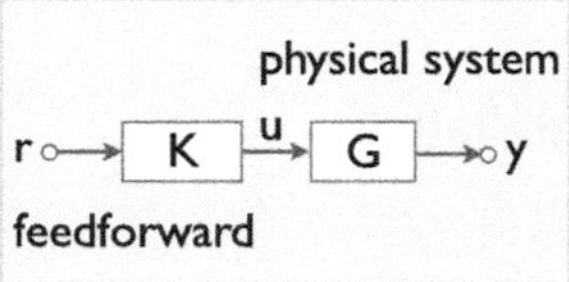	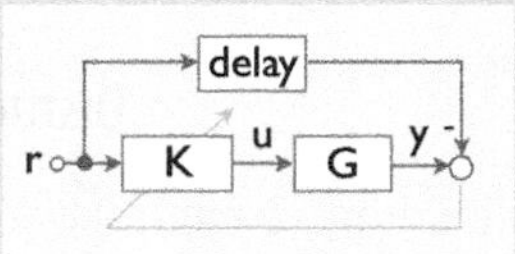
Identification	To invert G, we must know it.	Adaptive filters can be used to identify dynamics. Pass the input through both the system and a filter and adjust the latter until the two outputs match.
Robustness	G may vary over time.	Adaptive filters handle variations on a time scale fixed by the learning rate γ.
Realizability	G^{-1} may not be causal and may demand inputs beyond the limits of the controller range.	Delay the input to ensure causality. Low-pass filtering the reference reduces the required system input magnitudes.

to the error *at the output of the filter itself*.[9] The adaptive scheme reproduced at right uses the system output rather than the filter output, and the filter weights consequently diverge. However, in the reverse configuration, the LMS algorithm does converge to $\hat{K} = G^{-1}$. Unfortunately, that configuration cannot be implemented directly for control, since a feedforward controller must alter the signal before the system input, and not after. Problem 10.18 shows how to use $\hat{K}$ for control in a second step.

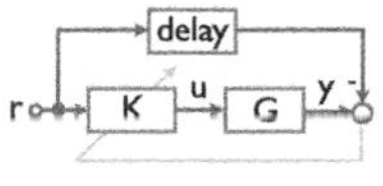

Diverges

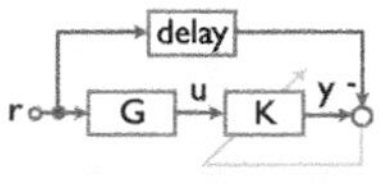

Converges

Example 10.14 (Identification of system inverse) We consider a system-filter cascade from reference r_k to input u_k to output y_k. The reference is fed into a discrete, first-order (IIR) system with unit delay, $u_k = (1-a)u_{k-1} + r_{k-1}$, where the delay would be present in a computer implementation. The transfer function of the system is $G(z) = \frac{a}{z-(1-a)}$. Its inverse is not causal, but we can make it so by delaying one time step. We set the delay to 1 unit, and $K(z)$ converges to

$$K(z) \to z^{-1}G(z)^{-1} = \frac{1}{a} - \left(\frac{1-a}{a}\right)z^{-1}. \tag{10.54}$$

At right, we see that $r_k \to y_{k+1}$ and that the first two FIR coefficients converge to 2 and -1 for $a = 0.5$. The reference $r_k \sim \mathcal{N}(0, 1)$ (gray), and the LMS $\gamma = 0.2$ (black).

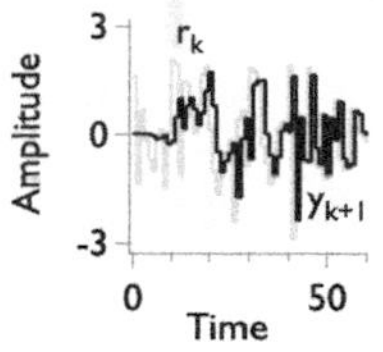

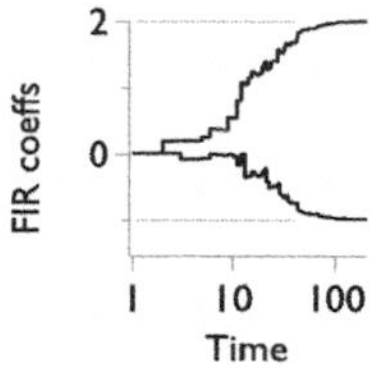

[9] The *differential steepest descent* strategy does work but is slow: pick a weight, increase it slightly, and accumulate a good estimate of the mean square error (MSE). Then decrease it and again measure the MSE, implying an estimate of one partial derivative. Repeat for all parameters; then step down the gradient.

In Example 10.14, we delayed the reference to make the desired FIR filter be causal. If we know the reference in advance – perhaps because it is periodic – then we can use the future signal to get an "instantaneous" (zero phase lag) response (Problem 10.19).

Disturbance Cancellation

FIR filters and adaptive feedforward algorithms have traditionally been a part of signal processing, a distinct discipline from control engineering. Yet many of the tasks of adaptive filtering are very similar to control problems. For example, we measure a noisy signal, $y_k = s_k + n_k$, and wish to recover the signal s_k. From the point of view of control, we could think of this as disturbance cancellation and could imagine generating "anti-noise" to counteract "noise." Here, we will follow the signal-processing language and imagery.

If we know the variances of both the signal and the noise, we can design an optimal filter that, when applied to y_k, produces the best estimate $\hat{s}_k$. This approach leads to the *Wiener filter* (cf. Problem 8.23 and Figure 10.3a). However, if you have extra information about the noise – e.g., you can measure a signal that is correlated with it – then you can do much better than the "optimal" filter that is based only on noise statistics.

Consider what happens when we can measure another signal $\bar{n}_k$ that is correlated with n_k but not with the signal (Figure 10.3b).[10] Let

$$\hat{s}_k = s_k + n_k - \hat{n}_k \,, \tag{10.55}$$

where $\hat{n}_k$ is the estimate of n_k produced by applying an FIR filter to the measured $\bar{n}_k$. Then

$$\left\langle \hat{s}^2 \right\rangle = \left\langle (s + n - \hat{n})^2 \right\rangle = \left\langle s^2 \right\rangle + \cancelto{0}{2\langle s(n - \hat{n})\rangle} + \left\langle (n - \hat{n})^2 \right\rangle \tag{10.56}$$

Since s_k is independent of both noise signals, minimizing $\langle \hat{s}^2 \rangle$ also minimizes the noise-estimation error.

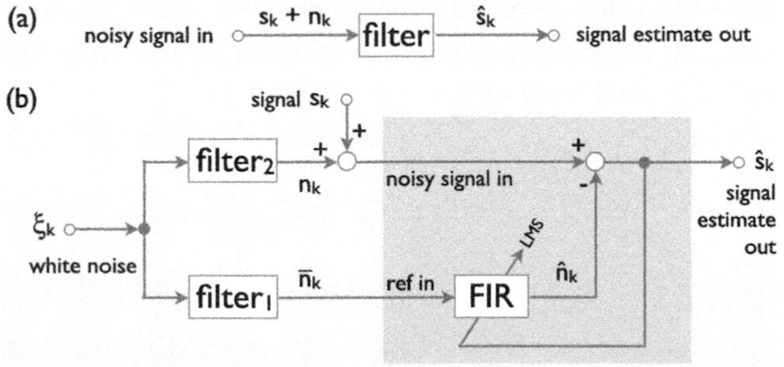

Fig. 10.3 Two ways to remove noise. (a) Conventional filtering of a noisy signal. (b) Adaptive noise cancellation based on a measurement of the noise that is correlated with the noise affecting the signal.

[10] We can generalize to the case where $\bar{n}_k$ is also correlated with the signal. As long as $\bar{n}_k$ is more strongly correlated with the noise than the signal, adaptive filtering will improve the signal estimate.

Example 10.15 (Noise cancellation) In adaptive noise-cancelling headphones, the speaker for each ear has a microphone on the outside that records directly the environmental noise but does not pick up the music being played in the headphone itself (Figure 10.3b). A second microphone near the ear canal picks up a close approximation to the sound the ear actually hears. Because environmental noise is passively low-pass filtered by the headphones themselves, the ear hears noise that is filtered from the outside microphone signal; however, we can use an adaptive filter instead to estimate the noise and to subtract off the estimate from the signal. In effect, the headphones add "antinoise" to the signal, so that the ear receives mainly the signal.

For simplicity, we assume that the outside microphone detects noise perfectly, with $\overline{n}_k = \xi_k$. We model the second, "headphone" microphone as having a low-pass response to outside noise ξ_k by $n_{k+1} = (1-a)n_k + a\xi_k$ (filter 2 in Figure 10.3b). The adaptive filter is based on an "ideal" FIR filter with $N+1$ parameters. Since the $\overline{n}_k$ elements are uncorrelated, the FIR approximation to the IIR "headphone filter" is[11]

$$\hat{n}^*_{k+1} = a\,\overline{n}_k + a(1-a)\,\overline{n}_{k-1} + a(1-a)^2\,\overline{n}_{k-2} + \cdots + a(1-a)^N\overline{n}_{k-N}\,. \tag{10.57}$$

Since we do not know a, we find the FIR coefficients adaptively. Let

$$\hat{n}_{k+1} = b_0^{(k)}\,\overline{n}_k + b_1^{(k)}\,\overline{n}_{k-1} + b_2^{(k)}\,\overline{n}_{k-2} + \cdots + b_N^{(k)}\overline{n}_{k-N}. \tag{10.58}$$

Using the LMS algorithm, Eq. (10.49), we update $\hat{\boldsymbol{\theta}}_k^{\mathsf{T}} = (b_0^{(k)}, b_1^{(k)}, \ldots, b_N^{(k)})$ based on $\boldsymbol{\varphi}_k^{\mathsf{T}} = (\overline{n}_k, \overline{n}_{k-1}, \ldots, \overline{n}_{k-N})$, which gives the results shown at right. At top is the direct output of the filter itself, where the adaptation begins at $k = 500$ and uses $\gamma = 0.005$. The white-noise source has $\xi = 1$, filtered by a first-order filter with $a = 0.3$. Since the signal-to-noise ratio is 0.1, the square-wave signal cannot be seen in the original signal. The subtraction, with 20 FIR coefficients, gives remarkable results. No filter based only on the noisy signal could do nearly as well. Note how the full set of frequencies contained in the square wave is preserved. Knowing the fundamental frequency is also not necessary. The bottom plot shows, for FIR coefficients $\{b_0, b_1, b_2, b_3\}$, that b_j converges to $a(1-a)^j$ (dotted lines).

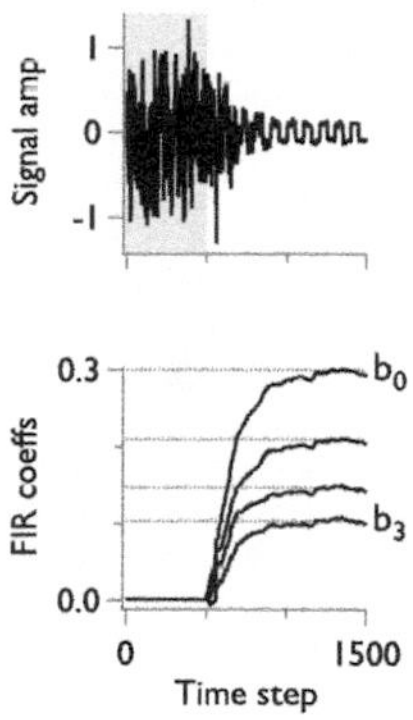

Not all noise-canceling headphones are adaptive. A cheaper solution uses only one microphone, with a fixed FIR filter computed at the factory, which is tuned to the "average ear" and "average fit" (the situation of Figure 10.3a).[12] Adaptive schemes are better because the way a headphone sits on the ear determines the fit and thus the filtering of outside noise. Either way – one microphone or two – the best results use all available information: if you can get any independent information about your noise, use it!

In summary, in this section and the previous, we have introduced two simple, somewhat idealized applications of adaptive feedforward techniques. When the full range

[11] If $\langle \overline{n}_k \overline{n}_\ell \rangle \neq 0$ when $k \neq \ell$, then the FIR coefficients will depend on the order of the expansion, just as the coefficients of a fit to a polynomial do. Expansions are simpler with orthogonal basis elements.

[12] An even-cheaper solution uses analog feedback to eliminate noise as a disturbance.

of complications of a real-life application are included – disturbances, contamination of the reference by the signal, etc. – the control structures become more complicated. Techniques include dithering (adding a random input) to make sure that the excitation is persistent and copying the system model to compute the controller off-line, rather than directly from the real control loop. (The off-line computation can often be done in real time, however.) See the references at the end of the chapter for guidance.

10.4 Optimal Adaptive Control

In adaptive control, we *learn* about the system while controlling it. Broadly, there are three basic approaches to the theory of learning:

- *Supervised learning*. For a given set of situations, you are told the best action to take. If you choose an action, you immediately learn whether some other would have been better. After being instructed about the training set, you try to respond to new situations.
- *Reinforcement learning*. You can evaluate the immediate cost of an action using a known function but must infer its long-term costs and consequences. No information is given as to whether some other action would have been better.
- *Unsupervised learning*. Without cost function or examples, you cluster, rank, or partition the data to simplify its description. For data collected as a matrix, singular value decomposition can reveal dominant "modes." Related techniques include principal component analysis and blind source separation.

Adaptive control falls mainly into the middle case of reinforcement learning: We know the immediate costs of a given control decision, via the cost function, but not much about the system being controlled. We must learn while doing. One distinction is that, traditionally, adaptive control focuses on a single (possibly long) trial, whereas reinforcement learning often allows for multiple training trials to learn the control problem.

Here, we consider adaptive control as a *multistep-decision*, optimal-control problem where, over time, we try to maximize performance while controlling a partially known dynamical system. At each step, we choose between a *greedy* action that *exploits* existing knowledge and a *probing* action that helps learn about (*explore*) the system, so that future control will improve. Optimizing the conflicting demands of exploitation versus exploration is a main theme of reinforcement learning (or *dual control*). From our brief consideration of adaptive control in the light of reinforcement learning will come four insights:

- *Probing control*. It can pay to choose controls that are not immediately optimal but that improve your knowledge of the system being controlled and increase future payoffs. Learning should thus be an active rather than a passive process.

- *Cautious control.* When we exploit existing knowledge, the remaining uncertainty in the dynamics translates into a cautious response, often with gains that are lower than would be used given perfect knowledge.
- *Bifurcations.* The optimal strategy can change suddenly, via bifurcations that have many aspects of phase transitions in physics.
- *Complexity.* Multistep optimization is hard, even in seemingly simple cases. Progress depends on finding suboptimal, heuristic solutions, often using problem-specific tricks.

Rather than surveying superficially a vast field, we first rephrase the learning / adaptive-control problem in the more general language of *dynamic programming* introduced in Chapter 7. We then present in detail a simple example that, while not very realistic, leads to most of the insights mentioned above (and is already hard enough). Finally, we look briefly at some suboptimal heuristic algorithms in the light of our example.

10.4.1 Adaptive Control and Dynamic Programming

We introduced the ideas of dynamic programming in our treatment of optimal control of deterministic systems (Section 7.4). Then, in Section 8.6, we generalized to continuous stochastic systems. For discrete stochastic systems, we proceed similarly:

- *System*: $\boldsymbol{x}_{k+1} = \boldsymbol{f}(\boldsymbol{x}_k, \boldsymbol{u}_k, \boldsymbol{v}_k)$, $k = 0, \ldots, N-1$.
- *Control constraints*: $\boldsymbol{u}_k \in U(\boldsymbol{x}_k)$.
- *Noise distribution*: $\boldsymbol{v}_k \sim p(\cdot|\boldsymbol{x}_k, \boldsymbol{u}_k)$.
- *Policies*: $\pi = \{\mu_0, \ldots, \mu_{N-1}\}$, with $\boldsymbol{u}_k = \boldsymbol{\mu}_k(\boldsymbol{x}_k)$ being rules that determine the control given the inputs. The $\boldsymbol{\mu}_k$ must give allowed inputs $\boldsymbol{u}_k \in U(\boldsymbol{x}_k)$ for all $\boldsymbol{x}_k$.
- *Policy cost* of π starting at $\boldsymbol{x}_0$: We define $J_\pi(\boldsymbol{x}_0) = \langle \varphi(\boldsymbol{x}_N) + \sum_{k=0}^{N-1} L[\boldsymbol{x}_k, \boldsymbol{\mu}_k(\boldsymbol{x}_k)] \rangle$.
- *Optimal cost function*: $J^*(\boldsymbol{x}_0) = \min_\pi J_\pi(\boldsymbol{x}_0)$.
- *Optimal policy* π^*: $J_{\pi^*} = J^*(\boldsymbol{x}_0)$.

For the diffusing particle, $\varphi(x_N) = x_N^2$ and $L[x_k, \mu_k(x_k)] = Ru_k^2$, and the goal is to determine the policy $u_k = \mu_k(x_k)$. The subtlety is that, for $R < R^*$, minimizing the cost at time step 0 by choosing $u_0 = 0$ (the *greedy* policy) is not the best strategy. Rather, we pay a little (choose $u_0 \neq 0$) in order to learn about the sign of the charge, and our reward is reduced future costs associated with the later choice u_1. At right, the dotted lines correspond to the path that minimizes total costs but has higher initial cost.

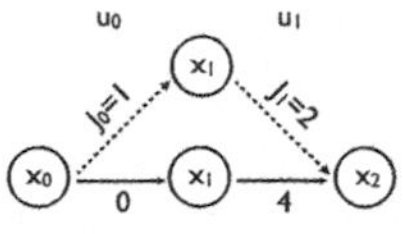

In Section 7.4, we saw that we can solve such problems by appealing to the *principle of optimality*, that the remaining part of an optimal path is also optimal. The principle of optimality leads to a discrete version of the Hamilton–Jacobi–Bellman equation for stochastic system discussed in Section 8.6. This recursive *Bellman equation for stochastic systems* is

$$J^*(\boldsymbol{x}_k) = \min_{\boldsymbol{u}_k \in U(\boldsymbol{x}_k)} \left(L[\boldsymbol{x}_k, \boldsymbol{u}_k] + \langle J[\boldsymbol{f}(\boldsymbol{x}_k, \boldsymbol{u}_k, \boldsymbol{v}_k)] \rangle \right) , \tag{10.59}$$

with terminal condition $J^*(\boldsymbol{x}_N) = \varphi(\boldsymbol{x}_N)$. Note that the various functions f, L, and so on can all depend on k; we can incorporate limited knowledge by adding the hyperparameters to an enlarged state vector $(\boldsymbol{x}_k, \boldsymbol{\theta}_k)$; and so on. The expectation in Eq. (10.59) is over $\boldsymbol{v}_k$ and is conditioned on the state value $\boldsymbol{x}_k$. Equation (10.59) is usually intractable unless the number of stages N is small.

10.4.2 Control a Particle with Unknown Electrical Charge

As an example of issues that can arise in the control of a discrete stochastic system, we consider the control of a charged, diffusing particle in one dimension. The problem will be simple, in having only two stages, but complex, in that parameter estimates will depend on the entire history (trajectory) of the system state.

The goal will be to minimize the diffusive motion of the charged particle, using the least amount of control effort.[13] The twist is that the sign of the particle's charge is unknown. Thus, when we apply an electric field to push the particle, we do not know, a priori, which way it will move. Of course, diffusion also makes the particle move unexpectedly, and both effects are at play. To further simplify the problem, we will assume that we know the magnitude of the charge (just not its sign).

In more detail, the equation of motion for the particle is

$$x_{k+1} = x_k + b u_k + v_k \,, \qquad x_0 = 0 \tag{10.60}$$

where the thermal noise is Gaussian and scaled so that $\langle v_k \, v_\ell \rangle = \delta_{k\ell}$ and where $b = \pm 1$. At time $k = 0$, nothing is known about the sign of the charge: $P_0(b = -1) = P_0(b = +1) = \frac{1}{2}$.

The goal of the optimal-control problem is to choose controls u_0 and u_1 to minimize

$$J = \left\langle x_2^2 + R\left(u_0^2 + u_1^2\right)\right\rangle \,. \tag{10.61}$$

In words: after two time steps, the particle should be as close to 0 as possible, while minimizing the control effort at times $k = 0$ and 1. Effort (work) is measured by u_0^2 and u_1^2, weighted by the constant R. There is no penalty on the value of the intermediate state x_1.

The general strategy will be to use the first control, u_0, to learn about the sign of the charge and the second, u_1, to exploit the information gained via u_0 to bring the particle back to the origin. Before launching into an analysis, let us consider two extreme cases: If work is very expensive, $R \gg 1$, any control will cost too much, and the optimal solution will be $u_0 = u_1 = 0$. Diffusion then implies a cost $J = 2$. Because the optimal control is independent of the observation x_1, it is a feedforward solution.

At the other extreme, $R \ll 1$, we can impose a very large input u_0 to learn with certainty the sign of the charge and then use u_1 to bring the particle back to the origin. Having assumed perfect measurements, we would then expect to miss the origin by the amount the particle diffuses in *one* time step, so that $J = 1$, half the value of the uncontrolled case. This is a feedback solution. But what happens for intermediate R?

[13] The example has similarities with Example 8.5 on "delayed choice."

We start by formalizing what we mean by "learning" the sign of the charge. Our knowledge of the charge at time k, implicitly conditioned on current and prior observations $x^k \equiv \{x_k, x_{k-1}, \ldots, x_0\}$, is expressed by $P(b|x^k) \equiv P_k(b)$, for $b = \pm 1$. It is convenient to parameterize these probabilities by defining a number β_k:

$$P_k(b = \pm 1) = \tfrac{1}{2}\,(1 \pm \beta_k)\ . \tag{10.62}$$

If the charge is surely positive, then $\beta_k = +1$. If surely negative, $\beta_k = -1$. If we have no idea, $\beta_k = 0$. Intermediate knowledge corresponds to $-1 < \beta_k < 1$.

Observations change our state of knowledge of the sign and hence about $P_k(b)$ and β_k. In Problem 10.20, we will see that

$$\beta_{k+1} \equiv \tanh\theta_{k+1}\,, \qquad \text{where} \qquad \theta_{k+1} = \theta_k + (x_{k+1} - x_k)\,u_k\,. \tag{10.63}$$

The θ_k variable may be viewed as an alternate parametrization of $P_k(b = 1)$ whose more complex form is justified by the simpler update law.[14] Note that $-\infty < \theta_k < +\infty$.

The update law in Eq. (10.63) is very intuitive: the greater the input, the more we learn about the sign of the charge. The favored sign depends on the relative signs of the input and the resulting displacement, $x_{k+1} - x_k$. Displacements and inputs are scaled by the thermal noise strength. With no input ($u_k = 0$), we learn nothing ($\theta_{k+1} = \theta_k$). Nonzero inputs will tend to drive θ toward $+\infty$ or $-\infty$ and the probabilities to 1 or 0. Finally, Eq. (10.63) also defines a dynamical system for θ_k. In the enlarged (x_k, θ_k) state space, the optimal-control problem is now a standard one for dynamics that are exactly known. Unfortunately, the equations are now nonlinear, as a result of the $x_k u_k$ term in the θ_k dynamics.

Working backwards from the end time (Chapter 7), we start at time $k = 1$, where the particle is at x_1 and our knowledge of the charge is given by θ_1 (or β_1). We then choose u_1 to optimize the intermediate *cost-to-go* function

$$J_1(x_1,\beta_1) = \min_{u_1} J_1(x_1,\beta_1,u_1) = \min_{u_1} \left\langle x_2^2 + Ru_1^2 \right\rangle_{b,\nu_1}\,, \tag{10.64}$$

where the subscripts on the angle brackets remind us that we have to average over *both* the noise and the parameters ($b = \pm 1$). In Problem 10.21, we will do this and find

$$J_1(x_1,\beta_1,u_1) = (1 + R)u_1^2 + x_1^2 + 2\beta_1 x_1 u_1 + 1\,. \tag{10.65}$$

Minimizing with respect to u_1, we have

$$u_1^* = -\frac{\beta_1 x_1}{1 + R} \qquad \text{and} \qquad J_1(x_1,\beta_1) = J_1(x_1,\beta_1,u_1^*) = \left(\frac{1 + R - \beta_1^2}{1 + R}\right) x_1^2 + 1\,. \tag{10.66}$$

Equation (10.66) reflects *cautious control*: the gain is directly proportional to our knowledge about the charge, as expressed by β_1. Complete certainty, $\beta_1 = \pm 1$, gives the largest gain magnitude. For total ignorance, $\beta_1 = 0$, the best strategy is to do nothing (set $u_1 = 0$). Increasing β_1^2 (learning) reduces the cost $J_1(x_1,\beta_1)$ from $x_1^2 + 1$ to

[14] In the literature on Bayesian statistics, parameters such as θ_k that characterize the prior distribution are known as *hyperparameters*. If prior and posterior have the same form, the Bayesian update is reduced to a parameter update law. The distributions are called *conjugate distributions* and the prior a *conjugate prior*.

$(\frac{R}{1+R})x_1^2 + 1$. Note that the optimal action violates the *certainty equivalence principle* used by most adaptive control algorithms, which is to pick the control as if parameter estimates were perfect (Section 10.2). Here, the principle translates to "set $\beta_1 = 1$" and use $u_1 = -\frac{x_1}{1+R}$. The *separation principle* – first estimate unknown parameters; then use the estimates in a control algorithm – still applies.

For the full two-stage problem, we pick a control u_0 at $k = 0$. At $k = 1$, we will know the result x_1 and can update our knowledge of the sign of the charge to $\beta_1 = \tanh[(x_1 - x_0)u_0]$. For any given outcome, we can then calculate the rest of the optimal path, as given by $J_1(x_1, \beta_1)$. We then calculate the total cost J for this particular outcome x_1. Of course, we then need to average over all possible outcomes x_1 (again, averaging over the noise and both possibilities for the charge b). Finally, we should repeat the procedure for each possible choice u_0, selecting the one that minimizes the full cost. More succinctly,

$$J^* = \min_{u_0} \left[R u_0^2 + \langle J_1(x_1, \beta_1) \rangle \right] . \tag{10.67}$$

Evaluating the averages in Eq. (10.67) leads to a disappointingly complex result (Problem 10.21). Fortunately, the problem simplifies in the two interesting limits.

For large R, we expect the optimal u_0 to be small, even zero. Thus, we Taylor expand about $u_0 = 0$ and find the approximate cost-to-go J_0 from $k = 0$ (Problem 10.21),

$$J_0(u_0) = 2 + \left((1+R) - \frac{3}{1+R} \right) u_0^2 + \left(\frac{4}{1+R} \right) u_0^4 + O(u_0^6) . \tag{10.68}$$

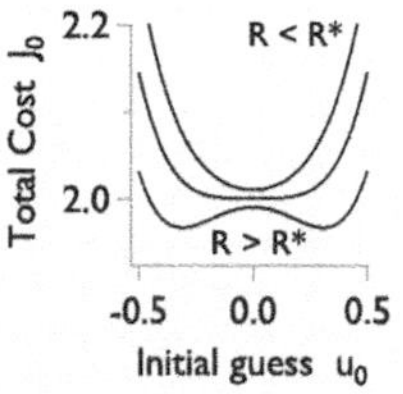

The form of J_0 in Eq. (10.68) is just that of the classic free energy in a symmetry-breaking, second-order phase transition.[15] Thus, we consider the cost-function parameter R to be a *control parameter* that plays the role of temperature. The analogous *order parameter* is u_0. Indeed, for $R \gg 1$, there is a single minimum at $u_0^* = 0$. At $R^* = \sqrt{3} - 1 \approx 0.7$, there is a bifurcation ("phase transition") as J_0 takes on the familiar double-well shape. See the plot at left for three values of R, with the curves displaced vertically for clarity.

For $R < R^*$, there are then two possible solutions, $u_0^*(R) = \pm\frac{1}{2}\sqrt{1 - R - \frac{1}{2}R^2}$. In physical units, $u_0 \sim \nu$, where the scale of thermal noise is defined by $\langle \nu_k^2 \rangle = \nu^2$. The cost $J^*(R)$ assuming optimal u_0^* is nonanalytic at the same R^*, where it goes below 2. See plots at left, where the bifurcation at R^* is indicated by a double arrow. Solid lines show the results of a numerical calculation based on Eq. (10.67). Dashed lines are Taylor series and asymptotic approximations.

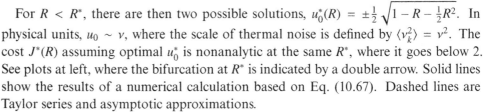

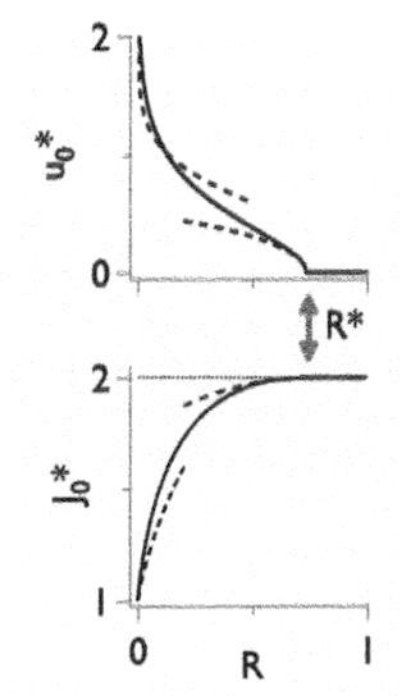

The $R \ll 1$ limit is also evaluated in Problem 10.21. The optimal choice u_0^* diverges as

$$u_0^* \sim \pm\sqrt{-\tfrac{1}{2}\ln R}, \quad \text{implying} \quad J^* \sim 1 + R \ln \frac{4}{R} . \tag{10.69}$$

[15] Although we do not have a thermodynamic system and "supercritical bifurcation" is the more precise terminology, we will use the language of phase transitions, which is more familiar to physicists.

As $R \to 0$, we find $|u_0^*| \to \infty$ and $J^* \to 1$, as claimed.[16] The two solutions in Eq. (10.69) diverge slowly with R: It does not take a huge displacement to be nearly sure of the sign of b (either it goes in the direction you intend or not). But when $R \to 0$ and control costs nothing, we choose $u_0^* = \pm\infty$ to be completely certain about the sign of the charge.

The two equivalent cases for u_0^* reflect the symmetry of the problem. Since we have no basis for choosing $u_0 > 0$ over $u_0 < 0$, we pick one at random in a kind of "participatory" *spontaneous symmetry breaking*. Half the time we will be right and half the time wrong. Either choice, after taking the ensemble average, leads to identical results (Problem 10.21).

To summarize, the deceptively simple example of the two-stage optimal control of a diffusing particle with unknown charge sign leads to several insights. Initially, it is favorable to pay the cost of probing the system ($u_0 = 0$ would minimize the immediate cost). In the second move, uncertainty leads to cautious control. The limits of small and large R can be understood from simple arguments, but not the bifurcation in crossing over from "do nothing" to "probe and then restore." Indeed, the bifurcation exists only because of the noise ("noise induced"): without noise, the best course of action is always to do nothing. It is striking how complex even a simple optimal-adaptive-control problem can be. Extending the calculation to more stages (time steps) leads to greater complexity: The bifurcation can be subcritical (like a first-order phase transition), with jumps in u_k as R is varied.

10.4.3 Heuristic Methods

The example of Section 10.4.2 showed how difficult it can be to find truly optimal solutions to nonlinear stochastic, adaptive problems. What to do in practice? We can simply accept the uncertainty and try to optimize the control given that uncertainty. That was the *robust-control* approach of Chapter 9. Here, it led to *cautious control*. Although the best solution would be based on the stochastic Bellman equation of the previous section, it is usually too hard to solve. Instead, we consider some suboptimal heuristic strategies.

Apart from potential instability, the major failing of standard adaptive routines is that a lack of persistent excitation can lead to control that is satisfactory under a given set of disturbances or reference signals but does not lead to convergence of the parameters being estimated. When conditions change, the response is poor and adaptation needs to start over. The optimal solution is "probing," changing the input so that it yields information about the dynamics that can be exploited later. But, since optimal probing is difficult to compute, we can try more heuristic algorithms to determine the probing signals.

16 The approximation in Eq. (10.69) is crude since u_0 does not become very large, even for extremely small values of R; however, the important point is that $u_0^* \to \pm\infty$ for $R \to 0$ and that the divergence is "reluctant." The minimum is also extremely broad: it does not matter much precisely what value you choose.

Since persistence is linked to frequency content, white noise (*dither*) is a simple input that is nearly always persistent and whose effect on the system being controlled is often minor. The noise reduces the immediate performance but improves parameter estimation. We can make the dither amplitude time dependent, increasing it when it is most needed and decreasing it when performance is most important. Of course, you then need more information about the situation (e.g., knowing when performance is particularly important). As always, the game in control is to incorporate all available side information.

In the control-theory literature, there have been various attempts to balance exploitation and exploration heuristically. For example, Filatov and Unbehauen (2004) introduce two conflicting optimization criteria, one to minimize the tracking error (or some other performance indicator), and the other to minimize the dynamical uncertainty (learn as much as possible about the system). Typically, the control that teaches you the most about the dynamics is much larger than the control that optimizes tracking, and the authors advocate a somewhat arbitrary balance between the two criteria.

Alternatively, the field of reinforcement learning deals with these exploitation-exploration trade-offs in a more general context than control. Two popular approaches (among many others, such as *policy search* and *approximate dynamical programming*) are

- *ϵ-greedy*. With probability ϵ, we explore and with probability $1 - \varepsilon$, we exploit. The case $\epsilon = 0$ corresponds to a standard "greedy" adaptive control algorithm that always chooses the best action, where "best" is estimated from prior stages.
- *Softmax*. We choose action i with probability P_i, which is a function of $\hat{Q}^{(i)}$, the estimated value of action i. One popular rule for choosing probabilities is according to a Boltzmann (Gibbs) distribution, $P_i \propto e^{\hat{Q}^{(i)}/T}$, where T is analogous to a temperature.

Such approaches are explored in Problem 10.22, in the context of a standard reinforcement-learning problem, the *n-armed bandit*. Imagine n slot machines ("bandits") with unknown average payout rates. If you knew the best machine, the optimal strategy would be to always play it. The problem is to decide how often to play different machines to improve payout-rate estimates. Although slot machines sound frivolous, the bandit problem has many important applications, for example in clinical trials (n possible treatments) and adaptive routing through a network (n routes to go from home to work).

One aspect of any learning system is whether the algorithm operates over reasonable time frames. A key concept is that algorithms should not grow more than polynomially with problem size. Leslie Valiant has emphasized the importance of such computational aspects in learning, introducing the notion of *probably approximately correct* (PAC) learning: "the number of computational steps is polynomially bounded and the errors are polynomially controlled." These algorithms, at least within a restricted space of problems, learn and improve in reasonable amounts of time (depending polynomially on problem size). In principle, such algorithms could be used for heuristic (or "theoryless") control algorithms. Similarly, the tools of reinforcement learning often have *model-free* versions where the dynamical equations are represented by neural networks that learn the functional form of equations directly from the data (Section 10.5).

To summarize, even the simplest situations of optimal adaptive control can lead to a very complicated analysis. In practice, start with fixed controllers, possibly designed to be robust against a given set of dynamical possibilities. If that is not good enough, the standard direct or indirect adaptive control methods from Sections 10.1 and 10.2, for deterministic and stochastic systems, respectively, should be your next options. If the system has short time scales, the feedforward approach of Section 10.3 is a good choice. For complex situations, try heuristic learning approaches.

10.5 Neural Networks

Adaptive control is intimately connected to *learning*, a vast topic with many applications to control. Reinforcement learning, introduced briefly above, is but one perspective among many. What if you could replace the tedious explicit modeling of dynamics with a kind of universal dynamical system that needs only be "trained" to have the desired behavior? If you could somehow create such a function and train it so that its output generates the required control signals, then you would have a *data-driven* controller even without explicit models for the system and controller dynamics.

Neural networks are a promising approach for such universal dynamics because they are based on a *universal function approximator*. Their design is inspired qualitatively by the neurons and architecture of the human brain. In a very simple version, "neurons" are nonlinear functions of weighted combinations of inputs that form a layer whose outputs are summed to form the system output. Given a set of input-output pairs for training, you adjust the weights to map the inputs to the outputs. The hope is that new data from the same general class as the training inputs will also map to desirable outputs. Such neural networks can work well for *classification* and *associative memories*, where a "suggestive" input signal causes the network to converge on the desired attractor ("content-addressable memory"). While classification is the principal use of neural networks, we focus here, instead, on their potential for modeling complex dynamical systems.

The literature on neural networks is, understandably, heavily influenced by neuroscience and efforts to model the brain. Since our interest is in neural networks as complex dynamical systems that can be trained to imitate other systems, we mostly just translate such terms into more familiar ones for dynamical systems (or ignore them).

A single, feedforward layer has but limited power, generally not enough to model interesting dynamical systems. But if one layer is fed to another hidden layer, and so on, the range and sophistication of representations increases dramatically: this is the

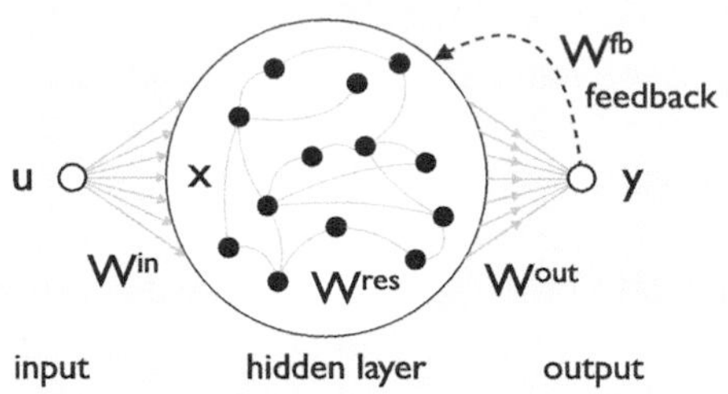

Fig. 10.4 Reservoir computer, with input(s) $\boldsymbol{u}$, the hidden layer nodes $\boldsymbol{x}$, and output(s) $\boldsymbol{y}$, along with the vectors and matrices for input coupling $\boldsymbol{W}^{\text{in}}$, output coupling $\boldsymbol{W}^{\text{out}}$, internal reservoir coupling $\boldsymbol{W}^{\text{res}}$, and optional feedback coupling $\boldsymbol{W}^{\text{fb}}$.

remarkably successful approach of *deep learning* for automatic language translation, image recognition, and other hard problems that until recently seemed beyond reach.

An alternative to this feedforward neural network is the feedback, or *recurrent* neural network, where the output of the network is fed back into itself. Such networks are typically much harder to train (and analyze). Below, we present one particular approach, *reservoir computing*, that avoids some of these difficulties and has the attractive possibility of being implemented using analog hardware.

10.5.1 Reservoir Computing

Figure 10.4 illustrates the layout of a reservoir computer, which is a special kind of recurrent neural network. Its discrete-time dynamics have the form,

$$\boldsymbol{x}_{k+1} = (1-\alpha)\boldsymbol{x}_k + \alpha \tanh\left(\boldsymbol{W}^{\text{res}}\,\boldsymbol{x}_k + \boldsymbol{W}^{\text{in}}\,\boldsymbol{u}_k + \boldsymbol{W}^{\text{fb}}\,\boldsymbol{y}_k + \boldsymbol{b}\right) \tag{10.70a}$$

$$\boldsymbol{y}_{k+1} = \boldsymbol{W}^{\text{out}}\,\boldsymbol{x}_{k+1}\,. \tag{10.70b}$$

Equation (10.70) is simply a special case of the nonlinear, discrete dynamical system $\boldsymbol{x}_{k+1} = \boldsymbol{f}(\boldsymbol{x}_k, \boldsymbol{u}_k)$ and $\boldsymbol{y}_{k+1} = \boldsymbol{h}(\boldsymbol{x}_{k+1})$. Here, $\boldsymbol{u}_k$ and $\boldsymbol{y}_k$ are the system inputs and outputs at time k. Only one of these is typically present in the state-dynamics equation: for signal generation, $\boldsymbol{W}^{\text{in}} = \boldsymbol{0}$; for control applications, $\boldsymbol{W}^{\text{fb}} = \boldsymbol{0}$.

The internal state vector $\boldsymbol{x}_k$ has a high dimension n, typically 100–1000. The $\tanh(\cdot)$ function makes the dynamics nonlinear. The precise form of the nonlinearity is not important, but function output values should be bounded for all input values of the state vector $\boldsymbol{x}$. The parameter α sets the time scale for state dynamics, while $\boldsymbol{b}$ is an n-dimensional constant bias vector that prevents the equivalent of dynamical "zeros." [17]

In traditional neural networks, all the elements of all the $\boldsymbol{W}$ matrices are considered as free parameters. For feedforward networks that lack internal state ($\boldsymbol{x}$) dynamics, optimizing these parameters so that a given input $\boldsymbol{u}$ leads to a desired output $\boldsymbol{y}$ is often

[17] Consider the case with an input u_k but no feedback. Then, every time u_k vanishes, the reservoir computer temporarily has no input. This tends to destabilize the reservoir coeffficents. Putting in the bias makes sure that there is always an input, even when the signal happens to vanish.

straightforward;[18] however, for recurrent networks, the optimization problem can be very hard. As discussed in Problem 10.23, the basic mathematical issue is that bifurcations can lead to discontinuities in the cost function that cause local optimization methods that try to go "down the gradient" to fail.

The reservoir-computing approach *fixes* the input coupling $\boldsymbol{W}^{\text{in}}$, the reservoir dynamics $\boldsymbol{W}^{\text{res}}$, and the possible feedback coupling $\boldsymbol{W}^{\text{fb}}$ as constant, random elements and trains *only* $\boldsymbol{W}^{\text{out}}$, the output coupling.[19] In the examples discussed here, we choose the matrix elements to be Gaussian random numbers, $\mathcal{N}(0,1)$, except that elements of $\boldsymbol{W}^{\text{res}}$ are divided by $\sqrt{n}$, to make the magnitude of its largest eigenvalue, the *spectral radius*, approximately 1.

The only elements that are "trained" are those comprising $\boldsymbol{W}^{\text{out}}$. Because the output coupling is linear, choosing these elements to minimize a cost function J of the form,

$$J\left[\boldsymbol{W}^{\text{out}}\right] = \sum_{k=1}^{N} \left\| \boldsymbol{W}^{\text{out}} \boldsymbol{x}_k - \boldsymbol{y}_k^{\text{target}} \right\|^2 + \beta \, \text{Tr}\left[(\boldsymbol{W}^{\text{out}})^{\mathsf{T}} \, \boldsymbol{W}^{\text{out}} \right], \tag{10.71}$$

is a linear problem that can be solved explicitly. Here, $\boldsymbol{y}_k^{\text{target}}$ is the target output signal for training, while β is a *regularization* parameter that tries to minimize the size of the elements of $\boldsymbol{W}^{\text{out}}$. A large β will reduce the size of $\boldsymbol{W}^{\text{out}}$ but track the target more loosely, while a small value of β will improve tracking, at the cost of larger elements of the output coupling, which can increase sensitivity to noise. (See Section A.8.8 online.)

To find the output weights $\boldsymbol{W}^{\text{out}}$, we first use the dynamical system to generate a sample of the desired output $\boldsymbol{y}$ and gather the time series of length N into an $N \times p$ matrix $\boldsymbol{Y}^{\text{target}}$, with p the number of outputs in $\boldsymbol{y}$. We then define random-element matrices $\boldsymbol{W}^{\text{res}}$ and $\boldsymbol{W}^{\text{fb}}$ and bias vector $\boldsymbol{b}$ to create a reservoir computer. Then, using Eq. (10.70a), we generate a time series of internal n-element states $\boldsymbol{x}$ that are gathered into an $N \times n$ matrix $\boldsymbol{X}$. Differentiating J with respect to the matrix $\boldsymbol{W}^{\text{out}}$then leads to (cf. Eq. A.226 online)

$$\boldsymbol{W}^{\text{out}} = \left(\boldsymbol{X}^{\mathsf{T}} \boldsymbol{X} + \beta \mathbb{I} \right)^{-1} \boldsymbol{X}^{\mathsf{T}} \boldsymbol{Y}^{\text{target}}, \tag{10.72}$$

where $\mathbb{I}$ is the $n \times n$ identity matrix. Equation (10.72) is easy and cheap to compute, even for $n = O(10^3)$. Moreover, we can compute the solution via the RLS algorithm from Section 10.2.2, allowing for online, real-time training of the reservoir network.

To understand how reservoir computers work, we discuss three examples below:

- *Signal generation*: a single reservoir computer can output three different signals, depending on the training input;
- *Feedforward control*: the reservoir learns the inverse of a given dynamical system;
- *Feedback control*: a reservoir can learn recursively and online the inverse of a given dynamical system; the training acts as a feedback controller.

[18] One method is *backpropagation* (MacKay, 2003, Section 39.2).
[19] In some variants to the basic scheme, minor tweaking of the other $\boldsymbol{W}$ matrices is allowed.

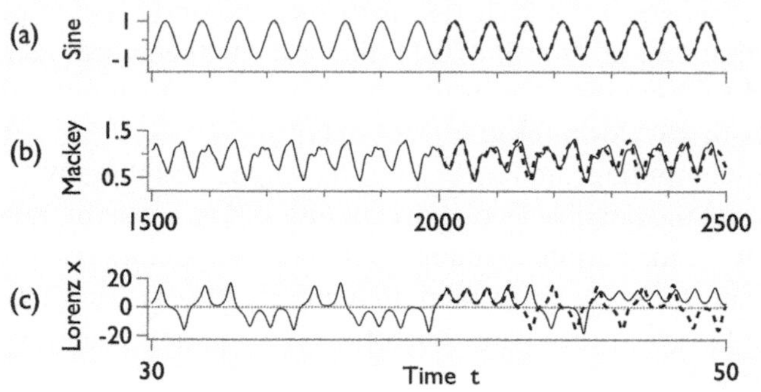

Fig. 10.5 Signal generation in a reservoir computer for (a) sine wave, (b) Mackey–Glass dynamical system, and (c) x-component of the Lorenz equations. The first 2000 time steps are used to train the system. Markers show "free running" signal. Reservoir parameters for all three cases: $n = 500, \alpha = 0.3, \beta = 1$. Time step $= 1$ in (a,b) and 0.02 in (c).

Signal Generation

For signal generation, $\boldsymbol{W}^{\text{in}} = \mathbf{0}$ (no input), but the output is fed back into the network via the $\boldsymbol{W}^{\text{fb}}$ matrix (a vector if there is but one output). Figure 10.5 shows that a single reservoir computer structure can be trained to give three different dynamical outputs. In part (a), the training signal is a sine wave; in part (b), it is the Mackey–Glass dynamical system; and in part (c), it is the x-element of the Lorenz equations.

In Figure 10.5, an identical reservoir computer is fed three different training signals for 2000 time steps.[20] Training is accomplished by using the desired dynamics y_k^{train} in place of the output y_k in the feedback in Eq. (10.70a). Then, starting at time step 2001, the reservoir output is substituted, and the network "free runs." Initially, all three systems track well the desired signal. The periodic sine wave does the best, then the Mackey–Glass, then the Lorenz. But the Mackey–Glass system is weakly chaotic, and the Lorenz system is even more strongly so; thus, any error on the initial conditions is magnified. Notice that the initial mistake of the Lorenz system is to "flip" too early to the other side of the attractor. For several oscillations, the value of $x(t)$ is very nearly the opposite of the target value.

We are thus not surprised that the chaotic signals diverge eventually from the reservoir output. But it is remarkable that even the strongly chaotic Lorenz dynamics can be tracked over many Lyapunov times. Remember that we have shown the same reservoir dynamical system three different training signals, and the system has learned each!

[20] The sine curve is $\sin t/10$. The Mackey–Glass system $x_{k+1} = 0.9x_k + 0.2x_{k-17}/(1 + x_{k-17}^{10})$ is a delay-difference equation. The Lorenz system is $\dot{x} = 10(y - x)$, $\dot{y} = x(28 - z) - y$, $\dot{z} = xy - (8/3)z$, integrated using a standard ODE solver and evaluated on a time step of 0.02. All inputs are normalized to have zero mean and unit variance.

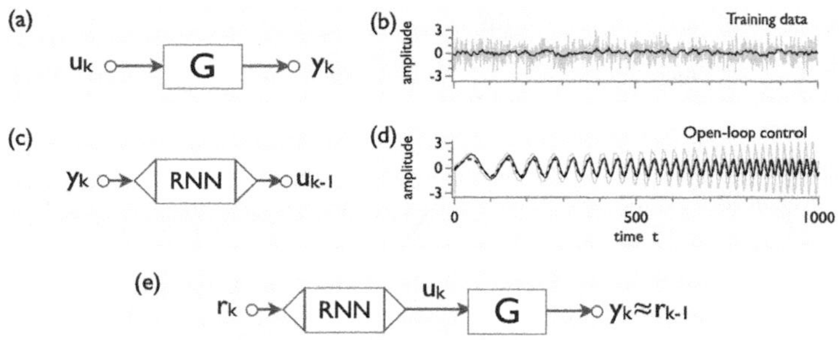

Fig. 10.6

Open-loop control of a first-order system using a reservoir computer. (a) Schematic of physical system. (b) Input-output training dynamics using random input for 1000 time steps. Input is gray line, output dark line. (c) Inverse training of the RNN. (d) Chirped-sine reference input leads to $y_k \approx r_{k-1}$. The output signal from the RNN and input signal to the physical system u_k is light gray. (e) Schematic for open-loop control. Reservoir parameters: $n = 100, \alpha = 1, \beta = 1$.

Feedforward Control

Section 10.5.1 showed that a reservoir computer can learn a dynamical system simply by recording input and output. Similarly, we can teach a system to learn the inverse of a dynamical system by reversing the training.

In Figure 10.6(a), we train a reservoir computer by first collecting input-output training pairs (u_k, y_k) on a physical system G for 1000 time steps.[21] The response of the first-order system, for random input, is shown in (b). The training (c) is done by using y_k as the input and the past value u_{k-1} as the target. This scheme is then tested in (d) by using a chirped-sine reference signal. Notice how the system can follow a wide range of frequencies, going from below to well above the low-pass cutoff frequency of the physical system. The input (light-gray line) grows as needed to supply the necessary output, even though such amplitudes were not supplied during the training period.

Feedback Control

We can extend the open-loop control scheme presented above to a closed-loop, learning control. For the open-loop scheme, we trained an RNN to create a system input u_{k-1} when presented with a desired outcome y_k. The training was done off-line. Here, we replace the "batch" off-line training with a recursive, online training where the matrix elements of the output $\mathbf{W}^{\text{out}}$ are adjusted in real time in response to an error signal created by the difference between the reference and the desired delayed output.

As illustrated in Figure 10.7(a), we implement such a scheme by sending d outputs $y^k_{k-d} \equiv \{y_{k-d}, \ldots, y_k\}$ to a reservoir computer (RNN), which is trained online by comparing the output of the RNN to the actual signal sent to the system, u_{k-d}. Then, in (b), the desired references are sent to a *copy* of the RNN, which has the same coefficients as in (a) – in particular the same $\mathbf{W}^{\text{out}}$ coefficients, which are being trained in real time. The RNN output u_k is sent to the physical system G. The output of this

[21] The system dynamics are $y_{k+1} = 0.9y_k + 0.1u_k$ and are trained using $u_k \sim \mathcal{N}(0, 1)$. The chirped-sine reference input is $r_k = \sin[2\pi k(1 + k/200)/200]$.

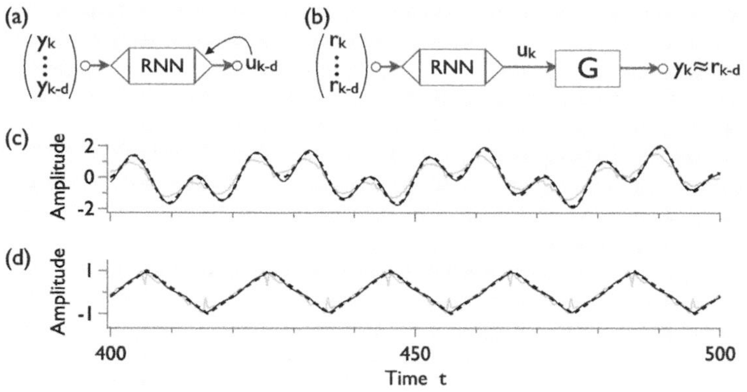

Fig. 10.7 Feedback control of second-order, underdamped dynamics, starting at $t = 0$. (a) Online training network. (b) Reference applied to same network, to track (c) quasiperiodic and (b) triangle-wave signals. Reference is dashed line; shifted system output is black line; system input is gray line. Reservoir parameters: $n = 300$, $\alpha = 1$, $P_0 = 1$, $d = 9$.

system is then the source of the signals used in (a), and so on through the loop. The target signal is now a delayed version, by d time steps, of the reference. Giving d inputs to the RNN gives better results for higher-order systems.[22]

Figure 10.7(c) and (d) shows that this scheme can actually work! The system is a discrete, underdamped harmonic oscillator ($\zeta = 0.1$), which is difficult to track because of its low damping. In (c), the tracking signal is a quasiperiodic signal, $\sin \omega t + \sin \pi\omega t$, that almost, but never quite, repeats. In (d), the signal is a triangle wave. Notice that in (c), the output (light gray line) gently "pulls back" to avoid an overshoot, whereas in (d), it "spikes" to approximate the triangle-wave "corner." [23]

To train the network in (a), we use the recursive-least-squares (RLS) algorithm from Section 10.2.2. Defining an observer $\boldsymbol{L}_k$ and an $n \times n$ covariance matrix $\boldsymbol{P}_k$, we replace the off-line ridge-regression algorithm of Eq. (10.72) with the RLS update equations. Adapting Eq. (10.26) gives an algorithm for updating the output $\boldsymbol{W}_k^{\text{out}}$ at time k,

$$\boldsymbol{L}_{k+1} \equiv \frac{\boldsymbol{P}_k \boldsymbol{x}_{k+1}}{1 + \boldsymbol{x}_{k+1}^{\mathsf{T}} \boldsymbol{P}_k \boldsymbol{x}_{k+1}}, \qquad \boldsymbol{P}_{k+1} = \left(\mathbb{I} - \boldsymbol{L}_{k+1} \boldsymbol{x}_{k+1}^{\mathsf{T}}\right) \boldsymbol{P}_k, \qquad \boldsymbol{W}_{k+1}^{\text{out}} = \boldsymbol{W}_k^{\text{out}} + \boldsymbol{L}_{k+1} \varepsilon_{k+1}, \tag{10.73}$$

where the *innovations* $\varepsilon_{k+1} = y_{k+1}^{\text{track}} - \boldsymbol{W}_k^{\text{out}} \boldsymbol{x}_{k+1}^{\mathsf{T}}$ is the difference between the observation and its prediction, with y_k^{track} the desired output of the RNN (here, the delayed input u_{k-d}). The covariance matrix is initialized by setting $\boldsymbol{P}_0 = P_0 \mathbb{I}$, where the constant P_0 is a regularization parameter that controls the size of the coefficients of the output coupling, $\boldsymbol{W}^{\text{out}}$.

[22] Recall from Section 5.4.2 that full control of a d-order linear system requires a d-order feedback law. The feedforward design considered above was first order and thus needed only a single input (and delay).

[23] The system is the zero-order-hold (ZOH) discretization of $\ddot{x} + 2\zeta\dot{x} + x = u(t)$, for $\zeta = 0.1$ and a discretization time step $\Delta t = 0.1$. The quasiperiodic input is given by $r_k = \sin(2\pi k/30) + \sin(2\pi k\pi/30)$. That is the frequency ratio is π. The triangle input has period 20. Since the resonant frequency of the system has period ≈ 6 time steps, most of the tracking-signal power is 3–5 times below the resonance frequency.

The State of the Art

The above examples demonstrate the potential of neural networks. The one describing online feedback tracking is tantalizingly close to what must be the dream of every control engineer: to force a system to track a desired reference by simply sending the reference to the system and having the controller learn iteratively from its actual performance.

However, the current state of the art is much more mixed than this optimistic assessment would indicate. The theoretical analysis of recurrent neural networks remains unsatisfactory. In creating the examples given above, I used the simplest, most uniform version of reservoir computing, with as generic (and as few) parameters as possible. You are encouraged to code these yourself to try them out. (Or see the book website for codes.)

You will then see that the system often fails to converge to the desired target, or takes far too long to do so. You can easily tweak the various parameters in the reservoir computer and make minor alterations to considerably improve the results for particular cases; the literature is full of such examples.[24] But that misses the point: the attraction of neural-network approaches is precisely in the possibility of generic dynamics. Rather than searching for tweaks that work for a specific case, it would be better to develop a deeper understanding of the dynamics so that stability can be designed into the structure of the network.

10.6 Summary

Adaptive control attempts to learn system dynamics while they are under control. Initial attempts at adaptive control focused on two overall approaches. *Direct methods* such as the local gradient approach known as the *MIT rule* perturb parameters to lower a cost function. Some varieties of these methods such as *model reference adaptive control* (MRAC) make intimate use of the supposed structure of the dynamical equations; others such as extremum-seeking control are model free. A second class, *indirect methods*, proceed by separately estimating unknown parameters and adjusting the control algorithm as a consequence. The estimation methods reformulate conventional least-squares fitting in a recursive algorithm adapted to real-time computation. These *recursive-least-squares* (RLS) methods are similar to the Kalman filter developed in Chapter 8.

These local methods are subject to two related challenges. The first is that in order to completely understand the dynamical behavior, the system needs to explore a wide range of dynamical conditions. Yet control objectives such as regulation probe a very limited set of conditions. In control-theory jargon, they lack *persistence*. It may then

[24] To discuss just one element, the reservoir matrix $\mathbf{W}^{\text{res}}$, there are claims that its spectral radius of $\mathbf{W}^{\text{res}}$ should be smaller than one, except in those cases where it should be greater than one. Sometimes, it should be sparse; sometimes a uniform distribution is better than a Gaussian, etc.

be that a system that is well-adjusted to one kind of condition (say, one particular set point), may be poorly adapted for other conditions. The second is that because even linear systems become nonlinear when parameters are adjustable, adaptation can easily lead to instability, even when methods such as *Lyapunov functions* are used that ensure stable nonadaptive dynamics. The instabilities are even more problematic than ordinary feedback instabilities because they may become apparent only in certain regions of state space. A system can run stably for a long time before suddenly "blowing up." Local methods can deal with both of these problems in a somewhat piecemeal fashion that produces algorithms that work under relatively well-defined conditions (e.g., first-order but not higher-order systems). A third kind of local method, *adaptive feedforward control*, combines feedforward FIR filters, which are inherently stable, with an overall, slow feedback loop that adjusts the filter parameters. This method works particularly well for high-speed calculations for systems that are either fixed or drift slowly. Noise-cancelling headphones are a prominent application.

A more principled approach to adaptation is to optimize the combined problem of learning and control over a longer time frame. This approach has various names, including *dual control* and *reinforcement learning*. Because the optimal-control *Bellman* equations are usually very difficult to solve, we focused on a simple example that illustrates some of the new features that can arise, including phase transitions in control strategy that depend on noise levels. A related powerful new approach combines reinforcement learning with *neural networks* to approximate the learned dynamics. Although much attention is currently focused on feedforward neural networks with several layers (*deep learning*), we presented the somewhat simpler (to formulate) recurrent neural network known as a *reservoir computer*, which has proven very successful for dynamical systems and for hardware implementation. The field is developing extremely quickly.

10.7 Notes and References

Model reference adaptive control (MRAC) dates at least to Whitaker et al. (1958). Our discussion of MRAC and recursive least-squares parameter estimation draws mainly from Åström and Wittenmark (2008), with some help from the slightly simpler presentation by Slotine and Li (1991) and the more advanced treatment of Ioannou and Fidan (2006). For control aspects of the X-15 disaster in 1967, see Dydek et al. (2010). Our presentation of extremum-seeking control follows Brunton and Kutz (2019). For details and proofs, see Ariyur and Krstic (2003).

The innovations approach to state estimation is due to Kailath (1968). Our presentation also draws from Åström (2006a). Its application to the adaptive determination of system parameters via correlation functions is due to Mehra (1970). The correlation-nulling technique was recently used to find the diffusion constant in a tracking problem by Wang and Moerner (2011) and has been extended to nonlinear systems by Berry and Sauer (2013). Kailath et al. (2000) stresses the generality of the innovations approach.

Adaptive signal processing owes much to the work of Widrow and Hoff, who developed the LMS algorithm in 1960. For applications to control problems, see Widrow and Walach (1996). For FIR filters and adaptive algorithms, see Hayes (1996) and then Sayed (2008).

Optimal adaptive control (Section 10.4) is often referred to as *dual control* in the control-theory literature. Our discussion follows generally Chapter 7 in Åström and Wittenmark (2008). See also Klenske and Hennig (2016). The diffusing-particle example is adapted from Åström and Helmersson (1986) and Tonk and Kappen (2010). For other noise-induced transitions, see Horsthemke and Lefever (2006). The dynamic programming presentation follows Bertsekas (2005a). The classic text for reinforcement learning is Sutton and Barto (2018). (See the first two chapters.) Recht (2019) gives an overview from the point of view of continuous control theory. For the computational perspective of PAC learning, developed since the 1980s, see Valiant (2013).

For overviews of neural networks, see MacKay (2003) or Coolen et al. (2005), and Goodfellow et al. (2016) for deep learning. For applications to control, see Narendra and Lewis (2001), Hagan et al. (2002), and Lewis and Ge (2005). Our presentation of reservoir computing follows Jaeger and Haas (2004) for the signal generation and feedforward examples and Sussillo and Abbott (2009) for online, recursive methods. Pathak et al. (2018) show how to use local spatial coupling to model large, spatially extended systems and have succeeded in making model-free predictions of the Kuramoto–Sivashinsky equation out to eight Lyapunov times, in the regime of spatiotemporal chaos. For tweaks to make reservoir computing work in specific cases, see Lukoševičius (2012). The idea of training for the delayed inverse of a dynamical system for feedforward is from Salmen and Plöger (2005).

Since the main requirement of reservoir computing is the need for a complex nonlinear dynamical system, one can implement the state dynamics in hardware. For a recent example for classification, based on telecom fiber optics, see Larger et al. (2017). The authors use an idea, introduced by Appeltant et al. (2011), wherein a single hardware node that has a recursive delay line acts as a kind of virtual spatial array, with one node per delay time. Similar results are possible using FPGA arrays (Haynes et al., 2015).

Neural networks and reservoir computing is an example of *data-driven* control of dynamical systems that works by encoding functional relations. A quite different data-driven approach has been championed by Michael Safonov over the past two decades. His idea is to consider an ensemble of possible controllers and to *falsify* (eliminate) those that cannot produce the desired set of input and output signals, for a given reference command (Safonov and Tsao, 1997). The notion of falsification derives from the philosophy of science of Karl Popper, who argued that scientific theories can never be proven right but can be proven wrong, or falsified (Popper, 1959).[25] Safonov uses

[25] Popper's claim is controversial. Imre Lakatos observed that scientists engage in *research programs*, which include both theory and experiment (Lakatos, 1968). One discordant observation usually will not rule

observed input-output pairs as "data" (with no need for a model linking them) and then rules out controllers that are incompatible with that data. We do not pursue this somewhat ad hoc approach here.

The deterministic adaptive control algorithms from Section 10.1 are implemented in the Matlab Adaptive Control Toolbox. An accompanying book (Ioannou and Fidan, 2006) gives examples of its use and discusses the theory behind the algorithms. There are many free and commercial implementations of adaptive filters, including the DSP System Toolbox in Matlab and noncommercial software to implement specific tasks available from the Matlab community pages. For neural networks and related machine-learning tasks, see the Matlab toolbox on Neural Network Control Systems, routines native in Mathematica since v11, and the open-source TensorFlow library. Many reinforcement learning algorithms have been benchmarked against classic control problems such as the pendulum-swingup or cart-pendulum problems. See the Gym toolkit (https://gym.openai.com/).

Problems

10.1 MRAC stability for a feedforward gain. Analyze the stability of a constant solution for the feedforward adaptive control of an underdamped oscillator presented in Eq. (10.8). That is, let $u_c = u_0 + \delta u_c(t)$, $\theta(t) = (k_m/k) + \delta\theta(t)$, and so on.

a. Define $\mu \equiv \gamma k k_m (u_0)^2$. Show that the gain perturbations $\delta\theta$ obey, to lowest order, $d_{ttt}(\delta\theta) + d_{tt}(\delta\theta) + d_t(\delta\theta) + \mu\,(\delta\theta) = -\gamma k_m u_0 (\delta y)$.
b. The Laplace transform of the $\delta\theta$ dynamics is $s^3 + s^2 + s + \mu = 0$. Deduce that stability of the MRAC system requires $0 < \mu < 1$. Hint: Look up the Routh–Hurwitz theorems, graph the roots, or just prove directly.

In the text, we use a square-wave input, not a constant. However, since the period of the square wave is long compared to the oscillation time scales, the square wave acts as a sequence of steady-state conditions, with "jump perturbations" at the start.

10.2 Normalized MRAC. Normalizing the model reference adaptive control algorithm can stabilize it for all operating amplitudes.

a. Simulate the MRAC system of Figure 10.2. Show stability for $\gamma = 0.1$ and $u_0 = 1$.
b. Confirm numerically that $u_0 > \sqrt{10}$ leads to instability for $\gamma = 0.1$.

out a dearly held theory; rather, some other effect "is to blame." For example, abnormalities in the motion of the planet Mercury were ascribed to the unobserved planet Vulcan for many years but are now taken as evidence that Newtonian gravity is falsified and General Relativity confirmed. Similarly, it is not obvious how "inconsistent" a controller should be with respect to input-output data before eliminating it.

c. The normalized MRAC algorithm replaces $e'(\theta) \to e'(\theta)/(\alpha + |e'(\theta)|^2)$, where α is a small constant. Find the normalized equation for $\dot{\theta}$ analogous to Eq. (10.5).

d. Explain how the algorithm works, using the result in Problem 10.1.

e. Reproduce the plot at right for $\gamma = 0.1$, $k = k_{\mathrm{m}} = a_{\mathrm{m}} = 1$, $a = 2$, $\alpha = 0.001$, and $u_0 = 10$. Without normalization, the response would be unstable.

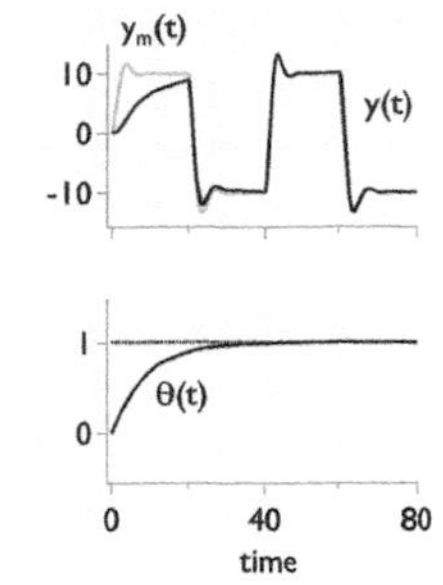

10.3 MRAC to stabilize a first-order system. Consider the system $\dot{y} = ay + u$, with $a > 0$ unknown. Let the desired stable dynamics be $\dot{y}_{\mathrm{m}} = a_{\mathrm{m}} y_{\mathrm{m}} + u_{\mathrm{c}}$, with $a_{\mathrm{m}} < 0$ and $u_{\mathrm{c}}(t)$ an arbitrary input function. Define the control $u = u_{\mathrm{c}} - \theta y$.

a. Show that the error $e = y - y_{\mathrm{m}}$ obeys $\dot{e} = a_{\mathrm{m}} e - (\theta - \theta^*) y$, with $\theta^* \equiv a - a_{\mathrm{m}}$.

b. Show that $V(t) = \frac{1}{2}[e^2 + \frac{1}{\gamma}(\theta - \theta^*)^2]$ is a Lyapunov function if $\dot{\theta} = \gamma y e$.

c. For $u_{\mathrm{c}}(t) = u_0 \neq 0$, show that the stationary solution $\theta(t) = \theta^*$. Why is this solution not valid for $u_0 = 0$?

d. Simulate the adaptive dynamics and show that the steady-state solution of Part (b) is reached in the long-time limit. The solution for parameters $\gamma = a = 1$ and $a_{\mathrm{m}} = -1$, initial conditions $y(0) = 2$, $y_{\mathrm{m}}(0) = \theta(0) = 0$, and input $u_{\mathrm{c}}(t) = 1$ should resemble the plots at right. Note that $\theta^* = 0$. What happens as $u_0 \to 0$?

e. Solve analytically for $u_0 = 0$ for $u_{\mathrm{c}} = y_{\mathrm{m}}(0) = 0$ and $\theta(0) = -a$. Then show that $y_{\mathrm{m}}(t) = 0$ and $\theta(t) = \sqrt{\gamma}\, y_0 \tanh\!\left(\sqrt{\gamma}\, y_0 t\right) - a$ and $y(t) = \sqrt{\gamma}\, y_0 \,\mathrm{sech}(\sqrt{\gamma}\, y_0 t)$. Notice that $\theta(\infty) = \sqrt{\gamma}\, y_0 - a$, but $y(\infty) = y_{\mathrm{m}}(\infty) = 0$: an input signal $u(t)$ that vanishes as $t \to \infty$ will not force $\theta \to \theta^*$, even though $y \to y_{\mathrm{m}}$.

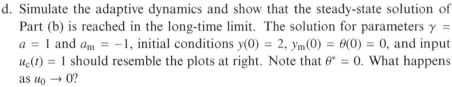

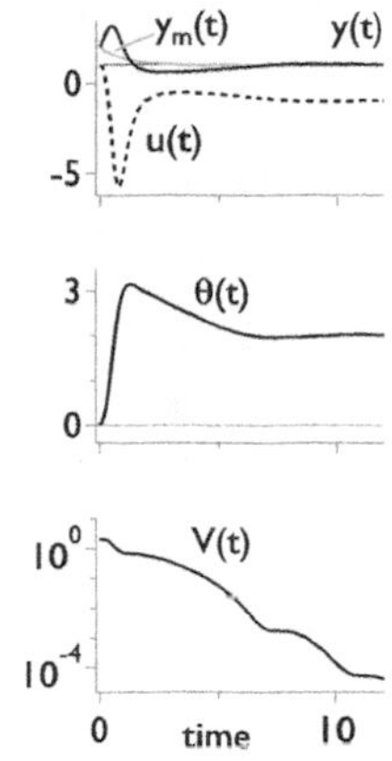

10.4 Lyapunov function for 2nd-order MRAC. In Example 10.3 the Lyapunov construction works if we observe the full state vector $\boldsymbol{e}_x$, or, equivalently, y and $\dot{y}$.

a. Find a Lyapunov function by showing that $\boldsymbol{Q} = \left(\begin{smallmatrix} 1 & 0 \\ 0 & 1 \end{smallmatrix}\right) \implies \boldsymbol{P} = \left(\begin{smallmatrix} \frac{3}{2} & \frac{1}{2} \\ \frac{1}{2} & 1 \end{smallmatrix}\right)$.

b. Show that the adaptation law for $\theta(t)$ becomes $\dot{\theta} = -\gamma(e + 2\dot{e})\, u_{\mathrm{c}}$.

c. Why might it be better to integrate the law for $\dot{\theta}$ in a practical implementation?

d. Verify by simulation that the resulting adaptive system is stable for step commands of any amplitude.

10.5 MRAC with unmodeled dynamics. Consider a Lyapunov control for a model system $y_{\mathrm{m}} = k_{\mathrm{m}} G_0(s) u_{\mathrm{c}}$, with $G_0(s) = \frac{1}{1+s}$ where the actual dynamics are given by $G_1(s) = \left(\frac{1}{1+s}\right)\left(\frac{1}{1+s/\alpha}\right)$, and the feedforward gain $\theta(t)$ is adjusted according to $\dot{\theta} = -\gamma\, u_{\mathrm{c}}\, e$, with $e = y - y_{\mathrm{m}}$ and command input $u_{\mathrm{c}}(t) = u_0 \cos \omega t$. Now analyze the stability of the θ dynamics by assuming a separation of time scales, $\gamma \ll \omega$, so that $\theta(t)$ evolves very slowly compared to the oscillation period of the forcing, $\tau = 2\pi/\omega$.

a. By averaging over the time τ, show that $\dot{\overline{\theta}} \approx -\gamma \overline{\theta} k \left(\overline{u_{\mathrm{c}}\, G_1\, u_{\mathrm{c}}}\right) + \gamma k_{\mathrm{m}} \left(\overline{u_{\mathrm{c}}\, G_0\, u_{\mathrm{c}}}\right)$, where the overline, $\overline{x} \equiv \frac{1}{\tau} \int_0^{\tau} \mathrm{d}t\, x(t)$, denotes averaging over one period.

b. Show that $\overline{\theta}(t)$ is unstable when $u_0^2 \,\mathrm{Re}\, G(\mathrm{i}\omega) < 0$.

c. Show that this happens for the example here when the input frequency $\omega > \sqrt{\alpha}$.

This *method of averaging* is widely used to analyze nonlinear oscillating systems.

10.6 Extremum-seeking control. Assuming that $J(\theta) - J_0 + \frac{1}{2}J''(\theta_0)(\theta - \theta_0)^2$ exactly, analyze Eqs. (10.18), with $k = 50$, $a = 0.2$, $\omega_0 = 2\pi \times 10$, $\omega_h = 2\pi$.

a. By averaging over the modulation period $2\pi/\omega_0$, show that $\dot{\hat{\theta}} \approx -\gamma J'(\hat{\theta})$, Hint: Expand $J[\theta(t)]$ to first order in a, filter DC terms, modulate, filter AC terms.

b. Solve Eqs. (10.18) numerically and confirm the plots of $J(t)$ and $\hat{\theta}(t)$ in Section 10.1.2. Investigate the effects of periodically modulating the coefficient J''.

10.7 Simplest LS. In Example 10.5, we wrote down the formulas for estimating the gain of a linear relationship $y_k = \theta u_k + \xi_k$. Here, we simplify further by choosing $u_k = 1$.

a. Interpret the formulas for $\hat{\theta}$, $\hat{\xi}$, and P in terms of elementary statistics.

b. Introduce a *forgetting factor* λ, with $0 < \lambda < 1$. Assume that observations last long enough that $(1 - \lambda) \gg \frac{1}{k}$. Show that $\hat{\theta}_k \approx (1 - \lambda)\sum_i \lambda^{k-i} y_i$.

c. Show that the recursive version for $\hat{\theta}_k$ is the moving average given in Eq. (10.39).

10.8 Recursive estimation of the noise strength. Derive Eq. (10.27). Reformulate the recursion relation as $\hat{\xi}_0^2 = 0$ and $\hat{\xi}_k^2 = \hat{\xi}_{k-1}^2 + \varepsilon_k^2$ for $1 \le k \le N_p + 1$. For $k \ge N_p + 2$, we have $\hat{\xi}_k^2 = (\frac{k-N_p-1}{k-N_p})\hat{\xi}_{k-1}^2 + (\frac{1}{k-N_p})\varepsilon_k^2$. Hint: Write the first few cases, for $N_p = 1$.

10.9 LS vs. RLS. The goal is to compare the ordinary least squares (LS) equations with their recursive (RLS) counterparts for $y_k = \theta u_k + \xi_k$. (Cf. Examples 10.5 and 10.6). To simplify the analysis, choose a step-function input, $u_k = 1$, and let $\xi^2 = 1$.

a. Show that the LS solution for k steps is $\hat{\theta}_k = \frac{1}{k}\sum y_k$, with variance $P_k = 1/k$.

b. Show that the RLS solution is the same, if you correctly choose P_1 and $\hat{\theta}_1$.

c. How does altering the step-input amplitude to $u_k = u$ affect the convergence?

d. How does choosing a pulse, $u_k = u$ for $k \le K$ and 0 otherwise, affect convergence?

e. (i) Write a simulation to illustrate that LS = RLS only if the RLS initial conditions are chosen correctly. (ii) Compare the convergence for step amplitudes $u = 1$ and $u = 10$. (iii) Show that $\hat{\theta}$ for a finite pulse gets "stuck" when the pulse ends.

10.10 Continuous-time recursive-least-squares (RLS). The continuous-time algorithm is simpler than the discrete-time case and illuminates its structure. For scalar output $y(t)$, the model is $y(t) = \boldsymbol{\varphi}^{\mathsf{T}}(t)\,\boldsymbol{\theta} + \xi_c(t)$, with Gaussian noise $\langle \xi_c(t)\,\xi_c(t') \rangle = \xi_c^2 \delta(t - t')$. If we scale $y \to y/\xi_c$ and $\boldsymbol{\varphi} \to \boldsymbol{\varphi}/\xi_c$, the loss function $\chi^2(\boldsymbol{\theta}) = \int_0^t \mathrm{d}t' \, [y(t') - \boldsymbol{\varphi}^{\mathsf{T}}(t')\,\boldsymbol{\theta}]^2$.

a. Show that $\hat{\boldsymbol{\theta}}(t) = \boldsymbol{P}(t) \int_0^t \mathrm{d}t' \, \boldsymbol{\varphi}(t')\, y(t')$, with $\boldsymbol{P}^{-1}(t) \equiv \int_0^t \mathrm{d}t' \, \boldsymbol{\varphi}(t')\, \boldsymbol{\varphi}^{\mathsf{T}}(t')$ minimizes χ^2. Note that $\boldsymbol{P}(t)$ is the covariance matrix for the estimate $\hat{\boldsymbol{\theta}}(t)$ (see Problem A.8.2 online).

b. Differentiating the equations for $\hat{\boldsymbol{\theta}}$ and $\boldsymbol{P}^{-1}$, derive an equivalent recursive algorithm. Hint: $\mathrm{d}_t \boldsymbol{I} = \mathrm{d}_t(\boldsymbol{P}\boldsymbol{P}^{-1}) = \boldsymbol{0}$. Show that $\mathrm{d}_t\hat{\boldsymbol{\theta}} = \boldsymbol{P}\boldsymbol{\varphi}\varepsilon$ and $\mathrm{d}_t\boldsymbol{P} = -\boldsymbol{P}\boldsymbol{\varphi}\boldsymbol{\varphi}^{\mathsf{T}}\boldsymbol{P}$, with $\varepsilon \equiv y - \boldsymbol{\varphi}^{\mathsf{T}}\hat{\boldsymbol{\theta}}$. The parameter estimate $\hat{\boldsymbol{\theta}}$ changes because of nonzero innovations ε, the difference between the prediction $\boldsymbol{\varphi}^{\mathsf{T}}\hat{\boldsymbol{\theta}}$ and the observation y.

c. Include a forgetting factor λ' by defining $\chi^2(\boldsymbol{\theta}) = \int_0^t \mathrm{d}t'\, \mathrm{e}^{-\lambda'(t-t')}\left[y(t') - \boldsymbol{\varphi}^{\mathsf{T}}(t')\,\boldsymbol{\theta}\right]^2$. Show that the only change to the RLS equations is to take $\dot{\boldsymbol{P}} = \lambda'\boldsymbol{P} - \boldsymbol{P}\boldsymbol{\varphi}\boldsymbol{\varphi}^{\mathsf{T}}\boldsymbol{P}$.

d. Derive the discrete RLS equations, Eq. (10.26). Hint: Apply the *Sherman–Morrison* matrix-inversion formula, Eq. (A.15 online), to $\mathrm{d}_t\boldsymbol{P}^{-1}$.

10.11 Persistent excitation. Input signals that are not persistent can bias parameter estimates. Estimate the parameters of the FIR filter $y_k = b_0 u_k + b_1 u_{k-1} + \xi_k$, where $\xi_k \sim \mathcal{N}(0, 1)$. Using nonrecursive least squares, find the asymptotic parameter estimates and associated covariance matrix for time steps $N \to \infty$ for a step input (what goes wrong?) and for a random input, $u_k \sim \mathcal{N}(0, 1)$. See Åström and Murray (2008).

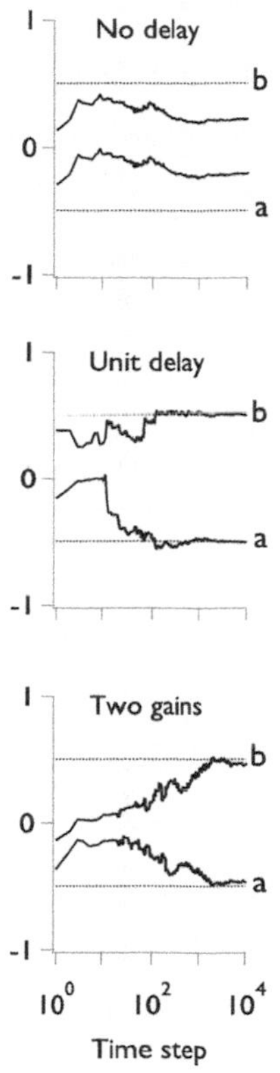

10.12 Identification in closed-loop systems. Investigate $y_{k+1} = -a y_k + b u_k + \xi_k$, from Example 10.7, with $a = -\frac{1}{2}$ and $b = \frac{1}{2}$ and $1 \le k \le N$. Compare two feedback laws: $u_k = -\kappa y_k$ (no delay) and $u_k = -\kappa y_{k-1}$ (unit delay).

a. Show that the closed-loop system is stable for $-1 < \kappa < 3$ for feedback without delay and for $-1 < \kappa < 2$ for feedback with delay. (As usual, delay limits gain.)

b. Show that the inverse covariance matrix $\boldsymbol{P}^{-1} = \boldsymbol{\Phi}^{\mathsf{T}}\boldsymbol{\Phi} \to N\begin{pmatrix} \langle y^2\rangle & -\langle y u\rangle \\ -\langle y u\rangle & \langle u^2\rangle \end{pmatrix}$.

c. For $u_k = -\kappa y_k$, show that $\boldsymbol{P}^{-1}$ is degenerate at large k and hence that the least-squares estimate will not converge.

d. For $u_k = -\kappa y_{k-1}$, show that $\boldsymbol{P}^{-1}$ is invertible if the closed-loop dynamics is stable.

e. For $u_k = -\kappa_1 y_k$ or $-\kappa_2 y_k$, with 50% probability, show that $\boldsymbol{P}^{-1}$ is invertible.

f. Simulate the system and reproduce the graphs at right, showing lack of identifiability for feedback with no delay, convergence with a delay, and also with no delay but two randomly alternating gains.

10.13 Colored-noise. For $x_k = \theta x_{k-1} + \xi_k + a\xi_{k-1}$, with $\langle \xi_k \xi_\ell \rangle = \delta_{k\ell}$,

a. Find $\langle x^2 \rangle$ and $\langle x\, x_{-1} \rangle$ by following the suggestions in Example 10.8.

b. Simulate the system with $\theta = 0.5$ and $a = -0.5$ and analyze by RLS on the raw and filtered signals. Make plots such as the one shown in Example 10.9.

c. For unknown a, implement the extended RLS scheme from Example 10.10. Show numerically that $\hat{\theta} \to \theta$ and $\hat{a} \to a$ unless $a = -\theta$. Why does that case not work?

10.14 Brownian particle with noisy observations. We seek to trap an overdamped particle diffusing in a liquid and measure its mobility and diffusion coefficients. In 1d, the position $x_k = x_{k-1} + \mu u_k + \nu_k$. The measurement $y_k = x_k + \xi_k$, reflecting the finite resolution of the microscope. The mobility is μ, the input u_k. The Gaussian noise terms have $\langle \nu_k\, \nu_\ell \rangle = \nu^2 \delta_{k\ell}$ and $\langle \xi_k \xi_\ell \rangle = \xi_0^2 \delta_{k\ell}$. The variance $\nu^2 = 2DT_{\mathrm{s}}$, with D the diffusion coefficient and T_{s} the sampling time.

a. For instantaneous observation-based feedback, $u_k = -\alpha_0 y_k$, eliminate x to find an equation for y and the noise. Find the correlation functions $\langle y^2\rangle$ and $\langle y y_{-1}\rangle$ and use them to show that the least-squares mobility estimate $\hat{\mu} = \mu\left[1 + \frac{(2-\alpha)\xi^2}{1+2\alpha\xi^2}\right]$, where $\alpha \equiv \mu\alpha_0$ and $\xi^2 = \xi_0^2/\nu^2$. The estimate is biased for $\xi^2 \neq 0$.

b. For $u_k = -\alpha_0 y_{k-1}$, show that $\hat{\mu} = \mu$ for all ξ. Why does delaying the feedback eliminate the bias? (You need not solve for $\langle y^2\rangle$, $\langle y y_{-1}\rangle$, and $\langle y y_{-2}\rangle$ to find $\hat{\mu}$.)

c. Use RLS to estimate μ recursively. Check that $\hat{\mu}$ converges to a biased value (calculate it) for no-delay feedback and to the correct value for unit delay. Include a recursive estimate of ν^2, based on Eq. (10.27). Why is $\hat{\nu}^2$ biased when $\hat{\mu} \neq \mu$?

d. For $\xi_k = 0$ and known μ, show numerically that choosing $\alpha \approx 0.47$ minimizes the position variance, with $\langle y^2\rangle_{\text{min}} \approx 2.4\,\nu^2$. For unknown μ, combine the RLS and simulation codes. Then try to adaptively control the particle variance by choosing the gain $(\alpha_0)_k = 0.47/\hat{\mu}_k$. Why does this algorithm fail? Propose a simple fix and then compare the observed variance with $\langle y^2\rangle_{\text{min}}$.

10.15 Colored noise subtleties. From the Wiener–Khintchine theorem, colored noise with temporal correlations has a nonflat power spectrum. A white-noise sequence ξ_k, with $\langle \xi_k \xi_\ell\rangle = \delta_{k\ell}$, that is filtered by a rational transfer function $C(z)$ has an output power spectrum density $\phi(\omega) = C(e^{i\omega T_s})C(e^{-i\omega T_s})$. Conversely, under reasonable conditions, a measured power spectrum can be approximated by the output of a linear system H driven by ideal white noise (Åström, 2006a). Use this representation to show that if a noise source characterized by $C(z^{-1}, a)$ in the z-domain has zeros at a which is outside the unit circle in the complex z-domain, then the power spectrum of $C(z^{-1}, a^{-1})$ matches that of the original function. Illustrate for $C(z^{-1}) = 1 - az^{-1}$.

10.16 Estimate mobility of a trapped particle by correlation methods. We use the correlation method to study the case of a diffusing particle with unknown mobility, from Example 10.12. Use steady-state statistics (i.e., assume that you analyze a long time series where any initial or final conditions can be neglected).

a. Set up the problem (e.g., define all estimators), including the effects of measurement noise ($y_k = x_k + \xi_k$, with $\langle \xi_k \xi_\ell\rangle = \xi^2 \delta_{k\ell}$).

b. State carefully the simplifications that occur if $\xi^2 = 0$.

c. Derive Eq. (10.47) for the innovation correlations, assuming $\xi^2 = 0$.

10.17 LMS and a variant. Explore the system from Example 10.13.

a. Implement the standard LMS adaptive filter, and reproduce the associated plots.

b. Let $u_k \sim \mathcal{N}(0, u^2)$. Keep $\gamma = 0.1$. What happens as the input amplitude u increases and why? Plot RMS convergence error versus time, for representative values of u.

c. The normalized LMS (n-LMS) algorithm is $\hat{\theta}_{k+1} = \hat{\theta}_k + \gamma\left(\frac{\varphi_k}{\alpha + \varphi_k^{\mathrm{T}}\varphi_k}\right)(y_k - \varphi_k^{\mathrm{T}}\hat{\theta}_k)$. Justify the algorithm when $\alpha = 0$. Why add the small parameter α?

d. Explore n-LMS convergence. Try changing the input level.

10.18 Feedforward control using LMS. In the simplest architecture for adaptive inverse feedforward control, shown at top right, the LMS algorithm does not converge.

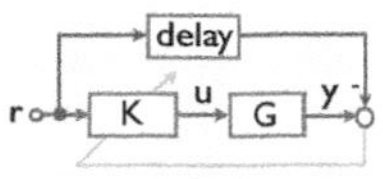

Diverges

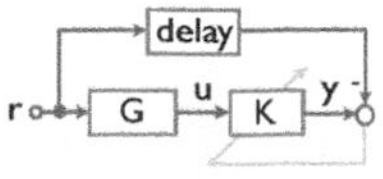

Converges

a. The reversed configuration, shown bottom right, does converge, as suggested by Example 10.14. Implement and make a version of the figures in that example.

b. Consider the more elaborate scheme depicted below. Why should it work?

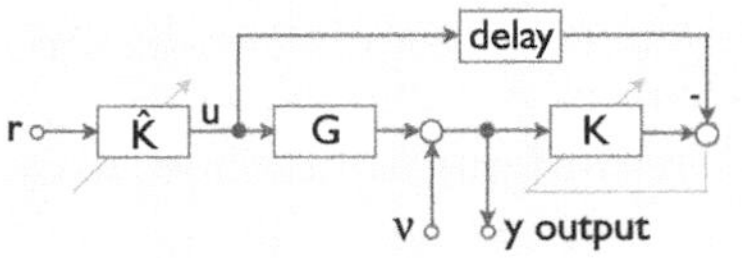

c. Simulate the above example, showing that you can control the system.

d. Add a disturbance $\nu \sim \mathcal{N}(0, \nu^2)$ and show that the above LMS scheme converges for small disturbances but not for large ones. An even more elaborate scheme – copy the system and find the controller off-line – is needed to both follow a reference and compensate for a disturbance. See Widrow and Walach (1996).

10.19 Feedforward with preview. Adaptive feedforward filters can have zero-phase-lag output. Of course, to obey causality, you need to know the reference signal in advance. If the desired waveform is periodic, then you do. We use the feedforward architecture from the previous problem, without disturbances. First, shift the desired reference N steps into the future. Then track this reference with a delay of N steps.

a. To understand the timing, let the G dynamics be a simple delay of Δ. Let the reference be a square wave. Your code should find that the required feedforward filter is zero, except for the "advanced weight," which is one.

b. Next, filter the reference to avoid aliasing. The figure below uses a binomial smoothing filter with $n + 1 = 51$ coefficients (or "taps"), with weights $2^{-n}\binom{n}{k}$. The filter is symmetric (acausal), to follow the reference with no phase shift. Use a square wave (dotted line) as input to generate the filtered reference (dashed line).

c. Input the filtered reference into an LMS algorithm, using the timing from (a). Reproduce the figure at right, for $y_k = (1 - a)y_{k-1} + au_k$, with $a = 0.1$ and a 21-tap LMS filter. The learning rate $\gamma = 0.05$, and $\Delta = 10$. The input (heavy line) leads the output so that the phase-shifted output is centered on the reference. Filtering the reference with more taps reduces the required amplitude for u.

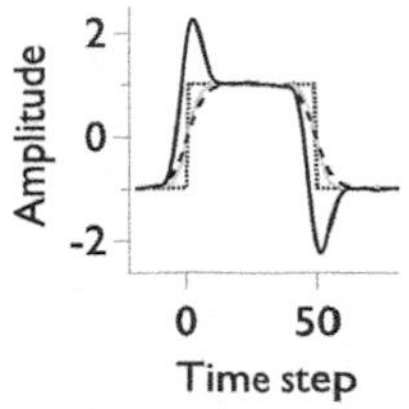

10.20 Learning an unknown charge. One step in the reinforcement-learning example of Section 10.4.2 is to parameterize the knowledge of the unknown charge gained from observations x_k and inputs u_k. In particular, given the diffusive dynamics of Eq. (10.60), use Bayes' theorem to prove the update law $\theta_{k+1} = \theta_k + (x_{k+1} - x_k)u_k$

stated in Eq. (10.63). Hint: Show that you can write $\frac{1}{2}(1+\beta_k) = \frac{\exp(b\theta_k)}{\exp(b\theta_k)+\exp(-b\theta_k)}$. You might want to start by showing the formula works for $k = 1$, given that $\theta_0 = 0$.

10.21 Control of a diffusing particle with unknown sign of charge.

a. Derive the single-stage optimization results in Eqs. (10.64) and (10.66) by averaging over both b and ν_1 and using the equation of motion for x_2.
b. Show that $J^* = 2+\min_{u_0}[(1+R)u_0^2-\frac{f_++f_-}{2(1+R)}]$, with $f_\pm = \langle(\nu_0 \pm u_0)^2 \tanh^2(\nu_0 \pm u_0)u_0\rangle_{\nu_0}$.
c. For large R, assume that u_0 is small. Taylor expand to show $f \approx 3u_0^2 - 4u_0^4$. Deduce Eq. (10.68) and the expression for u_0^*.
d. For small R, assume $u_0 \gg 1$ and show that $(u_0^2)^* \sim -\frac{1}{2}\ln R$, dropping constants and logarithmic corrections. Hint: Show, for $x \gg 1$, that $\tanh^2 x \sim 1 - 4\,\mathrm{e}^{-2|x|}$.
e. For $\beta_0 \neq 0$ (partial knowledge of the sign of the charge) and $x_0 \neq 0$ (biased initial position), show that the bifurcation is biased by the analog of an external field equal to $2\beta_0 x_0$. Why must β_0 and x_0 *both* be nonzero to have a finite field?

10.22 Multiarmed bandits. Consider n slot machines ("bandits"), each paying a reward $J_i \sim \mathcal{N}(Q^{(i)}, 1)$, with average payout $Q^{(i)} \sim \mathcal{N}(0, 1)$: On average, some bandits pay out, while others take money. Given an infinite number of trials, we could determine the payout of each bandit precisely and then play the best ever after. With a finite number of trials, we trade off finding the best bandit with having time to play it. Here, we play 10 bandits 1000 times, repeating for 2000 Monte Carlo trials. We explore different strategies, with the goal of reproducing the figure at left. The estimate of the average payout $\hat{Q}_k^{(i)}$ of bandit i at time k averages the payout over the $n_k^{(i)}$ times it has been observed at time step k. The estimated variance of bandit i at time k is $\hat{\sigma}_i{}^2 = 1/(\text{times played})$. The largest $\hat{Q}_k^{(i)}$ defines the "best bandit" at time k.

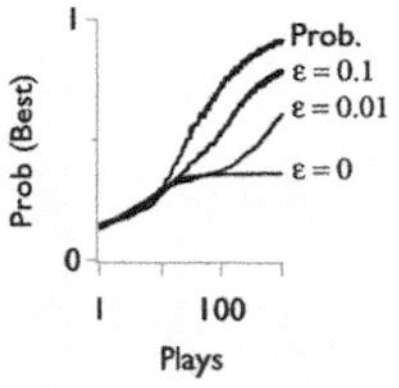

a. *ϵ-Greedy*. At each time step, pick the best bandit with probability $1 - \epsilon$ or another at random with probability ϵ. Three curves are shown here ($\epsilon = 0$, the greedy strategy, and $\epsilon = 0.01$ and 0.1). Notice that the $\epsilon = 0.01$ curve learns more slowly but will eventually surpass the $\epsilon = 0.1$ curve. Why? To maximize the fraction of times we play the best bandit at $k = 1000$, what is the best value of ϵ?
b. *Probabilistic*. At each time step and for each bandit, draw the random number $p_i \sim \mathcal{N}(\hat{Q}_i, \hat{\sigma}_i)$. Play the bandit with the highest p_i. Why might this be an effective strategy? (The rule differs from the typical Boltzmann rule but works better here.)

The optimal solution (Gittins) is complicated. Can you improve our heuristic one?

10.23 Training a recurrent neural network can be hard. Consider a toy RNN, $\dot{x} = -x + \tanh(wx - 1)$, with constant input ($u = -1$) and output $y(t) = x(t)$. Train it by tuning the parameter w, with the goal of minimizing the value of the steady state $x_{ss} \equiv x(t \to \infty)$. Show that the cost function $J(w) = \min_x \left(x_{ss}^2\right)$ has the form at right. Thus, since reasonable cost functions can have discontinuities, training algorithms that perturb coefficients locally can have problems (Doya, 1992).

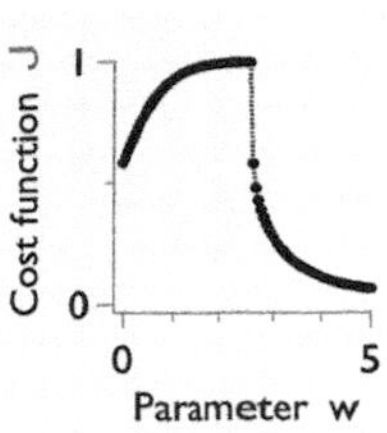

11 Nonlinear Control

Happy families are all alike; every unhappy family is unhappy in its own way.

Leo Tolstoy[1]

In previous chapters, we have discussed systematic ways to treat the linear equations arising in control problems. But the world is nonlinear: almost all dynamical systems obey nonlinear equations that must be treated by a variety of methods, each handling a certain class of nonlinearity in its own way. In Chapter 7, we presented a simple, robust, effective control strategy based on nonlinear feedforward, linear feedback, and event-triggered feedforward corrections. We also suggested ways for handling the particular type of nonlinearity created by the saturation of the control inputs. In this chapter, we consider several different approaches, focusing on nonlinear feedback algorithms that attempt to correct for perturbations that are too strong to be handled effectively by linear feedback algorithms. We also introduce new types of control – new goals – that are inherently nonlinear. Up to now, we have focused on regulation and tracking, which lend themselves well to heavily linear approaches. Other goals, such as synchronization and the creation of new types of motion (chaotic, etc.), have aspects that are more fundamentally nonlinear.

In Section 11.1, we start with the technique of *feedback linearization*, where the goal is to find a change of coordinates that transforms a nonlinear problem into a linear one. Then we may use familiar techniques to control the new system. A very simple heuristic method is *gain scheduling*, an algorithm for regulating nonlinear systems near a set point. The more systematic study of the conditions where linearization is possible leads to *geometric methods* for studying nonlinear dynamical systems. In Section 11.2, we return to the concept of *Lyapunov function*, introduced in Section 2.5.2, showing how it can be used to design a nonlinear control law. In Section 11.3.3, we discuss the *synchronization* of dynamical systems, as the simplest example of using control to create a collective motion. This is another way to view control: if a "control" dynamical system generates the desired state or output and a physical system synchronizes and follows that state or output, then we have controlled the physical system. More subtly, we may want to force two systems to behave identically, without specifying precisely how they behave at a given time. In Section 11.4, we discuss the control of *chaotic* dynamical systems. Part of our motivation is historical: beginning in 1990, the subject has attracted great interest from a wide group of physicists. We discuss how that work fits with the rest of our book. In addition, some of the specific methods we discuss give

[1] Opening line of *Anna Karenina*, Moscow, 1878.

good examples of how to leverage specific features of nonlinear dynamical systems to advantage in designing control laws.

11.1 Feedback Linearization

If nonlinear systems are hard and linear systems easy to solve, can we transform a nonlinear system into a linear one? Let us pursue this idea, starting from simple cases involving static transformations that are independent of the system dynamics and then proceeding to more elaborate cases. Previously, we have used *local linearization* to approximate a nonlinear system by a linear one in a small neighborhood about a (time-dependent) reference point in state space. Section 11.1.1 extends this idea to *gain scheduling*: we choose feedback parameters as a function of the desired reference point, using local linear dynamics and thereby ensuring a good linear controller under a wide range of operating conditions.

To go beyond such local approximations, we seek an exact transformation.[2] Starting from intuitive examples, we will develop a systematic approach based on concepts from differential geometry, Lie brackets in particular. The result nicely generalizes the Kalman rank condition on controllability (Section 4.1.1). The more general matrix we derive below will tell us whether we can transform a nonlinear dynamical system that is linear in the control variable (a *control affine* system) into a controllable linear system. Unfortunately, we will also see that such a change of variables is not sufficient to guarantee that the original system is controllable. Nonlinear systems come with their own challenges.

11.1.1 Gain Scheduling

Consider the nonlinear system $\dot{x}(t) = f(x,u)$, $y = h(x)$, for inputs u, outputs y, and internal states x. To regulate the system near a reference state x_0 (maintained by $u = u_0$), we linearize (Section 2.4): $\dot{x}(t) = Ax + Bu$ and $y = Cx$, with system matrices given by $A = \partial_x f$, $B = \partial_u f$, $C = \partial_x h$, with all derivatives evaluated at $\{x_0, u_0\}$. The control u is chosen to be a linear function of the state, $u = -Kx$. Since the gain matrix K depends on the system matrices $\{A, B, C\}$, we should choose it as a function of $y_0 = h(x_0)$. That is, based on x_0, we choose ("schedule") the gains $K(x_0)$.

As a feedforward operation, gain scheduling is simple and allows a quick response to changes in operating conditions; however, finding the optimal values of controller parameters for all anticipated conditions can be hard work. System dynamics must

[2] There is a superficial resemblance between feedback linearization and the theory of *normal forms* for dynamical systems (Guckenheimer and Holmes, 2002, Section 3.3). In both, the goal is to simplify nonlinear equations via a change of variables. Normal forms are used for autonomous systems with no control input, and the result is a (standard) *nonlinear equation*. Here, the control variable $u(t)$ gives an extra degree of freedom that can lead to complete linearization, not just the reduction to a standard nonlinear form.

be measured and may change. And even if you know the dynamics, the changing gains for different operating points may lead to large transients, even instability. Such complications rule out gain scheduling for all but rather simple situations.

Example 11.1 The response of a nonlinear dynamical system depends on the set point level. Consider a simplified description of temperature control (just a first-order relaxation here) with voltage control. But since the system is linear with respect to power, not voltage, as an input, the system is nonlinear. (Recall that power $P = V^2/R$.) Thus,

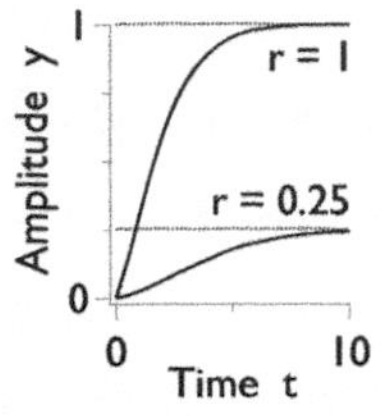

$$\dot{y} = -y + u^2\,, \qquad u = K_\mathrm{p}(r - y) + K_\mathrm{i} \int_0^t \mathrm{d}t' \, [r - y(t')]\,, \tag{11.1}$$

where we assume a PI controller. At left, we show step responses with amplitudes $r = 1$ and 0.25. The gains are fixed at $K_\mathrm{p} = K_\mathrm{i} = 0.5$. Because of the integral control, the system stabilizes at $y = r$. But notice that response speed also varies: the $r = 1$ step response is faster than the $r = 0.25$ response. Here, the effect of the nonlinearity is slight. In more complicated systems, the differences can be more dramatic. In Problem 11.2, we show how gain scheduling can keep the response speed independent of the set point r.

In this example, the nonlinearity is easy to account for exactly, simply by defining a new input $v = u^2$. This static transformation removes the nonlinearity, and we can simply design a linear controller, valid for all amplitudes, in terms of v. Then we transform back to u to compute the r-dependent output.

11.1.2 Static Linearization

An easy class of nonlinearity includes systems with linear evolution for the underlying state space but with static nonlinearities that affect either the input or output, or both. We can then use *pre-* and *postcompensators* to transform the input and outputs to new variables in terms of which the dynamics are linear. In particular, we look at dynamics of the form

$$\dot{\boldsymbol{x}} = \boldsymbol{A}\boldsymbol{x} + \boldsymbol{g}(u)\,, \quad y = h(\boldsymbol{x})\,, \tag{11.2}$$

with $\boldsymbol{g}(u)$ a nonlinear n-dimensional function of a scalar input $u(t)$ and $h(\boldsymbol{x})$ a nonlinear, scalar function of an n-dimensional state vector $\boldsymbol{x}$. Here, $\boldsymbol{A}$ is an $n \times n$ dynamical matrix. Assume that we can write $\boldsymbol{g}(u) = \boldsymbol{B}b(u)$, with $b(u)$ a *scalar* nonlinear function and $\boldsymbol{B}$ a constant n-dimensional vector (we restrict ourselves to SISO systems). If $b(\cdot)$ is invertible, then the *precompensator* $u = b^{-1}(v)$ linearizes the dynamics, with $\dot{\boldsymbol{x}} = \boldsymbol{A}\boldsymbol{x} + \boldsymbol{B}v$.

Similarly, assume that y can be written in the form $y = h(\boldsymbol{C}\boldsymbol{x})$, with $\boldsymbol{C}\boldsymbol{x} \equiv z$ a *scalar* quantity. Then, if $z = h^{-1}(y)$ exists, we will have converted the nonlinear dynamical relation $u \to y$ into a linear dynamical relation $v \to z$,

$$\dot{\boldsymbol{x}} = \boldsymbol{A}\boldsymbol{x} + \boldsymbol{B}v\,, \quad z = \boldsymbol{C}\boldsymbol{x}\,, \tag{11.3}$$

that can be controlled with linear feedback of the form $v = -K\hat{x}$, where $\hat{x}$ is an estimate of the state based on the transformed observations $z(t)$, as discussed in Chapters 4 and 8.

Example 11.2 (Temperature Control) In temperature control, there is a nonlinear relationship between power and voltage, $P = V^2/R$, where R is the heater resistance, V the voltage across the heater, and P the power dissipated by the load. As in Example 11.1, we model proportional-integral (PI) temperature control via $\dot{y} = -y + u^2$, with the controller $u = \sqrt{PR}$. A naive proportional-integral (PI) controller uses Eq. (11.1) for u_{naive}, while the feedback-linearization strategy uses $u^2 = u_{\text{naive}} \equiv v$. To implement this strategy, we have to invert the nonlinear function $u^2 = v$. Since only the positive inverse is physically relevant, we set $u = +\sqrt{v}$ and discard the negative root.

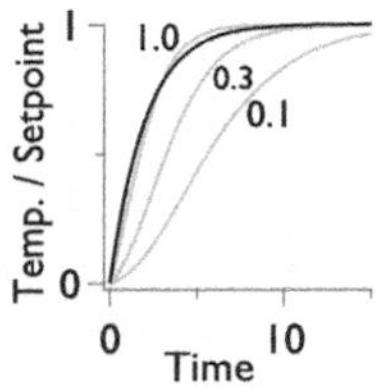

At right, we plot the normalized step response of the temperature, $y(t)/y(\infty)$. The light curves show the varying normalized response for set points of 0.1, 0.25, and 1.0 when using naive PI control, $K_p + K_i/s$, with $K_p = K_i = \frac{1}{2}$. The dark curve shows that all three set points give the *same* response for nonlinear PI control.

The relation $y = g(z)$ between temperature z and measured voltage, y, is also often nonlinear and can be dealt with using a similar strategy involving a postcompensator.

Example 11.3 (Remove a dead zone) DC motors are cheap and powerful but suffer from *stiction*, the static friction that must be overcome in order to move. At right (top) is a typical plot of rotation ω versus applied voltage V. For low voltages $|V| < V_0$, there is a *dead zone*: the motor is stuck and does not rotate. For $|V| \geq V_0$, the motor rotates with a linearly increasing angular velocity ω, with

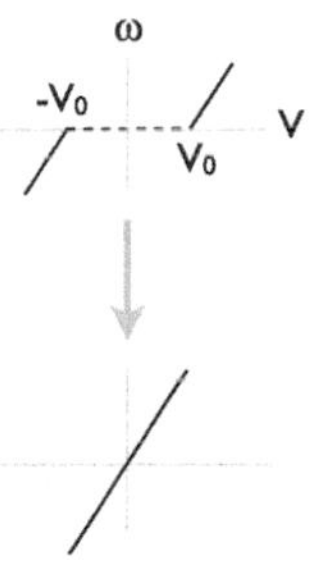

$$\omega = \begin{cases} k(V - V_0) & V > V_0 \\ 0 & -V_0 < V < +V_0 \\ k(V + V_0) & V < -V_0 \,. \end{cases} \tag{11.4}$$

A simple nonlinear transformation, $V(t) = u(t) + \text{sign}\,(u)V_0$, linearizes the relation to $\omega = ku$. The only difficulty is the need to measure V_0, which can change. An alternative, open-loop strategy, *pulse-width modulation* (PWM), sends a rapidly altering square wave of amplitude $V \gg V_0$. Varying the duty cycle varies ω. See Section 5.1.2 on dithering.

11.1.3 Canceling the Nonlinearity

The idea of transforming variables to linearize exactly a nonlinear system can work when the transformation of the input $u(t)$ involves the state variables $\boldsymbol{x}(t)$, too.

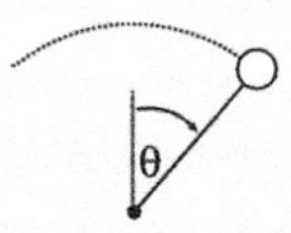

Example 11.4 (Stabilizing the nonlinear pendulum) For the undamped simple pendulum sketched at left,

$$\ddot{\theta} - \sin\theta = u\,, \quad u = -\sin\theta - K_{\rm p}\theta - K_{\rm d}\dot{\theta} \qquad \Longrightarrow \qquad \ddot{\theta} + K_{\rm d}\dot{\theta} + K_{\rm p}\theta = 0\,. \tag{11.5}$$

In Eq. (11.5), we transform the nonlinear pendulum dynamics via a nonlinear proportional-derivative (PD) control into a linear system with characteristic equation $s^2 + K_{\rm d}s + K_{\rm p} = 0$, which implies that we can choose the proportional gain $K_{\rm p}$ and derivative gain $K_{\rm d}$ to set the two poles $p_\pm = -\frac{1}{2}(K_{\rm d} \pm \sqrt{K_{\rm d}^2 - 4K_{\rm p}})$ at any desired position in the complex s-plane. Of course, the usual limitations affecting linear controller design apply (Section 3.7). See Problem 4.9 for more on the design of the linear controller. One general issue is that the angle of a pendulum, θ, can only take on values in a range of 2π, whereas the linear system in Eq. (11.5) allows θ to attain any value. Example 11.8 resolves this difficulty.

The substitution in Eq. (11.5) is easily generalized to nth-order equations,

$$y^{(n)} = f(y) + g(y)\,u\,, \tag{11.6}$$

by choosing $u = (v - f)/g$, which implies $y^{(n)} = v(t)$. This is an nth-order integrator that can be stabilized by choosing n feedback gains $v(t) = -K_0 - K_1\dot{y} - \ldots - K_{n-1}y^{(n-1)}$ to place the n poles as desired in the complex s-plane. Some caveats are the need to observe or estimate $y(t)$ and its n derivatives and the need to know the form of $f(y)$ and $g(y)$ exactly. Also, if $g(y) \approx 0$, then $u(t)$ becomes very large. Problem 11.4 discusses how to make the feedback more robust if the nonlinear dynamics are not precisely known.

Differentiating the Output

Let us now consider the *nonlinear state space* representation of Eq. (11.6),

$$\frac{\mathrm{d}}{\mathrm{d}t}\begin{pmatrix} x_1 \\ x_2 \\ \vdots \\ x_{n-1} \\ x_n \end{pmatrix} = \begin{pmatrix} x_2 \\ x_3 \\ \vdots \\ x_n \\ f(x_1) + g(x_1)u(t) \end{pmatrix}, \qquad y = x_1\,. \tag{11.7}$$

The special feature of Eq. (11.7) is that the nonlinearity $f(x)$ and control $u(t)$ affect the same component, x_n. What if the nonlinearity and control affect different terms? For

$$\frac{\mathrm{d}}{\mathrm{d}t}\begin{pmatrix} x_1 \\ x_2 \end{pmatrix} = \begin{pmatrix} x_2^2 \\ u \end{pmatrix}, \qquad y = x_1\,, \tag{11.8}$$

the idea of Eq. (11.7) will not work. But here is a trick: Simply differentiate the output variable $y(t)$ until the input variable $u(t)$ appears. This gives a closed system of equations for $y(t)$, and we can again choose $u(t)$ to cancel the nonlinearities.

Applying this trick to Eq. (11.8), we have $\dot{y} = \dot{x}_1 = x_2^2$ and $\ddot{y} = 2x_2\dot{x}_2 = 2x_2 u$. Thus,

$$u(t) = \frac{1}{2x_2}\left(-K_\mathrm{d}\dot{y} + K_\mathrm{p}[y_0 - y(t)]\right) \tag{11.9}$$

leads to the closed-loop dynamics of Eq. (11.5). Note that we need to estimate x_2 in the nonlinear feedback law. We should avoid the initial condition $x_2(0) = 0$.

The number of times you need to differentiate y before u appears is known as the *relative degree* of a dynamical system. As an example, for the pendulum,

$$\dot{x}_1 = x_2\,, \qquad \dot{x}_2 = -\sin x_1 + u\,, \qquad y = x_1\,, \tag{11.10}$$

the recipe gives a relative degree of two. In Problem 11.7, we will see that the relative degree generalizes the notion of pole excess (degree of the denominator minus the degree of the numerator) for rational transfer functions of linear systems. Here, the linear pendulum has a transfer function $G(s) = \frac{1}{1+s^2}$, which has a pole excess of two.

Internal Dynamics

One issue that can arise when forming an equation relating the input $u(t)$ to the output $y(t)$ is that part of the state-space dynamics is inaccessible. In linear systems (Chapter 4), we associated such internal dynamics with zeros. As a nonlinear example, consider

$$\frac{\mathrm{d}}{\mathrm{d}t}\begin{pmatrix} x_1 \\ x_2 \end{pmatrix} = \begin{pmatrix} x_2^3 + u \\ u \end{pmatrix}, \qquad y = x_1\,. \tag{11.11}$$

Differentiating the output $y(t)$ gives

$$\dot{y} = \dot{x}_1 = x_2^3 + u\,, \tag{11.12}$$

which contains the input. Thus, we choose the control law $u = -x_2^3 + v$, with $v = -x_1$ to stabilize the point $x_1 = 0$. The first state equation is then $\dot{x}_1 = -x_1$, which converges to zero. The second state equation, however, is

$$\dot{x}_2 = u = -x_2^3 - x_1\,, \tag{11.13}$$

which is a nonlinear differential equation that can be thought of as nonautonomous in the sense that $x_1(t)$ appears as an external driving term. In Problem 11.8, you will show that $x_2(t)$ also converges to 0 at long times. Conversely, if the second equation in Eq. (11.11) is modified to $\dot{x}_2 = -u$, then the internal dynamics is unstable.

Thus, the same issues of observability and controllability that we considered in Chapter 4 in the context of linear systems can be present in nonlinear systems, as well.

Zero Dynamics

Recall (Section 4.1.3, Eq. 4.24) that a transfer function $G(s)$ for a linear system with zeros,

$$\frac{b_0 s^k + b_1 s^{k-1} + \cdots b_k}{s^n + a_1 s^{n-1} + \cdots a_n} \implies y^{(n)} + a_1 y^{(n-1)} + \cdots a_n y = b_0 u^{(k)} + \cdots b_k u \,. \tag{11.14}$$

Equation (11.14) implies that the solutions $u(t)$ to the "numerator equation,"

$$b_0 u^{(k)} + b_1 u^{(k-1)} + \cdots + b_k u = 0 \,, \tag{11.15}$$

are "blocked" from exciting the dynamics of the output $y(t)$ in the sense that they lead to zero output. Equation (11.15) defines the *zero dynamics* of the original dynamical system. We then found that left-hand-plane (LHP) zeros corresponding to stable zero dynamics are relatively benign whereas right-hand-plane (RHP) zeros correspond to unstable dynamics that impose fundamental limitations on feedback performance.

This distinction between left- and right-hand-plane zeros in linear dynamics generalizes to internal dynamics. The idea is to first identify the internal-dynamics equations and then to solve for the $u(t)$ that forces the system output $y(t) = 0$ for all time. Because such an input could come from a disturbance, it cannot be corrected by feedback, since the output signal $y(t) = 0$. As long as the system is naturally stable to such disturbances, such a situation is likely tolerable. But if such a disturbance leads to instability in the internal state variables, then they will diverge.

To summarize our informal method of feedback linearization:

- differentiate the output y until the input u appears;
- choose u to cancel nonlinearities and create desired linear dynamics;
- test whether the resulting internal dynamics is stable.

In the next section, we adopt a more systematic approach that works for *control affine* systems, which are linear in the control variable u but nonlinear in the state vector $\boldsymbol{x}$.

11.1.4 Control-Affine Systems

Nonlinear, control-affine state-space systems have the form

$$\dot{\boldsymbol{x}} = \boldsymbol{f}(\boldsymbol{x}) + \boldsymbol{g}(\boldsymbol{x})\, u \,, \tag{11.16}$$

where the n-dimensional state vector $\boldsymbol{x}(t)$ is assumed to be observable, $u(t)$ is a scalar input, and $\boldsymbol{f}$, $\boldsymbol{g}$ are nonlinear functions. To linearize the dynamics, we seek a change of coordinates $z = \boldsymbol{T}(\boldsymbol{x})$ and a control $u = \alpha(\boldsymbol{x}) + \beta(\boldsymbol{x})v$ that leads to a canonical *linear* system,

$$\dot{z} = \boldsymbol{A}_c z + \boldsymbol{B}_c v \,, \tag{11.17}$$

where the $n \times n$ $\boldsymbol{A}_c$ and n-dimensional $\boldsymbol{B}_c$ are given by

$$\boldsymbol{A}_c = \begin{pmatrix} 0 & 1 & 0 & \cdots & & 0 \\ 0 & 0 & 1 & \cdots & & 0 \\ \vdots & \vdots & & & & \vdots \\ \vdots & \vdots & & \cdots & 0 & 1 \\ 0 & 0 & & \cdots & 0 & 0 \end{pmatrix}, \qquad \boldsymbol{B}_c = \begin{pmatrix} 0 \\ 0 \\ \vdots \\ 0 \\ 1 \end{pmatrix}. \tag{11.18}$$

The coordinate transformation $\boldsymbol{T}(\boldsymbol{x})$ should be a local *diffeomorphism*: one-to-one, onto, and continuously differentiable in a region about $\boldsymbol{x}$, with an inverse that is also continuously differentiable in a region about z. The transformation leads to

$$\dot{z} = \frac{\partial \boldsymbol{T}}{\partial \boldsymbol{x}} \dot{\boldsymbol{x}} = \frac{\partial \boldsymbol{T}}{\partial \boldsymbol{x}} \left[\boldsymbol{f}(\boldsymbol{x}) + \boldsymbol{g}(\boldsymbol{x})\, u \right], \tag{11.19}$$

with $\frac{\partial T}{\partial x}$ an $n \times n$ Hessian matrix.

To choose $\boldsymbol{T}(\boldsymbol{x})$ so that the dynamics have the form of Eq. (11.17), we follow our previous discussion (see Eq. 11.7) and try to write equations in the form

$$\frac{\mathrm{d}}{\mathrm{d}t} \begin{pmatrix} z_1 \\ z_2 \\ \vdots \\ z_{n-1} \\ z_n \end{pmatrix} = \begin{pmatrix} z_2 \\ z_3 \\ \vdots \\ z_n \\ \beta^{-1}\left[u - \alpha\right] \end{pmatrix}. \tag{11.20}$$

If we can, then $u = \alpha + \beta v$ is a *linearizing control law* and leads to Eq. (11.17), with z the *linearizing state*. Matching the terms in Eqs. (11.19) and (11.20) gives

$$\frac{\partial T_1}{\partial \boldsymbol{x}} \boldsymbol{f} = T_2, \quad \frac{\partial T_2}{\partial \boldsymbol{x}} \boldsymbol{f} = T_3, \quad \ldots, \quad \frac{\partial T_{n-1}}{\partial \boldsymbol{x}} \boldsymbol{f} = T_n, \quad \frac{\partial T_n}{\partial \boldsymbol{x}} \boldsymbol{f} = -\alpha\beta^{-1}, \tag{11.21a}$$

$$\frac{\partial T_1}{\partial \boldsymbol{x}} \boldsymbol{g} = 0, \quad \frac{\partial T_2}{\partial \boldsymbol{x}} \boldsymbol{g} = 0, \quad \ldots, \quad \frac{\partial T_{n-1}}{\partial \boldsymbol{x}} \boldsymbol{g} = 0, \quad \frac{\partial T_n}{\partial \boldsymbol{x}} \boldsymbol{g} = \beta^{-1} \neq 0. \tag{11.21b}$$

If we can find $T_1(\boldsymbol{x})$, then the recursion equations give $T_2, \ldots, T_n$, as well as α and β. Recall that the gradient $\frac{\partial T_1}{\partial x}$ is a row vector. Figure 11.1 illustrates the feedback-linearization setup via a block diagram.

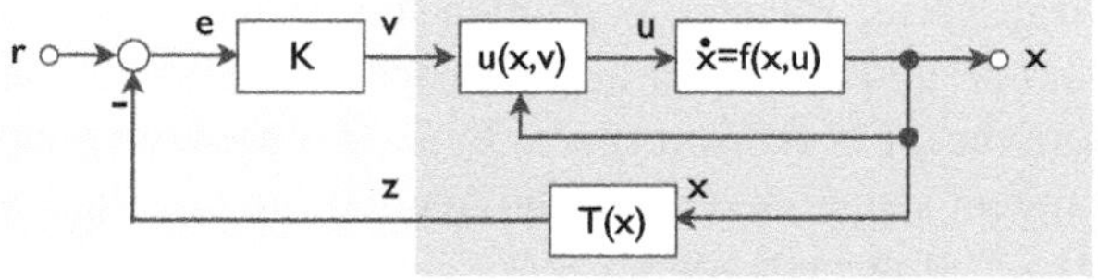

Fig. 11.1 Block diagram showing the feedback-linearization setup. The nonlinear elements are located in the darker-shaded region. Outside, the system behaves linearly.

Example 11.5 (Two-dimensional system) Consider a two-dimensional state space with a simple input:

$$\frac{\mathrm{d}}{\mathrm{d}t}\begin{pmatrix} x_1 \\ x_2 \end{pmatrix} = \begin{pmatrix} f_1(\boldsymbol{x}) \\ f_2(\boldsymbol{x}) \end{pmatrix} + \begin{pmatrix} 0 \\ 1 \end{pmatrix} u \,. \tag{11.22}$$

Intuitively, since only x_2 couples directly to the input, f_1 must depend on x_2 in order to control $x_1(t)$. We will see this below. From Eq. (11.21),

$$\frac{\partial T_1}{\partial \boldsymbol{x}}\boldsymbol{f} = \frac{\partial T_1}{\partial x_1} f_1 + \frac{\partial T_1}{\partial x_2} f_2 = T_2 \,, \qquad \frac{\partial T_1}{\partial \boldsymbol{x}}\boldsymbol{g} = \frac{\partial T_1}{\partial x_2}(1) = 0 \,. \tag{11.23}$$

Thus, $T_1 = T_1(x_1)$ and $\frac{\partial T_1}{\partial x_1} f_1 = T_2$. The second set of equations is

$$\frac{\partial T_2}{\partial x_1} f_1 + \frac{\partial T_2}{\partial x_2} f_2 = -\alpha\beta^{-1} \,, \qquad \frac{\partial T_2}{\partial x_2} = \beta^{-1} \,. \tag{11.24}$$

At this point, we make a guess and try the simplest possible function, $T_1(x_1) = x_1$. Then $T_2 = f_1$ and $\beta^{-1} = \frac{\partial f_1}{\partial x_2}$. From Eq. (11.24),

$$\frac{\partial f_1}{\partial x_1} f_1 + \frac{\partial f_1}{\partial x_2} f_2 = -\frac{\partial f_1}{\partial x_2}\alpha \,, \tag{11.25}$$

and the control $u = \alpha + \beta v$ implies that

$$u = -\left(\frac{\partial f_1}{\partial x_2}\right)^{-1}\left[f_1 \frac{\partial f_1}{\partial x_1} + f_2 \frac{\partial f_1}{\partial x_2} + v\right] = -f_2 + \left(\frac{\partial f_1}{\partial x_2}\right)^{-1}\left[v - f_1 \frac{\partial f_1}{\partial x_1}\right] . \tag{11.26}$$

In Eq. (11.26), if f_1 does not depend on x_2, then u will diverge. In Problem 11.9, you will check that the functions $\boldsymbol{T}$ and u do linearize Eq. (11.22).

Unfortunately, the ability to transform a nonlinear system into a controllable linear system does not always mean that you can reach arbitrary states in the original system. Consider $\dot{x}_1 = x_2^2$ and $\dot{x}_2 = u(t)$, a special case of Example 11.5. Even though there are coordinate changes that result in a linear, controllable system, it is still not possible to reach an arbitrary state from the origin. (See Problem 11.10.)

11.1.5 Geometrical Methods

The conditions of Eq. (11.21) have a geometric interpretation. Recall that a differential equation such as $\dot{\boldsymbol{x}} = \boldsymbol{f}(\boldsymbol{x})$ can be interpreted as a vector field $\boldsymbol{f}(\boldsymbol{x})$, which gives the rate $\boldsymbol{f}$ of change of the state vector $\boldsymbol{x}$ at a given point in state space.[3] The state vector $\boldsymbol{x}$ is an element of a *manifold* M, and the vector field $\boldsymbol{f}$ assigns an element of the *tangent space* to M at $\boldsymbol{x}$. We denote the set of tangent spaces by $T_{\boldsymbol{x}}M$. Often, we will be interested in subspaces to the tangent space. A *distribution* $\Delta(\boldsymbol{x})$ on the manifold M is a map that assigns a subspace to $T_{\boldsymbol{x}}M$ at each point $\boldsymbol{x} \in M$.

[3] For lack of space, we will not give the full development of the differential-geometric point of view, where a vector field is defined as a differential operator $f_i(\boldsymbol{x})\partial_i$, with $f_i(\boldsymbol{x})$ the component of $\boldsymbol{f}$ evaluated at $\boldsymbol{x}$ in a coordinate system with components x_i and $\partial_i = \frac{\partial}{\partial x_i}$.

Solutions to the differential equation can then be visualized as a "flow" in state space. Think of a fluid particle being swept along a path $\boldsymbol{x}(t)$ guided by $\boldsymbol{f}(\boldsymbol{x})$, as illustrated at right for the van der Pol equation introduced in Section 2.6 (with $\epsilon = 0.5$). The solution $\boldsymbol{x}(t)$ defines an *integral curve*.

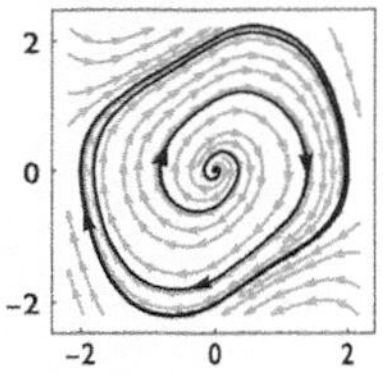

In this picture, $\frac{\partial T_1}{\partial \boldsymbol{x}}\boldsymbol{f}$ corresponds to the rate of change of the scalar quantity T_1, along the direction defined by the solution $\boldsymbol{x}(t)$. This is just the directional derivative along the direction defined by the vector field $\boldsymbol{f}(\boldsymbol{x})$. Such a concept is captured by the *Lie derivative*, which for a scalar field $h(\boldsymbol{x})$ and vector field $\boldsymbol{f}(\boldsymbol{x})$, is defined as

$$L_f h \equiv \frac{\partial h}{\partial \boldsymbol{x}} \boldsymbol{f}(\boldsymbol{x}) . \tag{11.27}$$

In terms of Lie derivatives, Eq. (11.21) becomes

$$L_f T_1 = T_2 , \quad L_f T_2 = T_3 , \quad \dots , \quad L_f T_{n-1} = T_n , \quad L_f T_n = -\alpha\beta^{-1} , \tag{11.28a}$$

$$L_g T_1 = 0 , \quad L_g T_2 = 0 , \quad \dots , \quad L_g T_{n-1} = 0 , \quad L_g T_n = \beta^{-1} \neq 0 . \tag{11.28b}$$

We can further simplify the equations by defining higher-order Lie derivatives as $L_f^{k+1} h \equiv L_f \left(L_f^k h \right)$ and also $L_f^0 h \equiv h$. Then, Eq. (11.28) implies

$$L_f^{k-1} T_1 = T_k , \qquad L_g L_f^k T_1 = 0 , \qquad k = 0, 1, \dots , n-2 . \tag{11.29}$$

Assuming that we can solve for $T_1(\boldsymbol{x})$, we can use the recursion relations Eq. (11.29) to solve for $T_2, \dots , T_n$. Then we find $\beta = (L_g T_n)^{-1}$ and $\alpha = -\frac{L_f T_n}{L_g T_n}$.

Lie Brackets

To see whether it is possible to solve for T_1, we introduce another geometrical concept, the *Lie bracket* of vector fields. Given two vector fields $\boldsymbol{f}(\boldsymbol{x})$ and $\boldsymbol{g}(\boldsymbol{x})$,

$$[\boldsymbol{f}, \boldsymbol{g}](\boldsymbol{x}) = \frac{\partial \boldsymbol{g}}{\partial \boldsymbol{x}} \boldsymbol{f} - \frac{\partial \boldsymbol{f}}{\partial \boldsymbol{x}} \boldsymbol{g} , \tag{11.30}$$

where $\frac{\partial \boldsymbol{f}}{\partial \boldsymbol{x}}$ is an $n \times n$ Jacobian matrix. The Lie bracket is itself another vector field. To motivate the definition of a Lie bracket, we return to the interpretation of a vector field as generating a flow. Let us consider four piecewise-constant flows, defined by letting the state $\boldsymbol{x}(t)$ evolve according to $\dot{\boldsymbol{x}} = \boldsymbol{f}(\boldsymbol{x})$ from time 0 to time ϵ. Then, from ϵ to 2ϵ, it evolves according to $\dot{\boldsymbol{x}} = \boldsymbol{g}(\boldsymbol{x})$. From 2ϵ to 3ϵ, it evolves according to $\dot{\boldsymbol{x}} = -\boldsymbol{f}(\boldsymbol{x})$. Finally, from 3ϵ to 4ϵ, it evolves according to $\dot{\boldsymbol{x}} = -\boldsymbol{g}(\boldsymbol{x})$. At the end of this circuit, depicted at right, it may happen that the state fails to return to its origin. In general, it will miss the origin by $\epsilon^2[\boldsymbol{f}, \boldsymbol{g}]$. The Lie bracket then assigns this displacement vector to each starting point $\boldsymbol{x}$. In Problem 11.12, you will establish this geometrical property, as well as others such as bilinearity, skewness, and the Jacobi identity.

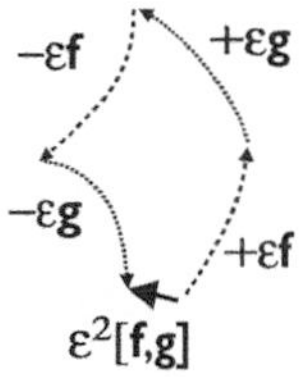

Lie brackets take two vector fields and turn them into a new vector field. At a given point in state space $\boldsymbol{x}$, is the new vector a linear combination of the old ones? More precisely, if

$$[\boldsymbol{f}, \boldsymbol{g}](\boldsymbol{x}) = \alpha_1(\boldsymbol{x})\boldsymbol{f} + \alpha_2(\boldsymbol{x})\boldsymbol{g} , \tag{11.31}$$

then the two vector fields f and g are *involutive*. More generally, a set of m vector fields f_k (for $k = 1, \ldots, m$) is involutive if

$$[f_i, f_j](x) = \sum_{k=1}^{m} \alpha_{ijk}(x) f_k(x). \tag{11.32}$$

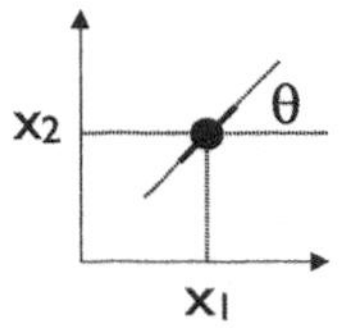

unicycle coordinates

Example 11.6 (Parking a unicycle) The Lie bracket of two vector fields can generate a "prohibited" motion that is outside the span of directly allowed motions. To illustrate this statement, consider the unicycle, which moves in the x_1-x_2 plane and has an orientation θ. Since $x_1 \in \mathbb{R}$ and $x_2 \in \mathbb{R}$, and since $\theta \in S^1$, the 1-dimensional "sphere," the state-space vector $(x_1, x_2, \theta)^{\mathsf{T}}$ is an element of the manifold $M = \mathbb{R}^2 \times S^1$. See left. The unicycle is capable of two motions: *drive*, a linear movement at fixed orientation θ, and *spin*, a rotation about a fixed point $\binom{x_1}{x_2}$. It is *not* allowed to skid, so that direct sideways motion (*slip*) is assumed to be not possible. Nevertheless, you can move a unicycle sideways by a careful combination of *drive* and *spin*. Let us see intuitively how such a motion is possible. As illustrated below, at left, you *drive* forward, *spin* counterclockwise, *drive* backward, and *spin* clockwise. These four operations comprise the Lie bracket. Here, we see that we do not return to our starting point but that there is a net slip to the right. In the diagram at left, $\theta = 90°$, so that an initial drive displacement $\binom{x_1}{x_2} = \binom{0}{1}$ generates a sideways displacement $\binom{1}{0}$. The result is not directly a prohibited motion, but proper combinations of *drive* and *spin* can approximate the prohibited motion arbitrarily closely.

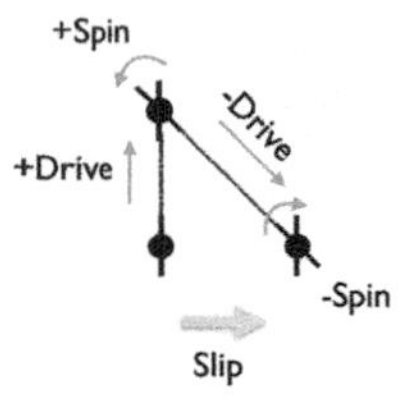

In Problem 11.13, we will see that the Lie bracket of *drive* and *spin* is not involutive, which amounts to a formal statement of our intuitive picture of parallel parking. Notice that the illustration at left is equivalent to the geometric illustration of the Lie bracket operation on two vector fields.

The case of cars is similar: You can always parallel park in a place that is ϵ larger than your car, as long as you are willing to make enough passes!

Involutivity is significant because it implies, via the *Frobenius theorem*, the property of *integrability*. The latter property asserts that a set of m vector fields $g_1, \ldots, g_m$ on $\mathbb{R}^n$ is integrable if there are $n - m$ scalar functions $h(x)_1, \ldots, h(x)_{n-m}$ that satisfy

$$\frac{\partial h_i}{\partial x} g_j(x) = 0. \tag{11.33}$$

Example 11.7 (Frobenius and foliation) To understand the Frobenius theorem, consider $x = (x_1\ x_2\ x_3)^{\mathsf{T}} \in \mathbb{R}^3$ and let the two vector fields g_1 and g_2 be given by

$$g_1 = \begin{pmatrix} 1 \\ 0 \\ 0 \end{pmatrix}, \qquad g_2 = \begin{pmatrix} 0 \\ 1 \\ 0 \end{pmatrix}, \qquad \Longrightarrow \qquad [g_1, g_2] = \mathbf{0}. \tag{11.34}$$

Because the Lie bracket is defined in terms of derivatives, it is zero for any two constant vector fields. The Frobenius theorem then asserts we can find a single $(3 - 2 = 1)$ function $h(x)$ that obeys

$$\frac{\partial h}{\partial x_1} = \frac{\partial h}{\partial x_2} = 0 \quad \Longrightarrow \quad h = h(x_3)\,. \tag{11.35}$$

Once we have found $h(x_3)$, then setting $h(x_3)$ = constant gives us a family of 2d vector fields that *foliate* $\mathbb{R}^3$. That is, the stack of 2d surfaces defined by the span of $\boldsymbol{f}$ and $\boldsymbol{g}$ fills space (in this case, globally; in general, locally). Here, the surfaces are all x_1-x_2 planes and we are simply stating that a stack of 2d parallel planes fills 3-space, as shown at right. We can choose any $h(x_3)$. Choosing $h = x_3$ means that the foliation is defined by x_3 = const. Any other function of x_3 would also be a foliation but with, in general, a nonuniform density of planes. In general, foliations are not planes but curved surfaces. Think of spherical onion shells or stackable potato chips, which foliate $\mathbb{R}^3 \setminus 0$ and $\mathbb{R}^3$, respectively.

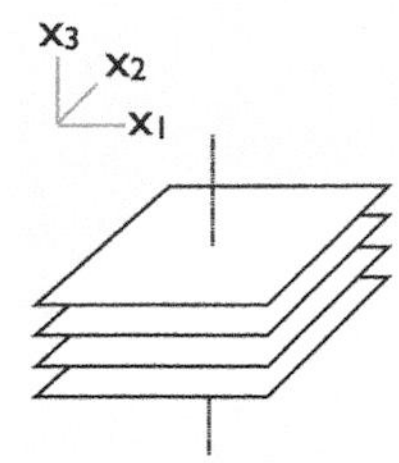

11.1.6 Controllability of Nonlinear Affine Systems

We can now begin to understand the connection between these differential-geometry ideas and controllability in dynamical systems and thereby extend the notions of controllability developed in Chapter 4. We start with a special case, the *driftless* affine system,

$$\dot{\boldsymbol{x}} = \sum_{i=1}^{m} \boldsymbol{g}_i(\boldsymbol{x})\, u_i(t)\,. \tag{11.36}$$

In Eq. (11.36), there is no dynamics beyond that induced directly by the control. Set the m control inputs to zero, and the system state remains constant. The unicycle (Example 11.6) is an example. Its position is fixed if $u_{\text{drive}} = 0$, as is its orientation if $u_{\text{spin}} = 0$.

The discussion of Example 11.6 implies that we can control the motion of the dynamical system on the manifold $M = \mathbb{R}^2 \times S^1$ using just the two controls for *drive* and *spin*.

On the other hand, if we construct the control system from Example 11.7,

$$\frac{\mathrm{d}}{\mathrm{d}t}\begin{pmatrix} x_1 \\ x_2 \\ x_3 \end{pmatrix} = \boldsymbol{g}_1 u_1(t) + \boldsymbol{g}_2 u_2(t) = \begin{pmatrix} u_1 \\ u_2 \\ 0 \end{pmatrix}, \tag{11.37}$$

it is clearly not controllable, as no action of $u_1(t)$ and/or $u_2(t)$ can alter the value of x_3.

More formally, for a driftless control-affine system such as Eq. (11.36), we can define a distribution $\Delta(\boldsymbol{x}) = \text{span}\ \{\boldsymbol{g}_i(\boldsymbol{x})\}$, for $i \in \{1, \ldots, m\}$. If the rank m of the $n \times m$ dimensional matrix $\{\boldsymbol{g}_i(\boldsymbol{x})\}$ is constant, then the distribution is said to have *constant rank*. Here, n is the state-space dimension (dimension of M), which also equals the dimension of the tangent space, $T_{\boldsymbol{x}}M$. If $m < n$, then the system is *underactuated*.

Next, we decompose an arbitrary control $\boldsymbol{u}(t)$ into a sequence of piecewise-continuous controls lasting a time T_s each, during which one u_i = const. and all others are 0. It is intuitive – and can be shown rigorously – that for small-enough T_s, we can find such a sequence that produces a state-space motion $\boldsymbol{x}(t)$ that approximates arbitrarily well the motion from an arbitrary continuous protocol involving all the $u_i(t)$'s simultaneously. Thus the link to Lie brackets, which involve a special sequence of four steps using two controls.

An underactuated system can still be controllable. Indeed, we need to look not only at direct displacements generated by each control $u_i(t)$ with its individual vector field $\boldsymbol{g}_i(\boldsymbol{x})$ but also at all possible Lie bracket combinations, including repeated brackets of the form $[\boldsymbol{g}_i, [\boldsymbol{g}_i, \boldsymbol{g}_j]]$, etc. If the Lie brackets are involutive – if all combinations of Lie brackets stay within a set whose rank is less than n – then the system is integrable and not fully controllable. To be fully controllable, vector fields and all combinations of their Lie brackets must span the full tangent space of the underlying state-space manifold.

Finally, we return to the feedback linearization of systems of the form $\dot{\boldsymbol{x}} = \boldsymbol{f} + \boldsymbol{g}\, u(t)$. Equation (11.29) gave a set of conditions that must be satisfied by a change of variables that linearizes the equations of motion. In Problem 11.14, we show that we can express these conditions as requiring that, in a region Ω about $\boldsymbol{x}$,

$$\text{the matrix } \boldsymbol{W}_c = \begin{pmatrix} \boldsymbol{g} & \mathrm{ad}_f \boldsymbol{g} & \cdots & \mathrm{ad}_f^{n-1} \boldsymbol{g} \end{pmatrix} \text{ has rank } n\,; \tag{11.38a}$$

$$\text{the set } \{\boldsymbol{g}\,,\ \mathrm{ad}_f \boldsymbol{g}\,,\ \ldots\,,\ \mathrm{ad}_f^{n-2} \boldsymbol{g}\} \text{ is involutive in } \Omega\,. \tag{11.38b}$$

Equation (11.38) introduces yet another notation, $\mathrm{ad}_f \boldsymbol{g} \equiv [\boldsymbol{f}, \boldsymbol{g}]$. The "ad" is short for *adjoint*, a concept that plays a key role in the theory of Lie algebras. In our case, it simply gives a convenient notation for repeated Lie brackets. Thus, $\mathrm{ad}_f^2 \boldsymbol{g} = [\boldsymbol{f}\,, [\boldsymbol{f}, \boldsymbol{g}]]$, and so on.

Equation (11.38a) generalizes the Kalman rank condition for controllability of the linear system $\dot{\boldsymbol{x}} = \boldsymbol{A}\boldsymbol{x} + \boldsymbol{B}u$ from Eq. (11.39), that the matrix

$$\boldsymbol{W}_c \equiv \begin{pmatrix} \boldsymbol{B} & \boldsymbol{A}\boldsymbol{B} & \boldsymbol{A}^2\boldsymbol{B} & \cdots & \boldsymbol{A}^{n-1}\boldsymbol{B} \end{pmatrix}, \tag{11.39}$$

has rank n. Equation (11.38b) is always satisfied for linear systems (Problem 11.15).

Equation (11.38) gives necessary and sufficient conditions for feedback linearization, but does NOT imply even local controllability. As already discussed, $\dot{x}_1 = x_2^2$, $\dot{x}_2 = u$ is linearizable but not controllable, as no $u(t)$ can make $x_1(t) < x_1(0)$. Unfortunately, there are no straightforward rules for determining controllability for general nonlinear systems. Still, the geometric approach introduced here is elegant and can be generalized to include nonlinear outputs $y = h(\boldsymbol{x})$ and address observability, stabilizability, and similar questions. More practically, it can give a straightforward way to tackle otherwise difficult control problems. See Problem 11.16 for a nice example.

11.2 Lyapunov-Based Design

Our second semisystematic method for designing controllers for nonlinear dynamics leverages Lyapunov's method for evaluating the stability of fixed points and limit cycles in nonlinear dynamical systems (Section 2.5.2). Recall that for an autonomous dynamical system $\dot{\boldsymbol{x}} = \boldsymbol{f}(\boldsymbol{x})$ with a fixed point at $\boldsymbol{x} = \boldsymbol{0}$, a Lyapunov function V must be positive definite ($V > 0$ for all $\boldsymbol{x} \neq \boldsymbol{0}$) and $\dot{V} = \frac{\partial V}{\partial x} \boldsymbol{f}$ should be negative definite (< 0) for asymptotic stability or negative semidefinite (≤ 0) for stability that may or may not be asymptotic.[4] For control systems, Lyapunov's method can be a valuable tool. Here, we give two ways that Lyapunov theory can help design a nonlinear controller. The first chooses the feedback form to give the closed-loop dynamics a Lyapunov function. The second takes advantage of assumed bounds on the nonlinear terms in the dynamics.

11.2.1 Control-Lyapunov Functions

If the dynamics have the form $\dot{\boldsymbol{x}} = \boldsymbol{f}(\boldsymbol{x}, \boldsymbol{u})$, we can try to choose a control law $\boldsymbol{u} = \boldsymbol{u}(\boldsymbol{x})$ such that the closed-loop dynamics $\dot{\boldsymbol{x}} = \boldsymbol{f}(\boldsymbol{x}, \boldsymbol{u}(\boldsymbol{x}))$ has a Lyapunov function. As discussed in Section 2.5.2, the main limitation of such Lyapunov-based design is that there is no general algorithm to construct Lyapunov functions, even with the extra degrees of freedom that $\boldsymbol{u}$ provides. Still, here are some cases where the idea works.

In Section 10.1.1 (adaptive control), we actually used this idea several times without really spelling out the general strategy. Here, we explore another example in some detail in order to understand better how to use Lyapunov functions to design nonlinear controllers. In Example 11.4, we discussed the control of a frictionless pendulum obeying $\ddot{\theta} - \sin\theta = u$, with the goal of stabilizing the up equilibrium, $\theta = \dot{\theta} = 0$. We advocated a strategy of "canceling out" the nonlinearity via a control law of the form $u = -\sin\theta - K_{\mathrm{p}}\theta - K_{\mathrm{d}}\dot{\theta}$. Such a controller treats the angle θ as a real number that can have any value. But the pendulum angle is periodic in 2π, implying that the controller should be a function with period 2π. From the theory of Fourier series (Appendix A.4.1 online), we know that such a function should be expressible as a sum of $\sin\theta$ and $\cos\theta$ and their harmonics.

Example 11.8 (Stabilizing the nonlinear pendulum, part 2) As in Example 11.4, we want to stabilize the up solution $\theta = \dot{\theta} = 0$ of the nonlinear pendulum, obeying $\ddot{\theta} - \sin\theta = u$. Consider the function

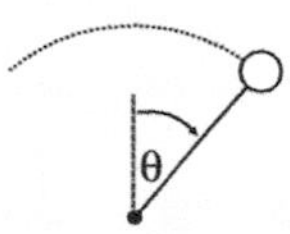

$$V = 2a \sin^2\left(\frac{\theta}{2}\right) + \frac{1}{2}\dot{\theta}^2 . \tag{11.40}$$

For the uncontrolled problem ($u = 0$), this choice of V is *not* a Lyapunov function; however, in Problem 11.17, you will show that choosing $a > 0$ and $K_{\mathrm{p}} = a + 1$ and setting

$$u(\theta, \dot{\theta}) = -K_{\mathrm{p}} \sin\theta - K_{\mathrm{d}}\dot{\theta}\cos^2\theta \tag{11.41}$$

[4] For $\dot{V} \leq 0$, recall the Krasovskii–LaSalle principle, that dynamical trajectories will end up on the invariant set.

does make V a (negative semidefinite) Lyapunov function that can stabilize the vertical solution to almost all perturbations. We can think of K_p and K_d as gains for a nonlinear proportional-derivative (PD) control scheme. Note that the 2π ambiguity affects θ, not $\dot{\theta}$.

As another example where the strategy of using Lyapunov functions to ensure control succeeds, Problem 11.18 discusses the use of Lyapunov functions to design a control that synchronizes two chaotic dynamical systems, even when there is a huge amount of noise present.

11.2.2 Bounds and Lyapunov Functions

Our second strategy depends on bounding the nonlinear terms in the dynamics. We will develop this idea in the context of nonlinear adaptive control, focusing on a generalization of the example from the beginning of Chapter 10. Let $\dot{x} = f(x) + u(x)$, with $f(x)$ a nonlinear function that has an unstable fixed point at $x = 0$ – i.e., $f(0) = 0$, $f'(0) > 0$. To stabilize, we add the variable proportional gain $u(x) = -k(t)x(t)$. The gain increases via $\dot{k} = \gamma x^2$, where γ sets the adaptation rate.

To show that the $x = 0$ solution is stabilized, consider a candidate Lyapunov function

$$V(x) = \frac{1}{2}x^2 + \frac{1}{2\gamma}(k - L)^2 , \tag{11.42}$$

where the constant L is to be determined. Differentiating V along solutions $x(t)$ gives

$$\dot{V} = x\dot{x} + \frac{1}{\gamma}(k - L)\dot{k} = x\,f(x) - kx^2 + (k - L)x^2 = x\,f(x) - Lx^2 . \tag{11.43}$$

Now, if $x\,f(x) < 0$, then V is automatically a Lyapunov function when $L > 0$. If not, $x\,f(x) \geq 0$, and our strategy will be to place a bound on $f[x(t)]$. In particular, we assume the *Lipschitz condition*, that $|f(x)| < \ell|x|$, for some constant ℓ. Visually, the function $f(x)$ should be contained within the shaded region shown at left.[5] In practice, systems where motion is confined to an attractor satisfy this condition. As a result, $x\,f(x) \leq |x\,f(x)| \leq |x|\,|f(x)| \leq \ell x^2$, and we can write Eq. (11.43) as

$$\dot{V} = x\,f(x) - Lx^2 \leq (\ell - L)x^2 . \tag{11.44}$$

Thus, for $L \geq \ell$, we can conclude that V is a Lyapunov function. Again, the basic idea is that the adaptive feedback law keeps increasing the proportional gain until the feedback term overwhelms the dynamics and stabilizes the desired fixed point, $x^* = 0$.

[5] The Lipschitz condition we impose is not quite the usual definition of *Lipschitz continuity*, which asks that $|f(x_1) - f(x_2)| < \ell|x_1 - x_2|$ for *all* x_1 and x_2. In terms of the picture given above right, we would need to be able to slide the shaded region along x and have $f(x)$ be contained for all x. This would occur if $|f'(x)| < \ell$, for all x, but this is clearly *not* the case above. Yet we allow such situations because a linear control $u = -kx$ with $k > \ell$ can still locally "overcome" the nonlinear term $f(x)$ near $x = 0$, no matter how steep it becomes.

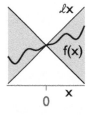

It is straightforward to change coordinates to $x' = x - x^*$ to stabilize another fixed point $x^* \neq 0$. Generalizing to an n-dimensional state vector $\boldsymbol{x}$ is possible but tricky.

Rather than developing this approach further, let us not forget the main lesson from our analysis of this example at the start of Chapter 10: *Having a Lyapunov function for a class of functions does not guarantee stability if the actual dynamics obey a different set of equations.* Here, disturbances plus an unmodeled time delay will increase the gain until there is an instability. As a workaround, we can simply cap the maximum allowable gain to $k = k_{\max}$. The dynamics will then increase the gain to the minimum value needed to stabilize or fail to stabilize at $k_{\max}$. If stabilization fails, it at least fails sooner rather than later, in a more predictable fashion.

11.3 Collective Dynamics

Up until now, we have mainly pursued two types of goals in control: stabilization of a system at some point $\boldsymbol{x}_0$ in state space (*regulation*) and follows a desired trajectory $\boldsymbol{x}_{\mathrm{d}}(t)$ (*tracking*). But, as pointed out in the book introduction (Section 1.2.1), there are other goals we can imagine. Here, we consider how feedback can create *collective dynamics* such as virtual potentials, virtual interactions, and synchronization. These goals all aim to create a new type of motion but do not specify the motion itself.

11.3.1 Virtual Potentials

The simplest way to obtain new motion is to define new dynamical rules. In an application taken from biophysics and statistical physics, imagine that you desire that a particle move in a potential $U(x)$ of your choice. For example, passive traps such as optical tweezers are not strong enough to trap single molecules but direct forces are.[6] We can thus use direct forces in a feedback loop to create a virtual harmonic potential that localizes the protein.

In other applications, particles are made to move in more complicated virtual potentials such as the double-well shape of Example 8.3, which was used to create a Szilard engine that converts information to work (Section 15.3.2).

All of the above applications share the need to make a particle obey dynamics of the form $\dot{\boldsymbol{x}} = \boldsymbol{f}(\boldsymbol{x}, t)$. For the above examples, the force can be written as the gradient of a potential, $\boldsymbol{f} = -\nabla U(\boldsymbol{x})$, but we could also generalize to time-dependent potentials or even cases where the force field does not derive from a potential.[7]

To create a virtual potential using feedback, we assume that we measure the particle position $\boldsymbol{x}_k$ at time kT_{s}. The measurement can be direct or use the techniques of filtering discussed in Chapter 8. Assuming that the force is well calibrated, then we simply

[6] The force from optical tweezers depends on polarizing the particle and has a magnitude $\sim r^3$, where r is the particle size. An external electric field, by contrast, creates a force $\sim r$. For small sizes, $r \gg r^3$. This method has been used to trap single dye molecules diffusing in water! See Fields and Cohen (2011).

[7] A simple example is a vortex field of the form $\boldsymbol{f} = f\hat{\theta}$ that tends to drive a particle around in a circle.

calculate the desired force as $\boldsymbol{f}_k = -\nabla U(\boldsymbol{x}_k)$ and apply it for the time interval T_s. As long as the update time (and any delays, which we here neglect) are short compared to the time scales of the dynamical system, the dynamics created will approximate accurately the motion of a particle in the desired potential.

Notice that the result of feedback here is to change the rules of motion. In the absence of feedback (and, for simplicity, assuming that no other forces are present), the particle obeys $\dot{\boldsymbol{x}} = \boldsymbol{0}$. With feedback, it obeys $\dot{\boldsymbol{x}} = \boldsymbol{f}(\boldsymbol{x}, t)$. But, unlike the cases of regulation and tracking, there is no explicit motion imposed – no setpoint or desired trajectory. Only the rules of motion are changed.

11.3.2 Virtual Interactions

We can generalize the ideas of Section 11.3.1 by progressing from virtual potentials to virtual interactions. Here, feedback is used to create new interactions between particles. That is, we now have a collection of N particles $\boldsymbol{x}_i(t)$ and their motion is specified, using feedback to create the interaction forces, so that $\dot{\boldsymbol{x}}_i(t) = \sum_j \boldsymbol{f}_{ij}(\boldsymbol{x}_j - \boldsymbol{x}_i, t)$. One implementation issue here is that we now need to supply local forces $\boldsymbol{r}_{ij}$ that affect each particle individually. For each particle, we would sum up the desired forces due to all the other interacting particles and then apply the vector sum to particle i and then quickly cycle through all the particles. The cycling and update, as before, must be done faster than the system dynamics.

What form should the virtual interactions take? Almost anything is possible: bizarre forms of distance dependence (perhaps oscillating with distance instead of decreasing monotonically); nonreciprocal forces, where the action of i on j is not simply the negative interaction from j on i expected from Newton's third law; interactions that single out special members (red particles interact with other red particles but not with green particles, etc.). The possibilities are endless.

Why might virtual interactions be interesting or useful? They can create new forms of collective dynamics (which may appear only in some parameter regimes and may arise via newly created phase transitions). For example, Khadka et al. (2018) and Bäuerle et al. (2018) used virtual interactions to create collective motions among active particles, where none would otherwise be present. And Barral et al. (2010) investigated the collective motion of sensory hair bundles by coupling a single physical hair bundle to two virtual clones, using stochastic simulations to generate the virtual interacting particles!

With virtual potentials and interactions, we can now envisage creating systematically a kind of *cyberphysics*, where information and physics are joined using feedback to create new kinds of dynamics that greatly expand the "limited set" that Nature has given us. The possibilities are rich and scarcely explored.

11.3.3 Synchronization

Synchronization is a way to make two different dynamical systems behave identically (or with some defined relationship). For example, given a "master" dynamical system, you would like another system to track it, whatever its output is. Alternatively, an outside system may be the master and your system the "slave" or follower. Or perhaps you do not care who leads, as long as the two systems are linked. Often, but not always, the desired motion is a limit cycle (periodic motion). Other kinds of collective motion that can be created include swarming and dancing (see below).

Synchronization is a *collective* phenomenon more characteristic of nonlinear dynamics than of control theory. But the control-theory view is equally valuable. Often, the control is open loop, where a control parameter is altered to "turn on" (or off) the synchronization.

Although the goals of synchronization might seem a bit abstract, synchronization is extremely important in both science and engineering. The natural world is full of systems that spontaneously synchronize. From the clapping of crowds to the flashing of fireflies to the swarming of birds and fish seeking to avoid a predator to the organized oscillations of cells in the heart, the biological world is full of systems that have synchronized spontaneously. And while the synchronized behavior of such systems might seem to imply an intelligent organizer, the behavior arises spontaneously, and similar features are seen in purely physical systems. Indeed, historically, the study of synchronization was initiated by Christiaan Huygens, who, in 1665, was "forced to stay in bed for a few days and made observations of two clocks," noticing a "wonderful effect that nobody could have thought of before." Huygens observed that two pendula hanging from the same wall spontaneously began swinging together. He found that nonlinearity plays an essential role in synching the clocks, allowing the period of one oscillator to adjust and match that of another dynamical system. Indeed, without some kind of nonlinearity, synchronization could not exist, as the chance that two independent oscillators have a rationally related frequency ratio is zero.[8] Since oscillators and nonlinearity are everywhere in physics, it is not surprising to find other instances, including Josephson junctions, chemical oscillators, and the like.

In engineering, there are many practical applications of synchronization, including techniques for locking different electrical oscillator circuits, laser and microwave arrays, and maintenance of electrical power networks. Biomedical applications include pacemakers and the mechanical entrainment of breathing in a respirator. And, most viscerally, synchronization plays a key role in music and dancing. Indeed, what is dancing but the movement of the body in time[9] to the "external forcing" of music? In sum, synchronization is one of the most common and important generic features of nonlinear dynamics.

Some key examples will give the flavor of the field. First, we recall from Section 7.6.2 that we can generalize the notion of observers to nonlinear systems. In fact, we can view the problem of using observations to infer underlying states as one of synchronizing the observer system to the physical system, with the observations serving as coupling.

[8] To be more precise, the set of rational numbers has measure zero among the reals. Since frequencies are real numbers, two randomly chosen oscillators will have frequencies that are irrationally related, implying unlocked, quasiperiodic motion.

[9] Or not.

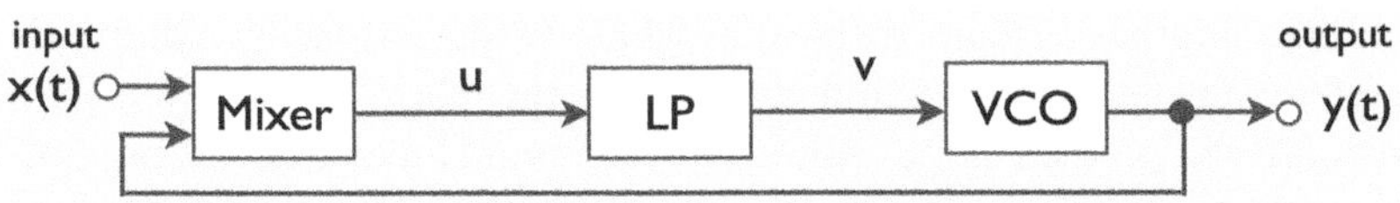

Fig. 11.2 Block diagram of a phase-locked loop.

Below, we first present the *phase-locked loop* (PLL), an electronic circuit that synchronizes a voltage with an external signal that is typically approximately sinusoidal but with drifting frequency and amplitude. In contrast to the nonlinear observer, the synchronization can work even if you do not know the dynamical system that generates the external signal. Although our analysis of PLLs will primarily use concepts from linear systems, we will see that the synchronization they produce is fundamentally a nonlinear phenomenon, suggesting that there should be deeper ways of analyzing the phenomenon. In Problem 11.20, we present another engineering example, with a somewhat similar flavor: the Pound–Drever–Hall technique for stabilizing the frequency of a laser against an external standard.

We then introduce a nonlinear view of synchronization via the Adler equation, a simple model for phase locking where an external system follows a control oscillator. An interesting example of Kuramoto shows that N oscillators, each with its own frequency ω_i, can spontaneously synchronize via a kind of phase transition, for strong coupling.

Phase-Locked Loops

The phase-locked loop originated in the 1930s as a way to tune a radio (i.e., synchronize to the carrier band of a radio signal). Figure 11.2 shows the basic elements. The goal is to synchronize to an external signal

$$x(t) = A(t)\, \sin\phi_{\mathrm{e}}(t)\,, \tag{11.45}$$

with $A(t)$ a slowly varying amplitude and $\phi_{\mathrm{e}}(t)$ a slowly varying phase. The main tool is a voltage-controlled oscillator (VCO), an electronic circuit that can produce a sinusoidal signal of controllable frequency. Specifically, we have

$$y(t) = 2B\, \cos\phi(t)\,, \qquad \dot{\phi} = \omega_0 + Kv(t)\,, \tag{11.46}$$

which is an oscillator of base frequency ω_0 that is altered by an control signal $v(t)$ that is coupled via a feedback gain K. Notice how the phase integrates $v(t)$.

The mixer multiplies the signals $x(t)$ and $y(t)$ together and serves as a phase detector. Indeed, using a trigonometric identity,[10] we have

$$u \equiv x(t)\, y(t) = 2AB \sin\phi_{\mathrm{e}}\, \cos\phi = AB\left[\sin(\phi_{\mathrm{e}} + \phi) + \sin(\phi_{\mathrm{e}} - \phi)\right], \tag{11.47}$$

[10] See also lock-in detection (Section 5.1.3) and the Pound–Drever–Hall technique (Problem 11.20).

which consists of a high-frequency component ($\approx 2\omega_0$) and a near-DC component that depends on the difference in phase. We send $u(t)$ into a low-pass filter (LP) that attenuates ("cancels") the $2\omega_0$ component. We will consider a simple one-pole filter described by $\dot{v} = -v/\tau + u(t)$. Finally, we define the phase difference between input and output signals, $\psi \equiv \phi - \phi_e$. Substituting, we find a pendulum equation,

$$\ddot{\psi} + \frac{1}{\tau}\dot{\psi} + KAB\sin\psi = -\ddot{\phi}_e - \frac{1}{\tau}\left(\dot{\phi}_e - \omega_0\right), \tag{11.48}$$

which is driven by the external signal and its frequency difference from the reference at ω_0.

Although it is typical to analyze Eq. (11.48) by linearizing around the phase-locked condition ($\psi = 0$), we recognize that the equations are fundamentally nonlinear, most obviously due to the $\sin\psi$ terms but also because the mixer multiplies x and y.

We begin by considering what happens when $x(t)$ is a sinusoid of frequency ω that is different from the reference frequency ω_0. Then $\phi_e = \omega t$. Equation (11.48) has a constant solution ψ_0 if $KAB\sin\psi_0 = \frac{\delta\omega}{\tau}$, with $\delta\omega \equiv \omega - \omega_0$. Since $\sin\psi \in (-1, 1)$, we must have

$$|\delta\omega| < KAB\tau. \tag{11.49}$$

This condition must be satisfied to have a time-independent solution. We should also look at the stability of such a solution. Linearizing about ψ_0 gives a transfer function,

$$G(s) = -\frac{s^2 + s/\tau}{s^2 + s/\tau + KAB\cos\psi_0}, \tag{11.50}$$

which implies poles (typically underdamped and therefore complex) at

$$s = \frac{1}{2}\left(-\frac{1}{\tau} \pm \sqrt{\frac{1}{\tau^2} - 4KAB\cos\psi_0}\right). \tag{11.51}$$

The decay time here is set by τ and the oscillation frequency by the KAB term.

Phase Locking as a Nonlinear Phenomenon

Why is synchronization so ubiquitous in nature? In brief: it is easy to perturb the phase of a limit cycle, keeping other features such as amplitude and waveform shape approximately constant. As a result, small perturbations that are otherwise unimportant can build up large phase shifts. If there is some preferred, "stable" phase relationship, then the phase shifts will likely find that value, resulting in synchronized motion.

But why are phase perturbations generically "softer" than other types? The answer traces back to notions of *spontaneous symmetry breaking*, one of the most important concepts from mid-twentieth-century physics. The fundamental laws of nature are believed to be invariant under time shifts. That is, they do not refer to an absolute time and take the same form under the transformation $t \to t + \tau$, with τ some arbitrary time shift. Yet solutions such as limit cycles break this symmetry. Indeed, a periodic limit cycle with $x(t) = x(t + nT)$ is invariant only under the translation $t \to nT$, with T the period of oscillation and n an arbitrary integer. But – and here is the main point

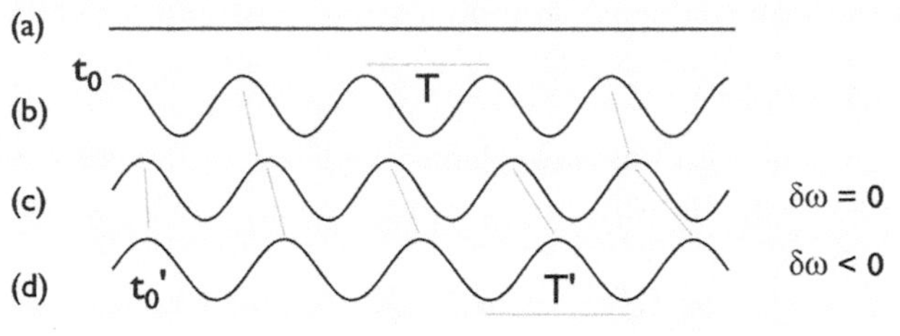

Fig. 11.3 Symmetry breaking and Goldstone modes. (a) Time-invariant solution; (b) limit cycle breaks time invariance; (c) constant phase shift; (d) small frequency shift.

– if $x(t)$ is a symmetry-breaking solution, then the original time-translation symmetry implies that there is a one-parameter family of equivalent but distinct solutions $x(t-t_0)$, parameterized by $0 < t_0 < T$.

Given this infinite family of solutions, a perturbation that shifts $t_0 \to t_0'$ must be neutral, with a zero relaxation rate (infinite relaxation time). The perturbation simply shifts the phase of the oscillator from one valid solution to another valid solution, implying that there will be no relaxation. As a further consequence, solutions that slowly perturb the phase (e.g., because the frequency is slightly different) relax slowly, because at each moment, the oscillations are nearly an exact solution.

Figure 11.3 illustrates these ideas, with the continuous invariance of (a) broken by the limit-cycle solution in (b). A uniform phase shift (c) leads to another solution, while a small frequency shift (d) leads to a slowly relaxing perturbation, a Goldstone mode.

More formally, from the Nambu–Goldstone theorem, symmetry breaking of a continuous symmetry leads to a "soft" mode. The eigenvalue s describing the growth rate of linear perturbations obeys $s(q) \to 0$ for perturbations that shift the wavenumber of the oscillator by q.[11] Less formally, it is easier to perturb the phase of a limit cycle than its amplitude.

With these ideas in mind, we can understand generically the effect of an external periodic perturbation on a limit cycle. If we write the instantaneous phase $\phi(t)$ of the oscillator, then an unperturbed oscillator obeys $\dot{\phi} = \omega_0$, with ω_0 its angular frequency. If the oscillator is perturbed, its phase will evolve faster or slower than normal. For convenience, we define the phase of the perturbed oscillator $\psi(t) = \phi - \omega_0 t$ relative to the phase of the unperturbed oscillator. Generically, the perturbed phase obeys a relation known as the *Adler* equation,

$$\dot{\psi} = -\nu + \varepsilon \sin \psi \,. \tag{11.52}$$

The first term, $-\nu$, represents the mean frequency shift produced by the perturbation. The second term represents the effect of perturbations, with ε a coupling constant measuring its strength. Generically, it must be periodic in 2π, since phases are defined

[11] For simplicity, we assume constant q. More generally, the wavenumber shifts can be local: $q = q(\boldsymbol{x})$.

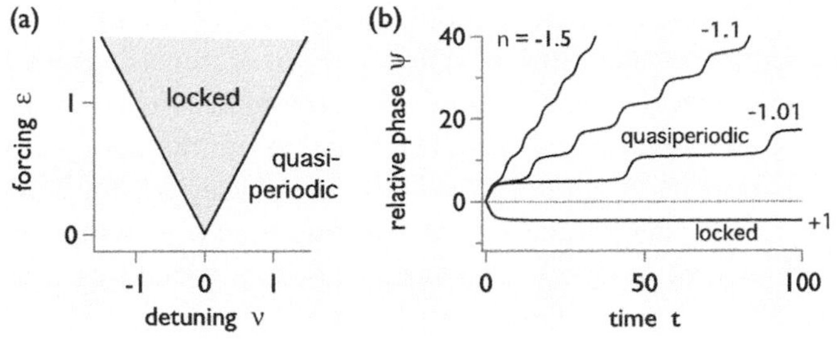

Fig. 11.4 Locked (synchronous) and quasiperiodic motion in the Adler equation. (a) Locked region in the forcing-detuning parameter plane. (b) Examples of locked and quasiperiodic time series, for $\varepsilon = 1$.

on the unit circle. Thus, the sine function is simply the simplest possible periodic perturbation.

Implicitly, the perturbation affects only the phase of the oscillator, not its amplitude nor waveform (mix of harmonics). This is justified by the ideas of symmetry breaking and soft modes discussed above for weak-enough perturbations. See Problem 11.21, on the forced van der Pol oscillator. Here, we focus on the implications of the Adler equation.

We first look for stationary solutions by imposing $\dot{\psi} = 0$, which implies $\varepsilon \sin\psi = \nu$. In other words, as illustrated in Figure 11.4, there is a range of frequencies, characterized by the detuning ν, for which ψ is constant. These are synchronized solutions and lie within a "tongue" in the ν-ε parameter plane, as shown in Figure 11.4a. Since $\sin\psi \in (-1, 1)$, the range is simply $(-\varepsilon, +\varepsilon)$. Recall that positive ν means the oscillator is running faster than the reference frequency ω_0, and negative ν means it is running slower. Figure 11.4b shows examples of time series in the locked and quasiperiodic regimes. Note that the quasiperiodic regime appears via a series of 2π *phase slips* that are increasingly spaced apart near the transition. Qualitatively, the oscillator locks for a time, then "loses it" and slips for a cycle, before temporarily relocking. See Pikovsky et al. (2001) for more details on this and other aspects of the transitions to synchrony.

Example 11.9 (Locking many oscillators) As an application of these ideas, we consider N oscillators with frequencies ω_k, all coupled to each other and governed by a phase equation of the form,

$$\frac{\mathrm{d}\phi_k}{\mathrm{d}t} = \omega_k + \frac{\varepsilon}{N}\sum_{j=1}^{N}\sin\big(\phi_j - \phi_k\big). \tag{11.53}$$

The coupling ε is global, as each oscillator perturbs all the others. For $\varepsilon < \varepsilon_c$, the oscillators run independently. For $\varepsilon > \varepsilon_c$, an increasing fraction will be synchronized. Below, we will see that keeping the total coupling $\sim 1/N$ makes ε_c independent of N.

To describe the state of synchronization, we introduce a complex *order parameter* $\mathcal{K} = K\,\mathrm{e}^{\mathrm{i}\Theta} = \frac{1}{N}\sum_j \mathrm{e}^{\mathrm{i}\phi_j}$. Intuitively, if all oscillators are perfectly synchronized, then $|\mathcal{K}| = K = 1$ and the variable Θ represents their collective phase. If no oscillators are

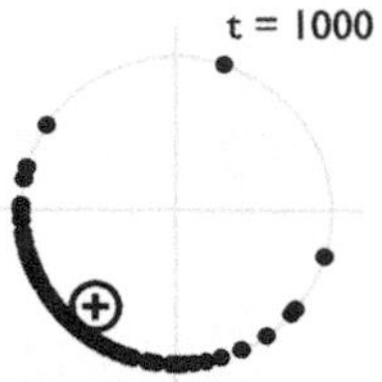

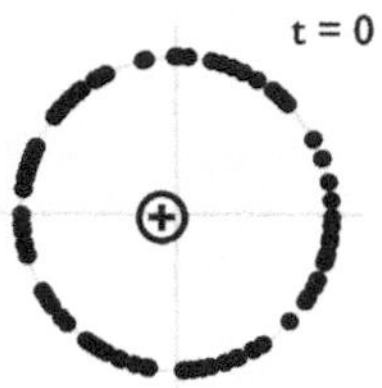

synchronized, then the phases ϕ_j are random, and $K \sim O(N^{-1/2})$, which vanishes in the thermodynamic limit, $N \to \infty$. Intermediate values of K imply that a fraction of the population is synchronized. The situation is illustrated at left, where black markers reflect the phase of an individual marker. The large central marker $\oplus$ shows the polar plot of the order parameter $\mathcal{K}$. Initially (at $t = 0$), the phases are randomly distributed and the order parameter magnitude $K \approx 0$. At long time ($t = 1000$), the interaction has led to partial synchronization, with K near 1 (as determined by the vector average of all the phases, where each oscillator contributes a vector from the origin to the unit circle).

Now let the distribution of oscillation frequencies be given by a function $g(\omega)$ that is peaked about ω_0. The critical coupling strength is then $\varepsilon_c = \frac{2}{\pi g(\omega_0)}$, and for $\varepsilon > \varepsilon_c$, the fraction of synchronized oscillators grows as $K \sim \sqrt{\varepsilon - \varepsilon_c}$ (Problem 11.22).

11.4 Controlling Chaos

Chaotic dynamical systems show sensitive dependence to initial conditions. Two nearby initial conditions will diverge exponentially, as illustrated in Figure 11.5 for the Lorenz equations (see Problem 11.18). The average exponential divergence rate is known as the *largest Lyapunov exponent*. Chaotic motion is characterized by a positive Lyapunov exponent, a notion that generalizes that of a positive eigenvalue in a linearly unstable system.[12]

Here, we will focus on dissipative chaotic systems.[13] Although the system is always locally unstable, dissipation implies that the state vector stays in a bounded region of state space. Is it possible to stabilize the dynamics through *small* variations of one or more of the system's control parameters? (Stabilization through large variations is less interesting, because we can usually find a stable region for some range of control parameters and because minimization of control effort is often an important consideration.)

Specialized techniques for the control of chaotic systems were introduced by physicists and mathematicians rather than by control theorists and, as a result, the methods (and the literature) have a somewhat different flavor. On the one hand, these methods leverage specific features of a (chaotic) dynamical system to do a better job than the more general-purpose algorithms that we mostly focus on. On the other hand, important aspects of the main ideas – and sometimes even the methods – are present in traditional approaches to control. In Section 11.4.1, we present the most famous of these algorithms (Ott et al., 1990), which, historically, created the subfield. In Section 11.4.2, we outline a couple more methods that are taken up in the problems.

[12] State spaces with n dimensions have n Lyapunov exponents. Only one need be positive for chaotic motion.

[13] The treatment of Hamiltonian systems requires somewhat different methods.

11.4.1 Ott–Grebogi–Yorke Algorithm

The first algorithm to stabilize a system with chaotic dynamics was given by Ott, Grebogi, and Yorke (OGY) in 1990. Their algorithm drew on several key features of chaotic motion:

- After initial transients, the system evolves on a *strange attractor* in phase space.
- Motion is ergodic on that attractor; a system repeatedly approaches any given point.
- Embedded in each strange attractor is a dense set of unstable periodic orbits.

The idea of OGY is to stabilize motion about an unstable orbit (or fixed point) x^* by waiting until the system brings the state vector x near to x^*. Since chaotic motion is ergodic, this will always happen, eventually. (The closer you want x to approach x^*, the longer you need to wait.) Once x is close enough to x^*, the control algorithm is activated. In effect, we change the attractor from a strange attractor to a simple fixed point.

For simplicity, assume discrete, one-dimensional dynamics $x_{k+1} = f(\lambda, x_k)$, where λ is a control parameter. In Chapter 5, we discretized continuous dynamical systems by sampling at a set of times kT_s. For nonlinear dynamics, it often makes sense to do a more sophisticated discretization called a *Poincaré section*, where the value is sampled each time it cuts a reference surface in the state space (see right). Sectioning turns an N-dimensional continuous-time dynamical system into an $N-1$-dimensional discrete-time system. In such a discretization, the intervals between times, $\Delta t_k \equiv t_{k+1} - t_k$, and so on will usually vary and be a function of both x_k and x_{k+1}. For Poincaré sections, a limit cycle of the continuous system corresponds to a fixed point of the discrete dynamical map, $x^* = f(x^*, \lambda)$.

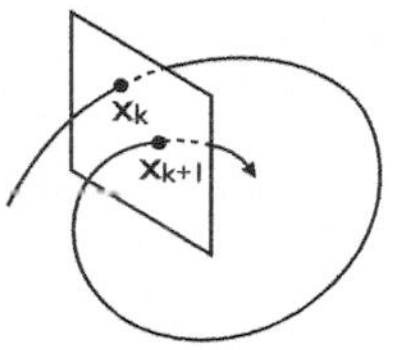

In the OGY method, when $x_k \approx x^*$, we can linearize about the fixed point:

$$x_{k+1} - x^* \approx (\partial_x f)(x_k - x^*) + (\partial_\lambda f)\, \Delta\lambda_k \,. \tag{11.54}$$

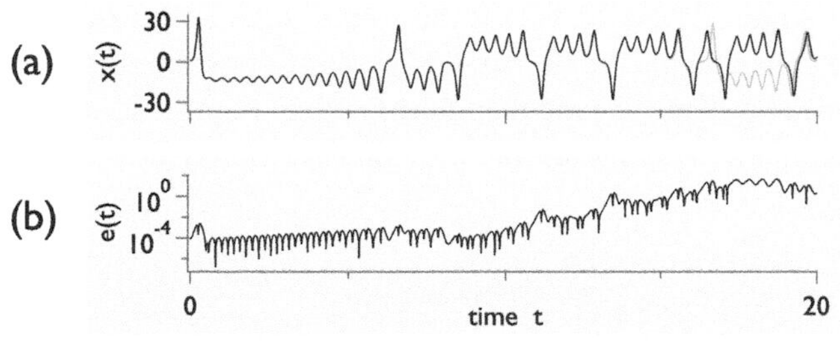

Fig. 11.5 Chaos in the Lorenz equations. (a) Two initial conditions that agree to within 10^{-4} diverge after $t \approx 17$. (b) Time-averaged difference between the two solutions grows approximately exponentially when they are close in state space. Calculated using the parameters from Problem 11.18a.

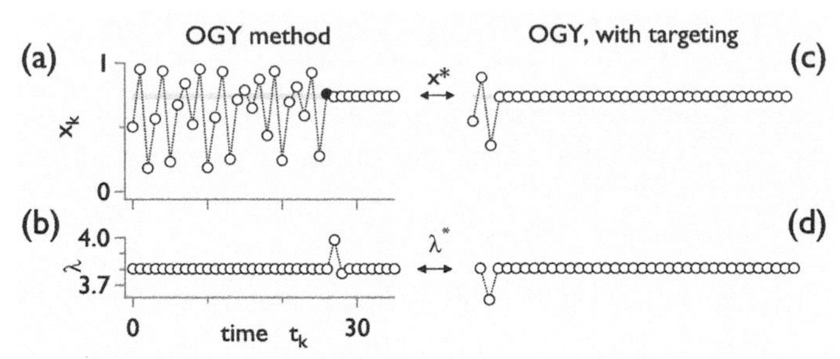

Fig. 11.6 OGY algorithm stabilizes logistic map. (a) Time series of x_k for $\lambda_0 = 3.8$. Control is initiated at $t = 0$ about unstable fixed point x^*. Shaded region shows the tolerance range $x^* \pm \varepsilon$ within which OGY algorithm is active. Here, the algorithm activates at $t = 27$. (b) Control parameter λ_k is adjusted by OGY algorithm to stabilize x^* at λ^*. (c,d) Like (a,b) but targeting directs the trajectory to the target region.

Then change λ in Eq. (11.54) so that $x_{k+1} = x^*$. That is, if $|x_k - x^*| < \varepsilon$, then set

$$\Delta\lambda_k = -\left(\frac{\partial_x f}{\partial_\lambda f}\right)(x_k - x^*). \tag{11.55}$$

Otherwise, fix λ for that iteration. Of course, using Eq. (11.55) to choose $\Delta\lambda_k$ will not make x_{k+1} precisely equal to x^*, since that choice is based upon the approximation in Eq. (11.54). But if the original distance $x_k - x^*$ is small enough, the system will quickly converge to x^*.

Example 11.10 (Logistic map) We illustrate the OGY algorithm on a simple chaotic dynamical system, the logistic map, $x_{k+1} = \lambda x_k(1 - x_k)$, which is a simple dynamical map that shows complex behavior as the control parameter λ is adjusted. At $\lambda = 3.8$, the system is normally chaotic. The goal will be to stabilize the system's motion about the normally unstable fixed point, given by $x^* = 1 - \lambda^{-1} \approx 0.74$. The OGY control algorithm is turned on at $t = 0$. Figure 11.6a shows a typical time series x_k. We wait until the natural motion of the system brings it to within a predefined tolerance $x^* \pm \varepsilon$ (at time step $k = 26$, black marker). The OGY algorithm is then activated at the next time step, as shown in Figure 11.6b. Note that the system state x_{26} is just slightly above the target value. We then change λ to $\lambda_k = \lambda + \Delta\lambda_k$ so as to position the fixed point of the modified dynamical system (logistic map with control parameter λ') *above* the point x_{26} so that the repulsive dynamics of the modified system pushes the point toward x^*. From Eq. (11.55),

$$\Delta\lambda_k = -\left[\frac{\lambda(1 - 2x^*)}{x^*(1 - x^*)}\right](x_k - x^*) \approx 9.3\,(x_k - x^*). \tag{11.56}$$

The values of λ_k are shown in Figure 11.6b. Notice how λ_{27} shoots up in response to x_{26}, which represents the first time the system has entered the tolerance zone $x^* \pm \varepsilon$. The system quickly settles down and stabilizes about the desired set point x^*. See Problem 11.23.

One problem with OGY is the need to wait for the system to wander near x^* before activating the control. If the tolerance zone is small, this can take a long time. To do better, *targeting* small corrections to the motion can drastically reduce the waiting time (Figure 11.6c,d). Here, given an initial condition x_0, we perturb λ to λ' at time step 1 and return it to λ at future time steps. Assume that we are allowed to set λ' within some range $\lambda \pm \Delta\lambda$. As we iterate forward, that range expands exponentially, eventually covering the target value x^*, maintained by $\lambda = \lambda^*$. If we know $f(x, \lambda)$ and if there is no noise, we can compute the value of control parameter λ^* that corresponds to x^*. Then the usual OGY algorithm keeps x near x^*. A more efficient alternative is to iterate forward from x_0 and *backwards* from $x^* \pm \varepsilon$ until the images intersect. Then construct a trajectory from x_0 to the vicinity of x^*.

Experimental noise or incomplete measurement of a multicomponent state vector $\boldsymbol{x}_k$ will make the above methods fail. A workaround is to compute a targeting perturbation at each time step. This and other tricks have led to many successful applications of OGY and targeting. One important application of targeting has been to direct spacecraft to distant parts of the solar system using very little fuel at just the right moments.

11.4.2 Other Chaos-Control Methods

The OGY method was the first of many algorithms. In Problem 11.18, we discuss the closely related issue of synchronizing two chaotic systems, a situation that is relevant to the problem of communicating messages. The basic idea is to choose a control that gives the coupled systems a Lyapunov function that enforces synchronization. In Problem 11.24, we discuss how a delayed signal can help stabilize limit cycles in continuous systems, while requiring much less information about the system than the OGY method does.

We can draw some general lessons from the field as a whole:

- If the desired motion is a solution to the equations of motion, very little control effort is needed to stabilize it. Strange attractors are special in having so many unstable solutions. The idea of focusing on "natural" (but unstable) dynamics that require only a small control effort should be contrasted with our discussion of feedback linearization. There, the algorithm often directly "canceled out" the nonlinearity, replacing it with nicer linear dynamics (Section 11.1.3). Typically much more control effort will be required.
- Unstable systems are more maneuverable than stable ones because you can use the instability mechanism to help push the state in the desired direction via small "tweaks" that again require little control effort. Strange attractors are special in that there can be unstable directions in state space nearly everywhere. By contrast, many systems, such as the inverted pendulum, have only a few unstable equilibria.
- Many of the methods that were developed independently by physicists have their counterparts in the control literature and are perhaps not as original as believed. To

give one example, the use of delayed signals to stabilize motion was already introduced in the control literature in the 1980s, where it was known as *repetitive control* (Costa-Castelló et al., 2005). One difference is that engineering applications of repetitive control have mostly been for linear systems, emphasizing the ability to track periodic reference signals and to compensate for periodic disturbances. Similarly, the notion that unstable systems can perform better because of greater maneuverability underlies many classic control designs, for example the Stealth fighter planes discussed in Section 3.5.2.

11.5 Summary

Most dynamical systems under control are nonlinear. Nonetheless, by (repeated) local linearization, control theory based on linear dynamics can be surprisingly successful. The possibility of superposition of solutions in linear systems leads to systematic principles of control. By contrast, nonlinear systems tend to require a number of different approaches, which each work best in particular situations. We began this chapter with a discussion of several methods that rely on coordinate transformations to effectively linear the dynamics. These methods range from gain scheduling a simple feedforward approach to a more systematic approach based on differential geometry. The latter leads to a multitude of new concepts – manifolds, tangent spaces, Lie derivatives, foliation, and more – and generalize the linear state-space approach notions of controllability and observability to nonlinear cases. As a rule, the most important results from the geometric theory are conceptual rather than practical. Another general set of ideas is based on notions of Lyapunov stability, where a Lyapunov function is sought for the closed-loop dynamics, including terms related to the controller. Finally, we showed how using nonlinear ideas could lead to new types of motion created by control, including synchronization (phase locking) and the creation or taming of chaotic motions.

11.6 Notes and References

The geometrical approach comes with a lot of baggage and requires a development beyond the scope of this book. Stone and Goldbart (2009) have a broad presentation of the required background, including notions of tangent spaces, Lie derivatives and brackets, the Frobenius theorem, and a general discussion of calculus on manifolds. Our discussion of the geometric approach to controllability, feedback linearization, and related issues is largely based on Slotine and Li (1991). The definitive presentation is by Isidori (1995). For a nice mix of history and pedagogy, see Brockett (2014).

An alternative approach to feedback linearization dates back to Koopman (1931), who proved that it is always possible to map a nonlinear dynamical system to an infinite-dimensional linear system. In favorable cases, a finite-dimensional approximation can be accurate. This *Koopman expansion* is particularly promising for *data-driven control*, which can be seen as an alternative version of the system-identification methods of Chapter 6. For an introduction to these methods, see Brunton and Kutz (2019).

Slotine and Li (1991) and Åström and Murray (2008) have accessible discussions of Lyapunov functions and their use in control problems. The main efforts are to extend the number of cases in which a negative semidefinite Lyapunov function still implies stability of a fixed point or limit cycle, to extend the theory to dynamical systems with explicit time dependence, and to make greater use of Lyapunov functions in the design process itself (e.g., to enforce convergence rates). The nearly 1000-page book by Haddad and Chellaboina (2008) gives a thorough and clear discussion of Lyapunov-based methods.

There is a vast physics literature on synchronization, starting from Huygens' discussion of pendulum clocks in the seventeenth century. For a general discussion of symmetry breaking and its consequences in the context of condensed matter physics, see Anderson (1984). Our presentation is based largely on Chapters 7 and 12 of Pikovsky et al. (2001). For a lively popular presentation, see Strogatz (2004). For phase-locked loops, we mostly follow the pleasant conceptual discussion in Pikovsky et al. (2001), filling in a few more details from a "control-centric" tutorial (Abramovitch, 2002).

Virtual potentials were introduced by Cohen (2005) and used by my group to investigate Landauer's principle (Section 15.3.6) in Jun et al. (2014). The term *cyberphysics* is a condensation of the similar notion of *cybernetical physics* (Fradkov, 2007).

For controlling chaos, the original presentation of the OGY method (Ott et al., 1990) is still good to read. For targeting trajectories, see Shinbrot (1999). The book by Abarbanel (1996) gives a good discussion of the experimental treatment of time-series data from nonlinear dynamical systems. Chapter 8, on "Control and Chaos," is a brief, clear introduction to the OGY method and to targeting. For a straightforward review of different methods and experiments, see Boccaletti et al. (2000). The most complete guide is the book by Schöll and Schuster (2008); however, it is long and repetitive.

Problems

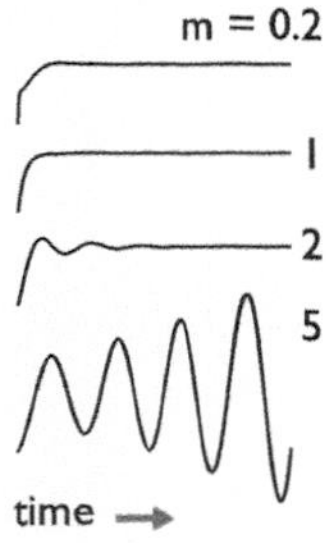

11.1 Gain scheduling for a linear system. Consider an oscillator with mass m and transfer function $G(s) = (ms^2 + s + 1)^{-1}$.

a. For $m = 1$, find a PID controller $K(s) = K_\text{p} + K_\text{i}/s + K_\text{d}s$ such that the closed-loop transfer function between reference and output $T(s) = \frac{1}{1+s}$.
b. Given $K(s)$ from (a), find the closed-loop transfer function $T_m(s)$ for arbitrary m.

c. Calculate the step response numerically for $m = 0.2$, 1, 2, and 5. (Compare at above right.) Note that the response is more robust for smaller than larger masses.
d. Show that the closed-loop response goes unstable at $m = 4$.
e. Given the mass m, design a "gain scheduled" controller $k_m(s)$ such that $T(s) = \frac{1}{1+s}$. This controller produces a nice step response for all (known) masses.

11.2 Gain scheduling for a nonlinear system. Analyzing a nonlinear system locally can improve control.

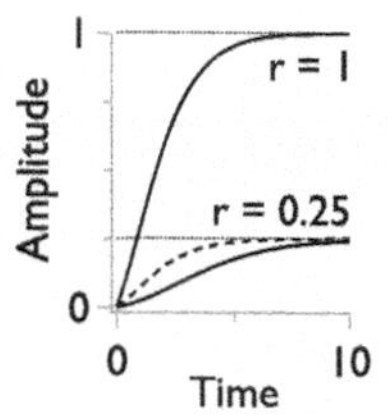

a. Integrate Eq. (11.1) numerically to reproduce the step plots shown. The PI gains are $K_p = K_i = 0.5$. Hint: Differentiate $u(t)$ and integrate the system $[y(t), u(t)]$.
b. Derive the linear system by expanding about the steady-state solution at r.
c. Show that choosing $K_p = K_i = 0.5/\sqrt{r}$ makes the transfer function of the resulting linear system between r and y equal to $\frac{1}{(1+s)^2}$.
d. Redo the numerical integration of the nonlinear system for this choice of gains and show that the response time is now independent of r (see dashed line at left).

11.3 Level control in a fluid tank.

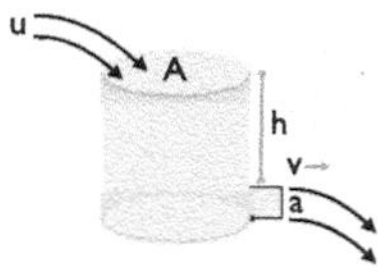

a. Using Bernouilli's Law, show that fluid exits a tank at a velocity $v^2 = 2gh$, where g is the gravitational acceleration and h is the fluid level above the outlet orifice. Conclude that the level of a tank with cross section A and outlet area a obeys $A\dot{h} = -a\sqrt{2gh} + u(t)$, where $u(t)$ is the added fluid flow volume/time. See left.
b. Design a simple nonlinear controller based on a sensor that measures $h(t)$ and controls $u(t)$. The controller should linearize the dynamics and converge to a desired reference height h_0. Adapted from Slotine and Li (1991).

11.4 Robustness and error cancellation. Consider the first-order system $\dot{x} = ax^2 + u$.

a. Choose a state-based control $u(t)$ that eliminates the nonlinearity and imposes linear dynamics with a pole at $s = -1$. Assume the state $x(t)$ is observable.
b. Assume now that you mistakenly estimate the parameter a as $\hat{a}$ and choose the feedback accordingly. Analyze the global stability of the closed-loop system.
c. Show that adding $-bx^3$ to the feedback can make $x = 0$ globally stable. Find the smallest value of b that stabilizes the origin.

11.5 Nonlinear integral control. In Chapter 3, we saw that integral control can stabilize a set point without offset. The same trick can work for nonlinear systems, too. Consider, as in Problem 11.4, the system $\dot{x} = ax^2 + u$. Assume that $a > 0$ but that its value is not known. Let the goal now be to stabilize $x(t)$ about the point $x_0 \neq 0$.

a. For proportional control $u = K_p(x_0 - x)$, find the equilibrium point x^* and determine the conditions for linear stability of the fixed point.

b. Add an integral control term, $K_i \int dt' \,(x_0 - x)$. Show that the set point x_0 is now a fixed-point solution. Determine its linear stability as a function of $\{a, x_0, K_p, K_i\}$. Does integral control improve the stability?

c. Reproduce the step responses at right for $K_p = 5$ and $K_i = \{0, 1, 5\}$.

d. Can integral control stabilize an arbitrary set point for a general nonlinear system?

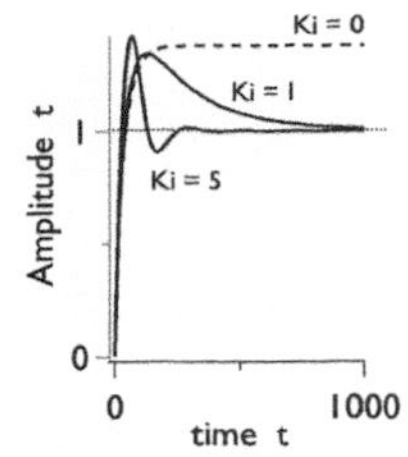

11.6 Stabilization in finite time. Nonlinear control can stabilize a system against perturbations in finite time, whereas linear feedback leads to exponential relaxation that takes an infinite amount of time to decay. As a simple example, consider $\dot{x} = x + u$, a one-dimensional system with unstable fixed point at $x = 0$. Let the feedback law be $u(t) = -k \,\mathrm{sign}[x(t)]$ for $|x| < 1$ and $-k\,x(t)$ for $|x| \geq 1$. See right, where $k = 2$.

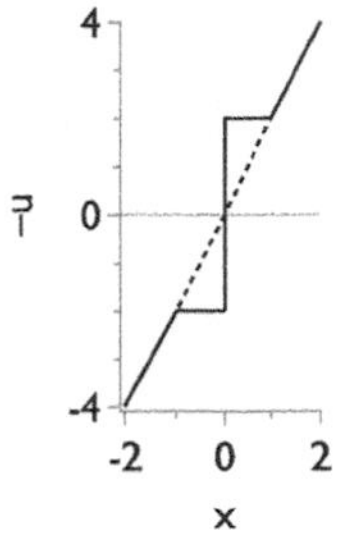

a. Show that a perturbation $x(0) = x_0$ returns in finite time to $x = 0$, for $k > 1$.

b. Compare the control effort, $\int_0^\infty dt\, u^2(t)$, for this system against that for $u = -kx$.

c. Show that you can replace the destabilizing $+x$ term with a nonlinear $f(x)$ satisfying $|f(x)| < \ell |x|$, for some $\ell > 0$ (Lipschitz condition), and $x = 0$ continues to be stable. Hint: Show that $V = \frac{1}{2}x^2$ is a Lyapunov function, using an inequality.

See Sun et al. (2017) for hints and applications to the control of complex networks.

11.7 Relative degree. Consider a linear system with transfer function $G(s)$,

$$G(s) = \frac{b_0 s^k + b_1 s^{k-1} + \cdots + b_k}{s^n + a_1 s^{n-1} + \cdots + a_n} \equiv \frac{b(s)}{a(s)} .$$

Show that the input $u(t)$ first appears after $n - k$ differentiations of the output $y(t)$. The relative degree is thus the difference between the numerator and denominator orders.

11.8 Internal dynamics. More on the simple example from Section 11.1.3.

a. Show that choosing $u = -x_2^3 - x_1$ stabilizes the $\left(\begin{smallmatrix}0\\0\end{smallmatrix}\right)$ solution to $\frac{d}{dt}\left(\begin{smallmatrix}x_1\\x_2\end{smallmatrix}\right) = \left(\begin{smallmatrix}x_2^3+u\\u\end{smallmatrix}\right)$.

b. Now alter $\dot{x}_2 = u$ to $\dot{x}_2 = -u$. Show that the internal dynamics is unstable.

11.9 Feedback linearization of a two-dimensional system. Complete Example 11.5 by deriving the equations of motion in the new coordinate system $z = \boldsymbol{T}(\boldsymbol{x})$. If the output is given as $y = x_1$, what is the relative degree?

11.10 Linearizable, but not reachable. Consider $\dot{x}_1 = x_2^2$ and $\dot{x}_2 = u(t)$.

a. Using the results from Example 11.5, find a nonlinear change of coordinates that makes this system linear and controllable.

b. Show that, nonetheless, you cannot reach all states starting from the origin, $\left(\begin{smallmatrix}0\\0\end{smallmatrix}\right)$. Qualitatively, what goes wrong?

c. Find the set of states that can be reached after a time T.

11.11 Involutive or not? Consider the vector fields, $\boldsymbol{f} = \begin{pmatrix}1\\0\\x_2\end{pmatrix}$, $\boldsymbol{g} = \begin{pmatrix}0\\-1\\\pm x_1\end{pmatrix}$. Show that choosing $+x_1$ gives an involutive pair but choosing $-x_1$ does not.

11.12 Lie brackets. Let $[f,g] \equiv \frac{\partial g}{\partial x} f - \frac{\partial f}{\partial x} g$, with $f(x)$ and $g(x)$ vector fields and $\frac{\partial f}{\partial x}$ and $\frac{\partial g}{\partial x}$ Jacobian matrices. Let $h(x)$ be a real-valued function, and recall that the Lie derivative $L_f h \equiv \frac{\partial h}{\partial x} f$. Then show the following:

a. *Geometrical interpretation*: Consider the following sequence of vector-field flows:

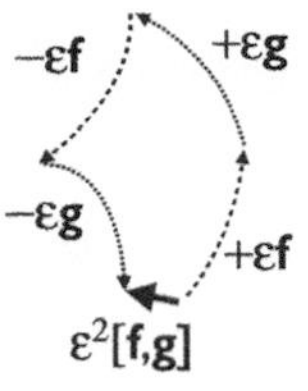

$$\dot{x} = \underbrace{+f(x)}_{0<t<\epsilon}, \quad \underbrace{+g(x)}_{\epsilon<t<2\epsilon}, \quad \underbrace{-f(x)}_{2\epsilon<t<3\epsilon}, \quad \underbrace{-g(x)}_{3\epsilon<t<4\epsilon}.$$

By Taylor expanding the state $x(t)$, show that the state fails to return to the origin by an amount $x(4\epsilon) - x(0) = \epsilon^2 [f,g] + O(\epsilon^3)$. (Caution: messy algebra.)

b. *Bilinearity*: $[\alpha_1 f_1 + \alpha_2 f_2, g] = \alpha_1 [f_1, g] + \alpha_2 [f_2, g]$.

c. *Skew commutivity*: $[f, g] = -[g, f]$.

d. *Jacobi identity*: $L_{[f,g]} h(x) = L_f L_g h(x) - L_g L_f h(x)$.

11.13 Parking a unicycle. The equations of motion are $\frac{d}{dt}\begin{pmatrix} x_1 \\ x_2 \\ \theta \end{pmatrix} u_1(t) \begin{pmatrix} \cos\theta \\ \sin\theta \\ 0 \end{pmatrix} + u_2(t) \begin{pmatrix} 0 \\ 0 \\ 1 \end{pmatrix}$ for the unicycle, where u_1 controls the *drive* and u_2 the *spin*. Show that the Lie bracket, [*drive*, *spin*] = *slip*, where *slip* is $\perp$ to *drive*.

11.14 Feedback linearization. Let us show that Eq. (11.38) gives necessary and sufficient conditions for a control affine system to be feedback linearizable.

a. Use the Jacobi identity (Problem 11.12d) to show, for a smooth function $h(x)$ and smooth vector fields $f(x)$ and $g(x)$ and arbitrary positive integer k, that

$$L_g h = L_g L_f h = \cdots = L_g L_f^k h = 0 \quad \Longleftrightarrow \quad L_g h = L_{\mathrm{ad}_f g} h = \cdots = L_{\mathrm{ad}_f^k g} h = 0.$$

b. Using the above result, show that Eqs. (11.28) and (11.29) imply,

$$\frac{\partial T_1}{\partial x} \mathrm{ad}_f^k g = 0 \text{ for } k = 0, 1, 2, \ldots, n-2 \quad \text{and} \quad \frac{\partial T_1}{\partial x} \mathrm{ad}_f^{n-1} g \neq 0.$$

c. Conclude that the vector fields $g, \mathrm{ad}_f g, \ldots, \mathrm{ad}_f^{n-1} g$ are linearly independent.

d. Using the Frobenius theorem, conclude that the vector fields are involutive.

11.15 Linear systems from a nonlinear point of view. By specializing the n-dimensional control-affine system of the form $\dot{x} = f + g\,u(t) \to Ax + Bu$, show that

$$W_c = \begin{pmatrix} g & \mathrm{ad}_f g & \cdots & \mathrm{ad}_f^{n-1} g \end{pmatrix} \quad \to \quad \begin{pmatrix} B & AB & A^2B & \cdots & A^{n-1}B \end{pmatrix}.$$

In both cases, the matrix W_c must have rank n. But the nonlinear condition refers to feedback linearizability, while the linear condition also implies controllability.

11.16 Flexible link. Consider controlling a pendulum via an elastic, "soft" torque such as that provided by a motor. The system may be used to model muscles, both real and artificial. In robotics jargon, it is known as a *single-link flexible joint*. With scaled units and simplified choices for spring constant and inertial moments, the equations of motion are $\ddot{\theta}_1 + \sin\theta_1 + (\theta_1 - \theta_2) = 0$ and $\ddot{\theta}_2 + (\theta_2 - \theta_1) = u(t)$, where θ_1 is the angle of the pendulum with respect to its down equilibrium, θ_2 is the angle of the input to the coupling, and u is the torque applied to the coupling.

a. Write these equations in control-affine form, $\dot{\boldsymbol{x}} = \boldsymbol{f}(\boldsymbol{x}) + \boldsymbol{g}(\boldsymbol{x})\,u(t)$.
b. Calculate the linearizability matrix $\boldsymbol{W}_{\mathrm{c}} = (\boldsymbol{g}\;\; \mathrm{ad}_f\,\boldsymbol{g}\;\; \mathrm{ad}_f^2\,\boldsymbol{g}\;\; \mathrm{ad}_f^3\,\boldsymbol{g})$, and verify that it has full rank. We thus can linearize the system exactly.
c. Find the change of variable $\boldsymbol{z} = \boldsymbol{T}(\boldsymbol{x})$ and control that linearizes the system in the form of Eq. (11.20). Hint: See Problem 11.14b. In particular, show that

$$u = -\left(\sin x_1(x_2^2 + \cos x_1 + 1) + (x_1 - x_3)(2 + \cos x_1)\right) + v(t)\,.$$

d. Find the inverse transformation $\boldsymbol{x} = \boldsymbol{T}^{-1}(\boldsymbol{z})$ and use it to confirm explicitly that the dynamics are linear in the new variables: $\dot{z}_1 = z_2,\;\; \dot{z}_2 = z_3,\;\; \dot{z}_3 = z_4,\;\; \dot{z}_4 = v$.

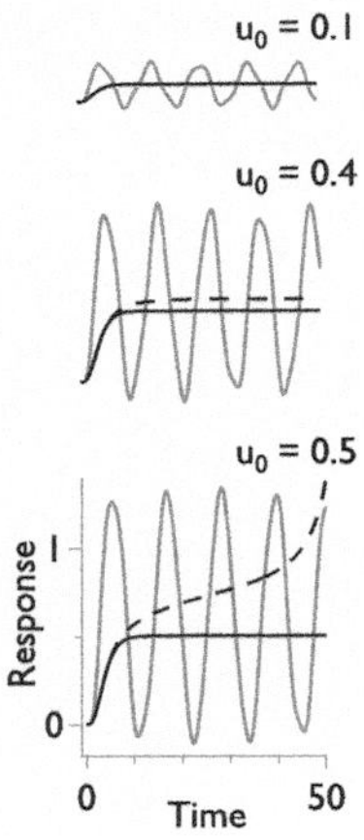

e. Reproduce the plots at right, for step inputs of amplitude 0.1, 0.4, and 0.5. The light oscillating trace shows the undamped, open-loop response. The heavy dark line uses the exact feedback linearization, assuming that the linear part of the control is set to have 4 poles at −1. You should find a gain (row) vector $\boldsymbol{K} = (1\;4\;6\;4)$. The dashed curve is based on the linearization of the system about $\theta_1 = \dot{\theta}_1 = \theta_2 = \dot{\theta}_2 = 0$. Its gain vector is designed so that the *linear* approximation has poles at −1. You should find $\boldsymbol{K}' = (-6\;-4\;3\;4)$. The approximate design matches the exact linearization very well for $u_0 = 0.1$ but goes unstable for large input torque, near $u_0 = 0.5$.

11.17 Global Stabilization of the nonlinear pendulum. For $\ddot{\theta} - \sin\theta = u$ and for the "up" position ($\theta = \dot{\theta} = 0$), consider the function $V = 2a\sin^2\left(\frac{\theta}{2}\right) + \frac{1}{2}\dot{\theta}^2$.

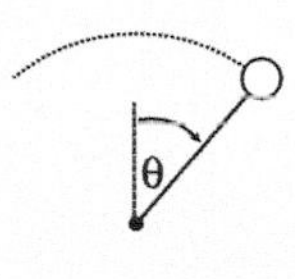

a. For the uncontrolled system ($u = 0$), show that V is not a Lyapunov function.
b. Show that choosing $u = -K_{\mathrm{p}}\sin\theta - K_{\mathrm{d}}\dot{\theta}\cos^2\theta$ can make V a Lyapunov function that stabilizes the top position. In particular, show that you need to choose $a > 0$, $K_{\mathrm{p}} = a + 1 > 1$, and $K_{\mathrm{d}} > 0$. What is the physical significance of these conditions?
c. What if there is friction in the pendulum, so that $\ddot{\theta} + \lambda\dot{\theta} - \sin\theta = u$?
d. Example 11.8 notes that $\dot{V}$ is negative semidefinite, not negative definite. Yet the up solution is stable to almost all perturbations. Consider a new term in the control algorithm $u \to u - \epsilon\theta$, with θ defined to be in the range of $(-\pi, \pi)$. There is a discontinuity of amplitude 2ϵ when crossing the down position, $\theta = \pm\pi$. Is such a term helpful? Is the modified V still a Lyapunov function?
e. Plot $\theta(t)$, $V(t)$, and $u(t)$ for a perturbation of the form $\theta(0) = 0$, $\dot{\theta}(0) = 2$, and $a = K_{\mathrm{d}} = 1$. Show that the controller recovers from a "kick" of almost 80°.

11.18 Sending "secret" messages by synchronized chaos. Section 10.5.1 introduced the Lorenz equations for state variables $\boldsymbol{x}^{\mathrm{T}} = (x, y, z)$. Here, think of this system as the *transmitter*, and set up a *receiver* with state variables $\boldsymbol{x}_{\mathrm{r}}^{\mathrm{T}} = (x_{\mathrm{r}}, y_{\mathrm{r}}, z_{\mathrm{r}})$ driven by $x(t)$ from the original system (but not in the $\dot{x}_{\mathrm{r}}$ equation). The two systems are

$$\underbrace{\begin{aligned} \dot{x} &= \sigma\,(y - x) \\ \dot{y} &= x\,(r - z) - y \\ \dot{z} &= xy - bz \end{aligned}}_{\text{transmitter}} \qquad\qquad \underbrace{\begin{aligned} \dot{x}_{\mathrm{r}} &= \sigma\,(y_{\mathrm{r}} - x_{\mathrm{r}}) \\ \dot{y}_{\mathrm{r}} &= x\,(r - z_{\mathrm{r}}) - y_{\mathrm{r}} \\ \dot{z}_{\mathrm{r}} &= xy_{\mathrm{r}} - bz_{\mathrm{r}} \end{aligned}}_{\text{receiver}}\,.$$

a. Simulate for $\sigma = 16$, $b = 4$, and $r = 45.6$. Demonstrate synchronization of the chaotic motion by plotting components of $\boldsymbol{x}_r(t)$ versus $\boldsymbol{x}(t)$. Also, plot the components of the error $\boldsymbol{e} \equiv \boldsymbol{x} - \boldsymbol{x}_r$, and show that they decay exponentially.
b. Show that $V = \frac{1}{2}(\frac{1}{\sigma}e_x^2 + e_y^2 + e_z^2)$ is a Lyapunov function for this dynamical system. Why does this explain the synchronization?
c. Cuomo and Oppenheim built analog electronic circuits corresponding to these equations and demonstrated synchronization experimentally (Strogatz, 2014). They then communicated "secret messages" by both analog and digital techniques. To understand the latter, let the "signal" $m(t)$ be a stream of 0's and 1's – we will use a square wave but a random bit sequence will work equally well. Use m to modulate the b coefficient. That is, let $b \to b(t) = b + a\,m(t)$. Take $a = 0.4$. Send the altered $x(t)$ signal to the receiver, compute x_r, and then plot the quantity $e_x = (x_r - x)^2$. Send e_x through a low-pass filter, and plot the result. Then apply a threshold: set the signal = 0 below a certain amplitude and 1 above it. Call this estimate $\hat{m}$ and compare to the original m (see below). Why does this scheme work?

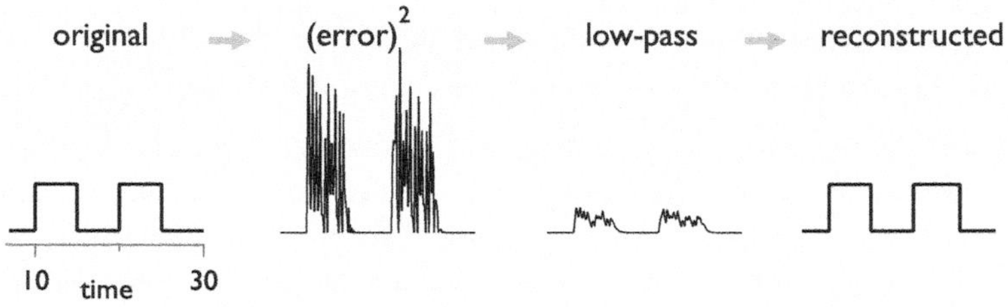

One note: although here the "carrier" of the message is chaotic and chaotic motion seems complicated, this problem shows that the message scheme is not inherently secure, in that any eavesdropper could achieve the same synchronization as the intended receiver. Whatever robustness exists in the detection scheme will be available to all, and any desired security will arise by encrypting the message itself. For a discussion of why using a chaotic carrier might nonetheless be useful, see Abarbanel (2008). Quantum communication methods, by contrast, can provide inherently secure protection against eavesdroppers (Mermin, 2007).

11.19 Backstepping. Here is a recursive way to start from a Lyapunov function for a simple system and "extend" it to a new Lyapunov function for an enlarged dynamical system. Consider a system of the form $\dot{\boldsymbol{x}} = \boldsymbol{f}(\boldsymbol{x}) + \boldsymbol{g}(\boldsymbol{x})\,v$, with $\dot{v} = u$. Here, the input variable is u, as usual, and v is an "extra" state variable, along with the n variables in x. Now pretend that v is the control variable for $\dot{\boldsymbol{x}} = \boldsymbol{f}(\boldsymbol{x}) + \boldsymbol{g}(\boldsymbol{x})\,v$, and imagine that we know a Lyapunov function $V_0(\boldsymbol{x})$ that stabilizes

$\boldsymbol{x} = \boldsymbol{0}$ for the system $\dot{\boldsymbol{x}} = \boldsymbol{f}(\boldsymbol{x}) + \boldsymbol{g}(\boldsymbol{x})\,\phi(\boldsymbol{x})$ for some appropriate "control" $v = \phi(\boldsymbol{x})$.

a. In the original problem, v is not directly controlled. Define a new variable $z = v - \phi(\boldsymbol{x})$. Find a u to make a new Lyapunov function, $V = V_0 + \frac{1}{2}z^2$.
b. Use backstepping to find a control u and Lyapunov function V for $\dot{x}_1 = x_1^2 + x_2$, $\dot{x}_2 = u$ that stabilizes $x_1 = x_2 = 0$.

Khalil (2001) extends this idea to a chain of dynamical equations (not restricted to the simple integrator discussed here).

11.20 Oscillator frequency stabilization. The *Pound–Drever–Hall* (PDH) technique is a way to stabilize the frequency of a tunable laser that is similar to *extremum-seeking control* (Section 10.1.2). It uses a standard PID loop based on an error signal proportional to the frequency shift between laser and reference cavity. The clever part is how the error signal is generated. Here, we discuss a simplified version of PDH that applies to a signal $y(t) = y(\omega)\,\mathrm{e}^{\mathrm{i}\omega t}$ + c.c., whose frequency $\omega(t)$ can drift in time, and a reference passive filter, whose response has a fixed resonance frequency ω_0. Our focus will be on generating an error signal proportional to the shift in frequency, $e(t) \propto \omega_0 - \omega(t)$. In keeping with the optical applications, we will assume that ω is so large that we cannot measure the time-domain signal $y(t)$ directly but must rather measure its time-averaged power, $P(\omega) \equiv \langle |y(\omega)|^2 \rangle$, which can drift slowly.

a. We begin with a naive strategy that seems like it should work but has flaws. As shown below, we pass the signal, $y(t)$, through a second-order filter with highly underdamped, resonant response. Select ω_0 so that the $\omega(t)$ is located on the side of the resonance, at or near the point of maximum slope. Using a Taylor expansion of the power variation $\delta P = P - P_0$, show that we expect $\delta P = \left(\frac{\partial P}{\partial P_0}\right)\delta P_0 + \left(\frac{\partial P}{\partial \omega}\right)\delta\omega$, to first order in variations of signal power (δP_0) and frequency ($\delta\omega$). We cannot distinguish between these variations without monitoring the signal power.
b. Now consider the PDH strategy, illustrated below. Set $\omega_0 \approx \omega(t)$ and add a deliberate phase modulation of amplitude a and frequency $\Omega \ll \omega_0/Q$, where Q is the enhancement factor of the resonance. Specifically, let the frequency be modulated so that $\omega(t) = \omega_0 + \delta\omega(t) + a\Omega\cos\Omega t$. Repeat the calculation from (a) and show, ignoring 2Ω and higher-frequency terms, that $\delta P = \left(\frac{\partial P}{\partial P_0}\right)\delta P_0 + \left(\frac{\partial^2 P}{\partial \omega^2}\delta\omega\right) a\Omega\cos\Omega t$, where $\left.\frac{\partial^2 P}{\partial \omega^2}\right|_{\omega_0}\delta\omega \equiv e(t)$ functions as an error signal. In the PDH strategy, we use a mixer to multiply the output of the filter $F(\omega)$ by $\cos\Omega t$ and a low-pass (LP) filter with cutoff frequency $\ll \Omega$. Explain why this isolates $e(t)$, the amplitude of the $\cos\Omega t$ term in δP, and why the effects of amplitude variation in the original signal now give a higher-order contribution.

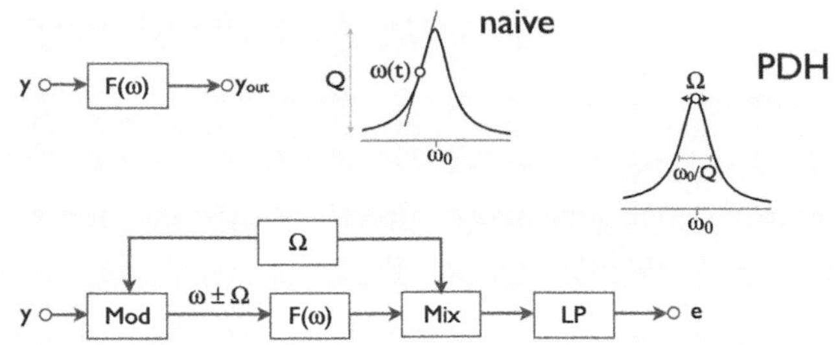

The real *Pound–Drever–Hall* technique not only adds the physics of laser cavities but also includes many subtleties, such as the advantages of working with a higher modulation frequency $\Omega > \omega_0/Q$, of working in reflection with an intensity minimum (as opposed to the maximum analyzed here), and the calculation of nonlinearities in the shape of the error function for larger frequency excursions (Black, 2001).

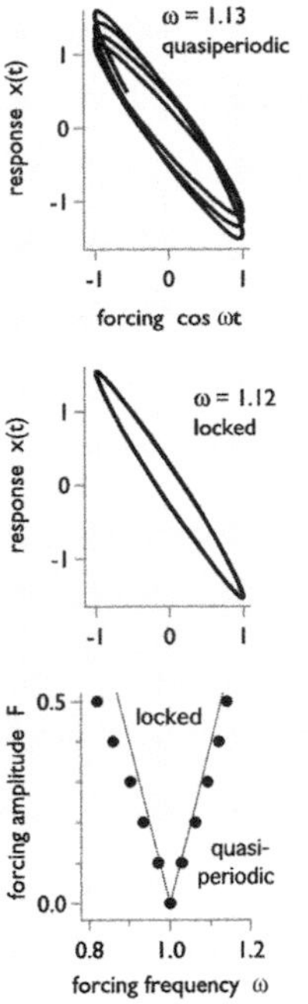

11.21 Synchronization of the van der Pol oscillator to periodic forcing. Consider $\ddot{x} - \varepsilon(1 - x^2)\dot{x} + \omega_0^2 x = F\cos\omega t$, a forced version of the van der Pol equation introduced in Problem 7.17, with $0 < \varepsilon \ll 1$. The forcing amplitude is weak ($F \lesssim \varepsilon$), and the detuning small ($\omega/\omega_0 = 1 + O(\varepsilon)$).

a. Assume a solution of the form $x(t) = A\,e^{i\omega t} + \text{c.c.}$, with A an amplitude that can vary on time scales ε^{-1}. Show that $\dot{A} = O(\varepsilon)$ and $\ddot{A} = O(\varepsilon^2)$. Then show that

$$\dot{A} = -i\left(\frac{\omega_0^2 - \omega^2}{2\omega}\right)A + \frac{1}{2}\varepsilon\left(1 - |A|^2\right)A - i\frac{F}{4\omega}.$$

Hint: Write $\ddot{x} + \omega^2 x = \cdots$ and include all other $O(\varepsilon)$ terms on the RHS. Substitute for x, multiply by $e^{-i\omega t}$, and average over one period.

b. Rewrite (a) to give a generic amplitude equation, $a'(\tau) = -i\nu a + (1 - |a|^2)a - if$, by rescaling and redefining quantities. Show that all terms are $O(1)$.

c. Write a in polar form, $a = r\,e^{i\phi}$, and show that $r' = (1 - r^2)r - f\sin\phi$, with $\phi' = -\nu - \frac{f}{r}\cos\phi$. Does ϕ obey an Adler equation?

d. Simulate the forced van der Pol equation and find conditions for synchronization. Compare to the asymptotic analysis based on the Adler equation. Use the parametric plot (Lissajous figure) between $x(t)$ and $\cos\omega t$ to test for synchronization. Reproduce the plots at left, where $\varepsilon = 0.1$, $\omega_0 = 1$, and (top two graphs) $f = 0.4$.

11.22 Synchronization of N globally coupled oscillators (Example 11.9). Consider N coupled oscillators obeying Eq. (11.53), with frequencies ω_k having symmetric distribution $g(\omega)$ and mean ω_0. Define the complex order parameter $\mathcal{K} = K\,e^{i\Theta}$, which can be viewed as a "mean field" that "externally" forces the oscillators.

a. Show that Eq. (11.53) can be rewritten as $\dot{\phi}_k = \omega_k + \varepsilon K\sin\left(\Theta - \phi_j\right)$.

b. Show that the synchronized oscillator frequency $\approx \omega_0$ and that $\psi_k \equiv \phi_k - \omega_0 t$ obeys the Adler equation $\dot{\psi}_k = \omega_k - \omega_0 - \varepsilon K\sin\psi_k$.

c. Argue that for $N \to \infty$, a self-consistent condition for the order parameter is $\mathcal{K} = K\,\mathrm{e}^{\mathrm{i}\Theta} = \int_{-\pi}^{\pi} \mathrm{d}\psi\, \mathrm{e}^{\mathrm{i}\phi}\, n_s(\psi)$, where $n_s(\phi)$ is the distribution of synchronized oscillators having phase ϕ. Why do only the *synchronous* (and not the asynchronous) oscillators contribute? Why is the integral over ψ?

d. Rewrite the self-consistent equations for $\mathcal{K}$ as $1 = \varepsilon \int_{-\pi/2}^{\pi/2} \mathrm{d}\psi \cos^2\psi\, g(\omega)$ and $0 = \varepsilon \int_{-\pi/2}^{\pi/2} \mathrm{d}\psi \cos\psi\, \sin\psi\, g(\omega)$, where $\omega = \omega_0 + \varepsilon K \sin\psi$.

e. Argue that the second equation is satisfied if the mean frequency is ω_0, as we guessed. Taylor expand the first equation about ω_0 and show that there is a phase transition at $\varepsilon_c = \frac{2}{\pi g(\omega_0)}$ from $K = 0$ to $K^2 = \frac{8 g(\omega_0)}{|g''(\omega_0)|\, \varepsilon_c^3}(\varepsilon - \varepsilon_c)$, as shown at right.

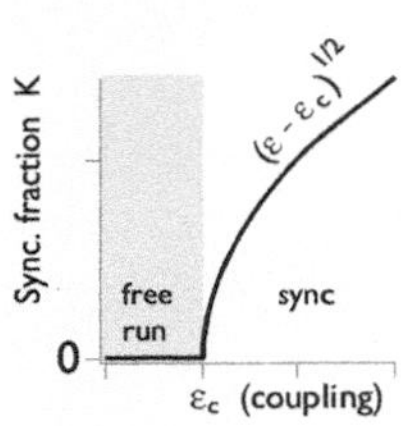

f. Integrate numerically the equations in (a), with $N = 10^5$ and $g(\omega) \sim \mathcal{N}(1, 0.1^2)$. Reproduce the order-parameter plot at right for $K(t)$. The values by the three curves indicate $\varepsilon/\varepsilon_c$. Show that the synchronization threshold $\varepsilon_c = \sqrt{8/\pi}\,\sigma$.

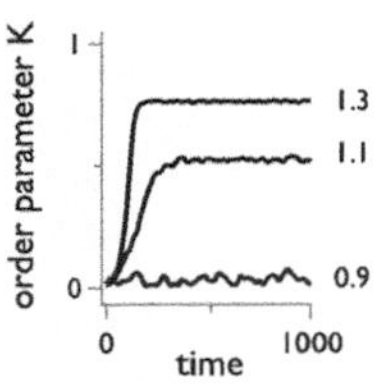

For details, see Kuramoto (1984) and also Pikovsky et al. (2001), Section 12.1.

11.23 Controlling chaos via the OGY method. Implement the OGY method and targeting to reproduce the plots in Figure 11.6. For targeting, write a function that expands the range $\lambda \pm \Delta\lambda$ and compute the range of possible images $x_1 \pm \Delta x_1$. Then count the number of further iterations (using λ as control parameter) to expand Δx_1 to a larger range Δx_n that includes the target x^*. You now have a function between the range $\lambda \pm \Delta\lambda$ and $x_n \pm \Delta x_n$ that includes x^*. Use this function in a root-finder routine to predict the value of the perturbation λ' that brings the system to x^* in n iterations. Verify that the number of required iterations grows logarithmically as the perturbation tolerance $\Delta\lambda$ is reduced. For plots, use $\varepsilon = 0.02$ and $x_0 = 0.5$.

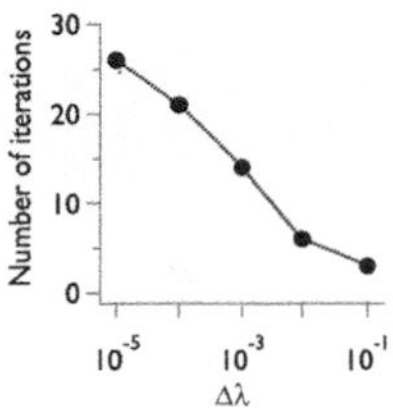

11.24 Time-delayed feedback for chaotic systems. Consider a control algorithm for a signal $y(t)$ that is based on $K[y(t) - y(t-\tau)]$, where K is a feedback gain and τ is the period of limit cycle that you wish to stabilize. The control signal will vanish when $y(t)$ is periodic with period τ. Use this idea to stabilize the Rössler equations, a canonical dynamical system exhibiting chaos. The equations for a three-dimensional state vector $(x\ y\ z)^{\mathsf{T}}$ are $\dot{x} = -y - z$, $\dot{y} = x + ay - K[y(t) - y(t-\tau)]$, and $\dot{z} = b + z(x - c)$.

a. Simulate a time series $y(t)$ for $0 \le t \le 300$, using $a = b = K = 0.2$ and $c = 5.7$. Start the control at $t = 100$. For $K = 0$ (no control), the motion should be chaotic.

b. The method needs the period τ. To find τ from the motion itself, calculate the mean-square error of the control signal (after transients have died away), as a function of τ. From the minima locations, find the periods to use in part (a).

See Pyragas and Pyragas (2011), who show how to tune τ adaptively. Note that using a delayed signal turns ordinary differential equations into *delay-differential equations*, whose infinite-dimensional state spaces are hard to analyze.

PART III

SPECIAL TOPICS

12 Discrete-State Systems

In this chapter, we discuss how discrete state spaces arise, how they can be described probabilistically using the Bayesian formalism developed in Chapter 8, and how they can be controlled using concepts of optimal control developed in Chapter 7.

The simplest nontrivial mathematical version of a discrete state space is a single state x that at each time can take on only two values, for example -1 and $+1$. Of course, quantum systems, such as particles with spin degrees of freedom, can have properties that are inherently discrete and low-dimensional. But even classical, continuous state spaces can often be well approximated by such a discrete system, via *coarse graining*. Think of a system that hops back and forth between two different configurations. For example, a protein in solution can hop back and forth between a looser, *unfolded state* (-1) and a compact, *folded state* $(+1)$, as sketched at right.[1]

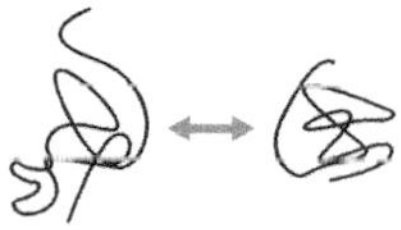

Why describe such systems using discrete state spaces?

- We can identify and isolate just the important degrees of freedom.
- Control theory and state estimation become much simpler, even when the equivalent continuum case would be intractable.
- Dynamics modeled on a computer must be discretized in both time and state.
- Recapitulating the essentials of control theory in this restricted context may clarify ideas that are harder to grasp when spread out over the entire book.

We will focus on systems where both the state space and time are discrete. Chapters 13 and 15 present examples where the state space is discrete but time is continuous.

12.1 Discrete-State Models

In this section, we describe the discrete versions of the continuous state-space models that we have used up to now, focusing on the discrete-time case. Models that are fully observable are known as *Markov chains*. If the state is not directly observable, then the model is known as a *hidden Markov model*. We discuss these and the *coarse-graining* procedure used to derive approximate discrete-state models from continuous ones.

[1] Other biological examples of two-state systems include *ion channels* that can be open or closed and *gene-transcription repressor sites* that can be occupied or not. More generally, "systems" such as gene transcription and *sensory receptors* can be active or not. See Phillips et al. (2012), chapter 7.

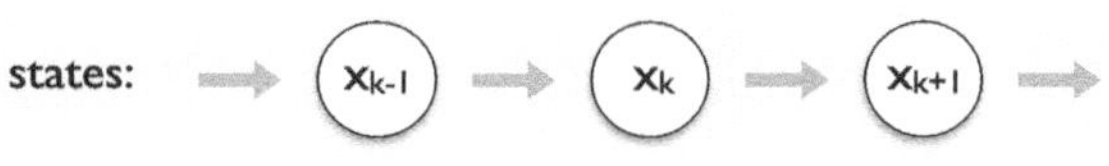

Fig. 12.1 Graphical representation of the Markov chain. State x_{k+1} depends only on x_k.

12.1.1 Markov Chains

Consider a system described at a discrete time k by a state x_k that can take on a finite number n values,[2] for example the set 1 to n. We can then define a probability distribution $P(x)$ over the set of allowed states, where $P(x_k = i)$ is the probability that the system is in state i at time k. The distribution is normalized by enforcing $\sum_{i=1}^{n} P(x_k = i) = 1$ or, more succinctly, $\sum_{x_k} P(x_k) = 1$. Our previous discussions of Markov dynamics (e.g., Section A.9.2 online) then apply almost without change, with the two chief differences being that integrals are replaced by discrete sums and that dynamics become simple lookup tables of probabilities. The Markov dynamics are completely specified via $P(x_{k+1}|x_k)$ and are known commonly as *Markov chains*.

Dynamics take the form of an $n \times n$ *transition matrix* $\boldsymbol{A}$ whose matrix elements are $A_{ij} \equiv P(x_{k+1} = i \,|\, x_k = j)$ and satisfy $0 \le A_{ij} \le 1$. That is, A_{ij} gives the $i \leftarrow j$ transition probability from j to i during one step. For a two-state system,

$$\boldsymbol{A} = \begin{pmatrix} 1 - a_0 & a_1 \\ a_0 & 1 - a_1 \end{pmatrix}. \tag{12.1}$$

Normalization implies that the sum along each column of $\boldsymbol{A}$ equals 1, $\sum_i P(x_{k+1} = i|x_k = j) = 1$. In words: From state j, the system must go to one of the n possible states, indexed by i. More generally, a matrix whose elements $A_{ij} \in [0, 1]$ and where sums along each column are all equal to one, $\sum_i A_{ij} = 1$ is known as a *stochastic matrix*.[3] At left is a diagram depicting the transitions of a two-state Markov model. By convention, only the transitions from one state to another are shown. The probability to stay in the same state is omitted, as normalization dictates its value.

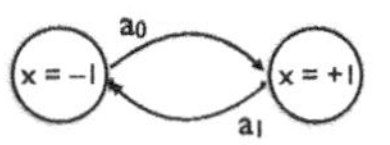

Let us define the n-dimensional *stochastic vector* $\boldsymbol{p}_k$, whose elements $p_k^{(j)} \equiv P(x_k = j)$ give the probability to be in state j at time k. Note that $0 \le p_k^{(j)} \le 1$ and $\sum_j p_k^{(j)} = 1$. Then,

$$p_{k+1}^{(i)} = \sum_{j=1}^{n} P(x_{k+1} = i, x_k = j) = \sum_{j=1}^{n} \underbrace{P(x_{k+1} = i \,|\, x_k = j)}_{A_{ij}} P(x_k = j) = \sum_{j=1}^{n} \boldsymbol{A}_{ij}\, p_k^{(j)}. \tag{12.2}$$

[2] The numerical values of states are somewhat arbitrary. Thus, a two-state system might take on values $\{\pm 1\}$ or $\{1, 2\}$ or $\{0, 1\}$. Values can also be nonnumerical: { Left, Right }, { Black, White }, etc.

[3] More precisely, a *left stochastic matrix*. In another commonly used convention, the *right stochastic matrix* $\boldsymbol{A}$ has elements $A_{ij} \equiv P(x_{k+1} = j|x_k = i)$ equal to the rate of $i \to j$ transitions. Sums across *rows* are then equal to one.

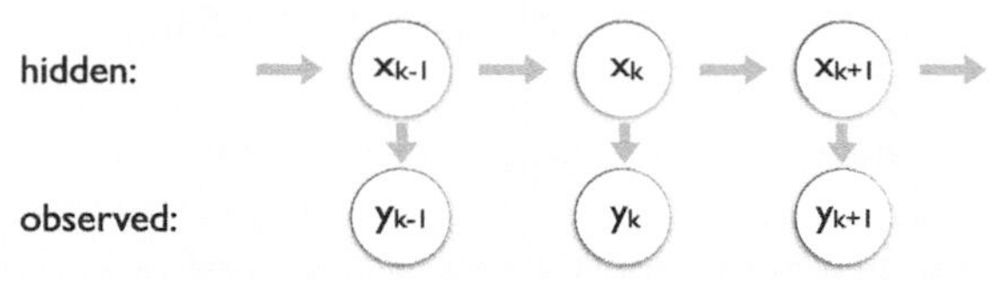

Fig. 12.2 HMM graphical structure. The states x_k form a Markov process that is not directly observable. The observations y_k depend only on x_k.

More compactly,[4]

$$p_{k+1} = A\, p_k \,. \qquad (12.3)$$

Formally, Eq. (12.3) is a linear difference equation of the type considered in Chapter 5, with solution $p_k = A^k\, p_0$. We are often interested in finding the steady-state distribution p, which satisfies $p = Ap$. One way is to repeatedly iterate Eq. (12.3); another is to note that the steady-state distribution of probabilities corresponds to the eigenvector associated with an eigenvalue equal to 1. A stochastic matrix must have at least one such eigenvalue. To see this, note that $A - \mathbb{I}$ is a matrix where sums along columns all equal zero. They are then linearly dependent, implying a zero determinant. Below, we focus on *ergodic* Markov models with a unique steady state where all states have nonzero probabilities. That is, we exclude situations where the system can be "stuck" in a given state (or subset of states).

Example 12.1 (Steady state for a two-state Markov model) For the two-state Markov model with transition matrix $A = \begin{pmatrix} 1-a_0 & a_1 \\ a_0 & 1-a_1 \end{pmatrix}$, the eigenvalues are 1 and $1 - (a_0 + a_1)$. The eigenvector corresponding to the unit eigenvalue gives the steady state, $p = \frac{1}{a_0+a_1} \begin{pmatrix} a_1 \\ a_0 \end{pmatrix}$, where the vector components are normalized to sum to unity. If $a_0 = a_1 = a$, then $p^* = \begin{pmatrix} 0.5 \\ 0.5 \end{pmatrix}$; by symmetry, both states are a priori equally probable.

Further properties of Markov chains and their steady states are explored in Problem 12.2.

12.1.2 Hidden Markov Models

Often, the states of a Markov chain are not directly observable; however, there may be *symbols* emitted that correlate with the underlying states. The combination is known as a *hidden Markov model* (HMM). The hidden states are also known as *latent variables*.

In the example of proteins that alternate between unfolded and folded states, the molecule itself is usually not directly observable. The configuration can be monitored, however, by attaching a small colloidal particle to one end of the protein and anchoring the other end to a surface. As the protein folds and unfolds, the bead moves up

[4] In the convention where A_{ij} denotes $i \to j$ transitions, $p_{k+1} = A^{\mathsf{T}} p_k$.

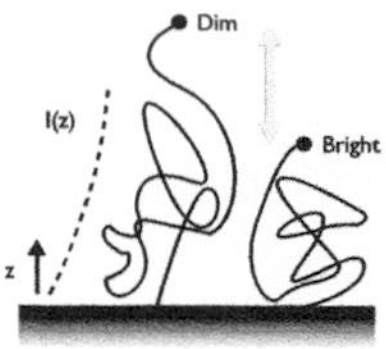

and down from the surface, as illustrated at left. We can illuminate the region near the surface using an evanescent wave via the technique known as total internal reflection microscopy. The intensity $I(z)$ of light scattered by the bead at height z from the surface will decrease exponentially as $I(z) \propto e^{-z/z_0}$, with $z_0 \approx 100$ nm.[5] The two states will then correspond to two different brightnesses (*Dim* and *Bright*), which we take to be two different symbols, as described above. But because light scattering is itself a stochastic process (as is the generation of photons in the laser used for the scattering), a protein may be in one state but emit the "wrong" symbol. We can describe such a situation by defining the observations $y_k = \pm 1$ and noting that they are related to the states probabilistically via the matrix $\boldsymbol{B}_{ij} \equiv P(y_k = i | x_k = j)$.

This leads to a symmetric *observation matrix* $\boldsymbol{B} = \left(\begin{smallmatrix} 1-b & b \\ b & 1-b \end{smallmatrix}\right)$.[6] In words, the probability that the observed symbol correctly reflects the state is $1 - b$. It is incorrect with probability b. Like the transition matrix $\boldsymbol{A}$, the matrix $\boldsymbol{B}$ is stochastic, with sums along columns that each equal 1. Here, sums across rows also equal 1, but only because of the symmetry between states. Figure 12.2 gives a graphical representation of an HMM.

12.1.3 Coarse Graining

A system described by a continuous state variable can sometimes be approximated by a finite-state Markov chain. For example, the protein that alternates between two configurations can be well approximated, in certain circumstances, by a two-state Markov process. Although that example would be complicated to derive from a microscopic description, we can explore a similar but simpler example. Specifically, we explore the reduction of overdamped motion in the double-well potential  to a two-state Markov model.[7]

Proteins are long polymers with many degrees of freedom. Assume that two distinct sets of configurations, *folded* and *unfolded*, exist in a high-dimensional configuration space. Consider the distance between the surface at one end and the attached scattering particle at the other end. As a function of this endpoint separation, we can calculate, at least conceptually, an effective free energy by averaging over all other coordinates. This preliminary coarse graining leads to a one-dimensional equation of motion for the coordinate $x(t)$,

$$\gamma \dot{x}(t) = x(t) - x(t)^3 + \nu(t) , \tag{12.4}$$

[5] Often, the bead is fluorescent, but a metal bead works, too, particularly if light excites its *plasmon resonance*.

[6] The number of observation symbols need not equal the number of internal states. With m symbols and n states, the matrix $\boldsymbol{B}$ is $m \times n$. The number m can be bigger or smaller than n, or even infinite (continuous observations; cf. Problem 12.4). If smaller, then there is the problem of observability (Chapter 4).

[7] If the energy barrier in a double-well potential is high, the phase (state) space splits into two disjoint regions, left and right, that can be described as *locally ergodic*. Recall that for an ergodic dynamical system, time averages are equivalent to ensemble averages, and a system passes arbitrarily close to any point in its phase (state) space. When energy barriers are high, global ergodicity can be broken, and the locally ergodic subregions are naturally associated with macroscopic states.

with $\langle v(t)\, v(t')\rangle = 2\gamma k_B T\, \delta(t - t')$ and with k_B Boltzmann's constant and T the absolute temperature. The fluctuation-dissipation relation constrains the magnitude of the noise. A simple physical realization of Eq. (12.4) is an overdamped colloidal particle moving in a one-dimensional potential in a fluid. The coefficient γ is then hydrodynamic damping.

Time Discretization

To begin, we discretize time, using a first-order forward Euler approximation for the derivative $\dot{x}$ (Section 5.3.4) and a short sampling time T_s. Introducing $x_k \equiv x(t = kT_s)$, we have

$$x_{k+1} = x_k + a\left(x_k - x_k^3\right) + v_k\,, \tag{12.5}$$

where $a = T_s/\gamma$. We will treat a as a control parameter that sets the size of the energy barrier E between the two local wells. The discrete noise $v_k \sim \mathcal{N}(0, v^2)$, where $v^2 = (2\gamma k_B T)\, T_s$.[8] This preliminary averaging is convenient for numerical illustrations of coarse graining. Rather than integrating the nonlinear stochastic Eq. (12.4), we simply time step its discrete approximation Eq. (12.5), which is easier. The time scale T_s is then effectively the microscopic time scale of the system.

State Discretization

Coarse graining, or discretizing state-space requires two steps:

- *Averaging*. We average the time series x_k over a time scale $t_{avg} = n_{avg} T_s$, where n_{avg} is an integer. The new time series is denoted $\bar{x}_k$, with the index k in steps of t_{avg}. Averaging divides dynamics into fast modes, which are lost, and slow modes, which are retained (see right). In the present context, choosing t_{avg} too short leads to spurious slow modes, which cause the reduced dynamics to be non-Markovian. Choosing t_{avg} too long loses some of the desired "signal."

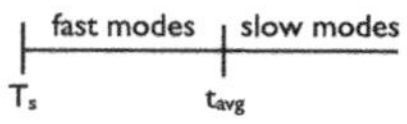

- *Classification*. The discretization itself is a nonlinear operation that maps sets of the continuous states to disjoint classes, which then constitute the discrete states. See right. For motion in a double-well potential, we can use the sign operation: $x_k \equiv \text{sgn}(\bar{x}_k)$. All states with $\bar{x}_k < 0$ are assigned to -1, and all states with $\bar{x}_k \geq 0$ are assigned to $+1$.

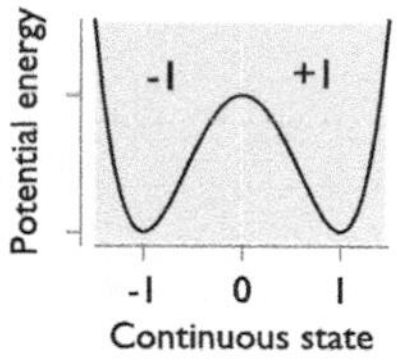

The nonlinear sign function performs a simple classification that is the key step in the coarse graining: whenever the (time averaged) state is on the right side of the potential ($\bar{x}_k > 0$), $x_k = +1$. It equals -1 when on the left side. More generally, this part of the coarse-graining operation requires a partition of a one- (or higher-) dimensional state space into a finite or countable discrete set, as shown at right.[9] Even

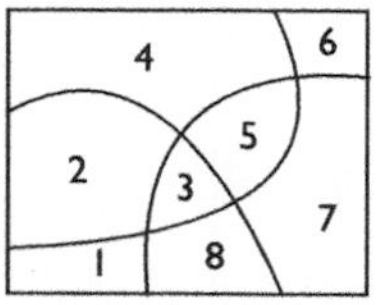

[8] As Problem 8.11 shows, nonlinear terms such as x_k^3 can make the dynamics formally unstable. The instability is an artifact of the discretization. Here, we will always choose parameter values that avoid this problem.

[9] The partitioning recalls Kadanoff's *block spin* approach to the renormalization group (Goldenfeld, 1992).

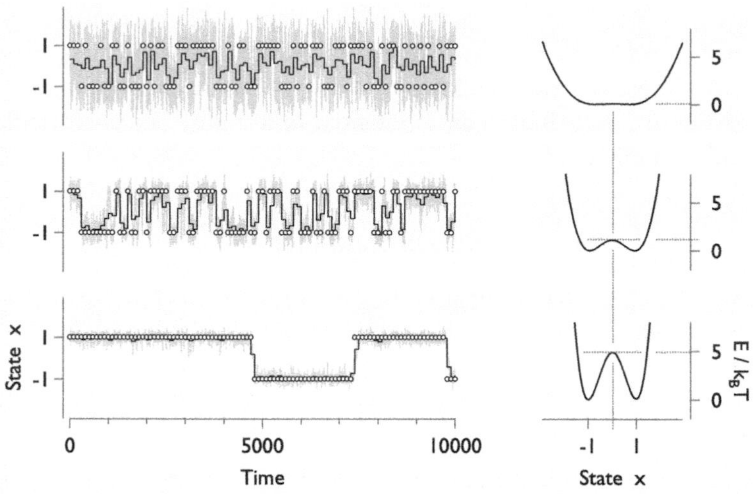

Fig. 12.3 Coarse graining requires high energy barriers ($E/k_BT \gg 1$). At left, 10^4 values of the continuous motion $x(t)$ are plotted in light gray; 100 discretized and averaged values $\bar{x}_k$ are plotted in black. The white markers denote the coarse-grained values $x_k = \text{sgn}(\bar{x}_k) = \pm 1$. Each time series has 10 000 points for x_k at time scale T_s and 100 coarse-grained points. At right are the potentials, $U(x)/k_BT = U_0 - \ln P(x)$, where $P(x)$ is estimated from the histogram of time series values, using 10^7 points per time series and $a = 0.1$. The constant U_0 is chosen to make the potential minimum $U = 0$. Noise amplitude from top to bottom: $\nu = 0.5, 0.2, 0.1$ $\Longleftrightarrow E/k_BT \approx 0.06, 1.1, 4.9$.

more generally, the different states need not be geometrically disjoint. More sophisticated ways of grouping and classifying states for such cases include *Gaussian mixture models* and *k-means clustering*.

Figure 12.3 gives the result of coarse graining of a double-well potential for three values of noise amplitude or, equivalently, three different values of energy barrier, E/k_BT. At top, $E/k_BT \ll 1$ and coarse graining amounts to an arbitrary partitioning of state space. At bottom, $E/k_BT \gg 1$ and coarse gaining captures the most important aspects of the continuous motion $x(t)$. In the marginal middle case, $E/k_BT \approx 1$ and coarse graining just starts to make sense: the time scale for local motion "within" a state starts to be less than the transit time to a different state. Coarse graining thus requires $t_{\text{transition}} \ll t_{\text{dwell}}$, a condition created, for example, by large energy barriers between local free energy minima.

Even if the physical requirements for creating macrostates are met, we still have to average the motion and apply a classification function in order to isolate the slow motion between states. Figure 12.4 shows a time series (top graph) of a particle in a double-well potential, with $E/k_BT \gtrsim 1$ set by choosing $a = 0.1$ and $\nu = 0.15$ in Eq. (12.5). We average N_ν values together and take the sign of the result. For $N_\nu = 1$, 10, noise in the transition region $x \approx 0$ leads to spurious state transitions. For $N_\nu = 1000$ and 10000, the real transitions are averaged over and lost. The factor $N_\nu = 100$ captures the reduced state dynamics well. Using N_ν to define an averaging time $t_{\text{averaging}}$, we can summarize the conditions for

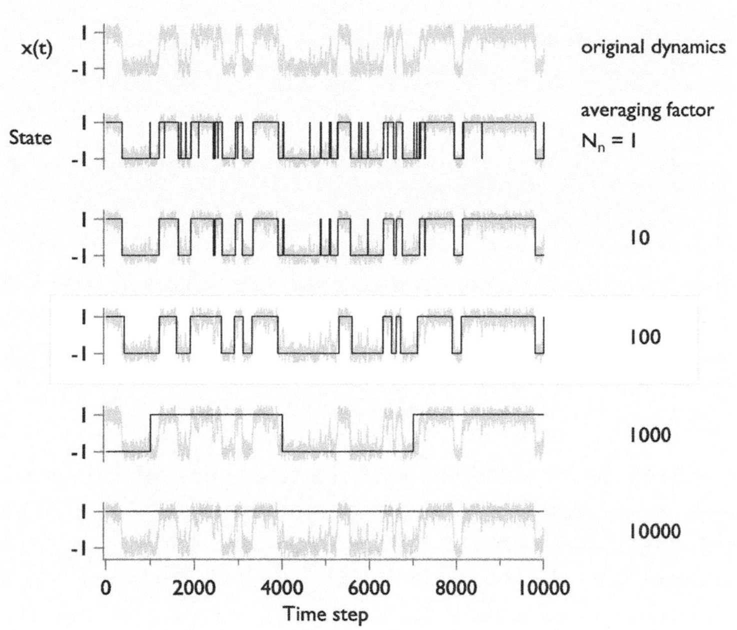

Fig. 12.4

Coarse graining requires $t_{\text{transition}} \ll t_{\text{averaging}} \ll t_{\text{dwell}}$. Original time series (light gray) is calculated for $\nu = 0.15$. Coarse graining (black curves) result from averaging N_ν points and applying the sign function to the result.

ideal coarse graining as

$$t_{\text{transition}} \ll t_{\text{averaging}} \ll t_{\text{dwell}} \,. \tag{12.6}$$

These conditions are only approximately met in the example shown in Figure 12.4, where $t_{\text{transition}} \approx 50$, $t_{\text{averaging}} = 100$, and $t_{\text{dwell}} \approx 500$. Problem 12.1 discusses how testing for Markov behavior leads to a more precise assessment of the success of coarse graining.

Observation Discretization

The difference between a Markov process and a hidden Markov process is illustrated in Figure 12.5. In the top graph, the observations, which can take continuous values, fall into disjoint sets that unambiguously identify the states. In the HMM case bottom, the observations overlap and do not unambiguously identify the system state. More precisely, the states $x_k = \pm 1$ give rise to observations $y_k^{(c)} \sim \mathcal{N}(\pm 1, \sigma)$.

We can also discretize the observation variables. If $y_k \equiv \text{sgn}\left(y_k^{(c)}\right)$, then

$$P(y_k = +1 | x_k = -1) = \frac{1}{\sqrt{2\pi\sigma^2}} \int_0^\infty \mathrm{d}y \, \mathrm{e}^{-\frac{(y+1)^2}{2\sigma^2}} \equiv b \,, \tag{12.7}$$

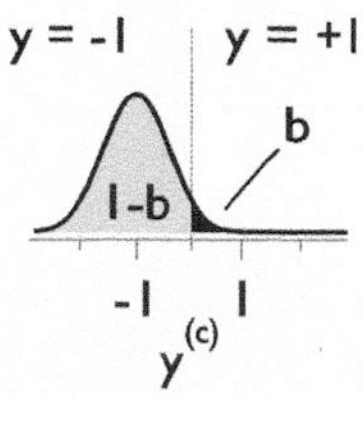

as illustrated at right. The constant $b = \frac{1}{2}\,\text{erfc}\left(1/\sqrt{2\sigma^2}\right)$.

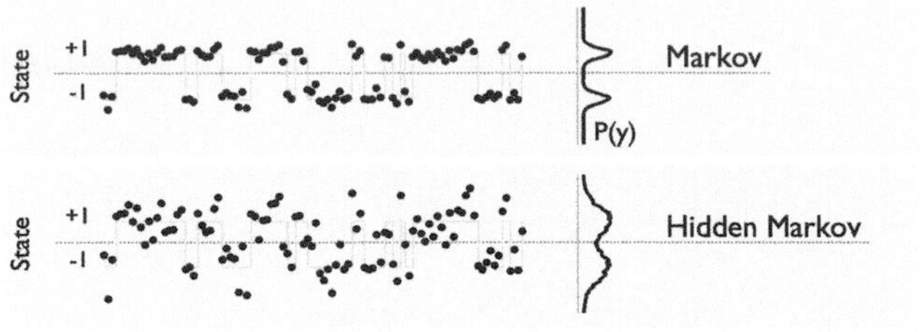

Fig. 12.5 Markov vs. hidden Markov models. Observations for a two-state Markov process (top) and a hidden Markov process (bottom). True state in light gray. Observations y_k are indicated by round markers and have Gaussian noise, with standard deviation $\sigma = 0.2$ (top) and 0.6 (bottom). Histograms of y_k are compiled from 10^4 observations, of which 100 are shown.

To summarize, the result of the combined coarse-graining operations is a two-state Markov model with states $x_k = \pm 1$ and transition matrix $\boldsymbol{A} = \begin{pmatrix} 1-a & a \\ a & 1-a \end{pmatrix}$ and a two-symbol observation model with symbols $y_k = \pm 1$ and $\boldsymbol{B} = \begin{pmatrix} 1-b & b \\ b & 1-b \end{pmatrix}$.

Note that discretizing the observations is not required. As Problem 12.4 shows, it is straightforward to infer discrete states from continuous observations. However, the analysis also shows that inferences based on discretized observations can sometimes be almost as accurate as those based on equivalent continuous observations.

12.2 Inferring States and Models

Given the output of a hidden Markov model (HMM), what can be inferred about the states and, if needed, the model of the dynamics and observations? We break this question into two parts: state estimation and then system identification.

12.2.1 State Estimation

Given the output of a hidden Markov model (HMM), what can be inferred about the states? We focus on three different but related types of inference:

1. *Filtering*, or $P(x_k|y^k)$. We estimate the probabilities for each state based on observations $y^k \equiv \{y_1, y_2, \ldots, y_k\}$ up to the present time k. Filtering is appropriate for real-time applications such as control.
2. *Smoothing*, or $P(x_k|y^N)$, for $N > k$. Smoothing uses data from the future as well as the past in the off-line postprocessing of N observations.
3. *Optimal path estimation*, or $\arg\max_{x^N} P(x^N|y^N)$ (MAP approximation). Ideally, we would calculate instead the entire distribution of paths through the states, $P(x^N|y^N)$, but evaluating the likelihood of all n^N state combinations is usually too hard and we

settle for finding the most likely state path. Here, n is the number of states that x_k may take.

Note the distinction between the most likely path, $\arg\max_{x^N} P(x^N|y^N)$, and the sequence of most likely states, $\cup_{k=1}^{N} \{ \arg\max_{x_k} P(x_k|y^N) \}$. To see this, consider speech recognition, where symbols are sounds and states are words. If you listen to each sound individually and infer the most likely word independently of the context, you might infer "wreck a nice beach," whereas the sounds heard together imply "recognize speech" (Murphy, 2012).

Finally, a note on formulas: The expressions for the filter, smoother, and related quantities that will be given here differ from their usual forms. One reason is that the HMM community uses different notation from the control community and from the information-theory community. (Our notation is closest to this last.) But also, because of historical accident, the tendency in the HMM literature is to use joint probabilities, such as $P(y^k, x_k = i)$, for which they use notation such as $\alpha_k(i)$, whereas we use conditional probabilities, such as $P(x_k = i|y^k)$, which are better behaved numerically. Our notation emphasizes the similarities between HMM and state-space models of dynamics; the formulas of one apply mostly to the other, with $\int \mathrm{d}x_k \leftrightarrow \sum_{x_k}$.

Filtering

We can derive the Bayesian filtering equations as in Section 8.3.2 using Bayes' Theorem, the Markov property $P(x_{k+1}|x^k, \cancel{y^k}) = P(x_{k+1}|x_k)$, and the memoryless property of observations $P(y_k|x^k, \cancel{y^{k-1}}) = P(y_k|x_k)$. Here, the "cancel" slash indicates the notion of *conditional independence* defined in Section 8.3.2: conditioning on x_k"blocks" the influence of all other variables. The x_{k-1} are blocked, too: the state at time $k+1$ depends only on the state at time k. The result matches Eq. (8.76), with integrals replaced by sums:

$$\begin{aligned} P(x_{k+1}|y^k) &= \sum_{x_k} P(x_{k+1}|x_k)\, P(x_k|y^k) && \text{predict} \\ P(x_{k+1}|y^{k+1}) &= \frac{1}{Z_{k+1}} P(y_{k+1}|x_{k+1})\, P(x_{k+1}|y^k) && \text{update}, \end{aligned} \tag{12.8}$$

with the normalizing constant ("partition function") given by

$$Z_{k+1} = P(y_{k+1}|y^k) = \sum_{x_{k+1}} P(y_{k+1}|x_{k+1})\, P(x_{k+1}|y^k). \tag{12.9}$$

The recursion relation runs in the forward direction ($k \to k+1$) and is equivalent to what the HMM literature calls the *forward algorithm*.

Figure 12.6 shows filtering in action for a symmetric, two-state, two-symbol hidden Markov model. The time series of observations y_k (markers) will disagree, on average, with the true state 30% of the time. The black line shows $P(x_k = 1|y^k)$. When that probability is below the dashed line at 0.5, the most likely state is 0. For the value of a used in the dynamic matrix ($a = 0.2$), the filter estimate $x_k^{(f)} = \arg\max_{x_k} P(x_k|y^k)$ is not significantly better at predicting the true state than the observation.

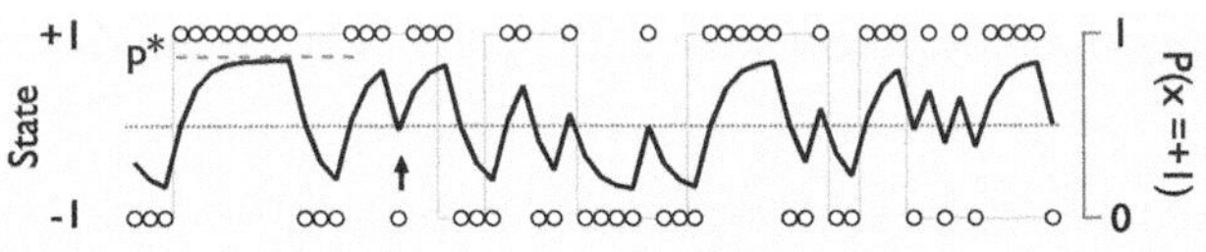

Fig. 12.6 Filtering for a symmetric, two-state, two-symbol hidden Markov model with $a = 0.2$ and $b = 0.3$. The light gray line shows true state, which is hidden. Markers show 60 observations. The heavy black line shows the probability that the state equals $+1$, given by $P(x_k = 1|y^k)$. The maximum confidence level $p^* \approx 0.85$ (dashed line).

But as $a \to 0$, the difference can be significant. For example, for $a = 0.01$, the filter estimate fails to match the true state only $\approx 8.6\%$ of the time, much lower than the 30% level using only the observation. Notice that whenever the state changes, the filter probability responds, with a time constant set by both observational noise (b) and dynamics (a).

Using the full probability estimates (black line) rather than simply the MAP (maximum a posteriori) estimate $x_k^{(f)}$ brings further advantages. When the filter is wrong, the two probabilities are often not that different. (See the case indicated by the arrow in the figure.) Thus, marginalizing (averaging) any prediction over all possibilities rather than just the most likely will improve estimates. Of course, a string of wrong symbols (e.g., the three to the left of the arrow) strengthens confidence in the (wrong) state.

In Figure 12.6, even after a string of $+1$ observations, the estimated probability that $x = +1$ never exceeds about 85% (and never goes below 15% for a long string of -1 observations). To explain this observation, we calculate $P(x_k = 1|y^k = 1)$, where the notation $y^k = 1$ means $\{y_1 = 1, y_2 = 1, \ldots, y_k = 1\}$. For $k \gg 1$, the maximum value of the state probability approaches a fixed point p^*. From Eq. (12.8),

$$\underbrace{P(x_k = 1|y^k = 1)}_{p^*} = \frac{1}{Z_k} \underbrace{P(y_k = 1|x_k = 1)}_{1-b} \sum_{x_{k-1}} P(x_k = 1|x_{k-1}) P(x_{k-1}|y^{k-1} = 1) . \tag{12.10}$$

Substituting for the matrix elements in Eq. (12.10), evaluating the normalization constant, and imposing the fixed point gives a quadratic equation for p^* whose relevant solution is

$$p^* = \frac{1 - 2b + a(4b - 3) + \sqrt{a^2 + (1 - 2a)(1 - 2b)^2}}{2(1 - 2a)(1 - 2b)} \tag{12.11}$$

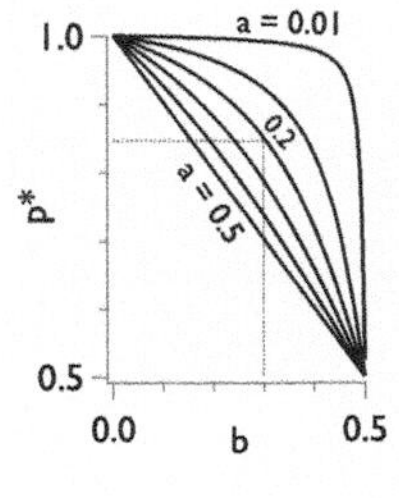

At left, we plot p^* as a function of the symbol error probability b, for $a = 0.01, 0.1, 0.2, 0.3, 0.4$, and 0.5. The light dotted lines show that $a = 0.2$ and $b = 0.3$ give $p^* \approx 0.852$, which agrees with the upper bound in Figure 12.6. In general, reducing the value of a implies longer dwell times for states and greater confidence limit p^* (Problem 12.5).

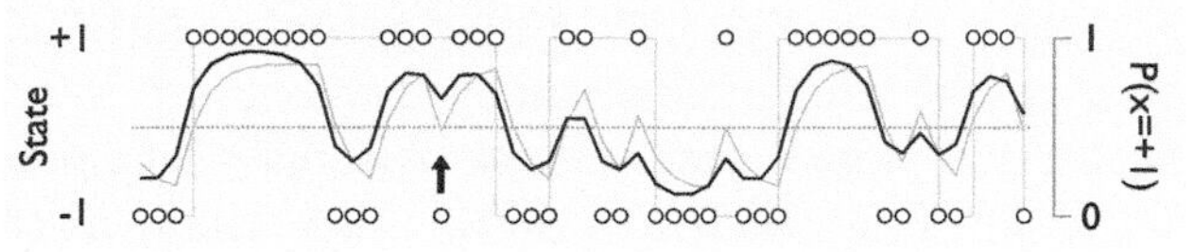

Fig. 12.7 Smoother estimates (black line) for two-state, two-symbol HMM with $a = 0.2$ and $b = 0.3$. Filter estimate is shown as a light-gray trace.

Smoothing

In Problem 8.20, we showed that future data can be used to make a better ("smoother") estimate of the state via the *backward* recursion relation from $k+1 \to k$,

$$P(x_k|y^N) = P(x_k|y^k) \sum_{x_{k+1}} \frac{P(x_{k+1}|x_k)\, P(x_{k+1}|y^N)}{P(x_{k+1}|y^k)} \,. \tag{12.12}$$

Again, the only difference here is that sums replace integrals. Notice that we first need to find $P(x_k|y^k)$ and $P(x_{k+1}|y^k)$ via the forward algorithm (filtering). Equation (12.12) then iterates back from the end time, $k = N$, back to the start, $k = 1$, to generate the probability distribution for the smoother estimate. In the HMM literature, this combination is known as the *forward-backward algorithm.*[10]

We can apply the smoother algorithm to the example of Section 12.2.1 and obtain results very similar to Figure 12.6. Indeed, in Figure 12.7, we plot the smoother estimate, with the filter estimate added as a light-gray trace. Despite their similarity, the differences are instructive: The filter always lag (reacts) to observations, whereas the smoother curve is statistically symmetric in time. Flipping the direction of time alters the overall form of the filter plot but not the smoother. The lag for the filter reflects causal response, whereas the zero-lag response of the smoother is acausal (Section 15.1).

The smoother estimates are more confident than the filter estimates, as they use more information. Look at the time step indicated by the arrow. The filter estimate is just barely mistaken, but the smoother estimate makes the correct call, aided by the three correct observations that come before and the three after. Indeed, repeating the calculation of the maximum confidence level using the smoother (Problem 12.5) shows that for $a = 0.2$, it rises from $p^* \approx 85\%$ to $q^* \approx 93\%$, as illustrated at right.

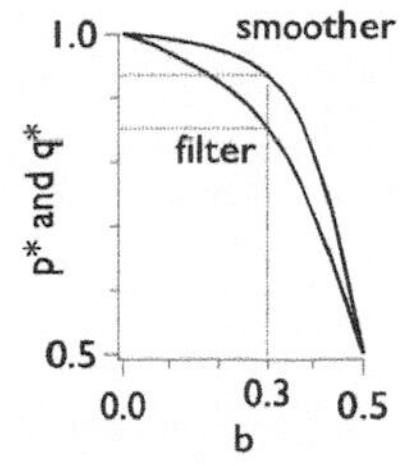

For continuous state spaces, the variance of the smoother estimate is lower than that of the filter estimate (Section 8.7), as extra information improves estimates. We can make a more systematic comparison here via the notion of conditional *Shannon entropy*, defined in Sections 15.2 and A.10.1. With $p_j \equiv P(x_k = j\,|\,y^k)$,

$$H(x_k|y^k) \equiv -\sum_{j=1}^{n} p_j \log p_j \,. \tag{12.13}$$

[10] The variant of the forward-backward algorithm used by the HMM community is closer to the naive smoother of Section 8.7. We use the same notation to describe both discrete and continuous state-space systems, whereas the different communities that historically studied each type of system used different notations.

For large-enough k, the average of $H(x_k|y^k)$ over y^k becomes independent of k. Averaging over a single long time series of observations then leads to $\langle H(x_k|y^k)\rangle \to H(x|\overleftarrow{y})$, where $\overleftarrow{y}$ denotes past and present observations. A similar definition holds for the smoother entropy, $H(x_k|y^N)$ and leads to a steady-state smoother entropy $H(x|\overleftrightarrow{y})$, where $\overleftrightarrow{y}$ includes both past and future observations. To characterize the performance of filtering and smoothing, we recall that for a two-state probability distribution, the entropy ranges from 1 bit (equal probabilities for each possibility) to 0 bits (certainty about each possibility).

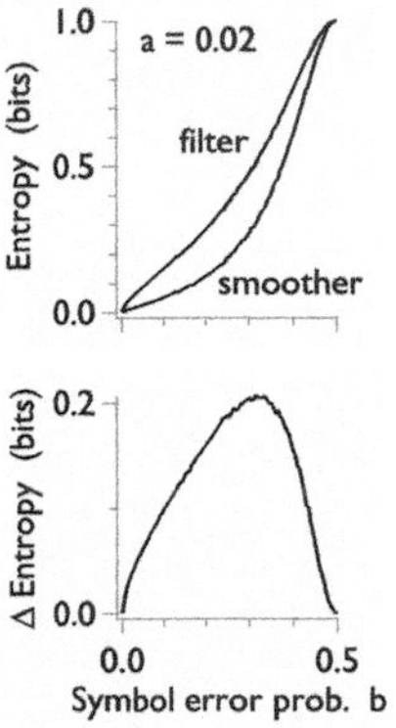

At left are the steady-state Shannon filter and smoother entropies for time series of length 10^5, as a function of b, the error rate in the observation matrix $\boldsymbol{B}$. The symmetric transition matrix $\boldsymbol{A}$ has parameter $a = 0.02$. At small values of a, the smoother has a greater advantage relative to the filter: when dwell times in each state are long, the information provided by averaging is more important. The lower graph plots the difference between filter and smoother entropies. For $b = 0$, the difference vanishes: with no noise, the observation perfectly determines the state, and there is no uncertainty about it afterwards. For $b = 0.5$, the observations convey no information, and $H(x|\overleftarrow{y}) = H(x|\overleftrightarrow{y}) = H(x) = 1$ bit and the difference is again zero. For intermediate values of b, the smoother entropy is lower than the filter entropy, analogous to the reduction in variance of the Kalman smoother relative to the Kalman filter. This also makes sense:

Extra information is most useful at moderate signal-to-noise ratios.

The extremes – complete certainty, where a single observation suffices, and complete ignorance, where observations are useless – are unaffected by extra information. This idea that information has greatest value for intermediate signal-to-noise ratios of measurements has been noticed in different contexts, for example in a model of cellular response to environmental signals (Sivak and Thomson, 2014). In general, if information is either perfect or worthless, then more is not better. But in between, extra information adds value.

Most Likely Path

The key to finding the optimal path is to recognize that finding an optimal path is equivalent to the shortest-path problem discussed in Section 7.4. To see this, we create a *trellis*, a directed acyclic graph illustrating the possible state paths through a discrete-state system. Figure 12.8 shows a trellis for a two-state Markov system.

The total likelihood of a path is a product of the form $L = P(x_0)\prod_{k=1}^{N} P(x_{k+1}|x_k)$ $P(y_k|x_k)$. Maximizing this likelihood is equivalent to minimizing the negative log likelihood, $-\ln L$, which is a sum. This equivalence means that finding the most likely path is equivalent to finding the shortest "distance" through the states, where the distance d_{ij} between nodes at k and $k+1$ is given by the negative log likelihood, $d_{ij} \equiv -\ln P(x_{k+1} = j|x_k = i) - \ln P(y_{k+1}|x_{k+1} = j)$. To account for initial conditions, we

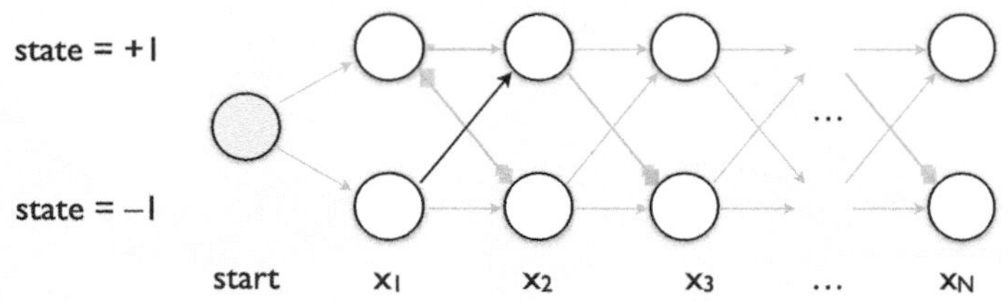

Fig. 12.8 Trellis graph for hidden Markov model. Top row of white circles represents being in state $+1$ at time k. Bottom row is equivalent, for the state -1. Darker circle represents a fictitious start state. The dark line has "length" equal to $-\ln P(x_2 = +1|x_1 = -1) - \ln P(y_2|x_2 = +1)$, where y_2 is the observed value at $k = 2$.

add a fictitious start node, whose distance to the states at x_1 are the a priori probabilities $P(x = -1)$ and $P(x = +1)$. For a symmetric transition matrix $\boldsymbol{A}$, these probabilities both equal 0.5.

Since finding the most likely path is a shortest-path problem, we can follow the procedure of Section 7.4 to solve the problem by dynamic programming, starting with a forward pass, then tracing back to record the most likely states. The recursive formulation reduces the number of calculations from $O[(2N)\,n^N]$ to $O(n^2N)$. For a two-state system with 1000 time steps, the reduction is from 10^{304} to 4000 calculations, a tremendous difference.

Why State Estimation Can Be a Hard Problem

We saw in Section 8.5 that nonlinear dynamics and non-Gaussian noise could lead to complexity in state estimation, but even a two-state Markov chain with two observation symbols – the simplest HMM – can show complex behavior. To see this, we plot in Figure 12.9 a *discord* order parameter $\mathcal{D} \equiv 1 - \langle y\,\hat{x}\rangle$, where y represents the time series of observations and $\hat{x}$ represents a state estimate (filter, smooth, or Viterbi).[11] A value $\mathcal{D} = 0$ means that the two time series agree (nearly) everywhere. A positive value means that they disagree, with $\mathcal{D} = 1$ corresponding to two independent time series.[12] As elsewhere, the numerical values assume that $\hat{x}, y \in \{-1, +1\}$.

As a start toward understanding the complex behavior displayed in Figure 12.9, we calculate, for the filter estimate, the critical value b_c that divides the $\mathcal{D} = 0$ and $\mathcal{D} > 0$ regimes. From Section 12.2.1, there is a maximum possible confidence level p^* in the estimate of a state, given by solving the relation $p^* = P(x_k = 1|y^k = 1)$ in terms of a and b. (See Eq. 12.11.) Then let $y_{k+1} = -1$. The discordant observation must lower the confidence in x_{k+1} to below $\frac{1}{2}$ in order for the filter estimate and observation to disagree. Thus,

$$P(x_{k+1} = 1|y_{k+1} = -1, y^k = 1) = \frac{1}{2}\,. \tag{12.14}$$

[11] This order parameter has nothing to do with the *quantum discord* order parameter that is used to distinguish between classical and quantum correlations (Zurek, 2003).

[12] We assume positive correlation between y and $\hat{x}$. Perfect anticorrelation implies $d = 2$.

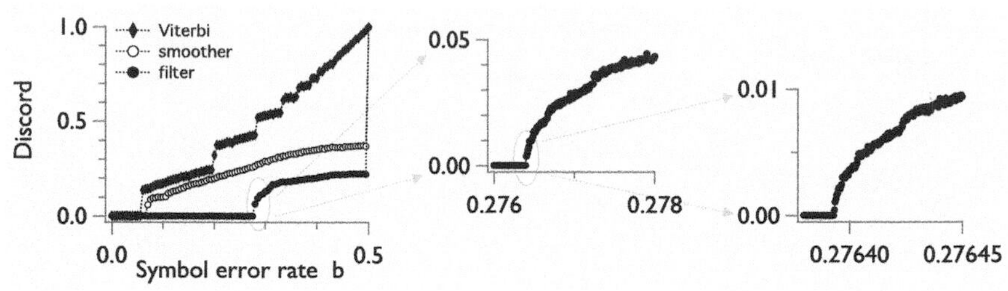

Fig. 12.9 Discord order parameter for two-state, two-symbol hidden Markov model with $a = 0.2$. At left, plots of discord between the observations and the Viterbi (most likely path), smoother, and filter state estimates. Center and right graphs enlarge the filter-based discord plot near the first bifurcation at $b \approx 0.27639$.

Writing this condition out explicitly, similar to Eq. (12.11), leads to a remarkably simple quadratic equation for b_c. Its relevant solution (Problem 12.6) is

$$b_c^{\text{filter}} = \frac{1}{2}\left(1 - \sqrt{1-4a}\right). \tag{12.15}$$

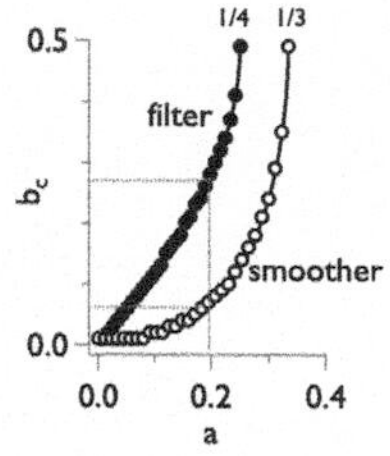

At right, we see that the thresholds of simulated data agree completely with Eq. (12.15). For example, with $a = 0.2$, the threshold is $b_c^{\text{filter}} = \frac{1}{2}(1 - \frac{1}{\sqrt{5}}) \approx 0.276393$, which agrees with the threshold shown at right in Figure 12.9 and also as shown by the light dashed lines in the margin illustration at left. Note that no parameters have been fit.

A similar calculation (Problem 12.6) gives the threshold for the smoother:

$$b_c^{\text{smoother}} = \frac{1}{2}\left(1 - \frac{\sqrt{(1+a)(1-3a)}}{1-a}\right). \tag{12.16}$$

The $a = 0.2$ case corresponds to $b_c^{\text{smoother}} = \frac{1}{2}(1 - \frac{\sqrt{3}}{2}) \approx 0.067$.

For the filter, $a > \frac{1}{4}$ implies that $\mathcal{D} = 0$ for all values of b. For the smoother, the corresponding critical value is $a > \frac{1}{3}$. The increase arises from the extra information supplied by the smoother that allows us to use memory – trusting our inference and not the observation – at shorter dwell times.

Yet there is clearly more to understand. The threshold for the Viterbi transition requires a different method. Moreover, there is clear evidence for a sequence of transitions. Finally, the singularities are sometimes continuous (filter estimate in Figure 12.9) but sometimes discontinuous (smoother and Viterbi estimates).

To make further progress, we note that the discontinuities in the discord "order parameter" are reminiscent of phase transitions. Indeed, we can map the two-state, two-symbol HMM onto a variant of the familiar Ising model from statistical physics. With x_k and $y_k = \pm 1$, we define $P(x_{k+1}|x_k) = \frac{e^{Jx_{k+1}x_k}}{2\cosh J}$ and $P(y_k|x_k) = \frac{e^{hy_kx_k}}{2\cosh h}$, with $J = \frac{1}{2}\ln(\frac{1-a}{a})$ and $h = \frac{1}{2}\ln(\frac{1-b}{b})$. We use these definitions to formulate a "Hamiltonian" $H = -\ln P(x^N, y^N)$,

$$H = -J\sum_{k=1}^{N} x_k\, x_{k+1} - h\sum_{k=1}^{N} y_k x_k\,, \tag{12.17}$$

where we have dropped constant terms that are independent of x_k and y_k. For $a < \frac{1}{2}$, the interaction term $J > 0$ is *ferromagnetic*: neighboring "spins" tend to align. The term h corresponds to an external field coupling constant. The field is of constant strength and, for $b < \frac{1}{2}$, has a sign is equal to the observation y_k. The picture is that a local, *quenched* field of strength hy_k tries to align its local spin. Notice that $h = 0$ for $b = \frac{1}{2}$; spins are independent of y_k; observations and states decouple. A further change of variables (gauge transformation), $z_k = y_k x_k$ and $\tau_k = y_k y_{k+1}$, gives

$$H(\tau, z) = -J \sum_k \tau_k \, z_k z_{k+1} - h \sum_k z_k \,, \tag{12.18}$$

which is a random-bond Ising model in a uniform external field h.

Starting in the late 1970s, both random-bond and random-field one-dimensional Ising chains were extensively studied as models of frustration in disordered systems such as spin glasses. In particular, Derrida et al. (1978) showed that the ground state at zero temperature has a countable infinity of transitions at $h = 2J/m$ for $m = 1, 2, \ldots, \infty$. Their solution used a transfer-matrix formalism that is equivalent to the factorization of the partition coefficient $Z = \prod_k Z_k$ given in Eq. (12.9).

The lowest-order transition, $h = 2J$, corresponds to a case where the external field at a site forces the local spin to align. In terms of the original HMM problem, the ground state corresponds to the most likely (Viterbi) path. Allahverdyan and Galstyan (2009) explicitly connect the sequence of transitions to the jumps in $\mathcal{D}$. In the regime where phase transitions occur, there are exponentially many paths that have likelihoods nearly identical to that of the most likely path. The most likely path is of little interest in such regimes, a fact that argues for a statistical-mechanics approach. For us, the main point is that the mapping of the two-state, two-symbol HMM to a random-field Ising model can help explain why HMM inference shows such rich behavior.

12.2.2 System Identification

So far, we have assumed that the transition matrix $\boldsymbol{A}$ and emission matrix $\boldsymbol{B}$ are known. If not, they can be estimated from the data. The equivalent problem for continuous-state dynamical systems is system identification (Chapter 6). For HMMs, it is often called *learning*, or *training*.

The general approach is to maximize the likelihood of the unknown quantities, grouped here into a single parameter vector $\boldsymbol{\theta}$. That is, we seek

$$\boldsymbol{\theta}^* = \arg\max_{\boldsymbol{\theta}} P(y^N|\boldsymbol{\theta}) = \arg\min_{\boldsymbol{\theta}} \left[-\ln P(y^N|\boldsymbol{\theta}) \right] . \tag{12.19}$$

It is better to compute $L(\boldsymbol{\theta}) \equiv -\ln P(y^N|\boldsymbol{\theta})$ because $P(y^N|\boldsymbol{\theta})$ decreases exponentially with N, leading to numerical underflow. The negative sign is a convention from least-squares curve fitting, where $\chi^2(\boldsymbol{\theta})$ is also proportional to the negative log likelihood of the data.

We can find the total likelihood $P(y^N \mid \boldsymbol{\theta})$ from the normalization condition in Eq. (12.8) and the chain rule for probabilities given in Problem A.6.1 online:

$$P\left(y^N\right) = \underbrace{\prod_{k=1}^{N} P\left(y_k | y^{k-1}\right)}_{\text{chain rule}} = \prod_{k=1}^{N} Z_k \,, \tag{12.20}$$

where $Z_1 \equiv P(y_1)$. Then

$$L(\boldsymbol{\theta}) = -\sum_{k=1}^{N} \ln \sum_{x_k} P\left(y_k | x_k\right) P\left(x_k | y^{k-1}\right), \tag{12.21}$$

where all right-hand-side terms depend also on $\boldsymbol{\theta}$. Since $L(\boldsymbol{\theta})$ is just a function of $\boldsymbol{\theta}$, we can use standard optimization routines to find the $\boldsymbol{\theta}^*$ that minimizes L. See Problem 12.7.

In the HMM literature, an alternate approach to finding $\boldsymbol{\theta}^*$ is based on the Expectation Maximization (EM), or Baum–Welch algorithm. In a two-step iteration, one finds $\boldsymbol{\theta}$ by maximum likelihood assuming that the hidden states x^N are known and then infers states x^N from the smoother algorithm assuming $\boldsymbol{\theta}$ is known. The algorithm converges locally and can be robust numerically but is slow relative to standard numerical optimizers based on Newton's Method and the like. Mostly, you should try standard optimizers first and EM if all else fails. EM algorithms can, however, be the starting point for recursive variants that allow for adaptation, as in Chapter 10.

12.3 Control

Finally, we can discuss the control of Markov models, fully observed and hidden. In the context of discrete state spaces, the control u_k influences the transition probability, which becomes $P(x_{k+1}|x_k, u_k)$ and is described by a time-dependent transition matrix $\mathbf{A}_k$. Much of the formalism of optimal control discussed in Chapter 7 then carries over and leads to similar results. In Chapter 15, we give some examples in the context of a discussion on thermodynamics and control. Here, we focus on a model problem, the choice of a path through state space towards a desired goal. Depending on the noise level, it is better to choose one route or the other. Because the allowed controls are discrete in this example, dynamic programming offers an effective, systematic approach.

Control theory for Markov chains is also known as *Markov Decision Processes* (MDP). Optimal-control protocols that minimize some cost function can be found using Bellman's dynamic programming, a general algorithm for problems involving sequential decisions introduced in Section 7.4. In this setting, control is viewed as a blend of state estimation and decision theory. The goal is to choose actions based on available information in order to minimize a cost function. Figure 12.10 is an *influence*

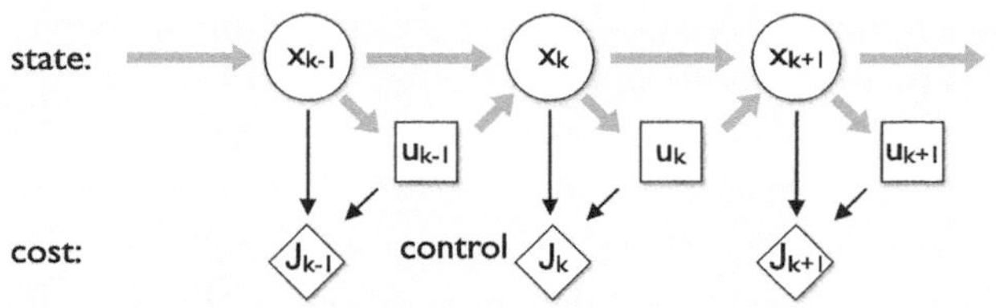

Fig. 12.10 Influence diagram for MDP. The states x_k form a Markov process that is controlled by u_k, which is calculated to minimize a cost J_k. Thin arrows denote that costs J_k depend on the current state x_k and current control u_k. The new state x_{k+1} depends on both x_k and u_k, as indicated by the thick arrows.

diagram that sketches the graphical structure of an MDP. States are denoted by round outlines, decisions by square ones, and utilities (or costs) by diamond-shaped ones.

Although the structure of Figure 12.10 seems complex, the state dynamics form a directed acyclic graph solvable by the Bellman equation (7.28) of Chapter 7. Recall that, for a deterministic dynamical system with state x_k and control u_k,

$$J^*(x_k) = \min_{\{u_k\}} \left[V(x_k, u_k) + J^*(x_{k+1}) \right] , \tag{12.22}$$

where $V(x_k, u_k) \equiv V_k$ is the cost of current time step, $J(x_k, u_k) = \sum_k V_k$ is the cost-to-go function (cost starting from time k), and $J^*(x_k) = \min_{u_k} J(x_k, u_k) \equiv J_k^*$ is the optimal cost-to-go function.

To adapt Eq. (12.22) to MDPs, we make the following extensions and alterations, inspired by the language of economics, where MDPs are particularly popular:

- We convert cost to *reward*, cost-to-go to *utility*, and *min* → *max*. We substitute the terms *decision* or *action* for *control*, and we use *policy* instead of *control law*.
- We discount the future by a factor γ, in order to keep total values finite and to account for the intuitive idea that we care less about far-off consequences than immediate ones.
- We generalize from a deterministic to a stochastic formalism and maximize the expected value of a given decision (control).

With these modifications, Eq. (12.22) becomes

$$J^*(x_k) = \max_{\{u_k\}} \left[V(x_k, u_k) + \gamma \sum_{x_{k+1}} P(x_{k+1}|x_k, u_k)\, J^*(x_{k+1}) \right] , \tag{12.23}$$

In steady state, $J^*(x_k) = J^*(x_{k+1}) = J^*(x)$, so that

$$J^*(x) = \max_{\{u\}} \left[V(x, u) + \gamma \sum_{x'} P(x'|x, u)\, J^*(x') \right] . \tag{12.24}$$

Equation (12.24) is nonlinear because of the max operation. A simple, if slow, way to solve Eq. (12.24) is by *value iteration*, where

$$J^*(x)^{(j+1)} = \max_{\{u\}} \left[V(x,u) + \gamma \sum_{x'} P(x'|x,u)\, J^*(x')^{(j)} \right]. \tag{12.25}$$

In Eq. (12.25), the indices j and $j+1$ denote iterations of the algorithm, *not* time. In particular, note that the Bellman equation, (12.22), is a backward iteration in time, whereas Eq. (12.25) is a forward computational iteration to find the fixed point. We can initialize the algorithm by $J^*(x)_0 = \min_{\{u\}} V(x,u)$.

Example 12.2 (Gridworld) A robot wanders a 2d grid, seeking a special termination state. Figure 12.11(a) shows a 2×3 example, with states labeled 1–6 and shaded termination state 4 at upper right. The other states have immediate value -1, except for the "bad" state 5, which has value -5 and is therefore to be avoided.

Stochastic dynamics are introduced as part of the control in Figure 12.11(b). At each time step, you choose a direction (N, E, S, W). Whatever the choice, you will go in the intended direction 80% of the time. But 10% of the time you will go left instead, and 10% of the time you will go right. You never go backwards. The outcomes are analogous to those of thermal noise on a drifting particle.

The results of solving numerical by value iteration, with $\gamma = 0.2$ and $\gamma = 0.9$, are shown in Figure 12.11(c) and (d). In solving Eq. (12.25), we note that the immediate reward $V(x)$ is a function only of the position (state x). The steady-state values $J^*(x)$ are shown in Figure 12.11(c). Notice that the termination state always has value $+10$, since there is no reward beyond the immediate reward $V(x = 3)$. Also, small γ implies

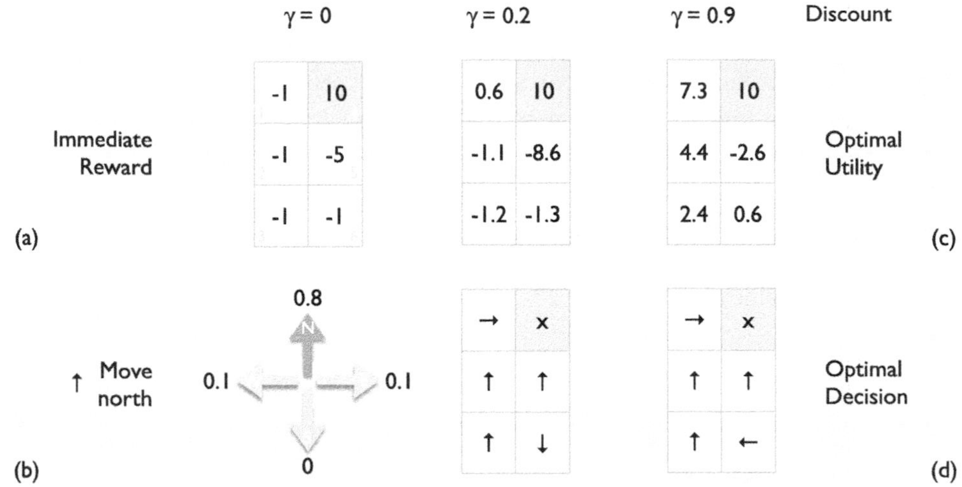

Fig. 12.11 Gridworld. (a) Immediate reward for being in states 1–6; (b) outcomes from an attempted move north; (c) optimal utility for discount $\gamma = 0.2$ and 0.9; (d) optimal decision for same discounts. Dark-shaded box denotes terminal state.

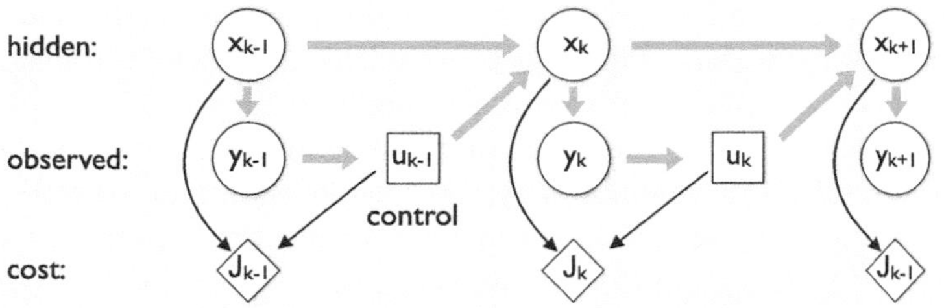

Fig. 12.12 Influence diagram for POMDP. The states x_k and observations y_k form a HMM that is controlled by u_k, which is calculated to minimize a cost J_k. Thin arrows denote that costs J_k depend on the current state x_k and current control u_k. The new state x_{k+1} depends on both x_k and u_k, as indicated by the thick arrows.

state values that are closer to the immediate reward $V(x)$, since the future has little value.

Figure 12.11(d) shows the optimal policies, given by $\text{argmax}_{\{u\}} \sum_{x'} P(x'|x,u)\, J^*(x')$. The optimal decision for state 6 is to go west when $\gamma = 0.9$. Perhaps surprisingly, it is to go south when $\gamma = 0.2$. Why head south? Heading in that direction, there is no chance of accidentally going north and falling into the bad state. The trade-off is that it takes 10 attempts to escape to state 3. This adds negative costs (those -1's add up) and delays the reward, which is worth less because of the discount. For small discounts, these costs are small and heading south is the best decision. When the future counts for more, it is better to head directly west and risk going to the bad state. For details, see Problem 12.8.

Finally, gridworld can be a model for a general path-integral problem where the goal is to maximize some quantity defined over an extended, unknown trajectory.

The analogous approach to optimal control on HMMs is known as *Partially Observable Markov Decision Processes* (POMDP). The structure of a POMDP is shown in Figure 12.12. In contrast to the MDP, we have only indirect information on the hidden states x_k via the observations y_k. The general approach is to average over state-estimation distributions. This increases the computations dramatically, motivating the development of more complicated suboptimal approximations, which are beyond our scope.

12.4 Summary

Coarse graining in space can convert a continuous state space into a discrete one. A similar, but independent coarse graining in time can convert continuous-time dynamics into discrete dynamics, as we already saw in Chapter 5. Since most of the ideas developed to describe the control of systems with continuous state spaces carry over to the description of discrete-state dynamics, this chapter may be seen as a recapitulation of ideas developed throughout the rest of the book. In particular, we discuss how the problem of state estimation carries over to the discrete-state and discrete-time case. Indeed, along with the Kalman filter, this is one of the few exactly solvable problems of state estimation. Discrete-state systems give simple examples of abrupt transition in estimation, which we already discussed more qualitatively for the continuous case in Chapter 8. Similarly, techniques of system identification and control design all carry over to the discrete case in a straightforward manner. Perhaps the biggest obstacle to carrying over ideas from the continuous to the discrete domain is that these topics were developed by separate communities, each with its own incommensurate jargon. We have tried to present the material in a way that emphasizes how similar the approaches really are.

12.5 Notes and References

Parts of this chapter are adapted from Bechhoefer (2015). For subtleties about stochastic equations and their discretizations, see van Kampen (2007). For an introduction to discrete-state Markov models, see the first chapter of Kelly (1979). Clustering and hidden Markov models are discussed from a physics perspective by Press et al. (2007), chapter 16.1 and 16.3, and from a machine-learning perspective by Murphy (2012). We have used several derivations by Durbin et al. (1998), which focuses on bioinformatics applications. Rabiner (1989) has been an influential tutorial, focused on speech recognition. Unfortunately, its notation differs from that used in Bayesian inference in general, not to mention control theory. Our presentation emphasizes the links with the usual state-estimation problem for continuous state spaces, an approach also adopted by Murphy (2012). Cappé et al. (2005) has a more complete and more mathematical treatment. Our notation is adapted from Särkkä (2013). *Numerical Recipes* (Press et al., 2007) has explicit codes for filtering, smoothing, and the Viterbi optimal-path methods.

The problem of finding the most likely path of an HMM was solved by Viterbi in the 1960s in the context of *error-correction codes* for information transmission and has greatly influenced technologies for reliable communication over noisy channels. MacKay (2003) gives a physicist-friendly treatment of coding theory.

EM for system identification, which we mostly do not recommend, is discussed in most of the standard HMM literature. Murphy (2012) has a good presentation, and Durbin et al. (1998) has some interesting proofs. For the adaptive version, see Krishnamurthy (2016).

Phase transitions in HMMs have been studied by Allahverdyan and Galstyan (2009), which focuses in particular on counting the number of paths with likelihoods very close to that of the most likely path. The change of variables to a random-bond Ising model is from Allahverdyan and Galstyan (2015).

Dynamical programming is introduced in Press et al. (2007) and treated extensively in Bertsekas (2005b). Jensen and Nielsen (2007) introduces decision theory and applies to Bayesian networks, which have a more complicated causal structure than the problems treated in this book. The robot-navigation (gridworld) example is adapted in this work; it is also commonly discussed in the reinforcement-learning literature (Sutton and Barto, 2018). Pearl (2009) develops a stimulating treatment of causality using this language. Krishnamurthy (2016) presents in detail MDPs and POMDPs and some nice applications.

Problems

12.1 Coarse graining. There are two steps: time averaging and then a nonlinear "classification." Here, we investigate the choice of time-averaging scale.

a. Write a code to make time series plots similar to those in Figure 12.3 for $x_{k+1} = x_k + a\left(x_k - x_k^3\right) + v_k$, with $v_k \sim \mathcal{N}(0, v^2)$. The control parameters are a and v.

b. Investigate the role of averaging time in coarse graining and reproduce Fig. 12.4.

c. Show that if the coarse-graining factor is too small, the resulting process is not Markov by using the update law, $p_{k+1} = \boldsymbol{A}\, p_k$ twice: $p_{k+2} = \boldsymbol{A}^2 p_k$. Compare with $p_{k+2} = (\boldsymbol{A}_2)\, p_k$, where $\boldsymbol{A}_2$ can be empirically estimated by looking at frequencies of the four different state combinations. Does $\boldsymbol{A}^2 = \boldsymbol{A}_2$? Average a simulated time series by a "coarse-graining factor" ; then compute the ratio of off-diagonal matrix elements $(\boldsymbol{A}^2)_{01} \,/\, (\boldsymbol{A}_2)_{01}$ as a function of v and coarse graining (see right).

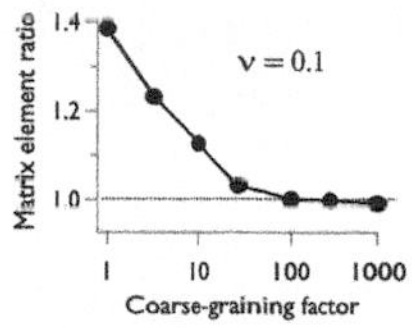

12.2 Equilibrium and steady states of a Markov chain. A steady state is *reversible* if the stochastic process forward in (discrete) time is indistinguishable from the backward process. The steady state in a reversible Markov chain is also termed an *equilibrium state*, as it is closely connected to the notion of thermodynamic equilibrium. In the following, the matrix element A_{ij} is the $j \to i$ transition probability.

a. *Detailed balance* for a homogeneous Markov chain is defined by $A_{ij}\, p_j = A_{ji}\, p_i$, for all i and j. Show that reversibility implies detailed balance, and vice versa. Hints: Sum i over all n states. Also, consider $P(x_{k+1} = i, x_k = j)$ and its time reversal, along with sequences of N elements and their time reversal.

b. Show that the steady state of the two-state Markov model in Example 12.1 obeys detailed balance and is hence an equilibrium state.

c. For a diagonalizable, stochastic transition matrix $\boldsymbol{A}$ with no zero entries, show that $\lim_{N\to\infty} \boldsymbol{A}^N = \boldsymbol{P}$, where each column of $\boldsymbol{P}$ is the steady-state distribution $\boldsymbol{p}$. Please interpret. Use the fact (or prove) that $\lambda = 1$ is the largest eigenvalue. Hints: $\boldsymbol{p}$ is a right eigenvector, and there is also a left eigenvector of all ones.

d. In order to use the detailed-balance condition for equilibrium, we have to first solve for $\boldsymbol{p}$. Kolmogorov derived a condition for equilibrium that depends only on the transition probabilities A_{ij}. In particular, a stationary Markov chain is reversible and obeys detailed balance if and only if

$$A_{\ell_1\ell_2} A_{\ell_2\ell_3} \dots A_{\ell_{N-1}\ell_N} A_{\ell_N\ell_1} = A_{\ell_N\ell_{N-1}} A_{\ell_{N-1}\ell_{N-2}} \dots A_{\ell_2\ell_1} A_{\ell_1\ell_N}$$

for *every* finite sequence of states $\ell_1, \ell_2, \dots, \ell_N$ for any length N. Show that detailed balance implies Kolmogorov's condition. The converse is trickier: consider a very long path, fix $\ell_1 = i$ and $\ell_N = j$, sum over all intermediate states $\ell_2, \dots, \ell_{N-1}$, and use the identity from (c). Kolmogorov's condition implies that clockwise and counterclockwise probability currents around a loop are equal for an equilibrium (reversible) state. Here, the $j \to i$ current is $J_{ij} = A_{ij} p_j - A_{ji} p_i$. In equilibrium, detailed balance implies $J_{ij} = 0$: the *only* way to have a nonequilibrium steady state is to have a loop with differing clockwise and counterclockwise probability currents. Finally, as a corollary of Kolmogorov's criterion, show that steady states must be reversible for *trees* – graphs with no loops. Cf. Kelly (1979).

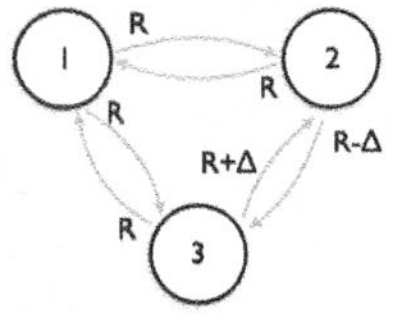

12.3 Kinetic proofreading. Many biological systems have error rates far below that predicted by the Boltzmann distribution of equilibrium thermodynamics. A simple *nonequilibrium* model can model this phenomenon. Consider the three-state, discrete-time Markov model depicted at left, with $\boldsymbol{A} = \begin{pmatrix} 1-2R & R & R \\ R & 1-2R-\Delta & R-\Delta \\ R & R+\Delta & 1-2R+\Delta \end{pmatrix}$.

a. Find the steady state of the Markov chain.
b. Using Kolmogorov's condition from Problem 12.2, show that the system is in equilibrium (reversible) if and only if $\Delta = 0$.
c. Show that choosing Δ controls the ratio $\frac{p_2}{p_1}$ to be in the range $(\frac{1}{3}, \frac{5}{3})$.
d. Find the nonequilibrium current $J(\Delta)$ circulating around the loop.
e. Compute the steady state when you convert the cycle into a linear chain by setting $A_{13} = A_{31} = 0$. For this chain, show that $\frac{p_2}{p_1}$ is independent of Δ.

The connection with kinetic proofreading is that adding a "useless" node 3 and creating a nonequilibrium cycle alters $\frac{p_2}{p_1}$ without changing A_{12} or A_{21}.

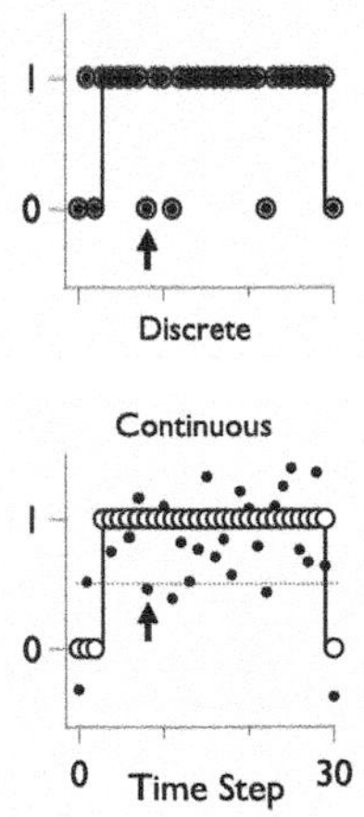

12.4 HMM with continuous observations. Consider a two-state, discrete-time, symmetric Markov model with hopping probability a and states $x_k = \pm 1$. Let the observations $y_k = x_k + \xi_k$ be continuous, with $\xi_k \sim \mathcal{N}(0, \sigma^2)$.

a. Write code to generate a hidden Markov sequence with both discrete and continuous observations. Match the noise variance ξ^2 to b, as in Eq. (12.7). Solve the filtering problem for continuous observations. Compare plots of a time series for $a = b = 0.1$. Plot the true state x_k, the observations (solid markers at left), and filter estimates (hollow markers). First, generate the continuous observations; then use them to form the discrete symbols via the sign operation.
b. Show that the filter performs slightly better using continuous observations. Hint: Look at the vertical arrows.
c. For $0 < b < 0.5$, compare the entropy with that of the equivalent HMM having two observation symbols, defined as in Eq. (12.7). You

should find something resembling the plots comparing filter and smoother in Section 12.2.1.

12.5 Confidence bound for state estimates.

a. For the filter, derive Eq. (12.11) and reproduce its associated plot.
b. For the smoother, show that the maximum confidence $q^* = \frac{1}{2}(1 + \frac{(1-a)(1-2b)}{\sqrt{a^2+(1-2a)(1-2b)^2}})$.

12.6 Phase transition in discord order parameter $\mathcal{D}$.

a. Using the argument given in Section 12.2.1, derive Eq. (12.15).
b. For the smoother case, let $y^{N\backslash k}$ refer to all observations except y_k. Explain why the condition to be imposed is now $P(x_k = 1|y_k = -1, y^{N\backslash k} = 1) = \frac{1}{2}$.
c. Show that $P(x_k = 1|y^{N\backslash k} = 1) = \frac{1}{Z}P(x_k = 1|y_{k+1}^N = 1)\,P(x_k = 1|y^{k-1} = 1)$.
d. Justify $P(x_k|y_{k+1}^N) = P(x_k|y^{k-1})$ and then derive Eq. (12.16).
e. Simulate the HMM and, for given a, scan b until $\mathcal{D} > 0.001$ in order to reproduce the numerical threshold data at right. Do for both filter and smoother.

12.7 Learning an HMM. Find the parameters a and b of a symmetric, two-state, two-symbol hidden Markov model. Write a program to generate an HMM time series of length N, given $a = 0.2$ and $b = 0.1$. Call a standard optimization program that can take the output series y^N and initial guesses for a and b and return estimates $\hat{a}$ and $\hat{b}$ based on minimizing the negative log likelihood function, Eq. (12.21). For given N, repeat enough times to estimate the mean and standard deviation of each parameter and then compute the relative error, versus N. Use the true values of a and b as initial guesses for the optimization. Reproduce the plot at right.

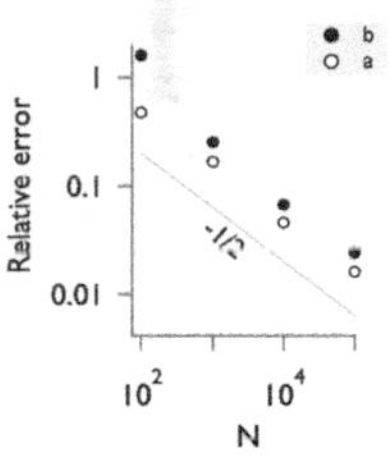

12.8 Gridworld. Code Example 12.2:

a. Create the 6×6 transition matrices $P(x'|x, u)$ for each of the four decisions $u = \{\text{N, E, S, and W}\}$. Each column should sum to one. If a move would leave gridworld, the system stays in its current state. Thus, for example, $P(x' = 1|x = 1, u = \text{N}) = 0.8 + 0.1 = 0.9$. You are in the NW corner trying to go N. You cannot go north, which adds 0.8 from the forward branch. You cannot go west, adding another 0.1 from the left branch. Include rules for the termination state, too.
b. Reproduce the tables of optimal utilities and policies with greater precision.
c. What happens if you always move forward and never go left, right, or back?

13 Quantum Control

Quantum control began with what now seem crude efforts. After the invention of the laser in the 1960s, an obvious strategy to pump a system into a desired excited state was to shine light with energy $\hbar\omega$ equal to the difference in energy levels between the initial and desired final states. Unfortunately, in real systems, this strategy can be horribly inefficient, as the deposited energy rapidly dissipates into other modes. Only after it was realized that exploiting quantum coherence can greatly improve efficiency and only after laser-pulse-shaping technology had improved enough to make use of that insight could real control be achieved. The ensuing rapid progress led to the 2012 Nobel Prize to Serge Haroche and David Wineland for "measuring and manipulation of individual quantum systems."

In this chapter, we will outline some basic aspects of the control of quantum systems. Our goal will be to understand which are similar to and which differ from the control of classical systems.[1]

We will not attempt to present the full formalism involving the quantum mechanics of open systems and continuous measurements, as it requires a higher technical level than the rest of the book, which mostly stops short of continuous-time stochastic processes. Also, there are good discussions for physicists available elsewhere. Still, the available book-length discussions and reviews on quantum control often have introductions to classical control that are too brief and too abstract to easily understand the physical implications. For their readers, I hope that this book can serve as a more easily digestible introduction to classical control and that this chapter can be a bridge to more advanced treatments.

In Section 13.1, we recall key ideas of quantum mechanics and introduce some basic concepts about dynamics and measurement. We focus on two-level systems, as they are the simplest setting to explore quantum ideas and, interpreted as *qubits*, they are quantum-mechanical analogs of classical bits for computation. Section 13.2, the heart of this chapter, presents three classes of control systems. The first is the analog of open-loop controllers, and the second is the analog of measurement-based closed-loop controllers. The third dispenses with explicit measurements, in somewhat the same way that measurements are only implicit in systems such as the steam-engine governor. However, the special role that measurement plays in quantum mechanics will imply a more fundamental difference between feedback that is or is not based on measurements than exists in classical control. In this section, we deliberately keep to

[1] Books on quantum control often contrast their subject with *classical control*. In this book, we reserve that term for the frequency-domain methods of Chapter 3.

the simplest possible examples. In Section 13.3, we discuss a physical example of a two-level system, a spin-$\frac{1}{2}$ particle in a magnetic field, and use it to calculate an optimal feedforward control.

13.1 Quantum Mechanics

We review some notions from quantum theory, focusing on states, dynamics, and measurement. I assume a basic knowledge of quantum physics (see Section 13.6 for references).

13.1.1 Quantum States

In Dirac notation, pure states of a quantum system can be represented by a unit vector $|\psi\rangle$, also known as the wavefunction.[2] States have components that are complex numbers and are *rays* in a complex Hilbert space $\mathcal{H}$. For wavefunctions, the global phase is irrelevant, so that both $|\psi\rangle$ and $e^{i\alpha}\,|\psi\rangle$ correspond to the same physical state. Moreover, the notion of "ray" also implies that the overall magnitude of $|\psi\rangle$ is unimportant. Our convention will be to normalize it to one.

In this chapter, we consider only finite-dimensional Hilbert spaces, focusing on the simplest case, the two-dimensional space $\mathcal{H}_2$. Quantum two-state (two-level) systems are extremely important and have many physical realizations, such as spin-$\frac{1}{2}$ particles (e.g., electrons), polarization states of a photon, hyperfine states of a trapped ion or atom, neighboring levels of a Rydberg atom, and presence or absence of a photon in a microcavity. Perhaps most important, they serve as model *qubits* that generalize notions of classical bits to quantum mechanical systems.[3] The topics of quantum computing and quantum information theory are often presented in terms of small sets of two-level systems.

Here, we discuss such systems abstractly, not worrying about their physical realization. Section 13.3 presents a physical example, a spin-$\frac{1}{2}$ particle in a uniform, external, time-dependent magnetic field, $\boldsymbol{B}(t)$. For background on classical two-state systems and their control, see Chapter 12.

For a two-level system, or qubit, the most general *pure state* is

$$|\psi\rangle = \alpha|0\rangle + \beta|1\rangle\,, \qquad |\alpha|^2 + |\beta|^2 = 1\,, \tag{13.1}$$

where $|0\rangle$ and $1\rangle$ are the two basis states and where the condition on the complex numbers α and β fixes the normalization. States also have a vector representation,

[2] Pure states are an important special case. More general states can be expressed as convex combinations of pure states and are usually described using the *density matrix* formulation of quantum mechanics.

[3] The term *qubit* is short for "quantum bit" and plays off, for no particularly good reason, the archaic unit of measure "cubit." Mermin (2007) has argued, with passion but not success, for "Qbit" as an alternate spelling.

$$|0\rangle \to \begin{pmatrix}1\\0\end{pmatrix}, \quad |1\rangle \to \begin{pmatrix}0\\1\end{pmatrix} \quad \Longrightarrow \quad |\psi\rangle \to \begin{pmatrix}\alpha\\ \beta\end{pmatrix}. \tag{13.2}$$

The basis states are eigenstates of the Pauli spin matrix σ_z. The $|0\rangle$ state is associated with eigenvalue $+1$, while $|1\rangle$ is associated with -1. These conventions constitute the *computational basis*.[4]

A nice way to visualize a qubit state is known as the *Bloch sphere*. Although a complex two-vector has four components, normalization to a unit vector fixes one and a global phase shift can fix a second. There remain just two free parameters, which we can take to be polar and azimuthal angles (θ, ϕ) indicating the orientation on a unit sphere. We thus rewrite the arbitrary state $|\psi\rangle$ as

$$|\psi\rangle = \cos\left(\tfrac{1}{2}\theta\right)|0\rangle + e^{i\phi}\sin\left(\tfrac{1}{2}\theta\right)|1\rangle, \tag{13.3}$$

where the angles are in the range $0 \le \theta \le \pi$ and $0 \le \phi \le 2\pi$. We then define the unit vector

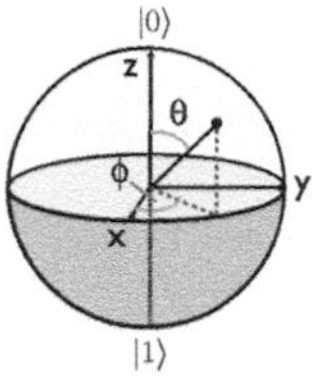

Bloch sphere

$$\hat{\boldsymbol{n}} = \begin{pmatrix}\sin\theta\cos\phi\\ \sin\theta\sin\phi\\ \cos\theta\end{pmatrix} \tag{13.4}$$

on a sphere in $\mathbb{R}^3$ with angles θ and ϕ, as depicted at left.

The state $|\psi\rangle$ then corresponds to $\hat{\boldsymbol{n}}$, with angles θ and ϕ, as depicted at left. In particular, $|0\rangle$ corresponds to the north pole and $|1\rangle$ to the south pole. Orthogonal states thus lie at the antipodes (opposite each other on the sphere). In Problem 13.2, we will see that unitary transformations are equivalent to rotations on the Bloch sphere, so that states $|\psi\rangle$ that start on the sphere remain on it.

13.1.2 Quantum Dynamics

States evolve according to the *Schrödinger equation*, a linear ordinary differential equation that is given by, for a time-independent Hamiltonian H,

$$\frac{d|\psi\rangle}{dt} = -\frac{i}{\hbar}H|\psi\rangle. \tag{13.5}$$

Its solution is

$$|\psi(t)\rangle = U(t)\,|\psi(0)\rangle, \qquad U(t) = e^{-iHt/\hbar}, \tag{13.6}$$

where $|\psi(0)\rangle$ is the wavefunction at the initial time 0 and $|\psi(t)\rangle$ the wavefunction at time t and where U is a *unitary transformation* satisfying $UU^\dagger = 1$, with the adjoint $U^\dagger = (U^\mathsf{T})^*$: the complex conjugate of the transpose is the inverse of U.

[4] For a physicist used to spin-$\frac{1}{2}$ particles where the ground state is spin down, the computational basis is a bit odd! Just remember to identify $|0\rangle$ with $|\uparrow\rangle$ and also $|1\rangle$ with $|\downarrow\rangle$. Assigning a $+1$ eigenvalue to the state $|0\rangle$ is widely used for qubits. See, for example, the influential text by Nielsen and Chuang (2000).

For qubits, unitary transformations are 2×2 complex matrices. In Problem 13.1, you will show that the most general unitary transformation on a qubit can be written as

$$U = \mathbb{I}\cos\varphi + \mathrm{i}\hat{\boldsymbol{m}}\cdot\boldsymbol{\sigma}\sin\varphi = \exp(\mathrm{i}\varphi\hat{\boldsymbol{m}}\cdot\boldsymbol{\sigma}), \tag{13.7}$$

where $\hat{\boldsymbol{m}}$ is a unit vector pointing in an arbitrary direction and $\boldsymbol{\sigma}$ is the vector of Pauli matrices, whose component matrices are[5]

$$\begin{aligned} \sigma_x &= \begin{pmatrix} 0 & 1 \\ 1 & 0 \end{pmatrix} & \sigma_y &= \begin{pmatrix} 0 & -\mathrm{i} \\ \mathrm{i} & 0 \end{pmatrix} & \sigma_z &= \begin{pmatrix} 1 & 0 \\ 0 & -1 \end{pmatrix} \\ &= |0\rangle\langle 1| + |1\rangle\langle 0| & &= \mathrm{i}|1\rangle\langle 0| - \mathrm{i}|0\rangle\langle 1| & &= |0\rangle\langle 0| - |1\rangle\langle 1|\,. \end{aligned} \tag{13.8}$$

Notice that the effect of σ_x is to swap $|0\rangle \leftrightarrow |1\rangle$. That is, $\sigma_x|0\rangle = |1\rangle$ and $\sigma_x|1\rangle = |0\rangle$. One can view this equivalently as a NOT logical operation on a qubit.

In Problem 13.2, we will explore the interpretation of U as a rotation of $|\psi\rangle$ on the Bloch sphere. The representation in Eq. (13.7) corresponds to a rotation by an angle φ about the axis $\hat{\boldsymbol{m}}$. The axis $\hat{\boldsymbol{m}}$ is distinct from the direction $\hat{\boldsymbol{n}}$ associated with the original Bloch state.

13.1.3 Quantum Measurement

In quantum mechanics, each observable O corresponds to a self-adjoint operator on elements of a Hilbert space $\mathcal{H}$. The adjoint $O^\dagger$ is defined via $\langle O^\dagger\psi_A|\psi_B\rangle = \langle\psi_A|O\psi_B\rangle$, and an operator is self-adjoint if $O = O^\dagger$. For finite-dimensional Hilbert spaces, self-adjoint operators are also known as *Hermitian*, and we will use the terms interchangeably. Hermitian operators can be written in a *spectral representation* as a sum over their eigenvalues o_n and eigenvectors $|n\rangle$,

$$O = \sum_n o_n|n\rangle\langle n|\,. \tag{13.9}$$

A measurement O on a state $|\psi\rangle = \sum_n c_n|n\rangle$ returns the eigenvalue o_n (corresponding to eigenstate $|n\rangle$ with probability $|\langle\psi|n\rangle|^2 = |c_n|^2$. The expected value of the measurement is $\langle\psi|O|\psi\rangle = \sum_n |c_n|^2 o_n$, where normalization implies $\sum_n |c_n|^2 = 1$.

The kind of measurement described above leaves the system in the corresponding eigenstate $|n\rangle$[6] and is known as a *von Neumann*, or *projective* measurement.[7] Notice that while the state evolution is deterministic, the measurement process introduces stochasticity. Also, the Schrödinger equation itself is linear, but the projection

[5] In this chapter, matrices are in normal font, while we continue to use bold font for vectors. Here, $\boldsymbol{\sigma}$ is a vector whose components are 2 × 2 matrices.

[6] Historically, the projection aspect of measurement has often been associated with the notion of wavefunction *collapse*. This soundness of and need for this interpretation has been vigorously debated. The basic experimental facts, however, are not in question, and "projection" can still serve as a convenient term, whether viewed as a collapse or not. Practical measurements often use intermediate, ancillary systems that lead to system states that are not simple projections onto an eigenstate, but those generalized measurements still suppose a projective measurement of the ancillary system.

[7] If the eigenvalues are degenerate, the projection is to the subspace spanned by the degenerate eigenvectors.

operation is nonlinear, since the wavefunction satisfies $\langle\psi|\psi\rangle = 1$ before and after the measurement.[8] Viewed from input to output, as control systems should be, quantum dynamics can be nonlinear.

Example 13.1 (Spin of a spin-$\frac{1}{2}$ particle) You prepare a spin-$\frac{1}{2}$ particle to be in the eigenstate $|0\rangle$ of the σ_z operator. Now you measure σ_x. To find the probability of getting $|+\rangle$ and $|-\rangle$, the eigenstates of the σ_x basis, we project the initial state $|0\rangle$ onto the basis states of σ_x. From Eq. (13.8), those states are

$$|+\rangle \equiv \frac{1}{\sqrt{2}}\left(|0\rangle + |1\rangle\right), \qquad |-\rangle \equiv \frac{1}{\sqrt{2}}\left(|0\rangle - |1\rangle\right), \tag{13.10}$$

with eigenvalues of $+1$ and -1, respectively. The probability of measuring $+1$ and leaving the system in the state $|+\rangle$ is then $|\langle 0|+\rangle|^2 = \frac{1}{2}$. Similarly, the probability of measuring -1 and leaving the system in $|-\rangle$ is $|\langle 0|-\rangle|^2 = \frac{1}{2}$.

13.1.4 Composite Systems

Above, we have described some properties of a single system. What if we had two such systems? In the simplest case, each state is independent. If the first system has state $|\psi\rangle_A$ and is an element of a Hilbert space $\mathcal{H}_A$ and the second has state $|\psi\rangle_B$ in $\mathcal{H}_B$, then the composite system state is $|\psi\rangle_A \otimes |\psi\rangle_B$ in $\mathcal{H}_{AB}$, given by the *tensor product* $\mathcal{H}_A \otimes \mathcal{H}_B$.

We can also use the tensor product to form operators on the composite space. Indeed, the operator $O_A \otimes O_B$ is defined so that O_A acts on the state $|\psi\rangle_A$ while O_B acts on the state $|\psi\rangle_B$. For the case of two qubits, the composite system has $2^2 = 4$ elements, and operators correspond to 4×4 matrices.

To have an operator that transforms only subsystem A, we can define $O_A \otimes \mathbb{I}_B$, where $\mathbb{I}_B$ is the identity operator on subsystem B. Similarly, $\mathbb{I}_A \otimes O_B$ operates only on subsystem B.

13.2 Three Types of Quantum Control

With the above background, we can understand the simplest versions of quantum control. We consider three different methods: open-loop control, measurement-based quantum feedback, and coherent quantum feedback control. Our goal is to understand where and how quantum control can differ from its classical counterparts. We will illustrate these three approaches using a typical task: to transform an initial state into a desired target state. Such *quantum state preparation* is crucial in many systems of great current interest, including cold atoms, trapped ions, superconducting qubits, and nitrogen-vacancy centers.

[8] If, after measurement, $|\psi\rangle = |n\rangle$, then the amplitude of this eigenstate will have changed from c_n to 1.

13.2.1 Open-Loop Quantum Control

In an ideal version of open-loop, or feedforward control, we know perfectly the dynamics of the system *and* its initial state. To transform this arbitrary but *known* initial state $|\psi\rangle$ into a desired target state (e.g., $|1\rangle$), it is easy to see that, for $|\psi\rangle = \alpha|0\rangle+\beta|1\rangle$, the operator

$$U = \begin{pmatrix} \beta & -\alpha \\ \alpha^* & \beta^* \end{pmatrix} \tag{13.11}$$

is both unitary and sends $|\psi\rangle \to |1\rangle$, as desired. We thus can find a unitary transformation that changes the (pure) state of a qubit in an arbitrary way. We can also use the Bloch-sphere representation to interpret U as a rotation on the Bloch sphere. See Problem 13.2.

To carry out such a transformation physically, we need to know the system Hamiltonian H, along with the initial state, characterized by α and β. These requirements are analogous to those for feedforward control of classical systems. Thus, while the equations of motion of quantum-mechanical systems have some peculiar features such as complex numbers, noncommutative operators, and the like, there is no fundamental difference – from the point of view of control – between quantum feedforward operations and their classical counterparts. Of course, good open-loop feedback techniques must account for and use quantum coherences. Below, in Section 13.3, we will explore how feedforward works in the more explicitly physical setting of a spin-$\frac{1}{2}$ particle in an external magnetic field.

13.2.2 Measurement-Based Quantum Feedback (MF)

Feedforward control requires the initial state of the system. If you do not know it, you can measure and apply feedback based on the result. We thus consider a modified control problem: start from an *unknown* initial state $|\psi\rangle = \alpha|0\rangle + \beta|1\rangle$ and transform it to $|1\rangle$.

A simple way to implement measurement-based quantum feedback (MF) in this case is to measure the observable σ_z. According to the postulates of quantum mechanics, the result of this *von Neumann measurement* will be one of the basis states: either the state $|0\rangle$ with probability $|\alpha|^2$ or the state $|1\rangle$ with probability $|\beta|^2 = 1 - |\alpha|^2$.

If the result happens to be -1, then the state must be $|1\rangle$: the measurement process has itself transformed the state from $|\psi\rangle$ to the desired state $|1\rangle$. We can stop: our job is done! If the result happens to be $+1$, then the state must be $|0\rangle$, and we can apply the unitary transformation $U = \sigma_x$, which swaps the states $|0\rangle$ and $|1\rangle$.

Thus, by applying the conditional unitary transformation

$$U = \begin{cases} \sigma_x, & \text{if measurement result} = |0\rangle \\ \mathbb{I}, & \text{if measurement result} = |1\rangle \end{cases} \tag{13.12}$$

we can use a single measurement-feedback pair to transform an initial unknown state $|\psi\rangle$ into the desired state $|1\rangle$. Notice that while the measurement is quantum mechanical (involving projection onto an eigenstate), the processing of the information and the actuation (via a control Hamiltonian) is classical. The decision as to which control action to take can be implemented on a classical computer, as can each of the actions themselves.

13.2.3 Coherent Quantum Feedback (CF)

In Section 15.3 on thermodynamics and control, we will argue that to understand the role of information in thermodynamics, we must explicitly model the measuring device and its interactions with the system of interest. We can imagine a quantum analog of such a device, where both the controller and system of interest are quantum systems that exchange quantum information and act in a "quantum" way that does not destroy correlations.

Let us again consider the problem of transforming an unknown state $|\psi\rangle = \alpha|0\rangle + \beta|1\rangle$ to $|1\rangle$. To control the system, we let it interact with a second qubit, which is initially in the known state $|0\rangle$. We then apply two unitary operations. The first, U_{corr}, "correlates" the two states and will play a role analogous to measurement. The second, U_{fb}, applies a feedback to the "measured state" and transforms it to the desired state, $|\,1\,\rangle$.

To fully describe two systems, we use their tensor product. The initial state of the joint controller and system is

$$|\Psi\rangle_0 = |0\rangle \otimes |\psi\rangle\,, \tag{13.13}$$

where the first state is that of the controller and the second that of the system. Because the states are written as a simple tensor product, they are uncorrelated, or *separable*. The first step, then, is to correlate the two systems:

$$U_{\text{corr}} = \mathbb{I} \otimes |0\rangle\langle 0| + \sigma_x \otimes |1\rangle\langle 1|\,, \tag{13.14}$$

where $\mathbb{I}$ is the identity operator and where σ_x swaps $|0\rangle \leftrightarrow |1\rangle$. The first argument of the tensor product is applied to the controller state and the second to the system state. The result is an *entangled state* that is not a simple tensor product and reflects correlations between the states of the system and controller:

$$|\Psi\rangle_{\text{corr}} = \Bigg(\underbrace{\mathbb{I} \otimes |0\rangle\langle 0| + \sigma_x \otimes |1\rangle\langle 1|}_{U_{\text{corr}}}\Bigg)\Bigg(|0\rangle \otimes \underbrace{\alpha|0\rangle + \beta|1\rangle}_{|\psi\rangle}\Bigg) = \alpha|0\rangle \otimes |0\rangle + \beta|1\rangle \otimes |1\rangle\,. \tag{13.15}$$

The second step is to apply the feedback operation $U_{\rm fb}$, which *disentangles* the subsystems:

$$|\Psi\rangle_{\rm fb} = \Bigg(\underbrace{|0\rangle\langle 0| \otimes \sigma_x + |1\rangle\langle 1| \otimes \mathbb{I}}_{U_{\rm fb}}\Bigg)\Bigg(\underbrace{\alpha|0\rangle \otimes |0\rangle + \beta|1\rangle \otimes |1\rangle}_{|\Psi_{\rm corr}\rangle}\Bigg)$$
$$= \Bigg(\underbrace{\alpha|0\rangle + \beta|1\rangle}_{|\psi\rangle}\Bigg) \otimes |1\rangle = |\psi\rangle \otimes |1\rangle \,. \tag{13.16}$$

The combined operation of $U_{\rm fb}\, U_{\rm corr}$ can be considered as a single unitary operator that maps $|0\rangle \otimes |\psi\rangle \to |\psi\rangle \otimes |1\rangle$. The important point is that we map the unknown system state $|\psi\rangle$ to the desired state $|1\rangle$. Unlike the previous case, no measurement has taken place; rather, we have created a quantum "supersystem" with an open-loop control that manipulates both subsystems in a way that makes the "system of interest" behave as desired.

Note that the original unknown state $|\psi\rangle$ has not disappeared but merely been shifted from the system of interest to the controller. Since we avoid measurement (and since we have assumed a closed system with no interactions coming from an outside world), the only allowed evolution is unitary. And unitary operators shift states but do not destroy them.[9]

How does coherent quantum feedback control compare to measurement-based quantum feedback control? Using a fully quantum system to control without measurement can have inherent advantages, as the set of possible unitary operators is restricted in measurement-based feedback schemes, which destroy coherences that quantum coherent systems preserve. The set of possibilities is greater with quantum coherent control, and sometimes that can translate into greater performance for a control task.

There are also practical differences: as with measurement-based control of classical systems, the measured quantity must be *amplified* in order to be processed by a (classical) computer to compute a feedback control. This amplification almost always slows the feedback loop to times that are longer than the natural time scales of the system under control. But once the measurement has been acquired by a computer, all subsequent processing is noise free. By contrast, the actions of a fully quantum controller are subject to interactions with the environment, which add noise.

13.3 Physical Example

The above overview of quantum control was in terms of the abstract language of qubits. In this section, we give a more physical (if still idealized) example of a qubit

9 Analogously, classical states are also conserved. There are classical and quantum Liouville Theorems.

and illustrate it with an open-loop control problem that uses optimal-control ideas from Chapter 7.

Our physical example is familiar from textbooks on quantum mechanics, a spin-$\frac{1}{2}$ particle in a controllable magnetic field, $\boldsymbol{B}(t)$. The goal is to prepare the system in a desired final state while minimizing the control effort. We proceed in stages:

1. State the problem.
2. Connect to optimal control.
3. Solve for the form of the controls $u(t)$.
4. Integrate the equations of motion forward in time.
5. Minimize the cost by choosing the free parameters in the control $u(t)$.

13.3.1 State the Problem

Assume a constant background field $B_0\,\hat{z}$ and a time-dependent horizontal field $\boldsymbol{B}_\perp(t)$ with components $B_x(t)$ and $B_y(t)$. Starting at time $t = 0$ from a pure state $|0\rangle$, we wish to drive $|\psi(\tau)\rangle$ as close as possible to the state $|1\rangle$ at time τ, while using the smallest field $\boldsymbol{B}_\perp(t)$. A cost functional that captures these desires is[10]

$$J\left[|\psi\rangle, u_x, u_y\right] = 1 - |\langle\psi(\tau)|1\rangle|^2 + \tfrac{1}{2}\eta \int_0^\tau \mathrm{d}t \left[u_x^2(t) + u_y^2(t)\right], \qquad (13.17)$$

where the controls $u_x(t) \equiv \gamma B_x(t)$ and $u_y(t) \equiv \gamma B_y(t)$, where γ is the gyromagnetic ratio (see below). To ease notation, we do not show, in the expression for J, that $|\psi\rangle$ is evaluated at the end time τ and that it depends on the control paths, $u_x(t)$ and $u_y(t)$ over the interval $[0, \tau]$.[11] In Eq. (13.17), the parameter η sets the cost of control: small values make $|\psi(\tau)\rangle \to |1\rangle$, while large values minimize the applied field, $\boldsymbol{B}_\perp(t)$. The role of the constant term $+1$ is to define the lowest possible cost to be $J = 0$.

The Hamiltonian for the system is $H = -\boldsymbol{\mu}\cdot\boldsymbol{B}$, with magnetic moment $\boldsymbol{\mu} = \gamma\boldsymbol{S}$, vector spin operator $\boldsymbol{S}$. The latter can be a positive or negative number, depending on the particle. For a spin-$\frac{1}{2}$ particle, $\boldsymbol{S} = \frac{1}{2}\hbar\boldsymbol{\sigma}$, where $\boldsymbol{\sigma}$ is a vector whose components are the three Pauli matrices. In terms of the Pauli matrices, the Hamiltonian is

$$H(t) = -\tfrac{1}{2}\gamma\hbar\left(B_0\,\sigma_z + B_x(t)\,\sigma_x + B_y(t)\,\sigma_y\right) \equiv -\tfrac{1}{2}\left(\omega_0\,\sigma_z + u_x(t)\,\sigma_x + u_y(t)\,\sigma_y\right), \qquad (13.18)$$

where $\omega_0 = \gamma B_0$ is the Larmor precession frequency, and we have scaled units so that $\hbar = 1$. (See Problem 13.3.) The state $|\psi(t)\rangle$ obeys the Schrödinger equation,

$$i\,\partial_t|\psi(t)\rangle = H\,|\psi(t)\rangle\,. \qquad (13.19)$$

[10] The magnitude squared of the overlap between pure states $|\psi(\tau)\rangle$ and $|1\rangle$, defined as $F \equiv |\langle\psi(\tau)|1\rangle|^2$ is often called the *fidelity* and satisfies $0 \le F \le 1$. By similar logic, its complement, $1 - F$, is the *infidelity*. See, for example, Schumacher and Westmoreland (2010), Section 7.5.

[11] Compare the form of J in Eq. (13.17) to Eq. (7.5) in our discussion of optimal control in Chapter 7.

13.3.2 Connect to Optimal Control

In short, we seek control inputs $u_x(t)$ and $u_y(t)$ that minimize the cost J in Eq. (13.17), while constraining $|\psi(t)\rangle$ to obey the Schrödinger equation. With the adjoint $|\lambda(t)\rangle$ as a Lagrange multiplier, we have

$$J' = \underbrace{1 - |\langle\psi(\tau)|1\rangle|^2}_{\text{infidelity}} + \underbrace{\tfrac{1}{2}\eta \int_0^\tau \mathrm{d}t \left(u_x^2 + u_y^2\right)}_{\text{cost of control}} + \underbrace{\mathrm{Re} \int_0^\tau \mathrm{d}t \, \langle\lambda| \left(-\mathrm{i}\partial_t + H\right) |\psi\rangle}_{\text{dynamical constraint}} . \qquad (13.20)$$

The main difference between Eq. (13.20) and the cost functions we considered in Chapter 7 is that the state $|\psi(t)\rangle$ is a normalized, complex vector (with J' nonetheless a real scalar). To extend the variations to complex numbers, we can write the real part as the sum of a complex term and its conjugate and then take variations with respect to $|\psi(t)\rangle$ and $|\psi^*(t)\rangle$. Normalization can be enforced by adding another Lagrange multiplier or, more simply, by observing that the variational equations imply the $|\psi^*(t)\rangle$ obeys the Schrödinger equation, which preserves normalization (see Problem 13.5a).

13.3.3 Solve for the Form of the Controls

In Section 7.2, we discussed how to use the calculus of variations to minimize cost functions such as J' in Eq. (13.20). The minimization (Problem 13.5) gives Euler–Lagrange equations, which lead to

$$u_x = \frac{1}{\eta} \mathrm{Re}\, \langle\lambda|\sigma_x|\psi\rangle , \qquad u_y = \frac{1}{\eta} \mathrm{Re}\, \langle\lambda|\sigma_y|\psi\rangle . \qquad (13.21)$$

The next step would normally be to substitute u_x and u_y back into the Schrödinger equation for the adjoint $|\lambda(t)\rangle$ and use the final condition at $t = \tau$ to integrate backwards to $t = 0$. Having solved for $u_x(t)$ and $u_y(t)$, we would then integrate the equation $|\psi(t)\rangle$ forward from $t = 0$ to τ. Here, we adopt a slightly different strategy: We differentiate each relation in Eq. (13.21) with respect to time and use the commutation relations to show that $\dot{u}_x = k\, u_y$ and $\dot{u}_y = -k\, u_x$, where $k = \omega_0 - \frac{1}{\eta}\mathrm{Re}\, \langle\lambda|\sigma_z|\psi\rangle$ is a constant, even though $|\lambda\rangle$ and $|\psi\rangle$ each vary in time. The controls then obey relations of the form,

$$u_x = B_1 \cos(\omega t - \theta_0) , \qquad u_y = B_1 \sin(\omega t - \theta_0) , \qquad (13.22)$$

where B_1 and ω are freely chosen. Since the initial state is about the z-axis and since u_x and u_y are in the xy plane, rotational symmetry allows us to set $\theta_0 = 0$.

13.3.4 Integrate the Equations of Motion Forward in Time

Having found that the optimal control for $u_x(t)$ and $u_y(t)$ is to use a rotating magnetic field, we can solve the equations of motion with the amplitude B_1 and rotational frequency ω as free parameters. This is a classic problem in quantum mechanics and is at the heart of techniques such as nuclear magnetic resonance (NMR) and ways to manipulate quantum gates in quantum computing. The result (see Problem 13.4) is that, in a frame of reference rotating with the horizontal field, the total applied field is effectively static. A spin-$\frac{1}{2}$ particle in a static field of magnitude B has magnetic moment expectation values that precess at a frequency γB about the field axis. Thus, the static effective field leads to precession at frequency $\omega_1 t$ about the axis defined by σ_a in the rotating frame. Back in the original frame, this new precession becomes a *nutation* on the original precession about $\hat{z}$. The quantitative result is that, starting from an initial state $|\psi(0)\rangle = |0\rangle$, the wave function evolves as

$$|\psi(t)\rangle = \mathrm{e}^{-\mathrm{i}\omega t/2}\left[\cos\left(\tfrac{1}{2}at\right) + \frac{\mathrm{i}(\omega_0+\omega)}{a}\sin\left(\tfrac{1}{2}at\right)\right]|0\rangle + \mathrm{e}^{\mathrm{i}\omega t/2}\left[\frac{\mathrm{i}\omega_1}{a}\sin\left(\tfrac{1}{2}at\right)\right]|1\rangle\,, \quad (13.23)$$

where $a = [(\omega_0+\omega)^2 + \omega_1^2]^{1/2}$ and where $\omega_0 = \gamma B_0$ defines the *Larmor frequency* (precession about $\hat{z}$) and $\omega_1 = \gamma B_1$ the *Rabi frequency* (nutation). The frequencies ω and ω_1 are considered to be free parameters, while ω_0 is given.

13.3.5 Minimize the Cost

Having used the calculus of variations to reduce an infinite-dimensional search for functions $u_x(t)$ and $u_y(t)$ to a search for two free parameters ω and ω_1, we finish by substituting the expression for $|\psi(t)\rangle$ into the cost function and performing standard minimization on the function $J(\omega,\omega_1)$. Recalling that the desired final function at time τ is $|\psi(\tau)\rangle = |1\rangle$, we can evaluate Eq. (13.17). Defining $a = a(\omega,\omega_1)$ as above, we find (Problem 13.5)

$$J(\omega,\omega_1) = 1 - \left(\frac{\omega_1}{a}\right)^2 \sin^2\left(\tfrac{1}{2}a\tau\right) + \tfrac{1}{2}\eta\omega_1^2\tau\,, \quad (13.24)$$

where η, ω_0, and τ are all given parameters.

We have reduced the optimization problem to that of minimizing a standard two-parameter function. In the limit $\eta \to 0$, we can choose ω and ω_1 to make $|\psi(\tau)\rangle = |1\rangle$ exactly, implying $J = 0$. For large η, the optimum cost is $J = 1$, which is achieved by setting $\omega_1 = 0$. Control is so expensive that it is best not to apply any at all.

In this example, there are no limits to the magnitude of the applied field, although η sets an approximate "soft bound." For hard limits, we can minimize J by applying the Pontryagin Minimum Principle (PMP) discussed in Section 7.5.2.

13.4 How Different Is Quantum Control?

This is an important, yet delicate question. On the one hand, there are clear conceptual differences between quantum and classical systems. Features such as the inevitability that the basic projective measurements alter the state of the system being measured and the consequences of coherence and superposition (the electron that cannot be said to pass through one slit or the other but must in some sense pass through both) – these are aspects that have no classical counterpart. On the other hand, if we begin to look in more detail at differences in practice, the distinctions become muddier. Put another way, one should be careful in claiming that a particular phenomenon is "quantum."[12]

One feature of quantum mechanics is that the state itself, $|\psi\rangle$, is not directly measurable, although it influences the results of measurements of observables, by setting the probabilities for different outcomes. For some authors, this distinguishes quantum systems conceptually from classical systems. Yet from the perspective of control theory, the distinction is not so clear in practice. Although it is true in principle that one can measure an internal state $\boldsymbol{x}$ of a classical dynamical system, this is almost never the case in practice.[13] For this reason, we have been careful to distinguish the observations $\boldsymbol{y}$ from the internal state $\boldsymbol{x}$ and have advanced different methods for reconstructing estimates of $\boldsymbol{x}$ at a current time t from a sequence of past observations $\boldsymbol{y}$. In other words, even in classical systems, the state is viewed as a set of underlying, hidden variables, and observations give information about the underlying state to the extent that they are correlated with it.

What about the measurement process, which seems so different in quantum physics? How does measurement-based quantum feedback compare with feedback control of classical systems? At first glance, they differ greatly, as the process of measuring the state of a quantum system can greatly perturb it. Indeed, in the example of Section 13.2.2, the perturbation is strong enough that sometimes the measurement itself supplies the entire needed control! In classical physics, it is typically assumed that the measurement process itself, while it may give a noisy result, does not itself affect the system state in any significant way. This difference is fundamental to quantum mechanics, where the projective action of measurement implies *back action* on the measured state. However, this clear conceptual difference is again less clear in practice:

1. The back action of quantum measurements can be reduced by replacing the direct von Neumann measurement with an indirect *weak* measurement. Instead of measuring directly the system of interest, one measures instead a second, *ancillary system* that is weakly coupled to it.
2. For feedback based on noisy classical measurements, noise will be injected into the feedback loop and perturb the physical system. Thus, in a feedback loop for a classical system, the process of measurement will typically affect the system state. The perturbations can be significant if measurement noise is large. Of course, quantum measurements may add "classical" noise, too, for example if their results are amplified.

[12] There is a long history of "quantum phenomena" that turn out to have classical counterparts. To cite just one, the *dynamical decoupling* method described in Section 13.5 has a classical counterpart that may be observed in a system as simple as two linearly coupled harmonic oscillators (Salerno and Carusotto, 2014).

[13] Recall, too, that the internal state of a linear system is defined only up to a coordinate transformation in state space, a fact that accentuates the difference between the state of a dynamical system and its observations.

3. Classical measurements need not create a back action, but often do. For example, a voltage measurement at finite impedance implies a current drain on the circuit being measured, an effect that can be important if the measured impedance is high.

These three scenarios all imply, in their own way, that the *practical* distinction between classical and quantum feedback control is more one of degree. Both quantum and classical measurements can end up affecting the system, although perhaps for different reasons. But, for control systems, noise is noise, no matter what its origin.

Our point is not to minimize the conceptual differences between quantum and classical systems, nor is it to focus on which specific features of quantum control are quantum or not. Rather, we draw the optimistic conclusion that the techniques to control classical systems often do apply to quantum systems and that those with an interest in quantum control can profitably take advantage of lessons learned from the control of classical counterparts.

13.5 Summary

In this chapter, we have introduced three types of quantum control: feedforward control, measurement-based feedback, and coherent feedback. We focused on the most basic examples, based on abstract two-level systems known as qubits. We have seen that the "weird and wonderful" features of quantum mechanics – the back action of measurements on the system measured, the possibility of coherent states and entanglement – give a distinctive twist to problems of control. Nonetheless, many of the ideas and issues from the use of feedback on classical systems apply to quantum control, as well. For example, we have seen that measurement noise is injected into a feedback loop, thereby affecting the system under control. And the notion of coherent quantum control has elements analogous to early control systems such as the steam-engine and governor developed by James Watt. Our hope is to build a bridge so that people familiar with control of classical systems can understand what changes in the quantum case and, conversely, that people whose primary interest is quantum control can understand the classical counterpart more deeply.

Inevitably, in such a brief introduction, we have left a great deal out. Many topics of interest to the quantum control community have classical counterparts. For example, the notions of controllability and observability in quantum systems are closely related to the geometric, Lie-bracket techniques presented in Chapter 11.

The range of experimental implementations achieved is far more impressive than the simple examples we have discussed. For example, H. Rabitz and collaborators have added a kind of feedback to quantum feedforward control. The idea is to guess a feedforward protocol, apply it, measure the quality of result, use that information to adjust the feedforward protocol, and repeat. They term this "adaptive feedback control," although it might be better termed a kind of *learning control*. The control is always open loop but is incorporated into an overall optimization strategy. Nonetheless, via techniques such as genetic algorithms for the optimization, impressive results have been achieved, such as the selective amplification of a nonlinear optical response for a chosen (high) harmonic.

Dynamical decoupling is another widely used open-loop control technique (Viola and Lloyd, 1998). The idea is to suppress the decoherence of a qubit due to environmental perturbations by applying strong periodic or random pulses on time scales fast compared to the bath, to disrupt the decoherence process. The implementation is a kind of *quantum bang-bang* control, in reference to the classical version discussed in Section 7.5.2.

Much of the work on quantum feedback has used weak measurements that give only a small amount of information about the system under control but also perturb it only slightly. Weak measurements are an instance of a more general notion of measurement that goes beyond the projective measurements discussed in this chapter. Using weak measurements, control can be implemented in a way that amounts to a continuous measurement; occasionally, it is accompanied by real-time control. A spectacular example of the latter comes from the group of S. Haroche (Sayrin et al., 2011), who stabilized the number of microwave photons in a superconducting cavity, using weak measurements. Their sensor was a beam of atoms that crossed the cavity. Each "atomic qubit" weakly interacts with the Fock state $|n\rangle$ of n photons. The information gained is used to adjust the classical microwave field (its photon number), closing the loop.

A recent experiment by Hirose and Cappellaro (2017) demonstrates nicely the power of quantum coherent control. Their system qubit is a defect in a diamond lattice consisting of a nitrogen atom substituted for carbon next to a lattice vacancy. This *NV center* qubit decoheres in $\approx 4\ \mu\text{s}$ because of interactions with a bath of nearby ^{13}C nuclei, which comprise about 1% of the carbon atoms in naturally occurring diamond. Using the ^{14}N nuclear spin of the nitrogen atom as an ancillary qubit that interacts with the NV center, Hirose and Cappellaro are able to maintain the coherence of the system qubit up to the decoherence time of the ancilla (3 ms), an extension of a factor of 1000 over the natural decay time.

We end with a brief historical note. All three types of quantum control have been realized on experimental systems, in roughly the same historical order as our exposition: coherent feedforward experiments date from nuclear magnetic resonance (NMR) experiments in the 1950s, and laser-pulse schemes to control chemical reactions began around 1990. Measurement-based feedback started in the 1990s (stimulated by the desire for error correction in quantum-computing schemes). Finally, both the theoretical proposal and first experimental realization of quantum coherent feedback date from 2000. Interestingly, this order is precisely the *reverse* of how classical control developed: the original steam engine and governor is analogous to coherent quantum feedback; the systems-theoretic approach starting in the 1920s is analogous to measurement-based feedback; and the full value of feedforward did not begin to be recognized until roughly the 1960s.

13.6 Notes and References

The prehistory of open-loop control of quantum systems dates back to early work on nuclear magnetic resonance (NMR). The first studies – e.g., Hahn and Purcell (1950) and Carr and Purcell (1954) – treated cases simple enough that the optimal solutions could be guessed. For more complex cases, one needs a more systematic approach. The open-loop control of quantum systems has been developed mainly by physical chemists, as described by Rice and Zhao (2000) and by Shapiro and Brumer (2012). Rabitz and his group introduced the optimal-control approach and, later, AFC, adaptive feedback control (Brif et al., 2010). For a nontechnical overview, see Walmsley and Rabitz (2003). An earlier review, Warren et al. (1993), is valuable for its frank discussion of why the first efforts to control quantum systems failed. D'Alessandro (2008) carefully relates the subject to standard control theory. Its chapters 1, 2, and 9 are particularly accessible, while chapters 3 and 4 are more mathematical, relating the geometric methods introduced in Section 11.1.5 to quantum open-loop control. In more complex settings (e.g., spin chains), even simple tasks such as the state-preparation problem of Section 13.3 lead to complex optimal-control problems, where the cost function, in the space of protocols, has an exponential number of attractors and strongly resembles that of a spin-glass energy surface (Day et al., 2019).

Gentler introductions to quantum measurement theory may be found in introductory books on quantum computing, such as Kaye et al. (2007), or quantum "systems," such as Schumacher and Westmoreland (2010). Ballentine (2015) gives a careful, clear presentation of the foundations of quantum mechanics and quantum information theory, forcefully arguing for an ensemble interpretation and for a "collapse-less" version of quantum mechanics. The *Qubist* interpretation also argues against wavefront collapse, from a Bayesian point of view (Fuchs et al., 2014). The problems and discussion of spin-$\frac{1}{2}$ particles in external fields draw from his chapter 12 (and D'Alessandro (2008)). Brun (2002) gives a good introduction to two-level quantum systems as qubits and as examples to understand continuous quantum measurements. The introduction to open systems is based in part on lecture notes by Preskill (2017), who is the source for the "weird and wonderful" quotation.

The original proposal for coherent quantum feedback as described here was by Lloyd (2000), a work that was rapidly followed by an experimental demonstration (Nelson et al., 2000). Wiseman and Milburn had earlier considered feedback without measurement from a somewhat different point of view. For a broad review of experimental work, see Zhang et al. (2017), which also has some good introductory summaries of basic techniques. Our example of coherent quantum feedback is taken from Section 3.1.

Jacobs (2014) gives a good, careful introduction to quantum measurement theory and control. The book by Wiseman and Milburn (2010), two of the pioneers in the field, covers similar ground at a slightly higher level. Nielsen and Chuang (2000) gives a simpler discussion of measurement theory. The presentation in these books focuses on aspects we have neglected: density matrices, open systems, and quantum notions of dissipation. Most of the discussion focuses on systems with continuous measurements, described by stochastic master equations and related techniques.

Problems

13.1 Most general unitary operator on a qubit. Show that

a. the most general two-dimensional Hermitian operator can be written $H = c_0\mathbb{I} + c_1\sigma_x + c_2\sigma_y + c_3\sigma_z$, where the c_i are real and the $\sigma_{x,y,z}$ are the Pauli matrices;
b. the associated unitary transformation can be written $U = \exp-(\mathrm{i}Ht/\hbar) = \mathrm{e}^{\mathrm{i}\varphi\,\hat{\boldsymbol{m}}\cdot\boldsymbol{\sigma}}$;
c. and $\mathrm{e}^{\mathrm{i}\varphi\,\hat{\boldsymbol{m}}\cdot\boldsymbol{\sigma}} = \cos\varphi\,\mathbb{I} + \mathrm{i}\sin\varphi\,(\hat{\boldsymbol{m}}\cdot\boldsymbol{\sigma})$, where $\hat{\boldsymbol{m}}$ is a 3d unit vector and $\boldsymbol{\sigma}$ is the 3-vector of Pauli matrices. Hint: Show that $(\hat{\boldsymbol{m}}\cdot\boldsymbol{\sigma})^2 = \mathbb{I}$, the 2×2 identity matrix.

13.2 Feedforward control of a qubit.

a. Show that if $|\alpha|^2 + |\beta|^2 = 1$, then $U = \begin{pmatrix} \beta & -\alpha \\ \alpha^* & \beta^* \end{pmatrix}$ is unitary and transforms the normalized state $|\psi\rangle = \alpha|0\rangle + \beta|1\rangle$ to the target state $|1\rangle$.
b. To interpret U as a rotation of $|\psi\rangle$ on the Bloch sphere, show that $R_z(\varphi) \equiv \mathrm{e}^{-\mathrm{i}\frac{\varphi}{2}\sigma_z}$ rotates $|\psi\rangle$ by φ about the z-axis on the Bloch sphere.
c. For a state $|\psi\rangle$ that lies in the x-z plane, show that $R_y(\varphi) \equiv \mathrm{e}^{-\mathrm{i}\frac{\varphi}{2}\sigma_y}$ rotates $|\psi\rangle$ by φ about the y-axis on the Bloch sphere.
d. Using (b) and (c), show that U can be decomposed into a rotation by $-\phi$ about the z-axis followed by a rotation $\pi - \theta$ about the y-axis. Write the two rotation matrices explicitly and show that the resulting U has the form supposed in (a).

13.3 Spin-$\frac{1}{2}$ particle in a constant external field. Consider a spin-$\frac{1}{2}$ particle in a constant field $B_0\,\hat{z}$, whose normalized state at $t = 0$ is given by $|\psi(0)\rangle = \alpha|0\rangle + \beta|1\rangle$.

a. Solve the Schrödinger equation to find $|\psi(t)\rangle$.
b. Show that $\langle\mu_x\rangle$ and $\langle\mu_y\rangle$ precess about $\hat{z}$ at a frequency $\omega_0 = \gamma B_0$.
c. Show that the field does no work on the particle (expected energy is constant).

13.4 Spin-$\frac{1}{2}$ particle in a rotating external field. Consider a spin-$\frac{1}{2}$ particle in a time-dependent field consisting of a stationary $\hat{z}$ component B_0 and a rotating horizontal component $\boldsymbol{B}_1(t) = B_1[\cos\omega t\,\hat{x} + \sin\omega t\,\hat{y}]$. Define $\omega_1 = \gamma B_1$.

a. Find the Hamiltonian in a frame that rotates about $\hat{z}$ at frequency ω by applying a rotation of angle ωt to the Hamiltonian.
b. Write the Schrödinger equation in coordinates that rotate with phase $-\omega t$ and thereby show that the field becomes static, with $\gamma\boldsymbol{B}_{\text{eff}} = -\frac{1}{2}(\gamma B_0 + \omega)\,\sigma_z + \frac{1}{2}\gamma B_1\sigma_x$.

c. Using (a), solve the Schrödinger equation in the rotating frame and transform back to the original frame of reference to get $|\psi(t)\rangle$. With $|\psi(0)\rangle = |0\rangle$, show

$$\langle 1|\psi(t)\rangle = \mathrm{e}^{\mathrm{i}\omega t/2}\left[\frac{\mathrm{i}\omega_1}{a}\sin\left(\tfrac{1}{2}at\right)\right], \qquad a \equiv \sqrt{(\omega_0+\omega)^2+\omega_1^2}\,.$$

d. Show that at resonance, $\omega = -\omega_0$, the average energy $E(t) = -\frac{1}{2}\hbar\omega_0 \cos\omega_1 t$. The rotating field thus pumps energy in and out of the system periodically.

13.5 Optimal control of a two-level system.

a. Derive Euler–Lagrange equations for $|\psi\rangle$, $|\lambda\rangle$, u_x, and u_y. Show that $|\psi\rangle$ and $|\lambda\rangle$ each obey Schrödinger equations and that $u_{\{x,y\}} = (1/\eta)\,\mathrm{Re}\,\langle\lambda|\sigma_{\{x,y\}}|\psi\rangle$.

b. Show that $\dot{u}_x = (\omega_0 - \frac{1}{\eta}\mathrm{Re}\,\langle\lambda|\sigma_z|\psi\rangle)u_y \equiv k\,u_y$, where k is constant in time. Similarly, show that $\dot{u}_y = -k\,u_x$ and hence, that the optimal control is of the form $u_x = A\cos\omega t$ and $u_y = A\sin\omega t$, where A and ω are free parameters.

c. Using the optimal controls, evaluate the cost function and confirm Eq. (13.24).

d. Show that $J(\omega,\omega_1)$ is minimized when $\omega = -\omega_0$ (resonance condition for the rotating field). Then minimize over the remaining variable, ω_1, and confirm the statements made in the text about the small- and large-η limits. Qualitatively, how would your conclusions change if J depended on $-(\langle\psi(\tau)|P_1|\psi(\tau)\rangle)^{1/2}$?

14 Networks and Complex Systems

More is different.

P. W. Anderson[1]

The examples in this book have been based on small systems, ones whose state-space dimension is typically less than ten. Since many theorems pertain to systems of arbitrary size, the restriction to small systems might seem a mere convenience.

In this chapter, we will see that large systems are not just more complicated versions of smaller systems but have their own character. Techniques that were successful on small systems may require learning too many parameters or supplying too much control effort. To make progress, we will introduce ideas such as mapping dynamical systems to graphs and then applying graph-theoretic methods.

There are also broader, deeper reasons to study networks and their control. Networks are a popular approach to the study of complex systems, with many examples: *physical* (the weather of a planet), *biological* (protein interactions within a cell), *technological* (the components of a computer), *economic* (agents in a market), *social* (actors in movies), or a *hybrid* (the Internet). Ideas from statistical physics, especially critical phenomena and phase transitions, have been important sources of inspiration and techniques. We shall touch on these aspects lightly, as they remain controversial and are perhaps irrelevant for control. That is, we can separate the pragmatic questions of how to control a system from the deeper ones of whether and why complex systems are organized in a particular way. For balance, we will also consider briefly a rather different perspective on complex systems that comes from engineering and control theory.

Recap of Linear Systems

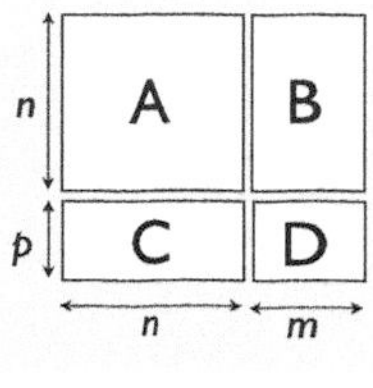

Because many of the issues in going from small to large systems are already present in linear, time-invariant systems, we concentrate on this case. In Chapters 3 and 4, we outlined the elementary theory of linear control in the frequency and time domains. The state-space formulation leads to dynamics of the form $\dot{\boldsymbol{x}} = \boldsymbol{A}\boldsymbol{x} + \boldsymbol{B}\boldsymbol{u}$, with output $\boldsymbol{y} = \boldsymbol{C}\boldsymbol{x} + \boldsymbol{D}\boldsymbol{u}$, where $\boldsymbol{x}$ is an n-dimensional state vector, $\boldsymbol{u}$ an m-dimensional input vector, and $\boldsymbol{y}$ a p-dimensional output vector of measurements or observations. The matrices $\boldsymbol{A}$, $\boldsymbol{B}$, $\boldsymbol{C}$, and $\boldsymbol{D}$ describe the dynamics, input and output coupling, and feedthrough, respectively. Because the dynamical equation has an explicit solution, $\boldsymbol{x}(t) = e^{\boldsymbol{A}t}\,\boldsymbol{x}(0) + \int_0^t dt'\, e^{\boldsymbol{A}(t-t')}\,\boldsymbol{B}\boldsymbol{u}(t')$, you might think that there is little else to say, beyond choosing the

[1] Title of article from *Science* **77**, 393–396 (1972).

linear feedback and observer matrices $\boldsymbol{K}$ and $\boldsymbol{L}$. In this chapter, we reconsider that conclusion, showing that the explicit solution may either be impossible to compute or impossible to implement on large systems. On a more positive note, casting control problems in terms of graph theory and networks leads to new strategies for control.

Controllability Revisited

We focus on the technical notion of *controllability*, introduced for linear systems in Chapter 4.1 and for nonlinear systems in Chapter 11.1.6. Because of the *duality* between input and output coupling (Section 4.1.2), the issues raised for controllability are also relevant to issues connected with observability.

Recall that a system is controllable if we can drive it from any initial state $\boldsymbol{x}_0$ to any desired state $\boldsymbol{x}_\tau$ at a given finite time τ. Recall, too, the *Kalman rank condition*: a linear system is controllable if

$$\text{rank}\; \boldsymbol{W}_\text{c} \equiv \begin{pmatrix} \boldsymbol{B} & \boldsymbol{AB} & \boldsymbol{A}^2\boldsymbol{B} & \cdots & \boldsymbol{A}^{n-1}\boldsymbol{B} \end{pmatrix} = n\,, \tag{14.1}$$

where $\boldsymbol{W}_\text{c}$ is the $n \times nm$ dimensional *controllability matrix*.

To see why Eq. (14.1) is not the end of the story, consider that a large system might have thousands or even millions of elements n. With an input for each variable, we can guarantee controllability, but what is the minimum number that we need? To find even a possible set of driver nodes (let alone the "best" set), we would typically have to consider all $O(2^n)$ possible combinations of control nodes, a task that becomes impractical for large n. Even worse, testing the rank condition requires knowing all the n^2 matrix elements of $\boldsymbol{A}$. But we have seen in Chapter 6 the challenges of measuring even a small number of parameters. A large system will almost certainly have many poorly known elements.[2]

Since the classic approach to controllability fails for large systems, we try an alternative based on the topological features of networks that requires only qualitative information about the existence and strength of connections. In Section 14.1, we give a brief introduction to networks and graphs, outlining some typical models of networks and their graphs. In Section 14.2, we show how dynamical systems have natural representations as graphs and networks and then introduce some canonical network models, including *random* and *scale-free* networks. Section 14.3 uses the graphical representation of dynamics to formulate the notion of *structural controllability*, which leads to a notion of controllability for networks that does not depend on knowing all the system's parameter values. In Section 14.4, we extend the graph approach and offer one solution to the question of whether a large network is controllable. In Section 14.5, we see that most networks that pass "in principle" controllability tests require so much control effort that they would be impossible to control in practice. Adding extra inputs beyond the minimum needed and adopting the more modest approach of control within a part of state space are practical responses. Finally, in

[2] The Kalman rank condition also requires the mn elements of $\boldsymbol{B}$, but since we are free to choose the inputs, we might hope to know better the values of the coupling constants.

Section 14.6.2, we present a rather different, "engineering" perspective on complex systems that contrasts sharply with visions based on statistical physics.

14.1 Networks in a Nutshell

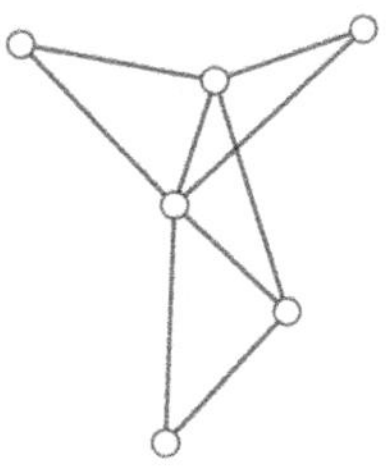

The overall goal of this chapter is to show that mapping dynamical systems to networks can be a useful approach to the study of complex systems and their control. We thus need to introduce network science and its mathematical language, *graph theory*.

Let us start with graph theory. A *graph* $G(V, E)$ is a structure consisting of a set of *vertices* V and edges E. In the *graph drawing* at right, we depict vertices as open circles and edges as lines connecting the vertices. Although it is natural to identify a graph with its drawing, the two really are distinct notions, with the former being an abstract definition using set theory and the latter a visual representation. Note that a vertex is connected to only some of the other vertices and may even be isolated (disconnected).

Loosely, a *network* is a graph with meaning. That is, networks have graphs whose vertices and edges have external significance. The jargon of networks differs from that of graphs. For example, vertices become *nodes*, edges *links*. In a physical example, each node could represent an electrical power generator, each link a transmission-line connection between two generators. In a social example, the nodes could be associated with actors, the links with appearing in the same movie. Mostly, in this chapter, we will use the language of network science, reserving graph-theory terminology for mathematical discussions.[3]

Networks are the appropriate language for empirical investigations. That is, we can define a set of nodes and links and construct the graph corresponding to this network. A natural question is then to ask what structure does this graph have or, more loosely, does the network have. To answer such questions, it is useful to refer to canonical examples of networks. Figure 14.1 shows graphs of four basic types, discussed below.

14.1.1 (Nearly) Regular Networks

These are networks that correspond to *regular graphs*, where all vertices have the same number of edges. Figure 14.1a shows a simple example, a circle graph where every vertex is connected to two edges. A key property is that the average distance between vertices is proportional to the number of nodes n. In particular, in Problem 14.1, you will show that $\langle d \rangle \sim \frac{1}{4}n$. Intuitively, the maximum path is $\frac{1}{2}n$, and the average path should be half that.

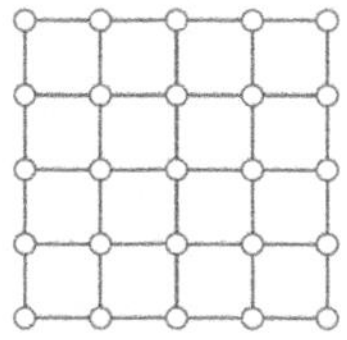

At right is a slightly more complicated example, a finite, two-dimensional *grid* graph. Now there are two classes of vertices, central ones connected to four edges and peripheral ones connected to three or two edges.[4] The average distance $\langle d \rangle \sim n^{1/2}$. In

[3] The terminology given here is common but not universal. For example, some graph-theory texts use *links* to denote edges between distinct vertices (Bondy and Murty, 1976).

[4] We will ignore the distinction between corners and other edge vertices.

(a) circle   **(b) random**   **(c) scale free**   **(d) star**

$$\langle d \rangle \sim \tfrac{1}{4}n \qquad \langle d \rangle \sim \frac{\ln n}{\ln\langle k\rangle} \qquad \langle d \rangle \sim \ln\ln n \qquad \langle d \rangle \sim 2$$

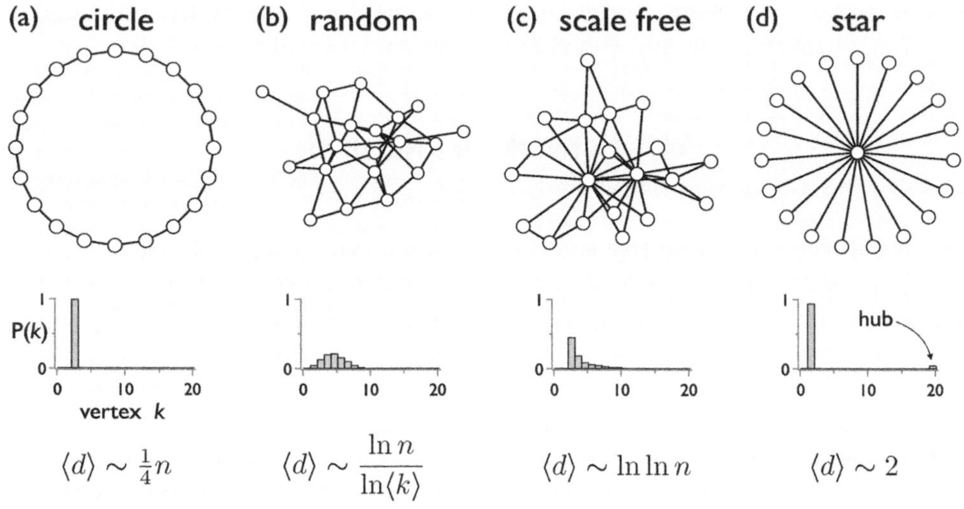

Fig. 14.1   Canonical network models, showing example graphs having $n = 20$ nodes each, the associated distribution of vertex numbers k, and the asymptotic scaling of the average path distance $\langle d\rangle$ for large n. Networks: (a) circle; (b) random (ErdŐs–Rényi); (c) scale-free (Barabási–Albert); (d) star.

the three-dimensional case (and applying to a model of materials), we would refer to volume and surface nodes (or atoms, in the case of a simple solid) and have $\langle d \rangle \sim n^{1/3}$.

Regular graphs are models for periodic lattices in materials. A one-dimensional regular graph with equal weights for links can model a one-dimensional periodic crystal. The "nearly regular" two-dimensional example can model a finite two-dimensional solid, with surface effects. Of course, we usually do not think about crystals as being a "material network," although a network enthusiast might!

14.1.2 Star Networks

If regular networks, which have only local connectivity, represent one extreme of the range of possibilities, *star networks* represent the other. See Figure 14.1d. As illustrated by the degree distribution $P(k)$, there are $n - 1$ *peripheral nodes* that have degree $k = 1$ and 1 *hub node* that has degree $n - 1$. For all network sizes n, the path lengths are either one or two. Conversely, for regular networks, the largest node degree is independent of n; however, for star networks, the largest node degree is proportional to n.

Star networks thus communicate between nodes in a fundamentally different way from a regular network. For regular networks, a perturbation to a node (or a message) spreads locally, and must cross order n nodes (or $n^{1/k}$ nodes in k dimensions). By contrast, in a star network, all nodes can communicate via the hub in two steps.

While many examples in physics correspond, at least implicitly, to cases of regular networks, transportation hubs might be arranged in a star.[5] Such star networks are extraordinarily efficient for communication, but they have an *Achilles' heel*: remove the hub and the entire network disintegrates into $n-1$ isolated nodes. By contrast, removing peripheral nodes has no effect on the path lengths between remaining nodes. Thus, star networks are robust to random attack but fragile to a targeted attack on the hub. The other big problem with a star topology is that a single node must handle *all* the traffic in the network.

14.1.3 Random Networks

Starting in 1959, Erdős and Rényi (ER) and, independently, Gilbert proposed models of random graphs. Gilbert's version is more commonly studied and is described by the notation $G(n,p)$, the set of graphs that have n nodes.[6] Each possible edge is included with a probability p that is independent of the existence of all other edges. Since there are $e_{\max} = n(n-1)/2$ possible edges, we can view a particular graph element of $G(n,p)$ as being drawn from a binomial distribution $B(e_{\max}, p)$ of $e_{\max}$ trials of success rate p. The expected number of edges k is thus in terms of mean ± standard deviation,

$$\langle e\rangle = e_{\max} p = \tfrac{1}{2}n(n-1)p \pm \sqrt{p(1-p)e_{\max}}\,. \tag{14.2}$$

And, since every edge links two vertices, the average node degree is given by

$$\langle k\rangle = \frac{2\langle e\rangle}{n} = \frac{2\frac{1}{2}n(n-1)p}{n} = (n-1)p \approx np\,, \tag{14.3}$$

with the last relation an approximation for $n \gg 1$.

Many examples of networks are large and *sparse*, with $n \gg 1$, $p \ll 1$, and $\langle k\rangle$ finite. In this limit, the binomial distribution is well approximated by the Poisson distribution,

$$P(k) = \frac{\langle k\rangle^k\, \mathrm{e}^{-\langle k\rangle}}{k!}\,. \tag{14.4}$$

As a result, node connectivity is typically in the relatively narrow range $\langle k\rangle \pm \sqrt{\langle k\rangle}$ (see Figure 14.1b), and the average degree is independent of the network size n.

In the sparse limit, the only adjustable parameter is $\langle k\rangle$. As it increases, the set of random graphs undergoes a series of abrupt *phase transitions* in behavior. For example, for

$$\langle k\rangle > \langle k\rangle_c \equiv \ln n\,, \tag{14.5}$$

or, equivalently, $p > p_c = \ln n/(n-1)$, the graph is always connected, in the $n \to \infty$ limit. For finite values of n, there is a sudden rise in the probability that nodes are all indirectly connected, a rise that becomes steeper for increasing n. See right and Problem 14.2.

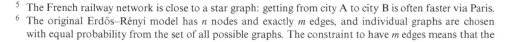

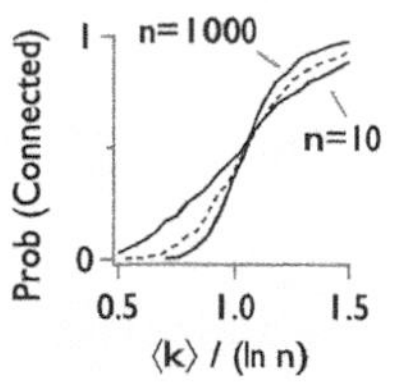

[5] The French railway network is close to a star graph: getting from city A to city B is often faster via Paris.
[6] The original Erdős–Rényi model has n nodes and exactly m edges, and individual graphs are chosen with equal probability from the set of all possible graphs. The constraint to have m edges means that the

This transition and others are closely related to those studied in *percolation theory*. Traditionally, percolation was studied in the context of nodes on a regular grid with links appearing with a given probability. In random networks, the graphs have a more complicated set of possible connections.

Finally, the average distance between nodes in a random network is (see Problem 14.2)

$$\langle d \rangle \approx \frac{\ln n}{\ln \langle k \rangle} \,. \tag{14.6}$$

The log-n scaling is referred to as the *small world phenomenon*, also known as *six degrees of separation*.[7] The dependence on n is qualitatively weaker than that of a regular network, where $\langle d \rangle \sim n^{1/D}$, where D is the dimension of the space (typically, 1, 2, or 3).

14.1.4 Scale-Free Networks

Scale-free networks (SF) have a degree distribution that is a power law $P(k) \sim k^{-\gamma}$, with *degree exponent* $2 < \gamma < 3$. Why "scale free"? If we change the scale of k, for example $k \to 10k'$ (measure edges in units of ten), then $P(k')$ has the same form as $P(k)$. Why the restricted range of γ? For $\gamma > 3$, the second moment (and variance) are finite. If so, the statistical properties of the network turn out to be very similar to those of the random (Erdős–Rényi) networks discussed in the previous section.[8] For $\gamma < 2$, the mean, $\langle k \rangle$, diverges. As Problem 14.3 shows, the only way to have such a mean is to allow multiple links between nodes. If we exclude such cases, there are no networks with $\gamma < 2$.

The range $2 < \gamma < 3$ is possible, yet quite different from the random-network case, whose degree distribution has a variance that is independent of system size. For scale-free networks, the variance diverges for $\gamma < 3$, and the node with largest degree has $k_{\max} = k_{\min} n^{1/(\gamma-1)}$. That is, $k_{\max}$ grows as a fractional power of n. As a result, nodes with large numbers of connections – hubs – are created, which is a qualitative distinction from random graphs, where the range of connection degree is sharply limited. For uncorrelated networks, the average path length $\langle d \rangle \sim \ln \ln n$, which is a distinctly weaker n dependence than the $\ln n$ result for random graphs. Since the latter is

presence of a given edge correlates with the absence of links in other edges. The edges are not independent. By the law of large numbers, the two models have similar properties for large n.

[7] The term originates from the play of the same name by John Guare and refers to a famous experiment by Stanley Milgram, who, in 1967, asked people in Wichita and Omaha to send a message to randomly chosen "target" recipients in Boston and Sharon, Massachusetts. The volunteers were asked to send a letter to someone who might know or know of someone, etc., who might know the target. It took, on average, six relays for the letters to reach their target via this *acquaintance network*. What do we expect? The population of the earth in 1967 was $n \approx 3.5 \times 10^9$. Inverting Eq. (14.6), $\langle k \rangle \approx n^{1/d} \approx 40$ acquaintances, on average, for each person. A recent study of Facebook friends determined that its $\langle d \rangle \approx 3.9$ (Barabási, 2016), implying a larger number of acquaintances (on Facebook).

[8] The *central limit theorem* has a similar qualification: Sums of large numbers of random variables tend to a Gaussian distribution, but only if the variances of the distributions involved are all finite.

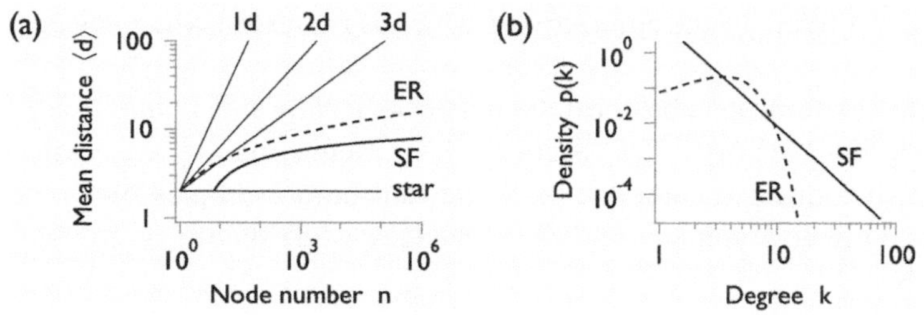

Fig. 14.2

Comparison of graph models. (a) Mean distance $\langle d \rangle$ for regular, random, scale-free, and star graphs. Asymptotic form is plotted. (b) Degree distributions for Erdős–Rényi (ER) and scale-free (SF) networks. The mean degree $\langle k \rangle = 4$ for both cases and corresponds to $k_{\text{min}} = 2$ for the SF case.

known as the *small-world* property, the $\ln \ln n$ scaling is known as the *ultrasmall-world* property.[9]

The wide range of node degrees implies that, on a randomly structured network, randomly removing nodes will have negligible effect until nearly all nodes are gone. But a *targeted attack* that systematically removes the largest remaining node can quickly destroy a network, reducing it to isolated nodes. In Section 14.6.2, we will revisit this conclusion, showing that scale-free networks with nonrandom, carefully chosen links can be robust to targeted attacks.

Figure 14.2 compares some key features of these classes of graphs. Part (a) shows the algebraic scaling of average path length $\langle d \rangle$ with node number n for regular graphs in one, two, and three dimensions – as n^1, $n^{1/2}$, and $n^{1/3}$, respectively. The random (ER) graph scales as $\ln n$, the scale-free (SF) as $\ln \ln n$. And finally, the maximum distance between nodes in a star graph distance is always two, independent of node numbers. Note the qualitative distinction between the regular graphs, where distances grow substantially with network size, and the other cases, which have small-world connectivity.

In Figure 14.2b, we compare the degree distributions between the ER and SF cases. Both are chosen to have $\langle k \rangle = 10$. We can see how the Poisson distribution of the ER case has a well-defined range, whereas the scale-free case generates increasingly large hubs as the network grows. (Remember that the larger the network is, the smaller the probability density that can be probed.) The regular graphs have degrees distributions concentrated on 2, 4, and 6 for 1, 2, and 3 dimensions, respectively, and are not shown.

14.1.5 Other Networks

There are many other types of network models, which we will mostly pass over. Here are two more examples: In *complete graphs*, every node is connected to every other node (see right). The mean distance between nodes $\langle d \rangle = 1$. In effect, every node is a hub. The network is the most robust possible, at the cost of needing $O(n^2)$ nodes.

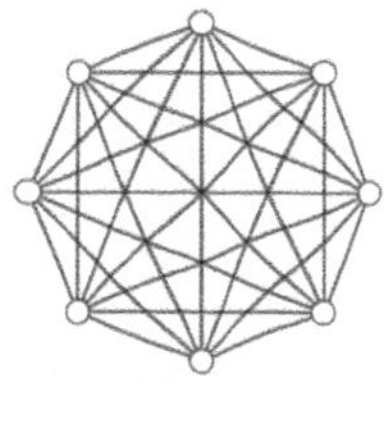

[9] Unlike the other cases, the ultrasmall-world scaling is tricky to derive. The case $\gamma = 3$ is special, with $\langle d \rangle \sim \ln n / (\ln \ln n)$. For derivations of all these results, see Cohen and Havlin (2003).

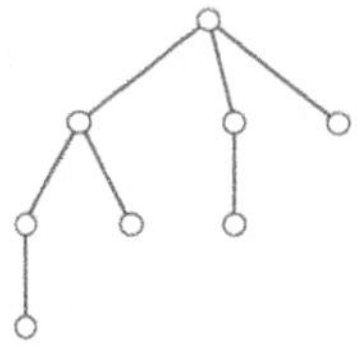

Here, robustness refers to the connectivity, which is preserved when any edge or node is removed.

At the other extreme in robustness are networks having *tree graphs* (see left), which are connected but lack *cycles* (cf. Section 14.3). Removing any edge will divide the graph into two connected components. Note that both examples shown here have 8 nodes. The complete graph has $8 \times 7/2 = 28$ edges, while the tree graph has $8 - 1 = 7$ edges.

14.2 From Dynamics to Graphs to Networks

Having familiarized ourselves with some typical classes of graphs, we turn to the use of graph theory and network science as tools for analyzing dynamical systems. Figure 14.3 illustrates a natural way to map a linear dynamical system to a graph. We identify the n elements $x_i(t)$ of the state vector $x(t)$ with nodes of a graph. A nonzero element A_{ij} of the dynamical matrix A then implies a directional link from node j to node i.[10] The zero elements correspond to node pairs with no link between them. In the language of graph theory, A is a *weighted adjacency matrix*.[11] Since the dynamical matrix of an LTI is usually asymmetric, a $j \to i$ link does not imply the existence of an $i \to j$ link. The graph representing the dynamics is thus directional (*digraph*). Because A_{ij} is a real number, so is the weight of the corresponding edge of the graph. Figure 14.3 applies these ideas to the case of two blocks connected by a spring.

(a)

$$\ddot{x}^{(1)} + x^{(1)} - x^{(2)} = u(t)$$
$$\ddot{x}^{(2)} + x^{(2)} - x^{(1)} = 0$$

(b)

$$\underbrace{\begin{pmatrix} 0 & 1 & 0 & 0 \\ -1 & 0 & 1 & 0 \\ 0 & 0 & 0 & 1 \\ 1 & 0 & -1 & 0 \end{pmatrix}}_{A} \underbrace{\begin{pmatrix} 0 \\ 1 \\ 0 \\ 0 \end{pmatrix}}_{B}$$

$$\dot{x} = Ax + Bu$$

(c)

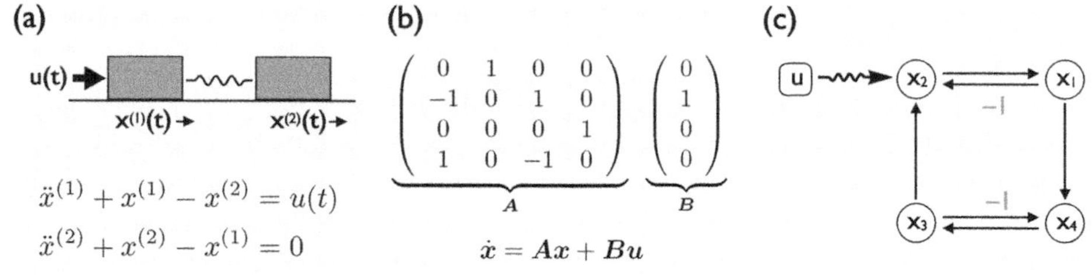

Fig. 14.3   Mapping an LTI system to a weighted digraph. (a) Two blocks connected by a spring and forced at one end. Dynamics described by two coupled second-order equations. (b) State-space representation with dynamical matrix A and input coupling B. (c) Graphical representation of the four internal states, x_i, and one input, $u(t)$. Unlabeled edges have weight $+1$.

[10] The convention in the graph-theory literature is for $A_{ij} = 1$ to represent a link from i to j, rather than the reverse convention adopted here and in most of the physics literature.

[11] The standard adjacency matrix sets all the nonzero elements of A equal to 1.

14.3 Structural Controllability

In the introduction to this chapter, we argued that the traditional definition of controllability based on the Kalman rank condition is not useful for large systems. The rank condition is based on the matrices $\boldsymbol{A}$ and $\boldsymbol{B}$ of a dynamical system in its state-space representation, and it is unlikely that all $n^2 + nm$ values will be known to high precision in a large system.

An approach introduced by Lin in 1974 is to replace controllability with the notion of *structural controllability*. We partition the links into two sets: those with precisely zero weight and those with nonzero weight. Two linear systems, $\{\boldsymbol{A}, \boldsymbol{B}\}$ and $\{\boldsymbol{A}', \boldsymbol{B}'\}$ then have the same *structure* if they have the same set of zero-weight elements. The weights that are nonzero can have arbitrary values, which need not be known. Since the nonzero weights are also what we associate with links in the graph derived from a dynamical system, we can say that two dynamical systems have the same structure if each has the same graph.

The system $\{\boldsymbol{A}, \boldsymbol{B}\}$ is then structurally controllable if there exists a controllable system $\{\boldsymbol{A}', \boldsymbol{B}'\}$ having the same structure as $\{\boldsymbol{A}, \boldsymbol{B}\}$. One can then show that systems having the same structure will be controllable for almost all possible values of the nonzero elements and that controllability will fail for at most a set of measure zero of matrix-element values.

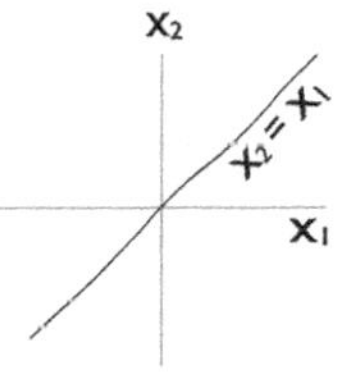

Figure 14.4 illustrates this definition of structural controllability. In (a), the graph corresponds to the equations $\dot{x}_1 = u(t)$ and $\dot{x}_2 = u(t)$. Since the difference $\dot{x}_1 - \dot{x}_2 = 0$, integration gives $x_2(t) = x_1(t) +$ const. Thus, the two solutions cannot be varied independently, being confined to a line in the x_1–x_2 plane (see right). In (b), we add self-interactions, so that

$$\dot{x}_1 = -\lambda_1 x_1 + u(t), \qquad \dot{x}_2 = -\lambda_2 x_2 + u(t). \tag{14.7}$$

We already discussed this system in Example 4.4 and Problem 4.1. With $\boldsymbol{A} = \begin{pmatrix} -\lambda_1 & 0 \\ 0 & -\lambda_2 \end{pmatrix}$ and $\boldsymbol{B} = \begin{pmatrix} 1 \\ 1 \end{pmatrix}$, the controllability matrix $\boldsymbol{W}_c = \begin{pmatrix} 1 & -\lambda_1 \\ 1 & -\lambda_2 \end{pmatrix}$, which has determinant $\lambda_1 - \lambda_2$. Generically, for all $\lambda_1 \neq \lambda_2$, the system is controllable, meaning that it is structurally controllable. Unless some symmetry imposes $\lambda_1 = \lambda_2$, the two will always be at least a little different. Notice that $\lambda_2 = 0$ and $\lambda_1 \neq 0$ implies det $\boldsymbol{W}_c = \lambda_1$, meaning that adding a single self-interaction node in Figure 14.4b already ensures controllability.

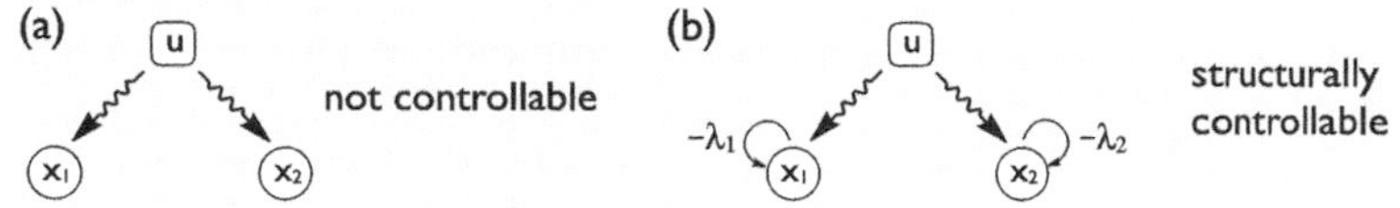

Fig. 14.4 Controllability examples. (a) Two nodes driven by one input are not controllable. (b) Adding self-interactions makes the system structurally controllable.

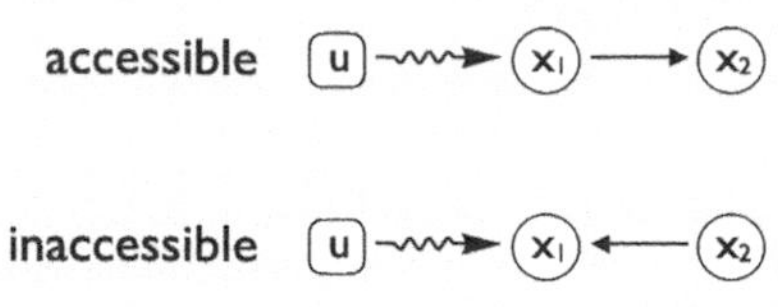

$A=\begin{pmatrix}0&0\\1&0\end{pmatrix},\quad B=\begin{pmatrix}1\\0\end{pmatrix}\quad\Longrightarrow\quad W_c=\begin{pmatrix}1&0\\0&1\end{pmatrix}$

$A=\begin{pmatrix}0&1\\0&0\end{pmatrix},\quad B=\begin{pmatrix}1\\0\end{pmatrix}\quad\Longrightarrow\quad W_c=\begin{pmatrix}1&0\\0&0\end{pmatrix}$

Fig. 14.5 Top: Node x_2 is accessible, and the system is controllable ($\det \mathrm{W_c} = 1$). Bottom: Node x_2 is inaccessible, and the system is not controllable ($\det \mathrm{W_c} = 0$).

Lin (1974) showed that structurally controllability fails when either *inaccessible nodes* or *dilations* are present. Figure 14.5 illustrates the first possibility. For a given node to be *accessible*, there must be a *directed path* to the node that starts from an input node (a node directly connected to an input) and proceeds by "following the arrows" along directed edges to the given node. The top line in Figure shows an example where a directed path exists from the input node x_1 to the given node x_2 and shows that it is controllable. In Problem 14.6, you will show that a directed chain of n nodes driven at its head remains controllable. Conversely, an inaccessible node is not reachable by any directed path from any input node and leads to a system that is not controllable (bottom line in Figure 14.5).

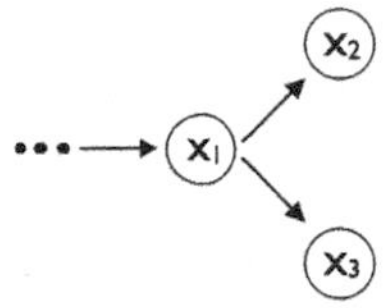

Dilations are the other graph property that leads to an uncontrollable system. Intuitively, a dilation is a branching where one node drives two or more others (see left). As in Figure 14.4a, one node (x_1) drives two (x_2, x_3) and cannot control their difference. Whether x_1 is driven by an input or *is* the input is immaterial. See Problem 14.7.

More formally, for any subset S of vertices V of a graph, there is a dilation if the *neighborhood* $T(S)$ has fewer elements than S does. Here, the neighborhood consists of the vertices that have an edge directed to an element of S. At left is a dilation: the set $S = \{x_2, x_3\}$ has two elements, but the neighborhood of S, which consists of x_1, has only one.

path

Lin's theorem is that an LTI system $\{\boldsymbol{A}, \boldsymbol{B}\}$ is structurally controllable if and only if its associated digraph $G(\boldsymbol{A}, \boldsymbol{B})$ has no inaccessible nodes or dilations. The important point is that these conditions depend only on the topology of the graph, not its weights. The detailed numerical evaluations required to evaluate the Kalman rank condition are unnecessary.

The version of Lin's theorem given above states what graphs of controllable networks *cannot* have, namely inaccessible nodes or dilations. A more positive version of the theorem states what the graph of a controllable system *can* have. We start with some definitions. All are for a directed graph and are illustrated at left and in Figure 14.6.

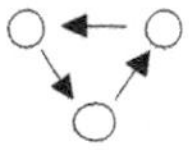

cycle

- *Elementary path*: a sequence of distinct, connected vertices.
- *Stem*: a directed, elementary path whose first vertex is connected to an input u.

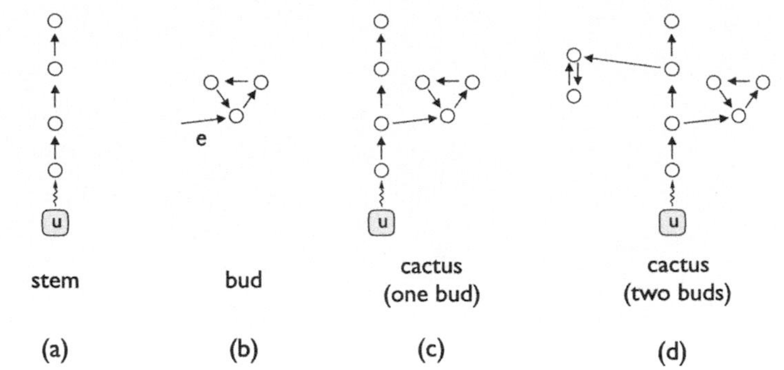

Fig. 14.6 Construction of a cactus. (a) Stem (directed path). (b) Bud – a cycle plus an extra inward edge e. (c) Cactus with one bud. (d) Cactus with two buds.

- *Elementary cycle* is an elementary path of length ≥ 2, whose last vertex links to the first.
- *Bud*: an elementary cycle that has an extra edge e that is directed to the cycle but whose head is not connected to a node.
- *Cactus*: union of stem and buds. The connection of the bud must not create a dilation. A stem is defined to be a cactus with zero buds.

Figure 14.6 shows a stem (a), a bud (b), and their union (c,d). In this figure, the graphs in (a), (c), and (d) are all cacti. The important point is that a cactus has no inaccessible nodes and no dilations. (See Problem 14.7.) A cactus thus corresponds to a controllable system, and the union of cacti is also controllable, provided that there is an independent input for each. Moreover, a cactus is a minimal controllable structure, in that removing any edge from a cactus will create either an inaccessible node or a dilation. (See Problem 14.8.)

14.4 Minimum Inputs Problem

Armed with the notions of structural controllability and its associated graph-theoretic jargon of cacti and the like, we can solve the problem of finding the minimum number of inputs needed to control a network. More precisely, we seek the *minimum driver node set* (MDNS) for a network. First, a couple of definitions: an *actuator node* is one that is directly connected to an input. The set of *driver nodes* is the subset of actuator nodes that share no input. This is the number of actual inputs needed, given by the number of columns in the input-coupling matrix $\boldsymbol{B}$.

Figure 14.7 illustrates an algorithm for determining which particular nodes should be connected to inputs. Part (a) shows a network with two disjoint pieces. Part (b) shows a *maximum matching* for the digraph. Here, a matching is a set of directed

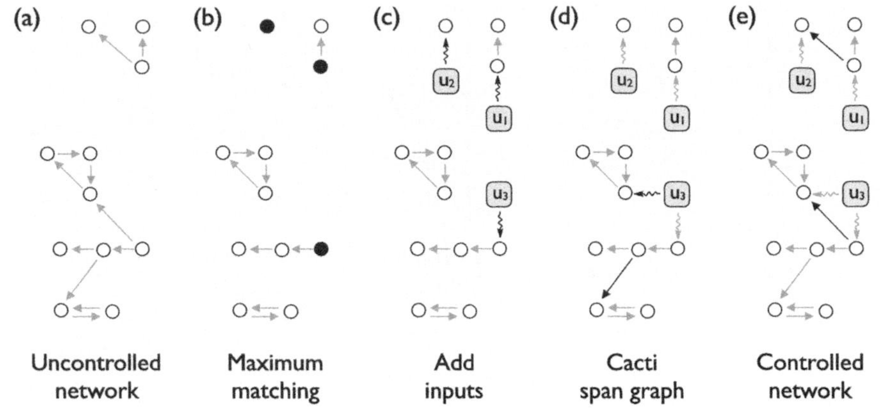

Fig. 14.7 Graphical algorithm to determine minimum inputs. (a) Original, uncontrolled network with $n = 11$ nodes. (b) Maximum matching decomposition into directed paths and cycles with $M = 8$ edges. Unmatched black nodes are the first vertex in a directed path. (c) Attach $11 - 8 = 3$ inputs (shaded) to these first vertices. (d) Add links to form cacti that span the original network. (e) Add remaining links from original network. Adapted from Liu and Barabási (2016).

edges that do not share common start or end vertices, and a maximum matching is the largest possible set fulfilling this condition. Directed paths are part of a matching, as are cycles, as both these objects are constructed from directed edges with different start and end vertices. Indeed, a maximum matching is, loosely, a decomposition of a network graph into paths and cycles. In Figure 14.7b, we can easily verify this decomposition visually. There are eight elements (edges). In general, the maximum matching is not unique.

In Figure 14.7b, the first nodes of each path, including the zero-length path corresponding to node x_2, are shaded black. In (c), we connect an independent input (u_1, u_2, and u_3) to each head of the path, forming three stems. In (d), we add links (in black) to create three separate cacti that span the graph (include all nodes). By computing the neighborhood numbers for each node, we verify that neither step (c) nor (d) creates a dilation. We have also found a minimal controllable structure: removing any link destroys the controllability of the system. In the final step, (e), we add in the remaining links from the original network (shown again in black). These links are redundant, since the network was already controllable in (d). Note the need for the horizontal input connection from u_3 in step (d); otherwise, we would create a dilation, destroying controllability.

The network in Figure 14.7a can thus be controlled using only three driver nodes (but four actuator nodes). The $n = 11$ nodes and $M = 8$ elements of this maximum matching imply $11 - 8 = 3$ driver nodes (independent inputs). Although this case seems simple, we would need to check $O(10^3)$ cases using the traditional Kalman rank condition in a "blind" search on this 11-node network. More generally, for n nodes and a maximum matching of M elements, the minimum driver node set (MDNS) has n_D elements, with

$$n_D = \max(n - M, 1) . \tag{14.8}$$

The maximum condition arises because there must be at least one driver node.

How does this solution to finding the minimum number of inputs scale to large networks? The hardest step is finding a maximum matching. Fortunately, there are well-developed graph-theoretic methods for this task. In particular, the Hopcroft–Karp algorithm requires $O(n^{5/2})$ computations, with n the number of nodes. This favorable polynomial-time scaling allows easy computations for large graphs.

14.4.1 Controllability of Large Networks

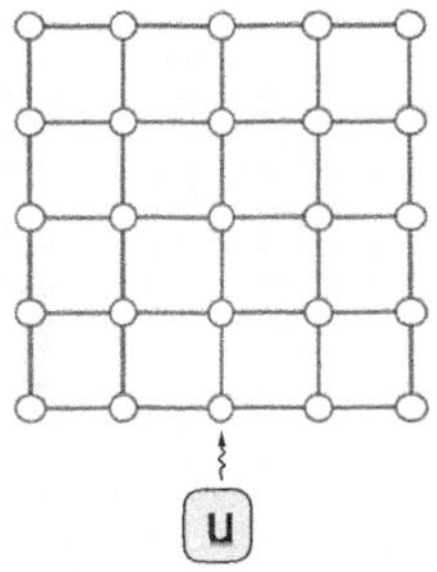

The above algorithm for finding the MDNS has been applied to both graph models and empirically derived networks. For example, a regular, undirected network in any dimension (such as the grid graph at right) is structurally controllable from a single node. Or, a star network with connections from the hub to the outer spokes would require $n - 1$ inputs, as controlling the hub can control only one outer node. Indeed, hubs – nodes with unusually high degree number k – are not as important to the overall controllability of a network as their dominant node degree might imply. Nodes with many edges can have many dilations.

For models of networks such as the ER and stochastic scale-free models, both numerical methods and analytical techniques from statistical physics (cavity method) can predict typical properties of large networks. If there are no self-interactions among nodes, then one must control independently a finite fraction of nodes. This implies, for large networks, that a large number of inputs is needed in order for the system to be controllable.

These conclusions have nuances. In accord with the limits of regular graphs (need only one input) and star graphs (need $n - 1$ inputs), dense homogeneous networks need fewer inputs than sparse inhomogeneous networks. In particular, for both ER and SF, the driver node density n_D/n (fraction of nodes needed to control the network) decreases exponentially for large values of the average degree $\langle k \rangle$. Also, any large network with uncorrelated degree distributions can be controlled by a single node if all nodes have in- and out-degrees greater than two. Thus, a relatively small connectivity for a highly connected network suffices for easy controllability. Nonetheless: large, sparse, inhomogeneous networks with generic edge weights need large numbers of independent control inputs.

Self-Interactions

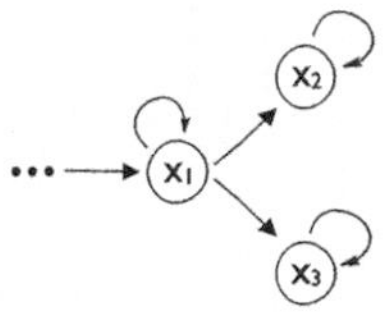

In the above analysis, self-interactions – nodes that interact with themselves – are assumed absent. What if that is not true? Imagine that every node in a network has a self-interaction (see right). In the spirit of structural controllability, the weight of each self-interaction, which corresponds to the diagonal elements of the interaction matrix, is different. If the interaction has first-order dynamics, then the network dynamics can be written as

$$\dot{x}_i = -\lambda_i x_i + \sum_{j=1}^{n} A^0_{ij} x_j + B_i u \,, \tag{14.9}$$

with the matrix A^0 capturing the interactions between nodes. (Alternatively, the vector of self-interactions can be absorbed into the matrix A.) Every node then belongs to its own neighborhood, ruling out dilations. If there are no inaccessible nodes, the network is automatically structurally controllable.

The seeming sensitivity of structural controllability to assumptions about just which edges have precisely zero weight is worrisome. A more fundamental critique, though, is to ask whether structural controllability – even if we did know how to calculate it – is the interesting quantity to consider. As discussed in Section 4.1.1, full-state controllability is a technical notion that often does not correspond to what one really wants. In particular, it neither implies the ability to stabilize about a given state nor to follow a given trajectory.

Perhaps most importantly, the notion of controllability does not address the control effort required. A system may pass the technical tests for controllability (or structural controllability), yet require a larger control magnitude $|u|$ (or control effort u^2) than can be delivered.

14.5 Control Effort

For physics-based control problems, the control effort $\mathcal{E}(\tau)$ is often proportional to the energy required to carry out the control. In nonphysical networks (biological, economic, social, etc.), $\mathcal{E}$ may not correspond to an energy but can still quantify the amount of control required. Linear control problems assume that the range of possible control values $\boldsymbol{u}(t)$ is unbounded, but all real control systems have limits on $\boldsymbol{u}(t)$.

From Problem 7.9, the input $\boldsymbol{u}(t)$ that minimizes control effort for trajectories in an n-dimensional state space that start at $\boldsymbol{x}(0) = \boldsymbol{x}_0$ and end at $\boldsymbol{x}(\tau) = \boldsymbol{x}_\tau$ is given, for linear, time-invariant dynamics, by $\boldsymbol{u}(t) = \boldsymbol{B}^{\mathsf{T}}\, \mathrm{e}^{\boldsymbol{A}^{\mathsf{T}}(\tau-t)}\, \boldsymbol{P}^{-1}(\tau)\, \Delta\boldsymbol{x}$ and corresponds to the effort

$$\mathcal{E}(\tau, \boldsymbol{x}_0, \boldsymbol{x}_\tau) = \int_0^T \mathrm{d}t\, \boldsymbol{u}^{\mathsf{T}}(t)\, \boldsymbol{u}(t) = \Delta\boldsymbol{x}^{\mathsf{T}} \boldsymbol{P}^{-1}(\tau)\, \Delta\boldsymbol{x}\,, \quad \boldsymbol{P}(\tau) = \int_0^T \mathrm{d}t\, \mathrm{e}^{\boldsymbol{A}t}\, \boldsymbol{B}\,\boldsymbol{B}^{\mathsf{T}}\, \mathrm{e}^{\boldsymbol{A}^{\mathsf{T}}t}\,, \tag{14.10}$$

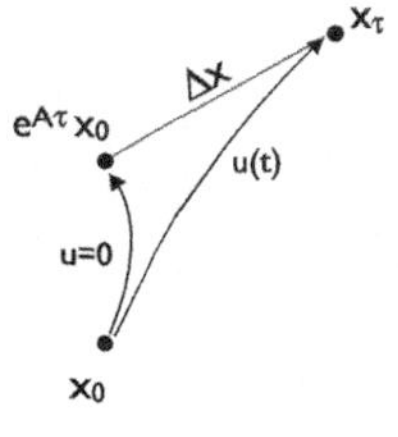

with $\boldsymbol{P}$ the control Gramian matrix and $\Delta\boldsymbol{x} = \boldsymbol{x}_\tau - \mathrm{e}^{\boldsymbol{A}\tau}\boldsymbol{x}_0$ the difference between the desired endpoint with control and the "natural" endpoint without (see left).

The control effort thus depends on the protocol time τ, the dynamics $\boldsymbol{A}$, and the initial and final positions in state space. Key points include how fast the protocol is relative to system time scales and whether the desired motion in state space is "along" or "against" the flow of the dynamical vector field (determined by $\boldsymbol{A}$).

To shape intuitions, consider the simple first-order system (Problem 14.10)

$$\dot{x} = \lambda x + u\,, \qquad x(0) = x_0\,,\ x(\tau) = x_\tau\,. \tag{14.11}$$

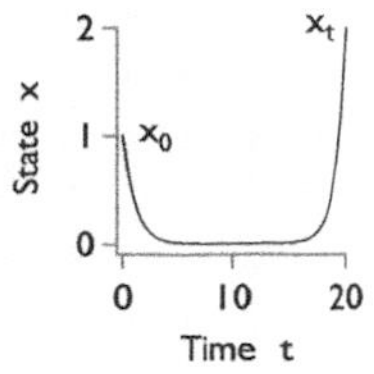

For fast protocols with $\tau \ll |\lambda|^{-1}$, the minimum control effort is $\mathcal{E} = (x_\tau - x_0)^2/\tau$. The τ^{-1} divergence was already noted in Problem 7.9 and depends quadratically on the distance moved. For slow protocols, with $\tau \gg |\lambda|^{-1}$, the effort is $\mathcal{E} = 2|\lambda| x_\tau^2$ for $\lambda < 0$ and $2\lambda x_0^2$ for $\lambda > 0$. For $\lambda < 0$, the optimal strategy is to "fall in" from x_0 toward the origin for free, wait, and then "pay" to go out to the destination x_τ a short time before the end of the protocol. See right (where $|\lambda| = 1$). For $\lambda > 0$, we pay to visit the origin but then receive a "free" ride to the destination x_τ. The trajectory turns out to be invariant under $\lambda \to -\lambda$. Thus, a large $|\lambda|$, whether positive or negative, leads to large required control efforts. As a corollary, systems that are neutrally stable, such as an undamped harmonic oscillator, require relatively low control effort because they are easily pushed around.

Having established some intuition for a simple system, we turn to larger systems. The directions in state space of maximum and minimum control effort are the eigenvectors corresponding to maximum and minimum eigenvalues of $\boldsymbol{P}^{-1}(\tau)$. Alternatively (and more simply), they are the directions corresponding to minimum and maximum eigenvalues[12] of $\boldsymbol{P}(\tau)$. Thus, the *condition number* of $\boldsymbol{P}$ describes the range of difficulty of the control problem. Gramians with large condition numbers will have directions that are much harder to control than those corresponding to the most favorable cases.

We next explore two mechanisms that lead to divergent values of $\mathcal{E}$. The first is near degeneracy in system parameters. The second is a consequence of network structure and is relevant for larger systems.

14.5.1 Near-Degeneracy and System Parameters

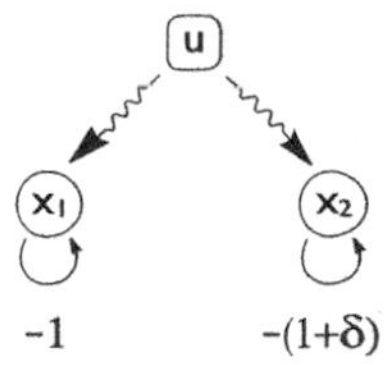

Loss of rank in the controllability and Gramian matrices due to symmetries that constrain the relative or absolute values of parameters and lead to uncontrollable dynamics can occur in many ways, making general statements problematic.[13] Nonetheless, it is intuitive that near-degeneracy will lead to large control efforts. Let us consider a simple example, two first-order systems with identical inputs and nearly identical rate constants (see Example 4.4, Figure 14.4, and Eq. 14.7). Written in a slightly different notation, the system is $\dot{x}_1 = -x_1 + u$ and $\dot{x}_2 = -(1+\delta)x_2 + u$ and is sketched at right. The system is not controllable for $\delta = 0$, when the equations at the two nodes become identical, but is so for any nonzero δ. Indeed, the controllability matrix $\boldsymbol{W}_c$ has determinant $-\delta$ and thus has rank 2 for $\delta \neq 0$ and rank 1 for $\delta = 0$. When $\delta \to 0$, the control effort $\mathcal{E} \sim \delta^{-2}$ (Problem 14.11). That is, when the system is *nearly* uncontrollable, the required control effort to move between states is large. This divergence, coupled with the finite range of u, implies a minimum "distance" from degeneracy (quantified by δ)

[12] The statement is equivalent to the *min-max* theorem of linear algebra (and related to *Rayleigh–Ritz* theory). To see why, expand an arbitrary target direction $\hat{\boldsymbol{n}} = \sum_i a_i \boldsymbol{e}_i$, where $\boldsymbol{e}_i$ are the eigenvectors of $\boldsymbol{P}$.

[13] I mean "problematic" in its literal sense of "posing a problem": I have not found a good discussion.

where the system becomes effectively uncontrollable, even though $\boldsymbol{W}_c$ still passes the Kalman rank test.

In our discussion of structural controllability, we leaned heavily on the idea of generic controllability, that only special combinations of parameters can create a degeneracy in the controllability matrix (or Gramian $\boldsymbol{P}$), causing loss of full rank. Because these parameter combinations have zero measure in the parameter space, a real system will never have exactly the same values and therefore is generically controllable. Only the structure of links with precisely zero weight can create noncontrollability.

But if parameters that lead to a nearly degenerate controllability or Gramian matrix imply systems that are effectively uncontrollable, the above reasoning fails. Stepping back, we can distinguish three types of parameters that appear in dynamical equations:

- *Accurate*: The values (or relations between values) are enforced by symmetries and their related conservation laws. For example, in the two-block, one-spring system of Figure 14.3a, the force terms in the two equations have identical magnitudes, $|x^{(2)} - x^{(1)}|$, because there is but one spring. (The opposite sign results from Newton's third law, a consequence of momentum conservation.) Other examples include the implications of Kirchhoff's voltage and current laws in the dynamical equations describing circuits.
- *Effectively accurate*: These parameters have values that are close enough to induce an effective loss of controllability. The example of two nominally identical first-order systems discussed above falls into this category.
- *Inaccurate*: These parameters are generically sufficiently different and, through the arguments underlying structural controllability, lead to generically controllable systems. For example, in a network of chemical reactions, each reaction has generically different rate constants and the rate constants are typically not accurately known. The ratios among parameters can span orders of magnitude.

The original theory of structural controllability distinguished only between inaccurate and one class of accurate parameters (links with precisely zero weight). This leads to the applications of graph theory discussed previously. But there is a need to incorporate more systematically and for large systems the effect of other types of accurate parameters and effectively inaccurate ones, as well. The problem remains open.

14.5.2 Large Networks and Ill-Conditioned Gramians

The second scenario that leads to divergent control effort originates in the structure of the network itself. Here, we show how large networks can have Gramian matrices whose *condition number* (ratio between largest and smallest singular value) diverges with network size. We start with a simple, two-dimensional example, which shows how easy it is to have large variation in the effort required to move in different directions through a state space. As with the one-dimensional problem, much more effort is required to go against the flow of the dynamics than to go with it.

Low-Dimensional Example

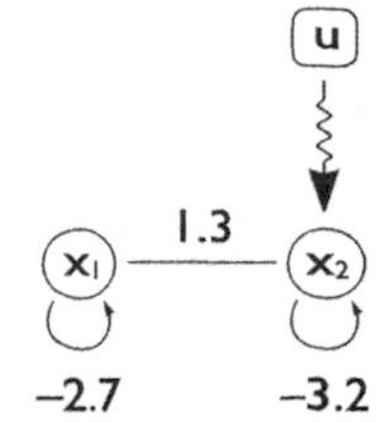

Consider the linear dynamical system at right, with state space matrices $\boldsymbol{A} = \left(\begin{smallmatrix} -3.2 & 1.3 \\ 1.3 & -2.7 \end{smallmatrix}\right)$, $\boldsymbol{B} = \left(\begin{smallmatrix} 0 \\ 1 \end{smallmatrix}\right)$. The eigenvalues of the dynamical matrix $\boldsymbol{A}$ are $\approx \{-4.3, -1.6\}$, implying that the origin is stable and that the two time scales are comparable (ratio ≈ 2.6). But the eigenvalues of the Gramian matrix (computed for a trajectory of length $\tau = 3$) are $\approx \{0.22, 0.0079\}$ and have a ratio ≈ 28, which is ten times larger. The light gray arrows in the plot at right depict the vector field of $\boldsymbol{A}$ that describes the system dynamics near the origin. The dashed and solid lines show trajectories corresponding to the minimum- and maximum-effort solutions. Both trajectories start at the origin and both end on the unit circle after a time $\tau = 3$. But the solid-line trajectory requires nearly 30 times more effort than the dashed-line one. In Problem 14.12, you will explore these issues and see how adding just one more dimension increases the condition number to nearly 1000.

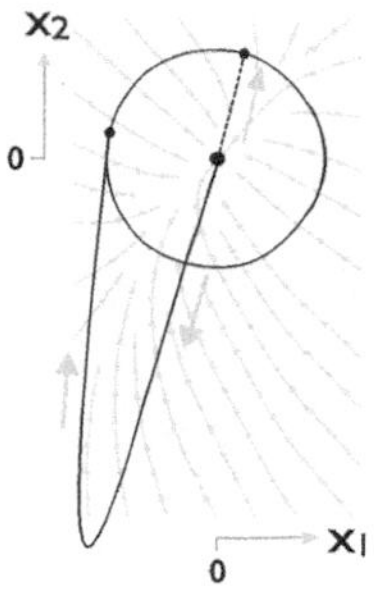

The trajectory illustrated in the above example is very nonlocal, traveling far outside the unit circle where the target state is located. This nonlocality is typical of minimum effort trajectories (cf. Problem 4.3). While you may prefer (and be able to find) a trajectory that stays local, you will pay more in control effort. Minimum-effort trajectories are useful for general discussions of control problems because they bound what is possible, even if they have other undesirable features.

Linear Chain

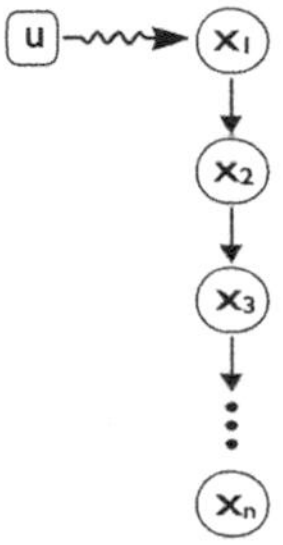

The above example shows that even the simplest, low-dimensional linear dynamical system can have control trajectories that require extremely large control effort and increase rapidly with the dimension n of state space. In fact, the increase is typically exponential. We can see how the exponential form arises in a simple example. Consider an n-element *chain* of one-dimensional dynamical systems with a single input at its head. In Problem 14.6, we will see that $\boldsymbol{W}_c$ is always the identity matrix, while Problem 14.13 shows that $\boldsymbol{P}$ (and thus $\boldsymbol{P}^{-1}$) have a condition number (ratio of largest to smallest eigenvalue) that diverges exponentially with n. Intuitively, the influence of $u(t)$ diminishes by the same factor at each successive link in the chain. Note that the benign behavior of $\boldsymbol{W}_c$ contrasts with its near-degeneracy in the example from Section 14.5.1.

14.5.3 Can We Control a Large System?

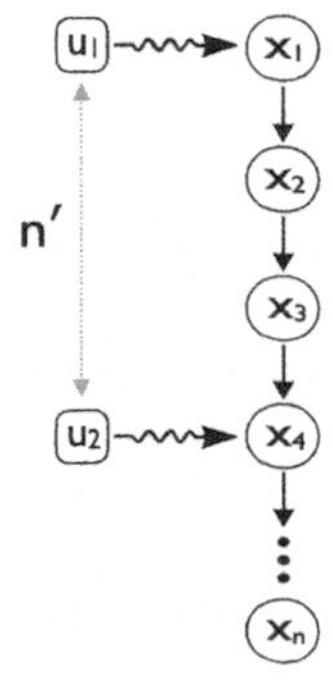

Large networks will thus generically require more control effort than can be delivered. They are impossible to control, at least in the technical sense of full state controllability. What to do? There are two basic strategies: use more inputs, or control fewer states.

More Inputs

In the linear chain, the key point was the number of "downstream" nodes. If we add an input every n' nodes, then the control effort will be reduced to approximately that of a chain that is n' elements long. (See right.) The second input, u_2, is not needed for controllability, but it reduces the condition number of the Gramian. Note that choosing (or *designing*) regular-spaced inputs nodes is much more effective than placing them at random. On a linear chain, this could easily be quantified, as random placement implies a Poisson distribution of control nodes (exponential distribution for the separations). There can then be relatively large gaps, which would dominate the control-effort requirements.

These ideas also apply to more complicated networks. In a study of scale-free networks, Yan et al. (2015) showed that $\mathcal{E}_{\max} \sim e^{n/n_D}$, where n is the number of nodes in the entire network and n_D the number of driver nodes (independent inputs). For a fully controlled network with $n = n_D$, the maximum effort $\mathcal{E}_{\max} \sim n^{1/(\gamma-1)}$ for degree distribution exponent γ. This ranges from $\mathcal{E}_{\max} \sim n^{1/2}$ to $\mathcal{E}_{\max} \sim n$ for the most relevant cases, where $2 < \gamma < 3$.

In short, adding inputs generally reduces the required control effort, and you may want (or need) more inputs than the minimum number imposed by structural controllability.

Fewer States

We can give up trying to control a system throughout its entire state space and ask, instead, for the ability to control a subspace. Because the condition number of the Gramian is large, some directions are easy to control and some are hard, as the low-dimensional examples above show. We can then imagine different strategies:

1. *Best case.* The simplest, most modest approach is to focus on the easiest control directions, whose associated control effort typically does *not* increase with network size. That is, the hardest directions become exponentially harder, while the easiest directions require roughly the same control effort. (We confirm this statement in Problem 14.12.)

   A simple way to find the easiest direction is the *power method*: multiply the Gramian matrix against an arbitrary unit vector and normalize the output vector. Iterating N times gives the largest eigenvector when $N \to \infty$. For a network of size n, the number of operations naively is $O(n^3 N)$. But since the dynamical matrices describing large networks are often sparse, the matrix multiplications can be fast. And because Gramian matrices are ill conditioned, the power method converges rapidly, aided by the large ratio between successive eigenvalues of the Gramian. (In Problem 14.12, $N = 3$ suffices.)

2. *Typical case.* A somewhat more ambitious goal is to be able to control to reach a typical, randomly chosen region of state space. In numerical studies, Sun and Motter (2013) showed that there is an abrupt *controllability transition* in random ER networks and in SF networks with $\gamma > 3$. For ER networks, the transition is abrupt, and the probability jumps from zero to one as the number of inputs is increased. For SF networks, the probability is continuous but has a discontinuity in the slope at the transition point. In the study of Sun and Motter (2013), the

criterion for practical controllability was the ability to control $\approx 20\%$ of the nodes, using randomly chosen driver nodes.

Thus, while large networks may generically require too much effort, adding more inputs than the bare minimum imposed by controllability tests or choosing a more modest goal than full state controllability can help.

14.6 Complex Systems

So far, we have focused on the different issues that arise when controlling large networks. In doing so, we used graph theory in two ways: as a tool to analyze issues such as controllability in the presence of uncertain coefficients and as a source of models (regular, star, random, and scale-free) to describe the connectivity of nodes in large systems. We have skirted, though, the most interesting and important questions: Are networks the appropriate description of complex systems? And, if so, what kinds of networks are found empirically? In this last section, we first discuss briefly these questions, which have inspired the new and influential discipline of *network science* over the last two decades. The perspective behind this interest has been that of statistical physics and its notions of criticality and universality. However, this way of looking at complex systems is controversial and, significantly, has been questioned by both engineers and biologists. To balance the physicists' perspective, we also discuss briefly an alternative viewpoint rooted in engineering and control theory.

14.6.1 Real Networks

We have introduced several models of networks, but are they relevant to the real world? Of course, regular grids are familiar in condensed-matter physics, serving as models for crystal lattices. Star graphs can be a good model for small networks, but it would clearly be an exceptional (and very fragile) situation to rely on a single hub. This leaves random and scale-free cases. The big surprise, dating from the late 1990s, when large-scale data on networks first began to become available, is that empirical networks tend to be much closer to scale-free (SF) models than to random (ER) models. This was truly a profound discovery.[14] Generically, most networks – whether informational (www, citations), technological (circuits, peer to peer), biological (protein interactions, metabolic), or social (actors, business relations) – have degree distributions that approximately follow a power law.

How might a scale-free network form? One simple scenario is the *Barabási–Albert* model, which has two elements: First, the size n of the network grows. Second, there

[14] Like many profound discoveries, there were precursors. To name but one, de Solla Price observed a power-law degree distribution in citation networks starting in the 1960s and even formulated a model based on preferential attachment (under a different name) in the 1970s (de Solla Price, 1976). However, it seems fair to say that the generality and profound implications of the fact that most large networks are not ER and show heavy degree-distribution tails was not recognized before roughly the year 2000.

is *preferential attachment*: new nodes link with existing nodes with a probability that is proportional to number of nodes possessed by the candidate node. In other words, the "rich get richer." Nodes with many links are likely to capture a higher fraction of the new network nodes.[15] Problem 14.4 shows that the model leads to $P(k) \sim k^{-3}$ for large k.

While extremely popular and influential, network science remains controversial. One reason is that although the networks associated with complex systems do have heavy tails, their nature and origin is much less clear. When the data are tested carefully, few networks show perfect power-law scaling. Whether the deviations are due to mundane or important complications, though, is debated. Problem 14.5 shows that distinguishing power-law distributions from alternatives such as lognormal distributions can be tricky.

14.6.2 Highly Optimized Tolerance

"Complex" can mean different things to people trained in different disciplines. Statistical physicists focus on *universal* features of models (e.g., the power-law scaling of degree distributions). They search for simple models that epitomize the universal phenomena. For example, preferential attachment shows how power laws can arise from the growth of the network, in a system that need not be "tuned" to a critical point. By contrast, power-law scaling is observed in equilibrium systems only when parameters are precisely tuned to a special set of values that are highly nongeneric points in parameter space.

Engineers put more emphasis on the implications of design (or evolution, for biological systems). Here, we offer a competing, sharply contrasting view of complex systems known as *highly optimized tolerance* (HOT) that comes from engineering ideas such as the robust, "worst-case" theory of control presented in Chapter 9.

Recall that robust control is an attempt to optimize control of systems with uncertain dynamics. Worst-case design methods ($\mathcal{H}_\infty$) follow the partition principle by implicitly partitioning the set of possible dynamics for the system under control into "normal" and "disaster" cases. As a simple example, the parameters describing a class of dynamical systems can be assumed to lie within some prescribed range (the "normal" cases). If a parameter is actually outside that range, then the instance is a "disaster." Robust control tries to optimize the behavior of an ensemble of dynamical systems assuming that we encounter the worst-possible member of the normal set of systems. The corollary is that the performance of systems outside the normal set may be very bad.

This partitioning leads to the *robust yet fragile* character of many complex systems. When subject to foreseen, normal perturbations or uncertainties, they function extremely well: they are robust. Yet they can fail when an unforeseen event occurs. A typical example is a modern airplane, which can function well in a wide variety of passenger loads, weather conditions, and so on. But disasters do occur, often resulting

[15] Merton (1968) dubbed this phenomenon the *Matthew effect*: "For to everyone who has, more shall be given." (Matthew 25:29, New American Standard Bible).

from unforeseen circumstances. To take an aerospace example, an early test of the Ariane 5 rocket ended in an explosion because a 64-bit integer representing the rocket's velocity was converted to a 16-bit integer and overflowed. The software had worked in previous rockets because they were slower, so that velocities had magnitudes that could be represented using 16 bits. A simple, unforeseen, "trivial" error destroyed the product of a $7 billion development program.

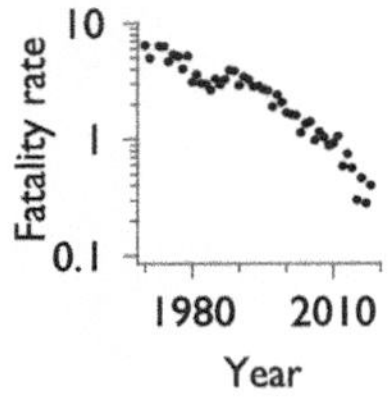

If an accident does occur, the modern engineering response is to thoroughly investigate its causes and make whatever alterations are needed to prevent that kind of disaster. By repeated cycles of disaster, investigation, and incremental fix, the system evolves to be more robust. The airline industry shows the power of this approach: The rate of fatalities per million flights (on commercial aircraft) has decreased twentyfold over the last fifty years. See right for data from the Aviation Safety Network (https://aviation-safety.net/statistics/).

HOT theory attempts to extend and formalize such ideas. It starts with the recognition that complex technological and biological systems are heterogeneous, composed of a wide variety of different parts. An airplane can have millions of separate parts, and most of them are different. Cells depend on *factories* that are agglomerations of proteins that carry out functions such as replication, transcription, and splicing. These factories have tens (even a hundred) different proteins, often present in but a single example. Further, these elements are *highly organized.* They are not self-assembled, nor are they randomly interacting. Rather, they are designed, whether by an engineering team or by the actions of evolution. The design is typically an iterative process and may include subsystems with feedback and control, designed according to the principles elaborated in this book.

The result is a system that acts to *deconstrain* the low- and high-level constraints that a system is subject to. That is, the individual components that a system is made from have various physical constraints. Mechanical systems obey Newton's equations, chemical systems the law of mass action, and so on. There are also system-level constraints, including functions that are desired, operating requirements involving both performance and safety, and so on. In practice, a common strategy in both engineering and biology is to organize subcomponents hierarchically and to develop *protocols* for manipulating them. These constrain the design space, so that variations on existing designs can be explored relatively efficiently. Successful technological systems hide the complexity during routine operation. You do not have to understand how a car works to drive nor how to program to use an app. (Such statements are not true at the beginning of a technology, when users must also be tinkerers.) Similarly, we are unaware of our immune system until we become sick.

The HOT theory view of complex systems contrasts with statistical-physics ideas inspired by critical phenomena. Rather than identical subunits interacting randomly and producing emergent behavior, HOT theory emphasizes heterogeneous subunits interacting according to selected rules. These rules evolve over repeated cycles where a problem (or disaster) occurs and then a local solution is implemented to prevent that type of problem from recurring. In industry, the continual improvement of a manufacturing process is known as *kaizen*, a Sino-Japanese term made popular by its role in the success of Japanese industry after World War II. In HOT theory, emergent phenomena such as power laws arise as a byproduct of heterogeneity but do not play a more fundamental role. We give a taste of the HOT approach to complex systems via a toy model of forest fires in Problem 14.14.

14.6.3 Control without Understanding?

The two views of complexity presented above are radically different. One is guided by statistical physics notions of criticality and universality, the other by engineering notions of design and robustness. Perhaps, for control, these differences are not so important. That is, it may not matter whether we understand complexity and its origins, nor even that we have found the "right" form of structure (degree distribution and other statistical quantities). Control strategies based on approximate dynamics that are approximately described as network structures can be successful. We can obtain good control from bad models if there are enough measurements to correct modeling errors. Proportional-integral-derivative (PID) control implicitly assumes the controller acts on a second-order system; nonetheless, it often works very well. Perhaps control designed for networks with power-law degree distributions may work well, even if the "true" distributions turn out to be lognormal.

14.7 Summary

The control of complex systems has led to great excitement: The paper by Liu et al. (2011) on maximum matching and driver nodes kicked off the recent surge in interest and received more than 2000 citations in eight years after publication. Older works such as Lin (1974) on structural controllability had attracted attention within a small group of specialists for nearly forty years before undergoing a surge in interest driven by its recognition as a precursor to current interests. And there is also no doubt that the surge of interest has demonstrated the weaknesses of 1960s notions of controllability: tests such as the Kalman rank condition simply cannot be applied to most large systems.

At the time of writing, the empirical utility of the ideas sketched here has been largely untested. Indeed, only one experimental effort has been reported. Yan et al. (2017) used ideas from structural controllability to identify which neurons in the nematode worm *C. elegans* are important for controlling various muscles. Notably, they correctly predicted that a new, previously untested neuron would lead to appreciable loss of motion control when the neuron was disabled.

On the broader questions of full state control, the early predictions that a large fraction of nodes is needed were clearly simplistic. The subsequent debate can be viewed more broadly as a search for the right question: what kinds of symmetries and network structures are relevant, what metric to impose on the control energy. Should the goal of control be to control all states, some target states, attractors, or something else? All these questions continue to be vigorously debated.

The interest in network science has been motivated as a possibly useful way to describe complex systems. The rich structure of nodes and connections, which involve nonequilibrium systems themselves, means that we should expect rich behavior in their networks. The application of graph theory and related notions from computer science, and the discovery of the peculiar structure of many real-world networks, with their heavy-tailed degree distributions, have spurred hope that statistical-physics notions of universality may be useful. The hope is hotly disputed – degree distributions that seem to follow power laws may have deviations and their origins may be murky – and applications to control theory are even less clear. The simplest analyses ignore (approximate) symmetries, and the control effort required may be huge. Conflicting alternative views – the highly optimized tolerance of engineers and modular systems of biologists – make important points.

And yet, amidst all the confusion, physics may have something to say. Consider the crystal: Its lattice of identical atoms with identical interactions can be viewed as a regular network. The special features of quantum mechanics ensure that atoms and interactions are truly identical, having states with identical properties and interactions of the same form. The interactions are typically local, sometimes limited to nearest neighbors.

What does it mean to control a crystal? We cannot – nor would we want to – control each of 10^{23} atoms in a macroscopic object. Rather, we identify a set of degrees of freedom of interest; the rest is noise. We may care about only the three translational degrees of motion, treating the solid as a rigid body. We may add rotations. We may include finite frequencies that describe bulk and surface vibrations. We may include defects and their modes. The formalism of condensed-matter physics – its broken symmetries, its emergent rigidity and soft modes, its recognition that "more is different" – gives us a way to describe the dynamics and control of a mix of identical and nonidentical elements of a network. It provides tools such as *mean-field theory*, which holds for complete graphs, among others. Condensed-matter theory is more than critical phenomena and can be extended in many ways. Although how to handle complex systems remains an open question, there is hope for a synthesis, even if "more" turns out to be more different than we currently imagine.

14.8 Notes and References

The year 1999 was key to the development of the ideas described in this chapter. Barabási and Albert (1999) pointed out that many important networks had degree distributions that were far from the Poisson distributions expected of randomly connected nodes. It also saw the first paper on the competing view of highly organized tolerance (HOT), sketched in Section 14.6.2, which emphasized design and optimization (Carlson and Doyle, 1999). And a group of biologists and physicists (Hartwell et al., 1999) called for the development of a "modular biology" to succeed molecular biology. Their program has been elaborated in the theory of *facilitated variation* (Gerhart and Kirschner, 2007), which describes the path from genetic to phenotypic variation, largely via changes in the regulatory function of genetic modules that have strongly conserved, yet weakly linked core components.

For a simple, relatively current introduction to network science, see Barabási (2016). We have left out important aspects, such as *degree correlations*, *clustering*, and *communities*. The preferential-attachment model has many precursors (Pólya, Yule, Gibrat, Zipf, Simon, Price, Merton) but did not attract much attention before the work of Barabási and Albert. Whether real networks conform to the scale-free hypothesis and have power-law distributions continues to be hotly debated. Broido and Clauset (2019) show that almost all networks fail rigorous statistical tests for power-law degree distributions. But the relevance of such tests is less clear, as they neglect correlations (Gerlach and Altmann, 2019) and because deviations can easily arise from "trivial" mechanisms such as corrections to scaling, and mixtures of different mechanisms. For more on HOT, see Willinger et al. (2004) and Alderson and Doyle (2010). The Ariane 5 disaster is described in Gleick (1996).

Network control is reviewed by Liu and Barabási (2016). The approach using networks and graphs arguably originates with Lin's paper on structural controllability (Lin, 1974). This work is a classic "late bloomer," known for 35 years to only a small group of specialists but now widely cited and recognized as a precursor to the approaches based on network science discussed in this chapter. For brevity, we have omitted discussion of a complementary notion of *strong structural controllability*, which asks for controllability for *all* non-zero parameter values corresponding to links among nodes (Mayeda and Yamada, 1979). Although such a notion of controllability of networks may be too strong, it does exclude the possibility that a system is uncontrollable because symmetries force special values of parameters. The "right" notion of controllability may lie somewhere in between ordinary and strong structural controllability (J.-C. Delvenne, private communication).

The connection between maximum matching and structural controllability was discovered several times independently, including Murota in the 80s (see Murota (2009)), and later by Commault et al. (2002). These papers attracted comparatively little attention before Liu et al. (2011), which applied these mathematical ideas from graph theory to real networks. The latter paper inspired the current burst of interest in these topics and was also the first to apply cavity methods to the controllability problem. We have left out some important aspects, such as the mapping between directed weighted graphs and undirected *bipartite* graphs, which is a key idea behind the algorithms to find maximum matchings.

The critique by Cowan et al. (2012) pointed out that a network with self-node dynamics can generically be controlled by a single input node and argued that real networks should then be generically controllable by a single node. Since they are not, other features such as control effort are more important than the technical notion of complete state controllability.

The role of low in- and out-degree nodes in controllability was discussed by Marchetti et al. (2014). Murota (2009) emphasized the role of the "two kinds of numbers" he termed "accurate" and "inaccurate." He did not consider explicitly the possibility of effectively accurate parameters, but they can be grouped together with the accurate parameters, as far as effective controllability is concerned. Murota has used *matroid theory*, a generalization of the notion of linear independence, to analyze systems that have both accurate (and effectively accurate) numbers and also inaccurate parameters. It is not clear whether his methods can be effective for large networks. For numerical methods involving sparse matrices, see Saad (2003). The discussion of control effort follows Sun and Motter (2013), Yan et al. (2012), and Yan et al. (2015). For an attempt to take advantage of the particular structure of a given $\boldsymbol{A}$ matrix, see Lindmark and Altafini (2018).

Because the field is in such a ferment, we cannot do justice to all the issues that have been raised. To mention just a few we have passed over, network structure can predict whether centralized or decentralized control is needed (Jia et al., 2013). Also, networks whose structure varies with time, *temporal networks*, can be much easier to control (Li et al., 2017). Along with the focus on controllability, there have been analogous studies of observability. For example, Yan et al. (2015) show that the large condition numbers of controllability Gramians imply that there can be similarly large condition numbers for observability Gramians. As a result, some directions in state space will have large estimation errors when estimating the underlying state from observations. Since sensors have minimum noise, these modes may be effectively unobservable and cause observability of the network as a whole to fail. More generally, we can ask where to *locate sensors* for best controllability. Further discussions of observability include Yang et al. (2012) and Liu et al. (2013), as well as the summary in Liu and Barabási (2016).

Finally, we have deliberately passed over the topic of controlling networks with nonlinear dynamics, as analyses are so tentative. Here are two interesting papers: (1) Cornelius et al. (2013) focus on controlling a network to steer its dynamics to a desired attractor by making small perturbations to the initial state. Since there are typically few attractor states in a nonlinear network, we can say, in statistical-physics language, that the attractors play the role of *coarse-grained macrostates*. (2) Kashima (2016) generalizes the definition of controllability Gramian to nonlinear systems, identifying it with the covariance matrix of randomly excited, but uncontrolled, state dynamics.

Problems

14.1 Deterministic graphs. Let us consider some properties of deterministic graphs.

a. Show that the average path length of a circle graph is $\langle d \rangle \sim \frac{1}{4}n$.
b. How does $\langle d \rangle$ of grid graphs scale with n if each node has $2k$ nearest neighbors?
c. Show that the average path length of a large star graph is $\langle d \rangle \sim 2$. (Hint: The paths from hub to periphery have negligible weight for large n).

14.2 Random graphs. Consider the properties of random Erdős–Rényi graphs with n nodes and a probability p for a link between any two nodes chosen at random. Work in the "Poisson" limit where the only parameter is $\langle k \rangle$, the average node degree.

a. Give a reasonable argument that the average path length $\langle d \rangle \sim \ln n / \ln k$. Hint: Consider the "fan out" of paths. Each node reaches roughly $\langle k \rangle$ nodes after paths of length one, $\langle k \rangle^2$ nodes after paths of length two, ...
b. Show that the graph becomes connected at $\langle k \rangle = \ln n$, for large n. Hints: Estimate the probability for a node to have no links and use $(1-p)^n \approx \mathrm{e}^{-np}$.

14.3 Scale-free graphs. Consider node distributions $P(k) = Z^{-1} k^{-\gamma}$, with $\gamma \geq 2$ and $k > k_{\min}$. Treat, for simplicity, the node degree as a continuous density, $p(k)$.

a. Find the normalization constant Z for the continuous probability density.
b. Show that changing the units $k \to ak'$ does not change the form of the distribution.
c. Show that $\langle k \rangle = \left(\frac{\gamma-1}{\gamma-2}\right) k_{\min}$.
d. For a finite network of n nodes, show that $k_{\max} = k_{\min}\, n^{\frac{1}{\gamma-1}}$.
e. Why then does $\gamma < 2$ imply multiple edges between node pairs?

14.4 Scaling in the Barabási–Albert model of preferential attachment. Consider a network that adds one node each time step. Let $n(k,t)$ be the number of nodes with degree k at time t and $p(k,t) = n(k,t)/n(t)$ the corresponding degree-node distribution at time t. Each new node adds m links. Each new link goes to an existing node, with the probability to connect to a node of degree k given by $\Pi(k) = k/\sum_j k_j = k/(2mt)$. For the denominator: at time t there are mt links, and each link connects two nodes.

a. Show that typically $\frac{k}{2} p(k,t)$ links are added to degree-k nodes at time t and that $(n+1)p(k,t+1) = np(k,t) + \left(\frac{k-1}{2}\right) p(k,t) - \left(\frac{k}{2}\right) p(k,t)$.
b. In the long-time limit, $p(k,t) \to p_k$. Show that the master equation in (a) becomes $p_k = \left(\frac{k-1}{2}\right) p_{k-1} - \left(\frac{k}{2}\right) p_k$.
c. Derive the continuum limit $p_k = -\frac{1}{2}\partial_k (k p_k)$ and verify that $p_k \sim k^{-3}$ is a solution.

14.5 Lognormal vs. power law. In practice, it is not easy to distinguish a power-law distribution from alternatives such as the lognormal distribution.

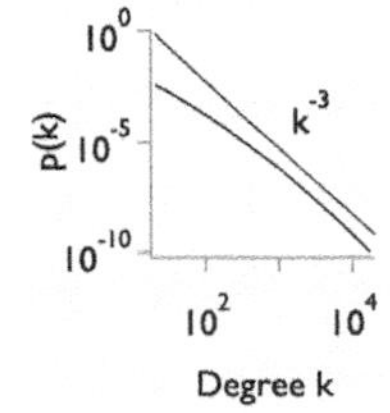

a. Show that a lognormal degree distribution $\ln k \sim N(\mu, \sigma^2)$ is "scale-free" in that a change of scale $k = ak'$ merely shifts the distribution on a log-log plot.
b. By creating a plot such as the one at left, show that you can find parameters in a lognormal distribution that are close to a given power law over some decades.

14.6 Controllability of an n-chain. Following an example from Sun and Motter (2013), we consider a chain of n one-dimensional systems with a single input at its head. The dynamics are that of an n-fold integrator: $\dot{x}_1 = u$, $\dot{x}_2 = x_1$, $\ldots$, $\dot{x}_n = x_{n-1}$.

a. Show that the controllability matrix $W_c = \mathbb{I}_n$, the n-dimensional identity matrix. The Kalman rank condition for controllability is thus satisfied.
b. A system has *strong* structural controllability if it is structurally controllable for all nonzero values of the weights. Show that the n-chain defined above has this property, allowing for arbitrary weights $b, a_{21}, a_{31}, \ldots a_{n1}$.

14.7 Dilation. Consider the graph at right of an LTI dynamical system.

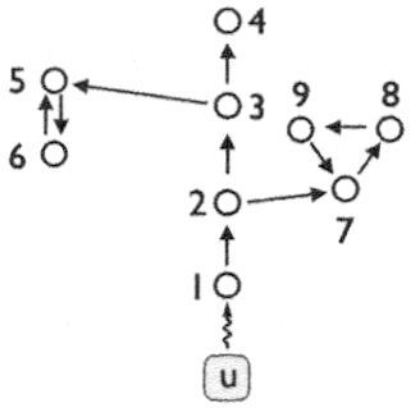

a. Show that the system is not controllable, for any values of the link weights.
b. Add a self-link of weight a_{22} to the node x_2. Show that the system is now controllable and has the strong structural controllability property.
c. Why does adding a self-interaction node remove the dilation?
d. Finally, add a second self-link of weight a_{33} to the node x_3. Show that the system is controllable but does not have the strong structural controllability property.

14.8 Cactus is a minimum controllable structure. Consider the cactus. At right is a cactus with two buds, reproduced from Figure 14.6d and then annotated. Show that removing *any* edge renders the resulting network uncontrollable.

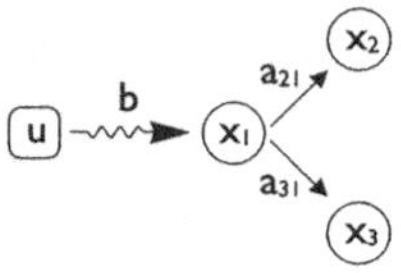

14.9 Self-interactions. By computing the determinant of the controllability matrix W_c, give a nongraphical proof that adding self-interactions generically implies that a system is controllable from one input. Hint: Set all nonself interactions equal to zero and all input couplings to one. (Justify these assumptions.)

14.10 Control effort in one dimension. For $\dot{x} = \lambda x + u$, with $x(0) = x_0$ and $x(\tau) = x_\tau$:

a. Show that the minimum control effort is $\mathcal{E} = 2\lambda(x_\tau - e^{\lambda\tau} x_0)^2/(e^{2\lambda\tau} - 1)$.
b. Deduce the short- and fast-protocol limits given in the text.
c. For $\tau^* = \lambda^{-1} \ln(x_\tau/x_0)$, the minimum effort $\mathcal{E} = 0$. What is going on?
d. Show that the minimum-effort trajectory $x(t)$ is identical for $\lambda \to -\lambda$.

14.11 Control effort diverges in a nearly uncontrollable system. Consider two first-order equations driven by a common input: $\dot{x}_1 = -x_1 + u, \quad \dot{x}_2 = -(1+\delta)x_2 + u$.

a. Calculate the Gramian $P(\tau)$ and show that its determinant $\sim \delta^2$, for $\delta \ll 1$. Argue that this implies that the control effort $\mathcal{E} \sim \delta^{-2}$.
b. Calculate numerically and then plot the minimum-effort trajectory connecting $\binom{0}{0} \to \binom{0}{1}$ for $0 \le t \le 1$. Plot, too, the control effort $\mathcal{E}$ as a function of δ.

14.12 Control effort can be sensitive to direction and dimension. Consider the linear system $A = \begin{pmatrix} -3.2 & 1.3 \\ 1.3 & -2.7 \end{pmatrix}$, $B = \binom{0}{1}$, with protocol duration $\tau = 3$ (Yan et al., 2015).

a. Find the eigenvalues of the system dynamics A and also the controllability (W_c) and Gramian matrices P; confirm the plots and numbers given in the text. Plot $u(t)$ for the minimum- and maximum-effort inputs.
b. Show that applying powers of the Gramian P to an arbitrary unit vector (normalizing at each step) gives the target direction requiring the least control effort, and powers of P^{-1} give the target direction requiring the most control effort.

c. Enlarge the system to three dimensions. For $\boldsymbol{A} = \begin{pmatrix} -3.2 & 1.3 & 1 \\ 1.3 & -2.7 & 0.7 \\ 1 & 0.7 & -2.2 \end{pmatrix}$, $\boldsymbol{B} = \begin{pmatrix} 0 \\ 1 \\ 0 \end{pmatrix}$, show that the ratio of largest to smallest control efforts ≈ 979.

d. Show that the efforts for the easiest direction are roughly the same in the $n = 2$ and $n = 3$ cases, whereas the efforts for the hardest direction differ significantly.

14.13 Effective controllability of an n**-chain**. (Continuation of Problem 14.6.)

a. Show that the control Gramian $\boldsymbol{P}(\tau)$ has elements $\boldsymbol{P}(\tau)_{ij} = \frac{(\tau)^{i+j-1}}{(i+j-1)(i-1)!(j-1)!}$.

b. Evaluate the eigenvalues of $\boldsymbol{P}$ numerically, for $n = 1, 2, \ldots, 20$, and show that the condition number increases exponentially as $\sim e^{8.4n}$.

14.14 Some like it HOT. Consider a simple model of a forest management in the face of forest fires. Plant trees on a square lattice of side N, with site probability ρ. The forest is subject to "sparks" that have a source above and left of the forest and fall on the ij site in the forest with probability $p_{ij} \equiv p_i p_j$, where $p_x \propto 2^{-(m_x + x/N)/\sigma_x^2}$. Choose $m_i = 1$, $\sigma_i = 0.4$, $m_j = 0.5$, and $\sigma_j = 0.2$. If a spark falls on an unoccupied site, there is no damage. If it falls on an occupied site, all contiguous occupied sites are burned. The goal is to maximize the yield $y(\rho)$, the average number of trees left after a fire.

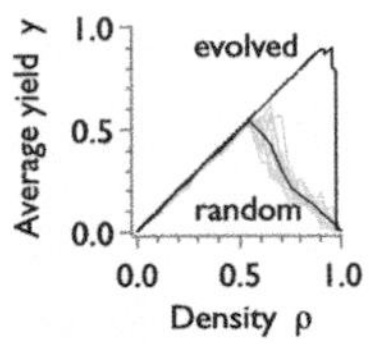

random, ρ=0.6

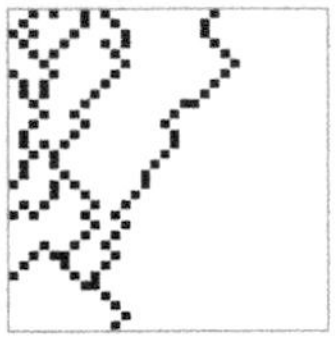

evolved, ρ=0.9

a. For randomly planted trees and $N \to \infty$, show that $y(\rho) = \rho - P_\infty^2$, where P_∞ is the probability that a lattice site is in the infinite *percolation* cluster.

b. Write code to find the yield plotted at left (top), for $N = 32$. Ten individual trials are shown in light gray, and the average in thick black. Hints: Find contiguous lattice sites of ones (the trees) by finding morphological components of a binary image and then counting their size. Find this number for every lattice site and average over the p_{ij} to determine the expected loss from a spark.

c. To increase the yield, *evolve* a design for planting trees as follows: When going from a density $\rho = n/N^2$ to $(n+1)/N^2$, explore D possibilities for where to plant the next tree. For each candidate position, calculate the average loss and then choose the position that minimizes the loss. $D = 1$ is equivalent to the random-forest case. The figure at left is for $D = N = 32$ (cf. thin black line for yield).

This simple design procedure naturally leads to *firebreaks* of unplanted sites that stop fires from spreading too far. See the lines of unplanted sites in black at left (bottom) The firebreaks become even more clearly organized if one optimizes over all free sites instead of only up to D sites. We emphasize that this organization arises from the repeated evolutionary cycle of trial planting and evaluation and not from any kind of self-organization. The distribution of fire sizes turns out to be approximately a power law, although that fact does not play an important role in the organization (*function*) of the tree-planting algorithm. For more, see Carlson and Doyle (2000).

15 Limits to Control

You Can't Always Get What You Want

Mick Jagger and Keith Richards[1]

Many factors can limit the performance of control systems: actuator range, feedback algorithm, controller architecture, and more. These are important practical limitations but are not fundamental: we can buy a bigger actuator, alter the controller, add feedforward. Yet some issues cannot be so easily bypassed. Such fundamental limits have three origins:

- *Causality*. The future cannot affect the past. Physical response is causal.
- *Information theory*. It provides a useful language and two important ideas:
  - Fundamental physical laws conserve the number of possible states;
  - information has value that can translate into performance.
- *Thermodynamics*. To control a system, it must be out of equilibrium. Control requires work, with fundamental performance trade-offs among energy, speed, and accuracy.

Applying these simple ideas to particular cases can be subtle.

15.1 Causal Limits

Many fundamental control limitations trace back to the requirement that response functions must be causal: the response $G(t)$ to a $\delta(t)$ input must be 0 for $t < 0$. For linear response in the frequency domain, the restrictions due to causality are not trivial. In Section 15.1.1, we review the Kramers–Kronig relations between the real and imaginary parts of a linear response function, which are a standard part of the physics curriculum. We give two derivations: one emphasizes the role of causality in the time domain, the other the analytic properties of response functions in the frequency domain.

Just as causality relates the real and imaginary parts of a linear response function, it also connects the magnitude (or gain) of a transfer function to its phase. In Section 15.1.2, we discuss and derive Bode's *gain-phase* law, which is less familiar to

[1] Title of song released as a single, July 4, 1969, and on the album *Let It Bleed*, December 1969.

physicists. One important difference between the Kramers–Kronig and Bode relations is that the former are *equalities*, while the latter is an *inequality*: there is a *minimum* phase lag consistent with causality, but the lag can be greater. This is the origin of non-minimum-phase (NMP) transfer functions.

It is likely no accident that physicists are more interested in the connections between real and imaginary parts of a response function while control engineers are more interested in the connections between magnitude and phase. The real part of the response function measures energy storage – for example, the energy of a charged capacitor is $U = \frac{1}{2}CV^2$, where the capacitance $C \sim \epsilon$, the dielectric response function. The imaginary part of the response function measures energy dissipation – for example, absorption in optics, viscosity in fluids.[2] By contrast, the magnitude of a response function measures the accuracy of response and appears, for example, in the cost functions of optimal control. And the phase response corresponds to a lag, which controls stability – for example, we have seen repeatedly the difficulties that the extra phase lags of non-minimum-phase systems create for control. Thus, the relative familiarity of the Kramers–Kronig relations in physics and the Bode relations in control simply reflects the differing concerns of the two domains. But the relations are intimately connected, and both can be relevant to both disciplines.

Lastly, in Section 15.1.3, we present another consequence of causality derived by Bode, known informally as the *waterbed effect*: it gives an integral constraint on the sensitivity function $S = \frac{1}{1+L}$, where L is the loop gain. It has many practical consequences.

15.1.1 Kramers–Kronig Relations

The Kramers–Kronig relations connect the real and imaginary parts of a causal linear response function. In our case, the response function $G(t)$ is the time-domain representation of the transfer function that gives the output $y(t)$ to an input $u(t)$:

$$y(t) = \int_0^{\infty} \mathrm{d}t'\, G(t-t')\, u(t') \equiv \int_{-\infty}^{\infty} \mathrm{d}t'\, \theta(t-t')F(t-t')\, u(t')\,, \tag{15.1}$$

where the step function, $\theta(t)$, enforces causality: only inputs from the past, $t' < t$, can affect y at time t. Thus,

$$G(t) = \theta(t)\, F(t)\,, \tag{15.2}$$

meaning that $G = F$ for positive times and is zero for negative times.

Before plunging into mathematics, consider a qualitative explanation of the relation between the real and imaginary parts of a causal response function. Figure 15.1a shows a typical causal response function (underdamped oscillator). (b) shows one of

[2] The fact that the real part of the response function measures energy storage while the imaginary part measures dissipation is a convention. The reverse convention is standard for electrical circuits, which measure the phase of voltage response with respect to the current. Thus, the response of a capacitor, which stores energy without dissipation, is usually taken as purely imaginary: $Z = (\mathrm{i}\omega C)^{-1}$.

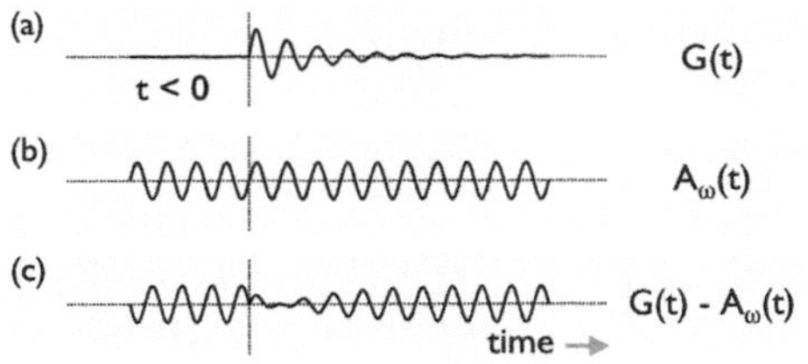

Fig. 15.1 Causality in the frequency domain. (a) Causal response function $G(t)$. (b) Removing a Fourier component $A_\omega(t)$ disrupts the balance of phases needed to cancel (c) the response for negative times.

its Fourier components. Removing it ("absorption") in (c) makes the function nonzero for $t < 0$.

Because Fourier components extend for all t, past and future, it is remarkable that in the original response all of the sines and cosines can sum to zero for all $t < 0$. This cancellation can happen only if there is a delicate balance of phases of each of the Fourier components. Removing one of them destroys this balance and leads to an acausal response, *unless the phases of the remaining components are all adjusted.* The Kramers–Kronig relations give precisely the adjustments that are needed. That is, absorption "in a causal way" at a single frequency implies altering the phases of all the remaining frequency components.

Fourier transforming the causal-response relation of Eq. (15.2), we find

$$G(\omega) = \frac{1}{2\pi}\,\theta(\omega) * F(\omega) = \int_{-\infty}^{\infty} \frac{d\omega'}{2\pi}\,\theta(\omega-\omega')F(\omega')$$
$$= -\,\mathrm{P}\int_{-\infty}^{\infty} \frac{d\omega'}{2\pi}\,\frac{\mathrm{i}\,F(\omega')}{\omega-\omega'} + \frac{F(\omega)}{2}, \qquad (15.3)$$

where we used the convolution theorem and the Fourier transform of the step function, $\theta(t)$,

$$\theta(\omega) = -\,\mathrm{P}\,\frac{\mathrm{i}}{\omega} + \pi\,\delta\,(\omega), \qquad (15.4)$$

with P the *principal value* of the integral (Appendix A.2 online).

Because G is causal, the function $F(t)$ can have arbitrary values for $t < 0$. If we use this freedom to define F to be even, so that $F(-t) = F(t)$, then the Fourier transform of F is pure real. Then Eq. (15.3) gives $F(\omega) = 2G'(\omega)$, with $G' \equiv \mathrm{Re}\,G$ and $G'' \equiv \mathrm{Im}\,G$, and

$$G''(\omega) = \frac{1}{\pi}\,\mathrm{P}\int_{-\infty}^{\infty} d\omega'\,\frac{G'(\omega')}{\omega'-\omega}. \qquad (15.5)$$

By contrast, choosing F to be odd, with $F(-t) = -F(t)$, makes $F(\omega)$ pure imaginary. Then $F = 2\,\mathrm{i}G''$ and

$$G'(\omega) = -\frac{1}{\pi}\,\mathrm{P}\int_{-\infty}^{\infty} d\omega'\,\frac{G''(\omega')}{\omega'-\omega} \qquad (15.6)$$

Since G'' is odd and G' even, we can simplify:

$$\int_{-\infty}^{\infty} d\omega' \frac{G''(\omega')}{\omega' - \omega} = \int_{-\infty}^{\infty} d\omega' \frac{G''(\omega')(\omega' + \omega)}{\omega'^2 - \omega^2}$$
$$= \int_{-\infty}^{\infty} d\omega' \frac{G''(\omega')\,\omega'}{\omega'^2 - \omega^2} = 2\int_{0}^{\infty} d\omega' \frac{G''(\omega')\,\omega'}{\omega'^2 - \omega^2}, \qquad (15.7)$$

and similarly for G', giving the positive-frequency Kramers–Kronig relations:[3]

$$G'(\omega) = -\frac{2}{\pi}\,\mathrm{P}\int_{0}^{\infty} d\omega' \frac{\omega'\,G''(\omega')}{\omega'^2 - \omega^2}, \qquad G''(\omega) = \frac{2\omega}{\pi}\,\mathrm{P}\int_{0}^{\infty} d\omega' \frac{G'(\omega')}{\omega'^2 - \omega^2}. \qquad (15.8)$$

Alternate derivation. The above derivation highlights the role of causality but obscures other aspects. Here is another derivation that connects causality to the analytic properties of the response function in the complex-s plane.[4] Let us define a *two-sided* Laplace transform,

$$G_{L2}(s) \equiv \int_{-\infty}^{\infty} dt\, G(t)\, e^{-st}\,. \qquad (15.9)$$

Because $G(t) = \theta(t)F(t)$ is a causal response function, the two-sided Laplace transform is identical to the ordinary one-sided Laplace transform. But because the integral in Eq. (15.9) extends from $t = -\infty$ to $+\infty$, $G_{L2}(s = i\omega) \equiv G(\omega)$ is also the Fourier transform of $G(t)$.

If, for $t \to \infty$, the response $F(t) \to 0$ quickly enough, such that $\int_0^\infty dt\, F(t)\,e^{-i\omega t}$ converges, then $G_{L2}(s)$ is *analytic* for all Re $s > 0$. For simplicity, we will assume that $G_{L2}(s)$ also has no poles on the imaginary axis (where $s = \pm i\omega_0$). Then $G_{L2}(s)$ is analytic in the right-hand part (RHP) of the complex s-plane, which includes the imaginary axis, Re $s = 0$. Note that requiring $F(t) \to 0$ at long times implies that the response is stable.

Since $G_{L2}(s)$ is analytic in the RHP, we can use *Cauchy's Integral Theorem* from complex analysis (see Appendix A.2 online) to write

$$\oint_{\gamma} ds' \frac{G_{L2}(s')}{s' - s} = 0\,, \qquad (15.10)$$

where the closed contour γ is depicted in Figure 15.2 and where $s = i\omega$. The semicircular indentation around $s = i\omega$ is necessary since there is a pole in the integrand in Eq. (15.10).

[3] In the physics literature, the terms in Eq. (15.8) usually have the opposite sign. This difference traces back to the convention of using $e^{-i\omega t}$ rather than $e^{+i\omega t}$ in the forward Fourier transform.

[4] In the physics literature, the argument is commonly phrased in terms of the complex ω plane. The two are rotated by 90°, so that our discussion of the analyticity properties in the left- and right-hand planes is equivalent to discussions in the physics literature of analyticity properties of the lower- and upper-half planes.

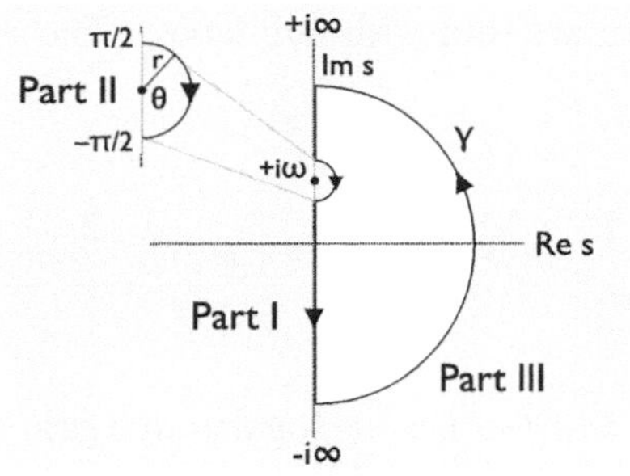

Fig. 15.2 Path γ for contour integral in Eq. (15.10).

The integral in Eq. (15.10) is divided into three parts, as labeled in Figure 15.2.

$$\text{Part I} = \mathrm{P}\int_{\infty}^{-\infty} \mathrm{d}\omega' \, \frac{G(\omega')}{\omega' - \omega}, \tag{15.11}$$

where $G(\omega) \equiv G_{L2}(s = \mathrm{i}\omega) = \int_0^\infty \mathrm{d}t \, F(t)\, \mathrm{e}^{-\mathrm{i}\omega t}$. We integrate from $+\infty$ to $-\infty$ because the convention in the Cauchy theorem is that the path γ goes counterclockwise in the complex plane.

Part II of the integral is the small semicircle with radius $r \to 0$. Writing $s' = \mathrm{i}\omega + r\,\mathrm{e}^{\mathrm{i}\theta}$, we can approximate $G_{L2}(s')$ by $G(\omega)$ and pull it out of the integral, leaving

$$\text{Part II} = G(\omega) \int_{\pi/2}^{-\pi/2} \mathrm{d}\theta \, \frac{\mathrm{i}\, r\, \mathrm{e}^{\mathrm{i}\theta}}{r\, \mathrm{e}^{\mathrm{i}\theta}} = -\mathrm{i}\,\pi\, G(\omega)\,. \tag{15.12}$$

Finally, we assume that $G_{L2}(s') \to 0$ fast enough for $|s'| \to \infty$ that Part III $\to 0$ as the contour radius $R \to \infty$. Then, from the Cauchy theorem, Parts I + II + III = 0, implying

$$G(\omega) = +\frac{\mathrm{i}}{\pi}\,\mathrm{P}\int_{-\infty}^{\infty} \mathrm{d}\omega' \, \frac{G(\omega')}{\omega - \omega'}\,. \tag{15.13}$$

Writing $G = G' + \mathrm{i}G''$ and isolating real and imaginary parts gives Eqs. (15.5) and (15.6). The integral $\frac{1}{\pi}\,\mathrm{P}\int_{-\infty}^{\infty} \mathrm{d}\omega' \, \frac{G(\omega')}{\omega - \omega'}$ is also known as the *Hilbert transform* of the function G.

The logic of this alternate derivation is that causality implies the analyticity of the Laplace transform of the transfer function in the RHP, which implies the Kramers–Kronig relations. Conversely, transfer functions with poles in the RHP do not obey the Kramers–Kronig relations. In our first derivation, the requirement for analyticity in the RHP was hidden in the expression for the Fourier transform in Eq. (15.4), which is well defined only if it is possible to close the contour in the RHP.

15.1.2 Bode's Gain-Phase Relation

The Kramers–Kronig relations are between the real and imaginary parts of a response function $G(s)$. Intuitively, there should also be relations between the magnitude (gain) and phase of $G(s)$. There are, but with subtleties. The *gain-phase relation* (Bode, 1945) is

$$\arg G(i\omega) = \frac{\pi}{2}\int_{-\infty}^{\infty} d\nu \, \frac{dM}{d\nu} f(\nu), \qquad f(\nu) \equiv \frac{2}{\pi^2} \ln \coth \frac{|\nu|}{2}, \tag{15.14}$$

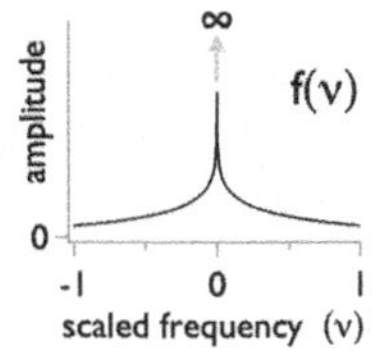

where $\nu = \ln(\frac{\omega'}{\omega})$ and $M = \ln |G(i\omega')|$ is the logarithmic gain and where the kernel $f(\nu)$ is approximately a delta function (see left) about $\nu = 0$, where $\omega' = \omega$. The kernel is normalized so that $\int_{-\infty}^{\infty} d\nu \, f(\nu) = 1$. Because the kernel has finite width, the argument, $\arg G$, depends on $dM/d\nu$ over a range of ν centered on $\nu = 0$. A delta function, by contrast, would pick out the value of $dM/d\nu$ at exactly $\nu = 0$.

A subtlety of the gain-phase relation is that it is more properly an *inequality*: As discussed in Section 3.6, a pure delay adds extra phase lag to a response function while leaving the gain unchanged. In other words, for a given gain $M(\nu)$, we can add an arbitrary amount of extra phase to $\arg G(i\omega)$. The gain-phase relations then give the *minimum* phase lag. Below, we will see that minimum-phase response functions must have all their poles and zeros in the left-hand side of the complex s-plane.

To understand the implications of the gain-phase relation intuitively, we consider a frequency response $G(i\omega) \sim (i\omega)^{-n}$, which holds at high frequencies for physical transfer functions. Recall that a low-pass filter has $n = 1$ and a harmonic oscillator $n = 2$. If this relation were to hold for *all* frequencies $\omega > 0$, then

$$\frac{dM}{d\nu} = \frac{d \ln |G|}{d \ln \omega} = -n, \tag{15.15}$$

and the phase delay is $-\frac{\pi}{2}n$. More generally, $n(\omega)$ is the *local* value of $\ln |G(i\omega)|$. In that case, we note that the kernel $f(\nu)$ illustrated above resembles a broadened delta function, with most of its weight near $\nu = 0$ ($\omega' = \omega$). If $n(\omega)$ is constant over about a decade of frequency centered on ω, then the Bode relation is, approximately,

$$\arg G(i\omega) \approx \frac{\pi}{2} \frac{d \ln |G(i\omega)|}{d \ln \omega} \approx -\frac{\pi}{2} n(\omega). \tag{15.16}$$

As a result, when frequency response is graphed on *Bode plots* with logarithmic frequency axes and a logarithmic magnitude axis, the phase lag is approximately the derivative of the magnitude curve times $\pi/2$. Figures 2.4 and 2.5 illustrate this equivalence.

Derivation

We consider a transfer function $G(s)$ with positive DC gain $G(0)$ and no RHP poles or zeros. The former restriction is needed, as a negative DC gain is more an overall conversion factor between input and output than an extra 180° phase shift. For the Bode theorem, $G(s) = \pm\left(\frac{1}{1+s}\right)$ are both minimum phase. The latter restriction (no RHP poles or zeros) will be necessary for minimum-phase response.

Let us now apply Kramers–Kronig to the logarithm of such a transfer function. Note that $\ln G(\mathrm{i}\omega) = \ln|G(\mathrm{i}\omega)| + \mathrm{i}\arg G(\mathrm{i}\omega)$, where $\arg G(\mathrm{i}\omega)$ is the phase of the complex number $G(\mathrm{i}\omega)$ at frequency ω. Then Eq. (15.8) gives

$$\arg G(\mathrm{i}\omega) = \frac{2\omega}{\pi}\,\mathrm{P}\int_0^\infty \mathrm{d}\omega'\,\frac{\ln|G(\mathrm{i}\omega')|}{\omega'^2-\omega^2}\,. \tag{15.17}$$

As Bode recognized, Eq. (15.17) becomes more intuitive after integrating by parts. First, we change variables: $\nu \equiv \ln(\frac{\omega'}{\omega})$, or $\omega' = \omega\,\mathrm{e}^\nu$, and $M(\nu) \equiv \ln|G(\mathrm{i}\omega')|$, giving

$$\arg G(\mathrm{i}\omega) = \frac{2}{\pi}\,\mathrm{P}\int_{-\infty}^\infty \mathrm{d}\nu\,\frac{\cancel{\omega} M(\nu)}{\cancel{\omega^2}\,(\mathrm{e}^{2\nu}-1)}\,\cancel{\omega}\,\mathrm{e}^\nu = \frac{1}{\pi}\,\mathrm{P}\int_{-\infty}^\infty \mathrm{d}\nu\,\frac{M(\nu)}{\sinh\nu}\,. \tag{15.18}$$

Because sinh is an odd function, only the odd part of $M(\nu)$ contributes in Eq. (15.18). Writing M as the sum of odd and even functions, $M = M_\mathrm{o} + M_\mathrm{e}$, we have

$$\begin{aligned}\mathrm{P}\int_{-\infty}^\infty \mathrm{d}\nu\,\frac{M(\nu)}{\sinh\nu} &= \lim_{\varepsilon\to0^+}\left[\int_{-\infty}^{-\varepsilon}\mathrm{d}\nu\,\frac{M_\mathrm{o}(\nu)+\cancel{M_\mathrm{e}(\nu)}}{\sinh\nu} + \int_\varepsilon^\infty \mathrm{d}\nu\,\frac{M_\mathrm{o}(\nu)+\cancel{M_\mathrm{e}(\nu)}}{\sinh\nu}\right]\\ &= 2\lim_{\varepsilon\to0^+}\int_\varepsilon^\infty \mathrm{d}\nu\,\frac{M_\mathrm{o}(\nu)}{\sinh\nu}\,,\end{aligned} \tag{15.19}$$

since the even contributions M_e cancel in the two integrals. Now integrate by parts, noting that $\int\frac{\mathrm{d}\nu}{\sinh\nu} = -\ln\coth\frac{\nu}{2}$. Then, Eq. (15.19) becomes

$$= 2\lim_{\varepsilon\to0^+}\left\{\int_\varepsilon^\infty \mathrm{d}\nu\,\frac{\mathrm{d}M_\mathrm{o}}{\mathrm{d}\nu}\ln\coth\frac{\nu}{2} - M_\mathrm{o}(\nu)\ln\coth\frac{\nu}{2}\bigg|_\varepsilon^\infty\right\}. \tag{15.20}$$

The boundary terms vanish:

1. As $\nu\to\infty$, $\ln\coth\frac{\nu}{2} \sim 2\,\mathrm{e}^{-\nu} = 2(\frac{\omega}{\omega'})$. As $\omega'\to\infty$, $|G(\mathrm{i}\omega')| \sim \omega'^{-n}$, since the response of physical transfer functions vanishes at infinite frequencies. Then, $\ln|G| \sim -n\ln\omega'$. Thus, $M_\mathrm{o}(\nu)\ln\coth\frac{\nu}{2} \sim \frac{\ln\omega'}{\omega'} \to 0$.
2. As $\nu\to0$, $\ln\coth\frac{\nu}{2} \approx -\ln\frac{\nu}{2}$. Since M_o is odd, $M_\mathrm{o} \sim \nu + O(\nu^3)$. And then $\nu\ln\nu\to0$.

Thus,

$$\arg G(\mathrm{i}\omega) = \frac{2}{\pi}\int_0^\infty \mathrm{d}\nu\,\frac{\mathrm{d}M_\mathrm{o}}{\mathrm{d}\nu}\ln\coth\frac{\nu}{2} = \frac{1}{\pi}\int_{-\infty}^\infty \mathrm{d}\nu\,\frac{\mathrm{d}M}{\mathrm{d}\nu}\ln\coth\frac{|\nu|}{2}\,. \tag{15.21}$$

In the last step, we used $2M_\mathrm{o} = M(\nu) - M(-\nu)$.

The relationship Eq. (15.17) between $\ln|G(\mathrm{i}\omega)|$ and $\arg G(\mathrm{i}\omega)$ holds only if $\ln G(s)$ has no poles in the RHP. But if $G(s) = 0$ in the RHP, the logarithm diverges, $\ln G$ has a pole, and the Bode relation, Eq. (15.21), does not hold. From Section 3.6, such transfer functions have extra phase delays. Transfer functions that obey the Bode relation are *minimum phase*, with poles and zeros in the left-hand side of the complex s-plane (LHP).

Consequences

Although the Bode gain-phase relations are less familiar to physicists than the Kramers–Kronig relations, they can be relevant. For example, they can play a role in the measurement of the optical properties of materials. If the light source is weak or incoherent, only the magnitude of the reflection coefficient can be measured. The Bode relation then determines the phase. Given the magnitude and phase, you can infer the index of refraction and absorption. If the response has RHP zeros, then you will need to take the extra phase shifts into account in order to correctly infer the absorption (Peiponen and Saarinen, 2009).

If the gain-phase relation is occasionally important in physics measurements, it is key to understanding the limitations of control. Chapter 3 introduced the notion of non-minimum-phase response. To recap, any phase lag is hard to control, and NMP functions with extra phase lags are harder still. The extra phase lags are associated with RHP zero or poles and with nonrational transfer functions such as pure delays, e^{-s}, and diffusion, $e^{-\sqrt{s}}$, which can both be approximated using rational functions. We can decompose an arbitrary $G(s)$ into the product of a minimum phase and an all-pass function, $G = G_{\mathrm{mp}}\, G_{\mathrm{ap}}$, where all-pass transfer functions have unit gain, $|G_{\mathrm{ap}}(\mathrm{i}\omega)| = 1$. By looking at simple cases of all-pass transfer functions, we saw that RHP zeros limit the maximum control bandwidth whereas RHP poles (unstable systems) impose a minimum bandwidth. Specifically, if there is an RHP zero at z, then stability requires a bandwidth $\lesssim 0.6\, z$; if there is an RHP pole at p, then stabilization requires a bandwidth $\gtrsim 1.7\, p$. See Problem 3.16. The best control strategy is to eliminate RHP poles and zeros by redesign of the system. If that is not possible, try to ensure $z \gg p$. See Problem 3.17.

15.1.3 Bode's Waterbed Effect

Causality and analyticity further constrain the sensitivity functions defined in Section 3.1. Recall that a SISO system that attempts to reject an output disturbance $v(s)$ has output $y(s) = S(s)\, v(s)$, where $S = \frac{1}{1+L}$ the sensitivity function (Eq. 3.24). The loop transfer function is $L = KG$, with K the controller and G the system. Robust performance requires $|S(\mathrm{i}\omega)| \ll 1$, as small values of $|S|$ reduce disturbance errors, including those generated by a model mismatch.

The loop transfer function L must go to zero at high frequencies, since the response G of physical systems is zero at high frequencies. As a result, $S \to 1$ as $\omega \to \infty$: it is impossible to compensate for disturbances to arbitrarily high frequencies. Actually, causality imposes even stronger constraints on the function $S(s)$. For stable systems whose L has relative degree ≥ 2,[5]

$$\int_0^{\infty} \mathrm{d}\omega \ln |S(\mathrm{i}\omega)| = 0\,. \tag{15.22}$$

[5] That is, $L(\mathrm{i}\omega) \sim \omega^{-n}$, with $n \geq 2$ as $\omega \to \infty$. See Problem 15.3 to understand the need for this condition.

Equation (15.22) implies that on a log-linear plot of $|S(i\omega)|$, the area below 0, where $|S| < 1$, is balanced by an equal area above 0, where $|S| > 1$. When $|S| < 1$, the control loop decreases the effect of disturbances, and the system is more *robust*. When $|S| > 1$, the control loop actually amplifies disturbances, and the system is more *fragile*. Indeed, $\ln |S(i\omega)|$ is sometimes called the *fragility*. See right for a sketch.

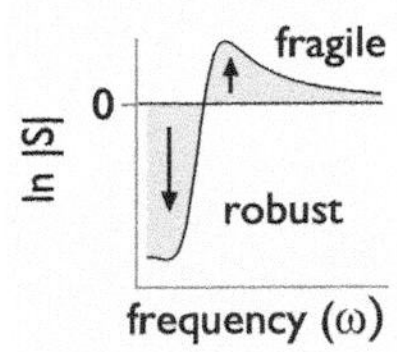

Equation (15.22) thus implies a kind of "conservation of fragility": frequencies where the system is robust to disturbances are "paid for" by frequencies where the system is fragile. This is the *waterbed effect*: push $|S|$ down at some frequencies, and it will pop up at others! As shown at right for the second-order system $G(s) = 1/(s^2 + s + 1)$ with proportional control $K_p = 1, 3, 9$, higher feedback gains increase both disturbance rejection and feedback bandwidth – and also system fragility. A controller can influence but not arbitrarily manipulate the frequency response of disturbance rejection.

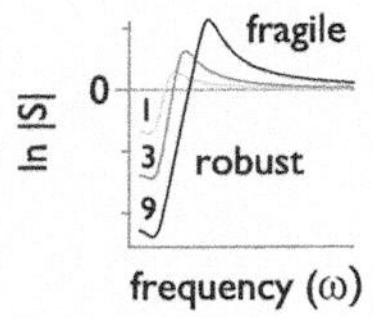

In principle, the fragile region in the waterbed effect can have very small amplitude, because the constraint is on the integral of $\ln |S|$ out to $\omega = \infty$. In practice, the magnitude of the closed-loop transfer function $|T(s)|$ (complementary sensitivity function) should be small beyond a closed-loop bandwidth ω_c (Section 3.1.3). Further, because of noise, we generally cannot make ω_c too big. Since $S + T = 1$, if $|T| \to 0$, then $|S| \to 1$ and $\ln |S| \to 0$. In words, restricting the bandwidth of T "bottles up" the fragile frequency range, creating a peak in the response function and enhancing the waterbed effect.[6]

Sensitivity constraints are even more severe when the open-loop system is unstable, as the integral in Eq. (15.22) will be positive, not just zero. The waterbed relation thus makes it clear that the difficulties in control associated with unstable poles and non-minimum-phase zeros are intrinsic and cannot be avoided via a better choice of controller. Recall the example of balancing an upside-down pendulum by hand (Example 3.4 and Problem 3.14), where the difficulty of the control task depended on the location of the pendulum position sensor. The transfer function from force input (moving one's hand) to pendulum top has no zeros, implying a minimum gain to stabilize the system. By contrast, the transfer function from input to the bottom of the pendulum has an NMP zero, making the control task very difficult, if not impossible. Again, although these control difficulties are intrinsic, the position of the sensor is not. Moving or adding more sensors can simplify enormously a control task.

Derivation

To derive Eq. (15.22), recall that stable, causal response functions $G(s)$ are analytic for Re $s > 0$. If the controller $K(s)$ is also minimum phase and stable, then L and S and $\ln S$ are all analytic in the RHP. Assume also that L has no poles or zeros on

[6] Another factor enhancing the waterbed effect is that the Poisson integral formula implies that the sensitivity trade-offs occur over an even tighter frequency range than that of the sensitivity integral (Doyle et al., 1992).

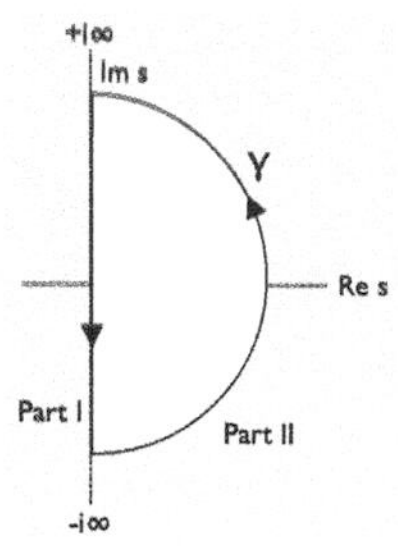

the imaginary axis $s = \mathrm{i}\omega$ itself. Then, since $\ln S(s)$ is analytic in the RHP, Cauchy's Theorem implies that

$$\oint_\gamma \mathrm{d}s \ln S(s) = 0 \qquad (15.23)$$

vanishes, where γ is the contour depicted at left. Break up the integral into two parts. Part I is evaluated from $s = +\mathrm{i}\omega$ to $s = -\mathrm{i}\omega$. In Problem 15.3, you will show that the real part of $\ln S(s)$ is an even and the imaginary part an odd function of ω,. Thus,

$$\text{Part I } = \mathrm{i}\int_R^{-R} \mathrm{d}\omega \ln S(\mathrm{i}\omega) = -2\,\mathrm{i}\int_0^R \mathrm{d}\omega \ln |S(\mathrm{i}\omega)| \,. \qquad (15.24)$$

If the relative degree n of L is at least 2 (i.e., $L \sim \omega^{-n}$ as $\omega \to \infty$, with $\nu \geq 2$), then Part II vanishes as the circle radius $R \to \infty$, and Eq. (15.22) follows from Eq. (15.23).

Extensions

There are many extensions and variants of the waterbed theorem. Here are three:

1. *Unstable poles*. In deriving the waterbed theorem, we assumed that the system to be controlled, $G(s)$, was stable. If not, the constraints on S are even more severe. Let G be unstable, with RHP poles $\{p_j\}$, and assume that the controller K does not try to cancel the poles of the system G – a bad strategy, as we saw in Section 3.5.3. The closed-loop system should be stable. The loop transfer function $L(s)$ inherits the unstable poles at $\{p_j\}$, meaning that $S = 0$ and $\ln S$ is not analytic at those positions. For $\nu > 1$, the waterbed integral becomes (Problem 15.3),[7]

$$\int_{-\infty}^{\infty} \frac{\mathrm{d}\omega}{2\pi} \ln |S(\mathrm{i}\omega)| = \sum_j p_j \,. \qquad (15.25)$$

   Since poles come in complex-conjugate pairs and since Re $p_j > 0$, the sum on the right-hand side of Eq. (15.25) is both real and positive, meaning that $|S|$ must be greater than 1 over an even greater range of frequencies. Defining the $\sum_j p_j$ to be the *degree of instability* of a system, we can interpret Eq. (15.25) as implying that systems having a higher degree of instability are harder to control. The statement sounds obvious but takes its power from the precise and fundamental limits that it summarizes.

2. *Waterbed theorem for T*. The complementary sensitivity function is $T = 1-S$, which should ≈ 1 for good tracking of reference signals. In Problem 15.5, we show

$$\int_{-\infty}^{\infty} \frac{\mathrm{d}\omega}{2\pi}\,\frac{1}{\omega^2} \ln |T(\mathrm{i}\omega)| = \sum_j \frac{1}{z_j} \,, \qquad (15.26)$$

   where z_j are the RHP zero positions. The right-hand side vanishes if L is minimum phase. Fast poles and slow zeros in the RHP both make a system hard to control.

[7] Because $\ln[S(\mathrm{i}\omega)]$ is an even function of ω, Eq. (15.25) is equivalent to $\int_0^\infty \mathrm{d}\omega \ln |S(\mathrm{i}\omega)| = \pi \sum_j p_j$. When there are no unstable open-loop poles, we recover Eq. (15.22). Problem 15.4 treats the case $\nu = 1$.

The factor ω^{-2} makes the integral in Eq. (15.26) converge, since $|T| \to 0$ at high frequencies.

3. *Discrete dynamics*. For $\nu \geq 1$,[8] the discrete waterbed theorem is

$$\int_{-\pi}^{\pi} \frac{d\omega}{2\pi} \ln |S(e^{i\omega})| = \sum_j \ln p_j \,, \tag{15.27}$$

where the sum is again over the unstable poles, which, in the discrete case, satisfy $|p_j| > 1$. Notice that because the integral is over a finite frequency range, the sensitivity peak in the fragile region must be finite. In the discrete case, the degree of instability is defined to be the *product* of unstable eigenvalues. Again, because poles come in complex-conjugate pairs (and $\ln(\cdot)$ is an analytic function), the sum in Eq. (15.27) is real. Thus, the discrete Bode sensitivity integral equals the log of the degree of instability.

Example 15.1 Consider stochastic, one-dimensional, discrete dynamics, with $\nu_k \sim \mathcal{N}(0, \nu^2)$, a noise term of variance ν^2 (e.g., from thermal fluctuations), obeying

$$x_{k+1} = a x_k + u_k + \nu_k \,, \qquad y_k = x_k \,, \qquad u_k = -K y_k \,, \tag{15.28}$$

where K is a proportional-feedback gain. The system transfer function $G(z) = \frac{y(z)}{u(z)} = \frac{1}{z-a}$ and loop gain $L = \frac{K}{z-a}$. The sensitivity function is

$$S(z) = \frac{1}{1+L} = \frac{1}{1+\frac{K}{z-a}} = \frac{z-a}{z-(a-K)} \,. \tag{15.29}$$

For simplicity, choose $K = a$, so that $S(z) = 1 - az^{-1}$. Then substituting $z = e^{i\omega}$ gives

$$\left|S(e^{i\omega})\right|^2 = \left|1 - a e^{-i\omega}\right|^2 = \left(1 + a^2 - 2a\cos\omega\right) . \tag{15.30}$$

From an identity in Problem 15.6, the Bode integral is

$$\int_{-\pi}^{\pi} \frac{d\omega}{2\pi} \ln |S(e^{i\omega})| = \begin{cases} 0 & |a| \leq 1 \\ \ln|a| & |a| > 1 \end{cases} \,, \tag{15.31}$$

which is consistent with the Bode theorem, Eq. (15.27).

Although we have assumed that the feedback gain $K = a$, Problem 15.7 shows that Eq. (15.31) holds for $K \neq a$, as long as K stabilizes the closed-loop system. That is, K must be in the range defined by $|a - K| < 1$. It is remarkable that the sensitivity integral can be independent of the feedback gain K, as illustrated in Figure 15.3: when $|a| < 1$ and the dynamics are stable, the positive areas of the log-sensitivity curve just balance the negative ones. But when $|a| > 1$ and the dynamics are unstable, the integral is positive.

---

[8] The $\nu = 1$ case is allowed – see Problem 15.6.

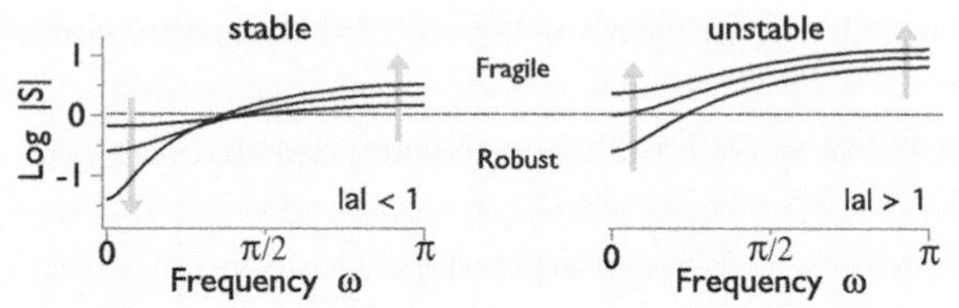

Fig. 15.3 Discrete system waterbed integrand, $\log_{10}|S(e^{i\omega})|$, for the system defined in Example 15.1. At left, the stable case, with curves for $a = 0.2, 0.5, 0.8$, and the integral over frequency is zero. At right, the unstable case, with curves for $a = 1.5, 2$, and 2.5, and the integral is positive. The arrows denote increasing a.

Living with the Waterbed Effect

The waterbed theorem constrains performance so strongly that you might question the value of adding feedback. If suppressing low-frequency disturbances necessarily amplifies high-frequency disturbances, and if disturbances actually have equal amplitudes at all frequencies, then the limitation is a serious one. And the postulate of disturbances at all frequencies might seem reasonable: Both white noise and sudden shocks do indeed have equal-amplitude components at all frequencies.

But what if only certain frequencies matter? Then you can design your control to suppress the relevant range, not worrying about what happens at other frequencies. Even better: a careful combination of active control and passive filtering can attenuate disturbances at *all* frequencies, despite the limitations of the waterbed effect.

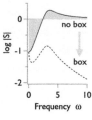

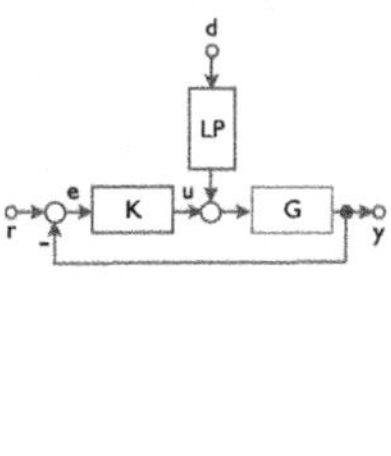

The trick is that *passive* filtering can partially isolate the system from its environment, ensuring that disturbances occur only in the limited range of frequencies where feedback is effective. Consider temperature control, where insulating a system by placing it in a box adds a low-pass filter that can remove the unwanted sensitivity peak. The illustration at left shows the sensitivity to disturbances of a simplified model representing temperature control. The waterbed effect leads to a sensitivity peak near $\omega \approx 3$ (continuous line). Putting the system in a box reduces the effective sensitivity function. Because the box suppresses the high-frequency and feeds back the low-frequency disturbances, the result is that *all* frequencies are attenuated (dashed line). In effect, the passive filter created by the box is analogous to the prefilters we have used for feedforward control (Section 3.4). See the block diagram at left, where the box is represented by a low-pass filter (LP) and Problem 15.8 for details.

Combining active feedback and passive filtering can thus outperform either technique alone. Because of the waterbed effect, an active component that suppresses disturbances at low frequencies will amplify them at high frequencies. But the passive component attenuates the high frequencies while leaving the low ones unaffected. Together, the two components attenuate all frequencies. These ideas have wide application. For example, for vibration control, place a device on a soft rubber base to passively isolate against high-frequency vibrations. Then add an active system for lower frequencies. Such methods are used for sensitive instruments such as atomic force, electron microscopes, and (in much more elaborate form) the LIGO gravitational wave detector. The main point is that the need to mix active and passive components traces back to fundamental issues in control.

Not every setting will allow such isolation, but many do. Regulation problems are amenable, as are input disturbances, which have a transfer function $G(s)\,S(s)$. The extra factor of $G(s)$ also filters the high-frequency response, although perhaps not as thoroughly as the box does for temperature control. An advantage of designing the isolation is that you can tune its bandwidth independently from that of the system. For example, in the temperature-control problem, we can choose the thickness ℓ and thermal diffusivity D of the wall of the box to set the low-pass cutoff frequency to match the feedback bandwidth of the internal system, $G(s)$. (See right.)

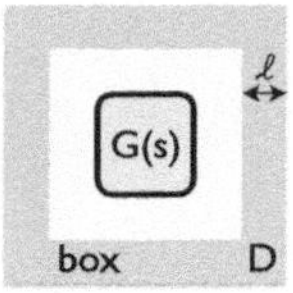

15.1.4 Acausal Control

Please read these lectures last week.

Sidney Coleman, 1970

Causality constrains transfer and response functions and thus limits the performance of feedback control. But what if control could use information about the future? Although the idea might seem crazy, it is important to realize that while the fundamental laws of physics are causal,[9] control systems can behave in ways that seemingly violate causality. Control systems are *open*, in the sense that they are a subsystem of the universe. And while the universe is governed by causal laws, subsystems need not be. Thus, although we tend to think that we cannot know the future of a signal, that is not always true. For example, we can know the future behavior of a reference signal that we impose on a system, since we are free to choose the reference signal however we want. For example, a mechanical stage, such as used in scanning probe microscopy, scans back and forth repeatedly. The desired position reference signal is periodic and thus known arbitrarily far into the future. Since we decide what the reference signal is to be, we can know $r(t)$ in advance.

If we know the future of a signal, then that future information can improve the control response. For example, the minimum delays implied by Bode's gain-phase relation are modified. To see this, consider a dynamical system whose response includes a delay τ. If we know the reference signal at least a time τ in the future, we can simply advance our signal by τ to compensate for the delay exactly. As shown at right for a first-order system with delay, the usual response (top) shows a delay τ relative to the input. In the acausal case, the step-function input is advanced by τ, and the response starts at the desired time.

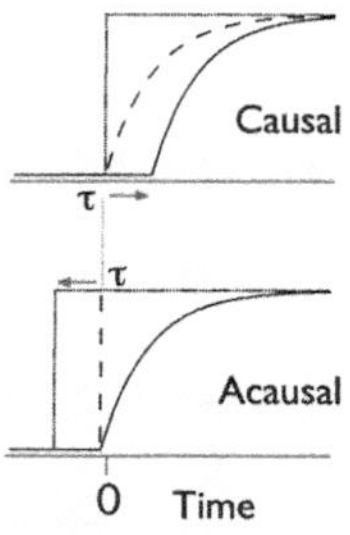

It is also possible to have information about future disturbances. For example, in controlling the temperature of a room, we can *anticipate* a warm disturbance in the morning, after sunrise, and a cool disturbance in the evening, after sunset. Controllers

[9] Many people have postulated the existence of particles, or even entire universes, whose arrow of time runs backwards (Carroll, 2010). There has never been any experimental confirmation of such ideas.

based on the anticipation of future information and that thus formally lack causality are *acausal*.[10]

To control an acausal system, recall from Chapter 3 that an NMP transfer function factors into the product of a minimum phase transfer function and an all-pass function (see Eq. 3.58). We control the minimum-phase part using whatever usual technique is appropriate for that and focus on techniques to control the all-pass function. In the simplest case, we have a function of the form of Eq. (3.57), $G = \frac{1-s}{1+s}$. Since G approximates a delay, its inverse approximates an advance.

Adaptive feedforward control is an example of acausal control of a reference signal. In Section 10.3.2, we showed that we could cancel periodic disturbances exactly, with no phase delay. The scheme worked because a periodic signal is implicitly known indefinitely far into the future.[11]

Disturbance feedforward compensation is another instance of acausal control. In Section 3.4.3, we showed that learning about disturbances before they reach the sample (or actuator) gives more time to correct them than does a causal controller that measures the disturbance at the point where they are to be corrected. Performance exceeds the ordinary causal limits because "advance warning" is designed into the controller structure.[12]

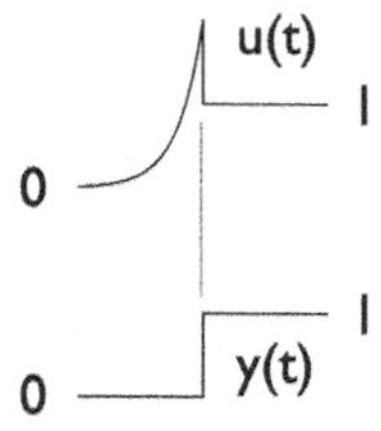

Example 15.2 (Acausal control of a first-order system) Let us consider the first-order, non-minimum-phase system $G(s) = \frac{1-\tau s}{1+\tau s}$. For $z > 0$, this system has a stable pole at $s = -\tau^{-1}$ and an unstable zero at $s = +\tau^{-1}$. The goal is to make the output $y(t)$ follow a step function, $\theta(t)$. The acausal feedforward control input illustrated at left, $u(t) = 2\,e^{t/\tau}$ for $t < 0$ and 1 afterwards, leads to the desired output (Problem 15.9). Although, in principle, you need to start the input infinitely far in advance, the required input decays exponentially with negative time. In practice, starting $\approx 10\tau$ in advance is enough.

Problem 15.10 gives another example where advance information can improve control.

15.2 Information-Theoretic Limits

We now turn to a different category of limits to control, those having to do with information. We have already seen that the available information can determine performance. The challenge is to quantify such intuitive concepts. Here, we investigate four aspects: first, we reformulate the frequency-domain Bode relations of the previous

[10] The terms *noncausal* and *preview* control are also used.

[11] *Iterative learning control* (ILC) is a technique for learning repeated control maneuvers adaptively. It can be very effective for controlling robot-arm maneuvers in a factory.

[12] Assuming the anticipations are correct. In everyday life, those who act on anticipation tend to be rewarded as farsighted when their guesses are correct and dismissed as foolish when not.

section in the time-domain. The time-domain versions apply to both time-varying and nonlinear systems. Equally important, information-theoretic quantities appear naturally. Second, since causality is important for feedback systems, we discuss how to extend basic information to define quantities that capture causal relations, applying it to the Bode relations. Third, we consider how to analyze dynamical systems in terms of information flows. One nice application is a better understanding of when feedback can (or cannot) improve measurements. Finally, we use information theory to give new insight into the feedback stabilization of an unstable system. For background on information theory, see Appendix A.10 online.

15.2.1 Bode Relations in the Time Domain

Let us now formulate a time-domain version of the waterbed theorem. Whereas the physical meaning of the sensitivity integral is somewhat opaque in the frequency domain, its time-domain formulation will turn out to be more intuitive.

We begin with the *entropy rate* $\mathcal{H}(X)$, which is the average entropy per observation of the random variable X,[13] when averaged over a very long time series $\{X_1, X_2, \dots, X_N\} \equiv X^N$, with $N \to \infty$. See Section A.10.2 online. The entropy rate $\mathcal{H}$ is also the average information gain per observation after transients have decayed. For *independent, identically distributed* (i.i.d.) variables X_i, the entropy $H(X^N) = \sum_k H(X_k) = NH(X)$, so that $\mathcal{H}(X) = H(X)$, the entropy of a single observation. Correlations between successive observations X_i reduce the information supplied by a single observation: $\mathcal{H}(X) < H(X)$. See Example A.25 online.

Now let us consider a discrete, linear, time-invariant, dynamical system described by the transfer function $S(z)$. In the Z-domain, the output $y(z) = S(z)\,\nu(z)$. Then, as we will see,

$$\mathcal{H}(Y) = \mathcal{H}(\nu) + \int_{-\pi}^{\pi} \frac{\mathrm{d}\omega}{2\pi} \ln \left|S(\mathrm{e}^{\mathrm{i}\omega})\right| , \tag{15.32}$$

where $\mathcal{H}(\nu)$ is the *entropy rate* of the input stochastic process ν and $\mathcal{H}(Y)$ is the entropy rate of the output stochastic process.[14]

To understand Eq. (15.32), we can view the transformation between the N-element input and output time series ν^N and Y^N as a linear change of real variables. From Eq. (A.267 online), the entropy (in nats) of the whole time series transforms as

$$H(Y^N) = H(\nu^N) + \ln |J| , \tag{15.33}$$

where J is the determinant of the Jacobian matrix of the transformation between ν and Y.

[13] When X is a continuous variable, $\mathcal{H}(X) = \lim_{N\to\infty} \sum_{k=1}^{N} H(X_k)/N$ is the rate of *differential entropy* increase, with $H(X) = -\int \mathrm{d}x\; p(x) \ln p(x)$. See Section A.10.1 online.

[14] We ought to denote the stochastic process ν by N, the capital of the Greek letter ν. But that conflicts with the length of a time series, and we abuse notation slightly by using ν instead.

To evaluate the Jacobian, we start with the time-domain relation $y_k = \sum_{n=0}^{k} S_n v_{k-n}$, with $S_n \in \mathbb{R}$. In matrix form, the N time-domain equations for the first N time-points are

$$\begin{pmatrix} y_0 \\ y_1 \\ \vdots \\ y_{N-1} \end{pmatrix} = \begin{pmatrix} S_0 & 0 & \cdots & 0 \\ S_1 & S_0 & \cdots & 0 \\ \vdots & & \ddots & \vdots \\ S_{N-1} & S_{N-2} & \cdots & S_0 \end{pmatrix} \begin{pmatrix} v_0 \\ v_1 \\ \vdots \\ v_{N-1} \end{pmatrix} \tag{15.34}$$

The determinant of the S-matrix is just $J = (S_0)^N$, since the matrix is lower triangular. Dividing the terms in Eq. (15.33) by N and taking the limit $N \to \infty$ gives

$$\mathcal{H}(Y) = \mathcal{H}(v) + \ln |S_0| \,. \tag{15.35}$$

The last step is to prove that $\ln |S_0| = \int_{-\pi}^{\pi} \frac{d\omega}{2\pi} \ln \left| S(e^{i\omega}) \right|$. If we assume that $v_\ell = 0$ for $\ell < 0$, then the Z-transform of the equations gives $Y(z) = S(z)v(z)$, with $S(z) = \sum_{k=0}^{\infty} S_k z^{-k}$. Notice the limit $|z| \to \infty$ gives $S(z) \to S_0$. We also assume that the dynamics between v_k and y_k are stable, so that both $S(z)$ and $S(z)^*$ are analytic on and outside the unit circle in the complex z-plane. Then

$$\begin{aligned} \int_{-\pi}^{\pi} \frac{d\omega}{2\pi} \ln \left| S(e^{i\omega}) \right|^2 &= \oint \frac{dz}{2\pi i z} \ln \left[S(z) S(z^*) \right] \\ &= \oint \frac{dz}{2\pi i z} \ln S(z) + \text{ c.c.} \\ &\to \oint \frac{dz}{2\pi i z} \ln S_0 + \text{ c.c.} \\ &= \ln |S_0|^2 \,. \end{aligned} \tag{15.36}$$

In Line 1, we substitute $z = e^{i\omega}$; the contour is the unit circle. In Line 2, we expand the integral into two pieces, one over $S(z)$ and the other over $S(z^*)$. The second integral is just the complex conjugate (c.c.) of the first, since $S(z)$ is analytic on the contour (and outside it). In Line 3, we enlarge the contour from the unit circle to one of radius $R \to \infty$. The value of the contour integral is unchanged because $S(z)$ is analytic outside the unit circle. And, in that limit, $S(z) \to S_0$. In the last line, $\ln S_0 + \ln S_0^* = \ln |S_0|^2$. Finally, expanding the log of the square, dividing by 2, and substituting into Eq. (15.35) gives Eq. (15.32). Notice that we did not place any restrictions on the stochastic process v_k.

Interpreting the sequence v_k as output disturbances to a closed-loop linear control system, y_k as the output, and $S(z)$ as the sensitivity function, we see that Eq. (15.32) contains the discrete Bode waterbed integral. Further, if the open-loop transfer function L associated with $S = \frac{1}{1+L}$ is stable and has relative degree ≥ 1, then Eq. (15.27) shows that the entropy rates of the disturbance and output are equal, $\mathcal{H}(Y) = \mathcal{H}(v)$.

Table 15.1 Finite-precision measurements y_k of $x_{k+1} = ax_k$.

	$a = 0.1$	$a = 10$
y_0	3.1416	3.1416
y_1	0.3142	31.4159
y_2	0.0314	314.1593
y_3	0.0031	3141.5926
y_4	0.0003	31415.9265
y_5	0.0000	314159.2654
y_6	0.0000	3141592.6536

When the class of allowed dynamics is enlarged to include unstable systems, we can combine Eq. (15.32) and the discrete waterbed relation Eq. (15.27) to write

$$\mathcal{H}(Y) = \mathcal{H}(v) + \sum_j \ln p_j , \tag{15.37}$$

where $\sum_j$ is over all the unstable poles of the linear dynamics.

To understand the information-theory interpretation of the Bode integral, we first look at a simple case, Example 15.3, where the dynamics have no stochastic input.

Example 15.3 Consider the discrete, deterministic dynamical system, $x_{k+1} = ax_k$. Let us write numbers in base 10 and assume that the output y_k is given by rounding x_k to m decimal places and truncating to be < 10, as would be done by a measuring instrument having finite precision and range.[15] If, for example, $x_0 = \pi = 3.14159265359\ldots$ and $m = 4$, then $y_0 = 3.1416$. Such a truncation is known as *coarse graining* and leads to the notion of *ϵ-entropy* – entropy defined by partitioning phase space at a scale ϵ. For the cases $a = 0.1$ and $a = 10$, the dynamics are shown in Table 15.1. When $a = 0.1$, we have $y_k = 0$ for $k \geq 5$. The entropy rate of the time series is thus zero. By contrast, when $a = 10$, each iteration reveals a new digit from the infinite-precision initial condition.[16] Thus, if we adopt for this example information units in base 10, the information rate is $\log_{10} a = \log_{10} 10 = 1$ "dit" (decimal digit),[17] as expected from Eq. (15.37).

Notice in Example 15.3 that the result is independent of the value of the precision m of the measurement, which need not actually be specified. (Transients are

[15] We could drop the condition that $0 \leq y_k < 10$, but then the output would not be a stationary process.

[16] Since the instrument presumably has a maximum (e.g., 9.9999) as well as a minimum measurement value (0.0000), we indicate the "overflow" digits in gray. These truncated digits are "remembered" in the collective entropy of the entire time series, as we do not "forget" information. The important point is that with unstable dynamics, new digits from the initial condition are continually revealed to the measurement instrument.

[17] Another name for the base-10 unit of information is *ban*, coined by Alan Turing and I. J. Good in 1940. The name came from the town of Banbury, which was near Bletchley Park, where they worked to decode German military communications during World War II. *Dit* is analogous to *bit* and seems more natural today.

excluded by the definition of rates.) The next example shows how stable dynamics (no entropy production) merely "shuffles" the information of a stochastic input in the time domain.

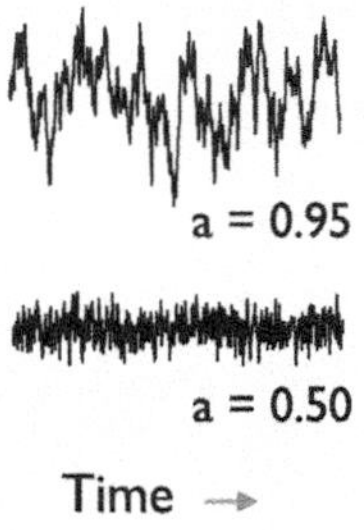

Example 15.4 Consider a diffusing particle in one dimension in a fluid, confined by a discrete feedback loop that implements proportional feedback. Let $x_{k+1} = ax_k + \nu_k$, with $\nu_k \sim \mathcal{N}(0, \nu^2)$ and $0 < a \equiv 1 - KT_s < 1$. The variance $\nu^2 = 2DT_s$ for diffusion coefficient D and time step T_s. From the waterbed theorem for discrete dynamics, the Bode integral $\int_0^{\pi} \mathrm{d}\omega \ln |S(e^{i\omega})| = 0$. Equation (15.32) then implies that the entropy rates for x_k and for ν_k are the same, for all feedback gains K (or a). See Problem 15.12. At left, we plot time series for $a = 0.95$ ("low feedback") and for $a = 0.50$ ("high feedback"). Above, low feedback gain leads to bigger excursions and greater information gain per observation. Below, high feedback gain leads to larger correlations among neighboring measurements, which decreases the information gain per observation. The two effects just balance, making the information rate independent of a, as long as the resulting system is stable.

Finally, Problem 15.11 raises a paradox in the interpretation of entropy rates. The important physical point is that the equality of entropy rates in Eq. (15.37) arises because output disturbances affect the output instantaneously and with unit gain.

15.2.2 Causality and Information Theory

The common tools of information theory express correlations, not causation. Because the mutual information $I(X;Y) = I(Y;X)$ is symmetric in its arguments, measuring Y tells you as much about X as measuring X tells about Y (Section A.10.4 online). By contrast, causality is an inherently asymmetric concept that is a crucial aspect of control theory, where systems *respond* in time to perturbations. Thus, an information-theoretic formulation of control theory ought to incorporate notions about causality.

Consider, then, two sets of random variables, X^N and Y^N. From the definition of conditional probability and from the chain rule for probabilities (Problem A.6.1 online),

$$P(X^N, Y^N) = P(Y^N|X^N)\, P(X^N) = \prod_{k=1}^{N} P(Y_k|Y^{k-1}, X^N)\, P(X_k|X^{k-1})\,. \tag{15.38}$$

Equation (15.38) holds for any two sets of N random variables. If they are sequences in time, then the first set of terms on the right-hand side depends on X^N, which contains past ($< k$), present ($= k$), *and future* ($> k$) values. If sequences are post-processed, the dependence on future values is not a problem. But in responding to a signal in real time, we would not normally have information about future values.[18]

[18] As we discussed in Section 15.1.4, many systems are open with respect to information. Outside information can lead to rational expectations about future values of a time series. Here, we assume that is not the case.

To impose causality and see the role of feedback more explicitly, we use an alternate decomposition of $P(X^N, Y^N)$:

$$P(X^N, Y^N) = \prod_{k=1}^{N} P(X_k, Y_k|X^{k-1}, Y^{k-1}) = \prod_{k=1}^{N} P(Y_k|Y^{k-1}, X^k)\, P(X_k|X^{k-1}, Y^{k-1})$$
$$\equiv P(Y^N||X^N)\, P(X^N||Y^{N-1}), \tag{15.39}$$

where we introduce the notion of *causal conditioning*. Compare the two notions:

$$P(Y^N|X^N) = \prod_{k=1}^{N} \underbrace{P(Y_k|Y^{k-1}, X^N)}_{\text{ordinary conditioning}} \quad \leftrightarrow \quad P(Y^N||X^N) \equiv \prod_{k=1}^{N} \underbrace{P(Y_k|Y^{k-1}, X^k)}_{\text{causal conditioning}}. \tag{15.40}$$

Each factor of the product replaces X^N with X^k, to respect causality. If the X^N are inputs and the Y^N outputs, then $P(Y^N||X^N)$ includes direct feedforward from the current input to the current output. Similarly, $P(X^N||Y^{N-1})$ represents feedback: the previous output influences the current input. Note that the conditioning in $P(X^N||Y^{N-1})$ is on Y^{k-1}, not Y^k.

Similarly, we can modify Eq. (A.258 online) to define *causally conditioned entropy*:

$$H(Y^N|X^N) = -\left\langle \log P(Y^N|X^N) \right\rangle = \sum_{k=1}^{N} H(Y_k|Y^{k-1}, X^N)$$

$$\text{suggests} \quad H(Y^N||X^N) \equiv -\left\langle \log P(Y^N||X^N) \right\rangle = \sum_{k=1}^{N} H(Y_k|Y^{k-1}, X^k). \tag{15.41}$$

Mutual information likewise becomes *directed information*:

$$I(X^N; Y^N) = H(Y^N) - H(Y^N|X^N) = \sum_{k=1}^{N} I(Y_k; X^N|Y^{k-1})$$

$$\text{suggests} \quad I(X^N \to Y^N) \equiv H(Y^N) - H(Y^N||X^N) = \sum_{k=1}^{N} I(Y_k; X^k|Y^{k-1}). \tag{15.42}$$

Conditioning on past and future events is a stronger constraint than conditioning only on the past and can only decrease entropy (Section A.10.4 online). Thus, $H(Y^N||X^N) \geq H(Y^N|X^N)$ and the associated rate $\mathcal{H}(Y||X) \geq \mathcal{H}(Y|X)$. Similarly, the extra information provided by future values of Y can only decrease the uncertainty about the X variables and increase the mutual information. In fact, from Eq. (15.39), you can show (Problem 15.14) that

$$I(X^N; Y^N) = \underbrace{I(X^N \to Y^N)}_{\text{feedforward}} + \underbrace{I(Y^{N-1} \to X^N)}_{\text{feedback}}. \tag{15.43}$$

The first term quantifies how much X^N can causally predict Y^N. The second term quantifies how much Y^{N-1} can causally predict X^N. Implicitly, $\ldots \to Y_{N-1} \to X_N \to Y_N \to \ldots$. In terms of rates, $\mathcal{I}(X; Y) = \mathcal{I}(X \to Y) + \mathcal{I}(Y^{(-1)} \to X)$. The last term is between the previous output and current input. Thus, $\mathcal{I}(X \to Y) \leq \mathcal{I}(X; Y)$, with equality if and only if there is no feedback from Y^N to X^N. Cf. Section A.10.5 online.

Example 15.5 (Mutual vs. directed information) Consider a simple delay, $y_{k+1} = x_k$, where x_k are i.i.d. variables, each having entropy $H(X)$. In Problem 15.15, we verify by direct calculation that the mutual-information rate $\mathcal{I}(X;Y) = \mathcal{I}(X \to Y) = H(X)$ per trial. On the other hand, $\mathcal{I}(Y \to X) = 0$, in accordance with Eq. (15.43). Thus, the directed-information rate correctly captures the idea that the process X drives ("causes") the process Y, and not *vice versa*. Because mutual information is symmetric in X and Y, all it can do is indicate that the two time series are correlated.

Let us use directed information to analyze the relation between a sequence of additive disturbances v_k and the corresponding output sequence y_k. Let $y_k = f(y^{k-1}, v^{k-1}) + v_k$, where f is a nonlinear function and v_k are autonomous random variables (independent of the stochastic process Y). From Eq. (15.42),

$$\begin{aligned} I(v^{N-1} \to Y^N) &= \sum_{k=1}^{N} I(Y_k; v^{k-1}|Y^{k-1}) = \sum_{k=1}^{N} H(Y_k|Y^{k-1}) - H(Y_k|v^{k-1}, Y^{k-1}) \\ &= \sum_{k=1}^{N} H(Y_k|Y^{k-1}) - H(v_k|v^{k-1}) = H(Y^N) - H(v^N)\,. \end{aligned} \tag{15.44}$$

In going from Line 1 to 2, note that Y^{k-1} is a deterministic function of v^{k-1}. That is, given the previous outputs y^{k-1} and disturbances v^{k-1}, the uncertainty in y_k is exactly the uncertainty in the random variable v_k.

We can define a directed-information rate $\mathcal{I}(v \to Y) = \lim_{N\to\infty} I(v^{N-1} \to Y^N)/N$ in analogy with the definitions of entropy and mutual-information rates given in Section A.10.4 online. If we restrict ourselves again to linear dynamics, then Eqs. (15.37) and (15.44) imply

$$\mathcal{I}(v \to Y) = \sum_j \ln p_j\,, \tag{15.45}$$

where $\sum_j$ is, again, over all the unstable poles of the linear dynamics. From our previous discussion, Eq. (15.45) implies that the information content of the output time series y^N is the causal contribution from the disturbance v_k plus the extra amount of the initial condition y_0 that is "revealed" by the unstable dynamics. In this information-theory version of Bode's waterbed relation, we see the explicit dependence on causality (dependence on directed rather than mutual information). Equation (15.45) can be extended to time-dependent and continuous dynamics. The analogous results for nonlinear systems are less clear.

Relations such as Eq. (15.45) place powerful constraints on the kinds of closed-loop dynamics that are achievable through feedback. But do they affect the performance that we really care about? Single-time quantities, such as the one-point Shannon differential entropy $H(X)$ of a system state – or even, more simply, the variance of X – may be of more interest.[19] For example, for temperature regulation, the size of typical temperature excursions may matter more than their temporal correlations. In the low-pass filter of Example 15.4, we saw that the two traces reproduced at right have identical entropy rates. But we may prefer the "better-regulated" behavior at bottom.

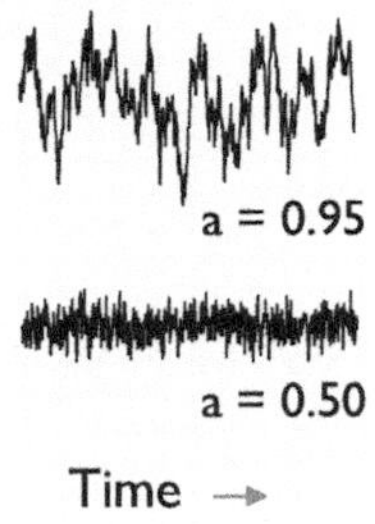

15.2.3 Does Feedback Improve Measurement?

Feedback is commonly used to "improve measurements," but does it? And, if so, why? Using both linear-systems and information-theory arguments, we will see that feedback does not help with linear systems but can with nonlinear ones.

In asking whether feedback "improves measurements," we should first define our terms. The traditional way is to consider *signals* and *noise* as two stochastic processes and to define terms such as signal-to-noise ratio (SNR) and bandwidth. Here, we add an information-theoretic criterion, maximizing the mutual-information rate $\mathcal{I}(U;Y)$,[20] where U is a stochastic process representing some desired signal (an input to a measuring device) and Y is a stochastic process representing the output of the measuring device. We also account for measurement noise, which we model as another stochastic process, Ξ.

Amplifiers with DC Gain

We begin by showing that a linear amplifier can boost the SNR of a constant (DC) signal relative to the noise produced by a measuring device that is after the amplifier. Let U be a stochastic process with realizations $u_k \sim \mathcal{N}(0,\sigma_u^2)$. We consider U to be a sequence of independent, mean-zero Gaussian random variables because we assume that all other deterministic aspects have been previously extracted, including temporal correlations. In a similar spirit, the measurement noise Ξ is a sequence ξ_k of independent Gaussian random variables distributed as $\mathcal{N}(0,\xi^2)$ and also independent of u_k.

Applying a noisy, linear amplifier to the constant signal gives $y_k = G_0 u_k + \xi_k$, where G_0 is the DC gain. The DC signal-to-noise ratio $\text{SNR}_0 \equiv G_0\sigma_u/\xi$ then clearly improves with gain. Here, as always in this book, we define SNR using amplitudes and not power.

Alternately, we can describe a linear amplifier in terms of entropy rates and channel capacity. For independent stochastic sequences, the entropy rate is just the entropy of one variable. Thus, for a sequence u_k of independent Gaussians, the entropy rate is $\mathcal{H}(U) = \log\sqrt{2\pi e\sigma_u^2}$. Similarly, for the measurements, $\mathcal{H}(Y) = \log\sqrt{2\pi e(G_0^2\sigma_u^2 + \xi^2)}$. Finally, from Eq. (A.294 online) and Problem A.10.11 online, the mutual information rate equals

[19] A state that fluctuates between two values, $\pm X_{\text{big}}$, would have low entropy but high variance.

[20] We look at mutual rather than directed information, assuming that the entire measurement record is available to make inferences about the input. Later measurements then do give information about earlier inputs.

$$\mathcal{I}(U;Y) = \frac{1}{2}\log\left(1 + \frac{G_0^2\sigma_u^2}{\xi^2}\right) = \frac{1}{2}\log\left(1 + \mathrm{SNR}_0^2\right), \tag{15.46}$$

which is also the channel capacity (A.306 online).

For small noise ($\xi \ll \sigma_u$), the output $\mathcal{H}(Y) \approx \mathcal{H}(U) + \log|G_0|$. Using a linear amplifier with gain improves the signal-to-noise ratio and increases the amount of information acquired about the signal (by $\approx \log|G_0|$ bits per measurement). Note that Eq. (15.46) gives the long-time average entropy per measurement. To convert to entropy per time, multiply by $f_s = 1/T_s$, where T_s is the sampling time.

Amplifiers with Frequency-Dependent Gain

Above, G_0 was a static gain that was independent of frequency. What if the amplifier gain depends on frequency? For example, the frequency dependence can reflect the amplifier bandwidth, which limits the transmitted information rate. We can immediately conclude that the signal-to-noise ratio depends on frequency: $\mathrm{SNR}(\omega) = |G(\omega)|\sigma_u/\xi$.

In information-theoretic terms, we use Eq. (A.298 online) and assume that the input $u(t)$ and measurement noise $\xi(t)$ are continuous-time, Gaussian processes with finite variances σ_u^2 and ξ^2. Then, since every frequency acts as an independent channel,

$$\mathcal{I}(U;Y) = \frac{1}{2}\int_{-\infty}^{\infty}\frac{d\omega}{2\pi}\log\left(1 + \frac{|G|^2(\omega)\,\sigma_u^2}{\xi^2}\right) = \frac{1}{2}\int_{-\infty}^{\infty}\frac{d\omega}{2\pi}\log\left(1 + \mathrm{SNR}^2(\omega)\right). \tag{15.47}$$

For example, an amplifier with DC gain G_0 and simple low-pass filter with cutoff-frequency ω_c has $|G|^2(\omega) = \frac{G_0^2}{1+\omega^2/\omega_c^2}$. Equation (15.47) then implies that

$$\mathcal{I}(U;Y) = \frac{\omega_c}{2}\left(\sqrt{1 + \frac{G_0^2\sigma_u^2}{\xi^2}} - 1\right) \to \frac{\omega_c G_0 \sigma_u}{2\xi} \equiv \frac{1}{2}\omega_c\,\mathrm{SNR}_0, \tag{15.48}$$

where SNR_0 represents the signal-to-noise ratio at DC. Here, $\mathrm{SNR}_0 \gg 1$ (Problem 15.16). Finite bandwidth limits information acquisition rates, just as a finite sampling rate limited the rate in the case of an amplifier with DC gain.[21] Note that Eq. (15.48) diverges as $\omega_c \to \infty$: an infinite bandwidth would imply infinite information acquisition rates.[22]

Feedback and Linear Systems

The argument of the last two sections shows that amplifiers can improve the SNR (and mutual information rate) of a noisy measurement by boosting the signal above the noise, at a given frequency. But feedback plays no role in such a picture. To include the effects of feedback, we first consider a stable system with input u_k but without

21 See Eq. (15.46) and the note in the following paragraph.

22 Equations (15.46) and (15.48) seem to contradict each other. Problem 15.17 shows that they are different limits of a more general result for finite transducer bandwidth ω_c and sampling time T_s. In Eq. (15.46), the bandwidth of the transducer G is tacitly assumed to greatly exceed the sampling frequency.

noise ($\xi_k = 0$). Equation (15.45) implies that $\mathcal{I}(U^N \to Y^N) = 0$, or $\mathcal{H}(Y) = \mathcal{H}(U)$, independent of whatever feedback we may use in a closed-loop controller. To understand why the information rates are the same in that case, see Problem 15.12. Here, an alternate linear-system analysis reaches the same conclusion when measurement noise is present.

Consider a linear dynamical system in the Laplace domain, $y_{\text{open}} = Gu + \xi$, where $G = G(s)$ is the transfer function of a dynamical system. Now, add a linear feedback signal to the desired measurement signal, $u \to u - Ky$. The block diagram then implies that $y_{\text{closed}} = (Gu + \xi)/(1 + GK)$. Comparing the two expressions gives

$$y_{\text{closed}} = \left(\frac{1}{1+GK}\right) y_{\text{open}} \,. \tag{15.49}$$

In the time domain, we can measure the signal y_{open} without using feedback at all, convolve with the time-domain transfer function $\mathcal{L}^{-1}(\frac{1}{1+GK})$, and obtain precisely the signal we would measure with feedback. (Or we can do the reverse.) As a result, adding linear feedback does not change the information that we can obtain from the signal.

Note that if our dynamical system includes a static gain $G(0) > 1$, then the information-acquisition rate for the output will be greater than that for the input by $\log |G(0)|$. But this gain is from amplifying the signal and is the same whether feedback is present or not.

More formally, for linear systems and Gaussian noise, the Linear-Quadratic-Gaussian (LQG) solution for the optimum state estimation (Kalman filter) derived in Chapter 8 leads to a covariance matrix $\boldsymbol{P}$ for state estimates that does not depend on the control. We can view the uncertainties in states as indicating the quality of measurements.

Feedback thus does not improve measurements for stable linear systems. But what about unstable systems? The same arguments will imply that no improvement in measurement results. However, an unstable system has states and outputs that grow exponentially, and any measurement must saturate. Feedback can keep values within a finite range and allow measurement by a linear system. Thus, *indirectly*, feedback improves the measurement of unstable linear systems. But if measurements can saturate, the original system is, in fact, nonlinear and should not be included in a discussion of linear systems.

Static Nonlinear Relation

A linear transducer maps each input to a unique output value, whereas a nonlinear transducer might map multiple inputs to the same output. Such degeneracy explains why nonlinearity in a measurement can reduce information rates. (The data-processing inequality, Eq. (A.291 online), shows that processing a stochastic variable by a function cannot *increase* information transfer.)

More generally, consider the static nonlinearity produced by a measuring apparatus, also known as a *transducer*. Thus, we generalize from the linear relation $y_k = Gu_k + \xi_k$

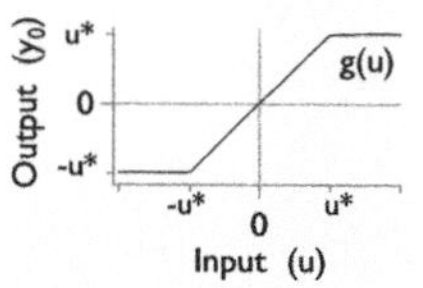

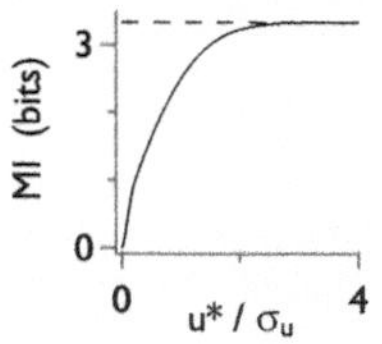

to the nonlinear relation $y_k = g(u_k) + \xi_k$, where $g(\cdot)$ often has a sigmoidal shape. Indeed, a standard analog-to-digital (A/D) converter has the piecewise linear response curve illustrated at left. The function $g(u)$ is linear in the range $|u| \leq u^*$ and saturates at values outside.[23]

Now let us use this piecewise, nonlinear transducer function to measure a sequence of values $u_k \sim \mathcal{N}(0, \sigma_u^2)$ with $\xi_k \sim \mathcal{N}(0, \xi^2)$. It is then easy to understand the limits of large and small u^*. For $u^*/\sigma_u \gg 1$, the output $y_k = u_k + \xi_k$, so that $\mathcal{I}(U;Y) = \frac{1}{2}\log\left(1 + \text{SNR}^2\right)$, with SNR$= \sigma_u/\xi$ (dotted line at left). This is just the result we found above in Eq. (15.46), with unit gain. For $u^*/\sigma_u \ll 1$, measurement noise dominates: $y_k = \xi_k$. Since the output y_k and input u_k are then independent, $\mathcal{I}(U;Y) = 0$. At left and in Problem 15.18, we see how the intermediate cases interpolate between these limits.

Given a nonlinear relation $y = g(u)$, how can we make the outputs y tell us the most about the inputs u? More formally, what distribution $p(u)$ maximizes the information transfer, $I(U;Y)$? From the definition of mutual information, Eq. (A.287 online),

$$I(U;Y) = H(Y) - H(Y|U) = H(Y) - H(\xi)\,. \tag{15.50}$$

The identity $H(Y|U) = H(\xi)$ holds for additive Gaussian white noise,[24] $y = g(u) + \xi$, with $\xi \sim \mathcal{N}(0, \xi^2)$. Intuitively, if the input u is known, then the only remaining source of uncertainty is the measurement noise ξ. More formally, the conditional entropy $H(Y|U)$ depends on $p(y|u) \sim \mathcal{N}[g(u), \xi^2]$. Since the entropy of a Gaussian random variable depends only on its variance (independent of its mean), we have $H(Y|U) = H(\xi) = \log\sqrt{2\pi \mathrm{e}\,\xi^2}$.

Here, we want to maximize $I(U;Y)$ over all input distributions $p(u)$, or, equivalently, over the corresponding output distribution, $p(y)$. Since $H(\xi)$ is independent of $p(y)$, we choose $p(y)$ to maximize $H(Y) = -\int \mathrm{d}y\, p(y) \log p(y)$ and then convert the $p(y)$ distribution into the corresponding distribution for the input, $p(u)$. We introduce a Lagrange multiplier λ to enforce normalization, $\int \mathrm{d}y\, p(y) = 1$. Dropping terms independent of $p(y)$, we have

$$\tilde{I}(U;Y) = \underbrace{-\int \mathrm{d}y\, p(y)\, \log p(y)}_{H(Y)} - \lambda \int \mathrm{d}y\, p(y)\,. \tag{15.51}$$

Then

$$\frac{\delta \tilde{I}[p(y)]}{\delta\,[p(y)]} = 1 + \log p(y) - \lambda = 0 \tag{15.52}$$

implies that the optimal $p(y)$ is constant, a uniform distribution in y. We can assume that $y \in [y_{\min}, y_{\max}]$, since any physical transducer must have a finite range of outputs. The distribution of outputs that maximizes information flow is then

[23] The measurement function of an A/D converter is actually slightly different, $y_k = g(u_k + \xi_k)$. The analysis becomes more difficult, but the conclusions are qualitatively the same.

[24] Indeed, it holds for white noise, in general.

simply

$$p(y) = \left(\frac{1}{y_{\max} - y_{\min}}\right) \quad \Longrightarrow \quad p(u) \approx \left(\frac{1}{y_{\max} - y_{\min}}\right) \left|\frac{dg}{du}\right| . \tag{15.53}$$

The approximation assumes low noise – i.e., that $y = g(u)+\xi \approx g(u)$ (cf. Problem 15.19). Intuitively, the probability to choose input u should be proportional to the local transducer sensitivity $g'(u)$. Alternatively, if $p(u)$ is given, then we can tune the sensitivity $g'(u)$.[25] Finally, an equivalent representation is found by integrating Eq. (15.53):

$$y = g(u) = y_{\min} + (y_{\max} - y_{\min}) \int_{u_{\min}}^{u} du'\, p(u') . \tag{15.54}$$

For the piecewise-linear $g(u)$ discussed above, the optimal input distribution is the uniform distribution (see right) $p(u) = p(y) = 1/(2u^*)$ over the range $(-u^*, u^*)$, with entropy $H[p(y)] = \log(y_{\max} - y_{\min})$ (see Eq. A.251 online). To find the maximum rate, we note that it corresponds to the case where there is no correlation between successive values of u:

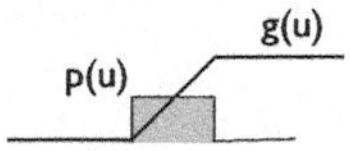

$$\mathcal{I}_{\max}(U;Y) = \log(y_{\max} - y_{\min}) - \log|\xi| = \log \mathrm{SNR} , \tag{15.55}$$

where $\mathrm{SNR} = |(y_{\max} - y_{\min})/\xi|$. For an n-bit A/D converter with a noise level of one bit,[26] the maximum possible information acquisition rate would be n bits per conversion.

Figure 15.4 shows how the optimal distribution balances conflicting requirements of sensitivity and SNR. The poor choice of input distribution in the left panel is centered on a region of low sensitivity, so that nearly all values of u are mapped to a narrow

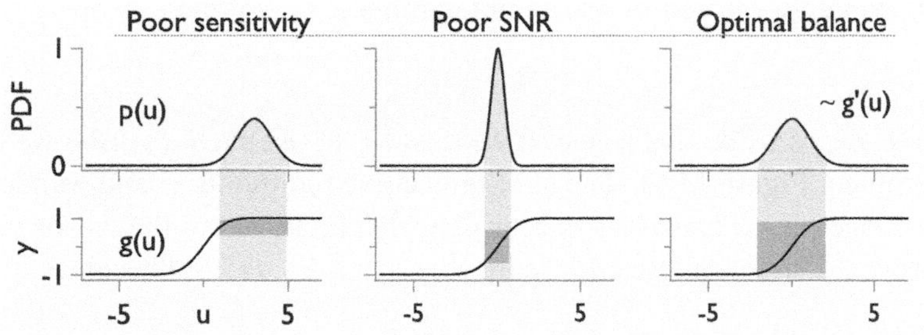

Fig. 15.4 Optimizing input distribution $p(u)$ for $y = g(u) = \mathrm{erf}(u/\sqrt{2})$. Left: $p(u) \sim \mathcal{N}(3, 1)$ has poor sensitivity. Center: $p(u) \sim \mathcal{N}(0, 0.4)$ has poor SNR. Right: $p(u) \sim \mathcal{N}(0, 1) \sim g'(u)$ has optimal information transfer. The light-shaded areas have width $\pm 2\sigma$. The heavier-shaded areas indicate the range of y in the output.

[25] This seems relevant at longer time scales and may be important in understanding evolution (Bialek, 2012).

[26] Quantization nonlinearities can be softened by adding Gaussian noise with $\sigma \approx 0.5$ bit (cf. Section 5.1.2 on *dithering*). The SNR for an A/D converter is also known as the *effective number of bits* (ENOB).

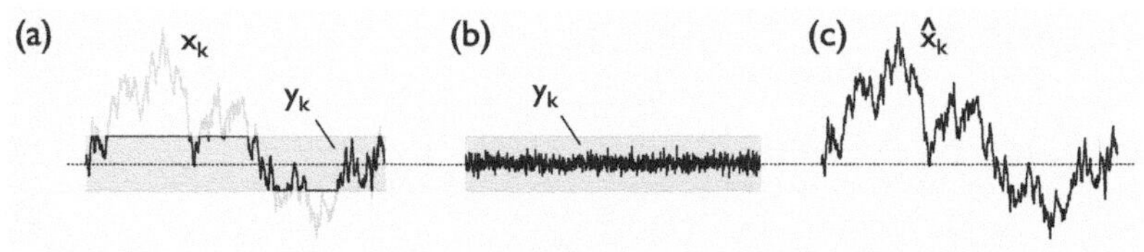

Fig. 15.5 Effect of feedback on information rates. (a) Signal x_k (gray) often exceeds the $\pm x_m$ range of the sensor (dark shaded region), clipping the open-loop measurement y_k. (b) Applying feedback keeps y_k within the sensor range. (c) Reconstructed $\hat{x}_k$. Sensor range $x_m = 5$. Times series length $N = 1000$.

range of y. The poor choice of input distribution in the center panel is too narrow: although centered on the region of highest sensitivity [maximum $g'(u)$], it uses too small a range of y. Finally, the optimal input distribution in the right panel is centered on the region of highest sensitivity and is wide enough to use the full range of output values, thereby maximizing the SNR.

Notice from Eq. (15.55) that $I_{max}(U;Y)$ does not depend on $g(u)$. The optimal input distribution adapts to the particular $g(u)$ that maximizes the information-transfer rate.

Dynamic Nonlinear Relation

Let x_k be a signal that we wish to measure. For the sake of definiteness, it is generated by

$$x_{k+1} = x_k + \nu_k, \quad \nu_k \sim \mathcal{N}(0,1). \tag{15.56}$$

In other words, our signal is a random walk. Its range is thus unbounded, growing $\sim \sqrt{2k}$. This range may reflect the actual extent of a signal or perhaps the effects of drift. Unfortunately, our sensor has a limited range $\pm x_m$, so that

$$y_k = x_k, \quad |x_k| < x_m, \tag{15.57}$$

and equal to $\pm x_m$ outside that range. If we then try to use such a sensor, we will get a result resembling Figure 15.5a. The light-gray curve represents x_k and wanders out of the sensor range $\pm x_m$. The darker trace is y_k, which is clipped to the sensor range.

To improve the situation, we add feedback, so that Eq. (15.56) becomes

$$x_{k+1} = x_k + u_k + \nu_k, \quad \nu_k \sim \mathcal{N}(0,1), \quad u_k = -Ky_k. \tag{15.58}$$

Because we ignore sensor noise, the optimal value of feedback is $K = 1$. As long as the feedback can keep x within the range $\pm x_m$, then $y_k = x_k$ and the closed-loop dynamics are simply $x_{k+1} = y_{k+1} = \nu_k$, which is the situation plotted in Figure 15.5b. Because the range $x_m/\nu = 5$ and because the feedback is noise free, there is little chance of clipping for y_k.

Having kept the signal within the sensor range, we can then reconstruct the original signal, generating $\hat{x}_k$ from the closed-loop measurement signal via the recursion relation $\hat{x}_{k+1} = \hat{x}_k + y_k$. Figure 15.5c shows the "perfect" reconstruction that results.

From the point of view of the information-theoretic view of control we have been developing, we observe that the mutual information rate $\mathcal{I}(Y;v)$ will approach zero for the open-loop situation. Over a long time, the signal will be almost always outside the range $\pm x_{\rm m}$, meaning that almost all correlation with the original signal will be lost. (The only information retained is whether it is too high or too low.) By contrast, the closed-loop signal allows perfect reconstruction and thus gives the information rate of the signal itself. Here, because we neglect sensor noise and because the signal is a random walk, it is formally infinite. Adding sensor noise and a cutoff to signal variations would make the rate finite.[27]

The above example actually resembles the situation of scanning probe microscopes (STMs and AFMs). They must keep a tip extremely close to a surface. If the tip is more than a few nanometers away, there is no signal (tunneling current for STM, cantilever deflection for AFM). If the tip is too close, it crashes into the surface, saturating the signal at some maximum value. The response curve thus resembles the one discussed here (except that the sensitivity is exponential). For all but the flattest samples, such instruments depend on a feedback loop to keep the tip in the proper range. Note that often the large range in an STM or AFM sample is due to trivial factors such as sample tilt. In such cases, it is common not to bother with a full reconstruction of the original signal but to simply use the measured signal as a kind of "high-pass" filter.

Although we might be tempted to push our analysis further, we would quickly run up against the "dirty secret" of information theory, which is that beyond basic examples such as a binary alphabet, Gaussian noise, or Markov process, relevant quantities such as entropy and mutual information rates can be difficult to calculate, even numerically. (See Problem A.10.9 in the Appendix online.) In our example, any clipping leads to non-Gaussian distributions.

Despite these practical limitations, information theoretic analyses shed light on the value of feedback, even when heuristic: we have seen how the local sensitivity governs the rate at which information can be extracted – measurements performed. Since linear systems have a sensitivity that is independent of the system state, feedback provides no advantages. But all systems are ultimately nonlinear, and then feedback can keep a system in a region of state space where the sensitivity and information throughput are high.

15.2.4 Information and Stabilization

Information-theoretic ideas give new understanding about the requirements to stabilize a dynamical system. Consider again the one-dimensional dynamics of Example 15.1:

[27] The feedback capacity theorem of information theory asserts that feedback cannot increase the capacity of a memoryless channel (Cover and Thomas, 2006, chapter 7). The theorem seems to contradict our statement that feedback can improve information flow rates in measurements with nonlinear response functions. However, the capacity of a channel is achieved by using an optimal input distribution. Our statement here is that feedback can alter a nonoptimal input distribution to be closer to the distribution that maximizes information transmission.

$$x_{k+1} = ax_k + u_k + \nu_k\,, \quad y_k = x_k\,, \quad u_k = -Ky_k\,, \quad \nu_k \sim \mathcal{N}(0, \nu^2)\,. \tag{15.59}$$

Define the average input power to be $P \equiv \langle u^2 \rangle$. What power is needed to stabilize the output? Of course, if $|a| < 1$, then the system is intrinsically stable, and choosing $K = 0$ implies $P = 0$. The interesting case is $|a| > 1$, where feedback is needed for stabilization.

We can find the feedback gain in the unstable case from the Linear-Quadratic-Gaussian (LQG) theory developed in Section 8.2; however, the problem is simple enough to do directly. Indeed, from Problem 15.7, we have $\langle y^2 \rangle = \frac{\nu^2}{1-(a-K)^2}$. Note that we implicitly assume that K stabilizes the dynamics; otherwise, you cannot define $\langle y^2 \rangle$ – exponentially growing values have no steady-state average.

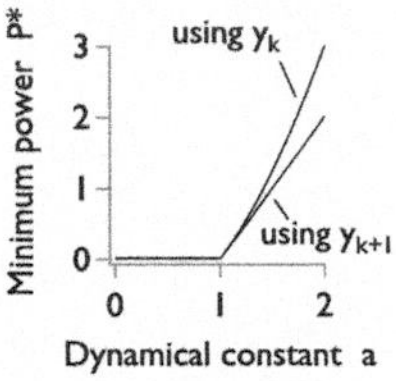

The power $P = \langle u^2 \rangle = K^2 \langle y^2 \rangle$ is then minimized by setting $\partial_K P = 0$, which implies

$$K^* = \frac{a^2 - 1}{a} \qquad \text{and} \qquad P^* = (a^2 - 1)\nu^2\,, \tag{15.60}$$

where K^* is the optimal gain and P^* the corresponding minimum power. See Problem 15.10 and the figure at left, which labels the $P^*(a)$ plot as "using y_k."

Let us rewrite the expression for P^* in Eq. (15.60) as $\ln |a| = \frac{1}{2} \ln \left(1 + \frac{P^*}{\nu^2}\right)$. The log of the degree of instability a appears in the Bode waterbed relation. It is equal to the expression for the channel capacity of an additive Gaussian white-noise communication channel for a signal of power P and noise ν^2. See Section A.10.5 online. Thus, at least in this simple case, the minimum power to stabilize a system is just that needed to transmit the "new information" produced by the dynamical system. The latter is now viewed as a communication channel that "transmits" information about the initial state of the system to the measured output.

More formally, to stabilize a system with a set of unstable poles at p_k, you need measurements that supply information at a rate that is at least $\sum_k \log p_k$. The result is known as the *data rate theorem*. For continuous systems, the corresponding requirement is $\sum_k p_k$. A natural conjecture is that the corresponding theorem for nonlinear dynamical systems would replace the p_k by positive Lyapunov exponents. Special cases have been studied.

The example explored above demonstrates the value of information. In Problem 15.10, we will see that knowing the output y_{k+1} above time k reduces the minimum power to stabilize from $a^2 - 1$ to $2(a - 1)$ (in units of ν^2). This new curve is plotted above left as the straight line (for $a > 1$). By contrast, using past information (e.g., y_{k-1} at time k) does not help.

15.3 Thermodynamic Limits

Section 15.2 revealed that control performance is limited by the information that can be acquired about the system under control. But there are also connections between information and thermodynamics. In the words of Rolf Landauer: "Information is physical." That is, information is always coded in physical systems that must obey the laws of physics. As we develop this notion, we will come to see that three seemingly disparate topics are intimately linked: feedback, information, and thermodynamics.

15.3.1 Does Feedback Violate the Second Law?

In 1871, at the dawn of statistical physics, James Clerk Maxwell proposed a troubling thought experiment:[28]

> [I]f we conceive a being whose faculties are so sharpened that he can follow every molecule in its course, such a being, whose attributes are still as essentially finite as our own, would be able to do what is at present impossible to us. For we have seen that the molecules in a vessel full of air at uniform temperature are moving with velocities by no means uniform, though the mean velocity of any great number of them, arbitrarily selected, is almost exactly uniform. Now let us suppose that such a vessel is divided into two portions, A and B, by a division in which there is a small hole, and that a being, who can see the individual molecules, opens and closes this hole, so as to allow only the swifter molecules to pass from A to B, and only the slower ones to pass from B to A. He will thus, without expenditure of work, raise the temperature of B and lower that of A, in contradiction to the second law of thermodynamics.

Figure 15.6 illustrates *Maxwell's demon*, the "being" with "sharpened faculties," as memorably dubbed by William Thomson (Lord Kelvin).[29] As imagined by Maxwell, the demon observes the state of the system (gas molecules in both chambers) and executes a policy: raise the door if a fast molecule approaches in A or a slow molecule approaches in B.[30] After many such exchanges, a temperature difference develops between A and B, which can be exploited to do useful work.

Recalling the Planck statement of the *second law of thermodynamics*

> It is impossible to construct an engine which will work in a complete cycle, and produce no effect except the raising of a weight and the cooling of a heat-reservoir.[31]

we see that unless there is an energetic cost to the repeated operation of the demon,[32] it will apparently violate the second law. To understand why that possibility aroused enormous concern, consider Arthur Eddington's 1928 pronouncement:

> The law that entropy always increases – the second law of thermodynamics – holds, I think, the supreme position among the laws of Nature. If someone points out to you that your pet theory of the universe is in disagreement with Maxwell's equations – then so much the worse for Maxwell's equations. If it is found to be contradicted by observation – well, these experimentalists do bungle things sometimes. But if your theory is found to be against the second law of thermodynamics I can give you no hope; there is nothing for it but to collapse in deepest humiliation.[33]

[28] J. C. Maxwell, *Theory of Heat*, Longmans, Green, and Co., 1871. The thought experiment was actually proposed informally in a December 1867, letter to Peter Guthrie Tait.

[29] Maxwell, a religious man, did not like the term.

[30] For simplicity, Figure 15.6 shows molecules with only two velocities: *fast* (darker, longer) and *slow* (lighter, shorter). As Maxwell knew, gas molecules in equilibrium have a continuous distribution of velocities.

[31] Planck (1903), p. 86. Clausius and Kelvin formulated equivalent alternate statements of the second law.

[32] As an example of a similar situation where there are other costs, *refrigerators* extract heat from cooler bodies, expelling it to the warmer environment. But, of course, they need external power to operate.

[33] Eddington (1928), p. 74.

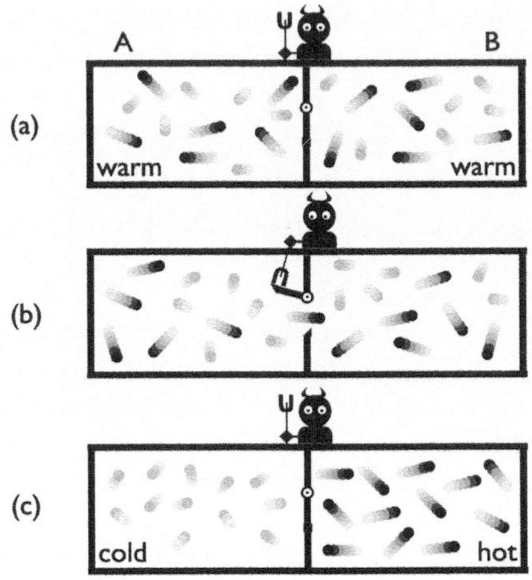

Fig. 15.6 Maxwell's demon creates a temperature difference. (a) Gas in equilibrium in chambers A and B. (b) Demon sees and then lets pass a fast molecule. (c) Fast molecules sorted to the right, slow ones to the left.

A year later, Leo Szilard simplified and sharpened Maxwell's thought experiment by realizing that it was sufficient to consider a "gas" with just one molecule inside a thermostatted box.[34] Indeed, in an ideal gas, all particles move independently, so that a single system of N particles can be equivalent to an ensemble of N single-particle systems.

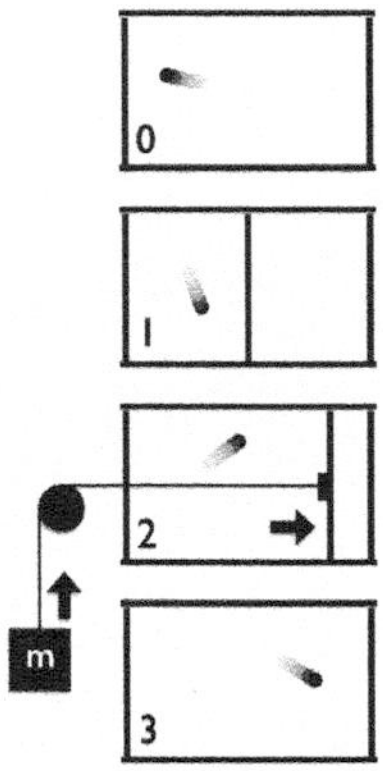

At left we show a cycle of a *Szilard engine*, which has a single molecule in a box whose walls are in equilibrium with a surrounding bath of temperature T. Starting from the initial state (0) of a single particle in a box of volume V, the cycle proceeds through three stages:

- *Measurement (1)*. We insert a partition that divides the box into two equal volumes, $V/2$. If the partition movement is frictionless, no work is done. We determine which side of the box the particle is in (L or R).
- *Feedback (2)*. Knowing which side contains the particle, we attach a cable plus weight and extract work by slowly moving the partition, letting the "gas" expand isothermally.
- *Reset (3)*. The partition, now at the edge of the box, is removed, again with no work.

To estimate the work that can be extracted, we use the ideal gas law for a single particle in equilibrium at temperature T, pressure p, and volume V, which is $pV = k_B T$,

[34] We use the anglicized form of Szilard's Hungarian name, as he spent the second half of his life in English-speaking countries.

with k_B Boltzmann's constant. From this law, the maximum work W_{ext} that can be extracted in an isothermal expansion is

$$W_{ext} = \int_{V/2}^{V} dV' \, \frac{k_B T}{V'} = k_B T \ln \frac{V}{V/2} = k_B T \ln 2 \,. \tag{15.61}$$

Notice that the volume V drops out of the expression for W_{ext}. What counts is the ratio, which reflects that a particle has been localized to one of two equally likely volumes.

Szilard's paper made the first link between physics and information theory.[35] In each cycle, we acquire one bit of information (determining which side of the box the particle is in) and use it to extract a work $W = k_B T \ln 2$ from the heat bath. There is explicit feedback, an action that depends on the result of the measurement: If we measure the particle on the left side, we attach the weight to that side, in order to extract work as the piston expands. The work is extracted by slowly raising the weight of mass m by a height $h = k_B T \ln 2/(mg)$, where g is the gravitational acceleration. Note that if we do not bother to attach the weight, the partition may still move, but no work will be extracted.

In short, using one bit of information, the Szilard engine can extract $k_B T \ln 2$ of work by cooling the outside bath and raising a weight, in apparent violation of Planck's statement of the second law.

15.3.2 Szilard Engine Based on Energy Levels

The above analysis of the Szilard engine follows its original presentation from 1929. Let us consider an alternate version based on a two-state system, where states are characterized only by their energy levels. Chapter 12 discussed how discrete-state systems can emerge from a suitable coarse-graining of system dynamics. There are many possible physical realizations of such systems, including molecules with two large-scale configurations and small capacitors where each electron arrival and departure may be monitored. One realization that is particularly simple and has been used for experiments testing many of the notions presented in this section is the Brownian particle moving in a double-well potential, as illustrated at right and discussed in detail in Section 12.1.3. The barrier is assumed high enough that the dwell time in a well greatly exceeds not only the time scale of dynamics within each well but also the time scales associated with any kind of protocol. That is, spontaneous thermal hops between states do not occur on any time scale of interest.

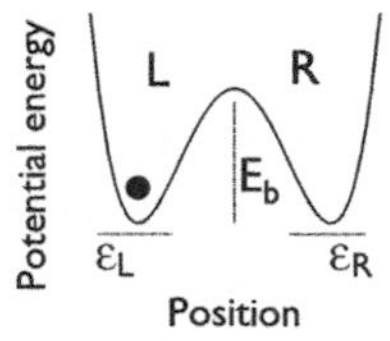

In fact, a physical picture, while useful for intuition, is not required for characterizing two-state systems. Since energies are relative, we can define the energy of one state to be 0 and the other to be $\epsilon(t)$, which is adjusted according to a slow, *quasistatic* protocol – so slowly, in fact, that they effectively stay in thermal equilibrium. If the energy

[35] Szilard's paper was actually the first application of information theory to a physical problem. Remarkably, it appeared twenty years before Shannon (1948).

difference is ϵ at time t, the state probabilities are given by the Boltzmann distribution $\pi_i = \frac{1}{Z}\,\mathrm{e}^{-\epsilon_i}$ (defined in Section 15.3.4, after Eq. 15.67),

$$\pi_1 = \frac{1}{1+\mathrm{e}^{-\epsilon}}\,, \qquad \pi_2 = \frac{1}{1+\mathrm{e}^{+\epsilon}}\,, \qquad \pi_1 + \pi_2 = 1\,. \tag{15.62}$$

In Eq. (15.62), energies are in units of $k_B T$, where k_B is Boltzmann's constant and T is the temperature of the bath in contact with the system. In the example of a Brownian particle, T is the temperature of the surrounding fluid. As ϵ is adjusted during a protocol, the equilibrium probabilities π_1 and π_2 adjust along with it. In Section 15.3.4, we will introduce a model for system dynamics that shows that the quasistatic limit is reached when the adjustments of ϵ are slow on the time scales defined by the system dynamics.

15.3.3 Work and Information

In the quasistatic limit, let us calculate the energy that can be transferred from the heat bath to system energy. We will call this the "work extracted from the heat bath." The reason we refer to this energy as "work" is that energy belonging to the system is available, at least in principle, for other purposes. For example, it could create mechanical motion (push a particle) or drive a chemical reaction. By contrast, energy that is "heat" belongs to the degrees of freedom of the bath, which are hidden and thus cannot be exploited for such purposes. This separation between useful macroscopic and useless microscopic degrees of freedom is at the heart of classical thermodynamics.

Consider a protocol where both energies are initially zero. We observe the system. If the observation implies that the system is in the L state, then ϵ_R is instantaneously increased to ϵ and then lowered very slowly (quasistatically) to zero. If the system is in the R state, ϵ_L is raised, instead. With perfect observations, there is no chance of choosing the wrong state. After the raised state energy is reduced to zero, another observation is made, repeating the cycle. Using Eq. (15.62), we find that the extracted work is given by

$$W = -\int_{\epsilon}^{0} \mathrm{d}\epsilon'\,\pi_2(\epsilon') = -\int_{\epsilon}^{0} \mathrm{d}\epsilon'\,\frac{\mathrm{e}^{-\epsilon'}}{1+\mathrm{e}^{-\epsilon'}} = +\ln\frac{2}{1+\mathrm{e}^{-\epsilon}}\,, \tag{15.63}$$

where the initial negative sign results because we count energy taken *from* the heat bath as positive. If the initial energy $\epsilon \to \infty$, the work $W \to \ln 2$, as with the "classic" version of the Szilard engine. In both cases, the "2" in the ln 2 term reflects the two states of the system: one bit of information has been acquired (which of two states is occupied). The logarithm in "ln 2" converts between the physical units of entropy, which is based on the natural logarithm (and k_B) and the bit, which is defined in terms of logarithms with base 2.

To strengthen the link between the information acquired and the work that can be extracted, let us generalize to the case where the observation of the state has a probability ξ of being incorrect.[36] Averaging the work over the possible observations, we have

$$W = \ln \frac{2}{1 + e^{-\epsilon}} - \xi\epsilon \,. \tag{15.64}$$

The extra term arises because if the observation is in error, you do work ϵ raising the energy level. This energy is transferred *to* the heat bath, reducing the net extraction of energy from the bath. The other term is unaltered: in the quasistatic limit, the system hops many times between the two states as the energy is lowered. The initial state does not matter.

With no observational error ($\xi = 0$), the optimal protocol maximizing work extraction was $\epsilon \to \infty$. Here, from solving $\partial_\epsilon W = 0$, the optimal $\epsilon^* = \ln[(1 - \xi)/\xi]$, which implies

$$W^* = \ln 2 + \xi \ln \xi + (1 - \xi) \ln(1 - \xi) \,. \tag{15.65}$$

We immediately recognize the Shannon entropy function for a two-state system, in nats, $H_2(\xi) = -\xi \ln \xi - (1-\xi) \ln(1 - \xi)$ (cf. Eq. A.247 online), and write $W^* = \ln 2 - H_2(\xi)$. The first term represents the uncertainty $H(X)$ in the state of the system before the measurement is done. The second, $H_2(\xi) = H(X|Y)$, is the uncertainty in the state *after* the measurement. Moreover, $H(X) - H(X|Y) = I(X;Y)$ is the mutual information between the two quantities (Section A.10.4 online). Thus, $W^* = I(X;Y)$. For all other choices of ϵ, the average work extracted will be smaller; as a result,

$$W \leq k_{\mathrm{B}} T \, (\ln 2) \, I \,, \tag{15.66}$$

in physical units, with I in *bits* (base 2 logarithms). A perfect measurement yields one bit of information about the state of the system and allows extraction of $k_{\mathrm{B}} T \ln 2$ of work. A useless measurement ($\xi = 0.5$) gives no information about the state of the system and no energy can be extracted from the heat bath. In between, a noisy measurement yields between zero and one bit and allows extraction of an intermediate amount of energy.

We thus have a precise link between the amount of information acquired and the maximum amount of work that can be extracted from a heat bath. The "paradox" of Maxwell's demon is now clear: The second law of thermodynamics states that in an isothermal system, $\Delta S \geq 0$. But the first law states that $-W - T\Delta S = \Delta U$, the change in internal energy of the system. (Again, the unusual sign for W arises because we are interested in the extracted work, the energy transferred from the heat bath.) Putting the two together and remembering that a cyclic protocol has $\Delta U = 0$, we see that the second law requires $W \leq 0$. Thus, a positive extraction $W > 0$ based on feedback would seem to violate the second law.

[36] We use the notation ξ as an analog of the notation for noise in a continuous observation, $y(t) = x(t) + \xi(t)$. Here, though, ξ is a number (a probability) and not a stochastic variable.

15.3.4 Stochastic Thermodynamics

The links between thermodynamics and information theory discussed above have been from the point of view of classical thermodynamics, where protocols are so slow that the system is at every moment in a thermodynamic equilibrium (quasistatic). In the last two decades, a new formalism, stochastic thermodynamics, has led to a detailed treatment of the thermodynamics of small systems far from equilibrium. It leads to deeper insight into the links among feedback, information, and thermodynamics.

Our setting is a "system" coupled to a single heat bath that is large, assumed not to be affected by anything the system can do, and thus describable using classical thermodynamics. The system can be out of equilibrium and in general is not characterized by a temperature. Typical examples are colloidal particles and biological molecules in an aqueous solution.

Implicitly, we adopt a *coarse-grained* point of view, where large numbers of molecules are described by a limited number of states and where the cumulative effects of atoms and molecules in the bath can be summarized by simple stochastic perturbations. The states are then characterized solely by their energy ϵ_i (assuming no chemical reactions, which can change particle numbers). Justifying such ideas from first principles, in either classical or quantum cases, is a long and subtle story (Section 12.1.3).

Systems can have a finite or an infinite number of degrees of freedom. For the latter, the Fokker–Planck equation describes how the probability density function $p(\boldsymbol{x}, t)$ evolves (cf. Chapter 8). Here, we also consider the equivalent description for the evolution of probability distributions over discrete states assumed to obey Markov dynamics. The evolution is then given by *master equations* that generalize the Markov chains studied in Chapter 12. Stochasticity enters because the bath causes stochastic transitions among states.

Master Equation

For a system with n discrete, nondegenerate states, the probability that the system is in state i at time t, given by $p_i(t)$, evolves as

$$\mathrm{d}_t p_i = \sum_j \left(W_{ij}\, p_j - W_{ji} p_i \right) \equiv \sum_j J_{ij} \,. \tag{15.67}$$

In Eq. (15.67), the W_{ij} are *transition rates* from state j to state i. The rates can be time dependent, for example when manipulated as part of a control process. The current $J_{ij} \equiv (W_{ij}\, p_j - W_{ji} p_i)$ is the net flux of transitions along the link from j to i. We also introduce the notation d_t for the time derivative of a *state function*. Later, we will use overdots for rates that are inexact differentials, corresponding to *path functions* that depend on the trajectory through state space and thus are not simple state functions.[37] Problem 15.21 relates this "physical version" of the master equation to an equivalent

[37] This notation for *rates* complements the traditional notation of d and đ for *differential changes*.

"mathematical version" given by $\mathrm{d}_t \boldsymbol{p} = \mathbb{W}\boldsymbol{p}$, where $\boldsymbol{p}$ is a vector with components p_i and where $\mathbb{W}$ is the *rate matrix*.

In thermodynamic equilibrium, the system is in equilibrium with the bath, has a well-defined temperature T, and is described by the *canonical ensemble*. The probabilities p_i are then denoted π_i and obey the *Boltzmann distribution*, $\pi_i = \frac{1}{Z}\,\mathrm{e}^{-\epsilon_i}$, where ϵ_i is the energy when the system is in state i, in units of $k_\mathrm{B}T$. The *partition function* $Z = \sum_i \mathrm{e}^{-\epsilon_i}$ is a numerical factor that ensures normalization, $\sum_i \pi_i = 1$.

In order for the master equation dynamics to be compatible with thermodynamic equilibrium, we must constrain the rates W_{ij}. One constraint comes from the principle of *microscopic reversibility*, which expresses, at the level of trajectories through system states, that the underlying microscopic dynamics are time reversible. For an arbitrary link between j and i, microscopic reversibility implies that transitions in both directions across the link must have equal rates (see right),

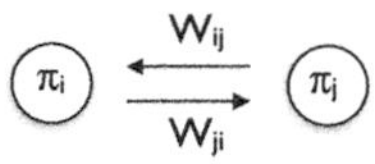

$$W_{ij}\pi_j = W_{ji}\pi_i\,, \tag{15.68}$$

a property known as *detailed balance*. Moreover, we must have

$$\frac{W_{ij}}{W_{ji}} = \mathrm{e}^{-(\epsilon_i-\epsilon_j)}\,. \tag{15.69}$$

This constraint ensures that the stationary distribution π_i obtained by setting $\mathrm{d}_t p_i = 0$ in the master equation is the desired equilibrium distribution.

Below, we add nonequilibrium *driving terms* to the transition rates. These create currents that do not obey detailed balance. They can be created, for example, by coupling a system to two heat baths (so that a current of heat runs through the system) or to a chemical reaction (which converts chemical potential into a particle current). For example, a nonequilibrium drive between nodes i and j can be modeled by assigning rates $W_{ij} = \mathrm{e}^{+f/2}$ and $W_{ji} = \mathrm{e}^{-f/2}$ (with f in units of $k_\mathrm{B}T$). Then $W_{ij}/W_{ji} = \mathrm{e}^{f}$, which superficially resembles Eq. (15.69) but is not connected with a detailed-balance condition.

First Law of Thermodynamics

For an ensemble of identically prepared systems, the expected value (mean) energy of our system is $E \equiv \langle\epsilon\rangle = \sum_i p_i\epsilon_i$. Taking a time derivative gives

$$\mathrm{d}_t E = \mathrm{d}_t\langle\epsilon\rangle = \underbrace{\sum_i p_i\,(\mathrm{d}_t\epsilon_i)}_{\dot{W}} + \underbrace{\sum_i (\mathrm{d}_t p_i)\,\epsilon_i}_{\dot{Q}} = \dot{W} + \dot{Q}\,, \tag{15.70}$$

a nonequilibrium version of the first law of thermodynamics. The $\dot{W}$ term in Eq. (15.70) corresponds to the path-dependent rate of work and reflects changes of the energy levels, $\mathrm{d}_t\epsilon_i$ at fixed state occupations p_i. The $\dot{Q}$ term corresponds to the path-dependent heat flow exchanged with the environment and reflects changes in state occupation levels $\mathrm{d}_t p_i$ at fixed energies ϵ_i. Signs are defined so that $\dot{W} > 0$ for manipulations of energy levels that increase the energy of the system and $\dot{Q} > 0$ for flows *from* the heat bath *to* the system.

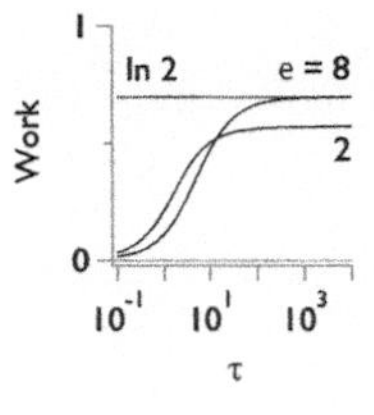

Example 15.6 (Szilard engine with a finite-time protocol) We can use the master-equation formalism and the first law to evaluate the work done in a finite-time protocol. Problem 15.22 generalizes the energy-level version of the Szilard engine to finite-time protocols, evaluating numerically a protocol that raises the unoccupied level to ϵ just after a measurement and lowers it linearly in a time τ. At left, $\epsilon = 2$ and $\epsilon = 8$, for the average energy extracted from the bath in the form of work. For long cycle times $\tau \gg 1$, the work tends to the quasistatic value $W \to \ln[2/(1+e^{-\epsilon})]$ found in Eq. (15.63).

Second Law of Thermodynamics

Another key state function that we can define is the *entropy* of the system,

$$S = -k_B \sum_i p_i \ln p_i = k_B \langle \ln(1/p) \rangle \,, \tag{15.71}$$

a measure of the average "surprise" in finding the system in state i. The definition in Eq. (15.71) is just the Gibbs entropy in equilibrium, where $p_i = \pi_i$ but applies to general nonequilibrium states of the system, too. We will usually scale S by k_B.

By evaluating the total time derivative $d_t S$, we can decompose changes in the system entropy into two path-dependent terms (see Problem 15.23):

$$d_t S = \dot{S}_i + \dot{S}_e \,, \tag{15.72}$$

where $\dot{S}_i \geq 0$ is the irreversible entropy production resulting from forcing the system out of thermal equilibrium and $\dot{S}_e = \dot{Q}/T$ is the net heat exchanged from the environment. In Problem 15.23, you derive Eq. (15.72) by showing that the time derivative of S is

$$d_t S = \frac{1}{2} \sum_{ij} J_{ij} \ln \frac{p_j}{p_i} \,, \tag{15.73}$$

which can be decomposed into

$$\dot{S}_i = \frac{1}{2} \sum_{ij} J_{ij} \ln \frac{W_{ij} p_j}{W_{ji} p_i} \quad \text{and} \quad \dot{S}_e = \dot{Q} = \frac{1}{2} \sum_{ij} J_{ij} \ln \frac{W_{ji}}{W_{ij}} \,. \tag{15.74}$$

In these equations, S is scaled by k_B and Q by $k_B T$. The current $J_{ij} = W_{ij} p_j - W_{ji} p_i$ is from j to i. Because $\dot{S}_i$ is the sum of terms of the form $(x - y) \ln(x/y)$, it must be nonnegative: $\dot{S}_i \geq 0$. It can be zero only if $J_{ij} = 0$, which is just the condition of detailed balance.

To understand the connection between $\dot{S}_{\mathrm{e}}$ and heat flow, notice that detailed balance and Eq. (15.69) implies that $\ln(W_{ji}/W_{ij}) = \epsilon_i - \epsilon_j$. Thus,

$$\dot{Q} = \frac{1}{2}\sum_{ij} J_{ij}\left(\epsilon_i - \epsilon_j\right). \tag{15.75}$$

The heat gained from the reservoir is given by the difference between the final and initial energy levels, times the number of net transitions per time. $\dot{Q}$ can have either sign. Problem 15.23 shows that this definition of $\dot{Q}$ is consistent with $\sum_i (\mathrm{d}_t p_i)\,\epsilon_i$ from Eq. (15.70).

With the above identifications, $\dot{S}_{\mathrm{i}} = (\mathrm{d}_t S - \dot{S}_{\mathrm{e}}) \geq 0$ is a nonequilibrium version of the second law. The terms $\mathrm{d}_t S - \dot{S}_{\mathrm{e}}$ represent the change in entropy of the system $\mathrm{d}_t S$ plus the change in the entropy of the heat bath $-\dot{S}_{\mathrm{e}}$. (The negative sign means that $\dot{S}_{\mathrm{e}} > 0$ when heat flows *into* the system and *out from* the bath.) The two contributions equal the entropy production of "the universe" (system plus bath), $\dot{S}_{\mathrm{i}}$, which is nonnegative.

Nonequilibrium Free Energy

For the canonical ensemble, a system can exchange energy with a heat bath at temperature T. At thermal equilibrium, such a system adopts the configuration that minimizes the (Helmholtz) free energy $F^{\mathrm{eq}} = E^{\mathrm{eq}} - TS^{\mathrm{eq}}$. We can define an analogous nonequilibrium free energy for ensemble stochastic thermodynamics as

$$F = E - TS = \sum_i p_i \epsilon_i - \sum_i p_i \ln p_i \,. \tag{15.76}$$

The temperature T disappears from the second term because we scale energies by $k_{\mathrm{B}}T$ and entropy by k_{B}. Differentiating gives

$$\mathrm{d}_t F = (\dot{W} + \dot{Q}) - (\dot{Q} + \dot{S}_{\mathrm{i}}) = \dot{W} - \dot{S}_{\mathrm{i}} \,. \tag{15.77}$$

Rewriting this expression as $\dot{W} - \mathrm{d}_t F \geq 0$ and integrating over a time interval gives

$$W - \Delta F \geq 0\,, \tag{15.78}$$

a different version of the second law. If the system is in equilibrium at start and end, then $W \geq \Delta F^{\mathrm{eq}}$. Thus, for a *reversible* transformation, $\Delta F^{\mathrm{eq}} = W$, and the increase in free energy can be used to undo the transformation, extracting $W_{\mathrm{extract}} = -W$. For an *irreversible* transformation, you extract less work.

Equation (15.78) also applies to cases where F is a nonequilibrium free energy. To understand its physical meaning, we first note the identity (Problem 15.24)

$$F - F^{\mathrm{eq}} = D(p\|\pi) \geq 0\,, \tag{15.79}$$

where $D(p\|\pi) = \sum_i p_i \ln(p_i/\pi_i)$ is the relative entropy between the nonequilibrium and equilibrium distributions p and π.[38] It vanishes for $p = \pi$ and is positive otherwise.

[38] The relative entropy is also known as the Kullback–Leibler divergence. Although asymmetric in p and π, it is "almost" a distance between the two probability distributions. For proof of nonnegativity, see Section A.10.2 online.

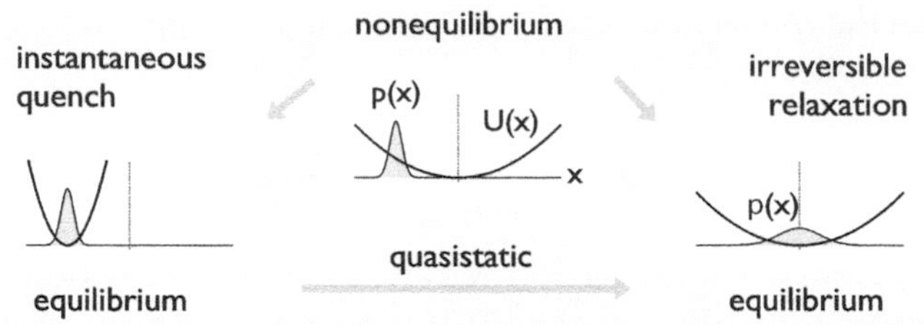

Fig. 15.7 Extracting work vs. irreversible relaxation. Adapted from Parrondo et al. (2015).

A nonequilibrium distribution $p \neq \pi$ will thus have a nonequilibrium free energy that *exceeds* F^{eq}. Equation (15.79) also applies to continuous state variables x with probability density functions $p(x)$ and $\pi(x)$, energies $\epsilon_i \rightarrow U(x)$, and $\sum_i \rightarrow \int dx$. (See Problem 15.24.)

To understand the meaning of the nonequilibrium free energy F, consider Figure 15.7, which might represent a Brownian particle in the harmonic trap produced by an optical tweezer. Initially, as shown at center, $p(x)$ differs from the thermodynamic equilibrium distribution $\pi(x)$, meaning that the system is out of equilibrium. Because $p(x) \neq \pi(x)$, the nonequilibrium free energy F is higher than the equilibrium value F^{eq}. In this case, the increase is due to a combination of higher energy (particle localized in a higher-energy region) and smaller entropy (distribution is narrower than the equilibrium distribution).

To understand how to extract work from a nonequilibrium distribution of states, we consider two possible protocols that start from the initial condition depicted at center. In the first, we do nothing and simply let the distribution relax to the distribution shown at right. The relaxation is an irreversible process, and the excess free energy is dissipated as heat in the bath. No work is extracted.

Alternatively, we can extract a work $W = F - F^{\mathrm{eq}}$ via the procedure going from center to left: Change the potential instantaneously, so that the particle distribution is the equilibrium distribution of the new potential and then quasistatically transform the potential back to its original form. In the optical-tweezer illustration, you can move the laser trap and increase its intensity ($\sim$ stiffness). The passage from left to right then depicts the quasistatic transformation back to the original equilibrium. Problem 15.24 shows that this protocol extracts $F - F^{\mathrm{eq}}$ of energy from the bath, the maximum allowed by the second law.

The above protocol is reversible: We could start at right, in equilibrium, quasistatically deform the potential to the state shown at left, and then instantaneously create our non-equilibrium state. The energetics of this process are exactly the reverse, and we will have prepared a nonequilibrium state with the *minimum* amount of work demanded by the second law. Conceptually, the combination of quench and quasistatic transformation plays a role in the transition between nonequilibrium states that simple quasistatic transformations play in equilibrium thermodynamics. Thus, the nonequilibrium free energy has a physical meaning entirely analogous to the conventional free energy.

15.3.5 Information Engines

The previous paragraph gives a protocol to create the nonequilibrium initial state depicted at the top of Figure 15.7. But there is an easier way: Imagine that the particle is initially in equilibrium with the potential, as shown at bottom right. Now, we *measure* the particle position. Just after the measurement, the distribution is the one shown at the top of the figure, with the width reflecting the noise of the measurement. But, as we have just argued, altering rapidly the potential to the form at lower left and quasistatically deforming to the original shape allows the recovery of the full amount of free energy generated by the original nonequilibrium distribution. Thus, a sequence of "simple" observations followed by feedback to exploit the information gained gives a cyclic protocol leading to the extraction of energy from the heat bath – all powered by information! This is an *information engine*.

We can analyze the Szilard engine from this point of view. In its original version, the system is a particle in a box whose volume is manipulated. The changes in volume lead to changes in entropy ΔS. In the second version, the volume associated with each system state is fixed, and one of the energies is altered instead. In both cases, there is no change in the average energy E, since the energies of the two states are equal at the time of observation. But the system entropy does change. When averaged over measurement results, it is simply

$$\Delta S = S(X|Y) - S(X) = -I(X;Y), \tag{15.80}$$

where $I(X;Y)$ is the mutual information between the system state and the observation, or more briefly the "mutual information of the measurement." The nonequilibrium free energy is then increased by $\Delta F_{\text{meas}} = +I(X;Y)$. If we isolate this part of the nonequilibrium free energy, then we can write a version of the second law that holds for feedback processes:

$$W - \Delta F \geq -I(X;Y), \tag{15.81}$$

where ΔF records other changes in nonequilibrium free energy. In a repeated cyclic protocol, for example, $\Delta F = 0$, so that $W_{\text{extract}} = -W \leq +k_{\text{B}}T(\ln 2)I$, again expressing I in bits and work in physical units. The Szilard engine, if operated slowly enough and without unnecessary dissipation (ideal pistons, etc.), meets this bound, extracting the full $k_{\text{B}}T(\ln 2)I$ of work from the heat bath each cycle.

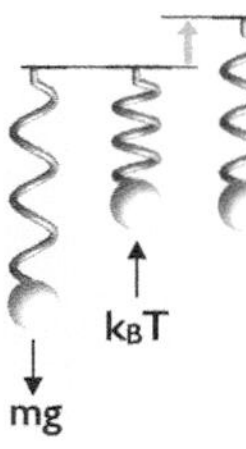

Example 15.7 (The information ratchet) At right is another type of device that converts heat to work, the *information ratchet*. A heavy bead is suspended from a spring. A chance thermal fluctuation pushes the bead up. After detection, the spring support is shifted up. For down fluctuations, the trap stays fixed. Repeated cycling then leads to a steady rise in the average height of the bead and a net work gain. (If the particle

is too heavy, the energy spent stretching the spring when moving up the trap center can exceed the gain in gravitational potential energy.) Experimentally, this situation has been implemented using a horizontal tweezer and a heavy particle in water, with overdamped dynamics. This is again a Maxwell demon: thermal energy in the bath raises the bead and is "captured" by feedback, leading to a steady extraction of useful energy from the bath.

15.3.6 Information Is Physical

By now, you should be worried. How can the mere act of observing a system throw it out of equilibrium? Moreover, heeding Eddington's warning, we cannot ignore these apparent violations of the second law of thermodynamics. Where is the hidden cost?

In fact, it took nearly a century and a half to understand fully the paradoxes first raised by Maxwell and Szilard. Starting in the 1960s, Rolf Landauer and later Charles Bennet made key contributions that were fully understood only in the last decade, via the formalism of stochastic thermodynamics. The main point, already recognized by Szilard, is that we cannot think of measurement as an abstract process. Rather, a measuring device must record the information, meaning that the information is encoded in a physical system obeying physical laws. In Landauer's pithy formulation, "Information is physical."

We thus turn to the physics of the measuring device. Its key features are *information-bearing degrees of freedom* (or "informational states"): macrostates that are long lived, so that the probability of a spontaneous thermal escape from the state is negligible on any relevant time scale. A particle in a double-well potential with high barrier $E_b \gg k_B T$ gives a paradigmatic example that is again illustrated at left. The particle will be in one of two macrostates, "left" or "right." The system can store one bit of information.

More generally, the phase space Γ of the measuring / memory device is divided into *ergodic regions* Γ_i, where dynamics are ergodic for the microstates within Γ_i, and each region is occupied with probability p_i. The global ergodicity of the system is broken. The memory then has a nonequilibrium free energy

$$F = \sum_i p_i F_i^{\text{eq}} + (\ln 2) H(p) , \tag{15.82}$$

where F_i^{eq} is the equilibrium free energy of the system in state i and $H(p) = -\sum_i p_i \log_2 p_i$ is the Shannon entropy of the informational states, in bits. Now, because of the high barriers separating informational states, the system will never spontaneously equilibrate under any relevant time scales. A memory device is thus characterized simply by the occupational probabilities p_i. In the simplest case, all of the macrostates are equivalent ($F_i^{\text{eq}} = F^{\text{eq}}$), and protocols that manipulate the memory always end with the potential in standard reference form. Then memory operations simply alter the probabilities p_i, and the nonequilibrium free energy changes by $\Delta H = H(p') - H(p)$.

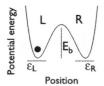

We can now consider the Szilard engine and measuring device as a joint system. For brevity, we will refer to them as the engine and memory, respectively. In calculating costs, it is important to consider a complete cycle of operation that ends in the same state as it began. We start with the two systems decoupled, with the engine in equilibrium ($p = \frac{1}{2}$ for each state) and the memory in the state $p_0 = 0$ and $p_1 = 1$. The latter choice is conventional, and the analysis can easily be done for other choices. The free energy of the joint system is $F(XY) = F(X) + F(Y)$, where $F(X)$ is the free energy of the engine and $F(Y)$ is the free energy of the memory, each in its initial state. Because the states X and Y are independent, there is no interaction energy.

To "make a measurement," we allow engine and memory to interact for some time and then separate the two subsystems. The memory state is then correlated with the engine state. The degree of this correlation is expressed by $I(X;Y)$ and ranges from 0 (no correlation at all) to 1 bit (perfect measurement). As discussed before, the measurement *raises* the free energy of the engine by $I(X;Y)$ (in units of $k_B T$). At the same time, the memory goes from a Shannon entropy of 0 (it started in state 1 with probability 1) to a Shannon entropy of ln 2 (storing one bit of information), which reduces the free energy $E - TS$. The measurement thus *lowers* the nonequilibrium free energy of the memory by $I(X;Y)$. The net change of the nonequilibrium free energy of the joint system is then zero, as the two contributions cancel. We can thus carry out this step for a work $W_{\text{meas}} \geq 0$, with the lower bound of zero reached when the measurement process is very slow.

The second step is to use the information gained to extract work from the engine, via the feedback protocol already described. The feedback – conditional action – consists of choosing the new potential so that the system is in equilibrium after the observation. In this step, we can extract $W_{\text{fb}} \leq I(X;Y)$ of work via the feedback operation.

But now, to close the cycle and return to the initial state of both the engine and the memory, we need a third step, as Landauer and Bennett recognized. We need to *reset* the memory back to its original state.[39] This requires performing $W_{\text{reset}} \geq I(X;Y)$ of work.

To summarize (and use physical units), we have shown that

$$\underbrace{W_{\text{fb}} \leq k_B T \ln 2}_{\text{engine}}, \qquad \underbrace{W_{\text{meas}} + W_{\text{reset}} \geq k_B T \ln 2}_{\text{memory}}, \qquad W_{\text{total}} \geq 0. \tag{15.83}$$

Thus, although the amount of work that we can extract from the engine can equal the information gained, the work needed to run the measuring device (memory) through a complete cycle will always exceed or at best equal this amount.

While the discussion above has put the cost of the memory operation entirely on the reset operation, that conclusion depends on our selection of $p = (0, 1)$ as a reference state for the memory. If the reference state of the memory is to be in thermal equilibrium, with $p = (\frac{1}{2}, \frac{1}{2})$, then the measurement process costs the price $I(X;Y)$ of correlating two formerly independent two-state systems. The correlation can be exploited to extract $I(X;Y)$ from the engine. At the end of the extraction, the memory

[39] The reset operation is also termed *erasing information*.

is still in the same equilibrium state, and only the correlation has been lost. Thus no reset is needed and this version puts all the cost on the measurement. But the sum of measurement and reset remains the same.[40]

Finally, let us return to the claim made at the beginning of this section that observations can change stochastic quantities by changing their probability distributions. I hope that this statement now seems obvious, even a bit trivial. In our everyday experience of the world, macroscopic objects are so big that thermal fluctuations are much smaller than any measurement noise. In such a limit, observations change the nonequilibrium free energy by negligible amounts, explaining why we do not meet Maxwell demons in everyday life.

15.3.7 Autonomous Control

In the last section, we considered a Szilard engine and its measuring device jointly and showed how the apparent violations of the second law provided by feedback are in fact consistent with thermodynamics. The costs of running the memory will always equal or exceed the work extracted from a heat bath. But a defect of our treatment is that physical details such as the nature of the interaction between the systems remain obscure. In this section, we analyze a simple four-state system using the master-equation formalism introduced in Section 15.3.4. Our goal will be to see how measurement and feedback can arise in a single physical system. In certain limits, it will make sense to say that one two-state subsystem "controls" the other via a "feedback loop."

Bipartite Systems

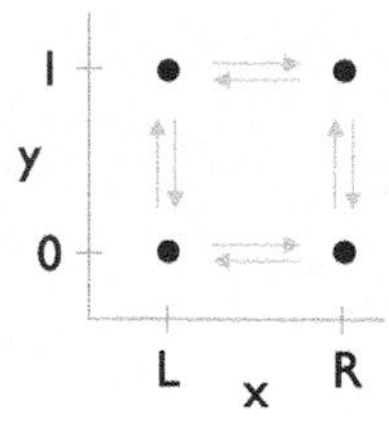

Our goal is to construct a master equation that models the dynamics of two independent subsystems, X and Y, that are coupled to each other yet retain their identity. Here, X represents the physical system of interest, and Y the measuring device. Our canonical example where X and Y are twin two-state systems is shown at left. Here, X represents the engine and can take values $x = \{L, R\}$ while Y represents a one-bit memory and can take on values $y = \{0, 1\}$. We will assume that the system is coupled to a single heat bath.

In the figure at left, the key simplifying feature, which we will draw upon repeatedly, is that the allowed transitions can only be horizontal or vertical on the diagram, never diagonal. In other words, at any moment in time, the state of X can change while that of Y is fixed, or vice versa. But the two systems are never allowed to change simultaneously. Roughly, for continuous-state systems, this requirement amounts to asserting the independence of noise sources that act on the system (process noise) from the measurement noise.

[40] Historically, the arbitrariness of the reference state led to much confusion. In hindsight, choosing $p = (0, 1)$ as a reference state for the memory might seem natural to those working in the computer industry (both Landauer and Bennett worked at IBM). But choosing a state of thermal equilibrium probably seems more natural to those interested in how biological molecules process information and carry out feedback.

We express this constraint via the structure of transition rates of a master equation,

$$\mathrm{d}_t p_{xy} = \sum_{x'y'} W_{xx'}^{yy'} p_{x'y'} - W_{x'x}^{y'y} p_{xy} \equiv \sum_{x'y'} J_{xx'}^{yy'}\,, \tag{15.84}$$

where $p_{xy}(t) \equiv p(x, y; t)$ is the joint probability for subsystem X to have state x and subsystem Y state y at time t. We can map from single- to double-index notation by $(p_1, p_2, p_3, p_4)^{\mathrm{T}} \to (p_{L0}, p_{R0}, p_{L1}, p_{R1})$. In the double-index notation, the current $J_{xx'}^{yy'}$ flows from $(x', y') \to (x, y)$. The bipartite structure is then created via choosing transition rates so that x or y changes at any given time but not both:

$$W_{xx'}^{yy'} = \begin{cases} W_{xx'}^{y} & x \neq x', y = y' \\ W_{x}^{yy'} & x = x', y \neq y' \\ 0 & \text{otherwise} \end{cases} \tag{15.85}$$

We can analyze the joint system using the formalism of Section 15.3.4. The entropy production decomposes as usual: the time derivative of the system entropy $\mathrm{d}_t S^{XY} = \dot{S}_{\mathrm{i}}^{XY} + \dot{S}_{\mathrm{e}}^{XY}$, the sum of irreversible entropy production and heat flow from reservoir to system:

$$\mathrm{d}_t S^{XY} = \frac{1}{2} \sum_{xy,x'y'} J_{xx'}^{yy'} \ln \frac{p_{x'y'}}{p_{xy}}\,, \tag{15.86}$$

and

$$\dot{S}_{\mathrm{i}}^{XY} - \frac{1}{2} \sum_{xy,x'y'} J_{xx'}^{yy'} \ln \frac{W_{xx'}^{yy'} p_{x'y'}}{W_{x'x}^{y'y} p_{xy}} \quad \text{and} \quad \dot{S}_{\mathrm{e}}^{XY} = \dot{Q}^{XY} = \frac{1}{2} \sum_{xy,x'y'} J_{xx'}^{yy'} \ln \frac{W_{x'x}^{y'y}}{W_{xx'}^{yy'}}\,. \tag{15.87}$$

Alternatively, we can view it as two coupled subsystems. The bipartite structure allows arbitrary functionals of current to be split into two parts. For example, the master equation itself can be rewritten in the form

$$\mathrm{d}_t p_{xy} = \sum_{x'y'} J_{xx'}^{yy'} = \sum_{x'} J_{xx'}^{y} + \sum_{y'} J_{x}^{yy'}\,, \tag{15.88}$$

where $J_{xx'}^{y} \equiv W_{xx'}^{y} p_{x'y} - W_{x'x}^{y} p_{xy}$ is the current from x' to x for fixed y and where $J_{x}^{yy'} \equiv W_{x}^{yy'} p_{xy'} - W_{x}^{y'y} p_{xy}$ is the current from y' to y for fixed x. We gather all the currents along x in one sum and those along y in the other.

Let us define the notion of *information flow* as the time derivative of the mutual information between the x and y states. The mutual information is given by

$$I^{XY} = \sum_{xy} p_{xy} \ln \frac{p_{xy}}{p_x p_y} \geq 0\,. \tag{15.89}$$

In Problem 15.25, you will show that $\mathrm{d}_t I^{XY} = \dot{I}^X + \dot{I}^Y$, with path-dependent rates

$$\dot{I}^X = \frac{1}{2} \sum_{xy,x'} J_{xx'}^{y} \ln \frac{p_{y|x}}{p_{y|x'}} \quad \text{and} \quad \dot{I}^Y = \frac{1}{2} \sum_{xy,y'} J_{x}^{yy'} \ln \frac{p_{x|y}}{p_{x|y'}}\,. \tag{15.90}$$

The rates $\dot{I}^X$ and $\dot{I}^Y$ quantify the rates of information exchange between the x and y subsystems. When $\dot{I}^Y > 0$, jumps in Y increase I. In the interpretation of X as the engine and Y as the memory, Y "learns" about X. Conversely, when $\dot{I}^Y < 0$, the memory erases information (resets to the initial state). When $\dot{I}^X < 0$, jumps in X decrease I. Then X "consumes" information by decorrelating the X and Y subsystems.

The key relation (Problem 15.25) is that the total entropy production can be decomposed into two other path-dependent rates, $\dot{S}_{\rm i}^{XY} = \dot{S}_{\rm i}^X + \dot{S}_{\rm i}^Y \geq 0$, with

$$\dot{S}_{\rm i}^X = {\rm d}_t S^X - \dot{S}_{\rm e}^X - \dot{I}^X \geq 0 \qquad \text{and} \qquad \dot{S}_{\rm i}^Y = {\rm d}_t S^Y - \dot{S}_{\rm e}^Y - \dot{I}^Y \geq 0\,. \tag{15.91}$$

Equation (15.91) gives substance to the claim that it is useful to decompose the whole system into X and Y subsystems, as it shows that $\dot{S}_{\rm i}^X$ and $\dot{S}_{\rm i}^Y$ are *both* nonnegative. The information-flow terms $\dot{I}^X$ and $\dot{I}^Y$ explicitly couple the two subsystems. Now, following Eq. (15.72), the terms ${\rm d}_t S^X - \dot{S}_{\rm e}^X$ in Eq. (15.91) are just what we would naively assume to be the entropy production $\dot{S}_{\rm i}^X$ of the X subsystem, if we did not know about Y. But because of the $\dot{I}^X$ term, this quantity can be negative, apparently violating the second law. We have thus created yet another Maxwell demon, where the exchange of information leads to a seeming violation of the second law. The new feature is that both system and measuring device are explicitly subsystems of a joint system that clearly does obey the second law.

For steady-state solutions, all time derivative terms are zero. Thus, ${\rm d}_t I^{XY} = \dot{I}^X + \dot{I}^Y = 0$, and $\dot{I}^Y = -\dot{I}^X \equiv \dot{I}$: the information flows between X and Y balance. Equation (15.91) then simplifies to

$$\dot{S}_{\rm i}^X = -\dot{S}_{\rm e}^X + \dot{I} \geq 0 \qquad \text{and} \qquad \dot{S}_{\rm i}^Y = -\dot{S}_{\rm e}^Y - \dot{I} \geq 0\,. \tag{15.92}$$

When $\dot{I} > 0$, subsystem Y learns about – senses – subsystem X. Moreover, Eq. (15.92) implies that $-\dot{S}_{\rm e}^Y = -\dot{Q}^Y \geq \dot{I}$. Since $-\dot{Q}^Y$ is the rate energy transferred from the system to the reservoir, sensing is an active, nonequilibrium process that pays in heat for the information gained. Conversely, $\dot{I} < 0$ means that the subsystem X can extract work from the bath at a rate up to $\dot{I}$. Adding the two contributions recovers the full entropy production $\dot{S}_{\rm i}^{XY} = -(\dot{S}_{\rm e}^X + \dot{S}_{\rm e}^Y) \geq 0$: the information terms cancel out, leaving just the rate of transfer of heat to the bath from each subsystem.

Szilard Engine and Measuring Device

The above formalism of bipartite systems and information flow is quite general and allows for an arbitrary number of states connected in complicated ways, with many different loops (*cycles*, in graph-theoretic language). Only the bipartite constraint that x or y must change individually is important. To understand better how autonomous control can arise in a bipartite system, we now specialize to the simple structure depicted in Figure 15.8, where four states represent the engine (x-projection) and memory (y-projection).

Because the joint system has but a single loop, its analysis simplifies (Problem 15.26). In the basic setup (a), each state has an energy $\epsilon = 0$ and the barriers are equal and symmetric. The symmetric system has four states, each with a steady-state probability

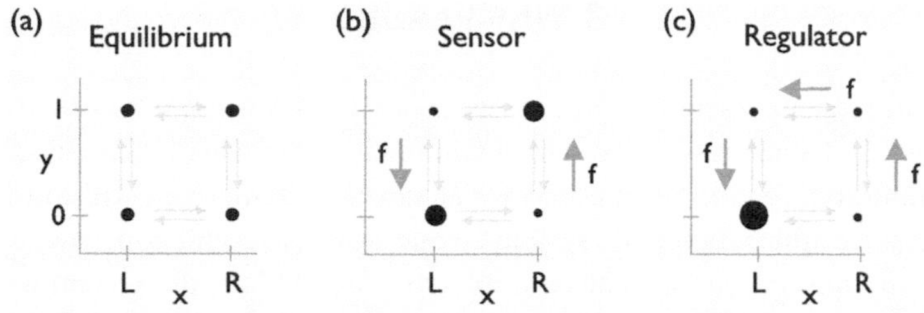

Fig. 15.8

2×2 Bipartite system. The engine has states L, R (x-axis), while the memory has states 0, 1. (a) Equilibrium setup. Equal areas of four states indicate steady-state probability $p = \frac{1}{4}$ for all four states. (b) Sensing limit. Nonequilibrium driving f correlates L and 0, R and 1. (c) Regulation limit. Adding a third nonequilibrium forcing stabilizes the system in one state.

$p = \frac{1}{4}$. The marginal probability p_L to be in the L state regardless of the state of the measuring device is $\frac{1}{2}$ and equals the marginal probability p_R.

Figure 15.8b adds nonequilibrium forces of magnitude f to the vertical transitions, as discussed in Section 15.3.4. As should be intuitive (and verified in Problem 15.26), the counterclockwise flow imposed by the two nonequilibrium contributions $L, 1 \to L, 0$ and $R, 0 \to R, 1$ bias the steady state to $L, 0$ and $R, 1$. For $f \gg 1$, the device functions as a sensor because L and 0 are correlated, as are R and 1. If the system is in the L state, it is also most likely in the 0 state. If in the R state, it is also most likely in the 1 state. Yet the transitions $L \leftrightarrow R$ are unperturbed, and we still find $p_L = p_R = \frac{1}{2}$.

Figure 15.8c has three driving potentials, all of magnitude f. The combined effect is to force the system to spend almost all its time in state $L, 0$. After the occasional $L, 0 \to R, 0$ transition, the three driving terms rapidly send the state "around the loop" until it returns to $L, 0$. The net result is to combine sensing and actuation, creating a feedback regulation of the state L. Problem 15.26 explores the performance and costs of this autonomous feedback "device" in more detail. It can "lower the entropy" of the system state from $\ln 2$ in equilibrium to a value approaching zero as $f \to \infty$. The device incorporates both engine and memory as subsystems, a view that is enforced both via the bipartite structure and the need to have $f \gg 1$,[41] which implies that the feedback control operates on a faster time scale than the system dynamics (the time scale for $L \leftrightarrow R$ transitions). The model thus gives us a starting point to imagine how systems such as the Watt governor or climate-change models of Chapter 1 might be analyzed as instances of feedback, even though they lack the explicit structure of sensors and controllers of textbook control systems.

[41] A bipartite structure is easy to visualize but not necessary in order to have Maxwell demons (Chétrite et al., 2019). For Kalman filters, the bipartite condition amounts to the independence of process and measurement noise sources. But Kalman filters can be defined when there is correlation between the two noise sources. After an appropriate coordinate transformation, such cases are equivalent to the bipartite case, with two uncorrelated noise sources.

15.3.8 Why Feedback?

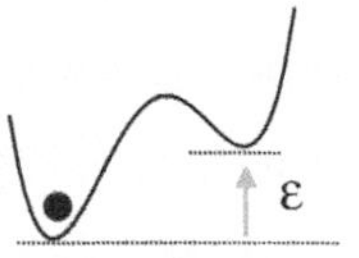

The analysis of bipartite systems of the previous section gives us an excuse to ask again, "Why feedback?"[42] That is, we have shown *how* feedback can arise in a physical system with no human intervention. But could we achieve the same goal without feedback? For example, to confine a two-state system to one state, we could construct a potential for which the desired outcome is the equilibrium state. At left, the system will spend a fraction of time $1-e^{-\epsilon}$ in the desired state, ≈ 1 for $\epsilon \gg 1$. After an initial cost to *prepare* the state, there is no further thermodynamic cost to *maintaining* it. The average cost tends thus tends to zero, in contrast to the feedback system, which dissipates energy every time it makes a correction, by forcing the system to go "around the loop." The key point is the ratio between initial and maintenance costs. If all you want to do is maintain a system in one state forever, then it is indeed better to make a system that has that state as its equilibrium. But perhaps the goal is for the system to follow a trajectory in the state space, for example *LRLRLR*..., a periodic alternation where the system should dwell in each state for a time τ before switching. In this case, in a system with damping, each switch has a dissipative cost, and now we can compare the switching cost per time against the steady-state cost rate of the four-state feedback system.

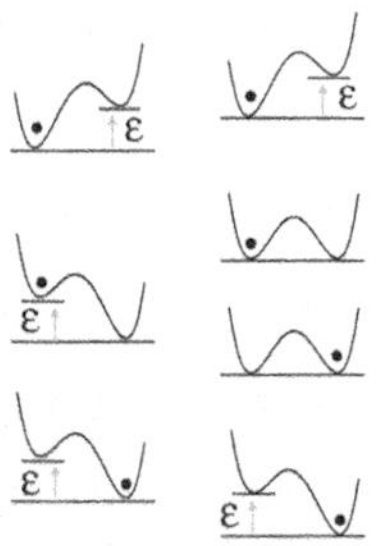

To see this trade-off more clearly, consider the nonautonomous control of a two-state system, where we can, from the outside, manipulate the energy levels (e.g., by tilting a double-well potential and manipulating its energy barrier). At left, we compare the two methods for imposing a state-space trajectory. The first, at left in the figure, is simply to switch the energy levels at a desired time. It will cost $\approx \epsilon/\tau$ after averaging the dissipation over the switch time τ.[43] The second method, shown at left in the figure, is to lower the second energy level to zero and wait for a spontaneous hop. Once the system is in the correct state, the other energy level is raised. Ignoring the costs of measurement and reset, we apparently can switch for free. But we then need to account for the costs of the controller, which include the fundamental costs of memory reset and also the much larger costs of running a camera, computer, lighting system, and so on.[44]

If we turn, however, to our four-state autonomous control system, we can explicitly calculate the costs of control. In Problem 15.26, we show that the rate of entropy production to run the bipartite regulator example is $3f/\tau_0$, where f is the nonequilibrium driving and τ_0 is the time scale for jumps in the unregulated system. Thus, when $\tau \lesssim \frac{1}{3}\tau_0\epsilon/f$, it pays to use feedback; for longer τ, perhaps not. The numerical details, of course, are less important than the idea that there are two regimes.

[42] Chapter 9 discusses robustness to uncertain dynamics as another motivation for feedback.

[43] If the occupation probability of the well that is raised ≈ 1 and if it is raised suddenly, the cost will be close to the energy difference ϵ. For a protocol executed every τ, you could design a more gradual protocol that would cost a fraction of ϵ on average each time. But the overall cost would still be $\propto \epsilon/\tau$.

[44] Another possibility is to rotate the potential out of the page, which can swap the states without changing energy levels. Assuming a drag force proportional to the imposed velocity also leads to a τ^{-1} cost.

Agility – the need or desire to change a set point rapidly – is one reason to use feedback, but there are others. As with linear differential equations, master-equation dynamics can be expressed as a sum of exponentially decaying modes. If the natural time constants are too slow, you can add more nodes and use feedback to speed up relaxation.

A biological example illustrates these issues. Bacteria such as *E. coli* often encounter fluctuating environments. In response, they can produce different phenotypes that grow fast in a favored environment. One can imagine two strategies: sense the environment continually and switch to the right phenotype as a feedback response; or switch stochastically (*bet hedging*), maintaining subpopulations of all phenotypes so that the population is never wiped out when the environment changes. Bacteria often adopt the latter strategy, which is cheaper when environmental changes are infrequent (Kussell and Leibler, 2005).

15.3.9 Regulating and Cooling

We conclude with an example of autonomous control for a continuous system where we can find the explicit mapping from an active controller to a nonequilibrium system with two temperature baths. At right is a current source and RC circuit, where the resistor is in equilibrium with a thermal bath at temperature T. If we take the Johnson–Nyquist voltage noise of the resistor into account, the equation for the currents gives

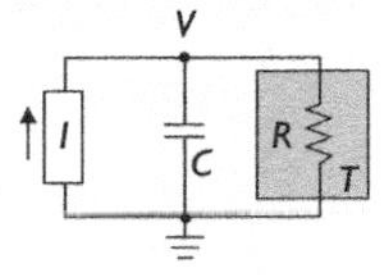

$$C\dot{V} = -\frac{V}{R} + I + \sqrt{\frac{2k_{\mathrm{B}}T}{R}}\,\eta(t)\,, \qquad \langle \eta(t)\,\eta(t_0)\rangle = \delta\,(t - t_0)\,, \tag{15.93}$$

with $\eta(t)$ a zero-mean Gaussian white-noise source.

Either by integrating Eq. (15.93) (Problem 15.27) or by appealing to Eq. (8.50), we can show that the fluctuating resistor voltage (Johnson–Nyquist noise), for $I = 0$, leads to

$$\langle V^2 \rangle = \frac{k_{\mathrm{B}}T}{C}\,. \tag{15.94}$$

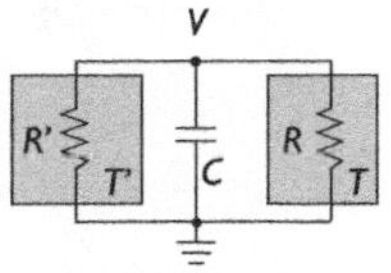

Now consider the circuit at right, which replaces the current source with a second resistor R', in thermal equilibrium with a second bath, at temperature T'. Let us first analyze this circuit from a purely physical point of view, defining a current

$$C\dot{V} = -\frac{V}{R} - \frac{V}{R'} + \sqrt{\frac{2k_{\mathrm{B}}T}{R}}\,\eta(t) + \sqrt{\frac{2k_{\mathrm{B}}T'}{R'}}\,\eta'(t)\,, \tag{15.95}$$

with each resistor having its own independent thermal noise source:

$$\langle \eta(t)\,\eta(t_0)\rangle = \langle \eta'(t)\,\eta'(t_0)\rangle = \delta\,(t - t_0)\,, \qquad \langle \eta(t)\,\eta'(t_0)\rangle = 0\,. \tag{15.96}$$

In Problem 15.27, you will show that the variance becomes

$$\langle V^2 \rangle = \frac{k_{\mathrm{B}}}{C}\left(\frac{R'}{R+R'}T + \frac{R}{R+R'}T'\right). \tag{15.97}$$

Equation (15.94) leads to an effective temperature for the capacitor,[45] given by $T_C \equiv C\langle V^2\rangle/k_B$ in the single-resistor case. In the two-resistor case, the temperature becomes

$$T_C = T + (T' - T)\left(\frac{R}{R + R'}\right), \tag{15.98}$$

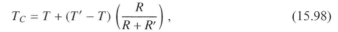

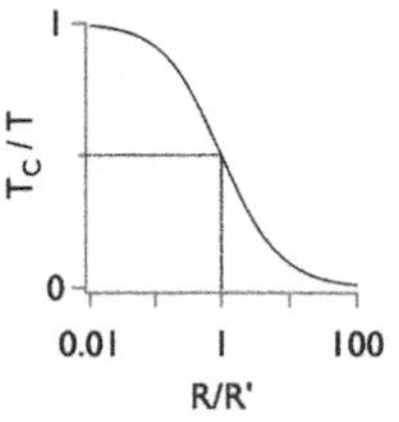

which interpolates between T (for $R' = \infty$) and T' (for $R' = 0$), as shown at left for $T' = 0$. The result should not be surprising: the capacitor is connected to two fluctuating resistors and ends up with voltage fluctuations that are intermediate between those produced by either resistor alone. Notice that it is essential to use a resistor at a lower temperature. If $T' = T$, then the capacitor temperature $T_C = T$, whatever the value of R'.

Now let us relate the two circuits shown above by choosing the current in Eq. (15.93) as

$$I(t) = -\frac{V}{R'} + \sqrt{\frac{2k_B T'}{R'}}\,\eta'(t) \equiv -\frac{1}{R'}y(t). \tag{15.99}$$

In Eq. (15.99), the circuit "measures" $y(t) = \left[V(t) + \sqrt{2k_B T' R'}\,[-\eta'(t)]\right]$ and then "applies" a current that is based on a negative proportional-feedback algorithm with gain $1/R'$.

Viewing the second resistor R' as an autonomous feedback controller, we see that measuring a signal and using the results in a feedback loop is equivalent to cooling the system by connecting it to a cold bath. Of course, we do not usually regulate a quantity by literally cooling it.[46] Rather, we resort to *active components*. For example, we can replace the cool resistor R' with a room-temperature op-amp circuit. The point is that active components are active – they draw power, much as a refrigerator draws power to cool.

The quantitative analysis of information flows, entropy production, and work rates can proceed as before. We refer the reader to the notes at the end of this chapter, as there remains debate about the most useful type of information-theoretic quantities to consider for continuous state variables. The important point is to see how feedback and control can arise "spontaneously" in a simple autonomous system and to see the need for that system to be out of thermodynamic equilibrium.

15.4 Summary

We have discussed three types of limits to control: those due to causality, to information flows and network structure, and to thermodynamics.

[45] The effective temperature, defined by the variance of voltage fluctuations, is that of the conduction electrons. The bulk of the capacitor remains at ambient temperature.

[46] Unless you are a low-temperature physicist!

Starting from the familiar Kramers–Kronig relations between the real and imaginary parts of a linear-response function, we derived Bode's gain-phase version, which gives the minimum phase lag of a causal response. Another consequence of causality is the Bode waterbed principle and its variants, which assert that reduced sensitivity at some frequencies (*robustness*) must be balanced by enhanced sensitivity at other frequencies (*fragility*). Finally, we briefly explore how lifting the constraints of causality can improve performance. The enterprise sounds silly until we realize that control systems are *open systems* as regards information. We often do know something about the future, either because there are commands that we know we will apply in the future or because we have early warning of incoming disturbances or even because of general knowledge of future disturbances ("the sun will set"). Understanding how the limits imposed by causality are softened in such situations is a topic that deserves more attention from physicists.

The classic constraints of causality for linear systems are expressed in the frequency domain. In the time domain, the waterbed principle has a close connection to information theory. For example, stable linear systems have output entropies that match those of their inputs. Looking at control theory from the point of view of information theory clarifies some basic questions: By directing system states in regions of high measurement sensitivity, feedback can improve measurement throughput. But linear systems have constant sensitivity, and feedback does not improve measurements in such cases.

More generally, the connections to information theory discussed here, along with the treatments of stochastic effects, system identification, and adaptive control, begin to flesh out the maxim for successful control that we presented in the Introduction to this book: *Exploit what you know; learn what you can.*

In a final section, we showed how control theory comes together with both information theory and thermodynamics to address issues that go back to the heroic age of thermodynamics and statistical physics. The Maxwell demon thought experiment, which has been realized in several laboratory experiments, forced physicists to confront questions such as whether learning something about a system allowed circumventing the second law of thermodynamics. Fortunately, Landauer's dictum that "information is physical" dissolves the paradoxes by accounting for the costs of running the measuring device, or memory.

The formalism of stochastic thermodynamics suggests many useful simple models. Perhaps the simplest control problem where the thermodynamics can be worked out explicitly is the case of twin two-state systems with bipartite dynamics. If time scales are appropriate, the two coupled subsystems can interact, so that one plays the role of Maxwell demon (feedback controller) and the other that of an engine. We can then begin to understand, at least in principle, how complex natural and engineered systems can show features of control systems that correspond to measurement and feedback. Thus, technologically important inventions such as James Watt's steam engine and natural systems such as the earth's climate may be seen as autonomous control systems that improve the performance of a subsystem, at the cost of greater dissipation in the whole.

Control systems are magical, yet subject to the laws of physics.

> We shall not cease from exploration
> And the end of all our exploring
> Will be to arrive where we started
> And know the place for the first time.
>
> T. S. Eliot[47]

15.5 Notes and References

Frequency-domain implications of causality and related issues are standard topics in control-theory books. Probably the most complete discussions are the book by Seron et al. (1997) and the review of Chen et al. (2019). See Goodwin et al. (2001) for an introduction. For further examples along the lines of Problems 3.16 and 3.17, see Åström (2006b). The time-domain derivation of the Kramers–Kronig relation unpacks the "two line" derivation of Hu (1989). For better and for worse, it avoids the discussion of analyticity in the complex plane given in our second derivation (Jackson, 1999, Chapter 7.10). The derivation of the gain-phase relation given here is, I believe, a simpler, clearer version of that given by Bode (1945) and follows Bechhoefer (2011). The physical interpretation of the meaning of the Kramers–Kronig relations as well as the inspiration for Figure 15.1 are due to Toll (1956). Toll also discusses the gain-phase version of the Kramers–Kronig relations but uses a form that differs from Bode's (and ours). Our discussion of acausal control (*preview control* in the engineering literature) follows Skogestad and Postlethwaite (2005), Section 5.7.4.

It is difficult to overstate the influence on engineers of Bode (1945), who explored systematically the implications of the analyticity due to causality. To this day, such consequences are not as well known to physicists as they should be. In addition, formulas similar to the Bode sensitivity integral constrain broadband impedance matching with passive components: The minimum bound of a reflection coefficient over a given frequency range is limited by the load impedance *(Bode–Fano limits)*. If the load is a parallel RC circuit, then $\int_0^\infty d\omega \ln 1/|\rho| \le \frac{\pi}{RC}$, where $\rho(\omega)$ is the reflection coefficient at frequency ω. If matching is attempted over a bandwidth $\Delta\omega = 2\pi\Delta f$ (see left), then $\rho_0 \ge \exp(-\frac{1}{2RC\Delta f})$. A recent application shows the practical impossibility of creating invisibility cloaks over the visible-light spectrum using passive materials (Monticone and Alù, 2016). Time-dependent (active) materials give more possibilities (Shlivinski and Hadad, 2018).

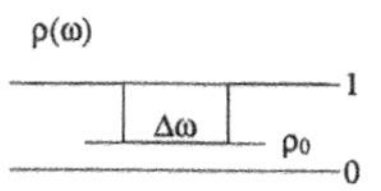

The connections between control theory and information theory is of great current interest. The effect of filtering a linear system on entropy rates, Eq. (15.32), was stated by Shannon (1948). Our derivation is a simplified, careful version of that found in

[47] Excerpt from "Little Gidding" from FOUR QUARTETS by T. S. Eliot. Copyright ©1942 by T. S. Eliot, renewed 1970 by Esme Valerie Eliot. Reprinted by permission of Houghton Mifflin Harcourt Publishing Company. All rights reserved.

Papoulis and Pillai (2002). Very few control-theory books have discussed the relations between information and control. One exception is Jacobs (1993), who noted that linear systems could not improve measurements. His *Principle of Neutrality* states, "A controlled object is said to be *neutral* if the rate of reduction of uncertainty about its state $\boldsymbol{x}$ is independent of the control $\boldsymbol{u}$." Touchette and Lloyd (2000, 2004) are influential works in the physics literature relating information and control. Lestas et al. (2010) apply an information-theoretic analysis to place general performance limits on feedback control of biological processes: The minimum standard deviation of the regulated species decreases very slowly with the number of molecules used in regulation (1/4 power).

Massey (1990) introduced *directed information* to capture aspects of causality that *mutual information* ignores. Causal conditioning was introduced by Kramer (2003). There are other quantities that capture different aspects of causality such as the transfer entropy of Schreiber (2000), nonanticipatory ϵ entropy, and more. Rivoire and Leibler (2011) discuss how the different types of problems in communications, control theory, finance, and biology all lead to different "natural" definitions of information-theoretic quantities. Our discussion of the connection between directed information rates and the Bode waterbed relation follows Elia (2004) and that between stabilization and information Liu and Elia (2009). See also Martins and Dahleh (2008) and then Nair et al. (2007) for overviews of the data rate theorem. Special cases of the conjectured relationship between the sum of positive Lyapunov exponents and the data rate theorem are given in Kawan (2013).

Trying to use feedback to improve measurements in linear systems is a Siren that still tempts heedless physicists. The idea seems particularly attractive for underdamped oscillators, as feedback can indeed alter damping (quality factors) and resonance frequencies. But information rates remain invariant. For example, Gavartin et al. (2012) claimed that feedback damping of a high-Q oscillator seemed to improve the oscillator-position measurements. However, Vinante et al. (2013) and Harris et al. (2013) showed that similar (indeed, better) results can be obtained by using an optimal filter.

Inspired by the seminal study of Laughlin (1981) on the neural response to light-intensity changes, Bialek (2012) discusses the optimization of information flow, translating the statistics of the environment into a design for an optimal response function. He argues that there is evidence that such optimized response has evolved in numerous biological systems. Appendix A.8 of his book introduces the issues in estimating entropy and mutual information.

Our brief presentation of stochastic thermodynamics discussed only the simplest version of that theory. More generally, state spaces can be continuous, dynamics can be underdamped, and there can be multiple baths that can exchange heat and also particle numbers, via chemical reactions. Further, we have focused just on mean values. Extending the formalism to stochastic trajectories leads to beautiful and important results, such as the various fluctuation relations. We follow Van den Broeck and Esposito (2015); see Seifert (2012) for a more advanced review and Sekimoto (2010) for physical insights. Shaw (1984) gives an early, insightful discussion of the relations among nonlinear dynamics, information, and thermodynamics. For background on master equations, see Kelly (1979) for mathematical details and van Kampen (2007) for physical aspects. Our discussion of the relations between stochastic thermodynamics, information, and feedback is largely based on Parrondo et al. (2015). Horowitz and Esposito (2014) compare a variety of proposed information-theoretic quantities that all lead to bounds on the work extraction rate of the form $\dot{W} \leq k_B T \dot{I}$, where I can be the mutual information, directed information, and so on. There does not yet seem to be a full consensus on which quantities to use and when.

A recent development is the *thermodynamic uncertainty relation* (TUR), whose generic form is Var $\langle O \rangle / \langle O \rangle^2 \geq 2/\langle \sigma \rangle$, with O an observable quantity and $\langle \sigma \rangle$ a dissipation rate including entropy production (Horowitz and Gingrich, 2019). For feedback control, $\langle \sigma \rangle$ includes a contribution from the mutual information between the system state and the measurement. Thus, the signal-to-noise ratio (SNR^2) of a regulated quantity is proportional to the overall thermodynamic cost of control (including the Landauer cost of measurements). The TUR has been proven as stated for nonequilibrium steady states; extensions to include feedback are still being investigated (Vu and Hasegawa, 2020).

Experiments on Maxwell's demon, information engines, and Landauer's principle have only recently been performed, making real what had been only long-standing thought experiments. See Toyabe et al. (2010), Bérut et al. (2012), Jun et al. (2014), and also the review by Ciliberto (2017). The gravitational information ratchet from Example 15.7 has been realized experimentally in my lab. A version based on flow was studied by Admon et al. (2018).

The circuit discussed at the end of the chapter was realized experimentally by Ciliberto et al. (2013). For its interpretation as a feedback controller, see Sandberg et al. (2014). See also Freitas et al. (2019) for a generalization to RLC circuits of arbitrary topology, with each resistor at a different temperature, and for a discussion of subtleties associated with our particular circuit of two resistors and one capacitors: Local heat currents diverge if fluctuations can occur at arbitrarily high frequency. In reality, fluctuations are cut off by stray inductances or capacitances that create low-pass filters in the circuit, as well as by quantum effects that cut off fluctuations when $\hbar\omega > k_B T$.

We end with a couple of comments about the application of control theory to biology. As a book devoted to teaching control theory to physicists, we have mostly avoided biological applications. Still, such applications are of great current interest, and physicists have been involved in those efforts. For a physicist, an immediate question might be whether control theory applied to biological systems is different in any fundamental sense from control theory applied to more familiar physical and engineering systems. The immediate answer, of course, is that the fundamental limits of causality, information theory, and thermodynamics sketched in this chapter apply to all systems, biological included. Nonetheless, the more important and successful applications of control theory in biology seem at least superficially quite different from their traditional forms (Del Vecchio and Murray, 2014).

One such application is stick (and human) balancing (Sections 3.5.2 and 5.4.2). Humans seem to use internal models of physical dynamics in order to predict or anticipate future events. Such internal models must be used in order to explain how we can balance sticks short enough that they would inevitably fall if controlled by traditional PID controllers. Moreover, there is strong evidence that such internal models are routinely used in other events. Although such model predictive control was developed independently by the engineering community, it plays a far more important role in physiological systems.

A second area of biology where control theory ends up looking quite different is in the control at the molecular level of the cell. Here, systems are *autonomous*: they look more like the steam-engine governor where the decomposition of a system into independent sensor, controller, and actuator *modules* is less clear-cut than in traditional engineering applications where each is a separate box wired together. Moreover, the numbers of molecules are often quite small, meaning that fluctuations are important and may follow non-Gaussian distributions (e.g., Poisson). In such a limit, general principles still hold. For example, perfect homeostasis (maintenance of some quantity in the face of steady-state perturbations) requires integral control (the internal model principle of Section 3.7.4). But the way integral control is implemented can look quite different (e.g., the *antithetical integral feedback* of Aoki et al. [2019]).

Problems

15.1 Causality and Kramers–Kronig for a first-order system. Consider a first-order, low-pass-filter system with transfer function $G(s) = \frac{1}{1+s}$. Using contour integration,

a. invert the Fourier transform of $G(i\omega)$ and verify that the impulse response function $G(t)$ (the Green function) is causal;
b. verify the Kramers–Kronig relations for G.

15.2 Sensitivity function for oscillator. Verify numerically that the waterbed integral of a second-order system with proportional feedback gain is zero. That is, show for $G(s) = \frac{1}{1+2\zeta\omega+\omega^2}$ and $K(s) = K_\mathrm{p}$ that $\int_0^\infty \mathrm{d}\omega \ln |S(i\omega)| = 0$. Plot for $K_\mathrm{p} = 1$ and $\zeta = 0.5$, and show its numerical integral = 0.

15.3 Bode's waterbed theorem. Derive Eqs. (15.22) and (15.25). For $S(s)$ analytic:

a. Show that Re $[\ln S(i\omega)]$ is an even function of ω and Im $[\ln S(i\omega)]$ is odd.
b. Fill in the missing steps leading to Eq. (15.24).
c. Show that Part II, evaluated along a circle of radius $R \to \infty$, vanishes if $L(s)$ is of relative order 2 or greater.
d. Deal with RHP poles using the contour at right, with a similar detour for each pole p_j. Evaluate the added contributions to prove Eq. (15.25).

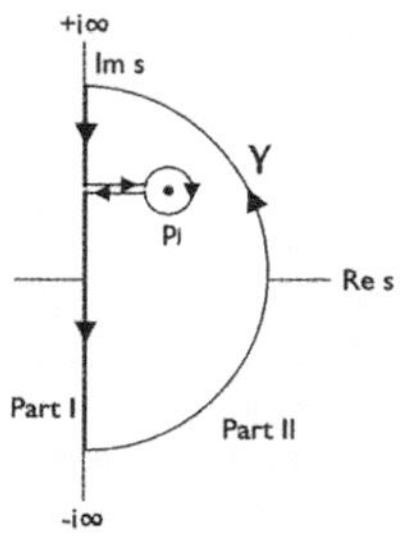

15.4 Waterbed theorem for relative degree 1 systems. For a stable, first-order loop transfer function $L(s)$ with $L(s) \to \alpha/s$ as $s \to \infty$, show that the Bode sensitivity integral is $\int_0^\infty \mathrm{d}\omega \ln |S(i\omega)| = -\frac{\pi}{2}\alpha$ and that, for degree ≥ 1, $\int_{-\infty}^{\infty} \frac{\mathrm{d}\omega}{2\pi} \ln |S(i\omega)| = \sum_j p_j - \frac{1}{2} \lim_{s\to\infty} s\,L(s)$. Hint: In deriving Eq. (15.22), the contribution of the big semicircle no longer vanishes. Uncertainties in the dynamics and delays mean that most practical systems have unmodeled high-frequency dynamics that are effectively higher order. The first-order case has relatively few practical consequences.

15.5 **Waterbed theorem for T.** Prove Eq. (15.26). Hint: Define $\bar{s} = 1/s$ and $\bar{L}(s) \equiv 1/L(1/s) = 1/L(\bar{s})$. From Åström and Murray (2008).

15.6 **Waterbed theorem for discrete dynamics.** Derive Eq. (15.27): for an open-loop transfer function $L(z)$ with relative degree $\nu \geq 1$ and unstable poles $|p_j| > 1$, we have $\int_{-\pi}^{\pi} \frac{d\omega}{2\pi} \ln|S(e^{i\omega})| = \sum_j \ln|p_j|$. Assume that the gain K of L stabilizes the closed-loop system. Hint: Use Jensen's relation, Eq. (A.51 online), to show $I(p) \equiv \int_{-\pi}^{\pi} \frac{d\omega}{2\pi} \ln\left|e^{i\omega} - p\right| = \ln|p|$ for $|p| > 1$ and 0 otherwise. Then write $L(z)$ in pole-zero form.

15.7 **One-dimensional, discrete dynamics: Bode's waterbed theorem.** For the dynamics of Example 15.1, reproduce the graphs in Figure 15.3 and show that the variance of observations is given by $\langle y^2 \rangle = \nu^2/[1-(a-K)^2]$.

15.8 **Temperature control and the waterbed theorem.** As a simplified temperature response let $G(s) = \frac{1}{(1+s)^2}$. Add a PID controller, $K(s) = K_p + \frac{K_i}{s} + K_d\left(\frac{s}{1+s/\omega_f}\right)$. Place the system in a box of thickness ℓ and thermal diffusivity D, whose temperature transfer function for high frequencies is approximately $G_{\text{box}}(s) = e^{-(\ell/\sqrt{D})\sqrt{s}}$ (see Problem 2.5). Use $\{D, \ell, K_p, K_i, K_d, \omega_f\} = \{1, 2, 10, 3, 7, 10\}$.

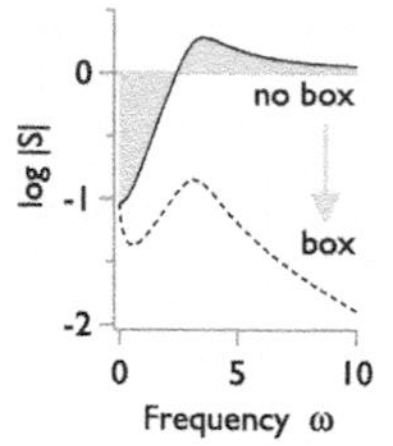

a. With no box, compute the sensitivity function $S = \frac{1}{1+L}$, with $L = K(s)G(s)$. Confirm the "no box" Bode integral plot at left.
b. Add the insulating box response and confirm the "box" plot at left.
c. Investigate the response to an impulse disturbance at the output, using a Padé approximant to the box transfer function or the inverse Laplace transform of $G_{\text{box}}(s)$: $G(x = \ell, t) = \frac{\ell}{\sqrt{4\pi D t^3}} \exp\left\{\left[-\frac{\ell^2}{4Dt}\right]\right\}$. Plot the disturbance as filtered by the box, $G(x = \ell, t)$, along with the closed-loop response that it provokes. Plot, too, the responses to a step input, showing the case of no control (just the system), PID control, and PID control augmented with a first-order feedforward filter between the reference signal and the controller input that eliminates the overshoot of the simple PID controller.

15.9 **Acausal control of a first-order system.** In reference to Example 15.2:

a. Show that the acausal feedforward input $u(t)$ given leads to the desired control. Hint: Solve the time-domain equations for $u(t < 0)$, then $u(t) \approx 0$, then $u(t) > 0$.
b. Explain physically (in words) how this solution works. How can a nonzero input $u(t)$ for $t < 0$ nonetheless produce a zero output? If that output is zero, where does the step response come from?

15.10 **Anticipating the future improves control.** Consider the system of Problem 15.7, for $|a| > 1$, with unstable uncontrolled dynamics. Scale $\nu^2 = 1$. What is the minimum power $P^* = \langle u^2(K^*) \rangle$ required to stabilize the system?

a. Use only current information. Assume $u_k = -K y_k$, and show that choosing $K^* = a - 1/a$ leads to a minimum power $P^* = a^2 - 1$. (See Section 15.2.4.)

b. Assume that somehow you know y_{k+1} at time k and choose $u_k = -K_0 y_k - K_1 y_{k+1}$. Show that choosing $K_0^* = K_1^* = a - 1$ minimizes the power, with $P_+^* = 2(|a| - 1)$. For unstable systems, note that $P_+^* < P^*$ (see right).

c. Show that using y_{k-1} does not help. That is, if $u_k = -k_0 y_k - k_1 y_{k-1}$, the best feedback gains are $k_0 = a - 1/a$ and $k_1 = 0$, leading again to $P_-^* = a^2 - 1$.

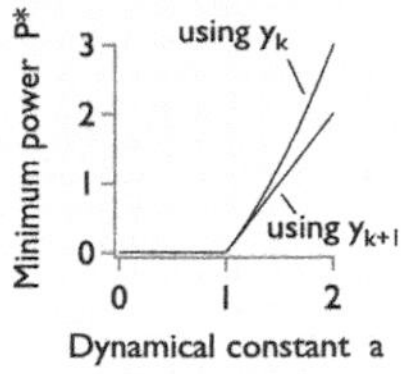

15.11 Entropy-rate paradox. Equation (15.37) claims that, for a stable open-loop linear dynamics, the entropy rate of the output, $\mathcal{H}(Y)$, equals the entropy rate of an output disturbance, $\mathcal{H}(\nu)$. Yet Eq. (A.264 online) claims that if $y_k = a\,\nu_k$, then $\mathcal{H}(Y) = \mathcal{H}(\nu) + \ln|a|$. Reconcile these two statements mathematically and physically.

15.12 Entropy rate of the output of a stable 1d system. Let $x_{k+1} = ax_k + \nu_k$, with $\nu_k \sim \mathcal{N}(0, \nu^2)$ and $|a| < 1$. Let the output $y_k = x_k$ (no measurement noise).

a. Show that the variance of the output is $\langle y_k^2 \rangle = \frac{\nu^2}{1-a^2}$.

b. By direct calculation in the time domain, show that the entropy rate $\mathcal{H}(Y) = \mathcal{H}(\nu)$, where the time series Y has realizations y_k and the series ν has realizations ν_k.

15.13 Causal conditioning. Prove that Eqs. (15.38) and (15.39) for ordinary and causal conditioning are equivalent. Hints: Use Bayes repeatedly and do $N = 2$ explicitly.

15.14 Directed information decomposition. Prove Eq. (15.43). Hint: Use Eq. (A.288 online).

15.15 Mutual vs. directed information. Using the chain rule, show the results for mutual and directed information claimed in Example 15.5.

15.16 Information rates for a finite-bandwidth, continuous system. Consider an amplifier that acts also as a low-pass filter, with transfer function $G(s) = G_0/(1+s)$, that is used as a transducer between a continuous input signal $u(t)$ and a continuous output signal $y(t)$. Let $y(s) = G(s)\,u(s) + \xi(s)$, with $\langle \xi(t)\,\xi(t') \rangle = \xi^2 \delta(t - t')$.

a. Using integration by parts, show that $\int_{-\infty}^{\infty} \frac{d\omega}{2\pi} \ln\left[\prod_i \left(\frac{\omega^2+a_i^2}{\omega^2+b_i^2}\right)\right] = \sum_i (a_i - b_i)$.

b. Show that $\mathcal{I}(U;Y)$ is given by Eq. (15.48).

15.17 Information rates in two different limits. In Section 15.2.3, the mutual information rate for a constant-gain amplifier, sampled at T_s, is given as $\mathcal{I}(U;Y) = \frac{1}{2T_s}\log(1 + \mathrm{SNR}_0^2)$. On the other hand, the rate for a continuously sampled amplifier of bandwidth ω_c is $\mathcal{I}(U;Y) = \frac{\omega_c}{2}\left(\sqrt{1 + \mathrm{SNR}_0^2} - 1\right)$. Calculate $\mathcal{I}(U;Y)$ for finite T_s and ω_c, and reconcile the two expressions. Assume $G(s) = \frac{G_0}{1+s/\omega_c}$.

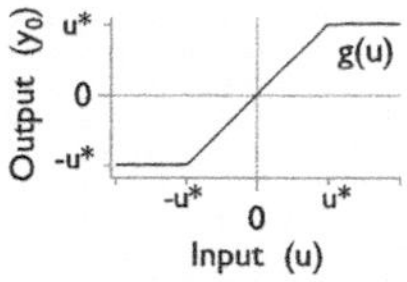

15.18 Nonlinearities can reduce information rates. Consider measuring the signal $u_k \sim \mathcal{N}(0, \sigma_u^2)$ with the nonlinear saturation function $y_0 = g(u)$ shown at right.

a. Show, for suitably defined a, that $p(y_0) = a\,[\delta(y_0 - u^*) + \delta(y_0 + u^*)] + \frac{1}{\sqrt{2\pi}}\,\mathrm{e}^{-y_0^2/2}$ for $|y_0| \le u^*$ and 0 otherwise. (Check that $p(y_0)$ is normalized, too.)

b. Then find the full distribution for $p(y)$ for $y = g(u) + \xi$ by convoluting with the noise distribution $\xi \sim \mathcal{N}(0, \xi^2)$. Do the convolution symbolically or numerically.

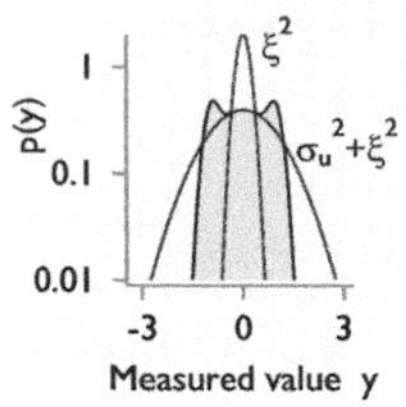

c. For $u^* = \sigma_u = 1$ and $\xi = 0.2$, confirm $p(y)$, left. Confirm, too, that the dashed lines show the limiting distributions: $\mathcal{N}(0, \xi^2)$ for $u^*/\sigma_u \ll 1$ and $\mathcal{N}(0, \sigma_u^2 + \xi^2)$ for $u^*/\sigma_u \gg 1$. Confirm, too, the information-rate plot in the text.

15.19 Information flow in the small-noise limit. Make the arguments about the small-noise limit that lead to Eq. (15.53) more precise.

15.20 Classic Szilard engine with noisy measurements. For the "energy" version of the Szilard engine discussed in the text, one can extract energy up to $k_B T \ln 2\, I(X; Y)$ for noisy measurements where the probability that the wrong state is observed is ξ. Here, we show that the same result occurs for the traditional version of the Szilard engine illustrated in Section 15.3.1. See Sagawa (2019).

a. Show that the average information gained by a single measurement y is $I(X; Y) = 1 - H_2(\xi)$, where $H_2(\xi)$ is the Shannon entropy function for two states, in bits.

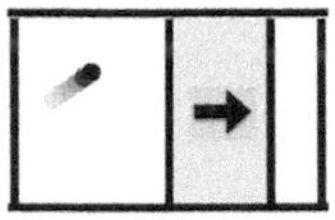

b. Extract an average work $k_B T \ln 2\, I(X; Y)$ as follows: If the measurement shows the particle on the left, move the partition from the center of the box to a position v chosen to maximize the extracted work. What if the measurement indicates that the particle is on the right? Explain intuitively this optimal protocol.

15.21 Two forms of the master equation. The master equation $\mathrm{d}_t p_i = \sum_j \left(W_{ij}\, p_j - W_{ji} p_i \right)$ has transition rates W_{ij} from state j to i.

a. Interpret this form of the master equation physically in terms of in and out currents. Can you give a graphical interpretation, too? (Cf. Section 15.3.4.)
b. An equivalent form of the master equation that is more convenient mathematically is given by $\mathrm{d}_t \boldsymbol{p} = \mathbb{W} \boldsymbol{p}$. For this form, why must each column of $\mathbb{W}$ sum to zero?
c. Express the rate matrix $\mathbb{W}$ in terms of the transition rates W_{ij}.
d. Relate $\mathrm{d}_t \boldsymbol{p} = \mathbb{W} \boldsymbol{p}$ to its discrete-time counterpart $\boldsymbol{p}_{k+1} = \boldsymbol{A} \boldsymbol{p}_k$ for a protocol of duration τ. (Hint: Exponentiate to find a condition between $\mathbb{W}$ and $\boldsymbol{A}$.)
e. Write Eq. (15.67) for a two-state system. What is its matrix $\mathbb{W}$?

15.22 Work extraction from a finite-time protocol. We explore numerically and analytically finite-time protocols of duration τ for a two-state Szilard engine.

a. Show that the master equation reduces to a single differential equation for $p(t)$, the probability to be in the initially unoccupied state, and $\epsilon(t)$, its energy level.
b. Integrate the master equation for the protocol $\epsilon(t) = \epsilon_0(1 - t/\tau)$ and evaluate the average work over the protocol. Confirm that $\langle W \rangle$ is maximized for $\tau \to \infty$.
c. Show numerically that the average power extracted, $\langle P \rangle \equiv \langle W \rangle / \tau$, is maximized for $\tau \to 0$. Given that, show analytically that $\epsilon_0 \approx 0.3$ maximizes $\langle P \rangle$.

15.23 Entropy production in stochastic thermodynamics.

a. Derive the entropy decomposition $\mathrm{d}_t S = \dot{S}_{\mathrm{i}} + \dot{S}_{\mathrm{e}}$ by showing (or identifying)

$$\mathrm{d}_t S = \frac{1}{2}\sum_{ij} J_{ij} \ln \frac{p_j}{p_i}, \quad \dot{S}_{\mathrm{i}} = \frac{1}{2}\sum_{ij} J_{ij} \ln \frac{W_{ij} p_j}{W_{ji} p_i}, \quad \dot{S}_{\mathrm{e}} = \frac{1}{2}\sum_{ij} J_{ij} \ln \frac{W_{ji}}{W_{ij}}.$$

b. Show that $\dot{Q} = T\dot{S}_{\mathrm{e}}$ is consistent with the definition $\dot{Q} = \sum_i (\mathrm{d}_t p_i)\epsilon_i$ used in the first law. For help on this problem, see Van den Broeck and Esposito (2015).

15.24 Nonequilibrium free energy

a. Show that $F(p) = -k_{\mathrm{B}}T \ln Z$ when p is the equilibrium distribution π.
b. Show that $F = F^{\mathrm{eq}} + D(p\|\pi)$, for relative entropy $D(p\|\pi) \equiv \sum_i p_i \ln(p_i/\pi_i) \geq 0$.
c. Generalize the result in (b) to one-dimensional continuous distributions, with $p_i \to p(x)$ and $\sum_i \to \int \mathrm{d}x$. Assume overdamped dynamics, so that $\epsilon_i \to U(x)$.
d. Show that the protocol of Figure 15.7 extracts a work $W_{\mathrm{extract}} = F - F^{\mathrm{eq}}$.
e. Show that the irreversible protocol dissipates heat $Q = F - F^{\mathrm{eq}}$ into the bath.

15.25 Bipartite system. Analyze aspects of their dynamics and thermodynamics.

a. Show that the information flow is given by Eq. (15.90).
b. Derive the decomposition of entropy given in Eq. (15.91). Give explicit expressions for all components and prove that $\dot{S}^X_{\mathrm{i}}$ and $\dot{S}^Y_{\mathrm{i}}$ are each nonnegative.
c. Imagine that, not knowing about System Y, you tried to define an X-only "entropy-production rate" $\sigma^X_{\mathrm{i}} = \frac{1}{2}\sum_{xy,x'} J^y_{xx'} \frac{W^y_{xx'} p_{x'}}{W^y_{x'x} p_x}$. Explain mathematically and physically how σ^X_{i} can be negative.

15.26 Four-state bipartite system. The system illustrated at right copies that of Figure 15.8 but relabels the states as $\{1, 2, 3, 4\}$, to simplify the analysis as a single joint system. In addition, we add up to four nonequilibrium driving potentials $f_{21}, f_{42}, f_{34}, f_{13}$ (all in units of $k_{\mathrm{B}}T$) going from $2 \to 1$, and so on. The $W_{ij} = 1$ for the "base" rates (light forward-backward arrows). When nonequilibrium driving is present, the rates are modified to $W_{21} = \mathrm{e}^{f_{21}/2}$ and $W_{12} = \mathrm{e}^{-f_{21}/2}$, so that $W_{21}/W_{12} = \mathrm{e}^{f_{21}}$, and so on.

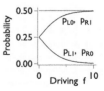

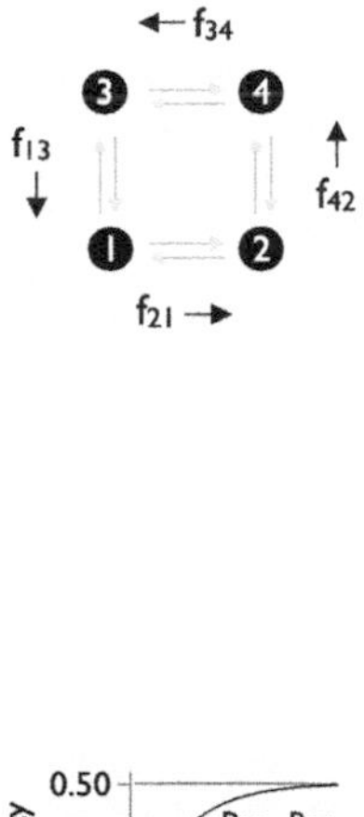

a. Write down the master equation for the joint bipartite system. Show that the steady-state solution has $p_i = \frac{1}{4}$ when all driving terms $f_{ij} = 0$.
b. When the driving terms are present, show that the entropy production rate is $\dot{S}^{XY}_{\mathrm{i}} = J(f_{21} + f_{42} + f_{34} + f_{13})$, where J is the current around the loop. Similarly, show that $\dot{S}^X_{\mathrm{i}} = J(f_{21} + f_{34})$ and $\dot{S}^Y_{\mathrm{i}} = J(f_{42} + f_{13})$, thus confirming $\dot{S}^{XY}_{\mathrm{i}} = \dot{S}^X_{\mathrm{i}} + \dot{S}^Y_{\mathrm{i}}$. Finally, show that the information flow is $\dot{I}^X = -\dot{I}^Y = J \ln(p_{RO}p_{LI} \,/\, p_{LO}p_{RI})$.
c. Consider the sensor case where $f_{13} = f_{42} = f$ and $f_{21} = f_{34} = 0$. Show that the steady-state probabilities states are $\frac{1}{4}(1 \pm \tanh \frac{1}{4}f)$, as plotted at right. Show, too, that the steady-state current around the loop is $J = \frac{1}{2}\tanh(\frac{1}{4}f)$. The total dissipation rate to run the sensor is then $\dot{S}^{XY}_{\mathrm{i}} = 2fJ \approx \frac{1}{4}f^2$ for $f \ll 1$ and $\approx f$ for $f \gg 1$. Show that the information flow $\dot{I}^Y = Jf \geq 0$, as expected for a sensor.

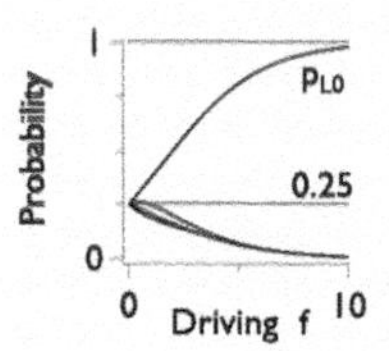

d. Consider the regulator case, where $f_{13} = f_{42} = f_{34} = f$ and $f_{21} = 0$. Show that the steady-state probabilities are as shown at left. Find J. Use the latter to find $\dot{S}_{\text{i}}^{XY}$ and the information flow. Show that for $f \gg 1$, $\dot{S}_{\text{i}}^{XY} \approx 3f$ and $\dot{I}^X \approx -f/2$.

15.27 Voltage fluctuations in an RCR circuit.

a. By finding $V(t)$ for arbitrary driving function $\eta(t)$, evaluate $\langle V(t)\, V(0)\rangle$ for an RC circuit and thereby derive Eq. (15.94).
b. Find $\langle V^2\rangle$ for an RC circuit using Eq. (8.50) for the evolution of the variance.
c. Generalize to the two-resistor case, Eq. (15.97).

References

Nothing is so dangerous as being too modern. One is apt to grow old-fashioned quite suddenly.

— Lady Markby, *An Ideal Husband*, Oscar Wilde

Abarbanel, H. D. I. 1996. *Analysis of Observed Chaotic Data*. Springer.

Abarbanel, H. D. I. 2008. Nonlinear communication strategies. Chap. 16, pages 349–368 of: Schöll, E., and Schuster, H. G. (eds.), *Handbook of Chaos Control*. Wiley-VCH.

Abarbanel, H. D. I. 2013. *Predicting the Future: Completing Models of Observed Complex Systems*. Springer.

Abramovitch, D. 2002. Phase-locked loops: A control centric tutorial. *Proc. American Control Conference*, **1**, 1–15.

Acerbi, C., and Tasche, D. 2002. Expected shortfall: A natural coherent alternative to value at risk. *Economic Notes by Banca Monte dei Paschi di Siena SpA*, **31**, 379–388.

Admon, T., Rahav, S., and Roichman, Y. 2018. Experimental realization of an information machine with tunable temporal correlations. *Phys. Rev. Lett.*, **121**, 180601.

Agüero, J. C., Yuz, J. I., Goodwin, G. C., and Delgado, R. A. 2010. On the equivalence of time and frequency domain maximum likelihood estimation. *Automatica*, **46**, 260–270.

Akaike, H. 1974. A new look at the statistical model identification. *IEEE Trans. Auto. Cont.*, **19**, 716–723.

Alderson, D. L., and Doyle, J. C. 2010. Contrasting views of complexity and their implications for network-centric infrastructures. *IEEE Trans. Syst. Man Cybern. A Syst. Hum.*, **40**, 839–852.

Alexander, R. M. 1996. *Optima for Animals*. Princeton University Press.

Allahverdyan, A., and Galstyan, A. 2009. On Maximum a Posteriori estimation of hidden Markov processes. *Proceedings of the Twenty-Fifth Conference on Uncertainty in Artificial Intelligence*.

Allahverdyan, A., and Galstyan, A. 2015. Active inference for binary symmetric Hidden Markov Models. *J. Stat. Phys.*, **161**, 452–466.

Anderson, B. D. O., and Moore, J. B. 2005. *Optimal Filtering*. Dover Publications.

Anderson, P. W. 1984. *Basic Notions of Condensed Matter Physics*. Addison-Wesley.

Andresen, B., Hoffmann, K. H., Nulton, J., Tsirlin, A., and Salamon, P. 2011. Optimal control of the parametric oscillator. *Eur. J. Phys.*, **32**, 827–843.

Andrieu, C., Doucet, A., and Holenstein, R. 2010. Particle Markov chain Monte Carlo methods. *J. Roy. Statist. Soc. B.*, **72, Part 3**, 269–342.

Ångström, A. J. 1861. Neue methode, das Wärmeleitungsvermögen der Körper zu bestimmen. *Annalen der Physik und Chemie*, **114**, 513–530.

Antenucci, F., Franz, S., Urbani, P., and Zdeborová, L. 2019. Glassy nature of the hard phase in inference problems. *Phys. Rev. X*, **9**, 011020.

Aoki, S. K., Lillacci, G., Gupta, A., Baumschlager, A., Schweingruber, D., and Khammash, M. 2019. A universal biomolecular integral feedback controller for robust perfect adaptation. *Nature*, **570**, 533–537.

Apalkov, V. M., Raikh, M. E., and Shapiro, B. 2004. Incomplete photonic band gap as inferred from the speckle pattern of scattered light waves. *Phys. Rev. Lett.*, **92**, 253902.

Appeltant, L., Soriano, M. C., der Sande, G. Van, Danckaert, J., Massar, S., Dambre, J., Schrauwen, B., Mirasso, C. R., and Fischer, I. 2011. Information processing using a single dynamical node as complex system. *Nat. Commun.*, **2**, 468.

Ariyur, K. B., and Krstic, M. 2003. *Real-Time Optimization by Extremum-Seeking Control*. Wiley-Interscience.

Arnold, V. I. 1989. *Mathematical Methods of Classical Mechanics*. 2nd edn. Springer.

Arulampalam, M. S., Maskell, S., Gordon, N., and Clapp, T. 2002. A tutorial on particle filters for online nonlinear/non-Gaussian Bayesian tracking. *IEEE Trans. Sig. Proc.*, **50**, 174–188.

Åström, K. J. 2006a. *Introduction to Stochastic Control Theory*. Dover Publications.

Åström, K. J. 2006b. Limitations on control system performance. *Eur. J. Control*, **6**, 2–20.

Åström, K. J. 2008. Event based control. Pages 127–147 of: Astolfi, A., and Marconi, L. (eds.), *Analysis and Design of Nonlinear Control Systems*. Springer.

Åström, K. J., and Furuta, K. 2000. Swinging up a pendulum by energy control. *Automatica*, **36**, 287–295.

Åström, K. J., and Hägglund, T. 2006. *Advanced PID Control*. ISA.

Åström, K. J., and Helmersson, A. 1986. Dual control of an integrator with unknown gain. *Comp. & Maths. with Appls.*, **12A**, 653–662.

Åström, K. J., and Kumar, P. R. 2014. Control: A perspective. *Automatica*, **50**, 3–43.

Åström, K. J., and Murray, R. M. 2008. *Feedback Systems: An Introduction for Scientists and Engineers*. Princeton University Press.

Åström, K. J., and Wittenmark, B. 1997. *Computer-Controlled Systems: Theory and Design*. 3rd edn. Prentice Hall.

Åström, K. J., and Wittenmark, B. 2008. *Adaptive Control*. 2nd edn. Dover Publications.

Athans, M., and Kendrick, D. 1974. Control theory and economics: A survey, forecast, and speculations. *IEEE Trans. Auto. Cont.*, **AC-19**, 518–524.

ATLAS. 2008. The ATLAS experiment at the CERN Large Hadron Collider. *J. Inst.*, **3**, 1–407.

Bailey, D. 2017. Not Normal: the uncertainties of scientific measurements. *R. Soc. Open Sci.*, **4**, 160600.

Bailey, D. 2018. Why outliers are good for science. *Significance*, **15**, 14–19.

Ballentine, L. E. 2015. *Quantum Mechanics: A Modern Development*. 2nd edn. World Scientific.

Barabási, A.-L. 2016. *Network Science*. Cambridge University Press.

Barabási, A.-L., and Albert, R. 1999. Emergence of scaling in random networks. *Science*, **286**, 509–512.

Barenblatt, G. I. 2003. *Scaling*. Cambridge University Press.

Barral, J., Dierkes, K., Lindner, B., Jülicher, F., and Martin, P. 2010. Coupling a sensory hair-cell bundle to cyber clones enhances nonlinear amplification. *PNAS*, **107**, 8079–8084.

Bateman, H. 1945. The control of an elastic fluid. *Bull. Amer. Math. Soc.*, **51**, 601–646.

Bauer, M., Barato, A. C., and Seifert, U. 2014. Optimized finite-time information machine. *J. Stat. Mech.*, P09010.

Bäuerle, T., Fischer, A., Speck, T., and Bechinger, C. 2018. Self-organization of active particles by quorum sensing rules. *Nat. Commun.*, **9**, 3232.

Becerra, V. M. 2010. *PSOPT Optimal Control Solver User Manual*.

Bechhoefer, J. 2005. Feedback for physicists: A tutorial essay on control. *Rev. Mod. Phys.*, **77**, 783–836.

Bechhoefer, J. 2011. Kramers-Kronig, Bode, and the meaning of zero. *Am. J. Phys.*, **79**, 1053–1059.

Bechhoefer, J. 2015. Hidden Markov models for stochastic thermodynamics. *New J. Phys.*, **17**, 075003.

Beiser, F. C. 1998. A romantic education: The concept of *Bildung* in early German romanticism. Pages 284–299 of: Rorty, A. O. (ed.), *Philosophers on Education: Historical Perspectives*. Routledge.

Beker, M. G., Bertolini, A., van den Brand, J. F. J., Bulten, H. J., Hennes, E., and Rabeling, D. S. 2014. State observers and Kalman filtering for high performance vibration isolation systems. *Rev. Sci. Instrum.*, **85**, 034501.

Bellon, L., Ciliberto, S., Boubaker, H., and Guyon, L. 2002. Differential interferometry with a complex contrast. *Opt. Comm.*, **207**, 49–56.

Bennett, S. 1979. *A History of Control Engineering: 1800–1930*. Peter Peregrinus LTD.

Bennett, S. 1993. *A History of Control Engineering: 1930–1955*. Peter Peregrinus LTD.

Bennett, S. 1996. A brief history of automatic control. *IEEE Contr. Syst. Mag.*, **16**, 17–25.

Bennett, S. 2002. Otto Mayr: Contributions to the history of feedback control. *IEEE Contr. Syst. Mag.*, **22**, 29–33.

Berg-Sørensen, K., and Flyvbjerg, H. 2004. Power spectrum analysis for optical tweezers. *Rev. Sci. Instrum.*, **75**, 594–612.

Berry, T., and Sauer, T. 2013. Adaptive ensemble Kalman filtering of non-linear systems. *Tellus*, **65**, 20331.

Bertsekas, D. P. 2005a. *Dynamic Programming and Optimal Control*. 3rd edn. Vol. 1. Athena Scientific.

Bertsekas, D. P. 2005b. *Dynamic Programming and Optimal Control*. 3rd edn. Vol. 2. Athena Scientific.

Bérut, A., Arakelyan, A., Petrosyan, A., Ciliberto, S., Dillenschneider, R., and Lutz, E. 2012. Experimental verification of Landauer's principle linking information and thermodynamics. *Nature*, **483**, 187–190.

Bialek, W. 2012. *Biophysics: Searching for Principles*. Princeton University Press.

Bissell, C. C. 2009. A History of Automatic Control. In: Nof, S. Y. (ed.), *Springer Handbook of Automation*. Springer.

Black, E. D. 2001. An introduction to Pound–Drever–Hall laser frequency stabilization. *Am. J. Phys.*, **69**, 79–87.

Black, H. S. 1977. Inventing the negative feedback amplifier. *IEEE Spectrum*, **14**, 55–60.

Block, D. J., Åström, K. J., and Spong, M. W. 2007. *The Reaction Wheel Pendulum*. Morgan and Claypool.

Boas, M. L. 2005. *Mathematical Methods in the Physical Sciences*. 3rd edn. Wiley.

Boccaletti, S., Grebogi, C., Lai, Y.-C., Mancini, H., and Maza, D. 2000. The control of chaos: theory and applications. *Phys. Rep.*, **329**, 103–197.

Bode, H. W. 1940. Relations between attenuation and phase in feedback amplifier design. *Bell. Syst. Tech. J.*, **19**, 421–454.

Bode, H. W. 1945. *Network Analysis and Feedback Amplifier Design*. New York: D. van Nostrand and Co.

Bondy, J. A., and Murty, U. S. R. 1976. *Graph Theory with Applications*. North-Holland.

Boyd, J. P. 2000. *Chebyshev and Fourier Spectral Methods*. 2nd edn. Dover Publications.

Bridgman, P. W. 1922. *Dimensional Analysis*. Yale University Press.

Brif, C., Chakrabarti, R., and Rabitz, H. 2010. Control of quantum phenomena: past, present, and future. *New J. Phys.*, 075008.

Brockett, R. 2014. The early days of geometric nonlinear control. *Automatica*, **50**, 2203–2224.

Broido, A. D., and Clauset, A. 2019. Scale-free networks are rare. *Nat. Commun.*, **10**, 1017.

Brun, T. A. 2002. A simple model of quantum trajectories. *Am. J. Phys.*, **70**, 719–737.

Brunton, S. L., and Kutz, J. N. 2019. *Data-Driven Science and Engineering*. Cambridge University Press.

Bryan, K., and Leise, T. 2013. Making do with less: An introduction to compressed sensing. *SIAM Review*, **55**, 547–566.

Bryson, A. E., and Ho, Y.-C. 1975. *Applied Optimal Control: Optimization, Estimation, and Control*. Taylor and Francis.

Buckingham, E. 1914. On physically similar systems; illustrations of the use of dimensional equations. *Phys. Rev.*, **4**, 345–376.

Candès, E. J., and Wakin, M. B. 2008. An introduction to compressive sampling. *IEEE Sig. Proc. Mag.*, **25**, 21–30.

Candès, E. J., Romberg, J., and Tao, T. 2006. Robust uncertainty principles: Exact signal reconstruction from highly incomplete Fourier information. *IEEE Trans. Inf. Theory*, **52**, 489–509.

Cappé, O., Moulines, E., and Rydén, T. 2005. *Inference in Hidden Markov Models*. Springer.

Carlson, J. M., and Doyle, J. 1999. Highly optimized tolerance: A mechanism for power laws in designed systems. *Phys. Rev. E*, **60**, 1412–1427.

Carlson, J. M., and Doyle, J. 2000. Highly optimized tolerance: Robustness and design in complex systems. *Phys. Rev. Lett.*, **84**, 2529–2532.

Carroll, S. 2010. *From Eternity to Here*. Penguin Group.

Chaikin, P. M., and Lubensky, T. C. 1995. *Principles of Condensed Matter Physics*. Cambridge, UK: Cambridge University Press.

Chaitin, G. 2006. The limits of reason. *Sci. Am.*, **294**, 74–81.

Chen, J., Fang, S., and Ishii, H. 2019. Fundamental limitations and intrinsic limits of feedback: An overview in an information age. *Ann. Rev. Control*, **47**, 155–177.

Chétrite, R., Rosinberg, M. L., Sagawa, T., and Tarjus, G. 2019. Information thermodynamics for interacting stochastic systems without bipartite structure. *J. Stat. Mech.*, 114002.

Ciliberto, S. 2017. Experiments in Stochastic Thermodynamics: Short History and Perspectives. *Phys. Rev. X*, **7**, 021051.

Ciliberto, S., Imparato, A., Naert, A., and Tanase, M. 2013. Heat flux and entropy produced by thermal fluctuations. *Phys. Rev. Lett.*, **110**, 180601.

Clayton, G. M., Tien, S., Leang, K. K., Zou, Q., and Devasia, S. 2009. A review of feedforward control approaches in nanopositioning for high-speed SPM. *ASME J. Dyn. Syst., Meas., Control*, **131**, 061101.

Cohen, A. E. 2005. Control of nanoparticles with arbitrary two-dimensional force fields. *Phys. Rev. Lett.*, **94**, 118102.

Cohen, R., and Havlin, S. 2003. Scale-free networks are ultrasmall. *Phys. Rev. Lett.*, **90**, 058701.

Cole, K. S., and Cole, R. S. 1941. Dispersion and absorption in dielectrics 1. Alternating current characteristics. *J. Chem. Phys.*, **9**, 341–351.

Coleman, S. 1970. Acausality. In: Zichichi, A. (ed.), *Subnuclear Phenomena*. Elsevier.

Commault, C., Dion, J.-M., and van der Woude, J. W. 2002. Characterization of generic properties of linear structured systems for efficient computations. *Kybernetika*, **38**, 503–520.

Conant, R. C., and Ashby, W. R. 1970. Every good regulator of a system must be a model of that system. *Int. J. Systems Sci.*, **1**, 89–97.

Cook, J. A., Komanovsky, I. V., McNamara, D., Nelson, E. C., and Prasad, K. V. 2007. Control, computing and communications: Technologies for the Twenty-First century Model T. *Proc. IEEE*, **95**, 334–355.

Coolen, A. C. C., Kühn, R., and Sollich, P. 2005. *Theory of Neural Information Processing Systems*. Oxford University Press.

Cooper, W. S. 1986. Use of optimal estimation theory, in particular the Kalman filter, in data processing and signal analysis. *Rev. Sci. Instrum.*, **57**, 2862–2869.

Cornelius, S. P., Kath, W. L., and Motter, A. E. 2013. Realistic control of network dynamics. *Nat. Commun.*, **4**, 1942.

Costa-Castelló, R., Nebot, J., and Griñó, R. 2005. Demonstration of the internal model principle by digital repetitive control of an educational laboratory plant. *IEEE Trans. Educ.*, **48**, 73–80.

Cover, T., and Thomas, J. 2006. *Elements of Information Theory*. 2nd edn. New York: John Wiley & Sons, Inc.

Cowan, N. J., Chastain, E. J., Vilhena, D. A., Freudenberg, J. S., and Bergstrom, C. T. 2012. Nodal dynamics, not degree distributions, determine the structural controllability of complex networks. *PLoS ONE*, **7**, e38398.

Crisan, D., and Rozovskii, B. (eds.). 2011. *The Oxford Handbook of Nonlinear Filtering*. Oxford University Press.

Cubitt, T. S., Eisert, J., and Wolf, M. M. 2012. Extracting dynamical equations from experimental data is NP hard. *Phys. Rev. Lett.*, **108**, 120503.

D'Alessandro, D. 2008. *Introduction to Quantum Control and Dynamics*. Chapman and Hall / CRC.

Daniels, B. C., and Nemenman, I. 2015. Automated adaptive inference of phenomenological dynamical models. *Nat. Commun.*, **6**, 8133.

Daum, F. 2005. Nonlinear filters: Beyond the Kalman filter. *IEEE A&E Systems Magazine*, **20**, 57–69.

Daum, F., and Huang, J. 2016. A plethora of open problems in particle flow research for nonlinear filters, Bayesian decisions, Bayesian learning and transport. In: Kadar, I. (ed.), *Signal Processing, Sensor/Information Fusion, and Target Recognition XXV*, vol. 9842. Proc. SPIE.

Davenport, W. B., and Root, W. L. 1958. *An Introduction to the Theory of Random Signals and Noise*. McGraw Hill.

Day, A. G. R., Bukov, M., Weinberg, P., Mehta, P., and Sels, D. 2019. Glassy phase of optimal quantum control. *Phys. Rev. Lett.*, **122**, 020601.

de Solla Price, D. 1976. A general theory of bibliometric and other cumulative advantage processes. *J. Am. Soc. Info. Sci.*, **27**, 292–306.

Del Vecchio, D., and Murray, R. M. 2014. *Biomolecular Feedback Systems*. Princeton University Press.

Derrida, B., Vannimenus, J., and Pomeau, Y. 1978. Simple frustrated systems: Chains, strips and squares. *J. Phys. C: Solid State Phys.*, **11**, 4749–4765.

Devasia, S. 2002. Should model-based inverse inputs be used as feedforward under plant uncertainty? *IEEE Trans. Auto. Cont.*, **47**, 1865–1871.

Diesenroth, M. P., Fox, D., and Rasmussen, C. E. 2015. Gaussian processes for data-efficient learning in robotics and control. *IEEE Trans. Patt. Anal. Machine Intell.*, **37**, 408–423.

Dobzhansky, T. 1973. Nothing in biology makes sense except in the light of evolution. *The Amer. Biol. Teacher*, **35**, 125–129.

Donelan, M. A., Hamilton, J., and Hui, W. H. 1985. Directional spectra of wind-generated waves. *Phil. Trans. Roy. Soc. A*, **315**, 509–562.

Donoho, D., and Tanner, J. 2009. Observed universality of phase transitions in high-dimensional geometry, with implications for modern data analysis and signal processing. *Phil. Trans. Roy. Soc. A*, **367**, 4273–4293.

Donoho, D. L. 2006. Compressed sensing. *IEEE Trans. Inf. Theory*, **52**, 1289–1306.

Doya, K. 1992. Bifurcations in the learning of recurrent neural networks. *Proc. IEEE Int. Symp. on Circuits and Systems*, 2777–2780.

Doyle, J. C. 1978. Guaranteed margins for LQG controllers. *IEEE Trans. Auto. Cont.*, **AC-23**, 756–757.

Doyle, J. C., Francis, B. A., and Tannenbaum, A. R. 1992. *Feedback Control Theory*. New York: Macmillan Publishing Co.

Duarte, M. F., Davenport, M. A., Takhar, D., Laska, J. N., Sun, T., Kelly, K. F., and Baraniuk, R. G. 2008. Single-pixel imaging via compressive sampling. *IEEE Sig. Proc. Mag.*, **25**, 83–91.

Durbin, R., Eddy, S., Krogh, A., and Mitchison, G. 1998. *Biological Sequence Analysis*. Cambridge University Press.

Dutton, K., Thompson, S., and Barraclough, B. 1997. *The Art of Control Engineering*. Harlow, England: Addison-Wesley.

Dydek, Z., Annaswamy, A., and Lavretsky, E. 2010. Adaptive control and the NASA X-15-3 flight revisited. *IEEE Contr. Syst. Mag.*, **30.3**, 32–48.

Eddington, A. S. 1928. *The Nature of the Physical World*. Cambridge University Press.

Eldar, Y. C., and Kutyniok, G. (eds.). 2012. *Compressed sensing: Theory and Applications*. Cambridge University Press.

Elia, N. 2004. When Bode meets Shannon: control-oriented feedback communication schemes. *IEEE Trans. Auto. Cont.*, **49**, 1477–1488.

Enßlin, T. A. 2013. Information field theory. *AIP Conf. Proc.*, **1553**.

Etchenique, R., and Aliaga, J. 2004. Resolution enhancement by dithering. *Am. J. Phys.*, **72**, 159–163.

Evensen, G. 2009. *Data Assimilation: The Ensemble Kalman Filter*. 2nd edn. Springer-Verlag.

Fields, A. P., and Cohen, A. E. 2011. Electrokinetic trapping at the one nanometer limit. *Proc. Natl. Acad. Sci. USA*, **108**, 8937–8942.

Filatov, N. M., and Unbehauen, H. 2004. *Adaptive Dual Control: Theory and Applications*. Springer.

Fishburn, P. C. 1970. *Utility Theory for Decision Making*. Research Analysis Corp.

Fleming, A. J., and Leang, K. K. 2014. *Design, Modeling and Control of Nanopositioning Systems*. Springer.

Foucart, S., and Rauhut, H. 2013. *A Mathematical Introduction to Compressive Sensing*. Birkhäuser.

Fradkov, A. L. 2007. *Cybernetical Physics: From Control of Chaos to Quantum Control*. Springer.

Francis, B. A., and Wonham, W. M. 1975. The internal model principle for linear multivariable regulators. *Appl. Math. & Opt.*, **2**, 170–194.

Frankel, F. C., and DePace, Angela H. 2012. *Visual Strategies: A Practical Guide to Graphics for Scientists & Engineers.*. Yale University Press.

Franklin, G. F., Powell, J. D., and Workman, M. L. 1998. *Digital Control of Dynamical Systems*. 3rd edn. Reading, MA: Addison-Wesley.

Freitas, N., Delvenne, J.-C., and Esposito, M. 2019. Stochastic and quantum thermodynamics of driven RLC networks. *arxiv.org*, 1906.11233. *Phys. Rev. X* **10**, 031005 (2020).

Frick, M., Gupta, S., and Bechhoefer, J. 2018. When geometry is irrelevant for heat diffusion: The transition from lumped element to field formulations. *Eur. J. Phys.*, **39**, 065104.

Friston, K. 2010. The free-energy principle: a unified brain theory? *Nat. Rev. Neurosci.*, **11**, 127–138.

Fuchs, C. A., Mermin, N. D., and Schack, R. 2014. An introduction to QBism with an application to the locality of quantum mechanics. *Am. J. Phys.*, **82**, 749–754.

Gajamohan, M., Merz, M., Thommen, I., and D'Andrea, R. 2012. The Cubli: a cube that can jump up and balance. In: *IEEE/RSJ Int. Conf. Intell. Robots Syst.*

Gamkrelidze, R. V. 1999. Discovery of the maximum principle. *J. of Dyn. and Contr. Syst.*, **5**, 437–451.

Gardiner, C. W. 2009. *Stochastic Methods: A Handbook for the Natural and Social Sciences*. 4th edn. Springer.

Gavartin, E., Verlot, P., and Kippenberg, T. J. 2012. A hybrid on-chip optomechanical transducer for ultrasensitive force measurements. *Nat. Nanotech.*, **7**, 509–514.

Geering, H. P. 2007. *Optimal Control with Engineering Applications*. Springer.

Gelfand, I. M., and Fomin, S. V. 1963, reprinted 2000. *Calculus of Variations*. Dover Publications.

Gerhart, J., and Kirschner, M. 2007. The theory of facilitated variation. *Proc. Natl. Acad. Sci. USA*, **104, suppl. 1**, 8582–8589.

Geri, M., Keshavarz, B., Divoux, T., Clasen, C., Curtis, D. J., and McKinley, G. H. 2018. Time-resolved mechanical spectroscopy of soft materials via optimally windowed chirps. *Phys. Rev. X*, **8**, 041042.

Gerlach, M., and Altmann, E. G. 2019. Testing statistical laws in complex systems. *Phys. Rev. Lett.*, **122**, 168301.

Germain, R. N. 2001. The art of the probable: System control in the adaptive immune system. *Science*, **293**, 240–245.

Gershenfeld, N. 2000. *The Physics of Information Technology*. Cambridge, UK: Cambridge University Press.

Gibson, J. D. 2014. *Information Theory and Rate Distortion Theory for Communications and Compression*. Morgan and Claypool.

Gilks, W. R., and Berzuini, C. 2001. Following a moving target – Monte Carlo inference for dynamic Bayesian models. *J. Roy. Statist. Soc. B.*, **63**, 127–146.

Gladwell, M. 2002. *The Tipping Point*. Back Bay Books.

Gleick, J. 1996. A bug and a crash. *New York Times Magazine*, December 1.

Goldenfeld, N. 1992. *Lectures on Phase Transitions and the Renormalization Group*. Reading, MA: Addison-Wesley.

Goldstein, H. 1980. *Classical Mechanics*. 2nd edn. Addison-Wesley.

Goodfellow, I., Bengio, Y., and Courville, A. 2016. *Deep Learning*. Massachusetts Institute of Technology Press.

Goodman, J. W. 2007. *Speckle Phenomena in Optics*. Roberts & Company.

Goodwin, G. C., Graebe, S. F., and Salgado, M. E. 2001. *Control System Design*. Upper Saddle River, NJ: Prentice Hall.

Gordon, N. J., Salmond, D. J., and Smith, A. F. M. 1993. Novel approach to nonlinear/non-Gaussian Bayesian state estimation. *IEE Proc.-F*, **140**, 107–113.

Graichen, K., Treuer, M., and Zeitz, M. 2007. Swing-up of the double pendulum on a cart by feedforward and feedback control with experimental validation. *Automatica*, **43**, 63–71.

Granger, C. W. J. 1980. Testing for causality. *J. Econ. Dyn. Contr.*, **2**, 329–352.

Gray, M., McClelland, D., Barton, M., and Kawamura, S. 1999. A simple high-sensitivity interferometric position sensor for test mass control on an advanced LIGO interferometer. *Optical and Quantum Electronics*, **31**, 571–582.

Gray, R. M., and Stockham, T. G. 1993. Dithered quantizers. *IEEE Trans. Inf. Theory*, **39**, 805–812.

Gross, C. G. 1998. Claude Bernard and the constancy of the internal environment. *The Neuroscientist*, **4**, 380–385.

Guckenheimer, J., and Holmes, P. 2002. *Nonlinear Oscillations, Dynamical Systems, and Bifurcations of Vector Fields*. Corrected 6th printing edn. Springer.

Gumbel, E. J. 1958. *Statistics of Extremes*. New York: Columbia University Press.

Haddad, W. M., and Chellaboina, V. 2008. *Nonlinear Dynamical Systems and Control*. Princeton University Press.

Hagan, M. T., Demuth, H. B., and Jesús, O. De. 2002. An introduction to the use of neural networks in control systems. *Int. J. Robust Nonlin. Control*, **12**, 959–985.

Hahs, D. W., and Pethel, S. D. 2011. Distinguishing anticipation from causality: Anticipatory bias in the estimation of information flow. *Phys. Rev. Lett.*, **107**, 128701.

Hansen, J., Sato, M., Kharecha, P., von Schuckmann, K., Beerling, D. J., Cao, J., Marcott, S., Masson-Delmotte, V., Prather, M. J., Rohling, E. J., Shakun, J., Smith, P., Lacis, A., Russell, G., and Ruedy, R. 2017. Young people's burden: requirement of negative CO2 emissions. *Earth Syst. Dynam.*, **8**, 577–616.

Harris, G. I., McAuslan, D. L., Stace, T. M., Doherty, A. C., and Bowen, W. P. 2013. Minimum requirements for feedback enhanced force sensing. *Phys. Rev. Lett.*, **111**, 103603.

Hartley, R. V. L. 1928. Transmission of information. *Bell. Syst. Tech. J.*, **7**, 535–563.

Hartwell, L. H., Hopfield, J. J., Leibler, S., and Murray, A. W. 1999. From molecular to modular biology. *Nature (London)*, **402**, C47–C52.

Hayes, M. H. 1996. *Statistical Digital Signal Processing and Modeling*. John Wiley & Sons, Inc.

Haykin, S. 2001. Kalman Filters. Chap. 1, pages 1–21 of: *Kalman Filter and Neural Networks*. John Wiley & Sons, Inc.

Haynes, N. D., Soriano, M. C., Rosin, D. P., Fischer, I., and Gautier, D. J. 2015. Reservoir computing with a single time-delay autonomous Boolean node. *Phys. Rev. Lett.*, **91**, 020810(R).

Haynes, W. M. (ed.). 2014. *CRC Handbook of Chemistry and Physics*. 95 edn. CRC Press.

Hirose, M., and Cappellaro, P. 2017. Coherent feedback control of a single qubit in diamond. *Nature*, **532**, 77–80.

Ho, Y.-C., and Lee, R. C. K. 1964. A Bayesian approach to problems in stochastic estimation and control. *IEEE Trans. Auto. Cont.*, **9**, 333–339.

Horowitz, J. M., and Esposito, M. 2014. Thermodynamics with continuous information flow. *Phys. Rev. X*, **4**, 031015.

Horowitz, J. M., and Gingrich, T. R. 2020. Thermodynamic uncertainty relations constrain non-equilibrium fluctuations. *Nat. Phys.* **16**, 15–20.

Horsthemke, W., and Lefever, R. 2006. *Noise-Induced Transitions: Theory and Applications in Physics, Chemistry, and Biology*. 2nd edn. Springer.

Hu, B. Y.-K. 1989. Kramers-Kronig in two lines. *Am. J. Phys.*, **57**, 821.

Ingalls, B. P. 2013. *Mathematical Modeling in Systems Biology: An Introduction*. Massachusetts Institute of Technology Press.

Insperger, T., and Milton, J. 2017. Stick balancing with feedback delay, sensory dead zone, acceleration and jerk limitation. *Procedia IUTAM*, **22**, 59–66.

Ioannou, P., and Fidan, B. 2006. *Adaptive Control Tutorial*. Philadelphia: SIAM Press.

Isermann, R., and Münchhof, M. 2011. *Identification of Dynamic Systems*. Berlin, Heidelberg: Springer Verlag.

Isidori, A. 1995. *Nonlinear Control Systems*. 3rd edn. Springer.

Isidori, A., and Byrnes, C. I. 1990. Output regulation of nonlinear systems. *IEEE Trans. Auto. Cont.*, **35**, 131–140.

Jackson, J. D. 1999. *Classical Electrodynamics*. 3rd edn. John Wiley & Sons, Inc.

Jacobs, K. 2014. *Quantum Measurement Theory and Its Applications*. Cambridge University Press.

Jacobs, O. L. R. 1993. *Introduction to Control Theory*. 2nd edn. Oxford University Press.

Jaeger, H., and Haas, H. 2004. Harnessing nonlinearity: Predicting chaotic systems and saving energy in wireless communication. *Science*, **304**, 78–80.

Jaynes, E. T. 2003. *Probability Theory: The Logic of Science*. Cambridge University Press.

Jensen, F. V., and Nielsen, T. D. 2007. *Bayesian Networks and Decision Graphs*. 2nd edn. Springer.

Jia, T., Liu, Y.-Y., Csóka, E., Pósfai, M., Slotine, J.-J., and Barabási, A.-L. 2013. Emergence of bimodality in controlling complex networks. *Nat. Commun.*, **4**, 2002.

Julier, S., and Uhlmann, J. 1997. A new extension of the Kalman filter to nonlinear systems. *Proc. SPIE*, **3068**, 182–193.

Jun, Y., Gavrilov, M., and Bechhoefer, J. 2014. High-precision test of Landauer's principle. *Phys. Rev. Lett.*, **113**, 190601.

Kailath, T. 1968. An innovations approach to least-squares estimation part I: Linear filtering in additive white noise. *IEEE Trans. Auto. Cont.*, **AC-13**, 646–655.

Kailath, T. 1980. *Linear Systems*. Englewood Cliffs, NJ: Prentice Hall.

Kailath, T., Sayed, A. H., and Hassibi, B. 2000. *Linear Estimation*. Prentice Hall.

Kalman, R. E. 1960a. A new approach to linear filtering and prediction problems. *ASME J. Basic Engineering*, **82**, 35–45.

Kalman, R. E. 1960b. On the general theory of control systems. Pages 481–492 of: *Proc. First IFAC Congress Automatic Control, Moscow*, vol. 1. London: Butterworth Scientific Publishers.

Kalman, R. E. 1963. The theory of optimal control and the calculus of variations. Chap. 16 of: Bellman, R. (ed.), *Mathematical Optimization Techniques*. University of California Press.

Kappen, H. J. 2005. Path integrals and symmetry breaking for optimal control theory. *J. Stat. Mech.*, P11011.

Kappen, H. J. 2011. Optimal control theory and the linear Bellman Equation. Chap. 17, pages 363–387 of: Barber, D., Cemgil, A. T., and Chiappa, S. (eds.), *Bayesian Time Series Models*. Cambridge University Press.

Kashima, K. 2016. Novel controllability Gramian for nonlinear network dynamics. *Sci. Rep.*, **6**, 27300.

Kawan, C. 2013. *Invariance Entropy for Deterministic Control Systems*. Springer.

Kaye, P., Laflamme, R., and Mosca, M. 2007. *An Introduction to Quantum Computing*. Oxford University Press.

Kelly, F. P. 1979. *Reversibility and Stochastic Networks*. Wiley.

Khadka, U., Holubec, V., Yang, H., and Cichos, F. 2018. Active particles bound by information flows. *Nat. Commun.*, **9**, 3864.

Khalil, H. K. 2001. *Nonlinear Systems*. 3rd edn. Prentice Hall.

Kim, K. H., and Qian, H. 2004. Entropy production of Brownian macromolecules with inertia. *Phys. Rev. Lett.*, **93**, 120602.

Kim, K. K., Shen, D. E., Nagy, Z. K., and Braatz, R. D. 2013. Wiener's polynomial chaos for the analysis and control of nonlinear dynamical systems with probabilistic uncertainties. *Control. Syst. Mag.*, **33**, 58–67.

Kitagawa, G. 1996. Monte Carlo filter and smoother for non-Gaussian nonlinear state space models. *J. Comp. Graph. Stat.*, **5**, 1–25.

Klenske, E. D., and Hennig, P. 2016. Dual control for approximate Bayesian reinforcement learning. *J. Mach. Learn. Res.*, **17**, 1–30.

Koopman, B. O. 1931. Hamiltonian systems and transformation in Hilbert space. *Proc. Natl. Acad. Sci. USA*, **17**, 315–318.

Kouvaritakis, B., and Cannon, M. 2016. *Model Predictive Control*. Springer.

Kramer, G. 2003. Capacity results for the discrete memoryless network. *IEEE Trans. Inf. Theory*, **49**, 4–21.

Krishnamurthy, V. 2016. *Partially Observed Markov Decision Processes: From Filtering to Controlled Sensing*. Cambridge University Press.

Krstic, M. 2009. *Delay Compensation for Nonlinear, Adaptive, and PDE Systems*. Birkhäuser.

Krzakala, F., Mézard, M., Sausset, F., Sun, Y. F., and Zdeborová, L. 2012. Statistical-physics-based reconstruction in compressed sensing. *Phys. Rev. X*, **2**, 021005.

Kuo, A. 1995. An optimal control model for analyzing human postural balance. *IEEE Trans. Biomed. Eng.*, **42**, 87–101.

Kuramoto, Y. 1984. *Chemical Oscillations, Waves and Turbulence*. Springer.

Kushner, H. J. 1962. On the differential equations satisfied by conditional probability densities of Markov processes, with applications. *SIAM J. Cont. A*, **2**, 106–119.

Kussell, E., and Leibler, S. 2005. Phenotypic diversity, population growth, and information in fluctuating environments. *Science*, **309**, 2075–2078.

Lakatos, I. 1968. Criticism and the methodology of scientific research programmes. *Proc. Arist. Soc.*, **69**, 149–186.

Lamont, C. H., and Wiggins, P. A. 2016. The development of an information criterion for change-point analysis. *Neural Comp.*, **28**, 594–612.

Landau, L. D., and Lifshitz, E. M. 1976. *Mechanics*. 3rd edn. Vol. 1. Butterworth-Heinemann.

Landau, L. D., Pitaevskii, L. P., Lifshitz, E. M., and Kosevich, A. M. 1986. *Theory of Elasticity*. 3rd edn. Butterworth-Heinemann.

Landauer, R. 1952. Path concepts in Hamilton-Jacobi theory. *Am. J. Phys.*, **20**, 363–367.

Larger, L., Baylón-Fuentes, A., Martinenghi, R., Udaltsov, V. S., Chembo, Y. K., and Jacquot, M. 2017. High-speed photonic reservoir computing using a time-delay-based architecture: Million words per second classification. *Phys. Rev. X*, **7**, 011015.

Laughlin, S. B. 1981. A simple coding procedure enhances a neuron's information capacity. *Z. Naturforsch*, **36c**, 910–912.

Leigh, J. R. 2004. *Control Theory*. 2nd edn. London: The Institution of Electrical Engineers.

Lemm, J. C. 2003. *Bayesian field theory*. Johns Hopkins University Press.

Lenhart, S., and Workman, J. T. 2007. *Optimal Control Applied to Biological Models*. Chapman and Hall / CRC.

Leong, Y. P., and Doyle, J. C. 2016. Understanding robust control theory via stick balancing. Pages 1508–1514 of: *Proc. IEEE Int. Conf. Decis. Control (CDC)*.

Lepschy, A. M. 1992. Feedback control in ancient water and mechanical clocks. *IEEE Trans. Educ.*, **35**, 3–10.

Lestas, I., Vinnicombe, G., and Paulsson, J. 2010. Fundamental limits on the suppression of molecular fluctuations. *Nature*, **467**, 174–178.

Lévine, J. 2009. *Analysis and Control of Nonlinear Systems*. Springer.

Levine, W. S. (ed.). 2011a. *The Control Handbook: Control System Applications*. 2nd edn. CRC Press.

Levine, W. S. (ed.). 2011b. *The Control Handbook: Control System Fundamentals*. 2nd edn. CRC Press.

Levine, W. S. (ed.). 2011c. *The Control Systems Handbook: Control System Advanced Methods*. 2nd edn. CRC Press.

Lewis, F. L., and Ge, S. S. 2005. Neural networks in feedback control systems. In: *Mechanical Engineer's Handbook*. John Wiley & Sons.

Lewis, F. L., Dawson, D. M., and Abdallah, C. T. 2004. *Robot Manipulator Control: Theory and Practice*. 2nd edn. Marcel Dekker, Inc.

Lewis, F. L., Vrabie, D. L., and Syrmos, L. 2012. *Optimal Control*. 3rd edn. John Wiley & Sons.

Li, A., Cornelius, S. P., Liu, Y.-Y., Wang, L., and Barabási, A.-L. 2017. The fundamental advantages of temporal networks. *Science*, **358**, 1042–1046.

Li, Y., and Bechhoefer, J. 2007. Feedforward control of a closed-loop piezoelectric translation stage for atomic force microscope. *Rev. Sci. Instrum.*, **78**, 013702.

Liberzon, D. 2012. *Calculus of Variations and Optimal Control Theory: A Concise Introduction*. Princeton University Press.

Lin, C. T. 1974. Structural controllability. *IEEE Trans. Auto. Cont.*, **19**, 201–208.

Lindley, D. V. 1987. The probability approach to the treatment of uncertainty in artificial intelligence and expert systems. *Stat. Sci.*, **2**, 17–24.

Lindmark, G., and Altafini, C. 2018. Minimum energy control for complex networks. *Sci. Rep.*, **8**, 3188.

Lipshitz, S. P., Pocock, M., and Vanderkooy, J. 1982. On the audibility of midrange phase distortion in audio systems. *J. Audio Eng. Soc.*, **30**, 580–595.

Liptser, R. S., and Shiryaev, A. N. 2000. *Statistics of Random Processes II: Applications*. Springer.

Liu, J., and Elia, N. 2009. Convergence of fundamental limitations in feedback communication, estimation, and feedback control over Gaussian channels. *arxiv.org*, 0910.0320.

Liu, N., Giesen, F., Below, M., Losby, J., Moroz, J., Fraser, A. E., McKinnon, G., Clement, T. J., Sauer, V., Hiebert, W. K., and Freeman, M. R. 2008. Time-domain control of ultrahigh-frequency nanomechanical systems. *Nat. Nanotech.*, **3**, 715–719.

Liu, Y.-Y., and Barabási, A.-L. 2016. Control principles of complex systems. *Rev. Mod. Phys.*, **88**, 035006.

Liu, Y.-Y., Slotine, J.-J., and Barabási, A.-L. 2011. Controllability of complex networks. *Nature*, **473**, 167–173.

Liu, Y.-Y., Slotine, J.-J., and Barabási, A.-L. 2013. Observability of complex systems. *Proc. Natl. Acad. Sci. USA*, **110**, 2460–2465.

Liutkus, A., Martina, D., Popoff, S., Chardon, G., Katz, O., Lerosey, G., Gigan, S., Daudet, L., and Carron, I. 2014. Imaging With Nature: Compressive imaging using a multiply scattering medium. *Sci. Rep.*, **4**, 5552.

Ljung, L. 1999. *System Identification: Theory for the User*. 2nd edn. Upper Saddle River, NJ: Prentice Hall.

Lloyd, S. 2000. Coherent quantum feedback. *Phys. Rev. A*, **62**, 022108.

Lüke, H. D. 1999. The origins of the sampling theorem. *IEEE Comm. Mag.*, **37**, 106–108.

Lukoševičius, M. 2012. A practical guide to applying echo state networks. Pages 659–686 of: Montavon, G., Orr, G. B., and Müller, K.-R. (eds.), *Neural Networks: Tricks of the Trade*. Lecture Notes in Computer Science, vol. 7700.

MacFarlane, A. G. C., and Karcanias, N. 1976. Poles and zeros of linear multivariable systems: a survey of the algebraic, geometric and complex-variable theory. *Int. J. Control*, **24**, 33–74.

Machta, B. B., Chachra, R., Transtrum, M. K., and Sethna, J. P. 2013. Parameter space compression underlies emergent theories and predictive models. *Science*, **342**, 604–607.

MacKay, D. J. C. 2003. *Information Theory, Inference, and Learning Algorithms*. Cambridge University Press.

Macleod, H. A. 2017. *Thin-Film Optical Filters*. 5th edn. CRC Press.

Mancini, R. 2002. *Op Amps for Everyone: Texas Instruments Guide*. 2nd edn. Newnes.

Mandelis, A. 2000. Diffusion waves and their uses. *Phys. Today*, **53**, 29–34.

Manneville, P. 2004. *Dynamique Non-linéaire appliquée au chaos et à son contrôle*. Tech. rept. Ecole Polytechnique (France).

Marchetti, G., Dall'Asta, L., and Bianconi, G. 2014. Network controllability is determined by the density of low in-degree and out-degree nodes. *Phys. Rev. Lett.*, **113**, 078701.

Marder, M. P. 2010. *Condensed Matter Physics*. Wiley.

Marriner, J. 2004. Stochastic cooling overview. *Nucl. Instr. Meth. Phys. A*, **532**, 11–18.

Martins, N. C., and Dahleh, M. A. 2008. Feedback control in the presence of noisy channels: "Bode-like" fundamental limitations of performance. *IEEE Trans. Auto. Cont.*, **53**, 1604–1615.

Massey, J. L. 1990. Causality, feedback, and directed information. Pages 303–305 of: *Proc. Intl. Symp. Info. Theory and Its Applications (ISITA-90)*.

Maybeck, P. S. 1979. *Stochastic Models, Estimation, and Control*. Vol. 1. New York: Academic Press.

Maybeck, P. S. 1982. *Stochastic Models, Estimation, and Control*. Vol. 2. Academic Press.

Mayeda, H., and Yamada, T. 1979. Strong structural controllability. *SIAM J. Cont. and Opt.*, **17**, 123–138.

Mayr, O. 1970. *The Origins of Feedback Control*. Massachusetts Institute of Technology Press.

McDonnell, M. D., Stocks, N. G., Pearce, C. E. M., and Abbott, D. 2008. *Stochastic Resonance: From Suprathreshold Stochastic Resonance to Stochastic Signal Quantization*. Cambridge University Press.

McGrayne, S. B. 2011. *The Theory That Would Not Die: How Bayes' Rule Cracked the Enigma Code, Hunted Down Russian Submarines, and Emerged Triumphant from Two Centuries of Controversy*. New Haven, CT: Yale University Press.

McNamee, D., and Wolpert, D. M. 2019. Internal models in biological control. *Ann. Rev. Control Robot. Auton. Syst.*, **2**, 339–364.

Mehra, R. K. 1970. On the identification of variances and adaptive Kalman filtering. *IEEE Trans. Auto. Cont.*, **AC-15**, 175–184.

Mehta, P., and. A. G. R. Day, C.-H. Wang, Richardson, C., Bukov, M., Fisher, C. K., and Schwab, D. J. 2019. A high-bias, low-variance introduction to machine learning for physicists. *Phys. Rep.*, **810**, 1–124.

Mermin, N. D. 2007. *Quantum Computer Science: An Introduction*. Cambridge University Press.

Merton, R. K. 1968. The Matthew effect in science. *Science*, **159**, 56–63.

Mesterton-Gibbons, M. 2009. *A Primer on the Calculus of Variations and Optimal Control Theory*. American Mathematical Society.

Miller, R. N., E. F. Carter, Jr., and Blue, S. T. 1999. Data assimilation into nonlinear stochastic models. *Tellus*, **51A**, 167–194.

Milton, J., Meyer, R., Zhvanetsky, M., Ridge, S., and Insperger, T. 2016. Control at stability's edge minimizes energetic costs: expert stick balancing. *J. R. Soc. Interface*, **13**, 20160212.

Mindell, D. A. 2002. *Between Mind and Machine: Feedback, Control, and Computing before Cybernetics*. Baltimore and London: The Johns Hopkins University Press.

Monticone, F., and Alù, A. 2016. Invisibility exposed: physical bounds on passive cloaking. *Optica*, **3**, 718–724.

Moore, B. C. 1981. Principal component analysis in linear systems: Controllability, observability, and model reduction. *IEEE Trans. Auto. Cont.*, **26**, 17–32.

Mora, T., and Nemenman, I. 2019. Physical limit to concentration sensing in a changing environment. *Phys. Rev. Lett.*, **123**, 198101.

Morari, M., and Zafirioiu, E. 1989. *Robust Process Control.* Upper Saddle River, NJ: Prentice Hall.

Morin, D. 2008. *Introduction to Classical Mechanics*. Cambridge University Press.

Mukherjee, S. 2018. My father's body, at rest and in motion. *The New Yorker*, **93**(43).

Müller, P., Quintana, F. A., Jara, A., and Hanson, T. 2015. *Bayesian Nonparametric Data Analysis*. Springer.

Munakata, T., and Kamiyabu, M. 2006. Stochastic resonance in the FitzHugh-Nagumo model from a dynamic mutual information point of view. *Eur. Phys. J. B*, **53**, 239–243.

Murota, K. 2009. *Matrices and Matroids for Systems Analysis*. Springer.

Murphy, K. P. 2012. *Machine Learning: A Probabilistic Perspective*. Massachusetts Institute of Technology Press.

Nagengast, A. J., Braun, D. A., and Wolpert, D. M. 2010. Risk-sensitive optimal feedback control accounts for sensorimotor behavior under uncertainty. *PLoS Comp. Biol.*, **6**, e1000857.

Naidu, D. S. 2002. *Optimal Control Systems*. Boca Raton, FL: CRC Press.

Nair, G. N., Fagnani, F., Zampieri, S., and Evans, R. J. 2007. Feedback control under data rate constraints: An overview. *Proc. IEEE*, **95**, 108–137.

Narendra, K. S., and Lewis, F. L. 2001. Introduction to the special issue on neural network feedback control. *Automatica*, **37**, 1147–1148.

Nelson, R. J., Weinstein, Y., Cory, D., and Lloyd, Seth. 2000. Experimental demonstration of fully coherent quantum feedback. *Phys. Rev. Lett.*, **85**, 3045–3048.

Nielsen, M. A., and Chuang, I. L. 2000. *Quantum Computation and Quantum Information*. Cambridge University Press.

Nørrelykke, S., and Flyvbjerg, H. 2011. Harmonic oscillator in heat bath: Exact simulation of time-lapse-recorded data and exact analytical benchmark statistics. *Phys. Rev. E*, **83**, 041103.

O'Keeffe, K. P., Hong, H., and Strogatz, S. H. 2017. Oscillators that sync and swarm. *Nat. Commun.*, **8**, 1504.

Oliva, A. I., Angulano, E., Denisenko, N., and Aguilar, M. 1995. Analysis of scanning tunneling microscopy feedback system. *Rev. Sci. Instrum.*, **66**, 3195–3203.

Oppenheim, A. V., and Schafer, R. W. 2010. *Discrete-Time Signal Processing*. 3rd edn. Upper Saddle River, NJ: Prentice Hall.

Ott, E., Grebogi, C., and Yorke, J. A. 1990. Controlling chaos. *Phys. Rev. Lett.*, **64**, 1196–1199.

Özbay, H. 2000. *Introduction to Feedback Control Theory*. Boca Raton: CRC Press.

Papoulis, A., and Pillai, S. Unnikrishna. 2002. *Probability, Random Variables, and Stochastic Processes*. 4th edn. McGraw Hill.

Parrondo, Juan M. R., Horowitz, Jordan M., and Sagawa, Takahiro. 2015. Thermodynamics of information. *Nature Phys.*, **11**, 131–139.

Pathak, J., Hunt, B., Girvan, M., Lu, Z., and Ott, E. 2018. Model-free prediction of large spatiotemporally chaotic systems from data: a reservoir computing approach. *Phys. Rev. Lett.*, **120**, 024102.

Pearl, J. 2009. *Causality: Models, Reasoning and Inference*. 2nd edn. Cambridge University Press.

Peiponen, K.-E., and Saarinen, J. J. 2009. Generalized Kramers-Kronig relations in nonlinear optical- and THz-spectroscopy. *Rep. Prog. Phys.*, **72**, 056401.

Pérez-Aparicio, R., Crauste-Thibierge, C., Cottinet, D., Tanase, M., Metz, P., Bellon, L., Naert, A., and Ciliberto, S. 2015. Simultaneous and accurate measurement of the dielectric constant at many frequencies spanning a wide range. *Rev. Sci. Instrum.*, **86**, 044702.

Phillips, R., Kondev, J., Theriot, J., and Garcia, H. 2012. *Physical Biology of the Cell*. 2nd edn. Garland Science.

Pikovsky, A., Rosenblum, M., and Kurths, J. 2001. *Synchronization: A Universal Concept in Nonlinear Sciences*. Cambridge, UK: Cambridge University Press.

Pintelon, R., and Schoukens, J. 2012. *System Identification: A Frequency Domain Approach*. 2nd edn. Wiley-IEEE Press.

Planck, M. 1903. *Treatise on Thermodynamics*. Longmans, Green, and Co.

Pontryagin, L. S., Boltyanskii, V. G., Gamkrelidze, R. V., and Mishchenko, E. F. 1964. *The Mathematical Theory of Optimal Processes*. New York: The Macmillan Co.

Popper, K. R. 1959. *The Logic of Scientific Discovery*. Hutchinson & Co., London, UK.

Preskill, J. 2017. *Quantum Computation*. Caltech, Physics 219 lecture notes.

Press, W. H., Plannery, B. P., Teukolsky, S. A., and Vetterling, W. T. 2007. *Numerical Recipes: The Art of Scientific Computing*. 3rd edn. Cambridge, UK: Cambridge University Press.

Pressé, S., Ghosh, K., Lee, J., and Dill, K. A. 2013. Principles of maximum entropy and maximum caliber in statistical physics. *Rev. Mod. Phys.*, **85**, 1115–1141.

Pyragas, V., and Pyragas, K. 2011. Adaptive modification of the delayed feedback control algorithm with a continuously varying time delay. *Phys. Lett. A*, **375**, 3866–3871.

Rabiner, L. R. 1989. A tutorial on Hidden Markov Models and selected applications in speech recognition. *Proc. IEEE*, **77**, 257–286.

Rasmussen, C. E., and Williams, C. K. I. 2006. *Gaussian Processes for Machine Learning*. Massachusetts Institute of Technology Press.

Rawlings, J. B., Mayne, D. Q., and Diehl, M. M. 2017. *Model Predictive Control: Theory, Computation, and Design*. 2nd edn. Nob Hill.

Recht, B. 2019. A tour of reinforcement learning: The view from continuous control. *Ann. Rev. Control Robot. Auton. Syst.*, **2**, 253–279.

Restrepo, J. M. 2008. A path integral method for data assimilation. *Physica D*, **237**, 14–27.

Reynolds, D. E. 2003. Coarse graining and control theory model reduction. *cond-mat/0309116*.

Reynolds, D. E. 2009. Construction of coarse-grained order parameters in nonequilibrium systems. *Phys. Rev. E*, **79**, 061107.

Rice, S. A., and Zhao, M. 2000. *Optical Control of Molecular Dynamics*. New York: Wiley-Interscience.

Risken, H. 1989. *The Fokker-Planck Equation: Methods of Solutions and Applications*. 2nd edn. Springer-Verlag.

Rissanen, J. 2007. *Information and Complexity in Statistical Modeling*. Springer.

Rivoire, O., and Leibler, S. 2011. The value of information for populations in varying environments. *J. Stat. Phys.*, **142**, 1124–1166.

Rossing, T. D., and Russell, D. A. 1990. Laboratory observation of elastic waves in solids. *Am. J. Phys.*, **58**, 1153–1162.

Routledge, R. 1900. *Discoveries and Inventions of the Nineteenth Century*. 13th edn. George Routledge and Sons, Ltd.

Rubio-Sierra, F. Javier, Vázquez, R., and Stark, R. W. 2006. Transfer function analysis of the micro cantilever used in atomic force microscopy. *IEEE Trans. Nanotech.*, **5**, 692–700.

Saad, Y. 2003. *Iterative Methods for Sparse Linear Systems*. Philadelphia: SIAM Press.

Safonov, M. G., and Tsao, T.-C. 1997. The unfalsified control concept and learning. *IEEE Trans. Auto. Cont.*, **42**, 843–847.

Sagawa, T. 2019. Second law, entropy production, and reversibility in thermodynamics of information. In: Lent, C., Orlov, A., W, W. Porod, and Snider, G. (eds.), *Energy Limits in Computation*. Springer, Cham.

Salerno, G., and Carusotto, I. 2014. Dynamical decoupling and dynamical isolation in temporally modulated coupled pendulums. *Euro. Phys. Lett.*, **106**, 24002.

Salmen, M., and Plöger, P. G. 2005. Echo state networks used for motor control. *Proc. ICRA*.

Sandberg, H., Delvenne, J.-C., Newton, N. J., and Mitter, S. K. 2014. Maximum work extraction and implementation costs for nonequilibrium Maxwell's demons. *Phys. Rev. E*, **90**, 042119.

Särkkä, S. 2013. *Bayesian Filtering and Smoothing*. Cambridge University Press.

Sayed, A. H. 2008. *Adaptive Filters*. Wiley-Interscience.

Sayrin, C., Dotsenko, I., Zhou, X., Peaudecerf, B., Rybarczyk, T., Gleyzes, S., Rouchon, P., Mirrahimi, M., Amini, H., Brune, M., Raimond, J.-M., and

Haroche, S. 2011. Real-time quantum feedback prepares and stabilizes photon number states. *Nature*, **477**, 73–77.

Schöll, E., and Schuster, H. G. (eds.). 2008. *Handbook of Chaos Control*. 2nd edn. New York: Wiley-VCH.

Schoukens, J., Pintelon, R., and Rolain, Y. 2012. *Mastering System Identification in 100 Exercises*. Wiley-IEEE Press.

Schreiber, T. 2000. Measuring information transfer. *Phys. Rev. Lett.*, **85**, 461–464.

Schulz, M. 2006. *Control Theory in Physics and Other Fields of Science*. Springer.

Schumacher, B., and Westmoreland, M. 2010. *Quantum Processes: Systems and Information*. Cambridge University Press.

Schwarz, G. 1978. Estimating the dimension of a model. *Ann. Stat.*, **6**, 461–464.

Seifert, U. 2012. Stochastic thermodynamics, fluctuation theorems and molecular machines. *Rep. Prog. Phys.*, **75**, 126001.

Sekimoto, K. 2010. *Stochastic Energetics*. Springer.

Seoane, L. F., and Solé, R. 2016. Multiobjective optimization and phase transitions. Pages 259–270 of: *Proceedings of ECCS 2014*. Springer.

Sepulchre, R., Drion, G., and Franci, A. 2019. Control across scales by positive and negative feedback. *Ann. Rev. Control Robot. Auton. Syst.*, **2**, 89–113.

Seron, M. M., Braslavsky, J. H., and Goodwin, G. C. 1997. *Fundamental Limitations in Filtering and Control*. Springer.

Shankar, R. 1995. *Basic Training in Mathematics: A Fitness Program for Science Students*. Springer.

Shannon, C. E. 1948. A mathematical theory of communication. *Bell. Syst. Tech. J.*, **27**, 379–423, 623–656.

Shannon, C. E. 1949. Communication in the presence of noise. *Proc. IRE*, **37**, 10–21.

Shannon, C. E. 1959. Coding theorems for a discrete source with a fidelity criterion. *IRE Int. Conv. Rec.*, **7**, 325–350.

Shannon, C. E., and Weaver, W. 1949. *The Mathematical Theory of Communication*. University of Illinois Press.

Shapiro, M., and Brumer, P. 2012. *Quantum Control of Molecular Processes*. 2nd edn. New York: Wiley-VCH.

Shaw, R. 1984. *The Dripping Faucet as a Model Chaotic System*. Aerial Press.

Shinbrot, T. 1999. Using chaotic sensitivity. Pages 157–180 of: Schuster, H. G. (ed.), *Handbook of Chaos Control*. New York: Wiley-VCH.

Shlivinski, A., and Hadad, Y. 2018. Beyond the Bode-Fano bound: Wideband impedance matching for short pulses using temporal switching of transmission-line parameters. *Phys. Rev. Lett.*, **121**, 204301.

Simon, H. A. 1956. Dynamic programming under uncertainty with a quadratic criterion function. *Econometrica*, **24**, 74–81.

Singer, N., and Seering, W. 1990. Preshaping command inputs to reduce system vibration. *ASME J. Dyn. Syst., Meas., Control*, **112**, 76–82.

Singh, T. 2010. *Optimal Reference Shaping for Dynamical Systems*. CRC Press.

Sivak, D. A., and Thomson, M. 2014. Environmental statistics and optimal regulation. *PLoS Comp. Biol.*, **10**, e1003826.

Sivia, D. S., and Skilling, J. 2006. *Data Analysis: A Bayesian Tutorial*. 2nd edn. Oxford University Press.

Sjöberg, J. 2005. *Descriptor Systems and Control Theory*. Tech. rept. LiTH-ISY-R-2688. Linköping University.

Skogestad, S., and Grimholt, C. 2012. The SIMC method for smooth PID controller tuning. In: Vilanova, R., and Visioli, A. (eds.), *PID Control in the Third Millennium*. Advances in Industrial Control. Springer.

Skogestad, S., and Postlethwaite, I. 2005. *Multivariable Feedback Control*. 2nd edn. Chichester, UK: John Wiley and Sons.

Slotine, J.-J. E., and Li, W. 1991. *Applied Nonlinear Control*. Prentice Hall.

Small, A., and Lam, K. S. 2011. Simple derivations of the Hamilton–Jacobi equation and the eikonal equation without the use of canonical transformations. *Am. J. Phys.*, **29**, 678–681.

Smith, O. J. M. 1957. Posicast control of damped oscillatory systems. *Proc. IRE*, **45**(1249–1255).

Smith, S. W. 1999. *The Scientist and Engineer's Guide to Digital Signal Processing*. 2nd edn. California Technical Publishing.

Solon, A. P., and Horowitz, J. M. 2018. Phase transitions in protocols minimizing work fluctuations. *Phys. Rev. Lett.*, **120**, 180605.

Sornette, D., and Ide, K. 2001. The Kalman-Lévy filter. *Physica D*, **151**, 142–174.

Souders, T. M., Flach, D. R., Hagwood, C., and Yang, G. L. 1990. The effects of timing jitter in sampling systems. *IEEE Trans. Instrum. and Meas.*, **39**, 80–85.

Stein, G. 2003. Respect the unstable. *Control. Syst. Mag.*, **23**, 12–25.

Stengel, R. F. 1994. *Optimal Control and Estimation*. Dover Publications.

Stepan, G. 2009. Delay effects in the human sensory system during balancing. *Phil. Trans. Roy. Soc. A*, **367**, 1195–1212.

Stone, M. 1977. An asymptotic equivalence of choice of model by cross-validation and Akaike's criterion. *J. Roy. Statist. Soc. B*, **38**, 44–47.

Stone, M., and Goldbart, P. 2009. *Mathematics for Physics: A Guided Tour for Graduate Students*. Cambridge University Press.

Stratonovich, R. L. 1960. Conditional Markov processes. *Theor. Prob. Appl.*, **5**, 156–178.

Strogatz, S. H. 2004. *Sync: How Order Emerges from Chaos in the Universe, Nature, and Daily Life*. Hyperion.

Strogatz, S. H. 2014. *Nonlinear Dynamics and Chaos*. 2nd edn. Westview Press.

Strunk, W., Jr. 1918. *The Elements of Style*. Ithaca, NY: W. P. Humphrey.

Sun, J., and Motter, A. E. 2013. Controllability transition and nonlocality in network control. *Phys. Rev. Lett.*, **110**, 208701.

Sun, Y.-Z., Leng, S.-Y., Lai, Y.-C., Grebogi, C., and Lin, W. 2017. Closed-loop control of complex networks: A trade-off between time and energy. *Phys. Rev. Lett.*, **119**, 198301.

Sussillo, D., and Abbott, L. F. 2009. Generating coherent patterns of activity from chaotic neural networks. *Neuron*, **63**, 544–557.

Sussmann, H. J., and Willems, J. C. 1997. 300 years of optimal control. *Control. Syst. Mag.*, **17**, 32–44.

Sutton, R. S., and Barto, A. G. 2018. *Reinforcement Learning: An Introduction*. 2nd edn. Massachusetts Institute of Technology Press.

Taleb, N. 2010. *The Black Swan: The Impact of the Highly Improbable*. 2nd edn. Random House.

Tang, E., and Bassett, D. S. 2018. Control of dynamics in brain networks. *Rev. Mod. Phys.*, **90**, 031003.

Taylor, J. R. 2005. *Classical Mechanics*. University Science Books.

Toll, J. S. 1956. Causality and the dispersion relation: Logical foundations. *Phys. Rev.*, **104**, 1760–1770.

Tomayko, J. E. 2000. *Computers take flight: A history of NASA's pioneering digital fly-by-wire project*. Tech. rept. The NASA History Series.

Tonk, S. R. H. M., and Kappen, H. J. 2010. *Optimal exploration as a symmetry breaking phenomenon*. https://repository.ubn.ru.nl/handle/2066/94186.

Touchette, H., and Lloyd, S. 2000. Information-theoretic limits of control. *Phys. Rev. Lett.*, **84**, 1156–1159.

Touchette, H., and Lloyd, S. 2004. Information-theoretic approach to the study of control systems. *Physica A*, **331**, 140–172.

Toyabe, S., Sagawa, T., Ueda, M., Muneyuki, E., and Sano, M. 2010. Experimental demonstration of information-to-energy conversion and validation of the generalized Jarzynski equality. *Nature Phys.*, **6**, 988–993.

Tufte, E. R. 1990. *Envisioning Information*. Graphics Press.

Tufte, E. R. 2001. *The Visual Display of Quantitative Information*. 2nd edn. Graphics Press.

Twiss, R. Q. 1955. Nyquist's and Thevenin's theorems generalized for nonreciprocal linear networks. *J. Appl. Phys.*, **26**, 599–602.

Valiant, L. 2013. *Probably Approximately Correct*. Basic Books.

Van den Broeck, C., and Esposito, M. 2015. Ensemble and trajectory thermodynamics: A brief introduction. *Physica A*, **418**, 6–16.

van Kampen, N. G. 2007. *Stochastic Processes in Physics and Chemistry*. 3rd edn. Elsevier.

van Leeuwen, P. J. 2010. Nonlinear data assimilation in geosciences: an extremely efficient particle filter. *Q. J. R. Meterol. Soc. B*, **136**, 1991–1999.

Van Trees, H. L., Bell, K. L., and Tian, Z. 2013. *Detection Estimation and Modulation Theory, Detection, Estimation, and Filtering Theory, Volume 1*. Wiley.

Vinante, A., Bonaldi, M., Marin, F., and Zendri, J.-P. 2013. Dissipative feedback does not improve the optimal resolution of incoherent force detection. *Nat. Nanotech.*, **8**, 470.

Viola, L., and Lloyd, S. 1998. Dynamical suppression of decoherence in two-state quantum systems. *Phys. Rev. A*, **58**, 2733–2744.

von der Linden, W., Dose, V., and von Toussaint, U. 2014. *Bayesian Probability Theory: Applications in the Physical Sciences*. Cambridge University Press.

von Neumann, J., and Morganstern, O. 1944. *Theory of Games and Economic Behavior*. Princeton University Press.

Vu, T. V., and Hasegawa, Y. 2020. Uncertainty relation in the presence of information measurement and feedback control. *J. Phys. A: Math. Theor.*, 075001.

Wallace, C. S., and Boulton, D. M. 1968. An information measure for classification. *Comput. J.*, **11**, 185–194.

Walmsley, I., and Rabitz, H. 2003. Quantum physics under control. *Phys. Today*, **56**, 43–49.

Wan, E. A., and van der Merwe, R. 2001. The Unscented Kalman Filter. Chap. 7, pages 221–280 of: Haykin, S. (ed.), *Kalman Filter and Neural Networks*. John Wiley & Sons, Inc.

Wang, Q., and Moerner, W. E. 2010. Optimal strategy for trapping single fluorescent molecules in solution using the ABEL trap. *Appl. Phys. B*, **99**, 23–30.

Wang, Q., and Moerner, W. E. 2011. An adaptative Anti-Brownian Electrokinetic Trap with real-time information on single-molecule diffusivity and mobility. *ACS Nano*, **5**, 5792–5799.

Warren, W. S., Rabitz, H., and Dahleh, M. 1993. Coherent control of quantum dynamics: The dream is alive. *Science*, **259**, 1581–1589.

Wasserman, L. 2004. *All of Statistics: A Concise Course in Statistical Inference*. Springer.

Waterfall, J. J., Casey, F. P., Gutenkunst, R. N., Brown, K. S., Myers, C. R., Brouwer, P. W., Elser, V., and Sethna, J. P. 2006. Sloppy-model universality class and the Vandermonde matrix. *Phys. Rev. Lett.*, **97**, 150601.

Wayman, E. 2012. Becoming human: The evolution of walking upright. *smithsonian-mag.com*.

Weaver, R. L., and Lobkis, O. I. 2001. Ultrasonics without a source: Thermal fluctuation correlations at MHz frequencies. *Phys. Rev. Lett.*, **87**, 134301.

Whitaker, H. P., Yamron, J., and Kezer, A. 1958. *Design of Model Reference Adaptive Control Systems for Aircraft, Report R-164*. Tech. rept. Massachusetts Institute of Technology.

Whittle, P. 1996. *Optimal Control: Basics and Beyond*. Wiley-Interscience.

Whittle, P. 2000. *Probability via Expectation*. 4th edn. Springer.

Whittle, P. 2002. Risk sensitivity, a strangely pervasive concept. *Macroeconomic Dynamics*, **6**, 5–18.

Widrow, B., and Walach, E. 1996. *Adaptive Inverse Control*. Prentice Hall.

Wiener, N. 1961. *Cybernetics or Control and Communication in the Animal and the Machine*. 2nd edn. Cambridge, MA: Massachusetts Institute of Technology Press.

Wilkinson, J. H. 1959. The evaluation of the zeros of ill-conditioned polynomials. Part I. *Numerische Mathematik*, **1**, 150–166.

Willems, J. C. 2007. The behavioral approach to open and interconnected systems: Modeling by tearing, zooming, and linking. *IEEE Contr. Syst. Mag.*, **27**, 46–99.

Willems, J. C. 2010. Terminals and ports. *IEEE Circuits and Systems Mag.*, **10**, 8–26.

Williams, G., Drews, P., Goldain, B., Rehg, J. M., and Theodorou, E. A. 2018. Information-theoretic model predictive control: Theory and applications to autonomous driving. *IEEE Trans. Robotics*, **34**, 1603–1622.

Willinger, W., Alderson, D., Doyle, J. C., and Li, L. 2004. More "normal" than normal: Scaling distributions and complex systems. Pages 130–141 of: Ingalls, R. G., Rossetti, M. D., Smith, J. S., and Peters, B. A. (eds.), *Proc. Winter Simul. Conf.*

Winter, D. A., Patla, A. E., Prince, F., Ishac, M., and Gielo-Perczak, K. 1998. Stiffness control of balance in quiet standing. *J. Neurophysiol.*, **80**, 1211–1221.

Wiseman, H. M., and Milburn, G. J. 2010. *Quantum Measurement and Control.* Cambridge University Press.

Wonham, W. M. 1968. On the separation theorem of stochastic control. *SIAM J. Cont.*, **6**, 312–326.

Yan, G., Ren, J., Lai, Y.-C., Lai, C.-H., and Li, B. 2012. Controlling complex networks: How much energy is needed? *Phys. Rev. Lett.*, **108**, 218703.

Yan, G., Tsekenis, G., Barzel, B., Slotine, J.-J., Liu, Y.-Y., and Barabási, A.-L. 2015. Spectrum of controlling and observing complex networks. *Nat. Phys.*, **11**, 779–786.

Yan, G., Vértes, P. E., Towlson, E. K., Chew, Y. L., Walker, D. S., Schafer, W. R., and A. 2017. Network control principles predict neuron function in the *Caenorhabditis elegans* connectome. *Nature*, **550**, 519–523.

Yang, Y., Wang, J., and Motter, A. E. 2012. Network observability transitions. *Phys. Rev. Lett.*, **109**, 258701.

Yi, T.-M., Huang, Y., Simon, M. I., and Doyle, J. 2000. Robust perfect adaptation in bacterial chemotaxis through integral feedback control. *Proc. Natl. Acad. Sci. USA*, **97**, 4649–4653.

Yuan, R.-S., Ma, Y.-A., Yuan, B., and Ao, P. 2014. Lyapunov function as potential function: A dynamical equivalence. *Chin. Phys. B*, **23**, 010505.

Zalta, E. N. (ed.). 2012. *Interpretations of Probability*.

Zhang, J., Liu, Y.-X., Jacobs, K., and Nori, F. 2017. Quantum feedback: Theory, experiments, and applications. *Phys. Rep.*, **679**, 1–60.

Znaimer, L., and Bechhoefer, J. 2014. Split PID control – two sensors can be better than one. *Rev. Sci. Instrum.*, **85**, 106105.

Zurek, W. H. 2003. Quantum discord and Maxwell's demons. *Phys. Rev. A*, **67**, 012320.

Index